# HEAT TRANSFER

Lindon C. Thomas

PRENTICE HALL, Englewood Cliffs, New Jersey 07632

Library of Congress Cataloging-in-Publication Data

Thomas, Lindon C.
Heat transfer / Lindon C. Thomas.
p. cm.
Includes bibliographical references (p. ) and index.
ISBN 0-13-384942-2
1. Heat--Transmission. 2. Heat--Convection. I. Title.
QC320.T493 1992
621.402'2--dc20 90-27484
CIP

Editorial/production supervision
and interior design: Virginia L. McCarthy
Acquisition editor: Doug Humphrey
Cover design: Joe DiDomenico
Pre-press buyer: Linda Behrens
Manufacturing buyer: David Dickey

To *Tisha* and our children,
*Sarah, Stephen, Mary, Mark,* and *Elissa.*



Printed in the United States of America
10 9 8 7 6 5 4 3 2 1

ISBN 0-13-384942-2

9 780133 849424 90000>

Prentice-Hall International (UK) Limited, *London*
Prentice-Hall of Australia Pty. Limited, *Sydney*
Prentice-Hall Canada Inc., *Toronto*
Prentice-Hall Hispanoamericana, S.A., *Mexico*
Prentice-Hall of India Private Limited, *New Delhi*
Prentice-Hall of Japan, Inc., *Tokyo*
Simon & Schuster Asia Pte. Ltd., *Singapore*
Editora Prentice-Hall do Brasil, Ltda., *Rio de Janeiro*

# CONTENTS

**PREFACE** vii

Acknowledgments *x*

**CHAPTER 1**

**INTRODUCTION** 1

1–1 Fundamental Laws *4*
1–2 Basic Transport Mechanisms and Particular Laws *5*
1–3 Analogy between Heat Transfer and the Flow of Electric Current *25*
1–4 Summary *28*
Review Questions *29*
Problems *29*

**CHAPTER 2**

**ONE-DIMENSIONAL HEAT TRANSFER** 34

2–1 Introduction *34*
2–2 Conduction Heat Transfer in a Plane Wall *34*
2–3 Conduction Heat Transfer in Radial Systems *58*
2–4 Variable Thermal Conductivity *67*
2–5 Internal Energy Sources *71*
2–6 Extended Surfaces *76*
2–7 Unsteady Heat-Transfer Systems *95*
2–8 Practical Solution Results *105*
2–9 Summary *108*
Review Questions *108*
Problems *109*

**CHAPTER 3**

**CONDUCTION HEAT TRANSFER: MULTIDIMENSIONAL** 118

3–1 Introduction *118*
3–2 General Fourier Law of Conduction *118*

3–3 Mathematical Formulations *119*
3–4 Differential Formulation *120*
3–5 Solution Techniques: Introduction *130*
3–6 Analytical Approaches *131*
3–7 Analogy Approaches *140*
3–8 Graphical Approaches *143*
3–9 Practical Solution Results *146*
3–10 Summary *180*
Review Questions *181*
Problems *182*

**CHAPTER 4**
**CONDUCTION HEAT TRANSFER: NUMERICAL APPROACH** 188
4–1 Introduction *188*
4–2 The Nodal Network *189*
4–3 Finite-Difference Approximations *190*
4–4 Numerical Analysis: Steady Systems *195*
4–5 Numerical Analysis: Unsteady Systems *224*
4–6 Summary *247*
Review Questions *248*
Problems *249*

**CHAPTER 5**
**RADIATION HEAT TRANSFER** 257
5–1 Introduction *257*
5–2 Physical Mechanism *257*
5–3 Thermal Radiation Properties *261*
5–4 Radiation Shape Factor *287*
5–5 Practical Thermal Analysis of Radiation Heat Transfer *296*
5–6 Solar Radiation *351*
5–7 Summary *357*
Review Questions *357*
Problems *358*

**CHAPTER 6**
**CONVECTION HEAT TRANSFER: INTRODUCTION** 364
6–1 Characterizing Factors Associated with Convection Systems *364*
6–2 The Boundary Layer *373*
6–3 Approaches to the Analysis of Convection Heat Transfer *374*
6–4 Characteristics of Internal Flow *381*
6–5 Summary *386*
Review Questions *387*
Problems *387*

**CHAPTER 7**
**CONVECTION HEAT TRANSFER: THEORETICAL ANALYSIS** 390
7–1 Introduction *390*
7–2 Laminar Flow Theory *390*

7–3 Turbulent Flow Theory *440*
7–4 Summary *455*
Review Questions *455*
Problems *456*

**CHAPTER 8**
**CONVECTION HEAT TRANSFER: PRACTICAL ANALYSIS—FORCED CONVECTION** 463

8–1 Introduction *463*
8–2 Internal Flow *463*
8–3 External Flow *502*
8–4 Flow Across Tube Banks *515*
8–5 Summary *535*
Review Questions *541*
Problems *542*

**CHAPTER 9**
**CONVECTION HEAT TRANSFER: PRACTICAL ANALYSIS—NATURAL CONVECTION** 549

9–1 Introduction *549*
9–2 Characterizing Parameters for Natural Convection *550*
9–3 External Natural Convection Flow *551*
9–4 Internal Natural Convection Flow *563*
9–5 Natural Convection Flow in Enclosed Spaces *570*
9–6 Combined Natural and Forced Convection *576*
9–7 Summary *577*
Review Questions *580*
Problems *580*

**CHAPTER 10**
**CONVECTION HEAT TRANSFER: PRACTICAL ANALYSIS—BOILING AND CONDENSATION** 583

10–1 Introduction *583*
10–2 Characterizing Parameters for Two-Phase Heat-Transfer Processes *584*
10–3 Boiling Heat Transfer *585*
10–4 Condensation Heat Transfer *605*
10–5 Summary *622*
Review Questions *624*
Problems *625*

**CHAPTER 11**
**CONVECTION HEAT TRANSFER: PRACTICAL ANALYSIS—HEAT EXCHANGERS** 628

11–1 Introduction *628*
11–2 Types of Heat Exchangers *628*
11–3 Evaluation and Design *641*
11–4 Overall Coefficient of Heat Transfer *641*

11–5 Heat Exchanger Analysis *652*
11–6 Summary *696*
Review Questions *698*
Problems *698*

**APPENDIXES** 704

A Mathematical Concepts *705*
B Dimensions, Units, and Significant Figures *707*
C Thermophysical Properties *710*
D Earth Temperature Data for Select U.S. Cities *735*
E Analytical Solutions for Conduction Heat Transfer *735*
F Numerical Computations *744*
G Mathematical Functions *745*
H Table for Radiation Functions *746*
I Viscous Dissipation: Mathematical Formulation for Laminar Flow *747*
J Film Condensation: Approximate Solution for Laminar Flow *750*

**NOMENCLATURE** 756

Fundamental Units *756*
Parameters *756*
Mathematical Operations and Functional Relations *761*

**GLOSSARY** 762

General *762*
Conduction *763*
Radiation *764*
Convection *767*

**REFERENCES** 774

**INDEX** 792

# PREFACE

Heat transfer has long been a part of the curricula of disciplines such as aerospace, chemical, environmental, mechanical, and nuclear engineering, as well as physics. The significance of heat transfer, having heightened in past years as a result of the worldwide energy problem, has become more widely recognized in other fields such as architectural, civil, electrical, and petroleum engineering.

*Heat Transfer* and the associated *mass transfer supplement* were developed for use in introductory undergraduate engineering heat-transfer courses that are taught at the junior or senior level. These volumes should also serve as a useful reference for graduate students and others interested in energy-related problems.

The text represents an extensive revision and refinement of *Fundamentals of Heat Transfer,* which was published by Prentice Hall in 1980. The text features (1) the thorough presentation of fundamental concepts in the context of simple one-dimensional analyses; (2) the analysis of multidimensional conduction heat-transfer processes, with special consideration given to the modern numerical finite-difference method; (3) the analysis of radiation heat transfer; and (4) the coverage of convection heat transfer, with the theory of convection followed by the practical analysis approach.

The presentation throughout the text generally involves the use of brief developments followed by examples. Elementary aspects of calculus and differential equations, which provide background for an introductory study of heat transfer, are reviewed in the Appendix. Fundamental concepts pertaining to the mechanisms of heat transfer are introduced in Chap. 1. One-dimensional analyses are developed in Chap. 2, with emphasis given to the importance of limiting criteria for common systems, such as fins, that are inherently multidimensional. Standard approaches for solving multidimensional conduction problems are considered in Chap. 3, with attention focused on the simple but accurate numerical finite-difference method in Chap. 4. Fundamentals of thermal radiation and the practical analysis of radiation heat transfer, which involves the use of the thermal network concept, are presented in Chap. 5. The topic of convection heat transfer is introduced in Chap. 6, with the theory of convection presented in Chap. 7. The practical solution approach, which involves the use of modern convection correlations and simple lumped formulations, is developed in Chaps. 8 through 11 for forced and natural convection, boiling and condensation, and heat exchangers, with attention given to both hydraulic and thermal aspects. Concerning the treatment of convection heat transfer, the format is such that the practical lumped approach of Chaps. 8 through 11 can be studied before or after (or independent of) the material in Chap. 7, which deals with the theory of convection.

The unique approach to the study of convection heat transfer developed in this text, which separates the theory of convection (Chap. 7) and the practical analysis approach (Chaps. 8–11), has been motivated by the fact that students often encounter difficulties in the study of this important topic. The traditional approach involves the presentation of the theory of convection, which is fairly complex and sometimes appears somewhat bewildering, interspersed with practical aspects. The topics generally covered in the theory of convection include (1) the development of the boundary layer (partial differential) equations for laminar and turbulent flows, (2) the development of the related integral equations, (3) the introduction of the similarity solution approach for laminar forced and natural convection external boundary layer flows, (4) the development of approximate integral solutions for laminar and turbulent external boundary layer flows, and more. These topics are important, but are not essential to the development of the practical analysis approach used in heat exchanger analysis and in the solution of most basic convection heat-transfer problems that engineers encounter in practice. The presentation of the practical analysis approach in a self-contained unit enables the student to gain a basic understanding of the key elements of evaluation and design of convection systems with or without the study of details of the theory of convection.

In this connection, the practical hydraulic and thermal analysis approach to analyzing heat exchangers and other internal flow processes involves the use of bulk-stream characteristics. This approach has traditionally been developed in the context of a defining equation for bulk-stream (or mixing-cup) temperature $T_b$. Unfortunately, this critical issue is frequently blurred by imprecise mathematical formulations. The approach taken in Chaps. 6 through 11 in this text features a less restrictive fundamental formulation, which involves the useful concept of bulk-stream enthalpy $\dot{H}_b$. The basic relationships among mass flow rate $d\dot{m}$, momentum rate $d\dot{M}$, enthalpy rate $d\dot{H}$, velocity, temperature, and thermophysical properties used to develop this more general perspective are introduced in Chap. 6. In addition to providing a basis for the development of a clearer and more general practical hydraulic and thermal analysis approach, the concept of enthalpy rate and momentum rate is used in Chap. 7 to develop what is believed to be a simpler, less tedious approach to the formulation of the differential and integral equations for boundary layer flow. Because of its versatility and relative simplicity, the approximate integral solution approach is featured in Chap. 7 for external flows.

The text contains sufficient material for coverage in two quarter courses. For use in a single semester or quarter course, the following material is suggested:

| | |
|---|---|
| 1–1 to 1–4 | Introduction |
| 2–1 to 2–9 | One-dimensional analyses |
| 3–1 to 3–6, 3–9, 3–10 | Conduction: Multidimensional |
| 4–1 to 4–3, selections from 4–4 to 4–6 | Conduction: Numerical Approach |
| 5–1 to 5–3–2, selections from 5–4 to 5–7 | Radiation |
| 6–1 to 6–5 | Convection: Introduction |
| 7–1, 7–2, 7–4 | Convection: Theory *(This material may be covered before or after sections in Chaps. 8 to 11, or may be omitted.)* |
| | Convection: Practical Analysis— |
| 8–1 to 8–5 | Forced Convection |
| 9–1 to 9–3, 9–7 | Natural Convection |
| 10–1, 10–2, selections from 10–3 to 10–5 | Boiling and Condensation |
| 11–1 to 11–4, selections from 11–5, 11–6 | Heat Exchangers |

This basic material can be coupled with selections from other sections in the text and the mass transfer supplement, depending on the specific objectives of the course, total number of contact hours, and background of the students.

The mass transfer supplement primarily deals with issues pertaining to heat and mass transfer in chemically nonhomogeneous substances, with emphasis placed on basic low mass-flux processes. This material is organized into three chapters. Chapter 12 provides a general introduction to the topic; Chap. 13 deals with the theory of diffusion mass transfer, which involves the formulation and solution of diffusion transport equations; and Chap. 14 presents the practical analysis approach to convection heat and mass transfer, which features the use of convection coefficients. As in the text itself, the organization is such that the practical analysis approach to convection heat and mass transfer featured in Chap. 14 can be studied before, after, or independent of the theoretical details developed in Chap. 13. In addition to providing insight into the analysis of low mass-flux processes, the presentation also provides a foundation for the analysis of moderate to high mass-flux (concentration and/or injection driven) processes.

Because the changeover to metric units continues in many countries, both the international system of units (SI) and the English engineering system are used, with the SI system being used in the body of the text and in approximately 80% of the examples and problems.

In regard to the examples, a consistent methodology is employed that involves the following format:

**EXAMPLE**

(statement)

**Solution**

*Objective* (concise statement of what is to be done)

*Schematic* (sketch of system showing all knowns and basic conditions)

*Assumptions/Conditions* (listing of assumptions and conditions not shown in schematic)

*Properties* (listing of pertinent thermophysical properties)

*Analysis* (presentation of mathematical formulation/solution, calculations, and comments)

It is believed that this systematic approach will aid the student in recognizing and understanding the specific concepts presented in each example.

It is my hope that *Heat Transfer* will contribute to the effective study and teaching of the challenging subject of heat transfer.

## ACKNOWLEDGMENTS

I wish to express my sincere appreciation to the many individuals who offered suggestions and criticism during the development of the text. Reviewers who provided important thoughtful and constructive input include Robert F. Boehm, University of Utah; Benjamin T. F. Chung and Jerry E. Drummond, University of Akron; Ronald D. Flack, University of Virginia; Ray Knight, Auburn University; Ekkehard Marschall, University of California at Santa Barbara; John McCutchen, Yuba Heat Transfer; Mohammad H. N. Naraghi, Manhattan College; Wilbert Pulkrabak, University of Wisconsin at Platteville; C. S. Reddy, Union College; Syed A. M. Said, King Fahd University of Petroleum and Minerals; Craig Saltiel, University of Florida Center for Advanced Study; John W. Sheffield, University of Missouri at Rolla; Richard L. Shilling, Brown Fintube Company; and E. M. Sparrow, University of Minnesota. I am grateful to these people for their willingness to undertake the sometimes thankless and difficult task of reviewing the work of a colleague.

Work on the text was initiated while I was on the faculty of King Fahd University of Petroleum and Minerals in Dhahran, Saudi Arabia. The consistent support provided by KFUPM for my professional activities and for the personal needs of my family contributed much in the early stages of the work and to the eventual completion of the project. Several of the faculty and administrative staff at KFUPM who influenced the development of the text through their suggestions, encouragement, or general support include Habib Abualhamayel, Alfred Guenkel, Kevin Loughlin, Mansour Nazer, and Syed Zubair.

The refinement of the manuscript and production of the book by Prentice Hall has been professionally administered by the senior managing editor, Doug Humphrey. His involvement in the project has been characterized by open-mindedness and objectivity, and an obvious determination to produce quality textbooks. It has been refreshing to deal with an editor who recognizes the potential long term value of his texts and resists pressures to rush or otherwise compromise the publication process.

In connection with the production of the text, I am pleased to acknowledge the able assistance of Mrs. Virginia McCarthy who served as production editor, Mr. James Tully who copy edited the manuscript, Dr. Tillie Amminger who proof read the galleys, and Mrs. Nazli Ishaq who typed early portions of the manuscript.

Finally, I am thankful to God who provided me with energy and good health, freedom and an education, encouraging parents, a loving and patient wife, and five wonderful children. I believe that what we are able to contribute professionally and otherwise is a result of what has been given to us by our creator and by those who love, encourage, teach, and support us, rather than something for which we alone should be credited.

Lindon C. Thomas

# CHAPTER 1

# INTRODUCTION

*Heat transfer* is defined as the transfer of energy across a system boundary caused solely by a temperature difference.

The study of heat transfer has long been a basic part of engineering curricula because of the significance of energy-related applications. For example, the transfer of heat in power plants from the energy source, be it fossil, nuclear, solar, or other, to the working fluid is one of the most basic processes in such systems. Similarly, the operation of refrigeration and air-conditioning units depend on the effective transfer of heat in condensers and evaporators. Other applications pertaining to environmental control, which are of particular interest currently, include the minimization of building-heat losses by means of improved insulating techniques and the use of supplemental energy sources, such as solar radiation, heat pumps, and fireplaces. Heat transfer is also very important in the operation of electrical machinery and transformers, and is often the controlling factor in the miniaturization of electronic systems.

Today the generation of electrical energy in power plants is primarily derived from fossil fuels—coal, natural gas, and oil. The cross section of a typical large coal-fired boiler used in power stations is shown in Fig. 1–1. In this system, high-pressure preheated water flowing in the vertical tubes surrounding the combustion chamber is heated by radiation and convection heat transfer, which results from the high temperatures produced by the combustion of coal. After separating the liquid and vapor in the upper steam drum, the vapor is further heated by a series of heat exchangers (known as superheaters), which operate at higher temperatures, and is then delivered to the steam turbine. The air supplied to the combustion chamber is first forced through a steam-heated coil in order to remove excess moisture and then through an exhaust gas/air preheater.

Although traditional fossil fuels such as coal and oil will be utilized for many years to come, a new era has been brought upon us by rising costs and environmental

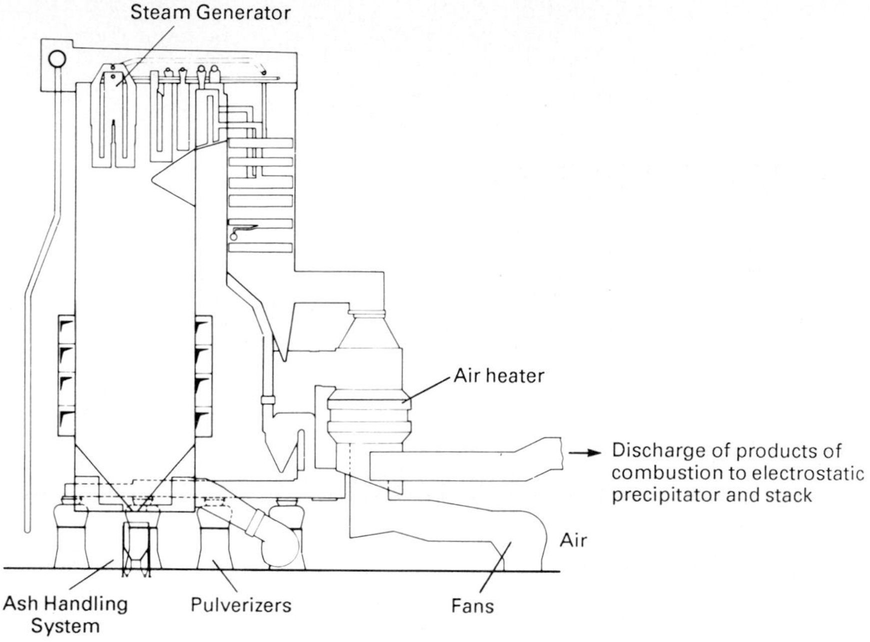
Steam Generator
Air heater
Discharge of products of combustion to electrostatic precipitator and stack
Air
Ash Handling System
Pulverizers
Fans

**FIGURE 1–1** Coal-fired steam boiler for electric power station. (Courtesy Babcock and Wilcox.)

considerations associated with the use of these resources. As solutions are sought to our energy problems, there is no doubt that heat transfer will be a key factor. The most obvious example is the generation of power by nuclear fission, which involves (1) the transfer of heat from the reactor core to fluid circulating in a primary loop, and (2) the transfer of energy by a heat exchanger to convert secondary water to steam to power an electric generator. Although safety and environmental concerns persist, the use of nuclear power continues to increase in many countries. Another example is the use of solar energy in heating, cooling, and energy generation. The schematic of a typical solar heating system is shown in Fig. 1–2(a). This arrangement is used in the Colorado State University Solar House shown in Fig. 1–2(b). In this and other solar energy applications, thermal radiation from the sun is captured in collectors, transferred to a working fluid such as water, and stored. The stored energy is then used for direct heating of the building during both day and night. Auxiliary electrical heating is provided for periods of inclement weather. Of course, the effective use of solar energy requires that the system be carefully insulated.

The primary objective in the analysis of most heat-transfer problems is either to (1) determine the temperature distribution within the system and the rate of heat transfer for specified operating conditions (the function of evaluation), or (2) prescribe the necessary configuration (size and shape) in order to accomplish a given heat-transfer rate and/or temperatures (the function of thermal design). Although emphasis

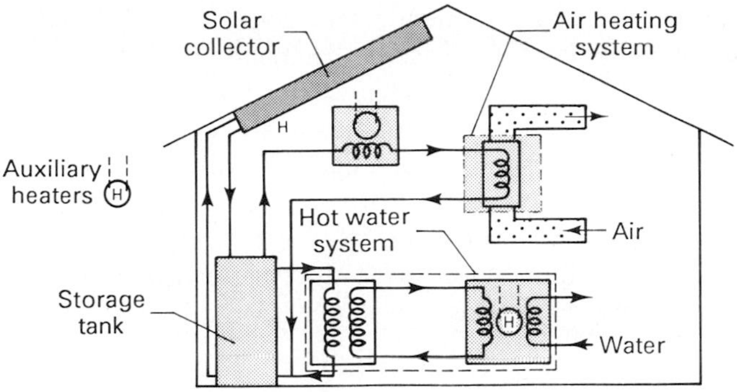


(a) Schematic of typical solar heating system.

(b) Solar house II in CSU Solar Village. (Courtesy of Solar Laboratory, Colorado State University.)

**FIGURE 1–2** Solar heating.

in our study will be placed on the evaluation function, the concepts of thermal (and hydraulic) analysis that will be developed provide the foundation for the actual design of systems involving heat transfer.

Because thermodynamics involves the study of heat and work for systems in equilibrium, a thermodynamic analysis can only provide us with predictions for the total quantity of heat transferred during a process in which a system goes from one equilibrium state (uniform temperature) to another. However, the length of time required for such processes to occur cannot be obtained by thermodynamics alone. On the other hand, the study of heat transfer involves a consideration of the mechanics of the transfer of thermal energy and is not restricted to equilibrium states. It is the science of heat transfer that enables us to perform the critical evaluation and design functions.

Analysis of heat-transfer processes requires the use of several *fundamental laws*, all of which are already familiar to us. These fundamental principles are of a general nature and are independent of the mechanism by which heat is transferred. In addition, *particular laws* pertaining to each of the mechanisms by which heat transfer can be accomplished must be satisfied. These fundamental and particular laws are reviewed in the following two sections, after which brief consideration is given to the very useful analogy between heat transfer and the flow of electric current. A review of basic mathematical concepts, dimensions, and units that pertain to our study of heat transfer is presented in Appendixes A and B.

## 1–1 FUNDAMENTAL LAWS

As in the case of thermodynamics, the *first law of thermodynamics* (conservation of energy) is the cornerstone of the science of heat transfer. This fundamental law takes the form

$$\text{rate of creation of energy} = 0$$

$$\Sigma\, \dot{E}_o - \Sigma\, \dot{E}_i + \frac{\Delta E_s}{\Delta t} = 0 \tag{1–1}$$

where $\dot{E}_i$ and $\dot{E}_o$ represent the rate of energy transfer into and out of the system, respectively, and $\Delta E_s/\Delta t$ is the rate of change in energy stored within the system.

Two other fundamental laws are required in the analysis of heat transfer in fluids that are in motion: (1) the *principle of conservation of mass* (for nonrelativistic conditions),

$$\text{rate of creation of mass} = 0$$

$$\Sigma\, \dot{m}_o - \Sigma\, \dot{m}_i + \frac{\Delta m_s}{\Delta t} = 0 \tag{1–2}$$

and (2) *Newton's second law of motion*, which is represented in terms of the $x$-component by

$$\text{rate of creation of momentum} = \text{sum of forces}$$

$$\Sigma\, \dot{M}_{o,x} - \Sigma\, \dot{M}_{i,x} + \frac{\Delta M_{s,x}}{\Delta t} = \Sigma\, F_x \tag{1–3}$$

where the momentum $M_x$ represents the product of the $x$-component of velocity $u$ and the mass $m$.

Three supplementary fundamental principles are required in the analysis of all heat-transfer processes: (1) *the second law of thermodynamics*, which provides us with the very critical conclusion that heat is transferred in the direction of decreasing temperature; (2) the principle of *dimensional continuity*, which requires that all equations be dimensionally consistent; and (3) *equations of state*, which provide information in equation, tabular, or graphical form pertaining to the thermodynamic properties at any state.

Concerning the thermodynamic properties, the symbols $U$ and $H$ ($H = U + PV$) will be used to designate the *internal energy* and *enthalpy* of a given mass $m$ of a substance. Information pertaining to the *specific internal energy* $e$ ($e = U/m$) and *specific enthalpy* $i$ ($i = H/m$) is tabulated for common substances (e.g., steam tables). In addition, $e$ and $i$ can be expressed in terms of the *constant-volume specific heat* $c_v$ and the *constant-pressure specific heat* $c_P$ by

$$c_v = \left.\frac{\partial e}{\partial T}\right|_v \qquad c_P = \left.\frac{\partial i}{\partial T}\right|_P \tag{1–4,5}$$

For important practical applications involving ideal fluids (i.e., ideal gases and incompressible liquids), $de$ and $di$ are given by

$$de = c_v\,dT \qquad di = c_P\,dT \tag{1–6,7}$$

or, for systems with constant mass $m$,

$$dU = mc_v\,dT \qquad dH = mc_P\,dT \tag{1–8,9}$$

## 1–2 BASIC TRANSPORT MECHANISMS AND PARTICULAR LAWS

Conduction and thermal radiation represent the two fundamental mechanisms by which heat transfer is accomplished. These heat-transfer mechanisms occur in both solids and fluids. Transfer of heat by conduction (and sometimes by thermal radiation) from a solid surface to a moving fluid is known as *convection heat transfer*. These three modes of heat transfer and the particular laws that govern these phenomena are introduced in the following sections.

### 1–2–1 Conduction Heat Transfer

From the thermodynamic view, *temperature* $T$ is a property that is an index of the kinetic energy possessed by the building-block particles of a substance (i.e., molecules, atoms, and electrons); the greater the agitation of these basic components of which matter is made, the higher the temperature. In this light, *conduction heat transfer* is the transfer of energy caused by physical interaction among molecular, atomic, and subatomic particles of a substance at different temperatures (level of kinetic energy). To expand upon this point, conduction in gases involves the collision and exchange of energy and momentum among molecules in continuous random motion. This same molecular transport mechanism occurs in liquids, but is complicated by the effects of molecular force fields, and can be augmented by the transport of free electrons in liquids that are good electrical conductors. On the other hand, conduction in solids occurs as a result of the movement of free electrons and vibrational energy in the atomic lattice structure of the material.

### *Fourier Law of Conduction*

On the basis of experimental observation, the rate of heat transferred by conduction in the $x$ direction through a finite area $A_x$ for the situation in which $T$ is a function only of $x$ can be expressed by

$$q_x = -kA_x \frac{dT}{dx} \tag{1–10}$$

where $A_x$ is normal to the direction of transfer $x$, and $k$ is the *thermal conductivity*. This equation was first used to analyze conduction heat transfer in 1822 by a French mathematical physicist named J. Fourier [1] and has come to be called the *Fourier law of conduction*. An example of a one-dimensional molecular conduction-heat-transfer problem for which this equation applies is illustrated in Fig. 1–3, which shows a plate with surface temperatures $T_1$ and $T_2$. Because no temperature differences occur in the $y$ and $z$ directions, $q_y$ and $q_z$ are both zero. For this case in which $T_1$ is greater than $T_2$, the temperature gradient is negative. (As shown in Chap. 2, the temperature distribution is linear in this application because $q_x$, $k$, and $A_x$ are all constants.)

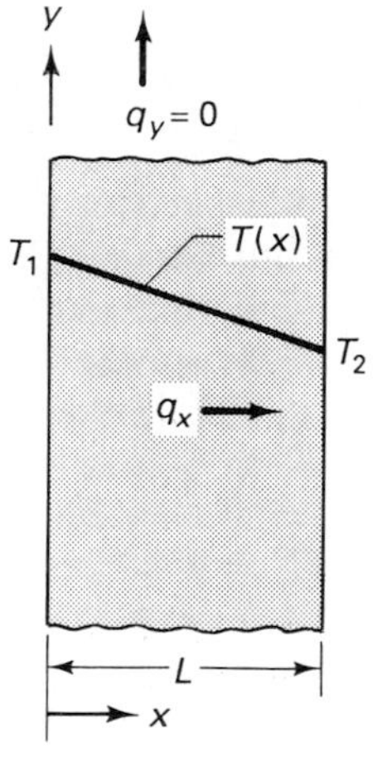


**FIGURE 1–3**
One-dimensional conduction heat transfer in a plate with $T_1 > T_2$; $q_y = 0$ and $q_z = 0$.

The consequence of the minus sign in Eq. (1–10) is that $q_x$ is positive for situations such as this in which the temperature gradient is negative. This result is consistent with the second law of thermodynamics, which stipulates that heat is transferred in the direction of decreasing temperature.

For situations in which the temperature is a function of time $t$ and one space variable, such as $x$, the Fourier law of conduction is written as

$$q_x = -kA_x \frac{\partial T}{\partial x} \tag{1–11}$$

where $q_x$ also is a function of $t$ and $x$. Expressions of the form of Eqs. (1–10) and (1–11) can also be written for conduction heat transfer in the $y$, $z$, or $r$ directions, as will be illustrated in Chap. 2. For applications in which the temperature $T$ is a function of more than one spatial dimension, the heat transfer in each direction must be accounted for. For example, for the case in which $T$ is a function of $x$ and $y$, we must write expressions for both $q_x$ and $q_y$. To accomplish this task, we utilize a more general form of the Fourier law of conduction. This general Fourier law of conduction is presented in Chap. 3, which deals with multidimensional conduction heat transfer.

***Thermal Conductivity*** The *thermal conductivity* $k$ is a thermophysical property of the conducting medium that represents the rate of conduction heat transfer per unit area for a temperature gradient of 1 °C/m (or 1 °F/ft). The units for $k$ are W/(m °C) [or Btu/(h ft °F)]. (Note that °C = 1 K and °F = 1 °R.) The thermal conductivities of various common substances are listed in Table 1–1 for standard atmospheric conditions. More extensive tabulations of thermal conductivities and other properties are given in Tables A–C–1 through A–C–5 of the Appendix and in references 2 through 6.

At room temperature, $k$ ranges from values in the hundreds for good conductors of heat such as diamond and various metals to less than 0.01 W/(m °C) for some gases. Materials with values of $k$ less than about 1 W/(m °C) are classified as insulators. As a rule of thumb, metals with good electrical conducting properties have higher thermal conductivities than do dielectric nonmetals or semiconductors. This is because the molecular interaction in good electrical conductors is enhanced by the movement of free electrons. Exceptions to this rule include dielectric crystals, such as diamond, sapphires, and quartz, and electric semiconductors, such as silicon and germanium. A second rule is that solid phases of materials generally have higher thermal conductivities than do liquid phases. An exception to this rule is bismuth, which has a higher thermal conductivity for the liquid phase than for the solid phase.

The variation of $k$ with temperature is shown in Fig. 1–4 for several representative substances and in Figs. A–C–1 through A–C–4 for various other common materials. The thermal conductivity of many of these substances varies by a factor of 10 or more for an order-of-magnitude change in temperature. On the other hand, the variation in $k$ with temperature for some materials over certain temperature ranges is small enough to be neglected. We also note that exceptionally high thermal conductivities occur among the solid materials that were judged to be good conductors at room temperature. For example, the thermal conductivity of aluminum reaches a maximum value of about 20,000 W/(m °C) at 10 K. This is over 100 times as large as the value that occurs at room temperature. Substances under low-temperature conditions that have such exceedingly high thermal conductivities are known as *superconductors*.

In homogeneous materials, $k$ can generally be assumed to be independent of direction (i.e., isotropic). However, some pure materials and laminates have thermal conductivities that are dependent upon the direction of heat flow. For example, the

**TABLE 1–1** Thermal conductivity of various substances at room temperature

| *Substance* | $k$ W/(m °C) | $k$ Btu/(h ft °F) |
|---|---|---|
| Metals | | |
| Silver | 420 | 240 |
| Copper | 390 | 230 |
| Gold | 320 | 180 |
| Aluminum | 200 | 120 |
| Silicon | 150 | 87 |
| Nickel | 91 | 53 |
| Chromium | 90 | 52 |
| Iron (pure) | 80 | 46 |
| Germanium | 60 | 35 |
| Carbon steel (0.5% C) | 54 | 31 |
| Nonmetallic Solids | | |
| Diamond, type 2A | 2300 | 1300 |
| Diamond, type 1 | 900 | 520 |
| Sapphire ($Al_2 0_3$) | 46 | 27 |
| Limestone | 1.5 | 0.87 |
| Glass (Pyrex 7740) | 1.0 | 0.58 |
| Teflon (Duroid 5600) | 0.40 | 0.23 |
| Brick, building | 0.69 | 0.399 |
| Plaster | 0.13 | 0.075 |
| Cork | 0.040 | 0.023 |
| Liquids | | |
| Mercury | 8.7 | 5.0 |
| Water | 0.6 | 0.35 |
| Freon F-12 | 0.08 | 0.046 |
| Gases | | |
| Hydrogen | 0.18 | 0.10 |
| Air | 0.026 | 0.015 |
| Nitrogen | 0.026 | 0.015 |
| Steam | 0.018 | 0.01 |
| Freon F-12 | 0.0097 | 0.0056 |

*Source*: From references 2 through 5.

thermal conductivity of wood is different for heat conduction across the grain than for heat transfer parallel to the grain. Other materials with such nonisotropic characteristics include crystalline substances, laminated plastics, and laminated metals. For an introduction to the topic of conduction heat transfer in nonisotropic materials, the textbook by Eckert and Drake [7] is recommended.

Heat-transfer applications involving the more familiar metallic conductors such as copper and aluminum and insulators such as rock wool and cork are known to us

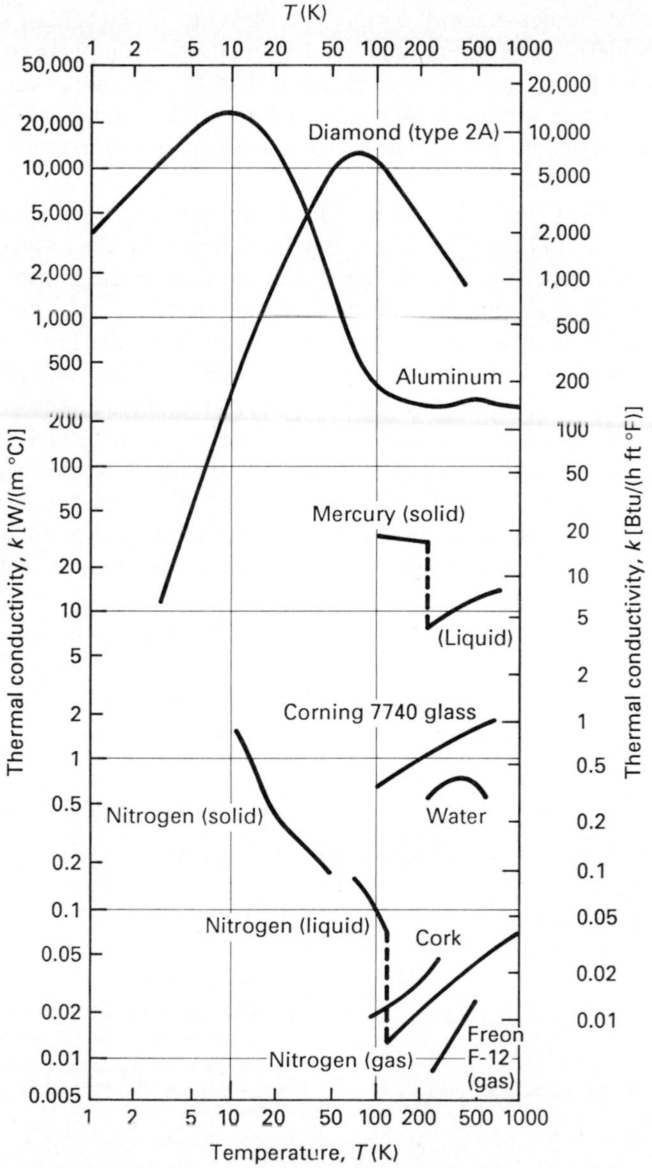


**FIGURE 1–4** Variation of thermal conductivity $k$ with temperature for representative substances. (From Touloukian et al. [2]. Used with permission.)

all. To name only a few, tubes made of copper, stainless steel, aluminum, and other metals are used in boilers, evaporators, and condensers to transmit energy from one fluid to another; metal pans are used to cook food; double glass panes are used to minimize heat loss through windows; and glass fibers, bricks, and other insulative materials are used to reduce building heat losses in the winter and heat gains in the summer.

Because of their high thermal conductivities and large electrical resistivities, silicon and diamond also find application, especially in the field of electronics. For example, silicon greases, pastes, and gaskets are often used in the construction of electronic systems in order to increase the rate of heat transfer while maintaining good electrical insulation between components. As another example, Fig. 1–5 shows a gold-plated diamond (type 2A) cube diode. Diodes such as these, ranging in size from below 0.1 mm to a few millimeters across, are used to generate high-frequency radio waves that relay telephone conversations and television broadcasts. These small diodes are characterized by very high power density operation, with operating temperatures in the range 150°C to 200°C. Because type 2A diamond has the highest thermal conductivity of all known materials in this temperature range, the diamond cube is used to remove the energy generated in the electronic semiconductor chip. Because of its effectiveness as a heat conductor, the diamond cube reduces the operating temperature of the diode, thereby increasing its lifetime and reliability.

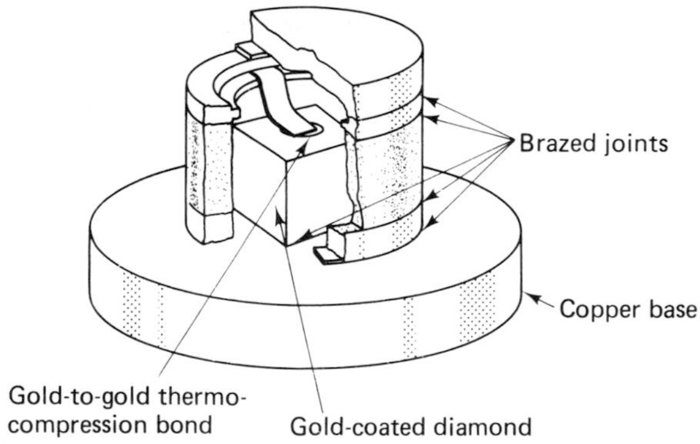
Brazed joints
Copper base
Gold-to-gold thermo-
compression bond
Gold-coated diamond

**FIGURE 1–5**
Microwave oscillator diode with diamond heat sink. (Courtesy of Bell Telephone Laboratories and D. Drukker & ZN. N.Y.)

### *Analysis of Conduction Heat Transfer*

Consideration is now given to the analysis of conduction heat transfer in solids or stationary fluids. The important topic of conduction heat transfer in moving fluids will be considered separately in Sec. 1–2–3, which deals with convection.

The general theoretical analysis of conduction-heat-transfer problems involves (1) the use of (a) the fundamental first law of thermodynamics and (b) the Fourier law of conduction (particular law) in the development of a mathematical formulation that represents the energy transfer in the system; and (2) the solution of the resulting system of equations for the temperature distribution. Once the temperature distribution is known, the rate of heat transfer is obtained by use of the Fourier law of conduction. The basic concepts involved in the theoretical analysis of conduction-heat-transfer problems will be presented in Chap. 2 in the context of fairly simple one-dimensional systems. These fundamentals will then be extended to multidimensional systems in Chap. 3.

***Conduction Shape Factor*** A simple practical approach to the analysis of basic steady-state conduction heat transfer problems has been developed that involves the use of an equation derived from the fundamental and particular laws. This practical equation for conduction heat transfer takes the form (for systems with uniform thermal conductivity)

$$q = kS(T_1 - T_2) \tag{1–12}$$

where $q$ is the rate of heat transfer conducted from a surface at temperature $T_1$ to a surface at $T_2$, and $S$ is known as the *conduction shape factor*; the unit for $S$ is m (or ft). The conduction shape factor $S$ is dependent upon geometry. Representative conduction shape factors are listed in Table 1–2 for several basic geometries. This practical approach to the analysis of conduction-heat-transfer problems is illustrated by several examples in this chapter. The theoretical basis for this simple method will be developed for one-dimensional systems in Chap. 2 and will be extended to more complex multidimensional systems in Chap. 3.

**TABLE 1–2** Conduction shape factors $S$

| *Geometry* | $S$ |
|---|---|
| Flat plate<br>Cross-sectional area $A$<br>Thickness $L$ | $A/L$ |
| Hollow cylinder<br>Radii $r_1$ and $r_2$<br>Length $L$ | $\dfrac{2\pi L}{\ln (r_2/r_1)}$ |
| Hollow sphere<br>Radii $r_1$ and $r_2$ | $\dfrac{4\pi r_1 r_2}{r_2 - r_1}$ |

**EXAMPLE 1–1**

Determine the rate of heat loss per unit area through a 10-cm-thick brick wall with surface temperatures of 15°C and 75°C.

**Solution**

*Objective* Determine the heat flux $q''$.

*Schematic* Plane wall.

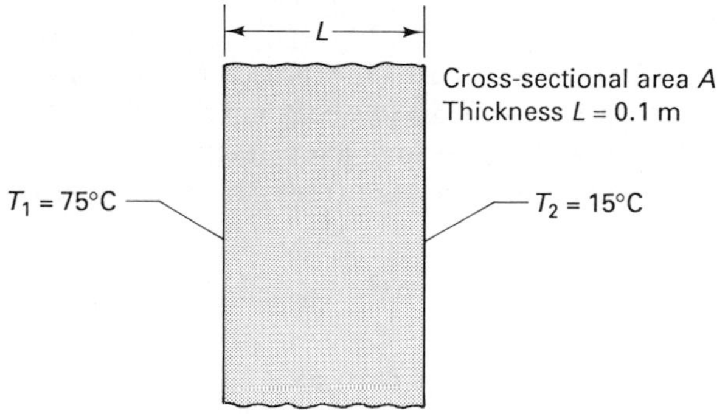


*Assumptions/Conditions*

steady-state
one-dimensional
uniform properties

*Properties* Building brick (Table 1–1): $k = 0.69$ W/(m °C).

*Analysis* Utilizing Eq. (1–12) with $S = A/L$, we write

$$q = kS(T_1 - T_2) = \frac{kA}{L}(T_1 - T_2)$$

or

$$q'' = \frac{q}{A} = \frac{k}{L}(T_1 - T_2) = \frac{0.69 \text{ W/(m °C)}}{0.1 \text{ m}}(75°\text{C} - 15°\text{C}) = 414 \text{ W/m}^2$$

### 1–2–2 Radiation Heat Transfer

The second fundamental mechanism of heat transfer involves the transfer of electromagnetic radiation emitted from a body as a result of vibrational and rotational movements of molecules, atoms, and electrons. As we have already noted, temperature is an index of the level of agitation of these microscopic particles. Because molecules and their components are continuously in motion, thermal radiation is always emitted by physical matter. Furthermore, the rate at which internal energy associated with the motion of molecules, atoms, and subatomic particles (indicated by temperature) is converted into thermal radiation increases with temperature. As discussed in Chap. 5, the manner in which thermal radiation, which encompasses ultraviolet (UV), visible light, and infrared (IR), is generated distinguishes it from other types of electromagnetic waves, such as $\gamma$ rays and X rays on the one hand, and microwaves and broadcasting waves on the other.

The medium through which thermal radiation passes can be a vacuum or a gas, liquid, or solid. Objects within the path absorb, reflect, and, if they are transparent, transmit incident thermal radiation. Electromagnetic radiation that is absorbed by matter is converted into internal energy, which can be stored, transferred by conduction, and converted back into electromagnetic radiation that is given off by the material itself. The absorptivity $\alpha$, reflectivity $\rho$, and transmissivity $\tau$ represent the fractions of incident thermal radiation that a body *absorbs*, *reflects*, and *transmits*, respectively. It follows that

$$\alpha + \rho + \tau = 1 \tag{1–13}$$

These properties are primarily dependent upon the temperature of the emitting source and the nature of the surface that receives the thermal radiation. To illustrate, most incoming thermal radiation is absorbed by a surface coated with lampblack paint (i.e., $\alpha \simeq 0.97$, $\rho \simeq 0.03$, $\tau \simeq 0$), but is reflected from the surface of a polished aluminum plate (i.e., $\alpha \simeq 0.1$, $\rho \simeq 0.9$, $\tau \simeq 0$). As another example, a thin glass plate will transmit most of the thermal radiation from the sun, but will absorb much of the thermal radiation emitted from the low-temperature interior of a building, such as a greenhouse.

Thermal radiation generally passes through gases such as air with no significant absorption taking place. Such gases with $\tau \simeq 1$ are known as *nonparticipating gases*. Gases and transparent liquids that absorb significant quantities of the thermal radiation are known as *participating fluids*. Carbon dioxide and water vapor are examples of participating gases and water is a participating liquid. Of course, many liquids, such as mercury, are opaque to thermal radiation.

### Stefan-Boltzmann Law

A body continually emits radiant energy in an amount that is related to its temperature and the nature of its surface. An object that absorbs all the radiant energy reaching its surface ($\alpha = 1$) is called a *blackbody*. Such ideal absorbers emit radiant energy at a rate that is proportional to the fourth power of the absolute temperature of the surface. The *Stefan-Boltzmann law* for blackbody thermal radiation takes the form

$$E_b = \sigma T_s^4 \tag{1–14}$$

where the *total emissive power* $E_b$ for a blackbody is the total rate of thermal radiation emitted by a perfect radiator per unit surface area, $\sigma$ is the *Stefan-Boltzmann constant*

[$\sigma = 5.670 \times 10^{-8}$ W/(m$^2$ K$^4$) $= 0.1714 \times 10^{-8}$ Btu/(h ft$^2$ °R$^4$)], and $T_s$ is the *absolute* surface temperature. The experimental basis for this famous fourth-power law was first established by the Austrian scientist J. Stefan in 1879. This discovery was followed in 1884 by a theoretical development by another Austrian, L. Boltzmann. This equation indicates that the thermal radiation energy content rapidly falls from a very substantial level for temperatures of the order of 500 K to relatively small values for common environmental temperatures of the order of 300 K.

For nonblackbody surfaces that absorb less than 100% of the incident radiant energy, the total emissive power $E$ (i.e., the rate of thermal radiation emitted per unit surface area) is generally expressed by

$$E = \epsilon E_b = \epsilon \sigma T_s^4 \tag{1–15}$$

where the *emissivity* $\epsilon$ lies between zero and unity. For example, the emissivities of polished aluminum and lampblack paint at room temperature are of the order of 0.1 and 0.96, respectively. The emissivities of various substances are listed in Table A–C–7 of the Appendix. The relationship between the emissivity $\epsilon$ and the absorptivity $\alpha$ will be considered in Chap. 5.

### *Radiation-Heat-Transfer Rate*

By definition, the rate of radiation heat transfer $q_R$ between two bodies is equal to the *net* rate of exchange of thermal radiation. For two infinite parallel blackbody plates that are separated by a vacuum or a nonparticipating gas, all the thermal radiation emitted from one surface reaches and is absorbed by the other body. The rate of thermal radiation leaving $A_s$ that is absorbed by $A_R$ is $A_s E_{bs}$, and the rate of radiant energy leaving $A_R$ that is absorbed by $A_s$ is $A_R E_{bR}$. With $A_s$ and $A_R$ being equal, it follows that the rate of radiation heat transfer $q_R$ from $A_s$ to $A_R$ is simply

$$q_R = A_s(E_{bs} - E_{bR}) \tag{1–16}$$

or

$$q_R = \sigma A_s(T_s^4 - T_R^4) \tag{1–17a}$$

Taking the opposite orientation, the rate of radiation heat transfer from surface $A_R$ to surface $A_s$ is

$$q_R = \sigma A_s(T_R^4 - T_s^4) \tag{1–17b}$$

A more general relation, which applies to other geometries, is given by

$$q_R = A_s F_{s-R}(E_{bs} - E_{bR}) = \sigma A_s F_{s-R}(T_s^4 - T_R^4) \tag{1–18}$$

where the *shape factor* (or *view factor*) $F_{s-R}$ represents the fraction of thermal radiation leaving surface $s$ that arrives directly on surface $R$. The shape factor varies from unity (for infinite parallel plates or enclosures) to zero. Further information pertaining to the development of Eq. (1–18) and to relations that enable us to determine $F_{s-R}$ for particular geometric configurations can be found in Chap. 5.

## EXAMPLE 1–2

The electrical cabinet shown in Fig. E1–2 is exposed to blackbody radiation with the surrounding walls, which are at 25°C. Determine the rate of radiation heat transfer if the cabinet temperature is 125°C.

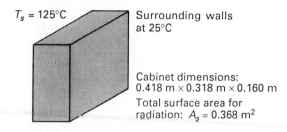

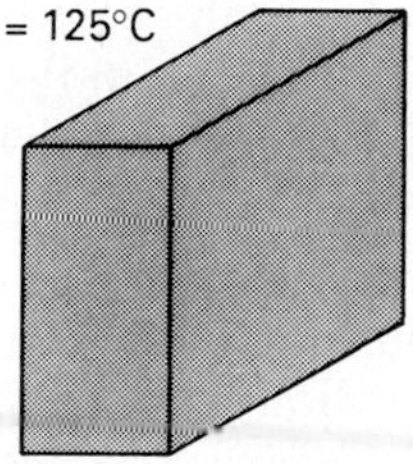
$T_s$ = 125°C
Surrounding walls at 25°C
Cabinet dimensions: 0.418 m × 0.318 m × 0.160 m
Total surface area for radiation: $A_s$ = 0.368 m²

**FIGURE E1–2**
Cabinet mounted on a vertical wall.

### Solution

*Objective* Determine the rate of radiation heat transfer $q_R$.

*Assumptions/Conditions*

steady-state
blackbody thermal radiation

*Analysis* The energy emitted per unit area from the cabinet and the walls is calculated first. Utilizing the Stefan-Boltzmann law, Eq. (1–14), we have

$$E_{bs} = \sigma T_s^4 = 5.67 \times 10^{-8} \frac{\text{W}}{\text{m}^2\,\text{K}^4} (398\ \text{K})^4 = 1420\ \text{W/m}^2$$

$$E_{bR} = \sigma T_R^4 = \sigma(298\ \text{K})^4 = 447\ \text{W/m}^2$$

The rate of blackbody radiation heat transfer from $A_s$ to the enclosure is given by Eq. (1–18), with the radiation shape factor $F_{s-R}$ equal to unity; that is,

$$q_R = A_s F_{s-R}(E_{bs} - E_{bR}) = 0.368\ \text{m}^2\,(1)\left(1420 \frac{\text{W}}{\text{m}^2} - 447 \frac{\text{W}}{\text{m}^2}\right) = 358\ \text{W}$$

If the space between the cabinet and the enclosure is evacuated, the total rate of heat transfer from the cabinet will be 358 W. By performing an energy balance on the cabinet for these conditions, we conclude that the power generated by the electrical system within the cabinet is 358 W. (If the cabinet is surrounded by air or another fluid, heat will also be removed by convection, which is the topic of the next section.)

## 1–2–3 Convection Heat Transfer

As we have already indicated, *convection* is the transfer of heat from a surface to a moving fluid. The conduction-heat-transfer mechanism always plays a primary role in convection. In addition, thermal radiation heat transfer (associated with absorbing and emitting fluids) and diffusion mass transfer (associated with chemically nonhomogeneous substances) are also sometimes involved.† In addition to the transfer of energy via the basic heat-transfer-mechanisms, convection heat transfer also involves the transfer of energy by macroscopic fluid motion. The process by which energy (or mass) is transferred by bulk-fluid motion is known as *advection*.

An introductory theoretical treatment of convection-heat-transfer processes is presented in Chap. 7. The theoretical analysis of convection requires that the fundamental laws of mass, momentum, and energy and the particular laws of viscous shear and conduction be utilized in the development of mathematical formulations for the fluid flow and energy transfer. The solution of these equations provides predictions for the velocity and temperature distributions within the fluid, after which calculations are developed for the rate of heat transfer into the fluid by the use of the Fourier law of conduction.

### *Newton Law of Viscous Stress*

In regard to the theoretical analysis of convection-heat-transfer processes, the particular law for viscous stress in Newtonian fluids can be written for simple steady one-dimensional shear flows as

$$\tau = \frac{dF}{dA} = \mu \frac{du}{dy} \tag{1–19}$$

where the viscosity $\mu$ is a property of the fluid with units kg/(m s) [or $lb_m$/(ft s)], $y$ the distance from the wall, $dA$ the differential area normal to $y$, $dF$ the differential shear force acting on the area $dA$, $\tau$ the shear stress, and $u$ the axial velocity. The Newton law of viscous stress is supported by experimental observation, kinetic theory, and statistical mechanics. The viscosity is often coupled with the density and written as $\nu = \mu/\rho$, where $\nu$ is called the *kinematic viscosity*; the units for $\nu$ are $m^2/s$ (or $ft^2/s$). The viscosity and kinematic viscosity of several Newtonian fluids are shown in Tables A–C–3 to A–C–5. The resistance to Newtonian fluid motion that results from shear stress is proportional to the viscosity, such that highly viscous fluids flow much less readily than do fluids with low viscosities such as water and air.

### *Newton Law of Cooling*

The engineer is generally primarily concerned about the rate of convection heat transfer rather than the temperature distribution within the fluid. Therefore, a practical approach

† Attention will be restricted to fluids in which thermal radiation can be neglected. The topic of diffusion mass transfer is considered in the accompanying *mass transfer supplement*.

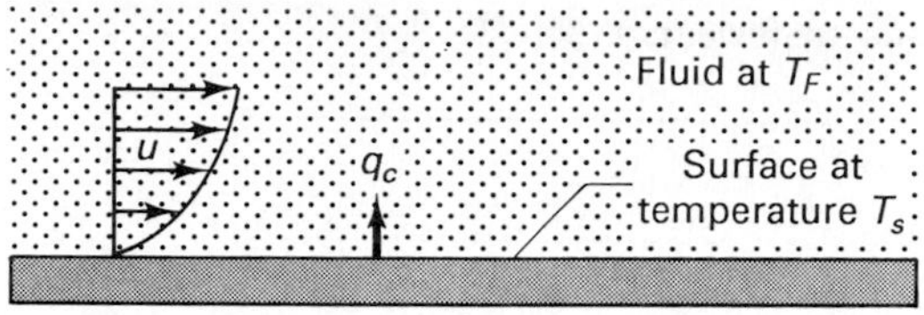


**FIGURE 1–6**
Convection heat transfer.

to the analysis of convection heat transfer from surfaces such as the flat plate shown in Fig. 1–6 has been developed which employs an equation of the form

$$q_c = \bar{h} A_s (T_s - T_F) \tag{1–20}$$

where $q_c$ is the rate of heat transferred from a surface at uniform temperature $T_s$ to a fluid with reference temperature $T_F$, $A_s$ is the surface area, and $\bar{h}$ is the *mean coefficient of heat transfer*; the units for $\bar{h}$ are W/(m$^2$ °C) [or Btu/(h ft$^2$ °F)]. Equation (1–20) is often referred to as the *Newton law of cooling* in honor of the British scientist Sir Isaac Newton.

A more general form of the Newton law of cooling, which applies to cases in which $T_s$ and/or $T_F$ are not uniform, is given by

$$dq_c = h_x \, dA_s \, (T_s - T_F) \tag{1–21}$$

where $dq_c$ represents the rate of heat transfer from a differential surface area $dA_s$ and $h_x$ is the *local coefficient of heat transfer*. The mean and local coefficients of heat transfer are related by

$$\bar{h} = \frac{1}{A_s} \int_{A_s} h_x \, dA_s \tag{1–22}$$

The Newton law of cooling in its simple or general form will be seen to be very useful in the analysis of heat-transfer processes involving convection combined with conduction and radiation in Sec. 1–2–4 and Chaps. 2 through 5, and in the evaluation and design of convection-heat-transfer systems in Chaps. 8 through 11.

***Coefficient of Heat Transfer*** Approximate ranges of $\bar{h}$ are shown in Table 1–3 for forced and natural convection in air and water. (For forced convection, the fluid motion is caused by mechanical means such as pumps and fans. On the other hand, natural convection is caused by temperature-induced density gradients within the fluid.†) The actual value of $\bar{h}$ depends upon the hydrodynamic conditions as well as on the thermodynamic and thermophysical properties of the fluid. The details of evaluating $h_x$ and $\bar{h}$ for standard fluid-flow systems will be considered in Chaps. 6 through 11.

† Natural convection can also occur in chemically nonhomogeneous mixtures as a result of density gradients caused by concentration gradients.

**TABLE 1–3** Convection-heat-transfer coefficients—range for representative applications

| *System* | $\bar{h}$ W/(m² °C) | $\bar{h}$ Btu/(h ft² °F) |
|---|---|---|
| Natural convection | | |
| Air | 5–30 | 0.9–5 |
| Water | 200–600 | 30–100 |
| Forced convection | | |
| Air | 10–500 | 2–100 |
| Water | $100–2 \times 10^4$ | $20–4 \times 10^3$ |
| Oil | $60–2 \times 10^3$ | 10–400 |
| Boiling water at 1 atm | $2 \times 10^3–5 \times 10^4$ | $300–9 \times 10^3$ |
| Condensation of steam | $5 \times 10^3–10^5$ | $900–2 \times 10^4$ |

## EXAMPLE 1–3

Determine the rate of convection heat transfer from the cabinet in Example 1–2 if it is surrounded by air at 25°C with a mean coefficient of heat transfer of 6.8 W/(m² °C).

### Solution

*Objective* Determine the rate of convection heat transfer $q_c$.

*Schematic* Cabinet.

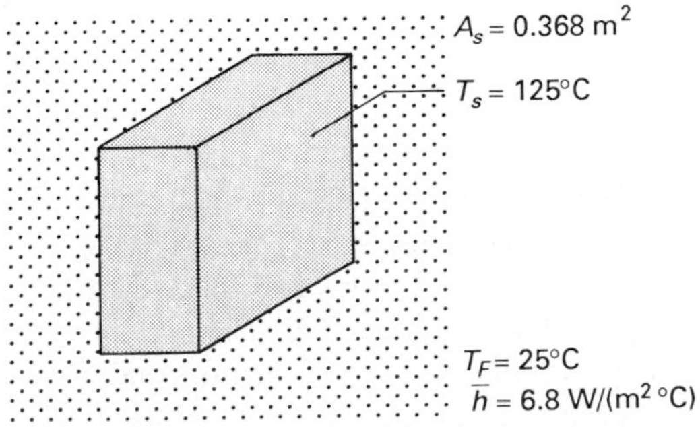
$A_s = 0.368$ m²
$T_s = 125$°C
$T_F = 25$°C
$\bar{h} = 6.8$ W/(m² °C)

*Assumptions/Conditions*

steady-state
convection and blackbody thermal radiation

*Analysis* Utilizing the Newton law of cooling, Eq. (1–20), we have

$$q_c = \bar{h}A_s(T_s - T_F) = 6.8 \frac{\text{W}}{\text{m}^2\ {}^\circ\text{C}} (0.368\ \text{m}^2)(125^\circ\text{C} - 25^\circ\text{C}) = 250\ \text{W}$$

As indicated in Example 1–4, this level of convection cooling contributes significantly to the total rate of heat transfer from the cabinet.

### 1-2-4 Combined Modes of Heat Transfer

Many heat-transfer processes encountered in practice involve combinations of conduction, thermal radiation, and convection. A situation in which combinations of these basic heat-transfer modes occur simultaneously is illustrated in Fig. 1–7. In this wall, which consists of two plates separated by a vacuum, heat is (1) convected from the fluid at $T_{F1}$ to plate 1, (2) conducted through plate 1, (3) radiated from plate 1 to plate 2, (4) conducted through plate 2, (5) and convected from plate 2 to the fluid at $T_{F2}$. Similar composite walls have been developed for the storage of cryogenic liquids, which consist of multiple layers of highly reflective materials that are separated by evacuated spaces. These *superinsulations* provide very efficient insulative walls with apparent thermal conductivities that are as low as $2 \times 10^{-5}$ W/(m °C).

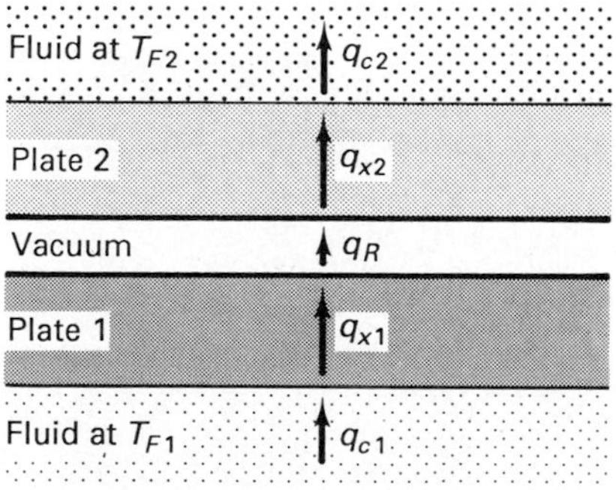


**FIGURE 1–7**
Heat-transfer process involving combined conduction, thermal radiation, and convection.

As illustrated by Examples 1–4 and 1–5, the analysis of basic multimode heat-transfer processes requires the use of the fundamental first law of thermodynamics and the particular laws for conduction, radiation, and convection.

**EXAMPLE 1–4**

An electronic cabinet made of anodized aluminum is cooled by natural convection and radiation. The surface area of the cabinet is 0.368 m², the temperature of the surrounding fluid and surface is 25°C, and $\bar{h}$ = 6.8 W/(m² °C). Estimate the rate of heat transfer from the cabinet if its surface temperature is to be maintained at 125°C.

**Solution**

*Objective* Determine the total rate of heat transfer $q$.

*Schematic* Cabinet.

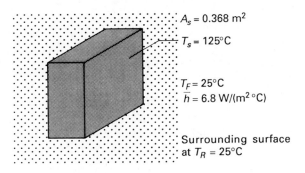

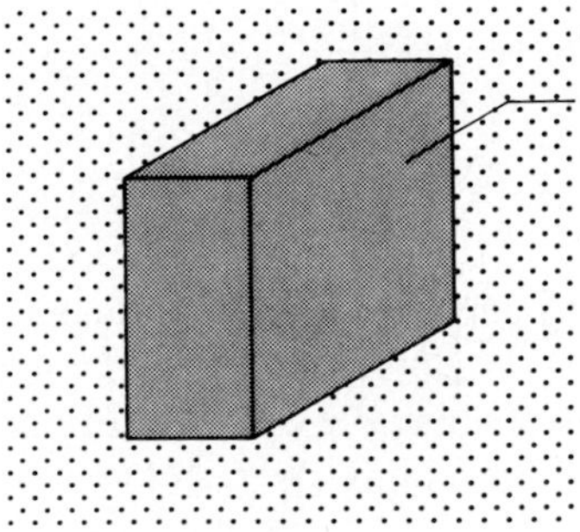
$A_s = 0.368$ m$^2$
$T_s = 125°C$
$T_F = 25°C$
$\bar{h} = 6.8$ W/(m$^2$ °C)
Surrounding surface
at $T_R = 25°C$

*Assumptions/Conditions*

steady-state

convection and blackbody thermal radiation

*Analysis* The total rate of heat transfer $q$ from the cabinet is

$$q = q_c + q_R$$

where $q_c$ is given by Eq. (1–20). Because the cabinet is made of black anodized aluminum, we will utilize the blackbody approximation given by Eq. (1–18). Referring to Examples 1–2 and 1–3, we obtain

$$q_R = A_s F_{s-R}(E_{bs} - E_{bR}) = \sigma A_s F_{s-R}(T_s^4 - T_R^4) = 358 \text{ W}$$

$$q_c = \bar{h} A_s (T_s - T_F) = 250 \text{ W}$$

such that the total rate of heat transfer $q$ is

$$q = 608 \text{ W}$$

Note that the radiation heat transfer accounts for a very significant 59% of the total. Whereas radiation is important for systems such as this involving natural convection, radiation is often not a factor in forced convection systems. For example, if the cabinet is cooled by forced air with $\bar{h} = 500$ W/(m$^2$ °C), only about 1.9% of the heat transfer is accomplished by radiation. (Radiation is also often insignificant in systems involving bright metal surfaces with low emissivities.)

## EXAMPLE 1–5

Liquid petroleum gas (LPG), which is widely used for domestic and industrial heating, consists of propane, butane, and mixtures of the two. LPG must be kept in liquid form because of its high volatility. The liquefication is often accomplished by refrigeration so that LPG can be stored at atmospheric pressure. Butane is stored at about 0°C and propane at −40°C.

With this brief background, we consider a cryogenic fluid that is to be stored in the 2-m-diameter spherical chamber shown in Fig. E1–5. A temperature of −40°C must be maintained by the refrigerant, which flows within the 1-cm-thick shell surrounding the inner storage chamber. The outer part of the vessel consists of a 10-cm-thick insulating material with $k = 0.60$ W/(m °C). The fluid surrounding the vessel is at 35°C with $\bar{h} = 150$ W/(m$^2$ °C). Determine the amount of refrigeration that is required to maintain the fluid at −40°C.

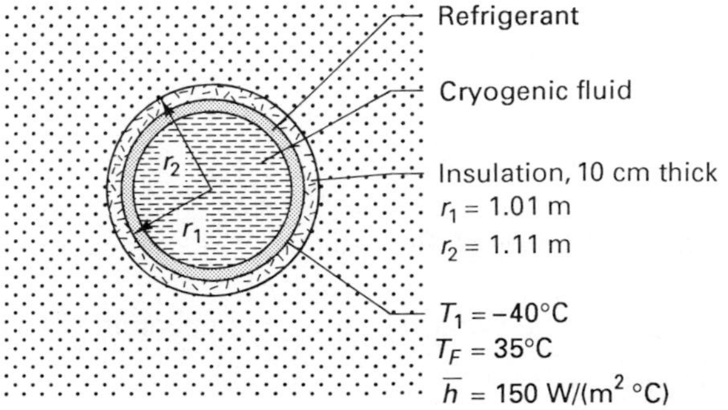
Refrigerant
Cryogenic fluid
Insulation, 10 cm thick
$r_1$ = 1.01 m
$r_2$ = 1.11 m
$T_1$ = −40°C
$T_F$ = 35°C
$\bar{h}$ = 150 W/(m$^2$ °C)
$r_2$
$r_1$

**FIGURE E1–5**
Spherical vessel.

### Solution

*Objective* Determine the total rate of heat transfer that must be compensated for by refrigeration.

*Assumptions/Conditions*

steady-state
one-dimensional
uniform properties

*Properties* Insulation: $k = 0.60$ W/(m °C).

*Analysis* By applying the first law of thermodynamics to the outer surface of the insulating wall, we obtain

$$\Sigma \dot{E}_i = \Sigma \dot{E}_o + \cancel{\frac{\Delta E_s}{\Delta t}} \tag{a}$$

$$q_c = q_r$$

The rate of heat transfer by convection $q_c$ is given by Eq. (1–20),

$$q_c = \bar{h}A_s(T_F - T_2) = \bar{h}4\pi r_2^2(35°\text{C} - T_2) \tag{b}$$

$$= 150 \frac{\text{W}}{\text{m}^2\,°\text{C}}(4\pi)(1.11\text{ m})^2(35°\text{C} - T_2) = 2320\frac{\text{W}}{°\text{C}}(35°\text{C} - T_2)$$

where the surface temperature is represented by $T_2$. The rate of heat transfer by radial conduction $q_r$ is given by Eq. (1–12) with $S = 4\pi r_1 r_2/(r_2 - r_1)$; that is,

$$q_r = kS(T_2 - T_1) = \frac{4\pi r_1 r_2 k}{r_2 - r_1}(T_2 + 40°\text{C}) = 84.5\frac{\text{W}}{°\text{C}}(T_2 + 40°\text{C}) \tag{c}$$

Similarly, the application of the first law of thermodynamics to the inside surface of the insulating wall gives

$$q_r = q_{\text{ref}} \tag{d}$$

such that the rate of heat transferred through the insulation by conduction $q_r$ must be taken out by refrigeration.

To determine $q_r$ or $q_c$, we first solve for the surface temperature $T_2$ by utilizing Eqs. (a) through (c).

$$84.5\frac{\text{W}}{°\text{C}}(T_2 + 40°\text{C}) = 2320\frac{\text{W}}{°\text{C}}(35°\text{C} - T_2)$$

$$T_2 = \frac{(2320\text{ W/°C})(35°\text{C}) - (84.5\text{ W/°C})(40°\text{C})}{2320\text{ W/°C} + 84.5\text{ W/°C}} = 32.4°\text{C}$$

Combining this result with Eqs. (c) and (d), we obtain

$$q_{\text{ref}} = q_r = kS(T_2 - T_1) = 84.5\frac{\text{W}}{°\text{C}}(32.4°\text{C} + 40°\text{C}) = 6120\text{ W}$$

Hence, approximately 6.12 kW of refrigeration is required.

As suggested in Sec. 1–2–4, superinsulation-type arrangements involving at least one evacuated space are often used in such cryogenic applications.

**EXAMPLE 1–6**

Although energy transfer by microwaves is generally not classified as thermal radiation because of the manner in which electromagnetic waves of this type are generated (see Chap. 5), microwave ovens are now widely and effectively used to cook many foods. As illustrated by Fig. E1–6, microwave cooking is made possible by the fact that energy supplied by electric current and converted to microwaves reflects from metal surfaces, passes through glass, ceramic, and plastic, but is readily absorbed and converted into internal energy by food molecules. Compare the heating (i.e., cooking) principles and characteristics of a microwave oven with conventional methods of cooking.

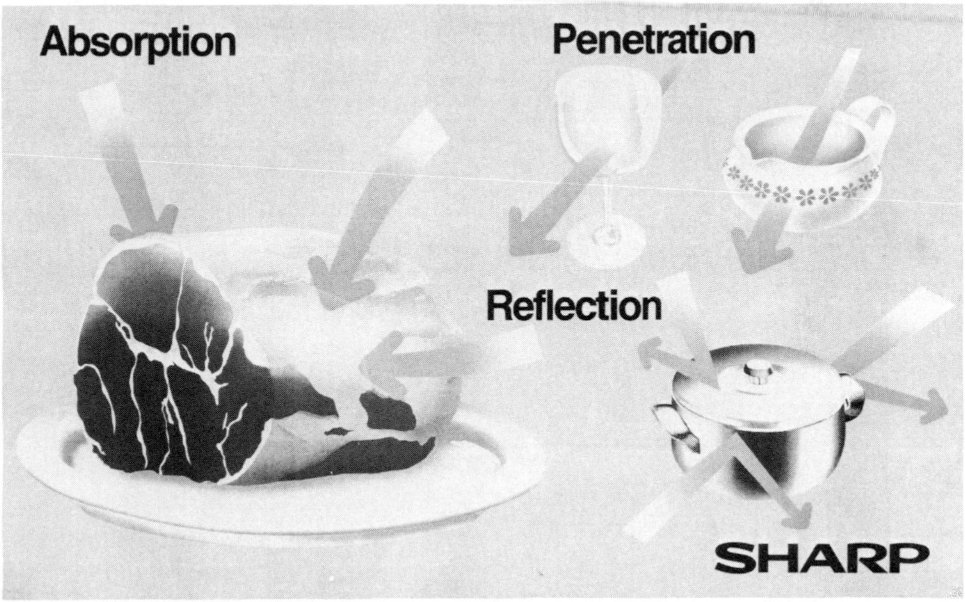


**FIGURE E1–6** Characteristics of microwaves. (Courtesy of Sharp Electronics Corporation.)

**Solution**

*Objective* Develop an understanding of the cooking principles of a microwave oven.

*Schematic* Heating principles and characteristics of range top, conventional oven, and microwave oven.

| Cooking Method | Heat Source | Characteristics |
|---|---|---|
| Range Top 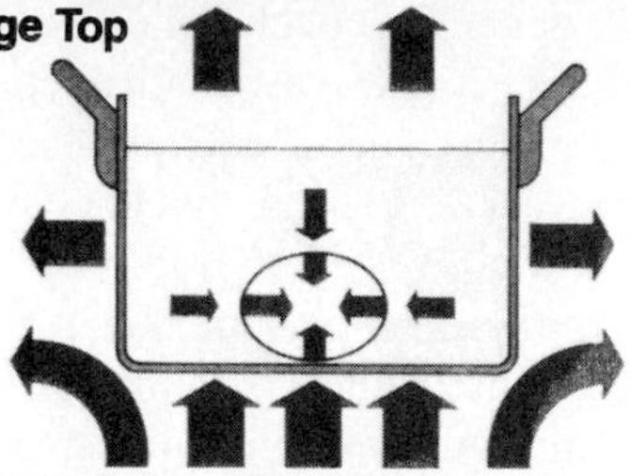 | Conduction heat | • Large energy loss<br>• Container and surrounding air are heated. |
| Conventional Oven 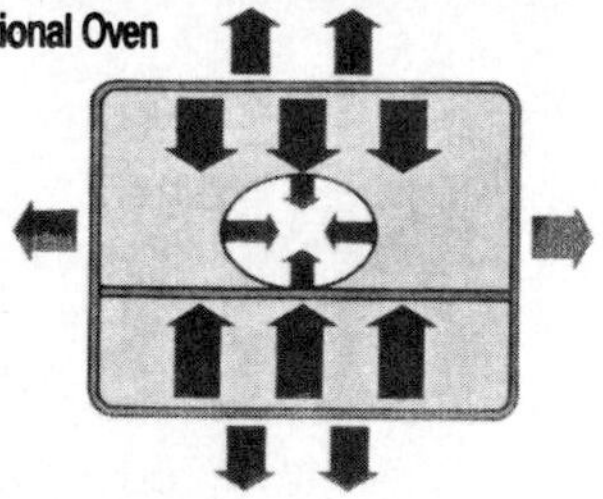 | Radiant heat | • Preheating required<br>• Container and surrounding air are heated. |
| Microwave Oven 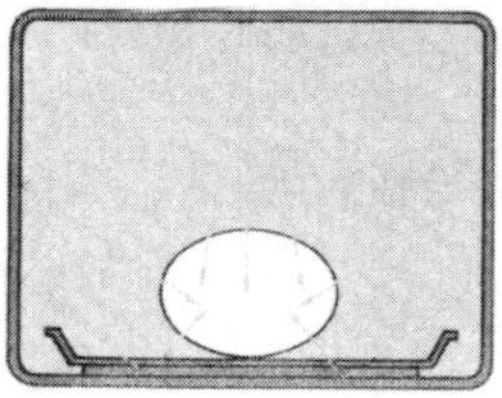 | Microwaves | • No preheating required<br>• Food heated directly for high thermal efficiency. |

*Analysis* Microwave ovens cook food by (1) the transfer of energy to the exposed surface of the food item via microwave radiation, (2) the absorption and conversion of microwave radiation into internal energy by the molecules of water, sugar, and fat near the surface of the item, and (3) transfer of energy into the interior of the item by conduction heat transfer. Because microwaves are reflected from metal surfaces and pass through glass and plastic cookware, the food is heated while the interior of the oven and the cooking tray stay relatively cool, with the result that no preheating is required and significantly greater cooking efficiency is achieved. On the other hand, conventional range top and oven cooking by means of conduction and thermal radiation involve large heat loss to the container and surrounding air and require oven preheating.

## 1–3 ANALOGY BETWEEN HEAT TRANSFER AND THE FLOW OF ELECTRIC CURRENT

The basic laws of conduction, thermal radiation, and convection heat transfer often lead to relationships for the heat transfer rate $q$ from a surface with area $A_s$ of the form†

$$q = \frac{\Delta T}{R} \tag{1–23}$$

where the *thermal resistance* $R$ is a function of the thermal and geometric characteristics. This equation is similar in form to the relationship for the flow of electric current $I_e$ through an electrical resistance $R_e$ with voltage drop $\Delta E_e$; that is,

$$I_e = \frac{\Delta E_e}{R_e} \tag{1–24}$$

A comparison of Eqs. (1–23) and (1–24) indicates that

$q$ is analogous to $I_e$,

$T$ is analogous to $E_e$,

and

$R$ is analogous to $R_e$.

The thermal resistance $R$ and the potential difference $\Delta T$ are dependent upon the heat-transfer mechanism considered. For example, Eq. (1–12) indicates that one-dimensional conduction heat transfer through a stationary medium with surface temperatures equal to $T_1$ and $T_2$ can be represented by a simple series circuit with resistance

$$R_k = \frac{1}{kS} \tag{1–25}$$

and potential difference $\Delta T = T_1 - T_2$. Similarly,

$$R_R = \frac{1}{\sigma A_s F_{s-R}(T_s + T_R)(T_s^2 + T_R^2)} \tag{1–26}$$

and $\Delta T = T_s - T_R$ for blackbody thermal radiation, and

† The corresponding relation for the rate of heat transfer $dq$ from a *differential area* $dA_s$ with finite temperature difference $\Delta T$ will be represented by

$$dq = \frac{\Delta T}{R'}$$

where $R'$ is the *local thermal resistance*.

$$R_c = \frac{1}{\bar{h}A_s} \tag{1-27}$$

and $\Delta T = T_s - T_F$ for convection. As illustrated in Examples 1–7 and 1–8, heat-transfer processes can be represented by electrical circuits. Such analogous electrical circuits often facilitate the solution of rather complex heat-transfer problems. As will be seen in Chaps. 2 through 4, electrical circuits can also be set up for unsteady and multidimensional systems.

### EXAMPLE 1–7

One hundred watts are generated by the flow of electric current in a 1-in.-diameter copper cable of 3-ft length. The surrounding air temperature is 75°F and $\bar{h}$ is equal to 40 Btu/(h ft$^2$ °F). Determine the surface temperature of the cable for negligible thermal-radiation losses.

#### Solution

*Objective* Determine the surface temperature $T_s$.

*Schematic* Electric conducting copper cable.

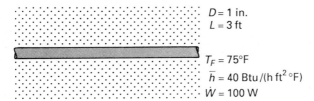

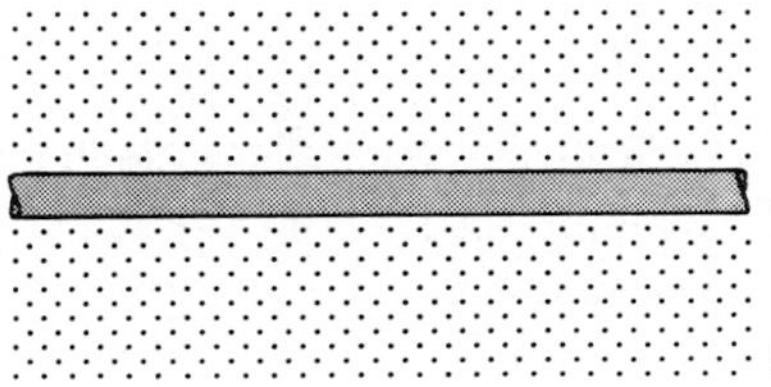

D = 1 in.
L = 3 ft
T_F = 75°F
h̄ = 40 Btu/(h ft² °F)
Ẇ = 100 W


*Assumptions/Conditions*

steady-state
one-dimensional
negligible thermal radiation

*Analysis* Applying the first law of thermodynamics to the cable, we have

$$q_c = \dot{W} = 100 \text{ W} = 341 \text{ Btu/h}$$

The analogous thermal circuit is shown in Fig. E1–7 for the case in which the radiation loss from the cable is negligible.

To obtain $T_s$, we write

$$q_c = \dot{W} = \frac{T_s - T_F}{R_c}$$

or

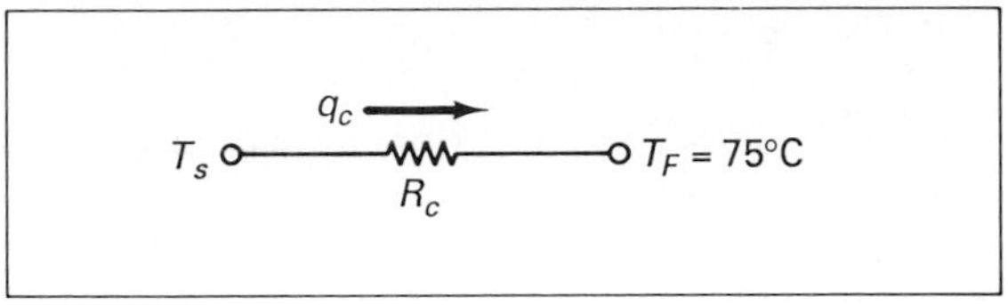


**FIGURE E1–7**
Thermal circuit; $R_c = 1/(\bar{h}A_s)$.

$$T_s = \dot{W}R_c + T_F = \frac{341 \text{ Btu/h}}{[40 \text{ Btu/(h ft}^2 \text{ °F)}][\pi(1 \text{ ft}/12)(3 \text{ ft})]} + 75°\text{F} = 85.8°\text{F}$$

## EXAMPLE 1–8

Because excessive temperature shortens the life of transistors, the heat transfer in transistors and compact integrated circuits (ICs, often called chips) that contain transistors, resistors, and capacitors is a critical design consideration. Consequently, manufacturers of transistors and ICs generally specify the maximum allowable internal junction temperature $T_J$ and/or case temperature $T_C$, and the thermal resistance from the junction to case $R_{JC}$ and from the case to ambient $R_{CA}$, assuming natural convection and radiative cooling. Whereas some silicon transistors and ICs can operate at temperatures as high as 200°C, the maximum temperature of components made of germanium is usually less than 100°C. In general, the life of a transistor or IC increases as its operating temperature decreases.

To illustrate, the thermal characteristics of the low-power silicon voltage regulator IC shown in Fig. E1–8a are specified by the manufacturer for an ambient temperature of 25°C as follows:

Maximum operating junction temperature: 150°C
Thermal resistance: $R_{JC} = 60$ °C/W
Thermal resistance: $R_{CA} = 90$ °C/W
Maximum power rating: 600 mW

We want to calculate the steady-state junction and case temperatures of this device at its maximum power output of 600 mW.

**FIGURE E1–8a** Integrated-voltage regulator (1/2 × 1 1/2 cm).

**Solution**

*Objective* Determine the temperature of the junction $T_J$ and case $T_C$ for steady-state conditions.

*Assumptions/Conditions*

steady-state
natural convection and radiation cooling

*Analysis* The thermal circuit for this system is shown in Fig. E1–8b. To calculate the junction temperature $T_J$ for 600-mW output, we write

$$q = \frac{T_J - T_F}{R_{JC} + R_{CA}}$$

$$T_J = q(R_{JC} + R_{CA}) + T_F = 0.6\text{ W}\left(60\frac{°\text{C}}{\text{W}} + 90\frac{°\text{C}}{\text{W}}\right) + 25°\text{C} = 115°\text{C}$$

This temperature is well within the safe operating range for the unit.

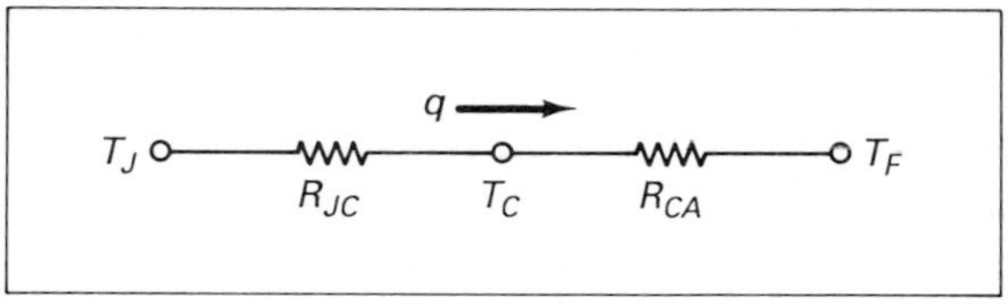


**FIGURE E1–8b**
Thermal circuit.

The case temperature $T_C$ at this maximum level of power dissipation is now calculated.

$$q = \frac{T_J - T_C}{R_{JC}}$$

$$T_C = T_J - qR_{JC} = 115°\text{C} - 0.6\text{ W}\left(60\frac{°\text{C}}{\text{W}}\right) = 79°\text{C}$$

## 1–4 SUMMARY

In this chapter we have reviewed the familiar fundamental laws upon which the science of heat transfer is based and we have introduced particular laws associated with conduction, thermal radiation, and convection heat transfer. As we have seen, the basic laws of conduction, thermal radiation, and convection provide the basis for a simple analogy between heat transfer and the flow of electric current and for the development of the practical analysis of basic heat-transfer problems. The principles introduced in this chapter will be used throughout the remainder of our study of heat transfer processes involving chemically homogeneous substances. The related topic of heat transfer and diffusion mass transfer in chemically nonhomogeneous substances is treated in references 7–9 and in the accompanying *mass transfer supplement*.

## REVIEW QUESTIONS

**1–1.** Define heat transfer.

**1–2.** Define temperature.

**1–3.** What is conduction heat transfer?

**1–4.** Write the Fourier law of conduction.

**1–5.** List several good conductors.

**1–6.** Write an expression for the rate of heat transfer $q$ through a plane wall with surface temperatures $T_1$ and $T_2$.

**1–7.** What is radiation heat transfer?

**1–8.** What is a blackbody?

**1–9.** Write the Stefan-Boltzmann law.

**1–10.** Write an expression for the rate of heat transfer $q$ between two blackbody parallel plates with surface temperatures $T_s$ and $T_R$.

**1–11.** What is convection heat transfer?

**1–12.** Write the Newton law of cooling.

**1–13.** Write the units for (a) thermal conductivity $k$ and (b) coefficient of heat transfer $\bar{h}$.

**1–14.** Write an expression for the rate of heat transfer $q$ in the form of an electrical analog.

**1–15.** Write relations for the thermal resistance $R$ associated with (a) conduction, (b) radiation, and (c) convection.

## PROBLEMS

**1–1.** One kilogram of helium at 300 K and 2 MPa contained in a piston and cylinder is heated at constant pressure to a temperature of 500 K. Determine the amount of heat that must be transferred. How does this value compare with the amount of heat that would be required to raise the temperature to 500 K in a constant-volume process?

**1–2.** Determine the rate of heat transfer per unit area through a 5 in.-thick brick wall with surface temperatures equal to 35°F and −20°F; $k = 0.399$ Btu/(h ft °F).

**1–3.** Determine the rate of heat transfer in a 5-cm-long hollow copper rod with radii equal to 1 cm and 5 cm. The surfaces at $r_1$ and $r_2$ are maintained at $T_1 = 25$°C and $T_2 = 75$°C, and the ends are insulated.

**1–4.** Determine the rate of heat transfer in the hollow cylinder of Prob. 1–3 if the surfaces at $r_1$ and $r_2$ are insulated and the ends are maintained at 25°C and 75°C, respectively.

**1–5.** Determine the rate of heat transfer in a copper spherical shell with radii equal to 3 cm and 6 cm if the surface temperatures are −120°C and 35°C.

**1–6.** An electrical device operating at 180°C with a 0.5-$m^2$ surface area is enclosed in a cabinet that has a surface temperature of 85°C. Determine the rate of radiation heat transfer between the two surfaces, assuming blackbody conditions.

**1–7.** A 1-mm-thick electrical heating plate generates 25 kW/$m^3$. One surface is insulated and the other is exposed to radiant exchange with a second surface at 70°C. Determine the surface temperature of the heater, assuming blackbody radiation.

**1–8.** A plate is exposed to a uniform heat flux of 100 Btu/(h $ft^2$) on one side. The other side exchanges thermal radiation with a surface at 0°F. Determine the temperature of the plate at the radiating surface.

**1–9.** A very thin copper plate coated with carbon black lies between two blackbody plates at temperatures of 1000°C and 50°C. Determine the temperature of the intermediate plate and the rate of heat transfer per unit area.

**1–10.** Assume that a fluid at 10°C passes over a plate with surface temperature equal to 100°C and with $\bar{h}$ equal to 150 W/($m^2$ °C). Determine the rate of heat transfer from 1 $m^2$ of the plate.

**1–11.** Water at atmospheric pressure and 30°F passes over a plate with 200°F surface temperature and $\bar{h}$ = 500 Btu/(h $ft^2$ °F). Sketch the analogous electrical circuit and determine the total quantity of heat transferred over a period of 30 s from a 1-$ft^2$ area.

**1–12.** Electricity passes through a 1-mm-diameter wire of 1 m length submerged in boiling water at a pressure of 20 psia. $\bar{h}$ is given as 4500 W/($m^2$ °C) and the resistance of the wire is 0.5 Ω. Determine the current required to maintain the wire surface at 115°C. Also determine the rate of heat transferred from the wire.

**1–13.** An electric current is passed through a wire 1 mm in diameter and 10 cm long. The wire is submerged in liquid water at atmospheric pressure and the current is increased until the water boils. For this situation $\bar{h}$ = 5000 W/($m^2$ °C). If the electric power supplied to the wire is 22 W, what is the surface temperature of the wire?

**1–14.** A very thin copper plate coated with carbon black exchanges radiant energy with a blackbody plate at 1000°C. The other surface is exposed to a fluid with $T_F$ equal to 50°C and $\bar{h}$ equal to 150 W/($m^2$ °C). Determine the rate of heat transfer per unit area and the surface temperature.

**1–15.** Determine the rate of radiation heat transfer between two concentric spheres with 2-cm and 10-cm diameters. The surfaces are coated with carbon black and maintained at 0°C and 150°C, respectively.

**1–16.** Determine the rate of radiation heat transfer per unit length between two very long blackbody concentric cylinders with 2-in. and 10-in. diameters. The surfaces are maintained at 100°F and 32°F, respectively.

**1–17.** Show that the thermal resistance associated with blackbody thermal radiation is given by $R_R = [\sigma A_s F_{s-R}(T_s^2 + T_R^2)(T_s + T_R)]^{-1}$.

**1–18.** Sketch the analogous electrical circuit for Example 1–4.

**1–19.** Solve Example 1–5 by the electrical analogy method.

**1–20.** Forced-convection air cooling with $\bar{h}$ = 500 W/($m^2$ °C) and radiation are used to maintain the cabinet surface of Example 1–4 at 125°C. Estimate the rate of heat transfer and the relative importance of radiation in this problem.

**1–21.** A house has a black tar roof that is well insulated on the lower surface. The upper surface is exposed to ambient air at 27°C with a convective heat-transfer coefficient of 10 W/($m^2$ °C). Calculate the temperature of the roof on a clear sunny day with an incident solar radiation flux of 500 W/$m^2$ and the ambient sky at an effective temperature of 50 K.

**1–22.** A thermocouple located at the center of a long circular duct with a 10-cm diameter has a spherical bead of 2-mm diameter. It indicates 185°C for a case in which the temperatures of the gas and duct are known to be 200°C and 140°C, respectively. Use this information to determine the convection coefficient for heat transfer between the gas and the bead, assuming radiating surfaces are black.

**1–23.** One hundred watts are generated by the flow of electric current in a 1-m-long, 1-cm-diameter cable that has been heavily oxidized. The surrounding air temperature is 50°C, $\bar{h}$ is 200 W/(m$^2$ °C), and blackbody-thermal-radiation exchange occurs between the cable and the surroundings at 50°C. Determine the surface temperature of the cable. Are the effects of thermal radiation important in this problem?

**1–24.** A power transistor is to be operated at 2 W. The manufacturer rating for the thermal resistance from the case to the ambient is 30 °C/W for natural convection cooling. Determine the temperature of the case if the air temperature is 25°C.

**1–25.** The junction-to-case thermal resistance of a germanium transistor is rated at 2 °C/W and the case-to-ambient thermal resistance is 30 °C/W. Determine the junction temperature if the power dissipation is 2 W and $T_F$ = 30°C. (Note that if the junction temperature of germanium transistors exceeds 100°C, burnout will often occur.)

**1–26.** Certain high-reliability silicon power transistors can operate at junction temperatures as high as 200°C without burning out. Determine the maximum power at which such a transistor can operate if the junction-to-case thermal resistance is 2.5 °C/W, the case-to-ambient thermal resistance is 75 °C/W, and $T_F$ = 25°C.

**1–27.** Estimate the level of importance of thermal radiation for the cable considered in Example 1–7.

**1–28.** Water beds have become very popular today. Because temperatures much above or below the skin temperature (about 90°F) cause discomfort, water beds are equipped with heaters that operate throughout the year to provide a temperature-controlled sleeping environment. Water bed heater manufacturers typically use 304 stainless steel electrical heating elements with about 36 Ω resistance. The average thermal resistance associated with heat loss from the warm water in a king-size mattress (84 in. × 72 in. × 6 in.) mounted on a bed and fitted with a comforter and two pillows has been reported to be about 0.03 °F h/Btu. Estimate the cost of heating a king-size water bed, assuming conditioned room air set points of 70°F for winter and 75°F for summer, water bed set points of 92°F for winter and 88°F for summer, and an electrical energy cost of $0.07/kWh.

**1–29.** According to the Tennessee Valley Authority, a family of six uses up to 150 gallons of hot water per day. As shown in Prob. 8–20, the annual energy requirement for heating this quantity of water from 55°F to 140°F is about 11,400 kWh/yr. The most common sources of energy used in residential and commercial water heaters are gas fuel and electric resistance elements. Based on tests conducted by GAMA (Gas Appliance Manufacturers Association), the operating cost of gas fuel models is about 50% of the cost of electric resistance models. Therefore, gas-fueled water heaters have generally been used when at all possible. Another less familiar method of providing hot water is to use a heat pump water heater (HPWH). Reliable HPWH units developed over the past decade provide coefficients of performance (COP) in the range of 2.1 to 3.6, depending on the operating conditions. Estimate the cost of providing an energy level of 11,400 kWh/yr to heat water by (a) electric resistance heating, and (b) electric heat pump, assuming a unit power cost of $0.07/kWh and a heat pump COP of 3.2. Also discuss the possible benefits of air cooling and dehumidification, which is provided by HPWH units.

How does the cost of operating an HPWH unit compare with the cost of gas heating? (Information pertaining to HPWH units can be obtained by contacting the electric power company or local heating and air-conditioning specialist.)

**1–30.** Two young people traveling in the early morning hours on a cold winter day were recently killed in a southwestern state when their car unexpectedly encountered ice on a bridge. Explain how it is possible for a bridge surface to ice over while the approaching roadway is ice free.

**1–31.** The ground near the surface of the earth acts as a heat sink during warm summer weather (as solar energy is collected and stored), and as a heat source during the winter. Referring to Fig. P1–31a, although the mean daily temperature distribution in the earth varies from a maximum during the summer to a minimum during the winter, the temperature beyond a depth of about 25 to 30 ft stays at a constant value (designated as the *annual mean earth temperature* $T_M$) throughout the year, regardless of the weather conditions. Soil temperature variation charts such as this are also characterized by the *amplitude of annual change in ground surface temperature* $\Delta T_s$, and the *phase constant* $t_0$ (i.e., the day of the year $T_s$ is a minimum). Earth temperature data for $T_M$, $\Delta T_s$, and $t_0$ are given in Table A–D–1 for selected cities in the U.S. Notice that relative to the outside air temperature $T_F$, the temperature several feet below the ground is cool during the summer and warm during the winter.

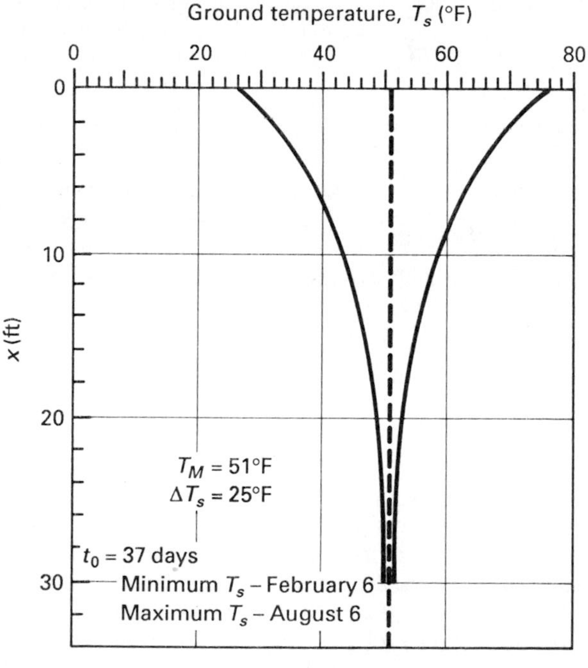


**FIGURE P1–31a**
Soil temperature variation chart: Chicago, IL (average soil conditions).

This favorable residual temperature difference provides the basis for developing ground source heat pump systems with heating and cooling efficiencies that are considerably higher than for air-to-air heat pump units. Ground source heat pump systems feature the use of buried plastic tubing through which liquid is circulated from the heat pump, as illustrated in Fig.

P1–31b. Explain briefly how such ground source heat pump systems operate and why high efficiencies can be expected.

**FIGURE P1–31b**
Typical horizontal loop ground source heat pump installation. (Courtesy of WaterFurnace International, Inc., Fort Wayne, IN).

**1–32.** Many skin disorders are either caused by or aggravated by exposure to ultraviolet and visible radiation from the sun. In this connection, the fact that the invisible ultraviolet rays are responsible for causing sunburn is well documented. The most serious consequence of long-term excessive exposure to ultraviolet radiation is the development of a malignant skin cancer known as *melanoma*. Although melanoma is generally fatal if left untreated, this type of cancer is completely curable when detected and removed in the early stages. List several means by which a person can guard against overexposure to ultraviolet solar radiation. Also describe the features of melanoma and list several warning signs.

# CHAPTER 2

# ONE-DIMENSIONAL HEAT TRANSFER

## 2–1 INTRODUCTION

The primary objective of this chapter is to establish the fundamental concepts involved in analyzing basic one-dimensional heat-transfer problems. These principles will first be demonstrated in the context of steady-state conduction heat transfer in a plane wall. One-dimensional analyses will then be developed for several other representative problems. This study will provide the foundation for our analysis of more complex multidimensional conduction, thermal radiation, and convection heat transfer in Chaps. 3 through 11.

## 2–2 CONDUCTION HEAT TRANSFER IN A PLANE WALL

We consider one-dimensional steady conduction heat transfer in the plane wall shown in Fig. 2–1(a). The classic differential approach to solving this problem is presented for several standard boundary conditions, after which an efficient but less general short method will be introduced.

### 2–2–1 Differential Formulation

To develop mathematical equations that represent the energy transfer within the wall, attention is focused upon the differential strip $A\ dx$ shown in Fig. 2–1(a). First, we recognize that $\Delta E_s/\Delta t$ is zero because of steady-state conditions, and conduction heat transfer in the $y$ and $z$ directions is zero because no temperature gradients occur in

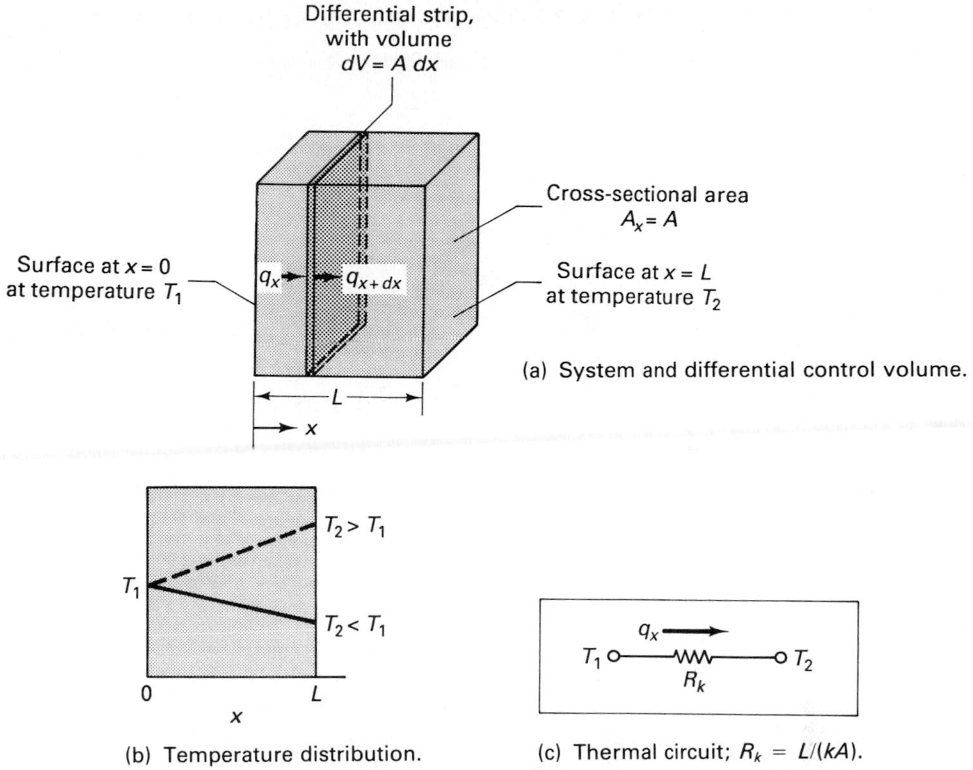


(a) System and differential control volume.

(b) Temperature distribution.

(c) Thermal circuit; $R_k = L/(kA)$.

**FIGURE 2–1** One-dimensional conduction heat transfer in a flat plate.

these directions. Thus, the temperature is a function of $x$ alone. With these points in mind, the application of the first law of thermodynamics given by Eq. (1–1) to the element $A\,dx$ gives

*Step 1*

$$q_x = q_{x+dx} \tag{2–1}$$

This equation indicates that $q_x$ does not change with $x$.

The next step in the development of the differential formulation requires the use of a basic relationship between $q_x$ and $q_{x+dx}$. This relationship is established on the basis of the definition of the derivative,

$$\frac{dq_x}{dx} = \lim_{\Delta x \to 0} \frac{q_{x+\Delta x} - q_x}{\Delta x} \tag{2–2}$$

By replacing $\Delta x$ by the infinitesimal quantity $dx$, this equation gives rise to

$$q_{x+dx} = q_x + \frac{dq_x}{dx}\,dx \tag{2–3}$$

Utilizing this expression, we eliminate $q_{x+dx}$ in Eq. (2–1) to obtain

*Step 2*
$$\frac{dq_x}{dx} = 0 \tag{2–4}$$

The quantity $q_x$ is now expressed in terms of the temperature distribution $T$ by introducing the Fourier law of conduction, Eq. (1–10); that is,

*Step 3*
$$\frac{d}{dx}\left(-kA_x\frac{dT}{dx}\right) = 0 \tag{2–5}$$

Because the cross-sectional area $A_x$ is independent of $x$ for this application ($A_x = A$), this equation reduces to

$$\frac{d}{dx}\left(k\frac{dT}{dx}\right) = 0 \tag{2–6}$$

For situations in which the thermal conductivity is essentially uniform, this expression takes the simple form

$$\frac{d^2T}{dx^2} = 0 \tag{2–7}$$

The differential formulation is completed by writing the boundary conditions. As the term implies, a boundary condition is a mathematical statement pertaining to the behavior of the dependent variable ($T$ in our case) at the system boundary. It should be recalled that the number of boundary conditions is equal to the highest-order derivative in an ordinary differential equation. Hence, two boundary conditions must be specified in order to solve steady one-dimensional conduction heat-transfer problems. For the moment, the surface temperatures are simply identified by $T_1$ and $T_2$,

$$T(0) = T_1 \qquad T(L) = T_2 \tag{2–8, 9}$$

The treatment of several common types of boundary conditions will be considered later in this section.

### 2–2–2 Solution

The solution to Eq. (2–7) is written on the basis of a simple double integration as

$$T = C_1x + C_2 \tag{2–10}$$

By applying the boundary conditions, the constants of integration $C_1$ and $C_2$ are written in terms of the surface temperatures as follows:

$$T_1 = C_2 \qquad T_2 = C_1L + C_2 \tag{2–11, 12}$$

or

$$C_1 = \frac{T_2 - T_1}{L} \tag{2–13}$$

Therefore, the temperature distribution takes the form

$$T = (T_2 - T_1)\frac{x}{L} + T_1$$

or

$$\frac{T - T_1}{T_2 - T_1} = \frac{x}{L} \tag{2–14}$$

This linear relationship is shown in Fig. 2–1(b) for situations in which $T_1 > T_2$ and $T_1 < T_2$. Because Eq. (2–14) does not involve $k$, we conclude that the temperature distribution is totally independent of the material (steel, wood, concrete, etc.).

An expression is obtained for the rate of heat transfer in the wall by utilizing the Fourier law of conduction together with Eq. (2–14); that is,

$$q_x = -kA\left.\frac{dT}{dx}\right|_x = \frac{kA}{L}(T_1 - T_2) \tag{2–15}$$

Consistent with Eq. (2–1), this equation indicates that the rate of heat transfer in the plate is independent of $x$.

Equation (2–15) provides the basis for the practical expressions given by Eqs. (1–12) and (1–23). For example, the thermal resistance $R_k$ is simply

$$R_k = \frac{L}{kA} \tag{2–16}$$

for a flat plate with uniform thermal conductivity. The thermal circuit associated with one-dimensional conduction heat transfer in a plate is shown in Fig. 2–1(c).

## EXAMPLE 2–1

Determine the temperature distribution and rate of heat transfer across a copper plate with cross-sectional area 1 m$^2$ and thickness 5 cm and with surface temperatures of 130°C and 15°C.

### Solution

*Objective* Determine the distribution in $T$ and $q_x$ for the plane wall shown in the schematic.

*Schematic* Temperature distribution in a plane wall.

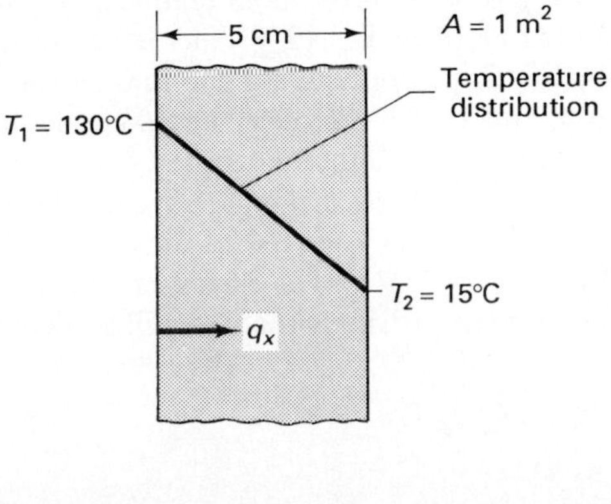

*Assumptions/Conditions*

steady-state
one-dimensional
uniform properties

*Properties* Copper at room temperature (Table A–C–1): $k = 386$ W/(m °C).

*Analysis* The temperature distribution is given by Eq. (2–14),

$$T = (15°\text{C} - 130°\text{C})\frac{x}{0.05\text{ m}} + 130°\text{C}$$

This linear distribution is shown in the schematic. The rate of heat transfer obtained from Eq. (2–15) is

$$q_x = 386\frac{\text{W}}{\text{m °C}}\left(\frac{1\text{ m}^2}{0.05\text{ m}}\right)(130°\text{C} - 15°\text{C}) = 888\text{ kW}$$

### 2–2–3 Boundary Conditions

For situations in which the surface temperatures $T_1$ and $T_2$ are known, the boundary conditions given by Eqs. (2–8) and (2–9), together with the differential equation, provide a complete mathematical model of the problem. However, for the many applications encountered in practice in which one or both surface temperatures are not known, the actual heat transfer at the boundaries must be accounted for. For such cases, the solutions for $T$ and $q$ developed above still apply, but $T_1$ and $T_2$ must be evaluated. Examples of other common types of boundary conditions include convection, thermal radiation, specified heat flux, and combinations of these. These standard types of boundary conditions will be considered in this section, as will another important type of boundary condition that involves interfacial conduction in composites.

Problems involving these other types of boundary conditions can be solved by any of several approaches. One way is to replace Eq. (2–8) and/or Eq. (2–9) by the formal mathematical statement of the boundary condition(s) and to then solve for the constants of integration. Thus, for a flat plate with uniform thermal conductivity, the boundary conditions are utilized to evaluate $C_1$ and $C_2$ in Eq. (2–10). In another somewhat simpler approach, Eqs. (2–8) and (2–9) are retained with the unknown surface temperatures $T_1$ and $T_2$ being determined by the use of the electrical analogy concept. For example, $R_k$ is given by Eq. (2–16) for a flat plate with uniform thermal conductivity. Both the formal and the thermal circuit approaches will be demonstrated in the sections that follow.

### *Standard Boundary Conditions*

For the case illustrated in Fig. 2–2(a), in which the surface at $x = 0$ is exposed to fluid with temperature $T_F$, the heat convected from the fluid is conducted into the wall. Consequently, the boundary condition at $x = 0$ is written as

$$q_c = q_x$$

$$\bar{h}[T_F - T(0)] = -k \left.\frac{dT}{dx}\right|_0 \tag{2–17}$$

where the surface temperature $T(0) = T_1$ is unknown. The other surface at $x = L$ is maintained at a known temperature $T_2$, such that the second boundary condition is given by Eq. (2–9).

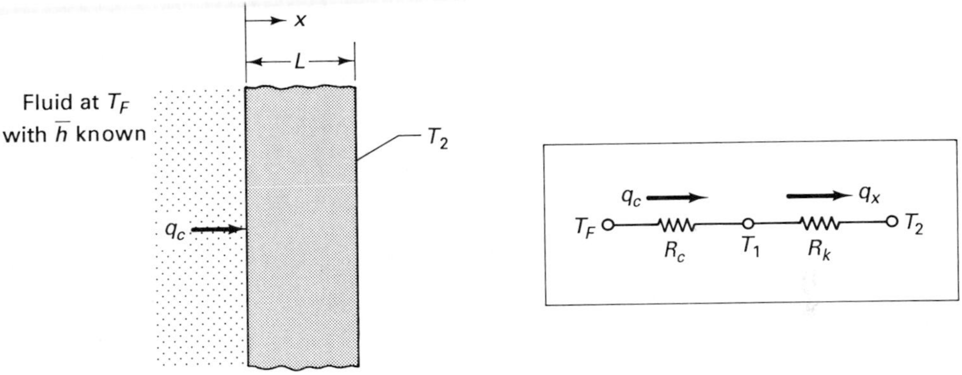


(a) System. (b) Thermal circuit.

**FIGURE 2–2** One-dimensional conduction heat transfer in a plane wall with convection at one surface.

At this point in our study, $\bar{h}$ is assumed to be known. Details concerning the evaluation of $\bar{h}$ are presented in Chaps. 6 through 11, which deal specifically with the study of convection heat transfer.

Substituting the solution for $T$ given by Eq. (2–10) into Eqs. (2–9) and (2–17), we obtain Eq. (2–12),

$$T_2 = C_1 L + C_2 \tag{2–12}$$

and

$$-kC_1 = \bar{h}(T_F - C_2) \tag{2–18}$$

The solution of these two equations for the two constants of integration gives rise to an expression for $T$ of the form

$$\frac{T - T_2}{T_F - T_2} = \frac{L - x}{L + k/\bar{h}} \tag{2–19}$$

Hence, the unknown surface temperature $T_1$ is given by

$$\frac{T_1 - T_2}{T_F - T_2} = \frac{L}{L + k/\bar{h}} = \frac{1}{1 + k/(\bar{h}L)} \tag{2–20}$$

With $T_1$ now known, the temperature distribution within the wall can also be conveniently expressed by Eq. (2–14),

$$\frac{T - T_1}{T_2 - T_1} = \frac{x}{L} \tag{2–14}$$

A more efficient approach to the solution of this convection-heat-transfer problem involves the use of the electrical analog. The series thermal circuit is shown in Fig. 2–2(b). An analysis of this elementary circuit indicates that the rate of heat transfer can be expressed as

$$q = \frac{T_F - T_2}{R_c + R_k} \tag{2–21}$$

where $R_c = 1/(\bar{h}A)$, $R_k = L/(kA)$, and $q = q_x = q_c$. This thermal circuit can also be solved for the surface temperature $T_1$ by writing

$$q = q_c = q_x$$

$$\frac{T_F - T_2}{R_c + R_k} = \frac{T_F - T_1}{R_c} = \frac{T_1 - T_2}{R_k} \tag{2–22}$$

By equating $q$ and $q_c$, we have

$$\frac{T_1 - T_F}{T_2 - T_F} = \frac{R_c}{R_c + R_k} = \frac{1}{1 + L\bar{h}/k} \tag{2–23}$$

Alternatively, by equating $q$ and $q_x$, we obtain Eq. (2–20), and by equating $q_c$ and $q_x$, we get

$$T_1 = \frac{T_F/R_c + T_2/R_k}{1/R_c + 1/R_k} = \frac{T_F\bar{h} + T_2k/L}{\bar{h} + k/L} \tag{2–24}$$

All three of these equations for $T_1$ are equivalent. Note that Eq. (2–24) can also be obtained directly from Eq. (2–17) by replacing $q_x$ by $(T_1 - T_2)/R_k$.

It should also be noted that the convective resistance is negligible for large values of $\bar{h}$. For this situation, Eq. (2–23) indicates that the surface temperature $T_1$ is simply

$$T_1 = T_F \tag{2–25}$$

Standard boundary conditions for one-dimensional steady-state convection, blackbody thermal radiation, and uniform heat flux are summarized in Table 2–1. Representative problems involving these boundary conditions are illustrated in Examples 2–2 through 2–4.

**TABLE 2–1** Standard boundary conditions for one-dimensional steady-state conduction heat transfer in a flat plate

| *Condition at boundary* | *Boundary condition—Mathematical statement* |
|---|---|
| Convection | |
| at $x = 0$ | $-k \left.\frac{dT}{dx}\right\|_0 = \bar{h}[T_F - T(0)]$ |
| at $x = L$ | $-k \left.\frac{dT}{dx}\right\|_L = \bar{h}[T(L) - T_F]$ |
| Blackbody thermal radiation | |
| at $x = 0$ | $-k \left.\frac{dT}{dx}\right\|_0 = \sigma F_{s-R}[T_R^4 - T(0)^4]$ |
| at $x = L$ | $-k \left.\frac{dT}{dx}\right\|_L = \sigma F_{s-R}[T(L)^4 - T_R^4]$ |
| Specified heat flux $q_s''$† | |
| at $x = 0$ | $-k \left.\frac{dT}{dx}\right\|_0 = q_s''$ |
| at $x = L$ | $-k \left.\frac{dT}{dx}\right\|_L = q_s''$ |
| Insulated surface, $q_s'' = 0$ | |
| at $x = 0$ | $-k \left.\frac{dT}{dx}\right\|_0 = 0$ |
| at $x = L$ | $-k \left.\frac{dT}{dx}\right\|_L = 0$ |

† $q_s''$ positive in $x$ direction.

## EXAMPLE 2–2

Two fluids are separated by a 2-in.-thick stainless steel plate [AISI 302] with an area of 10 ft$^2$. The fluid temperatures and mean coefficients of heat transfer are $T_{F1} =$ 50°F, $T_{F2} =$ 0°F, $\bar{h}_1 =$ 200 Btu/(h ft$^2$ °F), and $\bar{h}_2 =$ 150 Btu/(h ft$^2$ °F). Utilize the electrical analogy approach to determine the surface temperatures and the rate of heat transfer through the plate for negligible thermal radiation at the surfaces.

### Solution

***Objective*** Determine the surface temperatures $T_1$ and $T_2$ and $q$ for a plate with conditions shown in the schematic.

*Schematic* Two fluids separated by stainless steel plate.

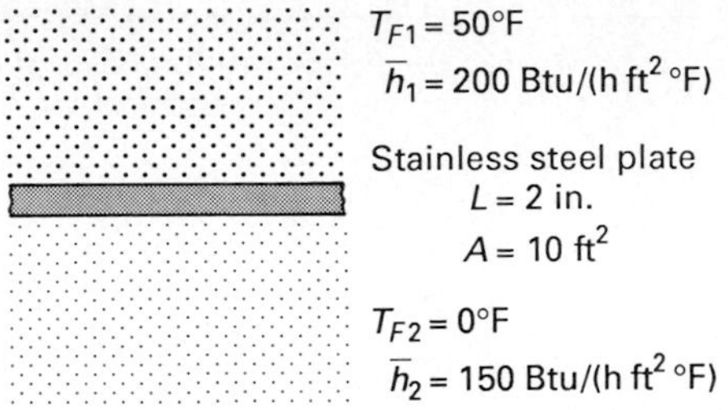

$T_{F1} = 50°F$
$\bar{h}_1 = 200$ Btu/(h ft$^2$ °F)
Stainless steel plate
$L = 2$ in.
$A = 10$ ft$^2$
$T_{F2} = 0°F$
$\bar{h}_2 = 150$ Btu/(h ft$^2$ °F)


*Assumptions/Conditions*

steady-state
one-dimensional
uniform properties
negligible thermal radiation

*Properties* Stainless steel AISI 302 near room temperature (Table A–C–1):

$$k = 15.1 \frac{\text{W}}{\text{m °C}} \frac{0.5778 \text{ Btu/(h ft °F)}}{\text{W/(m °C)}} = 8.72 \text{ Btu/(h ft °F)}$$

*Analysis* The thermal circuit for this arrangement is shown in Fig. E2–2. The thermal resistances are

$$R_{c1} = \frac{1}{\bar{h}_1 A} \qquad R_k = \frac{L}{kA} \qquad R_{c2} = \frac{1}{\bar{h}_2 A}$$

Solving for $q$, we obtain

$$q = \frac{T_{F1} - T_{F2}}{R_{c1} + R_k + R_{c2}} = \frac{(50°\text{F} - 0°\text{F})A}{\left[\frac{1}{200} + \frac{2}{8.72\,(12)} + \frac{1}{150}\right] \frac{\text{h ft}^2\ °\text{F}}{\text{Btu}}} = 16{,}200 \text{ Btu/h}$$

$$q'' = \frac{q}{A} = 1620 \text{ Btu/(h ft}^2)$$

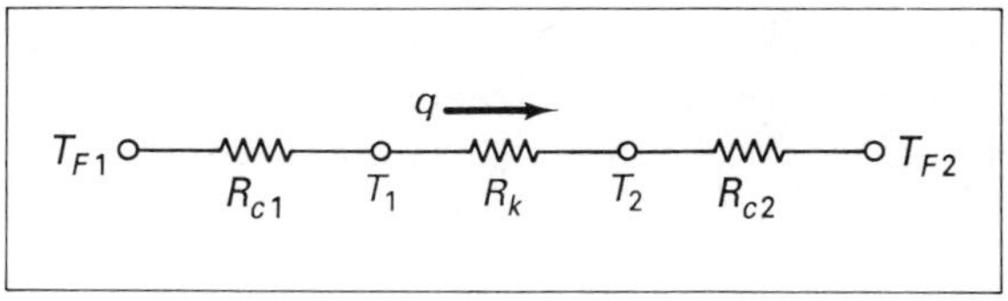

$q$
$T_{F1}$
$R_{c1}$
$T_1$
$R_k$
$T_2$
$R_{c2}$
$T_{F2}$


**FIGURE E2–2**
Thermal circuit.

The temperature distribution within the plate is given by Eq. (2–14),

$$\frac{T - T_1}{T_2 - T_1} = \frac{x}{L}$$

$T_1$ and $T_2$ are obtained as follows:

$$q = \frac{T_{F1} - T_1}{R_{c1}}$$

$$T_1 = T_{F1} - R_{c1}q = 50°\text{F} - \frac{16{,}200 \text{ Btu/h}}{[200 \text{ Btu/(h ft}^2\text{ °F)}](10 \text{ ft}^2)} = 41.9°\text{F}$$

$$q = \frac{T_2 - T_{F2}}{R_{c2}}$$

$$T_2 = T_{F2} + R_{c2}q = 0°\text{F} + \frac{16{,}200 \text{ Btu/h}}{[150 \text{ Btu/(h ft}^2\text{ °F)}](10 \text{ ft}^2)} = 10.8°\text{F}$$

## EXAMPLE 2–3

One surface of the plate shown in Fig. E2–3a is exposed to blackbody thermal radiation with $T_R = 1000°\text{C}$. The other surface is maintained at a temperature $T_2 = 15°\text{C}$. Obtain a solution for the unknown surface temperature at the radiating surface and the heat transfer rate for uniform property conditions.

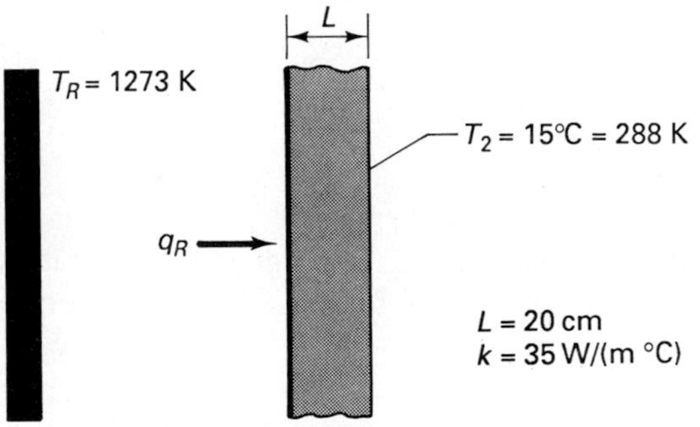
L
$T_R$ = 1273 K
$T_2$ = 15°C = 288 K
$q_R$
L = 20 cm
k = 35 W/(m °C)

**FIGURE E2–3a**
Plane wall with blackbody thermal radiation.

### Solution

*Objective* Determine the surface temperature $T_1$ and $q_x$ for a plate with conditions shown in Fig. E2–3a.

*Assumptions/Conditions*

steady-state
one-dimensional
uniform properties
blackbody thermal radiation

*Properties* Plate material: $k = 35$ W/(m °C).

*Analysis* The boundary condition at the surface which is exposed to thermal radiation is given by

$$q_R = q_x$$

$$\sigma F_{s-R} A_s (T_R^4 - T_1^4) = -kA \left.\frac{dT}{dx}\right|_0 \tag{a}$$

where $T(0)$ is denoted by $T_1$ and $A_s = A_x = A$. The thermal circuit for this problem is shown in Fig. E2–3b, where $R_R = [\sigma F_{s-R} A_s (T_R + T_1)(T_R^2 + T_1^2)]^{-1}$.

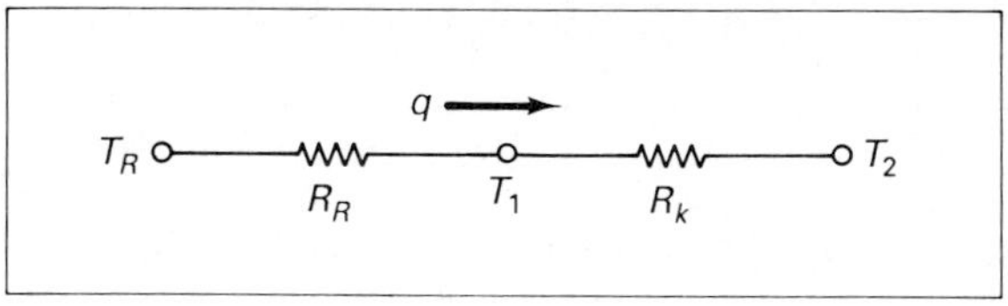


**FIGURE E2–3b**
Thermal circuit.

We have shown that the rate of heat transfer through the wall by conduction $q_x$ can be written as

$$q_x = \frac{kA}{L}(T_1 - T_2) \tag{b}$$

Thus, Eq. (a) reduces to

$$\sigma F_{s-R}(T_R^4 - T_1^4) = \frac{k}{L}(T_1 - T_2) \tag{c}$$

An iterative approach is suggested for solving this nonlinear equation for the unknown $T(0) = T_1$. This simple approach is outlined as follows: (1) rewrite Eq. (c) in the form (to four significant figures)

$$T_1 = T_2 + F_{s-R}\frac{\sigma L}{k}(T_R^4 - T_1^4) = \frac{T_2 + F_{s-R}(\sigma L/k)T_R^4}{1 + F_{s-R}(\sigma L/k)T_1^3}$$
$$= \frac{1139 \text{ K}}{1 + 3.24 \times 10^{-10}\, T_1^3/\text{K}^3} \tag{d}$$

(2) compute an approximate value of $T_1$ by substituting an assumed value of $T_1$ into the right-hand side of Eq. (d), (3) compute a new approximate value of $T_1$ by substituting the value computed in step 2 into the right-hand side of Eq. (d), (4) continue this procedure until the calculations for $T_1$ converge satisfactorily. With $T_1$ known, the rate of heat transfer can be calculated from Eq. (b).

For our particular problem, we start the iteration sequence with $T_1$ set equal to 1000 K. The convergence is slow, with the first several values of $T_1$ being

$$T_1^0 = 1000\text{ K} \qquad T_1^1 = 860.3\text{ K} \qquad T_1^2 = 944.2\text{ K} \qquad T_1^3 = 894.9\text{ K}$$

Continuing this iteration procedure, our calculations for $T_1$ converge to 913.4 K after 18 iterations.

The rate of heat transfer per unit area is

$$q_x'' = \frac{q_x}{A} = \frac{k}{L}(T_1 - T_2) = \frac{35\text{ W/(m °C)}}{0.2\text{ m}}(913.4\text{ K} - 288\text{ K})$$
$$= 1.094 \times 10^5\text{ W/m}^2 \tag{e}$$

To double-check, we calculate $q_R''$.

$$q_R'' = \frac{q_R}{A} = \sigma F_{s-R}(T_R^4 - T_1^4)$$
$$= 5.67 \times 10^{-8}\frac{\text{W}}{\text{m}^2\text{ K}^4}[(1273\text{ K})^4 - (913.4\text{ K})^4] \tag{f}$$
$$= 1.094 \times 10^5\text{ W/m}^2$$

The agreement between $q_x''$ and $q_R''$ assures us of a proper solution.

Returning briefly to the iterative solution for $T_1$, Eq. (d) can be written in the alternative forms

$$T_1 = 1139\text{ K} - 3.24 \times 10^{-10}\,T_1^4/\text{K}^3 \tag{g}$$

and

$$T_1 = \left(\frac{1139\text{ K} - T_1}{3.24 \times 10^{-10}/\text{K}^3}\right)^{1/4} \tag{h}$$

The use of Eq. (g) gives rise to a convergent iterative solution for $T_1$ of 913.4 K. But the convergence is much slower than for Eq. (d). On the other hand, the use of Eq. (h) produces iterative calculations for $T_1$ that diverge! A formal criterion for the iterative convergence or divergence of equations such as Eqs. (d), (g), and (h) is given in reference 1. However, it is generally a simple matter to perform a few calculations to determine whether convergence is being achieved.

It should also be mentioned that more efficient higher-order iterative schemes are available. For example, the Newton–Raphson method only requires three iteration steps. The use of this approach is suggested as an exercise.

### EXAMPLE 2–4

A 1-in.-thick carbon steel plate with a 1-ft$^2$ cross-sectional area is exposed to a uniform heat flux of 5000 Btu/(h ft$^2$). (This type of boundary condition can be achieved by heating the surface with an electrical heating plate or by thermal radiation with $T_R \gg T_s$.) The other surface of the plate is maintained at a temperature of 212°F. Determine the unknown surface temperature.

**Solution**

*Objective* Determine the surface temperature $T_1$ for a plate with conditions shown in the schematic.

*Schematic* Plane wall with uniform heat flux.

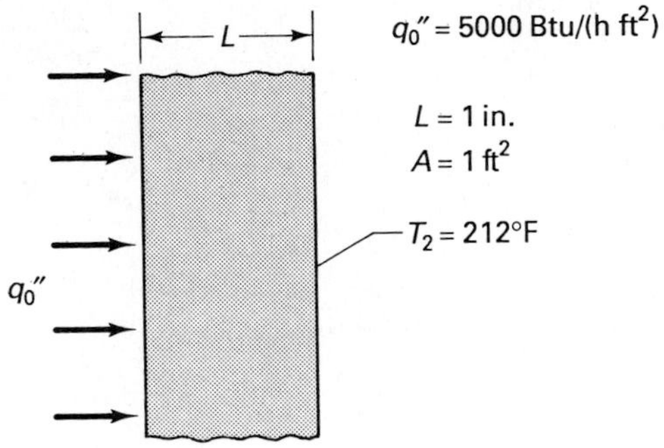
L
q₀″ = 5000 Btu/(h ft²)
L = 1 in.
A = 1 ft²
T₂ = 212°F
q₀″

*Assumptions/Conditions*

steady-state
one-dimensional
uniform properties

*Properties* Carbon steel (1% C) (Table A–C–1):

$$k = 43 \frac{\text{W}}{\text{m °C}} \frac{0.5778 \text{ Btu/(h ft °F)}}{\text{W/(m °C)}} = 24.8 \text{ Btu/(h ft °F)}$$

*Analysis* The thermal circuit for this problem is shown in Fig. E2–4, where $q_x = q_0$ and where $T(0) = T_1$ is unknown. Based on this simple circuit, we write

$$q_0'' = \frac{T_1 - T_2}{L/k}$$

such that $T_1$ is given by

$$T_1 = \frac{L}{k} q_0'' + T_2$$

Substituting into this equation, we obtain

$$T_1 = \frac{1 \text{ ft}/12}{24.8 \text{ Btu/(h ft °F)}} \frac{5000 \text{ Btu}}{\text{h ft}^2} + 212\text{°F} = 229\text{°F}$$

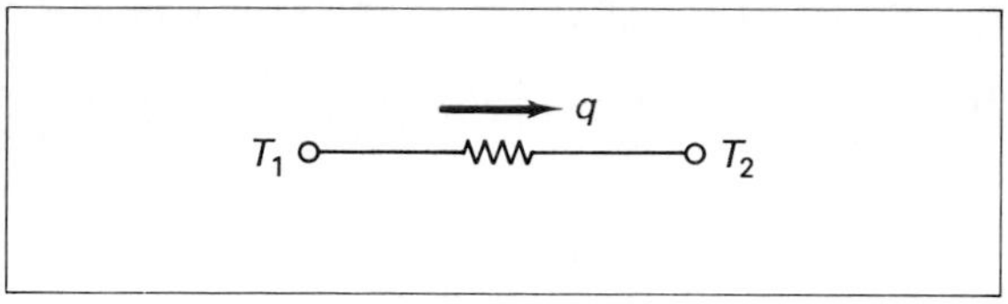


**FIGURE E2–4**
Thermal circuit.

### *Composite Walls*

We now consider the interfacial condition associated with conduction heat transfer within a wall that is composed of several layers of different materials. This situation is illustrated in Figs. 2–3(a) and 2–4(a) for two types of joints. We designate the temperature distributions in these two materials by $T_I$ and $T_{II}$. Because this composite actually involves two systems, materials I and II, we must write two boundary conditions for each material. Consequently, because the interface is a part of both materials, two interfacial boundary conditions are prescribed. Hence, for composite solids with perfect thermal contact, one interfacial boundary condition is

$$T_I(0) = T_{II}(0) \tag{2–26}$$

This situation in which the temperature distribution in the entire composite is continuous is illustrated in Fig. 2–3(b). The second boundary condition is written on the basis of the first law of thermodynamics, which states that the rate of energy conducted into the interface must be equal to the rate of energy conducted out; that is,

$$q_{Ix} = q_{IIx}$$

$$-k_I \frac{dT_I}{dx}\bigg|_0 = -k_{II}\frac{dT_{II}}{dx}\bigg|_0 \tag{2–27}$$

This statement is synonymous with the Kirchhoff law for electrical current flow at a junction. The thermal circuit for this composite is shown in Fig. 2–3(c).

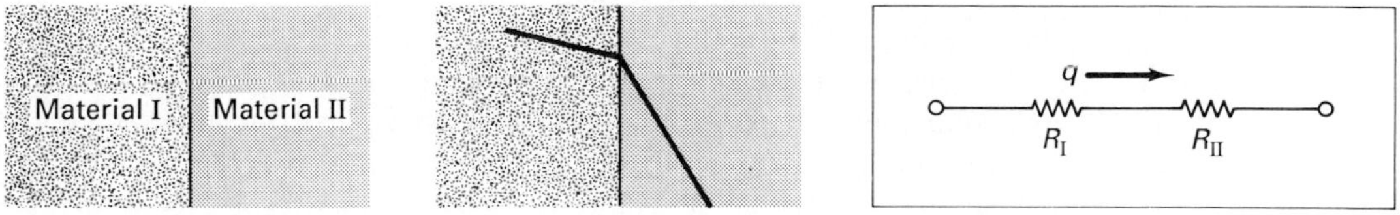


(a) System interface. (b) Representative temperature distribution. (c) Thermal circuit.

**FIGURE 2–3** Composite with perfect thermal contact.

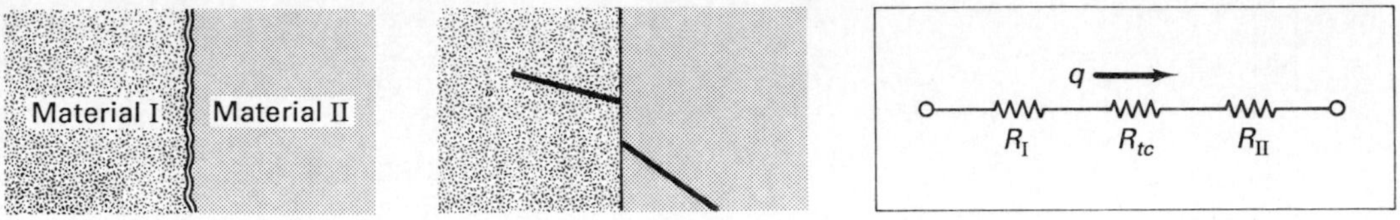


(a) System interface. (b) Representative temperature distribution. (c) Thermal circuit.

**FIGURE 2–4** Composite with imperfect thermal contact.

For situations in which an imperfect mechanical joint is made because of surface roughness, a discontinuity occurs in the temperature distribution at the interface, as shown in Fig. 2–4(b). For such cases, the interfacial temperatures in materials I and II are related empirically through an equation of the form

$$q_{tc} = h_{tc}A[T_I(0) - T_{II}(0)] \tag{2–28}$$

where $h_{tc}$ is called the *thermal contact coefficient*. The first law of thermodynamics still applies, such that we obtain the following two interfacial boundary conditions:

$$-k_I A \left.\frac{dT_I}{dx}\right|_0 = h_{tc}A[T_I(0) - T_{II}(0)] \tag{2–29}$$

$$h_{tc}A[T_I(0) - T_{II}(0)] = -k_{II} A \left.\frac{dT_{II}}{dx}\right|_0 \tag{2–30}$$

The thermal circuit for this type of problem is shown by Fig. 2–4(c); the thermal contact resistance $R_{tc}$ is equal to $1/(h_{tc}A)$. As $R_{tc}$ becomes small, this thermal circuit reduces to the circuit shown in Fig. 2–3(c).

The thermal contact coefficient $h_{tc}$ is dependent upon the material, surface roughness, contact pressure, and temperature. Experimental data for $h_{tc}$ are available in references 2 and 3 for standard materials such as aluminum, copper, and stainless steel. For example, $h_{tc}$ ranges from 5 to 50 kW/(m$^2$ °C) for various aluminum surfaces at 200°C with contact pressure between 1 and 30 atm. Thermal contact coefficients are generally much smaller for stainless steel [order of 3 kW/(m$^2$ °C)] and much larger for copper [order of 150 kW/(m$^2$ °C)].

A practical means of reducing the thermal contact resistance is to insert a material of good thermal conductivity between the two surfaces. Thermal greases containing silicon have been developed for this purpose. Thin soft metal foil can also be used for certain applications.

## EXAMPLE 2–5

A germanium power transistor is capable of operating at up to 5 W. The manufacturer's rating for the case-to-ambient thermal resistance of the transistor is 30 °C/W and the case temperature is not to exceed 80°C. In order to lower its operating temperature for a given power input, the transistor is to be mounted to a black anodized aluminum frame which serves as a heat sink, as shown in Fig. E2–5a. The frame provides an additional 1600-$mm^2$ surface area for cooling. To minimize the thermal contact resistance between the transistor and the heat sink and at the same time maintain proper electrical insulation, the surfaces are first cleaned and coated with a silicon grease (such as Dow-Corning Silicon Heat Sink Compound 340). A special electrical insulating gasket made of mica, beryllium oxide, or anodized aluminum is then inserted and the transistor is bolted tightly to the frame. As a rule of thumb, for proper thermal contact the thermal contact resistance is of the order of 0.5 °C/W.

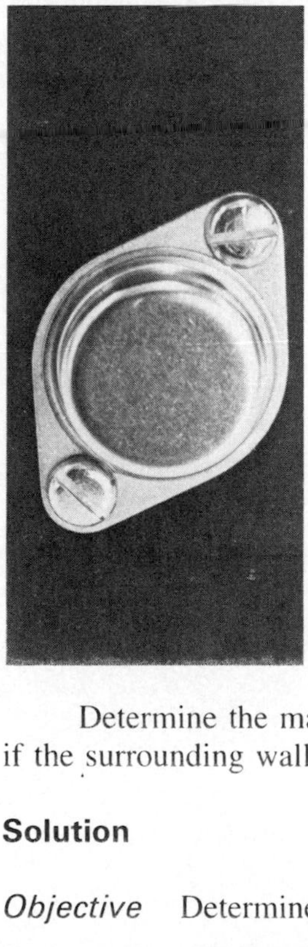

**FIGURE E2–5a**
Power amplifier mounted on aluminum frame.

Determine the maximum power at which the transistor can safely be operated if the surrounding walls and air are at 25°C and $\bar{h}$ = 10 W/($m^2$ °C).

## Solution

*Objective* Determine the maximum safe operating power of the transistor.

*Schematic* Germanium power transistor on anodized aluminum frame.

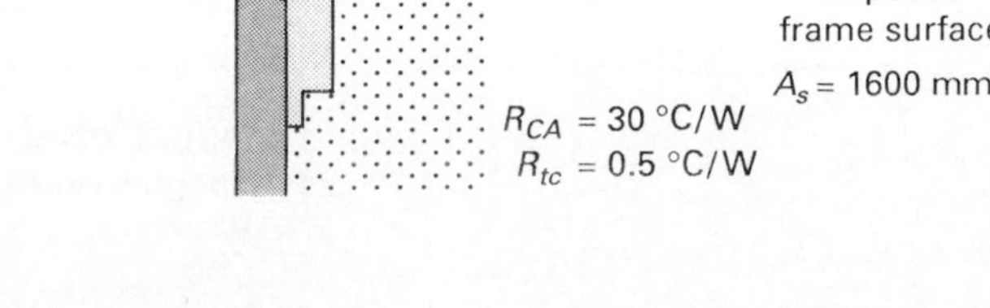

$T_C$ = 80°C
$T_F = T_R$ = 25°C
$\bar{h}$ = 10 W/($m^2$ °C)
Exposed
frame surface
$A_s$ = 1600 $mm^2$
$R_{CA}$ = 30 °C/W
$R_{tc}$ = 0.5 °C/W

*Assumptions/Conditions*

steady-state

convection and blackbody thermal radiation cooling

*Analysis* With the transistor case temperature $T_C$ equal to 80°C, the rate of heat transferred directly to the surroundings is

$$q_{CA} = \frac{T_C - T_F}{R_{CA}} = \frac{80°\text{C} - 25°\text{C}}{30\ °\text{C/W}} = 1.83\ \text{W}$$

In addition, heat is dissipated through the heat sink. The thermal circuit for the heat transfer through the heat sink (assuming negligible resistance to conduction) is shown in Fig. E2–5b. Assuming that blackbody conditions are approximated, $R_R$ is given by

$$R_R = \frac{1}{\sigma A_s F_{s-R}(T_s + T_R)(T_s^2 + T_R^2)}$$

where $T_s$ is the surface temperature of the frame. Setting $T_s$ equal to $T_C$ as a first approximation, we have

$$R_R = \frac{1}{\sigma(1600 \times 10^{-6}\ \text{m}^2)(1)(353\ \text{K} + 298\ \text{K})[(353\ \text{K})^2 + (298\ \text{K})^2]}$$
$$= 79.3\ °\text{C/W}$$

$R_c$ is given by

$$R_c = \frac{1}{\bar{h}A_s} = \frac{1}{[10\ \text{W/(m}^2\ °\text{C)}](1600 \times 10^{-6}\ \text{m}^2)} = 62.5\ °\text{C/W}$$

To calculate the equivalent resistance $R_{HS}$ for the heat transfer from the heat sink, we write

$$\frac{1}{R_{HS}} = \frac{1}{R_R} + \frac{1}{R_c} = \frac{1}{79.3\ °\text{C/W}} + \frac{1}{62.5\ °\text{C/W}}$$
$$R_{HS} = 35\ °\text{C/W}$$

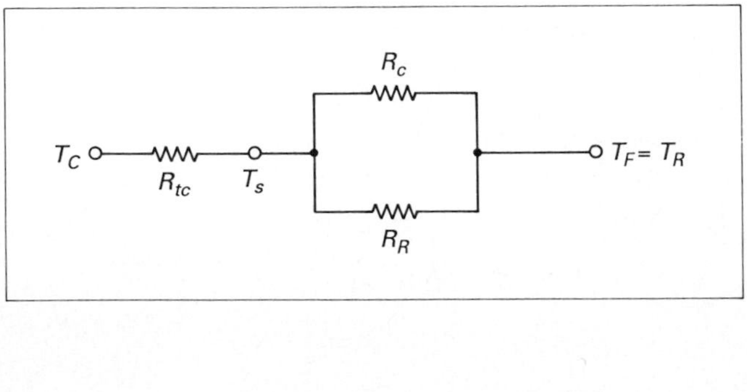


**FIGURE E2–5b** Thermal circuit.

It follows that the rate of heat transferred from the power amplifier through the heat sink is

$$q_{HS} = \frac{T_C - T_F}{R_{tc} + R_{HS}} = \frac{80°\text{C} - 25°\text{C}}{0.5\ °\text{C/W} + 35\ °\text{C/W}} = 1.55\ \text{W}$$

Based on this result, the temperature of the heat sink is calculated.

$$q_{HS} = \frac{T_C - T_s}{R_{tc}}$$

$$T_s = T_C - q_{HS}R_{tc} = 80°\text{C} - 1.55\ \text{W}\left(0.5\,\frac{°\text{C}}{\text{W}}\right) = 79.2°\text{C}$$

Because the difference between $T_s$ and $T_C$ is so small, our approximation for $R_R$ is quite adequate.

Thus, the total power that can be safely dissipated from the transistor/heat-sink unit is estimated to be

$$q = q_{CA} + q_{HS} = 1.83\ \text{W} + 1.55\ \text{W} = 3.38\ \text{W}$$

Without the heat sink, the transistor could only be operated at 1.83 W. For operation at the 5-W level, a larger more efficient heat sink would be required.

To compare the rate of heat transfer by radiation and convection from the heat sink, we write

$$q_R = \frac{T_s - T_F}{R_R} = \frac{79.2°\text{C} - 25°\text{C}}{79.3\ °\text{C/W}} = 0.683\ \text{W}$$

$$q_c = \frac{T_s - T_F}{R_c} = \frac{79.2°\text{C} - 25°\text{C}}{62.5\ °\text{C/W}} = 0.867\ \text{W}$$

Thus, radiation accounts for about 44% of the heat transfer from the heat sink.

The remainder of our study of composites will be restricted to cases in which the thermal contact resistance is assumed to be negligible. Three basic situations to be considered include series, parallel, and combined series–parallel arrangements.

***Series Arrangements*** A composite wall that is composed of two materials in series is shown in Fig. 2–5(a). For situations in which the surface temperatures are uniform, the heat transfer in composite walls that consist of two or more materials in series is always one-dimensional. The analogous electrical circuit for this two-wall series arrangement is shown in Fig. 2–5(b). The heat-transfer rate is simply

$$q_x = \frac{T_1 - T_3}{R_{\text{I}} + R_{\text{II}}} \tag{2–31}$$

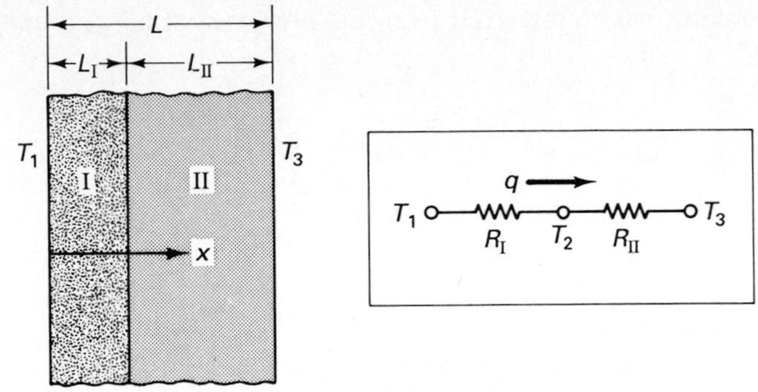


(a) System. (b) Thermal circuit.

**FIGURE 2–5** One-dimensional conduction heat transfer in a composite plane wall: series arrangement.

where $R_I = L_I/(k_I A)$ and $R_{II} = L_{II}/(k_{II} A)$. The unknown interfacial temperature $T_2$ can be expressed as

$$T_1 - T_2 = q_x R_I = (T_1 - T_3)\frac{R_I}{R_I + R_{II}} \tag{2–32}$$

Equations can also be written for the temperature profile within each material as

$$\frac{T_{Ix} - T_1}{T_2 - T_1} = \frac{x}{L_I} \qquad \frac{T_{IIx} - T_2}{T_3 - T_2} = \frac{x - L_I}{L_{II}} \tag{2–33, 34}$$

These equations both come directly from Eq. (2–14).

***Parallel Arrangements*** A simple parallel composite of two materials is shown in Fig. 2–6(a). For this situation, the heat transfer is one-dimensional and can be rep-

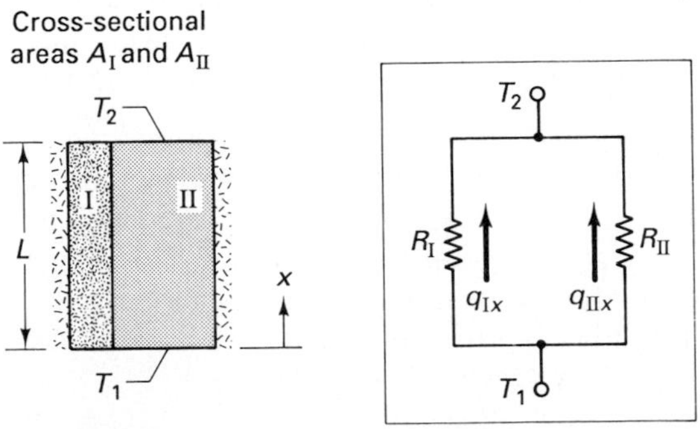


(a) System. (b) Thermal circuit.

**FIGURE 2–6** One-dimensional conduction heat transfer in a composite plane wall: parallel arrangement.

resented by the electrical circuit shown in Fig. 2–6(b). The solution for the total heat-transfer rate in this parallel network can be written as

$$q_x = \frac{T_{\text{I}} - T_2}{R_k} \tag{2–35}$$

where the equivalent parallel resistance is

$$\frac{1}{R_k} = \frac{1}{R_{\text{I}}} + \frac{1}{R_{\text{II}}} = \frac{k_{\text{I}}A_{\text{I}}}{L} + \frac{k_{\text{II}}A_{\text{II}}}{L} \tag{2–36}$$

This same result is achieved by recognizing that $q_x$ is equal to the sum of the heat-transfer rates in the individual materials; that is,

$$q_x = q_{\text{I}x} + q_{\text{II}x} = (T_1 - T_2)\left(\frac{1}{R_{\text{I}}} + \frac{1}{R_{\text{II}}}\right) \tag{2–37}$$

The temperature profile in each material is given by [from Eq. (2–14)]

$$T_{\text{I}x} - T_1 = (T_2 - T_1)\frac{x}{L} \qquad T_{\text{II}x} - T_1 = (T_2 - T_1)\frac{x}{L} \tag{2–38, 39}$$

The fact that these equations indicate that $T_{\text{I}x} = T_{\text{II}x}$ is consistent with the observation that this is a one-dimensional heat-transfer problem.

***Combined Series–Parallel Arrangements*** A composite wall that provides combined series and parallel paths for the heat transfer is illustrated in Fig. 2–7(a). Such heat-transfer problems are almost always two- or three-dimensional. However, approximate solutions to these types of problems can be obtained under certain conditions

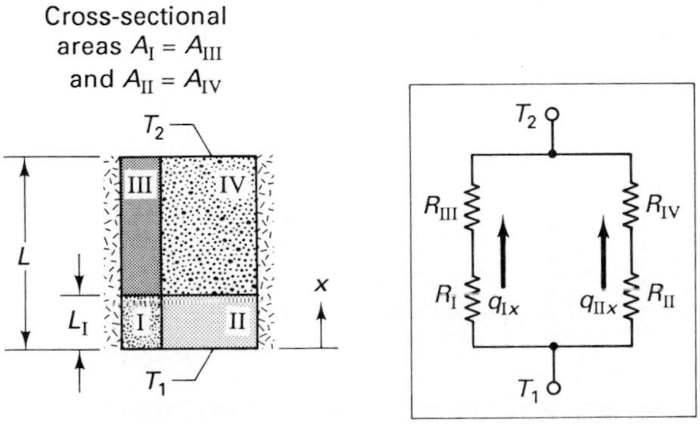


(a) System. (b) Approximate one-dimensional thermal circuit.

**FIGURE 2–7** Heat transfer in a composite plane wall: series-parallel arrangement.

by assuming one-dimensional heat transfer. For approximate one-dimensional heat transfer, the thermal circuit is given by Fig. 2–7(b). For this situation, the total rate of heat transfer can be expressed as

$$q_x = q_{\text{I}x} + q_{\text{II}x} \tag{2–40}$$

where $q_{\text{I}x}$ and $q_{\text{II}x}$ are given by

$$q_{\text{I}x} = q_{\text{III}x} = \frac{T_1 - T_2}{R_{\text{I}} + R_{\text{III}}} \qquad q_{\text{II}x} = q_{\text{IV}x} = \frac{T_1 - T_2}{R_{\text{II}} + R_{\text{IV}}} \tag{2–41, 42}$$

An approximate criterion for which a one-dimensional analysis is appropriate is given in Fig. 2–8 for a representative system involving four thermal resistances. This figure indicates that the heat transfer is essentially one-dimensional for situations in which $R_{\text{I}}/R_{\text{III}} \simeq R_{\text{II}}/R_{\text{IV}}$. In addition, this criterion is satisfied for small values of both $R_{\text{I}}/R_{\text{III}}$ and $R_{\text{II}}/R_{\text{IV}}$, and for large values of both $R_{\text{I}}/R_{\text{III}}$ and $R_{\text{II}}/R_{\text{IV}}$. For example, with $R_{\text{II}}/R_{\text{IV}} = 10$, the error is of the order of 1% for $R_{\text{I}}/R_{\text{III}} > 3$ and about 10% for

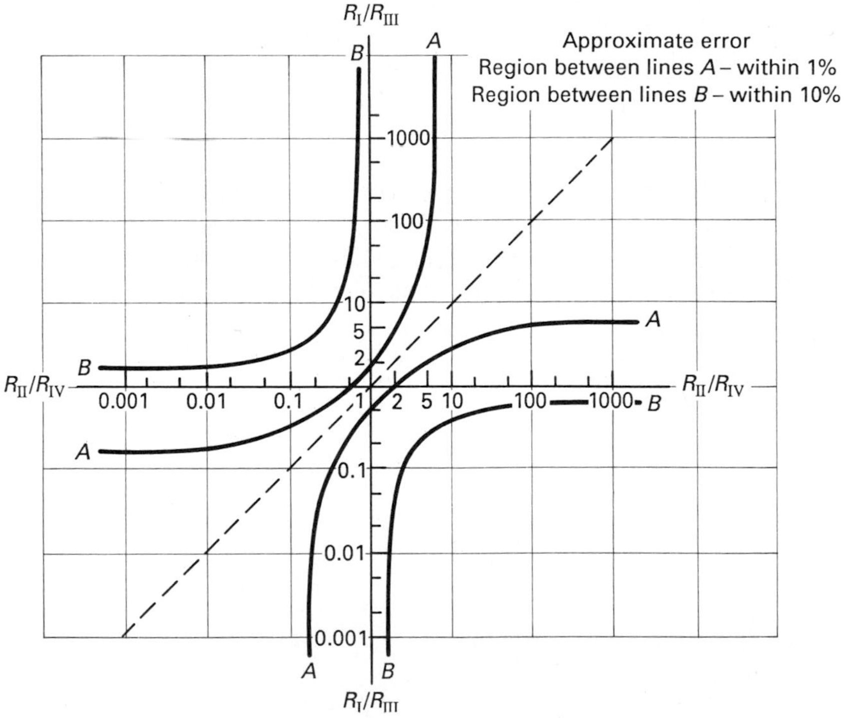


**FIGURE 2–8** Criterion for applicability of one-dimensional analysis to composite wall with four series-parallel thermal resistances.†

† This figure is based on a numerical finite difference solution for a problem that involves a parallel composite with convection at one surface. The numerical technique introduced in Chap. 4 can be used to develop criteria for other arrangements.

$R_I/R_{III} > 0.4$. For the case in which $R_I/R_{III}$ and $R_{II}/R_{IV}$ are both very small, the analogous electrical circuit essentially reduces to a simple parallel circuit involving $R_{III}$ and $R_{IV}$ alone. Similarly, for the case in which $R_I/R_{III}$ and $R_{II}/R_{IV}$ are very large, the circuit reduces to a parallel network which only involves $R_I$ and $R_{II}$. As indicated earlier, parallel heat-transfer systems involving only two thermal resistances are one-dimensional.

## EXAMPLE 2–6

One surface of the 1-m-thick composite plate shown in Fig. E2–6a is kept at 0°C. The other surface is exposed to a fluid with $T_F = 100°C$ and $\bar{h} = 1000$ W/(m² °C). Develop an approximate solution for the rate of heat transfer through this wall.

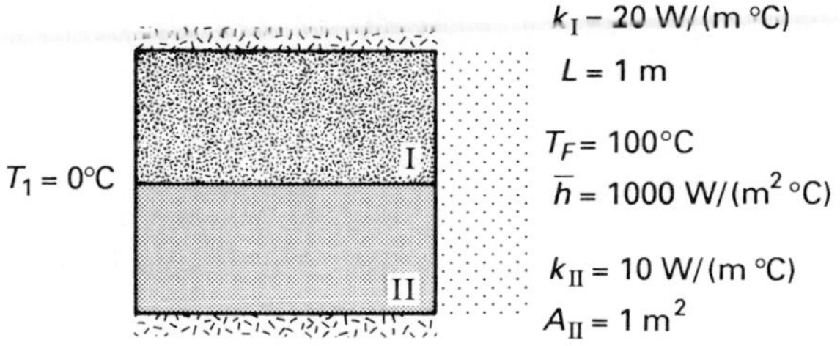
$k_I$ = 20 W/(m °C)
$L$ = 1 m
$T_F$ = 100°C
$\bar{h}$ = 1000 W/(m² °C)
$T_1$ = 0°C
I
II
$k_{II}$ = 10 W/(m °C)
$A_{II}$ = 1 m²

**FIGURE E2–6a**
Composite plane wall.

### Solution

*Objective* Estimate $q$ for the composite wall shown in Fig. E2–6a.

*Assumptions/Conditions*

steady-state
negligible two-dimensional effects
uniform properties in each material
negligible thermal contact resistance

*Properties*

Material I: $k_I = 20$ W/(m °C).
Material II: $k_{II} = 10$ W/(m °C).

*Analysis* Because this composite wall/fluid system provides both series and parallel paths, this heat-transfer problem is actually two-dimensional. Assuming for the moment that the two-dimensional effects are secondary, we sketch the approximate thermal circuit in Fig. E2–6b. Calculating $R_I$, $R_{II}$, $R_{III}$, and $R_{IV}$, we have

$$R_I = \frac{L}{k_I A_I} = \frac{1\ \text{m}}{[20\ \text{W/(m °C)}](1\ \text{m}^2)} = 0.05\ \text{°C/W}$$

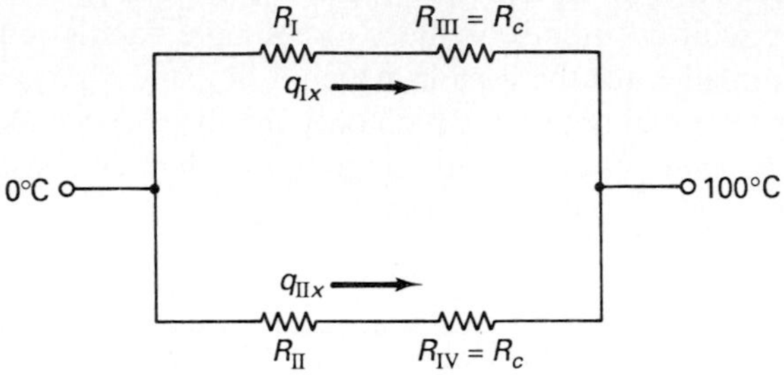


**FIGURE E2–6b** Thermal circuit.

$$R_{II} = \frac{L}{k_{II}A_{II}} = \frac{1 \text{ m}}{[10 \text{ W/(m °C)}](1 \text{ m}^2)} = 0.1 \text{ °C/W}$$

$$R_{III} = \frac{1}{\bar{h}A_{III}} = \frac{1}{[1000 \text{ W/(m}^2 \text{ °C)}](1 \text{ m}^2)} = 0.001 \text{ °C/W} \qquad R_{IV} = R_{III}$$

and

$$\frac{R_I}{R_{III}} = \frac{0.05 \text{ °C/W}}{0.001 \text{ °C/W}} = 50 \qquad \frac{R_{II}}{R_{IV}} = \frac{0.1 \text{ °C/W}}{0.001 \text{ °C/W}} = 100$$

Referring to Fig. 2–8, we find that the error for a one-dimensional analysis for this system is less than 1%.

The rate of heat transfer is therefore approximated by

$$q = \frac{T_1 - T_F}{R_I + R_{III}} + \frac{T_1 - T_F}{R_{II} + R_{IV}}$$

$$= (0\text{°C} - 100\text{°C})\left(\frac{1}{0.05 + 0.001} + \frac{1}{0.1 + 0.001}\right)\frac{\text{W}}{\text{°C}}$$

$$= -100\text{°C}\left(19.6\frac{\text{W}}{\text{°C}} + 9.9\frac{\text{W}}{\text{°C}}\right) = -2950 \text{ W}$$

### *Other Types of Boundary Conditions*

Several other types of boundary conditions are sometimes required in the analysis of heat-transfer problems. Examples include energy dissipation caused by relative interfacial motion, and a moving interface associated with phase change. Reference 5 is suggested as an introduction to the analysis of systems with moving boundaries.

### 2–2–4 Differential Formulation/Solution: Summary

To recapitulate, the development of the differential formulation/solution for conduction heat transfer involves (1) the application of the first law of thermodynamics to a differential element within the system, (2) the utilization of the definition of the derivative in relating the energy entering the differential element to the energy exiting, (3) the use of the appropriate particular law(s) (in this case the Fourier law of conduction), (4) a consideration of the conditions that exist at the boundaries, and (5) the solution of the resulting system of equations for the temperature distribution and the rate of heat transfer.

### 2–2–5 Short Method

The full differential formulation/solution approach outlined above can be utilized to analyze any steady one-dimensional conduction-heat-transfer problem. However, a much shorter method can be developed for one-dimensional problems in which $q_x$ is constant. This approach involves the direct integration of the one-dimensional Fourier law of conduction, Eq. (1–10). To develop this short method in the context of the flat-plate geometry, we merely separate and integrate Eq. (1–10) as follows:

$$\int \frac{q_x}{A_x}\,dx = -\int k\,dT \tag{2–43}$$

Because $q_x$ and $A_x$ are constant for this problem, this equation reduces to

$$\frac{q_x}{A}\int dx = -\int k\,dT \tag{2–44}$$

which can be integrated. For the case in which the thermal conductivity is uniform, Eq. (2–44) can be written as

$$\frac{q_x}{A}\int dx = -k\int dT \tag{2–45}$$

The integration of this equation from $x = 0$ to $L$ and $T = T_1$ to $T_2$ gives rise to Eq. (2–15),

$$q_x = \frac{kA}{L}(T_1 - T_2) \tag{2–15}$$

On the other hand, the integration from $x = 0$ to $x$ and $T = T_1$ to $T$ provides a relationship for the temperature distribution of the form

$$q_x = \frac{kA}{x}(T_1 - T) \tag{2–46}$$

Replacing $q_x$ from Eq. (2–15), we obtain Eq. (2–14),

$$\frac{T - T_1}{T_2 - T_1} = \frac{x}{L} \tag{2–14}$$

This short method can be generalized for any steady one-dimensional conduction heat-transfer system by writing

$$q_\xi \int \frac{d\xi}{A_\xi} = -\int k\, dT \tag{2–47}$$

where $\xi$ is equal to $x$, $y$, or $z$ in Cartesian coordinates and $r$ in cylindrical or spherical coordinates; $q_\xi$ is constant but $A_\xi$ can be a function of $\xi$. The integration from one boundary to the other provides us with an expression for $q_\xi$. The integration from one boundary to an intermediate location $\xi$ gives rise to a prediction for the temperature distribution. Because of its directness, this method will be used in the following sections whenever possible. However, the more general differential formulation/solution approach or numerical techniques will be required later when problems involving more complex situations are encountered.

## 2–3 CONDUCTION HEAT TRANSFER IN RADIAL SYSTEMS

Whereas the cross-sectional area $A_x$ for conduction heat transfer in a plane wall is constant, many situations are encountered in which the area through which the heat is transferred is dependent upon the spatial coordinate. For example, for radial conduction heat transfer in the hollow cylinder shown in Fig. 2–9(a), $A_r$ is equal to $2\pi rL$.

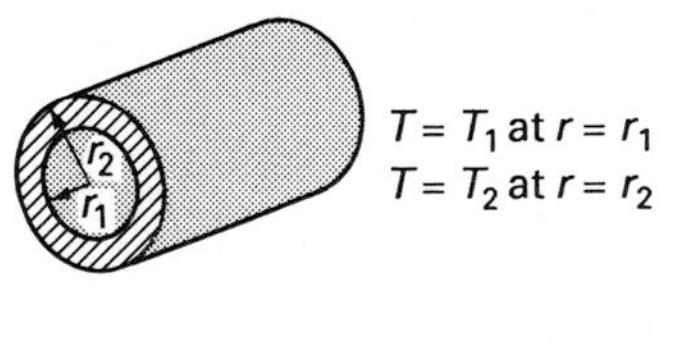


(a) System.

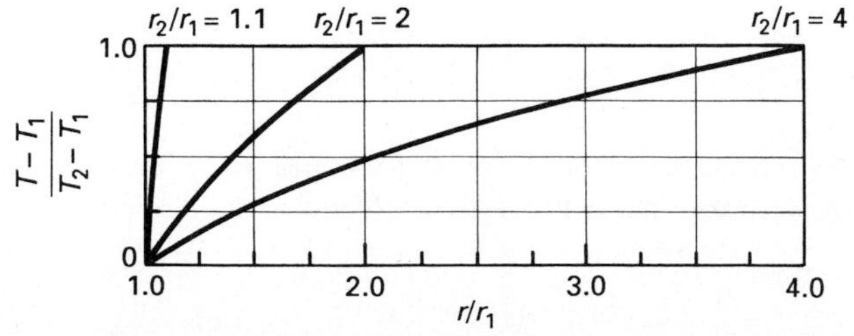


(b) Temperature distribution, Eq. (2–52).

**FIGURE 2–9** One-dimensional radial heat transfer in a hollow cylinder.

This classic problem is analyzed in this section, after which an interesting concept known as the critical radius will be introduced.

### 2–3–1 Analysis

For one-dimensional steady-state radial conduction heat transfer in a hollow cylinder, the Fourier law of conduction takes the form

$$q_r = -kA_r \frac{dT}{dr} \tag{2–48}$$

where $A_r = 2\pi rL$. Because $q_r = q_{r+dr}$ for this problem, the short method can be used to obtain the heat transfer by writing

$$q_r \int_{r_1}^{r_2} \frac{dr}{2\pi rL} = -k \int_{T_1}^{T_2} dT \tag{2–49}$$

for uniform thermal conductivity, or

$$q_r = \frac{2\pi Lk}{\ln (r_2/r_1)} (T_1 - T_2) \tag{2–50}$$

Thus, the thermal resistance to radial conduction in a hollow cylinder is

$$R_k = \frac{\ln (r_2/r_1)}{2\pi Lk} \tag{2–51}$$

Similar to the analysis of heat transfer in a plane wall, the temperature distribution in this hollow cylinder can be obtained by integrating from $r_1$ to $r$ and $T_1$ to $T$. The resulting expression for $T$ is given by

$$\frac{T - T_1}{T_2 - T_1} = \frac{\ln (r/r_1)}{\ln (r_2/r_1)} \tag{2–52}$$

This profile is shown in Fig. 2–9(b) for three values of $r_2/r_1$. It is observed that the temperature distribution is nearly linear for values of $r_2/r_1$ of the order of unity, but decidedly nonlinear for the larger values of $r_2/r_1$. This nonlinearity is caused by the variation in $A_r$ with respect to $r$.

For cases in which the surface temperatures are not specified, $T_1$ and $T_2$ are determined by utilizing the appropriate thermal boundary conditions. This point is illustrated in Example 2–7.

For one-dimensional radial heat transfer in a hollow sphere, $A_r$ is equal to $4\pi r^2$. Utilizing the short method, the rate of heat transfer in a hollow sphere with surface temperatures $T_1$ and $T_2$ is easily shown to be

$$q_r = \frac{4\pi r_1 r_2 k}{r_2 - r_1} (T_1 - T_2) \tag{2–53}$$

and the temperature profile is given by

$$\frac{T - T_1}{T_2 - T_1} = \frac{r - r_1}{r_2 - r_1}\frac{r_2}{r} \tag{2–54}$$

Note that the thermal resistance for this system is

$$R_k = \frac{r_2 - r_1}{4\pi r_1 r_2 k} \tag{2–55}$$

## EXAMPLE 2–7

Refrigerant flows in a 1.9-in.-O.D. copper tube with 0.281-in. wall thickness. The inside surface temperature is 5°F and the room temperature is 70°F. Determine the thickness of insulative pipe covering [$k_i$ = 0.428 Btu/(h ft °F)] required to reduce the heat gain to the pipe by 25% for the case in which forced convection heat transfer occurs with $\bar{h}$ = 10 Btu/(h ft$^2$ °F). Assume that the thermal radiation effects are negligible.

### Solution

*Objective* Determine the required insulation thickness.

*Schematic* Copper tube with insulation.

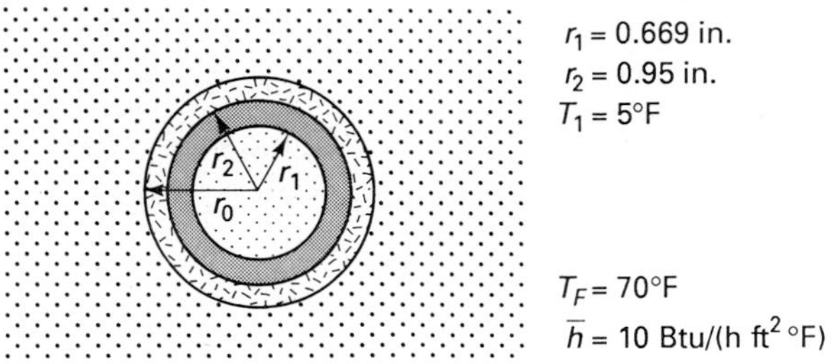
$r_1$ = 0.669 in.
$r_2$ = 0.95 in.
$T_1$ = 5°F
$T_F$ = 70°F
$\bar{h}$ = 10 Btu/(h ft$^2$ °F)
$r_2$
$r_1$
$r_0$

*Assumptions/Conditions*

steady-state
one-dimensional
uniform properties in each material
forced convection cooling
negligible thermal radiation

*Properties*

Copper at room temperature (Table A–C–1):

$$k = 386 \frac{\text{W}}{\text{m °C}} \frac{0.578 \text{ Btu/(h ft °F)}}{\text{W/(m °C)}} = 223 \text{ Btu/(h ft °F)}$$

Insulation: $k$ = 0.428 Btu/(h ft °F).

*Analysis* Referring to the thermal circuit shown in Fig. E2–7, the resistances are

$$R_{k1} = \frac{\ln(r_2/r_1)}{2\pi L k} = \frac{1}{L}\frac{\ln(0.95/0.669)}{2\pi}\frac{\text{h ft °F}}{223 \text{ Btu}} = \frac{2.50 \times 10^{-4}}{L}\frac{\text{h ft °F}}{\text{Btu}}$$

$$R_{k2} = \frac{\ln(r_0/r_2)}{2\pi L k_i} \qquad R_c = \frac{1}{\bar{h}A_s} = \frac{0.1}{A_s}\frac{\text{h ft}^2\text{ °F}}{\text{Btu}}$$

With no insulation, $q_r$ is given by

$$q_r = \frac{5\text{°F} - 70\text{°F}}{\dfrac{2.50 \times 10^{-4}}{L}\dfrac{\text{h ft °F}}{\text{Btu}} + \dfrac{0.1}{2\pi L(0.95 \text{ ft}/12)}\dfrac{\text{h ft}^2\text{ °F}}{\text{Btu}}} = q_c$$

$$\frac{q_r}{L} = \frac{-65\text{°F}}{(2.50 \times 10^{-4} + 0.201) \text{ h ft °F/Btu}} = -323 \text{ Btu/(h ft)}$$

Note that the thermal resistance $R_{k1}$ of the pipe wall is negligible.

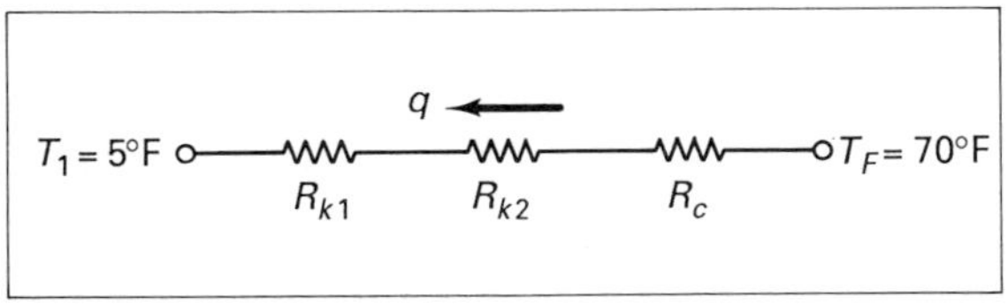


**FIGURE E2–7**
Thermal circuit.

A 25% reduction in the rate of heat transfer results in $q_r/L = -242$ Btu/(h ft). It follows that

$$q_r \simeq \frac{5\text{°F} - 70\text{°F}}{R_{k2} + R_c} = \frac{-65\text{°F}}{\dfrac{\ln(r_0/r_2)}{2\pi k_i L} + \dfrac{1}{\bar{h}2\pi r_0 L}}$$

$$\ln\left(\frac{r_0}{r_2}\right) + \frac{k_i}{\bar{h}r_0} = \frac{-65\text{°F}(2\pi k_i)}{q_r/L} = \frac{-65\text{°F}(2\pi)[0.428 \text{ Btu/(h ft °F)}]}{-242 \text{ Btu/(h ft)}} = 0.722$$

This nonlinear equation is solved for $r_0$ by iteration as follows:

$$\ln\left(\frac{r_0}{r_2}\right) = 0.722 - \frac{0.428}{10}\frac{\text{ft}}{r_0}$$

$$\frac{r_0}{r_2} = \exp\left(0.722 - \frac{0.0428}{r_0}\text{ ft}\right) \qquad \text{(a)}$$

Starting with an assumed value for $r_0/r_2$ of 2, we have

$$\frac{r_0}{r_2} = \exp\left[0.722 - \frac{0.0428}{2(0.95/12)}\right] = 1.57$$

Substituting this value back into Eq. (a), we obtain

$$\frac{r_0}{r_2} = 1.46$$

Continuing this iteration sequence, we arrive at

$$\frac{r_0}{r_2} = 1.4$$

after only three more steps. Thus, the thickness of insulation required to reduce the rate of heat loss by 25% is

$$\delta = 1.4r_2 - r_2 = 0.4(0.95 \text{ in.}) = 0.38 \text{ in.}$$

## EXAMPLE 2–8

Utilize the differential formulation approach to obtain expressions for the temperature distribution, rate of heat transfer, and thermal resistance for the cylindrical section shown in Fig. E2–8.

**FIGURE E2–8**
Cylindrical section.

### Solution

*Objective* Develop relations for $T$, $q$, and $R$ for this cylindrical section.

*Assumptions/Conditions*

steady-state
one-dimensional
uniform properties

*Analysis* Because no heat is transferred across the surfaces at $\theta = 0$ and $\theta_1$ and because $T_1$ and $T_2$ are uniform, the heat transfer in this system is one-dimensional. This one-dimensional ($r$ direction) conduction-heat-transfer problem can be analyzed by either the differential formulation approach or the short method.

The differential formulation is developed as follows:

*Step 1*

$$q_r = q_{r+dr}$$

*Step 2*

$$q_r = q_r + \frac{dq_r}{dr}\,dr \qquad \text{or} \qquad \frac{dq_r}{dr} = 0$$

*Step 3*

$$\frac{d}{dr}\left(-kA_r\frac{dT}{dr}\right) = \frac{d}{dr}\left(-k\theta_1 rL\frac{dT}{dr}\right) = 0$$

or, for uniform thermal conductivity,

$$\frac{d}{dr}\left(r\frac{dT}{dr}\right) = 0 \tag{a}$$

Finally, the boundary temperatures can be designated by $T(r_1) = T_1$ and $T(r_2) = T_2$.

Next, Eq. (a) is solved for the temperature distribution. Integrating, we have

$$r\frac{dT}{dr} = C_1$$

Continuing,

$$\int_{T_1}^{T} dT = \int_{r_1}^{r}\frac{C_1\,dr}{r} \qquad T - T_1 = C_1 \ln\left(\frac{r}{r_1}\right)$$

Utilizing the boundary condition at $r_2$, $C_1$ is given by

$$C_1 = \frac{T_2 - T_1}{\ln(r_2/r_1)}$$

and the temperature profile becomes

$$\frac{T - T_1}{T_2 - T_1} = \frac{\ln(r/r_1)}{\ln(r_2/r_1)}$$

To obtain an expression for the rate of heat transfer, we utilize the Fourier law of conduction as follows:

$$q_r = -k\theta_1 rL\frac{dT}{dr} = -k\theta_1 LC_1 = k\theta_1 L\frac{T_1 - T_2}{\ln(r_2/r_1)}$$

or

$$q_r = \frac{T_1 - T_2}{R_k}$$

where $R_k = \ln(r_2/r_1)/(\theta_1 kL)$. These results are seen to be identical to those which were obtained on the basis of the short method for the case of a hollow circular cylinder with $\theta_1 = 2\pi$.

### 2–3–2 Critical Radius

As shown in Example 2–7, convection and composite wall boundary conditions associated with geometries for which the area is not constant are handled in the same way that these complications were treated for the plane-wall geometry. For the case of a plane wall exposed to a fluid, an increase in the thickness of the wall results in an increase in the internal resistance $R_k = L/(kA)$ but does not change the surface resistance $R_c$. Hence, such an increase in the thickness of a plane wall always reduces the rate of heat transfer through the wall. Of course, a reduction in heat transfer is most easily accomplished by the use of an insulating material of low thermal conductivity. On the other hand, an increase in the wall thickness or the addition of an insulating material does not always bring about a decrease in the heat-transfer rate for geometries with nonconstant cross-sectional area.

To see this point, consider the hollow cylinder of radii $r_1$ and $r_2$ shown in Fig. 2–10(a). The inside surface is maintained at temperature $T_1$, and the temperature $T_F$ of the surrounding fluid is specified. The thermal circuit for this problem is shown in Fig. 2–10(b). Now we ask, what will be the effect of increasing the outside radius $r_2$, with $T_1$, $T_F$, and $\bar{h}$ held constant? The rate of heat transfer from the outside surface to the fluid is

$$q_r = \frac{T_1 - T_F}{R_k + R_c} = \frac{T_1 - T_F}{\dfrac{\ln (r_2/r_1)}{2\pi Lk} + \dfrac{1}{\bar{h}2\pi L r_2}} \tag{2–56}$$

An increase in $r_2$ is seen to increase $R_k$ but to decrease $R_c$. Therefore, the addition of material can either decrease or increase the rate of heat transfer, depending upon the change in the total resistance $R_c + R_k$ with $r_2$.

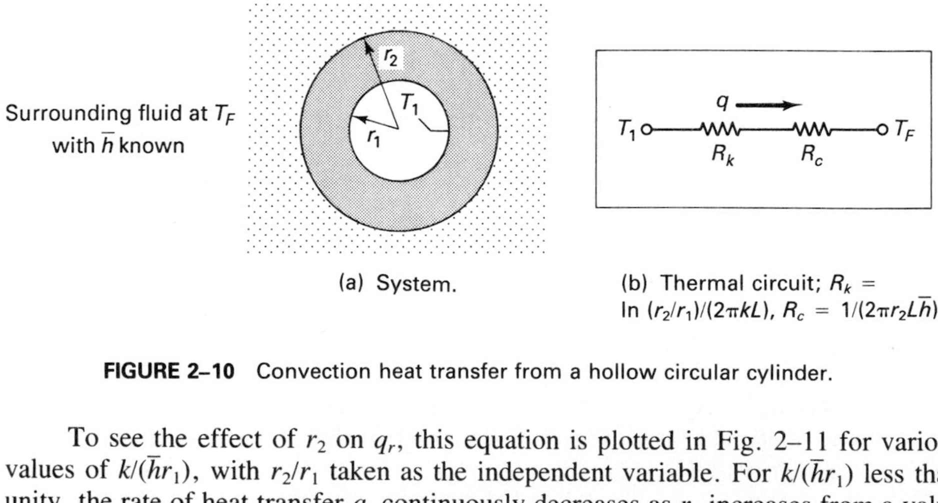


(a) System.

(b) Thermal circuit; $R_k = \ln (r_2/r_1)/(2\pi kL)$, $R_c = 1/(2\pi r_2 L\bar{h})$.

**FIGURE 2–10** Convection heat transfer from a hollow circular cylinder.

To see the effect of $r_2$ on $q_r$, this equation is plotted in Fig. 2–11 for various values of $k/(\bar{h}r_1)$, with $r_2/r_1$ taken as the independent variable. For $k/(\bar{h}r_1)$ less than unity, the rate of heat transfer $q_r$ continuously decreases as $r_2$ increases from a value of $r_1$. But, for $k/(\bar{h}r_1)$ greater than unity, $q_r$ increases to a maximum and then decreases. To determine the radius $r_c$ at which $q_r$ is maximized for $k/(\bar{h}r_1)$ greater than unity, we set $dq_r/dr_2$ equal to zero,

$$\left.\frac{dq_r}{dr_2}\right|_{r_c} = -(T_1 - T_F)2\pi Lk \left.\left(\frac{1}{r_2} - \frac{k}{\bar{h}r_2^2}\right)\right|_{r_c} = 0 \tag{2–57}$$

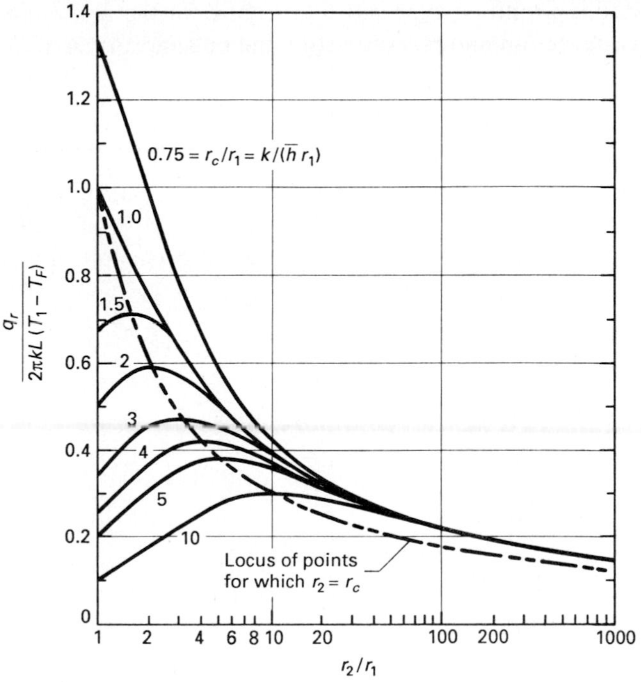
1.4
1.2
1.0
0.8
0.6
0.4
0.2
0
0.75 = r_c/r_1 = k/(h̄ r_1)
1.0
1.5
2
3
4
5
10
Locus of points
for which r_2 = r_c
q_r / 2πkL(T_1 − T_F)
1
2
4
6
8
10
20
100
200
1000
r_2/r_1

**FIGURE 2–11** Effect of increase in outside radius $r_2$ on heat transfer from a pipe.

with the result

$$r_c = \frac{k}{\bar{h}} \tag{2–58}$$

$r_c$ is referred to as the *critical radius*. If $r_2 > r_c$, the addition of material to the outside surface will decrease the rate of heat transfer, as in the case of Example 2–7, where $r_c = 0.514$ in. and $r_2 = 0.95$ in. This is often the case for conditions of forced convection for which $\bar{h}$ is large and $r_c$ is small. But if $r_2 < r_c$, the addition of material will increase the heat-transfer rate until $r_2 = r_c$, after which additional increases in $r_2$ will decrease $q_r$. This situation occurs more often for natural convection than for forced convection because of the low values of $\bar{h}$, especially in gases.

In contrast, if insulation is added to the inside surface, both $R_c$ and $R_k$ increase. Hence, the addition of insulation to the inside surface always reduces the heat-transfer rate and the critical radius concept has no significance.

Heat transfer in a sphere is affected by the addition of insulation to the outside or inside surface in much the same way as in a cylinder. However, for a sphere, the critical radius is given by

$$r_c = \frac{2k}{\bar{h}} \tag{2–59}$$

Likewise, a critical radius would be expected to be found for thermal radiation heat transfer or combined convection and radiation from the outside surface of a cylinder or sphere.

## EXAMPLE 2–9

Determine the thickness of insulative pipe covering [$k_i = 0.428$ Btu/(h ft °F)] required to reduce the convective heat loss from the 1.9-in.-O.D. pipe of Example 2–7 by 25% for the case in which natural convection cooling occurs with $\bar{h} = 3.6$ Btu/(h ft$^2$ °F).

### Solution

*Objective* Determine the required insulation thickness.

*Schematic* Copper tube with insulation.

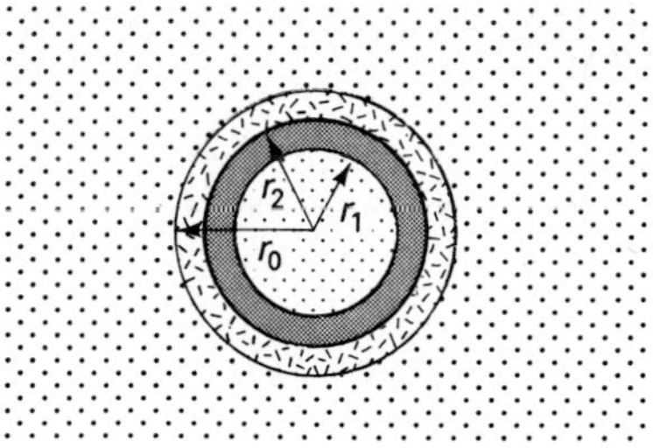
$r_2$
$r_1$
$r_0$
$r_1 = 0.669$ in.
$r_2 = 0.95$ in.
$T_1 = 5$°F
$T_F = 70$°F
$\bar{h} = 3.6$ Btu/(h ft$^2$ °F)

*Assumptions/Conditions*

steady-state
one-dimensional
uniform properties in each material
natural convection cooling
negligible thermal radiation

*Properties*

Copper tube (from Example 2–7): $k = 223$ Btu/(h ft °F).
Insulation: $k = 0.428$ Btu/(h ft °F).

*Analysis* The critical radius for this situation is

$$r_c = \frac{k_i}{\bar{h}} = \frac{0.428 \text{ Btu/(h ft °F)}}{3.6 \text{ Btu/(h ft}^2\text{ °F)}} = 0.119 \text{ ft} = 1.43 \text{ in.}$$

Thus, we have

$$\frac{r_c}{r_2} = \frac{1.43 \text{ in.}}{0.95 \text{ in.}} = 1.5$$

Because $r_c/r_2$ is greater than unity, the effect of insulation for $r_0$ less than $r_c$ will be to increase the rate of heat transfer $q_r$. In contrast, for the forced convection conditions of Example 2–7, $r_c$ is equal to 0.514, such that the effect of insulation is to reduce $q_r$ for all values of $r_0$.

Referring to Fig. 2–11, the dimensionless heat flux $q_r/[2\pi k_i L(T_1 - T_F)]$ increases from a value of 0.667 at $r_0/r_2 = 1$ to a maximum value of approximately 0.71 at $r_0/r_2$ equal to $r_c/r_2 = 1.5$, and then begins to fall toward zero. Note that the rate of heat transfer for $r_0/r_2 = 1$ is given by the simple Newton law of cooling,

$$q_r = 2\pi r_2 L\bar{h}(T_1 - T_F)$$

or

$$\frac{q_r}{2\pi k_i L(T_1 - T_F)} = \frac{\bar{h}r_2}{k_i} = \frac{r_2}{r_c} = \frac{1}{1.5} = 0.667$$

Our problem calls for a 25% reduction in the rate of heat transfer; that is,

$$\left.\frac{q_r}{2\pi k_i L(T_1 - T_F)}\right|_{r_0} = 0.5$$

Utilizing Fig. 2–11, we find that the dimensionless heat flux reaches this value for $r_0/r_2$ equal to approximately 5. Thus, the thickness of insulation is

$$\delta = r_0 - r_2 = 5(0.95 \text{ in.}) - 0.95 \text{ in.} = 3.8 \text{ in.}$$

## 2–4 VARIABLE THERMAL CONDUCTIVITY

As indicated in Chap. 1, the thermal conductivity of most materials is at least somewhat dependent upon temperature. The variation of thermal conductivity with temperature is shown in Fig. 2–12 for several common metals for the temperature range −100°C to 300°C. Whereas the assumption of uniform thermal conductivity is generally acceptable for problems involving small temperature differences, many situations are encountered in which the variation in $k$ with $T$ cannot be neglected. Referring to Fig. 2–12, we see that a linear approximation for $k$ can be utilized over limited temperature ranges for these materials; that is,

$$k(T) = k_0(1 + \beta_T T) \tag{2–60}$$

where $\beta_T$ is known as the *temperature coefficient of thermal conductivity*. The units for $\beta_T$ are 1/°C (or 1/°F).

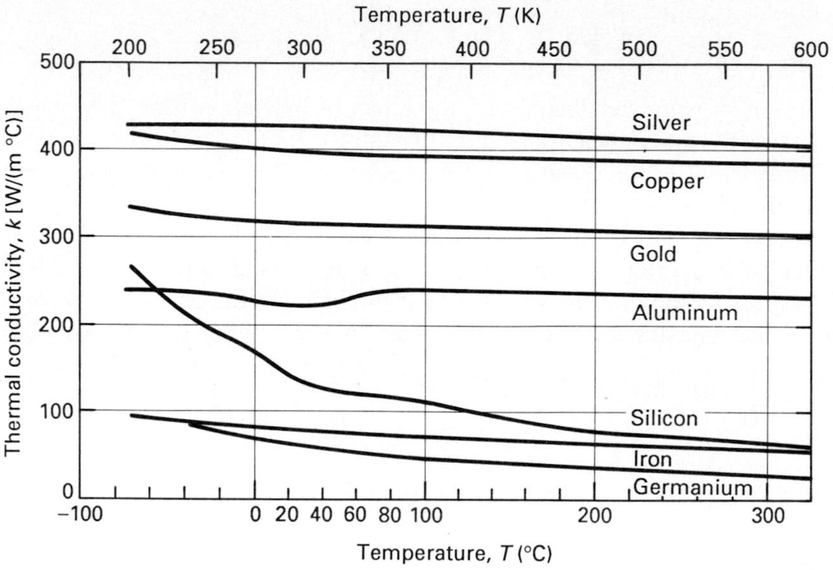


**FIGURE 2–12** Variation of thermal conductivity $k$ with temperature for several metals.

### 2–4–1 Analysis

Utilizing the short method, $k$ must be retained within the integral in Eq. (2–47) for cases in which the variation of the thermal conductivity with temperature is significant. By integrating from one boundary to the other, Eq. (2–47) gives

$$q_\xi \int_{\xi_1}^{\xi_2} \frac{d\xi}{A_\xi} = -\int_{T_1}^{T_2} k\, dT \tag{2–61}$$

where $\xi_1 = 0$ and $\xi_2 = L$ for a flat plate, and $\xi_1 = r_1$ and $\xi_2 = r_2$ for a hollow cylinder or sphere.

Introducing the mean thermal conductivity $\bar{k}$,

$$\bar{k} = \frac{1}{T_2 - T_1} \int_{T_1}^{T_2} k\, dT \tag{2–62}$$

the rate of heat transfer $q_\xi$ is given by

$$q_\xi = \frac{\bar{k}(T_1 - T_2)}{\displaystyle\int_{\xi_1}^{\xi_2} d\xi/A_\xi} \tag{2–63}$$

Based on this expression, equations can be written for the thermal resistances for flat plates, and hollow cylinders and spheres with variable thermal conductivity of the forms

$$R_k = \frac{L}{A\bar{k}} \quad \text{flat plate} \tag{2–64}$$

$$R_k = \frac{\ln (r_2/r_1)}{2\pi L\bar{k}} \quad \text{hollow cylinder} \tag{2–65}$$

$$R_k = \frac{r_2 - r_1}{4\pi r_1 r_2 \bar{k}} \quad \text{hollow sphere} \tag{2–66}$$

## EXAMPLE 2–10

The surfaces of a 10-cm-thick plate are maintained at 0°C and 100°C. The thermal conductivity varies with temperature according to $k = k_0(1 + \beta_T T)$ with $k = 50$ W/(m °C) at 0°C and $k = 100$ W/(m °C) at 100°C. Determine the temperature distribution and heat-transfer flux in the plate.

### Solution

*Objective* Determine $T$ and $q''$ for the plate.

*Schematic* Plate with variable thermal conductivity.

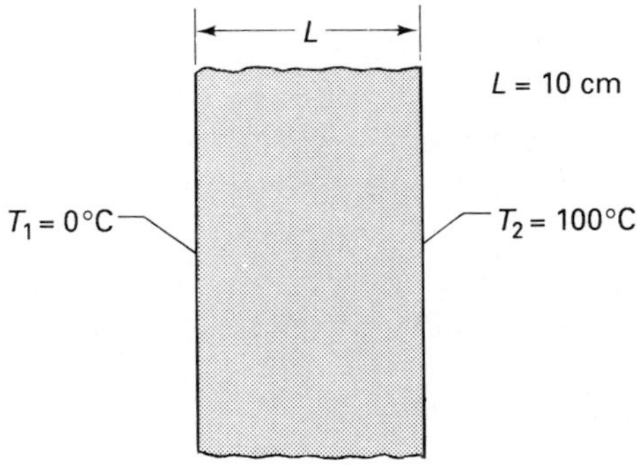


*Assumptions/Conditions*

steady-state
one-dimensional
variable thermal conductivity

*Properties* The thermal conductivity is

$$k = k_0 (1 + \beta_T T)$$

with $k$ specified at 0°C and 100°C. To evaluate $k_0$ and $\beta_T$, we set $k$ equal to 50 W/(m °C) at 0°C and 100 W/(m °C) at 100°C, with the result

$$k_0 = 50 \text{ W/(m °C)} \qquad \beta_T = 0.01/°\text{C}$$

*Analysis* To calculate the rate of heat transfer, we utilize Eq. (2–63), with the result

$$q = \frac{\bar{k}A}{L}(T_1 - T_2)$$

where

$$\bar{k} = \frac{1}{T_2 - T_1}\int_{T_1}^{T_2} k\,dT = \frac{k_0}{100°\text{C}}\int_{0°\text{C}}^{100°\text{C}} (1 + \beta_T T)\,dT$$

$$= \frac{k_0}{100°\text{C}}\left(T + \frac{\beta_T}{2}T^2\right)\Bigg|_{0°\text{C}}^{100°\text{C}} = \frac{k_0}{100°\text{C}}\left[100°\text{C} + \frac{10^{-2}}{2°\text{C}}(100°\text{C})^2\right] = 1.5k_0$$

Following through with the calculation, we have

$$q'' = 1.5\frac{k_0}{L}(T_1 - T_2) = \frac{1.5}{0.1\text{ m}}\left(50\frac{\text{W}}{\text{m °C}}\right)(0°\text{C} - 100°\text{C}) = -75\text{ kW/m}^2$$

To obtain the temperature distribution, the Fourier law of conduction is integrated from 0 to $x$ and from 0°C to $T$ as follows:

$$q = -kA\frac{dT}{dx} \qquad \frac{q}{A}\int_0^x dx = -\int_{0°\text{C}}^{T} k\,dT$$

$$\frac{q}{A}x = -k_0\left(T + \frac{\beta_T}{2}T^2\right)$$

where $q/A = q'' = -75\text{ kW/m}^2$. It follows that

$$T + \frac{\beta_T}{2}T^2 + 1.5(T_1 - T_2)\frac{x}{L} = 0$$

Solving this quadratic equation for $T$, we obtain

$$T = \frac{-1 \pm \sqrt{1 - 4(\beta_T/2)(1.5)(T_1 - T_2)(x/L)}}{\beta_T}$$

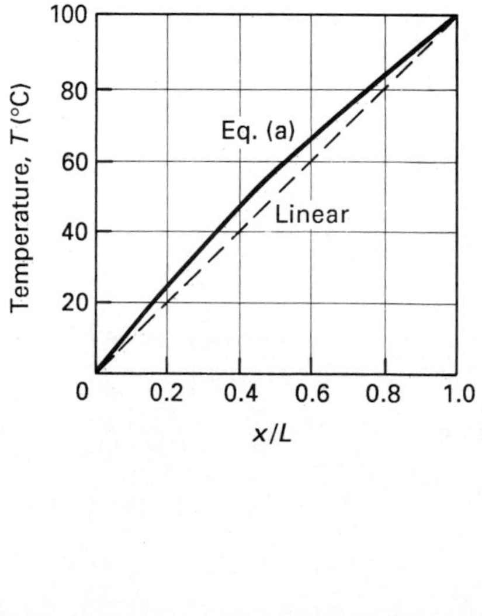


**FIGURE E2–10**
Temperature distribution in plane wall with variable thermal conductivity.

where the positive sign must be utilized to satisfy the boundary conditions. Substituting for $\beta_T$, $T_1$, and $T_2$, $T$ becomes

$$T = \frac{-1 + \sqrt{1 + 3(x/L)}}{0.01}\ ^\circ\text{C} \tag{a}$$

This temperature distribution is compared with the linear profile associated with uniform thermal conductivity in Fig. E2–10. Notice the distinct nonlinear behavior of $T$ for this case in which $k$ is a function of temperature.

## 2–5 INTERNAL ENERGY SOURCES

Heat-transfer systems involving internal energy sources include chemical and nuclear reactors in which energy is generated by chemical reaction or by the interaction of nuclear particles. Other important heat-transfer applications involving internal energy sources occur in electrical and electronic circuits. Examples include electrical resistance heaters made of nickel/chromium alloys, incandescent electric lamps with tungsten filaments, and semiconductor chips and transistors made of silicon or germanium. The dissipation of heat is also important in the operation of electric motors, generators, transformers, and relays.

The strength of an internal energy source is generally represented by the rate of energy generated per unit volume $\dot{q}$. For electrical and electronic systems, $\dot{q}$ is given by

$$\dot{q} = \frac{I_e^2 R_e}{V} = \frac{I_e^2 \rho_e}{A^2} \tag{2–67}$$

where the current $I_e$ is assumed to be uniform and $\rho_e$ $(= R_e A/L)$ is the *electrical resistivity*. The resistivity is generally a linear function of temperature; that is,

$$\rho_e = \rho_0(1 + \alpha_0 T) \tag{2–68}$$

where $\rho_0$ is the resistivity at a reference temperature such as 0°C, and $\alpha_0$ is the *temperature coefficient of resistance* at the reference temperature. Accordingly, the power dissipation per unit volume associated with the flow of electric current can be approximated by an equation of the form

$$\dot{q} = \dot{q}_0(1 + \alpha_0 T) \tag{2–69}$$

where $\dot{q}_0 = I_e^2 \rho_0/A^2$. Data for $\rho_0$ and $\alpha_0$ are tabulated in reference 6 for several common metals and alloys. For example, $\rho_0 = 1.8 \times 10^{-8}$ Ω m and $\alpha_0 = 0.004$/°C for copper wire, and $\rho_0 = 20 \times 10^{-8}$ Ω m and $\alpha_0 = 0.005$/°C for carbon steel. In comparison, the electrical resistivities of electrical insulators such as mica and electrical semiconductors such as germanium are of the order of $10^{15}$ Ω m and 0.4 Ω m, respectively. In general, $\alpha_0$ is quite small for metals and alloys, such that the internal energy generation can be assumed to be uniformly distributed (i.e., independent of the spatial coordinates) for moderate temperatures. However, the variation of $\dot{q}$ becomes an important factor for high-temperature operation.

### 2–5–1 Analysis

Application of the first law of thermodynamics to the problem of conduction heat transfer in a flat plate with internal energy generation (see Fig. 2–13) leads to

*Step 1*

$$\dot{q}\,dV + q_x = q_{x+dx} \tag{2–70}$$

(The energy generation can be considered as energy brought into the differential element.) Unlike the case of steady-state conduction heat transfer with no energy generation [represented by Eq. (2–1)], Eq. (2–70) indicates that $q_x$ is not equal to $q_{x+dx}$ for energy-generation problems. Therefore, the short method, which involves the mere integration of the Fourier law, cannot be utilized, unless the dependence of $q_x$ on $x$ is known. Hence, we proceed with the development of the differential formulation.

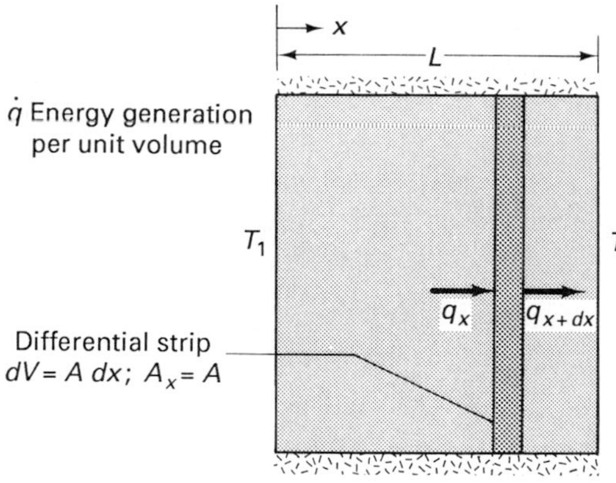


**FIGURE 2–13**
Conduction heat transfer in a plane wall with internal energy generation.

The substitution of Eq. (2–3) into Eq. (2–70) gives

*Step 2*

$$\dot{q}A\,dx = \frac{dq_x}{dx}\,dx \tag{2–71}$$

where $A_x$ is constant ($A_x = A$) and $dV = A\,dx$. With $q_x$ given by the Fourier law of conduction, Eq. (2–71) becomes

*Step 3*

$$\frac{d}{dx}\left(k\frac{dT}{dx}\right) + \dot{q} = 0 \tag{2–72}$$

For the case of uniform thermal conductivity and energy generation, this equation reduces to the form

$$\frac{d^2T}{dx^2} + \frac{\dot{q}_0}{k} = 0 \tag{2–73}$$

Notice that Eq. (2–73) is identical to Eq. (2–7) for $\dot{q}_0 = 0$.

Because this is a second-order differential equation, two boundary conditions are required; they are

$$T(0) = T_1 \qquad T(L) = T_2 \tag{2–74,75}$$

This simple differential equation can be solved by two integrations as follows:

$$\frac{dT}{dx} = -\frac{\dot{q}_0}{k}x + C_1 \qquad T = -\frac{\dot{q}_0}{k}\frac{x^2}{2} + C_1 x + C_2 \tag{2–76,77}$$

The boundary conditions require that $C_2 = T_1$ and $C_1 = (T_2 - T_1)/L + \dot{q}_0 L/(2k)$, such that Eq. (2–77) becomes

$$T - T_1 = (T_2 - T_1)\frac{x}{L} + \frac{\dot{q}_0}{k}\frac{L^2}{2}\left[\frac{x}{L} - \left(\frac{x}{L}\right)^2\right] \tag{2–78}$$

Focusing attention on the case in which symmetrical cooling takes place with $T_2 = T_1$, the temperature distribution is given by

$$T - T_1 = \frac{2\dot{q}_0 \ell^2}{k}\left[\frac{x}{L} - \left(\frac{x}{L}\right)^2\right] \tag{2–79}$$

or

$$T - T_1 = \frac{1}{2}\frac{\dot{q}_0 \ell^2}{k}\left[1 - \left(\frac{\xi}{L/2}\right)^2\right] \tag{2–80}$$

where $\ell = V/A_s = L/2$ and $\xi = x - L/2$. This temperature distribution is shown in Fig. 2–14 in terms of both $x$ and $\xi$. The maximum temperature within the plate for symmetrical cooling occurs at the center where $dT/dx = 0$.

Setting $x = L/2$ in Eq. (2–79) or $\xi = 0$ in Eq. (2–80), we have

$$T_{\max} = T_1 + \frac{\dot{q}_0}{k}\frac{\ell^2}{2} = T_1 + \frac{\dot{q}_0}{k}\frac{L^2}{8} \tag{2–81}$$

Because materials break down above certain temperatures, this maximum temperature represents a critical design consideration.

To obtain the rate of heat transfer for the case in which $T_2 = T_1$, we apply the Fourier law of conduction as follows:

$$q_x = -kA\left.\frac{dT}{dx}\right|_x = \dot{q}_0\frac{V}{2}\left(\frac{2x}{L} - 1\right) \tag{2–82}$$

Hence, the rates of heat transfer at the surfaces $x = 0$ and $x = L$ are

$$q_0 = -\dot{q}_0\frac{V}{2} \qquad q_L = \dot{q}_0\frac{V}{2} \tag{2–83,84}$$

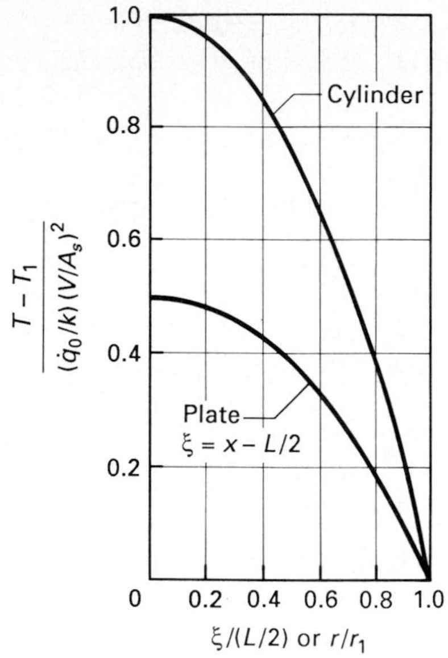


**FIGURE 2–14**
Temperature distributions for plate and solid circular cylinder with uniform energy generation and surface temperature $T_1$.

such that the total rate of energy generated within the plate $\dot{q}_0 V$ is transferred into the fluid from the two surfaces.

For the more general case in which $T_2 \neq T_1$, Eq. (2–78) is coupled with the Fourier law of conduction to obtain

$$q_x = \frac{kA}{L}(T_1 - T_2) + \frac{\dot{q}_0 V}{2}\left(\frac{2x}{L} - 1\right) \tag{2–85}$$

It follows that $q_0$ and $q_L$ are

$$q_0 = \frac{kA}{L}(T_1 - T_2) - \frac{\dot{q}_0 V}{2} \tag{2–86}$$

and

$$q_L = \frac{kA}{L}(T_1 - T_2) + \frac{\dot{q}_0 V}{2} \tag{2–87}$$

Thus, we see that the total rate of heat transfer is found by simply superimposing the heat transfer through a nongenerating plate with surface temperatures $T_1$ and $T_2$ upon the heat transfer in a generating plate with both surfaces at $T_1$.

Solutions are also available for the temperature distribution in heat-generating cylinders and spheres. For example, the temperature distribution in a solid circular cylinder with radius $r_1$, surface temperature $T_1$, and uniform heat generation takes the form

$$T - T_1 = \frac{\dot{q}_0 \ell^2}{k}\left[1 - \left(\frac{r}{r_1}\right)^2\right] \tag{2–88}$$

where $\ell = V/A_s = r_1/2$. This expression is shown in Fig. 2–14. The maximum temperature in the cylinder occurs at the center.

$$T_{max} = T_1 + \frac{\dot{q}_0 \ell^2}{k} = T_1 + \frac{\dot{q}_0}{k}\frac{r_1^2}{4} \tag{2–89}$$

## EXAMPLE 2–11

Four amperes of current flow in a 1-mm-diameter copper wire with resistivity equal to about 30 $\mu\Omega$ cm. The insulation is 2 mm thick and has a thermal conductivity of 0.05 W/(m °C). Determine the surface temperature of the insulation and the maximum operating temperature of the wire if the system is cooled by blackbody thermal radiation with $T_R = -200$°C and $F_{s-R} = 1.0$.

### Solution

*Objective* Determine the surface and centerline temperature of the wire.

*Schematic* Electric conducting copper wire with insulation.

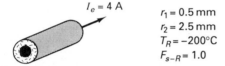

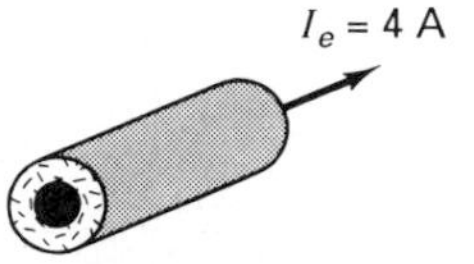

$I_e = 4$ A
$r_1 = 0.5$ mm
$r_2 = 2.5$ mm
$T_R = -200$°C
$F_{s-R} = 1.0$


*Assumptions/Conditions*

steady-state
one-dimensional
uniform internal energy generation
blackbody thermal radiation cooling

*Properties*

Copper at room temperature (Table A–C–1): $k = 386$ W/(m °C), $\rho_e = 30$ $\mu\Omega$ cm.
Insulation: $k = 0.05$ W/(m °C).

*Analysis* The power generated within the wire is represented by

$$\dot{W} = I_e^2 R_e = I_e^2 \rho_e \frac{L}{A}$$

such that

$$\frac{\dot{W}}{L} = (4\text{ A})^2(30 \times 10^{-6}\ \Omega\ 0.01\text{ m})\frac{4}{\pi(10^{-3}\text{ m})^2} = 6.11\text{ W/m}$$

Thus, the energy generated per unit volume is

$$\dot{q}_0 = \frac{\dot{W}}{LA} = \frac{6.11 \text{ W/m}}{\pi(10^{-3}\text{ m})^2/4} = 7.78 \times 10^6 \text{ W/m}^3$$

Because this energy must be radiated away,

$$\dot{W} = q_R = \sigma A_s F_{s-R}(T_s^4 - T_R^4)$$

or

$$T_s^4 = T_R^4 + 6.11L\frac{\text{W}}{\text{m}}\left(\frac{1}{\sigma A_s F_{s-R}}\right)$$

such that the surface temperature $T_s = T_2$ is found to be 288 K (=15.1°C).

To find the interfacial temperature $T_1$, we write

$$q_r = \frac{T_1 - T_s}{\dfrac{\ln (r_2/r_1)}{2\pi k_i L}} = q_R$$

$$T_1 = T_s + \frac{q_R}{L}\frac{\ln (r_2/r_1)}{2\pi k_i} = 15.1°\text{C} + 6.11\frac{\text{W}}{\text{m}}\frac{\ln (0.0025/0.0005)}{2\pi[0.05 \text{ W/(m °C)}]} = 46.4°\text{C}$$

Finally, utilizing Eq. (2–89), $T_{\text{max}}$ is calculated.

$$T_{\text{max}} = T_1 + \frac{\dot{q}_0}{k}\left(\frac{r_1}{2}\right)^2 = 46.4°\text{C} + \frac{7.78 \times 10^6 \text{ W/m}^3}{386 \text{ W/(m °C)}}\left(\frac{0.5 \times 10^{-3}\text{ m}}{2}\right)^2$$
$$= 46.4°\text{C} + 1.26 \times 10^{-3}\text{ °C} \simeq 46.4°\text{C}$$

Thus, we see that the temperature throughout the wire is essentially uniform.

## 2–6 EXTENDED SURFACES

Situations often arise in which means are sought for increasing the heat convected from a surface. A consideration of the Newton law of cooling, Eq. (1–20),

$$q_c = \bar{h}A_s(T_s - T_F) \tag{2–90}$$

suggests that $q_c$ can be increased by increasing $\bar{h}$, $T_s - T_F$, or $A_s$. As already indicated, $\bar{h}$ is a function of the geometry, fluid properties, and flow rate. The modulation of $\bar{h}$ through the control of these factors provides a means by which $q_c$ can be increased or decreased, as will be discussed in Chaps. 6 through 11. With regard to the effect of $T_s - T_F$ on the rate of heat transfer, difficulties are often encountered in automobile cooling systems in very hot weather because $T_F$ is too high. Concerning the third factor, which is the object of this section, the area of a surface that is exposed to the

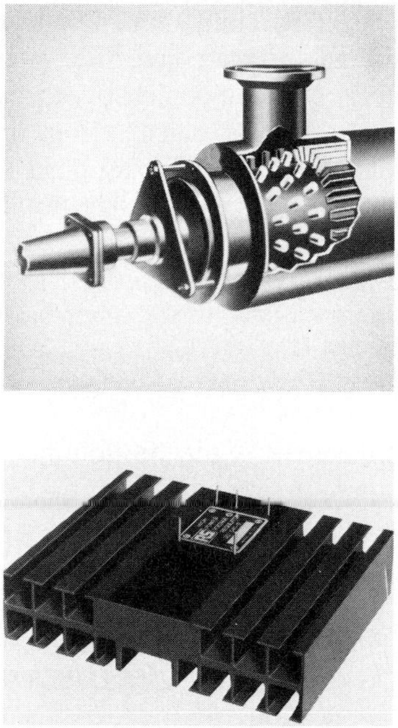

(a) Longitudinal fins and spines in a fired heater. Fluid flowing in the annulus is heated by a gas- or oil-fired flame inside the fintube. (Courtesy of Brown Fintube Company, Houston, Texas.)

(b) Fins on a high-power voltage regulator. (Courtesy of RS Components Ltd.)

**FIGURE 2–15** Typical applications of extended surfaces.

fluid is often "extended" by the use of fins or spines, as illustrated in Fig. 2–15. Familiar applications of such extended surface heat-transfer devices include automobile radiators, power transistors, and high-voltage electrical transformers. In addition, extended surfaces are commonly used to increase the rate of thermal radiation from surfaces.

Referring to the surface extension of the plane wall illustrated in Fig. 2–16,

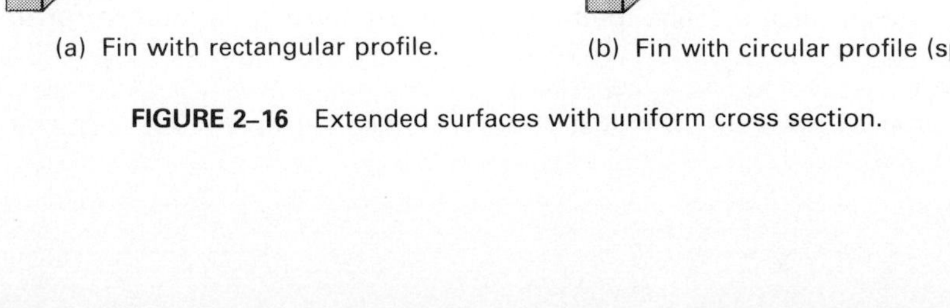

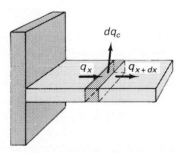

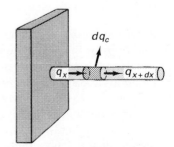


(a) Fin with rectangular profile. (b) Fin with circular profile (spine).

**FIGURE 2–16** Extended surfaces with uniform cross section.

heat is transferred from the wall into the fin itself by conduction and from the fin surface by convection. Hence, the decrease in the surface convection resistance $R_c$ brought about by the increase in surface area $A_s$ is accompanied by an increase in the conduction resistance $R_k$. In order for the rate of heat transfer from the wall to be increased by the use of a surface extension, the decrease in $R_c$ must be greater than the increase in $R_k$. As a matter of fact, the surface resistance must be the controlling factor ($R_k < R_c$, or preferably $R_k << R_c$) in practical fin applications.

In order to provide a guide for the design of extended surface systems, we introduce the *Biot number Bi*,

$$Bi = \frac{\bar{h}\ell}{k} \tag{2–91}$$

The characteristic length $\ell$ is equal to $V/A_s$, where $V$ is the fin volume. The Biot number is generally taken as a rule-of-thumb approximation for the ratio between the conduction resistance and the surface resistance; that is,

$$Bi \simeq \frac{R_k}{R_c} \tag{2–92}$$

Because $R_k$ must be considerably smaller than $R_c$ for a fin to be effective, we conclude that the Biot number should be small for fin applications. In practice, the following criterion is generally maintained in the design of convective surface extensions:†

$$Bi = \frac{\bar{h}\ell}{k} \lesssim 0.1 \tag{2–93}$$

In keeping with this criterion, we conclude that fins should generally be considered for applications involving small values of $\bar{h}$ and $\ell$ and large values of $k$. Referring back to Table 1–3 we see that likely fin applications occur for natural convection, forced convection of gases, and, to a lesser extent, forced convection of liquids. However, because of the very high values of $\bar{h}$ associated with two-phase fluids, fins are generally not useful for applications involving boiling or condensation.

## 2–6–1 Analysis

Returning to the fins with uniform cross section, which are shown in Fig. 2–16, the heat transfer in these systems is clearly two-dimensional. However, for situations in which the Biot number is much less than unity, the temperature is essentially a function of $x$ alone, such that an approximate one-dimensional analysis of the heat-transfer process can be developed. For values of the Biot number less than 0.1, the use of a one-dimensional analysis generally leads to an error less than about 10%, as will be demonstrated in Chap. 3. However, for values of the Biot number much greater than 0.1, the multidimensionality must be accounted for in the analysis. For

† An alternative less restrictive criterion given by $Bi \lesssim 1.0$ is preferred by some analysts.

such situations in which the Biot number is not small, the concept of using fins to increase the heat transfer must be questioned.

Assuming that the temperature distribution in the extended surfaces shown in Fig. 2–16 is essentially one-dimensional, the first law of thermodynamics is applied to the differential volume $A\ dx$, with the following result:

*Step 1*
$$q_x = q_{x+dx} + dq_c \tag{2–94}$$

*Step 2*
$$\frac{dq_x}{dx}\,dx + dq_c = 0 \tag{2–95}$$

where $q_c$ represents the rate of heat convected from the fin.

Whereas $q_x$ is given by the one-dimensional Fourier law of conduction, Eq. (1–10), $dq_c$ requires the use of the *general Newton law of cooling*, which is given by Eq. (1–21),

$$dq_c = h_x\,dA_s\,(T_s - T_F) \tag{2–96}$$

where $h_x$ is the *local* coefficient of heat transfer. Incidentally, the integration of this equation for situations in which $T_s$ and $T_F$ are uniform gives rise to the simpler form of the Newton law of cooling, Eq. (2–90), where $\bar{h}$ is defined by Eq. (1–22),

$$\bar{h} = \frac{1}{A_s}\int_{A_s} h_x\,dA_s \tag{2–97}$$

Utilizing the one-dimensional Fourier law of conduction and the general Newton law of cooling with $dA_s = p\,dx$, Eq. (2–95) takes the form

*Step 3*
$$\frac{d}{dx}\left(kA\frac{dT}{dx}\right)dx - h_x p\,dx\,(T - T_F) = 0 \tag{2–98}$$

where $T_s$ is set equal to $T$ in this approximate one-dimensional analysis. For our particular application, the cross-sectional area $A$, perimeter $p$, and $k$ are uniform such that Eq. (2–98) reduces to

$$\frac{d^2T}{dx^2} - \frac{h_x p}{kA}(T - T_F) = 0 \tag{2–99}$$

For the case in which the base temperature $T_0$ is specified, one boundary condition is written as

*Step 4*
$$T = T_0 \qquad \text{at } x = 0 \tag{2–100}$$

The second boundary condition is written by recognizing that the fin loses heat from the tip by convection; that is,

$$-k\frac{dT}{dx} = h_x(T - T_F) \qquad \text{at } x = L \tag{2–101}$$

where the coefficient $h_x$ at the tip is not necessarily equal to the coefficient along the perimeter. For the case in which the fin is very long, the temperature of the fin approaches $T_F$ as $x$ increases, such that Eq. (2–101) reduces to

$$\frac{dT}{dx} = T - T_F = 0 \qquad \text{as } x \to \infty \tag{2–102}$$

Of course, other boundary conditions can be written at $x = 0$ or at $x = L$, depending upon the dictates of the actual problem under consideration.

Equation (2–102) will now be used to obtain predictions for the temperature distribution and heat transfer in a very long fin. With $h_x$ approximated by $\bar{h}$ and with $T_F$ assumed to be uniform, Eq. (2–99) takes the form

$$\frac{d^2\psi}{dx^2} - m^2\psi = 0 \tag{2–103}$$

where $\psi = T - T_F$ and $m^2 = \bar{h}p/(kA)$. Recognizing that an exponential function satisfies Eq. (2–103), the substitution of

$$\psi = Ce^{cx} \tag{2–104}$$

into this differential equation gives

$$c^2 - m^2 = 0 \tag{2–105}$$

Hence, $c = \pm m$, and the solution is

$$\psi = C_1e^{mx} + C_2e^{-mx} \tag{2–106}$$

The constants $C_1$ and $C_2$ are evaluated on the basis of the boundary conditions. Based on Eq. (2–100), $\psi = T_0 - T_F$ at $x = 0$ such that

$$T_0 - T_F = C_1 + C_2 \tag{2–107}$$

As suggested above, the second boundary condition can be written in the form of Eq. (2–102) for cases in which $L$ is very long. For this situation, $\psi = 0$ as $x \to \infty$ and Eq. (2–106) gives

$$0 = \lim_{x\to\infty} (C_1e^{mx} + C_2e^{-mx}) \tag{2–108}$$

This equation requires that $C_1$ be equal to zero, such that $C_2$ is equal to $T_0 - T_F$ [from Eq. (2–107)]. Therefore, the solution for this case is

$$\psi = T - T_F = (T_0 - T_F)e^{-mx} \tag{2–109}$$

To obtain the total rate of heat transfer from the fin into the fluid $q_F$, we perform a lumped energy balance on the entire fin. It follows that $q_F = q_b$ where the rate of heat transfer at the base $q_b$ is obtained from the Fourier law of conduction,

$$q_b = -kA\left.\frac{dT}{dx}\right|_0 \tag{2–110}$$

With the temperature distribution given by Eq. (2–109), our prediction for $q_F$ becomes

$$q_F = q_b = \sqrt{\bar{h}pkA}\,(T_0 - T_F) \tag{2–111}$$

Parenthetically, this same result can also be obtained by equating $q_F$ to the total rate of heat convected from the surface,

$$q_F = \int dq_c = \int_0^\infty h_x p(T - T_F)\, dx \qquad (2\text{–}112)$$

where $h_x$ is again approximated by $\bar{h}$. Substituting for $T - T_F$ and integrating, we arrive at Eq. (2–111).

To determine the minimum fin length for which this solution applies, we merely require that $T$ be approximately equal to $T_F$ at $x$ equal to $L$; that is,

$$\frac{T_L - T_F}{T_0 - T_F} = e^{-mL} < \epsilon \qquad (2\text{–}113)$$

where $\epsilon$ is a small number. With $\epsilon$ equal to 0.01, $mL$ must be greater than a value of about 4.6 in order for our analysis to be reasonable.

For the case in which $mL$ is significantly less than 4.6, the heat transfer through the tip can be accounted for by the use of Eq. (2–101). The use of this more general boundary condition for ''short'' convecting fins gives rise to an expression for the temperature distribution of the form

$$\frac{T - T_F}{T_0 - T_F} = \frac{\cosh [m(L - x)] + [\bar{h}/(km)] \sinh [m(L - x)]}{\cosh (mL) + [\bar{h}/(km)] \sinh (mL)} \qquad (2\text{–}114)$$

To obtain $q_F$ for this type of fin, we apply the Fourier law of conduction as follows:

$$q_F = q_b = -kA \left.\frac{dT}{dx}\right|_0$$

$$= -kA \left\{ \frac{-m \sinh (mL) - m[\bar{h}/(km)] \cosh (mL)}{\cosh (mL) + [\bar{h}/(km)] \sinh (mL)} \right\} (T_0 - T_F)$$

$$= \sqrt{\bar{h}pkA} \left\{ \frac{\sinh (mL) + [\bar{h}/(km)] \cosh (mL)}{\cosh (mL) + [\bar{h}/(km)] \sinh (mL)} \right\} (T_0 - T_F) \qquad (2\text{–}115)$$

Other types of boundary conditions are sometimes encountered in fin applications. For example, for a fin with insulated tip, $T$ and $q_F$ are given by

$$\frac{T - T_F}{T_0 - T_F} = \frac{\cosh [m(L - x)]}{\cosh (mL)} \qquad (2\text{–}116)$$

and

$$q_F = \sqrt{\bar{h}pkA} \tanh (mL) (T_0 - T_F) \qquad (2\text{–}117)$$

Expressions can also be obtained for $T$ and $q_b$ (or $q_F$) for fins with specified tip temperature; that is,

$$T - T_F = (T_0 - T_F) \frac{\sinh [m(L - x)]}{\sinh (mL)} + (T_1 - T_F) \frac{\sinh (mx)}{\sinh (mL)} \qquad (2\text{–}118)$$

and

$$q_b = \frac{(T_0 - T_F)\sqrt{\bar{h}pkA}}{\sinh(mL)}\left[\cosh(mL) - \frac{T_1 - T_F}{T_0 - T_F}\right] \tag{2–119}$$

where $q_F \neq q_b$ (see Table 2–2 for $q_F$).

### 2–6–2 Fin Resistance

To express the rate of heat transfer from a fin in terms of a thermal resistance, we write

$$q_F = \frac{T_0 - T_F}{R_F} \tag{2–120}$$

For example, by introducing Eq. (2–111), the thermal resistance for a very long fin with small Biot number and negligible radiation effects is

$$R_F = \frac{1}{\sqrt{\bar{h}pkA}} \tag{2–121}$$

The thermal resistances of several standard fin geometries are summarized in Table 2–2. It should be noted that the manufacturers of fin units for electronic devices and other systems generally provide a rating for the thermal resistance of the unit, which accounts for conduction, natural convection, and radiation effects. Table 2–3 gives the thermal resistance rating for several standard anodized aluminum heat-sink fin units for electronic applications.

### 2–6–3 Fin Efficiency

Traditionally, the rate of heat transfer from surface extensions is generally presented in the literature in terms of the *fin efficiency* $\eta_F$. Fin efficiency is defined by

$$\eta_F = \frac{q_F}{q_{\max}} = \frac{q_F}{\bar{h}A_F(T_0 - T_F)} \tag{2–122}$$

where $q_{\max}$ is the rate of heat transfer for the idealistic situation in which the Biot number is equal to zero (i.e., $R_k \simeq 0$) and the entire surface area of the fin $A_F$ is at the base temperature $T_0$. Thus, $q_F$ is expressed in terms of $\eta_F$ by

$$q_F = \eta_F \bar{h} A_F (T_0 - T_F) \tag{2–123}$$

The fin efficiency is easily expressed in terms of the fin resistance by writing

$$\eta_F = \frac{T_0 - T_F}{R_F}\,\frac{1}{\bar{h}A_F(T_0 - T_F)} = \frac{1}{R_F\bar{h}A_F}$$

Thus

$$R_F = \frac{1}{\eta_F \bar{h} A_F} \tag{2–124}$$

**TABLE 2–2** Thermal resistance: Fins with small Biot number†

| *System* | *Thermal resistance $R_F$ and/or $R_b$* | *Comment* |
|---|---|---|
| Fin with uniform cross section: Very long with $Bi < 0.1$ | $\dfrac{1}{\sqrt{\bar{h}pkA}}$ | $R_F = R_b$ |
| 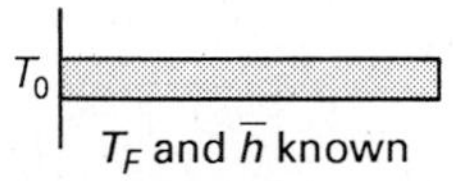$T_0$<br>$T_F$ and $\bar{h}$ known | | |
| Fin with uniform cross section: Insulated tip with $Bi < 0.1$ | $\dfrac{1}{\sqrt{\bar{h}pkA}\tanh(mL)}$ | $R_F = R_b$ |
| 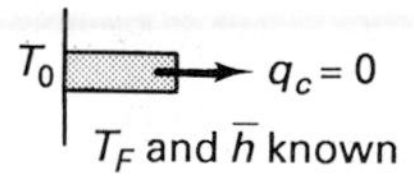$T_0$<br>$q_c = 0$<br>$T_F$ and $\bar{h}$ known | | |
| Fin with uniform cross section: Tip at $T_1$ with $Bi < 0.1$ | $\dfrac{\sinh(mL)}{\sqrt{\bar{h}pkA}\,[\cosh(mL) - (T_1 - T_F)/(T_0 - T_F)]}$ | $R_b$ |
| 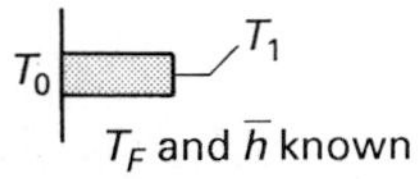$T_0$<br>$T_1$<br>$T_F$ and $\bar{h}$ known | $\dfrac{\sinh(mL)}{\sqrt{\bar{h}pkA}\,[\cosh(mL) - 1][1 + (T_1 - T_F)/(T_0 - T_F)]}$ | $R_F$ |
| Fin with uniform cross section: Convection from tip with $Bi < 0.1$ | $\dfrac{1}{\sqrt{\bar{h}pkA}\;\dfrac{\sinh(mL) + [\bar{h}/(mk)]\cosh(mL)}{\cosh(mL) + [\bar{h}/(mk)]\sinh(mL)}}$ | $R_F = R_b$ |
| 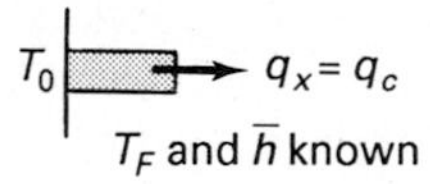$T_0$<br>$q_x = q_c$<br>$T_F$ and $\bar{h}$ known | | |
| Blackbody fin with convection and radiation and uniform cross section: Very long with $Bi < 0.1$ | $\left(\bar{h}pkA + \dfrac{2}{5}kA\sigma pF_{s-R}\dfrac{T_0^5 - 5T_0T_\infty^4 + 4T_\infty^5}{(T_0 - T_\infty)^2}\right)^{-1/2}$ | $R_F = R_b$ |
| 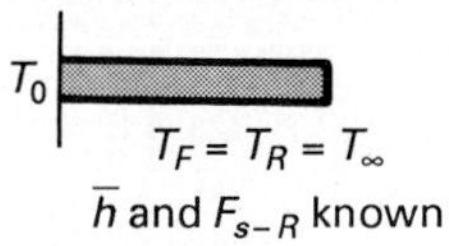$T_0$<br>$T_F = T_R = T_\infty$<br>$\bar{h}$ and $F_{s-R}$ known | | |

† $m^2 = \bar{h}p/(kA)$; $q_b = (T_0 - T_F)/R_b$.

**TABLE 2–3** Thermal resistance for natural convection and radiation from anodized aluminum fin heat-sink units†

| | *Description* | *Thermal resistance* $R_F$ |
|---|---|---|
|  | RS 401–778<br>Predrilled to accept T0–3 semiconductor case.<br>44.5 mm × 31.7 mm × 13.7 mm | 14 °C/W |
|  | RS 401–863<br>Predrilled to accept variety of plastic packaged semiconductor cases.<br>30 mm × 25 mm × 12.5 mm | 19 °C/W |
|  | RS 401–964<br>Predrilled to accept variety of plastic packaged semiconductor cases.<br>38 mm × 27 mm × 22.5 mm | 10.5 °C/W |
|  | RS 401–497<br>100 mm length, overall cross section 64.5 mm × 15 mm | 4 °C/W<br>(with fins vertical) |
|  | RS 401–403<br>100 mm length, overall cross section 123.8 mm × 26.7 mm | 2.1 °C/W<br>(with fins vertical) |
|  | RS 401–807<br>152 mm length, overall cross section 130 mm × 32 mm | 1.1 °C/W<br>(with fins vertical) |
|  | RS 401–958<br>115 mm length, overall cross section 120 mm × 120 mm | 0.5 °C/W<br>(with fins vertical) |

† Courtesy of RS Components Ltd.

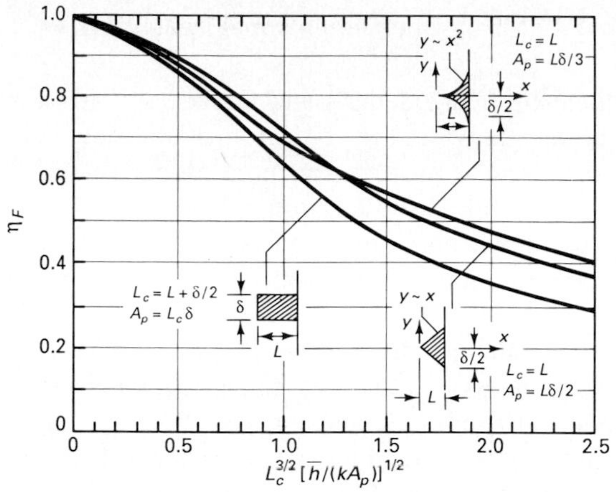


(a) Rectangular and triangular fins.

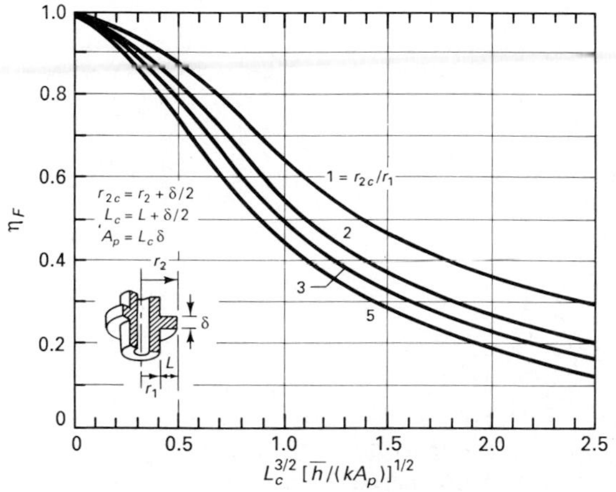


(b) Circumferential fins.

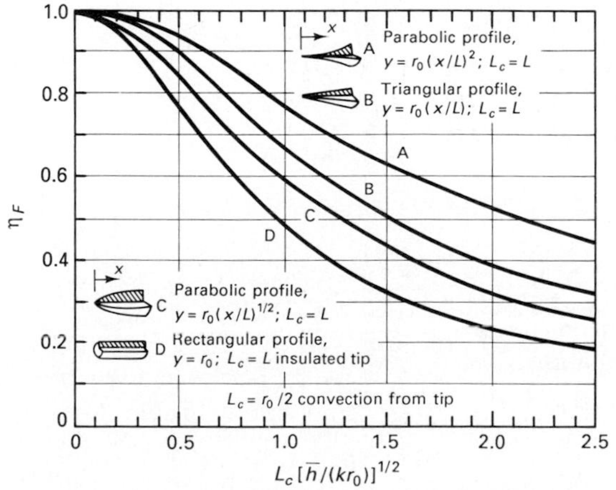


(c) Spines.

**FIGURE 2–17** Fin efficiencies for approximate one-dimensional heat transfer in various extended surfaces for small Biot number (Gardner [7]).

Relations for $\eta_F$ are given by Schneider [6] and Gardner [7] for a number of fin geometries. Fin efficiencies are shown in Fig. 2–17 for several standard-type fins. Notice that the parabolic and triangular profiles are more efficient than rectangular profiles. What is more, these nonuniform profiles contain less material than fins with rectangular profiles, which cuts down on weight and cost. Nevertheless, fins with uniform cross-sectional area are commonly employed in heat exchangers and other systems.

It should also be observed that $\eta_F$ decreases with fin length. This happens because the temperature of a fin approaches $T_F$ as $x$ increases. Of course, the local rate of heat transfer $dq_c$ convected from a fin decreases as $T - T_F$ falls. Since the Biot number $Bi$ is proportional to $\bar{h}/k$, these figures also indicate the importance of restricting the use of fins to applications in which the Biot number is small.

## EXAMPLE 2–12

A 1-cm-diameter, 3-cm-long carbon (1%) steel fin transfers heat from a wall at 200°C to a fluid at 25°C with $\bar{h}$ = 120 W/(m² °C). Determine the rate of heat transfer from the fin for the case in which the tip is insulated and thermal radiation effects are negligible.

### Solution

*Objective* Determine $q_F$.

*Schematic* Fin with insulated tip.

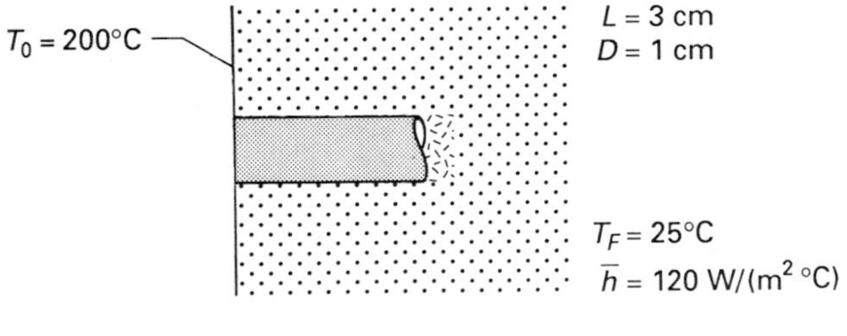

$T_0$ = 200°C
L = 3 cm
D = 1 cm
$T_F$ = 25°C
$\bar{h}$ = 120 W/(m² °C)


*Assumptions/Conditions*

steady-state
one-dimensional fin
negligible thermal radiation

*Properties* Carbon steel (1% C) at room temperature (Table A–C–1): $k$ = 43 W/(m °C).

*Analysis* We first compute the Biot number by writing

$$Bi = \frac{\bar{h}\ell}{k} = \frac{\bar{h}}{k}\frac{D}{4} = 0.00698$$

Because the Biot number is much less than 0.1, we are justified in using an approximate one-dimensional analysis.

For this situation in which the tip is insulated, $q_F$ is given by Eq. (2–117),

$$q_F = \sqrt{\bar{h}pkA}\tanh(mL)(T_0 - T_F)$$

where

$$m = \sqrt{\frac{\bar{h}p}{kA}} = 33.4/\mathrm{m}$$

and

$$\sqrt{\bar{h}pkA} = 0.113\ \mathrm{W/°C}$$

Following through with the calculation, we obtain

$$q_F = 15.1\ \mathrm{W}$$

### 2–6–4 Finned Systems

Referring to Fig. 2–18, to characterize the heat transfer from surfaces with one or more fins, we distinguish between the surface area of an individual fin $A_F$ and the area of the prime (i.e., unfinned) surface $A_p$ adjacent to the fin. The rate of heat convected per fin from the combined fin and prime surfaces is expressed in terms of the fin efficiency $\eta_F$ by

$$q_o = \bar{h}A_p(T_s - T_F) + \eta_F\bar{h}A_F(T_s - T_F) = \bar{h}(A_p + \eta_F A_F)(T_s - T_F) \quad (2\text{–}125)$$

where the coefficient of heat transfer over the fin and prime surfaces is approximated by the local mean coefficient of heat transfer $\bar{h}$. With the combined areas of the fin and prime surfaces represented by $A_o$ ($= A_p + A_F$), Eq. (2–125) is rearranged to obtain

$$q_o = \eta_o\bar{h}A_o(T_s - T_F) \quad (2\text{–}126)$$

where the *net surface efficiency* $\eta_o$† is

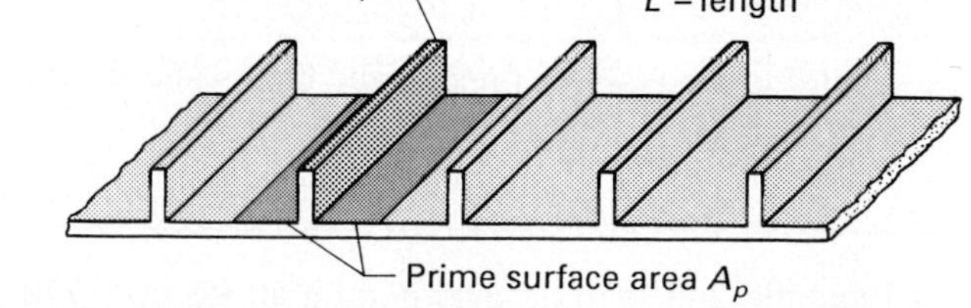


**FIGURE 2–18** Representative fin system with rectangular fins.

† $\eta_o$ is also commonly referred to as the *temperature effectiveness*.

$$\eta_o = \frac{A_p + A_F\eta_F}{A_o} = 1 - \frac{A_F}{A_o}(1 - \eta_F) \tag{2–127}$$

and $A_F/A_o$ is referred to as the *finned area fraction*. It follows that the resistance to convection heat transfer for a finned surface can be represented by

$$R_o = \frac{T_s - T_F}{q_o} = \frac{1}{\eta_o \bar{h} A_o} \tag{2–128}$$

per fin.

For applications in which the local mean coefficient of heat transfer $h$, the net surface efficiency $\eta_o$, and temperatures $T_s$ and $T_F$ are essentially uniform over the entire surface area $A_s$ of a finned surface with $N$ uniformly spaced fins, the total rate of heat transfer $q_c$ is simply written as

$$q_c = Nq_o = N[\eta_o \bar{h}\, A_o(T_s - T_F)] = \eta_o \bar{h} A_s(T_s - T_F) \tag{2–129}$$

such that the thermal resistance becomes

$$R_c = \frac{T_s - T_F}{q_c} = \frac{1}{\eta_o \bar{h} A_s} \tag{2–130}$$

On the other hand, for situations in which one or more of the four parameters $\bar{h}$, $\eta_o$, $T_s$, and $T_F$ change along the surface, as in the case of finned tube heat exchangers, more general relationships should be employed. To deal with this problem we represent the finned surface by an effective unfinned surface with equivalent surface area $A_s$ and effective perimeter $p = A_s/L$, and simply approximate the local heat flux $q_c''$ ($\equiv dq_c/dA_s$) by the local heat flux per fin $q_o''$; that is,

$$q_c'' = \frac{dq_c}{dA_s} = q_o'' = \frac{q_o}{A_o} \tag{2–131}$$

or

$$dq_c = q_o''\, dA_s = \eta_o \bar{h}(T_s - T_F)\, dA_s \tag{2–132}$$

Following this practical approach, the local convective resistance $R_c'$ for a finned surface can be approximated by

$$R_c' = \frac{T_s - T_F}{dq_c} = \frac{1}{\eta_o \bar{h}\, dA_s} \tag{2–133}$$

This relationship is useful in the analysis of finned tube banks and finned tube heat exchangers.

**EXAMPLE 2–13**

The power transistor of Example 2–5 is to be mounted on an RS 401–778 anodized aluminum heat-sink unit, as shown in Fig. E2–13a. Determine the maximum power that can be dissipated safely by this transistor for a case temperature of 80°C, ambient temperature of 25°C, and $\bar{h} = 10\ \mathrm{W/(m^2\,°C)}$, assuming that the thermal conductivity of anodized aluminum is about 200 W/(m °C). (This unit provides two fins plus an additional 800 $mm^2$ of prime surface area.)

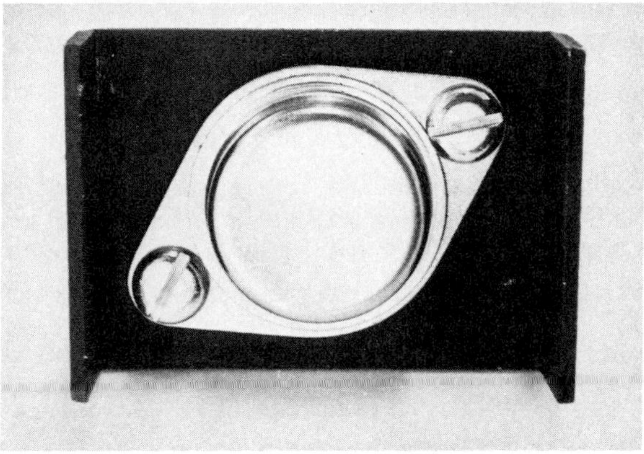

**FIGURE E2–13a**
Power transistor mounted on RS 401–778 heat sink. (Approximate fin dimensions: $L$ = 12 mm, $p$ = 64 mm, and $A$ = 31 $mm^2$.)

## Solution

*Objective* Determine the maximum operating power of the transistor.

*Schematic* Transistor/heat sink unit.

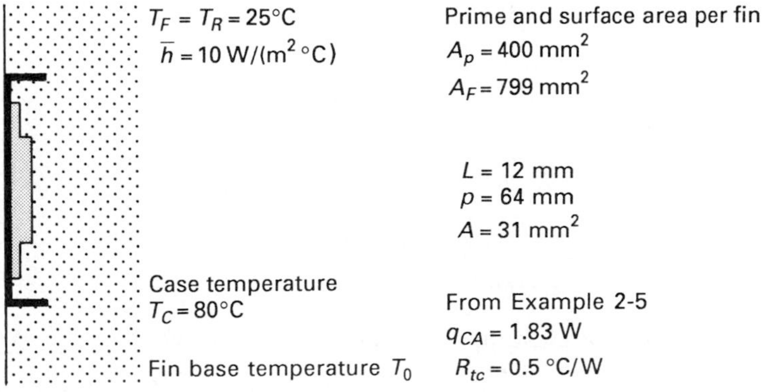
$T_F = T_R = 25°C$
$\bar{h} = 10\ W/(m^2\ °C)$
Prime and surface area per fin
$A_p = 400\ mm^2$
$A_F = 799\ mm^2$
$L = 12$ mm
$p = 64$ mm
$A = 31\ mm^2$
Case temperature
$T_C = 80°C$
From Example 2-5
$q_{CA} = 1.83$ W
$R_{tc} = 0.5$ °C/W
Fin base temperature $T_0$

*Assumptions/Conditions*

steady-state
short one-dimensional fins
uniform properties
convection and blackbody thermal radiation from heat sink

*Properties* Anodized aluminum: $k = 200$ W/(m °C).

*Analysis* Neglecting radiation effects for the moment and focusing attention on one of the two fins, we have

$$m = \sqrt{\frac{\bar{h}p}{kA}} = \sqrt{\frac{[10\ \text{W/(m}^2\ ^\circ\text{C)}](64 \times 10^{-3}\ \text{m})}{[200\ \text{W/(m}\ ^\circ\text{C)}](31 \times 10^{-6}\ \text{m}^2)}} = 10.2/\text{m}$$

$$mL = \frac{10.2}{\text{m}}(12 \times 10^{-3}\ \text{m}) = 0.122$$

Because $mL$ is much less than 4.6, we should account for the heat loss through the tip. Therefore, we employ Eq. (2–115),

$$q_F = \sqrt{\bar{h}pkA}\left\{\frac{\sinh(mL) + [\bar{h}/(km)]\cosh(mL)}{\cosh(mL) + [\bar{h}/(km)]\sinh(mL)}\right\}(T_0 - T_F)$$

where

$$\sqrt{\bar{h}pkA} = \left[10\frac{\text{W}}{\text{m}^2\ ^\circ\text{C}}(64 \times 10^{-3}\ \text{m})\left(200\frac{\text{W}}{\text{m}\ ^\circ\text{C}}\right)(31 \times 10^{-6}\ \text{m}^2)\right]^{1/2}$$

$$= 6.30 \times 10^{-2}$$

and

$$\frac{\bar{h}}{km} = \frac{10\ \text{W/(m}^2\ ^\circ\text{C)}}{[200\ \text{W/(m}\ ^\circ\text{C)}](10.2/\text{m})} = 4.90 \times 10^{-3}$$

It follows that

$$q_F = 7.95 \times 10^{-3}(T_0 - T_F)\ \text{W/}^\circ\text{C}$$

To calculate the fin efficiency, we write

$$\eta_F = \frac{q_F}{q_{\max}} = \frac{7.95 \times 10^{-3}(T_0 - T_F)\ \text{W/}^\circ\text{C}}{\bar{h}A_F(T_0 - T_F)}$$

Setting the surface area of each fin $A_F$ equal to 799 mm$^2$ and $\bar{h} = 10$ W/(m$^2$ °C), the fin efficiency is found to be $\eta_F = 99.5\%$.

After mounting the power transducer, the area $A_p$ of the primary surface of the fin unit, which is exposed to convective cooling, is approximately 400 mm$^2$ per fin. Therefore, the net surface efficiency $\eta_o$ can be computed as follows:

$$A_o = A_p + A_F = 400\ \text{mm}^2 + 799\ \text{mm}^2 = 1199\ \text{mm}^2$$

$$\eta_o = 1 - \frac{A_F}{A_o}(1 - \eta_F) = 1 - \frac{779}{1199}(1 - 0.995) = 99.7\%$$

Substituting this result into Eq. (2–129), the total rate of heat convected from the heat-sink fin unit with surface area $A_s = 2A_o = 2398$ mm$^2$ is

$$q_c = \eta_o \bar{h} A_s (T_0 - T_F) = 0.997 \left(10 \frac{\text{W}}{\text{m}^2\ ^\circ\text{C}}\right) (2398 \times 10^{-6}\ \text{m}^2)(T_0 - T_F)$$

$$= 0.0239\ (T_0 - T_F)\ \text{W/}^\circ\text{C}$$

Thus, the thermal resistance to convection for the heat-sink unit is

$$R_c = \frac{1}{0.0239\ \text{W/}^\circ\text{C}} = 41.8\ ^\circ\text{C/W}$$

(Note that $R_c$ for this heat-sink fin unit is about 49% lower than for the simple flat-plate frame system of Example 2–5.)

Assuming that the power transistor is properly attached to the heat sink (see Example 2–5), the thermal circuit for heat transfer through the heat sink is shown in Fig. E2–13b. It follows that the power dissipated through the heat sink is

$$q_{HS} = \frac{T_C - T_F}{R_{tc} + R_{HS}} = \frac{80^\circ\text{C} - 25^\circ\text{C}}{0.5\ ^\circ\text{C/W} + 41.8\ ^\circ\text{C/W}} = 1.3\ \text{W}$$

This coupled with the 1.83 W transferred directly from the transistor to the surroundings gives a total power dissipation rate of

$$q = q_T + q_{HS} = 1.83\ \text{W} + 1.3\ \text{W} = 3.13\ \text{W}$$

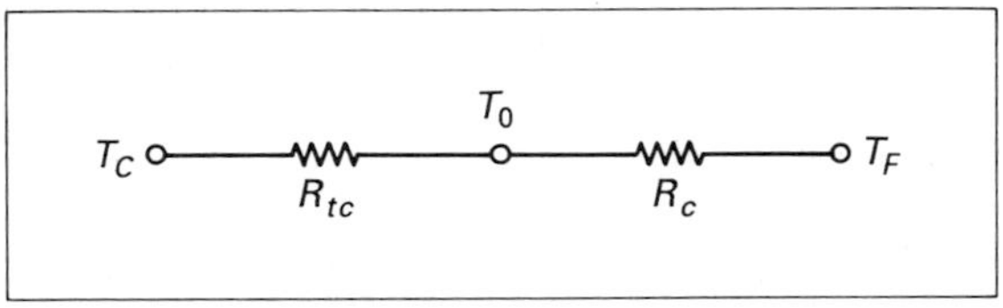


**FIGURE E2–13b**
Thermal circuit.

Because we have neglected the radiation losses, this will be a conservative estimate. The combined effects of convection and radiation heat transfer from a short fin such as this can be fairly easily analyzed by means of the numerical techniques introduced in Chap. 4. However, for design purposes, the manufacturers of heat-sink fin units generally specify the approximate thermal resistance for combined natural convection and radiation. Referring to Table 2–3, we find that the resistance of the fin unit in this example is rated by the manufacturer at 14 °C/W. But it should be noted that this rating is based on laboratory measurements in free air (at about 25°C) at an unspecified transistor frame temperature $T_C$. In actuality, $R_{HS}$ is strongly dependent upon $T_C$. Therefore, this value of $R_{HS}$ should be used as a rough estimate in design work. Replacing $R_c$ in Fig. E2–13b by this value, the rate of heat transfer from the heat sink becomes

$$q_{HS} = \frac{80^\circ\text{C} - 25^\circ\text{C}}{14.5\ ^\circ\text{C/W}} = 3.79\ \text{W}$$

such that the thermal limit on the total power dissipation from the transistor/heat-sink unit is

$$q = q_T + q_{HS} = 1.83 \text{ W} + 3.79 \text{ W} = 5.62 \text{ W}$$

Therefore, we conclude that the power transistor could be operated safely near its maximum power rating of 5 W.

## EXAMPLE 2–14

Determine the rate of heat transfer from the long anodized aluminum fin shown in Fig. E2–14. The base temperature $T_0$ is 80°C, the air and surrounding walls are 25°C, and the coefficient of heat transfer due to natural convection cooling is 10 W/(m$^2$ °C).

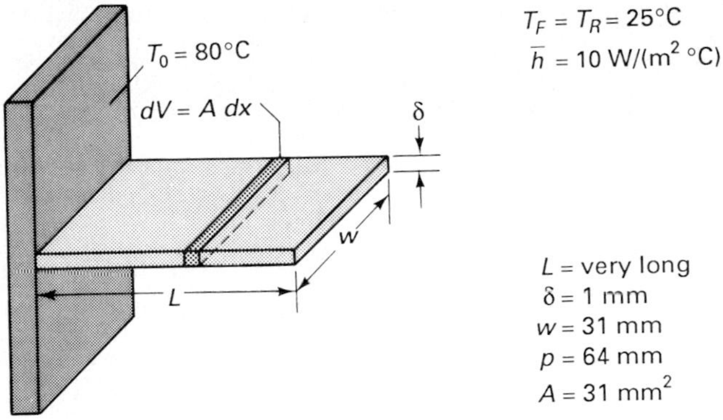
$T_0$ = 80°C
$dV = A\,dx$
δ
w
L
$T_F = T_R$ = 25°C
$\bar{h}$ = 10 W/(m$^2$ °C)
L = very long
δ = 1 mm
w = 31 mm
p = 64 mm
A = 31 mm$^2$

**FIGURE E2–14** Convecting and radiating fin.

### Solution

*Objective* Determine $q_F$ from this convecting and radiating fin.

*Assumptions/Conditions*

steady-state
long one-dimensional fin
uniform properties
natural convection and blackbody thermal radiation cooling
$F_{s-R} = 1$ for fin to surroundings

*Properties* Anodized aluminum (Example 2–13): $k = 200$ W/(m °C).

*Analysis* The heat transfer in this fin can be approximated by a one-dimensional analysis if the Biot number $R_k/R_s$ is very small. Assuming blackbody conditions, the Biot number for combined convection and radiation is approximated by

$$Bi = \frac{R_k}{R_s} = \frac{\ell}{k}[\bar{h} + \sigma F_{s-R}(T_s + T_R)(T_s^2 + T_R^2)]$$

where $\ell$ is set equal to $V/A_s$ for the fin and $F_{s-R} = 1$. As a conservative measure, $T_s$ and $T_R$ are both set equal to 353 K (= 80°C). The Biot number is then found to be $4.32 \times 10^{-5}$, such that a one-dimensional analysis can be safely used.

Referring to the lumped-differential element shown in Fig. E2–14, the application of the first law of thermodynamics gives

$$q_x = q_{x+dx} + dq_c + dq_R$$

It follows that

$$0 = \frac{dq_x}{dx}dx + dq_c + dq_R \tag{a}$$

Because the temperature along the fin is a function of $x$, we utilize the general Newton law of cooling as well as a generalized form of Eq. (1–18); that is,

$$dq_R = \sigma F_{dA_s - A_R}\, dA_s\, (T_s^4 - T_R^4)$$

where $F_{dA_s - A_R}$ represents the fraction of radiant energy leaving $dA_s$ that reaches $A_R$. Substituting these particular laws into Eq. (a), we obtain

$$\frac{d}{dx}\left(kA\frac{dT}{dx}\right)dx = h_x p\, dx\,(T - T_F) + \sigma p\, dx\, F_{dA_s - A_R}(T^4 - T_R^4)$$

or

$$\frac{d^2T}{dx^2} = \frac{h_x p}{kA}(T - T_F) + \frac{\sigma p F_{dA_s - A_R}}{kA}(T^4 - T_R^4) = m^2(T - T_F) + m_R^2(T^4 - T_R^4)$$

where $h_x$ is approximated by $\bar{h}$, $F_{dA_s - A_R}$ is approximated by $F_{s-R}$, $m^2 = \bar{h}p/(kA)$, and $m_R^2 = \sigma p F_{s-R}/(kA)$. The boundary condition at the base is

$$T = T_0 \qquad \text{at } x = 0$$

For the case in which $T_F$ and $T_R$ are both equal to $T_\infty$ and the fin is very long we also have

$$T = T_\infty \quad \text{or} \quad \frac{dT}{dx} = 0 \qquad \text{as } x \to \infty \tag{b}$$

The simplest way to solve this nonlinear system of equations for $T$ is to use the numerical finite-difference approach introduced in Chap. 4. However, we can obtain an analytical solution for the temperature gradient by making the substitution

$$\psi = \frac{dT}{dx} \tag{c}$$

This puts our nonlinear differential equation into the form

$$\frac{d\psi}{dx} = m^2(T - T_\infty) + m_R^2(T^4 - T_\infty^4) \qquad \text{(d)}$$

Based on Eq. (c), $dx = dT/\psi$. Therefore, Eq. (d) can be written as

$$\psi \frac{d\psi}{dT} = m^2(T - T_\infty) + m_R^2(T^4 - T_\infty^4)$$

We now separate the variables and integrate to obtain

$$\frac{\psi^2}{2} = m^2\left(\frac{T^2}{2} - T_\infty T\right) + m_R^2\left(\frac{T^5}{5} - T_\infty^4 T\right) + C_1$$

$$\psi = \frac{dT}{dx} = \pm\sqrt{2}\left[m^2\left(\frac{T^2}{2} - T_\infty T\right) + m_R^2\left(\frac{T^5}{5} - T_\infty^4 T\right) + C_1\right]^{1/2} \qquad \text{(e)}$$

where the negative sign is retained because the gradient is known to be negative for the case in which $T_0 > T$.

The constant $C_1$ can be evaluated by introducing the boundary condition given by Eq. (b); that is,

$$\left.\frac{dT}{dx}\right|_\infty = 0 = \sqrt{2}\left[m^2\left(\frac{T_\infty^2}{2} - T_\infty^2\right) + m_R^2\left(\frac{T_\infty^5}{5} - T_\infty^5\right) + C_1\right]^{1/2}$$

where $T(\infty) = T_\infty$. Hence, $C_1$ is given by

$$C_1 = m^2\frac{T_\infty^2}{2} + \frac{4}{5}m_R^2 T_\infty^5$$

and Eq. (e) becomes

$$\frac{dT}{dx} = -\sqrt{2}\left[m^2\left(\frac{T^2}{2} - T_\infty T + \frac{T_\infty^2}{2}\right) + m_R^2\left(\frac{T^5}{5} - T_\infty^4 T + \frac{4}{5}T_\infty^5\right)\right]^{1/2}$$

An expression can now be obtained for the rate of heat transfer from the fin by writing

$$q_F = -kA\left.\frac{dT}{dx}\right|_0$$

$$= kA\sqrt{2}\left[\frac{m^2}{2}(T_0^2 - 2T_\infty T_0 + T_\infty^2) + \frac{m_R^2}{5}(T_0^5 - 5T_\infty^4 T_0 + 4T_\infty^5)\right]^{1/2}$$

$$= \left[\bar{h}pkA(T_0 - T_\infty)^2 + \frac{2}{5}kA\sigma pF_{s-R}(T_0^5 - 5T_\infty^4 T_0 + 4T_\infty^5)\right]^{1/2}$$

or

$$q_F = \sqrt{q_c^2 + q_R^2}$$

where

$$q_c = \sqrt{\bar{h}pkA}\,(T_0 - T_\infty)$$

and

$$q_R = \left[\frac{2}{5}kA\sigma pF_{s-R}(T_0^5 - 5T_\infty^4T_0 + 4T_\infty^5)\right]^{1/2}$$

Calculating $q_c$ and $q_R$, we have

$$q_c = \left[10\,\frac{\text{W}}{\text{m}^2\,^\circ\text{C}}(64 \times 10^{-3}\,\text{m})\left(200\,\frac{\text{W}}{\text{m}\,^\circ\text{C}}\right)(31 \times 10^{-6}\,\text{m}^2)\right]^{1/2} \times (353^\circ\text{C} - 298^\circ\text{C}) = 3.46\,\text{W}$$

$$q_R = \left\{\frac{2}{5}\left(200\,\frac{\text{W}}{\text{m}\,^\circ\text{C}}\right)(31 \times 10^{-6}\,\text{m}^2)\left(5.67 \times 10^{-8}\,\frac{\text{W}}{\text{m}^2\,\text{K}^4}\right)(64 \times 10^{-3}\,\text{m})(1) \times [(353\,\text{K})^5 - 5(298\,\text{K})^4(353\,\text{K}) + 4(298\,\text{K})^5]\right\}^{1/2} = 2.94\,\text{W}$$

Thus, the total rate of heat transfer from the fin is

$$q_F = \sqrt{(3.46\,\text{W})^2 + (2.94\,\text{W})^2} = 4.54\,\text{W}$$

The fact that $q_F$ is about 31% greater than $q_c$ reinforces our earlier conclusion that radiation can play a very significant role in heat transfer from fins with near blackbody surfaces.

## 2–7 UNSTEADY HEAT-TRANSFER SYSTEMS

We now turn our attention to unsteady heat-transfer processes such as the cooling of a billet, which is illustrated in Fig. 2–19. The heat transfer in unsteady systems like this is actually multidimensional because the temperature within the body is a function

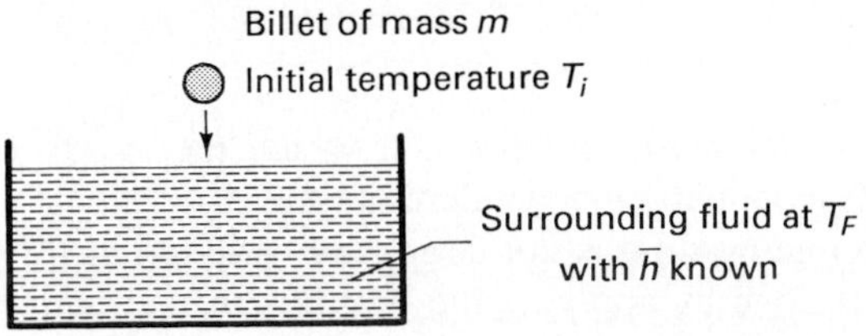
Billet of mass m
Initial temperature $T_i$
Surrounding fluid at $T_F$
with $\bar{h}$ known

**FIGURE 2–19**
Convective cooling of a billet initially at temperature $T_i$, which is dropped into a bath at temperature $T_F$.

of time $t$ and at least one space dimension. However, for problems such as this that involve convection and/or radiation, approximate lumped analyses can be utilized if the Biot number ($Bi = R_k/R_s$) is small (i.e., $Bi \lesssim 0.1$). Under these circumstances, the variation in temperature with the spatial coordinates will be very slight, such that the temperature can be taken as a function of time alone. Approximate lumped analyses are developed for representative unsteady convection heat-transfer processes in this section. Unsteady heat-transfer systems involving multidimensions and radiation will be considered in Chaps. 3 through 5.

### 2–7–1 Analysis

We consider the situation illustrated in Fig. 2–19 in which a body of mass $m$ with uniform initial temperature $T_i$ is suddenly exposed to an environmental temperature $T_F$. Applying the first law of thermodynamics to the lumped volume $V$, we obtain

$$q_c + \frac{\Delta E_s}{\Delta t} = 0 \qquad (2\text{–}134)$$

$$\bar{h}A_s(T - T_F) + \frac{dU}{dt} = 0 \qquad (2\text{–}135)$$

The internal energy $U$ can be expressed in terms of the specific heat at constant volume for this lumped system by

$$c_v = \left.\frac{\partial e}{\partial T}\right|_v = \left.\frac{\partial}{\partial T}\left(\frac{U}{m}\right)\right|_v = \left.\frac{1}{m}\frac{\partial U}{\partial T}\right|_v \qquad (2\text{–}136)$$

Since the volume is essentially constant for heat transfer in solids, $dU$ can be written as

$$dU = mc_v\,dT = \rho Vc_v\,dT \qquad (2\text{–}137)$$

Therefore, Eq. (2–135) takes the form

$$\frac{dT}{dt} + \frac{\bar{h}A_s}{\rho Vc_v}(T - T_F) = 0 \qquad (2\text{–}138)$$

The initial condition associated with this equation is

$$T = T_i \qquad \text{at } t = 0 \qquad (2\text{–}139)$$

This completes our formulation for situations in which the mass of the surrounding fluid is large and $T_F$ is essentially independent of time. However, for systems in which the mass of the surrounding fluid is not large, the variation in $T_F$ with time must be accounted for. This is done by applying the first law of thermodynamics to the surrounding fluid itself, as illustrated in Example 2–17.

For the case in which $T_F$ is constant, Eq. (2–138) can be written in the form

$$\frac{d\psi}{dt} + \frac{\bar{h}A_s\psi}{\rho V c_v} = 0 \tag{2–140}$$

where $\psi = T - T_F$. With $\bar{h}$ approximated by a constant, this homogeneous first-order equation can be separated and integrated to obtain

$$\int_{\psi_i}^{\psi} \frac{d\psi}{\psi} = \frac{-\bar{h}A_s}{\rho V c_v} \int_0^t dt \tag{2–141}$$

where $\psi_i = T_i - T_F$. Continuing, we have

$$\ln\left(\frac{\psi}{\psi_i}\right) = -\frac{\bar{h}A_s t}{\rho V c_v} \tag{2–142}$$

Hence, the temperature history is represented by

$$\frac{T - T_F}{T_i - T_F} = \exp\left(\frac{-\bar{h}A_s t}{\rho V c_v}\right) \tag{2–143}$$

The rate of heat convected from the surface at any instant $t$ can be written as

$$q_c = \bar{h}A_s(T - T_F) = \bar{h}A_s(T_i - T_F)\exp\left(\frac{-\bar{h}A_s t}{\rho V c_v}\right) \tag{2–144}$$

To obtain the total heat transferred from the surface over the time interval 0 to $t$, we simply integrate as follows:

$$\begin{aligned} Q_c &= \int_0^t q_c\,dt = \bar{h}A_s(T_i - T_F)\int_0^t \exp\left(\frac{-\bar{h}A_s t}{\rho V c_v}\right) dt \\ &= \rho V c_v(T_i - T_F)\left[1 - \exp\left(\frac{-\bar{h}A_s t}{\rho V c_v}\right)\right] \end{aligned} \tag{2–145}$$

The maximum total amount of energy that can be convected to the fluid is obtained by merely allowing $t$ to become very large; that is,

$$Q_{\max} = \rho V c_v(T_i - T_F) \tag{2–146}$$

Notice that $Q_{\max}$ is equal to the relative internal energy possessed by the body at time $t = 0$. The dimensionless ratio $Q_c/Q_{\max}$ is given by

$$\frac{Q_c}{Q_{\max}} = 1 - \exp\left(\frac{-\bar{h}A_s t}{\rho V c_v}\right) \tag{2–147}$$

The quantity $\rho V c_v/(\bar{h}A_s)$ appearing in Eq. (2–143) and associated equations is known as the *thermal time constant* $t_c$ of the system. The thermal time constant provides an indication of the length of time required for a system to approach thermal equilibrium; that is, the smaller the thermal time constant, the quicker the system response. Notice that $(T - T_F)/(T_i - T_F) = 0.368$ for $t = t_c$.

## EXAMPLE 2–15

An aluminum ball 0.5 cm in diameter initially at 250°C is dropped into a large tank of fluid (with $T_{sat} = 100°C$), which is maintained at 25°C. The mean coefficient of heat transfer is approximately equal to 3000 W/(m² °C) for boiling and 250 W/(m² °C) for nonboiling. Determine the temperature history of the ball and the instantaneous rate of convection heat transfer.

### Solution

*Objective* Determine the temperature and rate of heat transfer from the ball as a function of time.

*Schematic* Cooling of aluminum ball in extensive fluid bath.

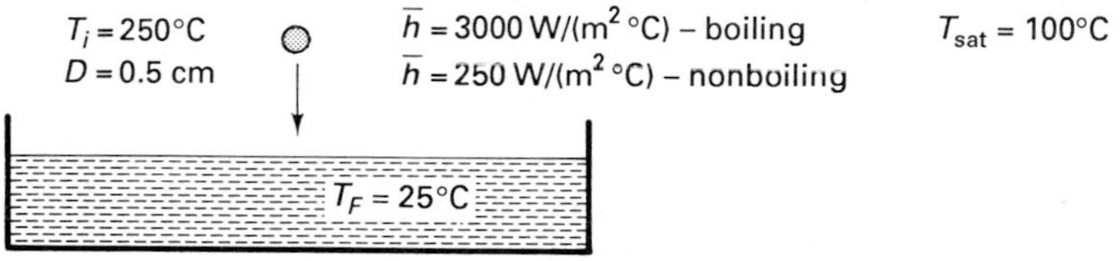

$T_i = 250°C$
$D = 0.5$ cm
$\bar{h} = 3000$ W/(m² °C) – boiling
$\bar{h} = 250$ W/(m² °C) – nonboiling
$T_{sat} = 100°C$
$T_F = 25°C$


*Assumptions/Conditions*

unsteady-state
negligible resistance to conduction
constant bath temperature with boiling occurring for ball temperature $\geq$ 100°C
constant properties of aluminum ball
negligible thermal radiation

*Properties* Aluminum at room temperature (Table A–C–1): $\rho = 2700$ kg/m³, $c_v = 0.896$ kJ/(kg °C), $k = 236$ W/(m °C).

*Analysis* Calculating the Biot number and assuming negligible radiation effects, we find that $Bi \ll 0.1$ for the entire process. Hence, the lumped differential formulation given by Eqs. (2–138) and (2–139) can be used.

The solution to Eq. (2–138) with the initial condition $T(0) = 250°C$ is given by Eq. (2–143) with $\bar{h} = 3000$ W/(m² °C); that is,

$$\frac{T - T_F}{T_i - T_F} = \frac{T - 25°C}{250°C - 25°C} = \exp\left(-\frac{\bar{h}A_s t}{\rho V c_v}\right) = \exp\left(-\frac{1.49t}{\text{s}}\right) \tag{a}$$

This equation applies to the period of time for which $T \geqslant 100°C$. To find the time $t_1$ at which boiling ceases, $T$ is set equal to 100°C in this equation, with the result

$$t_1 = \frac{s}{1.49} \ln\left(\frac{225}{75}\right) = 0.737 \text{ s}$$

The solution to Eq. (2–138) for the condition $T(t_1) = 100°C$ and $\bar{h} = 250$ W/(m$^2$ °C) is given by

$$\frac{T - 25°C}{100°C - 25°C} = \exp\left(-0.124\frac{t - t_1}{s}\right) \qquad \text{(b)}$$

Equations (a) and (b) are shown in Fig. E2–15.

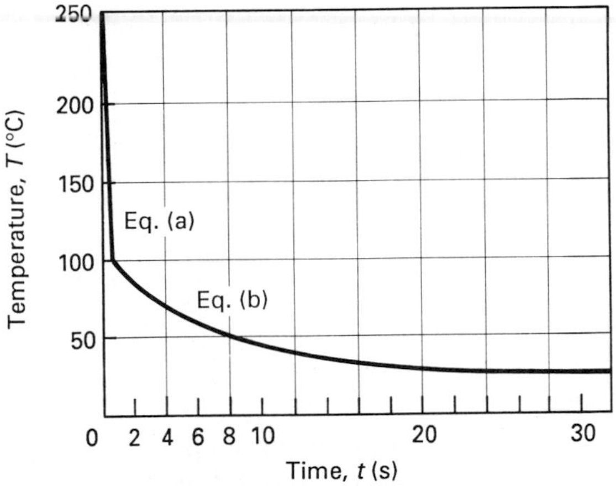


**FIGURE E2–15**
Temperature history.

To calculate the rate of convection from the surface, we write

$$q_c = \bar{h}A_s(T - T_F)$$

$$q_c'' = 3000 \frac{W}{m^2\ °C}(225°C)\exp\left(-\frac{1.49t}{s}\right)$$

for $t \leqslant 0.737$ s, and

$$q_c'' = 250 \frac{W}{m^2\ °C}(75°C)\exp\left(-0.124\frac{t - 0.737 \text{ s}}{s}\right)$$

for $t \geqslant 0.737$ s.

To estimate the levels of radiation heat transfer in this problem, one can easily calculate the instantaneous rate of thermal radiation emitted by a blackbody ball with temperature given by Eqs. (a) and (b). This calculation indicates that the maximum rate of radiation from the ball is less than 1% of the rate of convection heat transfer.

## EXAMPLE 2–16

Electric current of 5 A is suddenly passed through a 1-mm-diameter copper wire initially at temperature 25°C. Develop an expression for the instantaneous temperature of the wire for uniform internal energy generation ($\rho_e = 1.8 \times 10^{-8}\ \Omega$ m), constant surrounding fluid temperature $T_F$ of 25°C, coefficient of heat transfer $\bar{h}$ of 25 W/(m² °C), and negligible radiation effects.

### Solution

*Objective* Determine the temperature of the wire as a function of time.

*Schematic* Unsteady heat transfer from a wire; $I_e = 5$ A.

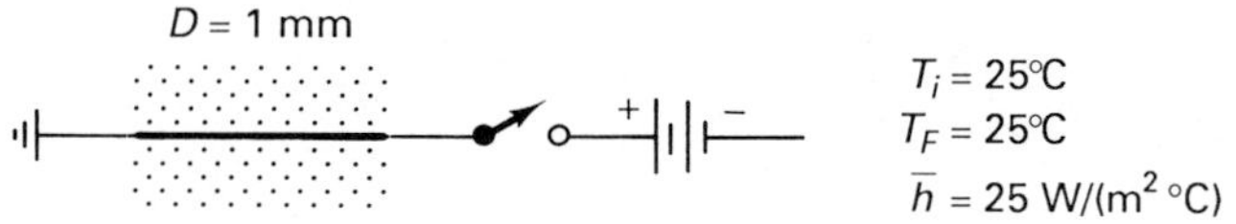
D = 1 mm
+
−
T_i = 25°C
T_F = 25°C
h̄ = 25 W/(m² °C)

*Assumptions/Conditions*

unsteady-state
constant bath temperature
uniform internal energy generation
constant properties of copper wire
negligible thermal radiation

*Properties* Copper at room temperature (Table A–C–1): $\rho = 8950$ kg/m³, $c_v = 0.383$ kJ/(kg °C), $k = 386$ W/(m °C), $\rho_e = 1.8 \times 10^{-8}\ \Omega$ m.

*Analysis* Because the Biot number for this system is much less than 0.1, an approximate one-dimensional formulation is developed. Utilizing the first law of thermodynamics, we obtain

$$\Sigma \dot{E}_o - \Sigma \dot{E}_i + \frac{\Delta E_s}{\Delta t} = 0$$

$$\bar{h}A_s(T - T_F) - \dot{q}_0 V + \rho V c_v \frac{dT}{dt} = 0$$

or

$$\frac{d\psi}{dt} + \frac{\bar{h}A_s\psi}{\rho V c_v} - \frac{\dot{q}_0}{\rho c_v} = 0 \tag{a}$$

where $\psi = T - T_F$. The initial condition is

$$\psi(0) = 0$$

To obtain the homogeneous solution $\psi_H$, we write

$$\frac{d\psi_H}{\psi_H} = -\frac{\bar{h}A_s}{\rho V c_v}\,dt$$

and integrate, with the result

$$\ln\left(\frac{\psi_H}{C_1}\right) = -\frac{\bar{h}A_s t}{\rho V c_v} \qquad \psi_H = C_1 \exp\left(\frac{-\bar{h}A_s t}{\rho V c_v}\right)$$

To obtain the particular solution $\psi_p$, we assume that

$$\psi_p = C_3 + C_4 t$$

Substituting this expression into Eq. (a),

$$C_4 + \frac{\bar{h}A_s}{\rho V c_v}(C_3 + C_4 t) = \frac{\dot{q}_0}{\rho c_v}$$

Hence, we see that $C_4 = 0$ and $C_3 = \dot{q}_0 V/(\bar{h}A_s)$; that is,

$$\psi_p = \frac{\dot{q}_0 V}{\bar{h}A_s}$$

Thus, $\psi$ takes the form

$$\psi = C_1 \exp\left(\frac{-\bar{h}A_s t}{\rho V c_v}\right) + \frac{\dot{q}_0 V}{\bar{h}A_s}$$

Utilizing the initial condition,

$$C_1 = -\frac{\dot{q}_0 V}{\bar{h}A_s}$$

and

$$\psi = T - T_F = \frac{\dot{q}_0 V}{\bar{h}A_s}\left[1 - \exp\left(\frac{-\bar{h}A_s t}{\rho V c_v}\right)\right]$$

To compute the rate of internal energy generation per unit volume $\dot{q}_0$, we write

$$\dot{q}_0 = \frac{I_e^2 \rho_e}{A^2} = \frac{(5\ \text{A})^2(1.8 \times 10^{-8}\ \Omega\ \text{m})}{[\pi(0.001\ \text{m})^2/4]^2} = 730\ \text{kW/m}^3$$

It follows that

$$\frac{\dot{q}_0 V}{\bar{h}A_s} = \frac{730\ \text{kW/m}^3}{25\ \text{W/(m}^2\ {}^\circ\text{C)}}\ \frac{0.001\ \text{m}}{4} = 7.3^\circ\text{C}$$

$$\frac{\bar{h}A_s}{\rho V c_v} = \frac{25\ \text{W/(m}^2\ {}^\circ\text{C)}}{(8950\ \text{kg/m}^3)[0.383\ \text{kJ/(kg}\ {}^\circ\text{C)}]}\ \frac{4}{0.001\ \text{m}} = 2.92 \times 10^{-2}/\text{s}$$

Hence, the instantaneous temperature of the wire is given by

$$T = 25°\text{C} + 7.3°\text{C}\left[1 - \exp\left(-\frac{2.92 \times 10^{-2}t}{\text{s}}\right)\right] \tag{b}$$

This equation is shown in Fig. E2–16. Notice that the wire temperature approaches a steady-state value of 32.3°C after about 160 s.

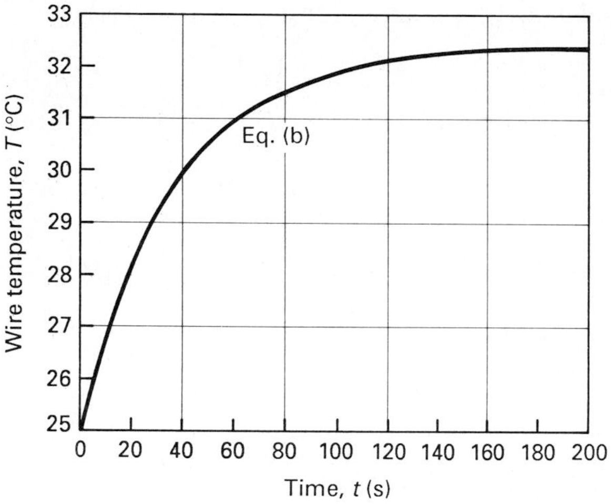


**FIGURE E2–16**
Temperature history.

The instantaneous convection heat-transfer flux is obtained by coupling the Newton law of cooling and Eq. (b).

$$q''_c = \bar{h}(T_s - T_F) = 25\,\frac{\text{W}}{\text{m}^2\,°\text{C}}\,(7.3°\text{C})\left[1 - \exp\left(-\frac{2.92 \times 10^{-2}t}{\text{s}}\right)\right]$$

$$= 183\,\frac{\text{W}}{\text{m}^2}\left[1 - \exp\left(-\frac{2.92 \times 10^{-2}t}{\text{s}}\right)\right]$$

The heat flux increases from zero at time zero to a maximum steady-state value of 183 $\text{W/m}^2$.

## EXAMPLE 2–17

The aluminum ball of Example 2–15 is dropped into a 0.0012-kg oil bath, which is initially at 25°C; the boiling point of the oil is 400°C and the properties are $\rho = 82.5$ $\text{kg/m}^3$, $c_v = 2.2$ kJ/(kg °C). The oil is well stirred such that $\bar{h} = 350$ W/($\text{m}^2$ °C) and the temperature throughout the bath is uniform, but time-dependent. The container is insulated. Determine the final steady-state temperature of the ball and oil and the time required to reach steady state for the case in which radiation effects are small.

**Solution**

*Objective* Determine the temperature and time to reach steady state.

*Schematic* Cooling of aluminum ball in finite oil bath.

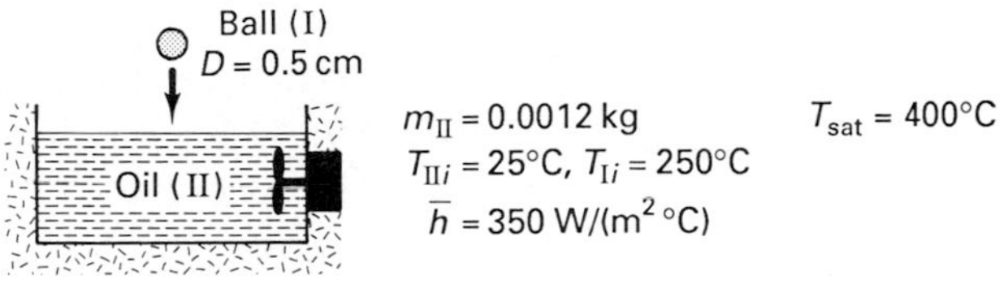
Ball (I)
D = 0.5 cm
Oil (II)
m_II = 0.0012 kg
T_IIi = 25°C, T_Ii = 250°C
h̄ = 350 W/(m² °C)
T_sat = 400°C

*Assumptions/Conditions*

unsteady-state
time dependent bath temperature
constant properties of the aluminum ball and oil
negligible thermal radiation

*Properties*

(I) Aluminum ball (Table A–C–1): $\rho = 2700$ kg/m$^3$, $c_v = 0.896$ kJ/(kg °C), $k = 236$ W/(m °C).

(II) Oil: $\rho = 82.5$ kg/m$^3$, $c_v = 2.2$ kJ/(kg °C).

*Analysis* We develop lumped energy balances on both the ball and the oil as follows:

*Ball* $$0 = \bar{h}A_s(T_\text{I} - T_\text{II}) + (\rho V c_v)_\text{I}\frac{dT_\text{I}}{dt} \tag{a}$$

*Oil* $$\bar{h}A_s(T_\text{I} - T_\text{II}) = (\rho V c_v)_\text{II}\frac{dT_\text{II}}{dt} \tag{b}$$

$$T_\text{I}(0) = T_{\text{I}i} = 250°\text{C} \qquad T_\text{II}(0) = T_{\text{II}i} = 25°\text{C} \tag{c,d}$$

These differential equations are put into the operator format

$$(D + K_\text{I})T_\text{I} = K_\text{I}T_\text{II} \tag{c}$$

$$(D + K_\text{II})T_\text{II} = K_\text{II}T_\text{I} \tag{f}$$

where $K_\text{I} = \bar{h}A_s/(\rho V c_v)_\text{I} = 0.174$/s and $K_\text{II} = \bar{h}A_s/(mc_v)_\text{II} = 1.04 \times 10^{-2}$/s. We now eliminate $T_\text{II}$ by combining Eqs. (e) and (f).

$$(D + K_\text{I})T_\text{I} = K_\text{I}K_\text{II}T_\text{I}\frac{1}{D + K_\text{II}}$$

or

$$\frac{d}{dt}\left(\frac{dT_\text{I}}{dt}\right) + (K_\text{I} + K_\text{II})\frac{dT_\text{I}}{dt} = 0$$

Separating the variables, we obtain

$$\frac{d(dT_\text{I}/dt)}{dT_\text{I}/dt} = -(K_\text{I} + K_\text{II})\,dt$$

A first integration gives

$$\ln\left(\frac{dT_\text{I}/dt}{C_\text{I}}\right) = -(K_\text{I} + K_\text{II})t$$

or

$$\frac{dT_\text{I}}{dt} = C_\text{I}\exp\left[-(K_\text{I} + K_\text{II})t\right]$$

A second integration gives

$$T_\text{I} = \frac{C_\text{I}}{K_\text{I} + K_\text{II}}\{1 - \exp\left[-(K_\text{I} + K_\text{II})t\right]\} + C_2$$

Referring to Eq. (a), we see that

$$\frac{dT_\text{I}}{dt} = -K_\text{I}(T_{\text{I}i} - T_{\text{II}i}) \qquad \text{at } t = 0$$

Thus, $C_\text{I} = -K_\text{I}(T_{\text{I}i} - T_{\text{II}i})$. To evaluate $C_2$, we employ the initial condition given by Eq. (c); that is, $C_2 = T_{\text{I}i}$. Hence, our solution for $T_\text{I}$ is

$$\frac{T_\text{I} - T_{\text{I}i}}{T_{\text{I}i} - T_{\text{II}i}} = \frac{K_\text{I}}{K_\text{I} + K_\text{II}}\{\exp\left[-(K_\text{I} + K_\text{II})t\right] - 1\}$$

The final steady-state temperature $T_{ss}$ of the system is

$$T_{ss} = 250°\text{C} - 225°\text{C}\frac{K_\text{I}}{K_\text{I} + K_\text{II}} = 37.7°\text{C}$$

The time for the temperature to reach 99% of its steady-state value is obtained by writing

$$0.01 = \exp\left[-(K_\text{I} + K_\text{II})t_{ss}\right]$$

$$t_{ss} = \frac{\ln 0.01}{-(K_\text{I} + K_\text{II})} = \frac{\ln 0.01}{-0.184} = 25.0 \text{ s}$$

### 2-7-2 Electrical Analogy

As suggested in Chap. 1, an analogy exists between unsteady one-dimensional heat transfer and the unsteady flow of electric current. For unsteady heat-transfer systems, we represent the thermal capacity $C$ ($\equiv mc_v$) of a mass by an electrical capacitor $C_e$. The electric current passing through a capacitor is proportional to the time rate of change of voltage,

$$I_e = -C_e \frac{dE_e}{dt} \qquad (2\text{–}148)$$

With this in mind, the thermal network for the problem under consideration is shown in Fig. 2–20. The thermal capacitor is initially charged at the potential $T_i$ with the switch in position 1. The process is then initiated by throwing the switch to position 2, with the energy stored in the capacitor being dissipated through the resistance. With the product $R_eC_e$ set equal to $R_cC$ [$= \rho c_v V/(\bar{h}A_s)$] (i.e., equal time constants) and with $E_{ei} - E_{e0}$ set equal to $T_i - T_F$, the flow of electric current in this circuit is perfectly analogous to the flow of heat in the thermal system. Hence, instantaneous measurements in $I_e$ and $E_e - E_{e0}$ correspond to the instantaneous rate of heat transfer $q$ and temperature difference $T - T_F$, respectively.

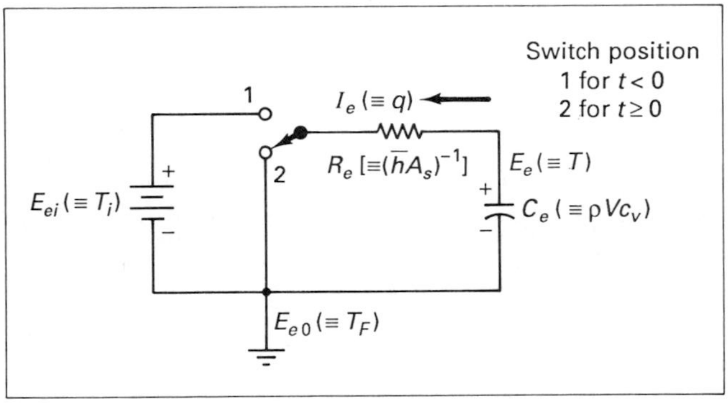


**FIGURE 2–20** Analogous electrical circuit for unsteady lumped-capacitance convection heat transfer.

## 2-8 PRACTICAL SOLUTION RESULTS

For systems without internal energy generation, we have found that the rate of heat transfer can be expressed in terms of practical equations taking either of the two formats,

$$q = \frac{\Delta T}{R} = kS\,\Delta T \qquad (2\text{–}149)$$

where $R$ is the *thermal resistance*, $S$ is the *conduction shape factor*, and $\Delta T$ is the *temperature difference* that characterizes the particular problem. The thermal resistances for the key steady one-dimensional nongenerating systems treated in this chapter are summarized in Table 2–4. Effects of convection, thermal radiation, and composite materials on steady-state heat transfer in these systems can be accounted for by the network approach. The thermal resistances for several fin units are given in Tables 2–2 and 2–3.

**TABLE 2–4** Thermal resistance: Steady one-dimensional systems

| *System* | *Thermal resistance* $R$ |
|---|---|
| Plane wall: Surfaces at $x = 0$ and $x = L$ at $T_1$ and $T_2$. 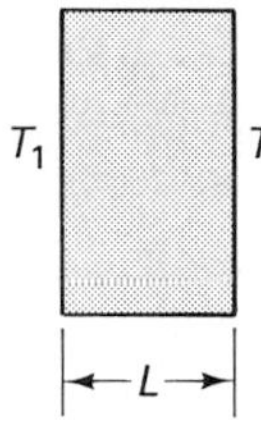  | $\frac{L}{kA}$ |
| Hollow Circular Cylinder: Surfaces at $r = r_1$ and $r = r_2$ at $T_1$ and $T_2$. 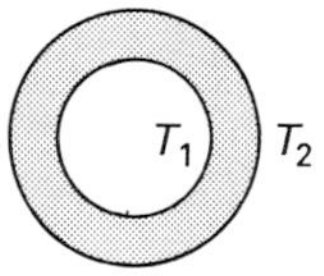  | $\frac{\ln (r_2/r_1)}{2\pi Lk}$ |
| Hollow Circular Cylinder Section: Surfaces at $r = r_1$ and $r = r_2$ at $T_1$ and $T_2$; surfaces at $\theta = 0$ and $\theta = \theta_1$ insulated. 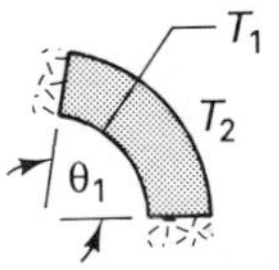  | $\frac{\ln (r_2/r_1)}{\theta_1 Lk}$ |
| Hollow Sphere: Surfaces at $r = r_1$ and $r = r_2$ at $T_1$ and $T_2$. 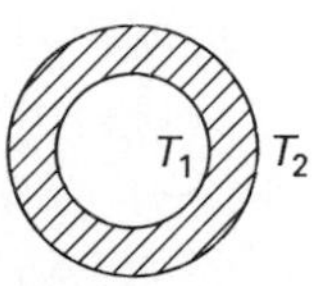  | $\frac{r_2 - r_1}{4\pi r_1 r_2 k}$ |

For situations involving nonuniform conditions, the local rate of heat transfer $dq$ is expressed in terms of a *local thermal resistance* $R'$ by

$$dq = \frac{\Delta T}{R'} \tag{2–150}$$

The local thermal resistance associated with convection heat transfer is of the form

$$R'_c = \frac{1}{h_x \, dA_s} \tag{2–151}$$

for unfinned surfaces [which follows from the general Newton law of cooling, Eq. (2–96)], and by Eq. (2–133),

$$R'_c = \frac{1}{\eta_o \bar{h} \, dA_s} \tag{2–133}$$

for finned surfaces. These relationships prove to be very useful in the analysis of heat exchangers.

In regard to the problem of heat transfer with internal energy generation, the maximum temperature $T_{max}$ is a very important design parameter. $T_{max}$ is given in Table 2–5 for two standard geometries.

**TABLE 2–5** $T_{max}$ for systems with uniform internal energy generation

| *System* | $T_{max}$ |
|---|---|
| Flat Plate 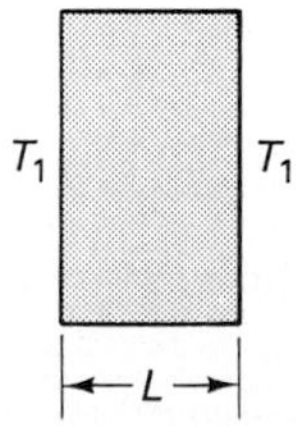  | $T_1 + \frac{\dot{q}_0}{k}\frac{L^2}{8}$ |
| Solid Circular Cylinder with radius $r_1$ 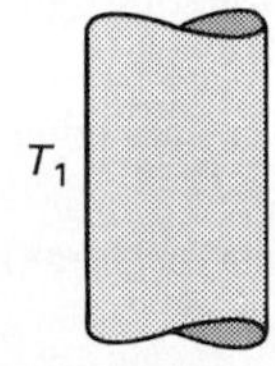  | $T_1 + \frac{\dot{q}_0}{k}\frac{r_1^2}{4}$ |

For unsteady convection systems with small values of Biot number we found that the instantaneous rate $q$ and total $Q$ are given by

$$q_c = hA_s(T_s - T_F)\exp\left(\frac{-\bar{h}A_st}{\rho Vc_v}\right) \tag{2–144}$$

$$Q_c = \rho Vc_v(T_i - T_F)\left[1 - \exp\left(\frac{-\bar{h}A_st}{\rho Vc_v}\right)\right] \tag{2–145}$$

## 2–9 SUMMARY

In this chapter we have dealt with classic one-dimensional heat-transfer systems. These include conduction in flat plates, hollow cylinders and spheres, and composites; conduction with convection and other types of boundary conditions; conduction with variable thermal conductivity; and conduction with internal energy generation. In addition, approximate analyses have been developed for steady heat transfer in extended surfaces and unsteady heat transfer in systems with small values of Biot number. The principles set forth in the chapter can generally be adapted to the analysis of any heat-transfer problem involving one independent variable for which solutions are not already available. These principles are also applied in the development of analyses for more complex multidimensional conduction heat-transfer systems, which are considered in Chaps. 3 and 4.

## ■ REVIEW QUESTIONS

**2–1.** What is meant by one-dimensional heat transfer?

**2–2.** Summarize the steps involved in the formal differential approach to determining the rate of heat transfer in a one-dimensional system.

**2–3.** How is the definition for derivative used in the development of the differential formulation for one-dimensional conduction heat transfer?

**2–4.** Write the differential formulation for steady one-dimensional heat transfer through a plane wall with specified surface temperatures $T_1$ and $T_2$.

**2–5.** Write the differential formulation for steady one-dimensional heat transfer through a plane wall with the surfaces exposed to fluids with temperatures $T_{F1}$ and $T_{F2}$.

**2–6.** Define the thermal contact coefficient. How can the thermal contact resistance be reduced?

**2–7.** Explain why the heat transfer in combined series–parallel composite wall arrangements is generally multidimensional.

**2–8.** Summarize the steps involved in the short method for analyzing one-dimensional conduction heat-transfer processes.

**2–9.** Write the Fourier law of conduction for one-dimensional heat transfer $q_r$ in (a) hollow cylinders, and (b) hollow spheres.

**2–10.** Define the critical radius.

**2–11.** Write relations for the thermal resistance $R_k$ for one-dimensional conduction in (a) flat plates, (b) hollow cylinders, and (c) hollow spheres.

**2–12.** Define the temperature coefficient of thermal conductivity.

**2–13.** Write the differential formulation for heat transfer in a hollow cylinder with uniform internal energy generation $\dot{q}$ and uniform surface temperatures $T_1$ and $T_2$.

**2–14.** What is the purpose of using fins?

**2–15.** Define the Biot number and explain its significance for heat transfer from extended surfaces.

**2–16.** Why is the *general* Newton law of cooling rather than the simple form of the law used in the analysis of heat transfer from extended surfaces?

**2–17.** Define fin resistance.

**2–18.** Define (a) fin efficiency, and (b) net surface efficiency. How are these two characteristics related?

**2–19.** What is the significance of the Biot number for unsteady heat transfer processes such as the one illustrated in Fig. 2–19?

**2–20.** Describe the analogy between heat transfer and the flow of electric current for unsteady processes in which the temperature is a function of time alone.

## PROBLEMS

**2–1.** Measurements for the axial temperature distribution in a 1-in.-diameter copper rod of 9-in. length are given in Fig. P2–1. The ends are at 345°F and 158°F, and the perimeter is insulated.

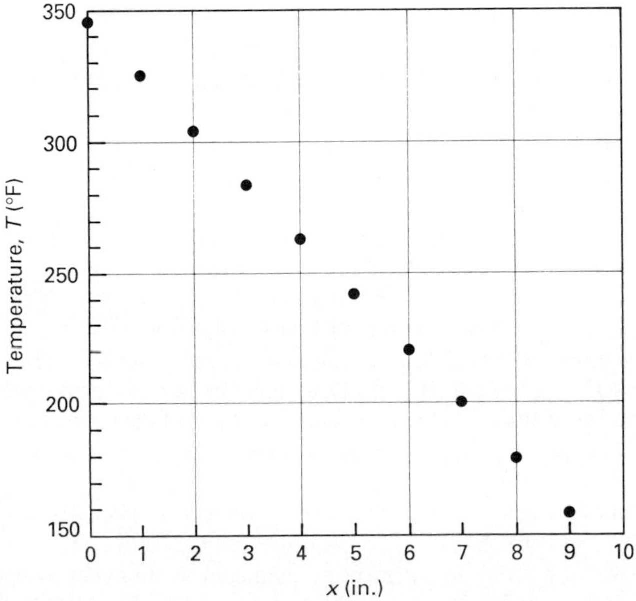


**FIGURE P2–1**

Compare the theoretical temperature distribution with these data and determine the rate of heat transfer.

**2–2.** One surface of a stainless steel plate [$k = 10$ W/(m K)] of 10-cm thickness is at a temperature of 80°C. The other surface is exposed to a uniform heat flux of 500 W/m$^2$. Determine the temperature distribution in the plate and the unknown surface temperature.

**2–3.** The surface temperatures of a glass window with dimensions $\frac{1}{32}$ in. by 1 ft by 1 ft are 100°F and 68°F. Determine the rate of heat conducted through the window and the total heat loss by conduction over a 12-h period.

**2–4.** An electrical heating plate of 1-mm thickness is sandwiched between two identical materials of 1-cm thickness and 0.5-m$^2$ cross-sectional area. The sides are insulated and $\dot{q}$ is equal to 500 kW/m$^3$. The surface temperatures of one plate are found to be equal to 85°C and 150°C. Determine the thermal conductivity of the material.

**2–5.** Determine the unknown temperature for Prob. 2–2 using the electrical analog approach.

**2–6.** Solve Prob. 2–3 using the electrical analog approach.

**2–7.** Determine the rate of heat conducted per unit area across a 1-mm-thick glass window with the air temperature being 100°F outside and 72°F inside. The coefficients of heat transfer are 50 Btu/(h ft$^2$ °F) and 10 Btu/(h ft$^2$ °F), respectively.

**2–8.** The inside of a furnace wall 1 ft thick constructed of fireclay brick is 1000°F. The room temperature is 50°F. Determine the rate of heat transfer through a 20-ft$^2$ area and the surface temperature if $\bar{h} = 150$ Btu/(h ft$^2$ °F).

**2–9.** A 10-cm-thick copper plate is exposed to a fluid at 5°C with $\bar{h} = 150$ W/(m$^2$ °C). The rate of heat transfer per unit area into the other surface of the plate is 1000 W/m$^2$. Determine the surface temperatures of the plate.

**2–10.** A 10-cm-thick plate [$k = 16$ W/(m °C)] coated with carbon black is exposed to radiation from a blackbody source at 200°C. The rate of heat transfer per unit area into the other surface of the plate is 1000 W/m$^2$. Determine the surface temperatures of the plate.

**2–11.** If the nonradiating surface of the plate described in Prob. 2–10 is maintained at 15°C, determine the unknown surface temperature and the rate of heat transfer per unit area.

**2–12.** A 10-cm-thick aluminum plate coated with carbon black is exposed to radiation on both sides from blackbody sources at 200°C and 2000°C. Determine the surface temperatures of the plate and the rate of heat transfer per unit area through the plate.

**2–13.** A furnace wall is constructed of 5-in.-thick brick. The radiating surfaces and the hot gases within the furnace are at a temperature of 2000°F with $\bar{h}$ equal to 500 Btu/(h ft$^2$ °F). The air temperature outside the furnace wall is at 85°F with $\bar{h}$ equal to 10 Btu/(h ft$^2$ °F). Determine the surface temperatures of the wall and the rate of heat transfer per unit area.

**2–14.** The inside surface temperature of a 5-cm-thick brick furnace wall is 1000°C. The outside air temperature is 30°C and $\bar{h} = 10$ W/(m$^2$ °C). (a) Determine the outside surface temperature. (b) Determine the thickness of fiber glass needed to reduce the surface temperature by 50%.

**2–15.** The general formula for the Newton–Raphson iteration method for solving an equation of the form $f(x) = 0$ is given in reference 1 as $x_{i+1} = x_i - f(x_i)/f'(x_i)$, where $i$ is the iteration index. Demonstrate that Eq. (d) in Example 2–3 can be solved in only three steps by this approach, starting with $T_1 = 1000$ K.

**2–16.** A composite wall consists of a series arrangement of aluminum (5 cm thick), copper (10 cm thick), and stainless steel (2 cm thick). The surface temperatures are 15°C and 100°C. Determine the rate of heat transfer per unit area, the temperature distribution, and the interfacial temperatures.

**2–17.** As illustrated in Fig. P2–17, the exterior wall of a typical home consists of a layer of common building brick, an air space created by 2 × 4 studs, and a layer of gypsum plaster. Estimate the rate of heat loss through such a wall with 30 ft length and 10 ft height for inside and outside surface temperatures of 72°F and 32°F, neglecting the effects of convection within the air space.

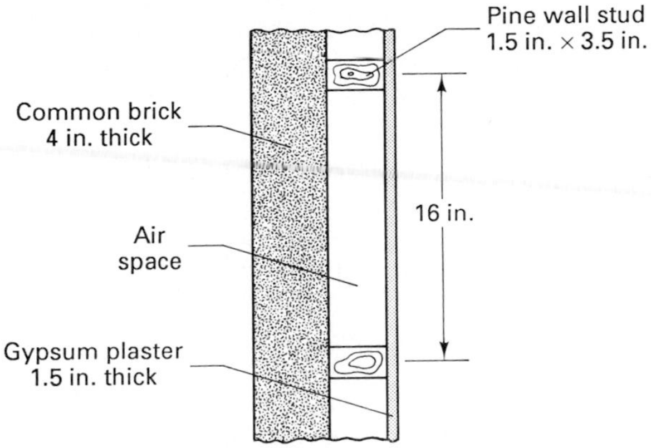
Pine wall stud
1.5 in. × 3.5 in.
Common brick
4 in. thick
16 in.
Air
space
Gypsum plaster
1.5 in. thick

**FIGURE P2–17**

**2–18.** Reconsider Prob. 2–17 for the case in which the air space is loosely packed with fiber glass insulation.

**2–19.** The wall of an ice chest consists of a 0.4-mm-thick low-carbon (1% C) steel plate, a 1.9-cm-thick layer of styrofoam insulation [$k$ = 0.035 W/(m °C)], and a 0.65-cm-thick fiber glass liner. Determine the rate of heat loss per unit area assuming that the coefficients of heat transfer are 850 W/($m^2$ °C) on the inside and 5 W/($m^2$ °C) on the outside and an ambient temperature of 27°C.

**2–20.** Two aluminum plates 10 cm thick are bolted together with a contact pressure of 15 atm. The surfaces of the composite plate are at temperatures 100°C and 0°C. Determine the temperature distribution and rate of heat transfer for the cases in which the interfacial surfaces are (a) polished and silicon grease is used; (b) ground with roughness of about 2.5 μm, for which case $h_{tc} \simeq$ 11.4 kW/($m^2$ °C).

**2–21.** A 10-cm-thick composite wall consists of a parallel arrangement of aluminum with 0.1-$m^2$ cross-sectional area and stainless steel with 0.5-$m^2$ cross-sectional area. The surface temperatures are 15°C and 100°C. Determine the rate of heat transfer and the temperature distribution in this one-dimensional system.

**2–22.** One surface of the composite wall of Prob. 2–21 is exposed to a fluid at 100°C with $\bar{h} = 5$ W/($m^2$ °C) and the other surface is held at 15°C. Estimate the rate of heat transfer in this two-dimensional system.

**2–23.** The thermal conductivity of a nonhomogeneous wall of thickness $L$ is given as a function of $x$. The surfaces at $x = 0$ and $x = L$ are maintained at $T_1$ and $T_2$. This system is a generalization of a series composite and is one-dimensional. Develop expressions for the rate of heat transfer $q_x$ for (a) $k = k_1 (1 + \gamma x)$, and (b) $k = k_1$ for $0 \leq x < L/2$ and $k = k_2$ for $L/2 < x \leq L$;

$k_1 = 15$ W/(m °C), $k_2 = 100$ W/(m °C), and $\gamma = 0.1$/m. Show that your solution to part (b) is equivalent to Eq. (2–31).

**2–24.** The thermal conductivity of a nonhomogeneous wall of thickness $L$ is given as a function of $y$. The surfaces at $x = 0$ and $x = L$ are held at $T_1$ and $T_2$. This system is a generalization of a parallel composite and is one-dimensional. Develop expressions for the rate of heat transfer for (a) $k = k_1(1 + \gamma y)$, and (b) for $k = k_1$ for $0 \leqslant y < w/2$ and $k = k_2$ for $w/2 < y \leqslant w$; $k_1 = 15$ W/(m °C), $k_2 = 100$ W/(m °C), and $\gamma = 0.1$/m. Show that your solution to part (b) is equivalent to Eq. (2–35).

**2–25.** The inside and outside surface temperatures of a hollow copper cylinder with 1-cm and 5-cm radii are 90°C and 15°C, respectively. Determine the rate of heat transfer. Also plot the temperature distribution and compare your results with the temperature distribution in a flat plate of 4 cm thickness.

**2–26.** Reconsider Prob. 2–25 for the case of a hollow sphere.

**2–27.** Saturated water at a pressure of 1 atm flows inside a hollow stainless steel cylinder of 5-cm and 6-cm radii. The outside air temperature is 50°C with $\bar{h}_1$ and $\bar{h}_2$ equal to 500 W/(m$^2$ °C) and 30 W/(m$^2$ °C), respectively. Determine the rate of heat transfer over a length of 1 m and determine the surface temperatures.

**2–28.** The inside surface temperature of a hollow aluminum sphere with $r_1$ and $r_2$ equal to 10 cm and 15 cm, respectively, is 50°C. The outside surface, which is coated with carbon black, exchanges radiant energy with the walls of an enclosure which are maintained at 500°C. Determine the outside surface temperature of the sphere and the rate of heat transfer.

**2–29.** The inside surface temperature of a hollow aluminum sphere of 1-in. and 6-in. radii is 0°F. The outside surface is exposed to air at a temperature of 50°F with $\bar{h}$ equal to 15 Btu/(h ft$^2$ °F). Determine the unknown surface temperature and the rate of heat transfer.

**2–30.** To reduce the rate of heat transfer in the sphere described in Prob. 2–29, a layer of fiber glass insulation is used. Determine the thickness of insulation required to reduce the heat loss by 50%. Is the critical radius of importance for this problem?

**2–31.** Steam at 150°C passes through a 2-cm-O.D. thin-walled tube that is insulated using a material with $k = 0.167$ W/(m °C). The outside air temperature is 35°C and $\bar{h}$ is equal to 6 W/(m$^2$ °C). Determine the thickness of the insulating material for which the heat loss is a maximum. What is the rate of heat transfer for this condition?

**2–32.** (a) Show that the differential equation for one-dimensional ($r$ direction) conduction heat transfer in a hollow sphere takes the form

$$\frac{d}{dr}\left(r^2 \frac{dT}{dr}\right) = 0$$

(b) Based on the differential-formulation approach, develop expressions for $T$ and $q_r$ for the case in which the surface temperatures at $r_1$ and $r_2$ are maintained at $T_1$ and $T_2$.

**2–33.** Develop the differential formulation for one-dimensional conduction heat transfer in a hollow cylinder with the temperature at $r_1$ maintained at $T_1$ and with the surrounding fluid at temperature $T_F$ with $\bar{h}$ known.

**2–34.** Show that the critical radius $r_c$ for convection from a hollow sphere is equal to $2k/\bar{h}$.

**2–35.** A spherical thin-walled metallic container is used to store liquid nitrogen at 1 atm, as shown in Fig. P2–35. The container has a diameter of 0.5 m and is covered with an evacuated reflective insulation wall composed of silica powder. The insulation is 25 mm thick, and its

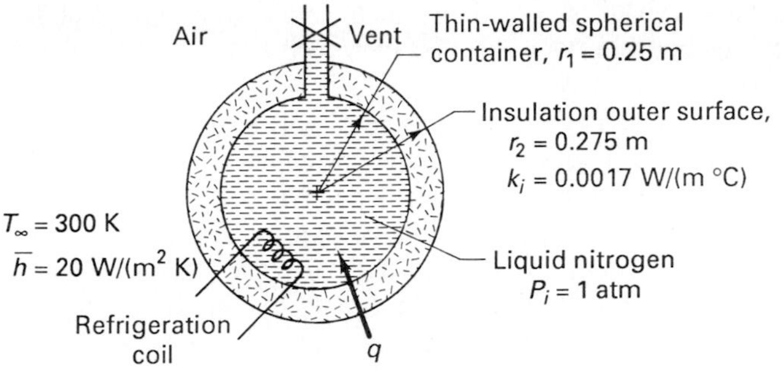


**FIGURE P2–35**

outer surface is exposed to ambient air at 300 K and 1 atm. The convection coefficient is known to be 20 W/(m$^2$ K). What rate of refrigeration is required if the container vent is closed? What is the rate of liquid boil-off if the vent is open and the refrigeration is shut off?

**2–36.** Develop a relation for the critical radius associated with the insulation of a hollow sphere with $T = T_1$ at $r = r_1$ and blackbody radiation with $F_{s-R}$ specified and $T_R \gg T_1, T_2$.

**2–37.** Determine the thickness of insulative pipe covering [$k_i = 0.428$ Btu/(h ft °F)] required to reduce the convective heat loss by 25% from a 1.9-in-O.D. pipe with surface temperature equal to 200°F. The surrounding fluid is at 70°F and $\bar{h} = 3.6$ Btu/(h ft$^2$ °F).

**2–38.** A 1-cm-diameter pipe with surface temperature of 200°C is surrounded by air at 20°C with $\bar{h} = 3$ W/(m$^2$ °C). Calculate the change in rate of heat loss if a 0.75-cm layer of fiber glass insulation is added.

**2–39.** Estimate the critical radius for blackbody radiation from a hollow cylinder [$k = 0.45$ W/(m °C)] with the inside surface at $r_1 = 1$ cm at a temperature $T_1 = 0$°C. The cylinder is surrounded by a vacuum with $F_{s-R} = 1$ and $T_R = 1000$ °C.

**2–40.** Determine the temperature distribution and rate of heat transfer through a 5-cm-thick wall in which the thermal conductivity varies linearly with temperature; that is, $k = k_0(1 + \beta T)$, where $k_0 = 50$ W/(m °C) and $\beta = 0.005$/°C. The surface temperatures are 15°C and 55°C. Utilize the short method.

**2–41.** Develop the differential formulation for Prob. 2–40 and solve these equations for the temperature distribution and the rate of heat transfer through the wall.

**2–42.** The wall described in Prob. 2–40 is exposed to a uniform heat flux equal to 500 W/m$^2$ on one surface and to a fluid at 50°C with $\bar{h}$ equal to 40 W/(m$^2$ °C) on the other side. Determine the surface temperatures of the wall.

**2–43.** Determine the heat transfer through a 10-cm-thick iron wall with surface temperatures equal to 2 K and 10 K. Refer to Fig. A–C–2 for the variation in thermal conductivity of iron at cryogenic temperatures.

**2–44.** Show that the energy equation for one-dimensional conduction heat transfer in a hollow sphere with internal energy generation takes the form

$$\frac{1}{r^2}\frac{d}{dr}\left(r^2\frac{dT}{dr}\right) + \frac{\dot{q}}{k} = 0$$

**2–45.** Develop a solution for the temperature distribution in a sphere with uniform internal energy generation and specified surface temperature.

**2–46.** For uniform energy generation in a flat plate with both surfaces maintained at $T_1$, $q_x$ can be easily deduced. With this point in mind, use the short method to solve for the temperature distribution in this system.

**2–47.** A 1-mm-diameter copper wire of 1 m length conducts 20 A of electric current. Determine the rate of heat transfer from the wire and determine the maximum temperature within the wire if the surrounding fluid temperature is 50°C and $\bar{h}$ is equal to 200 W/(m$^2$ °C).

**2–48.** A 1-mm-diameter aluminum wire is suspended in still air with an ambient temperature of 25°C and a convection heat-transfer coefficient of 10 W/(m$^2$ °C). The electrical resistance of this wire per unit length is 0.037 Ω/m. Determine the current that is required to produce wire temperatures of (a) 100°C, and (b) 200°C.

**2–49.** Show that the volume of a differential spherical shell is $dV = 4\pi r^2\, dr$.

**2–50.** If the wire described in Prob. 2–47 is insulated with 1 cm of material with a thermal conductivity of 0.1 W/(m °C), determine the temperature in the wire and the outside surface temperature of the insulation.

**2–51.** Heat transfer in electric motors is important for two reasons. First, energy lost through heat transfer obviously reduces the efficiency of the machine. Second, because of deterioration of the insulation with increasing temperature, the life expectancy of electric motors is shortened by overheating. (As a rule of thumb, the time to failure of organic insulation is halved for each 8° to 10°C rise.) Because of its importance in the operation of electric motors, the operating temperature should be carefully monitored. Describe how the measurement of the electrical resistance of a motor coil can be used in conjunction with Eq. (2–68) to estimate the temperature.

**2–52.** The one-dimensional idealization of temperature distribution in a circular fin with specified base temperature $T_0$ and small Biot number is shown in Fig. P2–52. Although the approximate one-dimensional approach to analyzing extended surfaces with small values of $Bi$ is generally quite adequate for design work, the actual temperature distribution in the vicinity of the base is distinctly multidimensional. This problem has been studied by Sparrow and Hennecke [8] and others. Show the nature of the actual temperature distribution in the vicinity of the base of this circular fin by sketching representative isotherms. How does the two-dimensionality of the temperature distribution affect the actual heat transfer from the fin and wall surface?

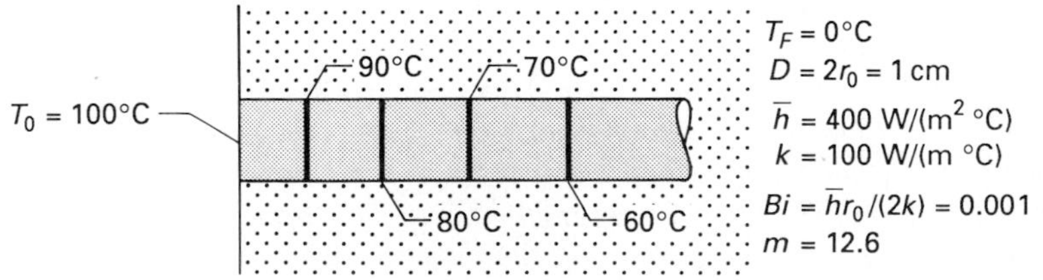


**FIGURE P2–52**

**2–53.** A 5-mm-diameter copper fin of 1 cm length is attached to a wall at 200°C. The air temperature is 5°C and $\bar{h}$ is equal to 25 W/(m$^2$ °C). Calculate the Biot number and determine the rate of heat transfer from this short fin, the temperature at the end, and the fin efficiency.

**2–54.** A 1-cm-diameter stainless steel rod connects two walls at 0°C and 25°C, which are 5 cm apart. The surrounding fluid temperature is 15°C and $\bar{h}$ is equal to 120 W/(m$^2$ °C). Determine the heat-transfer rate into the rod at each wall. What is the net rate of heat transfer between the fluid and the rod?

**2–55.** Develop Eqs. (2–114) and (2–115).

**2–56.** Develop Eqs. (2–118) and (2–119) and relations for $q_F$ and $q_L$.

**2–57.** One surface of a very thin plate [$k = 250$ W/(m °C)] has a surface temperature $T_s$ equal to 100°C. The other surface is exposed to a fluid with $T_F = 20$°C and $\bar{h} = 75$ W/(m$^2$ °C). A circular fin ($D = 2$ mm, $L = 1$ cm) is attached to the surface. Determine the rate of heat transfer from the fin, and the number of fins required per square meter of primary surface in order to increase the rate of heat transfer by 100%.

**2–58.** Air and water are separated by a thin plane wall made of carbon steel. It is proposed to increase the heat transfer rate between these two fluids by adding straight rectangular carbon steel fins $\frac{1}{8}$ in. thick, 1 in. long, and spaced 1 in. between centers, to the wall. What percent increase in heat transfer can be realized by adding fins to (a) the air side and (b) the water side of the plane wall? The air and water side coefficients may be taken as 40 and 250 Btu/(h ft$^2$ °F), respectively.

**2–59.** Assuming that the emissivity of the fin in Example 2–12 is equal to 0.2, show that the effects of thermal radiation from the fin are small.

**2–60.** Estimate the rate of heat transfer in cylindrical fins with diameter of 2 cm and length 5 cm which are exposed to a convection environment with $\bar{h}$ equal to 20 W/(m$^2$ °C), for two fin materials (copper and 1.5% carbon steel); $T_0 = 100$°C and $T_F = 0$°C. Utilize the one-dimensional assumption in these calculations. Also calculate the Biot number and comment on the appropriateness of the one-dimensional analysis for these fins.

**2–61.** A 2-cm-diameter 10-cm-long copper heating rod is connected to two surfaces which are at 50°C. The energy generation uniformly within the rod is equal to 100 W. The surrounding fluid is also at 50°C with $\bar{h} = 150$ W/(m$^2$ °C). Determine the temperature distribution in the rod and the rate of heat transfer to the fluid and to the walls.

**2–62.** Estimate the increase in heat-transfer rate which can be obtained from a cylindrical wall by using one aluminum pin-shaped fin per square inch, each having a diameter of $\frac{3}{16}$ in. and a length of 1 in. Assume that the heat-transfer coefficient is 25 Btu/(h ft$^2$ °F), $T_0 = 600$°F, and $T_F = 70$°F.

**2–63.** Estimate the operating case temperature of the power amplifier (without heat sink) of Example 2–5 for a power output of 3.4 W. Assume that $T_R = T_F = 25$°C and $\bar{h} = 10$ W/(m$^2$ °C).

**2–64.** Silicon high-power transistors with TO-3 case are to operate in the range 3 to 30 W. The thermal resistances are rated at $R_{JC} = 2$ °C/W and $R_{CA} = 38$ °C/W. The junction temperature must not exceed 150°C. Estimate the maximum power that a transistor can be operated at if the ambient temperature is 25°C and it is to be mounted (a) without a heat sink, (b) on a RS 401–778 heat sink, and (c) on a RS 401–403 heat sink.

**2–65.** The high-power voltage regulator unit shown in Fig. 2–15(b) is to operate at 30 W. Its thermal resistance when standing upright in free air is rated by the manufacturer at 1.3 °C/W. Determine the maximum air temperature if the case temperature is to be held to 80°C.

**2–66.** In order to operate a silicon power transistor with a TO-3 case at its maximum power rating of 115 W, the large RS 401–958 heat-sink fin unit is to be used. Determine the operating temperature of the case and the junction temperature for this application, assuming $R_{JC} =$ 1 °C/W, $R_{CA} = 38$ °C/W, and the ambient temperature is 25°C.

**2–67.** The case-to-ambient thermal resistance for a silicon transistor housed in a TO-5 case is rated at 150 °C/W. Determine the power at which the case temperature will reach an operating temperature of 175°C for an ambient temperature of 25°C.

**2–68.** Determine the power at which the case temperature of the transistor of Prob. 2–67 will reach the maximum value if the transistor is mounted on the RS 401–964 heat-sink fin unit.

**2–69.** Determine the maximum power at which a germanium transistor can be operated and the case temperature if the junction-to-case thermal resistance is rated at 2 °C/W. The transistor is housed in a TO-3 case with case-to-ambient thermal resistance of 30 °C/W. The air temperature is 25°C and the junction temperature must not exceed 100°C.

**2–70.** Determine the maximum power dissipation for Prob. 2–69 if the transistor is mounted on a RS 401–497 heat-sink fin unit.

**2–71.** A copper ball of 1 in. diameter initially at 75°F is dropped into an ice bath which is maintained at 32°F. The bath is stirred such that the coefficient of heat transfer is 250 Btu/(h ft$^2$ °F). Determine the temperature history of the ball and the length of time required to reach steady state. Also determine the rate of heat transfer to the fluid as a function of time.

**2–72.** Show that the effects of thermal radiation are small for the unsteady process in Example 2–15. Assume that the emissivity of the surface is 0.2.

**2–73.** The sole plate of a household iron has a surface area of 0.05 m$^2$ and is made of AISI 316 stainless steel with a total mass of 1.5 kg. The iron is rated at 500 W and is initially at the ambient temperature, which is 30°C. The coefficient of heat transfer is 10 W/(m$^2$ °C). After the iron is turned on, how long will it take to reach 110°C?

**2–74.** A wire of diameter $D$ initially at a temperature of 30°C is suddenly exposed to a fluid with sinusoidal temperature given by $T_F = [100 + 50 \sin (2\pi t)]$°F with $\bar{h}$ equal to 5 Btu/(h ft$^2$ °F). Use the lumped analysis approach to develop a relation for the temperature of the wire.

**2–75.** A 1-mm-diameter copper constantan thermocouple junction is to be used for temperature measurement in a gas stream. The convection coefficient between the junction surface and the gas is known to be 400 W/(m$^2$ °C). If the junction is at 25°C and is placed in a gas stream that is at 200°C, how long will it take for the junction to reach 199°C? What is the time constant?

**2–76.** The electrical circuit shown in Fig. P2–76 is reported to be analogous to an unsteady lumped thermal system. However, this circuit is clearly different from the circuit shown in Fig. 2–20. How is it that both circuits can be used to represent the behavior of the same thermal system?

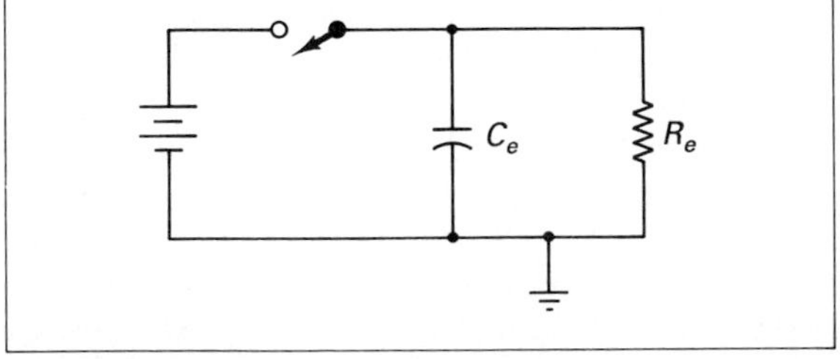


**FIGURE P2–76**

**2–77.** A well mixed fluid (1) contained in a thin walled metallic vessel (2) are both initially at a uniform temperature $T_i$. At time $t = 0$ the system is suddenly immersed in a surrounding fluid at temperature $T_F$. Sketch an electric circuit that is analogous to this two-lump thermal system.

**2–78.** Measurements for the axial temperature distribution in the copper rod shown in Fig. P2–78a are given in Fig. P2–78b. To estimate the significance of two-dimensional effects, develop an approximate one-dimensional analysis for the axial temperature distribution and compare your results with the data. Estimate the maximum percent difference between one-dimensional theory and the data.

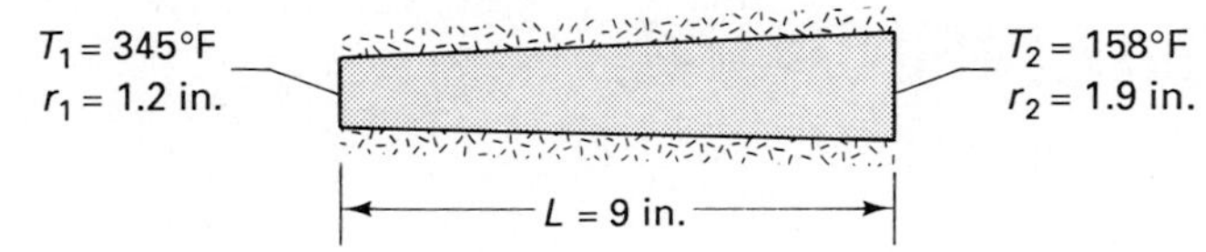


**FIGURE P2–78a**

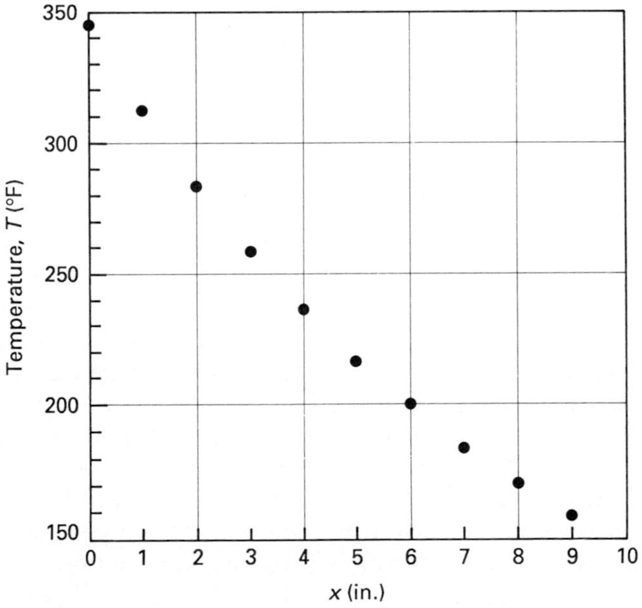


**FIGURE P2–78b**

# CHAPTER 3

# CONDUCTION HEAT TRANSFER: MULTIDIMENSIONAL

## 3–1 INTRODUCTION

As we have seen, one-dimensional heat transfer sometimes occurs in flat-plate, cylindrical, and spherical geometries. We have also seen that the heat transfer in multidimensional thermal systems such as series–parallel composites, extended surfaces, and unsteady convective or radiative heating and cooling of bodies can sometimes be approximated by one-dimensional analyses. But many heat-transfer systems encountered in practice are distinctly multidimensional and cannot be approximated by one-dimensional mathematical models. Therefore, we now turn our attention to the analysis of multidimensional conduction heat transfer. Our first order of business will be to present the general Fourier law of conduction for multidimensional conduction heat transfer. Differential formulations and solutions will then be developed for representative systems. In addition, practical solution results will be given to aid in the design of standard multidimensional conduction-heat-transfer systems. Throughout our study, attention will be restricted to systems involving isotropic materials.

## 3–2 GENERAL FOURIER LAW OF CONDUCTION

As indicated in Chap. 1, a more general form of the Fourier law must be utilized in the analysis of heat transfer when more than one spatial dimension is involved. For such multidimensional problems, the Fourier law of conduction takes the form (in Cartesian coordinates; see Fig. 3–1)

$$\boldsymbol{q}'' = q''_x\,\mathbf{i} + q''_y\,\mathbf{j} + q''_z\,\mathbf{k} = -k\left(\frac{\partial T}{\partial x}\mathbf{i} + \frac{\partial T}{\partial y}\mathbf{j} + \frac{\partial T}{\partial z}\mathbf{k}\right) \tag{3–1}$$

where the heat flux vector $\boldsymbol{q}''$ is the resultant heat transfer per unit area, with components

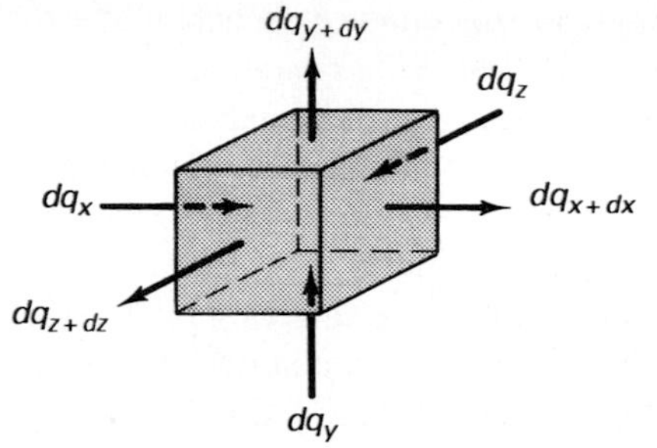


**FIGURE 3–1**
Differential control volume for rectangular solid; $dA_x = dy\,dz$, $dA_y = dx\,dz$, $dA_z = dx\,dy$.

represented by†

$$q''_x = \frac{dq_x}{dA_x} = -k\frac{\partial T}{\partial x} \qquad (3\text{–}1a)$$

$$q''_y = \frac{dq_y}{dA_y} = -k\frac{\partial T}{\partial y} \qquad (3\text{–}1b)$$

and

$$q''_z = \frac{dq_z}{dA_z} = -k\frac{\partial T}{\partial z} \qquad (3\text{–}1c)$$

The terms $dq_x$, $dq_y$, and $dq_z$ represent the rates of heat transfer through the differential areas $dA_x = dy\,dz$, $dA_y = dx\,dz$, and $dA_z = dx\,dy$, respectively. It follows that the total rate of heat transfer through a finite area, say $A_x = w\delta$, can be written as

$$q_x = \int_{A_x} -k\frac{\partial T}{\partial x}\,dA_x = \int_0^{\delta}\int_0^{w} -k\frac{\partial T}{\partial x}\,dy\,dz \qquad (3\text{–}2)$$

Notice that the term $\partial T/\partial x$ must be retained within the integral(s) in Eq. (3–2) for situations in which $T$ is a function of two or three spatial dimensions. However, for the case in which $T$ and $k$ are functions of $t$ and/or $x$ only, Eq. (3–2) reduces to the simple form of the Fourier law given by Eq. (1–11),

$$q_x = -kA_x\frac{\partial T}{\partial x} \qquad (3\text{–}3)$$

## 3–3 MATHEMATICAL FORMULATIONS

Of the various methods that have been developed for the mathematical modeling of energy transfer in conduction-heat-transfer systems, the differential and finite-difference formulations are the most frequently used. The differential formulation approach,

† The general Fourier law of conduction is also commonly represented by the vector equation

$$\mathbf{q}'' = -k\,\nabla T$$

and the tensor equation

$$q''_j = -k\frac{\partial T}{\partial x_j}$$

which was introduced in Chap. 2, is used in this chapter. Whereas the differential formulation approach provides the basis for establishing exact analytical solutions to many fundamental problems, the finite-difference approach lends itself to the analysis of the more complex problems often encountered in practice for which exact solutions are difficult, if not impossible to obtain. The finite-difference approach is featured in Chap. 4.

Two other methods commonly employed in the analysis of conduction heat transfer involve the use of finite-element and integral formulations. The finite-element approach, which is the newest of these four methods, has gained in popularity over the last few years. However, the development of finite-element formulations is generally more involved than finite-difference formulations. Therefore, this approach will not be developed in our present study. The simpler integral approach will be introduced in the context of the theoretical analysis of convection heat transfer in Chap. 7.

## 3–4 DIFFERENTIAL FORMULATION

Differential formulations are developed in this section for multidimensional conduction heat transfer in Cartesian and standard radial coordinate systems. The differential formulation for a multidimensional system is developed by following the same steps outlined in Sec. 2–2–4 for one-dimensional systems. The boundary conditions to be considered in this chapter include the basic type, which were introduced in Sec. 2–2–3 (i.e., specified temperature, convection, blackbody thermal radiation, specified heat flux, and conduction in composites).

### 3–4–1 Cartesian Coordinate System

Consider heat transfer in the rectangular solid shown in Fig. 3–1. The initial temperature is $T_i$ and the internal energy generation per unit volume is given by $\dot{q}$. The conditions at the six surfaces are, of course, prescribed by the boundary conditions. Applying the first law of thermodynamics to the differential element $dx\ dy\ dz$, we obtain

*Step 1*

$$dq_x + dq_y + dq_z + \dot{q}\ dV = dq_{x+dx} + dq_{y+dy} + dq_{z+dz} + \frac{\partial e}{\partial t}\ dm \qquad (3\text{–}4)$$

where $\Delta E_s/\Delta t = dm\ \partial e/\partial t = \rho\ dV\ c_v\ \partial T/\partial t$ with $dV = dx\ dy\ dz$. (Note that differential rates of heat are transferred through the differential areas $dA_x$, $dA_y$, and $dA_z$). Utilizing the definition of the partial derivative, we have [see Eq. (A–A–6) in the Appendix]

$$dq_{x+dx} = dq_x + \frac{\partial(dq_x)}{\partial x}\ dx \qquad (3\text{–}5)$$

plus similar expressions relating $dq_y$ and $dq_z$ to $dq_{y+dy}$ and $dq_{z+dz}$. Thus, Eq. (3–4) becomes

*Step 2*
$$\dot{q}\,dV = \frac{\partial(dq_x)}{\partial x}\,dx + \frac{\partial(dq_y)}{\partial y}\,dy + \frac{\partial(dq_z)}{\partial z}\,dz + dV\,\rho c_v \frac{\partial T}{\partial t} \tag{3–6}$$

Introducing the Fourier law of conduction, we have

*Step 3*
$$\frac{\partial}{\partial x}\left(k\,dA_x \frac{\partial T}{\partial x}\right) dx + \frac{\partial}{\partial y}\left(k\,dA_y \frac{\partial T}{\partial y}\right) dy + \frac{\partial}{\partial z}\left(k\,dA_z \frac{\partial T}{\partial z}\right) dz + \dot{q}\,dV = dV\,\rho c_v \frac{\partial T}{\partial t} \tag{3–7}$$

Dividing through by the differential volume, this equation reduces to

$$\frac{\partial}{\partial x}\left(k \frac{\partial T}{\partial x}\right) + \frac{\partial}{\partial y}\left(k \frac{\partial T}{\partial y}\right) + \frac{\partial}{\partial z}\left(k \frac{\partial T}{\partial z}\right) + \dot{q} = \rho c_v \frac{\partial T}{\partial t} \tag{3–8}$$

for general variable property conditions, or

$$\frac{\partial^2 T}{\partial x^2} + \frac{\partial^2 T}{\partial y^2} + \frac{\partial^2 T}{\partial z^2} + \frac{\dot{q}}{k} = \frac{1}{\alpha}\frac{\partial T}{\partial t} \tag{3–9}$$

for constant thermal conductivity, where the *thermal diffusivity* $\alpha$ is equal to $k/(\rho c_v)$.

The formulation is completed by writing

$$T = T_i(x,y,z) \qquad \text{at } t = 0 \tag{3–10}$$

plus six conditions in $x$, $y$, and $z$. These boundary conditions are expressed in the same way as those that were developed for one-dimensional systems in Chap. 2, except for the fact that partial derivatives are used instead of total derivatives. This point is illustrated in Examples 3–1 through 3–3.

With $\dot{q} = 0$, Eq. (3–9) reduces to the *Fourier equation*,

$$\frac{\partial^2 T}{\partial x^2} + \frac{\partial^2 T}{\partial y^2} + \frac{\partial^2 T}{\partial z^2} = \frac{1}{\alpha}\frac{\partial T}{\partial t} \tag{3–11}$$

For conditions of steady state, this equation reduces further to the *Laplace equation*,

$$\frac{\partial^2 T}{\partial x^2} + \frac{\partial^2 T}{\partial y^2} + \frac{\partial^2 T}{\partial z^2} = 0 \tag{3–12}$$

If we have steady-state conditions with internal energy generation, Eq. (3–9) reduces to the *Poisson equation*,

$$\frac{\partial^2 T}{\partial x^2} + \frac{\partial^2 T}{\partial y^2} + \frac{\partial^2 T}{\partial z^2} + \frac{\dot{q}}{k} = 0 \tag{3–13}$$

The steady one-dimensional form of Eq. (3–9) or Eq. (3–13) takes the form

$$\frac{d^2T}{dx^2} + \frac{\dot{q}}{k} = 0 \tag{3–14}$$

This equation was solved for various conditions in Chap. 2.

## EXAMPLE 3–1

Write the differential formulation for the energy transfer in the rectangular plate shown in Fig. E3–1.

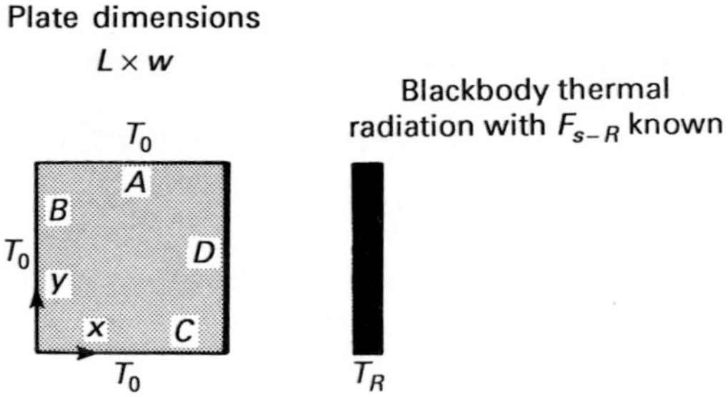
Plate dimensions
L × w
Blackbody thermal
radiation with $F_{s-R}$ known
$T_0$
A
B
$T_0$
D
y
x
C
$T_0$
$T_R$

**FIGURE E3–1**
Conduction in rectangular plate with blackbody thermal radiation.

### Solution

*Objective* Write the energy equation and boundary conditions for this two-dimensional plate.

*Assumptions/Conditions*

steady-state
two-dimensional
uniform properties
conduction with blackbody thermal radiation

*Analysis* Because heat is conducted from radiating surface $D$ to surfaces $A$ and $C$ as well as to surface $B$, the temperature distribution within this plate is a function of both $x$ and $y$. Thus, the two-dimensional Laplace equation applies; that is,

$$\frac{\partial^2 T}{\partial x^2} + \frac{\partial^2 T}{\partial y^2} = 0$$

The boundary conditions are

$$T = T_0 \quad \text{at } x = 0 \qquad -k\frac{\partial T}{\partial x} = \sigma F_{s-R}(T^4 - T_R^4) \quad \text{at } x = L$$

$$T = T_0 \quad \text{at } y = 0 \qquad T = T_0 \quad \text{at } y = w$$

Because of symmetry, either one of the $y$ conditions can be replaced by

$$\frac{\partial T}{\partial y} = 0 \qquad \text{at } y = \frac{w}{2}$$

## EXAMPLE 3–2

The flat plate shown in Fig. E3–2 is initially at a temperature $T_i$. It is suddenly exposed to a fluid at temperature $T_F$ with $h$ known. Write the differential formulation for the energy transfer, assuming uniform properties.

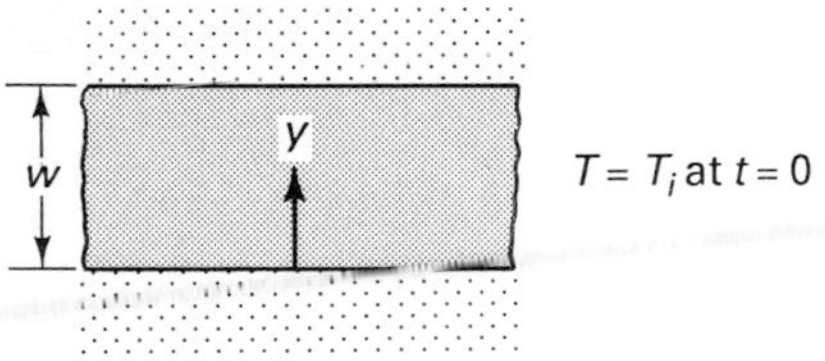


**FIGURE E3–2**
Unsteady conduction in plane wall; surfaces at $y = 0$ and $y = w$ suddenly exposed to a fluid at $T_F$ with $h$ known.

### Solution

*Objective* Write the energy equation and boundary conditions for this plane wall.

*Assumptions/Conditions*

unsteady-state
one spatial dimension
uniform coefficient of heat transfer
uniform properties

*Analysis* The temperature distribution for this problem is a function of both distance $y$ and time $t$. The energy equation for this unsteady conduction heat-transfer system takes the form

$$\alpha \frac{\partial^2 T}{\partial y^2} = \frac{\partial T}{\partial t}$$

This is the one-dimensional form of the Fourier equation. The initial and boundary conditions are

$$T = T_i \qquad \text{at } t = 0$$

$$h(T_F - T) = -k\frac{\partial T}{\partial y} \qquad \text{at } y = 0$$

$$-k\frac{\partial T}{\partial y} = h(T - T_F) \qquad \text{at } y = w$$

Note that because of symmetry, either one of the $y$ conditions can be replaced by

$$\frac{\partial T}{\partial y} = 0 \qquad \text{at } y = \frac{w}{2}$$

Strictly speaking, the coefficient of heat transfer $h$ is a function of time for problems such as this. However, approximate solutions are generally obtained by taking the average steady-state value of $h$. For the case in which $h$ is very large, the boundary conditions reduce to

$$T = T_F \qquad \text{at } y = 0 \text{ and } y = w$$

For steady-state conditions, the energy equation reduces to

$$\frac{\partial^2 T}{\partial y^2} = 0$$

such that

$$T = T_F$$

The rate of heat transfer through the plate is zero for this case.

Another limiting condition occurs for small values of the Biot number. As indicated in Chap. 2, the temperature is nearly independent of location for values of the Biot number less than 0.1, such that the energy equation can be approximated by

$$hA_s(T - T_F) + \rho V c_v \frac{dT}{dt} = 0$$

with

$$T = T_i \qquad \text{at } t = 0$$

## EXAMPLE 3–3

Write the formal differential formulation for energy transfer in the composite plate shown in Fig. E3–3.

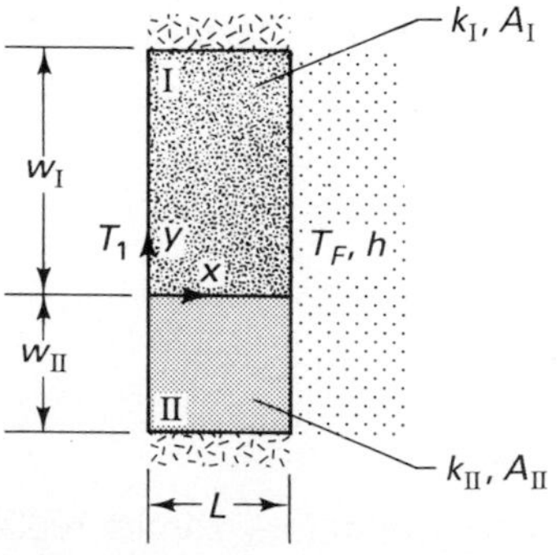


**FIGURE E3–3**
Composite system with convection at one surface.

**Solution**

*Objective* Write the energy equations and boundary conditions for this composite system.

*Assumptions/Conditions*

steady-state
two-dimensional
uniform coefficient of heat transfer
uniform properties in each material
negligible thermal conduction resistance

*Analysis* Even though the surfaces at $y = w_{\mathrm{I}}$ and $y = -w_{\mathrm{II}}$ are insulated, the temperature distribution is two-dimensional because the system involves a combination of series and parallel heat-transfer paths. The energy equation is therefore written for each material as follows:

$$\frac{\partial^2 T_{\mathrm{I}}}{\partial x^2} + \frac{\partial^2 T_{\mathrm{I}}}{\partial y^2} = 0 \qquad \text{for } 0 \le y \le w_{\mathrm{I}}$$

for material I, and

$$\frac{\partial^2 T_{\mathrm{II}}}{\partial x^2} + \frac{\partial^2 T_{\mathrm{II}}}{\partial y^2} = 0 \qquad \text{for } -w_{\mathrm{II}} \le y \le 0$$

for material II. Because we have two second order partial differential equations in $x$ and $y$, we require eight boundary conditions. The boundary conditions can be represented by

$$T_{\mathrm{I}} = T_{\mathrm{II}} \qquad -k_{\mathrm{I}}\frac{\partial T_{\mathrm{I}}}{\partial y} = -k_{\mathrm{II}}\frac{\partial T_{\mathrm{II}}}{\partial y} \qquad \text{at } y = 0$$

$$\frac{\partial T_{\mathrm{I}}}{\partial y} = 0 \qquad \text{at } y = w_{\mathrm{I}}$$

$$\frac{\partial T_{\mathrm{II}}}{\partial y} = 0 \qquad \text{at } y = -w_{\mathrm{II}}$$

$$T_{\mathrm{I}} = T_1 \qquad T_{\mathrm{II}} = T_1 \qquad \text{at } x = 0$$

$$-k_{\mathrm{I}}\frac{\partial T_{\mathrm{I}}}{\partial x} = h(T_{\mathrm{I}} - T_F) \qquad -k_{\mathrm{II}}\frac{\partial T_{\mathrm{II}}}{\partial x} = h(T_{\mathrm{II}} - T_F) \qquad \text{at } x = L$$

These equations represent the general mathematical formulation for a combined series–parallel composite with perfect thermal contact at the boundary. The conditions for which these equations can be approximated by the simple one-dimensional formulation are considered in Sec. 2–2–3 (see Example 2–6).

### 3-4-2 Radial Coordinate Systems

Attention is given next to systems that lend themselves to cylindrical or spherical coordinates.

The differential control volume for a cylindrical coordinate system is shown in Fig. 3–2. The cylindrical coordinates $r$ and $\phi$ are related to $x$ and $y$ by

$$x = r\cos\phi \qquad y = r\sin\phi \tag{3–15a,b}$$

The general Fourier law of conduction for heat transfer in the $r$, $\phi$, and $z$ directions takes the form

$$\boldsymbol{q}'' = q_r''\,\mathbf{i_r} + q_\phi''\,\mathbf{j_\phi} + q_z''\,\mathbf{k_z} = -k\left(\frac{\partial T}{\partial r}\mathbf{i_r} + \frac{1}{r}\frac{\partial T}{\partial \phi}\mathbf{j_\phi} + \frac{\partial T}{\partial z}\mathbf{k_z}\right) \tag{3–16}$$

where $\mathbf{i_r}$, $\mathbf{j_\phi}$, and $\mathbf{k_z}$ are unit vectors, the heat flux components are given by

$$q_r'' = \frac{dq_r}{dA_r} = -k\frac{\partial T}{\partial r} \tag{3–16a}$$

$$q_\phi'' = \frac{dq_\phi}{dA_\phi} = -k\frac{1}{r}\frac{\partial T}{\partial \phi} \tag{3–16b}$$

$$q_z'' = \frac{dq_z}{dA_z} = -k\frac{\partial T}{\partial z} \tag{3–16c}$$

and $dA_r = r\,d\phi\,dz$, $dA_\phi = dr\,dz$, and $dA_z = r\,d\phi\,dr$.

Consider, for example, the cylindrical body shown in Fig. 3–3. The initial temperature is given by $T_i$ and the conditions at the boundaries $r = r_0$, $z = 0$, and

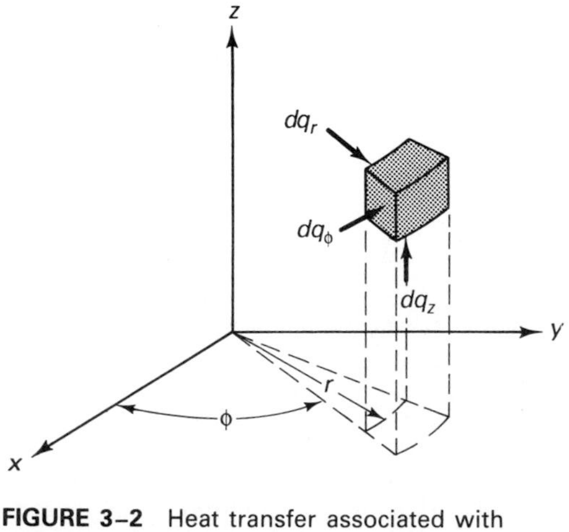


**FIGURE 3–2** Heat transfer associated with a differential-control volume for a cylindrical system; $dA_r = r\,d\phi\,dz$, $dA_z = r\,d\phi\,dr$, $dA_\phi = dr\,dz$, $dV = r\,d\phi\,dr\,dz$.

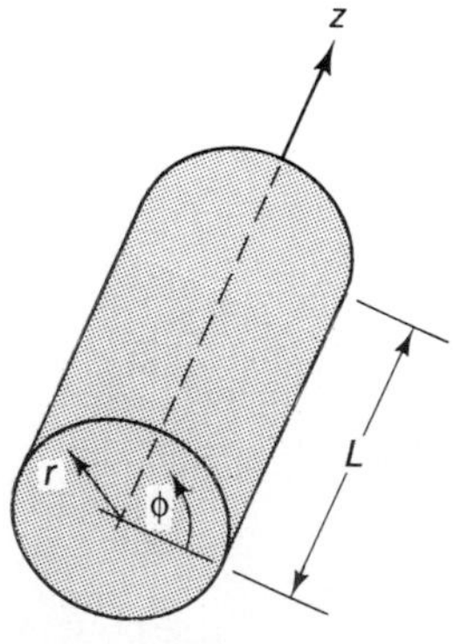


**FIGURE 3–3** Cylindrical body.

$z = L$ are assumed to be known. Applying the first law of thermodynamics to a differential volume $dV$ ($= r\,d\phi\,dr\,dz$) such as that shown in Fig. 3–2, we obtain

*Step 1*

$$dq_r + dq_\phi + dq_z + \dot{q}\,dV = dq_{r+dr} + dq_{\phi+d\phi} + dq_{z+dz} + \frac{\partial e}{\partial t}\,dm \tag{3–17}$$

where $\Delta E_s/\Delta t = dm\,\partial e/\partial t = \rho\,dV\,c_v\,\partial T/\partial t$. Utilizing the definition for the partial derivative, Eq. (3–17) takes the form

*Step 2*

$$\dot{q}\,dV = \frac{\partial}{\partial r}(dq_r)\,dr + \frac{\partial}{\partial \phi}(dq_\phi)\,d\phi + \frac{\partial}{\partial z}(dq_z)\,dz + \rho\,dV\,c_v\frac{\partial T}{\partial t} \tag{3–18}$$

Then,

*Step 3*

$$\frac{\partial}{\partial r}\left(kr\,d\phi\,dz\frac{\partial T}{\partial r}\right)dr + \frac{\partial}{\partial \phi}\left(k\,dr\,dz\frac{1}{r}\frac{\partial T}{\partial \phi}\right)d\phi + \frac{\partial}{\partial z}\left(kr\,d\phi\,dr\frac{\partial T}{\partial z}\right)dz + \dot{q}\,dV = \rho\,dV\,c_v\frac{\partial T}{\partial t} \tag{3–19}$$

In simplifying Eq. (3–19), the first term must be handled carefully when differentiating because $dA_r$ ($= r\,d\phi\,dz$) is a function of $r$. Taking note of this point, Eq. (3–19) reduces to

$$\frac{1}{r}\frac{\partial}{\partial r}\left(kr\frac{\partial T}{\partial r}\right) + \frac{1}{r^2}\frac{\partial}{\partial \phi}\left(k\frac{\partial T}{\partial \phi}\right) + \frac{\partial}{\partial z}\left(k\frac{\partial T}{\partial z}\right) + \dot{q} = \rho c_v\frac{\partial T}{\partial t} \tag{3–20}$$

or, for uniform properties,

$$\frac{1}{r}\frac{\partial}{\partial r}\left(r\frac{\partial T}{\partial r}\right) + \frac{1}{r^2}\frac{\partial^2 T}{\partial \phi^2} + \frac{\partial^2 T}{\partial z^2} + \frac{\dot{q}}{k} = \frac{1}{\alpha}\frac{\partial T}{\partial t} \tag{3–21}$$

The formulation is completed by writing

$$T = T_i(r, \phi, z) \qquad \text{at } t = 0 \tag{3–22}$$

together with six conditions in $r$, $\phi$, and $z$.

The steady one-dimensional form of Eq. (3–21),

$$\frac{1}{r}\frac{d}{dr}\left(r\frac{dT}{dr}\right) + \frac{\dot{q}}{k} = 0 \tag{3–23}$$

was solved in Chap. 2 for several conditions.

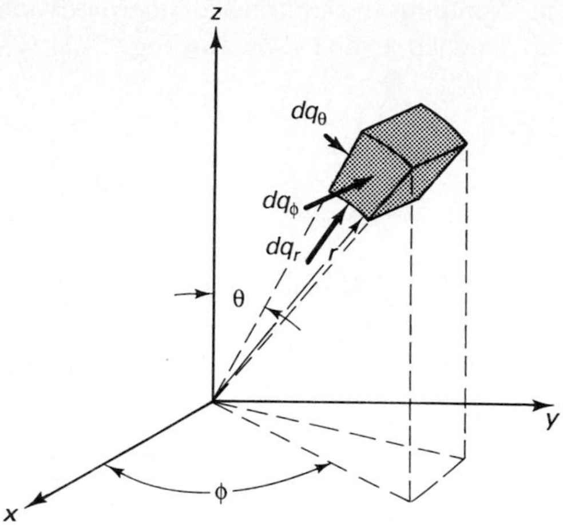
z
$dq_\theta$
$dq_\phi$
$dq_r$
r
θ
y
ϕ
x

**FIGURE 3–4**
Heat transfer associated with a differential control volume for a spherical system; $dA_r = r^2 \sin\theta\, d\theta\, d\phi$, $dA_\phi = r\, d\theta\, dr$, $dA_\theta = r \sin\theta\, d\phi\, dr$, $dV = r^2 \sin\theta\, d\phi\, d\theta\, dr$.

Referring to Fig. 3–4, the unsteady three-dimensional energy equation for conduction heat transfer in spherical geometries is given by

$$\frac{1}{r^2}\frac{\partial}{\partial r}\left(kr^2\frac{\partial T}{\partial r}\right) + \frac{1}{r^2\sin^2\theta}\frac{\partial}{\partial\phi}\left(k\frac{\partial T}{\partial\phi}\right) + \frac{1}{r^2\sin\theta}\frac{\partial}{\partial\theta}\left(k\sin\theta\frac{\partial T}{\partial\theta}\right) + \dot{q} = \rho c_v\frac{\partial T}{\partial t} \tag{3–24}$$

The derivation of this equation follows the pattern established in the formulation of the energy equation for cylindrical systems.

## EXAMPLE 3–4

Write the differential formulation for steady heat transfer in the cylindrical fin with circular cross section shown in Fig. E3–4.

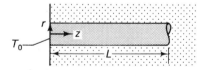

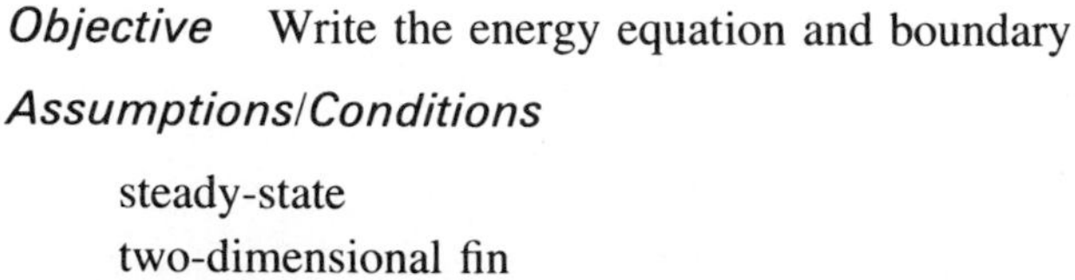
Fluid surrounding fin
at $T_F$ with $\bar{h}$ known
r
z
$T_0$
L

**FIGURE E3–4**
Circular fin.

### Solution

*Objective* Write the energy equation and boundary conditions for this fin.

*Assumptions/Conditions*

steady-state
two-dimensional fin
uniform properties
uniform coefficient of heat transfer

*Analysis* The differential formulation for this two-dimensional problem takes the form

$$\frac{\partial^2 T}{\partial z^2} + \frac{1}{r}\frac{\partial}{\partial r}\left(r\frac{\partial T}{\partial r}\right) = 0$$

$$T = T_0 \qquad \text{at } z = 0 \qquad -k\frac{\partial T}{\partial z} = \bar{h}(T - T_F) \qquad \text{at } z = L$$

$$\frac{\partial T}{\partial r} = 0 \qquad \text{at } r = 0 \qquad -k\frac{\partial T}{\partial r} = \bar{h}(T - T_F) \qquad \text{at } r = r_0$$

As indicated in Chap. 2, for small values of the Biot number the variation of temperature with $r$ is very slight, such that an approximate one-dimensional formulation can be utilized. The approximate one-dimensional energy equation is written as

$$\frac{d^2 T}{dz^2} - \frac{\bar{h}p}{kA}(T - T_F) = 0$$

(The axial coordinate is commonly designated by $x$ instead of $z$.)

## EXAMPLE 3–5

A spherical ball initially at temperature $T_i$ is suddenly exposed to a convecting fluid at temperature $T_F$ with $h$ known. Develop the differential formulation.

### Solution

*Objective* Write the energy equation and boundary conditions for this spherical ball.

*Schematic* Differential volume for unsteady heat transfer in a sphere.

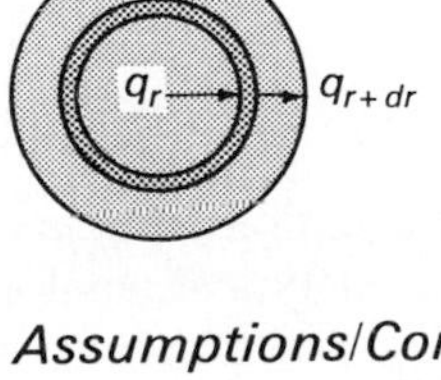


*Assumptions/Conditions*

unsteady-state
one spatial dimension
uniform coefficient of heat transfer
uniform properties

*Analysis* Referring to the schematic, the energy balance gives

$$\textit{Step 1} \qquad q_r = q_{r+dr} + \frac{\Delta E_s}{\Delta t}$$

(where $dV = 4\pi r^2\,dr$).

$$\textit{Step 2} \qquad q_r = q_r + \frac{\partial q_r}{\partial r}\,dr + \rho\,dV\,c_v\,\frac{\partial T}{\partial t}$$

$$\textit{Step 3} \qquad 0 = \frac{\partial}{\partial r}\left(-kr^2\frac{\partial T}{\partial r}\right) + r^2\rho c_v\frac{\partial T}{\partial t}$$

For uniform properties, this equation reduces to

$$\frac{1}{r^2}\frac{\partial}{\partial r}\left(r^2\frac{\partial T}{\partial r}\right) = \frac{1}{\alpha}\frac{\partial T}{\partial t}$$

or

$$\frac{\partial^2 T}{\partial r^2} + \frac{2}{r}\frac{\partial T}{\partial r} = \frac{1}{\alpha}\frac{\partial T}{\partial t}$$

The initial-boundary conditions are

$$T = T_i \qquad \text{at } t = 0$$

$$\frac{\partial T}{\partial r} = 0 \qquad \text{at } r = 0$$

$$-k\frac{\partial T}{\partial r} = h(T - T_F) \qquad \text{at } r = r_0$$

As indicated in Chap. 2, for small values of the Biot number, the temperature is essentially independent of $r$ such that the energy equation can be approximated by

$$\frac{dT}{dt} + \frac{hA}{\rho V c_v}(T - T_F) = 0$$

## 3–5 SOLUTION TECHNIQUES: INTRODUCTION

Several techniques are available for the solution of the equations produced by our differential formulation. These solution techniques include analytical, analogical, and graphical methods, all of which are introduced in the following sections of this chapter. The differential formulation can also be solved by the numerical finite-difference approach, as will be shown in Chap. 4.

Solutions obtained by these various methods for standard two- and three-dimensional conduction-heat-transfer problems are available in formula and chart form as a convenience for design calculations. These practical results will be presented in Sec. 3–9.

## 3–6 ANALYTICAL APPROACHES

Among the various methods that can be utilized to develop analytical solutions for multidimensional heat-transfer problems, the *separation-of-variables* technique and the *approximate integral* approach are the simplest. The classical separation-of-variables approach is introduced in this section. Although the integral approach can be used to solve many conduction-heat-transfer problems, this popular method will be introduced in Chap. 7 in the context of convection heat transfer. Other analytical solution techniques for multidimensional conduction-heat-transfer problems are available in the literature.

### 3–6–1 Separation-of Variables Method

The *separation-of-variables* technique is particularly well suited to the solution of steady and unsteady multidimensional conduction-heat-transfer problems that can be represented by linear equations. To introduce this solution concept, we consider the steady-state two-dimensional conduction-heat-transfer problem illustrated in Fig. 3–5.

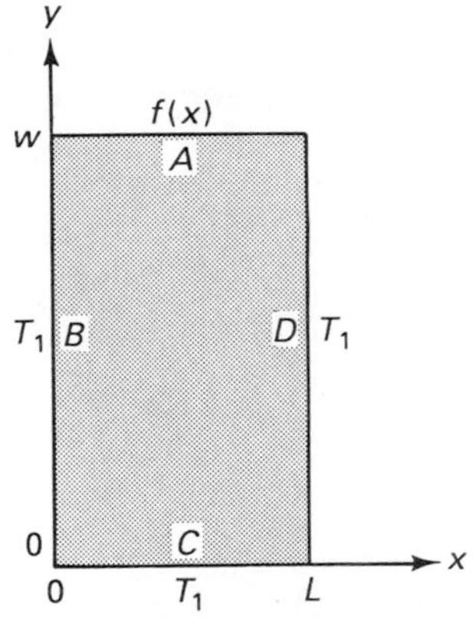


**FIGURE 3–5**
Steady two-dimensional conduction heat transfer in a rectangular plate.

The differential formulation for this problem for the case in which $k$ is uniform, $\dot{q} = 0$, $\partial T/\partial z = 0$, and $\partial T/\partial t = 0$ is written as

$$\frac{\partial^2 T}{\partial x^2} + \frac{\partial^2 T}{\partial y^2} = 0 \tag{3–25}$$

$$T = T_1 \quad \text{at } x = 0 \qquad T = T_1 \quad \text{at } x = L \qquad \text{(3–26,27)}$$

$$T = T_1 \quad \text{at } y = 0 \qquad T = f(x) \quad \text{at } y = w \qquad \text{(3–28,29)}$$

The use of the separation-of-variables method requires that only one nonhomogeneous term appear in the linear differential formulation. Therefore, we eliminate three of the nonhomogeneous boundary conditions by utilizing the substitution $\psi = T - T_1$. This simple substitution transforms Eqs. (3–25) through (3–29) to

$$\frac{\partial^2 \psi}{\partial x^2} + \frac{\partial^2 \psi}{\partial y^2} = 0 \tag{3–30}$$

$$\psi = 0 \quad \text{at } x = 0 \qquad \psi = 0 \quad \text{at } x = L \tag{3–31,32}$$

$$\psi = 0 \quad \text{at } y = 0 \qquad \psi = F(x) \quad \text{at } y = w \tag{3–33,34}$$

where $F(x) = f(x) - T_1$. Here $x$ is referred to as the homogeneous direction and $y$ is the nonhomogeneous direction.

The solution to Eqs. (3–30) through (3–34) is developed in Appendix E–1 by assuming a product solution of the form

$$\psi(x,y) = X(x)\,Y(y) \tag{3–35}$$

where $X(x)$ is a function of $x$ and $Y(y)$ is a function of $y$. The general solution is

$$T - T_1 = \sum_{n=1}^{\infty} c_n \frac{\sinh(n\pi y/L)}{\sinh(n\pi w/L)} \sin\frac{n\pi x}{L} \tag{3–36}$$

where

$$c_n = \frac{2}{L}\int_0^L F(x)\sin\frac{n\pi x}{L}\,dx \tag{3–37}$$

For the case in which $f(x)$ is equal to a uniform temperature $T_2$, we have

$$c_n = \frac{2}{L}\int_0^L (T_2 - T_1)\sin\frac{n\pi x}{L}\,dx = -\frac{2}{L}(T_2 - T_1)\frac{L}{n\pi}[\cos(n\pi) - 1]$$

$$= (T_2 - T_1)\frac{4}{n\pi} \qquad n = 1, 3, 5, \ldots \tag{3–38}$$

$$= 0 \qquad n = 2, 4, 6, \ldots$$

Hence, Eq. (3–36) reduces to

$$\frac{T - T_1}{T_2 - T_1} = \frac{2}{\pi}\sum_{n=1}^{\infty} \frac{1 - (-1)^n}{n} \frac{\sinh(n\pi y/L)}{\sinh(n\pi w/L)} \sin\frac{n\pi x}{L} \tag{3–39}$$

This series is uniformly convergent (see Kreyszig [1]) in the region $0 \le y < w$ for all values of $x$. Isotherms obtained on the basis of Eq. (3–39) are shown in Fig. 3–6.

For the case in which $f(x) = T_1 + T_2 \sin(\pi x/L)$ [i.e., $F(x) = T_2 \sin(\pi x/L)$], we find that

$$c_1 = T_2 \qquad c_n = 0 \qquad n = 2, 3, 4, \ldots \tag{3–40}$$

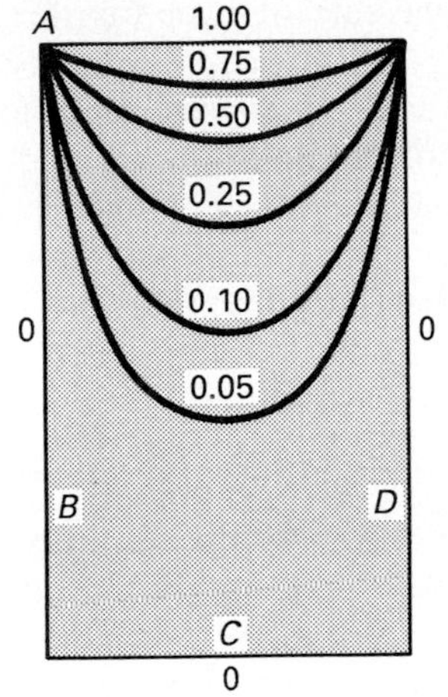


**FIGURE 3–6**
Dimensionless isotherms $(T - T_1)/(T_2 - T_1)$ for steady two-dimensional conduction heat transfer in a rectangular plate; $f(x) = T_2$ and $L/w = 0.6$.

such that the temperature distribution is given by

$$T - T_1 = T_2 \frac{\sinh (\pi y/L)}{\sinh (\pi w/L)} \sin \frac{\pi x}{L} \tag{3–41}$$

Once the temperature distribution for the problem of interest is known, the rate of heat transfer can be obtained by utilizing the appropriate particular law. To illustrate, we select the situation for which $f(x) = T_1 + T_2 \sin (\pi x/L)$. To obtain the local rate of heat transfer in the $y$ direction, the general Fourier law of conduction is utilized; that is,

$$dq_y = -k\, dA_y \frac{\partial T}{\partial y} = -k\delta\, dx\, T_2 \frac{\pi}{L} \frac{\cosh (\pi y/L)}{\sinh (\pi w/L)} \sin \frac{\pi x}{L} \tag{3–42}$$

The total rate of heat transfer $q_y$ is now obtained by integrating from $x$ equal to 0 to $L$ as follows:

$$\begin{aligned} q_y &= \int_0^L -k \frac{\partial T}{\partial y} \delta\, dx = -k\delta T_2 \frac{\pi}{L} \frac{\cosh (\pi y/L)}{\sinh (\pi w/L)} \left( -\frac{L}{\pi} \cos \frac{\pi x}{L} \right)\Bigg|_0^L \\ &= -2k\delta T_2 \frac{\cosh (\pi y/L)}{\sinh (\pi w/L)} \end{aligned} \tag{3–43}$$

Thus, the rate of heat transfer across surface $A$ at $y = w$ is

$$q_A = -2k\delta T_2 \coth \frac{\pi w}{L} \tag{3–44}$$

and the rate of heat transfer across surface $C$ at $y = 0$ is

$$q_C = -2k\delta T_2 \frac{1}{\sinh (\pi w/L)} \tag{3–45}$$

Similar expressions can be developed for the rate of heat transfer in the $x$ direction.

As illustrated in Examples 3–6 and 3–7, the same approach is used to develop expressions for the rate of heat transfer for the more complex steady and unsteady systems in which the temperature field is expressed in terms of infinite series.

The separation-of-variables approach has been utilized in the solution of a large number of standard steady and unsteady conduction heat-transfer problems. The results of some of these analyses will be presented in Sec. 3–9.

## EXAMPLE 3–6

Consider the two-dimensional problem shown in Fig. 3–5 for the case in which $f(x) = T_2 = 100°C$, $T_1 = 0°C$, $k = 100$ W/(m °C), and $\delta = 1$ cm. The solution to this problem for uniform property conditions is given by Eq. (3–39). First, demonstrate that this equation satisfies the energy equation. Then utilize this result to obtain predictions for the rates of heat transfer across surfaces $A$ and $C$ of the plate for the case in which $L = w$.

### Solution

*Objective* Demonstrate the validity of Eq. (3–39) and use this equation to determine $q_A$ and $q_C$.

*Schematic* Steady two-dimensional heat transfer in a rectangular plate.

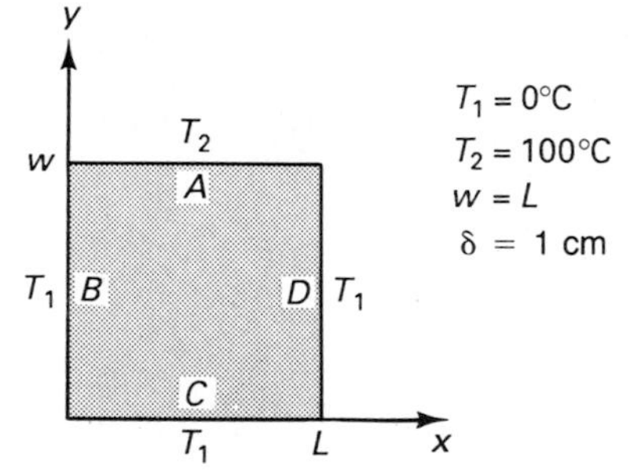


*Assumptions/Conditions*

steady-state

two-dimensional

uniform properties

*Properties* $k = 100$ W/(m °C)

*Analysis* As already mentioned, this infinite series converges for all values of $x$ and $y$ within the domain of the body. Predictions for the temperature at various locations within the body are shown in Fig. 3–6. Because Eq. (3–39) is uniformly convergent in the region $0 \le y < w$ for all values of $x$, this equation can be differentiated term by term in this domain.

To demonstrate that Eq. (3–39) satisfies the energy equation, Eq. (3–25), we obtain $\partial^2 T/\partial x^2$ and $\partial^2 T/\partial y^2$ as follows:

$$\frac{\partial T}{\partial x} = (T_2 - T_1)\frac{2}{\pi}\sum_{n=1}^{\infty}\frac{1-(-1)^n}{n}\frac{\sinh(n\pi y/L)}{\sinh(n\pi w/L)}\frac{n\pi}{L}\cos\frac{n\pi x}{L}$$

$$\frac{\partial^2 T}{\partial x^2} = (T_2 - T_1)\frac{2}{\pi}\sum_{n=1}^{\infty}\frac{1-(-1)^n}{n}\frac{\sinh(n\pi y/L)}{\sinh(n\pi w/L)}\left(\frac{-n^2\pi^2}{L^2}\right)\sin\frac{n\pi x}{L}$$

and

$$\frac{\partial T}{\partial y} = (T_2 - T_1)\frac{2}{\pi}\sum_{n=1}^{\infty}\frac{1-(-1)^n}{n}\frac{n\pi\cosh(n\pi y/L)}{L\sinh(n\pi w/L)}\sin\frac{n\pi x}{L}$$

$$\frac{\partial^2 T}{\partial y^2} = (T_2 - T_1)\frac{2}{\pi}\sum_{n=1}^{\infty}\frac{1-(-1)^n}{n}\left(\frac{n\pi}{L}\right)^2\frac{\sinh(n\pi y/L)}{\sinh(n\pi w/L)}\sin\frac{n\pi x}{L}$$

Substituting these results for $\partial^2 T/\partial x^2$ and $\partial^2 T/\partial y^2$ into Eq. (3–25),

$$\frac{\partial^2 T}{\partial x^2} + \frac{\partial^2 T}{\partial y^2} = 0$$

we find that this equation is indeed satisfied.

To determine the local rate of heat transfer in the $y$ direction, we apply the general Fourier law of conduction, with the result

$$dq_y = -k\,dA_y\frac{\partial T}{\partial y}$$

$$= -k(T_2 - T_1)\left\{\frac{2}{L}\sum_{n=1}^{\infty}[1-(-1)^n]\frac{\cosh(n\pi y/L)}{\sinh(n\pi w/L)}\sin\frac{n\pi x}{L}\right\}\delta\,dx$$

Integrating with respect to $x$, the total rate of heat transfer over the region $0 \le x \le L$ becomes

$$q_y = k\delta(T_2 - T_1)\frac{2}{L}\sum_{n=1}^{\infty}[1-(-1)^n]\frac{\cosh(n\pi y/L)}{\sinh(n\pi w/L)}\frac{L}{n\pi}[\cos(n\pi) - 1] \qquad \text{(a)}$$

$$= -k\delta(T_2 - T_1)\frac{4}{\pi}\sum_{n=1}^{\infty}\frac{1-(-1)^n}{n}\frac{\cosh(n\pi y/L)}{\sinh(n\pi w/L)}$$

Setting $y$ equal to zero, the total rate of heat transfer across face $C$ is given by

$$q_C = -k\delta(T_2 - T_1)\frac{4}{\pi}\sum_{n=1}^{\infty}\frac{1-(-1)^n}{n}\frac{1}{\sinh(n\pi)}$$

where $w = L$ for the square body. The infinite series

$$\frac{4}{\pi}\sum_{n=1}^{\infty}\frac{1-(-1)^n}{n}\frac{1}{\sinh(n\pi)}$$

converges fairly rapidly to approximately 0.221. Hence, $q_C$ is given by

$$q_C = -0.221k\delta(T_2 - T_1) = -22.1 \text{ W}$$

To obtain the rate of heat transfer across face $A$, we allow $y$ to approach $w$ in Eq. (a).

$$q_A = -k\delta(T_2 - T_1)\frac{4}{\pi}\sum_{n=1}^{\infty}\frac{1-(-1)^n}{n}\coth(n\pi)$$

The infinite series

$$\sum_{n=1}^{\infty}\frac{1-(-1)^n}{n} = 2\left(1 + \frac{1}{3} + \frac{1}{5} + \frac{1}{7} + \ldots\right)$$

is divergent [1,2]. This, together with the fact that coth $(n\pi)$ is always greater than unity, indicates that $q_A$ is unbounded. This rather startling result stems from the fact that we have the drastic situation in which the two surfaces $A$ and $B$ at different temperatures $T_2$ and $T_1$ are in direct contact. If we were to attempt to set up an electrical analog for this problem, we would have burn out because of shorting at the corner sections.

## EXAMPLE 3–7

The surfaces of a flat plate initially at uniform temperature $T_i$ are suddenly brought to a temperature $T_1$. The differential formulation takes the form

$$\alpha\frac{\partial^2\psi}{\partial x^2} = \frac{\partial\psi}{\partial t}$$

$$\psi = T_i - T_1 \qquad \text{at } t = 0$$

$$\psi = 0 \qquad \text{at } x = 0$$

$$\psi = 0 \qquad \text{at } x = L$$

where $\psi = T - T_1$. These equations can be solved by means of the separation-of-variables solution approach [3–5], with the result†

$$\frac{T - T_1}{T_i - T_1} = \frac{2}{\pi}\sum_{n=1}^{\infty}\frac{1-(-1)^n}{n}\exp(-n^2\pi^2\alpha t/L^2)\sin\frac{n\pi x}{L}$$

which is uniformly convergent for $t > 0$ over the entire plate $0 \leq x \leq L$. Develop expressions for the instantaneous rate of heat transfer from the plate and the total heat transferred over a period of time $t$.

† This system of equations can also be conveniently solved by means of the approximate integral technique and the Laplace-transform method.

**Solution**

*Objective* Develop relations for $q_0$ and for the total heat transferred over the period 0 to $t$.

*Schematic* Unsteady one-dimensional heat transfer in a flat plate.

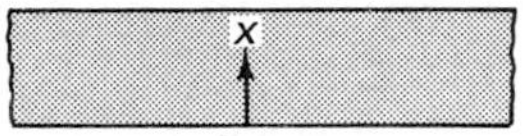


Uniform temperature = $T_i$
Surfaces suddenly brought to $T_1$

*Assumptions/Conditions*

unsteady-state
one spatial dimension
uniform properties

*Analysis* Taking $q_x$ in the $x$ direction as positive,

$$q = -q_0 + q_L$$

and because of symmetry

$$q = -2q_0$$

Introducing the Fourier law of conduction,

$$q_0 = -kA_x \left.\frac{\partial T}{\partial x}\right|_0$$

$$= -kA(T_i - T_1)\frac{2}{\pi}\sum_{n=1}^{\infty}\frac{1-(-1)^n}{n}\exp\left(-n^2\pi^2\alpha t/L^2\right)\frac{n\pi}{L}\cos\left.\frac{n\pi x}{L}\right|_0 \quad \text{(a)}$$

$$= -kA(T_i - T_1)\frac{2}{L}\sum_{n=1}^{\infty}[1-(-1)^n]\exp\left(-n^2\pi^2\alpha t/L^2\right)$$

To obtain an expression for the accumulated heat transfer $Q_0$ over the period 0 to $t$, we write

$$Q_0 = \int_0^t q_0\,dt = -kA(T_i - T_1)\frac{2}{L}\sum_{n=1}^{\infty}[1-(-1)^n]\frac{-L^2}{n^2\pi^2\alpha}\exp\left(-n^2\pi^2\alpha t/L^2\right)\Big|_0^t$$

$$= -kA(T_i - T_1)\frac{2L}{\alpha\pi^2}\sum_{n=1}^{\infty}\frac{1-(-1)^n}{n^2}\left[1-\exp\left(-n^2\pi^2\alpha t/L^2\right)\right] \quad \text{(b)}$$

The accumulated heat transfer for the entire plate $Q$ is

$$Q = -2Q_0$$

The infinite series in Eqs. (a) and (b) are both convergent. These infinite series are generally evaluated by digital computer, except for large values of $\alpha t/L^2$ or $\alpha\tau/L^2$ for which cases hand calculations can be quickly made. Calculations for $T$, $q$, and $Q$ for this problem are given in chart form in Sec. 3–9.

## EXAMPLE 3–8

Solutions to heat-transfer problems involving more than one nonhomogeneous term can be developed by applying the *principle of superposition* if the equations are linear (see Appendix A). In this approach, solutions to more involved problems are obtained by summing up or superimposing solutions of sets of simpler problems. To demonstrate this important superposition concept, consider steady two-dimensional conduction heat transfer in the rectangular solid shown in Fig. 3–5 for the case in which surfaces $A$ and $D$ are both at $T_2$.

### Solution

*Objective* Use the superposition principle to develop a solution for the temperature distribution in the rectangular plate shown in the schematic.

*Schematic* Steady two-dimensional heat transfer in a rectangular plate.

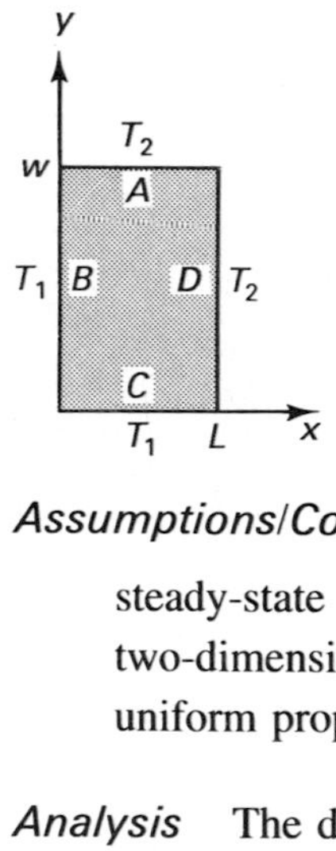


*Assumptions/Conditions*

steady-state
two-dimensional
uniform properties

*Analysis* The differential formulation for this problem can be written as

$$\frac{\partial^2 \psi}{\partial x^2} + \frac{\partial^2 \psi}{\partial y^2} = 0 \tag{a}$$

$$\psi = 0 \quad \text{at } x = 0 \qquad \psi = T_2 - T_1 \quad \text{at } x = L \tag{b,c}$$

$$\psi = 0 \quad \text{at } y = 0 \qquad \psi = T_2 - T_1 \quad \text{at } y = w \tag{d,e}$$

where the use of $\psi$ ($\equiv T - T_1$) has created homogeneous conditions at $x = 0$ and $y = 0$. This mathematical formulation is represented by Fig. E3–8a. However, we are left with two nonhomogeneous boundary conditions. Because the use of the separation-of-variables technique requires that only one nonhomogeneous condition occur, we turn to the principle of superposition to achieve a solution.

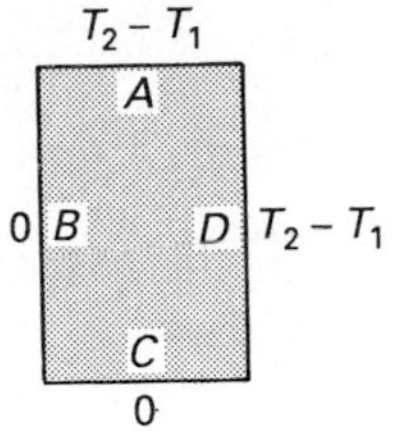


**FIGURE E3-8a**
Rectangular plate with two nonhomogeneous boundary conditions.

Utilizing the superposition concept, $\psi$ is assumed to be equal to the sum of two simpler solutions; that is,

$$\psi = \psi_I + \psi_{II} \tag{f}$$

Our differential formulation can be written in terms of $\psi_I$ and $\psi_{II}$ by merely substituting Eq. (f) into Eqs. (a) through (e).

$$\frac{\partial^2 \psi_I}{\partial x^2} + \frac{\partial^2 \psi_{II}}{\partial x^2} + \frac{\partial^2 \psi_I}{\partial y^2} + \frac{\partial^2 \psi_{II}}{\partial y^2} = 0$$

$$\psi_I + \psi_{II} = 0 \quad \text{at } x = 0 \qquad \psi_I + \psi_{II} = T_2 - T_1 \quad \text{at } x = L$$

$$\psi_I + \psi_{II} = 0 \quad \text{at } y = 0 \qquad \psi_I + \psi_{II} = T_2 - T_1 \quad \text{at } y = w$$

Based on this system of equations, the following two simpler problems are formulated such that the differential formulation for $\psi$ ($= \psi_I + \psi_{II}$) remains satisfied:

$$\frac{\partial^2 \psi_I}{\partial x^2} + \frac{\partial^2 \psi_I}{\partial y^2} = 0 \qquad \frac{\partial^2 \psi_{II}}{\partial x^2} + \frac{\partial^2 \psi_{II}}{\partial y^2} = 0$$

$$\psi_I = 0 \quad \text{at } x = 0 \quad \psi_{II} = 0$$

$$\psi_I = 0 \quad \text{at } x = L \quad \psi_{II} = T_2 - T_1$$

$$\psi_I = 0 \quad \text{at } y = 0 \quad \psi_{II} = 0$$

$$\psi_I = T_2 - T_1 \quad \text{at } y = w \quad \psi_{II} = 0$$

The formulations for $\psi_I$ and $\psi_{II}$ are represented by Figs. E3-8b and E3-8c. Note that Fig. E3-8a is equivalent to superimposing Fig. E3-8b on Fig. E3-8c. The solution for $\psi_I$ is given by Eq. (3-39),

$$\psi_I = T_I - T_1 = (T_2 - T_1)\frac{2}{\pi}\sum_{n=1}^{\infty}\frac{1-(-1)^n}{n}\frac{\sinh(n\pi y/L)}{\sinh(n\pi w/L)}\sin\frac{n\pi x}{L}$$

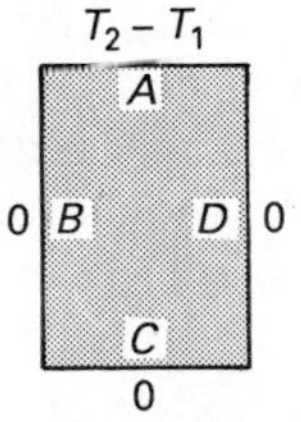


**FIGURE E3-8b**
Rectangular plate with one nonhomogeneous boundary condition at $y = w$.

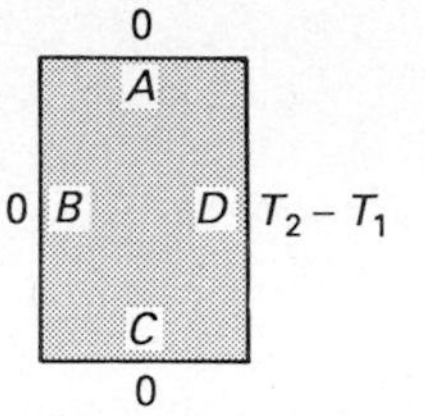


**FIGURE E3–8c**
Rectangular plate with one nonhomogeneous boundary condition at $x = L$.

By inspection, we see that the solution for $\psi_{II}$ can be obtained from Eq. (3–39) by merely interchanging $y$ and $x$ and $w$ and $L$; that is,

$$\psi_{II} = T_{II} - T_1 = (T_2 - T_1)\frac{2}{\pi}\sum_{n=1}^{\infty}\frac{1-(-1)^n}{n}\frac{\sinh(n\pi x/w)}{\sinh(n\pi L/w)}\sin\frac{n\pi y}{w}$$

Returning to Eq. (f), our full solution is

$$T - T_1 = \psi = \psi_I + \psi_{II} =$$

$$(T_2 - T_1)\frac{2}{\pi}\sum_{n=1}^{\infty}\frac{1-(-1)^n}{n}\left[\frac{\sinh(n\pi y/L)}{\sinh(n\pi w/L)}\sin\frac{n\pi x}{L} + \frac{\sinh(n\pi x/w)}{\sinh(n\pi L/w)}\sin\frac{n\pi y}{w}\right]$$

The principle of superposition can be utilized to solve a wide range of steady and unsteady heat-transfer problems, as long as the differential equation and boundary-initial conditions are linear.

## 3–7 ANALOGY APPROACHES

Analogies have been found to exist between heat transfer and (1) the flow of electric current, (2) fluid flow, and (3) membrane behavior, which provide useful tools for developing predictions for the temperature distribution and rate of heat transfer in multidimensional systems. Because of its prominence in heat-transfer work, the electrical analogy will be presented in this section. (An introduction to the fluid flow and membrane analogies is given by Schneider [6].)

### 3–7–1 Electrical Analogy

As shown in Chaps. 1 and 2, electrical circuits can be designed in which the flow of current simulates the rate of heat transfer in one-dimensional systems. The electrical/heat-transfer analogy developed to this point is summarized in Table 3–1. However, the analogy between current flow and heat transfer is more far reaching in that it extends to multidimensional electric fields and thermal systems. The differential and numerical finite-difference formulations provide the basis for this broader electrical analogy.

**TABLE 3–1** Electrical analogy

| *Electrical* | *Thermal* |
|---|---|
| $E_e$ Voltage (V) | $T$ Temperature (°C) |
| $I_e$ Current (A) | $q$ Rate of heat transfer (W) |
| $R_e$ Resistance (Ω) | $R$ Thermal resistance (°C/W) |
| $C_e$ Capacitance (F) | $C$ Thermal capacitance (J/°C) |

To expand upon this point, we first compare the Fourier equation (in Cartesian coordinates), Eq. (3–11),

$$\frac{\partial^2 T}{\partial x^2} + \frac{\partial^2 T}{\partial y^2} + \frac{\partial^2 T}{\partial z^2} = \frac{1}{\alpha}\frac{\partial T}{\partial t} \tag{3–46}$$

with the following differential equation that governs the distribution of voltage in an electrically conducting multidimensional system:

$$\frac{\partial^2 E_e}{\partial x^2} + \frac{\partial^2 E_e}{\partial y^2} + \frac{\partial^2 E_e}{\partial z^2} = \frac{R_e C_e}{L^2}\frac{\partial E_e}{\partial t} \tag{3–47}$$

Based on the similarity between these two equations, we conclude that an electrical field can be set up within an electrical conductor which corresponds to the thermal field in the modeled heat-transfer problem, with equipotential lines and orthogonal paths of electric current flow in the voltage field being representative of isotherms and paths of heat flow, respectively. These relations provide the basis for an experimental method known as the *analog field plotter*, and a resistance and capacitance (*R/C*) *network approach*, which is introduced in Chap. 4.

### *Analog Field Plotter*

A convenient experimental arrangement by which the electric field within steady multidimensional systems can be measured is shown in Fig. 3–7. For many two-dimensional systems, electrical conducting paper, thin strips of certain metals such as Inconel, or a shallow saline bath can be patterned after the conduction system of interest. Such experimental arrangements are known as *analog field plotters*. In regard to boundary conditions, a uniform surface temperature is modeled by maintaining a uniform voltage at the surface, and insulated surfaces correspond to surfaces that are not connected to voltage sources. Lines of constant voltage, which correspond to isotherms, are then found by utilizing a millivoltmeter. Using the resulting measured system of equipotential lines, orthogonal current flow lines can be sketched in. (The flow lines can sometimes be determined by reversing the electrical boundary conditions.) The resulting network of equipotential lines and current flow lines represent the isotherms and heat-flow paths in the analogous heat-transfer system. The usefulness

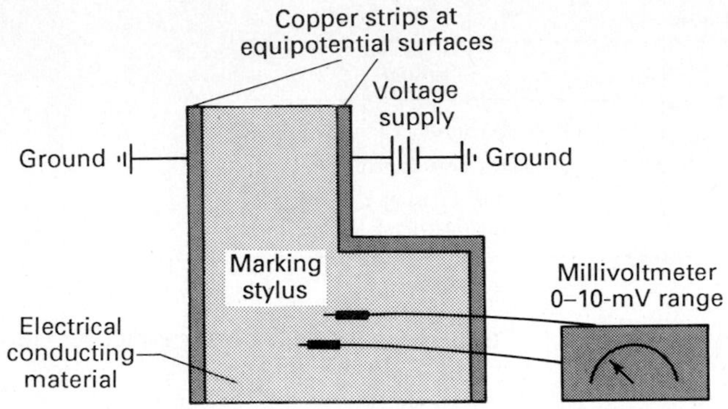


**FIGURE 3–7** Representative analog field plotter arrangement.

of this information will be seen in the next section when we consider the graphical solution technique.

## EXAMPLE 3–9

In Chap. 2 it was concluded that one-dimensional heat transfer occurs in composite walls consisting of only two dissimilar sections in parallel. It was also concluded that the heat transfer in composite walls consisting of combined series–parallel resistances is not one-dimensional. Demonstrate the validity of these conclusions by the use of an analog field plotter.

### Solution

*Objective* Develop a laboratory experiment using the analog field plotter in order to show that heat transfer in composite walls is (1) one-dimensional for two sections in parallel, and (2) two-dimensional for sections arranged in series/parallel.

*Assumptions/Conditions*

steady-state
one- and two-dimensional systems
uniform properties

*Analysis* Electrical circuits that are analogous to conduction heat transfer in (a) a simple parallel composite and (b) a series–parallel composite were constructed of strips of 0.635 mm Inconel (No. 600 cold rolled), as shown in Figs. E3–9a(i) and (ii). Voltages were then applied to the ends of these two composite strips and the equipotential lines shown in the two sketches were measured. These lines are clearly one-dimensional in Fig. E3–9a(i) and two-dimensional in Fig. E3–9a(ii). Therefore, we conclude that the heat transfer in simple parallel composites consisting of two thermal resistances is one-dimensional, but that the heat transfer in combined series–parallel composites is two-dimensional.

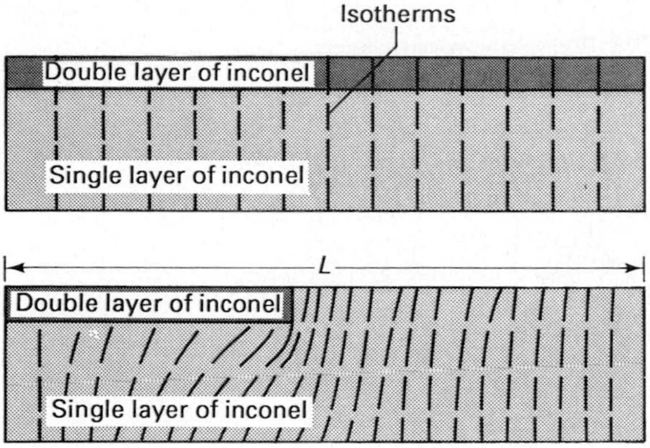


(i) Parallel arrangement.

(ii) Series-parallel arrangement.

**FIGURE E3–9a**
Composite wall (double layer – 5 mm wide, total wall width – 32 mm).

It follows that the heat transfer in the parallel composite can be represented by the one-dimensional thermal circuit shown in Fig. E3–9b. Representing the electrical resistivity of Inconel by $\rho_e$, the electrical resistance of the single layers and double layers are given by ($\delta = 0.635$ mm)

$$R_{e1} = \frac{\rho_e L}{\delta(0.027\ \text{m})} \qquad R_{e2} = \frac{\rho_e L}{2\delta(0.005\ \text{m})} = 2.7\, R_{e1}$$

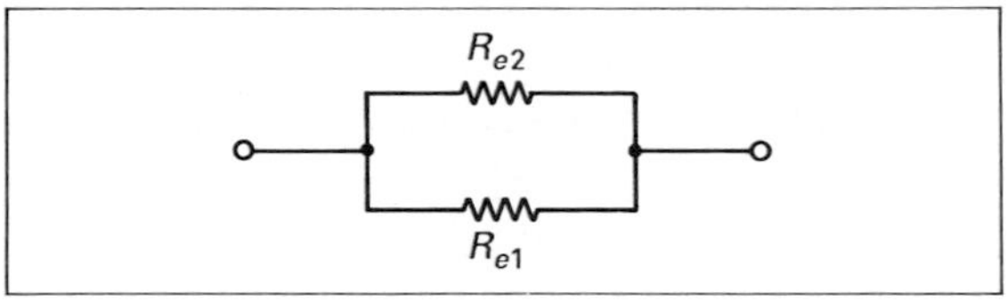


**FIGURE E3–9b**
Thermal circuit for parallel arrangement.

Because of the two dimensionality of the system shown in Fig. E3–9a(ii), the heat transfer in composites of this type can only be approximated by a one-dimensional thermal circuit when the criterion introduced in Sec. 2–2–4 of Chap. 2 is satisfied.

## 3–8 GRAPHICAL APPROACHES

Graphical solution techniques have been developed for both steady and unsteady multidimensional conduction-heat transfer problems. The use of these approaches in analyzing two-dimensional conduction heat transfer provides further insight into these fairly complex processes. The graphical approach is presented in this section for steady two-dimensional systems.

### 3–8–1 Steady Two-Dimensional Systems

The key to the graphical approach to solving steady two-dimensional problems is the fact that isotherms and heat-flow lines are orthogonal (perpendicular). This point is reflected in the Fourier law of conduction itself and was touched upon in our brief

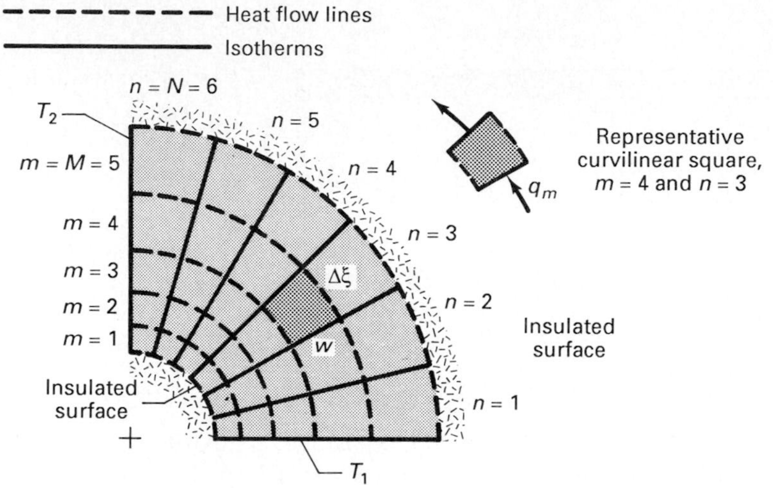


**FIGURE 3–8** Cylindrical section with $r_1 = 2$ cm and $r_2 = 7.4$ cm.

study of the electrical analogy. To illustrate this point, isotherms and heat-flow lines are sketched in Fig. 3–8 for a fairly simple two-dimensional problem, such that a network of curvilinear squares is constructed.

Because heat is transferred along the $M$ paths formed by adjacent heat-flow lines, the heat transfer in a single heat-flow lane is essentially one-dimensional (with respect to the curved coordinates, which follow the path $\xi$ taken by individual heat-flow lanes) and is given by the following form of the Fourier law:

$$q_\xi = -kA_\xi \frac{dT}{d\xi} \tag{3–48}$$

The rate of heat transfer across each curvilinear square within the $m$th heat-flow path can be approximated by

$$q_m = -kL \frac{w}{\Delta\xi} \Delta T = -\frac{\Delta T}{R_n} \tag{3–49}$$

where $\Delta T$ is the mean temperature drop, $\Delta\xi$ the mean length, $w$ the mean width, and $L$ the mean depth associated with the $n$th curvilinear square. Because $\Delta\xi = w$ for full curvilinear squares, the quantity $R_n$ is equal to $1/(kL)$.

Assuming uniform thermal conductivity $k$ and plate length $L$, the rate of heat transfer in the $m$th heat-flow path can also be expressed in terms of the total temperature drop $T_1 - T_2$ by

$$q_m = \frac{T_1 - T_2}{\sum_{n=1}^{N_m} R_n} = \frac{kL}{N_m}(T_1 - T_2) \tag{3–50}$$

where $N_m$ is the number of curvilinear squares in the $m$th heat-flow lane. The total rate of heat transfer in the system can now be obtained by summing the rates for each heat-flow path; that is,

$$q = kL(T_1 - T_2) \sum_{m=1}^{M} \frac{1}{N_m} \tag{3–51}$$

for uniform $k$ and $L$. For simple problems in which the same number of full curvilinear squares $N$ occur in each heat-flow path, Eq. (3–51) reduces to

$$q = kL \frac{M}{N} (T_1 - T_2) \tag{3–52}$$

where $N$ and $M$ take on integer values.

Suggestions for developing freehand plots of curvilinear networks have been developed by Bewley [7] and are summarized by Kreith and Black [8]. Although the development of such sketches is an art, one can be guided by experimental electrical analogy measurements.

## EXAMPLE 3–10

Referring to the hollow quarter cylinder shown in Fig. 3–8 with $L = 1$ m, $r_1 = 2$ cm, $r_2 = 7.4$ cm, and $k = 125$ W/(m °C), determine the rate of heat transfer for (a) $T_1 = 150$°C and $T_2 = 35$°C; and (b) $T_1 = 150$°C and surface at $\theta = 0$ rad exposed to a convecting fluid with $T_F = 35$°C and $\bar{h} = 100$ W/(m² °C).

### Solution

*Objective* Determine $q$ within the cylinder section for cases (a) and (b).

*Schematic* Cylindrical sections.

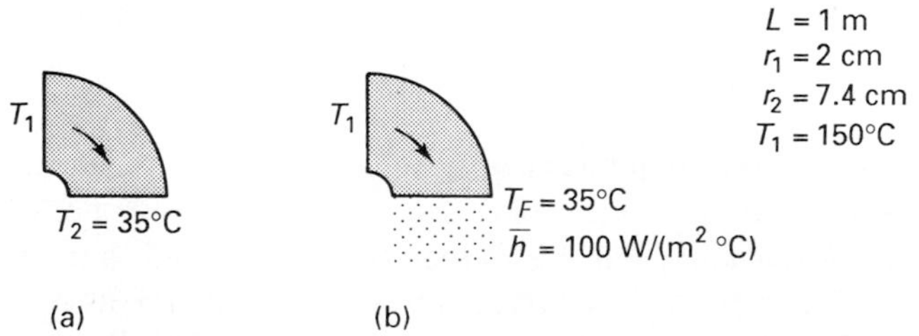
L = 1 m
r1 = 2 cm
r2 = 7.4 cm
T1 = 150°C
T1
T2 = 35°C
(a)
T1
TF = 35°C
h = 100 W/(m2 °C)
(b)

*Assumptions/Conditions*

steady-state
two-dimensional
uniform properties

*Properties* $k = 125$ W/(m °C).

*Analysis* Referring to Fig. 3–8, in which a network of curvilinear squares has already been developed, we see that $N = 6$ and $M = 5$.

(a) The rate of heat transfer for the case in which the surface temperatures are specified is given by Eq. (3–52),

$$q = kL\frac{M}{N}(T_1 - T_2) = 125\frac{\text{W}}{\text{m °C}}(1\text{ m})\left(\frac{5}{6}\right)(150\text{°C} - 35\text{°C}) = 12{,}000\text{ W}$$

Note that the equivalent thermal resistance for this problem is

$$R_k = \frac{N}{MkL} = \frac{1}{104}\frac{\text{°C}}{\text{W}}$$

(b) For the system with convection at one surface, we sketch the analogous electrical circuit in Fig. E3–10. The rate of heat transfer through this circuit is

$$q = \frac{150\text{°C} - 35\text{°C}}{\dfrac{1}{104}\dfrac{\text{°C}}{\text{W}} + \dfrac{\text{m}^2\text{ °C}}{100\text{ W}(0.054\text{ m})(1\text{ m})}} = 590\text{ W}$$

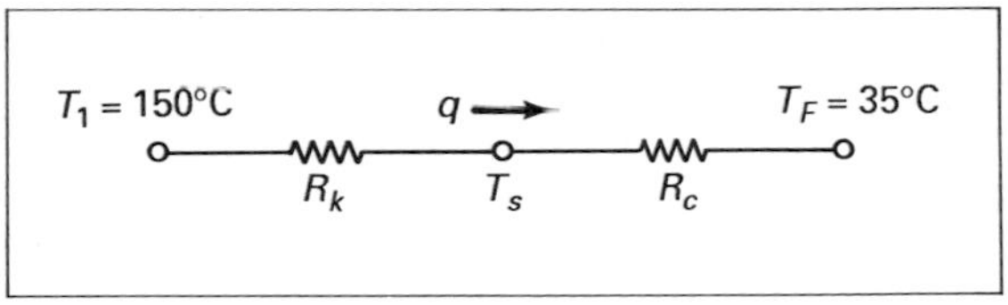

$T_1$ = 150°C
$q$
$T_F$ = 35°C
$R_k$
$T_s$
$R_c$


**FIGURE E3–10**
Thermal circuit; $R_c = 1/(\bar{h}A_s)$.

In addition, the surface temperature is obtained by writing

$$T_s = T_F + R_c q = 35\text{°C} + \frac{590\text{ W}}{5.4\text{ W/°C}} = 144\text{°C}$$

## 3–9 PRACTICAL SOLUTION RESULTS

The several approaches introduced in this chapter have been utilized to develop design equations for the temperature distribution and heat transfer for a number of standard conduction-heat-transfer problems that are commonly encountered in practice. Representative practical solution results are presented in this section. This useful design information is cast in the form of (1) tables for thermal resistance $R$, and (2) charts and analytical relations for unsteady one-dimensional processes.

### 3–9–1 Steady-Multidimensional Heat-Transfer Systems

The thermal resistance $R$ for several representative steady two- and three-dimensional conduction-heat-transfer systems are given in Tables 3–2 and 3–3. A comprehensive summary of conduction shape factors $S$ [$R = 1/(kS)$] is given by Hahne and Grigull [12].

**TABLE 3–2** Thermal resistance: Steady two-dimensional systems

| *System* | *Thermal resistance* $R = 1/(kS)$ |
|---|---|
| Hollow Circular Cylinder Section: Surfaces at $\theta = 0$ and $\theta = \theta_1$ at $T_1$ and $T_2$; surfaces at $r = r_1$ and $r = r_2$ insulated. 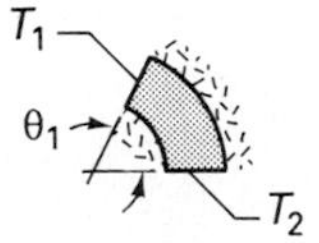  | $\dfrac{\theta_1/(Lk)}{\ln(r_2/r_1)}$ |
| Circular Cylinder Buried Horizontally in Semi-infinite Medium: $L >> r_1$. 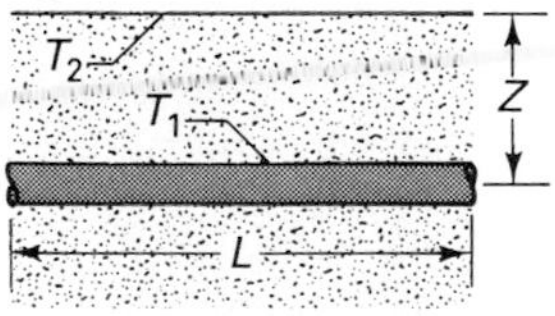  | $\dfrac{\cosh^{-1}(Z/r_1)}{2\pi Lk}$ |
| Circular Cylinder Buried Vertically in Semi-infinite Medium: $L >> r_1$. 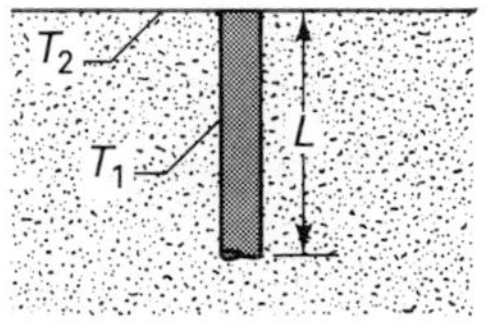  | $\dfrac{\ln(2L/r_1)}{\pi Lk}$ |
| Two Circular Cylinders Buried Horizontally in Infinite Medium: $L >> r_1, r_2$. 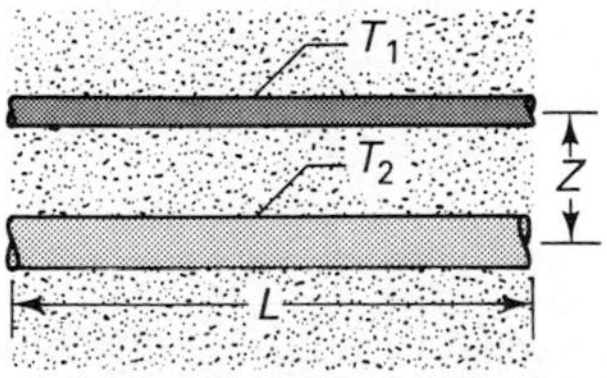  | $\dfrac{\cosh^{-1}\dfrac{Z - r_1 - r_2}{2r_1r_2}}{2\pi Lk}$ |
| Sphere Buried in Semi-inifinite Medium. 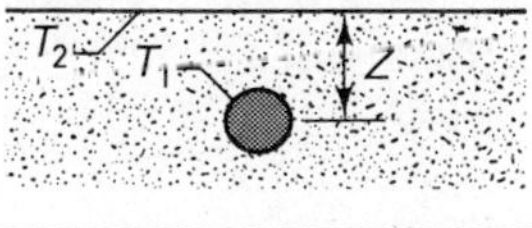  | $\dfrac{1 - r_1(2Z)}{4\pi r_1 k}$ |

*Source*: Summarized from references 9 through 11.

*Note*: $\cosh^{-1}\dfrac{x}{a} = \ln\dfrac{x + \sqrt{x^2 - a^2}}{a}$ for $x/a \leqslant 1$

**TABLE 3–3** Thermal resistance: Steady three-dimensional systems

| *System* | *Thermal resistance* $R = 1/(kS)$ |
|---|---|
| Circular Cylinder Buried Horizontally in Semi-infinite Medium: $L$ short.<br>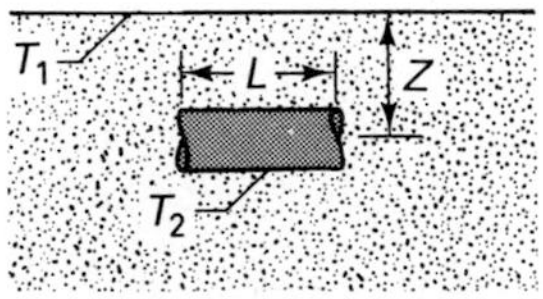<br> | $\dfrac{\ln (L/r_1) - \ln [L/(2Z)]}{2\pi Lk}$ |
| Two Plane Walls with Edge Section: Inside dimension greater than $\delta$.<br>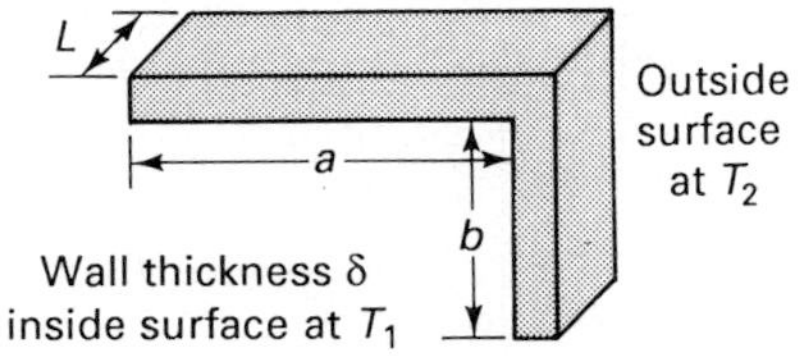<br> | $\dfrac{1}{\left(\dfrac{aL}{\delta} + \dfrac{bL}{\delta} + 0.54L\right)k}$ |
| Corner Section of Three Plane Walls: Inside dimensions greater than $\delta$.<br>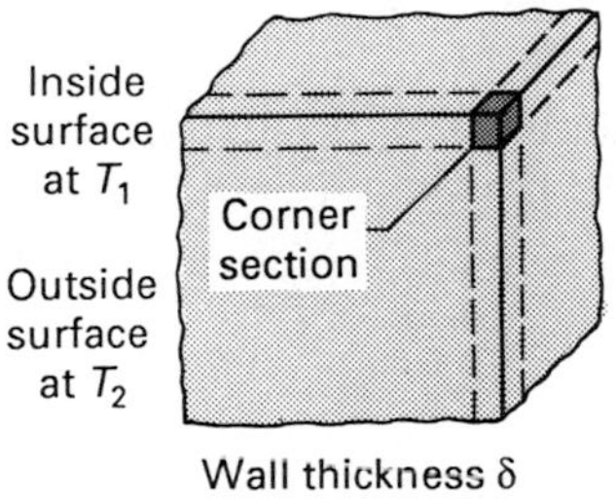<br> | $\dfrac{1}{0.15\delta k}$ |

*Source*: Summarized from references 9 through 11 and Chap. 3.

## EXAMPLE 3–11

Saturated steam at atmospheric pressure is passed through a long thin-walled horizontal pipe of 5 in. diameter, which is buried in the earth at a depth of 4 ft. Given an annual mean earth surface temperature of 51°F, estimate the mean rate of heat loss per unit length of pipe in the Chicago area for a soil thermal conductivity of 0.75 Btu/(h ft °F).

### Solution

*Objective* Estimate the annual mean value of $q'$.

*Schematic* Steady-state indealization of thin-walled pipe buried in the earth.

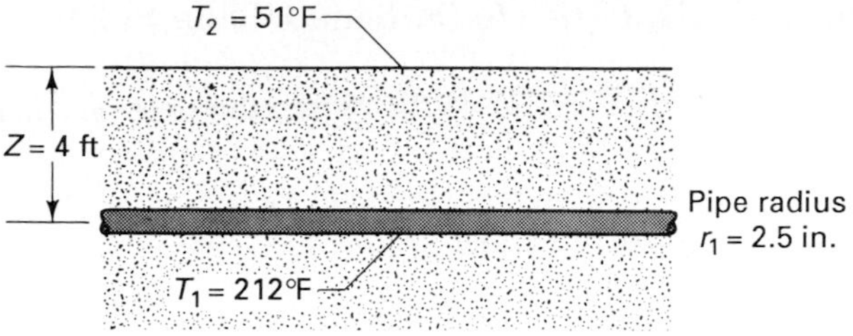


*Assumptions/Conditions*

steady-state
two-dimensional
uniform properties

*Properties* Earth: $k = 0.75$ Btu/(h ft °F).

*Analysis* According to the ASHRAE *Handbook of Fundamentals* [13], the annual mean earth temperature $T_M$ is approximately constant for all depths up to about 200 ft, with the annual variation in daily mean surface temperature generally of the order of ±25°F in Chicago (see Table A–D–1). Therefore, although the temperature distribution within the earth changes over the course of a year, the mean rate of heat loss from the pipe can be approximated by assuming steady-state conditions with the earth surface temperature set equal to 51°F and the pipe temperature set equal to 212°F.

Referring to Table 3–2, we find that the thermal resistance $R$ for this steady-state idealization of the actual process is given by

$$R = \frac{\cosh^{-1}(Z/r_1)}{2\pi L k} = \frac{\ln[(Z + \sqrt{Z^2 - r_1^2})/r_1]}{2\pi L k}$$

$$= \frac{\ln\{[4 + \sqrt{4^2 - (2.5/12)^2}]/(2.5/12)\}}{2\pi L[0.75 \text{ Btu/(h ft °F)}]} = \frac{0.774}{L}\frac{\text{h ft °F}}{\text{Btu}}$$

It follows that the rate of heat transfer becomes

$$q = \frac{T_1 - T_2}{R} = \frac{212\text{°F} - 51\text{°F}}{0.774/L}\frac{\text{Btu}}{\text{h ft °F}}$$

or

$$q' = \frac{q}{L} = 208 \text{ Btu/(h ft)}$$

### 3–9–2 Unsteady One-Dimensional Heat Transfer in Semi-Infinite Solids

Because of its importance in many practical applications, the unsteady temperature distribution and heat transfer in a semi-infinite solid with a change in conditions imposed at the surface has been extensively studied. Referring to Fig. 3–9 and assuming uniform properties, the energy equation for this type problem is given by

$$\frac{\partial^2 T}{\partial x^2} = \frac{1}{\alpha}\frac{\partial T}{\partial t} \tag{3–53}$$

Useful analytical solutions to this equation have been developed for four basic situations. The first three of these fundamental problems involve a uniform initial temperature distribution $T_i$, with an instantaneous change in the boundary condition imposed at $t = 0$. These three conditions include (1) a sudden step change in surface temperature $T_s$, (2) a sudden application of a constant heat flux $q_0''$, and (3) a sudden exposure of the surface to convection with $T_F$ and $h$ both constant.† The fourth fundamental problem involves the thermal response of a semi-infinite solid to a periodic change in the surface temperature. The initial-boundary conditions and solution results [i.e., temperature distribution $T(x,t)$, surface-heat flux $q_s''(t)$, and/or surface temperature $T_s(t)$] for these four basic cases are given in Table 3–4.

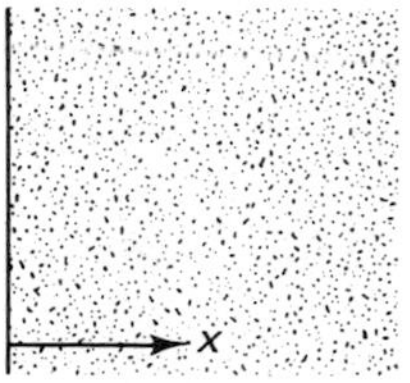


**FIGURE 3–9**
Semi-infinite solid with a change in boundary condition at $x = 0$.

It should be observed that these solutions involve the *error function* erf $X$, which is defined by the integral relation

$$\text{erf } X = \frac{2}{\sqrt{\pi}}\int_0^X \exp(-\beta^2)\, d\beta \tag{3–54}$$

This important mathematical function is tabulated as a function of $X$ in Appendix G–1. The *complementary error function* erfc $X$ is related to the error function by

$$\text{erfc } X = 1 - \text{erf } X = \frac{2}{\sqrt{\pi}}\int_X^\infty \exp(-\beta^2)\, d\beta \tag{3–55}$$

The temperature distribution for a semi-infinite solid initially at $T_i$ with surface suddenly exposed to convection (i.e., case iii) is shown in Fig. 3–10. It should be observed that the limiting curve associated with $h = \infty$ corresponds to a sudden step change in temperature (i.e., case i).

† $\bar{h}$ is represented by $h$ for situations in which the coefficient of heat transfer is uniform over the surface.

**TABLE 3–4** Solution results: Unsteady heat transfer in a semi-infinite solid with uniform properties.

| *Boundary condition* | *Solution results* |
|---|---|
| Case i: Step change in surface temperature<br>$T(x,0) = T_i$<br>$T(0,t) = T_0$<br>$T(\infty,t) = T_i$<br>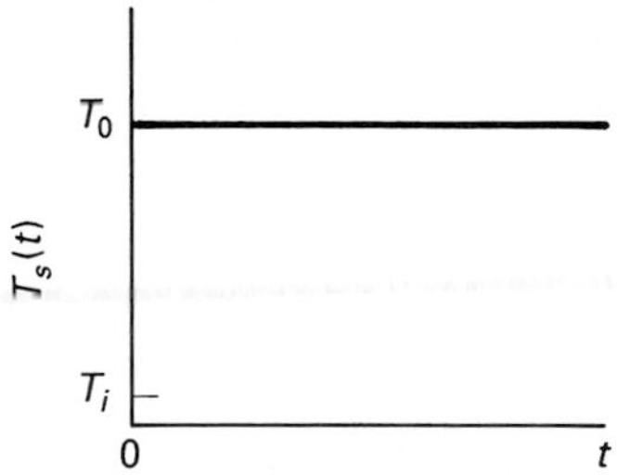<br> | $$\frac{T(x,t) - T_0}{T_i - T_0} = \text{erf}\left(\frac{x}{2\sqrt{\alpha t}}\right)$$<br>$$q_s''(t) = \frac{k(T_0 - T_i)}{\sqrt{\pi\alpha t}}$$<br>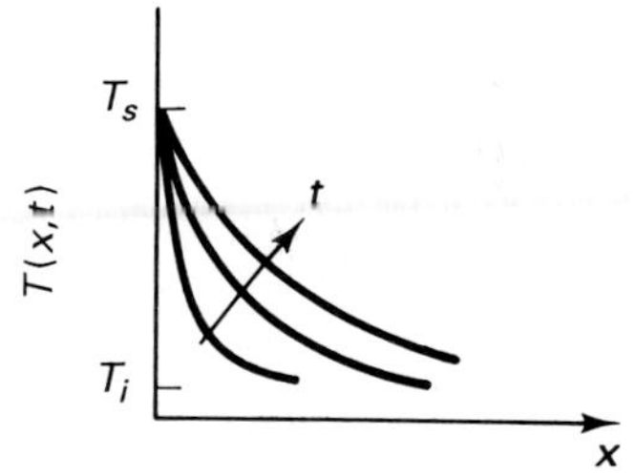<br> |
| Case ii: Constant surface-heat flux<br>$T(x,0) = T_i$<br>$q_s'' = -k\frac{\partial T}{\partial x}\Big\vert_{x=0} = q_0''$<br>$T(\infty,t) = T_i$<br>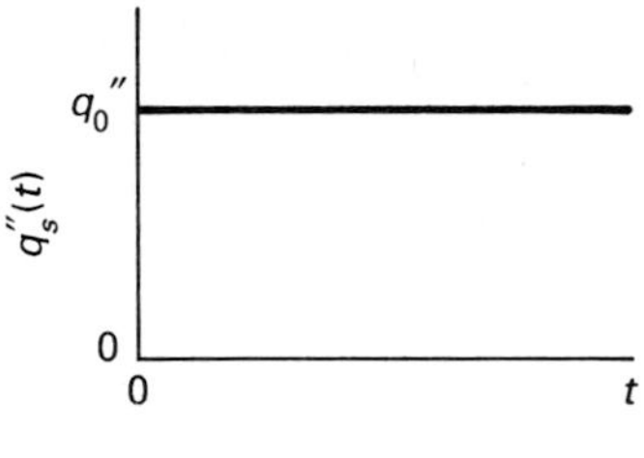<br> | $$T(x,t) - T_i = 2\frac{q_0''}{k}\sqrt{\frac{\alpha t}{\pi}}\exp\left(\frac{-x^2}{4\alpha t}\right) - \frac{q_0''x}{k}\text{erfc}\left(\frac{x}{2\sqrt{\alpha t}}\right)$$<br>$$T_s(t) - T_i = 2\frac{q_0''}{k}\sqrt{\frac{\alpha t}{\pi}}$$<br><br> |

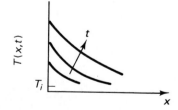

*Continued on next page*

**TABLE 3–4** (*Continued*)

| *Boundary condition* | *Solution results* |
|---|---|
| Case iii: Surface convection<br>$T(x,0) = T_i$<br>$-k\left.\frac{\partial T}{\partial x}\right\vert_{x=0} = h[T_F - T(0,t)]$<br>$T(\infty,t) = T_i$ | $\frac{T(x,t) - T_i}{T_F - T_i} = \operatorname{erfc}\left(\frac{x}{2\sqrt{\alpha t}}\right) - \left[\exp\left(\frac{hx}{k} + \frac{h^2\alpha t}{k^2}\right)\right]\left[\operatorname{erfc}\left(\frac{x}{2\sqrt{\alpha t}} + \frac{h\sqrt{\alpha t}}{k}\right)\right]$<br>$\frac{T_s(t) - T_i}{T_F - T_i} = 1 - \exp\left(\frac{h^2\alpha t}{k^2}\right)\operatorname{erfc}\left(\frac{h\sqrt{\alpha t}}{k}\right)$<br>$q''_s(t) = h(T_F - T_i)\exp\left(\frac{h^2\alpha t}{k^2}\right)\operatorname{erfc}\left(\frac{h\sqrt{\alpha t}}{k}\right)$<br>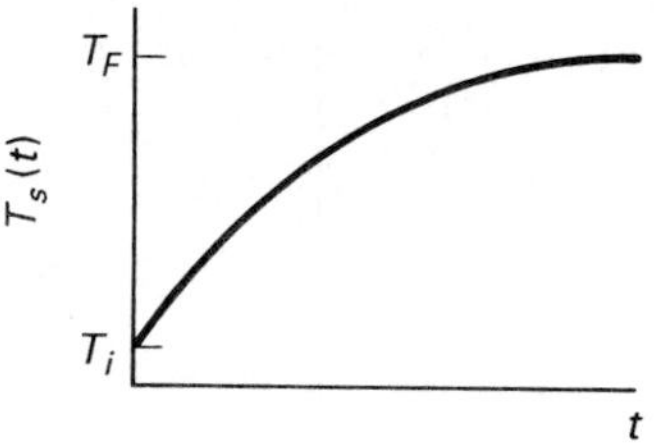<br><br>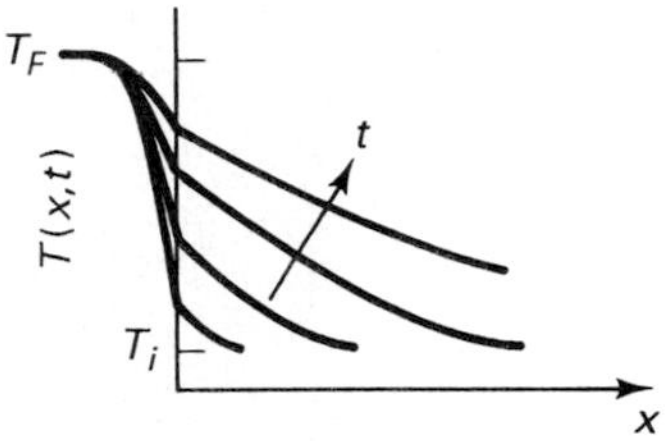<br> |
| Case iv: Periodic surface temperature<br>$T(0,t) = T_M - \Delta T_s \cos\left[\frac{2\pi}{\tau}(t - t_0)\right]$<br>$T(x,t) = T(x,t+n\tau);\ n = 0,1,2,3,\ldots$<br>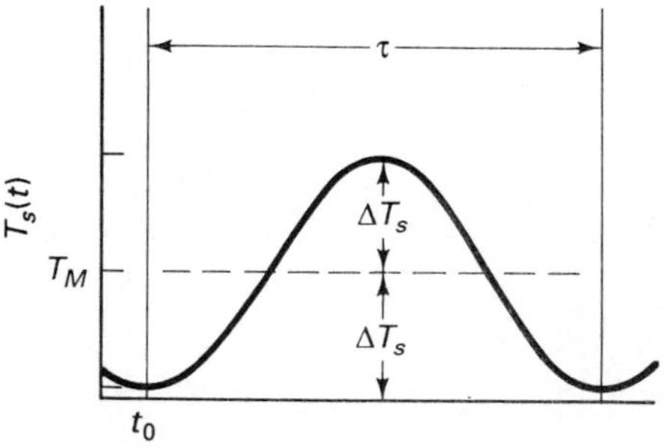<br> | $T(x,t) = T_M - \Delta T_s \exp\left(-x\sqrt{\frac{\pi}{\alpha\tau}}\right) \times \cos\left[\frac{2\pi}{\tau}(t - t_0) - x\sqrt{\frac{\pi}{\alpha\tau}}\right]$<br>$q''_s(t) = \frac{k\,\Delta T_s}{\sqrt{\alpha\tau/\pi}}\left\{-\cos\left[\frac{2\pi}{\tau}(t - t_0)\right] + \sin\left[\frac{2\pi}{\tau}(t - t_0)\right]\right\}$<br>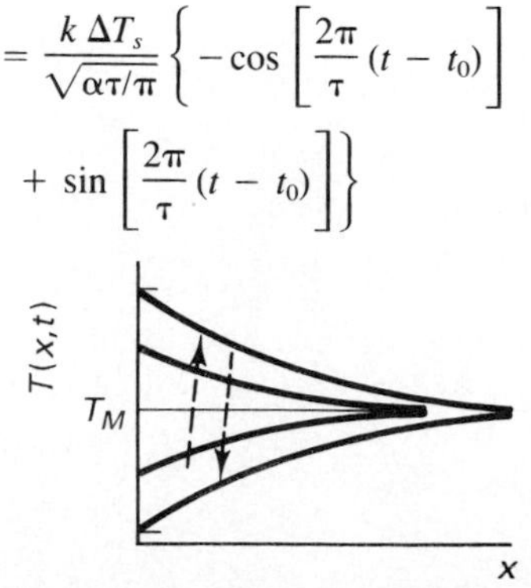<br> |

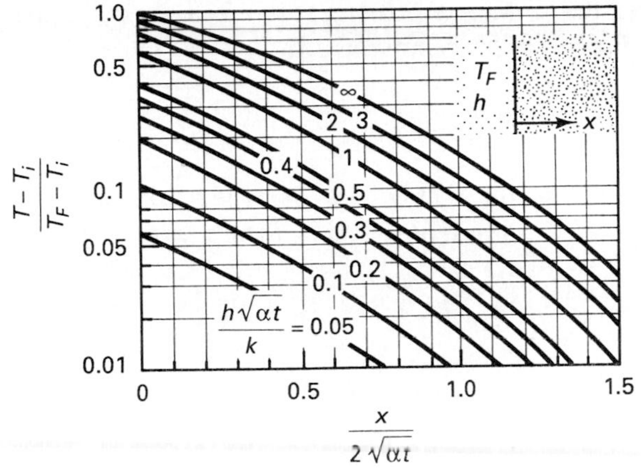


**FIGURE 3–10**
Chart for instantaneous temperature distribution in a semi-infinite solid with surface suddenly exposed to a fluid. (From Schneider [6]. Used with permission.)

## EXAMPLE 3–12

Referring to Table 3–4, the temperature distribution in a semi-infinite solid initially at uniform temperature $T_i$ with surface temperature suddenly changed to $T_0$ is

$$\frac{T(x,t) - T_0}{T_i - T_0} = \text{erf}\left(\frac{x}{2\sqrt{\alpha t}}\right)$$

for uniform properties. Use this relation to develop expressions for the instantaneous rate $q_s(t)$ and total accumulative $Q_s$ heat transfer at the surface.

### Solution

*Objective* Develop relations for $q_s(t)$ and $Q_s$.

*Schematic* Unsteady one-dimensional heat transfer in a semi-infinite solid.

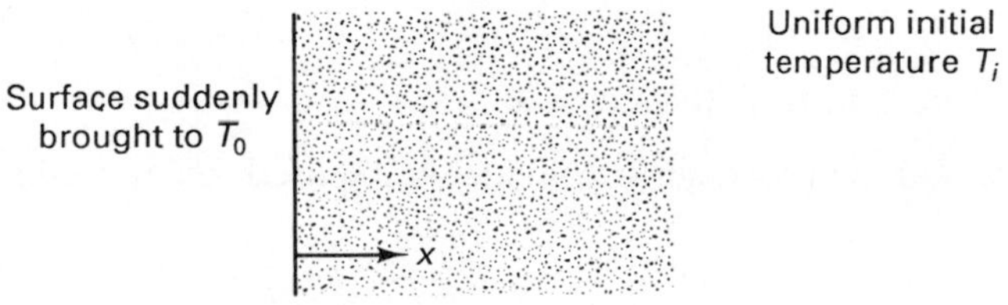


*Assumptions/Conditions*

unsteady-state
one spacial dimension
uniform properties

*Analysis* Making use of the defining equation for the error function given by Eq. (3–54), the temperature distribution is represented by

$$\frac{T(x,t) - T_0}{T_i - T_0} = \frac{2}{\sqrt{\pi}} \int_0^{x/(2\sqrt{\alpha t})} e^{-\beta^2}\, d\beta$$

Using this relation together with the Fourier law of conduction, the instantaneous rate of heat transfer within the semi-infinite solid is written as

$$q_x(t) = -kA \left.\frac{\partial T}{\partial x}\right|_x = -kA(T_i - T_0) \int_0^{x/(2\sqrt{\alpha t})} e^{-\beta^2}\, d\beta$$

The differentiation of this integral is accomplished by employing the Leibnitz rule (Appendix A), with the result

$$q_x(t) = -kA(T_i - T_0)\,\frac{2}{\sqrt{\pi}}\,\frac{e^{-x^2/(4\alpha t)}}{2\sqrt{\alpha t}} = \frac{kA(T_0 - T_i)}{\sqrt{\pi \alpha t}}\, e^{-x^2/(4\alpha t)}$$

Setting $x = 0$, we obtain

$$q_s(t) = \frac{kA(T_0 - T_i)}{\sqrt{\pi \alpha t}}$$

which is equivalent to the relation listed in Table 3–4 (case i) for $q_s''(t)$.

The total accumulative heat transfer $Q_s$ is obtained by writing

$$Q_s = \int_0^t q_s(t)\, dt = \int_0^t \frac{kA(T_0 - T_i)}{\sqrt{\pi \alpha t}}\, dt = 2kA(T_0 - T_i)\sqrt{\frac{t}{\pi \alpha}}$$

## EXAMPLE 3–13

One surface of a thick aluminum slab initially at 75°F is suddenly exposed to a fluid with $T_F = 250$°F and $h = 3000$ Btu/(h ft$^2$ °F). Determine the instantaneous surface temperature and heat flux after a time of 1 s.

### Solution

*Objective* Determine $T_s$ and $q_s''$ for $t = 1$ s.

*Schematic* Thick slab with one surface suddenly exposed to convection.

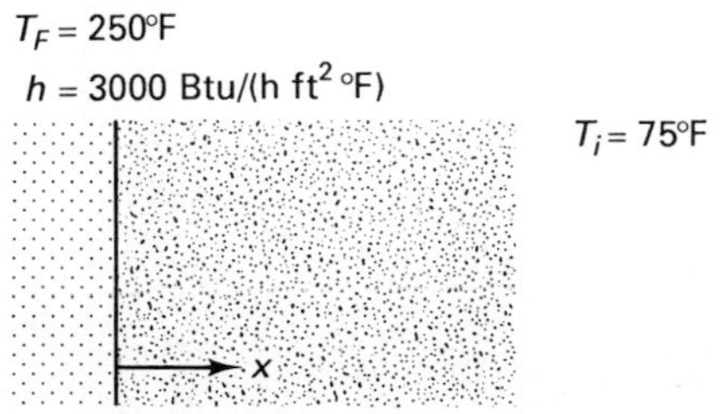
$T_F$ = 250°F
$h$ = 3000 Btu/(h ft$^2$ °F)
$T_i$ = 75°F
$x$

*Assumptions/Conditions*

unsteady-state
one spacial dimension
uniform properties

*Properties* Aluminum (Table A–C–1):

$$k = 236 \frac{\text{W}}{\text{m °C}} \frac{0.578 \text{ Btu/(h ft °F)}}{\text{W/(m °C)}} = 136 \text{ Btu/(h ft °F)}$$

$$\alpha = 9.75 \times 10^{-5} \frac{\text{m}^2}{\text{s}} \frac{10.8 \text{ ft}^2\text{/s}}{\text{m}^2\text{/s}} = 1.05 \times 10^{-3} \text{ ft}^2\text{/s}$$

*Analysis* Since the slab is thick and only one surface is exposed to convection, this system can be modeled as a semi-infinite solid. Referring to case iii of Table 3–4, the instantaneous surface temperature and heat flux are given by

$$\frac{T_s - T_i}{T_F - T_i} = 1 - \exp\left(\frac{h^2 \alpha t}{k^2}\right) \text{erfc}\left(\frac{h\sqrt{\alpha t}}{k}\right) \tag{a}$$

and

$$q''_s = h(T_F - T_i) \exp\left(\frac{h^2 \alpha t}{k^2}\right) \text{erfc}\left(\frac{h\sqrt{\alpha t}}{k}\right) \tag{b}$$

Setting $t = 1$ s, we have

$$\frac{h\sqrt{\alpha t}}{k} = 3000 \frac{\text{Btu}}{\text{h ft}^2 \text{°F}} \frac{\text{h ft °F}}{136 \text{ Btu}} \left[\frac{1.05 \times 10^{-3} \text{ ft}^2 \text{ (1 s)}}{\text{s}}\right]^{1/2} = 0.715$$

$$\frac{h^2 \alpha t}{k} = 0.715^2 = 0.511$$

and, using Table A–G–1,

$$\text{erfc}\left(\frac{h\sqrt{\alpha t}}{k}\right) = 1 - \text{erf}\,(0.715) = 1 - 0.688 = 0.312$$

Substituting into Eqs. (a) and (b), we obtain

$$\frac{T_s - T_i}{T_F - T_i} = 1 - 0.312 \exp\,(0.511) = 0.480$$

or

$$T_s = (250 \text{ °F} - 75\text{°F})(0.480) + 75\text{°F} = 159\text{°F}$$

and

$$q''_s = 3000 \frac{\text{Btu}}{\text{h ft}^2 \text{ °F}} (250\text{°F} - 75\text{°F})(0.312) \exp\,(0.511)$$

$$= 2.73 \times 10^5 \text{ Btu/(h ft}^2\text{)}$$

for $t = 1$ s.

It should be noted that Fig. 3–10 could be used instead of Eq. (a) to evaluate $T_s$. Following this practical but somewhat less accurate approach, the surface heat flux could be computed by use of the Newton law of cooling,

$$q_s'' = h(T_s - T_F)$$

## EXAMPLE 3–14

The daily mean temperature distribution within the region near the surface of the earth at any time of the year is commonly approximated by the solution for a semi-infinite solid with a periodic boundary condition of the form

$$T_s(t) = T_M - \Delta T_s \cos\left[\frac{2\pi}{\tau}(t - t_0)\right] \qquad \text{at } x = 0 \tag{a}$$

where the period $\tau$ is set equal to 365 days, $T_M$ is the annual mean earth temperature, $\Delta T_s$ is the amplitude of annual variation in surface soil temperature, and $t_0$ is the phase constant. The solution for the temperature distribution for this situation is given in Table 3–4 as (case iv)

$$T(x,t) = T_M - \Delta T_s \exp\left(-x\sqrt{\frac{\pi}{\alpha\tau}}\right)\cos\left[\frac{2\pi}{\tau}(t - t_0) - x\sqrt{\frac{\pi}{\alpha\tau}}\right] \tag{b}$$

Earth data for $T_M$, $\Delta T_s$, and $t_0$ are listed in Appendix Table A–D–1 for selected cities in the U.S. Use this relation to estimate the depth of the *freezing line* (i.e., the minimum soil depth at which freezing will not occur over the course of a year under normal conditions) in the Chicago area, assuming damp, heavy soil.

### Solution

*Objective* Approximate the freeze line depth $x_{fl}$ in Chicago, Illinois.

*Schematic* Heat transfer near the surface of the earth.

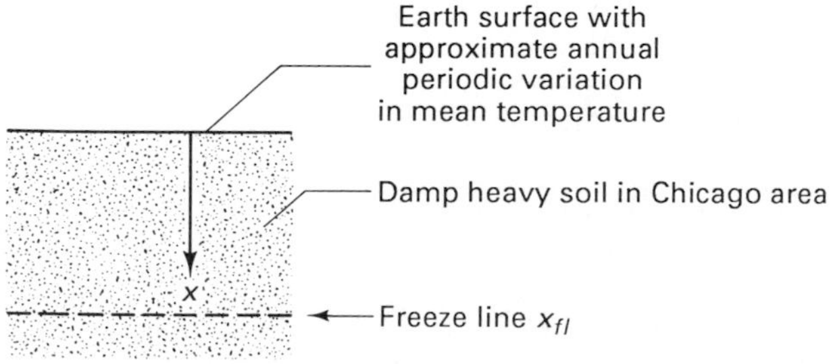


*Assumptions/Conditions*

unsteady-state
one spatial dimension
uniform properties

*Properties* Soil (damp heavy) (Table A–C–2): $k = 0.75$ Btu/(h ft °F), $\alpha = 0.6$ ft$^2$/day.

*Analysis* The distribution in minimum temperature, which is associated with Eq. (b), is obtained by setting the cosine term equal to unity; that is,

$$T(x,t)_{\min} = T_M - \Delta T_s \exp\left(-x\sqrt{\frac{\pi}{\alpha\tau}}\right)$$

Rearranging, this expression is put into the form

$$x = \sqrt{\frac{\alpha\tau}{\pi}} \ln\left[\frac{\Delta T_s}{T_M - T(x,t)_{\min}}\right]$$

Setting $T(x,t)_{\min} = 32°\text{F}$ and substituting for the various parameters and earth temperature data for Chicago (i.e., $T_M = 51°\text{F}$ and $\Delta T_s = 25°\text{F}$), the depth of the freezing line becomes

$$x_{fl} = \sqrt{\frac{(0.6\ \text{ft}^2/\text{day})(365\ \text{day})}{\pi}} \ln\left(\frac{25°\text{F}}{51°\text{F} - 32°\text{F}}\right) = 2.29\ \text{ft}$$

Thus, under normal weather conditions the freezing line depth for damp heavy soil in the Chicago area is about 2.29 ft. Adding a 20% safety factor to allow for variations in weather, water pipes buried below 2.75 ft should be safe from freezing for this soil and location.

### 3–9–3 Unsteady One-Dimensional Heat Transfer in Flat Plates, Circular Cyclinders, and Spheres

We now turn our attention to unsteady one-dimensional heat transfer in the flat plate, circular cylinder, and spherical geometries pictured in Fig. 3–11. Solutions for the temperature distribution and heat transfer in these geometries have been developed for various boundary conditions.

Focusing attention on the case in which a flat plate of width $2L$ initially at uniform temperature $T_i$ is suddenly subjected to convection, the mathematical formulation is given by (for uniform properties)

$$\frac{\partial^2 T}{\partial x^2} = \frac{1}{\alpha}\frac{\partial T}{\partial t} \tag{3–56}$$

$$T = T_i \qquad \text{at } t = 0$$

$$-k\frac{\partial T}{\partial x} = h(T - T_F) \qquad \text{at } x = L$$

and

$$\frac{\partial T}{\partial x} = 0 \qquad \text{at } x = 0$$

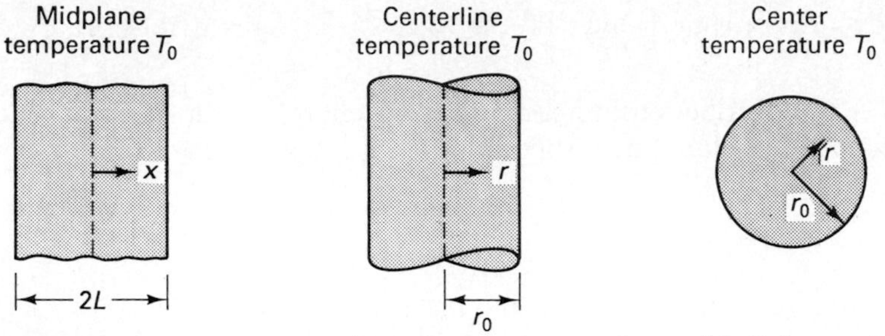
Midplane temperature $T_0$
Centerline temperature $T_0$
Center temperature $T_0$
x
r
r
$r_0$
2L
$r_0$

(a) Infinite plate of thickness $2L$; $\ell_0 = L$.

(b) Infinite cylinder of radius $r_0$; $\ell_0 = r_0$.

(c) Sphere of radius $r_0$; $\ell_0 = r_0$.

**FIGURE 3–11** One-dimensional solids with a change in boundary conditions at surface; characteristic length $\ell_0$.†

or

$$-k\frac{\partial T}{\partial x} = h(T_F - T) \qquad \text{at } x = -L$$

It should also be observed that this formulation also applies to a flat plate of thickness $L$ with surface at $x = 0$ insulated and surface at $x = L$ exposed to a convective fluid.

The exact solution to this system of equations for the case in which $h$ is constant, which can be obtained by the separation-of-variables method introduced in Sec. 3–6–1, is given in Appendix E–2. The general solution is in the form of an infinite series, with the number of significant terms in the series dependent upon a dimensionless time known as the *Fourier number Fo* ($Fo = \alpha t/L^2 = \alpha t/\ell_0^2$). As pointed out by Heisler [14], for values of $Fo$ greater than 0.2, which applies to 80% to 90% of the process period, the solution for the temperature distribution is approximated with an error of less than 1% by truncating the second- and higher-order terms in the series. The resulting approximate solution for dimensionless temperature $\Theta$ is

$$\Theta = \frac{T - T_F}{T_i - T_F} = \Theta_0 \cos\left(\gamma_1 \frac{x}{L}\right) \tag{3–57}$$

where the dimensionless midplane temperature $\Theta_0$ is

$$\Theta_0 = \frac{T_0 - T_F}{T_i - T_F} = C_1 \exp\left(-\gamma_1^2\, Fo\right) \tag{3–58}$$

and the coefficients $\gamma_1$ and $C_1$ are given in terms of Biot number $Bi_0$ ($Bi_0 = Bi = hL/k$) in Table 3–5. It follows that the surface temperature $T_s$ and instantaneous rate of convection heat transfer $q_c$ are given by

† Whereas $\ell = V/A_s$ is generally used as the characteristic length in the approximate lumped analysis approach of Chap. 2, $\ell_0$ is used in the formal solution approach considered in this section. Note that $\ell$ and $\ell_0$ are equivalent for the flat plate, but differ for the circular cylinder and sphere.

$$\Theta_s = \frac{T_s - T_F}{T_i - T_F} = \Theta_0 \cos \gamma_1 \tag{3–59}$$

and

$$q_c = hA_s(T_s - T_F) = hA_s(T_i - T_F)\, \Theta_0 \cos \gamma_1 \tag{3–60}$$

**TABLE 3–5** Heisler relations: Unsteady one-dimensional convection cooling of a plane wall

Coefficients $C_1$ and $\gamma_1$ as function of $Bi_0 = Bi = hL/k$

| $Bi_0$ | $\gamma_1$ (rad) | $C_1$ |
|---|---|---|
| 0.01 | 0.0998 | 1.0017 |
| 0.02 | 0.1410 | 1.0033 |
| 0.03 | 0.1732 | 1.0049 |
| 0.04 | 0.1987 | 1.0066 |
| 0.05 | 0.2217 | 1.0082 |
| 0.06 | 0.2425 | 1.0098 |
| 0.07 | 0.2615 | 1.0114 |
| 0.08 | 0.2791 | 1.0130 |
| 0.09 | 0.2956 | 1.0145 |
| 0.10 | 0.3111 | 1.0160 |
| 0.15 | 0.3779 | 1.0237 |
| 0.20 | 0.4328 | 1.0311 |
| 0.25 | 0.4801 | 1.0382 |
| 0.30 | 0.5218 | 1.0450 |
| 0.4 | 0.5932 | 1.0580 |
| 0.5 | 0.6533 | 1.0701 |
| 0.6 | 0.7051 | 1.0814 |
| 0.7 | 0.7506 | 1.0919 |
| 0.8 | 0.7910 | 1.1016 |
| 0.9 | 0.8274 | 1.1107 |
| 1.0 | 0.8603 | 1.1191 |
| 2.0 | 1.0769 | 1.1795 |
| 3.0 | 1.1925 | 1.2102 |
| 4.0 | 1.2646 | 1.2287 |
| 5.0 | 1.3138 | 1.2402 |
| 6.0 | 1.3496 | 1.2479 |
| 7.0 | 1.3766 | 1.2532 |
| 8.0 | 1.3978 | 1.2570 |
| 9.0 | 1.4149 | 1.2598 |
| 10.0 | 1.4289 | 1.2620 |
| 20.0 | 1.4961 | 1.2699 |
| 30.0 | 1.5202 | 1.2717 |
| 40.0 | 1.5325 | 1.2723 |
| 50.0 | 1.5400 | 1.2727 |
| 100.0 | 1.5552 | 1.2731 |

Other solution relations

$$\Theta_0 = C_1 \exp(-\gamma_1^2 Fo)$$

$$\Theta = \Theta_0 \cos(\gamma_1 x/L)$$

$$\Theta_s = \Theta_0 \cos \gamma_1$$

$$q_c = hA_s(T_i - T_F)\Theta_s$$

$$\frac{Q}{Q_{max}} = 1 - \Theta_0 \frac{\sin \gamma_1}{\gamma_1}$$

The coefficients $C_1$ and $\gamma_1$ are formally given by Eq. (A–E–2–3),

$$\gamma_1 \tan \gamma_1 = Bi_0$$

and Eq. (A–E–2–1),

$$C_1 = \frac{4 \sin \gamma_1}{2\gamma_1 + \sin(2\gamma_1)}$$

The Heisler relations are within 1% of the exact solution for $Fo > 0.2$.

To obtain an expression for the total accumulated heat transfer $Q$ from the surface, we write†

$$Q = \Sigma E_o - \Sigma E_i = -\Delta E_s = -[E_s(t) - E_s(0)]$$

$$= -\int_V \rho c_v (T - T_i)\, dV = -\int_0^L \rho c_v A (T - T_i)\, dx \tag{3–61}$$

$$= \rho c_v A (T_i - T_F) \int_0^L (1 - \Theta)\, dx = \rho c_v V (T_i - T_F)\left(1 - \Theta_0 \frac{\sin \gamma_1}{\gamma_1}\right)$$

or

$$\frac{Q}{Q_{max}} = 1 - \Theta_0 \frac{\sin \gamma_1}{\gamma_1} \tag{3–62}$$

where $Q_{max} = \rho c_v V (T_i - T_F)$.

These solution results for dimensionless midplane temperature $\Theta_0$, dimensionless temperature distribution $\Theta$, and total accumulated heat transfer $Q$ are represented by the convenient Heisler [14] charts shown in Fig. 3–12. With $Fo$ and $Bi_0$ specified, Fig. 3–12(a) can be used to evaluate $T_0$; Fig. 3–12(b) can be used to evaluate the temperature $T$ at any location off the midplane, such as the surface temperature $T = T_s$ at $x/L = 1$; and Fig. 3–12(c) can be used to evaluate $Q$.

The mathematical formulations for energy transfer from a circular cylinder and sphere, respectively, which are initially at uniform temperature $T_i$ and suddenly exposed to a convecting fluid, are represented by (for uniform properties)

$$\frac{1}{r}\frac{\partial}{\partial r}\left(r \frac{\partial T}{\partial r}\right) = \frac{1}{\alpha}\frac{\partial T}{\partial t} \tag{3–63}$$

for a circular cylinder, and

$$\frac{1}{r^2}\frac{\partial}{\partial r}\left(r^2 \frac{\partial T}{\partial r}\right) = \frac{1}{\alpha}\frac{\partial T}{\partial t} \tag{3–64}$$

for a sphere, with initial and boundary conditions of the form

$$\begin{aligned} T &= T_i && \text{at } t = 0 \\ \frac{\partial T}{\partial r} &= 0 && \text{at } r = 0 \\ -k\frac{\partial T}{\partial r} &= h(T - T_F) && \text{at } r = r_0 \end{aligned} \tag{3–65}$$

† $Q$ can also be represented by

$$Q = \int_0^t q_c\, dt$$

However, since Eq. (3–60) is restricted to the time domain for which $Fo > 0.2$, this alternative formulation is not useful in the present analysis.

for both geometries. The exact solution to these equations is given in Appendix E–2 for constant values of $h$. Approximate solutions, which are applicable for values of the Fourier number $Fo$ ($Fo = \alpha t/r_0^2 = \alpha t/\ell_0^2$) greater than 0.2, are listed in Tables 3–6 and 3–7 and are represented graphically by the Heisler charts shown in Figs. 3–13 and 3–14. With $Fo$ and the $r_0$ based Biot number $Bi_0$ specified, these solution results enable us to evaluate $\Theta_0$, $\Theta$, $q_c$, and $Q$.

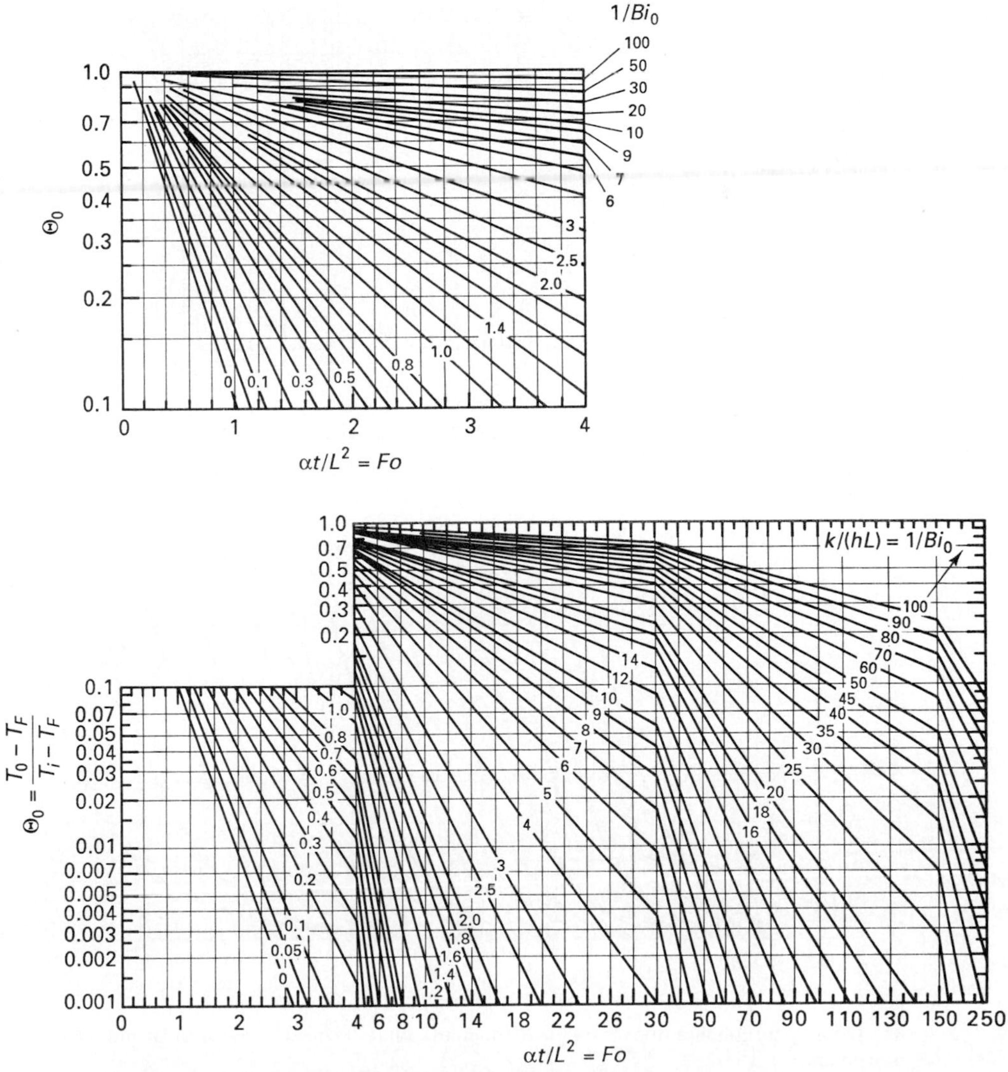


(a) Instantaneous midplane temperature $T_0$ (Heisler [14]).

**FIGURE 3–12** Heisler charts: Unsteady one-dimensional convection cooling of a plane wall; $Bi_0 = Bi = hL/k$ and $Fo = \alpha t/L^2$.

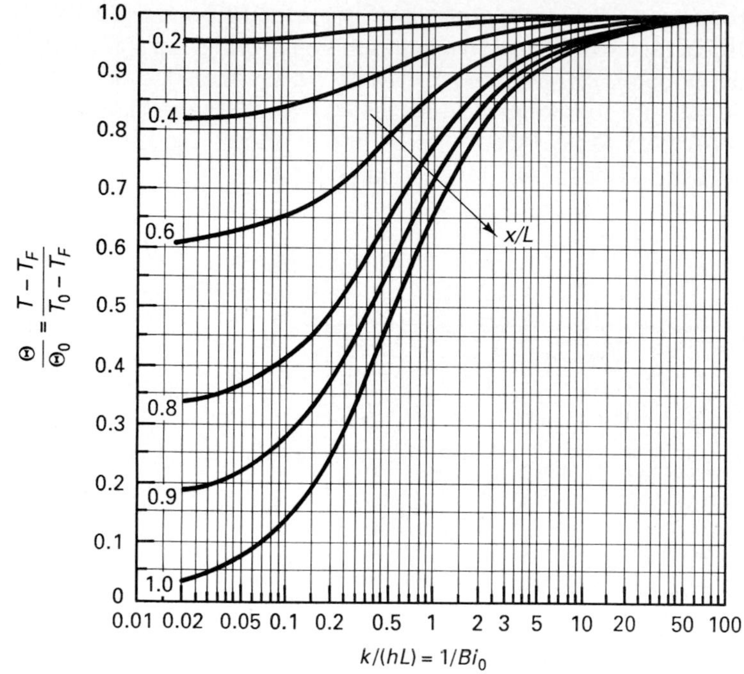


(b) Instantaneous temperature distribution for plane wall in terms of $T_0$ (Heisler [14]).

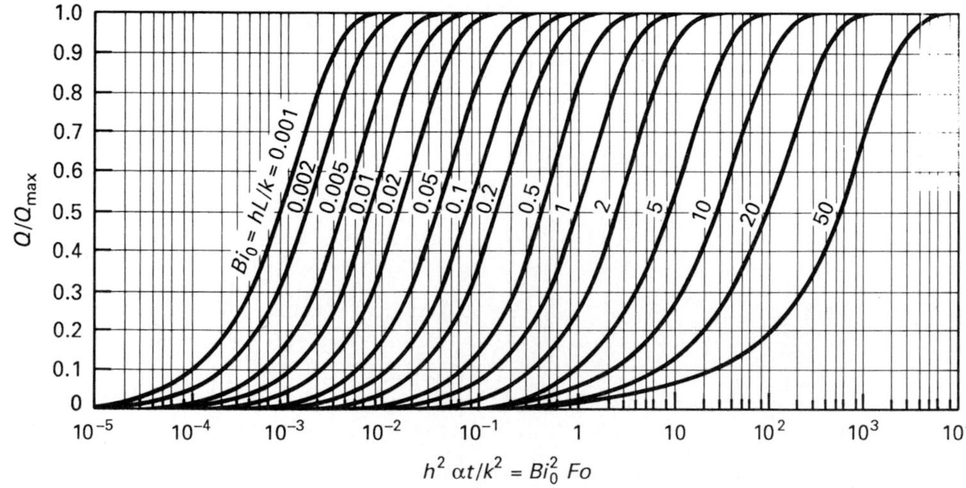


(c) Total accumulative heat transfer $Q$ for plane wall. (From Grober and Grigull [15]. Used with permission.)

**FIGURE 3–12** *(Continued)*

**TABLE 3–6** Heisler relations: Unsteady one-dimensional convection cooling of an infinite circular cylinder

Coefficients $C_1$ and $\gamma_1$ as function of $Bi_0 = hr_0/k$

| $Bi_0$ | $\gamma_1$ (rad) | $C_1$ |
|---|---|---|
| 0.01 | 0.1412 | 1.0025 |
| 0.02 | 0.1995 | 1.0050 |
| 0.03 | 0.2439 | 1.0075 |
| 0.04 | 0.2814 | 1.0099 |
| 0.05 | 0.3142 | 1.0124 |
| 0.06 | 0.3438 | 1.0148 |
| 0.07 | 0.3708 | 1.0173 |
| 0.08 | 0.3960 | 1.0197 |
| 0.09 | 0.4195 | 1.0222 |
| 0.10 | 0.4417 | 1.0246 |
| 0.15 | 0.5376 | 1.0365 |
| 0.20 | 0.6170 | 1.0483 |
| 0.25 | 0.6856 | 1.0598 |
| 0.30 | 0.7465 | 1.0712 |
| 0.4 | 0.8516 | 1.0932 |
| 0.5 | 0.9408 | 1.1143 |
| 0.6 | 1.0185 | 1.1346 |
| 0.7 | 1.0873 | 1.1539 |
| 0.8 | 1.1490 | 1.1725 |
| 0.9 | 1.2048 | 1.1902 |
| 1.0 | 1.2558 | 1.2071 |
| 2.0 | 1.5995 | 1.3384 |
| 3.0 | 1.7887 | 1.4191 |
| 4.0 | 1.9081 | 1.4698 |
| 5.0 | 1.9898 | 1.5029 |
| 6.0 | 2.0490 | 1.5253 |
| 7.0 | 2.0937 | 1.5411 |
| 8.0 | 2.1286 | 1.5526 |
| 9.0 | 2.1566 | 1.5611 |
| 10.0 | 2.1795 | 1.5677 |
| 20.0 | 2.2881 | 1.5919 |
| 30.0 | 2.3261 | 1.5973 |
| 40.0 | 2.3455 | 1.5993 |
| 50.0 | 2.3572 | 1.6002 |
| 100.0 | 2.3809 | 1.6015 |

Other solution relations

$$\Theta_0 = C_1 \exp(-\gamma_1^2 Fo)$$

$$\Theta = \Theta_0 J_0(\gamma_1 r/r_0)$$

$$\Theta_s = \Theta_0 J_0(\gamma_1)$$

$$q_c = hA_s(T_i - T_F)\Theta_s$$

$$\frac{Q}{Q_{max}} = 1 - 2\,\Theta_0 \frac{J_1(\gamma_1)}{\gamma_1}$$

$J_0$ – Bessel function of first kind, zero order

$J_1$ – Bessel function of first kind, first order

Representative values of $J_0$ and $J_1$ are tabulated in Table A–G–2.

The coefficients $C_1$ and $\gamma_1$ are formally given by Eq. (A–E–2–6),

$$\gamma_1 J_1(\gamma_1) = Bi_0 J_0(\gamma_1)$$

and Eq. (A–E–2–5),

$$C_1 = \frac{2}{\gamma_1} \frac{J_1(\gamma_1)}{J_0^2(\gamma_1) + J_1^2(\gamma_1)}$$

The Heisler relations are within 1% of the exact solution for $Fo > 0.2$.

Notice in Figs. 3–12(b), 3–13(b), and 3–14(b) that the temperature $T$ throughout the plane wall, circular cylinder, or sphere is only slightly dependent on location within the body for small values of the Biot number $Bi_0$ (or $Bi$). It follows that the unsteady lumped analysis developed in Chap. 2 can be used for such conditions; that is,

$$\Theta_0 = \frac{T_0 - T_F}{T_i - T_F} = \exp\left(-\frac{hA_s t}{\rho c_v V}\right) = \exp\left(-\frac{ht}{\rho c_v \ell}\right) \tag{3-66}$$

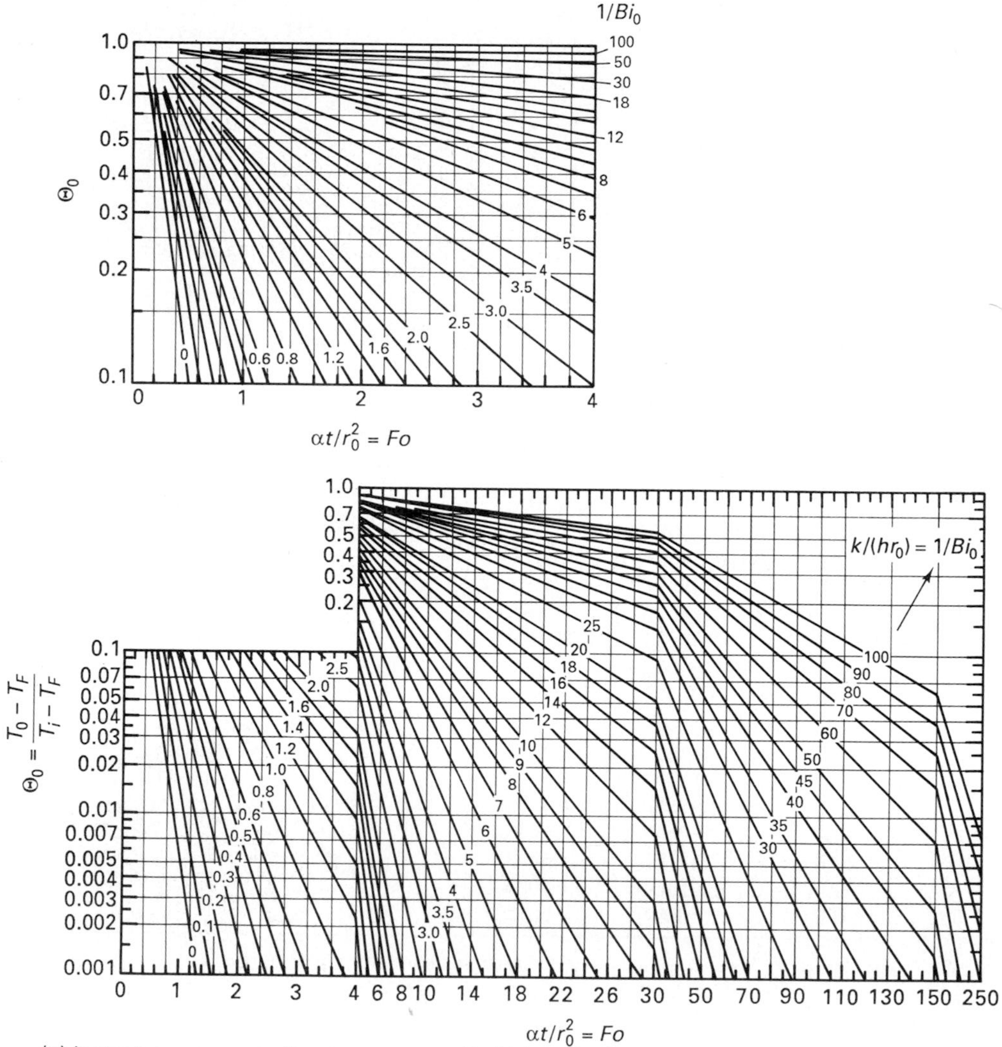

(a) Instantaneous centerline temperature $T_0$ (Heisler [14]).

**FIGURE 3–13** Heisler charts: Unsteady one-dimensional convection cooling of an infinite cylinder; $Bi_0 = hr_0/k$ and $Fo = \alpha t/r_0^2$.

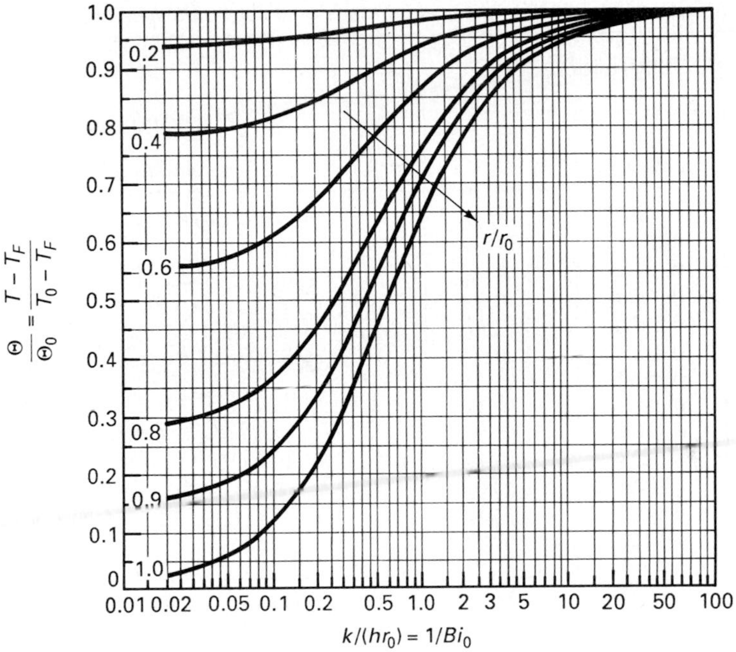


(b) Instantaneous temperature distribution for infinite cylinder in terms of $T_0$ (Heisler [14].

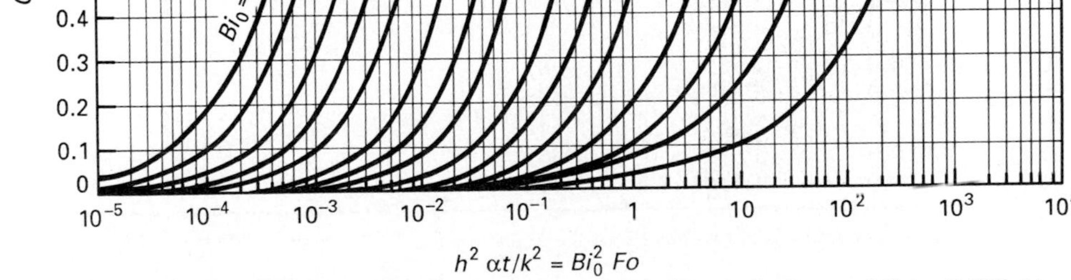

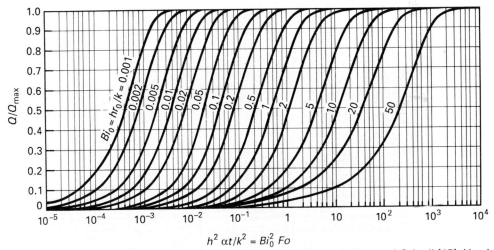


(c) Total accumulative heat transfer $Q$ for infinite cylinder. (From Grober and Grigull [15]. Used with permission.)

**FIGURE 3–13** *(Continued)*

**TABLE 3–7** Heisler relations: Unsteady one-dimensional convection cooling of a sphere

Coefficients $C_1$ and $\gamma_1$ as function of $Bi_0 = hr_0/k$

| $Bi_0$ | $\gamma_1$ (rad) | $C_1$ |
|---|---|---|
| 0.01 | 0.1730 | 1.0030 |
| 0.02 | 0.2445 | 1.0060 |
| 0.03 | 0.2989 | 1.0090 |
| 0.04 | 0.3450 | 1.0120 |
| 0.05 | 0.3852 | 1.0149 |
| 0.06 | 0.4217 | 1.0179 |
| 0.07 | 0.4550 | 1.0209 |
| 0.08 | 0.4860 | 1.0239 |
| 0.09 | 0.5150 | 1.0268 |
| 0.10 | 0.5423 | 1.0298 |
| 0.15 | 0.6608 | 1.0445 |
| 0.20 | 0.7593 | 1.0592 |
| 0.25 | 0.8448 | 1.0737 |
| 0.30 | 0.9208 | 1.0880 |
| 0.4 | 1.0528 | 1.1164 |
| 0.5 | 1.1656 | 1.1441 |
| 0.6 | 1.2644 | 1.1713 |
| 0.7 | 1.3525 | 1.1978 |
| 0.8 | 1.4320 | 1.2236 |
| 0.9 | 1.5044 | 1.2488 |
| 1.0 | 1.5708 | 1.2732 |
| 2.0 | 2.0288 | 1.4793 |
| 3.0 | 2.2889 | 1.6227 |
| 4.0 | 2.4556 | 1.7201 |
| 5.0 | 2.5704 | 1.7870 |
| 6.0 | 2.6537 | 1.8338 |
| 7.0 | 2.7165 | 1.8674 |
| 8.0 | 2.7654 | 1.8921 |
| 9.0 | 2.8044 | 1.9106 |
| 10.0 | 2.8363 | 1.9249 |
| 20.0 | 2.9857 | 1.9781 |
| 30.0 | 3.0372 | 1.9898 |
| 40.0 | 3.0632 | 1.9942 |
| 50.0 | 3.0788 | 1.9962 |
| 100.0 | 3.1102 | 1.9990 |

Other solution relations

$$\Theta_0 = C_1 \exp(-\gamma_1^2 Fo)$$

$$\Theta = \Theta_0 \frac{\sin(\gamma_1 r/r_0)}{\gamma_1 r/r_0}$$

$$\Theta_s = \Theta_0 \frac{\sin \gamma_1}{\gamma_1}$$

$$q_c = hA_s(T_i - T_F)\Theta_s$$

$$\frac{Q}{Q_{max}} = 1 - 3\,\Theta_0 \frac{\sin \gamma_1 - \gamma_1 \cos \gamma_1}{\gamma_1^3}$$

The coefficients $C_1$ and $\gamma_1$ are formally given by Eq. (A–E–2–9),

$$1 - \gamma_1 \cot \gamma_1 = Bi_0$$

and Eq. (A–E–2–8),

$$C_1 = \frac{4(\sin \gamma_1 - \gamma_1 \cos \gamma_1)}{2\gamma_1 - \sin(2\gamma_1)}$$

The Heisler relations are within 1% of the exact solution for $Fo > 0.2$.

from Eq. (2–143). The error in this equation is within about 10% for $Bi < 0.1$. It also follows that the instantaneous rate $q_c$ and accumulative $Q$ heat transfer for $Bi < 0.1$ can be approximated by Eq. (2–144),

$$\frac{q_c}{hA_s(T_i - T_F)} = \exp\left(-\frac{hA_st}{\rho c_v V}\right) \tag{3–67}$$

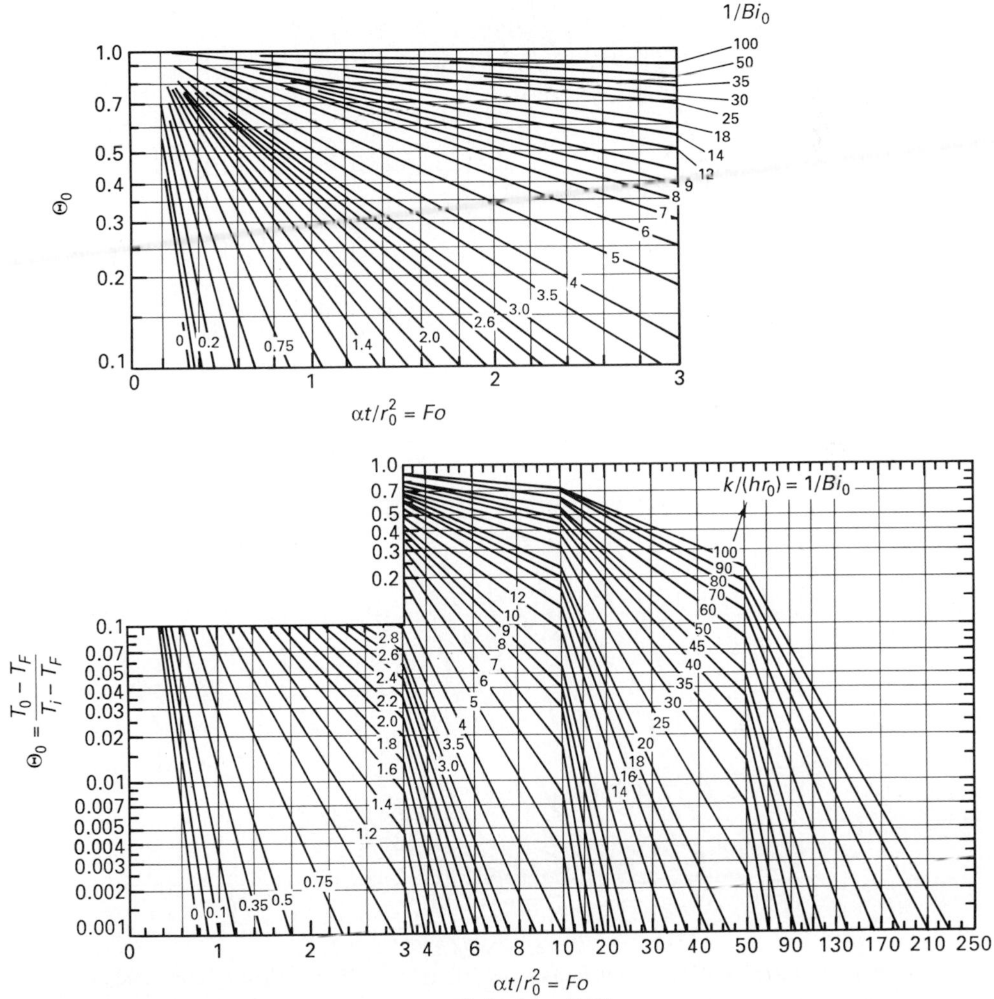


(a) Instantaneous midplane temperature $T_0$ (Heisler [14]).

**FIGURE 3–14** Heisler charts: Unsteady one-dimensional convection cooling of a sphere; $Bi_0 = hr_0/k$ and $Fo = \alpha t/r_0^2$.

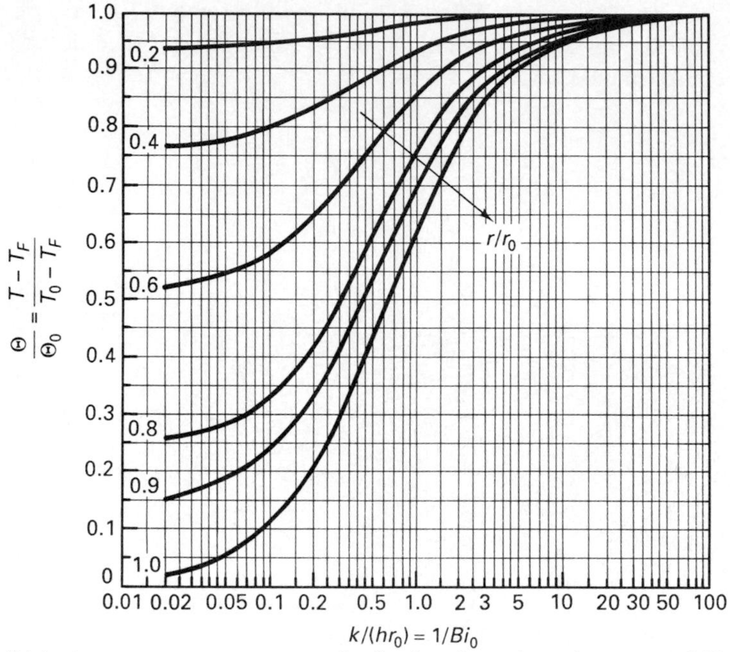


(b) Instantaneous temperature distribution for sphere in terms of $T_0$ (Heisler [14]).

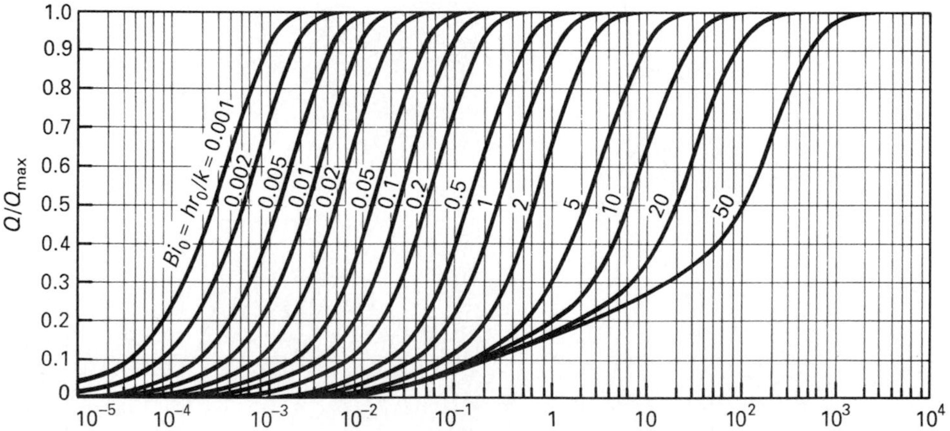


(c) Total accumulative heat transfer $Q$ for sphere. (From Grober and Grigull [15]. Used with permission.)

**FIGURE 3–14** *(Continued)*

and Eq. (2–147),

$$\frac{Q}{Q_{\max}} = 1 - \exp\left(-\frac{hA_s t}{\rho c_v V}\right) \tag{3-68}$$

These equations are written in terms of $Fo$ and $Bi_0$ by simply expressing $hA_s t/(\rho c_v V)$ as

$$\frac{hA_s t}{\rho c_v V} = \frac{h\ell_0}{k}\frac{\alpha t}{\ell_0^2}\frac{\ell_0}{\ell} = Bi_0\, Fo\,\frac{\ell_0}{\ell} \tag{3-69}$$

Because the Heisler relations and charts given in Tables 3–5 to 3–7 and Figs. 3–12 to 3–14 are based on approximate solutions, which are restricted to situations for which the Fourier number $Fo$ is not too small (i.e., $Fo = \alpha t/\ell_0^2 > 0.2$), these practical solution results can be used for all but the first 10% to 20% of an unsteady convective cooling or heating process. To obtain calculations for the small values of time $t$ for which $Fo < 0.2$, the exact solution given in Appendix E–2 can be used. Alternatively, the solution for small values of time during which the temperature in the body interior (i.e., in the vicinity of the midplane, centerline, or center) is not significantly influenced by the change in surface condition can be approximated by the use of the solution results for a semi-infinite solid.

Finally, we should recognize that limiting results obtained from Figs. 3–12(a), 3–13(a), and 3–14(a) for large values of Biot number $Bi_0$ (i.e., $1/Bi_0 \approx 0$) correspond to the case in which the surface is suddenly brought to a constant temperature $T_s = T_F$. Graphical solutions for the instantaneous rate $q_s$ and total accumulative $Q$ heat transfer at the surface of a plane wall, circular cylinder, and sphere for this important boundary condition are given in Fig. 3–15.

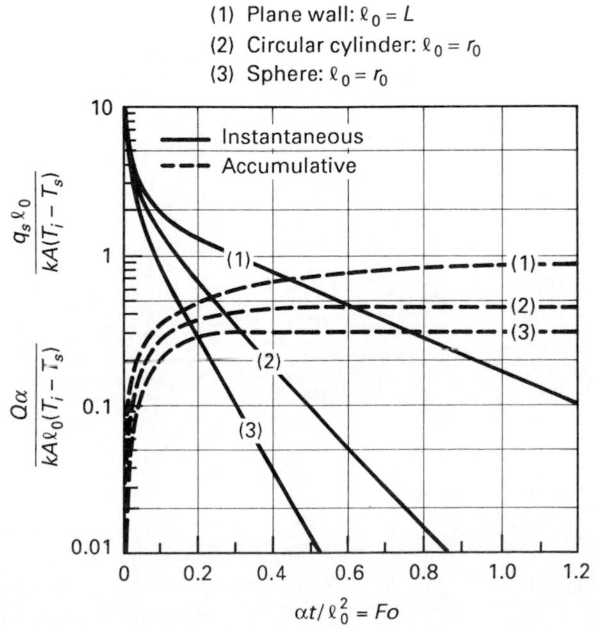


**FIGURE 3–15**
Unsteady one-dimensional heat transfer in plane wall, circular cylinder, and sphere with surface suddenly brought to a constant temperature $T_0$. (From Schneider [6]. Used with permission.)

## EXAMPLE 3–15

A 2-in.-thick aluminum plate initially at 75°F is suddenly exposed to convection with $T_F = 250°F$ and $h = 3000$ Btu/(h ft² °F). Determine the temperature at the surface after (1) 100 s, (2) 10 s, and (3) 1 s.

### Solution

*Objective* Determine $T_s$ for $t = 100$ s, $t = 10$ s, and $t = 1$ s.

*Schematic* Flat plate with both surfaces suddenly exposed to convection.

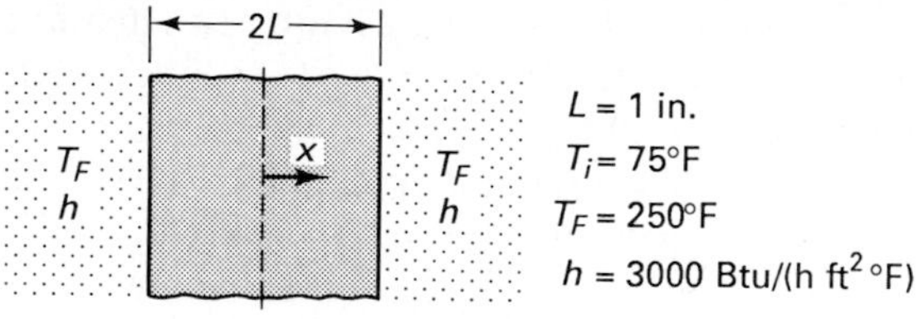

2L
x
$T_F$
h
$T_F$
h
L = 1 in.
$T_i$ = 75°F
$T_F$ = 250°F
h = 3000 Btu/(h ft² °F)


*Assumptions/Conditions*

unsteady-state
one spatial dimension
uniform properties

*Properties* Aluminum (Table A–C–1): $k = 136$ Btu/(h ft °F), $\alpha = 10.5 \times 10^{-4}$ ft²/s.

*Analysis* As a first step we compute the Biot number $Bi_0$.

$$Bi_0 = Bi = \frac{hL}{k} = 3000\,\frac{\text{Btu}}{\text{h ft}^2\text{ °F}}\,\frac{\text{ft}}{12}\,\frac{\text{h ft °F}}{136\text{ Btu}} = 1.84 = \frac{1}{0.544}$$

Since $Bi \gg 0.1$, the approximate lumped analysis of Chap. 2 is inappropriate. However, the temperature distribution for this problem can be determined by use of the Heisler relations or charts given by Table 3–5 or Fig. 3–12, providing that the Fourier number $Fo$ is not too small (i.e., $Fo \nless 0.2$). To compute $Fo$, we write

$$Fo = \frac{\alpha t}{L^2} = 10.5 \times 10^{-4}\,\frac{\text{ft}^2}{\text{s}}\,\frac{t}{(\text{ft}/12)^2} = 0.151\ t/\text{s}$$

such that

$$Fo = 0.151(100) = 15.1 \qquad \textit{case 1}$$

for $t = 100$ s,

$$Fo = 0.151(10) = 1.51 \qquad \textit{case 2}$$

for $t = 10$ s, and

$$Fo = 0.151(1) = 0.151 \qquad \textit{case 3}$$

for $t = 1$ s. Thus, the Heisler relations and charts can be used with confidence for cases 1 and 2, but provide limited accuracy for case 3.

*Case 1*: $t = 100$ s

Referring to the Heisler chart given by Fig. 3–12(a) and setting $Fo = 15.1$ and $1/Bi_0 = 0.544$, we find that the dimensionless midplane temperature $\Theta_0$ is very small; that is,

$$\Theta_0 = \frac{T_0 - T_F}{T_i - T_F} \ll 0.001 \qquad T_0 \simeq T_F = 250°\text{F}$$

Using this result together with Fig. 3–12(b), we conclude that

$$T_s \simeq T_F = 250°\text{F}$$

Thus, the temperature throughout the entire plate is essentially at 250°F for $t = 100$ s. This result is confirmed by use of the Heisler relations, which indicate $\gamma_1 = 1.04$, $C_1 = 1.17$, $\Theta_0 = 1.05 \times 10^{-7}$, and $\Theta_s/\Theta_0 = 0.506$.

*Case 2*: $t = 10$ s

Setting $Fo = 1.51$ and following the approach taken in case 1, we obtain

$$\Theta_0 = \frac{T_0 - T_F}{T_i - T_F} = 0.23$$

or

$$T_0 = 0.23(75°\text{F} - 250°\text{F}) + 250°\text{F} = 210°\text{F}$$

from Fig. 3–12(a), and

$$\frac{\Theta_s}{\Theta_0} = \frac{T_s - T_F}{T_0 - T_F} = 0.5$$

or

$$T_s = 0.5(210°\text{F} - 250°\text{F}) + 250°\text{F} = 230°\text{F}$$

from Fig. 3–12(b). These values are in good agreement with the Heisler relations, which indicate $\gamma_1 = 1.04$, $C_1 = 1.17$, $\Theta_0 = 0.231$, and $\Theta_s/\Theta_0 = 0.506$.

*Case 3*: $t = 1$ s

Since $Fo < 0.2$ for case 3, we turn to the exact solution given by Eq. (A–E–2–1) in the Appendix, which indicates

$$\Theta_0 = \frac{T_0 - T_F}{T_i - T_F} = \sum_{n=1}^{\infty} C_n \exp(-\gamma_n^2 Fo) \cos \gamma_n$$

where

$$C_n = \frac{4 \sin \gamma_n}{2\gamma_n + \sin(2\gamma_n)} \tag{a}$$

from Eq. (A–E–2–2). Referring to Table A–E–2–1, the first three eigenvalues $\gamma_n$, which correspond to $Bi_0 = 1.84$, are

$$\gamma_1 = 1.04 \qquad \gamma_2 = 3.61 \qquad \gamma_3 = 6.55$$

Substituting these values into Eq. (a), we obtain

$$C_1 = 1.17 \qquad C_2 = -0.224 \qquad C_3 = 0.0784$$

Using these results and setting $Fo = 0.151$, $\Theta_0$ becomes

$$\begin{aligned}\Theta_0 &= (1.17)\exp[-(1.04)^2(0.151)]\cos(1.04)\\ &\quad + (-0.224)\exp[-(3.61)^2(0.151)]\cos(3.61)\\ &\quad + (0.0784)\exp[-(6.55)^2(0.151)]\cos(6.55) + \cdots\\ &= 0.503 + 0.0279 + 0.000116 + \cdots \simeq 0.531\end{aligned} \tag{b}$$

It follows that

$$T_s = 0.531(75°\text{F} - 250°\text{F}) + 250°\text{F} = 157°\text{F}$$

Equation (b) indicates that the third- and higher-order terms in the exact solution can be neglected, but that the second term contributes about 5.3% to the solution. Thus, the error resulting from the use of the Heisler relations for this problem would be about 5.3%. Of course, we would expect larger errors in the Heisler relations for smaller values of $Fo$.

## EXAMPLE 3–16

A brass sphere 50 cm in diameter initially at 80°C is placed in a cooling fluid with $T_F = 15°\text{C}$ and $h = 500\ \text{W/(m}^2\ °\text{C)}$. Determine the length of time required for the center of the sphere to cool to 30°C.

### Solution

*Objective* Determine time $t$ required for $T_0 = 30°\text{C}$.

*Schematic* Convection cooling of brass sphere.

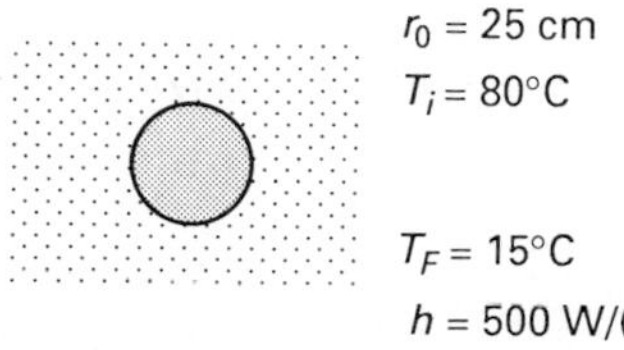
$r_0$ = 25 cm
$T_i$ = 80°C
$T_F$ = 15°C
$h$ = 500 W/(m² °C)

*Assumptions/Conditions*

unsteady-state
one spatial dimension
uniform properties

*Properties* Brass (Table A–C–1): $k = 111$ W/(m °C), $\alpha = 3.41 \times 10^{-5}$ m²/s.

*Analysis* First we calculate the Biot number $Bi_0$.

$$Bi_0 = \frac{hr_0}{k} = \frac{[500 \text{ W/(m}^2 \text{ °C)}](0.25 \text{ m})}{111 \text{ W/(m °C)}} = 1.126 = \frac{1}{0.888}$$

We also note that $Bi = (hr_0/3)/k = 0.375$. Because $Bi > 0.1$, we will utilize the Heisler relations/charts instead of the approximate lumped analysis approach developed in Chap. 2.

The dimensionless center temperature $\Theta_0$ is

$$\Theta_0 = \frac{T_0 - T_F}{T_i - T_F} = \frac{30\text{°C} - 15\text{°C}}{80\text{°C} - 15\text{°C}} = 0.23$$

Referring to Fig. 3–14(a), we estimate $Fo = \alpha t/r_0^2 = 0.7$. For better accuracy, the Heisler relations listed in Table 3–7 can be used to obtain $\gamma_1 = 1.63$, $C_1 = 1.30$, and

$$\Theta_0 = C_1 \exp(-\gamma_1^2 Fo)$$

Solving for $Fo$, we obtain

$$Fo = \frac{1}{\gamma_1^2} \ln \frac{C_1}{\Theta_0} = \frac{1}{1.63^2} \ln \frac{1.30}{0.23} = 0.653$$

Using this more reliable value for $Fo$, the time $t$ is given by

$$t = \frac{0.653(0.25 \text{ m})^2}{3.41 \times 10^{-5} \text{ m}^2/\text{s}} = 1190 \text{ s}$$

As a point of interest, this value is 50% greater than the value indicated by Eq. (2–143), which is based on the lumped analysis approach.

### 3–9–4 Unsteady Two- and Three-Dimensional Heat-Transfer Systems

As we have seen, the temperature charts given by Figs. 3–10, 3–12, and 3–13 are applicable to unsteady one-dimensional heat transfer in semi-infinite solids, infinite plates, and infinite circular cylinders, as indicated in Table 3–8. Notice that the dimensionless temperature distribution $\Theta$ $[(T - T_F)/(T_i - T_F)]$ in these three unsteady one-dimensional systems is represented by $S(x,t)$, $P(x,t)$, and $C(r,t)$, respectively. These fundamental solutions can be combined in the form of products to obtain

**TABLE 3–8** Unsteady one-dimensional temperature charts applicable to multidimensional systems

| *Geometry* | *Temperature chart* | *Notation for dimensionless temperature distribution* $\Theta = \dfrac{T - T_F}{T_i - T_F} = 1 - \dfrac{T - T_i}{T_F - T_i}$ |
|---|---|---|
| (i) Semi-infinite solid<br>Uniform $T_i$<br>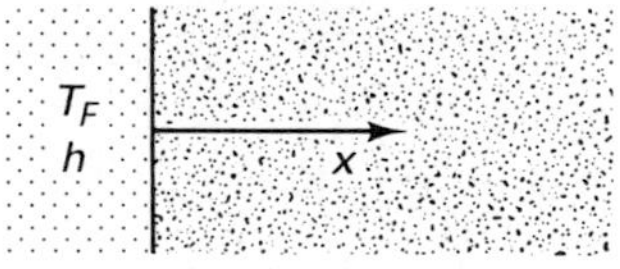 | Fig. 3–10 | $S(x,t)$ |
| (ii) Infinite plate<br>Uniform $T_i$<br>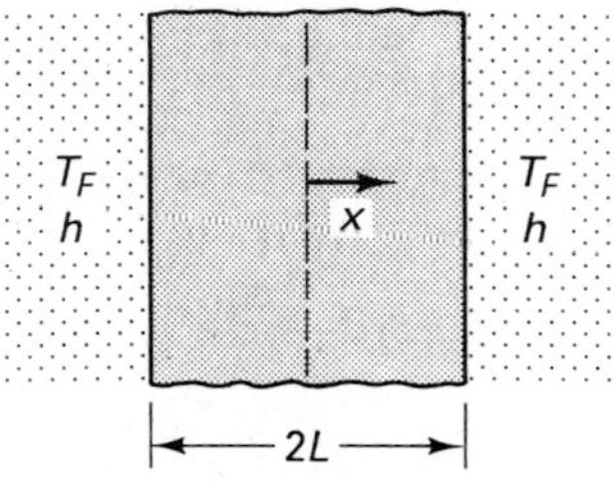 | Fig. 3–12 | $P(x,t)$ |
| (iii) Infinite cylinder<br>Uniform $T_i$<br>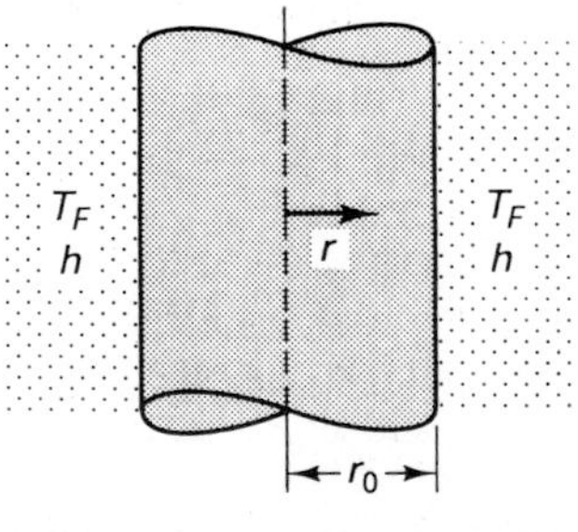 | Fig. 3–13 | $C(r,t)$ |

solutions for the unsteady temperature distribution in semi-infinite plates, short cylinders, and the other multidimensional systems shown in Fig. 3–16. For example, the temperature distribution is represented by

$$\Theta = \frac{T(x,y,t) - T_F}{T_i - T_F} = S(x,t)\, P(y,t) \tag{3–70}$$

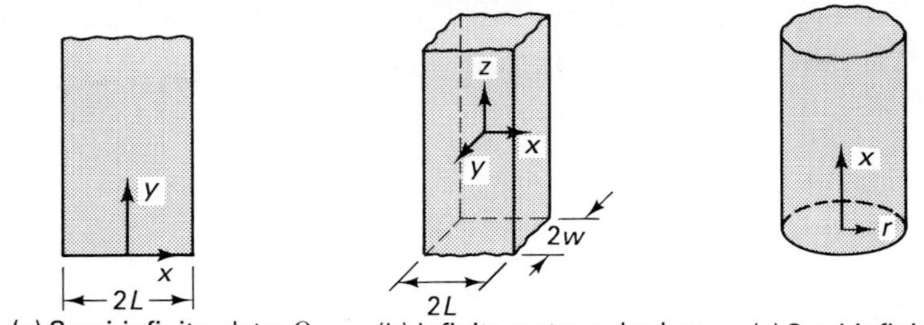

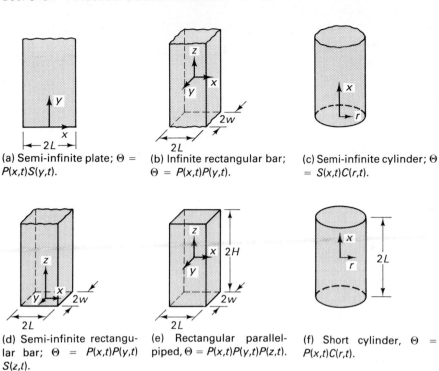

(a) Semi-infinite plate; $\Theta = P(x,t)S(y,t)$.

(b) Infinite rectangular bar; $\Theta = P(x,t)P(y,t)$.

(c) Semi-infinite cylinder; $\Theta = S(x,t)C(r,t)$.

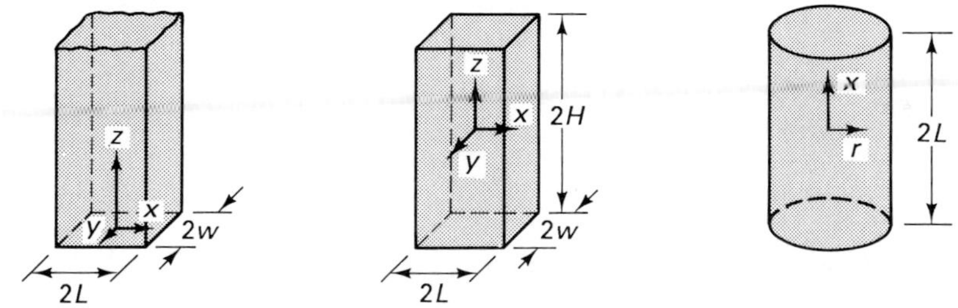


(d) Semi-infinite rectangular bar; $\Theta = P(x,t)P(y,t)S(z,t)$.

(e) Rectangular parallelpiped, $\Theta = P(x,t)P(y,t)P(z,t)$.

(f) Short cylinder, $\Theta = P(x,t)C(r,t)$.

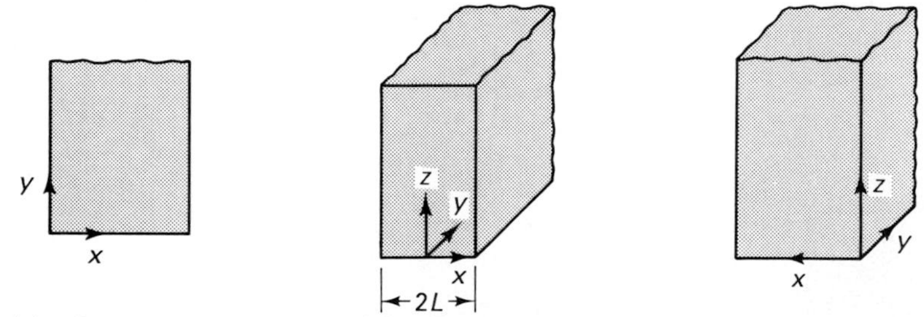


(g) One-quarter infinite solid; $\Theta = S(x,t)S(y,t)$.

(h) One-quarter infinite plate; $\Theta = P(x,t)S(y,t)S(z,t)$.

(i) One-eighth infinite solid; $\Theta = S(x,t)S(y,t)S(z,t)$.

**FIGURE 3–16** Temperature distributions for unsteady multidimensional systems with uniform initial temperature $T_i$ and convection boundary conditions expressed as products of unsteady one-dimensional solutions $S(\zeta,t)$, $P(\zeta,t)$, and $C(r,t)$; $\zeta = x, y, z$, or $r$.

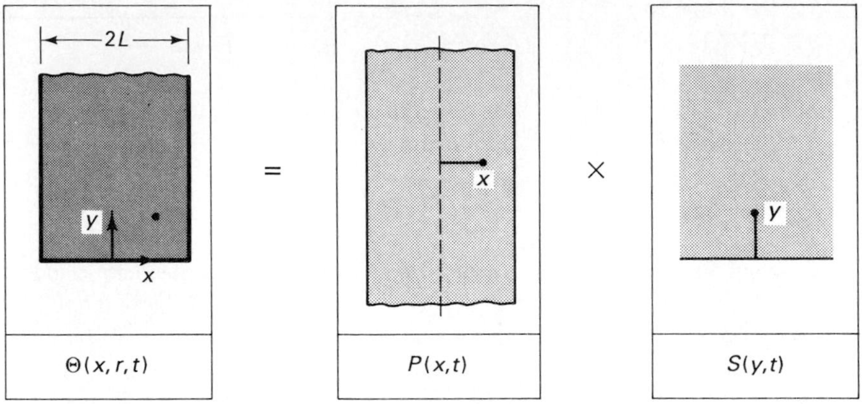


**FIGURE 3–17** Product solution scheme for unsteady two-dimensional temperature distribution in a semi-infinite plate.

as shown in Fig. 3–17 for a semi-infinite plate, and

$$\Theta = \frac{T(x,r,t) - T_F}{T_i - T_F} = S(x,t)\,C(r,t) \tag{3–71}$$

as shown in Fig. 3–18 for a short cylinder. This practical solution approach is based on the separation-of-variables method, which is illustrated in Example 3–17.

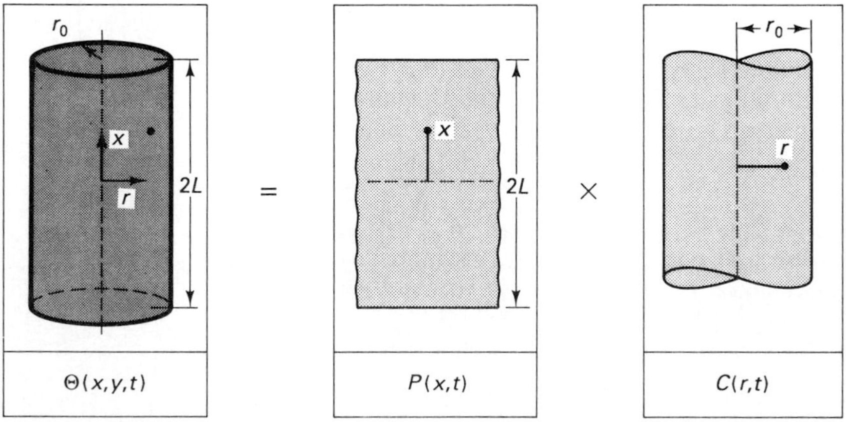


**FIGURE 3–18** Product solution scheme for unsteady two-dimensional temperature distribution in a short cylinder.

## EXAMPLE 3–17

The short circular cylinder shown in Fig. E3–17a, which is initially at a uniform temperature $T_i$, is suddenly exposed to a fluid with $h$ and $T_F$ specified. Demonstrate how the instantaneous temperature at any location within the cylinder can be determined by the use of Heisler charts.

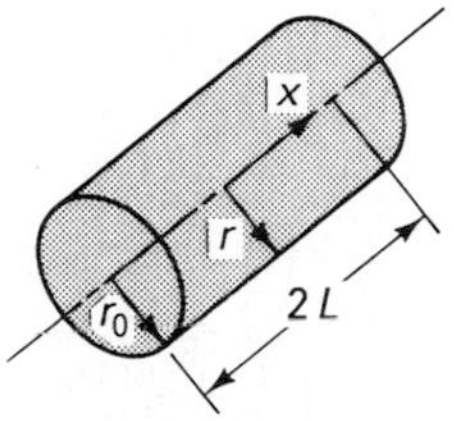
x
r
r0
2L
Fluid at $T_F$ with $h$ known
$T(x,r,0) = T_i$

**FIGURE E3–17a**
Convective cooling or heating of short circular cylinder.

### Solution

*Objective* Show how the Heisler charts can be used for this unsteady two-dimensional heat-transfer problem.

*Assumptions/Conditions*

unsteady-state
two spatial dimensions
uniform properties

*Analysis* The differential formulation for this multidimensional system takes the form [for $\Theta = (T - T_F)/(T_i - T_F)$]

$$\frac{1}{r}\frac{\partial}{\partial r}\left(r\frac{\partial \Theta}{\partial r}\right) + \frac{\partial^2 \Theta}{\partial x^2} = \frac{1}{\alpha}\frac{\partial \Theta}{\partial t}$$

$$\Theta = 1 \qquad \text{at } t = 0$$

$$\frac{\partial \Theta}{\partial r} = 0 \qquad \text{at } r = 0 \qquad -k\frac{\partial \Theta}{\partial r} = h\Theta \qquad \text{at } r = r_0$$

$$k\frac{\partial \Theta}{\partial x} = h\Theta \qquad \text{at } x = -L \qquad -k\frac{\partial \Theta}{\partial x} = h\Theta \qquad \text{at } x = L$$

These equations can be reduced to two simpler problems by assuming the product solution

$$\Theta(x,r,t) = C(r,t)\,P(x,t)$$

Using this substitution, we obtain

$$\frac{1}{r}\frac{\partial}{\partial r}\left(r\frac{\partial C}{\partial r}\right) = \frac{1}{\alpha}\frac{\partial C}{\partial t} \qquad \frac{\partial^2 P}{\partial x^2} = \frac{1}{\alpha}\frac{\partial P}{\partial t}$$

$$C = 1 \quad \text{at } t = 0 \qquad P = 1 \quad \text{at } t = 0$$

$$\frac{\partial C}{\partial r} = 0 \quad \text{at } r = 0 \qquad k\frac{\partial P}{\partial x} = hP \quad \text{at } x = -L$$

$$-k\frac{\partial C}{\partial r} = hC \quad \text{at } r = r_0 \qquad -k\frac{\partial P}{\partial x} = hP \quad \text{at } x = L$$

The solutions to these two unsteady one-dimensional problems are represented by the Heisler charts given in Figs. 3–12 and 3–13. Therefore, we see that the solution to our multidimensional problem is equal to the product of the solutions to simpler unsteady one-dimensional problems. A simple geometric perspective of this practical result is represented by Fig. E3–17b, which shows a short circular cylinder resulting from the intersection of a flat plate and a circular cylinder.

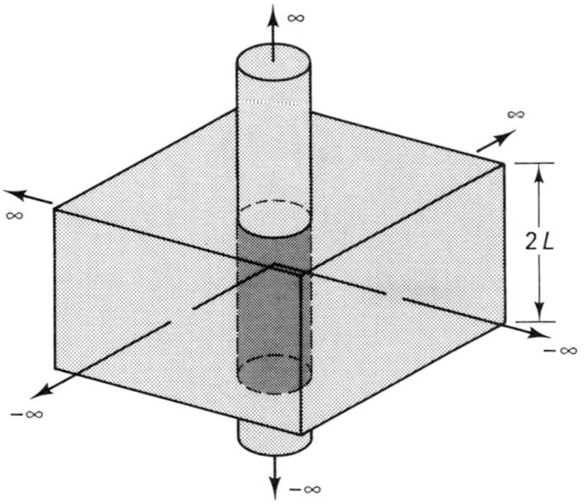


**FIGURE E3–17b**
Intersection of flat plate and circular cylinder.

## EXAMPLE 3–18

In manufacturing a cylindrical stainless steel (AISI 302) disk 10 cm in diameter and 8 cm thick, the piece is quenched in an oil bath with $h = 400 \text{ W/(m}^2 \text{ K)}$ from an initial temperature of 700 K to 300 K. Determine the minimum and maximum temperatures within the piece after a time of 10 minutes.

### Solution

*Objective* Determine $T_{\min}$ and $T_{\max}$ for $t = 10$ min.

*Schematic* Convective cooling of short circular cylinder.

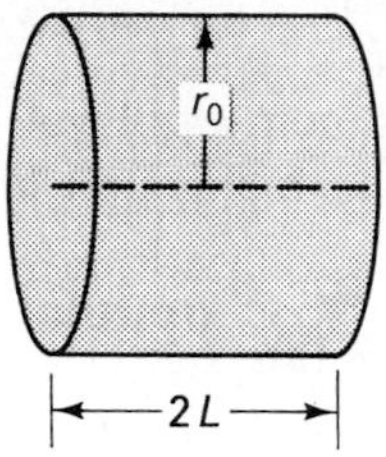


$L = 0.04$ m
$r_0 = 0.05$ m
$T_i = 700$ K

Surrounding fluid
$T_F = 300$ K
$h = 400$ W/(m$^2$ K)

*Assumptions/Conditions*

unsteady-state
two spatial dimensions
uniform properties

*Properties* Stainless steel (AISI 302) at average temperature of 500 K (Table A–C–1): $\rho = 8050$ kg/m$^3$, $k = 18.6$ W/(m K), $c_v = 0.536$ kJ/(kg K); the thermal diffusivity $\alpha$ is calculated by writing

$$\alpha = \frac{18.6 \text{ W/(m K)}}{(8050 \text{ kg/m}^3)[536 \text{ J/(kg K)}]} = 4.31 \times 10^{-6} \text{ m}^2\text{/s}$$

*Analysis* As a first step, we calculate the value of Biot number $Bi$ for the piece.

$$Bi = \frac{h\ell}{k} = \frac{hV}{kA_s} = \frac{h}{k}\frac{\pi r_0^2 (2L)}{2\pi r_0 (2L) + 2\pi r_0} = \frac{h}{k}\frac{r_0 L}{2L + r_0}$$

$$= \frac{400 \text{ W/(m}^2\text{ K)}}{18.6 \text{ W/(m K)}}\frac{0.05 \text{ m } (0.04 \text{ m})}{0.08 \text{ m} + 0.05 \text{ m}} = 0.331$$

Since the value of $Bi$ is significantly greater than 0.1, we should account for the variation of temperature within the cylinder.

As we have seen, the dimensionless temperature distribution $\Theta(r,x,t)$ in a short cylinder such as this can be represented by the product of dimensionless temperature distributions in an infinite cylinder $C(r,t)$ and an infinite plate $P(x,t)$; that is (see Fig. 3–16f),

$$\Theta(r,x,t) = C(r,t)\, P(x,t)$$

where

$$C(r,t) = \frac{T(r,t) - T_F}{T_i - T_F} \qquad P(x,t) = \frac{T(x,t) - T_F}{T_i - T_F}$$

It is obvious that at any instant of time the temperature is a maximum at the center of the piece and is a minimum along the circumference at both ends. The calculations required to determine $T_{\min}$ and $T_{\max}$ for $t = 10$ min are summarized as follows:

*Infinite Cylinder*

$$\frac{\alpha t}{r_0^2} = \frac{(4.31 \times 10^{-6}\ \text{m}^2/\text{s})(600\ \text{s})}{(0.05\ \text{m})^2} = 1.03$$

$$Bi_{0,C} = \frac{hr_0}{k} = \frac{[400\ \text{W}/(\text{m}^2\ \text{K})](0.05\ \text{m})}{18.6\ \text{W}/(\text{m K})} = 1.076 = \frac{1}{0.929}$$

$$C(0,t) = \left.\frac{T_0 - T_F}{T_i - T_F}\right|_{\text{Cylinder}} = 0.26 \qquad \text{from Fig. 3–13a}$$

$$C(r_0,t) = 0.62(0.26) = 0.161 \qquad \text{from Fig. 3–13b}$$

*Infinite Plate*

$$\frac{\alpha t}{L^2} = \frac{(4.31 \times 10^{-6}\ \text{m}^2/\text{s})(600\ \text{s})}{(0.04\ \text{m})^2} = 1.62$$

$$Bi_{0,P} = \frac{hL}{k} = \frac{[400\ \text{W}/(\text{m}^2\ \text{K})](0.04\ \text{m})}{18.6\ \text{W}/(\text{m K})} = 0.860 = \frac{1}{1.16}$$

$$P(0,t) = \left.\frac{T_0 - T_F}{T_i - T_F}\right|_{\text{Plate}} = 0.35 \qquad \text{from Fig. 3–12a}$$

$$P(L,t) = 0.70(0.35) = 0.245 \qquad \text{from Fig. 3–12b}$$

*Resultant Short Cylinder*

Maximum temperature

$$\Theta_{\text{max}} = C(0,t)\,P(0,t) = 0.26(0.35) = 0.114$$

$$T_{\text{max}} = 0.114(700\ \text{K} - 300\ \text{K}) + 300\ \text{K} = 346\ \text{K}$$

Minimum temperature

$$\Theta_{\text{min}} = C(r_0,t)\,P(L,t) = 0.161(0.245) = 0.394$$

$$T_{\text{min}} = 0.394(700\ \text{K} - 300\ \text{K}) + 300\ \text{K} = 316\ \text{K}$$

## 3–10 SUMMARY

In this chapter we have introduced the differential formulation and related analytical, analogical, and graphical solution concepts that are commonly used in the analysis of basic steady and unsteady conduction heat-transfer in isotropic materials involving two or more independent variables. These formulation and solution methods provide a basis for establishing an understanding of the physical nature of a problem. Furthermore, analytical methods are useful in establishing the effect of variations of the parameters on the solution, and in the development of criteria for limiting solution results.

Basic solution results in the form of analytic relations and charts are summarized in Sec. 3–9. These results provide the basis for a practical approach to analyzing many standard multidimensional conduction-heat-transfer problems involving convection and specified wall temperature or heat flux boundary conditions.

As we have seen, the analytical, analogical, and graphical solution methods all have their place in the science of heat transfer. However, numerical methods are generally required in the analysis of more complex problems. Therefore, special attention is given to basic numerical methods in the next chapter.

As indicated, our study has been restricted to isotropic media. The analysis of conduction heat transfer in anisotropic materials is introduced by Eckert and Drake [16].

## REVIEW QUESTIONS

**3–1.** Write the Fourier law of conduction for unsteady two-dimensional $(x,y,t)$ systems.

**3–2.** Write the Fourier law of conduction using cylindrical coordinates.

**3–3.** Define thermal diffusivity.

**3–4.** Write the following multidimensional forms of the energy equation: (a) Fourier equation, (b) Laplace equation, and (c) Poisson equation.

**3–5.** What kind of conduction-heat-transfer problems can be solved by the product solution method?

**3–6.** Assuming that the temperature distribution $T(x,y)$ within the rectangular plate of Fig. 3–5 is known, explain how the rate of heat transfer across face $A$ can be determined.

**3–7.** Explain how the principle of superposition can be used to solve steady two-dimensional conduction-heat-transfer problems.

**3–8.** Define the error function.

**3–9.** What is a semi-infinite solid?

**3–10.** Explain the operation of an analog field plotter.

**3–11.** Define the conduction shape factor.

**3–12.** Define the Fourier number.

**3–13.** Explain how the use of the thermal resistance listed in Table 3–2 for a circular cylinder buried horizontally in a semi-infinite medium would be expected to be in error when used to determine the average heat loss from a buried pipeline.

**3–14.** What are the Heisler charts? What type of one-dimensional problems can be solved by the use of these charts?

**3–15.** Explain how the one-dimensional Heisler charts can be use to solve two- and three-dimensional problems.

## PROBLEMS

**3–1.** Write differential formulations for the convecting fin shown in Fig. P3–1 for the conditions listed below.

(a) $Bi < 0.1$ tip insulated (c) $Bi > 0.1$ tip at $T_1$
(b) $Bi > 0.1$ tip insulated (d) $Bi > 0.1$ convection from tip

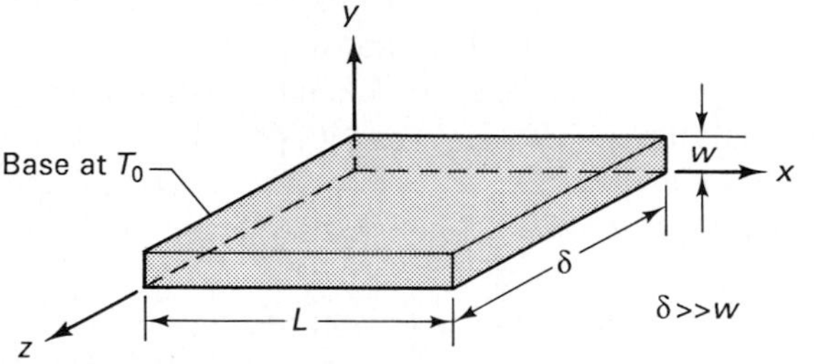


FIGURE P3–1

**3–2.** Write differential formulations for the convecting fin shown in Fig. P3–2 for the conditions listed below.

(a) $Bi < 0.1$ tip insulated (c) $Bi > 0.1$ tip at $T_1$
(b) $Bi > 0.1$ tip insulated (d) $Bi > 0.1$ convection from tip

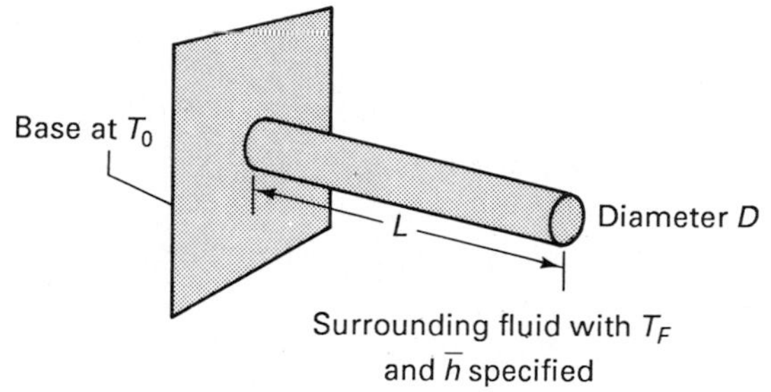


FIGURE P3–2

**3–3.** Assuming that the fin shown in Fig. P3–2 is exposed to blackbody radiation with an enclosure at $T_R$ rather than convection, write differential formulations for the conditions listed below.

(a) $Bi < 0.1$ tip insulated
(b) $Bi > 0.1$ tip insulated
(c) $Bi > 0.1$ thermal radiation from tip

**3–4.** Write the differential formulation for the systems shown in Fig. P3–4.

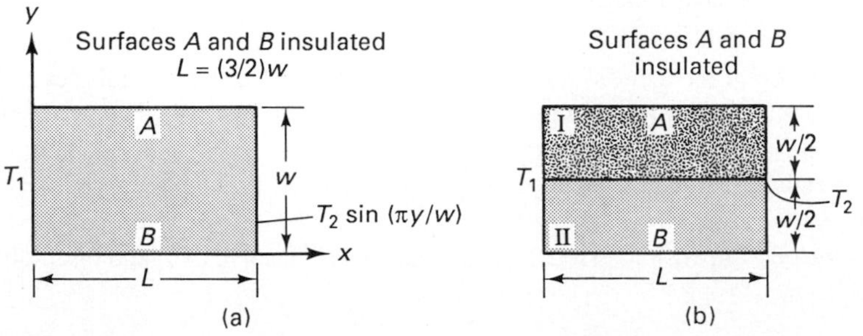


FIGURE P3–4

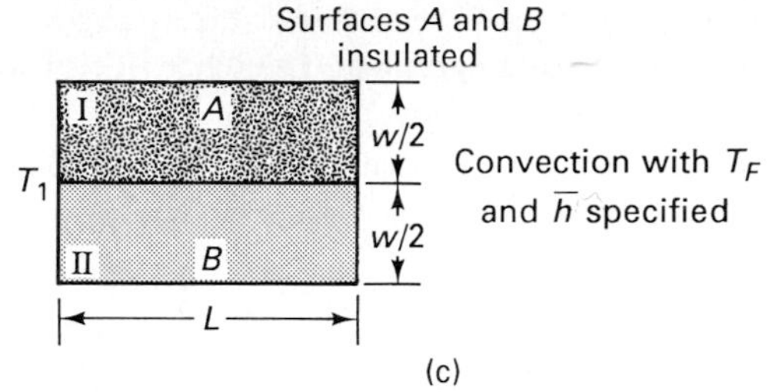


**FIGURE P3–4** *(Continued)*

**3–5.** A fin with a 1-cm-diameter circular cross section and length $L$ initially at $T_0$ is suddenly exposed to a fluid at $T_F$ with $h$ known. Develop the differential formulation for small-Biot-number conditions.

**3–6.** A sphere initially at $T_i$ is suddenly exposed to blackbody thermal radiation with its surroundings at $T_R$. The shape factor $F_{s-R}$ is equal to unity. Develop the differential formulation.

**3–7.** Write the differential formulation for the composite system shown in Fig. P3–7.

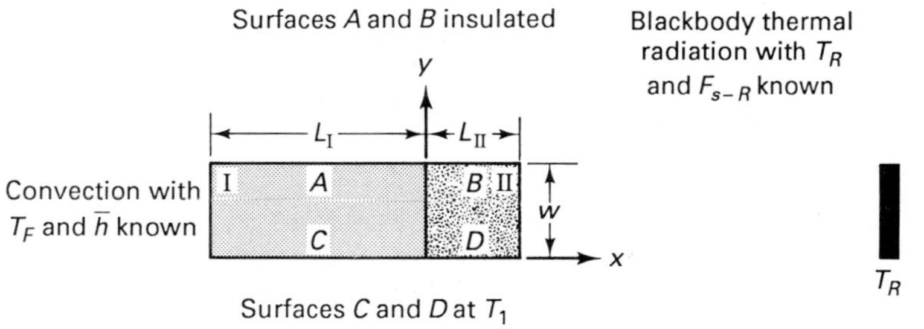


**FIGURE P3–7**

**3–8.** (a) Show $dq_{\phi+d\phi} = dq_\phi + [\partial(dq_\phi)/\partial\phi]d\phi$ for the cylindrical coordinate system. (b) Write the following equations in cylindrical coordinates: (i) Fourier equation; (ii) Laplace equation; (iii) Poisson equation.

**3–9.** Transform Eq. (3–13) into cylindrical coordinates by utilizing the coordinate transformation given by Eq. (3–15).

**3–10.** The spherical coordinate system is shown in Appendix A–2. Write the Fourier law of conduction in terms of the spherical coordinates $r,\phi,\theta$.

**3–11.** Develop the energy equation for unsteady multidimensional conduction heat transfer in a sphere.

**3–12.** A fin is attached to the surface of a flat plate as shown in Fig. P3–12. The plate and fin are made of different materials. Write the differential formulation for this problem for the case in which the plate thickness is significant and the fin Biot number is of the order of unity.

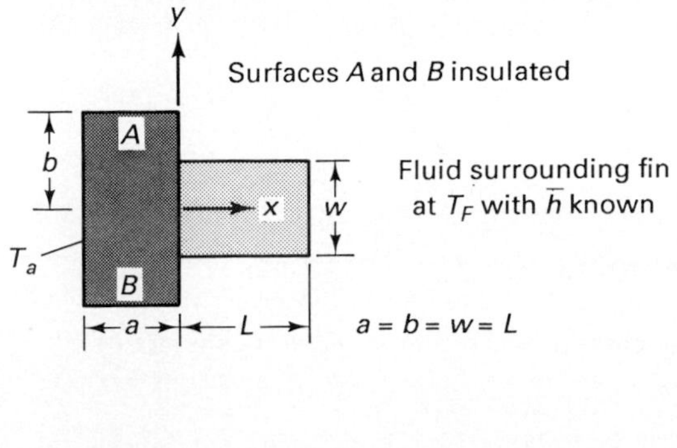


**FIGURE P3–12**

**3–13.** A plate is initially at temperature $T_i$. One side of the plate is then suddenly exposed to a fluid with temperature $T_F$ and coefficient of heat transfer specified. The other side of the plate is maintained at $T_i$. Write the differential formulation for this problem.

**3–14.** Develop the differential formulation for the cylindrical system shown in Fig. P3–14.

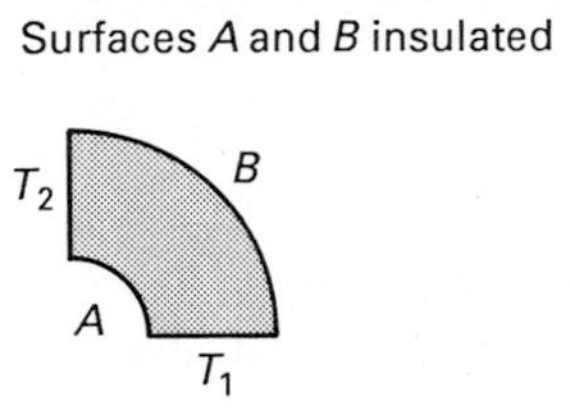
Surfaces A and B insulated
T2
B
A
T1

**FIGURE P3–14**

**3–15.** Write the differential formulation for the system shown in Fig. P3–15.

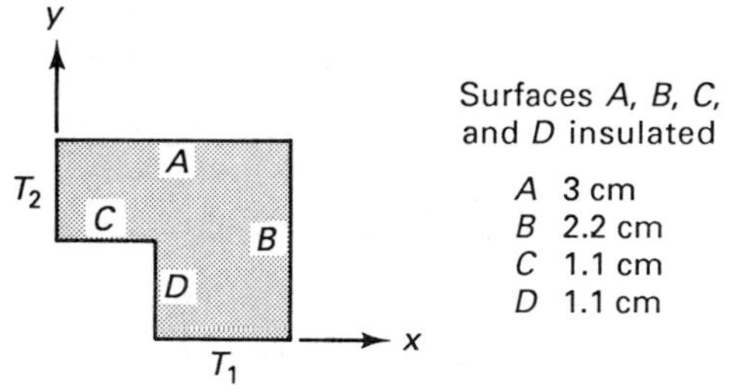
y
T2
A
C
B
D
T1
x
Surfaces A, B, C, and D insulated
A 3 cm
B 2.2 cm
C 1.1 cm
D 1.1 cm

**FIGURE P3–15**

**3–16.** Develop the differential formulation for heat transfer in the tapered rod shown in Fig. P3–16.

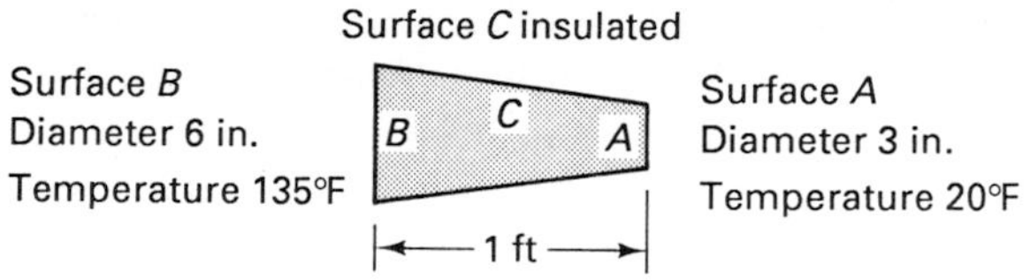
Surface C insulated
Surface B
Diameter 6 in.
Temperature 135°F
B
C
A
Surface A
Diameter 3 in.
Temperature 20°F
1 ft

**FIGURE P3–16**

**3–17.** Write the solution for temperature distribution in a rectangular plate with the surface at $x = 0$ maintained at $T_2$ and the other three surfaces at $T_1$.

**3–18.** Write the solution for temperature distribution in the rectangular plate shown in Fig. P3–18.

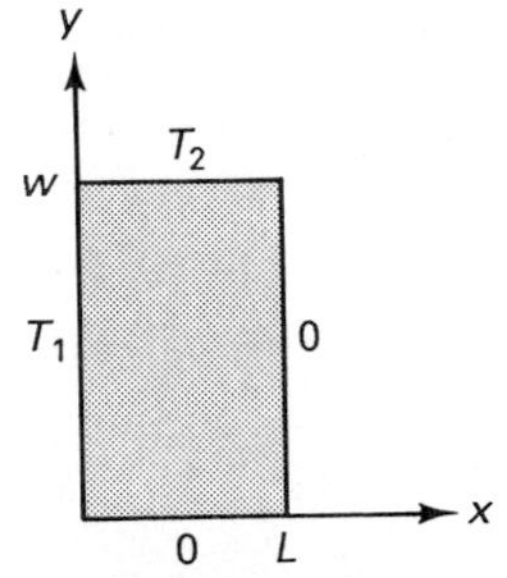
y
T2
W
T1
0
x
0
L

**FIGURE P3–18**

**3–19.** Evaluate $c_n$ in Eq. (3–37) for the case in which $f(x) = T_1 + T_2 \sin(\pi x/L)$.

**3–20.** Develop expressions for $q''_x$ and $q''_L$ for steady-state heat transfer in the rectangular solid of Fig. 3–5 for the case in which $f(x) = T_1 + T_2 \sin(\pi x/L)$.

**3–21.** A solid 10-cm-diameter aluminum sphere is initially at 100°C. Its surface is then suddenly changed to 0°C. (a) Write the differential formulation for this heat transfer problem. (b) The solution for the temperature distribution in this problem is given by

$$\frac{T - T_1}{T_i - T_1} = \frac{2}{\pi}\sum_{n=1}^{\infty}\frac{(-1)^{n+1}}{n}\exp(-n^2\pi^2\alpha t/r_0^2)\frac{\sin(n\pi r/r_0)}{r/r_0}$$

(i) Show that this equation satisfies the differential equation, and (ii) develop expressions for the instantaneous rate $q$ and accumulative total $Q$ heat transfer from the surface of the sphere. (This series is uniformly convergent for $t > 0$ and $0 \leq r \leq r_0$.)

**3–22.** Develop an approximate graphical solution for the rate of heat transfer in the hollow square steel rod shown in Fig. P3–22.

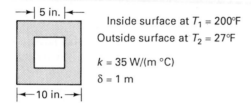

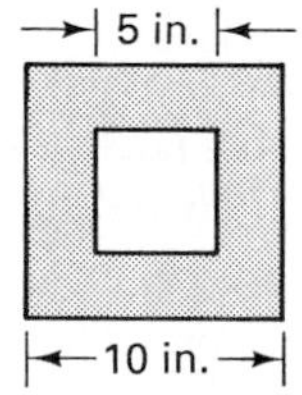


Inside surface at $T_1 = 200°F$

Outside surface at $T_2 = 27°F$

$k = 35$ W/(m °C)

$\delta = 1$ m

**FIGURE P3–22**

**3–23.** Use the graphical approach to estimate the rate of heat transfer per unit length through a 10-cm × 7.5-cm block [$k$ = 125 W/(m °C)] with a centered 2.4-cm diameter hole running along the axis. The inner and outer surface temperatures are maintained at 100°C and 30°C, respectively.

**3–24.** Utilize the graphical approach to determine the rate of heat transfer in a cylindrical quarter section such as the one shown in Fig. 3–8 for $r_1$ = 1 cm, $r_2$ = 10 cm, $T_1$ = −50°C, $T_2$ = 250°C, and $k$ = 25 W/(m °C).

**3–25.** A 2-cm-thick copper plate initially at 100°C is suddently exposed to a cooling fluid at 0°C with a very high coefficient of heat transfer. Develop a graphical solution for the temperature distribution and rate of heat transfer from the surface. Utilize a grid spacing of $L/10$.

**3–26.** The buildings in a complex are heated by steam-fed from a central boiler plant through 3 miles of 6-in.-diameter bare pipe buried 3 ft below the surface. The pipe wall temperature is 340°F, the mean ground surface temperature is 75°F, and the thermal conductivity of the soil is 1.2 Btu/(h ft °F). Develop a conservative estimate of the rate of heat loss to the ground.

**3–27.** A small cubic furnace 50 × 50 cm on the inside is constructed of fireclay brick [$k$ = 1.04 W/(m °C)] with a wall thickness of 10 cm. The inside of the furnace is maintained at 500°C and the ambient air temperature is 27°C with $h$ = 5 W/(m$^2$ °C). Determine the outside surface temperature and the rate of heat transfer through the walls. (See Holman [17], Example 3–2; Kreith and Bohn [18], Example 2–9.)

**3–28.** A 7-cm-diameter spherical heat source is buried 20 cm below the surface of a large box of sand. The surface of the sand is 27°C and 3.4 W are required to maintain the surface of the heater at 40°C. Determine the thermal conductivity of the sand.

**3–29.** A 5-cm-diameter water pipe maintained at 30°C is buried in a midwestern state at a depth of 2 m in heavy damp soil with ground surface temperature equal to 0°C. Calculate the maximum rate of heat loss per unit length of pipe. Explain why the actual rate of heat transfer is likely to be lower. How would one make a more accurate calculation?

**3–30.** A 0.2-m-diameter sphere maintained at 22°C is buried at a depth of 2.5 m in light dry soil. Determine the maximum rate of heat gain by the sphere if the surface of the ground is at 40°C. Explain why the actual heat gain is likely to be lower.

**3–31.** A Heisler chart for accumulated heat transfer for unsteady heat transfer in a long solid cylinder, which is suddenly exposed to a fluid, is given in Fig. 3–13. Show that the simple one-dimensional solution for $Q$ is in close agreement with the Heisler curves for Biot number $Bi$ less than 0.1.

**3–32.** A long steel bar (10 cm × 10 cm) initially at 500°C is suddenly placed in a cooling fluid at 15°C with $h = 1000$ W/(m$^2$ °C), $k = 35$ W/(m °C), and $\alpha = 1.5 \times 10^{-5}$ m$^2$/s. Utilize the Heisler charts and the principle of superposition to determine the length of time for the centerline temperature to reach 50°C.

**3–33.** A large copper plate of 1 in. thickness initially at 98°F is suddently immersed in a well-stirred fluid kept at 5°F with $h = 75$ Btu/(h ft$^2$ °F). Determine the time required for the surface of the plate to reach 30°F. Also determine the accumulative heat transfer over this period of time.

**3–34.** A concrete wall of 2 ft thickness is initially at a uniform temperature of 1100°F. The wall is suddenly exposed on both sides to a convecting fluid with $T_F = 100$°F and $h = 5$ Btu/(h ft$^2$ °F). Determine the temperature at the center of the slab after 5 h and 20 h. Assume $k = 0.694$ Btu/(h ft °F), and $\alpha = 8.6 \times 10^{-6}$ ft$^2$/s.

**3–35.** The surface of a very thick slab of iron initially at 200°C is suddenly exposed to a fluid with $T_F = 10$°C and $h = 75$ W/(m$^2$ °C). Determine the temperature and rate of heat transfer at the surface after 5 minutes.

**3–36.** Solve Prob. 3–35 for the case in which the slab is 8 cm thick, with both surfaces exposed to the fluid.

**3–37.** A long 5-in.-diameter copper cylinder initially at $-100$°F is suddenly exposed to a fluid with $T_F = 27$°F and $h = 100$ Btu/(h ft$^2$ °F). Determine the centerline and surface temperatures at increments of 60 s for a 10-min period of time. Determine the length of time for the cylinder to essentially reach steady-state conditions.

**3–38.** A semi-infinite 10-cm-diameter copper cylinder initially at 200°C is suddenly exposed to a fluid with $T_F = 50$°C and $h = 250$ W/(m$^2$ °C). Calculate the temperatures at the axis and the surface of the cylinder 5 cm from the end 10 minutes after the process has begun.

**3–39.** Given that the burn threshold of human skin is about 65°C, estimate the length of time the finger can be exposed to a candle flame at 800°C [assuming $h \approx 100$ W/(m$^2$ °C)] without suffering injury. (See Lienhard [19], Example 5.5.)

**3–40.** Roller bearings are to be heat-treated in a continuous type of process in which they are passed through a furnace on a conveyor chain. The bearings are cylindrical in shape, with a diameter of 2 in. The bearings are placed in a line on the conveyor such that with regard to heat transfer they form an infinitely long cylinder. The furnace temperature is at 1400°F, and the roller bearings are initially at 80°F. What is the minimum time that each cylinder should remain in the furnace if the centerline temperature should reach at least 1000°F? Take the properties of the bearings to be the same as those for carbon steel (1.5% C). (See Janna [20], Example 6.5.)

**3–41.** Develop an analytical solution for the temperature distribution in the rectangular solid shown in Fig. 3–5 for the case in which both surfaces $A$ and $D$ are maintained at $T_2$, with surfaces $B$ and $C$ at $T_1$. Also develop an expression for the rate of heat transfer across the base at $x = 0$.

**3–42.** The semi-infinite rectangular solid shown in Fig. P3–42 has a base temperature of $T_2$, with $T = T_1$ at $x = 0$ and $x = L$. (a) Utilize the separation-of-variables approach to show that the temperature distribution takes the form

$$\frac{T - T_1}{T_2 - T_1} = \frac{4}{\pi} \sum_{n=0}^{\infty} \frac{1}{2n+1} \exp\left[-(2n+1)\pi y/L\right] \sin \frac{(2n+1)\pi x}{L}$$

(b) Develop an expression for the rate of heat transfer $q_y$ and show that $q_y \to 0$ as $y \to \infty$, and $q_y \to \infty$ as $y \to 0$.

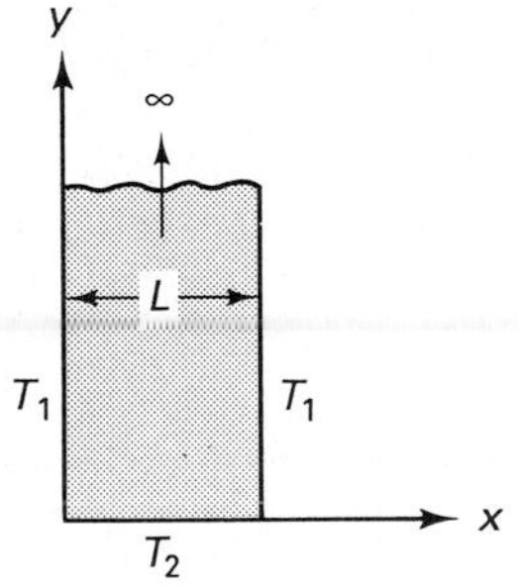


**FIGURE P3–42**

**3–43.** The very long plate of thickness $w$ shown in Fig. P3–43 is initially at uniform temperature $T_i$. Its surfaces are then suddenly brought to a temperature $T_1$. (a) Utilize the product solution approach to show that the unsteady temperature distribution in the plate takes the form

$$\frac{T - T_1}{T_i - T_1} = \frac{4}{\pi} \sum_{n=0}^{\infty} \frac{1}{2n+1} \exp\left[-(2n+1)^2\pi^2\alpha t/w^2\right] \sin \frac{(2n+1)\pi y}{w}$$

(This series is uniformly convergent for $t > 0$ and $0 \leq y \leq w$.) (b) Develop expressions for $q$ and $Q$. (To check your solution for $q$ as $t \to \infty$, note that $\Sigma_{n=0}^{\infty} 1/(2n+1)^2 = \pi^2/8$.)

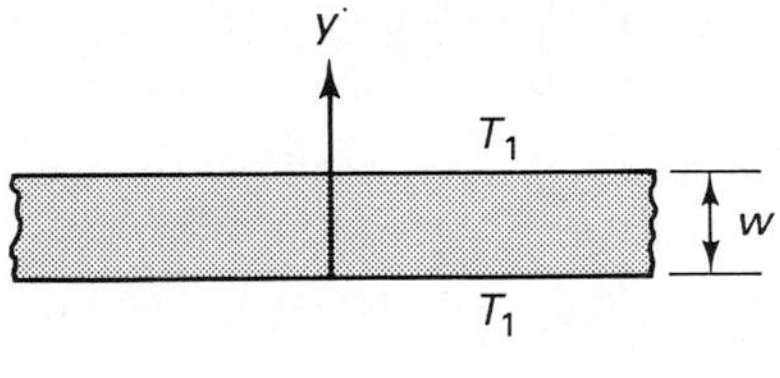


**FIGURE P3–43**

# CHAPTER 4

# CONDUCTION HEAT TRANSFER: NUMERICAL APPROACH

## 4–1 INTRODUCTION

As discussed in Chap. 3, numerical methods provide the basis for analyzing the more complex problems for which other approaches are inadequate. Conduction-heat-transfer problems that fall into this category generally include those involving nonlinear boundary conditions, temperature-dependent properties, and complex geometries. Numerical methods are also commonly used in the analysis of systems involving radiation and convection.

The *finite-difference* method and the *finite-element* method are the two basic approaches generally used in numerical analysis. The finite-difference method is featured in this chapter and is thoroughly discussed in reference 1. The finite-element method is presented in references 1 and 2.

The finite-difference approach involves the use of (1) nodal networks, (2) finite-difference approximations for derivatives in space and time, (3) standard energy conservation formulation concepts, and (4) computer solution of systems of algebraic nodal equations. The nodal network and finite-difference approximations are introduced in Secs. 4–2 and 4–3, after which finite-difference formulation and solution concepts are introduced in the context of steady systems in Sec. 4–4 and unsteady systems in Sec. 4–5. The basic principles are introduced in the framework of two-dimensional rectangular systems with uniform properties and internal energy generation, and are extended to other multidimensional conduction-heat-transfer systems in several examples.

## 4–2 THE NODAL NETWORK

In the finite-difference approach to the analysis of conduction heat transfer in a rectangular solid such as is shown in Fig. 4–1, we designate a number of discrete *nodal points* at which the temperature is to be approximated. These nodal points are established by subdividing the entire system into subvolumes, with the distance between adjacent nodes represented by $\Delta x$ or $\Delta y$. Each subvolume is treated as a lumped subsystem, with the temperature of a node assumed to represent the *average* temperature of its subvolume. The $x$ and $y$ location of each node within the *nodal network* is given by $(m-1)\Delta x$ and $(n-1)\Delta y$, respectively; the values of $m$ and $n$ take on integer values with $m$ ranging from 1 to $M$, and $n$ taking values from 1 to $N$.

The temperature at node $(m,n)$ is designated by $T_{m,n}$ for steady-state processes. To extend the representation to unsteady systems, we simply designate the nodal temperature by $T^{\tau}_{m,n}$, where the time index $\tau$ takes on integer values 0, 1, 2, . . . , and is defined in terms of the time increment $\Delta t$ by $t = \tau \, \Delta t$.

The mechanics of developing a nodal network is quite straightforward. We first sketch in horizontal and vertical construction lines that are $\Delta y$ and $\Delta x$ apart, with $\Delta y$ commonly set equal to $\Delta x$. These construction lines also include the boundaries of the system. The nodes are then located at all intersections of the construction lines. The subvolumes are formed by sketching in horizontal and vertical lines that lie

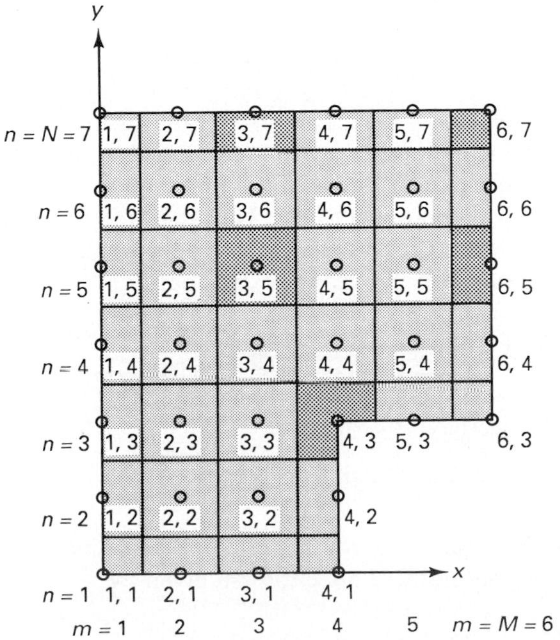


**FIGURE 4–1**
Representation of a rectangular plate by network of subvolumes and nodes. Shading indicates representative interior and exterior nodes. Plate thickness is $\delta$.

halfway between adjacent construction lines. This procedure will produce the desired system of nodes and subvolumes. As we will see, the accuracy of a finite-difference representation is dependent upon the number of nodes, represented by $Z_s$, employed in the nodal network. Whereas a small number of nodes (*coarse grid*) can sometimes be used to obtain useful estimates, a relatively large number of nodes (*fine grid*) is generally required to develop accurate solutions.

## 4–3 FINITE-DIFFERENCE APPROXIMATIONS

The numerical finite-difference approach to analyzing conduction-heat-transfer processes features the use of finite-difference approximations for derivatives in temperature with respect to space and, in the case of unsteady processes, time $t$.

### 4–3–1 Spatial Derivatives

The formal defining equation for the partial derivative $\partial T/\partial x$ is given by Eq. (A–A–4),

$$\frac{\partial T}{\partial x} = \lim_{\Delta x \to 0} \frac{T(x+\Delta x,y) - T(x,y)}{\Delta x} \quad \text{at } x,y \tag{4–1}$$

By permitting the spatial increment $\Delta x$ to remain finite, we obtain the well known finite-difference approximation

$$\frac{\partial T}{\partial x} = \frac{T(x+\Delta x,y) - T(x,y)}{\Delta x} \quad \text{at } x,y \tag{4–2}$$

Referring to Example 4–1, this equation can also be established by truncating second- and higher-order terms in the *Taylor series expansion*. Equation (4–2) is referred to as a *first-order* approximation because the order of the first term which is neglected in the Taylor series expansion is proportional to $\Delta x$. This first-order finite-difference approximation is expressed in terms of the nodal network for steady conditions by writing the *forward difference*,

$$\frac{\partial T}{\partial x} = \frac{T_{m+1,n} - T_{m,n}}{\Delta x} \quad \text{at } m,n \tag{4–3}$$

or the *backward difference*,

$$\frac{\partial T}{\partial x} = \frac{T_{m,n} - T_{m-1,n}}{\Delta x} \quad \text{at } m,n \tag{4–4}$$

both of which approximate the gradient *at the nodal point* $(m,n)$ itself. The corresponding relations for the gradient in the $y$-direction are given by

$$\frac{\partial T}{\partial y} = \frac{T_{m,n+1} - T_{m,n}}{\Delta y} \quad \text{at } m,n \tag{4–5}$$

or

$$\frac{\partial T}{\partial y} = \frac{T_{m,n} - T_{m,n-1}}{\Delta y} \qquad \text{at } m,n \qquad (4\text{–}6)$$

As shown in Example 4–1, the Taylor series expansion gives rise to a *second-order* finite-difference approximation for the gradient *at the interface between two subvolumes* of the form

$$\frac{\partial T}{\partial x} = \frac{T_{m+1,n} - T_{m,n}}{\Delta x} \qquad \text{at } m + 1/2,n \qquad (4\text{–}7)$$

and

$$\frac{\partial T}{\partial x} = \frac{T_{m,n} - T_{m-1,n}}{\Delta x} \qquad \text{at } m - 1/2,n \qquad (4\text{–}8)$$

The error in these more accurate central-difference approximations is proportional to $\Delta x^2$. Similar second-order approximations for gradients in the $y$-direction are given by

$$\frac{\partial T}{\partial y} = \frac{T_{m,n+1} - T_{m,n}}{\Delta y} \qquad \text{at } m,n + 1/2 \qquad (4\text{–}9)$$

and

$$\frac{\partial T}{\partial y} = \frac{T_{m,n} - T_{m,n-1}}{\Delta y} \qquad \text{at } m,n - 1/2 \qquad (4\text{–}10)$$

It should also be noted that a finite-difference approximation can be established for the second derivative $\partial^2 T/\partial x^2$ at a nodal point $(m,n)$ by writing

$$\frac{\partial^2 T}{\partial x^2} = \frac{\partial T/\partial x_{m+1/2,n} - \partial T/\partial x_{m-1/2,n}}{\Delta x} \qquad \text{at } m,n \qquad (4\text{–}11)$$

Substituting Eqs. (4–7) and (4–8) into this relation, we obtain

$$\frac{\partial^2 T}{\partial x^2} = \frac{T_{m+1,n} - 2\,T_{m,n} + T_{m-1,n}}{\Delta x^2} \qquad \text{at } m,n \qquad (4\text{–}12)$$

which can be shown to have an error of the order of $\Delta x^2$. The corresponding second-order finite-difference approximation for $\partial^2 T/\partial y^2$ is written as

$$\frac{\partial^2 T}{\partial y^2} = \frac{T_{m,n+1} - 2\,T_{m,n} + T_{m,n-1}}{\Delta y^2} \qquad \text{at } m,n \qquad (4\text{–}13)$$

As we have noted, the finite-difference approximations given by Eqs. (4–3)–(4–13) apply to steady-state systems. These relations are readily extended to unsteady systems by merely introducing the time index $\tau$ as a superscript for the nodal temperatures.

### 4–3–2 Time Derivatives

Following the approach introduced in the development of spatial derivatives, basic *first-order* approximations can be written for the time derivatives in terms of the *forward-time difference*,

$$\frac{\partial T}{\partial t} = \frac{T_{m,n}^{\tau+1} - T_{m,n}^{\tau}}{\Delta t} \qquad \text{at } m,n,\tau \qquad (4\text{–}14)$$

or the *backward-time difference*,

$$\frac{\partial T}{\partial t} = \frac{T_{m,n}^{\tau} - T_{m,n}^{\tau-1}}{\Delta t} \qquad \text{at } m,n,\tau \qquad (4\text{–}15)$$

Both of these finite-difference approximations are commonly used in the numerical solution of heat-transfer problems. However, care must be taken in the use of Eq. (4–14) to assure that the resulting systems of equations produce a stable solution. The issue of *stability* is considered in Sec. 4–5–2.

### 4–3–3 Summary

To summarize, finite-difference approximations have been presented in this section for spatial gradients in temperature at the subvolume interfaces and for spatial (first and second) and time derivatives at the individual nodes. Although these relations provide the basis for developing simple and accurate finite-difference solutions for many problems, it should be noted that more accurate higher-order finite-difference approximations can be developed for these derivatives by use of the Taylor series expansion. However, in addition to the question of accuracy, we must also concern ourselves with the issue of stability when analyzing unsteady conduction heat transfer problems.

### EXAMPLE 4–1

Utilize the Taylor theorem [3–6] to evaluate the error of standard finite-difference approximations for $\partial T/\partial x$ in the context of steady two-dimensional conditions.

#### Solution

*Objective* Determine the error associated with basic finite-difference approximations for $\partial T/\partial x$.

*Assumptions/Conditions*

steady-state
two-dimensional

*Analysis* Based on the Taylor theorem, $T$ at $x + \Delta x$ can be expanded in terms of $T$ at $x$ by

$$T(x+\Delta x,y) = T(x,y) + \Delta x \frac{\partial T}{\partial x} + \frac{1}{2!}\Delta x^2 \frac{\partial^2 T}{\partial x^2} + \frac{1}{3!}\Delta x^3 \frac{\partial^3 T}{\partial x^3} + \frac{1}{4!}\Delta x^4 \frac{\partial^4 T}{\partial x^4} + \cdots \quad \text{(a)}$$

Rearranging this equation, we obtain

$$\frac{\partial T}{\partial x} = \frac{T(x+\Delta x,y) - T(x,y)}{\Delta x} + O(\Delta x) \qquad \text{at } x,y \quad \text{(b)}$$

where $O(\Delta x)$ designates terms containing first- and higher-order powers of $\Delta x$. This equation provides the basis for the *forward-difference* approximation,

$$\frac{\partial T}{\partial x} = \frac{T_{m+1,n} - T_{m,n}}{\Delta x} \qquad \text{at } m,n \quad \text{(c)}$$

which has an error of the order of $\Delta x$. Hence, the error is reduced by approximately one-half by halving the increment $\Delta x$.

Similarly, $T(x-\Delta x,y)$ can be written as

$$T(x-\Delta x,y) = T(x,y) - \Delta x \frac{\partial T}{\partial x} + \frac{1}{2!}\Delta x^2 \frac{\partial^2 T}{\partial x^2} - \frac{1}{3!}\Delta x^3 \frac{\partial^3 T}{\partial x^3} + \frac{1}{4!}\Delta x^4 \frac{\partial^4 T}{\partial x^4} - \cdots \quad \text{(d)}$$

which leads to

$$\frac{\partial T}{\partial x} = \frac{T(x,y) - T(x-\Delta x,y)}{\Delta x} + O(\Delta x) \qquad \text{at } x,y \quad \text{(e)}$$

Thus, we conclude that the *backward-difference* approximation,

$$\frac{\partial T}{\partial x} = \frac{T_{m,n} - T_{m-1,n}}{\Delta x} \qquad \text{at } m,n \quad \text{(f)}$$

also has an error of the order of $\Delta x$.

A higher-order approximation can be developed for $\partial T/\partial x$ by subtracting Eq. (d) from Eq. (a); that is,

$$\frac{\partial T}{\partial x} = \frac{T(x+\Delta x,y) - T(x-\Delta x,y)}{2\,\Delta x} + O(\Delta x^2) \qquad \text{at } x,y \quad \text{(g)}$$

This equation gives rise to the *central-difference* approximation,

$$\frac{\partial T}{\partial x} = \frac{T_{m+1,n} - T_{m-1,n}}{2\,\Delta x} \qquad \text{at } m,n \quad \text{(h)}$$

which has an improved accuracy since the error is of the order of $\Delta x^2$.

Equation (h) provides the basis for writing finite-difference approximations for the temperature gradient at the interface between adjacent subvolumes given by Eq. (4–7),

$$\frac{\partial T}{\partial x} = \frac{T_{m+1,n} - T_{m,n}}{\Delta x} \qquad \text{at } m + 1/2, n \tag{i}$$

and Eq. (4–8),

$$\frac{\partial T}{\partial x} = \frac{T_{m,n} - T_{m-1,n}}{\Delta x} \qquad \text{at } m - 1/2, n \tag{j}$$

both of which have an error proportional to $\Delta x^2$.

## EXAMPLE 4–2

Write second-order finite-difference approximations for the two-dimensional $(x,y)$ Fourier law of conduction.

### Solution

*Objective* Write second-order finite-difference approximations for $\Delta q_x$ and $\Delta q_y$.

*Schematic* Finite-difference subvolume.

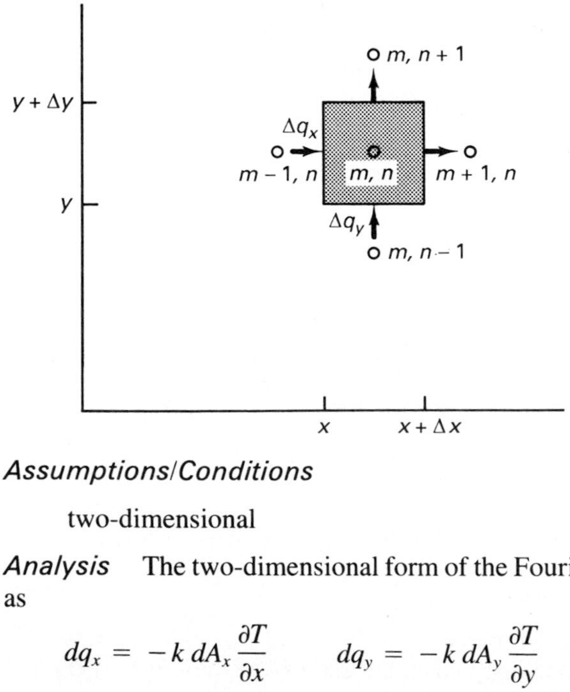


*Assumptions/Conditions*

two-dimensional

*Analysis* The two-dimensional form of the Fourier law of conduction can be written as

$$dq_x = -k\, dA_x \frac{\partial T}{\partial x} \qquad dq_y = -k\, dA_y \frac{\partial T}{\partial y}$$

Referring to the schematic, we see that the differential areas are approximated by $dA_x = \delta\ \Delta y$ and $dA_y = \delta\ \Delta x$, where $\Delta y = \Delta x$. The gradient $\partial T/\partial x$ at $x$ can be approximated by the simple second-order difference (with respect to the face at $x$) given by Eq. (4–8),

$$\left.\frac{\partial T}{\partial x}\right|_x = \frac{T_{m,n} - T_{m-1,n}}{\Delta x}$$

Similarly, $\partial T/\partial y$ at $y$ can be approximated by Eq. (4–10),

$$\left.\frac{\partial T}{\partial y}\right|_y = \frac{T_{m,n} - T_{m,n-1}}{\Delta y}$$

Thus, the steady two-dimensional Fourier law can be approximated by second-order finite-difference equations of the form

$$dq_x = \Delta q_x = -k\delta(T_{m,n} - T_{m-1,n})$$

and

$$dq_y = \Delta q_y = -k\delta(T_{m,n} - T_{m,n-1})$$

Similarly, second-order approximations can be written for $dq_{x+dx}$ and $dq_{y+dy}$ of the form

$$dq_{x+dx} = \Delta q_{x+\Delta x} = -k\delta(T_{m+1,n} - T_{m,n})$$

and

$$dq_{y+dy} = \Delta q_{y+\Delta y} = -k\delta(T_{m,n+1} - T_{m,n})$$

These relations will be used in the development of numerical formulations by means of an energy balance method in the sections that follow.

## 4–4 NUMERICAL ANALYSIS: STEADY SYSTEMS

Once the nodal network is established, finite-difference approximations can be used together with energy conservation principles to develop algebraic nodal equations for each of the $Z_s$ nodes. Basic concepts pertaining to the formulation and solution of the nodal equations are introduced in this section for steady conduction-heat-transfer processes.

### 4–4–1 Finite-Difference Formulation

To develop nodal equations that represent the energy transfer within the system, we want to distinguish between *interior nodes* and *exterior nodes*.

### *Interior Nodal Equations*

Nodal equations can be developed for a representative interior node $(m,n)$ by (1) the discretization of the applicable differential energy equation, or (2) the development of an energy balance for the corresponding subvolume. Following the *discretization method*, we first write the applicable differential energy equation for steady two-dimensional conduction heat transfer with uniform properties and internal energy generation,

$$\frac{\partial^2 T}{\partial x^2} + \frac{\partial^2 T}{\partial y^2} + \frac{\dot{q}}{k} = 0 \tag{4–16}$$

[from Eq. (3–13)]. Substituting the second-order finite-difference approximations for $\partial^2 T/\partial x^2$ and $\partial^2 T/\partial y^2$ given by Eqs. (4–12) and (4–13), this equation takes the form

$$\frac{T_{m+1,n} - 2\,T_{m,n} + T_{m-1,n}}{\Delta x^2} + \frac{T_{m,n+1} - 2\,T_{m,n} + T_{m,n-1}}{\Delta y^2} + \frac{\dot{q}}{k} = 0 \tag{4–17}$$

or, with $\Delta y$ set equal to $\Delta x$,

$$T_{m+1,n} + T_{m-1,n} + T_{m,n+1} + T_{m,n-1} - 4\,T_{m,n} + \frac{\dot{q}}{k}\Delta x^2 = 0 \tag{4–18}$$

which represents the nodal equation for any interior node $(m,n)$. With $m$ and $n$ permitted to take on the integer values associated with each interior node, we obtain a system of algebraic equations that represent the finite-difference equivalent of the original partial differential equation, Eq. (4–16).

To introduce the alternative energy balance method for developing the nodal equations, we apply the first law of thermodynamics to an interior subvolume such as the one shown in Fig. 4–2, with the result

$$\Delta q_x + \Delta q_y + \dot{q}\,\Delta V = \Delta q_{x+\Delta x} + \Delta q_{y+\Delta y} + \cancel{\frac{\Delta E_s}{\Delta t}} \tag{4–19}$$

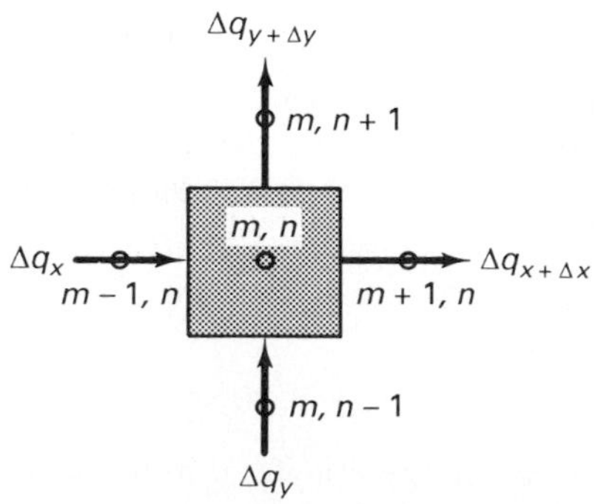


**FIGURE 4–2**
Representative interior node associated with rectangular plate of Fig. 4–1.

where $\Delta E_s/\Delta t = 0$ for steady-state conditions and $\Delta V = \delta\,\Delta x\,\Delta y$. Using the second-order finite-difference approximation for the Fourier law of conduction (see Example 4–2), Eq. (4–19) becomes

$$-k\delta\,\Delta y\,\frac{T_{m,n}-T_{m-1,n}}{\Delta x}-k\delta\,\Delta x\,\frac{T_{m,n}-T_{m,n-1}}{\Delta y}+\dot{q}\,\Delta V$$
$$=-k\delta\,\Delta y\,\frac{T_{m+1,n}-T_{m,n}}{\Delta x}-k\delta\,\Delta x\,\frac{T_{m,n+1}-T_{m,n}}{\Delta y} \qquad (4\text{–}20)$$

which gives rise to Eq. (4–17) and reduces to Eq. (4–18) for $\Delta y = \Delta x$.

### *Exterior Nodal Equations*

Representative exterior nodes associated with the rectangular solid of Fig. 4–1 are shown in Fig. 4–3. In addition to being exposed to energy transfer at the boundaries, these nodes are characterized by subvolumes that are smaller in size than the interior

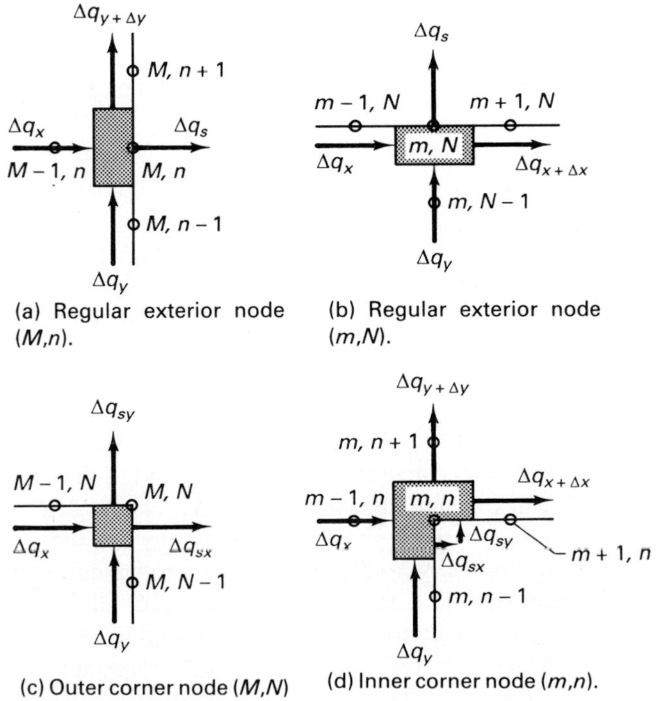


**FIGURE 4–3** Representative exterior nodes associated with rectangular plate of Fig. 4–1.

subvolumes. In order to characterize the energy transfer associated with exterior nodes such as these, we will make use of the *energy balance method*.

Referring to the regular exterior node shown in Fig. 4–3(a), we apply the first law of thermodynamics to obtain

$$\Delta q_x + \Delta q_y + \dot{q}\,\Delta V = \Delta q_s + \Delta q_{y+\Delta y} + \cancel{\frac{\Delta E_s}{\Delta t}} \tag{4–21}$$

where $\Delta E_s/\Delta t = 0$, $\Delta V = \delta\,\Delta x\,\Delta y/2$, and $\Delta q_s$ is the rate of heat transfer out of the subvolume surface. Employing the second-order finite-difference approximation for the Fourier law of conduction, Eq. (4–21) gives rise to

$$-k\delta\,\Delta y\,\frac{T_{M,n} - T_{M-1,n}}{\Delta x} - k\delta\,\frac{\Delta x}{2}\,\frac{T_{M,n} - T_{M,n-1}}{\Delta y} + \dot{q}\,\Delta V$$
$$= \Delta q_s - k\delta\,\frac{\Delta x}{2}\,\frac{T_{M,n+1} - T_{M,n}}{\Delta y} \tag{4–22}$$

Rearranging this equation with $\Delta q_s = \delta\,\Delta y\,q_s''$ and $\Delta y = \Delta x$, we write

$$2\,T_{M-1,n} + T_{M,n+1} + T_{M,n-1} - 4\,T_{M,n} + \left(\frac{\dot{q}}{k} - \frac{2}{\Delta x}\frac{q_s''}{k}\right)\Delta x^2 = 0 \tag{4–23}$$

Similar expressions can be developed for exterior nodes along the other surfaces and at the corners.

For convection, thermal radiation, or specified heat flux boundary conditions, the unknown exterior nodal temperatures satisfy equations such as Eq. (4–23) for steady-state conditions with $q_s''$ $(= \Delta q_s/\Delta A_s)$ specified by

$$q_s'' = h(T_{m,n} - T_F) \qquad \text{convection} \tag{4–24}$$

$$q_s'' = \sigma F_{s-R}(T_{m,n}^4 - T_R^4) \qquad \text{blackbody thermal radiation} \tag{4–25}$$

$$q_s'' = f(x,y) \qquad \text{specified heat flux} \tag{4–26}$$

$$q_s'' = 0 \qquad \text{insulated surface} \tag{4–27}$$

Once the surface temperature is calculated, Eqs. (4–24) and (4–25) provide the means by which the rate of heat transfer from the surface $\Delta q_s$ can be determined for convection and blackbody thermal radiation boundary conditions. On the other hand, for specified wall-temperature boundary conditions, the exterior nodal temperatures are known *a priori*, with nodal equations such as Eq. (4–23) providing the basis for calculating the rate of heat transfer from the surface. For example, for an isothermal boundary condition at the surface $x = (M-1)\Delta x$, the nodal temperatures along the surface are given by $T_{M,n} = T_{M,n+1} = T_{M,n-1} = T_0$, such that Eq. (4–23) reduces to

$$q_s'' = \frac{k}{\Delta x}(T_{M-1,n} - T_0) + \frac{\Delta x}{2}\,\dot{q} \tag{4–28}$$

Thus, after $T_{M-1,n}$ has been determined, the rate of heat transfer across the surface can be calculated.

### Nodal Equations: Summary

To recap, the full numerical finite-difference formulation for steady conditions consists of the $Z_s$ algebraic equations produced at the interior and exterior nodes. Representative second-order nodal equations are given by Eq. (4–18) and Eqs. (4–23)–(4–27). Standard interior and exterior nodal equations are summarized in Table 4–1 for steady two-dimensional systems. Similar equations can be readily developed for steady one-dimensional and three-dimensional systems by use of the discretization method and/or the energy balance method.

### Accuracy

The differential formulation is generally the standard against which the numerical finite-difference formulation of a problem is compared. The error introduced by using finite-difference approximations of the derivatives decreases toward zero as the increments approach infinitesimal proportions. Consequently, the error of a finite-difference formulation for steady conduction heat transfer can be reduced by decreasing the volume $\Delta V$ ($= \delta\, \Delta x\, \Delta y$). For instance, the error associated with the second-order nodal equations listed in Table 4–1 is proportional to $\Delta x^2$. Thus, a 50% reduction in $\Delta x$ would be expected to reduce the error by a factor of about four.

However, the use of smaller volumetric increments also brings about an increase in the number of subvolumes $Z_s$ and algebraic equations required to represent a system. Of course, the length of time required to obtain a solution increases as the number of equations increases. Consequently, a compromise concerning element size must be made on the basis of the accuracy required and the cost of computation.

It should also be mentioned that improved accuracy can be achieved by the use of higher-order finite-difference approximations for derivatives. This point is illustrated in Prob. 4–2.

The error associated with the finite-difference increment $\Delta x$ (and $\Delta t$ for unsteady systems) is commonly referred to as the *truncation error* or the *discretization error*. As we turn our attention to the development of solutions, it should also be noted that errors are introduced at each step in calculating nodal temperatures as a result of the use of a finite number of *significant figures*. Whereas the truncation error decreases with decreasing grid spacing, the *round-off error* increases. Fortunately, the number of significant figures that can be carried by modern digital computers is quite large, such that round-off error normally only becomes a factor in cases involving the use of extremely small increments (i.e., excessively large numbers of nodal equations).

The question remains: What grid size $\Delta x$ must be utilized to achieve reasonable accuracy? The most reliable way to obtain an accurate solution is to actually solve the problem for a number of successively smaller increments in $\Delta x$. As long as the solution continues to converge, the round-off error can be assumed to be secondary. For example, with $L/\Delta x$ represented by $M$, solutions can be obtained for $M = 2, 3,$

**TABLE 4–1** Summary of nodal equations for steady two-dimensional systems with internal heat generation

| *Type node* | *Nodal equation* |
|---|---|
| 1. Interior node<br>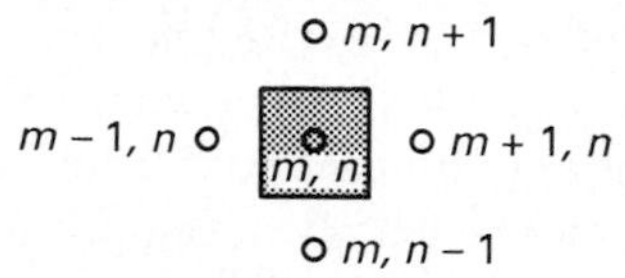m, n + 1<br>m − 1, n<br>m, n<br>m + 1, n<br>m, n − 1 | $T_{m+1,n} + T_{m-1,n} + T_{m,n+1} + T_{m,n-1} - 4T_{m,n} + \frac{\dot{q}}{k}\Delta x^2 = 0$ (1) |
| 2. Exterior nodes | |
| a. General equations | |
| i. Regular exterior node $(M,n)$<br>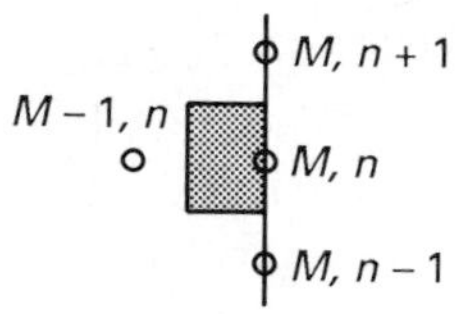M, n + 1<br>M − 1, n<br>M, n<br>M, n − 1 | $2T_{M-1,n} + T_{M,n+1} + T_{M,n-1} - 4T_{M,n} + \frac{\dot{q}}{k}\Delta x^2 - 2\frac{q_s''}{k}\Delta x = 0$ (2i) |
| ii. Outer corner node $(M,N)$<br>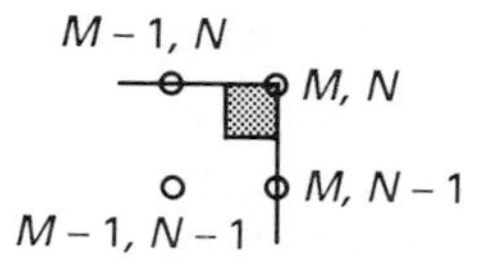M − 1, N<br>M, N<br>M, N − 1<br>M − 1, N − 1 | $2T_{M-1,N} + 2T_{M,N-1} - 4T_{M,N} + \frac{\dot{q}}{k}\Delta x^2 - 4\frac{q_s''}{k}\Delta x = 0$ (2ii) |
| iii. Inner corner node $(m,n)$<br>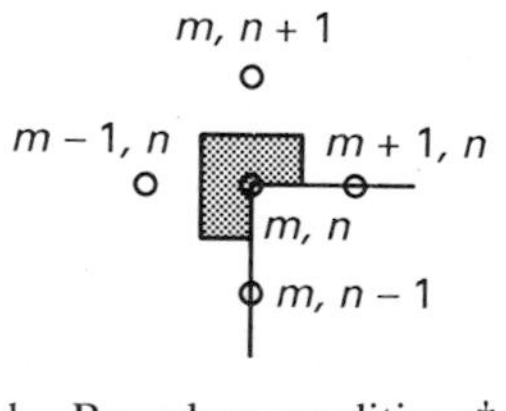m, n + 1<br>m − 1, n<br>m + 1, n<br>m, n<br>m, n − 1 | $\frac{2}{3}(T_{m+1,n} + 2T_{m-1,n} + 2T_{m,n+1} + T_{m,n-1} - 6T_{m,n}) + \frac{\dot{q}}{k}\Delta x^2 - \frac{4}{3}\frac{q_s''}{k}\Delta x = 0$ (2iii) |
| b. Boundary conditions† | |
| i. Specified temperature | $T_{m,n}$ known |
| ii. Specified heat flux | $q_s''$ known |
| iii. Insulated surface | $q_s'' = 0$ |
| iv. Convection | $q_s'' = h(T_{m,n} - T_F)$ |
| v. Blackbody thermal radiation | $q_s'' = \sigma F_{s-R}(T_{m,n}^4 - T_R^4)$ |

† These boundary conditions apply to each type of exterior node.

4, . . . . Since the finite-difference formulation for steady systems is stable, a value of $M$ can be selected for which adequate convergence has been achieved, providing that the limiting point at which round-off error becomes significant has not been reached.

Alternatively, the accuracy of a finite-difference formulation can sometimes be estimated by using the formulation to obtain the numerical solution for simplified conditions or similar problems for which exact solutions are available. However, it must be emphasized that this approach does not guarantee the same solution accuracy for the problem being solved and for the similar problems with known solutions.

## EXAMPLE 4–3

Develop a numerical finite-difference formulation for two-dimensional heat transfer in a square with 5-cm-long sides if three faces are maintained at 30°C and the other face exchanges blackbody thermal radiation with a surface at 500°C with $F_{s-R} = 0.25$. Utilize a grid spacing with $L/\Delta x = 3$. The thermal conductivity $k$ and thickness $\delta$ of the square are assumed to be known. (The differential formulation for this problem is given in Example 3–1.)

### Solution

*Objective* Develop the finite-difference nodal equations for energy transfer in this rectangular plate.

*Schematic* Finite-difference grid for rectangular plate.

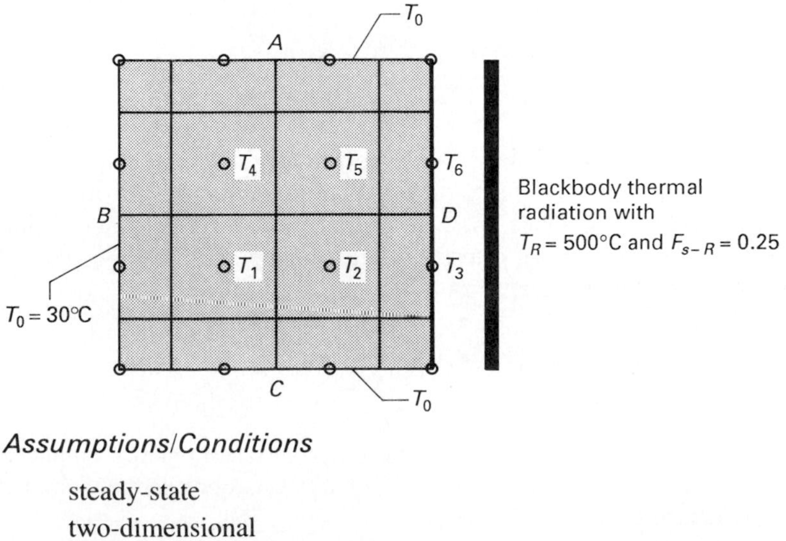


*Assumptions/Conditions*

steady-state
two-dimensional
uniform properties
blackbody thermal radiation

*Analysis* The finite-difference grid with $L/\Delta x = 3$ is shown in the schematic. This network involves the six unknown nodal temperatures $T_1$, $T_2$, $T_3$, $T_4$, $T_5$, and $T_6$. However, because of symmetry, we know that $T_4 = T_1$, $T_5 = T_2$, and $T_6 = T_3$. Therefore, we need only develop the nodal equations for $T_1$, $T_2$, and $T_3$. These nodal equations are developed as follows:

*Interior nodes at $T_1$ and $T_2$* $\quad \Delta q_x + \Delta q_y = \Delta q_{x+\Delta x} + \Delta q_{y+\Delta y}$

$$-k\,\Delta y\,\delta\,\frac{T_{m,n} - T_{m-1,n}}{\Delta x} - k\,\Delta x\,\delta\,\frac{T_{m,n} - T_{m,n-1}}{\Delta y} = -k\,\Delta y\,\delta\,\frac{T_{m+1,n} - T_{m,n}}{\Delta x}$$

or

$$T_{m-1,n} + T_{m,n-1} + T_{m+1,n} = 3\,T_{m,n}$$

We therefore have

$$T_0 + T_0 + T_2 = 3T_1 \tag{a}$$

for the node at $T_1$, and

$$T_1 + T_0 + T_3 = 3T_2 \tag{b}$$

for the node at $T_2$.

*Exterior Node at $T_3$* $\quad \Delta q_x + \Delta q_y = \Delta q_R$

$$-k\,\Delta y\,\delta\,\frac{T_3 - T_2}{\Delta x} - k\,\frac{\Delta x}{2}\,\delta\,\frac{T_3 - T_0}{\Delta y} = \sigma\,\Delta y\,\delta F_{s-R}(T_3^4 - T_R^4)$$

or

$$T_2 + \frac{1}{2}T_0 + \frac{\sigma\,\Delta y\,F_{s-R}T_R^4}{k} = \frac{3}{2}T_3 + \frac{\sigma\,\Delta y\,F_{s-R}T_3^4}{k} \tag{c}$$

Because Eqs. (a) through (c) can be solved for the temperatures $T_1$, $T_2$, and $T_3$, this completes our numerical finite-difference formulation for the temperature distribution with $L/\Delta x = 3$. Once these temperatures are known, calculations can be developed for the rate of heat transfer at any surface. To illustrate, the rate of heat transfer from radiating surface $D$ is

$$q_R = 2\left[\sigma\,\Delta y\,\delta F_{s-R}(T_3^4 - T_R^4) + \sigma\,\frac{\Delta y}{2}\,\delta F_{s-R}(T_0^4 - T_R^4)\right]$$

To obtain the rate of heat transfer at the other surfaces, energy balances must be developed at each node along the surface of interest. For example, the rate of heat transfer at surface $C$ is

$$q_C = -k\delta(T_1 - T_0 + T_2 - T_0) - \frac{k\delta}{2}(T_3 - T_0) + \sigma\delta\,\frac{\Delta y}{2}(T_0^4 - T_R^4)$$

As we shall see momentarily, the rather course grid spacing used in this illustration gives rise to an error of the order of 10%. To obtain higher levels of accuracy, a finer-grid mesh must be used.

### 4–4–2 Finite-Difference Solutions

The finite-difference formulation consists of nodal equations for each interior node and for those exterior nodes for which the temperature is unspecified. For the moment, we represent these equations for a steady two-dimensional system as follows:

$$i = 1 \qquad a_{11}T_1 + a_{12}T_2 + \cdots + a_{1j}T_j + \cdots + a_{1Z}T_Z = C_1 \tag{4–29–1}$$

$$i = 2 \qquad a_{21}T_1 + a_{22}T_2 + \cdots + a_{2j}T_j + \cdots + a_{2Z}T_Z = C_2 \tag{4–29–2}$$

$$\vdots$$

$$i = j \qquad a_{j1}T_1 + a_{j2}T_2 + \cdots + a_{jj}T_j + \cdots + a_{jZ}T_Z = C_j \tag{4–29–j}$$

$$\vdots$$

$$i = Z \qquad a_{Z1}T_1 + a_{Z2}T_2 + \cdots + a_{Zj}T_j + \cdots + a_{ZZ}T_Z = C_Z \tag{4–29–Z}$$

where $Z$ is equal to $Z_s$ minus the number of external nodes for which the temperature is specified, and $i$ represents the row and $j$ the column; the equations are generally ordered such that the matrix of elements $a_{ij}$ are diagonally dominant (i.e., $|a_{jj}| > |a_{ij}|$ for $i = 1, 2, \ldots, Z, i \neq j$). This system of equations must be solved for the $Z$ unknown nodal temperatures. Once the solution for the temperature distribution is determined, the rate of heat transfer within the body or at its surface can be obtained.

The mechanics involved in producing finite-difference calculations for the nodal temperatures and rate of heat transfer are introduced in Examples 4–4 and 4–5 in the context of rather coarse grids, which result in small numbers of equations that are solved by hand. As would be expected, the accuracy of such coarse grid formulations is often inadequate. Since the number of equations required to accurately model multidimensional heat-transfer systems usually range from 10 or so to many thousands, more efficient computer solution techniques are generally used. Both iterative and direct numerical methods are available for the systematic solution of simultaneous algebraic equations. Several of the more simple and commonly used methods are discussed in the following sections.

**EXAMPLE 4–4**

A 1-cm-diameter 3-cm-long carbon (1%) steel fin transfers heat from the wall of a heat exchanger at 200°C to a fluid at 25°C with $\bar{h} = 120$ W/(m² °C). Develop a numerical finite-difference solution for the rate of heat transfer for the case in which the tip is insulated and thermal radiation effects are negligible. Utilize a grid spacing of $\Delta x = L/4$.

### Solution

*Objective* Use the finite-difference approach with $\Delta x = L/4$ to develop an approximate solution for $q_F$.

*Schematic* Circular fin with insulated tip.

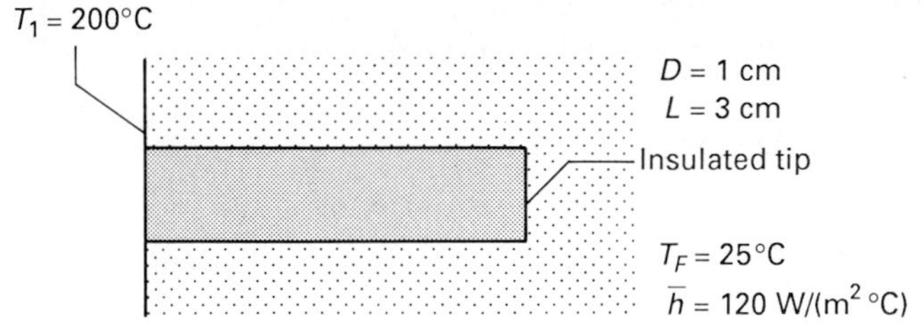


*Assumptions/Conditions*

steady-state
one-dimensional fin
uniform properties
negligible thermal radiation

*Properties* Carbon steel (1% C) at room temperature (Table A–C–1): $k = 43$ W/(m °C).

*Analysis* As shown in Example 2–12, the Biot number is equal to 0.00698, such that the heat transfer in this two-dimensional system can be approximated by a one-dimensional analysis.

A one-dimensional finite-difference grid is shown in Fig. E4–4 for $\Delta x = L/4$. Because $T_1$ is specified, this network only involves four unknown nodal temperatures. Therefore, four algebraic equations must be written to solve for $T_2$, $T_3$, $T_4$, and $T_5$. Focusing attention on node (2), an energy balance is developed by writing

*Node* (2) $\quad q_x = q_{x+\Delta x} + \Delta q_c$

$$-kA\,\frac{T_2 - T_1}{\Delta x} = -kA\,\frac{T_3 - T_2}{\Delta x} + \bar{h}p\,\Delta x\,(T_2 - T_F)$$

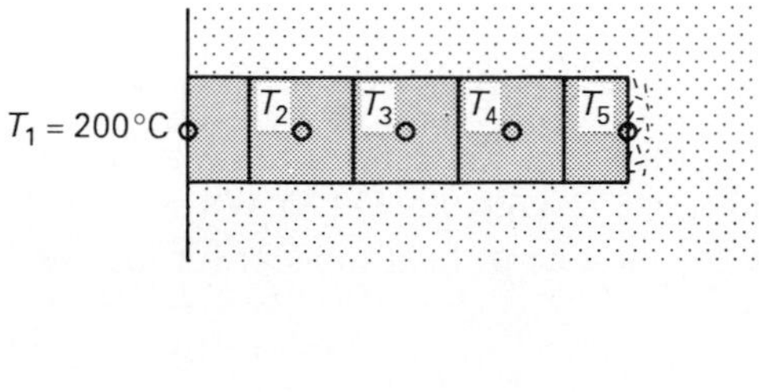


**FIGURE E4–4**
Finite-difference grid for fin with small Biot number; $\Delta x = L/4 = 0.0075$ m.

where $\bar{h}p\,\Delta x^2/(kA) = 0.0628$ for this problem. Rearranging, we obtain

$$T_1 - \left(2 + \frac{\bar{h}p\,\Delta x^2}{kA}\right) T_2 + T_3 + \frac{\bar{h}p\,\Delta x^2}{kA} T_F = 0$$

Similar equations are written for nodes (3) and (4).

$$\textit{Node}\,(3) \qquad T_2 - \left(2 + \frac{\bar{h}p\,\Delta x^2}{kA}\right) T_3 + T_4 + \frac{\bar{h}p\,\Delta x^2}{kA} T_F = 0$$

$$\textit{Node}\,(4) \qquad T_3 - \left(2 + \frac{\bar{h}p\,\Delta x^2}{kA}\right) T_4 + T_5 + \frac{\bar{h}p\,\Delta x^2}{kA} T_F = 0$$

The energy balance for node (5) takes the form

$$\textit{Node}\,(5) \qquad q_x = \Delta q_c$$

$$-kA\frac{T_5 - T_4}{\Delta x} = \bar{h}p\frac{\Delta x}{2}(T_5 - T_F)$$

or

$$T_4 - \left(1 + \frac{\bar{h}p}{kA}\frac{\Delta x^2}{2}\right) T_5 + \frac{\bar{h}p}{kA}\frac{\Delta x^2}{2} T_F = 0$$

Substituting for the various parameters, our four nodal equations are summarized as follows:

$$200^\circ\text{C} - 2.06T_2 + T_3 + 1.57\ ^\circ\text{C} = 0 \tag{a}$$

$$T_2 - 2.06T_3 + T_4 + 1.57\ ^\circ\text{C} = 0 \tag{b}$$

$$T_3 - 2.06T_4 + T_5 + 1.57\ ^\circ\text{C} = 0 \tag{c}$$

$$T_4 - 1.03T_5 + 0.785^\circ\text{C} = 0 \tag{d}$$

These equations can be easily solved by standard algebraic manipulation. For example, Eqs. (a) and (b) can be combined to eliminate $T_2$ and Eqs. (c) and (d) can be coupled to remove $T_5$. The solution of the resulting two equations gives

$$T_3 = 155^\circ\text{C} \qquad T_4 = 145^\circ\text{C}$$

$T_2$ and $T_5$ are then found from Eqs. (a) and (d); that is,

$$T_2 = 173^\circ\text{C} \qquad T_5 = 141^\circ\text{C}$$

By comparing these calculations for the nodal temperatures with the exact analytical solution given by Eq. (2–116), we find a maximum error of only about 2%.

To obtain an approximation for the rate of heat transfer from the fin, we develop an energy balance at node (1) by writing

*Node* (1) $q_F = q_{\Delta x/2} + \Delta q_c$

$$q_F = -kA\frac{T_2 - T_1}{\Delta x} + \bar{h}p\frac{\Delta x}{2}(T_1 - T_F) = 14.6 \text{ W}$$

Referring back to Example 2–12, we find that the analytical solution for the rate of heat transfer $q_F$ is

$$q_F = 15.1 \text{ W}$$

Thus, the error in our finite-difference calculations for $q_F$ is about 3.3%.

To improve the accuracy of our solution, a finer grid can be utilized. [To get a better feel for the effect of grid size on the accuracy of finite-difference solutions, it is suggested that this rather simple problem be solved for $\Delta x = L$, $\Delta x = L/2$, and $\Delta x = L/3$.

## EXAMPLE 4–5

Develop an approximate numerical solution for the rate of heat transfer across surfaces $A$, $B$, $C$, and $D$ of the rectangular plate shown in Fig. E4–5a by utilizing a grid with $\Delta x = L/3$. Also evaluate the accuracy of the solution.

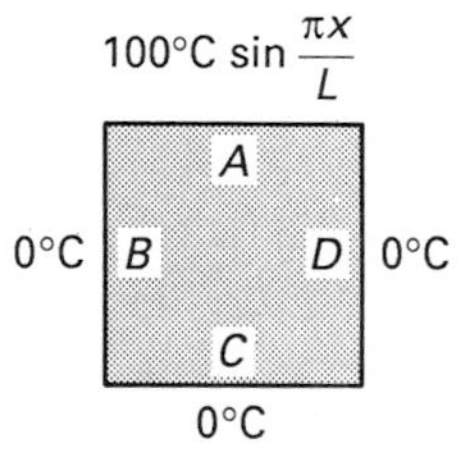
100°C sin πx/L
A
0°C
B
D
0°C
C
0°C

$k$ = 100 W/(m °C)
$\delta$ = 1 cm
Square $w = L$ = 10 cm

**FIGURE E4–5a**
Two-dimensional heat transfer in rectangular plate.

### Solution

*Objective* Use the finite-difference approach with $\Delta x = L/3$ to develop an approximate solution for $q_A$, $q_B$, $q_C$, and $q_D$. Evaluate the accuracy of the calculations.

*Assumptions/Conditions*

steady-state
two-dimensional
uniform properties

*Properties* $k = 100$ W/(m °C).

*Analysis* The finite-difference grid for this problem is shown in Fig. E4–5b for $\Delta x = L/3$. This grid produces 16 nodes, but only four of the nodal temperatures are unknown; these are $T_1$, $T_2$, $T_3$, and $T_4$. Referring to Fig. E4–5b and utilizing Eq. (1) in Table 4–1, or by developing energy balances on each interior node, the nodal equations for this system are written as

*Node* (2,2) $\quad -4T_1 + T_2 + T_3 = 0 \quad$ (a)

*Node* (2,3) $\quad T_1 - 4T_2 + T_4 + 86.6°\text{C} = 0 \quad$ (b)

*Node* (3,2) $\quad T_1 - 4T_3 + T_4 = 0 \quad$ (c)

*Node* (3,3) $\quad + T_2 + T_3 - 4T_4 + 86.6°\text{C} = 0 \quad$ (d)

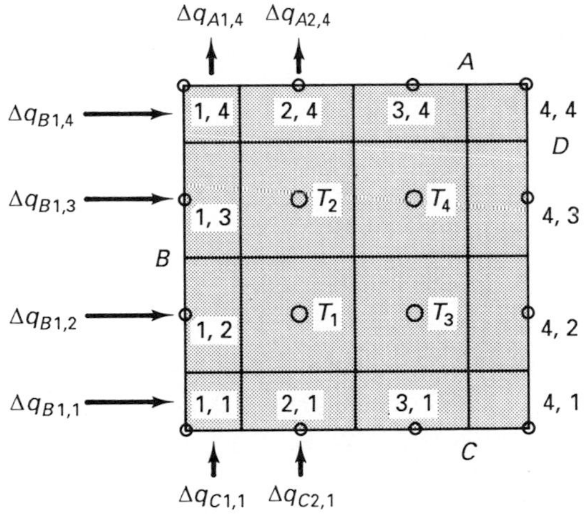


**FIGURE E4–5b**
Finite-difference grid for rectangular plate.

These four equations can be solved for the unknown nodal temperatures. However, our computational work can be eased by recognizing that system symmetry requires $T_3 = T_1$ and $T_4 = T_2$. Consequently, we really only have two unknown nodal temperatures to concern ourselves with. Utilizing this fact, Eqs. (a) and (b) become

*Node* (2,2) $\quad -3T_1 + T_2 = 0 \quad$ (e)

*Node* (2,3) $\quad T_1 - 3T_2 + 86.6°\text{C} = 0 \quad$ (f)

Equations (e) and (f) are easily solved to obtain

$$T_1 = \frac{1}{8}(86.6°\text{C}) = 10.8°\text{C} \qquad T_2 = \frac{3}{8}(86.6°\text{C}) = 32.4°\text{C}$$

A comparison of these predictions with the exact analytical solution developed in Sec. 3–6 reveals an error of 15.3% in $T_1$ and 8.06% in $T_2$.

Referring to Fig. E4–5b, the rate of heat transfer across the heating surface is

$$q_A = 2(\Delta q_{A1,4} + \Delta q_{A2,4})$$

To approximate the rate of heat transfer at nodes (1,4) and (2,4), energy balances are written for these nodes. Considering node (2,4) first, we write

$$\Delta q_{A2,4} = \Delta q_x + \Delta q_y = -k\delta \frac{\Delta y}{2} \frac{86.6°\text{C} - 0°\text{C}}{\Delta x} - k\delta\, \Delta x \frac{86.6°\text{C} - T_2}{\Delta y}$$

Setting $T_2 = 32.4°\text{C}$ and utilizing the values of $k$ and $\delta$ for this problem (i.e., $k\delta = 1\ \text{W/°C}$), this equation reduces to

$$\Delta q_{A2,4} = -k\delta \left( \frac{86.6°\text{C} - 0°\text{C}}{2} + 86.6°\text{C} - 32.4°\text{C} \right) = -97.5\ \text{W}$$

The negative sign indicates that heat is transferred into the surface at node (2,4).

The energy balance for the corner node (1,4) is written as

$$\Delta q_{B1,4} + \Delta q_y = \Delta q_{A1,4} + \Delta q_x$$

where

$$\Delta q_x = -k\delta \frac{\Delta y}{2} \frac{86.6°\text{C} - 0°\text{C}}{\Delta x} = -43.3\ \text{W}$$

$$\Delta q_y = -k\delta \frac{\Delta x}{2} \frac{0°\text{C} - 0°\text{C}}{\Delta y} = 0\ \text{W}$$

Thus, we have the result

$$\Delta q_{A1,4} = \Delta q_{B1,4} + 43.3\ \text{W} \tag{i}$$

where $\Delta q_{B1,4}$ is unknown. Because we have one equation but two unknowns, both $\Delta q_{A1,4}$ and $\Delta q_{B1,4}$ are indeterminant. About the best that we can do in this situation is to simply neglect the corner node, that is, assume that $\Delta q_{A1,4}$ is negligible.

Summing up, $q_A$ is approximated by

$$q_A = 2(0\ \text{W} - 97.5\ \text{W}) = -195\ \text{W}$$

The exact solution for $q_A$ is found from Eq. (3–44), which is developed in Sec. 3–6, to be $-200$ W. Thus, the error in $q_A$ is only 2.5%. However, upon closer inspection we find that the exact solutions for $\Delta q_{A1,4}$ and $\Delta q_{A2,4}$ are $-13.4$ W and $-86.6$ W, respectively. The error in $\Delta q_{A2,4}$ is a substantial 12.6%, with the contribution of $\Delta q_{A1,4}$ and $\Delta q_{A4,4}$ being about 13% of $q_A$. These errors are of the order of the errors in the predictions for the temperature distribution. It just so happens that these errors compensate each other. We are not always so fortunate.

As seen in this example, the inherent indeterminant nature of the heat transfer through corner nodes for surfaces with specified temperatures is one of the main sources of error in this type of numerical finite-difference formulation. (This type of error does not occur for convection, thermal radiation, or specified heat flux boundary conditions.) This error can be reduced by decreasing the grid size. (This feat is sometimes accomplished by using variable grid spacing, with the smallest subvolumes utilized in the region where the accuracy is most critical.)

Following the pattern established in approximating $q_A$, the following calculations are obtained for $q_B$ and $q_C$:

$$q_B = -86.5 \text{ W} \qquad q_C = -21.6 \text{ W}$$

Because of symmetry, we write

$$q_D = -q_B = 86.5 \text{ W}$$

To check our calculations, we perform an energy balance on the entire system.

$$q_A = q_B + q_C - q_D = (-86.5 - 21.6 - 86.5) \text{ W} = -194.6 \text{ W}$$

Because this value is within 0.3% of the direct calculation for $q_A$ ($= -195$ W), we conclude that no calculation errors have been made. This small difference is a consequence of the use of three significant figures in our calculations and should not be confused with the actual error that results from the use of finite-difference approximations. Such an energy balance cannot be used to determine the accuracy of a numerical finite-difference analysis!

As indicated in Sec. 4–4–1, the error associated with our finite-difference approximations should be of the order of $\Delta x^2$. In checking back over the calculations for the temperatures and rates of heat transfer, we find that the average error is about 11%. Thus, we would expect that a grid spacing of $\Delta x/L = 0.1$ would be required to reduce the errors to the order of 1%. This is indeed the case, as is shown in Example 4–8.

### *Iterative Methods*

The most efficient methods of developing solutions to conduction-heat-transfer problems requiring large numbers of nodes (50 or more) generally involve the use of iterative techniques. Iterative techniques have been devised for the simultaneous solution of systems of equations of the form of Eqs. (4–29–1) through (4–29–$Z$), which are similar to the simple iterative scheme introduced in Chap. 2. This approach involves the development of successive approximations for the unknowns that converge toward the exact solution.

The *Gauss-Seidel method* is one of the more popular iterative schemes for solving systems of algebraic equations. In this approach, Eqs. (4–29–1) through (4–29–$Z$) are rewritten as follows:

$$T_1 = \frac{1}{a_{11}}[C_1 - (a_{12}T_2 + a_{13}T_3 + \cdots + a_{1j}T_j + \cdots + a_{1Z}T_Z)] \tag{4–30–1}$$

$$T_2 = \frac{1}{a_{22}}[C_2 - (a_{21}T_1 + a_{23}T_3 + \cdots + a_{2j}T_j + \cdots + a_{2Z}T_Z)] \tag{4–30–2}$$

$$\vdots$$

$$T_j = \frac{1}{a_{jj}}[C_j - (a_{j1}T_1 + \cdots + a_{jj-1}T_{j-1} + a_{jj+1}T_{j+1} + \cdots + a_{jZ}T_Z)] \tag{4–30–j}$$

$$\vdots$$

$$T_Z = \frac{1}{a_{ZZ}}[C_Z - (a_{Z1}T_1 + a_{Z2}T_2 + \cdots + a_{Zj}T_j + \cdots + a_{ZZ-1}T_{Z-1})] \tag{4–30–Z}$$

By introducing approximations for the unknown temperatures ($T_1 = T_1^0$, $T_2 = T_2^0$, $T_j = T_j^0$, etc.) into Eqs. (4–30–1) through (4–30–$Z$) new first-order calculations are developed for the unknowns, $T_1^1$, $T_2^1$, and so on. The first-order calculations are then substituted into Eqs. (4–30–1) through (4–30–$Z$) to produce second-order calculations. Assuming convergence, this procedure is continued until the $k$th iteration produces sufficient accuracy for all values of $T_j$; that is,

$$\left|\frac{T_j^{k+1} - T_j^k}{T_j^k}\right| < \epsilon \tag{4–31}$$

where $k = 1,2, \ldots$, and $\epsilon$ is a small number that determines the accuracy of the solution of Eqs. (4–29). In this connection, convergence of the Gauss-Seidel method is guaranteed if the equations are ordered such that the elements $a_{ij}$ are diagonally dominant. (Formal convergence criteria for the Gauss-Seidel method is presented by Kreyszig [3].)

The mechanics involved in the use of the Gauss-Seidel iterative method is illustrated in Example 4–6. Notice that new temperature values are utilized in the iteration pattern as soon as they are calculated. This approach is adapted to digital computation in Example 4–6 and in several other examples.

Although the Gauss-Seidel method is always convergent when applied to diagonally dominant equations, the convergence can be relatively slow. Iterative techniques that provide faster convergence are discussed in references 4 and 5. It should also be mentioned that a hand-calculation iterative scheme known as the *relaxation method* is available in the literature.

## EXAMPLE 4–6

Develop a more accurate numerical solution for the rate of heat transfer from the fin of Example 4–4 by employing the Gauss-Seidel iteration method.

### Solution

*Objective* Use the Gauss-Seidel iteration method to develop accurate calculations for $q_F$.

*Schematic* Circular fin with insulated tip.

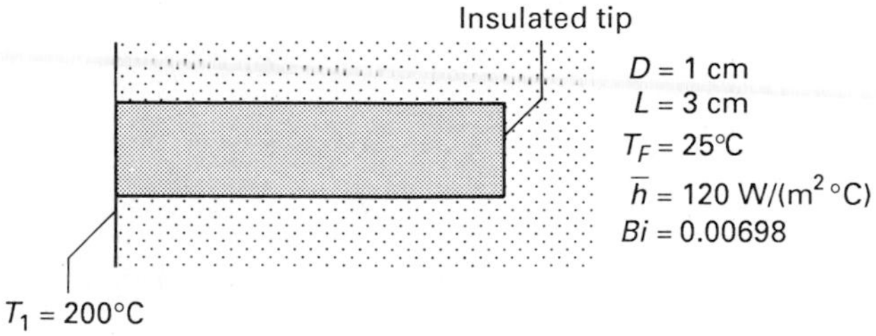
Insulated tip
D = 1 cm
L = 3 cm
T_F = 25°C
h̄ = 120 W/(m² °C)
Bi = 0.00698
T₁ = 200°C

*Assumptions/Conditions*

steady-state
one-dimensional fin
uniform properties
negligible thermal radiation

*Properties* Carbon steel (1% C) at 200°C: $k = 43$ W/(m °C).

*Analysis* A finite-difference grid network is shown in Fig. E4–6a for $M$ nodes. By applying the first law of thermodynamics to each node, we obtain

$$m = 2,3, \ldots, M-1 \qquad q_x = q_{x+\Delta x} + \Delta q_c$$

$$T_{m-1} - \left(2 + \frac{\bar{h}p}{kA}\Delta x^2\right) T_m + T_{m+1} + \frac{\bar{h}p}{kA}\Delta x^2 T_F = 0$$

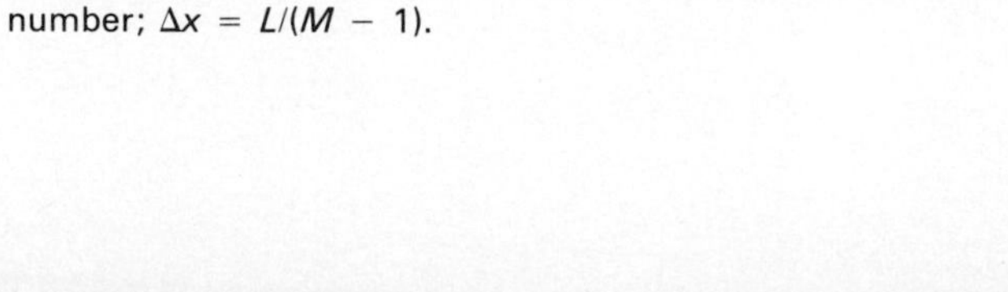
1
2
3
m − 1
m
m + 1
M − 1
M

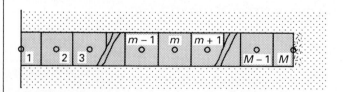

**FIGURE E4–6a** Finite-difference grid for fin with small Biot number; $\Delta x = L/(M-1)$.

and $m = M$ $\quad q_x = \Delta q_c$

$$T_{M-1} - \left(1 + \frac{\bar{h}p}{kA}\frac{\Delta x^2}{2}\right) T_M + \left(\frac{\bar{h}p}{kA}\frac{\Delta x^2}{2}\right) T_F = 0$$

These $M-1$ equations are written in the Gauss-Seidel format as

$m = 2,3, \ldots, M-1$

$$T_m = \frac{1}{2 + \bar{h}p\,\Delta x^2/(kA)}\left(T_{m-1} + T_{m+1} + \frac{\bar{h}p}{kA}\Delta x^2\,T_F\right) \tag{a}$$

$m = M$

$$T_M = \frac{1}{1 + \bar{h}p\,\Delta x^2/(2kA)}\left(T_{M-1} + \frac{\bar{h}p}{kA}\frac{\Delta x^2}{2}T_F\right) \tag{b}$$

For example, for the $M = 5$ grid of Example 4–4, we have (to four significant figures)

$$T_2 = \frac{1}{2.063}(200°\text{C} + T_3 + 1.57°\text{C})$$

$$T_3 = \frac{1}{2.063}(T_2 + T_4 + 1.57°\text{C})$$

$$T_4 = \frac{1}{2.063}(T_3 + T_5 + 1.57°\text{C})$$

$$T_5 = \frac{1}{1.031}(T_4 + 0.785°\text{C})$$

Starting our iterative calculations with $T_m = 200°\text{C}$, first-round calculations are obtained as follows:

$$T_2 = \frac{1}{2.063}(200°\text{C} + 200°\text{C} + 1.57°\text{C}) = 194.7°\text{C}$$

$$T_3 = \frac{1}{2.063}(194.7°\text{C} + 200°\text{C} + 1.57°\text{C}) = 192.1°\text{C}$$

$$T_4 = \frac{1}{2.063}(192.1°\text{C} + 200°\text{C} + 1.57°\text{C}) = 190.8°\text{C}$$

$$T_5 = \frac{1}{1.031}(190.8°\text{C} + 0.785°\text{C}) = 185.8°\text{C}$$

A second iteration gives

$$T_2 = \frac{1}{2.063}(200°\text{C} + 192.1°\text{C} + 1.57°\text{C}) = 190.8°\text{C}$$

$$T_3 = \frac{1}{2.063}(190.8°\text{C} + 190.8°\text{C} + 1.57°\text{C}) = 185.7°\text{C}$$

$$T_4 = \frac{1}{2.063}(185.7°\text{C} + 185.8°\text{C} + 1.57°\text{C}) = 180.8°\text{C}$$

$$T_5 = \frac{1}{1.031}(180.8°\text{C} + 0.785°\text{C}) = 176.1°\text{C}$$

The iteration calculations for $T_2$, $T_3$, $T_4$, and $T_5$ converge within about 2% of the values obtained in Example 4–4 after 22 iterations. But because of the slow convergence, hand calculation is impractical for this problem. Therefore, in order to develop more accurate finite-difference solutions with larger values of $M$, numerical digital computation is now employed.

Utilizing BASIC computer language, Eqs. (a) and (b) are written as

```
T(J,2)=(T(J−1,1)+T(J+1,1)+CC1*TF)/(2+CC1)
```

for J = $m$ = 2,3, . . . , $M-1$, and

```
T(M,2)=(T(M−1,1)+CC1/2*TF)/(1+CC1/2)
```

where CC1=H*P*DX*DX/(K*A), H = $\bar{h}$, P = $p$, DX = $\Delta x$, K = $k$, and TF = $T_F$. The second index represents the new (index 2) and old (index 1) iteration values for the nodal temperatures. The rate of heat transfer $q_F$ from the fin is obtained by performing an energy balance on node (1); $q_F$ is expressed in BASIC by

```
QF= −K*A/DX*(T(2,2)−T(1,2))+H*P*DX/2*(T(1,2)−TF)          (j)
```

A simple flowchart and BASIC program are given in Figs. E4–6b and E4–6c, respectively, which solves the $M-1$ nodal equations by the Gauss-Seidel method. The program also calculates the rate of heat transfer from the fin. This BASIC program requires that the temperature at node $M$ satisfy the specified iterative convergence criterion $\epsilon$ (designated by CHK in the program). (Logic that permits us to check and satisfy convergence for all nodal temperatures is developed in Example 4–8.)

Calculations for the rate of heat transfer from the fin are shown in Fig. E4–6d for $M$ = 3,4, . . . , 20, and for $\epsilon = 10^{-2}$, $10^{-3}$, $10^{-4}$, and $10^{-5}$. Note that the calculations are dependent upon both $M$ and $\epsilon$! The calculations for $q_F$ essentially converge to approximately 15.1 W for $M \geq 6$ (i.e., $\Delta x < 0.006$ m) and $\epsilon = 10^{-5}$. This value is in agreement with the analytical solution developed in Example 2–13.

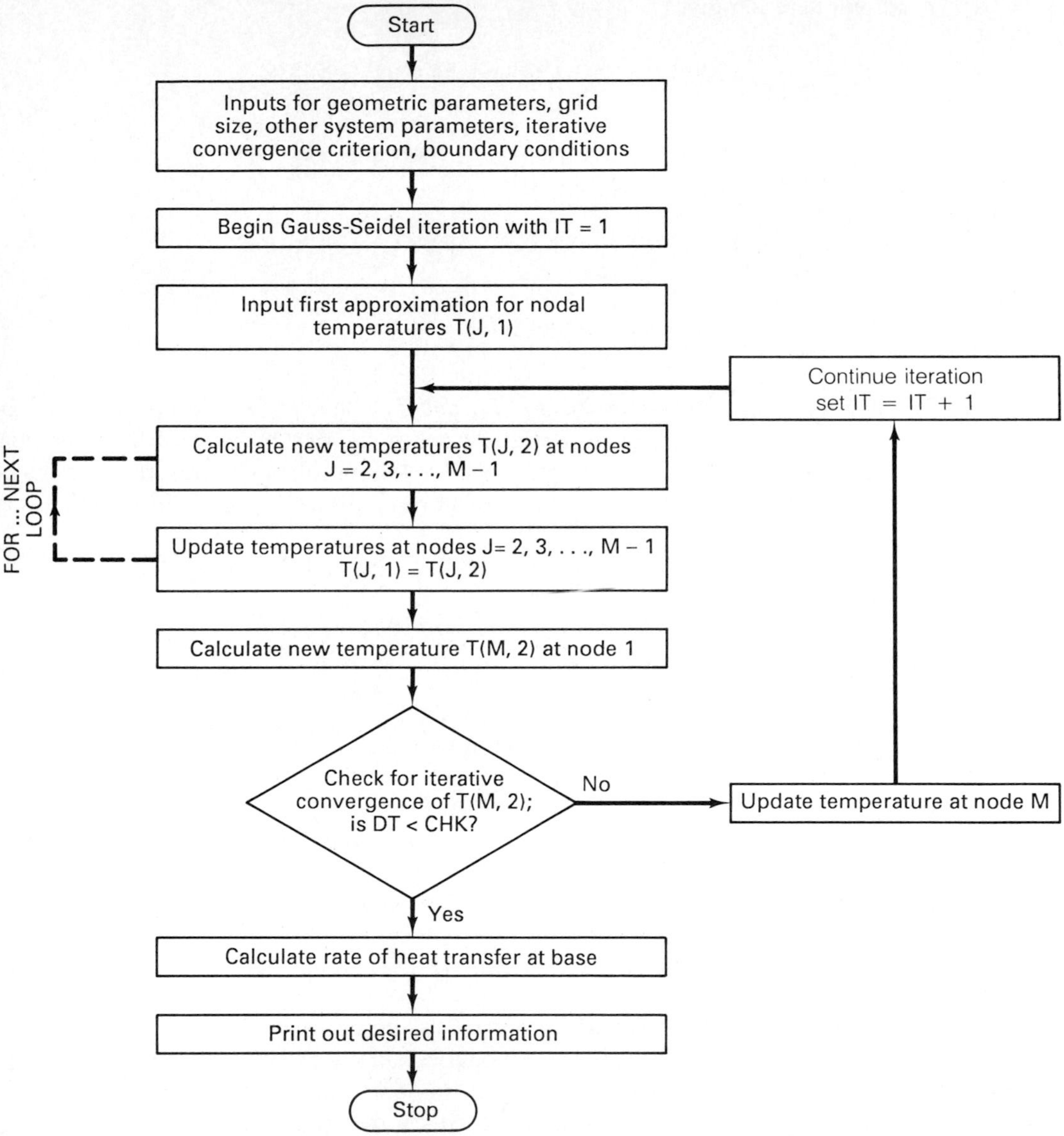


**FIGURE E4–6b** Program flowchart for Example 4–6.

```
                                                Comments
                              Inputs for geometric parameters
   L=.03
   D=.01
   A=3.1416*D^2/4
   P=3.1416*D
                                           Inputs for grid size
   FOR M=3 TO 20
   M1=M-1
   DX=L/M1
                                          Inputs for properties
   K=43
                               Inputs for other system parameters
   H=120
   CC1=H*P*DX^2/(K*A)
   T1=200
   TF=25
                         Input for iterative convergence criterion
   CHK=.00001
                                  Input for boundary conditions
   T(1,1)=T1
   T(1,2)=T1
                         Begin Gauss Seidel iteration with IT = 1
   IT=1
                    Input first approximation for nodal temperatures
   FOR J=2 TO M
   T(J,1)=T1
            Calculate new temperatures T(J,2) at nodes J = 2,3, . . . , M - 1
   NEXT J
2  FOR J=2 TO M1
   T(J,2)=(T(J-1,1)+T(J+1,1)+CC1*TF)/(2+CC1)
                         Update temperatures at J = 2,3, . . . , M - 1
   T(J,1)=T(J,2)
                        Calculate new temperature T(M,2) at node M
   NEXT J
   T(M,2)=(T(M-1,1)+CC1/2*TF)/(1+CC1/2)
                        Check for iterative convergence of T(M,2)
   DT=ABS((T(M,2)-T(M,1))/T(M,1))
   IF DT<CHK GOTO 4
                                    Update temperature at node M
   T(M,1)=T(M,2)
                                              Continue iteration
   IT=IT+1
   GOTO 2
                          Calculate rate of heat transfer at base
4  QF=-K*A/DX*(T(2,2)-T(1,2))+H*P*DX/2*(T(1,2)-TF)
                                   Print out desired information
   LPRINT USING "####.##"; M, IT, T(M,2), QF
   NEXT M
   STOP
   END
```

**FIGURE E4–6c** BASIC program for Example 4–6.

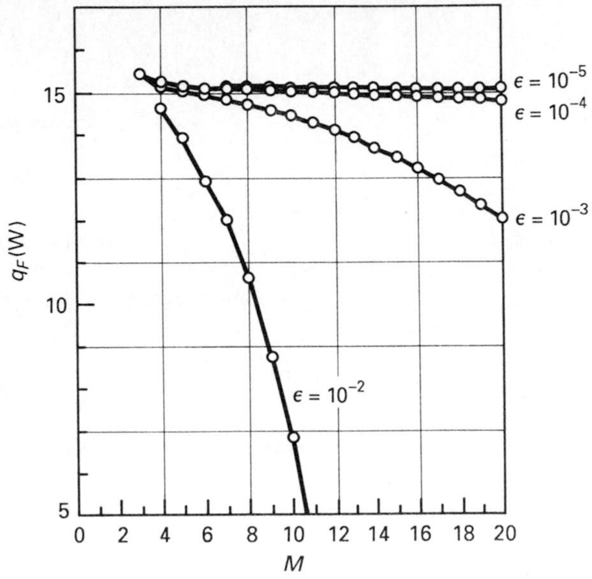


**FIGURE E4–6d** Numerical calculations for $q_F$.

## EXAMPLE 4–7

Determine the rate of heat transfer and temperature distribution for the anodized aluminum fin shown in Fig. E4–7a. The base is at 80°C, the air and surrounding walls are at 25°C, and the coefficient of heat transfer due to natural convection cooling is 10 W/(m² °C).

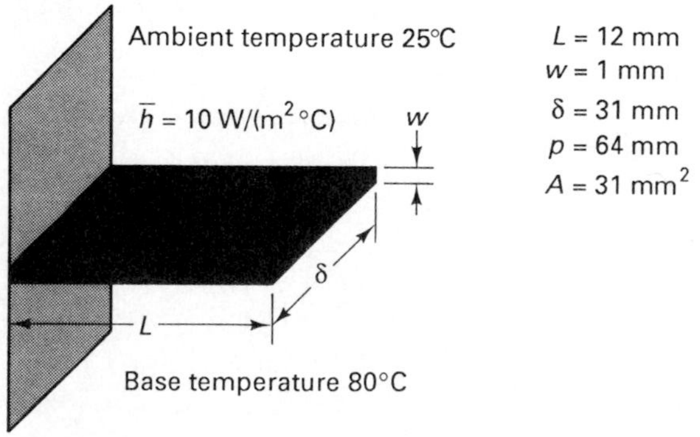


**FIGURE E4–7a**
Convecting and radiating fin.

### Solution

*Objective* Calculate $q_F$ and $T(x)$.

*Assumptions/Conditions*

steady-state
one-dimensional fin
uniform properties
convection and blackbody thermal radiation cooling
$F_{s-R} = 1$ from fin to surroundings

*Properties* Anodized aluminum (Example 2–13): $k = 200$ W/(m °C).

*Analysis* As shown in Example 2–14, a one-dimensional analysis is warranted for this problem because the Biot number is much less than 0.1. Because of the complexity of the problem, we will utilize the numerical approach.

Utilizing the finite-difference grid of Fig. E4–6a in Example 4–6, the nodal equations are given by

$m = 2, 3, \ldots, M-1 \qquad q_x = q_{x+\Delta x} + \Delta q_c + \Delta q_R$

$$-kA\frac{T_m - T_{m-1}}{\Delta x} = -kA\frac{T_{m+1} - T_m}{\Delta x} + \bar{h}p\,\Delta x\,(T_m - T_F) + \sigma p\,\Delta x\,F_{s-R}(T_m^4 - T_R^4) \tag{a}$$

$m = M \qquad q_x = \Delta q_c + \Delta q_R + q_c + q_R$

$$-kA\frac{T_M - T_{M-1}}{\Delta x} = \bar{h}p\frac{\Delta x}{2}(T_M - T_F) + \sigma p F_{s-R}\frac{\Delta x}{2}(T_M^4 - T_R^4) + \bar{h}A(T_M - T_F) + \sigma A F_{s-R}(T_M^4 - T_R^4) \tag{b}$$

where $q_c$ and $q_R$ account for the heat transfer from the tip. To put Eqs. (a) and (b) into the Gauss-Seidel iteration form, these equations are written as

$m = 2, 3, \ldots, M-1$

$$T_m = \frac{T_{m-1} + T_{m+1} + CC_1\,T_F + CR_1\,T_R^4}{2 + CC_1 + CR_1\,T_m^3} \tag{c}$$

$m = M$

$$T_M = \frac{T_{M-1} + (CC_1/2 + CC_2)T_F + (CR_1/2 + CR_2)T_R^4}{1 + CC_1/2 + CC_2 + (CR_1/2 + CR_2)T_M^3} \tag{d}$$

where $CC_1 = \bar{h}p\,\Delta x^2/(kA)$, $CC_2 = \bar{h}\,\Delta x/k$, $CR_1 = \sigma p F_{s-R}\,\Delta x^2/(kA)$, and $CR_2 = \sigma F_{s-R}\,\Delta x/k$. Notice that $T_m$ appears on both sides of these nonlinear equations. (A similar iterative formulation was developed in Example 2–3 for one-dimensional radiation heat transfer from a flat plate.)

Equations (c) and (d) are written in BASIC language as follows:

$$T(J,2)=(T(J-1,1)+T(J+1,1)+CC1*TF+CR1*TR**4)/(2+CC1+CR1*T(J,1)**3) \tag{e}$$

for J = 2,3, . . . , $M-1$, and

$$T(M,2)=(T(M-1,1)+(CC1/2+CC2)*TF+(CR1/2+CR2)*TR**4)/(1+CC1/2+CC2)+(CR1/2+CR2)*T(M,1)**3) \tag{f}$$

for J = $M$, where CC1 = H*P*DX*DX/(K*A), CC2 = H*DX/K, CR1 = SIGMA*P*DX*DX*FSR/(K*A), and CR2 = SIGMA*DX*FSR/K. By applying the first law of thermodynamics to node (1), an expression is obtained for the rate of heat transfer $q_F$ from the fin of the form

$$QF=-K*A/DX*(T(2,2)-T(1,2))+H*P*DX/2*(T(1,2)-TF)+SIGMA*P*DX/2*FSR*(T(1,2)**4-TR**4) \tag{g}$$

The BASIC program presented in Example 4–6 is employed to solve these equations by utilizing the following inputs for geometric parameters, properties, and other system parameters:

*Inputs for geometric parameters*

L = 12 E−3
P = 64 E−3
A = 31 E−6

*Inputs for properties*

K = 200

*Inputs for other system parameters*

H = 10
SIGMA = 5.67 E−8
FSR = 1
CC1 = H*P*DX*DX/(K*A)
CC2 = H*DX/K
CR1 = SIGMA*P*DX*DX*FSR/(K*A)
CR2 = SIGMA*DX*FSR/K
T1 = 353
TF = 298
TR = 298

Of course, the equations for T(J,2), T(M,2), and QF must be specified in accordance with Eqs. (e), (f), and (g).

Calculations are shown in Fig. E4–7b for the rate of heat transfer from the fin for $M = 3, 4, \ldots, 20$ and $\epsilon = 10^{-5}$ and $10^{-7}$. The calculations for $q_F$ converge to approximately 1.02 W for $M \geq 3$ (i.e., $\Delta x \lesssim 0.006$ m) and $\epsilon = 10^{-7}$. To see the effect of radiation, the program is also run with the radiation terms eliminated. The calculations for $q_F$ without radiation converge to about 0.575 W. Thus, the thermal radiation contributes 43.6% to the total rate of heat transfer.

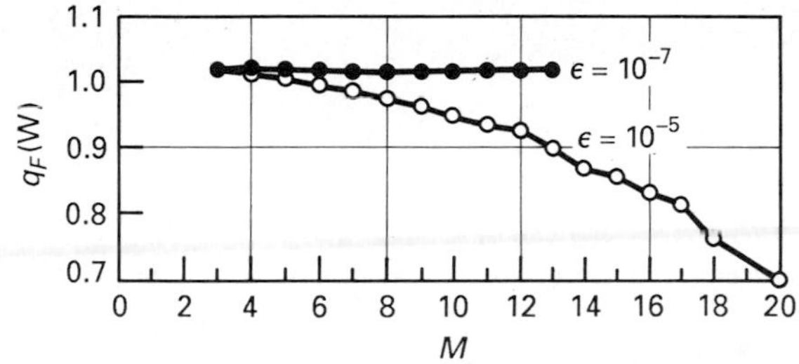


**FIGURE E4–7b**
Numerical calculations for $q_F$.

Assuming that no errors have been made in our analysis and program inputs, we can be reasonably sure of an accurate solution, since the calculations for $q_F$ converge for increasing $M$ and decreasing $\epsilon$. However, to further reinforce our confidence in the solution accuracy, calculations are made for $q_F$ for longer fin lengths to enable us to compare the results with the analytical solution of Example 2–14. The comparison of numerical calculations for $q_F$ for fins of various lengths with the analytical result for a long fin is made in Fig. E4–7c. Notice that the numerical calculations approach the analytical solution of Example 2–14 for lengths of the order of 0.3 m and greater.

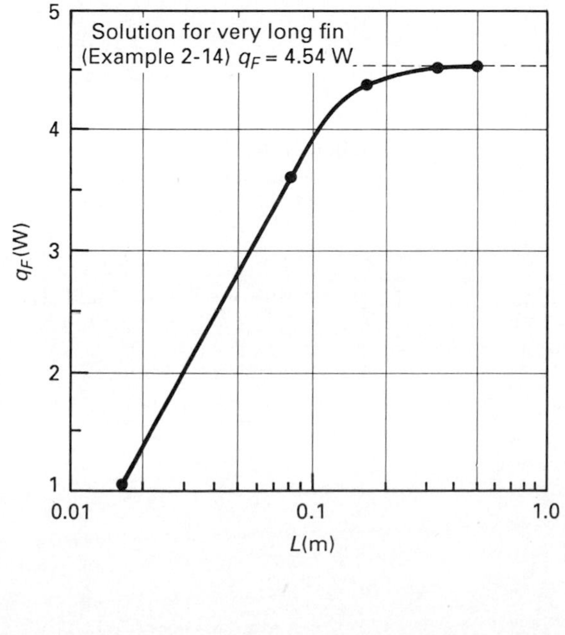


**FIGURE E4–7c**
Numerical calculations for $q_F$ for various lengths.

The temperature distribution in the fin is shown in Table E4–7. The temperature drop along this short fin is seen to be very small, such that the fin efficiency is quite high.

**TABLE E4–7** Numerical calculations for temperature distribution in fin

| $m$ | $T_m$(K) |
|---|---|
| 1 | 353.0 |
| 2 | 352.7 |
| 3 | 352.5 |
| 4 | 352.3 |
| 5 | 352.1 |
| 6 | 351.9 |
| 7 | 351.8 |
| 8 | 351.73 |
| 9 | 351.67 |

## EXAMPLE 4–8

Develop a BASIC program for determining the temperature distribution and rate of heat transfer across face $A$ for the rectangular plate shown in Fig. E4–8a. Use the Gauss-Seidel iteration method and a grid network with $\Delta x = L/M$, where $M$ can be any integer value. Then run the program to obtain an accurate numerical solution for (*a*) $T_1 = 0°\text{C}$ and $f(x) = 100°\text{C} \sin(\pi x/L)$, and (b) $T_1 = 0°\text{C}$ and $T_2 = 100°\text{C}$.

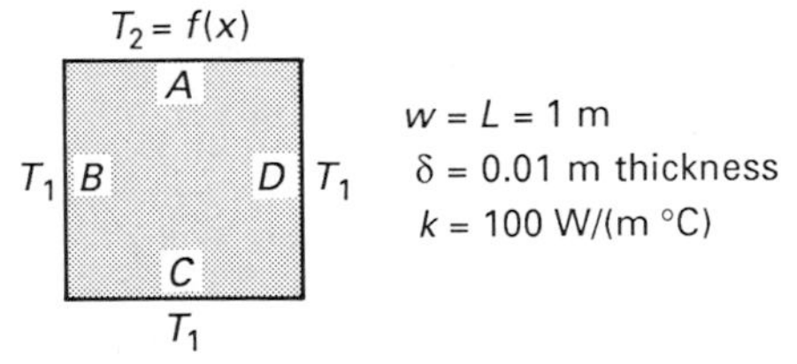

$T_2 = f(x)$
A
$T_1$ B
D $T_1$
C
$T_1$
$w = L = 1$ m
$\delta = 0.01$ m thickness
$k = 100$ W/(m °C)


**FIGURE E4–8a**
Two-dimensional heat transfer in square plate.

### Solution

*Objective* Develop accurate finite-difference solutions for $T_{m,n}$ and $q_A$.

*Assumptions/Conditions*

steady-state
two-dimensional
uniform properties

*Properties* $k = 100$ W/(m °C).

*Analysis* A finite-difference grid with $M = 6$ is shown in Fig. E4–8b for this conduction-heat-transfer problem. The interior nodal equations take the general form

$$4\,T_{m,n} = T_{m+1,n} + T_{m-1,n} + T_{m,n+1} + T_{m,n-1}$$

with $m = 2, 3, \ldots, M-1$, and $n = 2, 3, \ldots, N-1$. Our formulation consists of $Z\,[= (M-2)(N-2)]$ such equations plus the boundary conditions. Using BASIC computer language, this system of equations is put into the interative form

$$\text{T(J,I,2)} = (\text{T(J+1,I,1)} + \text{T(J}-\text{1,I,1)} + \text{T(J,I+1,1)} + \text{T(J,I}-\text{1,1)})/4$$

where $\text{J} = m = 2, 3, 4, \ldots, M-1$, and $\text{I} = n = 2, 3, 4, \ldots, N-1$. The third index represents the new (index 2) and old (index 1) iteration values for the nodal temperatures. Using this format, the rate of heat transfer $q_A$ is represented by

$$\text{QA} = \text{K*D*(T(J,N}-\text{1,2)} + \text{T(J}-\text{1,N,2)/2}$$
$$+ \text{T(J+1,N,2)/2} - \text{2*T(J,N,2))}$$

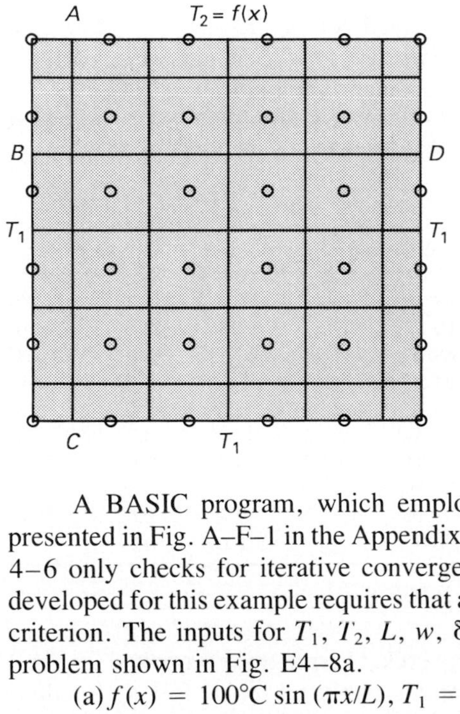


**FIGURE E4–8b**
Finite-difference grid for square plate.

A BASIC program, which employs the Gauss-Seidel iteration technique, is presented in Fig. A–F–1 in the Appendix. Whereas the program developed in Example 4–6 only checks for iterative convergence of one nodal temperature, the program developed for this example requires that all nodal temperatures satisfy the convergence criterion. The inputs for $T_1$, $T_2$, $L$, $w$, $\delta$, and $k$ are specified in accordance with the problem shown in Fig. E4–8a.

(a) $f(x) = 100°\text{C} \sin(\pi x/L)$, $T_1 = 0°\text{C}$: Calculations for $q_A$ obtained by running this program on a digital computer are shown in Fig. E4–8c for values of $M = 2, 3, 4, \ldots, 20$ and for $\epsilon = 10^{-1}$ and $10^{-4}$. Again we notice that the accuracy of the numerical solution depends upon both $M$ and $\epsilon$! The calculations for $q_A$ are seen to converge to approximately $-200.7$ W for $M$ equal to 9 with $\epsilon = 10^{-4}$. The exact solution given by Eq. (3–44) is $-200$ W, such that the error is only 0.35%.

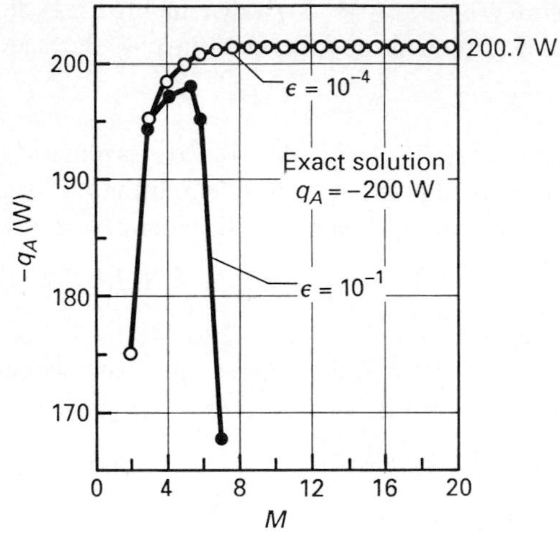


**FIGURE E4–8c**
Numerical calculations for $q_A$; $T_2 = 100 \sin(\pi x/L)$ °C.

Calculations are also easily obtained for $q_B$, $q_C$, and $q_D$. For example, the calculations for $q_C$ converge to approximately $-17.5$ W at a value of $M$ equal to 13. The exact solution for $q_C$ is $-17.3$ W, such that the error is about 1.5%.

Calculations are shown in Table E4–8 for the nodal temperatures obtained for $M = 9$. The accuracy of these predictions is extremely good, with a minimum error of about 0.1%.

**TABLE E4–8** Numerical calculations for temperature distribution in square plate†

| | | | | | | | | | |
|---|---|---|---|---|---|---|---|---|---|
| 0.0 | 34.20 | 64.28 | 86.60 | 98.48 | 98.48 | 86.6 | 64.28 | 34.20 | 0.0 |
| 0.0 | 24.16 | 45.41 | 61.18 | 69.57 | 69.57 | 61.1 | 45.41 | 24.16 | 0.0 |
| 0.0 | 17.03 | 32.01 | 43.13 | 49.04 | 49.04 | 43.1 | 32.01 | 17.03 | 0.0 |
| 0.0 | 11.96 | 22.48 | 30.28 | 34.44 | 34.44 | 30.2 | 22.48 | 11.96 | 0.0 |
| 0.0 | 8.33 | 15.65 | 21.09 | 23.98 | 23.98 | 21.0 | 15.65 | 8.33 | 0.0 |
| 0.0 | 5.70 | 10.72 | 14.44 | 16.42 | 16.42 | 14.4 | 10.72 | 5.70 | 0.0 |
| 0.0 | 3.77 | 7.08 | 9.54 | 10.84 | 10.84 | 9.5 | 7.08 | 3.77 | 0.0 |
| 0.0 | 2.28 | 4.29 | 5.78 | 6.58 | 6.58 | 5.7 | 4.29 | 2.28 | 0.0 |
| 0.0 | 1.08 | 2.02 | 2.73 | 3.10 | 3.10 | 2.7 | 2.02 | 1.08 | 0.0 |
| 0.0 | 0.0 | 0.0 | 0.0 | 0.0 | 0.0 | 0.0 | 0.0 | 0.0 | 0.0 |

† $T_2 = 100$°C $\sin(\pi x/L)$

(b) $T_2 = f(x) = 100$°C, $T_1 = 0$°C: The program presented in Fig. A–F–1 is easily run for $T_2 = 100$°C. Calculations produced for $q_A$ and $q_C$ for this condition are shown in Fig. E4–8d for $M = 2, 3, 4, \ldots, 20$ and $\epsilon = 10^{-4}$. For this problem, $q_C$ converges to approximately $-22.2$ W, but convergence is nowhere in sight for $q_A$. Referring back to Example 3–6, the exact solution for $q_C$ is $-22.1$ W, but $q_A$ is actually unbounded. Thus, our finite-difference solution is consistent with the exact analytical solution.

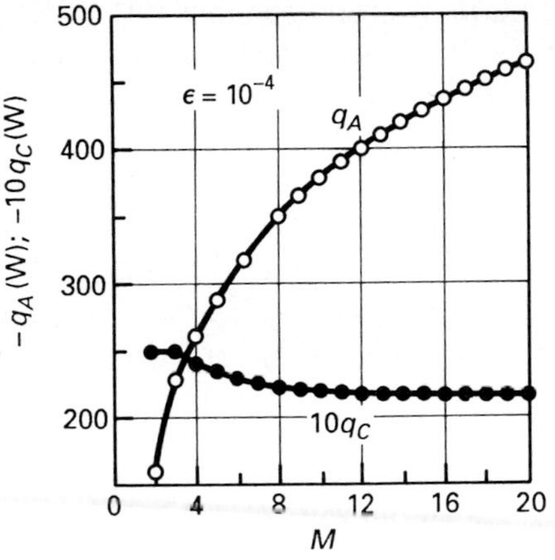


**FIGURE E4-8d**
Numerical calculations for $q_A$ and $q_C$; $T_2 = 100$ °C.

### *Direct Methods*

The most commonly used direct methods for solving systems of linear equations include matrix and Gaussian elimination methods [3–6].

In the Gaussian elimination method, the nodal equations, Eqs. (4–29–1) through (4–29–Z), are put into the form

$$a_{11}T_1 + a_{12}T_2 + \cdots + a_{1j}T_j + \cdots + a_{1Z}T_Z = C_1 \tag{4–32–1}$$

$$a'_{22}T_2 + \cdots + a'_{2j}T_j + \cdots + a'_{2Z}T_Z = C'_2 \tag{4–32–2}$$

$$\vdots$$

$$a'_{jj}T_j + \cdots + a'_{jZ}T_Z = C'_j \tag{4–32–j}$$

$$\vdots$$

$$a'_{ZZ}T_Z = C'_Z \tag{4–32–Z}$$

This arrangement is obtained by first dividing Eq. (4–29–1) by $a_{11}$. The resulting equation is used to algebraically eliminate $T_1$ from each of the remaining equations. For example, by combining Eqs. (4–29–1) and (4–29–2), we obtain

$$\begin{aligned} a'_{2j} &= a_{2j} - a_{1j}a_{21}/a_{11} \qquad j = 1,2, \ldots, Z \\ C'_2 &= C_2 - C_1 a_{21}/a_{11} \end{aligned} \tag{4–33}$$

The second equation is then used to eliminate $T_2$ from the remaining equations. This elimination procedure is continued until the final equation is obtained. This technique produces a solution for the unknown nodal temperature $T_Z$ of the form

$$T_Z = \frac{C'_Z}{a'_{ZZ}} \tag{4–34}$$

The other unknowns can then be calculated in reverse order (i.e., $T_{Z-1}, T_{Z-2}, \ldots, T_j, \ldots, T_1$). This elimination procedure is simplified by the fact that a given nodal temperature $T_j$ appears in no more than four equations for conduction problems involving two dimensions, such as $x$ and $y$.

A number of standardized programs are available that employ matrix and Gaussian elimination methods. However, except for applications involving relatively few nodal equations, these direct methods are generally less efficient than iterative methods.

### *Unsteady Analysis Method*

Because our next topic deals with the numerical solution of unsteady multidimensional conduction-heat-transfer problems, we merely mention here that steady-state problems are sometimes solved by means of unsteady analyses which are carried through to the steady-state limit.

## 4–5 NUMERICAL ANALYSIS: UNSTEADY SYSTEMS

Basic finite-difference formulation and solution concepts associated with the numerical analysis of unsteady conduction-heat-transfer processes are considered in this section.

### 4–5–1 Nodal Equations

To develop a numerical finite-difference formulation for unsteady conduction heat transfer in a rectangular solid with internal energy generation, the system is subdivided into $Z_s$ subvolumes, with the temperature at each node designated by $T^{\tau}_{m,n}$. Using the energy balance method, which was introduced in the previous section, the first law of thermodynamics is applied to each interior and exterior node at an instant of time $t$. To illustrate, the development of an energy balance on an interior subvolume such as the one shown in Fig. 4–2 gives

$$\Delta q_x + \Delta q_y + \dot{q}\,\Delta V = \Delta q_{x+\Delta x} + \Delta q_{y+\Delta y} + \frac{\Delta E_s}{\Delta t} \tag{4–35}$$

where $\Delta E_s/\Delta t$ is no longer equal to zero and the heat-transfer rates are specified by unsteady finite-difference approximations for the Fourier law of conduction.

Based on our previous study, we know that $\Delta E_s/\Delta t$ can be expressed in terms of the specific heat and instantaneous temperature difference by

$$\frac{\Delta E_s}{\Delta t} = \rho c_v \, \Delta V \frac{\Delta T}{\Delta t} \tag{4-36}$$

where $\Delta T/\Delta t$ represents the finite-difference approximation for $\partial T/\partial t$. This relationship for $\Delta E_s/\Delta t$ together with the second-order finite-difference approximation for the Fourier law of conduction are substituted into Eq. (4–35), to obtain

$$\begin{aligned} -k\delta \, \Delta y \frac{T^{\tau}_{m,n} - T^{\tau}_{m-1,n}}{\Delta x} - k\delta \, \Delta x \frac{T^{\tau}_{m,n} - T^{\tau}_{m,n-1}}{\Delta y} + \dot{q} \, \Delta V \\ = -k\delta \, \Delta y \frac{T^{\tau}_{m+1,n} - T^{\tau}_{m,n}}{\Delta x} - k\delta \, \Delta x \frac{T^{\tau}_{m,n+1} - T^{\tau}_{m,n}}{\Delta y} + \rho c_v \, \Delta V \frac{\Delta T}{\Delta t} \end{aligned} \tag{4-37}$$

Setting $\Delta y = \Delta x$ and rearranging, this equation is put into the form

$$\frac{\alpha}{\Delta x^2}\left(T^{\tau}_{m+1,n} + T^{\tau}_{m-1,n} + T^{\tau}_{m,n+1} + T^{\tau}_{m,n-1} - 4\,T^{\tau}_{m,n} + \frac{\dot{q}}{k}\Delta x^2\right) = \frac{\Delta T}{\Delta t} \tag{4-38}$$

Similar nodal equations can be written for the exterior subvolumes. For example, the nodal equation for a regular exterior node at $m = M$ with surface heat flux $q''_s$ is given by

$$\frac{\alpha}{\Delta x^2}\left(2\,T^{\tau}_{M-1,n} + T^{\tau}_{M,n+1} + T^{\tau}_{M,n-1} - 4\,T^{\tau}_{M,n} + \frac{\dot{q}}{k}\Delta x^2 - 2\frac{q''_s}{k}\Delta x\right) = \frac{\Delta T}{\Delta t} \tag{4-39}$$

It should be observed that these unsteady nodal equations and the corresponding relations for corner nodes can be obtained from the steady nodal equations given in Table 4–1 by merely replacing the right-hand side by $\Delta T/\Delta t$ and introducing the time index $\tau$ as a superscript for the nodal temperatures.

To complete the formulation, $\Delta T/\Delta t$ is usually expressed in terms of the *forward-time difference* [see Eq. (4–14)],

$$\frac{\Delta T}{\Delta t} = \frac{T^{\tau+1}_{m,n} - T^{\tau}_{m,n}}{\Delta t} \tag{4-40}$$

or the *backward-time difference* [see Eq. (4–15)],

$$\frac{\Delta T}{\Delta t} = \frac{T^{\tau}_{m,n} - T^{\tau-1}_{m,n}}{\Delta t} \tag{4-41}$$

both of which are *first-order* approximations.

The accuracy of the finite-difference formulations resulting from the use of Eqs. (4–38) and (4–39) and Eqs. (4–40) and (4–41) is controlled by component errors that are proportional to $\Delta x^2$ and $\Delta t$. Whereas accuracy requirements for limiting steady-state conditions can generally be used to establish an effective grid spacing

$\Delta x$, the time increment $\Delta t$ normally must be determined by trial-and-error testing. However, in the case of the forward-time difference formulation, the selection of $\Delta t$ is restricted by stability considerations.

Because the forward-time difference formulation produces algebraic equations that can be solved by straightforward explicit computational techniques, this approach will be emphasized in our study. Brief consideration will also be given to the alternative implicit finite-difference method, which is developed by use of the backward-time difference for $\Delta T/\Delta t$.

### 4–5–2 Explicit Method

Utilizing the forward-time difference, Eq. (4–38) takes the form

$$\frac{\alpha}{\Delta x^2}\left(T^{\tau}_{m+1,n} + T^{\tau}_{m-1,n} + T^{\tau}_{m,n+1} + T^{\tau}_{m,n-1} - 4\,T^{\tau}_{m,n} + \frac{\dot{q}}{k}\Delta x^2\right) = \frac{T^{\tau+1}_{m,n} - T^{\tau}_{m,n}}{\Delta t} \tag{4–42}$$

or

$$T^{\tau+1}_{m,n} = \frac{\alpha\,\Delta t}{\Delta x^2}\left(T^{\tau}_{m+1,n} + T^{\tau}_{m-1,n} + T^{\tau}_{m,n+1} + T^{\tau}_{m,n-1} + \frac{\dot{q}}{k}\Delta x^2\right) + \left(1 - 4\frac{\alpha\,\Delta t}{\Delta x^2}\right)T^{\tau}_{m,n} \tag{4–43}$$

This forward-time difference interior nodal equation expresses the nodal temperature $T^{\tau+1}_{m,n}$ at time $(\tau+1)\,\Delta t$ in terms of the nodal temperature distribution $T^{\tau}_{m,n}$, $T^{\tau}_{m+1,n}$, $T^{\tau}_{m-1,n}$, $T^{\tau}_{m,n+1}$, and $T^{\tau}_{m,n-1}$ at the earlier instant of time $\tau\,\Delta t$. Because the nodal temperature distribution is known at some initial instant of time, this type of equation is *explicit* in that the unknown nodal temperatures at the next instant of time can be calculated directly.

The explicit nodal equation for a regular exterior node at $m = M$ is given by

$$T^{\tau+1}_{M,n} = \frac{\alpha\,\Delta t}{\Delta x^2}\left(2\,T^{\tau}_{M-1,n} + T^{\tau}_{M,n+1} + T^{\tau}_{M,n-1} + \frac{\dot{q}}{k}\Delta x^2 - 2\frac{q''_s}{k}\Delta x\right) + \left(1 - 4\frac{\alpha\,\Delta t}{\Delta x^2}\right)T^{\tau}_{M,n} \tag{4–44}$$

Setting $q''_s = h(T^{\tau}_{m,n} - T_F)$ for the case of convection, this equation becomes

$$T^{\tau+1}_{M,n} = \frac{\alpha\,\Delta t}{\Delta x^2}\left(2\,T^{\tau}_{M-1,n} + T^{\tau}_{M,n+1} + T^{\tau}_{M,n-1} + \frac{\dot{q}}{k}\Delta x^2 + 2\frac{h\,\Delta x}{k}T_F\right) + \left[1 - \frac{\alpha\,\Delta t}{\Delta x^2}\left(4 + 2\frac{h\,\Delta x}{k}\right)\right]T^{\tau}_{M,n} \tag{4–45}$$

Equations (4–43) and (4–44) and other standard explicit nodal equations are listed in Table 4–2.

**TABLE 4-2** Summary of explicit nodal equations for unsteady two-dimensional systems with internal heat generation

| *Type node* | *Nodal equation* |
|---|---|
| 1. Interior node 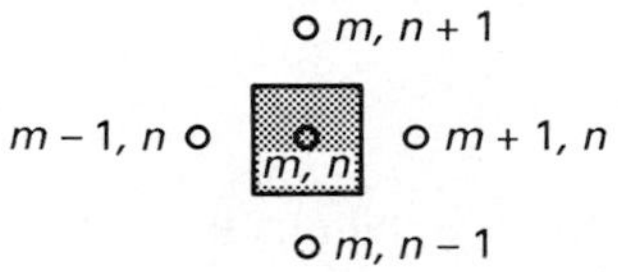 m, n + 1<br>m − 1, n<br>m, n<br>m + 1, n<br>m, n − 1 | $T_{m,n}^{\tau+1} = \frac{\alpha\,\Delta t}{\Delta x^2}\left(T_{m+1,n}^{\tau} + T_{m-1,n}^{\tau} + T_{m,n+1}^{\tau} + T_{m,n-1}^{\tau} + \frac{\dot{q}}{k}\Delta x^2\right) + \left(1 - 4\frac{\alpha\,\Delta t}{\Delta x^2}\right)T_{m,n}^{\tau}$ (1) |
| 2. Exterior nodes | |
| a. General equations | |
| i. Regular exterior node $(M,n)$ 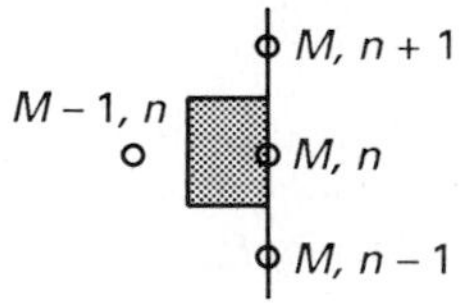 M, n + 1<br>M − 1, n<br>M, n<br>M, n − 1 | $T_{M,n}^{\tau+1} = \frac{\alpha\,\Delta t}{\Delta x^2}\left(2\,T_{M-1,n}^{\tau} + T_{M,n+1}^{\tau} + T_{M,n-1}^{\tau} + \frac{\dot{q}}{k}\Delta x^2 - 2\frac{q_s''}{k}\Delta x\right) + \left(1 - 4\frac{\alpha\,\Delta t}{\Delta x^2}\right)T_{M,n}^{\tau}$ (2i) |
| ii. Outer corner node $(M,N)$ 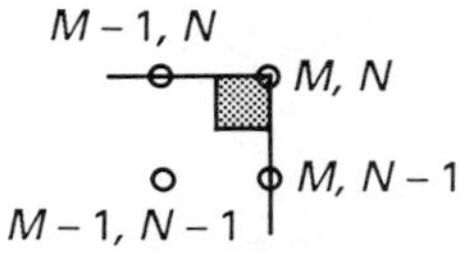 M − 1, N<br>M, N<br>M, N − 1<br>M − 1, N − 1 | $T_{M,n}^{\tau+1} = \frac{\alpha\,\Delta t}{\Delta x^2}\left(2\,T_{M-1,N}^{\tau} + 2\,T_{M,N-1}^{\tau} + \frac{\dot{q}}{k}\Delta x^2 - 4\frac{q_s''}{k}\Delta x\right) + \left(1 - 4\frac{\alpha\,\Delta t}{\Delta x^2}\right)T_{M,N}^{\tau}$ (2ii) |
| iii. Inner corner node $(m,n)$ 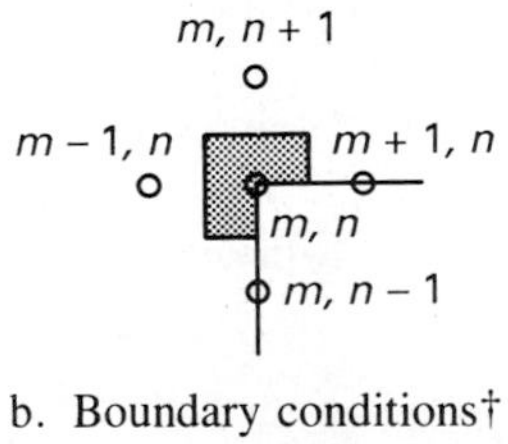 m, n + 1<br>m − 1, n<br>m + 1, n<br>m, n<br>m, n − 1 | $T_{m,n}^{\tau+1} = \frac{\alpha\,\Delta t}{\Delta x^2}\left[\frac{2}{3}\left(T_{m+1,n}^{\tau} + 2\,T_{m-1,n}^{\tau} + 2\,T_{m,n+1}^{\tau} + T_{m,n-1}^{\tau}\right) + \frac{\dot{q}}{k}\Delta x^2 - \frac{4}{3}\frac{q_s''}{k}\Delta x\right] + \left(1 - 4\frac{\alpha\,\Delta t}{\Delta x^2}\right)T_{m,n}^{\tau}$ (2iii) |
| b. Boundary conditions† | |
| i. Specified temperature | $T_{m,n}^{\tau}$ known |
| ii. Specified heat flux | $q_s''$ known |
| iii. Insulated surface | $q_s'' = 0$ |
| iv. Convection | $q_s'' = h(T_{m,n}^{\tau} - T_F)$ |
| v. Blackbody thermal radiation | $q_s'' = \sigma F_{s-R}[(T_{m,n}^{\tau})^4 - T_R^4]$ |

† These boundary conditions apply to each type of exterior node.

## EXAMPLE 4–9

A plate of 3 mm thickness is initially at 50°C. One side of the plate is suddenly exposed to blackbody thermal radiation with $F_{s-R} = 0.5$ and $T_R = 500°C$. The other side of the plate is maintained at 50°C. Develop an explicit finite-difference formulation for this unsteady problem with $\Delta x = L/3$.

### Solution

*Objective* Develop explicit finite-difference nodal equations for temperature using $\Delta x = L/3$.

*Schematic* Finite-difference grid for plane wall.

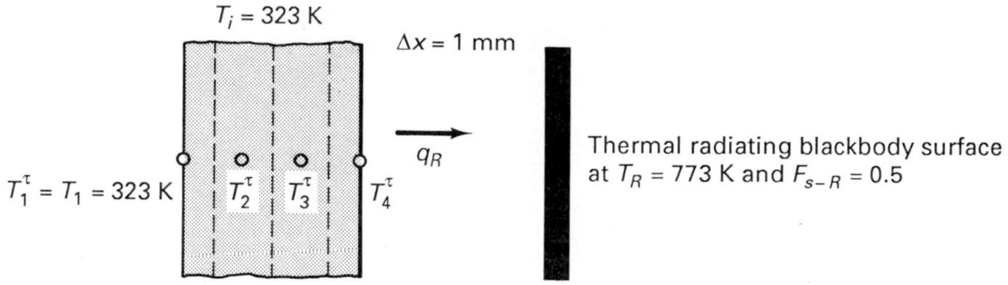


*Assumptions/Conditions*

unsteady-state
one spatial dimension
uniform properties
blackbody thermal radiation

*Analysis* The finite-difference grid network for this problem is shown in the schematic. Note that we have the three unknowns: $T_2^\tau$, $T_3^\tau$, and $T_4^\tau$. The energy balance on each node with unknown temperature is developed as follows, assuming uniform properties:

*Interior Nodes m = 2 and 3* $\qquad q_x = q_{x+\Delta x} + \dfrac{\Delta E_s}{\Delta t}$

$$-kA\frac{T_m^\tau - T_{m-1}^\tau}{\Delta x} = -kA\frac{T_{m+1}^\tau - T_m^\tau}{\Delta x} + \rho A\,\Delta x\,c_v\frac{T_m^{\tau+1} - T_m^\tau}{\Delta t}$$

or

$$T_m^{\tau+1} - T_m^\tau = \frac{\alpha\,\Delta t}{\Delta x^2}(T_{m+1}^\tau + T_{m-1}^\tau - 2\,T_m^\tau)$$

Hence, we obtain

$$T_2^{\tau+1} - T_2^\tau = \frac{\alpha\,\Delta t}{\Delta x^2}(T_3^\tau + T_1^\tau - 2\,T_2^\tau) \tag{a}$$

and

$$T_3^{\tau+1} - T_3^\tau = \frac{\alpha\,\Delta t}{\Delta x^2}(T_4^\tau + T_2^\tau - 2\,T_3^\tau) \tag{b}$$

*Exterior Node* $m = M = 4$ $\qquad q_x = q_R + \dfrac{\Delta E_s}{\Delta t}$

$$-kA\frac{T_4^\tau - T_3^\tau}{\Delta x} = Aq_R'' + \rho A\frac{\Delta x}{2}c_v\frac{T_4^{\tau+1} - T_4^\tau}{\Delta t}$$

or

$$T_4^{\tau+1} - T_4^\tau = \frac{2\alpha\,\Delta t}{\Delta x^2}\left(-\frac{\Delta x\,q_R''}{k} + T_3^\tau - T_4^\tau\right) \tag{c}$$

where $q_R'' = \sigma F_{s-R}[(T_4^\tau)^4 - T_R^4]$. To complete the formulation, we write the initial condition as

$$T_2^0 = T_3^0 = T_4^0 = T_i$$

Utilizing this initial condition, Eqs. (a) through (c) can be solved for $T_2^1$, $T_3^1$, and $T_4^1$. These values of $T_m^\tau$ at $\tau = 1$ can, in turn, be utilized to obtain $T_m^2$. Following through in this fashion, calculations can be developed for $T_m^\tau$ for successively larger and larger values of $\tau$.

Once $T_m^\tau$ is known, predictions can be developed for the rate of heat transfer. For example, the heat transfer flux at the radiating surface is simply

$$q_R'' = \sigma F_{s-R}[(T_4^\tau)^4 - T_R^4]$$

To obtain the instantaneous rate of heat transfer at the other face, we merely perform an energy balance on exterior node (1); that is,

$$q_0 = q_{\Delta x/2} + \frac{\Delta E_s}{\Delta t} = -kA\frac{T_2^\tau - T_1^\tau}{\Delta x}$$

where $\Delta E_s/\Delta t$ is zero because $T_1^\tau = T_1$ is independent of time.

### Stability Considerations

In connection with the specification of the finite-difference increments $\Delta t$ and $\Delta x$ (and $\Delta y$, which has been set equal to $\Delta x$), these increments must be selected such that the nodal calculations do not violate the physical requirement represented by the *second law of thermodynamics*. Otherwise, the resulting finite-difference "solution" will become unstable and blow up after a number of time steps have been taken. The physical basis for this instability is illustrated in Example 4–10. References 7 and 8 provide mathematical details pertaining to the stability and convergence of numerical solutions.

Focusing attention on systems with $\dot{q} = 0$, stability is assured by merely requiring that the coefficients associated with the term $T^{\tau}_{m,n}$ for all nodal equations within the nodal network be equal to or greater than zero. The resulting *stability criterion* for explicit finite-difference nodal equations takes the form

$$\frac{\alpha\,\Delta t}{\Delta x^2} \leq \frac{1}{X} \tag{4–46}$$

with $X$ given in Table 4–3 for interior and representative exterior nodes and boundary conditions. With $\Delta x$ specified, $\Delta t$ can be *no larger* than $\Delta x^2/(\alpha X)$. This limitation prohibits the use of larger increments of time, regardless of accuracy and computational considerations.

Referring to Table 4–3, we observe that the minimum value of $X$ for unsteady two-dimensional systems with $\dot{q} = 0$ is 4. However, since Eq. (4–46) must be satisfied for all nodes, $X$ must be greater than 4 for convection and blackbody thermal radiation. In this connection, with $\alpha\,\Delta t/\Delta x^2$ set equal to $\frac{1}{4}$ for specified wall temperature or specified wall heat flux boundary conditions, the interior nodal equation given by Eq. (4–43) reduces to

$$T^{\tau+1}_{m,n} = \frac{1}{4}(T^{\tau}_{m+1,n} + T^{\tau}_{m-1,n} + T^{\tau}_{m,n+1} + T^{\tau}_{m,n-1}) \tag{4–47}$$

for $\dot{q} = 0$. Utilizing this value of $\alpha\,\Delta t/\Delta x^2$, we see that $T^{\tau}_{m}$ is equal to the arithmetic average of the four surrounding nodal temperatures at $t = \tau\,\Delta t$.

**TABLE 4–3** Stability parameter for unsteady two-dimensional explicit nodal equations for $\dot{q} = 0$

| *Type node* | $X$ |
|---|---|
| Interior | 4 |
| Exterior | |
| Specified temperature | 4 |
| Specified heat flux | 4 |
| Convection | |
| Regular | $4 + 2\,h\,\Delta x/k$ |
| Outer corner | $4 + 4\,h\,\Delta x/k$ |
| Inner corner | $4 + \frac{4}{3}h\,\Delta x/k$ |
| Blackbody thermal radiation | |
| Regular | $4 + 2\,\sigma F_{s-R}\,\Delta x\,(T^{\tau}_{m,n})^3/k$ |
| Outer corner | $4 + 4\,\sigma F_{s-R}\,\Delta x\,(T^{\tau}_{m,n})^3/k$ |
| Inner corner | $4 + \frac{4}{3}\sigma F_{s-R}\,\Delta x\,(T^{\tau}_{m,n})^3/k$ |

**TABLE 4–4** Stability parameter for unsteady one-dimensional explicit nodal equations for $\dot{q} = 0$

| *Type node* | $X$ |
|---|---|
| Interior | 2 |
| Exterior | |
| Specified temperature | 2 |
| Specified heat flux | 2 |
| Convection | $2 + 2\,h\,\Delta x/k$ |
| Blackbody thermal radiation | $2 + 2\,\sigma F_{s-R}\,\Delta x\,(T^{\tau}_{m,n})^3/k$ |

The stability criterion given by Eq. (4–46) also applies to explicit nodal equations for unsteady one-dimensional and three-dimensional systems with $\dot{q} = 0$; the values of $X$ associated with one-dimensional nodal equations are listed in Table 4–4. With $\alpha\,\Delta t/\Delta x^2$ set equal to $\frac{1}{2}$ in the case of unsteady one-dimensional systems with $\dot{q} = 0$, the interior nodal equations are written as

$$T^{\tau+1}_{m} = \frac{1}{2}(T^{\tau}_{m+1} + T^{\tau}_{m-1}) \tag{4–48}$$

The fact that $T^{\tau+1}_{m}$ is equal to the average of the two adjacent nodal temperatures at the preceding increment of time provides the basis for a graphical approach to solving unsteady one-dimensional conduction-heat-transfer problems. This Schmidt graphical method is described in detail by Jacob [9].

Stability criterion can also be established for systems with $\dot{q} \neq 0$. In this regard, the values of $X$ listed in Tables 4–3 and 4–4 provide a conservative criterion for $\dot{q} > 0$.

As a consequence of the stability requirements, the explicit computational approach developed above generally requires the use of relatively small increments in time. However, the fact that this finite-difference method produces direct calculations of future nodal temperatures sometimes compensates for the restriction in $\Delta t$.

## EXAMPLE 4–10

Demonstrate that Eq. (4–43) violates the second law of thermodynamics for values of $\alpha\,\Delta t/\Delta x^2 > \frac{1}{4}$ and $\dot{q} = 0$.

### Solution

*Objective* Show that the forward-time difference nodal equation,

$$T^{\tau+1}_{m,n} = \frac{\alpha\,\Delta t}{\Delta x^2}(T^{\tau}_{m+1,n} + T^{\tau}_{m-1,n} + T^{\tau}_{m,n+1} + T^{\tau}_{m,n-1}) + \left(1 - 4\frac{\alpha\,\Delta t}{\Delta x^2}\right)T^{\tau}_{m,n}$$

violates the second law of thermodynamics for $\alpha\,\Delta t/\Delta x^2 > \frac{1}{4}$.

*Assumptions/Conditions*

unsteady-state
two spatial dimensions
no internal energy generation

*Analysis* To demonstrate this point, we consider the case in which the sides of a square plate ($L \times L \times \delta$) initially at 110°F are suddenly brought to 100°F at time $t = t_1$. To simplify our test case, we utilize a grid with $\Delta x = L/2$, as shown in Fig. E4–10. At the instant $t = \tau\ \Delta t = 1\ \Delta t$, the nodal temperatures surrounding node $(m,n)$ are given by

$$T^{\tau}_{m-1,n} = T^{\tau}_{m+1,n} = T^{\tau}_{m,n-1} = T^{\tau}_{m,n+1} = 100°\text{F}$$

but

$$T^{\tau}_{m,n} = 110°\text{F}$$

for $\tau = 1$. We now want to know the temperature at node $(m,n)$ at the next increment of time. Therefore, we utilize Eq. (4–43) to obtain

$$T^{\tau+1}_{m,n} = \frac{\alpha\ \Delta t}{\Delta x^2}\,400°\text{F} + \left(1 - 4\frac{\alpha\ \Delta t}{\Delta x^2}\right) 110°\text{F} \tag{a}$$

We select a value of $\Delta t$ such that $\alpha\ \Delta t/\Delta x^2$ is greater than $\frac{1}{4}$, say

$$\frac{\alpha\ \Delta t}{\Delta x^2} = \frac{1}{3}$$

Substituting this value into Eq. (a), we have

$$T^{\tau+1}_{m,n} = \frac{400°\text{F}}{3} + \left(1 - \frac{4}{3}\right) 110°\text{F} = 96.7°\text{F}$$

Because $T^{\tau+1}_{m,n}$ is less than $T^{\tau}_{m,n}$, we conclude that heat has been conducted out of node $(m,n)$ to the surrounding nodes. However, the fact that $T^{\tau+1}_{m,n}$ is actually less than 100°F indicates that heat has been transferred in directions of increasing temperature during part of the $\Delta t$ time increment. This result is a violation of the second law of thermodynamics, which requires that heat cannot be transferred from a low-temperature system to a high-temperature system without the input of work.

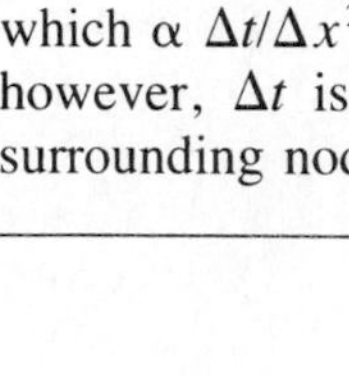


**FIGURE E4–10**
Finite-difference grid for square plate; $\Delta x = L/2$ and $\tau = 1$.

By examining Eq. (a), it can be seen that the use of any time increment for which $\alpha\ \Delta t/\Delta x^2$ is greater than $\frac{1}{4}$ will bring about a violation of the second law. If, however, $\Delta t$ is selected such that $\alpha\ \Delta t/\Delta x^2 \leq \frac{1}{4}$, $T^{\tau+1}_{m,n}$ does not fall below the surrounding nodal temperatures and no violation occurs.

### *Numerical Solutions*

With the initial temperature distribution specified and with $\Delta t$ and $\Delta x$ specified in accordance with necessary stability and accuracy requirements, the explicit interior and exterior nodal equations provide the means by which the temperature distribution can be calculated at the next increment of time, 1 $\Delta t$. The temperature distribution at 1 $\Delta t$ can then be used as an input to calculate the distribution at 2 $\Delta t$. This calculation procedure can be continued to obtain the temperature distribution over the number of time increments desired, or until the steady-state condition is approached. The use of the explicit numerical finite-difference method is illustrated in Example 4–11.

## EXAMPLE 4–11

A plate [$k = 50$ W/(m °C), $\alpha = 2 \times 10^{-5}$ m²/s] of 4 mm thickness is initially at 0°C. One side of the plate is then suddenly brought to a temperature of 100°C, with the other side maintained at 0°C. Develop an accurate explicit finite-difference solution for the rate of heat transfer from the hot surface.

### Solution

*Objective* Use the explicit finite-difference approach to obtain an accurate solution for $q_L$.

*Schematic* Unsteady heat transfer in a plane wall.

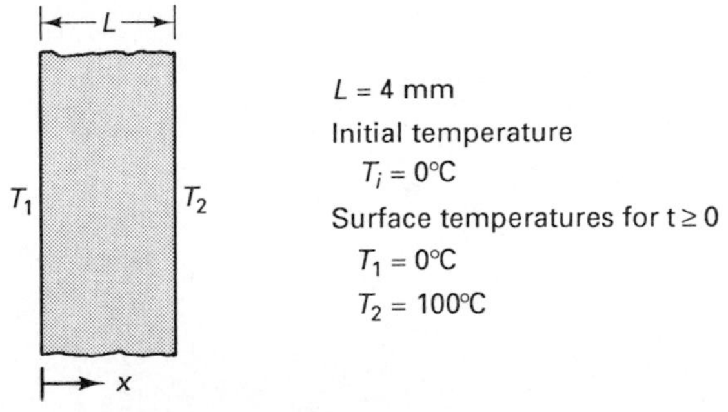
L
T1
T2
x
L = 4 mm
Initial temperature
Ti = 0°C
Surface temperatures for t ≥ 0
T1 = 0°C
T2 = 100°C

*Assumptions/Conditions*

unsteady-state
one spatial dimension
uniform properties

*Properties* $k = 50$ W/(m °C), $\alpha = 2 \times 10^{-5}$ m²/s.

*Analysis* To get a feeling for the explicit solution technique, we first utilize the finite-difference grid with $\Delta x = L/4 = 1$ mm shown in Fig. E4–11a. Notice that this grid produces only three unknown nodal temperatures ($T_2^\tau$, $T_3^\tau$, and $T_4^\tau$). An explicit finite-difference energy balance is developed for one of these three interior nodes ($m$) by writing

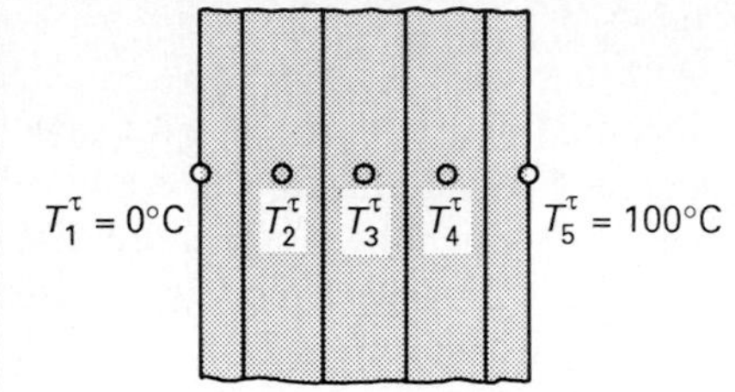
$T_1^\tau$ = 0°C
$T_2^\tau$
$T_3^\tau$
$T_4^\tau$
$T_5^\tau$ = 100°C

**FIGURE E4–11a**
Finite-difference grid for plane wall.

$$q_x = q_{x+\Delta x} + \frac{\Delta E_s}{\Delta t}$$

$$-kA\,\frac{T_m^\tau - T_{m-1}^\tau}{\Delta x} = -kA\,\frac{T_{m+1}^\tau - T_m^\tau}{\Delta x} + \rho A\,\Delta x\,c_v\,\frac{T_m^{\tau+1} - T_m^\tau}{\Delta t} \tag{a}$$

The solution for $T_m^{\tau+1}$ is

$$T_m^{\tau+1} = \frac{\alpha\,\Delta t}{\Delta x^2}(T_{m-1}^\tau + T_{m+1}^\tau) + \left(1 - 2\,\frac{\alpha\,\Delta t}{\Delta x^2}\right)T_m^\tau \tag{b}$$

To maintain stability, the coefficients associated with the $T_m^\tau$ term for each nodal equation must be equal to or greater than zero. Therefore, we require

$$1 - 2\alpha\frac{\Delta t}{\Delta x^2} \geq 0 \qquad \text{or} \qquad \frac{\alpha\,\Delta t}{\Delta x^2} \leq \frac{1}{2}$$

Setting $\alpha\,\Delta t/\Delta x^2 = \frac{1}{2}$, Eq. (b) becomes

$$T_m^{\tau+1} = \frac{1}{2}(T_{m-1}^\tau + T_{m+1}^\tau)$$

and $\Delta t$ is given by

$$\Delta t = \frac{1}{2}\frac{\Delta x^2}{\alpha} = \frac{1}{2}\frac{(10^{-3}\text{ m})^2}{2 \times 10^{-5}\text{ m}^2/\text{s}} = 0.025\text{ s}$$

To summarize, our three nodal equations are

$$T_2^{\tau+1} = \frac{1}{2}(T_1^\tau + T_3^\tau) = \frac{1}{2}(0°\text{C} + T_3^\tau) \tag{c}$$

$$T_3^{\tau+1} = \frac{1}{2}(T_2^\tau + T_4^\tau) \tag{d}$$

$$T_4^{\tau+1} = \frac{1}{2}(T_3^\tau + T_5^\tau) = \frac{1}{2}(T_3^\tau + 100°\text{C}) \tag{e}$$

The solution to these equations after the first increment of time ($t = \Delta t$) is

$$T_2^1 = \frac{1}{2}(0°C + 0°C) = 0°C$$

$$T_3^1 = \frac{1}{2}(0°C + 0°C) = 0°C$$

$$T_4^1 = \frac{1}{2}(0°C + 100°C) = 50°C$$

These results are then substituted back into Eqs. (c) through (e) to obtain predictions for $T_2^2$, $T_3^2$, and $T_4^2$. This procedure is continued as the solution is built up for increasing time steps. The predictions for $T_m^\tau$ are summarized in Table E4–11 for the number of time steps required to reach steady conditions. Note that the final steady-state profile is linear, which is consistent with the simple one-dimensional analysis developed in Chap. 2.

**TABLE E4–11** Numerical calculations for $T_m^\tau$

| $\tau$ | $T_2^\tau$ | $T_3^\tau$ | $T_4^\tau$ |
|---|---|---|---|
| 0 | 0. | 0. | 50. |
| 1 | 0. | 25. | 50. |
| 2 | 12.5 | 25. | 62.5 |
| 3 | 12.5 | 37.5 | 62.5 |
| 4 | 18.8 | 37.5 | 68.8 |
| 5 | 18.8 | 43.8 | 68.8 |
| 6 | 21.9 | 43.8 | 71.9 |
| 7 | 21.9 | 46.9 | 71.9 |
| 8 | 23.4 | 46.9 | 73.4 |
| 9 | 23.4 | 48.4 | 73.4 |
| 10 | 24.2 | 48.4 | 74.2 |
| 11 | 24.2 | 49.2 | 74.2 |
| 12 | 24.6 | 49.2 | 74.6 |
| 13 | 24.6 | 49.6 | 74.6 |
| 14 | 24.8 | 49.6 | 74.8 |
| 15 | 24.8 | 49.8 | 74.8 |
| 16 | 24.9 | 49.8 | 74.9 |
| 17 | 24.9 | 49.9 | 74.9 |
| 18 | 25.0 | 49.9 | 75.0 |
| 19 | 25.0 | 50.0 | 75.0 |
| 20 | 25.0 | 50.0 | 75.0 |

Observing that no change occurs in the nodal temperatures (to three significant figures) for $\tau \geq 19$, the length of time required to reach steady-state conditions is approximated by

$$t_{ss} = \tau_{ss}\,\Delta t = 19(0.025 \text{ s}) = 0.475 \text{ s}$$

It should be mentioned that the calculations are generally stopped when the percent change in all nodal temperatures falls below a specified steady-state convergence criterion, SSC. For instance, with SSC set equal to 0.005, the calculations would be terminated at $\tau = 15$.

To determine the rate of heat transfer from the surface $x = L$ at time $t$, we apply the first law of thermodynamics to node (5), with the result

$$-kA\,\frac{T_5^\tau - T_4^\tau}{\Delta x} = q_L + \frac{\Delta E_s}{\Delta t}$$

or

$$q_L = -\frac{kA}{\Delta x}(T_5^\tau - T_4^\tau) - \rho A\,\frac{\Delta x}{2}\,c_v\left(\frac{T_5^{\tau+1} - T_5^\tau}{\Delta t}\right) \tag{f}$$

where $T_5^{\tau+1} = T_5^\tau = 100°\text{C}$. Thus, to predict heat-transfer flux $q_L''$ at any instant $t$, the calculations for $T_m^\tau$ are substituted into Eq. (f). For example, at $t = 0.275$ s (i.e., $\tau = 11$), we obtain

$$q_L'' = \frac{k}{\Delta x}(74.2°\text{C} - 100°\text{C}) = -1.29 \text{ MW/m}^2$$

As we have seen in Examples 4–6 through 4–8, one way to assure a reasonably accurate numerical finite-difference solution is to compare the solutions for smaller and smaller subvolume sizes. Because of the obvious computational involvement in producing solutions for smaller values of $\Delta x$ and $\Delta t$, a simple BASIC program is developed for calculating the nodal temperatures and heat-transfer rate at the surface. The general nodal equation given by Eq. (b) is written in BASIC computer language as

```
T(J,2) = (T(J + 1,1) + T(J − 1,1))*S + (1 − 2 *S) *T(J,1)
```

where J $= m = 2, 3, 4, \ldots, M-1$, and $S = \alpha\,\Delta t/\Delta x^2$. The second index designates the time stations $\tau + 1$ (index 2) and $\tau$ (index 1). The flowchart and BASIC program are given in Figs. E4–11b and E4–11c. In addition to calculating the instantaneous nodal temperatures for any specific grid space $\Delta x$ and time increment $\Delta t$, this program calculates the heat flux at $x = L$. The instantaneous heat flux $q_L''$ is obtained from Eq. (f),

$$q_L'' = -\frac{k}{\Delta x}(T_5^\tau - T_4^\tau)$$

or

```
QFL = −K/DX*(T(M,2) − T(M − 1,2))
```

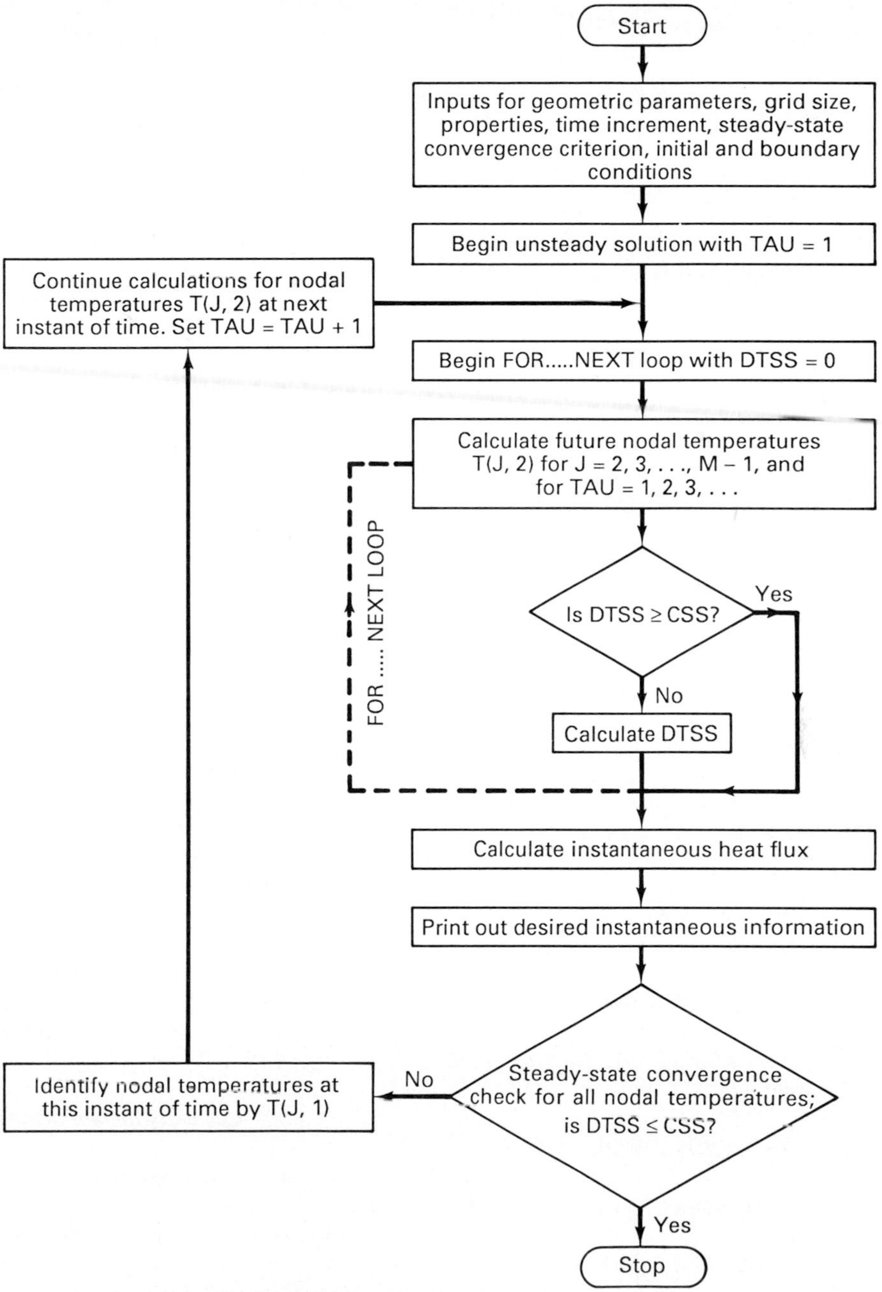


**FIGURE E4–11b** Program flowchart for Example 4–11.

```
                                                    Comments
                              Innputs for geometric parameters
   L=.004
                                           Inputs for grid size
   M=10
   M1=M-1
   DX=L/M1
                                         Inputs for properties
   K=50
   ALPHA=.00002
                                      Input for time increment
   S=.5
   DTIME=S*DX^2/ALPHA
                     Input for steady-state convergence criterion
   CSS=.0001
                                  Inputs for initial conditions
   T1=0
   FOR J=1 TO M STEP 1
                                 Inputs for boundary conditions
   T(J,1)=T1
   NEXT J
   T(1,1)=T1
   T(1,2)=T(1,1)
   TM=100
   T(M,1)=TM
   T(M,2)=T(M,1)
                             Begin unsteady solution with τ=1
   TAU=1
                                  Begin do loop with DTSS=0
2  DTSS=0
                       Calculate future nodal temperatures T(J,2)
   FOR J=2 TO M1
   T(J,2)=(T(J+1,1)+T(J-1,1))*S+(1-2*S)*T(J,1)
                  Check for convergence to steady state for all nodes
   IF DTSS>CSS GOTO 3
                                                Calculate DTSS
   DTSS=ABS ((T(J,2)-T(J,1))/T(J,2))
3  NEXT J
                               Calculate instantaneous heat flux
   QFL=-K/DX*(T(M,2)-T(M1,2))
                     Print out desired instantaneous information
   LPRINT USING "#######.#";M, TAU, QFL
                       Steady-state convergence check for all nodes
   IF DTSS<CSS GOTO 50
                     Identify current nodal temperatures by T(J,1)
   For J=2 TO M1
   T(J,1)=T(J,2)
   NEXT J
```

**FIGURE E4–11c** BASIC program for Example 4–11.

***Calculate future nodal temperatures T(J,2)***

```
      TAU=TAU+1                                                    32
      GOTO 2                                                       33
50    STOP                                                         34
      END                                                          35
```

*Note*: A dimension statement must be used for $M>10$. For example, for $10<M\leq 100$ we may write **DIM** T(100,2).

**FIGURE E4–11c** (*Continued*)

Calculations obtained for $q_L''$ by running this program are shown in Fig. E4–11d (to four significant figures) as a function of $M$ and $t$ with $\alpha\ \Delta t/\Delta x^2 = 0.5$. Note that as $M$ increases, both $\Delta x$ and $\Delta t$ become smaller. Based on this result, we conclude that a reasonably accurate solution is obtained for $M \gtrsim 10$, but considerable error occurs for $M$ much less than 10.

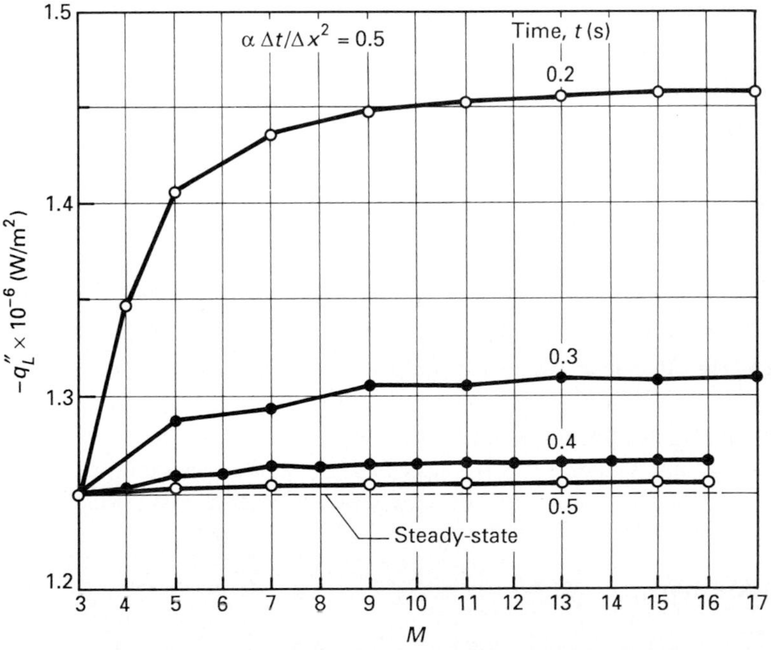


**FIGURE E4–11d** Numerical calculations for $q_L''$; $\alpha\ \Delta t/\Delta x^2 = 0.5$.

To see the effect of $\alpha\ \Delta t/\Delta x^2$ on our numerical solution, predictions are shown in Fig. E4–11e for the heat flux $q_L''$ at surface $x = L$ versus $\tau$ for $\alpha\ \Delta t/\Delta x^2$ equal to 0.4, 0.5, and 0.6, with $M = 10$. For $\alpha\ \Delta t/\Delta x^2 \leq 0.5$, the solution is stable and converges to the proper steady-state value. However, for $\alpha\ \Delta t/\Delta x^2 = 0.6$, the numerical solution is seen to be unstable and divergent. This result underscores the importance of establishing the appropriate stability criteria for explicit finite-difference formulations of unsteady problems.

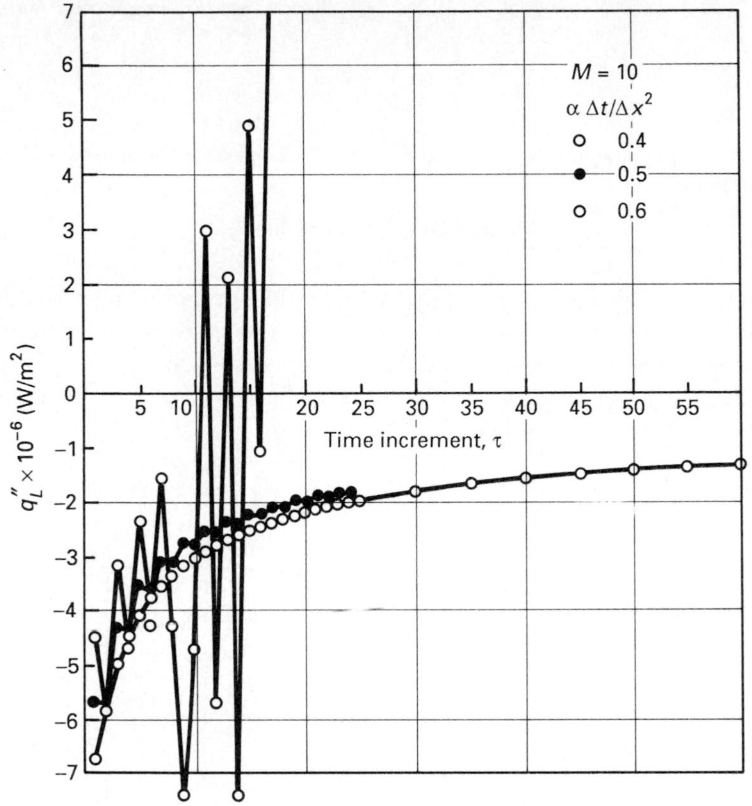


**FIGURE E4–11e** Numerical calculations for $q''_L$; several values of $\alpha\ \Delta t/\Delta x^2$.

This computer program can be modified in order to print out the temperature distribution at any value of $\tau$. This task is suggested as an exercise.

## 4–5–3 Implicit Method

The alternative backward-time difference formulation is commonly used for problems in which the computational time becomes a factor. The substitution of the backward-time difference given by Eq. (4–41) into Eq. (4–38) gives rise to an interior nodal equation of the form

$$T^{\tau}_{m,n} - T^{\tau-1}_{m,n} = \frac{\alpha\,\Delta t}{\Delta x^2}\Big(T^{\tau}_{m+1,n} + T^{\tau}_{m-1,n} + T^{\tau}_{m,n+1} + T^{\tau}_{m,n-1} - 4\,T^{\tau}_{m,n} + \frac{\dot{q}}{k}\Delta x^2\Big) \qquad (4\text{–}49)$$

or

$$T^{\tau+1}_{m,n} - T^{\tau}_{m,n} = \frac{\alpha\,\Delta t}{\Delta x^2}\left(T^{\tau+1}_{m+1,n} + T^{\tau+1}_{m-1,n} + T^{\tau+1}_{m,n+1} + T^{\tau+1}_{m,n-1} - 4\,T^{\tau+1}_{m,n} + \frac{\dot{q}}{k}\,\Delta x^2\right) \tag{4–50}$$

Similar equations can also be written for the exterior nodes. This backward-time-difference-type equation is implicit because the nodal temperature $T^{\tau+1}_{m,n}$ at any instant $(\tau+1)\Delta t$ is given in terms of unknown nodal temperatures at that same instant of time as well as at the preceding instant $\tau\,\Delta t$. Thus, like the situation encountered in our solution of steady-state problems, the entire system of $Z$ nodal equations must be solved simultaneously to develop calculations for $T^{\tau+1}_{m,n}$ However, this method has the advantage of being stable for all values of $\Delta t$. Consequently, larger increments in $t$ can be utilized in this approach, such that less computer time is sometimes required than is necessary in the explicit method. Of course, it should be realized that the use of larger time increments results in larger discretization errors.

**EXAMPLE 4–12**

Develop an implicit finite-difference solution for the unsteady problem of Example 4–11.

**Solution**

*Objective* Use the implicit finite-difference approach to obtain an accurate solution for $q_L$.

*Schematic* Unsteady heat transfer in a plane wall.

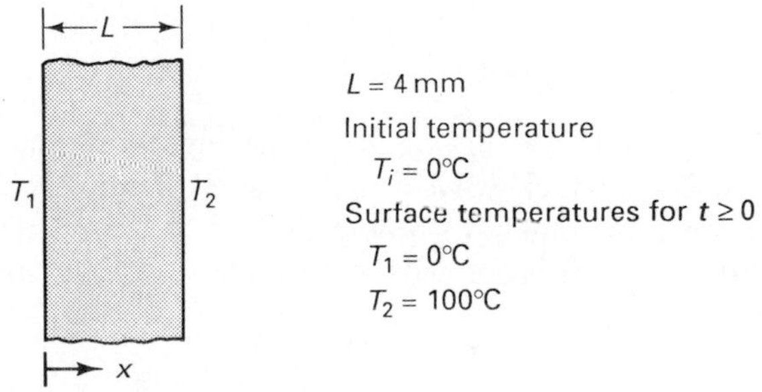
L
T1
T2
x
L = 4 mm
Initial temperature
Ti = 0°C
Surface temperatures for t ≥ 0
T1 = 0°C
T2 = 100°C

*Assumptions/Conditions*

unsteady-state
one spatial dimension
uniform properties

*Properties* $k = 50$ W/(m °C), $\alpha = 2 \times 10^{-5}$ m²/s.

*Analysis* The finite-difference grid for an implicit formulation is identical to the grid for an explicit formulation. Applying the first law of thermodynamics to interior node $(m)$, we write

$$q_x = q_{x+\Delta x} + \frac{\Delta E_s}{\Delta t}$$

$$-kA\frac{T_m^\tau - T_{m-1}^\tau}{\Delta x} = -kA\frac{T_{m+1}^\tau - T_m^\tau}{\Delta x} + \rho\,\Delta V\,c_v\frac{T_m^\tau - T_m^{\tau-1}}{\Delta t}$$

$$T_{m-1}^\tau + T_{m+1}^\tau - \left(2 + \frac{\Delta x^2}{\alpha\,\Delta t}\right)T_m^\tau + \frac{\Delta x^2}{\alpha\,\Delta t}T_m^{\tau-1} = 0$$

or

$$T_{m-1}^{\tau+1} + T_{m+1}^{\tau+1} - \left(2 + \frac{\Delta x^2}{\alpha\,\Delta t}\right)T_m^{\tau+1} + \frac{\Delta x^2}{\alpha\,\Delta t}T_m^{\tau} = 0$$

where $\Delta E_s/\Delta t$ has been approximated by the backward difference, rather than the forward difference used in Example 4–11.

For this problem in which both surface temperatures are specified, $m$ takes values of 2, 3, . . . , $M-1$, such that a total of $M-2$ unknown nodal temperatures occur at each time station $\tau = 1, 2, 3, \ldots$ . However, unlike the explicit formulation of Example 4–11, these equations must be solved simultaneously for each value of $\tau$. Therefore, the direct and indirect methods introduced in Sec. 4–4–2 can be utilized.

Because of its simplicity, the Gauss-Seidel iterative method is utilized. Our general nodal equation is therefore put into the form

$$T_m^{\tau+1} = \frac{1}{2 + \dfrac{\Delta x^2}{\alpha\,\Delta t}}\left(T_{m-1}^{\tau+1} + T_{m+1}^{\tau+1} + \frac{\Delta x^2}{\alpha\,\Delta t}T_m^\tau\right)$$

where $m = 2, 3, \ldots, M-1$, and $\tau = 1, 2, 3, \ldots$ . A finite-difference BASIC program is given in Fig. A–F–2, which solves this equation iteratively for $T_2^{\tau+1}$, $T_3^{\tau+1}$, $T_4^{\tau+1}$, . . . , $T_{M-1}^{\tau+1}$ at each increment of time $\tau = 1, 2, 3, \ldots$ This program is patterned after the BASIC program presented in Example 4–8. The solution for $q''_L$ is given by

$$q''_L = -\frac{k}{\Delta x}(T_M^\tau - T_{M-1}^\tau)$$

Calculations produced by this program for the rate of heat transfer per unit area from the surface at $x = L$ with $\alpha\,\Delta t/\Delta x^2$ equal to 0.5, 1, and 2, and $M = 10$ are compared in Fig. E4–12 with the explicit calculations of Example 4–11. Although much larger increments in $\Delta t$ can be used in the implicit approach, a price is paid in accuracy of the solution for small values of $t$.

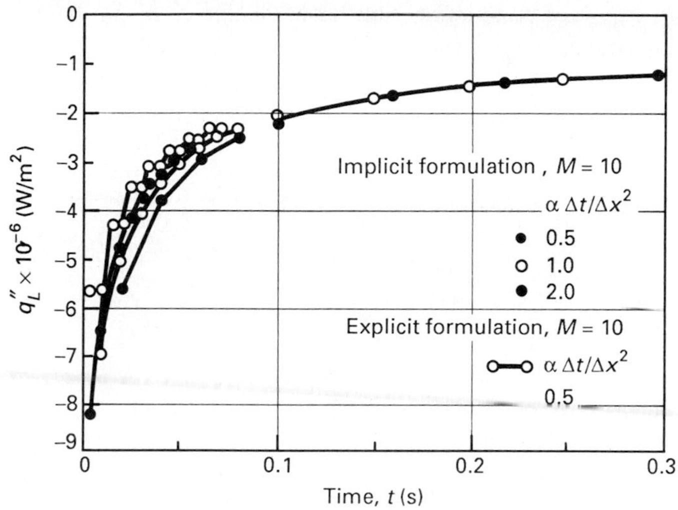


**FIGURE E4–12** Numerical calculations for $q''_L$.

## 4–5–4 R/C Network Formulation

To complete our study of the numerical approach, we want to introduce the popular *R/C* network representation of the finite-difference formulations. To develop this perspective, we first consider the energy balance for the interior node shown in Fig. 4–4; that is,

$$\dot{q}\,\Delta V + \Delta q_x + \Delta q_y - \Delta q_{x+\Delta x} - \Delta q_{y+\Delta y} = \frac{\Delta E_s}{\Delta t} \tag{4–51}$$

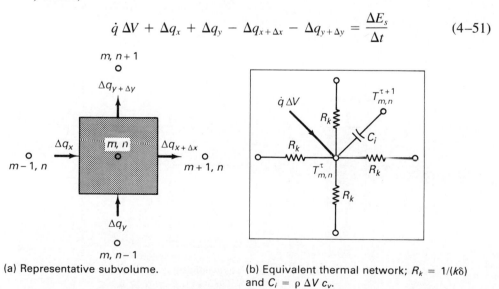

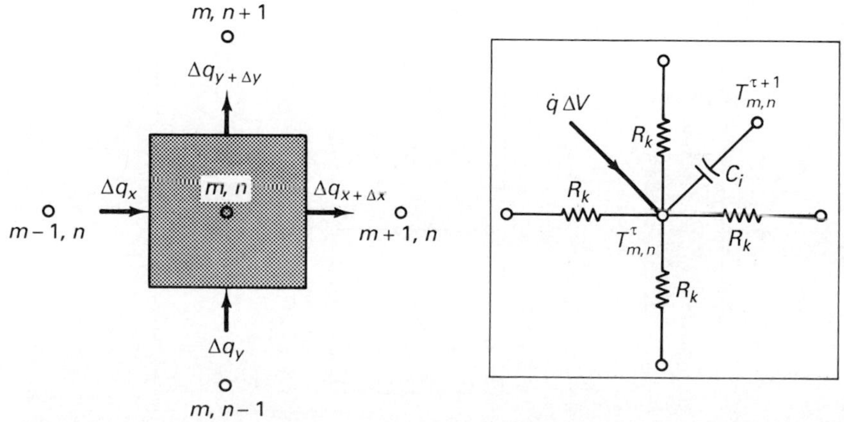


(a) Representative subvolume.

(b) Equivalent thermal network; $R_k = 1/(k\delta)$ and $C_i = \rho\,\Delta V\,c_v$.

**FIGURE 4–4** Interior finite-difference node.

Utilizing the Fourier law of conduction, the heat-transfer rates can be expressed in terms of thermal resistances by writing

$$\Delta q_x = -k\delta\,\Delta y\,\frac{T^{\tau}_{m,n} - T^{\tau}_{m-1,n}}{\Delta x} = \frac{T^{\tau}_{m-1,n} - T^{\tau}_{m,n}}{1/(\delta k)}$$

$$\Delta q_{x+\Delta x} = -k\delta\,\Delta y\,(T^{\tau}_{m+1,n} - T^{\tau}_{m,n}) = \frac{T^{\tau}_{m,n} - T^{\tau}_{m+1,n}}{1/(\delta k)}$$

and

$$\Delta q_y = \frac{T^{\tau}_{m,n-1} - T^{\tau}_{m,n}}{1/(\delta k)} \qquad \Delta q_{y+\Delta y} = \frac{T^{\tau}_{m,n} - T^{\tau}_{m,n+1}}{1/(\delta k)}$$

Utilizing this result and a forward-difference approximation for $\partial T/\partial t$, Eq. (4–51) can be rewritten in the explicit form

$$\dot{q}\,\Delta V + \sum \frac{T^{\tau}_i - T^{\tau}_{m,n}}{R_i} = C_i\,\frac{T^{\tau+1}_{m,n} - T^{\tau}_{m,n}}{\Delta t} \tag{4–53}$$

where $R_i = R_k = 1/(\delta k)$, $C_i = \rho\,\Delta V\,c_v$, and $T^{\tau}_i$ represents the temperatures of the nodes that surround the $(m,n)$ node. Equation (4–53) is represented in terms of an analogous electrical circuit in Fig. 4–4(b).

Similarly, the energy balance for the exterior node shown in Fig. 4–5(a) can be written as

$$\dot{q}\,\Delta V + \Delta q_x + \Delta q_y - \Delta q_s - \Delta q_{y+\Delta y} = \frac{\Delta E_s}{\Delta t} \tag{4–54}$$

where

$$\Delta q_x = \frac{T^{\tau}_{M-1,n} - T^{\tau}_{M,n}}{1/(\delta k)} \qquad \Delta q_y = \frac{T^{\tau}_{M,n-1} - T^{\tau}_{M,n}}{2/(\delta k)}$$

$$\Delta q_{y+\Delta y} = \frac{T^{\tau}_{M,n} - T^{\tau}_{M,n+1}}{2/(\delta k)} \tag{4–55}$$

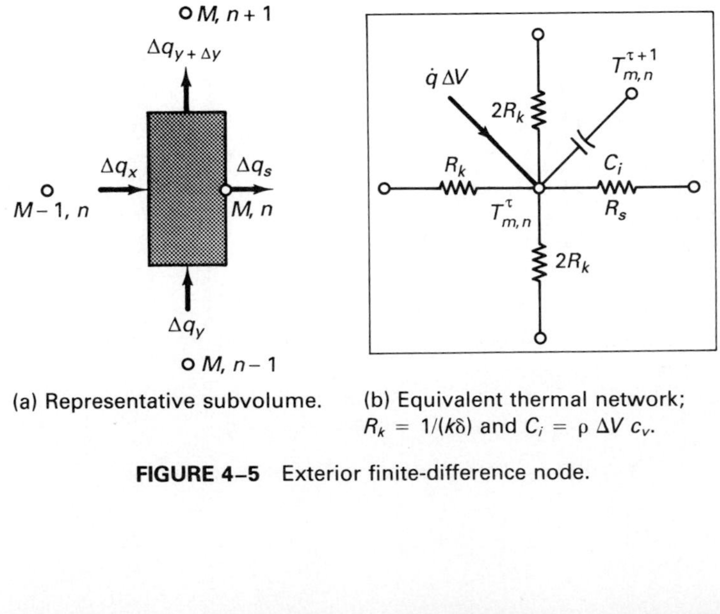


(a) Representative subvolume. (b) Equivalent thermal network; $R_k = 1/(k\delta)$ and $C_i = \rho\,\Delta V\,c_v$.

**FIGURE 4–5** Exterior finite-difference node.

The rate of heat transfer $\Delta q_s$ from the surface by conduction, convection, or thermal radiation, can also be written in terms of a thermal resistance, such that Eq. (4–54) can be put into the form of Eq. (4–53). The thermal circuit for this exterior node is shown in Fig. 4–5(b). Notice that $R_i$ is dependent upon the geometry of the subvolume and the boundary condition.

In summary, we find that Eq. (4–53) applies to both interior and exterior nodes and to both steady and unsteady multidimensional heat-transfer problems. In effect, this equation states that the summation of the currents flowing into the $(m,n)$ node from the surroundings is equal to the flow of current into the capacitor. For steady-state conditions $T_{m,n}^{\tau+1} = T_{m,n}^{\tau}$, and no current flows to the capacitor. For this case, the summation of current flow into the $(m,n)$ node is zero.

It should be mentioned that Eq. (4–53) must satisfy stability requirements which restrict $\Delta t$. For example, for $\dot{q} = 0$, $\Delta t$ must be less than or equal to $C_i/\Sigma(1/R_i)$. If the restrictions imposed on $\Delta t$ become too severe, an implicit formulation can be developed by merely utilizing a backward difference in $\partial T/\partial t$; that is,

$$\dot{q}\,\Delta V + \sum \frac{T_i^{\tau} - T_{m,n}^{\tau}}{R_i} = C_i \frac{T_{m,n}^{\tau} - T_{m,n}^{\tau-1}}{\Delta t} \qquad (4\text{–}56)$$

Equations (4–53) and (4–56) provided the basis for building up electrical networks that are analogous to steady and unsteady multidimensional conduction-heat-transfer problems. To illustrate, a finite-difference grid and an electrical network are shown in Fig. 4–6 for unsteady conduction in a plane wall.

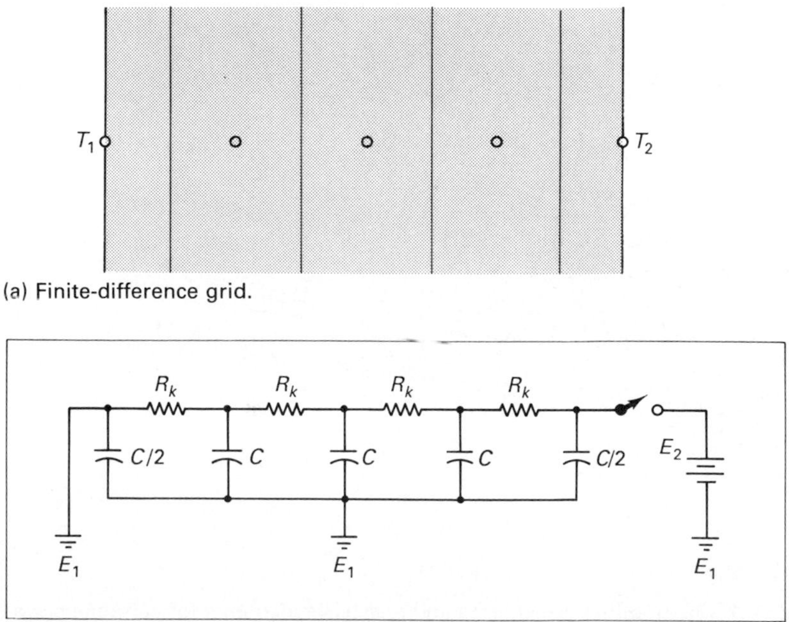


(a) Finite-difference grid.

(b) $R/C$ network; $R_k = \Delta x/(kA)$ and $C = \rho A c_v \Delta x$.

**FIGURE 4–6** Unsteady conduction heat transfer in a plane wall.

The $R/C$ network approach also lends itself to the analysis of three-dimensional rectangular, cylindrical, and spherical systems. The volume and resistance elements for the three standard coordinate systems are represented in Fig. 4–7 and Table 4–5. To generalize the notation, the node is represented by $(l,m,n)$, with the plus and minus signs associated with the resistance subscripts used to designate location of the resistance relative to the node.

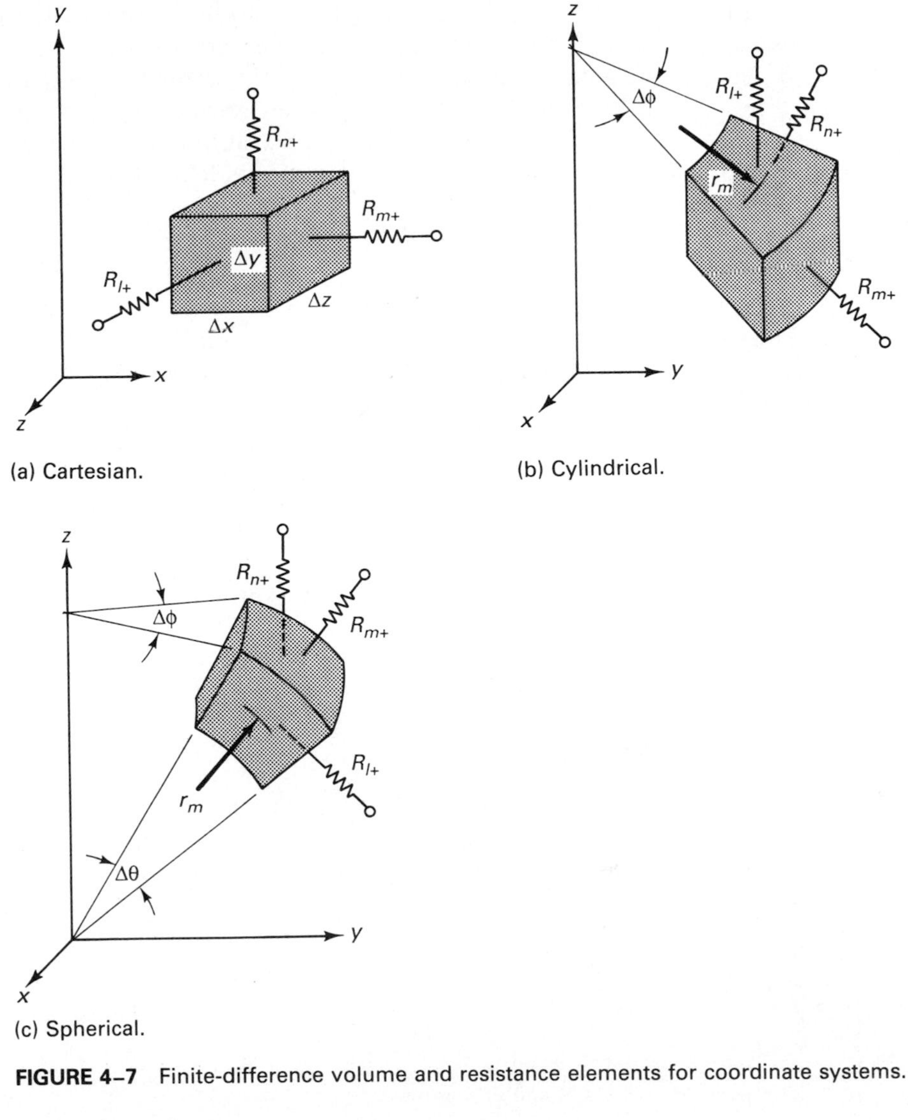


**FIGURE 4–7** Finite-difference volume and resistance elements for coordinate systems.

**TABLE 4–5** Internal element resistances and volumes for Cartesian, cylindrical, and spherical coordinate systems

| | *Cartesian* | *Cylindrical* | *Spherical* |
|---|---|---|---|
| *Volume element* $\Delta V$ | $\Delta x\ \Delta y\ \Delta z$ | $r_m\ \Delta r\ \Delta\phi\ \Delta z$ | $r_m^2 \sin\theta\ \Delta r\ \Delta\phi\ \Delta\theta$ |
| $R_{m+}$ | $\dfrac{\Delta x}{\Delta y\ \Delta z\ k}$ | $\dfrac{\Delta r}{(r_m + \Delta r/2)\ \Delta\phi\ \Delta z\ k}$ | $\dfrac{\Delta r}{(r_m + \Delta r/2)^2 \sin\theta\ \Delta\phi\ \Delta\theta\ k}$ |
| $R_{m-}$ | $\dfrac{\Delta x}{\Delta y\ \Delta z\ k}$ | $\dfrac{\Delta r}{(r_m - \Delta r/2)\ \Delta\phi\ \Delta z\ k}$ | $\dfrac{\Delta r}{(r_m - \Delta r/2)^2 \sin\theta\ \Delta\phi\ \Delta\theta\ k}$ |
| $R_{n+}$ | $\dfrac{\Delta y}{\Delta x\ \Delta z\ k}$ | $\dfrac{r_m\ \Delta\phi}{\Delta r\ \Delta z\ k}$ | $\dfrac{\Delta\phi \sin\theta}{\Delta r\ \Delta\theta\ k}$ |
| $R_{n-}$ | $\dfrac{\Delta y}{\Delta x\ \Delta z\ k}$ | $\dfrac{r_m\ \Delta\phi}{\Delta r\ \Delta z\ k}$ | $\dfrac{\Delta\phi \sin\theta}{\Delta r\ \Delta\theta\ k}$ |
| $R_{l+}$ | $\dfrac{\Delta z}{\Delta x\ \Delta y\ k}$ | $\dfrac{\Delta z}{r_m\ \Delta\phi\ \Delta r\ k}$ | $\dfrac{\Delta\theta}{\sin(\theta + \Delta\theta/2)\ \Delta r\ \Delta\phi\ k}$ |
| $R_{l-}$ | $\dfrac{\Delta z}{\Delta x\ \Delta y\ k}$ | $\dfrac{\Delta z}{r_m\ \Delta\phi\ \Delta r\ k}$ | $\dfrac{\Delta\theta}{\sin(\theta - \Delta\theta/2)\ \Delta r\ \Delta\phi\ k}$ |
| *Nomenclature for increments* | $x,m$<br>$y,n$<br>$z,l$ | $r,m$<br>$\phi,n$<br>$z,l$ | $r,m$<br>$\phi,n$<br>$\theta,l$ |

## 4–6 SUMMARY

In this chapter we have considered the numerical finite-difference approach to the solution of conduction-heat-transfer problems. This approach involves the formulation of systems of nodal equations by use of nodal networks, finite-difference approximations for differential terms, and discretization or energy balance methods. The nodal equations provide the basis for developing solutions for the nodal temperatures and heat fluxes. Since the grid space and time increment (for unsteady processes) are selected in accordance with accuracy requirements, finite-difference formulations generally result in fairly large numbers of nodal equations. In addition, in the case of explicit formulations for unsteady processes, stability criterion must also be satisfied.

Both iterative and direct solution approaches are available, with iterative techniques such as the Gauss-Seidel method generally providing the most effective solution approach. Because of the widespread use of digital computers, which can quickly solve the large numbers of algebraic equations often produced by finite-difference formulations, the numerical finite-difference solution approach is today a primary method of solution of multidimensional heat-transfer problems.

The finite-difference formulation concepts have been developed in this chapter in the context of conduction heat transfer in rectangular systems. Nodal equations are presented in Tables 4–1 and 4–2 for standard interior and exterior nodes associated with such simple geometries. In addition, *R/C* network finite-difference volume and resistance elements have been presented for standard three-dimensional Cartesian, cylindrical, and spherical systems. The formulation of finite-difference nodal equations for rectangular, cylindrical, and spherical geometries proves to be quite straightforward. Although finite-difference nodal equations can also be developed for complex geometries involving irregular boundaries, the finite-element method [1,2] is generally more suitable for such applications. Two general finite-element formulation approaches include (1) discretization methods, which involve the use of variational calculus or the method of weighted residuals, and (2) energy balance methods (also referred to as control-volume finite-element methods CVFEM). In this connection, a very attractive CVFEM, which is similar in nature to the energy balance finite-difference method, has recently been presented by Baliga and Patankar [10] for steady two-dimensional conduction and related problems.

The numerical concepts introduced in this chapter are also widely used in the analysis of radiation and convection heat transfer. In connection with the solution of the fluid flow and energy-transfer equations associated with convection systems, it should be noted that stability becomes an important issue when dealing with finite-difference approximations for terms such as $u\ \partial u/\partial x$ and $u\ \partial T/\partial x$, which appear in the fundamental conservation equations [4].

## ■ REVIEW QUESTIONS

**4–1.** State the primary advantage of the numerical finite-difference method relative to analytical methods for solving multidimensional conduction-heat-transfer problems.

**4–2.** In connection with Ques. 4–1, describe several disadvantages of the numerical finite-difference method relative to analytical methods.

**4–3.** Write finite-difference approximations for the *general* Fourier law of conduction using Cartesian coordinates.

**4–4.** Write second-order finite-difference approximations for the Fourier law of conduction for unsteady two-dimensional $(x,y,t)$ systems.

**4–5.** What is the significance of the Taylor theorem in the development of finite-difference approximations?

**4–6.** Explain how to set up a finite-difference grid for a two-dimensional rectangular solid with specified surface temperature.

**4–7.** How is the finite difference approach used to determine the heat flux at the surface of a rectangular solid with specified surface temperature?

**4–8.** How is the finite-difference approach used to determine the surface temperature of a rectangular solid with specified surface heat flux?

**4–9.** Write forward- and backward-time difference approximations for the time derivative $\partial T/\partial t$.

**4–10.** What is the difference between explicit and implicit finite-difference formulations for unsteady heat-transfer problems?

**4–11.** How can one be assured of obtaining a reasonably accurate numerical solution?

**4–12.** How does the issue of stability affect the development of numerical finite-difference solutions for conduction-heat-transfer problems?

**4–13.** What is the Gauss-Seidel method?

**4–14.** What is the Gaussian elimination method?

**4–15.** Describe the R/C network formulation method.

## PROBLEMS

**4–1.** Referring to the steady two-dimensional system shown in Fig. P4–1, once the nodal temperatures are known the surface heat flux $q''_s$ at node (0) can be calculated by use of the *discretization method* or the *energy balance method*. (a) Use the energy balance method to express $q''_s$ in terms of the nodal temperatures. (b) Use the discretization method to obtain an alternative first-order approximation for $q''_s$ in terms of the nodal temperatures.

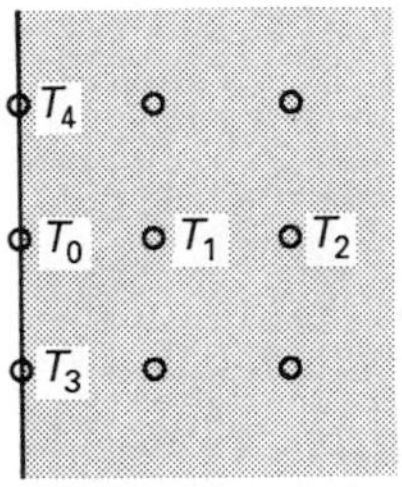


FIGURE P4–1

**4–2.** Referring to Prob. 4–1, a more accurate discretization approximation can be obtained by utilizing the following three-point formula:

$$\left.\frac{\partial T}{\partial x}\right|_0 = \frac{-3T_0 + 4T_1 - T_2}{2\,\Delta x}$$

(a) Use this formula to obtain a finite-difference approximation for $q''_0$.
(b) Use the Taylor theorem to show that the error in this approximation is of the order of $\Delta x^2$.

**4–3.** Set up the finite-difference grid for the radiating fin with rectangular cross section shown in Fig. P4–3 for the conditions listed below. Identify the unknown nodal temperatures and write their nodal equations. Also write an expression for the rate of heat transfer at the base.

(a) $Bi < 0.1$, $\Delta x = L$ tip insulated
(b) $Bi < 0.1$, $\Delta x = L/2$ radiation from tip
(c) $Bi < 0.1$, $\Delta x = L/3$ radiation from tip
(d) $Bi > 0.1$, $\Delta x = L$ tip insulated
(e) $Bi > 0.1$, $\Delta x = L/2$ radiation from tip

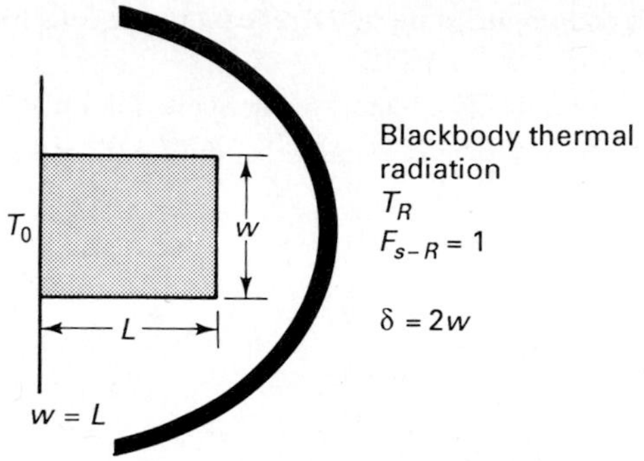
Blackbody thermal
radiation
$T_R$
$F_{s-R} = 1$
$\delta = 2w$
$T_0$
$w$
$L$
$w = L$

FIGURE P4–3

**4–4.** Set up the finite-difference grid for the convecting fin shown in Fig. P4–4 for the conditions listed below. Identify the unknown nodal temperatures and write their nodal equations. Write an expression for the rate of heat transfer at the base.

| | | |
|---|---|---|
| (a) $Bi < 0.1$, | $\Delta x = L$ | tip insulated |
| (b) $Bi < 0.1$, | $\Delta x = L$ | tip not insulated |
| (c) $Bi < 0.1$, | $\Delta x = L/2$ | tip insulated |
| (d) $Bi < 0.1$, | $\Delta x = L/3$ | tip insulated |
| (e) $Bi < 0.1$, | $\Delta x = L/4$ | perimeter insulated with convection from tip |
| (f) $Bi > 0.1$, | $\Delta x = L/2$ | tip insulated |
| (g) $Bi > 0.1$, | $\Delta x = L/4$ | tip not insulated |

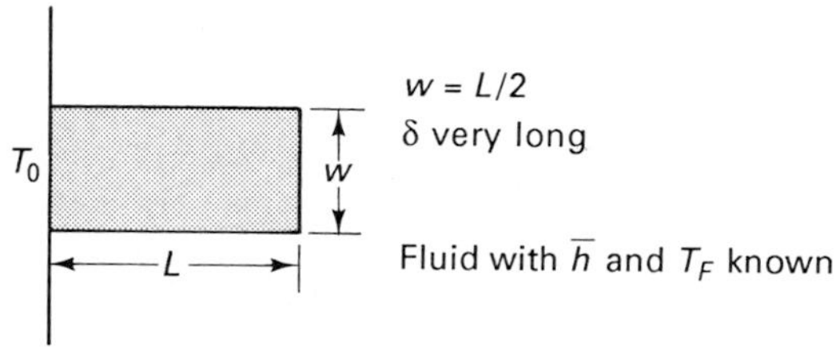
$w = L/2$
$\delta$ very long
$T_0$
$w$
$L$
Fluid with $\bar{h}$ and $T_F$ known

FIGURE P4–4

**4–5.** Develop the nodal equation given by Eq. (2iii) in Table 4–1.

**4–6.** To complement the nodal equations presented in Table 4–1, the nodal equation associated with the interior node near a curved boundary shown in Fig. P4–6 is represented by

$$\frac{2}{a+1}T_{m+1,n} + \frac{2}{b+1}T_{m,n-1} + \frac{2}{a(a+1)}T_1 + \frac{2}{b(b+1)}T_2 - \left(\frac{2}{a} + \frac{2}{b}\right)T_{m,n} + \frac{\dot{q}}{k}\Delta x^2 = 0$$

Use the energy balance method to develop this equation.

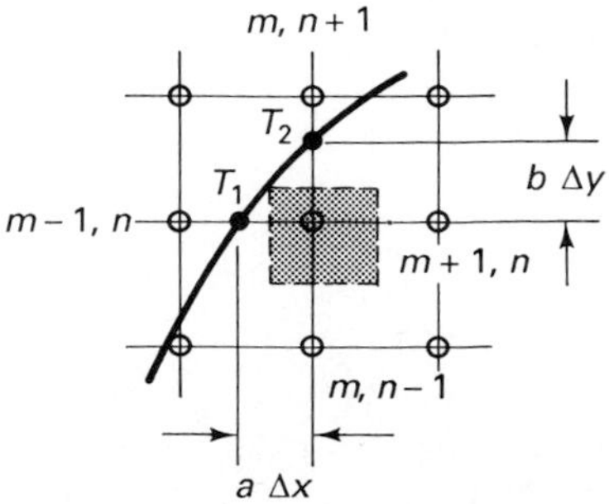
$m, n+1$
$T_2$
$T_1$
$b\,\Delta y$
$m-1, n$
$m+1, n$
$m, n-1$
$a\,\Delta x$

FIGURE P4–6

**4–7.** Develop a numerical finite-difference formulation with $\Delta x = L/2$ for the square plate shown in Fig. P4–7. Then solve for the unknown nodal temperature and develop predictions for the rates of heat transfer from each surface. (Note that the rate of heat transfer through the surfaces of each corner node is indeterminant.) (More accurate numerical solutions for this system are the subject of Examples 4–5 and 4–8.)

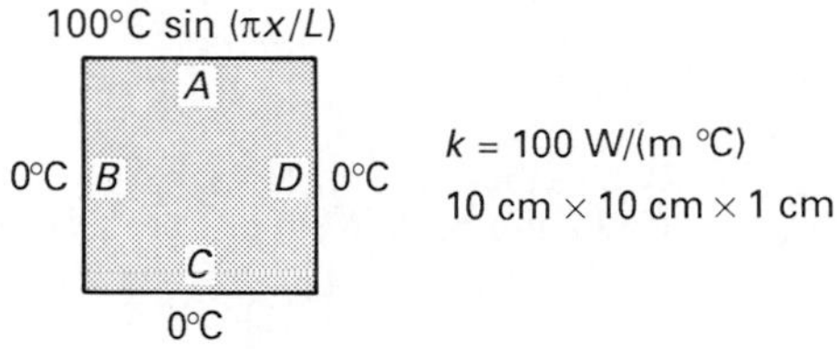
100°C sin (πx/L)
A
0°C B
D 0°C
C
0°C
k = 100 W/(m °C)
10 cm × 10 cm × 1 cm

**FIGURE P4–7**

**4–8.** Develop a numerical finite-difference formulation with $\Delta x = L/2$ for the square plate shown in Fig. P4–8. Then solve for the unknown nodal temperatures and develop predictions for the rates of heat transfer from surfaces $A$ and $B$. (Note that the heat transfer in corner nodes involving convection is determinant.)

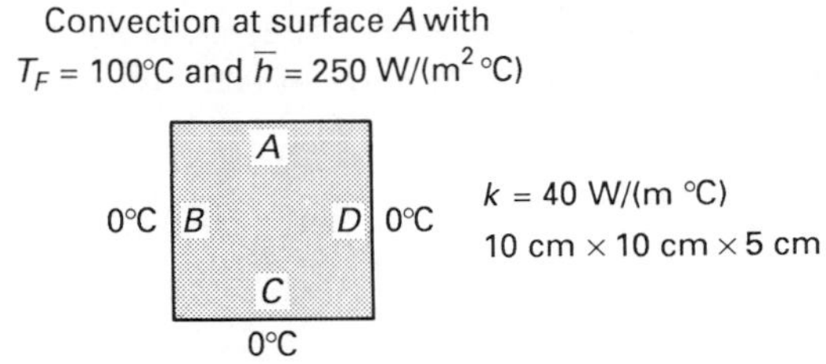
Convection at surface A with
T_F = 100°C and h̄ = 250 W/(m² °C)
A
0°C B
D 0°C
C
0°C
k = 40 W/(m °C)
10 cm × 10 cm × 5 cm

**FIGURE P4–8**

**4–9.** Develop a numerical finite-difference formulation with $\Delta x = L = 1$ cm for the rectangular plate shown in Fig. P4–9. Then solve for the unknown nodal temperature and develop predictions for the rates of heat transfer from each surface. (Note that the heat transfer through the surfaces of corner nodes involving radiation is determinant. The differential formulation for this problem is given in Example 3–1.)

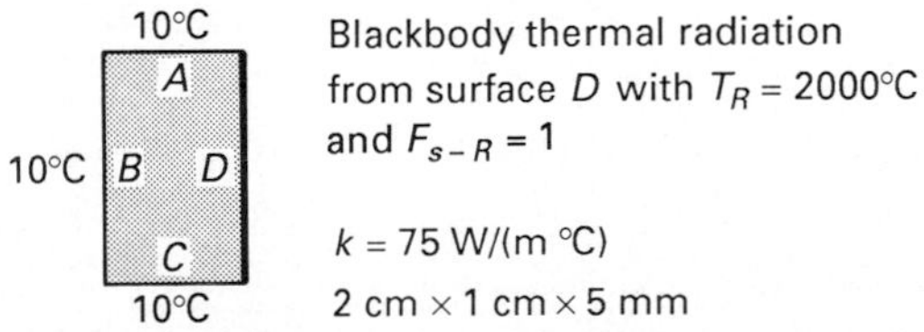
10°C
A
10°C B
D
C
10°C
Blackbody thermal radiation
from surface D with T_R = 2000°C
and F_s−R = 1
k = 75 W/(m °C)
2 cm × 1 cm × 5 mm

**FIGURE P4–9**

**4–10.** Develop a numerical finite-difference formulation with $\Delta x = L/2$ for the square plate shown in Fig. P4–10. Then solve for the unknown nodal temperatures and develop calculations for the rates of heat transfer from each surface.

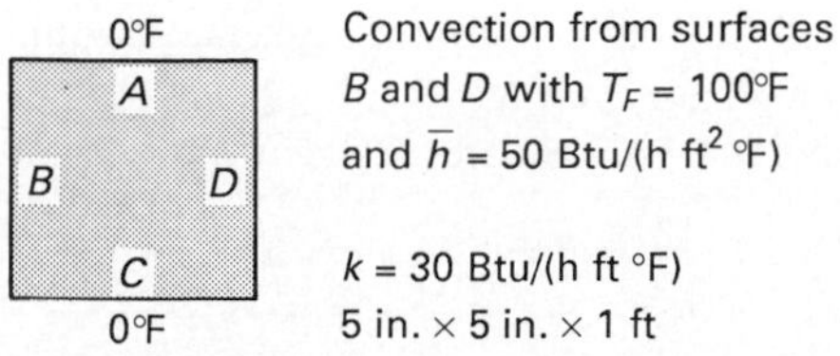
0°F
A
B
D
C
0°F
Convection from surfaces
B and D with T_F = 100°F
and h̄ = 50 Btu/(h ft² °F)
k = 30 Btu/(h ft °F)
5 in. × 5 in. × 1 ft

**FIGURE P4–10**

**4–11.** Develop a numerical finite-difference formulation with $\Delta x = 1$ cm for the fin unit shown in Fig. P4–11. Also develop calculations for the rate of heat transfer from surface $A$.

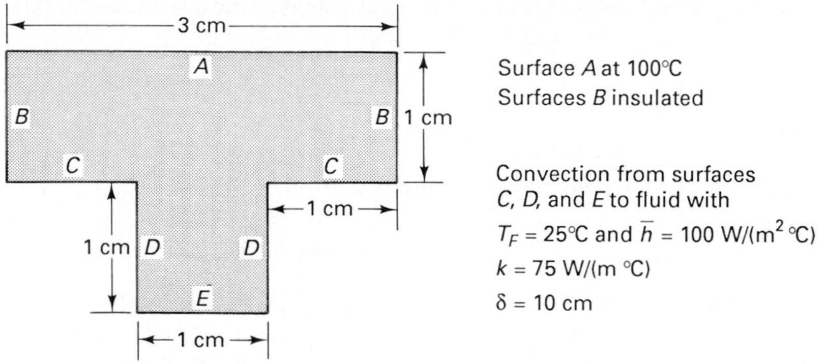
3 cm
A
B
B
1 cm
C
C
1 cm
1 cm
D
D
E
1 cm
Surface A at 100°C
Surfaces B insulated
Convection from surfaces
C, D, and E to fluid with
$T_F$ = 25°C and $\bar{h}$ = 100 W/(m$^2$ °C)
k = 75 W/(m °C)
δ = 10 cm

**FIGURE P4–11**

**4–12.** (a) Develop a numerical finite-difference formulation with $\Delta x = L/2$ for the square plate shown in Fig. P4–12. Then solve for the unknown nodal temperatures and develop calculations for the rates of heat transfer from each surface. (b) Solve this problem with $\Delta x = L$.

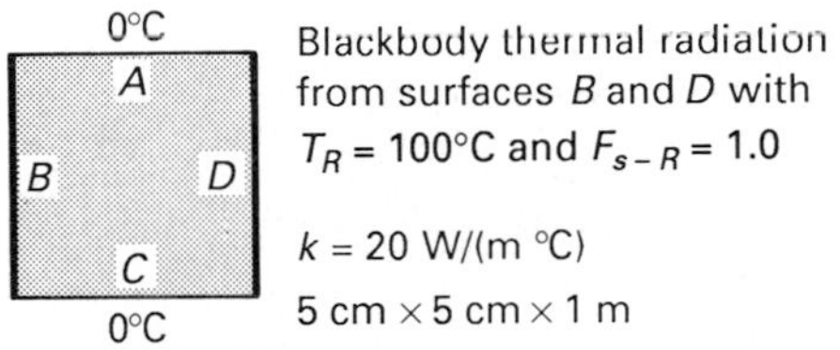
0°C
A
B
D
C
0°C
Blackbody thermal radiation
from surfaces B and D with
$T_R$ = 100°C and $F_{s-R}$ = 1.0
k = 20 W/(m °C)
5 cm × 5 cm × 1 m

**FIGURE P4–12**

**4–13.** Develop a numerical finite-difference formulation with $\Delta x = L/2$ for the square plate shown in Fig. P4–13. Then solve for the unknown nodal temperatures and develop calculations for the rates of heat transfer from each surface.

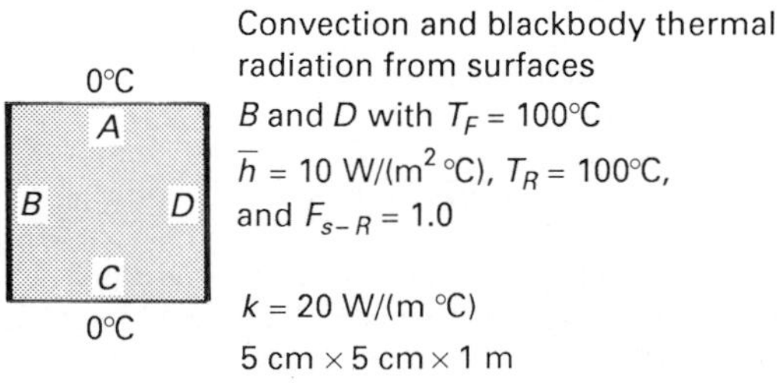
Convection and blackbody thermal
radiation from surfaces
B and D with $T_F$ = 100°C
$\bar{h}$ = 10 W/(m$^2$ °C), $T_R$ = 100°C,
and $F_{s-R}$ = 1.0
0°C
A
B
D
C
0°C
k = 20 W/(m °C)
5 cm × 5 cm × 1 m

**FIGURE P4–13**

**4–14.** Solve Prob. 4–8 for the case in which 0.5 MW/m$^3$ is uniformly generated within the body.

**4–15.** Develop a numerical finite-difference formulation with $\Delta x = L$ for the square plate shown in Fig. P4–15. Then solve for the unknown nodal temperatures and develop calculations for the rates of heat transfer from each surface.

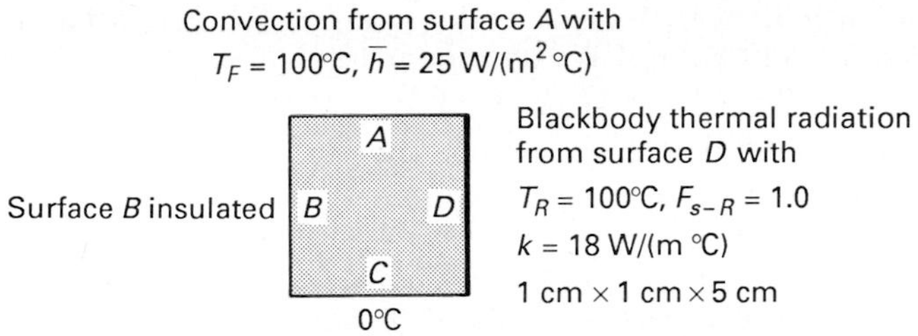
Convection from surface A with
$T_F$ = 100°C, $\bar{h}$ = 25 W/(m$^2$ °C)
A
Blackbody thermal radiation
from surface D with
Surface B insulated
B
D
$T_R$ = 100°C, $F_{s-R}$ = 1.0
k = 18 W/(m °C)
C
1 cm × 1 cm × 5 cm
0°C

**FIGURE P4–15**

**4–16.** Develop a numerical finite-difference formulation with $\Delta x = 10$ cm for the composite plate shown in Fig. P4–16. Then solve for the unknown nodal temperatures and develop calculations for the rates of heat transfer from the two surfaces.

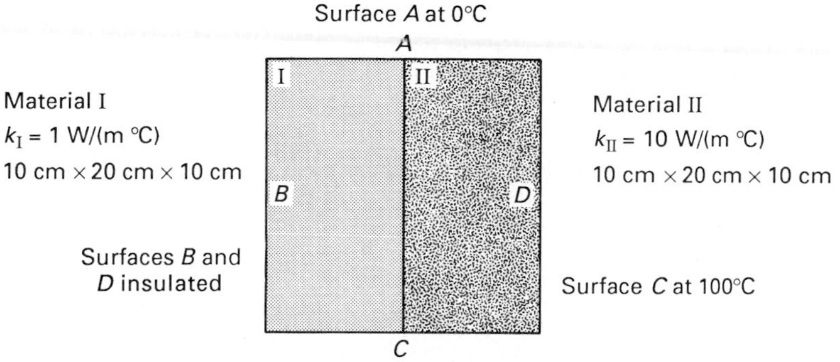
Surface A at 0°C
A
I
II
Material I
$k_I$ = 1 W/(m °C)
10 cm × 20 cm × 10 cm
Material II
$k_{II}$ = 10 W/(m °C)
10 cm × 20 cm × 10 cm
B
D
Surfaces B and
D insulated
Surface C at 100°C
C

**FIGURE P4–16**

**4–17.** Develop a numerical finite-difference formulation with $\Delta x = 10$ cm for the composite plate shown in Fig. P4–17. Then solve for the unknown nodal temperatures and develop calculations for the rates of heat transfer from the two surfaces.

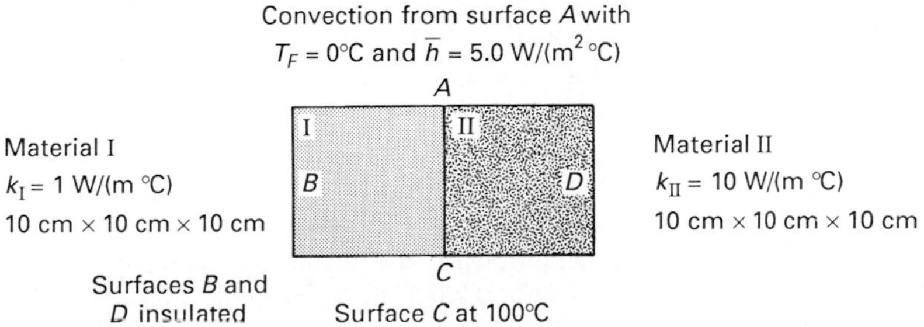
Convection from surface A with
$T_F$ = 0°C and $\bar{h}$ = 5.0 W/(m$^2$ °C)
A
I
II
Material I
$k_I$ = 1 W/(m °C)
10 cm × 10 cm × 10 cm
Material II
$k_{II}$ = 10 W/(m °C)
10 cm × 10 cm × 10 cm
B
D
C
Surfaces B and
D insulated
Surface C at 100°C

**FIGURE P4–17**

**4–18.** A 1-cm-diameter, 3-cm-long steel fin [$k = 43$ W/(m °C)] transfers heat from the wall of a heat exchanger at 200°C to a fluid at 25°C with $\bar{h} = 120$ W/(m$^2$ °C). Develop a numerical finite-difference formulation with $\Delta x = L/2$. Then solve for the unknown nodal temperatures and develop a prediction for the rate of heat transfer from the fin. (The development of more accurate numerical solutions for this system is the subject of Examples 4–4 and 4–6.)

**4–19.** Develop a finite-difference formulation for heat transfer in a quarter cylindrical section with surfaces at $\phi = 0$ and $\phi = \pi/2$ maintained at $T_A$ and $T_B$, surfaces at $r_i$ and $r_o$ insulated, and $\Delta r = (r_o - r_i)/2$ and $\Delta\phi = \pi/6$.

**4–20.** Develop a numerical finite-difference formulation for steady one-dimensional heat transfer in a 4-mm-diameter wire [$k = 19$ W/(m °C)] with uniform heat generation of 500 MW/m$^3$ and surface temperature of 200°C. Use a grid spacing of $\Delta r = 0.5$ mm.

**4–21.** Show that the solution to the nodal equations given in Prob. 4–20 for heat transfer in a circular cylinder is given by $T_1 = 229$°C, $T_2 = 226$°C, $T_3 = 221$°C, and $T_4 = 212$°C. Also show that these values are within about 1% of the exact analytical solution.

**4–22.** A 1-by-2-cm ceramic strip [$k = 3.0$ W/(m °C), $\rho = 1600$ kg/m$^3$, and $c_v = 0.8$ kJ/(kg °C)] is embedded in a high-thermal-conductivity material, as shown in Fig. P4–22, so that the sides are maintained at a constant temperature of 900°C. The bottom surface of the ceramic is insulated, and the top surface is exposed to a convection and radiation environment at $T_F = 50$°C with $h = 50$ W/(m$^2$ °C), and the radiation heat loss is calculated from (see Chap. 5) $q = \sigma A_s \epsilon (T_s^4 - T_R^4)$, where $\epsilon = 0.7$ and $T_R = 50$°C. Solve for the steady-state temperature distribution of the nodes shown and the rate of heat loss. (See Holman [11], Example 3–8.)

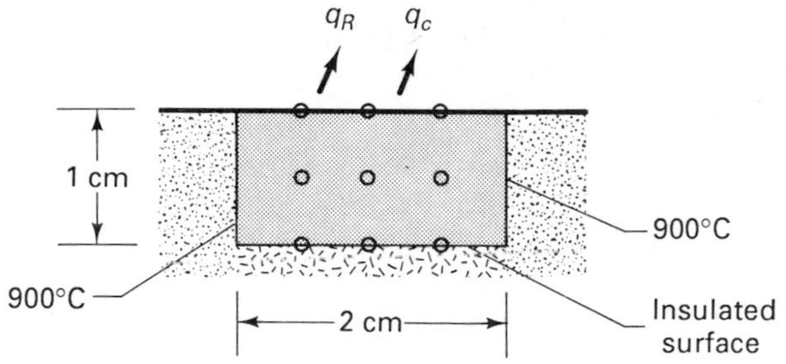

**FIGURE P4–22**

**4–23.** Modify the BASIC program presented in Example 4–8 to determine the temperature distribution and rate of heat transfer at face $A$ for the case in which face $A$ is exposed to blackbody thermal radiation with a surface $R$ at 1000°C with $F_{A-R} = 0.5$.

**4–24.** Utilize the BASIC program presented in Example 4–8 to obtain the temperature distribution and the rate of heat transfer at face $A$ for the case in which the boundary temperatures are given as follows: Face $A$, convection with $T_F = 100$°C, $h = 50$ W/(m$^2$ °C); face $B$, 0°C; face $C$, insulated; and face $D$, 0°C.

**4–25.** Develop an accurate numerical solution to Prob. 4–17.

**4–26.** A 5-in.-long 1-in.-diameter blackbody fin [$k = 100$ Btu/(h ft °F)] extends from a wall at 1000°F. Develop a numerical finite-difference formulation and computer program for the temperature distribution and rate of heat transfer at the base for the case in which the surroundings are at $T_R = 250$°F with $F_{s-R} = 0.4$, radiation occurs at tip, and convection is negligible.

**4–27.** A large industrial furnace is supported on a long column of fireclay brick, which is 1-m-by-1-m on a side. During steady-state operation, installation is such that three surfaces of the column are maintained at 500 K while the remaining surface is exposed to an airstream for which $T_F = 300$ K and $h = 10$ W/(m$^2$ °C). Develop a numerical solution for the two-dimensional temperature distribution in the column and the heat rate to the airstream per unit length of column. Use a grid with $\Delta x = \Delta y = 0.25$ m. (Incropera and DeWitt [12]; Example 4–3—matrix inversion method; Example 4–4—Gauss-Seidel method.)

**4–28.** Referring to the unsteady one-dimensional system shown in Fig. P4–28, write finite-difference approximations for $q''_0$ using (a) the energy balance method, (b) a first-order discretization formula, and (c) a second-order discretization formula.

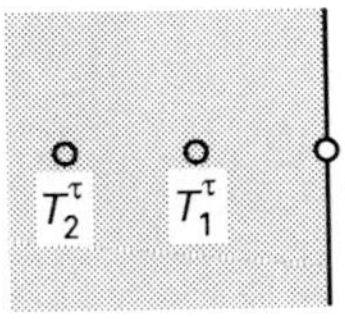


**FIGURE P4–28**

**4–29.** With $\Delta x^2/(\alpha\,\Delta t)$ set equal to 6 in an explicit finite-difference formulation for unsteady three-dimensional conduction heat transfer with $\dot{q} = 0$, demonstrate that the temperature $T_{m,n,l}^{\tau+1}$ (where $z = l\,\Delta z$) is equal to the arithmetic average of the 6 surrounding nodal temperatures at the instant $t = \tau\,\Delta t$.

**4–30.** Develop the nodal equation given by Eq. (2ii) in Table 4–2.

**4–31.** Develop the stability criterion given in Table 4–3 for convection at an inner corner node.

**4–32.** Develop the stability criterion given in Table 4–4 for an exterior node with blackbody thermal radiation.

**4–33.** Develop a numerical finite-difference formulation with $\Delta y = w$ for the plate of Example 3–2, where $T_i = 0°\text{C}$, $T_F = 100°\text{C}$, $h = 50\ \text{W/(m}^2\ °\text{C)}$, $k = 35\ \text{W/(m °C)}$, $\alpha = 2 \times 10^{-5}\ \text{m}^2\text{/s}$, and $w = 1$ cm. Also calculate the instantaneous heat flux at the surface at $t = 1$, 2, and 3 s.

**4–34.** Repeat Prob. 4–33 for $\Delta y = w/2$.

**4–35.** The fin of Prob. 4–18 is initially at 25°C. The base is then suddenly brought to 200°C. Develop a numerical finite-difference formulation with $\Delta x = L/2$. Then solve for the unknown nodal temperature and develop predictions for the instantaneous rate of heat transfer from the fin.

**4–36.** The fin of Example 2–12 is initially at 25°C. The base is then suddenly brought to a temperature of 200°C. Develop a numerical finite-difference formulation with $\Delta x = L/2$. Then solve for the unknown nodal temperatures and develop predictions for the instantaneous rate of heat transfer from the fin.

**4–37.** The square plate of Prob. 4–8 is initially at 0°C. Surface $A$ is then suddenly exposed to convection with $T_F = 100°\text{C}$ and $h = 25\ \text{W/(m}^2\ °\text{C)}$. Develop a numerical finite-difference formulation with $\Delta x = L/2$ and $\alpha = 2 \times 10^{-5}\ \text{m}^2\text{/s}$. Then solve for the unknown nodal temperatures and the instantaneous rate of heat transfer at surface $A$.

**4–38.** Develop a finite-difference formulation for a spherical ball that is suddenly exposed to convection.

**4–39.** Develop a numerical finite-difference formulation with $\Delta r = r_0/3$ for the spherical ball of Example 3–5.

**4–40.** A long rectangular structural member whose cross section is shown in Fig. P4–40 is to be heated during a manufacturing process. Because of space limitations, a heater can only be mounted on one surface, with the other surfaces being exposed to air at 30°C with $h = 75\ \text{W/(m}^2\ °\text{C)}$. The structural member has a thermal conductivity of 25 W/(m °C) and a thermal diffusivity of $2.5 \times 10^{-5}\ \text{m}^2\text{/s}$. It is originally at a uniform temperature of 50°C, with the heated surface being maintained at 300°C by the heater. Determine the length of time that the heater must remain in place for the minimum temperature of the opposite face to reach 120°C. (Kreith and Black [13]; Example 3–10.)

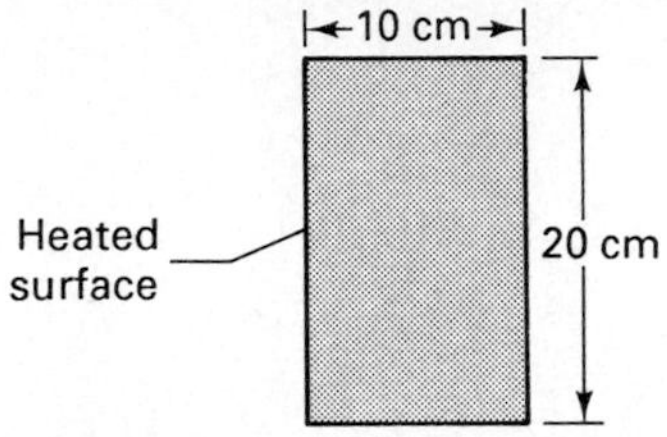


**FIGURE P4–40**

**4–41.** Modify the BASIC program of Example 4–11 to produce calculations for the temperature at 0.2 s, 0.3 s, 0.4 s, and 0.5 s for increments of 0.5 mm.

**4–42.** Develop a computer solution for the temperature distribution and instantaneous rate of heat transfer in the plate of Example 4–11 for the case in which one side is exposed to a fluid with $T_F = 100°C$ and $h = 150$ W/(m$^2$ °C) and the other side is kept at 0°C.

**4–43.** Develop an accurate numerical finite-difference solution for Prob. 4–33. Also determine the length of time required to reach steady state.

# CHAPTER 5

# RADIATION HEAT TRANSFER

## 5–1 INTRODUCTION

*Radiation heat transfer* is defined as the transfer of energy across a system boundary by means of an electromagnetic mechanism which is caused solely by a temperature difference. Some of the more basic and practical aspects of thermal radiation were introduced in Chap. 1. We now want to take a closer look at the physical mechanism, properties, and geometric factors associated with thermal radiation. In addition, modeling concepts are developed and design information is presented in this chapter which provide the basis for practical thermal analysis of radiation heat transfer. Practical thermal analyses will be developed for blackbody and diffuse nonblackbody surfaces, with special consideration given to unsteady systems, participating medium, combined mode processes, and solar radiation.

## 5–2 PHYSICAL MECHANISM

Whenever a charged particle undergoes acceleration, energy possessed by the particle is converted into a form of energy known as *electromagnetic radiation*. Electromagnetic radiation includes cosmic rays, gamma rays, X rays, ultraviolet radiation, visible light, infrared radiation, microwaves, broadcasting waves, and ultrasonic electrical waves. Electromagnetic radiation can be produced by various means, depending on the type of charged particles that are involved in the process. To illustrate, $\gamma$ rays are produced by fission of nuclei or by radioactive disintegration, X rays by the bombardment of metals with high-energy electrons, microwaves by special types of electron tubes (klystrons, magnetrons, or traveling wave-tubes), and radio waves by the excitation of certain crystals or by the flow of alternating current through electric conductors. It should also be noted that astronomical sources provide significant quantities of electromagnetic radiation that range from extremely short-wavelength cosmic rays to long-wavelength radio waves. Consequently, the reception and evaluation of X-ray, $\gamma$-ray, ultraviolet, visible, infrared, and radio-wave radiation are critical elements of modern astronomy.

Of particular interest to us is electromagnetic radiation that is produced by vibrational and rotational movements of atoms and molecules of a substance and/or by changes in the atomic energy levels of the least strongly bound electrons. Because the level of energy associated with the fluctuating motion of these small oscillators is indicated by temperature, the resulting electromagnetic radiation is referred to as *thermal radiation*. Thermal radiation heat transfer represents the exchange of thermal radiation between bodies at different temperature, with each body (1) converting internal energy into outflowing electromagnetic waves, and (2) absorbing incoming electromagnetic waves, which are converted into internal energy.

### 5-2-1 The Electromagnetic Spectrum

All the various types of electromagnetic waves are characterized by a frequency $\nu$ and a propagation velocity in free space (a vacuum or transparent medium) equal to the speed of light $c$. The speed of light $c$ in a gas, liquid, or solid is related to the speed of light in a vacuum $c_0$ ($= 3 \times 10^8$ m/s) by the *index of refraction* $n = c_0/c$. (The index of refraction of air and most gases is essentially unity, but for liquids and solids such as water and glass it is of the order of 1.5.) The wavelength $\lambda$ is defined in terms of $\nu$ and $c$ by

$$\lambda = \frac{c}{\nu} \tag{5-1}$$

The wavelength $\lambda$ is generally expressed in terms of the *micrometer* $\mu m = 10^{-6}$ m, *nanometer* nm $= 10^{-9}$ m, or *angstrom* Å $= 10^{-10}$ m. Whereas the propagation velocity and the wavelength of a radiant beam depend on the medium, the frequency $\nu$ depends only on the radiating source and is independent of the substance through which it is transmitted.

In addition to being described by continuous waves with characteristic wavelength $\lambda$ (or frequency $\nu$) and velocity $c$, electromagnetic radiation is also generally perceived as discrete packets of energy known as *quanta* or *photons*. This concept was first proposed in 1900 by Max Planck in the context of his *quantum theory*. Briefly, Planck related the *photon energy* $e$ to the frequency $\nu$ by $e = h\nu = ch/\lambda$, where $h = 6.625 \times 10^{-34}$ J s is *Planck's constant*. Notice that according to this celebrated theory the energy associated with electromagnetic radiation is inversely proportional to wavelength. It should be noted that this perspective is compatible with properties of electromagnetic radiation that pertain to its absorption by matter and its ability to cause photobiologic effects.

The various types of electromagnetic radiation are characterized according to wavelength or frequency by the *electromagnetic spectrum*, which is shown in Fig. 5–1. Notice that the spectrum of electromagnetic radiation ranges from the very short-wavelength

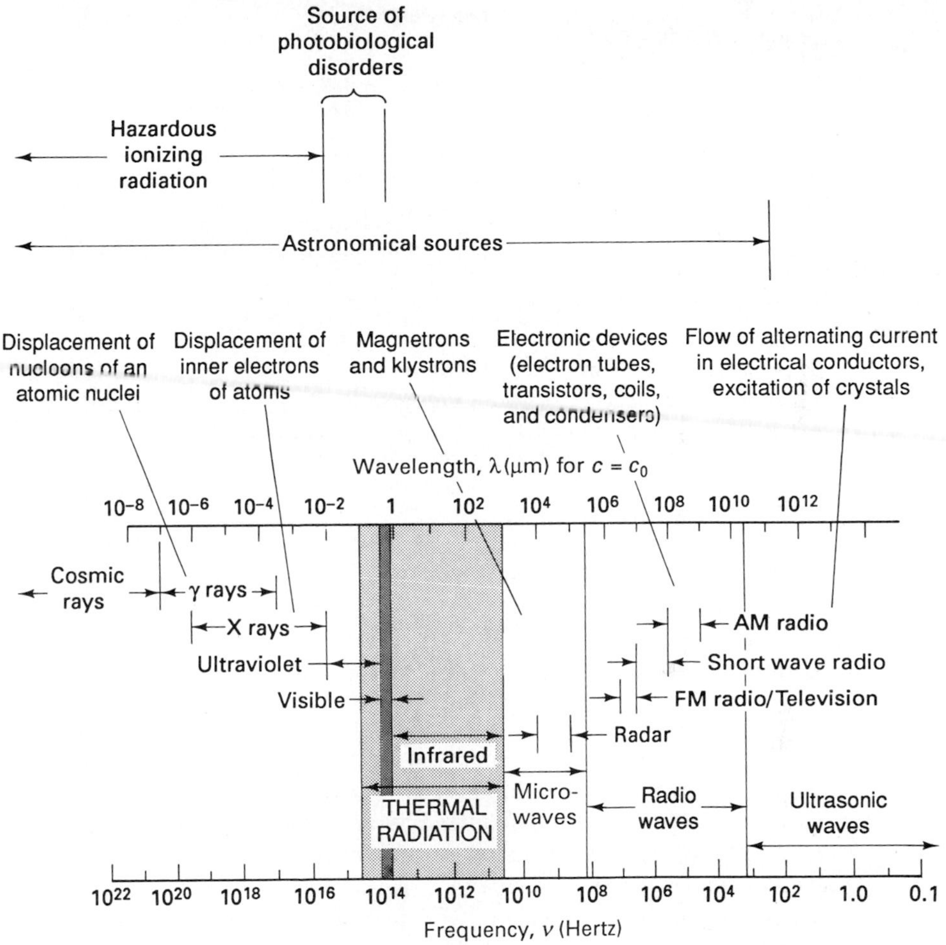


**FIGURE 5–1** Electromagnetic frequency/wavelength spectrum.

cosmic-ray, γ-ray, and X-ray phenomena, through the intermediate-wavelength thermal radiation, to the long-wavelength microwaves, radio waves, and ultrasonic waves.

### *Thermal Radiation*

For practical purposes, the thermal radiation wavelength band may be considered to extend from 0.1 to 1000 μm, which includes ultraviolet (0.1 to 0.38 μm), visible (0.38 to 0.76 μm), and infrared (0.76 to 1000 μm) regions. This part of the electromagnetic spectrum is focused upon in Fig. 5–2. Although the *solar radiation* band essentially lies in the heart of this thermal radiation region, astronomical sources provide thermal (as well as nonthermal) radiation with both shorter (UVC and vacuum UV) and longer (microwave and radio wave) wavelengths.

| Ultraviolet | | Visible light | | Infrared | |
|---|---|---|---|---|---|
| Vacuum UV | 0.10 – 0.20 μm | Violet | 0.40 – 0.44 μm | Near | 0.76 – 1.50 μm |
| UVC | 0.20 – 0.29 μm | Blue | 0.44 – 0.49 μm | Middle | 1.50 – 5.60 μm |
| UVB | 0.29 – 0.32 μm | Green | 0.49 – 0.54 μm | Far | 5.60 – 1000 μm |
| UVA | 0.32 – 0.40 μm | Yellow | 0.54 – 0.60 μm | | |
| | | Orange | 0.60 – 0.63 μm | | |
| | | Red | 0.63 – 0.76 μm | | |

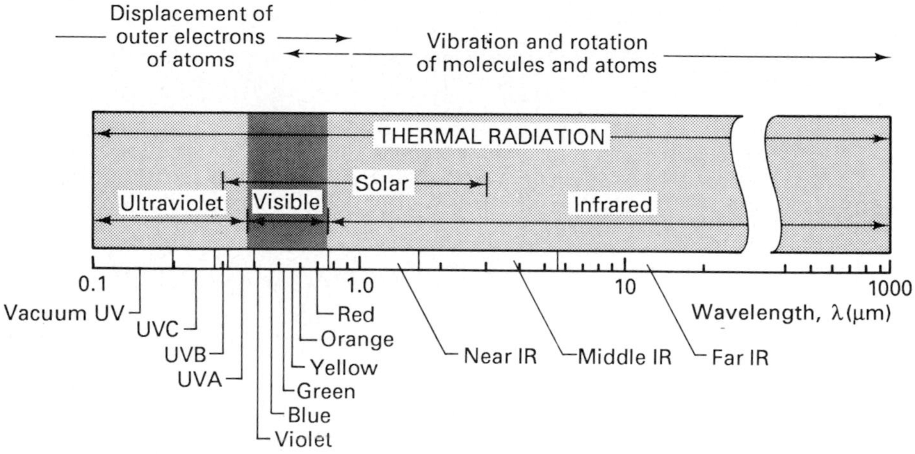

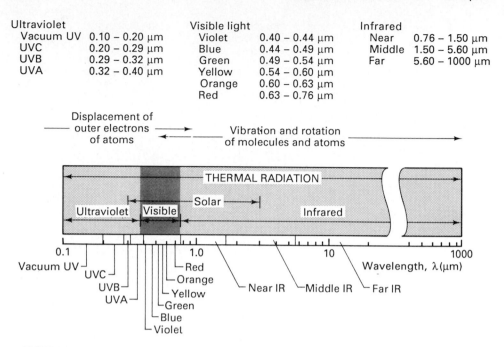

**FIGURE 5–2** Electromagnetic spectrum for thermal radiation wavelength band ($c = c_0$).

***Ultraviolet Radiation (UV)*** The part of the ultraviolet radiation region which forms the low-wavelength boundary of the thermal radiation spectrum is mainly produced by changes in the atomic energy level that occurs when outer electrons of an atom are displaced. Ultraviolet radiation includes UVA (also called *black light*) (0.32–0.40 μm), UVB (also called *sunburn radiation*) (0.29–0.32 μm), UVC (0.20–0.29 μm), and vacuum UV (0.1–0.2 μm). The UVC (which causes sunburn and kills microorganisms) and vacuum UV components of solar radiation are completely absorbed by the ozone $O_3$ layer in the stratosphere. Furthermore, the ozone layer and atmosphere absorb large amounts of the UVB radiation, such that only about 2 to 3% of the radiation contained in terrestrial sunlight is in the UVA and UVB range. However, as portions of the ozone layer near the Antarctic and Arctic are destroyed by fluorochloromethanes (CFCs) from aerosols and refrigerants and other causes, significant increases in the amounts of UVB can be expected to reach the earth's surface. Whereas UVB is the primary natural cause of sunburn, strong evidence indicates that skin cancers such as malignant melanoma, which is sometimes lethal, and basal cell carcinoma result from long-term excessive exposure to UVB and UVA. It should also be noted that ultraviolet radiation can be produced by artificial light sources. For example, low-pressure mercury vapor lamps are commonly used to produce UVC radiation, which is used to destroy certain types of bacteria and to sterilize foodstuffs and medical equipment. In addition, fluorescent tubes are available that emit little or no UVB and shorter wavelength radiation. Such fluorescent UVA lamps are commonly used for artificial tanning. However, evidence concerning the relation between skin cancer and other photobiological disorders has caused dermatologists to issue a strong warning against the use of artificial UVA for cosmetic tanning.

***Visible Radiation (Light)*** The term *light* is of course used to describe the visible portion of the electromagnetic spectrum (0.4–0.76 μm). Light consisting of narrow wavelength bands makes up the colors of the visible spectrum; that is, violet (0.40–0.44 μm), blue (0.44–0.49 μm), green (0.49–0.54 μm), yellow (0.54–0.60 μm), orange (0.60–0.63 μm), and red (0.63–0.76 μm). The summation of all visible wavelengths is referred to as *white light*. Visible and other electromagnetic radiation emanates naturally from the sun and stars and is produced artificially by devices such as fluorescent tubes consisting of a low-pressure mercury discharge source housed in a thin quartz glass tube coated on the inside with phosphor; incandescent lamps, which are usually designed with a coiled tungsten filament contained in an evacuated or inert-gas-filled glass envelope; lasers, carbon and xenon arcs, and low- or high-pressure mercury vapor lamps. The biological process of *photosynthesis* in which light is absorbed and transformed into chemical energy by green plants, algae, and certain bacteria ultimately provides the energy required by all living organisms.

***Infrared Radiation (IR)*** Thermal radiation in the infrared region is primarily associated with molecules or lattice vibrations. All bodies at temperatures above absolute zero emit infrared radiation. In this connection, hot solid bodies generally emit far more infrared energy than visible and UV radiation. Other sources of infrared radiation include emissions of electronic discharges in gases and some lasers (e.g., $CO_2$, $\lambda = 10.6$ μm; neodymium, $\lambda = 1.06$ μm). Infrared radiation is very important in many temperature-sensing applications. Unlike visible and ultraviolet radiation, infrared (and microwave) radiation does not cause biological effects (either adverse or beneficial), except for producing heat, which makes possible the maintenance of temperatures that enable vital metabolic processes to occur.

#### *Microwave Radiation*

Although electromagnetic radiation produced by microwave tubes, known as *magnetrons*, is generally not classified as thermal radiation because of the method of generation, microwaves with wavelengths in the range $10^3$ to $10^5$ μm are reflected by metal, pass through glass and plastic, and are absorbed and converted into internal energy by food (water, sugar, and fat) molecules. As indicated in Chap. 1, this transport and absorption characteristic of microwaves is now widely used in microwave ovens, which are capable of efficiently cooking many foods.

## 5-3 THERMAL RADIATION PROPERTIES

The exchange of thermal radiation between surfaces is a function of (1) surface emissions properties; (2) surface absorption, reflection, and transmission properties; and (3) properties of the medium that lies within the path of the thermal radiation. Each of these issues is considered in this section.

## 5–3–1 Surface Emission Properties

The rate of thermal radiation emitted by a body is dependent upon the surface temperature $T_s$, the nature of the surface, and the electromagnetic radiation wavelength $\lambda$ or frequency $\nu$. The effect of each of these factors on the emission of thermal radiation must be considered.

### *Total Emissive Power*

The effect of emitter surface temperature $T_s$ on the rate of thermal radiation emitted by a body is seen by examining the *total emissive power* $E$, which was introduced in Chap. 1. Because the total emissive power represents the total rate of thermal radiation emitted per unit surface area (i.e., thermal radiation flux) over all wavelengths, it is sometimes designated by $E_{0\to\infty}$ instead of $E$. However, for convenience, we will retain the symbol $E$.

As indicated in Chap. 1, a surface that emits the maximum possible thermal radiation at any given temperature is called a *blackbody*. The total emissive power for thermal radiation in a vacuum from such ideal emitting blackbody surfaces is given by the *Stefan-Boltzmann law*,

$$E_b = \sigma T_s^4 \tag{5–2}$$

where the Stefan-Boltzmann constant $\sigma$ is equal to $5.67 \times 10^{-8}$ W/(m$^2$ K$^4$). Referring back to Sec. 1–2–2, we are reminded that the blackbody thermal radiation flux ranges from very significant levels for source temperatures of the order of 1000 K and above to quite small and often negligible quantities for normal environmental temperatures.

Although some surfaces and geometrical configurations approach ideal emitting conditions, perfect blackbody surfaces do not exist. (The nature of ideal blackbody thermal radiating surfaces will be explored further in Sec. 5–3–3.) The total emissive power of real nonblackbody surfaces is expressed in terms of $E_b$ by

$$E = \epsilon E_b \tag{5–3}$$

where the *emissivity* $\epsilon$ ranges from zero to unity. (Because $\epsilon$ accounts for the thermal radiation emitted over all wavelengths into the entire hemispherical space above a surface, it is also commonly referred to as the total hemispherical emissivity.) It is important to note that $\epsilon$ is a property which is dependent only on the nature of the surface and its temperature $T_s$. The emissivity is given for common surfaces in Table 5–1, Fig. 5–3, and in Table A–C–7 of the Appendix; very comprehensive tabulations of radiation properties are available in *Thermophysical Properties of Matter* by Touloukian et al. [1–3] and in *Thermal Radiation Properties Survey* by Gubareff et al. [4]. Referring to Table 5–1, we observe that blackbody conditions are approached by surfaces coated with lampblack paint. On the other hand, metals have emissivities that range from very low values for polished surfaces to fairly high values for surfaces that have been oxidized or anodized. However, modifying terms such as polished, commercial finish, oxidized, anodized, and so on, which are used to describe the

**TABLE 5-1** Emissivities of representative surfaces

| *Surface* | *Emissivity*, $\epsilon$ | *Temperature*, $T$(K) |
|---|---|---|
| Aluminum | | |
| Polished | 0.04 | 500 |
| Anodized | 0.94 | 310 |
| Brass | | |
| Polished | 0.07 | 320 |
| Dull | 0.22 | 320 |
| Copper | | |
| Polished | 0.041 | 340 |
| Slightly polished | 0.12 | 320 |
| Polished, lightly tarnished | 0.05 | 320 |
| Dull | 0.15 | 320 |
| Oxidized at 1030 K | 0.50 | 590 |
| Nickel, polished | 0.09 | 270 |
| Silver, polished | 0.02 | 300 |
| Stainless steel 18-8, polished | 0.25 | 310 |
| Tungsten, polished | 0.33 | 3400 |
| Asphalt | 0.93 | 310 |
| Glass, Pyrex | 0.88 | 420 |
| Parsons black paint | 0.98 | 240 |
| Lampblack paint | 0.96 | 310 |

*Source*: Based on data primarily from Touloukian and DeWitt [1,2].

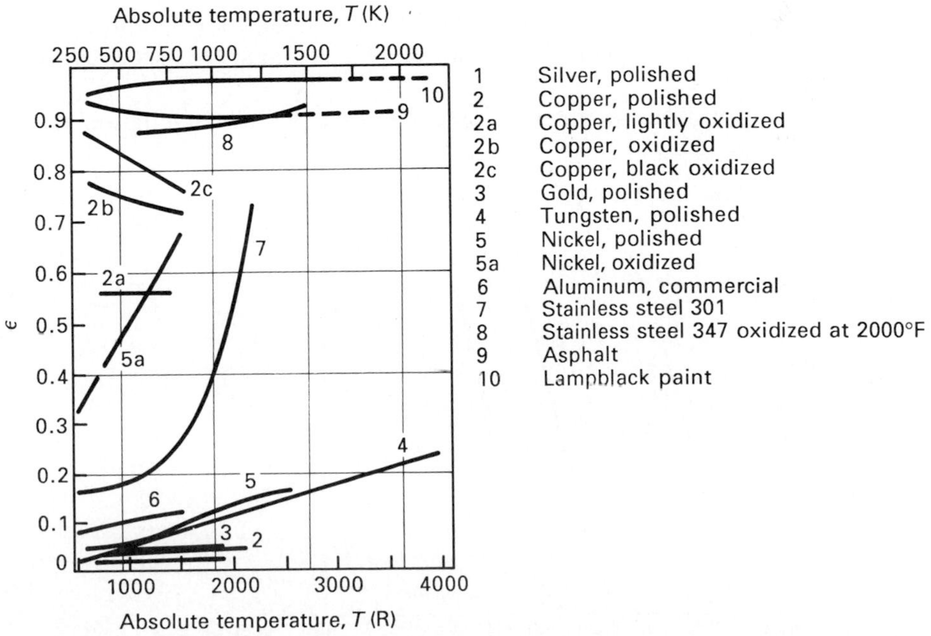
Absolute temperature, $T$ (K)
250 500 750 1000 1500 2000
0.9
0.8
0.7
0.6
0.5
0.4
0.3
0.2
0.1
0
$\epsilon$
1000 2000 3000 4000
Absolute temperature, $T$ (R)
10
9
8
2c
2b
7
2a
5a
4
5
6
3
2
1 Silver, polished
2 Copper, polished
2a Copper, lightly oxidized
2b Copper, oxidized
2c Copper, black oxidized
3 Gold, polished
4 Tungsten, polished
5 Nickel, polished
5a Nickel, oxidized
6 Aluminum, commercial
7 Stainless steel 301
8 Stainless steel 347 oxidized at 2000°F
9 Asphalt
10 Lampblack paint

**FIGURE 5-3** Dependence of emissivity $\epsilon$ of various surfaces on surface temperature. (Based on data from Touloukian and DeWitt [2] and Gubareff et al. [4].)

nature of a surface, are very subjective. As pointed out by Sparrow and Cess [5], because of the ambiguity of such terms, it is unwise to assume that radiative property values reported in literature apply with high precision to other similarly described materials.

Although the total emissive power $E$ of real surfaces is less than $E_b$, $E$ always increases with emitter surface temperature. This point is reinforced by Fig. 5–3, which shows the variation of $\epsilon$ with temperature. By multiplying $\epsilon$ by $E_b$ $(= \sigma T_s^4)$, we find that $E$ increases with increasing temperature, even for the materials for which $\epsilon$ itself decreases.

### *Subtotal Emissive Power*

To see the effect of wavelength on thermal radiation, we consider the energy flux emitted from a surface over wavelengths from zero to $\lambda$. This *subtotal emissive power* $E_{0\to\lambda}$ is shown in Fig. 5–4 as a function of $\lambda$ for a blackbody at several temperatures. $E_{b,0\to\lambda}$ increases from zero at small values of $\lambda$ and approaches the total emissive power $E_b$ as $\lambda$ becomes large. Consistent with the electromagnetic spectrum shown in Figs. 5–1 and 5–2, we find that the significant contribution to the thermal radiation for these temperatures occurs within wavelengths of about 0.1 and 100 μm. For solar radiation, which has an effective blackbody source temperature of roughly 5800 K, the wavelength band essentially lies between 0.30 and 3.0 μm, with about 98% of the energy associated with $\lambda < 3.0$ μm. On the other hand, the wavelength range for a surface temperature of 400 K is mainly between 3.0 and 40 μm. As a matter of fact, less than 1% of the thermal radiation emitted at environmental temperatures below 400 K is contained in the part of the electromagnetic spectrum for which $\lambda <$ 3.0 μm.

A more general representation of the subtotal emissive power for a blackbody

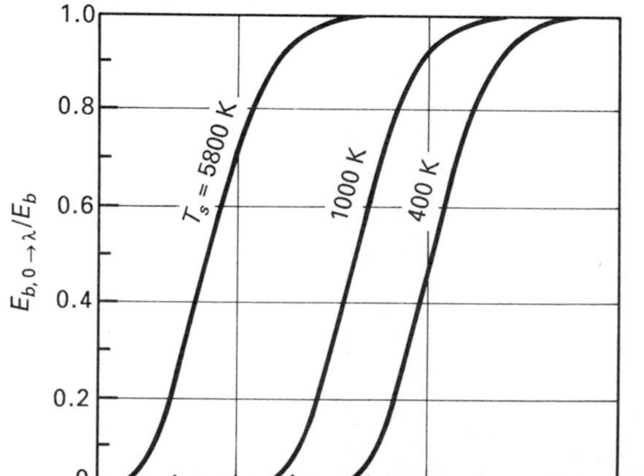


**FIGURE 5–4**
Subtotal emissive power $E_{b,0\to\lambda}$.

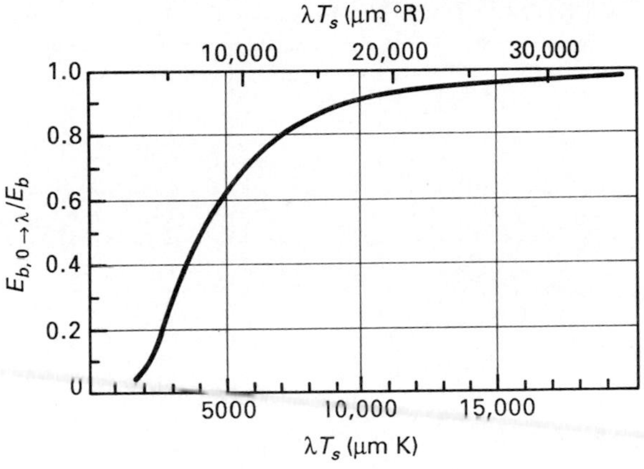


**FIGURE 5–5**
General representation of subtotal emissive power $E_{b,0\to\lambda}$.

that applies to the entire temperature range is given in Fig. 5–5 and Table A–H–1 of the Appendix in terms of $E_{b,0\to\lambda}/E_b$ versus $\lambda T_s$.

### *Monochromatic Emissive Power*

To complete the picture concerning the effect of wavelength on thermal radiation emitted from a surface, we introduced the *monochromatic emissive power* $E_\lambda$, which is defined as the thermal radiation flux emitted per unit wavelength $d\lambda$. This important thermal radiation emission property is related to the subtotal emissive power $E_{0\to\lambda}$ by

$$E_\lambda = \frac{dE_{0\to\lambda}}{d\lambda} \qquad \text{or} \qquad E_{0\to\lambda} = \int_0^\lambda E_\lambda \, d\lambda \tag{5–4,5}$$

As $\lambda$ becomes large, it follows that

$$E = E_{0\to\infty} = \int_0^\infty E_\lambda \, d\lambda \tag{5–6}$$

***Blackbody Thermal Radiation*** Theoretical predictions based on quantum theory were developed for the monochromatic emissive power for blackbody thermal radiation by M. Planck in 1901. The famous *Planck law* [6] is written as

$$E_{b\lambda} = \frac{C_1}{\lambda^5\{\exp\,[C_2/(\lambda T_s)] - 1\}} \tag{5–7}$$

where $C_1 = 3.743 \times 10^8$ W μm$^4$/m$^2$ and $C_2 = 1.439 \times 10^4$ μm K. This equation is shown in Fig. 5–6 for several temperatures, with $\lambda$ taken as the independent variable. For each temperature, $E_{b\lambda}$ is seen to increase from zero at low wavelengths to a peak, and to then gradually fall back toward zero.

The peak in $E_{b\lambda}$ increases and shifts to shorter wavelengths as the temperature

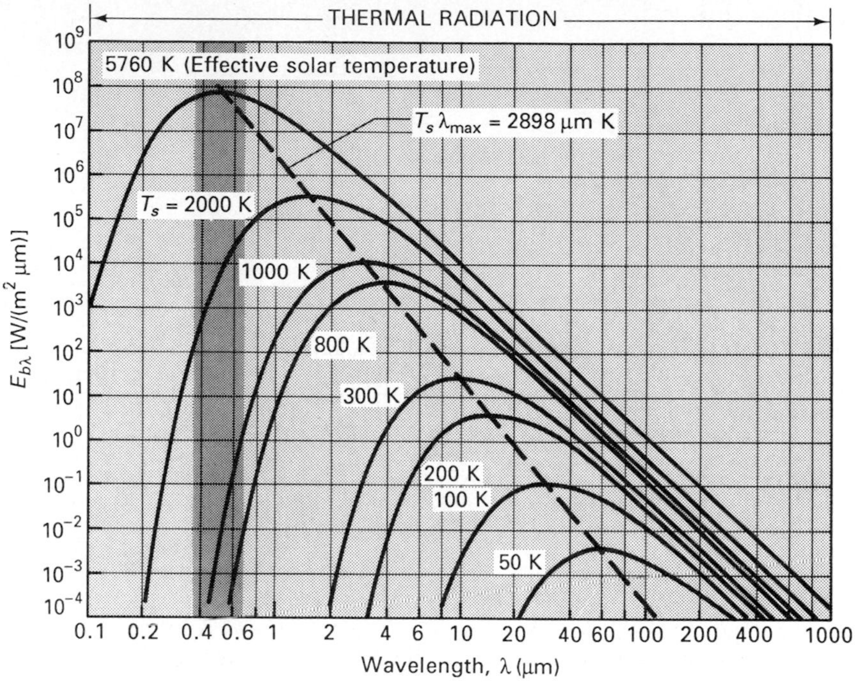


**FIGURE 5–6** Monochromatic emissive power of a blackbody for several temperatures.

increases. The wavelength at which the peak occurs is given as a function of temperature by *Wien's displacement law*,

$$T_s \lambda_{max} = 2898 \ \mu\text{m K} \tag{5–8}$$

This shift in $\lambda_{max}$ and increase in $E_{b\lambda}$ with increasing temperature is responsible for the familiar change in color of heat-treated steel, which goes from a dull red at around 700°C to bright red, then to bright yellow, and finally becomes glowing white at approximately 1300°C. Referring to Fig. 5–6, we observe that little thermal radiation emitted by low-temperature blackbodies lies in the portion of the electromagnetic spectrum from 0.38 to 0.76 μm that is visible to the eye. However, as the temperature increases, more and more of the thermal radiation falls within the visible range, thus producing this array of color.

## EXAMPLE 5–1

For practical purposes, the sun is generally considered to be a blackbody radiator with an effective temperature of about 5800 K. Determine the fraction of energy emitted by the sun that falls in the visible region of the electromagnetic spectrum.

**Solution**

*Objective* Determine the fraction of solar radiation that is visible.

*Assumptions/Conditions*

solar irradiation is equivalent to blackbody emission at 5800 K

*Properties* Sun at 5800 K: $\epsilon = 1$.

*Analysis* The wavelength band for solar radiation that is visible to the eye is 0.38 to 0.76 μm. Using Table A–H–1, we are able to determine the fraction of solar radiation with wavelength in the bands 0 to 0.38 μm and 0 to 0.76 μm; that is,

$$\frac{E_{b,0\rightarrow\lambda_1}}{E_b} = 0.102$$

for $\lambda_1 T_s = 0.38\ \mu\text{m}\ (5800\ \text{K}) = 2200\ \mu\text{m K}$, and

$$\frac{E_{b,0\rightarrow\lambda_2}}{E_b} = 0.55$$

for $\lambda_2 T_s = 0.76\ \mu\text{m}\ (5800\ \text{K}) = 4410\ \mu\text{m K}$. The difference between these two values gives the fraction of solar radiation falling in the visible spectrum; that is,

$$\frac{E_{b,\lambda_1\rightarrow\lambda_2}}{E_b} = 0.55 - 0.102 = 0.448$$

Therefore, approximately 44.8% of extraterrestrial solar radiation is visible to the human eye, with about 10.2% lying in the ultraviolet region and 45% in the infrared region. However, it should be noted that the amount of solar radiation actually reaching the surface of the earth is diminished by absorption within the stratosphere and atmosphere. Depending on environmental conditions, atmospheric pollution, latitude, season, time of day, and other factors, the total solar radiation reaching the earth's surface normally consists of approximately 60% IR, 37% visible light, and 3% UV.

***Thermal Radiation Emitted from Real Surfaces*** The monochromatic emissive power $E_\lambda$ of polished copper, anodized aluminum, and a blackbody are compared in Fig. 5–7. The ratio $E_\lambda/E_{b\lambda}$ at any given wavelength is known as the *monochromatic emissivity* $\epsilon_\lambda$:

$$\epsilon_\lambda = \frac{E_\lambda}{E_{b\lambda}} \tag{5–9}$$

Like the emissivity $\epsilon$, $\epsilon_\lambda$ is a property that is dependent on the surface alone. The monochromatic emissivities of these two real surfaces are shown in Fig. 5–8. Note

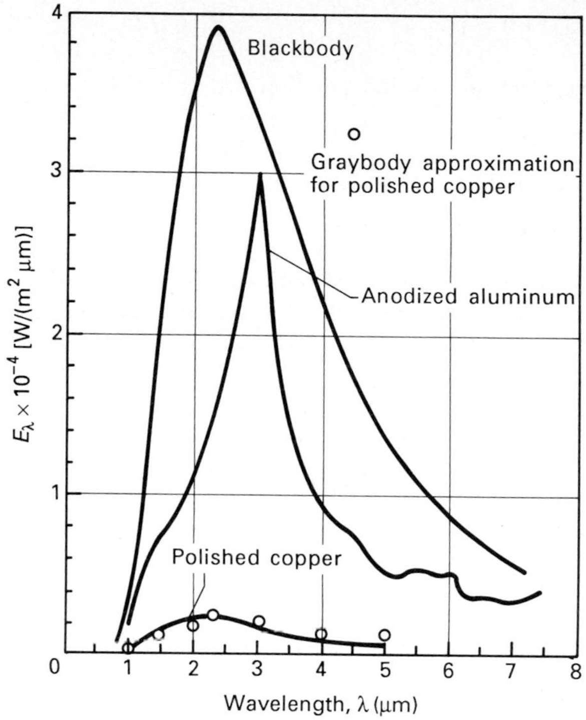


**FIGURE 5–7**
Monochromatic emissive power for blackbody and representative real surfaces at 1240 K.

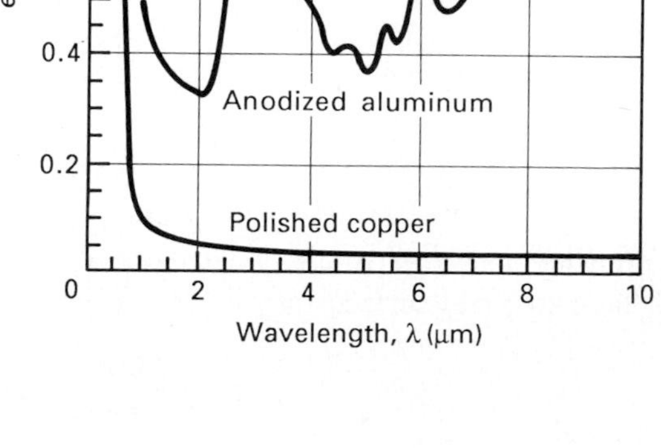

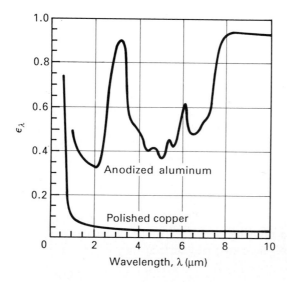


**FIGURE 5–8**
Monochromatic emissivity of polished copper and anodized aluminum (Dunkle et al. [7] and Seban [8]).

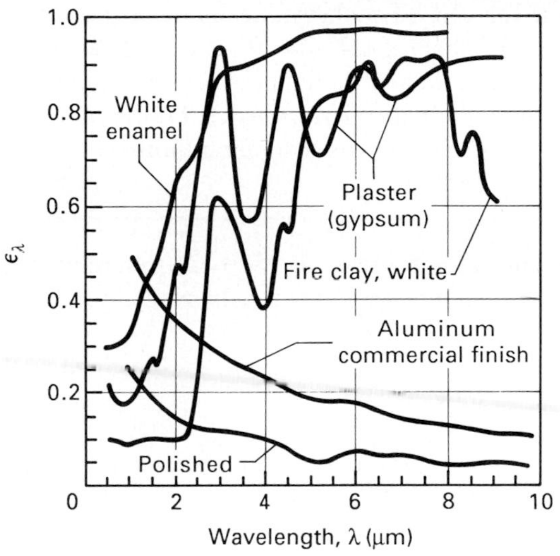


**FIGURE 5–9**
Monochromatic emissivity of several materials (Dunkle et al. [7] and Sieber [9]).

that $\epsilon_\lambda$ of these surfaces are less than unity and vary rather irregularly with $\lambda$. To at least some extent, all real surfaces exhibit these same characteristics. To reinforce this point, the monochromatic emissivities of several other materials at room temperature are shown in Fig. 5–9. An impressive listing of data for $\epsilon_\lambda$ is given in references 1–4 for many types of surfaces. Data in these references indicate that $\epsilon_\lambda$ is essentially independent of $T_s$ for many substances. For example, $\epsilon_\lambda$ is nearly independent of surface temperature for metals such as polished copper, polished iron, and tungsten and for nonmetals such as carbon, Pyrex, and certain carbides. However, large changes occur in $\epsilon_\lambda$ with $T_s$ for many nonmetallic substances, such as aluminum oxide ($Al_2O_3$).

The relationship between the emissivity $\epsilon$ and the monochromatic emissivity $\epsilon_\lambda$ is obtained by writing

$$\epsilon = \frac{E}{E_b} = \frac{\int_0^\infty E_\lambda \, d\lambda}{\int_0^\infty E_{b\lambda} \, d\lambda} = \frac{\int_0^\infty \epsilon_\lambda E_{b\lambda} \, d\lambda}{\int_0^\infty E_{b\lambda} \, d\lambda} \tag{5–10}$$

Note that for the case in which $\epsilon_\lambda$ is independent of $\lambda$, Eq. (5–10) reduces to

$$\epsilon = \epsilon_\lambda \tag{5–11}$$

Surfaces that satisfy this equation are known as *graybodies*. Referring to Fig. 5–8, we see that $\epsilon_\lambda$ of the polished copper surface exhibits approximate graybody behavior for wavelengths greater than about 2. A graybody approximation for the monochromatic emissive power $E_{g\lambda}$ for real surfaces such as this is given by

$$E_{g\lambda} = \epsilon E_{b\lambda} \tag{5–12}$$

This graybody approximation for the monochromatic emissive power of polished copper is shown in Fig. 5–7. Notice that $E_{g\lambda}$ follows the same wavelength-dependence pattern as $E_{b\lambda}$. The anodized aluminum and the other substances shown in Fig. 5–9 are observed to exhibit distinct nongraybody characteristics. Furthermore, the polished copper surface is nongray for wavelengths less than about unity.

**EXAMPLE 5–2**

The filament of an incandescent lamp operates at 2500 K. Assuming approximate graybody characteristics, determine the fraction of radiant energy emitted by the filament that falls in the visible spectrum.

**Solution**

*Objective* Determine the fraction of thermal radiation emitted by a graybody at 2500 K that produces light.

*Assumptions/Conditions*

graybody thermal radiation

*Properties* Lamp (graybody): $\epsilon_\lambda = \epsilon =$ constant.

*Analysis* Because the spectral distribution of radiant energy emitted by a graybody is the same as for a blackbody, we utilize Table A–H–1 to solve this problem. Following the pattern established in Example 5–1, we write

$$\frac{E_{b,0\to\lambda_1}}{E_b} = \frac{E_{g,0\to\lambda_1}}{E_g} = 0.00021$$

for $\lambda_1 T_s = 0.38\ \mu\text{m}\ (2500\ \text{K}) = 950\ \mu\text{m K}$, and

$$\frac{E_{b,0\to\lambda_2}}{E_b} = \frac{E_{g,0\to\lambda_2}}{E_g} = 0.0522$$

for $\lambda_2 T_s = 0.76\ \mu\text{m}\ (2500\ \text{K}) = 1900\ \mu\text{m K}$. Therefore, the fraction of energy that falls in the visible part of the spectrum is

$$\frac{E_{g,\lambda_1\to\lambda_2}}{E_g} = 0.0522 - 0.00021 = 0.052$$

For this operating temperature, only 5.2% of the radiant energy dissipated by the filament produces light, with most of the remaining energy producing infrared heating.

### *Thermal Radiation Intensity*

Referring to Fig. 5–10, the *intensity I* is defined as the total rate of thermal radiation emitted per unit solid angle $d\omega$ and per unit area normal to the direction $\phi,\theta$. (The intensity is sometimes defined in the literature in terms of the total radiant energy

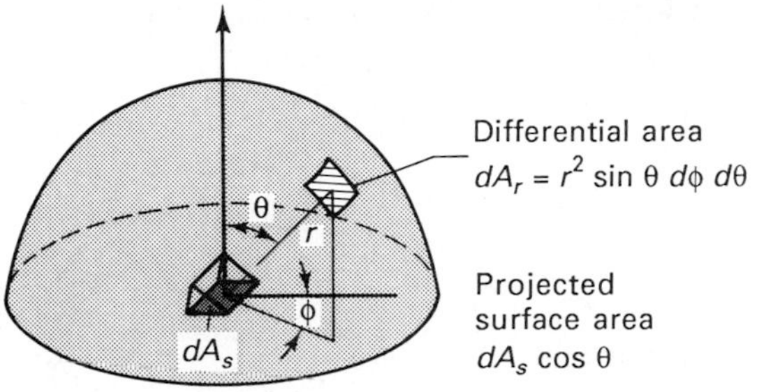
Differential area
$dA_r = r^2 \sin\theta\, d\phi\, d\theta$
θ
r
ϕ
$dA_s$
Projected
surface area
$dA_s \cos\theta$

**FIGURE 5–10**
Geometric perspective for concept of intensity.

emitted, reflected, and transmitted from a surface.) For a surface area $dA_s$, the projected area is simply $dA_s \cos\theta$. The solid angle $d\omega$ is defined by

$$d\omega = \frac{dA_r}{r^2} \tag{5–13}$$

where the area $dA_r$ of the hemispherical surface element is shown in Fig. 5–10. The intensity $I$ is written in terms of the total emissive power $E$, which is the rate of energy emitted per unit surface area $dA_s$, as

$$I = \frac{dE}{\cos\theta\, d\omega} \tag{5–14}$$

Rearranging this equation and assuming no variations in the emissive properties with the azimuthal angle $\phi$, we have

$$\frac{dE}{d\omega} = E_\theta = I\cos\theta \tag{5–15}$$

where $E_\theta$ is the directional emissive power.

An important aspect of ideal blackbody emitting surfaces is that the intensity $I$ is the same in all directions. Surfaces that exhibit this characteristic are said to be *diffuse* emitters. Many real surfaces such as industrially rough surfaces approach diffuse conditions. Utilizing Eq. (5–15), $I$ is equal to the directional emissive power normal to the surface $E_0$ for diffuse surfaces. For this case, Eq. (5–15) reduces to

$$E_\theta = E_0 \cos\theta \tag{5–16}$$

which is known as the *Lambert cosine law*.

The *directional emissivity* $\epsilon_\theta$,

$$\epsilon_\theta = \frac{E_\theta}{E_{b\theta}} = \frac{I}{I_b} \tag{5–17}$$

is shown in Fig. 5–11 for several real surfaces. Notice that the nonconductors are essentially diffuse for $\theta$ less than about 40 degrees but violate the Lambert cosine law for larger angles, with $\epsilon_\theta$ falling to very small values. The metallic surfaces obey the Lambert cosine law over about the same range in $\theta$, but $\epsilon_\theta$ increases quite sharply before falling to zero at 90 degrees. Consistent with these findings, a hot metallic

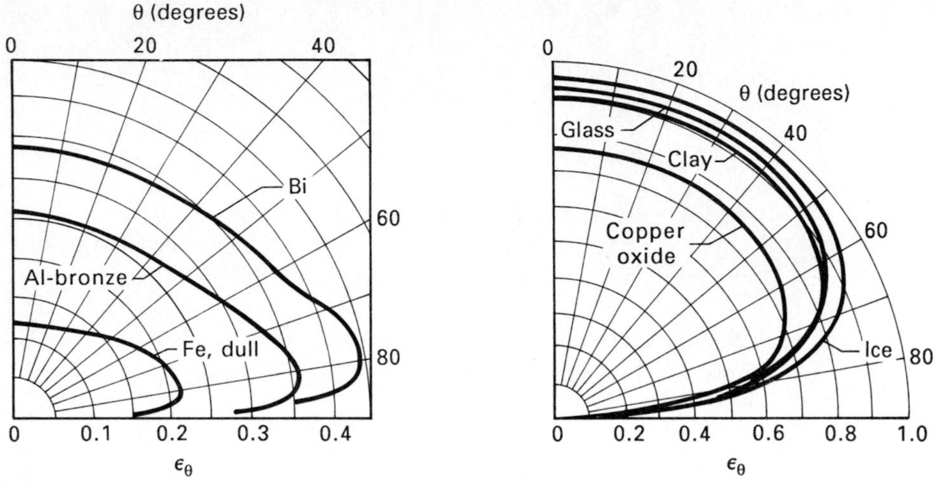


(a) Metal surfaces. (b) Electric nonconducting surfaces.

**FIGURE 5–11** Distribution of directional emissivity for several surfaces (Schmidt and Eckert [10]).

sphere will appear brighter near the base ($\theta \simeq 80$ degrees) than at the center $\theta \simeq 0$ degrees). The opposite holds true for a nonmetallic sphere. These directional effects are generally accounted for in design by utilizing the emissivity $\epsilon$, which accounts for the radiant energy emitted into the entire hemispherical space above the surface. The emissivity $\epsilon$ is usually set equal to some fraction of $\epsilon_0$. For example, $\epsilon/\epsilon_0 \simeq 1.2$ for bright metallic surfaces and $\epsilon/\epsilon_0 \simeq 0.96$ for nonconductors.

## EXAMPLE 5–3

Develop an expression for the total emissive power $E$ of a surface in terms of the intensity $I$.

### Solution

*Objective* Express $E$ in terms of $I$.

*Analysis* Based on Eqs. (5–13) and (5–14), we write

$$dE = I \frac{\cos\theta}{r^2}\, dA_r$$

Referring to Fig. 5–10, $dA_r$ is equal to $(r\,d\theta)(r\sin\theta\,d\phi)$, such that $dE$ becomes

$$dE = I\cos\theta\sin\theta\,d\phi\,d\theta$$

To obtain $E$, we integrate over the hemisphere (i.e., $0 \leq \phi \leq 2\pi$, $0 \leq \theta \leq \pi/2$).

$$E = \int_0^{\pi/2}\int_0^{2\pi} I\cos\theta\sin\theta\,d\phi\,d\theta$$

For a diffuse surface, $I$ is uniform and is brought outside the integrals, such that

$$E = I \int_0^{\pi/2} \int_0^{2\pi} \cos\theta \sin\theta \, d\phi \, d\theta \qquad \text{(a)}$$

By employing the double-angle trigonometric formula,

$$\sin(a + b) = \sin a \cos b + \cos a \sin b$$

we obtain

$$E = \pi I$$

### 5–3–2 Surface Irradiation Properties

#### *Total Irradiation Properties*

As illustrated in Fig. 5–12, thermal radiation incident upon a surface is absorbed, reflected, and transmitted through the body. The *absorptivity* $\alpha$, *reflectivity* $\rho$, and *transmissivity* $\tau$ were defined in Chap. 1. These total hemispherical surface irradiation properties account for the fractions of incident thermal radiation flux $G$ at *all* wavelengths over the entire hemisphere above a surface that are absorbed, reflected, and transmitted. The incoming thermal radiation flux $G$ is called the *irradiation*. Referring to Fig. 5–12, the thermal irradiation received by the surface is distributed as follows:

| | |
|---|---|
| Thermal radiation flux absorbed | $\alpha G$ |
| Thermal radiation flux reflected | $\rho G$ |
| Thermal radiation flux transmitted | $\tau G$ |
| Total irradiation | $G$ |

The relationship between these surface irradiation properties is given by Eq. (1–13),

$$\alpha + \rho + \tau = 1 \qquad (5\text{–}18)$$

Except for a few materials such as glass, rock salt, and other inorganic crystals, most solids are essentially opaque, with $\tau$ equal to zero.

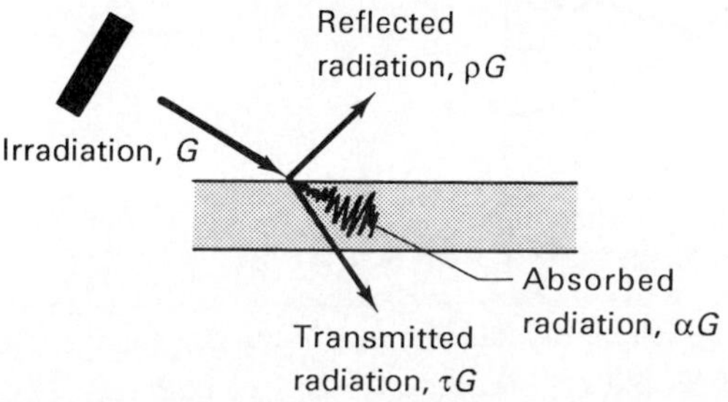
Reflected radiation, ρG
Irradiation, G
Absorbed radiation, αG
Transmitted radiation, τG

**FIGURE 5–12**
Absorption, reflection, and transmission of thermal radiation incident on a surface.

Although all real surfaces reflect and/or transmit at least some thermal radiation, the concept of an ideal blackbody surface which absorbs all incident irradiation (i.e., $\alpha = 1, \rho = 0, \tau = 0$) is extremely important. As we have already seen, it is for such ideal absorbing and emitting surfaces that the pioneering theoretical studies by Stefan, Boltzmann, Planck, and others were developed. Moreover, the radiative performance of ideal blackbody surfaces provides a standard against which the performance of real surfaces can be compared. The term "blackbody" stems from the observation that surfaces which absorb nearly all of the thermal radiation in the visible part of the electromagnetic spectrum are black in color as a result of the absence of reflected light. The eye is a very good indicator of reflected visible thermal radiation, but is totally insensitive to the reflection of thermal radiation outside this narrow wavelength spectrum. It just so happens that surfaces that appear black in color generally are also good absorbers of thermal radiation outside the visible range.

The total hemispherical absorption, reflection, and transmission properties are dependent upon the nature and temperature $T_R$ of the emitting source and upon the character of the receiving surface. The importance of $T_R$ and the type of surface are shown in Fig. 5–13. In this figure, the absorptivity $\alpha$ is shown as a function of emitter source temperature $T_R$ for several common materials which are at room temperature. Notice that the white fire clay, which would be judged by the eye to be a poor absorber of thermal radiation, is actually a very good absorber ($\alpha \gtrsim 0.8$) of thermal radiation which is emitted from sources with temperatures below 500 K. However, white fire clay reflects most of the incoming solar radiation ($\alpha \simeq 0.1$) which is associated with an effective blackbody source temperature of approximately 5800 K.

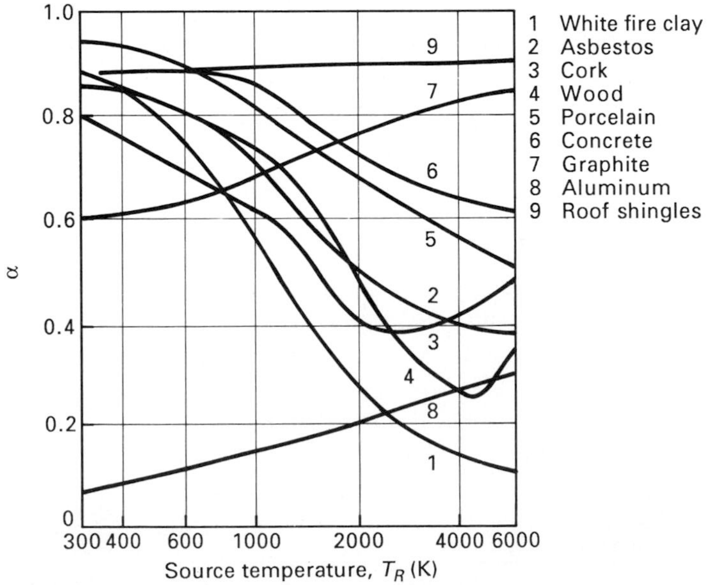


**FIGURE 5–13** Dependence of absorptivity $\alpha$ on source temperature $T_R$ of incident thermal radiation (Sieber [9]).

Data for $\alpha$, $\rho$, and $\tau$ are available in references 1 through 4 and elsewhere for many surfaces. However, these data are generally restricted to situations involving irradiation from ideal emitting sources at one or two values of $T_R$. For example, considerable data are available for surfaces that receive solar radiation and radiation from blackbody sources at 300 K. But aside from the calculations by Sieber [9], which are shown in Fig. 5–13, relatively little information is available in the literature on the general effect of $T_R$ on the surface irradiation properties. To circumvent this problem, $\alpha$ can sometimes be evaluated on the basis of tabulated data for $\epsilon$ by the use of *Kirchhoff's law* for thermal radiation. The limiting form of Kirchhoff's law indicates that $\alpha$ and $\epsilon$ are equal for thermal radiation exchange between a surface $s$ and a *blackbody or graybody* $R$ under conditions of *thermal equilibrium* (i.e., $T_R = T_s$). This form of Kirchhoff's law is developed in the context of a blackbody enclosure in Example 5–5. Other forms of this important law are considered in the next section.

## EXAMPLE 5–4

No perfect blackbody surface has been found to exist. Referring to Table A–C–7 in the Appendix, we see that emissivities of the order of 0.95 to 0.98 are common for near-blackbodies such as surfaces coated with flat black paint. However, geometrical blackbodies can be constructed which perform even closer to the ideal. Show that blackbody conditions are approached by a small hole in the wall of a large cavity with opaque partially absorbing isothermal surface.

### Solution

*Objective* Show that the radiation characteristics within a large cavity with opaque partially absorbing isothermal surface approach those of a blackbody.

*Schematic* Large cavity with uniform temperature $T_R$.

*Analysis* A spherical cavity with a small opening in its wall is shown in the schematic. If we trace the path of an incident ray of thermal radiation entering the cavity, we find that the ray is reflected within the interior of the cavity many times, with a part of the energy being absorbed each time. When the reflected ray eventually reaches the opening and escapes, its energy content is extremely small. Because nearly all of the thermal radiation entering the cavity is absorbed, the radiation leaving the hole approaches that of a blackbody at $T_R$. Furthermore, blackbody radiation is approximated within such a cavity, regardless of whether the surface is highly reflective or absorbing.

## EXAMPLE 5–5

Referring to the large isothermal enclosure $A_R$ containing a small body $A_s$ shown in Fig. E5–5, develop a relation for $\alpha_s$ in terms of $\epsilon_s$ for the body.

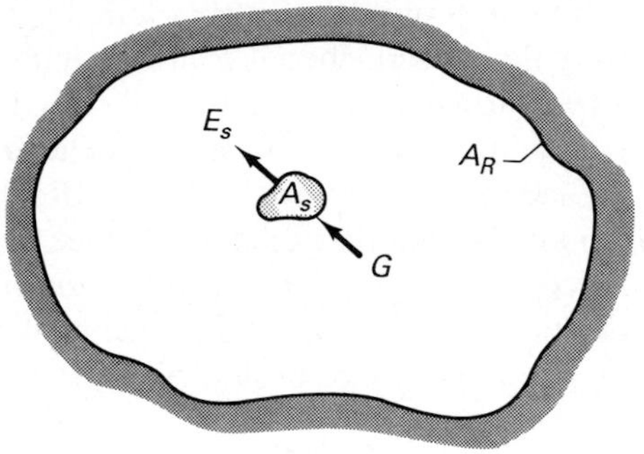


**FIGURE E5–5**
Small body contained in a large isothermal enclosure.

### Solution

*Objective* Develop a relation between $\alpha_s$ and $\epsilon_s$ for a small body contained in a large isothermal enclosure.

*Assumptions/Conditions*

surface $A_R$ is opaque
effect of surface $A_s$ on irradiation from $A_R$ is negligible

*Analysis* Assuming that the walls of the enclosure are opaque and that the body has no appreciable effect on the irradiation $G$, the enclosure forms a blackbody cavity with $G$ represented by

$$G = E_b(T_R) = \sigma T_R^4$$

For steady-state conditions, thermal equilibrium must exist within the enclosure such that the temperature $T_s$ of the body is equal to $T_R$. Thus, the net rate of energy transfer from the body must be equal to zero, and the energy balance becomes

$$\alpha_s G A_s - E_s(T_s)\, A_s = 0$$

or

$$\frac{\alpha_s G}{E_s(T_s)} = \frac{\alpha_s\, E_b(T_s)}{\epsilon_s\, E_b(T_s)} = \frac{\alpha_s}{\epsilon_s} = 1$$

Therefore, we conclude that $\alpha_s = \epsilon_s$, or simply

$$\alpha = \epsilon$$

for thermal radiation exchange between a surface $s$ and a blackbody $R$ under conditions of thermal equilibrium, which represents the limiting form of Kirchhoff's law.

### *Monochromatic Irradiation Properties*

Spectral surface irradiation properties are now defined which account for the effect of wavelength $\lambda$. The *monochromatic absorptivity* $\alpha_\lambda$ is the fraction of incident thermal radiation with wavelength $\lambda$ that is absorbed by a surface. Similarly, the *monochromatic reflectivity* $\rho_\lambda$ and the *monochromatic transmissivity* $\tau_\lambda$ represent the fractions of incoming thermal radiation with wavelength $\lambda$ that are reflected and transmitted, respectively. These important spectral surface irradiation properties are related by

$$\alpha_\lambda + \rho_\lambda + \tau_\lambda = 1 \tag{5–19}$$

The relationship between the monochromatic absorptivity $\alpha_\lambda$ of a receiving surface and its monochromatic emissivity $\epsilon_\lambda$ is given by Kirchhoff's law, which indicates

$$\alpha_\lambda = \epsilon_\lambda \tag{5–20}$$

for systems in which (1) the irradiation is diffuse, or (2) the surface is diffuse. The first of these conditions is approximately satisfied for many practical arrangements, and the second condition is approached for electrically nonconducting materials. This relation and its more general specular (i.e., directional) form, which is applicable to nondiffuse as well as diffuse thermal radiation, is developed by Planck [6] and Siegel and Howell [11].

In addition to the considerable amount of data available for $\epsilon_\lambda$ and $\alpha_\lambda$, extensive tabulations of data have also been developed for $\rho_\lambda$ and $\tau_\lambda$. Again, references 1 through 4 are very good sources of data for these spectral surface irradiation properties.

The monochromatic transmissivity $\tau_\lambda$ is shown in Fig. 5–14(a) and (b) for ordinary window glass and type 2A diamond. It is observed that glass transmits thermal radiation quite well in the low-wavelength visible range of the electromagnetic

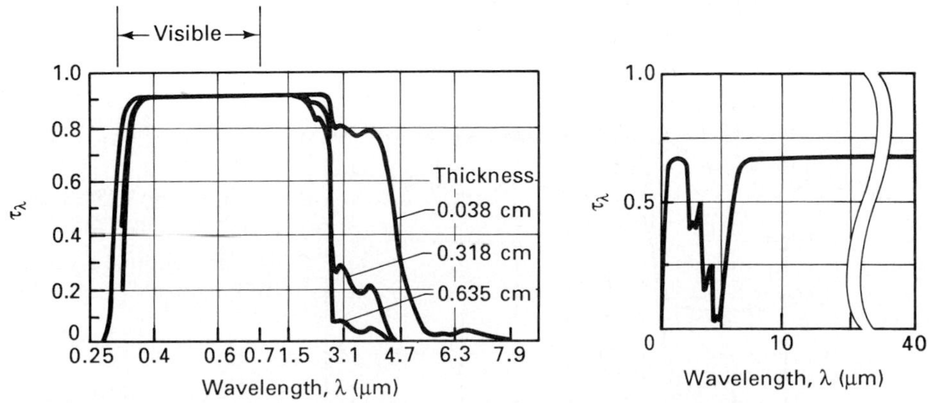


(a) Glass (with 0.02 $Fe_2O_3$) at room temperature (Dietz [12]).

(b) Type 2A diamond at room temperature (Seal [13]).

**FIGURE 5–14** Monochromatic transmissivity $\tau_\lambda$.

spectrum. However, in the longer-wavelength infrared part of the spectrum ($\lambda \gtrsim 2.6$ μm), the glass is nearly opaque to thermal radiation, with most of the energy being absorbed and reflected. As we have already mentioned, this nongraybody characteristic of glass is responsible for the *greenhouse effect*, in which glass is transparent to short-wavelength irradiation from the sun but is essentially opaque to longer-wavelength thermal radiation emitted by the low-temperature interior of an enclosure. Some plastic films such as polyethylene have similar characteristics.

On the other hand, the type 2A diamond is transparent throughout much of the infrared region (6 μm $\lesssim \lambda \lesssim$ 40 μm), as well as in visible range. Because of its transparency across such a broad range of wavelengths and because of its great strength, type 2A diamond was utilized in the development of two 18.2-mm-diameter windows for the 1978 Pioneer Venus Space Mission. The function of the windows was to protect the infrared radiometer equipment from the extremely hostile Venusian environment. The conditions withstood by the diamond window included "a 4-month journey through the cold and vacuum of space, entry decelerations to 565*g*, searing heat (the Venusian surface is red hot), crushing pressure (100 times that of the earth's atmosphere), and a highly corrosive atmosphere containing sulfuric acid and other aggressive gases" [13].

Representative measurements are shown in Fig. 5–15 for the monochromatic reflectivity $\rho_\lambda$ for several surfaces. These surfaces are essentially opaque, such that $\alpha_\lambda$ is equal to $1 - \rho_\lambda$. Notice that $\rho_\lambda$ of the pure aluminum surface is quite high for all wavelengths. But $\rho_\lambda$ for aluminum surfaces with certain coatings such as lead sulfide fall to much smaller values for wavelengths below 3 μm. This selective characteristic of certain types of metallic surfaces is very important in solar applications.

In order to express the absorptivity $\alpha$ in terms of $\alpha_\lambda$ (or $\epsilon_\lambda$) we first introduce the *monochromatic irradiation* $G_\lambda$, which is defined as the irradiation flux per unit wavelength. The total irradiation $G$ and the monochromatic irradiation $G_\lambda$ are themselves related by

$$G = \int_0^\infty G_\lambda \, d\lambda \tag{5–21}$$

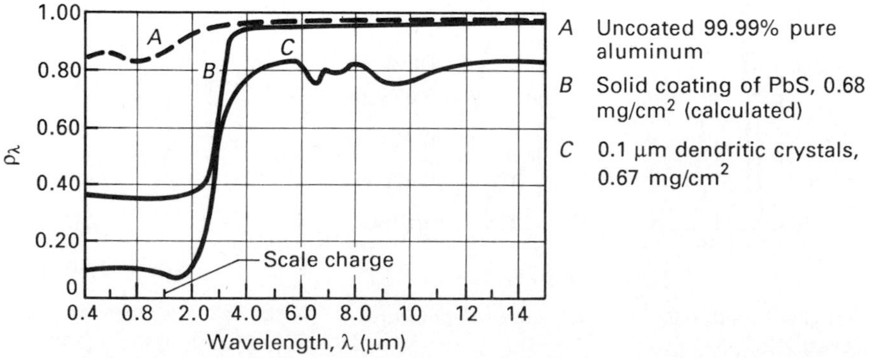
1.00
0.80
0.60
0.40
0.20
0
ρλ
A
B
C
Scale charge
0.4 0.8 2.0 4.0 6.0 8.0 10 12 14
Wavelength, λ (μm)
A Uncoated 99.99% pure aluminum
B Solid coating of PbS, 0.68 mg/cm² (calculated)
C 0.1 μm dendritic crystals, 0.67 mg/cm²

**FIGURE 5–15** Monochromatic reflectivity $\rho_\lambda$ for lead sulfide coatings on aluminum substrates (Williams et al. [14]).

The absorptivity is now formally defined by the equation

$$\alpha = \frac{\int_0^\infty \alpha_\lambda G_\lambda \, d\lambda}{\int_0^\infty G_\lambda \, d\lambda} \tag{5–22}$$

For the many practical situations in which $\alpha_\lambda = \epsilon_\lambda$, this expression becomes

$$\alpha = \frac{\int_0^\infty \epsilon_\lambda G_\lambda \, d\lambda}{\int_0^\infty G_\lambda \, d\lambda} \tag{5–23}$$

This equation provides a basis for the evaluation of $\alpha$ in terms of $\epsilon$ for several situations. First, for approximate graybody conditions, $\epsilon_\lambda$ is essentially independent of wavelength and equal to $\epsilon$, such that Eq. (5–23) reduces to the *graybody* $\alpha$ *approximation*

$$\alpha = \epsilon \tag{5–24}$$

where both $\alpha$ and $\epsilon$ are evaluated at $T_s$. Although very few materials exist for which $\epsilon_\lambda$ is nearly constant for all wavelengths, some substances exhibit approximate graybody characteristics over significant parts of the spectrum. The key issue when using this graybody $\alpha$ approximation is that $\epsilon_\lambda$ (and $\alpha_\lambda$) be essentially uniform in the wavelength range where there are appreciable amounts of both emitted and incident radiation. For example, referring to Fig. 5–8, this graybody $\alpha$ approximation could be used for a polished copper surface with $T_s$ and $T_R$ both less than approximately 1000 K, for which only 10% of the radiant energy is in the wavelength band 0 to 2 μm, but should not be used for situations in which $T_R$ is much greater than 1000 K.

For nongraybody surfaces, the use of Eq. (5–23) in evaluating $\alpha$ requires that the monochromatic irradiation $G_\lambda$ be specified. For situations involving blackbody or graybody radiation sources, $G_\lambda$ in Eq. (5–23) can be replaced by $E_{b\lambda}$; that is,

$$\alpha = \frac{\int_0^\infty \epsilon_\lambda(T_s)\, E_{b\lambda}(T_R) \, d\lambda}{\int_0^\infty E_{b\lambda}(T_R) \, d\lambda} \tag{5–25}$$

By comparing Eqs. (5–10) and (5–25), we see that the absorptivity is approximately equal to the emissivity for an approximate blackbody or graybody source with $T_R$ equal to $T_s$, which represents the limiting form of Kirchhoff's law. However, it should be noted that the rate of radiation heat transfer is zero for such isothermal conditions. It follows that the most practical consequence of the limiting form of Kirchhoff's law is its application to cases in which $T_s$ and $T_R$ are *not* equal. The first and most obvious extension of this law applies to systems in which the difference between $T_s$ and $T_R$ is small. For such cases, we simply utilize the *isothermal* $\alpha$ *approximation*

$$\alpha = \epsilon(T_s) \tag{5–26}$$

which is equivalent to the graybody $\alpha$ approximation given by Eq. (5–24). However, because of the strong dependence of the surface irradiation properties on the characteristics of the incident thermal radiation for nongraybody surfaces, the following *nonisothermal* $\alpha$ *approximation* is generally preferred:

$$\alpha = \epsilon(T_R) \tag{5–27}$$

This equation is obtained from Eq. (5–25) by merely setting $\epsilon_\lambda(T_s)$ equal to $\epsilon_\lambda(T_R)$. As we have already noted, this restriction on $\epsilon_\lambda$ is essentially satisfied by many metals and certain nonmetallic substances. [For thermal radiation from a blackbody or graybody source incident on a metallic surface with $T_R$ low enough to exclude significant radiation in the near-infrared, visible, or ultraviolet ranges, some investigators prefer an $\alpha$ approximation of the form $\alpha = \epsilon(\sqrt{T_s T_R})$.]

For nongraybody surfaces for which $\epsilon_\lambda(T_s)$ is not approximately equal to $\epsilon_\lambda(T_R)$, Eq. (5–25) can be integrated numerically, provided that $\epsilon_\lambda(T_s)$ is known. This approach was used by Sieber [9] to obtain the absorptivities shown in Fig. 5–13.

To evaluate the graybody (or isothermal) and nonisothermal $\alpha$ approximations, we consider data for commerical-finish aluminum surfaces. Referring back to Fig. 5–9, we see that the aluminum surface exhibits nongraybody characteristics. A comparison of the data for $\epsilon$ shown in Fig. 5–3 with the values of $\alpha$ given in Fig. 5–13 reveals a very close agreement between the two. This result reinforces our confidence in the usefulness of the nonisothermal $\alpha$ approximation. On the other hand, the very fact that $\alpha$ for aluminum shown in Fig. 5–13 ranges from a value less than 0.1 for a source temperature $T_R$ of 300 K to 0.3 for $T_R$ equal to 6000 K alerts us to the limitation of the graybody or isothermal $\alpha$ approximation for this material. For that matter, except for applications involving small temperature differences, the use of the graybody $\alpha$ approximation would lead to serious error in the analysis of radiation from surfaces 1 through 8 listed in Fig. 5–13.

For situations in which most of the irradiation upon a surface originates from a single approximate blackbody or graybody source, the nonisothermal approximation is fairly easy to employ. For example, in the heating of a small metal ball in a furnace, $\alpha$ can readily be estimated by using Eq. (5–27). However, in applications involving multiple nongraybody surfaces and complex reflection patterns, the nonisothermal $\alpha$ approximation is less reliable and is not so easily administered, since the irradiation upon a surface originates from various sources. Consequently, the simple graybody $\alpha$ approximation is often utilized, at least as a first cut.

### *Directional Effects*

Finally, mention should be made of the directional characteristics of irradiation surface properties. Two limiting types of reflection are illustrated in Fig. 5–16. The reflection is *diffuse* if the intensity of the reflected thermal radiation is constant for all angles of irradiation and reflection. On the other hand, if the angle of reflection is equal to the angle of incidence, the reflection is *specular*. Although real surfaces have neither diffuse nor specular irradiation surface properties, many common surfaces can be

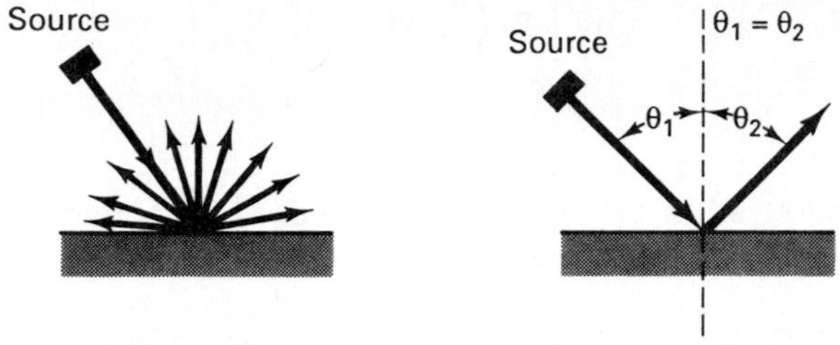


(a) Diffuse reflection. (b) Specular reflection.

**FIGURE 5–16**
Types of reflection.

placed in one or the other of these categories for purpose of design. For example, industrially rough surfaces essentially possess diffuse properties, with polished and smooth surfaces exhibiting near-specular characteristics.

## EXAMPLE 5–6

Assuming that a glass plate transmits 90% of the incident thermal radiation in the wavelength range from 0.29 to 2.70 μm (see Fig. 5–14), is opaque outside this range, and reflects 5% for all wavelengths, determine the total transmissivity $\tau$ of the glass for solar radiation.

### Solution

*Objective* Determine $\tau$ for this glass plate.

*Schematic* Transmission of solar radiation through a glass plate.

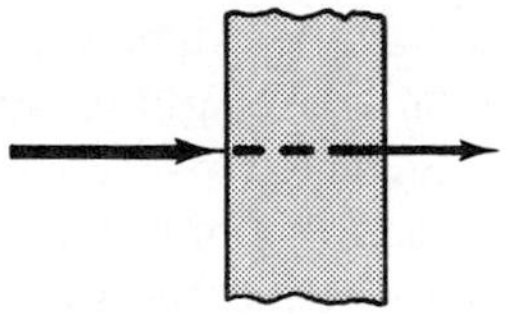

*Properties*

$$\tau_\lambda = 0.9 \qquad \text{for } 0.29\ \mu\text{m} < \lambda < 2.7\ \mu\text{m}$$
$$\tau_\lambda = 0 \qquad \text{for } \lambda \leq 0.29\ \mu\text{m},\ \lambda \geq 2.7\ \mu\text{m}$$

*Assumptions/Conditions*

solar irradiation is equivalent to blackbody emission at 5800 K

*Analysis* Assuming that the sun behaves as a blackbody radiator at 5800 K, we utilize Table A–H–1 to calculate the fraction of solar radiation that falls in the wavelength band from 0.29 to 2.70 μm. Using Table A–H–1, we obtain

$$\frac{E_{b,0\rightarrow\lambda_1}}{E_b} = 0.0254$$

for $\lambda_1 T_R = 0.29\ \mu m\ (5800\ K) = 1680\ \mu m\ K$, and

$$\frac{E_{b,0\rightarrow\lambda_2}}{E_b} = 0.972$$

for $\lambda_2 T_R = 2.70\ \mu m\ (5800\ K) = 15{,}700\ \mu m\ K$. Thus, the fraction of the solar radiation falling in the wavelength band for which $\tau_\lambda$ is 0.9 is

$$\frac{E_{b,\lambda_1\rightarrow\lambda_2}}{E_b} = 0.972 - 0.0254 = 0.947$$

It follows that the total transmissivity $\tau$ is

$$\tau = 0.947(0.9) = 0.852$$

To calculate the total absorptivity, we write

$$\alpha = 1 - \rho - \tau = 1 - 0.05 - 0.852 = 0.098$$

### 5–3–3 Thermal Radiation Properties of Gases

We are reminded that a vacuum provides the ideal medium for the transfer of thermal radiation from one surface to another. Thermal radiation passes through such evacuated spaces at the speed of light. Of course, in most practical situations at least some fluid, often in the form of a gas, lies in the path of the thermal radiation. Elementary gases such as oxygen ($O_2$), hydrogen ($H_2$), nitrogen ($N_2$), and dry air, which have symmetrical molecular structures, are essentially transparent to thermal radiation at low to moderate temperatures. Hence, the presence of such *nonparticipating gases* can generally be ignored. But, other polyatomic gases, such as water vapor ($H_2O$), carbon dioxide ($CO_2$), sulfur dioxide ($SO_2$), and various hydrocarbons absorb and emit significant amounts of thermal radiation. Gases such as these are known as *participating gases*. For example, water vapor contained at 1 atm pressure and 100°C between two plates 1 m apart would emit as much as 55% of the energy that would be emitted by a blackbody at 100°C with the same surface area as the plates. Carbon dioxide at the same temperature and pressure would emit about 20%. But carbon monoxide, which is diatomic, would only emit about 3% of the energy that would be emitted by an ideal radiator. Hence, the presence of participating gases between thermal radiating surfaces can have very important effects.

Thermal radiation emission and absorption characteristics of participating gases are generally much more complicated than for opaque solids. Whereas the emission and absorption of thermal radiation for opaque materials are surface phenomena, the thickness, shape, surface area, pressure, and temperature distribution can all affect thermal radiation in gases. Representative measurements for $\alpha_\lambda$ are shown in Fig. 5–17 for carbon dioxide in terms of the wave number $1/\lambda$. The absorbing spectrum of this important gas is seen to consist of distinct narrow bands. These absorbing and emitting patterns are typical of participating gases.

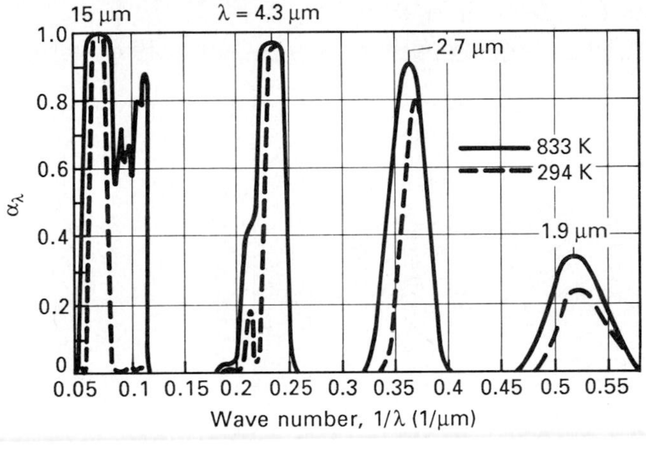


**FIGURE 5–17** Monochromatic absorptivity $\alpha_\lambda$ for carbon dioxide (Edwards [15]).

Concerning the more practical information pertaining to the total properties, which account for all wavelengths and all directions in which thermal radiation passes, Hottel and Egbert [16] have developed the charts shown in Fig. 5–18(a) and (b) for water vapor and carbon dioxide total emissivities $\epsilon$. Note that $\epsilon$ increases with increasing *mean beam length* $L_e$ and partial pressure. The mean beam length, which accounts for all possible directions the thermal radiation may take, is listed in Table 5–2 for several standard systems. For situations in which $L_e$ has not been evaluated, it is generally approximated by $L_e = 3.6V/A_s$. The data in Fig. 5–18(a) and (b) were obtained for a total pressure of 1 atm. Correction charts are available in Hottel [17] for systems under other total pressures. For systems involving combustion, both water vapor and carbon dioxide are present. For such situations, the total gas emissivity is simply equal to the sum of the emissivities for each component, minus a small correction factor that accounts for the mutual emission that takes place between the two gases. Corrections that account for this mutual emission factor are presented in Hottel [17]. This small difference can generally be neglected as a first approximation.

In regard to the total absorptivity $\alpha$ of participating gases, Hottel [17] has developed approximate graybody correlations for water vapor and carbon dioxide which take the forms

$$\alpha_w = \epsilon_w(T_s)\left(\frac{T_m}{T_s}\right)^{0.45} \qquad \alpha_c = \epsilon_c(T_s)\left(\frac{T_m}{T_s}\right)^{0.65} \tag{5–28a,b}$$

where $T_m$ is the mean temperature of the gas and $T_s$ the surface temperature. $\epsilon_w$ and $\epsilon_c$ are evaluated from Fig. 5–18 with the parameters $P_wL_e$ and $P_cL_e$ replaced by $P_wL_eT_s/T_m$ and $P_cL_eT_s/T_m$. For a mixture of water vapor and carbon dioxide, the absorptivity of the mixture is simply equal to the sum of $\alpha_w$ and $\alpha_c$, minus a small correction.

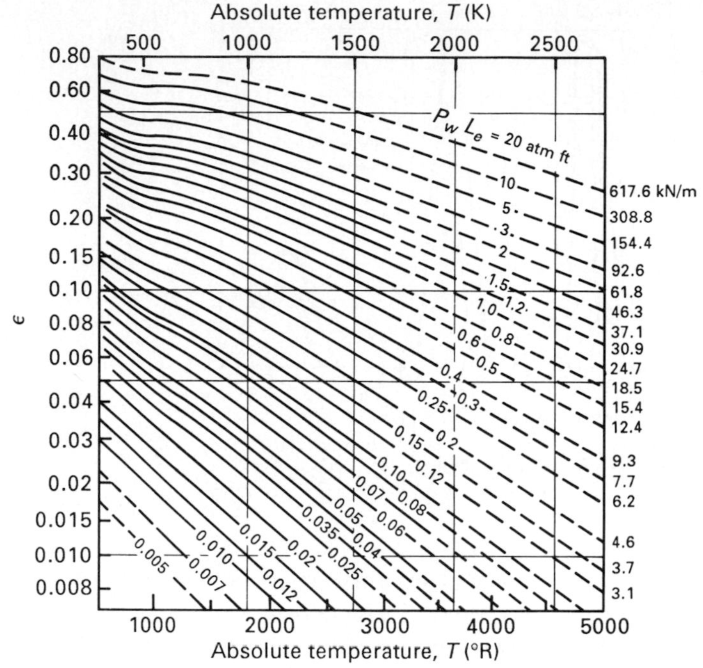


(a) Water vapor.

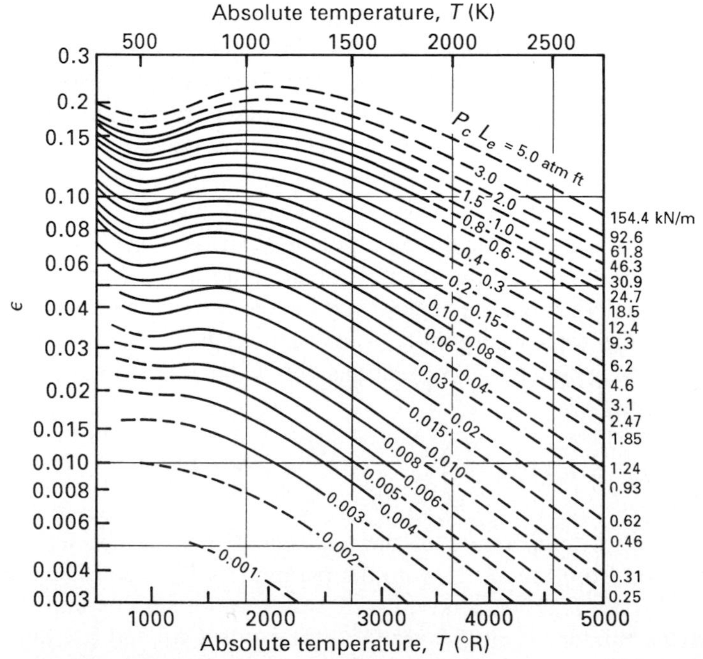


(b) Carbon dioxide.

**FIGURE 5–18** Emissivities of gases at 1 atm total pressure (Hottel and Egbert [16]).

**TABLE 5–2** Mean beam length $L_e$ for radiation from entire gas volume

| *Gas volume* | *Characteristic dimension* | $L_e$ |
|---|---|---|
| Volume between two infinite planes | Separation distance $L$ | $1.8L$ |
| Cube; radiation to any face | Edge $L$ | $0.60L$ |
| Sphere | Diameter $D$ | $0.65D$ |
| Circular cylinder with $L = D$; radiation to entire surface | Diameter $D$ | $0.60D$ |
| Circular cylinder, with semi-infinite length; radiation to entire base | Diameter $D$ | $0.65D$ |

*Source:* From Hottel [17] and Eckert and Drake [18].

It should be mentioned that in the high-temperature process of combustion which produces nonluminous products such as $H_2O$ and $CO_2$, clouds of carbon particles radiate intensely at short wavelengths in the visible-light region of the electromagnetic spectrum. This visible thermal radiation which is emitted during combustion is what is referred to as the *flame*. The total emissivities of luminous flames range from values of the order of 0.2 for gaseous hydrocarbon fuels to almost unity for fuels such as oil which are burned under conditions of large carbon/hydrogen ratios. In engineering applications, the emission from luminous flames generally must be determined experimentally.

## EXAMPLE 5–7

The walls of a cubical furnace 1 m on a side are maintained at 500 K. The products of combustion at 1 atm and 1500 K consist of 20% $CO_2$, 15% $H_2O$, and 65% $N_2$ (on molar basis). Assuming that the walls are essentially black, estimate the gas emissivity $\epsilon_m$ and absorptivity $\alpha_m$.

### Solution

*Objective* Approximate $\epsilon_m$ and $\alpha_m$ for the products of combustion within the furnace.

*Schematic* Cubical furnace with blackbody surfaces.

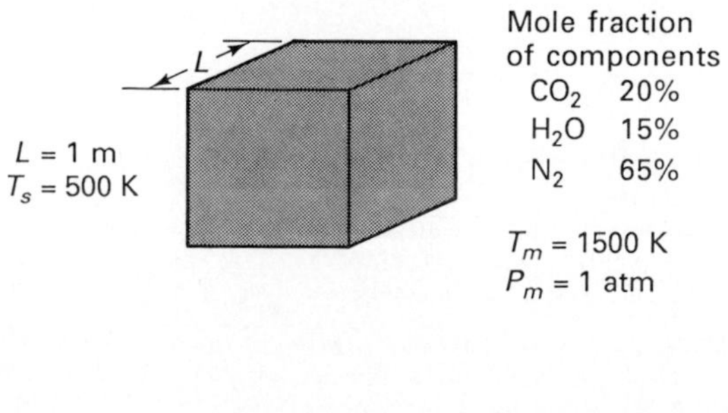
L
L = 1 m
Ts = 500 K
Mole fraction
of components
CO2 20%
H2O 15%
N2 65%
Tm = 1500 K
Pm = 1 atm

*Assumptions/Conditions*

blackbody surface irradiation
enclosure contains participating gases $H_2O$ and $CO_2$

*Properties* Walls: $\epsilon = 1$. The properties of the gases are considered in the analysis.

*Analysis* Assuming that the nitrogen is essentially nonparticipating, we focus our attention on the contributions of the water ($w$) and carbon-dioxide ($c$) to the thermal radiation properties of the mixture. The ideal gas law ($PV = mRT = n\bar{R}T$) indicates that the partial pressure of an ideal gas in a mixture is equal to the product of the total pressure of the mixture and the mole fraction of the gas. It follows that

$$P_w = 0.15(1 \text{ atm}) = 0.15 \text{ atm}$$
$$P_c = 0.2(1 \text{ atm}) = 0.2 \text{ atm}$$

Referring to Table 5–2, the mean equivalent beam length $L_e$ is given by

$$L_e = 0.6L = 0.6(1 \text{ m}) = 0.6 \text{ m}$$

Thus, the pressure/length parameters are

$$P_w L_e = 0.15 \text{ atm } (0.6 \text{ m}) = 0.09 \text{ atm m} = 0.295 \text{ atm ft}$$
$$P_c L_e = 0.2 \text{ atm } (0.6 \text{ m}) = 0.12 \text{ atm m} = 0.393 \text{ atm ft}$$

Making use of Fig. 5–18, the emissivities $\epsilon_w$ and $\epsilon_c$ are approximated by

$$\epsilon_w = 0.08 \qquad \epsilon_c = 0.094$$

for a mean gas temperature of 1500 K. Thus, the total emissivity of the gas is approximated by

$$\epsilon_m = \epsilon_w + \epsilon_c = 0.08 + 0.094 = 0.174$$

By reference to Hottel [17], we find that the correction for $\epsilon_m$, which accounts for the mutual emission of the two participating gases, is of the order of $-10\%$. Therefore, as a first approximation, we are safe in assuming $\epsilon_m \simeq 0.174$.

To estimate the absorptivities, we calculate the modified pressure/length parameters.

$$P_w L_e \frac{T_s}{T_m} = 0.295 \text{ atm ft} \frac{500 \text{ K}}{1500 \text{ K}} = 0.0983 \text{ atm ft}$$

$$P_c L_e \frac{T_s}{T_m} = 0.393 \text{ atm ft} \frac{500 \text{ K}}{1500 \text{ K}} = 0.131 \text{ atm ft}$$

Evaluating the emissivities at $T_s$, we have

$$\epsilon_w(T_s) = 0.097 \qquad \epsilon_c(T_s) = 0.075$$

The absorptivities are now calculated by utilizing Eqs. (5–28a) and (5–28b).

$$\alpha_w = \epsilon_w(T_s)\left(\frac{T_m}{T_s}\right)^{0.45} = 0.097\left(\frac{1500\ \text{K}}{500\ \text{K}}\right)^{0.45} = 0.159$$

$$\alpha_c = \epsilon_c(T_s)\left(\frac{T_m}{T_s}\right)^{0.65} = 0.075\left(\frac{1500\ \text{K}}{500\ \text{K}}\right)^{0.65} = 0.153$$

Referring to Hottel [17], we find the correction for mutual absorption is less than 1%. It follows that the total absorptivity of the gas is about

$$\alpha_m = \alpha_w + \alpha_c = 0.159 + 0.153 = 0.312$$

## 5–4 RADIATION SHAPE FACTOR

As a final step toward our objective of developing practical predictions for radiation heat transfer, attention is turned to important geometric aspects of thermal radiation exchange between surfaces.

The *radiation shape factor* $F_{s-R}$ introduced in Chap. 1 is defined as the fraction of thermal radiation leaving a diffuse surface $A_s$ that passes through a nonparticipating medium to surface $A_R$. $F_{s-R}$ is also commonly referred to as *view factor*, *configuration factor*, and *shape factor*. Because we will be dealing with systems involving more than one source, it is convenient to replace the subscript $R$ by the source surface identification index $j$, where $j = 1, 2, 3, \ldots, N$. Thus, the radiation shape factor will generally be denoted by $F_{s-j}$, except for systems involving only two surfaces, for which case we will write $F_{1-2}$.

To illustrate, we consider the two simple systems shown in Fig. 5–19(a) and (b). The radiation shape factors $F_{1-2}$ and $F_{2-1}$ for the infinitely long parallel-plate system of Fig. 5–19(a) are both unity, because all the energy leaving either surface reaches the other surface. For the concentric-sphere arrangement shown in Fig. 5–19(b), $F_{1-2}$ is also equal to unity. But, because only a fraction of the energy leaving the outer spherical surface $A_2$ reaches the inner surface $A_1$, $F_{2-1}$ is less than unity.

We now turn our attention to fundamental principles that apply to radiation shape factors and design curves for several standard geometries.

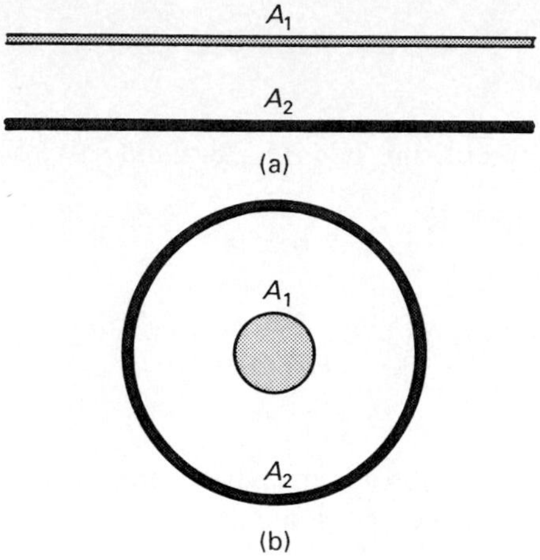


(a) Parallel plates; $F_{1-2} = 1$, $F_{2-1} = 1$.

(b) Concentric spheres; $F_{1-2} = 1$, $F_{2-1} < 1$.

**FIGURE 5–19**
Radiation systems.

## 5–4–1 Law of Reciprocity

The general relationship between $F_{s-j}$ and $F_{j-s}$ is given by the *reciprocity law*,

$$A_s F_{s-j} = A_j F_{j-s} \tag{5–29}$$

This relationship is developed in Sparrow and Cess [5] and Siegel and Howell [11] on the basis of geometric considerations. Based on this principle, we see that $F_{2-1}$ for the spherical system shown in Fig. 5–19(b) is

$$F_{2-1} = \frac{A_1}{A_2} F_{1-2} = \left(\frac{r_1}{r_2}\right)^2 \tag{5–30}$$

Note that $F_{2-1}$ approaches zero as $r_1/r_2$ decreases, and approaches unity as $r_1$ approaches $r_2$.

## 5–4–2 Summation Principles

Another important geometric concept pertains to the radiation shape factors from surface $s$ to $N$ surfaces forming an enclosure. This *first summation principle* is written as

$$\sum_{j=1}^{N} F_{s-j} = 1 \tag{5–31}$$

For example, for the three-surface enclosure shown in Fig. 5–20, we have

$$F_{1-2} + F_{1-3} = 1 \tag{5–32}$$

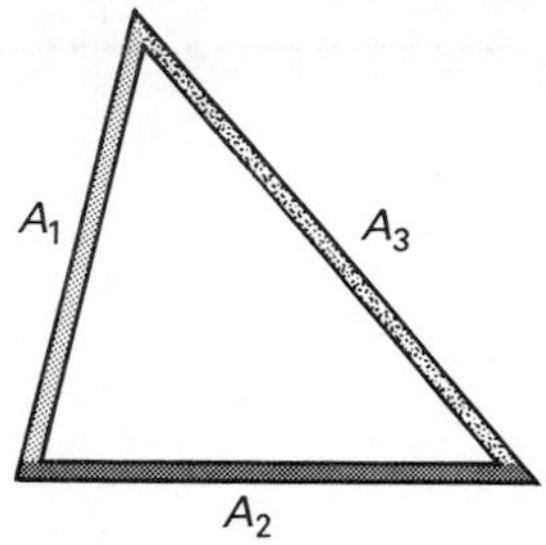


**FIGURE 5–20**
Trisurface enclosure.

This principle also requires that

$$F_{2-1} + F_{2-3} = 1 \qquad F_{3-1} + F_{3-2} = 1 \tag{5–33,34}$$

For systems involving concave surfaces, we must include the term $F_{s-s}$, which accounts for the fraction of thermal radiation leaving surface $s$ that is directly incident upon itself. Thus, for the spherical system of Fig. 5–19(b), we write

$$F_{2-1} + F_{2-2} = 1 \tag{5–35}$$

A *second summation principle* states that the total radiation shape factor is equal to the sum of its parts. To illustrate, the radiation shape factor from surface $A_1$ to the combined surfaces of $A_2$ and $A_3$ in Fig. 5–20 is

$$F_{1-(2,3)} = F_{1-2} + F_{1-3} \tag{5–36}$$

By employing the law of reciprocity, this equation is put into the useful form

$$(A_2 + A_3)F_{(2,3)-1} = A_2F_{2-1} + A_3F_{3-1} \tag{5–37}$$

## 5–4–3 Design Curves

Relationships have been developed for radiation shape factors for a great many geometries. The theoretical approach to this task involves the use of the intensity concept for diffuse surfaces, as illustrated in Example 5–9.

Because the theoretical evaluation of radiation shape factors is generally quite involved, standard design curves are heavily relied upon in practice. Design equations for radiation shape factors are given in Tables 5–3 and 5–4 and Figs. 5–21, 5–22, and 5–23 for several arrangements that are commonly encountered. More comprehensive listings of radiation shape factors are given in references 11, 17, 20–22. As shown in Example 5–8, information on radiation shape factors for standard geometries such as those given in Figs. 5–21 and 5–22 can sometimes be extended to other geometrical arrangements of practical interest by utilizing the reciprocity and summation principles.

**TABLE 5–3** Radiation shape factors for two-dimensional geometries

| *Geometry* | *Relation* |
|---|---|
| Parallel Plates with Midlines Connected by Perpendicular<br>$W_1 = w_1/L,\ W_2 = w_2/L$<br>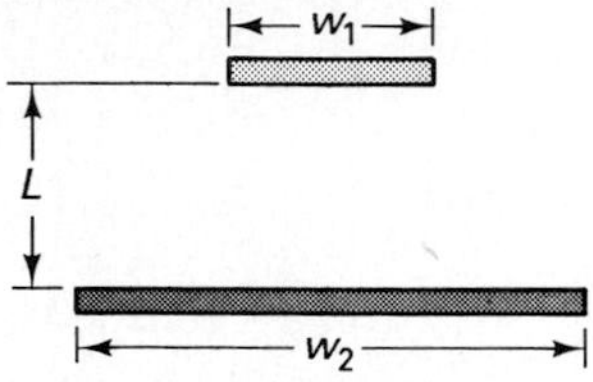 | $$F_{1-2} = \frac{[(W_1 + W_2)^2 + 4]^{1/2} - [(W_2 - W_1)^2 + 4]^{1/2}}{2W_1}$$ |
| Inclined Parallel Plates with Equal Width and a Common Edge<br>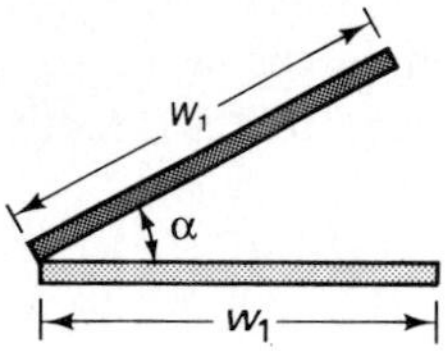 | $$F_{1-2} = 1 - \sin\left(\frac{\alpha}{2}\right)$$ |
| Perpendicular Plates with a Common Edge<br>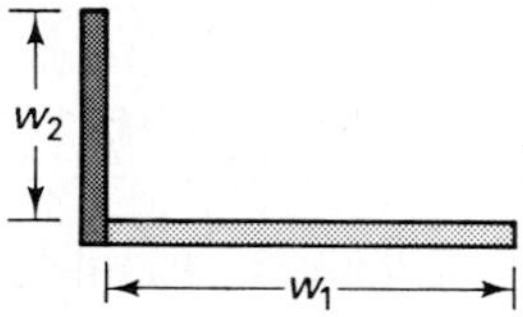 | $$F_{1-2} = \frac{1 + (w_2/w_1) - [1 + (w_2/w_1)^2]^{1/2}}{2}$$ |
| Three-Sided Enclosure<br>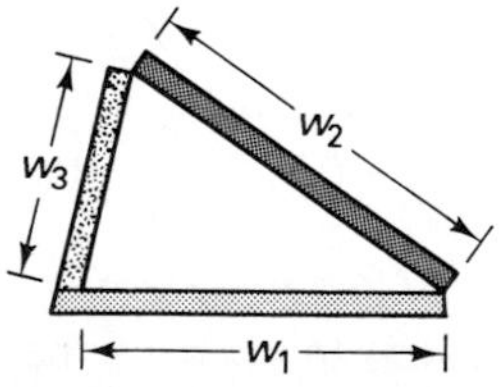 | $$F_{1-2} = \frac{w_1 + w_2 - w_3}{2w_1}$$ |
| Parallel Cylinders of Different Radius<br>$R = r_2/r_1,\ S = s/r_1$<br>$C = 1 + R + S$<br>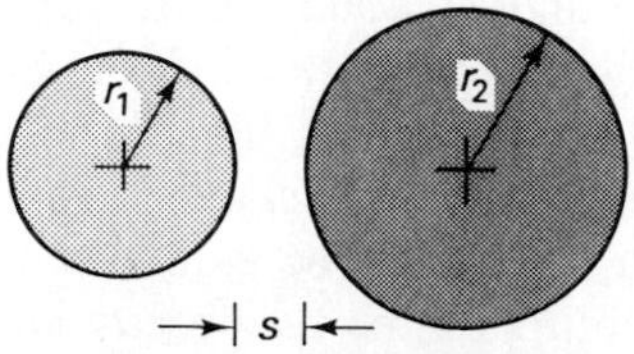 | $$F_{1-2} = \frac{1}{2\pi}\left\{\pi + [C^2 - (R + 1)^2]^{1/2} - [C^2 - (R - 1)^2]^{1/2} + (R - 1)\cos^{-1}\left(\frac{R}{C} - \frac{1}{C}\right) - (R + 1)\cos^{-1}\left(\frac{R}{C} + \frac{1}{C}\right)\right\}$$ |

*Sources*: Howell [21] and Hamilton and Morgan [22].

**TABLE 5-4** Radiation shape factors for three-dimensional geometries

| *Geometry* | *Relation* |
|---|---|
| Aligned Parallel Rectangles<br>$\bar{X} = X/Z$, $\bar{Y} = Y/Z$<br>(See Fig. 5-21)<br>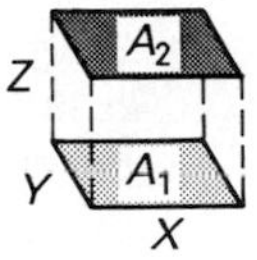 | $F_{1-2} = \frac{2}{\pi\bar{X}\bar{Y}}\left\{\ln\left[\frac{(1+\bar{X}^2)(1+\bar{Y}^2)}{1+\bar{X}^2+\bar{Y}^2}\right]^{1/2} + \bar{X}(1+\bar{Y}^2)^{1/2}\tan^{-1}\left[\frac{\bar{X}}{(1+\bar{Y}^2)^{1/2}}\right] + \bar{Y}(1+\bar{X}^2)^{1/2}\tan^{-1}\left[\frac{\bar{Y}}{(1+\bar{X}^2)^{1/2}}\right] - \bar{X}\tan^{-1}\bar{X} - \bar{Y}\tan^{-1}\bar{Y}\right\}$ |
| Perpendicular Rectangles with a Common Edge<br>$H = Z/X$, $W = Y/X$<br>(See Fig. 5-22)<br>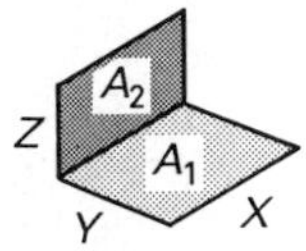 | $F_{1-2} = \frac{1}{\pi W}\left(W\tan^{-1}\frac{1}{W} + H\tan^{-1}\frac{1}{H} - (H^2+W^2)^{1/2}\tan^{-1}\frac{1}{(H^2+W^2)^{1/2}} + \frac{1}{4}\ln\left\{\frac{(1+W^2)(1+H^2)}{1+W^2+H^2}\left[\frac{W^2(1+W^2+H^2)}{(1+W^2)(W^2+H^2)}\right]^{W^2} \times \left[\frac{H^2(1+H^2+W^2)}{(1+H^2)(H^2+W^2)}\right]^{H^2}\right\}\right)$ |
| Coaxial Parallel Disks<br>$R_1 = r_1/L$, $R_2 = r_2/L$<br>$S = 1 + (1+R_2^2)/R_1^2$<br>(See Fig. 5-23)<br>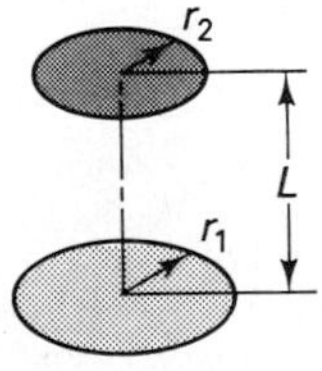 | $F_{1-2} = \frac{1}{2}\{S - [S^2 - 4(r_2/r_1)^2]^{1/2}\}$ |

*Sources*: Mackey et al. [19], Howell [21], and Hamilton and Morgan [22].

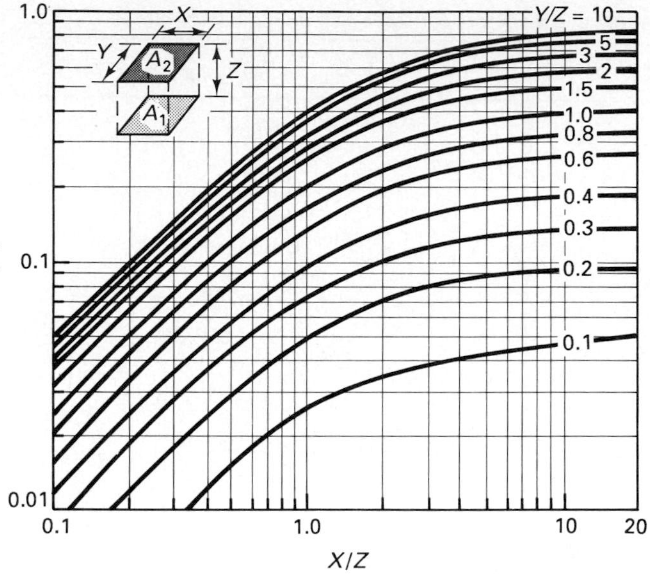


**FIGURE 5–21** Radiation shape factor for parallel plates.

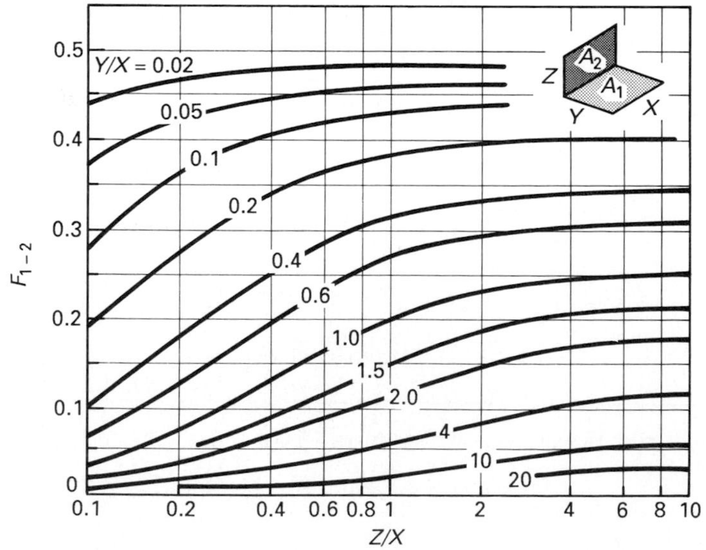


**FIGURE 5–22** Radiation shape factor for perpendicular rectangles with a common edge.

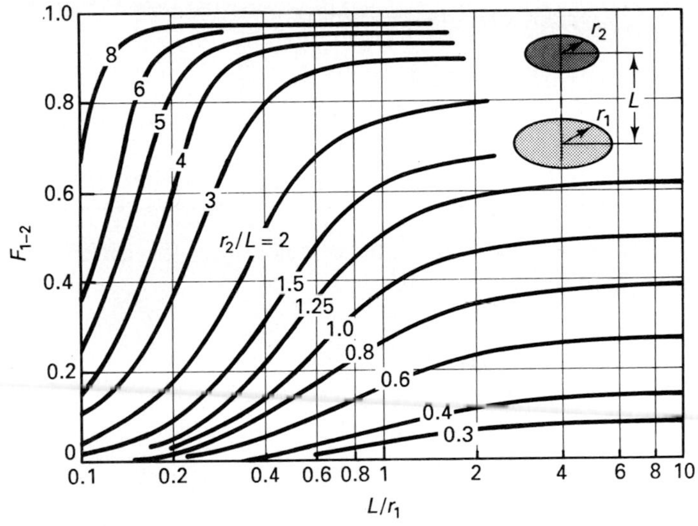


**FIGURE 5–23** Radiation shape factor for coaxial parallel disks.

## EXAMPLE 5–8

Determine the radiation shape factor for surfaces $A_1$ and $A_2$ shown in Fig. E5–8.

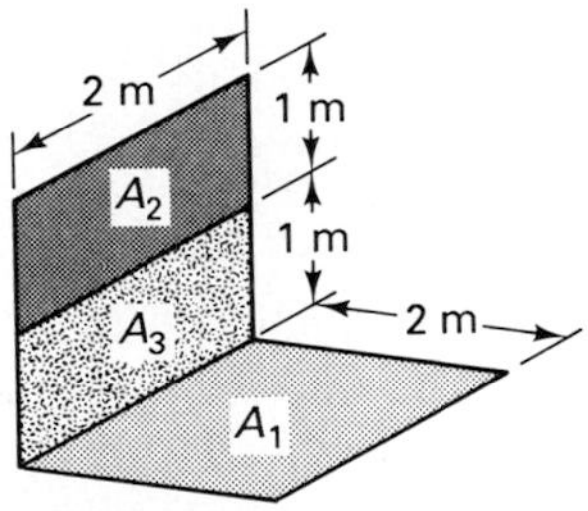


**FIGURE E5–8**
Trisurface system.

### Solution

*Objective* Determine $F_{1-2}$ and $F_{2-1}$.

*Analysis* According to the second summation principle, we write

$$F_{1-(2,3)} = F_{1-2} + F_{1-3}$$

or

$$F_{1-2} = F_{1-(2,3)} - F_{1-3}$$

Referring to Fig. 5–22, $F_{1-3} = 0.15$ and $F_{1-(2,3)} = 0.2$, such that

$$F_{1-2} = 0.2 - 0.15 = 0.05$$

Using the principle of reciprocity, $F_{2-1}$ becomes

$$F_{2-1} = \frac{A_1}{A_2} F_{1-2} = \frac{2}{1}(0.05) = 0.1$$

## EXAMPLE 5–9

Develop an expression for the radiation shape factor from the differential element $dA_1$ to the finite disk $A_2$ shown in Fig. E5–9.

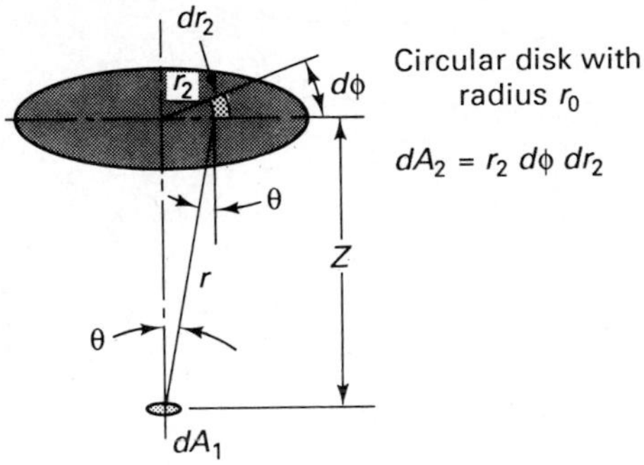
$dr_2$
$r_2$
$d\phi$
Circular disk with radius $r_0$
$dA_2 = r_2\, d\phi\, dr_2$
$\theta$
$Z$
$r$
$\theta$
$dA_1$

**FIGURE E5–9**
Parallel circular disks.

### Solution

*Objective* Determine $F_{dA_1-A_2}$.

*Assumptions/Conditions*

diffuse surfaces

*Analysis* Referring to Fig. E5–9, the total rate of thermal radiation diffusely emitted by surface element $dA_1$ is $E_1\, dA_1$. To determine the portion of this energy that reaches surface $A_2$, we return to the definition of intensity given by Eq. (5–14),

$$I_1 = \frac{dE_1}{\cos\theta\, d\omega}$$

or

$$dE_1 = I_1 \cos\theta\, d\omega$$

where $\theta$ is the angle between the normal to surface $dA_1$ and the line $r$ drawn between the area elements $dA_1$ and $dA_2$. The solid angle $d\omega$ is given by Eq. (5–13),

$$d\omega = \frac{dA_r}{r^2}$$

where $dA_r$ is the differential area normal to the line $r$; that is,

$$dA_r = dA_2 \cos\theta$$

Thus, the thermal radiation flux emitted from $dA_1$ that reaches $dA_2$ is

$$dE_1 = I_1 \cos^2\theta \frac{dA_2}{r^2}$$

where $I_1 = E_1/\pi$ for diffuse conditions (see Example 5–3).

The radiation shape factor $F_{dA_1-dA_2}$ is simply equal to the ratio $dE_1/E_1$; that is,

$$F_{dA_1-dA_2} = \frac{1}{\pi}\cos^2\theta \frac{dA_2}{r^2}$$

To obtain $F_{dA_1-A_2}$, we must integrate over the entire surface $A_2$.

$$F_{dA_1-A_2} = \frac{1}{\pi}\int_{A_2} \cos^2\theta \frac{dA_2}{r^2}$$

Noting that

$$dA_2 = r_2\, d\phi\, dr_2$$

we obtain

$$F_{dA_1-A_2} = \frac{1}{\pi}\int_0^{r_0}\int_0^{2\pi} \cos^2\theta \frac{r_2\, d\phi\, dr_2}{r^2} \tag{a}$$

The quantities $r$ and $\theta$ are expressed in terms of $r_2$ by

$$r^2 = Z^2 + r_2^2 \qquad \cos\theta = \frac{Z}{(Z^2 + r_2^2)^{1/2}}$$

Thus, Eq. (a) takes the form

$$F_{dA_1-A_2} = \frac{1}{\pi}\int_0^{r_0}\int_0^{2\pi} \frac{Z^2 r_2\, d\phi\, dr_2}{(Z^2 + r_2^2)^2} = 2Z^2\int_0^{r_0} \frac{r_2\, dr_2}{(Z^2 + r_2^2)^2}$$

Setting $Z^2 + r_2^2 = \xi$, we continue the integration process,

$$F_{dA_1-A_2} = Z^2\int_{Z^2}^{Z^2+r_0^2} \frac{d\xi}{\xi^2} = -\frac{Z^2}{\xi}\Bigg|_{Z^2}^{Z^2+r_0^2}$$

$$= Z^2\left(\frac{1}{Z^2} - \frac{1}{Z^2 + r_0^2}\right) = \frac{r_0^2}{Z^2 + r_0^2} \tag{b}$$

Notice that $F_{dA_1-A_2}$ appropriately approaches unity as the radius $r_0$ of the disk becomes large.

## 5–5 PRACTICAL THERMAL ANALYSIS OF RADIATION HEAT TRANSFER

Now that we have a basic understanding of the physical mechanism, properties, and geometric factors pertaining to the thermal radiation phenomenon, we are in a position to develop predictions for the rate of heat transfer (i.e., *net* rate of exchange of thermal radiation) that occurs in diffuse thermal radiating systems. Therefore, we turn our attention to the practical analysis of radiation heat transfer.

To illustrate the perspective that we will be taking, consider the radiating body in Fig. 5–24. The thermal radiation heat transfer $q_R$ from surface $A_s$ to other surrounding surfaces will be analyzed in sections to follow for cases in which the surface temperature $T_s$ and thermal radiation properties are uniform over $A_s$. But we must recognize that the rate of thermal radiation heat transfer $q_R$ from surface $A_s$ must be in balance with changes in the rate of storage and the rates of energy transfer into and out of the body through other surfaces and by other means. That is, the energy transfer associated with this body must satisfy the first law of thermodynamics,

$$\Sigma \dot{E}_i = q_R + \frac{\Delta E_s}{\Delta t} \tag{5–38}$$

where $\Sigma \dot{E}_i$ is the net rate of energy transfer into the body, not including $q_R$. For steady-state conditions, $\Delta E_s/\Delta t$ is zero with $T_s$ being independent of time, such that Eq. (5–38) reduces to

$$\Sigma \dot{E}_i = q_R \tag{5–39}$$

This equation simply indicates that the rate of thermal radiation heat transfer from surface $A_s$ with steady surface temperature $T_s$ must be replaced from an outside source, such as by the electrical generation of power within the body. If, on the other hand, a net rate of thermal radiation is received by surface $A_s$ such that $q_R$ is negative, this same rate of energy must be transferred from the body into a heat sink if steady-state conditions are to be maintained. For example, energy can be taken out of the back surface of the body by radiation or convection.

For unsteady conditions, $T_s$ is time-dependent and $q_R$ must satisfy Eq. (5–38). That is, $q_R$ must be in balance with the rate of energy brought into a body $\Sigma \dot{E}_i$ and the rate of change in energy stored within the body $\Delta E_s/\Delta t$. In this regard, for bodies with negligible thermal resistance to conduction (i.e., small thermal radiation Biot number $Bi_R$), the temperature throughout the body is essentially uniform. For such

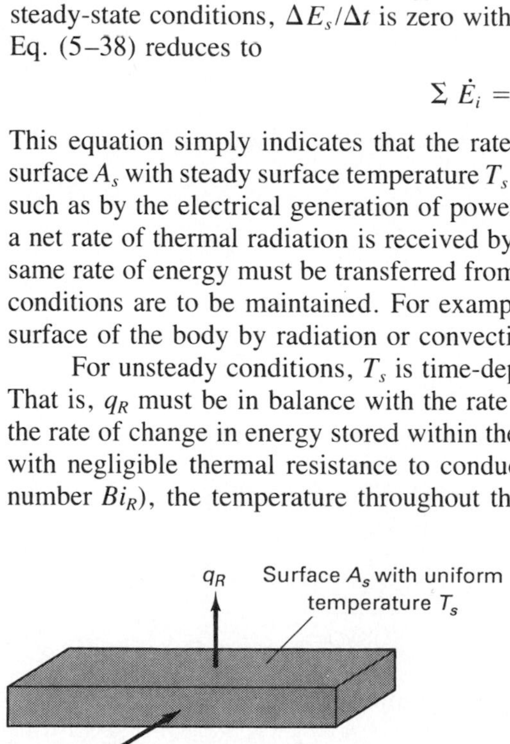
$q_R$
Surface $A_s$ with uniform temperature $T_s$
$\Sigma \dot{E}_i$

**FIGURE 5–24**
Thermal radiating body; $\Sigma \dot{E}_i$ represents the net rate of energy transfer into body, excluding radiation heat transfer $q_R$.

bodies, a lumped approximation can be made for $\Delta E_s/\Delta t$, with Eq. (5–38) taking the simpler form (for constant properties)

$$\Sigma \dot{E}_i = q_R + mc_v \frac{dT_s}{dt} \tag{5–40}$$

Keeping in mind that the energy transfer associated with the body of each radiating surface must satisfy the first law of thermodynamics, we now move on to the development of practical thermal analyses for the radiation heat transfer between two or more diffuse thermal radiating surfaces with uniform properties and uniform heating. To begin with, we focus on radiating surfaces in which conduction and convection are not significant. In addition to analyzing ideal blackbody and diffuse nonblackbody thermal radiating systems under steady-state conditions with a non-participating medium, consideration is given to systems involving unsteady conditions, participating medium, combined modes, and solar radiation. Approaches to analyzing more complex systems involving surfaces with nondiffuse characteristics, nonuniform properties, and nonuniform heating are introduced in references 5 and 11.

To simplify our notation, the subscript $R$ for radiation heat transfer will be omitted throughout the remainder of this chapter, except for cases involving combined modes of heat transfer. Accordingly, the rate of radiation heat transfer from surface $A_s$ to a second surface $A_j$ will be designated by $q_{s-j}$. In addition, the total radiation-heat-transfer rate from surface $A_s$ to $N$ surfaces of an enclosure will be represented by $q_s$.

## 5–5–1 Blackbody Thermal Radiation

The analysis of radiation heat transfer in systems involving ideal blackbody surfaces is quite straightforward because the energy transfer is direct, with no reflections. In this section we will consider basic two-surface and multisurface blackbody systems. The analysis of these ideal systems provides a foundation for the practical analysis of real nonblackbody surfaces, which is considered in Sec. 5–5–3. In addition, the results of this analysis provide us with a new viewpoint concerning the analogy between the flow of electric current and radiation heat transfer.

### *Bisurface Systems*

Representative blackbody systems consisting of two surfaces are shown in Fig. 5–25(a) and (b). Whereas no thermal radiation enters or leaves the closed system of Fig. 5–25(a), thermal radiation is propagated through the open spaces or "windows" of the radiatively open system of Fig. 5–25(b).

To determine the radiation heat transfer rate between two blackbody surfaces $A_1$ and $A_2$ in radiatively open or closed systems, we must recognize that the rate of thermal radiation leaving $A_1$ that reaches and is absorbed by $A_2$ is $A_1F_{1-2}E_{b1}$. Similarly, the rate of thermal radiation emitted by $A_2$ that is absorbed by $A_1$ is $A_2F_{2-1}E_{b2}$. It

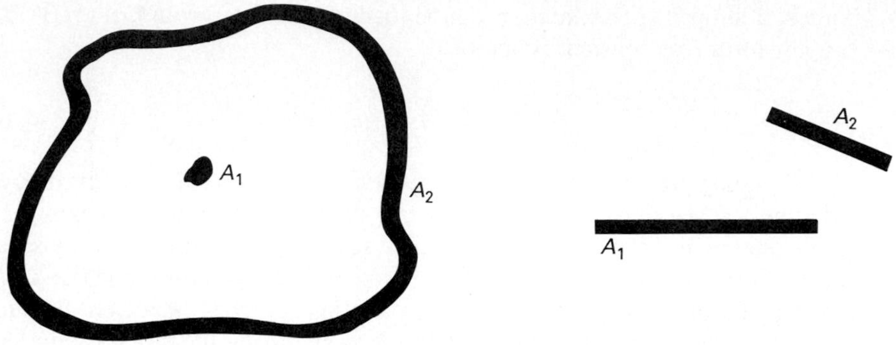


(a) Closed system. (b) Open system.

**FIGURE 5–25** Bisurface thermal radiation blackbody systems.

follows that the rate of radiation heat transfer $q_{1-2}$ from $A_1$ to $A_2$ in bisurface systems is equal to the net exchange of thermal radiation between $A_1$ and $A_2$; that is,

$$q_{1-2} = A_1F_{1-2}E_{b1} - A_2F_{2-1}E_{b2} \tag{5–41}$$

Utilizing the principle of reciprocity, which states that $A_1F_{1-2}$ and $A_2F_{2-1}$ are equal, this equation can be written as

$$q_{1-2} = A_1F_{1-2}(E_{b1} - E_{b2}) \tag{5–42}$$

Finally, introducing the Stefan-Boltzmann law, Eq. (5–2), we have

$$q_{1-2} = A_1F_{1-2}\sigma(T_1^4 - T_2^4) \tag{5–43}$$

### *Multisurface Systems*

To determine the rate of radiation heat transfer from blackbody surface area $A_s$ to $N$ blackbody surfaces in radiatively closed or open systems such as those shown in Fig. 5–26(a) and (b), we must account for the exchange of thermal radiation between surface $A_s$ and each of the $N$ surfaces. The net rate of thermal radiation exchange between $A_s$ and one of the surfaces $A_j$ is simply

$$q_{s-j} = A_sF_{s-j}E_{bs} - A_jF_{j-s}E_{bj} = A_sF_{s-j}(E_{bs} - E_{bj}) \tag{5–44}$$

It follows that the total rate of radiation heat transfer $q_s$ from $A_s$ to all $N$ blackbody surfaces is given by

$$q_s = \sum_{j=1}^{N} A_sF_{s-j}(E_{bs} - E_{bj}) \tag{5–45}$$

With $E_{bs}$ and $E_{bj}$ specified by the Stefan-Boltzmann law, we obtain

$$q_s = \sum_{j=1}^{N} A_sF_{s-j}\sigma(T_s^4 - T_j^4) \tag{5–46}$$

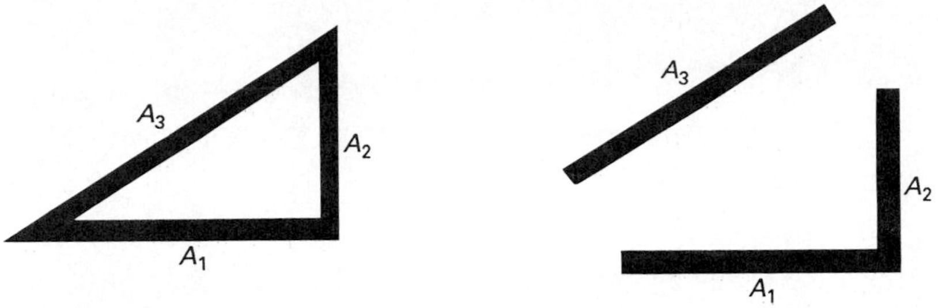


(a) Closed system. (b) Open system.

**FIGURE 5–26** Trisurface thermal radiation blackbody systems.

## EXAMPLE 5–10

Demonstrate the validity of the principle of reciprocity by analyzing the radiation heat transfer between two blackbodies $A_1$ and $A_2$ which are at the same temperature.

### Solution

*Objective* Deduce the principle of reciprocity, $A_1F_{1-2} = A_2F_{2-1}$.

*Schematic* Thermal radiation between two blackbodies.

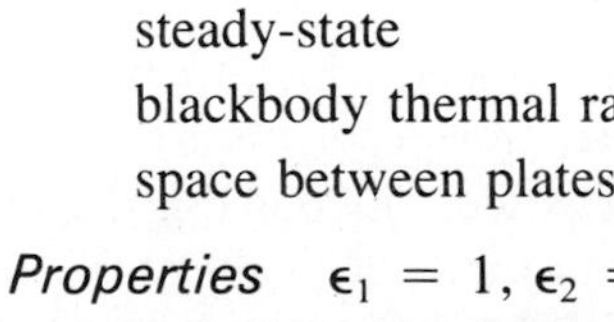


*Assumptions/Conditions*

steady-state
blackbody thermal radiation
space between plates is evacuated

*Properties* $\epsilon_1 = 1$, $\epsilon_2 = 1$.

*Analysis* The rate of radiation heat transfer between two blackbodies is given by Eq. (5–41),

$$q_{1-2} = A_1F_{1-2}E_{b1} - A_2F_{2-1}E_{b2}$$

Of course, $q_{1-2}$ must be zero and $E_{b1}$ is equal to $E_{b2}$, since both surfaces are at the same temperature. It follows that

$$(A_1F_{1-2} - A_2F_{2-1})E_{b1} = 0$$

Because

$$E_{b1} = E_{b2} = \sigma T_1^4$$

the factor $(A_1F_{1-2} - A_2F_{2-1})$ must be zero. Thus, the reciprocity law

$$A_1F_{1-2} = A_2F_{2-1}$$

must be satisfied.

This same reasoning can be applied to any two blackbody surfaces $A_s$ and $A_j$, with the result that

$$A_sF_{s-j} = A_jF_{j-s}$$

A more rigorous derivation of the principle of reciprocity is developed in references 5 and 11, which is based solely on geometric considerations.

### *Thermal Radiation Networks*

As we have seen in Chaps. 1 through 3, the practical thermal analysis of many basic heat-transfer processes can be facilitated by use of the standard electrical analogy concept in which temperature $T$ is taken to be analogous to electrical potential $E_e$. Based on this electrical analogy of the *first kind*, the radiation heat-transfer resistance $R_R$ between two blackbody surfaces is

$$R_R = \frac{1}{A_sF_{s-j}\sigma(T_s + T_j)(T_s^2 + T_j^2)} \tag{5-47}$$

This analogy concept is indeed useful for systems such as the ones shown in Fig. 5–25(a) and (b), which involve only two thermal radiating surfaces with known temperatures. However, because $R_R$ is a function of temperature, the utility of this approach decreases as the number of thermal radiating surfaces with unknown temperature increases.

An alternative electrical analogy has been developed which more readily lends itself to the analysis of the more complex thermal radiation problems involving multiple surfaces and unknown surface temperatures. In this electrical analogy approach the total emissive power $E$ is taken to be analogous to electrical voltage $E_e$. Equations (5–42) and (5–45) provide the basis for this powerful radiation-heat-transfer analysis tool. Simply put, the radiation-heat-transfer resistance $R_{s-j}$ between surfaces $A_s$ and $A_j$ based on this electrical analogy of the *second kind* is

$$R_{s-j} = \frac{1}{A_sF_{s-j}} \tag{5-48}$$

To illustrate, analogous electrical circuits of the second kind (which we will refer to as *thermal radiation networks*) are utilized in Examples 5–11 through 5–13 for blackbody systems involving two, three, and four surfaces.

## EXAMPLE 5–11

Referring to Fig. E5–11a, the blackbody surfaces $A_1$ and $A_2$ are at 27°C and 500°C, respectively. Determine the rate of radiation heat transfer between $A_1$ and $A_2$. Also sketch the thermal circuits of the first and second kinds.

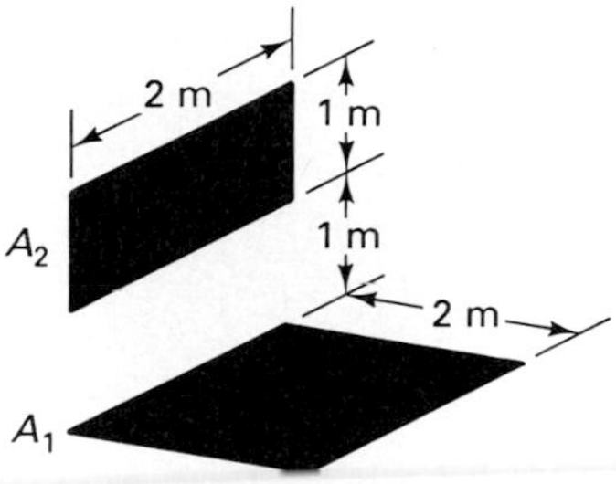


**FIGURE E5–11a**
Open thermal radiating blackbody system.

### Solution

*Objective* Determine $q_{1-2}$ and sketch the thermal radiation circuit and network.

*Assumptions/Conditions*

steady-state
blackbody thermal radiation
surrounding space is evacuated

*Properties* $\epsilon_1 = 1$, $\epsilon_2 = 1$.

*Analysis* The rate of radiation heat transfer is given by Eq. (5–42),

$$q_{1-2} = A_1 F_{1-2}(E_{b1} - E_{b2})$$

Referring back to Example 5–8, we see that $F_{1-2} = 0.05$. Calculating $q_{1-2}$, we obtain

$$q_{1-2} = 4 \text{ m}^2 (0.05)\left(5.67 \times 10^{-8} \frac{\text{W}}{\text{m}^2 \text{ K}^4}\right)[(300 \text{ K})^4 - (773 \text{ K})^4]$$

$$= 4(0.05)(5.67)(3^4 - 7.73^4) \text{ W} = -3960 \text{ W}$$

Thermal circuits of the first and second kinds are shown in Figs. E5–11b and E5–11c. The thermal resistances $R_R$ and $R_{1-2}$ are given by

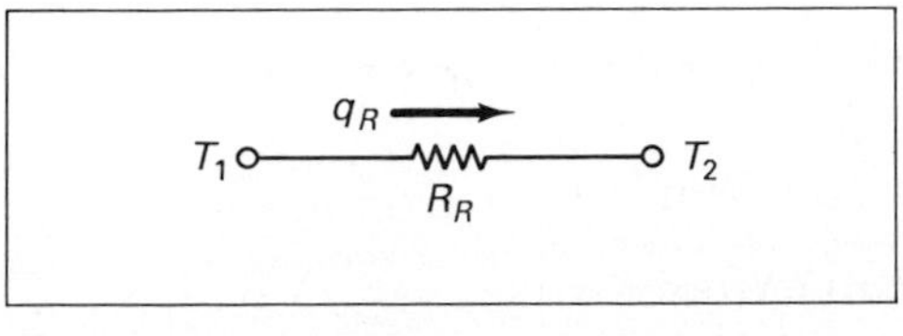


**FIGURE E5–11b**
Thermal circuit of first kind.

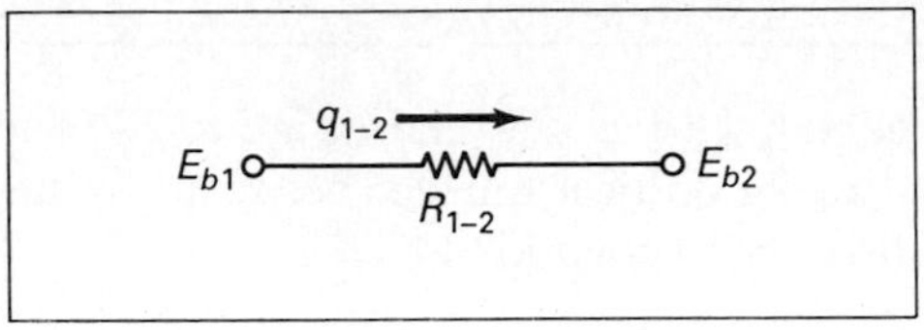


**FIGURE E5–11c**
Thermal radiation network.

$$R_R = \frac{1}{A_1F_{1-2}\sigma(T_1 + T_2)(T_1^2 + T_2^2)} = 0.12 \text{ K/W}$$

and

$$R_{1-2} = \frac{1}{A_1F_{1-2}} = 5/\text{m}^2$$

## EXAMPLE 5–12

Utilize the thermal-radiation-network approach to evaluate the rate of radiation heat transfer between each of the three blackbody surfaces of the cylindrical enclosure shown in Fig. E5–12a.

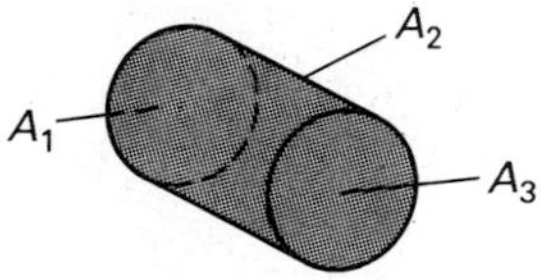


$A_1$ at $T_1 = 227°C$
$A_2$ at $T_2 = 200°C$
$A_3$ at $T_3 = 50°C$
$L = 1$ m, $D = 1$ m

**FIGURE E5–12a**
Cylindrical blackbody enclosure.

### Solution

*Objective* Determine $q_{1-2}$, $q_{1-3}$, and $q_{2-3}$.

*Assumptions/Conditions*

steady-state
blackbody thermal radiation
enclosure is evacuated

*Properties* $\epsilon_1 = 1$, $\epsilon_2 = 1$, $\epsilon_3 = 1$.

*Analysis* The thermal radiation network is shown in Fig. E5–12b, where

$$E_{b1} = \sigma T_1^4 = 5.67 \times 10^{-8} \frac{\text{W}}{\text{m}^2\text{ K}^4} (500 \text{ K})^4 = 3540 \text{ W/m}^2$$

$$E_{b2} = \sigma T_2^4 = \sigma(473 \text{ K})^4 = 2840 \text{ W/m}^2$$

$$E_{b3} = \sigma T_3^4 = \sigma(323 \text{ K})^4 = 617 \text{ W/m}^2$$

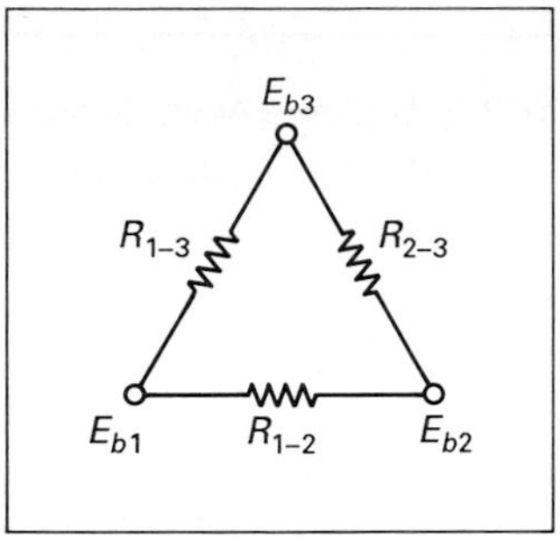


**FIGURE E5–12b**
Thermal radiation network.

Referring to Fig. 5–23, $F_{1-3}$ is equal to 0.17. Utilizing the summation principle, we have

$$F_{1-2} = 1 - F_{1-3} = 0.83 \qquad F_{3-2} = 0.83$$

The thermal resistances are

$$R_{1-2} = \frac{1}{A_1F_{1-2}} = \frac{1}{[\pi(1\text{ m})^2/4](0.83)} = 1.53/\text{m}^2$$

$$R_{1-3} = \frac{1}{A_1F_{1-3}} = 7.47/\text{m}^2$$

$$R_{2-3} = \frac{1}{A_2F_{2-3}} = \frac{1}{A_3F_{3-2}} = 1.53/\text{m}^2$$

The rates of radiation heat transfer flowing in this network are now easily calculated.

$$q_{1-2} = \frac{E_{b1} - E_{b2}}{R_{1-2}} = \frac{(3540 - 2840)\text{ W/m}^2}{1.53/\text{m}^2} = 458\text{ W}$$

$$q_{1-3} = \frac{E_{b1} - E_{b3}}{R_{1-3}} = \frac{(3540 - 617)\text{ W/m}^2}{7.47/\text{m}^2} = 391\text{ W}$$

$$q_{2-3} = \frac{E_{b2} - E_{b3}}{R_{2-3}} = \frac{(2840 - 617)\text{ W/m}^2}{1.53/\text{m}^2} = 1450\text{ W}$$

The total rate of radiation heat transfer from each surface is

$$q_1 = q_{1-2} + q_{1-3} = 849\text{ W}$$

$$q_2 = q_{2-3} + q_{2-1} = q_{2-3} - q_{1-2} = 992\text{ W}$$

$$q_3 = q_{3-1} + q_{3-2} = -q_{1-3} - q_{2-3} = -1841\text{ W}$$

and we note that

$$q_1 + q_2 + q_3 = 0$$

which is consistent with requirements of the first law of thermodynamics.

## EXAMPLE 5–13

The blackbody plates shown in Fig. E5–13a are located in a very large blackbody enclosure $A_4$ which is at 27°C. The back side of each plate is insulated, and 1 kW is electrically generated in plate 3. Sketch the thermal radiation network for this problem, and determine the temperature of surface $A_3$ and the rate of radiation heat transfer between $A_1$ and $A_3$.

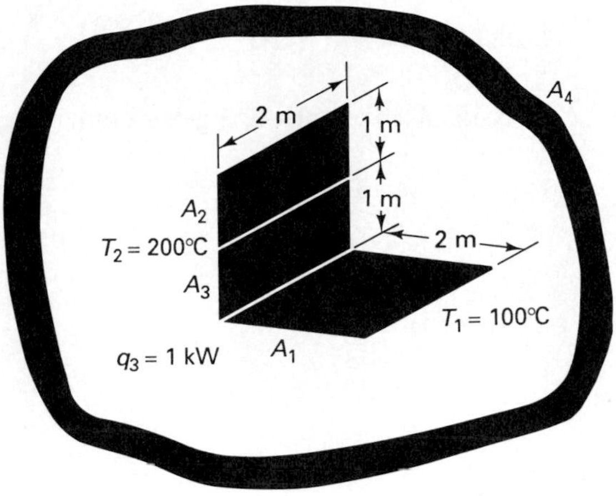


**FIGURE E5–13a**
Closed thermal radiating system.

### Solution

*Objective* Sketch the thermal radiation network and determine $T_3$ and $q_{1-3}$.

*Assumptions/Conditions*

steady-state
blackbody thermal radiation
enclosure surface $A_4$ is very large
surrounding space is evacuated

*Properties* $\epsilon_1 = 1$, $\epsilon_2 = 1$, $\epsilon_3 = 1$, $\epsilon_4 = 1$.

*Analysis* Referring back to Example 5–8, we have

$$F_{1-2} = 0.05 \qquad F_{1-3} = 0.15$$

By reciprocity,

$$F_{3-1} = \frac{A_1}{A_3} F_{1-3} = 2(0.15) = 0.3$$

$$F_{2-1} = \frac{A_1}{A_2} F_{1-2} = 2(0.05) = 0.1$$

It is clear that $F_{2-3} = 0$. By employing the summation principle, we are able to write

$$F_{1-4} = 1 - F_{1-2} - F_{1-3} = 1 - 0.05 - 0.15 = 0.8$$

$$F_{2-4} = 1 - F_{2-1} - F_{2-3} = 1 - 0.1 - 0 = 0.9$$

$$F_{3-4} = 1 - F_{3-1} - F_{3-2} = 1 - 0.3 - 0 = 0.7$$

The thermal radiation network for this four-surface system is shown in Fig. E5–13b, where

$$R_{1-3} = \frac{1}{A_1F_{1-3}} = \frac{1}{4\ \text{m}^2\ (0.15)} = 1.67/\text{m}^2 \quad E_{b1} = \sigma(373\ \text{K})^4 = 1100\ \text{W/m}^2$$

$$R_{1-2} = \frac{1}{A_1F_{1-2}} = \frac{1}{4\ \text{m}^2\ (0.05)} = 5/\text{m}^2 \quad E_{b2} = \sigma(473\ \text{K})^4 = 2840\ \text{W/m}^2$$

$$R_{1-4} = \frac{1}{A_1F_{1-4}} = \frac{1}{4\ \text{m}^2\ (0.8)} = 0.313/\text{m}^2 \quad E_{b4} = \sigma(300\ \text{K})^4 = 459\ \text{W/m}^2$$

$$R_{2-3} = \frac{1}{A_2F_{2-3}} = \frac{1}{2\ \text{m}^2\ (0)} = \infty/\text{m}^2$$

$$R_{3-4} = \frac{1}{A_3F_{3-4}} = \frac{1}{2\ \text{m}^2\ (0.7)} = 0.714/\text{m}^2$$

$$R_{2-4} = \frac{1}{A_2F_{2-4}} = \frac{1}{2\ \text{m}^2\ (0.9)} = 0.556/\text{m}^2$$

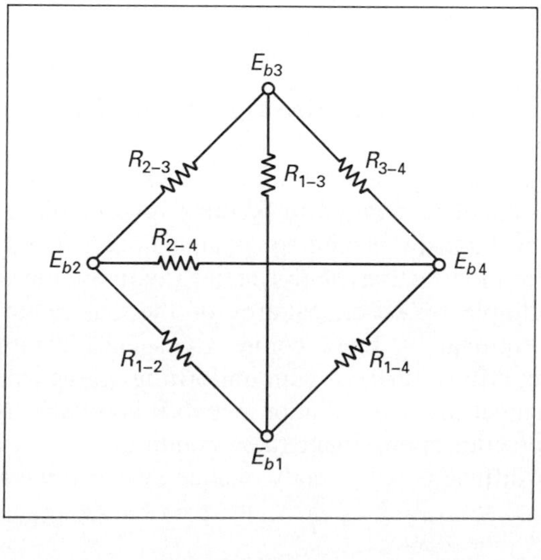


**FIGURE E5–13b**
Thermal radiation network.

To determine the unknown potential $E_{b3}$, we perform an energy balance at surface $A_3$.

$$q_3 = 1\text{ kW} = \frac{E_{b3} - E_{b1}}{R_{1-3}} + \frac{E_{b3} - E_{b2}}{R_{2-3}} + \frac{E_{b3} - E_{b4}}{R_{3-4}}$$

Making substitutions and solving for $E_{b3}$, we obtain

$$E_{b3} = \frac{1\text{ kW} + \dfrac{E_{b1}}{R_{1-3}} + \dfrac{E_{b2}}{R_{2-3}} + \dfrac{E_{b4}}{R_{3-4}}}{\dfrac{1}{R_{1-3}} + \dfrac{1}{R_{2-3}} + \dfrac{1}{R_{3-4}}}$$

$$= \frac{(1000 + 1100/1.67 + 2840/\infty + 459/0.714)\text{ W}}{(1/1.67 + 1/\infty + 1/0.714)\text{ m}^2} = 1150\text{ W/m}^2$$

To calculate $T_3$, we write

$$E_{b3} = \sigma T_3^4$$

$$T_3 = \left[\frac{1150\text{ W/m}^2}{5.67 \times 10^{-8}\text{ W/(m}^2\text{ K}^4)}\right]^{1/4} = 377\text{ K} = 104°\text{C}$$

The rate of radiation heat transfer between $A_1$ and $A_3$ is

$$q_{1-3} = \frac{E_{b1} - E_{b3}}{R_{1-3}} = \frac{(1100 - 1150)\text{W/m}^2}{1.67/\text{m}^2} = -29.9\text{ W}$$

To check our solution, the summation $q_1 + q_2 + q_3 + q_4$ can be shown to be equal to zero.

### 5-5-2 Nonblackbody Thermal Radiation from Diffuse Opaque Surfaces

As we have seen, some real surfaces exhibit approximate blackbody characteristics, but most thermal radiation systems encountered in practice involve distinctly nonblackbody surfaces. Radiation heat transfer in common opaque nonblackbody systems is complicated by the occurrence of reflections. For systems involving more than one nonblackbody surface, the multiple reflection patterns of thermal radiation often become extremely complex. Fortunately, these complexities can be rather easily overcome for systems involving diffuse surfaces with uniform properties and heating conditions by the use of a practical thermal radiation network approach. Numerous real surface systems found in practice approximate these conditions. To develop this practical analysis approach for diffuse nonblackbody opaque systems, we first introduce the concept of radiosity.

### *Radiosity*

Referring to Fig. 5–27, we define the *radiosity* $J_s$ as the rate of thermal radiation emitted and reflected per unit surface area from $A_s$.

$$J_s = E_s + \rho_s G_s \tag{5–49}$$

Recall that the irradiation $G_s$ is the thermal radiation flux incident on the surface $A_s$.

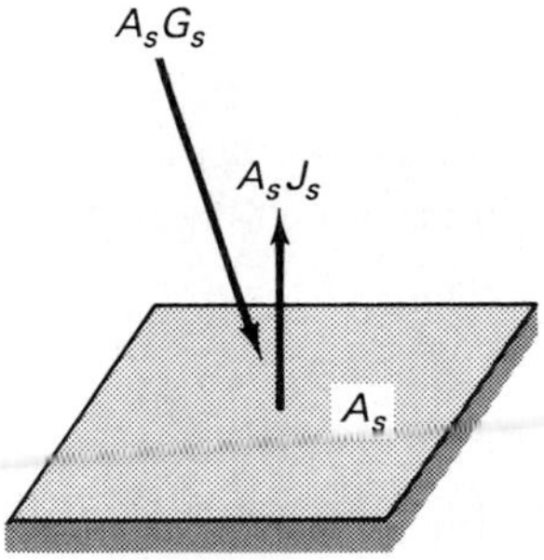


**FIGURE 5–27**
Thermal radiating surface $A_s$: concept of radiosity.

The total rate of radiation heat transfer from an opaque surface $A_s$ is simply equal to the difference between the rate of thermal radiation leaving, $A_sJ_s$, and the rate of thermal radiation coming in, $A_sG_s$; that is,

$$q_s = A_s(J_s - G_s) \tag{5–50}$$

To eliminate the irradiation $G_s$, we combine Eqs. (5–49) and (5–50).

$$q_s = A_s\left[J_s - \frac{1}{\rho_s}(J_s - E_s)\right] = \frac{A_s}{\rho_s}[E_s - (1 - \rho_s)J_s] \tag{5–51}$$

Since the body is opaque,

$$\rho_s = 1 - \alpha_s \tag{5–52}$$

and we have

$$q_s = \frac{A_s\alpha_s}{\rho_s}\left(\frac{\epsilon_s}{\alpha_s}E_{bs} - J_s\right) = \frac{A_s\alpha_s}{1 - \alpha_s}\left(\frac{\epsilon_s}{\alpha_s}E_{bs} - J_s\right) \tag{5–53}$$

Based on this equation, the total rate of radiation heat transfer $q_s$ can be represented by a thermal radiation network element as shown in Fig. 5–28; the node potentials are $(\epsilon_s/\alpha_s)E_{bs}$ and $J_s$, and the thermal resistance $R_s$ is

$$R_s = \frac{1 - \alpha_s}{A_s\alpha_s} = \frac{\rho_s}{A_s\alpha_s} \tag{5–54}$$

As indicated in Sec. 5–3, the total absorptivity $\alpha_s$ of a surface is dependent upon the sources from which the irradiation $G_s$ originates. Because of the reflection

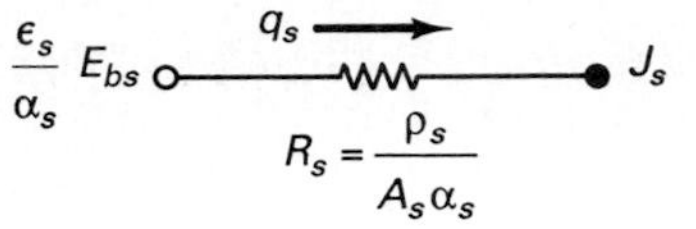


**FIGURE 5–28**
Thermal radiation network element for $q_s$.

patterns associated with nonblackbody surface systems, special care must be taken in the evaluation of $\alpha_s$ for nongraybody conditions, as illustrated in several examples in the next section. For systems with surfaces that exhibit approximate graybody characteristics, $\alpha_s$ is independent of the source of the various components of incoming irradiation, and $\alpha_s$ can be set equal to $\epsilon_s$. For this limiting condition, we write

$$q_s = \frac{A_s\epsilon_s}{1 - \epsilon_s}(E_{bs} - J_s) \tag{5–55}$$

The thermal radiation network element for this simple situation is shown in Fig. 5–29. The node potentials for this case are simply $E_{bs}$ and $J_s$, and $R_s$ is given by

$$R_s = \frac{1 - \epsilon_s}{A_s\epsilon_s} = \frac{\rho_s}{A_s\epsilon_s} \tag{5–56}$$

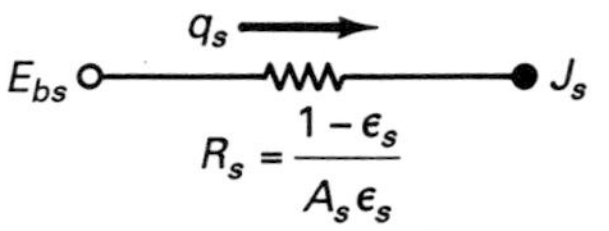


**FIGURE 5–29**
Thermal radiation network element for $q_s$: graybody conditions.

Equation (5–53) and its limiting form Eq. (5–55), together with the corresponding thermal radiation network elements given in Figs. 5–28 and 5–29, provide the key building block for the development of a simple practical analysis of radiation heat transfer. However, before we turn our attention to the completion of our practical analysis approach for bisurface and multisurface systems, it should be mentioned that another approach to handling Eqs. (5–49) and (5–50) that is favored by some is to eliminate the radiosity $J_s$. This alternative method is introduced in Example 5–14.

## EXAMPLE 5–14

The radiation heat transfer $q_s$ associated with graybody surfaces is given in terms of $E_{bs}$ and radiosity $J_s$ by Eq. (5–55). Develop an alternative formulation for $q_s$ by eliminating $J_s$ instead of $G_s$ in Eqs. (5–49) and (5–50).

### Solution

*Objective* Develop an alternative relationship for $q_s$ for graybody surfaces.

*Assumptions/Conditions*

diffuse opaque graybody surfaces

*Analysis* Starting with Eqs. (5–49) and (5–50),

$$J_s = E_s + \rho_s G_s \qquad q_s = A_s(J_s - G_s)$$

we eliminate $J_s$, to obtain

$$q_s = A_s(E_s + \rho_s G_s - G_s) = A_s(E_s - \alpha_s G_s) \tag{a}$$

The irradiation $G_s$ falling on surface $A_s$ is expressed in terms of the radiosities of the surrounding surfaces by

$$A_s G_s = \sum_{j=1}^{N} A_j F_{j-s} J_j$$

Utilizing the principle of reciprocity, this equation takes the form

$$A_s G_s = \sum_{j=1}^{N} A_s F_{s-j} J_j$$

or

$$G_s = \sum_{j=1}^{N} F_{s-j} J_j \tag{b}$$

The total rate of thermal radiation absorbed by surface $A_s$ is

$$A_s \alpha_s G_s = \sum_{j=1}^{N} \alpha_{sj} A_s F_{s-j} J_j \tag{c}$$

where $\alpha_{sj}$ is the absorptivity component associated with the $J_j$ source. Substituting this result into Eq. (a), we have

$$q_s = A_s \left( E_s - \sum_{j=1}^{N} \alpha_{sj} F_{s-j} J_j \right) \tag{d}$$

Although this equation does not conveniently lend itself to electrical analogy representation, as does Eq. (5–53), it does provide a basis for the analytical or numerical solution of more complex thermal radiation systems. For the simple limiting case of graybody conditions, Eq. (d) reduces to

$$q_s = A_s \epsilon_s \left( E_{bs} - \sum_{j=1}^{N} F_{s-j} J_j \right) \tag{e}$$

As in the case of Eq. (5–53), Eqs. (d) and (e) apply to systems with uniform emission, reflection, and irradiation over each surface.

As a point of interest, it is noted that the total absorptivity $\alpha_s$ at surface $A_s$ can be obtained by combining Eqs. (b) and (c); that is,

$$\alpha_s = \frac{\sum_{j=1}^{N} \alpha_{sj} F_{s-j} J_j}{\sum_{j=1}^{N} F_{s-j} J_j} \tag{f}$$

### Bisurface Systems

We consider radiation heat transfer between two opaque diffuse surfaces $A_1$ and $A_2$. For uniform radiosities, the rate of thermal radiation leaving $A_1$ that reaches $A_2$ is $A_1F_{1-2}J_1$, and from $A_2$ to $A_1$ is $A_2F_{2-1}J_2$. Therefore, the thermal radiation-heat-transfer rate $q_{1-2}$ from surface $A_1$ to $A_2$ is

$$q_{1-2} = A_1F_{1-2}J_1 - A_2F_{2-1}J_2 \tag{5–57}$$

or, since $A_1F_{1-2} = A_2F_{2-1}$,

$$q_{1-2} = A_1F_{1-2}(J_1 - J_2) \tag{5–58}$$

This equation provides the basis for the thermal radiation network element shown in Fig. 5–30.

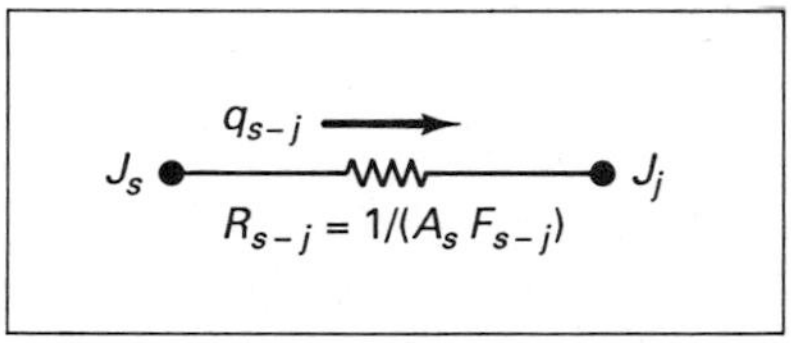


**FIGURE 5–30** Thermal radiation network element between potentials $J_s$ and $J_j$.

To close our two-surface system analysis, we couple the nodes $J_1$ and $J_2$ to the surface potentials $(\epsilon_1/\alpha_1)E_{b1}$ and $(\epsilon_2/\alpha_2)E_{b2}$ by using the thermal radiation network element given in Fig. 5–28. The resulting thermal radiation network is shown in Fig. 5–31. Because Eq. (5–58) was developed for surfaces with uniform radiosities, this thermal radiation network only applies to systems with uniform emission, reflection, and irradiation over each surface. The requirement that the reflected radiation flux be uniform is strictly satisfied only for symmetrical systems, such as infinite parallel plates or concentric spheres, with each surface having a uniform temperature and uniform thermal radiation properties. However, this convenient thermal radiation network approach is often applied to nonsymmetrical systems as a first approximation.

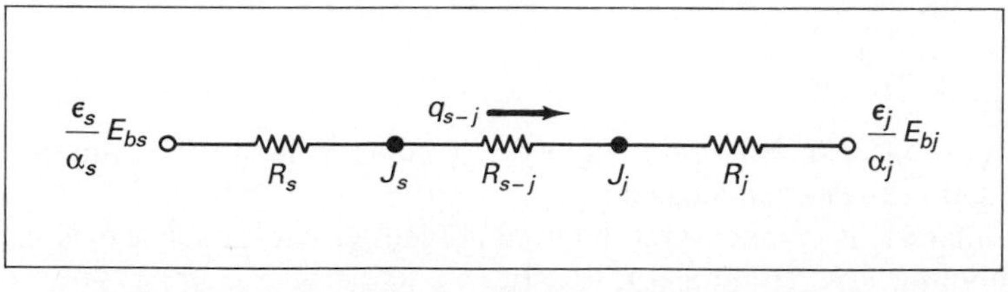


**FIGURE 5–31** Thermal radiation network for bisurface system.

The rate of radiation heat transfer $q_{1-2}$ associated with this thermal radiation network is given by

$$q_{1-2} = \frac{(\epsilon_1/\alpha_1)E_{b1} - (\epsilon_2/\alpha_2)E_{b2}}{R_1 + R_{1-2} + R_2} = \frac{(\epsilon_1/\alpha_1)E_{b1} - (\epsilon_2/\alpha_2)E_{b2}}{\dfrac{1-\alpha_1}{A_1\alpha_1} + \dfrac{1}{A_1F_{1-2}} + \dfrac{1-\alpha_2}{A_2\alpha_2}} \tag{5–59}$$

For blackbody conditions, $R_1$ and $R_2$ are both zero, such that the thermal radiation network reduces to the one shown in Fig. E5–11c, and Eq. (5–59) reduces to Eq. (5–42). For approximate graybody conditions, Eq. (5–59) simplifies to

$$q_{1-2} = \frac{E_{b1} - E_{b2}}{\dfrac{1-\epsilon_1}{A_1\epsilon_1} + \dfrac{1}{A_1F_{1-2}} + \dfrac{1-\epsilon_2}{A_2\epsilon_2}} \tag{5–60}$$

## EXAMPLE 5–15

Consider exchange of radiation heat transfer between the small surface $A_1$ and an enclosure $A_2$ shown in Fig. E5–15a. Demonstrate that the enclosure approximates blackbody conditions as $A_2$ becomes large.

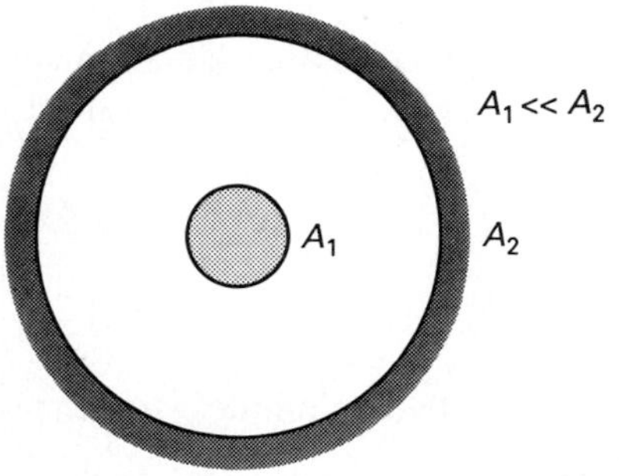


**FIGURE E5–15a**
Closed thermal radiating system.

### Solution

*Objective* Show that blackbody conditions are approached within this enclosure as $A_2$ becomes large or as $A_1$ becomes small.

*Assumptions/Conditions*

steady-state
diffuse opaque surfaces
enclosure is evacuated

*Analysis* The thermal radiation network for this system is shown in Fig. E5–15b, where

$$R_1 = \frac{\rho_1}{A_1\alpha_1} \qquad R_{1-2} = \frac{1}{A_1F_{1-2}} \qquad R_2 = \frac{\rho_2}{A_2\alpha_2}$$

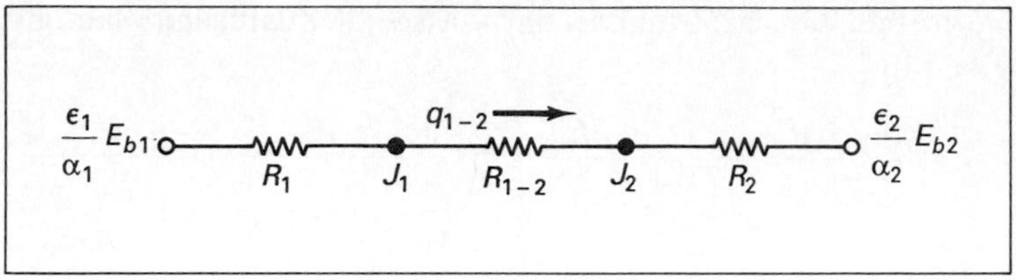


**FIGURE E5–15b** Thermal radiation network.

For very large values of $A_2$, most of the energy incident upon $A_2$ originates from itself, unless the body is a perfect reflector. Consequently, the nonisothermal $\alpha$ approximation reduces to the graybody $\alpha$ approximation with $\epsilon_2/\alpha_2$ equal to unity. Further, the resistance $R_2$ approaches zero as $A_2$ increases. For these conditions, the radiosity $J_2$ is essentially equal to $E_{b2}$, such that blackbody conditions are approached at the surface of the enclosure.

Incidentally, the thermal radiation network for this system implicitly accounts for thermal radiation from surface $A_2$ to itself.

## EXAMPLE 5–16

Develop an expression for the net rate of exchange of thermal radiation between two diffuse nonblackbody infinite parallel plates $A_1$ and $A_2$ without the use of radiosity.

### Solution

*Objective* Develop a relation for $q_{1-2}$ for this parallel plate system.

*Schematic* Nonblackbody thermal radiation between infinite parallel plates.

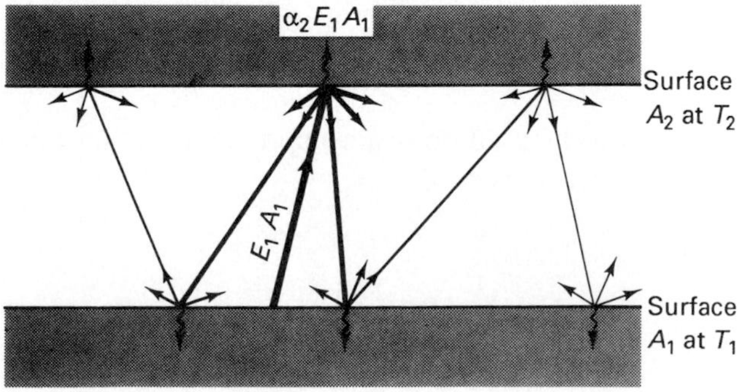


*Assumptions/Conditions*

steady-state
diffuse nonblackbody surfaces
space between plates is evacuated

*Analysis* The reflection and absorption pattern of thermal radiation originally emitted from surface $A_1$ is shown in the schematic. The energy rate content of this thermal radiation before absorption and reflection is $E_1A_1$. Of this amount of radiation power, the quantity $\alpha_2E_1A_1$ is absorbed by $A_2$ and $\rho_2E_1A_1$ is reflected back toward $A_1$. Of this reflected energy rate, the quantity $\rho_1(\rho_2E_1A_1)$ is rereflected toward surface $A_2$. Surface $A_2$ absorbs $\alpha_2(\rho_1\rho_2E_1A_1)$ of this rate of energy and reflects the remainder, $\rho_2(\rho_1\rho_2E_1A_1)$. The total rate of thermal radiation originating from surface $A_1$ that is absorbed by $A_2$ is written as

$$\alpha_2E_1A_1[1 + \rho_1\rho_2 + (\rho_1\rho_2)^2 + \cdots + (\rho_1\rho_2)^n + \cdots]$$

or

$$\alpha_2E_1A_1 \sum_{n=0}^{\infty} (\rho_1\rho_2)^n$$

In a similar fashion, we can show that the total rate of thermal radiation originating from surface $A_2$ that is absorbed by $A_1$ is

$$\alpha_1E_2A_2 \sum_{n=0}^{\infty} (\rho_1\rho_2)^n$$

It follows that the net rate of thermal radiation exchange between $A_1$ and $A_2$ becomes

$$q_{1-2} = (\alpha_2E_1A_1 - \alpha_1E_2A_2)\left[\sum_{n=0}^{\infty} (\rho_1\rho_2)^n\right]$$

where $A_1 = A_2$.

Referring to reference 23, the series $\Sigma_{n=0}^{\infty} x^n$ converges to $1/(1 - x)$ for $x$ less than unity. Thus, the rate of radiation heat transfer is

$$q_{1-2} = \frac{A_1(\alpha_2E_1 - \alpha_1E_2)}{1 - \rho_1\rho_2}$$

Introducing the identities $E_1 = \epsilon_1E_{b1}$, $E_2 = \epsilon_2E_{b2}$, $\rho_1 = 1 - \alpha_1$, and $\rho_2 = 1 - \alpha_2$, we obtain

$$q_{1-2} = \frac{A_1\left(\dfrac{\epsilon_1}{\alpha_1}E_{b1} - \dfrac{\epsilon_2}{\alpha_2}E_{b2}\right)}{\dfrac{1}{\alpha_1} + \dfrac{1}{\alpha_2} - 1}$$

which is equivalent to Eq. (5–59) for $A_1 = A_2$ and $F_{1-2} = 1$. For graybody conditions, we have

$$q_{1-2} = \frac{A_1(E_{b1} - E_{b2})}{\dfrac{1}{\epsilon_1} + \dfrac{1}{\epsilon_2} - 1}$$

This equation is equivalent to Eq. (5–60) for this infinite parallel-plate geometry.

The fact that this direct approach is consistent with the radiosity-based network approach should reinforce our confidence in the more efficient network method. Further, because this direct method becomes extremely unwieldly for multisurface systems, the concept of radiosity is the key to developing practical thermal analyses for radiation processes.

## EXAMPLE 5–17

Surface $A_2$ of the system shown in Fig. E5–17a is a graybody with emissivity of 0.56 and surface $A_1$ is a blackbody. Determine the rate of radiation heat transfer between $A_1$ and $A_2$ if $T_1 = 27°C$ and $T_2 = 500°C$.

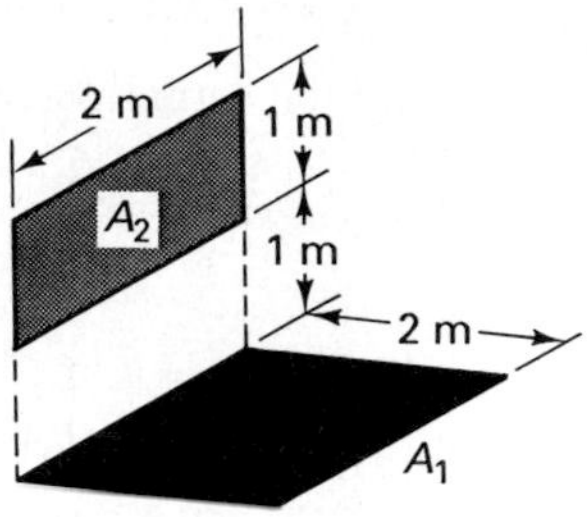


**FIGURE E5–17a**
Open thermal radiating system.

### Solution

*Objective* Determine $q_{2-1}$.

*Assumptions/Conditions*

steady-state
blackbody and diffuse opaque graybody surfaces
surrounding space is evacuated

*Properties* $\epsilon_1 = 1$, $\epsilon_2 = 0.56$. Because surface $A_2$ is a graybody, we also have

$$\alpha_2 = \epsilon_2 = 0.56$$

*Analysis* The radiation shape factor is given in Example 5–11 as

$$F_{1-2} = 0.05$$

Assuming diffuse radiation, the approximate thermal radiation network for this nonsymmetrical bisurface system is shown in Fig. E5–17b, where $E_{b1} = 459\ \text{W/m}^2$, $E_{b2} = 20{,}200\ \text{W/m}^2$, and

$$R_{1-2} = \frac{1}{A_1 F_{1-2}} = 5/\text{m}^2$$

$$R_2 = \frac{\rho_2}{A_2\alpha_2} = \frac{1 - 0.56}{2\ \text{m}^2\,(0.56)} = 0.393/\text{m}^2$$

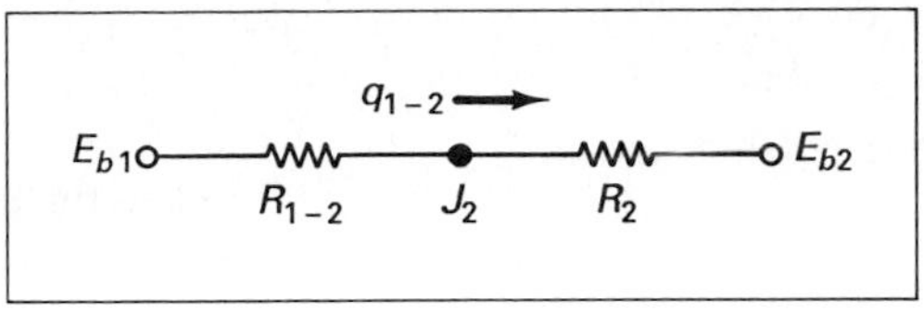
$q_{1-2}$
$E_{b1}$
$R_{1-2}$
$J_2$
$R_2$
$E_{b2}$

**FIGURE E5–17b**
Thermal radiation network.

It follows that the rate of radiation heat transfer $q_{2-1}$ is approximately

$$q_{2-1} = -q_{1-2} = \frac{(20{,}200 - 459)\ \text{W/m}^2}{(5 + 0.393)/\text{m}^2} = 3660\ \text{W}$$

This result lies about 7.5% below the value of 3960 W obtained in Example 5–11 for blackbody radiation from both surfaces.

To improve the accuracy of our analysis for this nonsymmetrical system, surface $A_2$ can be subdivided into smaller areas.

## EXAMPLE 5–18

Surface $A_2$ of the system shown in Fig. E5–17a is made of oxidized nickel and surface $A_1$ is a blackbody. Determine the rate of radiation heat transfer between $A_1$ and $A_2$ if $T_1 = 27°\text{C}$ and $T_2 = 500°\text{C}$.

### Solution

*Objective* Determine $q_{2-1}$.

*Schematic* Open thermal radiating system.

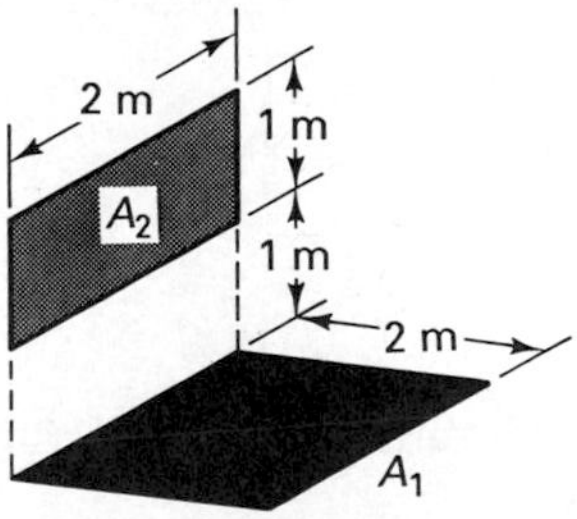
2 m
1 m
$A_2$
1 m
2 m
$A_1$

*Assumptions/Conditions*

steady-state
blackbody and diffuse opaque graybody surfaces
surrounding space is evacuated

*Properties*

$\epsilon_1 = 1$.

Oxidized nickel (Fig. 4–3): $\epsilon = 0.63$ at 500°F, $\epsilon = 0.32$ at 27°F.

Therefore, we take

$$\epsilon_2 = 0.63$$

*Analysis* Noting that all the thermal radiation from $A_1$ to $A_2$ is emitted by blackbody $A_1$, which is at a temperature of 27°C, we employ the nonisothermal $\alpha$ approximation to obtain

$$\alpha_2 = 0.32$$

As we have already seen, $F_{1-2} = 0.05$.

The thermal radiation network for this system is shown in Fig. E5–18 with $(\epsilon_2/\alpha_2)E_{b2}$ and $R_2$ given by

$$\frac{\epsilon_2}{\alpha_2} E_{b2} = \frac{0.63}{0.32}\left(20{,}200 \frac{\text{W}}{\text{m}^2}\right) = 39{,}800 \text{ W/m}^2$$

$$R_2 = \frac{\rho_2}{A_2\alpha_2} = \frac{1 - 0.32}{2 \text{ m}^2 (0.32)} = 1.06/\text{m}^2$$

The rate of radiation heat transfer is written as

$$q_{2-1} = \frac{(\epsilon_2/\alpha_2) E_{b2} - E_{b1}}{R_{1-2} + R_2} = \frac{(39{,}800 - 459) \text{ W/m}^2}{(5 + 1.06)/\text{m}^2} = 6490 \text{ W}$$

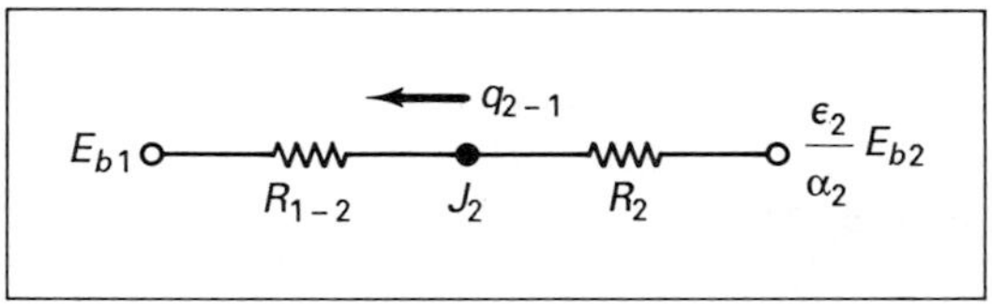


**FIGURE E5–18**
Thermal radiation network.

As a point of interest, the graybody $\alpha$ approximation gives

$$\alpha_2 = 0.63 \qquad R_2 = \frac{\rho_2}{A_2\alpha_2} = \frac{1 - 0.63}{2 \text{ m}^2 (0.63)} = 0.294/\text{m}^2$$

The resulting prediction for the rate of radiation heat transfer is

$$q_{2-1} = \frac{E_{b2} - E_{b1}}{R_{1-2} + R_2} = \frac{(20{,}200 - 459) \text{ W/m}^2}{(5 + 0.294)/\text{m}^2} = 3730 \text{ W}$$

which is a substantial 40% below the value obtained by utilizing the preferred nonisothermal $\alpha$ approximation.

As in Example 5–17, the accuracy of our analysis for this nonsymmetrical system can be improved by breaking surfaces $A_1$ and $A_2$ into smaller areas.

### *Multisurface Systems*

Following the pattern of our analysis of bisurface systems, the net rate of thermal radiation between opaque diffuse surfaces $A_s$ and $A_j$ of a multisurface system with uniform radiosities is

$$q_{s-j} = A_s F_{s-j} J_s - A_j F_{j-s} J_j \tag{5–61}$$

or, based on the principle of reciprocity,

$$q_{s-j} = \frac{J_s - J_j}{R_{s-j}} \qquad R_{s-j} = \frac{1}{A_s F_{s-j}} \tag{5–62,63}$$

Thus, we have a general thermal radiation network element such as the one shown in Fig. 5–30 for each surface combination.

The radiosity $J_s$ associated with each surface is linked to its surface potential $(\epsilon_s/\alpha_s)E_{bs}$ by means of the element shown in Fig. 5–28. To illustrate, thermal radiation networks are shown in Figs. 5–32 and 5–33 for representative three-surface and four-surface diffuse systems. For blackbody surfaces, $R_s$ is zero, and these thermal radiation

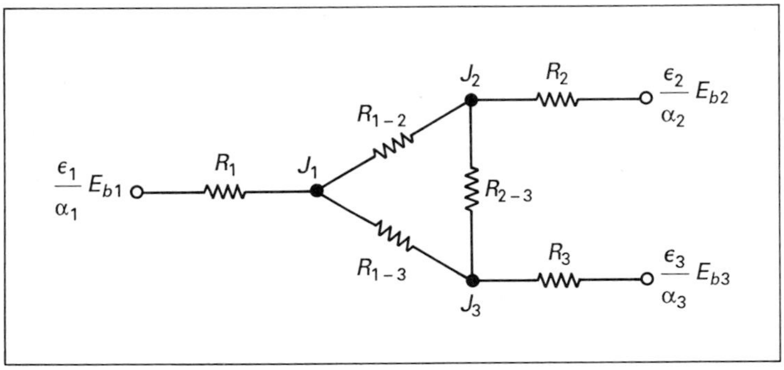


**FIGURE 5–32** Thermal radiation network for trisurface system.

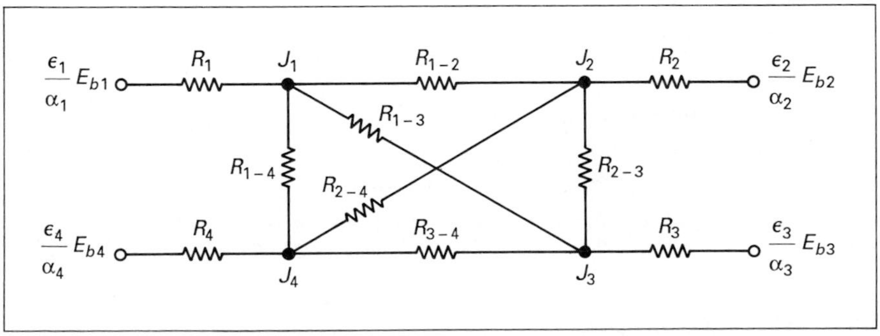


**FIGURE 5–33** Thermal radiation network for four-surface system.

networks reduce to the networks shown in Examples 5–12 and 5–13. And, for graybody conditions, $\alpha_s$ is set equal to $\epsilon_s$. As in the case of bisurface systems, these multisurface thermal radiation networks strictly only apply to symmetrical systems, but are commonly used as a first estimate for nonsymmetrical systems.

Referring back to the development of Eq. (5–53),

$$q_s = \frac{A_s\alpha_s}{1 - \alpha_s}\left(\frac{\epsilon_s}{\alpha_s} E_{bs} - J_s\right) \tag{5–53}$$

which is the basis for the network element of Fig. 5–28, we are again reminded that the rate of radiation heat transfer $q_s$ under steady-state conditions must be in balance with the net rate of energy entering the body, $\Sigma\, \dot{E}_i$. Certain situations occur in practice in which the only significant energy transfer to or from a body is caused by thermal radiation. Two rather special cases of practical importance that fall into this category are thermal radiation shields and reradiating surfaces.

***Thermal Radiation Shields*** As suggested in Chap. 1, the heat transfer through a wall can be greatly reduced by utilizing a composite construction that consists of layers of highly reflective materials separated by evacuated spaces. The thin plates or shells utilized in such superinsulative composite walls are known as *thermal radiation shields*. Referring to Fig. 5–34(a), we see that under steady-state conditions the rate of radiation heat transfer $q_{1-s}$ from enclosure surface $A_1$ to the single thermal radiation shield is equal to the rate of radiation heat transfer $q_{s-2}$ from the shield to enclosure surface $A_2$. Although such highly reflective surfaces have specular characteristics, the total radiation in this symmetrical system and in an infinite parallel-plate system can be treated as diffuse. A thermal radiation network is shown for this

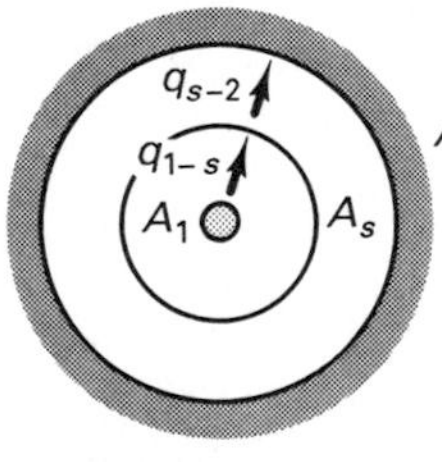


(a) Concentric sphere system.

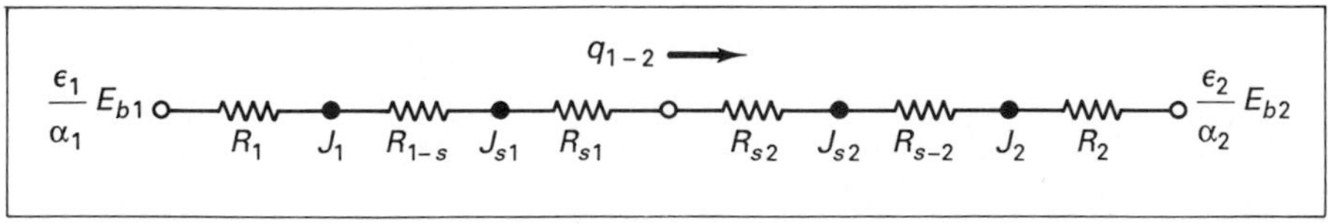


(b) Thermal radiation network.

**FIGURE 5–34** Thermal radiation shield.

arrangement in Fig. 5–34(b). Note that the thermal radiation shield is assumed to be very thin with negligible resistance to conduction. Based on this thermal radiation network, the rate of radiation heat transfer through the shield from surface $A_1$ to $A_2$ is

$$q_{1-2} = \frac{(\epsilon_1/\alpha_1)E_{b1} - (\epsilon_2/\alpha_2)E_{b2}}{\dfrac{\rho_1}{A_1\alpha_1} + \dfrac{1}{A_1F_{1-s}} + \dfrac{\rho_{s1}}{A_s\alpha_{s1}} + \dfrac{\rho_{s2}}{A_s\alpha_{s2}} + \dfrac{1}{A_sF_{s-2}} + \dfrac{\rho_2}{A_2\alpha_2}} \tag{5–64}$$

with $F_{1-s}$ and $F_{s-2}$ both equal to unity. Because the shield surfaces are highly reflective, the thermal radiation reaching surface $A_1$ primarily originates from surface $A_1$ itself. Therefore, the nonisothermal $\alpha$ approximation essentially reduces to the graybody $\alpha$ approximation (i.e., $\epsilon_1/\alpha_1 \simeq 1$). The same is true for surface $A_2$, with $\epsilon_2/\alpha_2 \simeq 1$. As a final simplification, the radiation properties of the shield can generally be taken as the average of its two surfaces; that is,

$$\alpha_s = \frac{\alpha_{s1} + \alpha_{s2}}{2} \qquad \rho_s = \frac{\rho_{s1} + \rho_{s2}}{2} = 1 - \alpha_s \tag{5–65a,b}$$

Introducing these simplifications, $q_{1-2}$ is given by

$$q_{1-2} = \frac{E_{b1} - E_{b2}}{\dfrac{1-\epsilon_1}{A_1\epsilon_1} + \dfrac{1}{A_1} + \dfrac{1-\epsilon_2}{A_2\epsilon_2} + \dfrac{2-\epsilon_s}{A_s\epsilon_s}} \tag{5-66}$$

For $N$ shields, it is a simple matter to show that $q_{1-2}$ becomes

$$q_{1-2} = \frac{E_{b1} - E_{b2}}{\dfrac{1-\epsilon_1}{A_1\epsilon_1} + \dfrac{1}{A_1} + \dfrac{1-\epsilon_2}{A_2\epsilon_2} + \displaystyle\sum_{j=1}^{N} \frac{1}{A_{sj}}\frac{2-\epsilon_{sj}}{\epsilon_{sj}}} \tag{5–67}$$

For a parallel-plate arrangement with each reflective shield having approximately the same radiation properties, this equation reduces to

$$q_{1-2} = \frac{A_1(E_{b1} - E_{b2})}{\dfrac{1-\epsilon_1}{\epsilon_1} + 1 + \dfrac{1-\epsilon_2}{\epsilon_2} + \dfrac{2-\epsilon_s}{\epsilon_s}N} = \frac{A_1(E_{b1} - E_{b2})}{\dfrac{1}{\epsilon_1} + \dfrac{1}{\epsilon_2} - 1 + \dfrac{2-\epsilon_s}{\epsilon_s}N} \tag{5–68}$$

Taking one step further, for the case in which the radiation properties of all the surfaces are approximately equal, we have

$$q_{1-2} = \frac{\epsilon A_1(E_{b1} - E_{b2})}{(2-\epsilon)(1+N)} \tag{5–69}$$

An examination of Eqs. (5–67) through (5–69) reveals that the rate of radiation heat transfer becomes small as (1) the reflectivity of the shield increases toward unity (i.e., as $\alpha_s$ or $\epsilon_s$ falls toward zero), and (2) the number of shields increases.

## EXAMPLE 5–19

A blackbody plate at 0°F exchanges radiation with a parallel stainless steel 301 plate at 1500°F. Determine the percent reduction in heat transfer if a thin polished aluminum plate with emissivity and absorptivity approximately equal to 0.08 is placed between these two large plates.

### Solution

*Objective* Determine the percent change in $q_{1-2}$ that is caused by the use of the radiation shield $A_s$.

*Schematic* Thermal radiation exchange between plates with radiation shield.

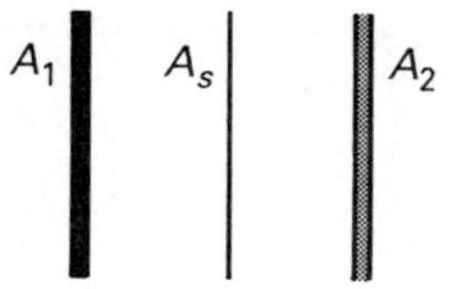


Surface
1 Blackbody plate at 0°F
2 Stainless steel (301) plate at 1500°F
s Polished aluminum plate

*Assumptions/Conditions*

steady-state
blackbody and diffuse opaque graybody surfaces
space between plates is evacuated

*Properties*

$\epsilon_1 = 1.$
Polished aluminum: $\epsilon_s = \alpha_s = 0.08.$
Stainless steel (301) at 1500°F (Fig. 4–3): $\epsilon_2 = 0.5.$

*Analysis* For the case in which no radiation shield is used, we employ the non-isothermal α approximation,

$$\alpha_2 = \epsilon_2(T_1) \simeq 0.16$$

The thermal radiation network for the parallel plates without the shield is shown in Fig. E5–19a, where

$$E_{b1} = \sigma T_1^4 = 0.171 \times 10^{-8} \frac{\text{Btu}}{\text{h ft}^2\ {}^\circ\text{R}^4}(460^\circ\text{R})^4 = 76.6\ \text{Btu/(h ft}^2)$$

$$E_{b2} = \sigma T_2^4 = \sigma(1960^\circ\text{R})^4 = 2.52 \times 10^4\ \text{Btu/(h ft}^2)$$

$$\frac{\epsilon_2}{\alpha_2} E_{b2} = \frac{0.5}{0.16}\left(2.52 \times 10^4 \frac{\text{Btu}}{\text{h ft}^2}\right) = 7.88 \times 10^4\ \text{Btu/(h ft}^2)$$

$$R_{1-2} = \frac{1}{A_1 F_{1-2}} = \frac{1}{A_1}$$

$$R_2 = \frac{\rho_2}{A_2\alpha_2} = \frac{1 - 0.16}{A_1(0.16)} = \frac{5.25}{A_1}$$

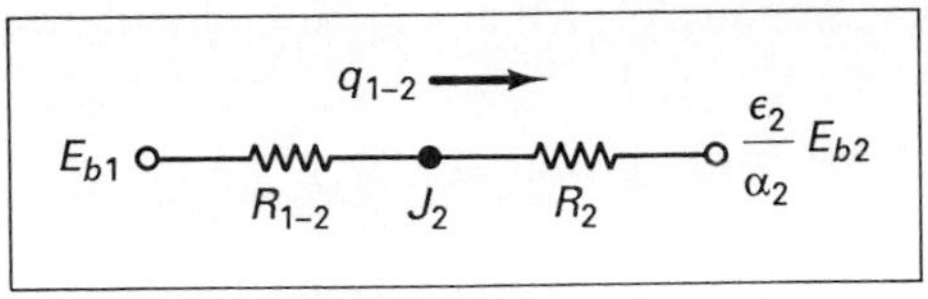


**FIGURE E5–19a**
Thermal radiation network for parallel plates.

Calculating the rate of radiation heat transfer $q_{2-1}$, we obtain

$$q_{2-1} = \frac{(\epsilon_2/\alpha_2)E_{b2} - E_{b1}}{R_{1-2} + R_2} = \frac{(7.88 \times 10^4 - 76.6)\ \text{Btu/(h ft}^2)}{(1 + 5.25)/A_1}$$

$$q''_{2-1} = 1.26 \times 10^4\ \text{Btu/(h ft}^2)$$

The thermal radiation network for the situation in which a thin plate lies between $A_1$ and $A_2$ is shown in Fig. E5–19b, where

$$R_{1-s} = \frac{1}{A_1 F_{1-s}} - \frac{1}{A_1}$$

$$R_{s1} = R_{s2} = \frac{\rho_s}{A_s \alpha_s} = \frac{1 - 0.08}{A_s(0.08)} = \frac{11.5}{A_1}$$

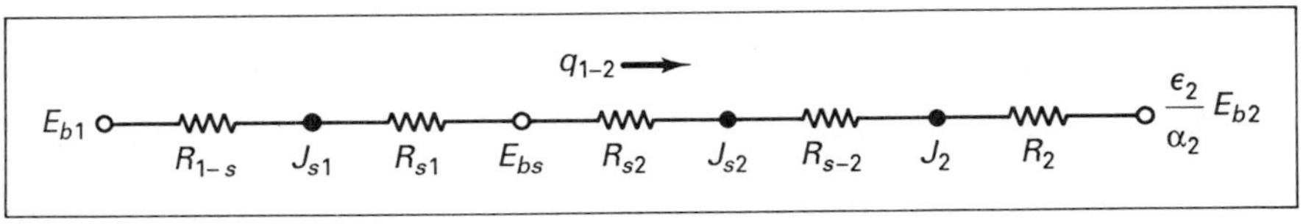


**FIGURE E5–19b** Thermal radiation network for parallel plates with shield.

As we have already noted, the emissivity of the stainless steel plate at 1500°F is approximately 0.5. Because most of the radiation incident upon $A_2$ actually originates at surface $A_2$ itself, we evaluate $\alpha_2$ at a source temperature of 1500°F, as a first estimate; that is,

$$\alpha_2 = \epsilon_2(T_2) = 0.5$$

It follows that

$$R_2 = \frac{\rho_2}{A_2 \alpha_2} = \frac{1 - 0.5}{A_1(0.5)} = \frac{1}{A_1}$$

The rate of radiation heat transfer is then

$$q_{2-1} = \frac{E_{b2} - E_{b1}}{R_{1-s} + R_{s1} + R_{s2} + R_{s-2} + R_2} = \frac{(2.52 \times 10^4 - 76.6)\ \text{Btu/(h ft}^2)}{(1 + 11.5 + 11.5 + 1 + 1)/A_1}$$

$$q''_{2-1} = 966\ \text{Btu/(h ft}^2)$$

Thus, based on this approximate analysis, we have a reduction in the radiation heat transfer of about 92%.

To determine the approximate temperature of the radiation shield, we write

$$q_{2-1} = \frac{E_{bs} - E_{b1}}{R_{1-s} + R_{s1}}$$

$$E_{bs} = E_{b1} + (R_{1-s} + R_{s1})q_{2-1} = [76.6 + (1 + 11.5)966] \text{ Btu/(h ft}^2)$$

$$\sigma T_s^4 = 12{,}200 \text{ Btu/(h ft}^2) \qquad T_s = 1630°\text{R} = 1170°\text{F}$$

Our analysis can be refined by evaluating the absorptivity at surface $A_2$ at a temperature which better represents the actual source of radiation. For example, the radiosity $J_{s2}$ represents the radiant energy incident upon $A_2$. Therefore, we could evaluate $\alpha_2$ at the temperature of a blackbody with total emissive power equal to $J_{s2}$.

***Reradiating Surfaces*** In the furnace arrangement of Fig. 5–35(a), the plate on the insulated floor is heated to a steady-state temperature which is totally governed by the thermal radiation from the walls at $T_2$ and ceiling at $T_1$. Under steady-state conditions, the total rate of radiation heat transfer $q_s$ from the plate to its enclosure

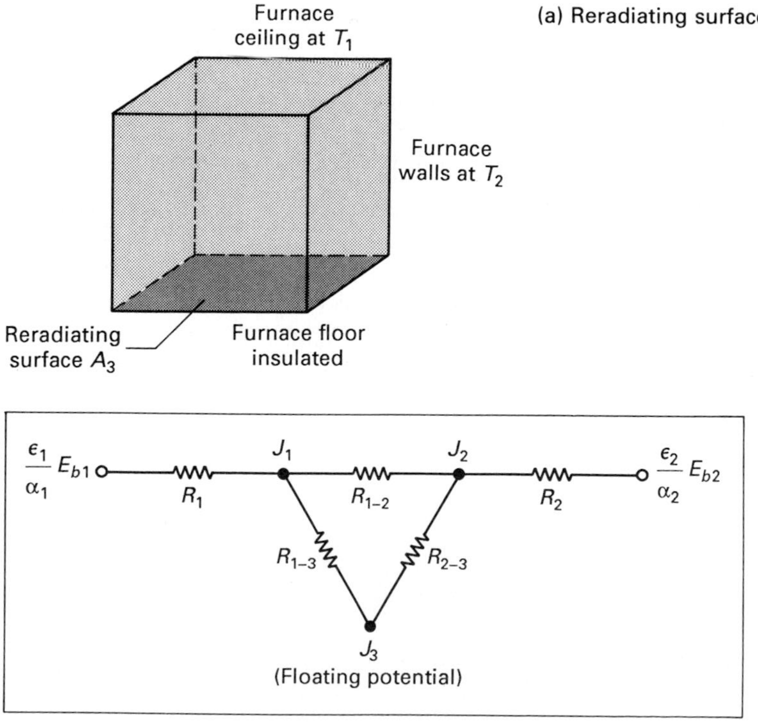


(a) Reradiating surface in furnace.

(b) Thermal radiation network.

**FIGURE 5–35** Reradiation.

is clearly zero. Bodies such as this with $q_s$ equal to zero are said to have *reradiating surfaces*.

For the usual case in which the reflectivity $\rho_s$ of an irradiating surface is not equal to unity, the thermal resistance $R_s$ given by Eq. (5–54) or Eq. (5–56) is finite. Thus, with $q_s$ set equal to zero in Eq. (5–55), we conclude that for irradiating surfaces with diffuse characteristics, the floating potential $E_{bs}$ must be equal to the radiosity $J_s$. This conclusion gives rise to the thermal radiation network shown in Fig. 5–35(b) for the system in Fig. 5–35(a). Solving this network for $q_{1-2}$, we obtain

$$q_{1-2} = \frac{(\epsilon_1/\alpha_1)E_{b1} - (\epsilon_2/\alpha_2)E_{b2}}{\dfrac{\rho_1}{A_1\alpha_1} + \dfrac{\rho_2}{A_2\alpha_2} + R_R} \tag{5–70}$$

where

$$R_R = \left[A_1F_{1-2} + \frac{1}{1/(A_1F_{1-3}) + 1/(A_2F_{2-3})}\right]^{-1} \tag{5–71}$$

Incidentally, for polished metallic surfaces such as copper (refer back to Fig. 5–8), the absorptivity $\alpha_s$ and emissivity $\epsilon_s$ are very small, with $\rho_s$ being almost equal to unity. The thermal resistance $R_s$ for such highly reflective surfaces is very large. If we neglect directional aspects, the thermal radiation network for a near-perfect reflector with specified temperature will be exactly the same as for a reradiating surface with floating potential. But, in reality, important nondiffuse specular characteristics of polished surfaces enter into the picture for many geometric arrangements that compromise the accuracy of Eqs. (5–62) and (5–63).

## EXAMPLE 5-20

Outline the steps required for determining the temperature of the reradiating plate surface $A_3$ shown in Fig. 5–35(a) if the temperature $T_1$ of the ceiling is 1000°C and the temperature $T_2$ of the walls is 500°C. Radiation from the interior of the furnace exhibits graybody conditions with emissivity equal to 0.8 and the plate is made of commercial aluminum.

### Solution

***Objective*** Summarize the steps required for determining $T_3$.

***Schematic*** Reradiating surface in furnace.

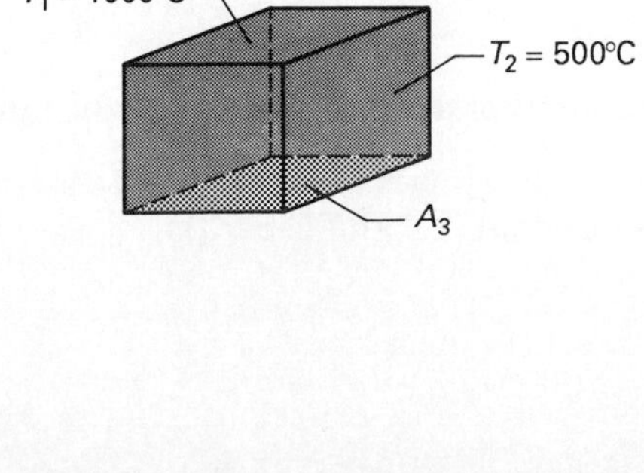

$T_1$ = 1000°C
$T_2$ = 500°C
$A_3$

*Assumptions/Conditions*

steady-state
diffuse opaque graybody surfaces
surface $A_3$ is reradiating
furnace is evacuated

*Properties* $\epsilon_1 = 0.8$, $\epsilon_2 = 0.8$.

*Analysis* First, the radiation properties $\epsilon_3$ and $\alpha_3$ of surface $A_3$ can be obtained from Fig. 5–3 and by utilizing either the nonisothermal or graybody $\alpha$ approximation. According to the graybody $\alpha$ approximation, $\alpha_1 = \epsilon_1 = 0.8$ and $\alpha_2 = \epsilon_2 = 0.8$.

Second, the radiation shape factors can be evaluated by utilizing Fig. 5–21 and the principles of reciprocity and summation.

Third, referring to Fig. 5–35, which represents the approximate thermal radiation network for this problem, $q_{1-2}$ can be calculated from Eq. (5–70), $J_1$ and $J_2$ can be calculated by writing

$$q_1 = q_{1-2} = \frac{(\epsilon_1/\alpha_1)E_{b1} - J_1}{R_1}$$

$$q_1 = -q_2 = \frac{J_2 - (\epsilon_2/\alpha_2)E_{b2}}{R_2}$$

$q_{1-3}$ can be calculated from

$$q_{1-3} = q_1 - \frac{J_1 - J_2}{R_{1-2}}$$

and $E_{b3}$ $(= J_3)$ can be obtained from

$$q_{1-3} = \frac{J_1 - E_{b3}}{R_{1-3}} = \frac{J_1 - \sigma T_3^4}{R_{1-3}}$$

### Radiation Factor

In looking back over the analyses developed for diffuse thermal radiation between opaque surfaces, we observe that the rate of radiation heat transfer $q_{s-j}$ between two surfaces $A_s$ and $A_j$ can be expressed in the compact form

$$q_{s-j} = A_s \mathscr{F}_{s-j}\left(\frac{\epsilon_s}{\alpha_s} E_{bs} - \frac{\epsilon_j}{\alpha_j} E_{bj}\right) \tag{5–72}$$

We will refer to $\mathscr{F}_{s-j}$ as the *radiation factor*. For approximate graybody conditions, Eq. (5–72) reduces to

$$q_{s-j} = A_s \mathscr{F}_{s-j}(E_{bs} - E_{bj}) = A_s \mathscr{F}_{s-j}\sigma(T_s^4 - T_j^4) \tag{5–73}$$

For example, for thermal radiation exchange in a bisurface system, $\mathcal{F}_{1-2}$ is given by

$$\mathcal{F}_{1-2} = \frac{1}{\dfrac{\rho_1}{\alpha_1} + \dfrac{1}{F_{1-2}} + \dfrac{A_1}{A_2}\dfrac{\rho_2}{\alpha_2}} \tag{5–74}$$

This equation reduces to

$$\mathcal{F}_{1-2} = \frac{1}{\dfrac{1-\epsilon_1}{\epsilon_1} + \dfrac{1}{F_{1-2}} + \dfrac{A_1}{A_2}\dfrac{1-\epsilon_2}{\epsilon_2}} \tag{5–75}$$

for approximate graybody conditions, and to

$$\mathcal{F}_{1-2} = F_{1-2} \tag{5–76}$$

for the limiting blackbody case.

For purposes of design, $\mathcal{F}_{s-j}$ is listed in Table 5–5 for several practical arrangements. For the more complex geometries for which $\mathcal{F}_{s-j}$ is not tabulated, the thermal radiation network should be solved systematically by numerical or analytical techniques. The systematic solution of thermal radiation problems is discussed in the following section.

**TABLE 5–5** Radiation factor $\mathcal{F}_{1-2}$

| | $\mathcal{F}_{1-2}$ | |
|---|---|---|
| *System* | *General* | *Graybody conditions* |
| Surfaces $A_1$ and $A_2$ | $\left(\dfrac{\rho_1}{\alpha_1} + \dfrac{1}{F_{1-2}} + \dfrac{A_1}{A_2}\dfrac{\rho_2}{\alpha_2}\right)^{-1}$ | $\left(\dfrac{\epsilon_1 - 1}{\epsilon_1} + \dfrac{1}{F_{1-2}} + \dfrac{A_1}{A_2}\dfrac{\epsilon_2 - 1}{\epsilon_2}\right)^{-1}$ |
| Infinite parallel plates | $\left(\dfrac{\rho_1}{\alpha_1} + 1 + \dfrac{\rho_2}{\alpha_2}\right)^{-1}$ | $\left(\dfrac{1}{\epsilon_1} + \dfrac{1}{\epsilon_2} - 1\right)^{-1}$ |
| Concentric cylinders or spheres | $\left(\dfrac{\rho_1}{\alpha_1} + 1 + \dfrac{D_1}{D_2}\dfrac{\rho_2}{\alpha_2}\right)^{-1}$ | $\left(\dfrac{1-\epsilon_1}{\epsilon_1} + 1 + \dfrac{A_1}{A_2}\dfrac{1-\epsilon_2}{\epsilon_2}\right)^{-1}$ |
| Small body $A_1$ inside large body $A_2$ | $\left(\dfrac{\rho_1}{\alpha_1} + 1\right)^{-1}$ | $\epsilon_1$ |
| Blackbody surfaces $A_1$ and $A_2$ | $F_{1-2}$ | $F_{1-2}$ |
| Surfaces $A_1$ and $A_2$ with with one radiation shield | | $\left(\dfrac{1-\epsilon_1}{\epsilon_1} + 1 + \dfrac{A_1}{A_2}\dfrac{1-\epsilon_2}{\epsilon_2} + \dfrac{A_1}{A_s}\dfrac{2-\epsilon_s}{\epsilon_s}\right)^{-1}$ |
| Parallel plates with $N$ radiation shields; $\epsilon_{sj} = \epsilon_s$ | | $\left(\dfrac{1-\epsilon_1}{\epsilon_1} + 1 + \dfrac{1-\epsilon_2}{\epsilon_2} + \dfrac{2-\epsilon_s}{\epsilon_s}N\right)^{-1}$ |

### *Systematic Solution Approach*

To develop a systematic solution for the radiation heat transfer in a thermal radiation network, an energy balance can be made on each $J$ node. For the moment, we focus our attention on systems with opaque diffuse surfaces. Referring to the $J_s$ node shown in Fig. 5–36, we have our building-block equation,

$$q_s = \frac{A_s \alpha_s}{\rho_s}\left(\frac{\epsilon_s}{\alpha_s} E_{bs} - J_s\right) \tag{5–53}$$

where $\rho_s = 1 - \alpha_s$. Based on the first law of thermodynamics, $q_s$ must also satisfy

$$q_s = \sum_{j=1}^{N} A_s F_{s-j}(J_s - J_j) \tag{5–77}$$

To obtain a nodal equation for $J_s$ for the case in which the surface temperature $T_s$ is known, we combine Eqs. (5–53) and (5–77), with the result

$$\frac{\alpha_s}{\rho_s} J_s + \sum_{j=1}^{N} F_{s-j}(J_s - J_j) = \frac{\epsilon_s}{\rho_s} E_{bs} \tag{5–78}$$

On the other hand, if the rate of radiation heat transfer $q_s$ from a surface is specified with $T_s$ being unknown, then Eq. (5–77) serves as the nodal equation for $J_s$. A nodal equation of one of these types can be written for each of the $N$ surfaces. For example, for the three-surface system of Fig. 5–37, we write

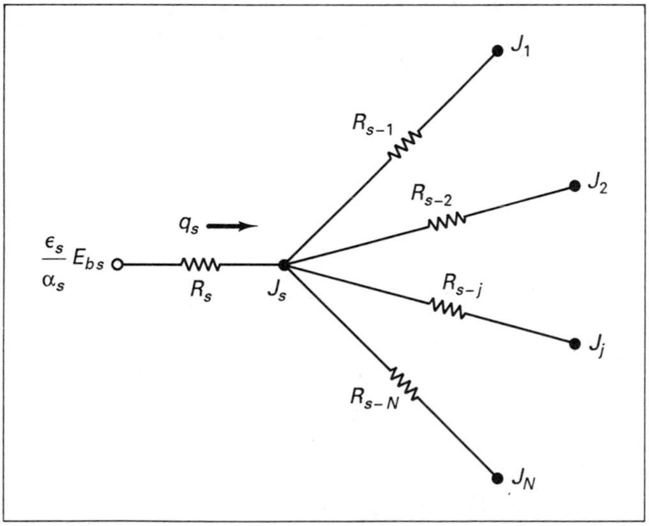


**FIGURE 5–36** Segment of thermal radiation network involving $J_s$ node.

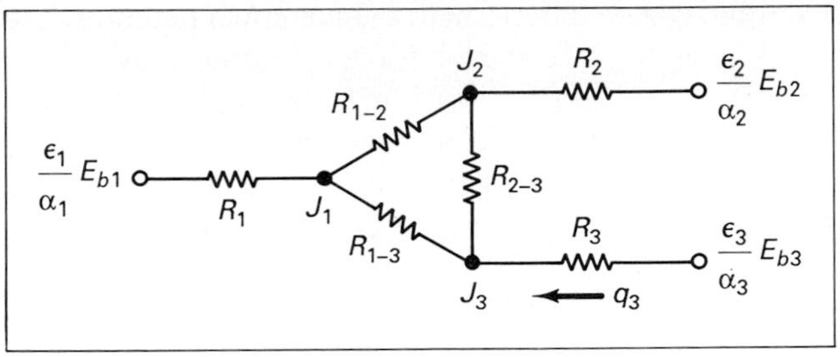


**FIGURE 5–37** Thermal radiation network for trisurface system with $T_1$, $T_2$, and $q_3$ specified.

$$\left(\frac{\alpha_1}{\rho_1} + F_{1-2} + F_{1-3}\right)J_1 - F_{1-2}J_2 - F_{1-3}J_3 = \frac{\epsilon_1}{\rho_1} E_{b1} \quad (5\text{–}79\text{–}1)$$

$$-F_{2-1}J_1 + \left(\frac{\alpha_2}{\rho_2} + F_{2-1} + F_{2-3}\right)J_2 - F_{2-3}J_3 = \frac{\epsilon_2}{\rho_2} E_{b2} \quad (5\text{–}79\text{–}2)$$

$$-F_{3-1}J_1 - F_{3-2}J_2 + (F_{3-1} + F_{3-2})J_3 = \frac{q_3}{A_3} \quad (5\text{–}79\text{–}3)$$

In general, for $N$ surfaces the nodal equations are written as

$$a_{11}J_1 + a_{12}J_2 + \cdots + a_{1j}J_j + \cdots + a_{1N}J_N = C_1 \quad (5\text{–}80\text{–}1)$$

$$a_{21}J_1 + a_{22}J_2 + \cdots + a_{2j}J_j + \cdots + a_{2N}J_N = C_2 \quad (5\text{–}80\text{–}2)$$

$$\vdots$$

$$a_{j1}J_1 + a_{j2}J_2 + \cdots + a_{jj}J_j + \cdots + a_{jN}J_N = C_j \quad (5\text{–}80\text{–j})$$

$$\vdots$$

$$a_{N1}J_1 + a_{N2}J_2 + \cdots + a_{Nj}J_j + \cdots + a_{NN}J_N = C_N \quad (5\text{–}80\text{–N})$$

The $N$ unknowns $J_1, J_2, \ldots, J_j, \ldots, J_s, \ldots, J_N$ in these $N$ equations can be solved by the analytical or numerical techniques which were introduced in Chap. 3. For example, the Gauss-Seidel approach can be used to obtain a hand calculation or computer solution. Or, hand calculation and computer solutions can be affected by powerful matrix methods. For situations in which the graybody $\alpha$ approximation can be utilized, the solutions are quite straightforward. However, for nongraybody conditions, $\alpha_s$ is expressed in terms of the unknown radiosities $J_j$, such that this approach requires iteration. For this situation, $\alpha_s$ can be specified by Eq. (f) in Example 5–14.

Once the radiosities have been determined, the unknown radiation-heat-transfer rate $q_s$ or temperature $T_s$ can be determined for each surface by employing Eq. (5–53).

As an alternative approach, the equation developed in Example 5–14 can be used. But here again, because of the dependence of the various components of absorptivity $\alpha_{sj}$ on the radiosities $J_j$, this approach also involves iteration, unless graybody conditions are assumed.

## EXAMPLE 5–21

The following information is available for the very long three-surface graybody enclosure shown in Fig. E5–21:

$$T_1 = 200°\text{C} \qquad \epsilon_1 = 0.2$$

$$T_2 = 27°\text{C} \qquad \epsilon_2 = 0.7$$

$$q_3'' = 1 \text{ kW/m}^2 \qquad \epsilon_3 = 0.5$$

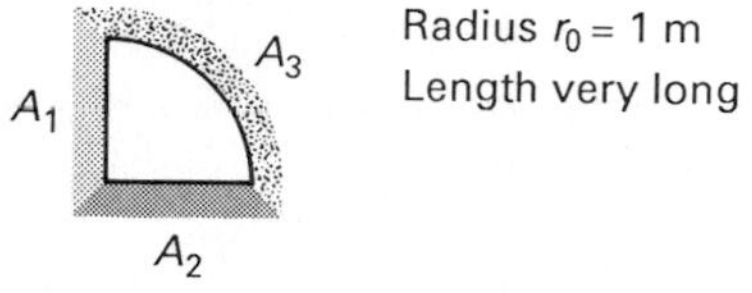
A3
A1
A2
Radius $r_0$ = 1 m
Length very long

**FIGURE E5–21**
Trisurface graybody enclosure.

Determine the radiation-heat-transfer fluxes from surfaces $A_1$ and $A_2$ and the temperature of surface $A_3$.

### Solution

*Objective* Determine $q_1''$, $q_2''$, and $T_3$.

*Assumptions/Conditions*

steady-state
diffuse opaque graybody surfaces
enclosure is evacuated

*Properties* $\epsilon_1 = 0.2$, $\epsilon_2 = 0.7$, $\epsilon_3 = 0.5$.

*Analysis* The approximate thermal radiation network for this nonsymmetrical system is given in Fig. 5–37.

We first consider the geometric aspects of the problem. As seen in Prob. 5–15, $F_{1-2}$ is equal to 0.293. Based on the reciprocity and summation principles, we write

$$F_{2-1} = \frac{A_1}{A_2} F_{1-2} = 1(0.293) = 0.293$$

$$F_{1-3} = 1 - F_{1-2} = 1 - 0.293 = 0.707$$

$$F_{3-1} = \frac{A_1}{A_3} F_{1-3} = \frac{4}{2\pi}(0.707) = 0.45$$

$$F_{3-2} = \frac{A_2}{A_3} F_{2-3} = 0.45$$

Because $A_3$ is concave, we also have

$$F_{3-3} = 1 - F_{3-1} - F_{3-2} = 0.1$$

The total emissive powers $E_{b1}$ and $E_{b2}$ are written as

$$E_{b1} = \sigma T_1^4 = 5.67 \times 10^{-8} \frac{\text{W}}{\text{m}^2\,\text{K}^4}(473\text{ K})^4 = 2.84\text{ kW/m}^2$$

$$E_{b2} = \sigma T_2^4 = \sigma(300\text{ K})^4 = 0.459\text{ kW/m}^2$$

And, because we are dealing with opaque graybody surfaces, the absorptivities and reflectivities are given by

$$\alpha_1 = \epsilon_1 = 0.2 \qquad \rho_1 = 1 - \alpha_1 = 0.8$$

$$\alpha_2 = \epsilon_2 = 0.7 \qquad \rho_2 = 1 - \alpha_2 = 0.3$$

$$\alpha_3 = \epsilon_3 = 0.5 \qquad \rho_3 = 1 - \alpha_3 = 0.5$$

The nodal equations at the three radiosity nodes are given by Eqs. (5–79–1), (5–79–2), and (5–79–3). Substituting for the various potentials, radiation shape factors, and radiation properties, we obtain

$$\left(\frac{0.2}{0.8} + 0.293 + 0.707\right)J_1 - 0.293J_2 - 0.707J_3 = \frac{0.2}{0.8}\left(2.84\frac{\text{kW}}{\text{m}^2}\right)$$

$$-0.293J_1 + \left(\frac{0.7}{0.3} + 0.293 + 0.707\right)J_2 - 0.707J_3 = \frac{0.7}{0.3}\left(0.459\frac{\text{kW}}{\text{m}^2}\right)$$

$$-0.45J_1 - 0.45J_2 + (0.45 + 0.45)J_3 = 1\text{ kW/m}^2$$

or

$$1.25J_1 - 0.293J_2 - 0.707J_3 = 0.71\text{ kW/m}^2$$

$$-0.293J_1 + 3.33J_2 - 0.707J_3 = 1.07\text{ kW/m}^2$$

$$-0.45J_1 - 0.45J_2 + 0.9J_3 = 1\text{ kW/m}^2$$

**Solution**

*Objective* Determine the instantaneous plate temperature $T_1$, heat transfer rate $q''_R$, and time $t_{ss}$ required to reach steady state.

*Schematic* Plate heated in an oven—all blackbody surfaces.

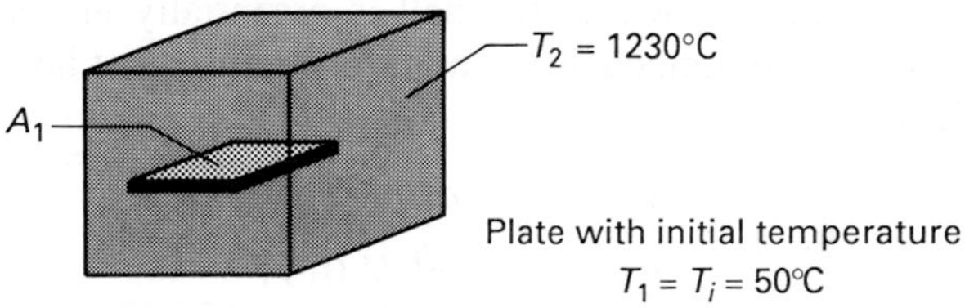
$T_2$ = 1230°C
$A_1$
Plate with initial temperature
$T_1 = T_i = 50$°C

*Assumptions/Conditions*

unsteady-state
blackbody surfaces
enclosure is evacuated

*Properties* All surfaces: $\epsilon = 1$.

*Analysis* Assuming that the thermal radiation Biot number is less than 0.1, the differential formulation for this problem is given by Eqs. (5–83) and (5–84).

$$q_{1-2} + \frac{dU}{dt} = 0$$

$$\frac{dT_1}{dt} + \frac{\sigma A_1 \mathcal{F}_{1-2}}{\rho V c_v}(T_1^4 - T_2^4) = 0$$

$$T_1 = T_i = 323 \text{ K}$$

where $\mathcal{F}_{1-2} = F_{1-2} = 1$ for blackbody radiation.

Separating the variables in this nonlinear first-order differential equation, we write

$$\frac{dT_1}{T_1^4 - T_2^4} = \frac{dT_1}{(T_1^2 - T_2^2)(T_1^2 + T_2^2)} = -\frac{\sigma A_1 \mathcal{F}_{1-2}}{\rho V c_v} dt$$

The left-hand side of this relationship can be broken into partial fractions as follows:

$$\frac{dT_1}{(T_1^2 - T_2^2)(T_1^2 + T_2^2)} = \left(\frac{C_1}{T_1^2 - T_2^2} + \frac{C_2}{T_1^2 + T_2^2}\right) dT_1$$

or

$$1 = C_1 T_1^2 + C_1 T_2^2 + C_2 T_1^2 - C_2 T_2^2$$

where $C_1$ and $C_2$ must be evaluated. By equating the coefficients associated with $T_1$, we obtain

$$0 = C_1 + C_2$$

We also have

$$1 = (C_1 - C_2)T_2^2$$

Hence, $C_1$ and $C_2$ are given by

$$C_1 = -C_2 \qquad C_1 = \frac{1}{2T_2^2}$$

Therefore, our differential equation takes the simpler form

$$\frac{dT_1}{2T_2^2(T_1^2 - T_2^2)} - \frac{dT_1}{2T_2^2(T_1^2 + T_2^2)} = -\frac{\sigma A_1 \mathcal{F}_{1-2}}{\rho V c_v}\, dt$$

This equation can be integrated with the help of integration tables. The result is

$$\frac{1}{4T_2^3} \ln \left| \frac{T_1 - T_2}{T_1 + T_2} \right| - \frac{1}{2T_2^3} \tan^{-1} \frac{T_1}{T_2} = -\frac{\sigma A_1 \mathcal{F}_{1-2}}{\rho V c_v} t + C_3$$

Applying the initial condition, we have

$$C_3 = \frac{1}{2T_2^3} \left( \frac{1}{2} \ln \left| \frac{T_i - T_2}{T_i + T_2} \right| - \tan^{-1} \frac{T_i}{T_2} \right)$$

The solution now can be written as

$$\frac{1}{2} \left( \ln \left| \frac{T_1 - T_2}{T_1 + T_2} \right| - \ln \left| \frac{T_i - T_2}{T_i + T_2} \right| \right) - \left( \tan^{-1} \frac{T_1}{T_2} - \tan^{-1} \frac{T_i}{T_2} \right) = -2T_2^3 \frac{\sigma A_1 \mathcal{F}_{1-2}}{\rho V c_v} t \qquad \text{(a)}$$

Setting $\mathcal{F}_{1-2} = 1$, $A_1 = 2\ \text{m}^2$, $V = 10^{-3}\ \text{m}^3$, $T_2 = 1500$ K, $\rho = 8950$ kg/m³, and $c_v = 0.383$ kJ/(kg °C), Eq. (a) reduces to the form

$$t = \frac{\text{s}}{0.223} \left[ \left( \tan^{-1} \frac{T_1}{1500\ \text{K}} - 0.212 \right) - \frac{1}{2} \left( \ln \left| \frac{T_1 - 1500\ \text{K}}{T_1 + 1500\ \text{K}} \right| + 0.438 \right) \right] \qquad \text{(b)}$$

This equation is solved for $t$ by taking $T_1$ as the independent variable. The resulting predictions for $T_1$ are shown in Fig. E5–22.

*Assumptions/Conditions*

steady-state
blackbody surfaces
enclosure contains participating gases $H_2O$ and $CO_2$

*Properties* Walls: $\epsilon = 1$.

*Analysis* The gas emissivity and absorptivity are shown in Example 5–7 to be approximately

$$\epsilon_m = 0.174 \qquad \alpha_m = 0.312$$

The total emissive powers $E_{bs}$ and $E_{bm}$ are given by

$$E_{bs} = \sigma T_s^4 = 5.67 \times 10^{-8} \frac{\text{W}}{\text{m}^2\ \text{K}^4} (500\ \text{K})^4 = 3.54 \times 10^3\ \text{W/m}^2$$

$$E_{bm} = \sigma T_m^4 = \sigma(1500\ \text{K})^4 = 2.87 \times 10^5\ \text{W/m}^2$$

The radiation heat transfer between the burning gases and the surface of the furnace are calculated by utilizing Eq. (5–86).

$$q_{m-s} = -q_{s-m} = A_s F_{s-m}(\epsilon_m E_{bm} - \alpha_m E_{bs})$$

$$= 6\ \text{m}^2\,(1)[0.174(2.87 \times 10^5) - 0.312(3.54 \times 10^3)] \frac{\text{W}}{\text{m}^2} = 293\ \text{kW}$$

*Gray Gas/Multisurface Enclosure* For multisurface blackbody enclosures, we have equations similar to Eq. (5–85) for each surface. For example, for two surfaces $A_s$ and $A_j$ we have

$$q_{s-m} = A_s F_{s-m}(\alpha_{ms} E_{bs} - \epsilon_m E_{bm}) \tag{5–87}$$

and

$$q_{j-m} = A_j F_{j-m}(\alpha_{mj} E_{bj} - \epsilon_m E_{bm}) \tag{5–88}$$

where the second subscript on the absorptivity of the gas medium designates the source of the irradiation.

In addition, an expression must be developed for the rate of thermal radiation $q_{s-j}$ transmitted through the medium from surface $s$ to surface $j$. The rate of thermal radiation leaving surface $A_s$ that reaches $A_j$ is $A_s F_{s-j} E_{bs} \tau_{ms}$ and the rate from surface $A_j$ to $A_s$ is $A_j F_{j-s} E_{bj} \tau_{mj}$. Therefore, the rate of thermal radiation from $A_s$ to $A_j$ is

$$q_{s-j} = A_s F_{s-j}(E_{bs}\tau_{ms} - E_{bj}\tau_{mj}) \tag{5–89}$$

If the surface temperatures $T_s$ and $T_j$ are not too different, $\tau_{ms}$ and $\tau_{mj}$ can be represented by an average value $\tau_m$, such that Eq. (5–89) becomes

$$q_{s-j} = A_s F_{s-j} \tau_m (E_{bs} - E_{bj}) \tag{5–90}$$

Following through, $\alpha_{ms}$ and $\alpha_{mj}$ in Eqs. (5–87) and (5–88) can be set equal to $\alpha_m$. Otherwise, Eq. (5–89) can be left in its present form and the distinction between $\alpha_{ms}$ and $\alpha_{mj}$ in Eqs. (5–87) and (5–88) can be retained.

Equations (5–87) through (5–90) provide the basis for the thermal radiation network representation of multisurface blackbody systems involving a participating gray gas. For example, the parallel-plate system in Fig. 5–38(a) is represented by the network shown in Fig. 5–38(b) for the case in which $\tau_{m1} \simeq \tau_{m2} \simeq \tau_m$ and $\alpha_{m1} \simeq \alpha_{m2} \simeq \alpha_m$. Based on this thermal network, we see that the total radiation-heat-transfer rate $q_m$ from the medium is

$$q_m = \frac{(\epsilon_m/\alpha_m)E_{bm} - E_{b1}}{R_{1-m}} + \frac{(\epsilon_m/\alpha_m)E_{bm} - E_{b2}}{R_{2-m}} \qquad (5\text{–}91)$$
$$= A_1F_{1-m}\alpha_m\left(\frac{\epsilon_m}{\alpha_m}E_{bm} - E_{b1}\right) + A_2F_{2-m}\alpha_m\left(\frac{\epsilon_m}{\alpha_m}E_{bm} - E_{b2}\right)$$

The rate $q_m$ must also satisfy the first law of thermodynamics,

$$\Sigma\,\dot{E}_i = q_m + \frac{\Delta E_s}{\Delta t} \qquad (5\text{–}92)$$

which, for steady-state conditions, reduces to

$$q_m = \Sigma\,\dot{E}_i \qquad (5\text{–}93)$$

If no energy is transferred to the medium from external sources (i.e., $\Sigma\,\dot{E}_i = 0$), then $q_m$ is zero under steady-state conditions and the node $E_{bm}$ becomes a simple floating point. Under these passive equilibrium conditions, the same rate of energy

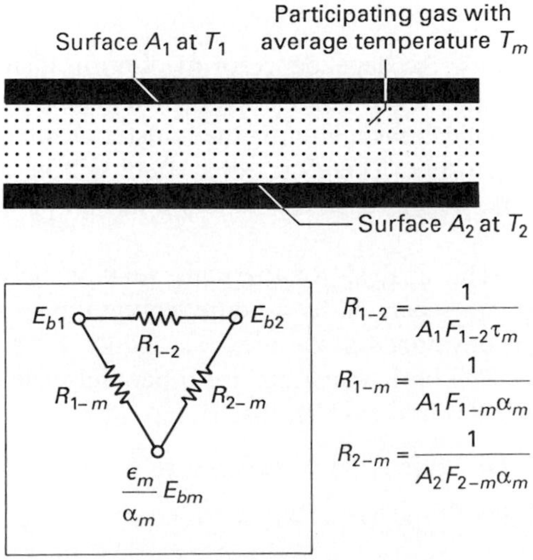
Surface $A_1$ at $T_1$
Participating gas with average temperature $T_m$
Surface $A_2$ at $T_2$
$E_{b1}$
$E_{b2}$
$R_{1-2}$
$R_{1-m}$
$R_{2-m}$
$\frac{\epsilon_m}{\alpha_m} E_{bm}$

$$R_{1-2} = \frac{1}{A_1F_{1-2}\tau_m}$$
$$R_{1-m} = \frac{1}{A_1F_{1-m}\alpha_m}$$
$$R_{2-m} = \frac{1}{A_2F_{2-m}\alpha_m}$$

(a) Parallel-plate system.

(b) Thermal radiation network.

**FIGURE 5–38**
Radiation heat transfer in a bisurface blackbody system with participating gas.

is emitted by the medium as is absorbed. The solution to the thermal network for this simple case gives rise to

$$q_{1-2} = A_1 \left[ F_{1-2}\tau_m + \frac{\alpha_m}{\dfrac{1}{F_{1-m}} + \dfrac{A_1}{A_2F_{2-m}}} \right] (E_{b1} - E_{b2}) \tag{5–94}$$

Notice that for a nonparticipating gas with $\alpha_m = 0$, this equation reduces to Eq. (5–42),

$$q_{1-2} = A_1F_{1-2}(E_{b1} - E_{b2}) \tag{5–42}$$

***Near-Blackbody Enclosures*** For situations in which the emissivity of the wall of a single enclosure is of the order of 0.8 and larger, Hottel [17] has shown that the net rate of radiation heat transfer can be approximated by multiplying Eq. (5–86) by the factor $(\epsilon_s + 1)/2$.

$$q_{s-m} = A_sF_{s-m}(\alpha_m E_{bs} - \epsilon_m E_{bm}) \frac{\epsilon_s + 1}{2} \tag{5–95}$$

But it should be emphasized that this approximation is only valid for near-blackbody surfaces.

For enclosures with low emittance surfaces, the gray gas method is not appropriate. For problems of this type, the radiation properties of the gas must be obtained experimentally, or more comprehensive analyses must be developed that account for the band-absorption characteristics of the gas. Higher-order analyses are discussed in references 5, 11, and 24.

### *Transparent Solids*

As indicated in the preceding sections, the dependence of medium irradiation properties on the source is sometimes important in dealing with participating gases. This factor is very critical in important applications involving the transfer of solar radiation through a glass medium into an enclosure. Therefore, this effect will be accounted for as we consider the practical solution approach for radiation in transparent solids.

***Blackbody Bisurface System*** The analysis of thermal radiation between two blackbody surfaces $A_1$ and $A_2$ that are separated by a nonreflecting transparent solid medium is identical to the analysis developed in the previous section for a gray gas/multisurface blackbody enclosure. The basic equations for a parallel-plate bisurface system are taken from Eqs. (5–87) through (5–89); that is,

$$q_{1-m} = A_1F_{1-m}(\alpha_{m1}E_{b1} - \epsilon_m E_{bm}) \tag{5–96}$$

$$q_{2-m} = A_2F_{2-m}(\alpha_{m2}E_{b2} - \epsilon_m E_{bm}) \tag{5–97}$$

$$q_{1-2} = A_1F_{1-2}(E_{b1}\tau_{m1} - E_{b2}\tau_{m2}) \tag{5–98}$$

where

$$\alpha_{m1} + \tau_{m1} = 1 \qquad \alpha_{m2} + \tau_{m2} = 1 \tag{5–99,100}$$

In order to develop a thermal radiation network for the important case in which $\tau_{m1}$ and $\tau_{m2}$ are significantly different, we rearrange Eqs. (5–96) through (5–98) as follows:

$$q_{1-m} = A_1F_{1-m}\frac{\alpha_{m1}E_{b1} - \epsilon_m E_{bm}}{E_{b1} - E_{bm}}(E_{b1} - E_{bm}) \tag{5–101}$$

$$q_{2-m} = A_2F_{2-m}\frac{\alpha_{m2}E_{b2} - \epsilon_m E_{bm}}{E_{b2} - E_{bm}}(E_{b2} - E_{bm}) \tag{5–102}$$

$$q_{1-2} = A_1F_{1-2}\frac{E_{b1}\tau_{m1} - E_{b2}\tau_{m2}}{E_{b1} - E_{b2}}(E_{b1} - E_{b2}) \tag{5–103}$$

These equations provide the basis for the thermal radiation network shown in Fig. 5–39.

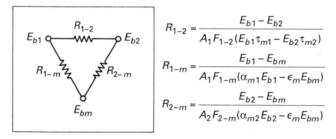

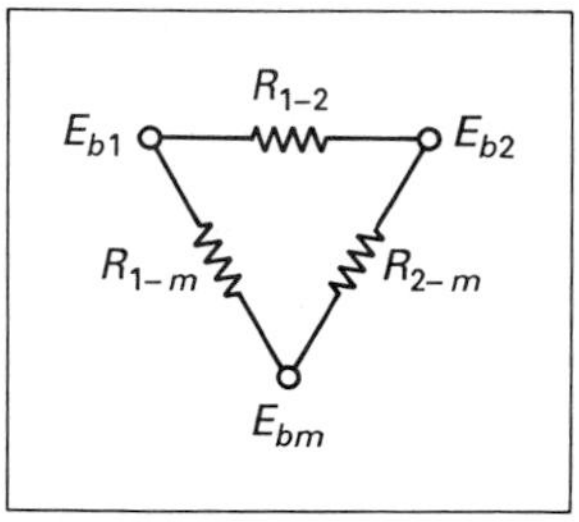


$$R_{1-2} = \frac{E_{b1} - E_{b2}}{A_1F_{1-2}(E_{b1}\tau_{m1} - E_{b2}\tau_{m2})}$$

$$R_{1-m} = \frac{E_{b1} - E_{bm}}{A_1F_{1-m}(\alpha_{m1}E_{b1} - \epsilon_m E_{bm})}$$

$$R_{2-m} = \frac{E_{b2} - E_{bm}}{A_2F_{2-m}(\alpha_{m2}E_{b2} - \epsilon_m E_{bm})}$$

**FIGURE 5–39** Thermal radiation network for blackbody bisurface system with transparent solid medium.

The absorptivities and transmissivities can be approximated by the method developed in Example 5–13. Of course, for situations in which the surface temperatures $T_1$ and $T_2$ are of the same order of magnitude, $\tau_{m1} \simeq \tau_{m2}$ (and $\alpha_{m1} \simeq \alpha_{m2}$) such that this network reduces to the thermal radiation network shown in Fig. 5–38(b). However, for important applications involving the transfer of solar or high-temperature radiation through glass into enclosures, say from $A_1$ to $A_2$, we have a maximum difference between $\tau_{m1}$ and $\tau_{m2}$, with the glass being essentially transparent to the high-temperature radiation ($\tau_{m1} \simeq 1$, $\alpha_{m1} \simeq 0$) and nearly opaque to the energy emitted within the enclosure ($\tau_{m2} \simeq 0$, $\alpha_{m2} \simeq 1$). By referring to Fig. 5–39 and by reexamining Eqs. (5–96) through (5–98), we see that this combination of glass properties allows a large net rate of thermal radiation to be transmitted through the glass.

By performing an energy balance on the $E_{bm}$ node, we have (neglecting convection)

$$q_{1-m} = q_{m-2} \tag{5–104}$$

That is, the rate of radiation heat transfer between $A_1$ and the glass is equal to the net rate of radiant exchange between the glass and $A_2$. It follows that the total rate of radiation heat transfer from $A_1$ is given by

$$q_1 = q_{1-2} + q_{1-m} \tag{5–105}$$

The rate of radiation heat transfer $q_{1-2}$ transmitted directly through the glass can be evaluated by using Eq. (5–103). To evaluate $q_{1-m}$, we must solve for $\epsilon_m E_{bm}$. This is done by utilizing Eq. (5–104). Substituting for $q_{1-m}$ and $q_{m-2}$ in this equation, we obtain

$$A_1F_{1-m}(\alpha_{m1}E_{b1} - \epsilon_m E_{bm}) = A_2F_{2-m}(\epsilon_m E_{bm} - \alpha_{m2}E_{b2})$$

$$\epsilon_m E_{bm} = \frac{A_1F_{1-m}\alpha_{m1}E_{b1} + A_2F_{2-m}\alpha_{m2}E_{b2}}{A_1F_{1-m} + A_2F_{2-m}} \tag{5–106}$$

Employing the principle of reciprocity and recognizing that $F_{m-1} = 1$ and $F_{m-2} = 1$, Eq. (5–106) reduces to

$$\epsilon_m E_{bm} = \frac{\alpha_{m1}E_{b1} + \alpha_{m2}E_{b2}}{2} \tag{5–107}$$

The substitution of this expression for $\epsilon_m E_{bm}$ into Eq. (5–101) or Eq. (5–102) gives

$$\begin{aligned} q_{1-m} = q_{m-2} &= A_mF_{m-1}\left(\alpha_{m1}E_{b1} - \frac{\alpha_{m1}E_{b1} + \alpha_{m2}E_{b2}}{2}\right) \\ &= A_m\left(\frac{\alpha_{m1}E_{b1} - \alpha_{m2}E_{b2}}{2}\right) \end{aligned} \tag{5–108}$$

Returning to Eq. (5–105), $q_1$ becomes

$$q_1 = A_1F_{1-2}(E_{b1}\tau_{m1} - E_{b2}\tau_{m2}) + A_m\left(\frac{\alpha_{m1}E_{b1} - \alpha_{m2}E_{b2}}{2}\right) \tag{5–109}$$

Based on geometric considerations, $F_{1-2}$ can be set equal to $F_{1-m}$, such that $A_1F_{1-2} = A_1F_{1-m} = A_mF_{m-1} = A_m$. Thus, our final expression for $q_1$ takes the form

$$\begin{aligned} q_1 &= A_m\left(E_{b1}\tau_{m1} - E_{b2}\tau_{m2} + \frac{\alpha_{m1}E_{b1} - \alpha_{m2}E_{b2}}{2}\right) \\ &= \frac{A_m}{2}[E_{b1}(1 + \tau_{m1}) - E_{b2}(1 + \tau_{m2})] \end{aligned} \tag{5–110}$$

This rate of thermal radiation heat transfer from the high-temperature source is equal to the total rate of thermal radiation received by $A_2$, $-q_2$.

## EXAMPLE 5–24

Determine the rate of radiation heat transfer from the interior of a furnace with surface area of 1 m$^2$ at 2000°C through a 0.1-m$^2$ glass plate to the interior of a room with a 10-m$^2$ surface area at 27°C. The glass properties are specified as $\tau_\lambda = 0.90$ for

$0.29\ \mu m < \lambda < 2.7\ \mu m$, $\tau_\lambda = 0$ outside this range, and $\rho_\lambda = 0$ for all wavelengths. Assume that the walls of the furnace and room can be approximated as blackbodies.

**Solution**

*Objective* Determine $q_{1-2}$.

*Schematic* Furnace with glass window.

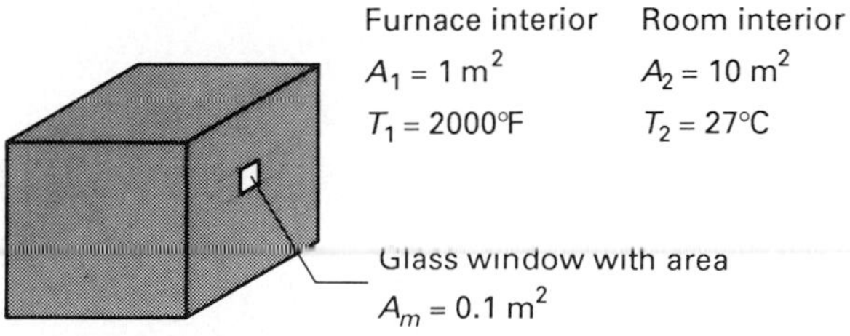
Furnace interior
$A_1 = 1\ m^2$
$T_1 = 2000°F$
Room interior
$A_2 = 10\ m^2$
$T_2 = 27°C$
Glass window with area
$A_m = 0.1\ m^2$

*Assumptions/Conditions*

steady-state
glass window in blackbody enclosure
enclosure is evacuated

*Properties*

Furnace and room walls: $\epsilon = 1$.
Glass:

$$\tau_\lambda = 0.9 \quad \text{for } 0.29\ \mu m < \lambda < 2.7\ \mu m$$
$$\tau_\lambda = 0 \quad \text{for } \lambda \leqslant 0.29\ \mu m,\ \lambda \geqslant 2.7\ \mu m$$
$$\rho_\lambda = 0 \quad \text{for all } \lambda$$

*Analysis* Following the pattern established in Example 5–6, we obtain $E_{b,\lambda_1\to\lambda_2}/E_b$ for both source temperatures in the wavelength range $0.29\ \mu m < \lambda < 2.7\ \mu m$. For the 2270 K temperature source, the corresponding $\lambda T$ range is $659\ \mu m\ K < \lambda T < 6140\ \mu m\ K$. Referring to Table A–H–1, we find

$$\frac{E_{b,\lambda_1\to\lambda_2}}{E_b} = 0.748 - 0.17 \times 10^{-7} \simeq 0.748$$

It follows that

$$\tau_{m1} = 0.9(0.748) = 0.673$$

For the 300 K source, the $\lambda T$ range is from 87 to 810 $\mu m$ K and

$$\frac{E_{b,\lambda_1\to\lambda_2}}{E_b} \simeq 0.738 \times 10^{-4}$$

such that $\tau_{m2}$ is essentially zero.

To obtain the total rate of radiation heat transfer we employ Eq. (5–110).

$$q_1 = A_m\left(E_{b1}\tau_{m1} - E_{b2}\tau_{m2} + \frac{\alpha_{m1}E_{b1} - \alpha_{m2}E_{b2}}{2}\right)$$

where $\tau_{m1} = 0.673$, $\tau_{m2} = 0$, $\alpha_{m1} = 0.327$, $\alpha_{m2} = 1$, and

$$E_{b1} = \sigma T_1^4 = 5.67 \times 10^{-8}\,\frac{\text{W}}{\text{m}^2\,\text{K}^4}\,(2270\ \text{K})^4 = 1.51 \times 10^6\ \text{W/m}^2$$

$$E_{b2} = \sigma T_2^4 = \sigma(300\ \text{K})^4 = 459\ \text{W/m}^2$$

Substituting into Eq. (5–110), we calculate

$$q_1 = 0.1\ \text{m}^2\left[1.51 \times 10^6\,(0.673) - 0 + \frac{0.327(1.51 \times 10^6) - 1(459)}{2}\right]\frac{\text{W}}{\text{m}^2}$$

$$= 1.26 \times 10^5\ \text{W}$$

For purpose of comparison, the radiation-heat-transfer rate is calculated for an opening with no glass plate.

$$q_{1-2} = A_m(E_{b1} - E_{b2}) = 0.1\ \text{m}^2\,(1.51 \times 10^6 - 459)\,\frac{\text{W}}{\text{m}^2} = 1.51 \times 10^5\ \text{W}$$

Thus, we find that the glass plate reduces the rate of radiation heat transfer by about 16%.

***Effects of Diffuse Nonblackbody Surfaces and Reflecting Medium*** Of course, the transfer of thermal radiation through a transparent solid medium generally occurs in the context of nonblackbody surfaces. In addition, reflection by the surfaces of the medium is sometimes significant. For example, the reflectivity of glass is usually of the order of 0.1. These complexities are approximately accounted for in Examples 5–25 and 5–26 by utilizing the radiosity concept.

### EXAMPLE 5–25

Develop a thermal radiation network for diffuse nonblackbody parallel plates which are separated by a nonreflecting transparent solid medium.

#### Solution

*Objective* Develop a thermal radiation network for this parallel plate system.

*Schematic* Parallel plates separated by transparent plate.

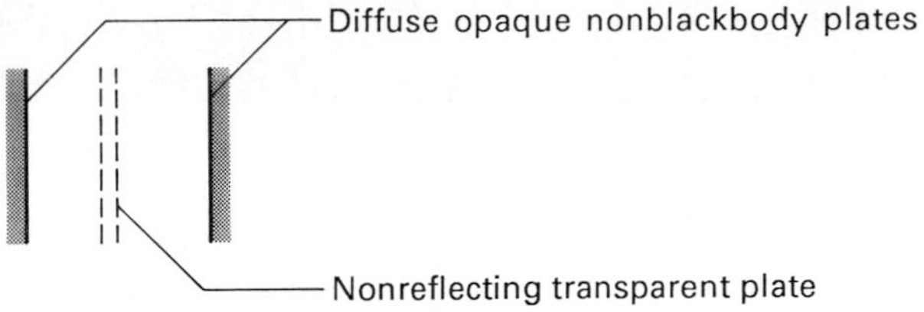
Diffuse opaque nonblackbody plates
Nonreflecting transparent plate

*Assumptions/Conditions*

steady-state
enclosed spaces are evacuated

*Analysis* To analyze radiation heat transfer between two diffuse surfaces $A_1$ and $A_2$ which are separated by a transparent solid medium, we utilize the concepts that have been introduced in the previous two sections. First, the net rate of radiation heat transfer from each surface is expressed in terms of radiosity by relationships of the form of Eq. (5–53); that is,

$$q_1 = \frac{A_1\alpha_1}{\rho_1}\left(\frac{\epsilon_1}{\alpha_1}E_{b_1} - J_1\right) \qquad q_2 = \frac{A_2\alpha_2}{\rho_2}\left(\frac{\epsilon_2}{\alpha_2}E_{b_2} - J_2\right)$$

A second set of equations is written for the net rate of radiation heat transfer between each surface and the medium $A_m$. By recognizing that $q_{1-m}$ must be equal to the difference between (1) the energy leaving $A_1$ that is absorbed by the medium, $\alpha_{m1}(A_1F_{1-m}J_1)$, and (2) the thermal radiation which is emitted by the medium and reaches $A_1$, $A_mF_{m-1}\epsilon_mE_{bm}$, we have

$$q_{1-m} = \alpha_{m1}(A_1F_{1-m}J_1) - A_mF_{m-1}\epsilon_mE_{bm} = A_1F_{1-m}(\alpha_{m1}J_1 - \epsilon_mE_{bm})$$

Similarly, $q_{2-m}$ takes the form

$$q_{2-m} = A_2F_{2-m}(\alpha_{m2}J_2 - \epsilon_mE_{bm})$$

Finally, the rate of radiation heat transfer between $A_1$ and $A_2$ is equal to the difference between (1) the rate of thermal radiation leaving $A_1$ that reaches $A_2$, $A_1F_{1-2}J_1\tau_{m1}$, and (2) the rate from $A_2$ to $A_1$, $A_2F_{2-1}J_2\tau_{m2}$; that is,

$$q_{1-2} = A_1F_{1-2}(J_1\tau_{m1} - J_2\tau_{m2}) \tag{a}$$

These equations provide the basis for the thermal radiation network shown in Fig. E5–25, where

$$R_{1-2} = \frac{J_1 - J_2}{A_1F_{1-2}(J_1\tau_{m1} - J_2\tau_{m2})}$$

$$R_{1-m} = \frac{J_1 - E_{bm}}{A_1F_{1-m}(\alpha_{m1}J_1 - \epsilon_mE_{bm})}$$

$$R_{2-m} = \frac{J_2 - E_{bm}}{A_2F_{2-m}(\alpha_{m2}J_2 - \epsilon_mE_{bm})}$$

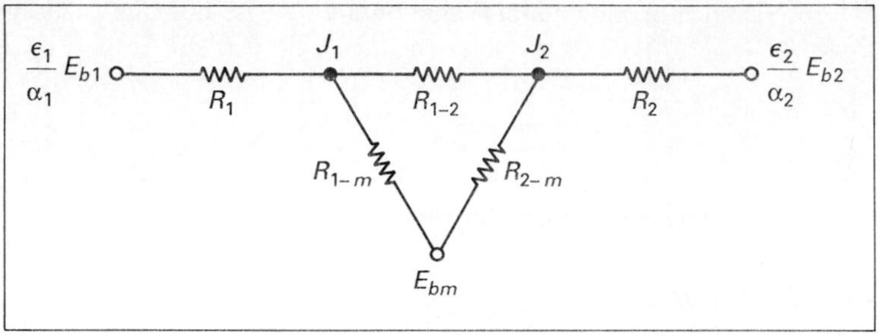


**FIGURE E5–25** Thermal radiation network.

Note that this network reduces to the one shown in Fig. 5–39 for blackbody surfaces $A_1$ and $A_2$.

## EXAMPLE 5–26

Develop a thermal radiation network for two diffuse nonblackbody parallel plates which are separated by a reflecting transparent solid medium.

### Solution

*Objective* Develop a thermal radiation network for this parallel plate system.

*Schematic* Parallel plates separated by reflecting/transparent plate.

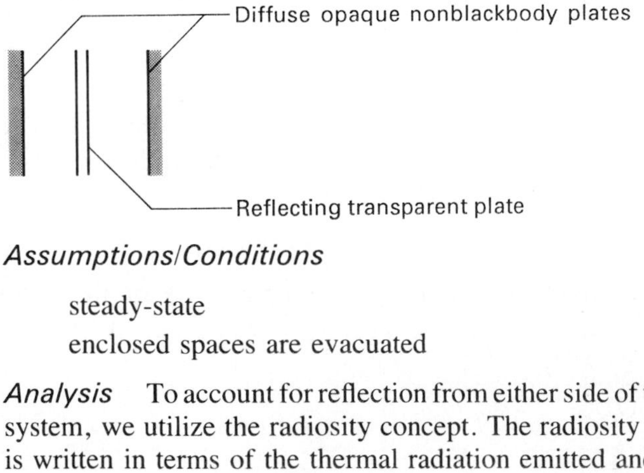


*Assumptions/Conditions*

steady-state
enclosed spaces are evacuated

*Analysis* To account for reflection from either side of the solid medium in a bisurface system, we utilize the radiosity concept. The radiosity $J_{m1}$ from one side of the solid is written in terms of the thermal radiation emitted and reflected as

$$J_{m1} = \epsilon_m E_{bm} + \rho_{m1} G_{m1} \tag{a}$$

where $G_{m1}$ is the irradiation from the surface $A_1$. It is important to note that the thermal radiation transmitted through the solid medium is *not* included in our defining equation for radiosity, but rather is treated separately. The net rate of radiation heat transfer from one side of the solid medium $q_{m1}$ (not including energy transmitted through the solid) is expressed by

$$q_{m1} = A_m(\epsilon_m E_{bm} - \alpha_{m1}G_{m1}) \quad \text{(b)}$$

Combining Eqs. (a) and (b) to eliminate $G_{m1}$, we have

$$q_{m1} = A_m\left[\epsilon_m E_{bm} - \frac{\alpha_{m1}}{\rho_{m1}}(J_{m1} - \epsilon_m E_{bm})\right] = \frac{A_m}{\rho_{m1}}[E_{bm}\epsilon_m(\rho_{m1} + \alpha_{m1}) - \alpha_{m1}J_{m1}]$$

$$= \frac{A_m}{\rho_{m1}}(1 - \tau_{m1})\left(\epsilon_m E_{bm} - \frac{\alpha_{m1}J_{m1}}{1 - \tau_{m1}}\right)$$

Similarly, an equation can be written for $q_{m2}$ of the form

$$q_{m2} = \frac{A_m}{\rho_{m2}}(1 - \tau_{m2})\left(\epsilon_m E_{bm} - \frac{\alpha_{m2}J_{m2}}{1 - \tau_{m2}}\right)$$

To obtain an expression for the net rate of thermal radiation from surface $A_1$ to the solid medium excluding the energy transmitted through the medium, we take the difference between (1) the rate of nontransmitted thermal radiation leaving $A_1$ that reaches the medium, $A_1F_{1-m}J_1(1 - \tau_{m1})$, and (2) the rate of nontransmitted thermal radiation leaving the medium that reaches $A_1$, $A_mF_{m-1}J_{m1}$. That is,

$$q_{1-m} = A_1F_{1-m}J_1(1 - \tau_{m1}) - A_mF_{m-1}J_{m1} = A_1F_{1-m}[J_1(1 - \tau_{m1}) - J_{m1}]$$

$$= A_1F_{1-m}(1 - \tau_{m1})\left(J_1 - \frac{J_{m1}}{1 - \tau_{m1}}\right)$$

Similarly, we write

$$q_{2-m} = A_2F_{2-m}(1 - \tau_{m2})\left(J_2 - \frac{J_{m2}}{1 - \tau_{m2}}\right)$$

To account for the thermal radiation transmitted through the medium, we utilize Eq. (a) in Example 5–25,

$$q_{1-2} = A_1F_{1-2}(J_1\tau_{m1} - J_2\tau_{m2})$$

Finally, the rate of radiation heat transfer from each surface is given by use of Eq. (5–53).

$$q_1 = \frac{A_1\alpha_1}{\rho_1}\left(\frac{\epsilon_1}{\alpha_1}E_{b1} - J_1\right) \qquad q_2 = \frac{A_2\alpha_2}{\rho_2}\left(\frac{\epsilon_2}{\alpha_2}E_{b2} - J_2\right)$$

These expressions permit us to construct the thermal radiation network shown in Fig. E5–26, where

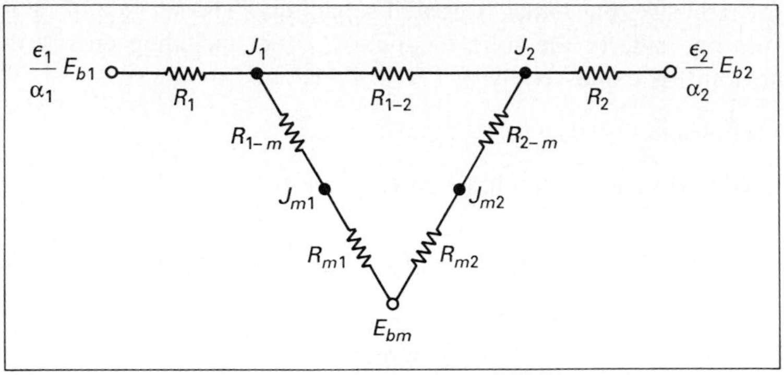


**FIGURE E5–26** Thermal radiation network.

$$R_{1-2} = \frac{J_1 - J_2}{A_1 F_{1-2}(J_1\tau_{m1} - J_2\tau_{m2})}$$

$$R_{1-m} = \frac{J_1 - J_{m1}}{A_1 F_{1-m}(1 - \tau_{m1})\left(J_1 - \dfrac{J_{m1}}{1 - \tau_{m1}}\right)}$$

$$R_{m1} = \frac{J_{m1} - E_{bm}}{\dfrac{A_m}{\rho_m}(1 - \tau_{m1})\left(\epsilon_m E_{bm} - \dfrac{\alpha_{m1} J_{m1}}{1 - \tau_{m1}}\right)}$$

with similar relations for $R_{2-m}$ and $R_{m2}$.

## 5–5–5 Thermal Radiation Systems with Combined Modes

Most thermal radiation problems encountered in practice involve at least one of the other two modes of heat transfer. For example, radiation heat transfer in a furnace includes conduction heat transfer through the walls and within bodies being heated as well as convection heat transfer within and without the furnace. Another example is the transfer of heat through superinsulative walls, which involves all three modes.

Practical analyses were developed in Chaps. 1 through 4 for mixed-mode blackbody radiation systems by utilizing the thermal resistance $R_R$. This practical approach is extended to more realistic one-dimensional nonblackbody combined-mode systems in Examples 5–27 and 5–28. For more complex multidimensional mixed-mode systems, which involve nonuniform surface temperatures, numerical solution techniques are required.

## EXAMPLE 5-27

A lightly oxidized thin copper plate with a 1-m$^2$ surface area is mounted horizontally outdoors with the lower surface insulated. Determine the temperature of the plate on a clear winter night for which the air temperature is 1°C and $\bar{h}$ is 10 W/(m$^2$ °C). (*Note*: According to Duffie and Beckman [25], the sky can be considered as a blackbody radiator with temperature given by

$$T_{sky} = T_{air} - A$$

where $A$ = 20°C in the winter and $A$ = 6°C in the summer.)

### Solution

*Objective* Determine the plate temperature $T_s$.

*Schematic* Thin copper plate (lightly oxidized).

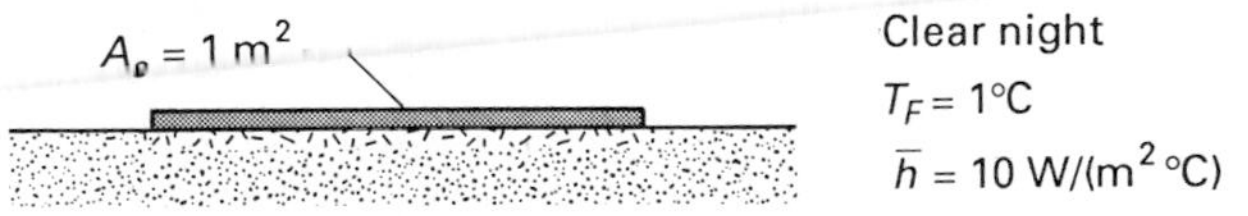
$A_s$ = 1 m$^2$
Clear night
$T_F$ = 1°C
$\bar{h}$ = 10 W/(m$^2$ °C)

*Assumptions/Conditions*

steady-state
diffuse opaque graybody plate
sky considered to be a blackbody radiator

*Properties*

Copper, lightly oxidized (Fig. 5-3): $\epsilon_s = 0.56$.

Sky: $\epsilon_{sky} = 1$.

*Analysis* Focusing attention on the radiation heat transfer, the thermal radiation network is given in Fig. E5-27a, where

$$T_R = T_{air} - 20°\text{C} = 254 \text{ K}$$

$$E_{bR} = \sigma T_R^4 \qquad R_{s-R} = \frac{1}{A_s F_{s-R}} = \frac{1}{A_s}$$

Utilizing the nonisothermal or graybody approximation, we have

$$\alpha_s = \epsilon_s = 0.56$$

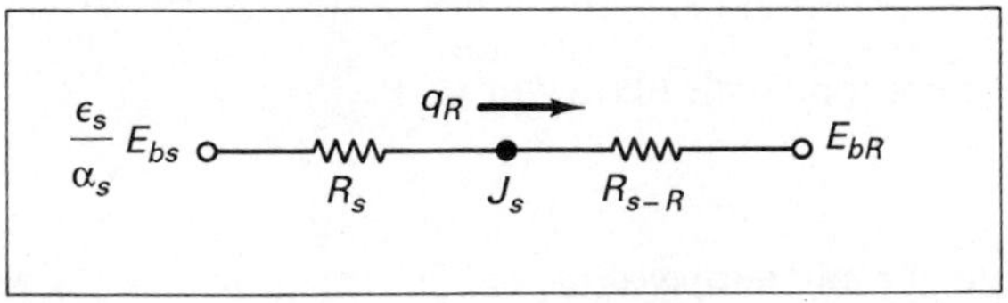
$\frac{\epsilon_s}{\alpha_s}$ $E_{bs}$
$q_R$
$R_s$
$J_s$
$R_{s-R}$
$E_{bR}$

**FIGURE E5-27a**
Thermal radiation network.

Therefore, $R_s$ is given by

$$R_s = \frac{\rho_s}{A_s\alpha_s} = \frac{1 - 0.56}{A_s(0.56)} = \frac{0.786}{A_s}$$

The rate of radiation heat transfer is now expressed in terms of the unknown surface potential $T_s$ by

$$q_R = \frac{(\epsilon_s/\alpha_s)E_{bs} - E_{bR}}{R_s + R_{s-R}} = \frac{E_{bs} - E_{bR}}{(0.786 + 1)/A_s} = \frac{\sigma A_s(T_s^4 - T_R^4)}{1.79} \tag{a}$$

where $T_R = 254$ K.

Turning to the overall problem, a mixed-mode thermal circuit is shown in Fig. E5–27b. We obtain $R_R$ by rearranging Eq. (a) as follows:

$$q_R = \left[\frac{\sigma A_s}{1.79}(T_s + T_R)(T_s^2 + T_R^2)\right](T_s - T_R) = \frac{T_s - T_R}{R_R}$$

$$R_R = \frac{1.79}{\sigma A_s(T_s + T_R)(T_s^2 + T_R^2)}$$

Of course, $R_c$ is simply equal to $1/(\bar{h}A_s)$.

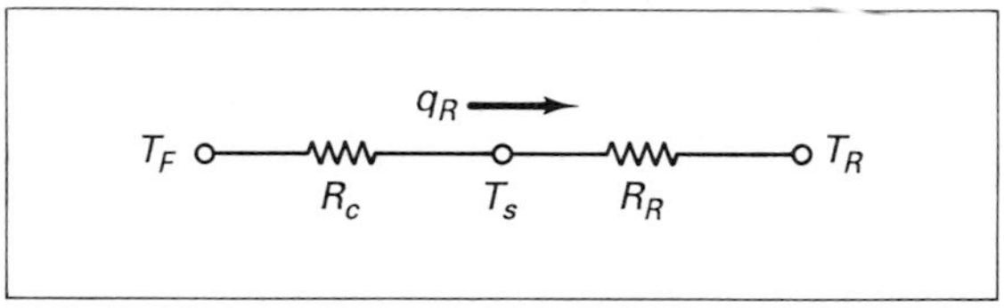


**FIGURE E5–27b**
Thermal circuit.

To determine the unknown surface temperature $T_s$, an energy balance is performed on the $T_s$ node.

$$q_c + q_R = 0 \qquad \bar{h}A_s(T_s - T_F) + \frac{\sigma A_s}{1.79}(T_s^4 - T_R^4) = 0$$

$$T_s = T_F - \frac{\sigma}{1.79\bar{h}}(T_s^4 - T_R^4) = 274\text{ K} - \frac{3.17 \times 10^{-9}}{\text{K}^3}[T_s^4 - (254\text{ K})^4]$$

$$= 287\text{ K} - \frac{3.17 \times 10^{-9}}{\text{K}^3}T_s^4$$

This equation is solved by iteration, with the result that

$$T_s \simeq 270\text{ K} = -3°\text{C}$$

Thus, we find that although the air temperature is 1°C, the surface temperature is below freezing because of radiation to the sky.

## EXAMPLE 5–28

Determine the rate of heat transfer through a superinsulative spherical wall which consists of an inner surface at $-40°C$ with a radius 1.01 m, a 1-mm evacuated space, and a 10-cm-thick layer of insulation [$k = 0.60$ W/(m °C)]. The surrounding fluid is at 35°C with $\bar{h} = 150$ W/(m$^2$ °C). The inner radiative surface is constructed of a highly reflective material with $\rho = 0.9$ while the other surface is a blackbody.

### Solution

*Objective* Determine $q_{1-2}$.

*Schematic* Superinsulative spherical wall.

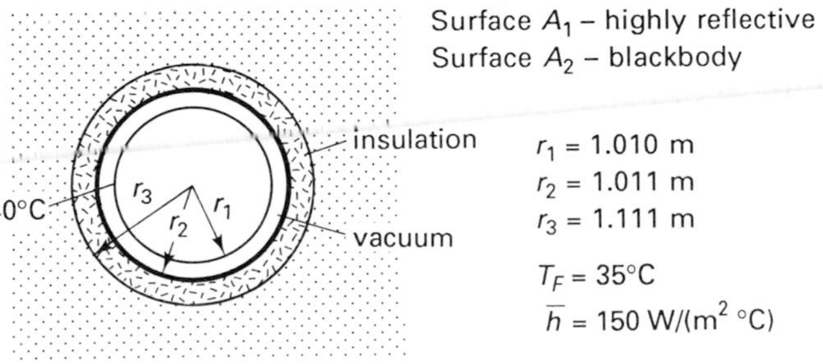
Surface $A_1$ – highly reflective
Surface $A_2$ – blackbody
insulation
vacuum
$T_1 = -40°C$
$r_3$
$r_2$
$r_1$
$r_1 = 1.010$ m
$r_2 = 1.011$ m
$r_3 = 1.111$ m
$T_F = 35°C$
$\bar{h} = 150$ W/(m$^2$ °C)

*Assumptions/Conditions*

steady-state
one-dimensional

*Properties*

Insulation: $k = 0.6$ W/(m °C).
Surfaces: $\rho_1 = 0.9$, $\alpha_2 = 1$.

*Analysis* The thermal circuit for this mixed-mode radiation, conduction, and convection system is shown in Fig. E5–28a. The thermal resistances for conduction and convection are given by

$$R_k = \frac{r_3 - r_2}{4\pi r_3 r_2 k} = \frac{(1.111 - 1.011)\ \text{m}}{4\pi(1.111\ \text{m})(1.011\ \text{m})[0.6\ \text{W/(m °C)}]} = 0.0118\ \text{°C/W}$$

$$R_c = \frac{1}{\bar{h}A_s} = \frac{1}{[150\ \text{W/(m}^2\ \text{°C)}](4\pi)(1.111\ \text{m})^2} = 4.30 \times 10^{-4}\ \text{°C/W}$$

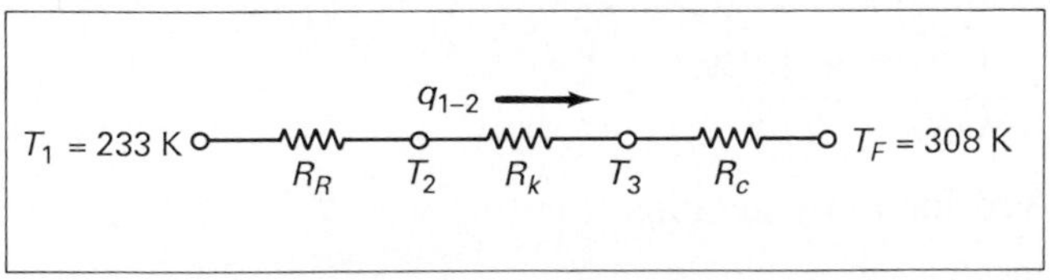
$q_{1-2}$
$T_1 = 233$ K
$R_R$
$T_2$
$R_k$
$T_3$
$R_c$
$T_F = 308$ K

**FIGURE E5–28a** Thermal circuit.

To determine the resistance $R_R$ for thermal radiation, we utilize the thermal radiation network shown in Fig. E5–28b, where

$$E_{b2} = \sigma T_2^4 \qquad R_1 = \frac{\rho_1}{A_1\alpha_1} = \frac{0.9}{A_1(0.1)} = \frac{9}{A_1} \qquad R_{1-2} = \frac{1}{A_1F_{1-2}} = \frac{1}{A_1}$$

and, assuming that $\epsilon_1 \simeq \alpha_1$,

$$\frac{\epsilon_1}{\alpha_1} E_{b1} \simeq E_{b1} = \sigma T_1^4$$

It follows that the rate of radiation heat transfer $q_{1-2}$ can be written as

$$q_{1-2} = \frac{E_{b1} - E_{b2}}{9/A_1 + 1/A_1} = \frac{A_1(E_{b1} - E_{b2})}{10}$$

Rearranging this expression, we have

$$q_{1-2} = \frac{[A_1\sigma(T_1 + T_2)(T_1^2 + T_2^2)](T_1 - T_2)}{10}$$

or

$$q_{1-2} = \frac{T_1 - T_2}{R_R} \tag{a}$$

where $A_1 = 4\pi(1.01 \text{ m})^2$, $T_1 = 233$ K, and

$$R_R = \frac{10}{A_1\sigma(T_1 + T_2)(T_1^2 + T_2^2)}$$

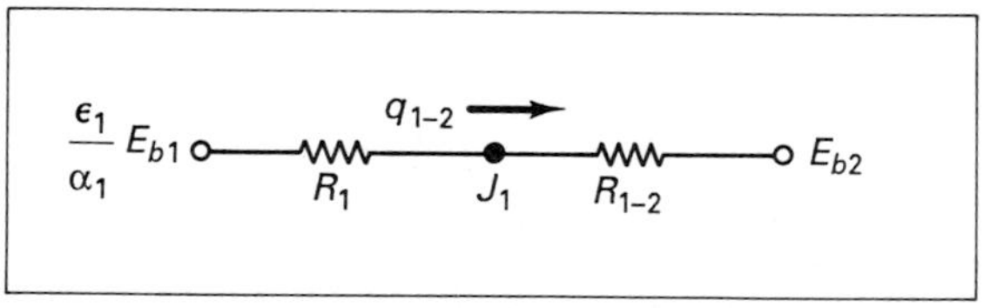


**FIGURE E5–28b**
Thermal radiation network.

Returning to the circuit in Fig. E5–28a, we perform an energy balance on the $T_2$ node, with the result that

$$\frac{233 \text{ K} - T_2}{R_R} = \frac{T_2 - 308 \text{ K}}{(1.18 \times 10^{-2} + 4.3 \times 10^{-4}) \text{ K/W}}$$

Rearranging, we obtain

$$T_2 = \frac{233 \text{ K}/R_R + 2.52 \times 10^4/\text{W}}{81.8 \text{ W/K} + 1/R_R}$$

This equation is solved for $T_2$ by iteration. Starting with $T_2 = 290$ K,

$$R_R = 0.191 \text{ K/W} \qquad T_2 = 304 \text{ K}$$

A second iteration gives

$$R_R = 0.174 \text{ K/W} \qquad T_2 \simeq 303 \text{ K}$$

Substituting this result into Eq. (a), we obtain

$$q_{1-2} = \frac{233 \text{ K} - 303 \text{ K}}{0.174 \text{ K/W}} = -402 \text{ W}$$

As a point of interest, we note that for blackbody radiation at surface $A_1$, the resistance $R_R$ reduces to

$$R_R = \frac{1}{A_1\sigma(T_1 + T_2)(T_1^2 + T_2^2)}$$

which is a factor of 10 less than for the case in which a highly reflective surface is used. For the blackbody situation, we find

$$T_2 - 270 \text{ K} \qquad q_R = -2.22 \times 10^4 \text{ W}$$

This heat loss is greater by a factor of 55 than for the case in which one reflective surface is employed. This result gives a good indication of why highly reflective surfaces are used in superinsulative walls.

## 5-6 SOLAR RADIATION

### 5-6-1 The Solar Resource

The sun is an essentially spherical body ($r_0 \simeq 0.695 \times 10^6$ km) of extremely high-temperature matter. Within its inner core ($r \lessapprox 0.23r_0$), temperatures of the order of $8 \times 10^6$ K to $40 \times 10^6$ K are maintained by a continuous fusion process in which mass is converted into energy. This fusion process produces X-ray and $\gamma$-ray electromagnetic radiation that emanates from the high-density core. In the low-density gaseous region between $0.7r_0$ and $r_0$, energy is also transported by convection. The temperature is believed to drop from about 130,000 K to 5000 K across this convective zone. The *photosphere*, which makes up the outer layer of the convection zone, is essentially opaque and well defined. Evaluations of thermal radiation received from the sun, ranging from visible to long wavelength radio waves, indicate that three gaseous layers lie outside the photosphere, with temperatures increasing from about 5000 K within the inner layer to as high as $10^6$ K in the outer layer.

Solar radiation consists of energy emitted by the various layers, with the major contributions being provided by the photosphere. Although the radiation emitted by the sun originates from the various temperature zones, for practical purposes the sun can be considered as a blackbody radiator at an effective temperature of about 5800 K. A more accurate value for the effective temperature of the sun according to

Thekaekara [26] is 5762 K. Based on Thekaekara's estimate, the effective total emissive power of the sun is calculated to be about

$$E_{\text{sun}} = \sigma T_{\text{sun}}^4 = 5.67 \times 10^{-8} \frac{\text{W}}{\text{m}^2\,\text{K}^4}(5762\ \text{K})^4 = 6.25 \times 10^7\ \text{W/m}^2$$

The approximate monochromatic emissive power $E_{b\lambda}$ of the sun calculated by the use of the Planck law, Eq. (5–7), is shown in Fig. 5–6.

Of course, only a small fraction of this enormous solar radiation flux reaches the outer fringes of the earth's atmosphere, and an even smaller portion reaches the surface of the earth itself. Based on direct measurements of solar irradiation in the outer reaches of earth's atmosphere, which have been obtained by high-altitude aircraft, balloons, and spacecraft, the monochromatic extraterrestrial solar irradiation $G_{s\lambda}$ reaching the earth's atmosphere for the mean earth–sun distance of about $1.50 \times 10^8$ km is given by the standard NASA curve shown in Fig. 5–40. This solar irradiation consists of very short wavelength $\gamma$ rays, X rays, and ultraviolet rays, as well as thermal radiation. The total extraterrestrial solar irradiation $G_s$ reaching the earth's atmosphere for the mean earth–sun distance is approximately 1350 W/m$^2$. Strictly speaking, $G_s$ varies by about $\pm 3\%$ as a result of changes in the earth–sun distance and the conditions on the sun. However, for practical purposes $G_s$ can be taken as a constant. For this reason $G_s$ is generally referred to as the *solar constant.*

The solar irradiation flux is further attenuated by the atmosphere before it reaches the surface of the earth. The $\gamma$-ray and X-ray components of the solar irradiation are absorbed in the ionosphere by nitrogen, oxygen, and other materials, and nearly all of the UVC and much of the UVB is absorbed by the ozone layer in the stratosphere.

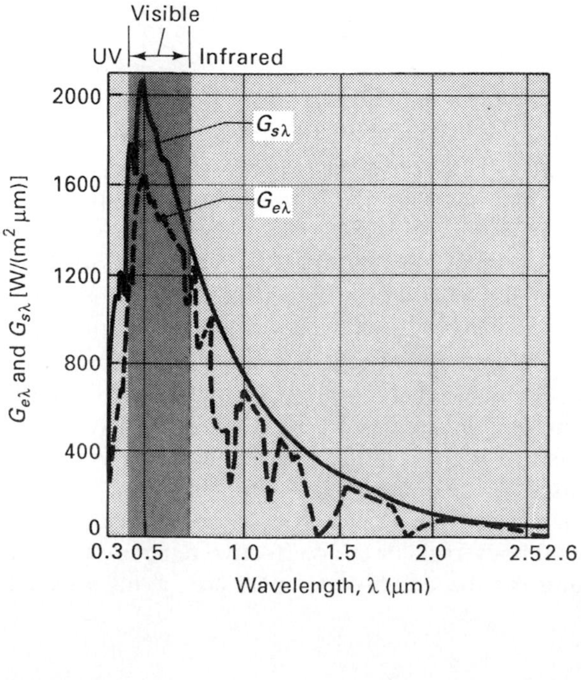


**FIGURE 5–40**
Representative monochromatic extraterrestrial solar irradiation $G_{s\lambda}$ and monochromatic solar irradiation $G_{e\lambda}$ (Thekaekara [26]).

Absorption also contributes to the attenuation of longer wavelength radiation, with $O_2$ and $O_3$ affecting visible light, and water vapor and $CO_2$ affecting the infrared. Of course, atmospheric pollutants also contribute to the absorption of solar radiation. In addition to absorption, molecular scattering in the atmosphere results in the redirection of much of the incoming solar radiation. According to the work of J. W. S. Rayleigh, the degree of scattering is inversely proportional to $\lambda^4$. This phenomenon is known as the *Rayleigh effect*. It follows that the influence of scatter increases as wavelength decreases according to the progression IR < Visible < UVA < UVB < UVC. In this connection, it has been estimated that as much as 50% of the UVB radiation reaching the atmosphere is scattered to space [27]. It follows that only about 3.5% of the visible light and even less infrared radiation entering the atmosphere is scattered to space.

The geometric factors associated with irradiation can be handled with the aid of celestial geometry. However, the lack of meteorological information generally makes it impractical to base predictions for the terrestrial solar resource at any location on extraterrestrial solar irradiation. Rather, the terrestrial solar irradiation is usually measured over a reasonable length of time at the location of interest. Data of this type have been collected at a number of stations across the United States and Europe for many years. Numerous other stations are being set up to measure the solar resource throughout the world. Such data are provided on hourly, daily, and annual bases, with the annual solar irradiation data being of the most practical value in the design of solar systems. Representative experimental measurements for the monochromatic solar irradiation $G_{e\lambda}$ reaching the earth's surface for a very clear atmosphere are shown in Fig. 5–40. The irregularities in $G_{e\lambda}$ are caused by absorption of water vapor, carbon dioxide, and oxygen. Notice that the solar radiation reaching the surface of the earth primarily consists of thermal radiation with wavelength ranging from 0.3 to 2.5 μm.

To illustrate the importance of weather conditions and time of day, the total terrestrial solar irradiation $G_e$ measured in the springtime at Madison, Wisconsin, is shown in Fig. 5–41 for both clear and cloudy days.

Concerning the location of a receiving surface, the solar irradiation is a function of the geometrical relation between the surface and the sun, which is continuously changing on both a daily and annual basis. One of the most important practical aspects involving location is the total number of hours of daytime per year. Another important factor pertains to the orientation of the surface itself. Of course, the maximum irradiation flux on a receiving surface at a given location on the earth is obtained by continual adjustment of the surface orientation to maintain normal solar incidence throughout the daylight hours. But such adjustment requires rotation about two axes, which is generally not practical, except for high-temperature power generation stations. Based on the analysis of much experimental data, researchers in the area of solar radiation have recommended that for maximum annual collection, a receiver should be oriented toward the equator with a slope approximately equal to the latitude $\phi$. For best winter collection the slope should be about $\phi + 10$ degrees, and for optional summer irradiation the slope should be approximately $\phi - 10$ degrees.

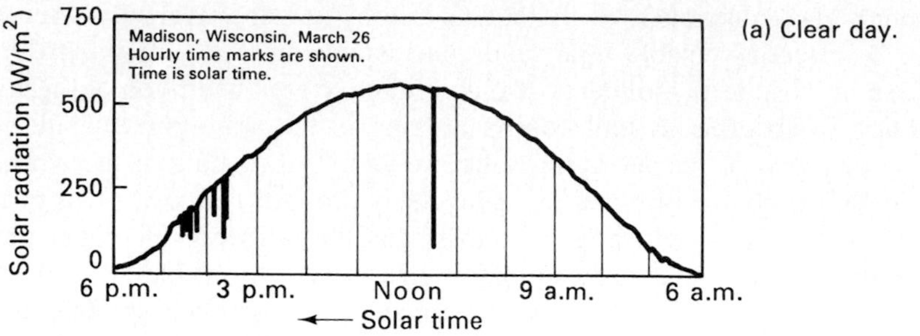


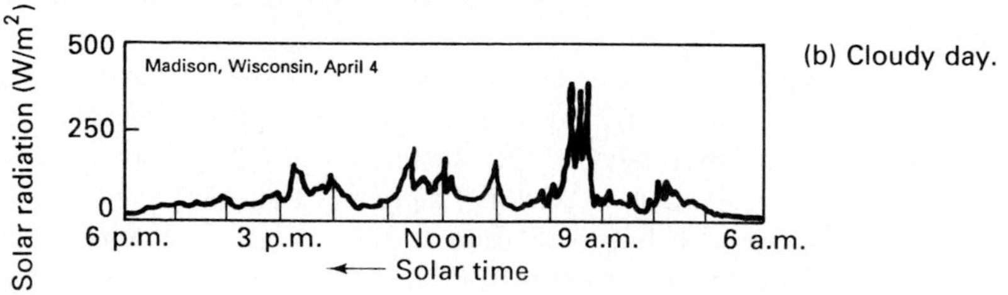


**FIGURE 5–41** Total solar irradiation on a horizontal surface vs. time. (From Duffie and Beckman [25]. Used with permission.)

The solar radiation reaching the surface of the earth is absorbed and reflected. Some of the energy that is absorbed is immediately reemitted, with the remainder being stored thermally in the land masses and oceans, chemically in vegetation through the process of photosynthesis, and mechanically in form of wind and eventually reemitted from water vapor as it condenses in the upper atmosphere.

## EXAMPLE 5–29

Assuming that the sun radiates as a blackbody at 5760 K, develop a prediction for the extraterrestrial irradiation $G_s$ on the atmosphere of the earth.

### Solution

*Objective* Calculate the extraterrestrial irradiation $G_s$.

*Schematic* Earth-sun geometric relationship (not to scale).

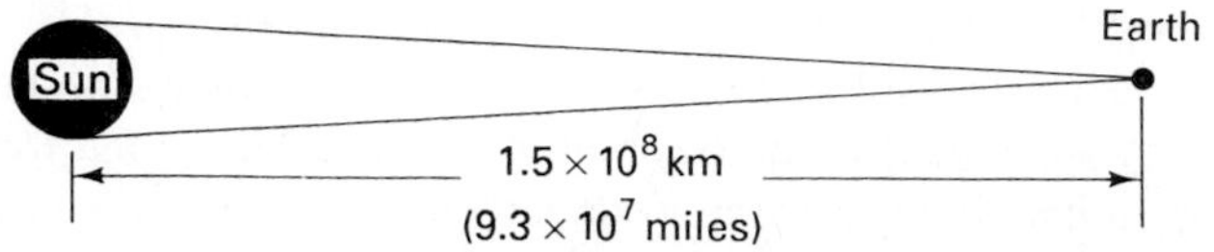

*Assumptions/Conditions*

solar irradiation is equivalent to blackbody emission at 5760 K

*Properties* Sun at 5760 K: $\epsilon = 1$.

*Analysis* Referring to the geometric relationship between the earth and the sun shown in the schematic, the solar irradiation $G_s$ reaching the earth's outer atmosphere is given by

$$G_s\, dA_e = A_{sun} F_{A_{sun}-dA_e} E_{sun} \tag{a}$$

where the differential area $dA_e$ is normal to the line $r$ drawn between the earth and the sun. Based on reciprocity, we write

$$G_s = F_{dA_e - A_{sun}} E_{sun} \tag{b}$$

By visualizing the differential area $dA_e$ and sun as two parallel disks, $F_{dA_e - A_{sun}}$ can be approximated by Eq. (b) of Example 5–9,

$$F_{dA_e - A_{sun}} = \frac{r_0^2}{Z^2 + r_0^2} \tag{c}$$

where the distance $Z$ between the earth and sun is about $1.5 \times 10^8$ km and the radius of the sun $r_0$ is about $0.695 \times 10^6$ km. Combining Eqs. (b) and (c) and employing the Stefan-Boltzmann law, we have

$$G_s = \frac{\sigma T_{sun}^4}{(Z/r_0)^2 + 1} = \frac{[5.67 \times 10^{-8}\ \text{W/(m}^2\ \text{K}^4)](5760\ \text{K})^4}{(1.5 \times 10^8/0.695 \times 10^6)^2 + 1} = 1340\ \text{W/m}^2$$

This value of the radiation constant $G_s$ lies between the standard value of 1353 $\text{W/m}^2$ proposed in 1971 by Thekaekara and Drummond [28] and an earlier standard of 1322 $\text{W/m}^2$.

## 5–6–2 Solar-Energy Systems

A variety of solar-energy systems have been devised to satisfy applications ranging from the heating of water, the heating and cooling of residential and commercial buildings, and the operation of desalination plants, to the production of power on a commercial basis and the energizing of electronic equipment in remote locations on land, sea, and in space. Several basic types of solar-energy systems include solar-thermal systems, ocean thermal systems, wind systems, photovoltaic devices, as well as others. We will restrict our attention to solar-thermal systems.

### *Solar-Thermal Systems*

All solar systems that receive, collect, store, and utilize thermal radiation directly from the sun fall under the heading of solar thermal. These systems consist of two

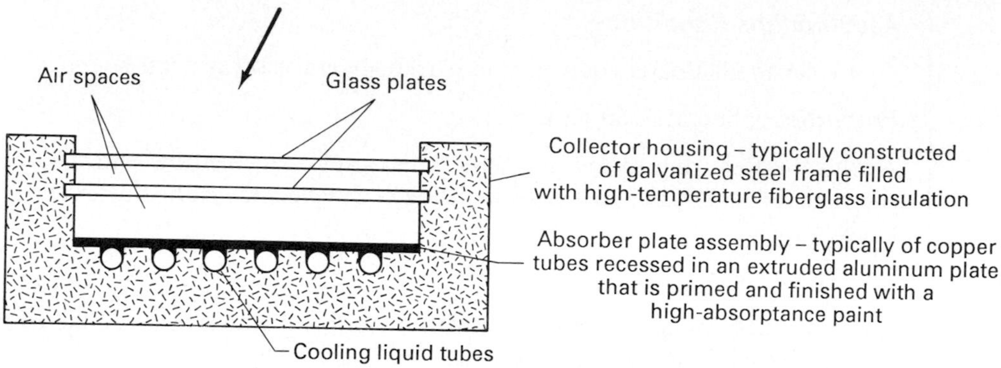
Air spaces
Glass plates
Collector housing – typically constructed of galvanized steel frame filled with high-temperature fiberglass insulation
Absorber plate assembly – typically of copper tubes recessed in an extruded aluminum plate that is primed and finished with a high-absorptance paint
Cooling liquid tubes

**FIGURE 5–42** Flat plate collector cross-section.

basic components: a collector and a storage unit. Of course, these solar components must be interfaced with conversion devices (i.e., air-conditioning units and engines), loads, auxiliary energy supplies, and controls to obtain a total energy system.

The solar collector is actually a thermal radiation heat exchanger in that such devices transfer solar energy to a fluid. The cross section of a typical flat-plate collector arrangement is illustrated in Fig. 5–42. The simple flat-plate collector consists of a near-black solar energy absorbing surface, which collects and transfers the solar energy to a fluid; a transparent cover, which minimizes radiation and convection losses to the atmosphere; and back insulation, which minimizes conduction losses. Flat-plate collectors are utilized in applications requiring moderate temperatures up to about 100°C above ambient. Applications involving the use of flat-plate collectors include water and environmental heating, refrigeration, and seawater desalination.

It should be noted that concentrating collectors are capable of operating at much higher temperatures than flat-plate collectors. Whereas single-axis systems can operate at temperatures up to 315°C, systems with double-axis focusing are able to attain temperatures as high as 3600°C.

Because of the intermittent nature of solar radiation, the storage unit is an essential part of most solar systems. Simply put, the procedure for storing energy in thermal solar-energy systems consists of heating a fluid or solid which is confined in a well-insulated space. For example, hot water is often circulated from the collector into a heavily insulated tank. For systems that utilize air as the working fluid, the hot air is sometimes passed from the collector into a porous bed of stone, which receives and stores the energy. However, simple methods such as these are generally restricted to storage periods of only a few days, at best, because of imperfect thermal insulation and size limitations. Other, more sophisticated methods, which involve materials that store energy by change of phase from solid to liquid, have been found to reduce the storage volume and extend the length of time energy can be stored.

## 5–7 SUMMARY

As we have seen, the physical mechanism of thermal radiation heat transfer involves the transport of energy between bodies at different temperature by means of electromagnetic waves that are associated with the body temperature, that travel at the speed of light, and that are characterized by wavelength or frequency. The thermal radiation spectrum essentially encompasses ultraviolet radiation (0.1–0.38 μm), visible light (0.38–0.76 μm), and infrared radiation (0.76–1000 μm). Basic concepts pertaining to thermal emission and irradiation properties, in addition to geometric factors, were introduced, and practical thermal analyses were developed for thermal radiation transfer in ideal or near-ideal systems involving diffuse surfaces with uniform and known thermal radiation properties $\epsilon$, $\alpha$, $\rho$, and $\tau$, uniform heating conditions (including uniform emission, reflection, and emission), and nonparticipating or gray gases. The use of this basic approach in analyzing problems involving solar radiation, combined modes, and unsteady conditions was also introduced. As we have seen, the concept of *radiosity* is the key to the practical analysis of radiation heat transfer. Either the thermal radiation network approach or the alternative approach of Example 5–14 can be used.

The radiosity concept and other approaches, such as the Monte Carlo method, are used in the analysis of nonideal systems involving surfaces with nondiffuse characteristics, complex spectral properties, and nonuniform heating. Introductions to the analysis of these types of nonideal systems, as well as systems involving multidimensional combined modes and participating gases, are presented by Siegel and Howell [11] and Sparrow and Cess [5]. With regard to solar radiation, which was briefly touched upon in Sec. 5–6, this timely topic is more thoroughly introduced in references 25, 29, and 30.

## ■ REVIEW QUESTIONS

**5–1.** What is electromagnetic radiation?

**5–2.** Electromagnetic radiation is characterized by (a) frequency, (b) wavelength, and (c) propagation velocity. Define these characteristics and indicate which ones are independent of the substance through which the radiation is transmitted.

**5–3.** What is the index of refraction?

**5–4.** What is thermal radiation?

**5–5.** What is solar radiation? Does it contain components of microwave and radio-wave radiation?

**5–6.** List several important biological effects of (a) visible light and (b) ultraviolet radiation.

**5–7.** Explain how it is possible for an individual to get a sunburn on a cloudy day, even though the skin actually feels cool.

**5–8.** What are microwaves and what role do they play in heat-transfer processes that occur in microwave ovens?

**5–9.** Examine a microwave oven door and explain how it is possible to see through the window during the cooking process without being harmed by microwave radiation.

**5–10.** Define (a) absorptivity, (b) reflectivity, and (c) transmissivity.

**5–11.** What is (a) total emissive power, (b) subtotal emissive power, and (c) monochromatic emissive power?

**5–12.** What can be concluded on the basis of Planck's law concerning the effect of wavelength on the energy associated with electromagnetic radiation?

**5–13.** Define (a) emissivity and (b) monochromatic emissivity.

**5–14.** What is a graybody?

**5–15.** Define (a) intensity and (b) total irradiation.

**5–16.** What is the greenhouse effect? How does the greenhouse effect apply to the earth's atmosphere?

**5–17.** State the following approximations for absorptivity $\alpha$: (a) graybody, (b) isothermal, and (c) nonisothermal.

**5–18.** What is a participating gas?

**5–19.** Define radiation shape factor.

**5–20.** Write two important summation principles pertaining to radiation shape factor.

**5–21.** Write the law of reciprocity.

**5–22.** Write a relation for the rate of heat transfer between two blackbody surfaces.

**5–23.** Define radiosity.

**5–24.** Define the radiation-heat-transfer resistances $R_{s-j}$ and $R_s$, which are used in the thermal radiation network approach.

**5–25.** Define (a) thermal radiation shield and (b) reradiating surface.

**5–26.** Explain why the sky appears blue and the sun appears yellow.

**5–27.** What are the main factors responsible for attenuating the irradiation from the sun that reaches the surface of the earth?

**5–28.** What is the solar constant?

**5–29.** What is a solar-thermal system?

**5–30.** What are the main components of a solar collector?

## ■ PROBLEMS

**5–1.** Calculate the total emissive power of a blackbody at (a) 25 K; (b) 100°F; (c) 1000°C; (d) 10,000°R.

**5–2.** Referring to Example 5–1, estimate the fraction of energy emitted by a blackbody at 1000°C that falls in the visible region.

**5–3.** The monochromatic emissivity of a polished copper surface is approximately 0.8 for $\lambda < 2\ \mu m$ and 0.5 for $\lambda \geq 2\ \mu m$. Calculate the emissivity $\epsilon$ for this surface at (a) 500 K and (b) 1000 K.

**5–4.** The monochromatic emissivity of a diffuse surface at 1500 K is given as $\epsilon_\lambda = 0.7$ for $0 < \lambda < 1.6\ \mu m$ and $\epsilon_\lambda = 0.4$ for $1.6 < \lambda < 4\ \mu m$. Determine the emissivity $\epsilon$ and the total emissive power $E$.

**5–5.** The interior surface of the blackbody cavity shown in the schematic of Example 5–4 is at a uniform temperature of 2500 K. With reference to the energy emitted from the cavity, determine (a) the total emissive power, (b) the wavelengths above and below which 1% of the emission occurs, and (c) the maximum monochromatic emissive power and the wavelength at which it occurs.

**5–6.** Estimate the total emissive power for a type 18–8 stainless steel surface at (a) 0°C; (b) 500°F; (c) 500°C.

**5–7.** A 10-cm square glass plate is used as a window in an oven door. The monochromatic transmissivity of the glass is listed as $\tau_\lambda = 0.4$ for $0.1 \le \lambda \le 4$ μm and $\tau_\lambda = 0$ outside this range. What is the rate of radiant energy transmitted from the oven if the interior is at 500°F, assuming blackbody conditions?

**5–8.** A glass plate transmits 90% of thermal radiation between 0.2 and 4 μm and is essentially opaque outside this wavelength band. Determine the thermal radiation flux transmitted through the glass from blackbody sources at (a) 500 K and (b) 1000 K.

**5–9.** Type 2A diamond was used in the development of 18.2-mm diameter windows for the 1978 Pioneer Venus Space Mission. Compare the approximate radiant energy transmitted through such a window with that of glass for blackbody source temperature of 1000 K.

**5–10.** Estimate the absorptivity $\alpha$ of a type 18–8 stainless steel surface $A_s$ at 500°C that is exposed to radiation from a blackbody surface at (a) 300 K and (b) 1000 K.

**5–11.** Estimate the emissivity $\epsilon$ and absorptivity $\alpha$ of a gas at 1 atm and 1500°C that consists of 40% $CO_2$ (molar basis) and 60% $N_2$. The parallel-plate blackbody walls are at 500°C, with $L_e$ equal to 1 m.

**5–12.** Demonstrate that water vapor contained between parallel plates 1 m apart at 1 atm pressure and 100°C emits as much as 55% of the energy that would be emitted by a blackbody at 100°C with the same surface area as the plates.

**5–13.** Determine the radiation shape factor from the walls of a cubical furnace with sides of 1 m length to a 4-cm² window located in the center of one of the walls.

**5–14.** Determine the radiation shape factor from the walls of a 10-cm-long, 5-cm-diameter pipe to one end which is open.

**5–15.** Demonstrate that the radiation shape factor $F_{1-2}$ for the enclosure shown in Fig. E5–21 is equal to 0.293.

**5–16.** Determine the radiation shape factors $F_{1-3}$ and $F_{1-4}$ for the system shown in Fig. P5–16.

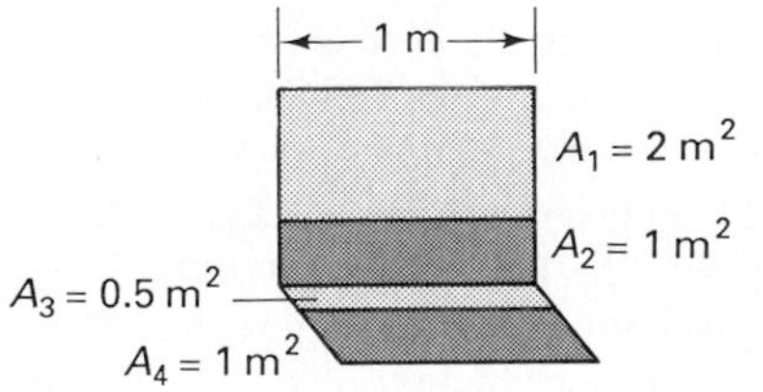


**FIGURE P5–16**

**5–17.** Determine the shape factor from the side to one end of a circular cylinder with length $L$ equal to diameter $D$.

**5–18.** Utilize the double-angle formula to perform the required integration in Eq. (a) of Example 5–3.

**5–19.** Determine the radiation-heat-transfer rate between two concentric blackbody spheres with radii $r_1 = 10$ cm and $r_2 = 25$ cm, and surface temperatures $T_1 = 1000°C$ and $T_2 = 30°C$.

**5–20.** Determine the rate of radiation heat transfer per unit length between two long concentric blackbody cylinders with radii $r_1 = 2$ in. and $r_2 = 10$ in., and surface temperatures $T_1 = -40°F$ and $T_2 = 100°F$. Sketch the thermal radiation network.

**5–21.** Determine the rate of radiation heat transfer between each surface of a long equilateral triangular duct for blackbody conditions with each side being 10 cm in length, $T_1 = 500°C$, $T_2 = 25°C$, and $T_3 = 200°C$. Sketch the thermal radiation network.

**5–22.** If the trisurface enclosure of Fig. 5–26(a) represents a very long triangular duct with $A_1 = A_2 = 100$ ft$^2$, determine the total radiation-heat-transfer rate from surfaces $A_1$ and $A_2$ and the temperature of surface $A_3$ for the case in which each surface is black and $T_1 = 70°F$, $T_2 = 300°F$, and $q_3'' = 100$ Btu/(h ft$^2$).

**5–23.** Referring to the open radiation system of Fig. 5–26(b), determine the rate of radiation heat transfer between each surface for blackbody conditions with $A_1 = 4$ m$^2$, $A_2 = 2$ m$^2$, $A_3 = 5$ m$^2$, $F_{1-2} = 0.2$, $F_{1-3} = 0.4$, $F_{2-3} = 0.3$, $T_1 = 500°C$, $T_2 = 25°C$, and $T_3 = 200°C$. Explain why the total rate of heat transfer from each surface cannot be determined.

**5–24.** Assuming that the open trisurface system of Prob. 5–23 is surrounded by a blackbody at 0 K, determine the total rate of radiation heat transfer from each surface. Sketch the thermal radiation network for this system.

**5–25.** Check the solution to Example 5–13 by showing that $q_1 + q_2 + q_3 + q_4 = 0$.

**5–26.** Sketch the approximate thermal radiation network for Example 5–11 for the case in which surface $A_1$ is a graybody with $\epsilon_1 = 0.6$. Then determine $q_{1-2}$.

**5–27.** Sketch the approximate thermal radiation network for Example 5–11 for the case in which $A_1$ and $A_2$ are both graybodies with $\epsilon_1 = 0.6$ and $\epsilon_2 = 0.1$. Determine $q_{1-2}$.

**5–28.** Sketch the approximate thermal radiation network for Example 5–12 for the case in which $A_2$ is a graybody with $\epsilon_2 = 0.8$. Determine the total rate of radiation heat transfer from $A_2$.

**5–29.** Sketch the thermal radiation network for two infinite parallel plates at temperatures of 100°C and 500°C; $A_1$ is a blackbody and $A_2$ is a graybody with $\epsilon_2 = 0.8$. Also determine the radiation heat transfer flux $q''_{1-2}$.

**5–30.** Estimate the total rate of radiation heat transfer from $A_2$ in Example 5–12 for the case in which surfaces $A_1$ and $A_2$ are graybodies with $\epsilon_1 = 0.2$ and $\epsilon_2 = 0.5$.

**5–31.** High-pressure steam at 250°C is carried in a long 0.8-m-diameter pipe encased in a second pipe with an inside diameter of 1 m. The annular space is evacuated and the emissivites of the inner and outer pipe surfaces are 0.4 and 0.3, respectively. Determine the rate of heat loss per unit length if the temperature of the outer surface is 35°C.

**5–32.** A tungsten filament of 0.1 mm diameter and 20 cm length is used in a light bulb. The electrical resistivity is given as 5.5 μΩ cm. Assuming that the heat generated is transferred from the light bulb by thermal radiation and that the emissivity of the tungsten is about 0.32 (see Table A–C–7), determine the temperature of the filament for voltages of (a) 115 V and (b) 220 V. Is the filament likely to melt for either of these cases?

**5–33.** Equation (5–58),

$$q_{1-2} = A_1F_{1-2}(J_1 - J_2)$$

is strictly applicable to symmetrical arrangements. To expand upon this point, refine the approximate solution for the nonsymmetrical system of Example 5–17 by breaking $A_2$ into two 1-m$^2$ areas.

**5–34.** Equation (5–43),

$$q_{1-2} = A_1F_{1-2}\sigma(T_1^4 - T_2^4) = A_1F_{1-2}(E_{b1} - E_{b2})$$

is applicable to nonsymmetrical as well as symmetrical blackbody systems. To see this point, repeat Prob. 5–33 for the case in which $A_1$ and $A_2$ are blackbodies.

**5–35.** Develop a thermal radiation network for a concentric spherical graybody system that accounts for the thermal radiation which leaves the outer surface and reaches itself. (It is suggested that this concave surface be broken into a number of smaller surfaces.) Then show that this more general network reduces to the simple thermal radiation network given in Example 5–15.

**5–36.** Determine the rate of radiation heat transfer between the concentric spheres of Prob. 5–19 for the case in which a thin radiation shield with radius of 15 cm is utilized. The emissivity and absorptivity of the shield material are both equal to 0.2.

**5–37.** Solve Prob. 5–36 for the case in which the shield is made of polished copper.

**5–38.** A thin radiation shield of 5 in. diameter is placed between the two concentric cylinders of Prob. 5–20. Determine the rate of radiation heat transfer and temperature of the shield for the case in which the emissivity and absorptivity of the shield are equal to 0.1.

**5–39.** A blackbody plate at 2000°F exchanges radiation with an oxidized nickel plate at 500°F. Determine the percent decrease in the rate of radiation heat transfer for the case in which a thin polished aluminum plate is placed between these two bodies.

**5–40.** Referring to Prob. 5–39, determine the number of polished aluminum shields required to reduce the rate of heat transfer by 99%.

**5–41.** A blackbody 1-m-diameter spherical surface at 0°C exchanges radiation with a concentric 2-m-diameter blackbody sphere at 500°C. Determine the percent reduction in heat transfer if a polished aluminum shield at 1.5 m with $\epsilon \approx \alpha \approx 0.05$ is placed between these two bodies.

**5–42.** Refine the solution for Example 5–19 by evaluating $\alpha_2$ at the temperature of a blackbody with total emissive power equal to $J_{s2}$.

**5–43.** Both surfaces $A_1$ and $A_2$ of the system shown in Fig. E5–17a are constructed of oxidized nickel. Determine the rate of thermal radiation heat transfer between $A_1$ and $A_2$.

**5–44.** A cryogenic storage system consists of a spherical container of 1-ft diameter, an evacuated space, a 1.5-ft-ID concentric outer sphere, and a 5-in. layer of fiberglass. Both spheres are made of polished aluminum with $\epsilon \simeq 0.03$. Calculate the rate of heat gain by radiation if liquid oxygen at −297°F is stored and the temperature of the inside surface of the outer sphere is found to be 30°F.

**5–45.** Referring to Example 5–20, calculate the rate of heat transfer $q_1$ in this reradiating system if $A_1 = 1$ m$^2$ and $A_2 = 4$ m$^2$.

**5–46.** Referring to Example 5–21, determine the radiation heat transfer fluxes $q_1''$, $q_2''$, and $q_3''$ if surface $A_3$ is a blackbody at 500°C.

**5–47.** An anodized aluminum surface can be taken as a graybody with $\epsilon = 0.92$ for long wavelength thermal radiation associated with moderate to low temperatures (see Fig. 5–8). Reevaluate $q_R$ in Example 2–14 by utilizing this information.

**5–48.** Solve Example 2–14 for the case in which polished aluminum with $\alpha \approx \epsilon \approx 0.04$ is used.

**5–49.** Solve Example 4–7 for the case in which $\alpha$ and $\epsilon$ of the anodized aluminum are equal to 0.92.

**5–50.** Develop a computer program to solve Example 2–14 for the temperature distribution in radiating fins. Also compute the rate of heat transfer for comparison with the analytical solution.

**5–51.** A furnace cavity is in the form of a 15-cm cube that is open at the top. The sides and bottom are maintained at 1200°C and 1500°C, respectively, by electric heating. Determine the amount of electric power required, assuming blackbody conditions and an ambient temperature of 30°C.

**5–52.** The floor $A_1$ and ceiling $A_3$ of a 1-m$^3$ cubical enclosure are made of lightly oxidized copper and are at temperatures $T_1 = 100°C$ and $T_3 = 300°C$. The other walls are reradiating surfaces. Determine the temperature of the reradiating surfaces.

**5–53.** The base of a small cylindrical furnace cavity with 2 cm diameter and 3 cm length is maintained at 1000°C by electrical heating, with the side wall insulated. The emissivity of the furnace surfaces is about 0.6 and the ambient temperature is 20°C. Determine the rate of radiation heat transfer between the cavity and the room and the temperature of the insulated wall, assuming that the radiosity over the furnace surfaces is essentially uniform.

**5–54.** Referring to Prob. 5–53, determine the rate of radiation heat transfer between the cavity and the surroundings if the wall of the cavity is also maintained at 1000°C.

**5–55.** Referring to Prob. 5–53, in reality the radiosity varies over the heated surface. To account for this variation, develop a numerical solution for the rate of radiation heat transfer between the cavity and the room by breaking the length of the cylinder into 3 equal subvolumes.

**5–56.** Calculate the radiation Biot number for the thin plate of Example 5–22 for the case in which the thermal conductivity is equal to 250 W/(m °C).

**5–57.** Solve Example 5–22 for the case in which a thin polished copper plate initially at 50°C is placed in the oven.

**5–58.** On a cold clear night, air at 0°F passes over the surface of an electrically heated graybody radiator with $\epsilon = 0.7$ that is exposed to radiation from the sky. The mean convection coefficient $\bar{h}$ is equal to 7.5 Btu/(h ft$^2$ °F) and the sky temperature can be approximated by $T_{sky} = (T_F - 20)°C$. Determine the rate of heat transfer per unit area for a radiator surface temperature of 100°F.

**5–59.** The walls of a 5-ft-diameter spherical blackbody furnace are maintained at 500°F. The gas within the furnace consists of 25% $CO_2$ and 75% nitrogen (molar basis) at 1 atm pressure and 1000°F. Evaluate the radiation properties of the gas and calculate the rate of heat transfer to the walls.

**5–60.** A gas turbine combustion chamber is 1 ft in diameter. The products of combustion are at 2000°F and 1 atm, and have a molar composition of 10% $CO_2$ and 20% $H_2O$. Determine the rate of radiation heat transfer assuming a very long combustion chamber with a wall temperature of 1000°F.

**5–61.** Two graybody plates at $T_1 = 1000$ K and $T_2 = 500$ K with $\epsilon_1 = 0.4$ and $\epsilon_2 = 0.8$ are separated by a nonreacting gas with experimental measurements indicating $\epsilon_m = 0.2$ and $\tau_m = 0.8$. Use the radiation network method to determine the heat flux between the two plates and the temperature of the gas.

**5–62.** A glass plate receives a solar radiation flux of 500 W/m$^2$. The glass properties are given as $\tau_\lambda = 0.92$ for $0.3\ \mu m < \lambda < 3\ \mu m$, $\tau_\lambda = 0$ outside this range, and $\rho_\lambda = 0$ for all wavelengths. Determine the rate of solar radiation transmitted through the glass.

**5–63.** Determine the rate of radiation heat transfer to the interior of a room with surface area of 20 $m^2$ at 27°C. A solar radiation flux of 500 W/$m^2$ enters through a glass plate with a 1-$m^2$ area. The glass properties are given in Prob. 5–62. Assume that the walls of the room are black.

**5–64.** Solve Prob. 5–63 for the case in which the interior wall is a graybody with emissivity of 0.1.

**5–65.** Solve Prob. 5–63 for the case in which the reflectivity of the glass at all wavelengths is approximately 0.1.

**5–66.** A very thin stainless steel 301 plate with uniform temperature of 81°F exchanges radiant energy with a blackbody plate at 1000°F. A cool fluid passes over the other side of the stainless steel plate with $\bar{h}$ equal to 250 Btu/(h ft$^2$ °F). Determine the temperature $T_F$ of the fluid.

**5–67.** Show that the resistance to conduction within the radiation shield of Example 5–19 is negligible.

**5–68.** Solve Example 5–19 for the case in which the shield is 1-cm-thick with $k = 30$ W/(m °C), and emissivity and absorptivity equal to 0.8.

**5–69.** Two parallel blackbody plates at $T_1 = 600$ K and $T_2 = 1000$ K are separated by a plate with $k \simeq 35$ W/(m °C), $\epsilon = 0.8$, and $\alpha = 0.8$. Determine the rate of radiation heat transfer for plate thicknesses of (a) 1 mm and (b) 10 cm.

**5–70.** To develop an expression for the radiation shape factor between the two bodies shown in Fig. P5–70, attention is first focused on the exchange of blackbody thermal radiation between the differential areas $dA_1$ and $dA_2$. Following the pattern established in the solution of Example 5–9, show that

$$F_{dA_1-A_2} = \int_{A_2} \frac{\cos\theta_1 \cos\theta_2}{\pi r^2}\, dA_2 \qquad F_{1-2} = \frac{1}{A_1}\int_{A_1} F_{dA_1-A_2}\, dA_1$$

and

$$F_{dA_2-A_1} = \int_{A_1} \frac{\cos\theta_1 \cos\theta_2}{\pi r^2}\, dA_1 \qquad F_{2-1} = \frac{1}{A_2}\int_{A_2} F_{dA_2-A_1}\, dA_2$$

Then demonstrate the validity of the principle of reciprocity.

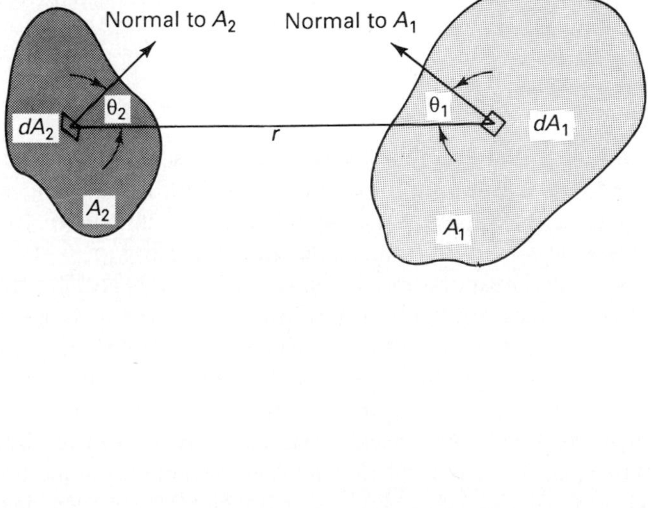


FIGURE P5–70

# CHAPTER 6

# CONVECTION HEAT TRANSFER: INTRODUCTION

To introduce the subject of convection heat transfer, we consider the main features that distinguish the various convection systems from one another, the concept of the boundary layer, the approaches to the analysis of convection heat transfer that are available to us, and characteristics of internal flow.

## 6–1 CHARACTERIZING FACTORS ASSOCIATED WITH CONVECTION SYSTEMS

Convection flow systems can be categorized according to several basic factors. These several categories are now introduced.

### 6–1–1 Forced and Natural Convection

Forced convection and natural convection are two basic mechanisms by which fluid motion can be produced. *Forced convection* represents fluid flow that is caused by mechanical devices such as pumps, fans, or compressors. Examples of forced convection systems include the forced-draft air cooler illustrated in Fig. 6–1, as well as gas turbines, condensers, and evaporators in steam power plants and in refrigeration units, and oil and gas pipelines, to name only a few.

*Natural convection* refers to fluid motion that is caused by temperature- (or concentration-) induced density gradients within the fluid. Natural convection (or free-convection) flow of air over a steam pipe is represented in Fig. 6–2. Notice that the less-dense air near the steam pipe rises while the heavier cool air falls. Other familiar examples of thermal driven natural convection flow include circulation through fireplaces and the cooling of electronic devices. Concentration driven natural con-

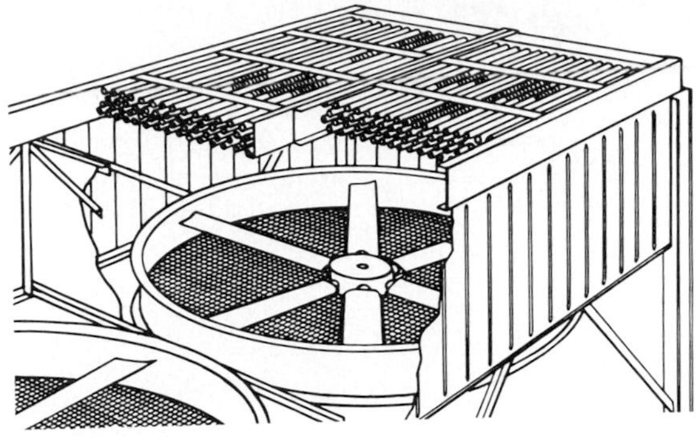

**FIGURE 6–1**
Cutaway view of forced-draft air cooler. (Courtesy of Yuba Heat Transfer Corporation.)

vection occurs in interfacial mass transfer processes such as evaporation from a vertical porous wet surface.

In practice, many convection-heat-transfer systems involve both the forced and natural convection mechanisms. The topic of combined natural and forced convection is considered in Chap. 9.

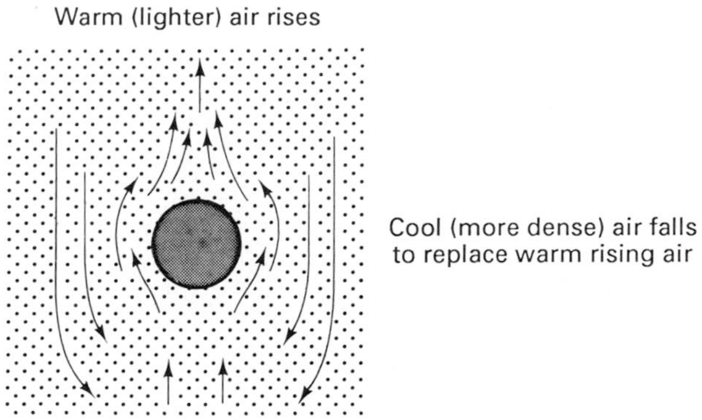
Warm (lighter) air rises
Cool (more dense) air falls to replace warm rising air

**FIGURE 6–2**
Natural convection flow of air over a heated steam pipe.

## 6–1–2 Internal and External Flow

Examples of practical internal and external convection flow systems are illustrated in Figs. 6–3 through 6–5. Flow in tubes, channels, annuli, and heat exchangers are examples of internal flows. External flows involve such geometries as flat plates, wing foils, cylinders, and so forth.

It is important to note that the velocity and temperature distributions are generally functions of axial location $x$ in both external and internal flow fields. Such fields are said to be hydrodynamically and thermally developing. However, under certain conditions the velocity and temperature distributions are geometrically similar in the streamwise $x$-direction. Situations for which *similar flow* fields occur include flow in

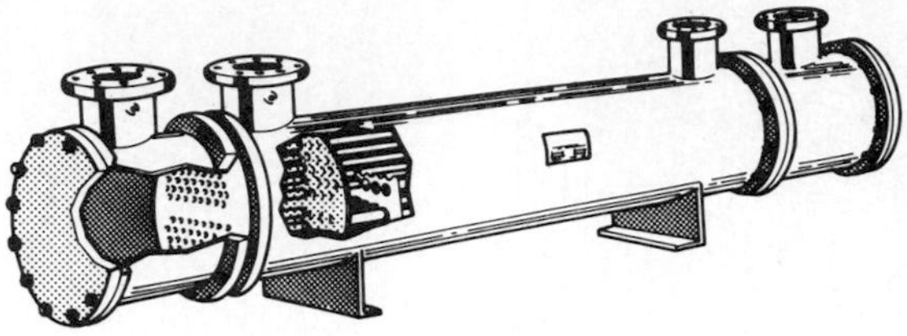

**FIGURE 6–3** Typical industrial shell-and-tube heat exchanger—single pass construction on shell side and tube side. (Courtesy of Enerquip.)

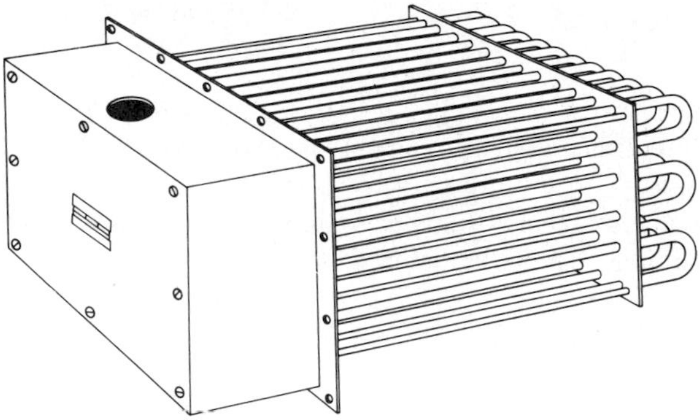

**FIGURE 6–4** Forced-air duct heater. (Courtesy of Heatube Company.)

**FIGURE 6–5** External flow over wing foil. (Courtesy of H. Werle, ONERA, Paris.)

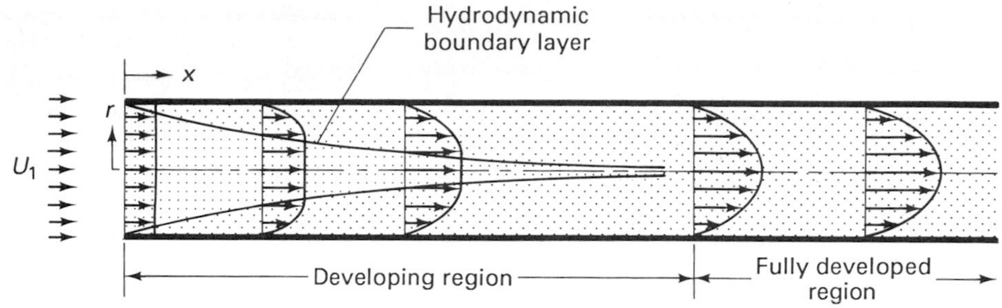


**FIGURE 6–6** Typical axial velocity distributions for laminar tube flow.

a tube in the region well downstream from the entrance, flow over a flat plate with uniform free-stream velocity, flow over wedge-shaped bodies, and natural convection flow over a vertical isothermal surface. To illustrate, Fig. 6–6 shows representative distributions in axial velocity $u$ for tube flow. The velocity profiles are seen to be developing in the entrance region and are fully developed (unchanging) in the region downstream. Further consideration is given to fully developed internal flows in Secs. 6–4 and 7–2–1 and to similar external flows in Secs. 7–2–2 and 7–2–3.

### 6–1–3 Time and Space Dimensions

As in the analysis of conduction-heat-transfer systems, covection-heat-transfer processes are categorized according to the time and space dimensions that the temperature and velocity distributions are dependent upon. In our study, attention will be focused upon steady one- and two-dimensional processes such as occur in a circular tube with uniform wall temperature or uniform wall heat flux. However, one should be aware of the importance of more complex multidimensional processes involving unsteady operation and nonuniform heating. For example, unsteady effects are important in the start-up of a boiler and in pulsating flows, and variations in the surface temperature or heat flux around the perimeter of a tube could be important if the tube is radiantly heated on only one side.

### 6–1–4 Laminar and Turbulent Flow

Laminar forced or natural convection flows exist when individual elements of fluid follow smooth streamline paths, whereas the flow is considered turbulent when the movement of elements of fluid is unsteady and random in nature. This important distinction between laminar and turbulent flow is demonstrated in Fig. 6–7, in which the path followed by fluid is marked by dye. In both cases, the main flow is from left to right. It should be noted that the random fluctuations associated with turbulent flow are superimposed upon the main flow.

One of the simplest ways of experimentally determining whether the flow is laminar or turbulent is to utilize very small electrical heating probes, such as those

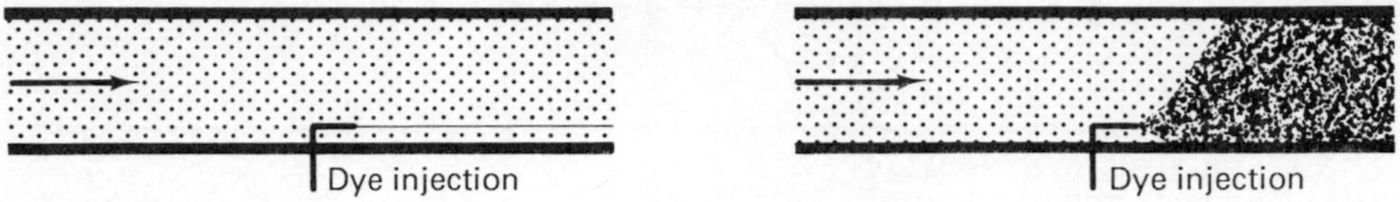


(a) Laminar. (b) Turbulent.

**FIGURE 6–7** Typical dye streak patterns for channel or tube flow.

shown in Fig. 6–8, which can be mounted within a flow stream or flush with the surface of a wall. These *anemometer* probes are maintained at an essentially constant temperature by controlling the instantaneous electrical current flow. The instantaneous bridge voltage for a flush-mounted probe is shown in Fig. 6–9 for laminar and turbulent channel flow of liquid mercury. The unsteady character of the turbulent condition is clearly indicated by this signal. Similar signals are obtained from standard probes mounted in the flow stream. When properly calibrated, the signal from a standard anemometer probe can be used to determine the approximate velocity at the location of the probe.

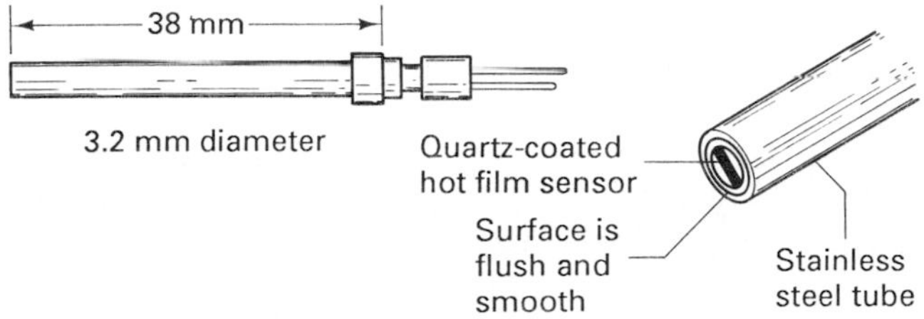


(a) Flush surface probe.

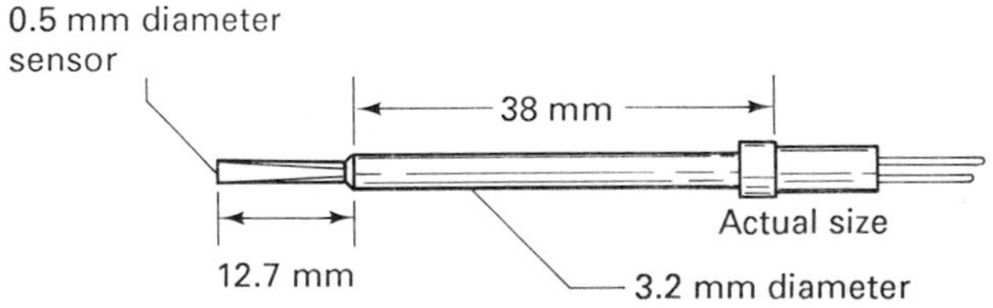


(b) Standard probe with cylindrical sensor.

**FIGURE 6–8** Anemometer probes. (Courtesy of TSI Inc.)

## 6–1–5 Boundary Conditions

The two simplest thermal boundary conditions encountered in convection heat transfer are the uniform wall heat flux and the uniform wall temperature conditions. A more general situation is also frequently encountered in heat exchangers involving convection between two fluids separated by a wall. For situations such as this, the

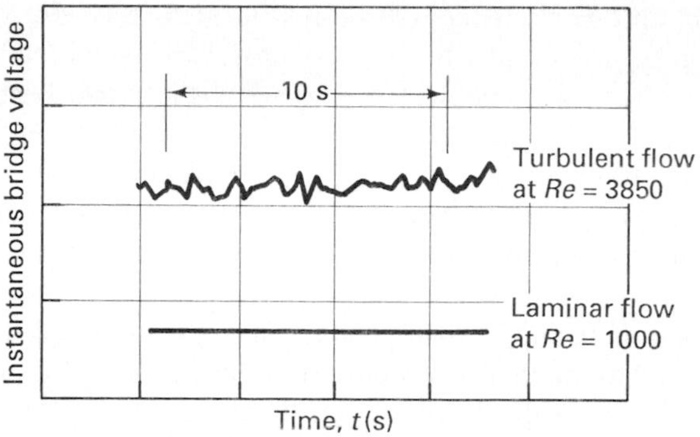


**FIGURE 6–9**
Signals from flush-mounted anemometer probe for channel flow of liquid mercury; Reynolds number $Re$ is defined by Eq. (6–28).

temperatures or heat fluxes at the surfaces are not known *a priori*. These three thermal boundary conditions are illustrated in Fig. 6–10. In this connection, a uniform wall heat flux condition can be experimentally achieved by the generation of energy in the wall itself by the flow of electric current. A uniform wall temperature boundary condition is approximated for the case in which heat is convected from a saturated fluid at constant temperature and very high coefficient of heat transfer through a thin metallic wall into the fluid of interest. As mentioned in Sec. 6–1–3, more complex boundary conditions are sometimes encountered in practice, which involve variations in wall temperature or heat flux around the system perimeter.

For momentum transfer, we can use the nonslip condition $u = 0$ at a stationary wall for most fluids. Of course, if the wall itself possesses an axial velocity $u_0$, then $u = u_0$ at the wall. For nonporous walls, the transverse velocity $v$ is zero at the wall. But for transpired flows $v_0 \neq 0$ since fluid actually passes through the wall.

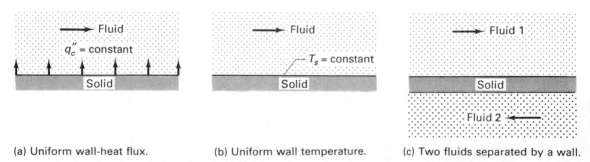

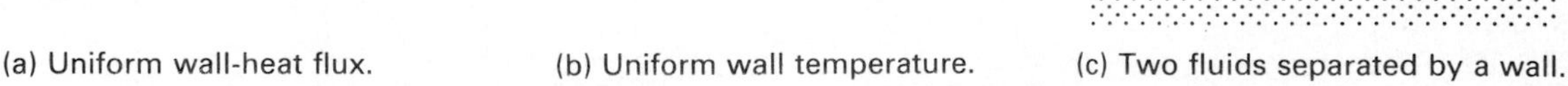


**FIGURE 6–10** Three common types of convection boundary conditions.

## 6–1–6 Type of Fluid

The thermophysical properties of a fluid are in general dependent on its chemical composition, temperature, pressure, and phase. Focusing attention on chemically homogeneous fluids, the most prominent thermophysical properties associated with forced and natural convection heat transfer include density $\rho$, viscosity $\mu$, specific enthalpy $i$, specific heat $c_P$, and thermal conductivity $k$.

Our study of convection heat transfer will be restricted to processes in which the effects of compressibility are small—that is, on processes in which pressure-induced changes in the density can be neglected. Liquids usually can be treated as incompressible. In addition, compressibility effects can generally be neglected for gas flow processes that operate in the low-Mach-number subsonic range. However, compressibility effects must be accounted for in processes involving highly compressed liquids and supersonic gas flow in which significant pressure related changes in density occur.

In regard to the properties $\mu$ and $k$, many of the fluids encountered in practice can be classified as being Newtonian and isotropic. Such fluids satisfy the following fluid-stress and conduction-heat-transfer laws for two-dimensional ($x$, $r$ or $x$, $y$) conditions:

$$\tau = -\mu \frac{\partial u}{\partial r} \qquad \text{or} \qquad \tau = \mu \frac{\partial u}{\partial y} \tag{6–1a,b}$$

$$q_r'' = -k \frac{\partial T}{\partial r} \qquad \text{or} \qquad q_y'' = -k \frac{\partial T}{\partial y} \tag{6–2a,b}$$

Whereas most fluids are essentially isotropic, some important fluids such as blood exhibit distinct non-Newtonian characteristics. In our study of the fundamentals of convection heat transfer, we will concentrate entirely on Newtonian and isotropic fluids.

The *specific enthalpy i*, defined by

$$i = e + \frac{P}{\rho} \tag{6–3}$$

accounts for the internal energy ($e$) and flow energy ($P/\rho$) associated with mass entering or exiting a control volume. Referring to Fig. 6–11, the rate of enthalpy $d\dot{H}$ entering a differential control volume is expressed in terms of $i$ and the mass flow rate $d\dot{m}$ by†

$$d\dot{H} = i\, d\dot{m} \tag{6–4}$$

With $d\dot{m}$ expressed in terms of density $\rho$, velocity $u$, and cross-sectional area $dA$ by

$$d\dot{m} = \rho u\, dA \tag{6–5}$$

(a) Mass flow rate. (b) Rate of enthalpy.

**FIGURE 6–11** Mass flow rate $d\dot{m}$ and rate of enthalpy $d\dot{H}$ for fluid with velocity $u$ and density $\rho$ entering a differential control volume with cross-sectional area $dA$.

† The rate of momentum $d\dot{M}$ associated with mass flow rate $d\dot{m}$ is given by $d\dot{M} = u\, d\dot{m}$.

Eq. (6–4) becomes

$$d\dot{H} = i\rho u \, dA \tag{6–6}$$

Information pertaining to the variation in $i$ with pressure $P$ and temperature $T$ is given in the *Steam Tables* [1] and other references for many common substances. The specific enthalpy is expressed in terms of quality $X$ for two-phase fluids by

$$i = Xi_{fg} + i_f = Xi_g + (1 - X)i_f \tag{6–7}$$

where $i_{fg}$ $(= i_g - i_f)$ is the *latent heat of vaporization*. The specific enthalpy of saturated liquid $i_f$ and that of saturated vapor $i_g$ are tabulated as a function of $T$ or $P$. As a matter of practicality, $e$ and $i$ are generally expressed in terms of specific heats $c_v$ and $c_P$ for single-phase fluids by Eqs. (1–4) and (1–5),

$$c_v = \left.\frac{\partial e}{\partial T}\right|_v \qquad c_P = \left.\frac{\partial i}{\partial T}\right|_P \tag{6–8,9}$$

in general, or

$$c_v = \frac{de}{dT} \qquad c_P = \frac{di}{dT} \tag{6–10,11}$$

for ideal gases, or

$$c_v = c_P = \frac{di}{dT} \tag{6–12}$$

for incompressible liquids. It should be noted that the specific enthalpy $i$ for ideal fluids (i.e., ideal gases and incompressible liquids) can be expressed in terms of $c_P$ as

$$i = \int_{T_R}^{T} c_P \, dT + i_R \tag{6–13}$$

or, for uniform $c_P$,

$$i = c_P(T - T_R) + i_R \tag{6–14}$$

where $i_R$ is the value of the specific enthalpy at the reference temperature $T_R$. In this connection, it is the value of relative specific enthalpy $i - i_R$ that is tabulated in the *Steam Tables* [1].

The dimensionless *Prandtl number Pr*,

$$Pr = \frac{\mu c_P}{k} \tag{6–15}$$

is a key parameter in characterizing convection in single-phase fluids. The Prandtl numbers of several common fluids are given in Tables A–C–3 through A–C–5 in the Appendix. Referring to Fig. 6–12, the Prandtl number ranges from very small values for liquid metals to very large values for highly viscous liquids such as oil. The Prandtl number for gases is generally of the order of unity.

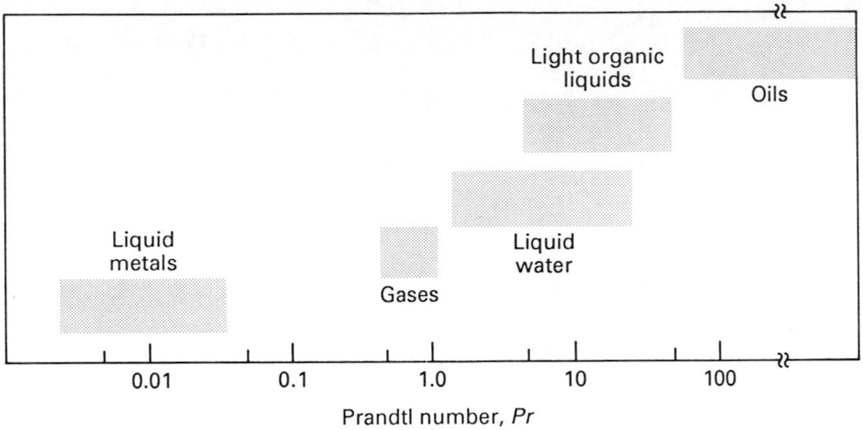


**FIGURE 6–12** Prandtl-number spectrum of common type fluids.

The performance of convection processes is also dependent upon whether the fluid exists as a single-phase gas or liquid, or as a multiple-phase substance. Of course, boiling and condensation are the most common types of multiple-phase convection processes. Both of these two-phase convection-heat-transfer processes are of great technological importance.

Convection heat transfer processes involving chemically nonhomogeneous fluids with interfacial mass transfer include evaporation such as occurs in body perspiration, natural- and forced-draft cooling towers, drying, humidification, and psychrometry; ablation cooling; transpiration cooling; and combustion. The analysis of convection heat transfer processes of this type requires the modeling of species diffusion mass transfer, which is caused on a molecular scale by concentration gradients of two or more of the components. Although the topic of convection heat and mass transfer in chemically nonhomogeneous systems can be quite complex and is a specialization in the field of chemical engineering, practical solution approaches have been developed which are in common use in the engineering profession. The practical solution approach to convection heat and mass transfer is introduced in the *mass transfer supplement*.

## 6–1–7 Other Factors

Numerous other factors are encountered in convection heat transfer that characterize the particular problem being considered. For example, for situations in which the temperature variation is not large, the properties are generally assumed to be uniform. However, for cases in which the properties vary noticeably with temperature, this temperature dependence must be accounted for. Further, factors such as kinetic energy, energy dissipation, body forces, axial conduction, thermal radiation, and three-dimensional effects can be neglected under certain conditions, but are sometimes quite important. In analyzing any convection-heat-transfer problem, care must be taken to ensure that all the significant factors are accounted for.

## 6–2 THE BOUNDARY LAYER

For flow over surfaces such as the flat plate shown in Fig. 6–13, the relative velocity is brought to zero at the wall. As the fluid proceeds downstream, the influence of the surface on the velocity reaches further and further into the fluid because of viscous shearing forces. For most practical applications, the region in which the viscous effects are significant is confined to a distinct layer near the surface known as the *hydrodynamic boundary layer* (HBL). In practice, the thickness of the HBL, represented by $\delta$, is generally designated as the distance from the wall at which the velocity differs by some small percentage $\epsilon$ (say 1%) from the free-stream or centerline velocity.

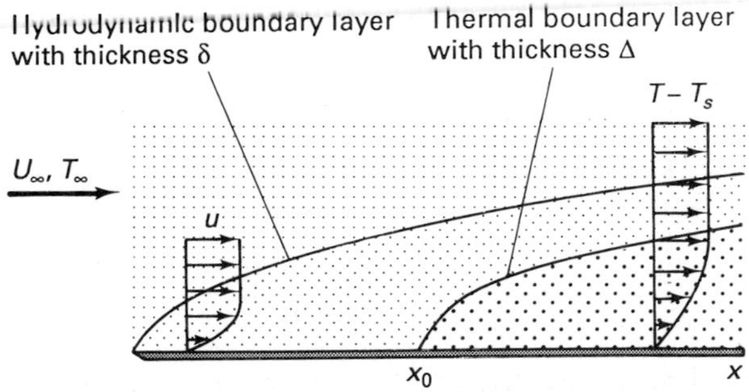


**FIGURE 6–13** Development of hydrodynamic and thermal boundary layers for flow over a flat plate with heating in the region $x \geq x_0$.

Similar to the development of the hydrodynamic boundary layer, as fluid flows over a body the temperature of the fluid in contact with the wall is brought to the surface temperature $T_s$. As the fluid moves downstream from the point at which heating or cooling is initiated, the effect of the wall on the temperature of the fluid penetrates deeper into the fluid, causing the development of a thermal boundary layer (TBL) such as illustrated in Fig. 6–13. The thickness of the TBL is represented by $\Delta$.

The concept of the boundary layer was introduced in 1904 by Prandtl [2] in the context of basic flows of the type shown in Figs. 6–6 and 6–13 for which the boundary layers are very thin or are confined by the system geometry. The approximations proposed by Prandtl for boundary-layer flows of this type provided the impetus for early developments in boundary-layer theory. For future reference, these classic *boundary-layer approximations* are given in the context of two-dimensional flow with $x$-component velocity $u$ and $y$-component velocity $v$ by

$$u \gg v \qquad \frac{\partial u}{\partial y} \gg \frac{\partial u}{\partial x}, \frac{\partial v}{\partial y}, \frac{\partial v}{\partial x} \qquad \text{for the HBL} \qquad (6\text{–}16)$$

and

$$\frac{\partial T}{\partial y} \gg \frac{\partial T}{\partial x} \quad \text{for the TBL} \tag{6–17}$$

Prandtl's pioneering work is generally recognized as one of the most important achievements in the modern developments of viscous fluid flow and convection heat transfer.

As illustrated by Fig. 6–13, for external forced convection flow the boundary layers continue to grow as $x$ increases, with the velocity and temperature at the outer edge of the boundary layer being equal to the free-stream values $U_\infty$ and $T_\infty$. Thus, the reference velocity $U_F$ and reference temperature $T_F$ for external forced-convection flow are generally set equal to $U_\infty$ and $T_\infty$, respectively. (The distinguishing features of natural convection boundary layers are considered in Chaps. 7 and 9.) On the other hand, for internal flows such as the one shown in Fig. 6–6, the thickness of the boundary layers that form around the perimeter of the confining surface grow as $x$ increases, until coming together, with the reference velocity and temperature generally specified in terms of bulk-stream values. The bulk-stream characteristics associated with internal flow are introduced in Sec. 6–4.

Referring back to the defining relation for the Prandtl number $Pr$, Eq. (6–15), by multiplying the numerator and denominator by density, we obtain

$$Pr = \frac{\mu}{\rho}\frac{\rho c_P}{k} = \frac{\nu}{\alpha} = \frac{\text{kinematic viscosity}}{\text{thermal diffusivity}} \tag{6–18}$$

Written in this form, the Prandtl number can be seen to represent the *relative effectiveness of molecular transport of momentum and energy within the hydrodynamic and thermal boundary layers*.

The hydrodynamic characteristics associated with forced-convection boundary layer flow are generally expressed in terms of the *Reynolds number* $Re_L$,

$$Re_L = \frac{U_F \mathbf{L}}{\nu} \tag{6–19}$$

where $\mathbf{L}$ is a reference length (such as $D$, $D_H$, $x$, or $L$). The Reynolds number actually represents the *ratio between the inertia and viscous forces* acting upon the fluid. This important dimensionless parameter provides an index that indicates whether the flow is likely to be laminar or turbulent. For example, with $U_F = U_\infty$ and $\mathbf{L} = x$ for forced convection flow over a flat plate with uniform free-stream velocity, the flow is normally laminar for $Re_x \lessapprox 5 \times 10^5$ and turbulent for $Re_x \gtrapprox 5 \times 10^5$.

## 6–3 APPROACHES TO THE ANALYSIS OF CONVECTION HEAT TRANSFER

The engineering analysis of convection-heat-transfer processes can be achieved by means of theoretical or practical approaches, with the practical approach sometimes being supplemented by dimensional analysis.

### 6–3–1 Theoretical Analysis

Convection heat transfer is generally more complex than conduction in a solid or in a stationary fluid because of the superimposed effects of fluid motion. Aside from this complicating factor, the basic concepts involved in the treatment of heat transfer from a solid/fluid interface to a fluid are the same as the treatment of the heat transfer within a solid. Similar to the analysis of conduction heat transfer, the complete theoretical solution of a convection-heat-transfer problem requires the development of a mathematical formulation that represents the energy transport within the fluid itself. In addition, because energy is transported by fluid movement, a mathematical formulation must be developed for the fluid motion. These formulations involve the application of the fundamental conservation principles for mass, momentum, and energy in the context of differential control volumes and generally take the form of systems of partial differential equations.

Strictly speaking, it is the formulation and solution of such equations of motion and energy that provide the means of obtaining predictions for the velocity and temperature distributions within a fluid and the wall shear stress or pressure drop and convection heat transfer. The formulation and solution of the equations of motion and energy will be presented for several convection processes in Chap. 7, which pertains to the theoretical analysis of convection.

### 6–3–2 Practical Analysis

The engineer is often primarily interested in knowing the pressure drop or drag and the rate of heat transfer (or the surface temperature) and is not always concerned with the details of the velocity and temperature distributions within the fluid. Therefore, a much simpler lumped-analysis approach has been developed that provides a means for obtaining practical engineering calculations. This practical analysis approach involves the use of coefficients of friction or drag and heat transfer and application of the fundamental conservation principles to lumped and lumped/differential control volumes, which generally result in much simpler mathematical formulations than are obtained in the theoretical approach. For situations in which relations for the coefficients of friction and heat transfer are available, the practical lumped-analysis approach will be seen to produce calculations for the wall shear stress (or pressure drop) and the rate of heat transfer (or the surface temperature) for prescribed fluid-flow/heating conditions. However, the velocity and temperature distributions within the fluid cannot be obtained by the lumped approach. If the velocity and temperature profiles are required or if information pertaining to the coefficients of friction and heat transfer for the specific problem of interest is not available, the full theoretical analysis referred to above must be developed or experimental measurements and empirical correlations must be obtained.

#### *Coefficients of Friction and Heat Transfer*

The *local* wall shear stress $\tau_s$ and heat flux $q_c''$ are traditionally expressed in terms of coefficients of friction and heat transfer.

The *local Fanning friction factor* $f_x$ is defined by†

$$\tau_s = \rho U_F^2 \frac{f_x}{2} \tag{6–20}$$

where $U_F$ is a reference velocity that depends on the geometry. The *mean Fanning friction factor* $\bar{f}$ over an area of surface $A_s$ is defined as

$$\bar{f} = \frac{1}{A_s} \int_{A_s} f_x \, dA_s \tag{6–21}$$

Both $f_x$ and $\bar{f}$ are dimensionless coefficients. Related dimensionless coefficients that account for entrance, exit, and component pressure losses in internal flow systems are introduced in Chaps. 8 and 11.

The *local coefficient of heat transfer* $h_x$ is defined by the *general Newton law of cooling*, Eq. (1–21),

$$q_c'' = \frac{dq_c}{dA_s} = h_x(T_s - T_F) \tag{6–22}$$

(The heat flux from the wall is given in terms of the distribution in temperature $T$ by Eq. (6–2b) at $y = 0$.) The mean coefficient of heat transfer $\bar{h}$ is defined in terms of $h_x$ by

$$\bar{h} = \frac{1}{A_s} \int_{A_s} h_x \, dA_s \tag{6–23}$$

The coefficients of heat transfer [with dimensions W/(m$^2$ °C) or Btu/(h ft$^2$ °F)] are usually expressed in terms of the dimensionless Nusselt number $Nu_L$ (and $\overline{Nu_L}$),

$$Nu_L = \frac{h_x L}{k} \qquad \overline{Nu_L} = \frac{\bar{h} L}{k} \tag{6–24a,b}$$

or dimensionless *Stanton number* $St$ (and $\overline{St}$),

$$St = \frac{Nu_L}{Re_L \, Pr} = \frac{h_x}{\rho c_P U_F} \qquad \overline{St} = \frac{\overline{Nu_L}}{Re_L \, Pr} = \frac{\bar{h}}{\rho c_P U_F} \tag{6–25a,b}$$

where $\rho$, $c_P$, and $k$ are properties of the fluid. It should be noted that the Nusselt number represents the *ratio of convection heat transfer for fluid in motion to conduction heat transfer for a motionless layer of fluid*. On the other hand, the Stanton number, which combines the Nusselt number, Reynolds number, and Prandtl number, indicates the *relative magnitude of the actual convection heat flux and the enthalpy energy flux capacity of the fluid flow*.

As we shall see in the next several chapters, theoretical and empirical relations are available for coefficients of friction and heat transfer for many standard convection

† The Fanning friction factor is also commonly represented by $C_f$. The product $4f$ is referred to as the *Darcy friction factor* $\lambda$,

$$\lambda = 4f$$

processes. These relations are generally expressed in terms of dimensionless parameters effectively representing the geometry, flow, and fluid characteristics. For example, the key independent dimensionless parameters for forced convection include the Nusselt number (or Stanton number), the Prandtl number $Pr$, and the Reynolds number $Re_L$.

As an aside, it should be mentioned that Eq. (6–22) actually appears to have first been proposed by Fourier (1768–1830), many years after the death of Newton (1643–1727) [4]. However, it was Newton's earlier work with the concept of temperature that laid the groundwork for the development of the concept of heat transfer in the late eighteenth century and the eventual framing of Eq. (6–22). Strictly speaking, the heat flux $q_c''$ is not directly proportional to the temperature difference $T_s - T_F$ for all convection heat-transfer processes. For example, $q_c''$ is a nonlinear function of $T_s - T_F$ for natural convection, boiling and condensation, and forced convection with large temperature differences, such that $h_x$ is a function of $T_s - T_F$. However, $h_x$ is independent (or at least approximately so) of $T_s - T_F$ for many practical forced-convection processes involving mild to moderate temperature differences, such that Eq. (6–22) effectively separates the key variables $q_c''$ and $T_s - T_F$. As we shall see, the coefficient of heat transfer has been assimilated into the modern practical engineering approach to the analysis of heat exchangers and other convection processes. As we get further into the study of convection heat transfer, additional benefits and occasional liabilities associated with the use of coefficients of friction $f_x$ and heat transfer $h_x$ will be uncovered. Mostly for the better, but sometimes for the worse, these coefficients of friction and heat transfer "have been assimilated irrevocably into the engineering literature and will no doubt be with us forever."†

### 6–3–3 Dimensional Analysis

For cases that must be dealt with empirically, dimensional analysis is commonly used to determine the pertinent dimensionless parameters. This approach is based on the principle of dimensional continuity, which requires that all terms in every equation be dimensionally consistent. To properly apply the dimensional-analysis approach to convection, we must know what dimensional parameters the coefficients of friction and heat transfer depend upon. These parameters can generally be established on the basis of a thorough physical understanding of the problem.

The formal theoretical principle used in dimensional analysis is known as the *Buckingham $\pi$ theorem* [5]. The Buckingham $\pi$ theorem indicates that the number $N$ of *independent* dimensionless groups $\pi_i$ which are associated with a physical phenomenon, is equal to the total number of significant dimensional parameters $I$ minus the number of fundamental dimensions $J$, which are required to define the dimensions of all the $I$ parameters, with the relationship among the various dimensionless groups taking the form

$$\pi_1 = \text{fn}\,(\pi_2, \pi_3, \ldots, \pi_i, \ldots, \pi_N) \tag{6–26}$$

† This statement is a generalization of an argument made by White [3] pertaining to coefficients of friction.

The mechanics involved in the general dimensional analysis approach are discussed by Buckingham [5], Bridgeman [6], and Langhaar [7], and introductions to the use of this tool in analyzing convection-heat-transfer and fluid-mechanics processes are given by Kreith and Bohn [8], Lienhard [9], and Fox and McDonald [10]. The use of this approach is illustrated in Example 6–1 in the context of a basic forced-convection system.

## EXAMPLE 6–1

Assuming that the coefficient of heat transfer $\bar{h}$ for forced convection flow in a tube with uniform wall temperature is dependent upon the geometric, hydrodynamic, and thermal parameters listed in Table E6–1, utilize the dimensional-analysis approach to determine the key dimensionless groups for this problem.

**TABLE E6–1** Forced convection parameters—flow in a tube

| *Dimensional parameter* | *Symbol* | *Dimensions* |
|---|---|---|
| Geometric | | |
| Axial distance | $x$ | $[L]$ |
| Diameter | $D$ | $[L]$ |
| Hydrodynamic | | |
| Entrance velocity | $U_1$ | $[L/t]$ |
| Density | $\rho$ | $[m/L^3]$ |
| Viscosity | $\mu$ | $[m/(tL)]$ |
| Thermal | | |
| Thermal conductivity | $k$ | $[mL/(t^3\,T)]$ |
| Specific heat | $c_P$ | $[L^2/(t^2\,T)]$ |
| Coefficient of heat transfer | $\bar{h}$ | $[m/(t^3\,T)]$ |

### Solution

*Objective* Use the dimensional-analysis approach to obtain the primary dimensionless groups pertaining to the coefficient of heat transfer $\bar{h}$ for forced convection flow in a tube with uniform wall temperature $T_0$.

*Schematic* Flow in a tube with uniform wall temperature $T_0$.

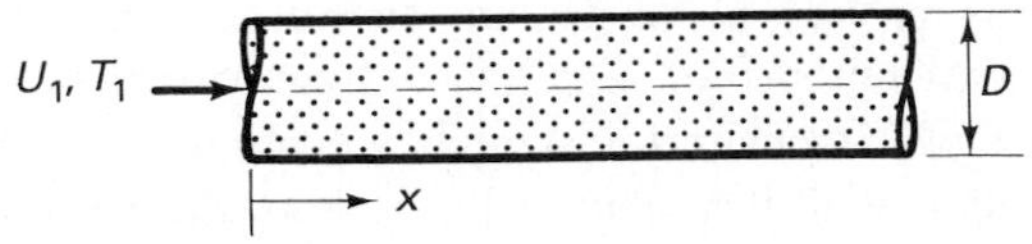


*Assumptions/Conditions*

forced convection

uniform properties

standard conditions (i.e., steady-state; two-dimensional; ideal, Newtonian, isotropic, single-phase, and chemically homogeneous fluid; incompressible flow; nonporous walls; and negligible buoyancy, kinetic energy, viscous dissipation, external body forces, axial conduction, and thermal radiation)

*Properties* Assume that fluid properties $\rho$, $\mu$, $k$, and $c_P$ are specified.

*Analysis* Referring to Table E6–1, we see that this problem involves eight dimensional parameters ($I = 8$) and four fundamental dimensions ($J = 4$). Therefore, according to the Buckingham $\pi$ theorem,

$$N = I - J = 8 - 4 = 4$$

which indicates four dimensionless groups. Following the procedure outlined by various authors [5–9], to determine the dimensionless groups $\pi_i$ (where $i = 1, 2, 3, 4$), we write

$$\pi_i = U_1^a \mu^b \rho^c k^d c_P^e \bar{h}^f D^g x^h$$

Substituting for the dimensions of each parameter and noting that the $\pi_i$ groups are dimensionless, we have

$$1 = \left[\frac{L}{t}\right]^a \left[\frac{m}{Lt}\right]^b \left[\frac{m}{L^3}\right]^c \left[\frac{mL}{t^3T}\right]^d \left[\frac{L^2}{t^2T}\right]^e \left[\frac{m}{t^3T}\right]^f [L]^g[L]^h$$

Because the summation of the exponents of each fundamental dimension must be equal to zero, we conclude that

$$\begin{aligned} b + c + d + f &= 0 \quad \text{for mass } m & \text{(a)} \\ a - b - 3c + d + 2e + g + h &= 0 \quad \text{for length } L & \text{(b)} \\ -a - b - 3d - 2e - 3f &= 0 \quad \text{for time } t & \text{(c)} \\ -d - e - f &= 0 \quad \text{for temperature } T & \text{(d)} \end{aligned}$$

These four equations must be satisfied for each dimensionless group. But, because we have eight unknowns, the values of four of the exponents must be specified for each of the four dimensionless groups $\pi_1$, $\pi_2$, $\pi_3$, and $\pi_4$. (The fact that these dimensionless groups must be independent simply means that no group can be expressed as the product of any combination of the other groups.) To help in selecting these four exponents for each dimensionless group, we will purposefully seek one thermal group ($\pi_1$) involving $\bar{h}$ and $D$, one thermophysical property group ($\pi_2$), one hydrodynamic group ($\pi_3$), and one geometric group ($\pi_4$).

To obtain dimensionless thermal group $\pi_1$, we set $f$ equal to unity, and the hydrodynamic and geometric exponents $a$, $b$, and $h$ equal to zero. Solving Eqs. (a) through (d) with these inputs, we obtain $c = 0$, $d = -1$, $e = 0$, and $g = 1$. Therefore, this dimensionless thermal group is

$$\pi_1 = \frac{\bar{h}D}{k}$$

which is the mean *Nusselt number Nu*.

To formulate a thermophysical property group, $a, f, g$, and $h$ are set equal to zero. Equations (a) through (d) then give rise to $b = e$, $c = 0$, and $d = -e$. Setting the common exponent $e$ equal to unity, we obtain

$$\pi_2 = \frac{\mu c_P}{k}$$

which is recognized as the *Prandtl number Pr*.

To obtain a dimensionless hydrodynamic group, the thermal exponents $d$, $e$, and $f$ and the geometric exponent $h$ are set equal to zero. It follows that $a = 1$, $b = -1$, $c = 1$, and $g = 1$, such that

$$\pi_3 = \frac{\rho D U_1}{\mu}$$

which is the *Reynolds number Re*.

Finally, to develop a dimensionless geometric group, we leave $g$ unspecified with $h = 1$, $b = 0$, $e = 0$, and $f = 0$. Substituting these values into Eqs. (a) through (d), we obtain $a = 0$, $c = 0$, $d = 0$, and $g = -1$. Hence, we have the natural dimensionless geometric grouping

$$\pi_4 = \frac{x}{D}$$

Based on these results, we can expect correlations for the coefficient of heat transfer associated with forced convection in smooth tubes to take the general form

$$\frac{\bar{h}D}{k} = \overline{Nu} = \text{fn}\,(Re, Pr, x/D) \tag{e}$$

And this is indeed the case, as we shall see in Chap. 8, which presents design equations for the coefficient of heat transfer.

It should be noted that the actual functional relationship between $\overline{Nu}$ and $Re$, $Pr$, and $x/D$, which is dependent upon the specific operating conditions, can often be determined on the basis of experimental data. In this connection, Eq. (e) reduces the data correlation problem from one involving the eight variables $U_1$, $\mu$, $\rho$, $k$, $c_P$, $\bar{h}$, $D$, and $x$ to one with only the four dimensionless groups, $\overline{Nu}$, $Re$, $Pr$, and $x/D$. To obtain an empirical correlation for $\bar{h}$ in terms of $U_1$, $\mu$, $\rho$, $k$, $c_P$, $D$, and $x$, it would be necessary to vary each of these seven parameters at least three or four times. Assuming that we obtained four data points for each of these seven variables, we would require $4^7$, or about 16,400 individual measurements! On the other hand, to obtain a correlation for $\overline{Nu}$ in terms of $Re$, $Pr$, and $x/D$ with as much information would only require $4^3$, or 64, data points. Thus, although dimensional analysis cannot be used to determine the functional relationship between a parameter such as $\bar{h}$ and the other system parameters, this approach can be utilized to greatly simplify the correlation of experimental data to produce empirical correlations for design.

It should be noted that the dimensionless parameters that characterize a convection heat-transfer process can also often be obtained by the theoretical approach introduced in Sec. 6–3–1. This more comprehensive approach will be developed in Chap. 7 for several basic convection processes.

Because of their common use and significance, basic dimensionless parameters that are used to characterize convection-heat-transfer processes are summarized for forced convection, natural convection, and boiling and condensation in Tables 8–9, 9–8, and 10–4. In this connection, Arpaci and Larsen [11] provide an excellent discussion of the interpretation of dimensionless groups for convection heat transfer.

## 6–4 CHARACTERISTICS OF INTERNAL FLOW

To establish a framework for the theoretical and practical analysis of internal flow, we now introduce the key bulk-stream characteristics and discuss the nature of developing and fully developed flow in the context of passages with uniform cross-sectional area $A$ and impermeable walls, such as the circular tube shown in Fig. 6–14. (The concepts presented in this section are extended to flow through tube banks and heat-exchanger cores in Chaps. 8 and 11.) For arrangements of this type, flow enters with a uniform velocity $U_1$, the mass flow rate $\dot{m}$ is constant, and heating is initiated at some distance $x_0$ from the entrance.

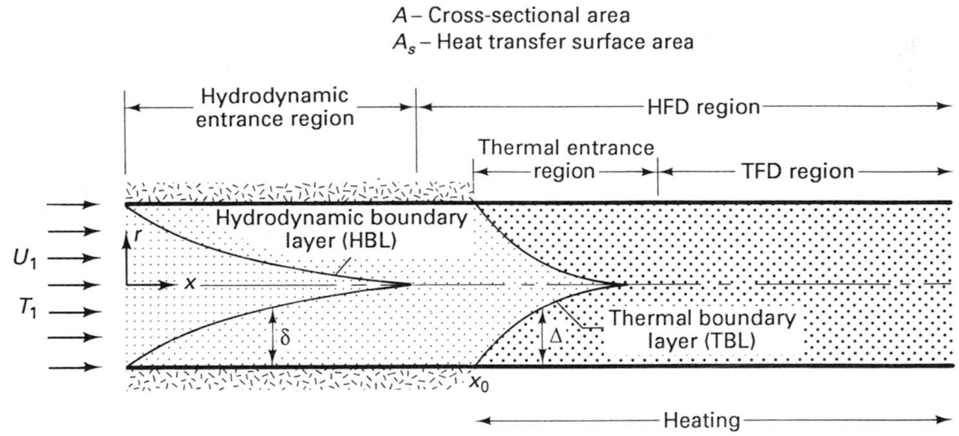


**FIGURE 6–14** Hydrodynamic and thermal entrance regions and fully developed regions associated with flow in a circular tube.

The reference length $L$ for internal flows is generally set equal to the *hydraulic diameter*† $D_H$, which is defined by

$$D_H = 4\frac{A}{p_w} \tag{6–27}$$

† The hydraulic radius $r_H$ also commonly appears in the literature; $r_H$ is related to $D_H$ by $D_H = 4r_H$.

where $p_w$ is the wetted perimeter. For tubular passages with $p_w$ equal to the heat-transfer surface perimeter $p$, Eq. (6–27) reduces to

$$\frac{D_H}{L} = 4\frac{A}{pL} = 4\frac{A}{A_s} \tag{6–28}$$

where $A_s$ ($= pL$) is the heat-transfer surface area. The hydraulic diameters of several common passages are given as follows:

*Circular Tube*

$$D_H = 4\frac{\pi D^2/4}{\pi D} = D \tag{6–29}$$

*Circular Annulus*

$$D_H = \frac{4\pi(D_o^2 - D_i^2)/4}{\pi(D_o + D_i)} = D_o - D_i \tag{6–30}$$

*Parallel Plates*
(width $w$, depth $Z$)

$$D_H = 4\frac{wZ}{2Z} = 2w \tag{6–31}$$

### 6–4–1 Hydrodynamic Entrance and Fully Developed Regions

As indicated in Fig. 6–14, the hydrodynamic boundary layer thickness $\delta$ for flow in a tube grows as $x$ increases, until it reaches the centerline. Beyond the axial location at which this occurs, the growth of the boundary layer is constrained and the shape of the velocity profile becomes independent of $x$; that is,

$$\frac{\partial u}{\partial x} = 0 \tag{6–32}$$

for constant mass flow rate $\dot{m}$ and constant properties. In this region, the flow is *hydrodynamically fully developed* (HFD). The region upstream of the HFD region is known as the *hydrodynamic entrance region*.

Internal flows are generally characterized in terms of the *bulk-stream mass flux G*,

$$G = \frac{\dot{m}}{A} = \frac{1}{A}\int_{\dot{m}} d\dot{m} = \frac{1}{A}\int_A \rho u\, dA = \frac{\rho}{A}\int_A u\, dA \tag{6–33}$$

or the *bulk-stream velocity* $U_b$,

$$U_b = \frac{\dot{m}}{\rho A} = \frac{G}{\rho} = \frac{1}{A}\int_A u\, dA \tag{6–34}$$

for uniform density. For uniform property flow in systems with constant mass flow rate, $U_b$ is constant and is equal to the entering velocity $U_1$.

Setting $L = D_H$ and $U_F = U_b$ in Eq. (6–19) and designating the Reynolds number for internal flow by $Re$ rather than $Re_D$, we write

$$Re = \frac{U_b D_H}{\nu} \quad \text{or} \quad Re = \frac{G D_H}{\mu} \qquad (6\text{–}35\text{a,b})$$

Using this defining relation for $Re$, the flow is generally laminar for values of $Re$ less than about 2000, and the flow is turbulent for values of $Re$ greater than this value. In the turbulent region, the flow is usually fully turbulent for $Re \gtrsim 10^4$ and transitional turbulent for $2000 \lesssim Re \lesssim 10^4$.

The Fanning friction factor for internal flow is defined by Eq. (6–20) with $U_F = U_b$; that is,

$$\tau_s = \rho U_b^2 \frac{f_x}{2} \qquad (6\text{–}36)$$

The mean coefficient of friction over the length 0 to $x$ for two-dimensional flow is given by

$$\bar{f} = \frac{1}{x}\int_0^x f_x\, dx \qquad (6\text{–}37)$$

It should be noted that the hydrodynamic entrance region is characterized by a local friction factor $f_x$ that varies with $x$. However, $f_x$ is independent of $x$ in the HFD region for systems with uniform mass flow rate $\dot{m}$ and uniform properties. To see this point, we couple the Newtonian shear law, Eq. (6–1), with the defining equation for $f_x$, Eq. (6–36), and the defining equation for HFD flow, Eq. (6–32), as follows:

$$\tau_s = \rho U_b^2 \frac{f_x}{2} = \mu \left.\frac{\partial u}{\partial y}\right|_0$$
$$\frac{d\tau_s}{dx} = \frac{\rho U_b^2}{2}\frac{df_x}{dx} = \mu \frac{\partial}{\partial y}\left.\left(\frac{\partial u}{\partial x}\right)\right|_0 = 0 \qquad (6\text{–}38)$$

Hence, $\tau_s$ is constant and

$$f_x = f = \text{constant} \qquad (6\text{–}39)$$

in the HFD region. The symbol $f$ will be consistently used in our study to designate the Fanning friction factor for HFD flow.

### 6–4–2 Thermal Entrance and Fully Developed Regions

Because the fluid is heated or cooled as it flows downstream, its enthalpy or energy content changes with $x$. To characterize this change, we define the bulk-stream enthalpy rate $\dot{H}_b$ as the total rate of enthalpy transferred through the cross-sectional area $A$ at any axial location $x$. Referring to Fig. 6–15, $\dot{H}_b$ is expressed in terms of the distribution in specific enthalpy $i$ at $x$ by

$$\dot{H}_b = \int_{\dot{m}} i\, d\dot{m} = \int_A i\rho u\, dA = \rho \int_A iu\, dA \qquad (6\text{–}40)$$

for uniform density.

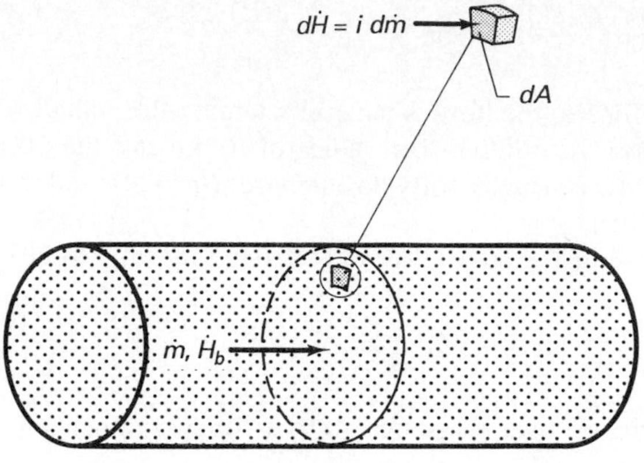


**FIGURE 6–15**
Representation of bulk-stream enthalpy rate $\dot{H}_b$ for flow in a tube with mass flow rate $\dot{m}$.

To express $\dot{H}_b$ in terms of the temperature distribution $T$ over the cross section, we make use of Eq. (6–14) for ideal fluids; that is,

$$i = c_P(T - T_R) + i_R \tag{6–41}$$

for uniform specific heat $c_P$, where the subscript $R$ represents the reference state. Combining this equation with Eq. (6–40), we have

$$\dot{H}_b = \rho \int_A [c_P(T - T_R) + i_R]u\,dA = \rho c_P \int_A uT\,dA + \dot{m}(i_R - c_P T_R) \tag{6–42}$$

The local average thermal energy state of the fluid over the entire cross section $A$ is characterized by the *bulk-stream temperature* $T_b$, which is formally defined for uniform property flow as the average value of $T$ for which Eq. (6–42) is satisfied; that is,

$$\dot{H}_b = \rho c_P \int_A uT_b\,dA + \dot{m}(i_R - c_P T_R) = \dot{m}[c_P(T_b - T_R) + i_R] \tag{6–43}$$

We are now in a position to establish the gradient $d\dot{H}_b/dT_b$, which is essential to the practical thermal analysis approach. Based on Eq. (6–43), $d\dot{H}_b/dT_b$ is given by

$$\frac{d\dot{H}_b}{dT_b} = \frac{d}{dT_b}\{\dot{m}[c_P(T_b - T_R) + i_R]\} \tag{6–44}$$

or, since $\dot{m}$ $(= GA)$ is independent of $T_b$, and since $T_R$, $i_R$, and $c_P$ are constant,

$$d\dot{H}_b = \dot{m}c_P\,dT_b \tag{6–45}$$

Finally, to express $T_b$ in terms of $T$, we equate Eqs. (6–42) and (6–43), with the result

$$\dot{m}[c_P(T_b - T_R) + i_R] = \rho c_P \int_A uT\,dA + \dot{m}(i_R - c_P T_R)$$

or

$$T_b = \frac{\rho}{\dot{m}} \int_A uT\, dA = \frac{1}{AU_b} \int_A uT\, dA \tag{6–46}$$

which is the defining equation for the bulk-stream temperature associated with uniform property flow. Since $T_b$ represents the temperature that would be measured if the fluid flowing through the cross-sectional area were collected and mixed in a cup, this parameter is also commonly referred to as the *mixing cup temperature*. Note that whereas $T_b$ is a function of $x$ for heating or cooling, its counterpart $U_b$ is constant for flow in passages with uniform properties, cross-sectional area, and mass flow rate.

The thickness $\Delta$ of the TBL grows with respect to $x$ until it becomes constrained by the system geometry (see Fig. 6–14). Beyond the axial location at which $\Delta$ becomes a constant, the shape of the temperature profile becomes independent of $x$ for uniform wall temperature, uniform wall heat flux, and certain other boundary conditions; that is,†

$$\frac{\partial}{\partial x}\left(\frac{T - T_s}{T_b - T_s}\right) = 0 \tag{6–47}$$

which is the defining equation for *thermal fully developed flow* (TFD).

The coefficient of heat transfer $h_x$ for internal flow is defined by Eq. (6–22) with $T_F = T_b$.

$$q_c'' = \frac{dq_c}{dA_s} = h_x(T_s - T_b) \tag{6–48}$$

The mean coefficient of heat transfer over the length 0 to $x$ is defined by

$$\bar{h} = \frac{1}{x} \int_0^x h_x\, dx \tag{6–49}$$

As will be seen, $\bar{h}$ is primarily used in the analysis of problems involving uniform wall-temperature conditions and in heat-exchanger applications.

Whereas the coefficients $h_x$ and $\bar{h}$ are functions of $x$ in the thermal entrance region, the TFD region is characterized by a coefficient $h_x$ that is independent of $x$ for uniform mass flow rate and uniform properties. To demonstrate this point, we combine the Fourier law of conduction, Eq. (6–2b), with the general Newton law of cooling, Eq. (6–48), to obtain

$$q_c'' = h_x(T_s - T_b) = -k \left.\frac{\partial T}{\partial y}\right|_0 \tag{6–50}$$

† For situations in which the wall temperature varies around the perimeter, $T_s$ is replaced by the peripheral mean wall temperature.

where $k$ is the thermal conductivity of the fluid and $\partial T/\partial y|_0$ is the temperature gradient within the fluid at $y = 0$. Rearranging this equation and introducing the defining equation for TFD flow, Eq. (6–47), we have

$$h_x = -k\frac{\partial}{\partial y}\left(\frac{T}{T_s - T_b}\right)\bigg|_0 = k\frac{\partial}{\partial y}\left(\frac{T - T_s}{T_b - T_s}\right)\bigg|_0$$
$$\frac{dh_x}{dx} = k\frac{\partial}{\partial y}\left[\frac{\partial}{\partial x}\left(\frac{T - T_s}{T_b - T_s}\right)\right]\bigg|_0 = 0 \tag{6–51}$$

Hence, we see that

$$h_x = h = \text{constant} \tag{6–52}$$

where $h$ specifically designates the coefficient of heat transfer for TFD conditions.

As indicated earlier, the coefficient of heat transfer is generally expressed in terms of the Nusselt number or Stanton number. To distinguish between $h$, $h_x$, and $\bar{h}$, when employing these dimensionless parameters, we will use the following notation for internal flow:

$$Nu = \frac{hD_H}{k} \qquad (Nu)_x = \frac{h_x D_H}{k} \qquad \overline{\mathrm{Nu}} = \frac{\bar{h}D_H}{k}$$
$$St = \frac{h}{\rho c_P U_b} \qquad (St)_x = \frac{h_x}{\rho c_P U_b} \qquad \overline{St} = \frac{\bar{h}}{\rho c_P U_b} \tag{6–53}$$

## 6–5 SUMMARY

In this chapter we discussed features that characterize convection-heat-transfer processes and the concept of the boundary layer. We also gave brief consideration to the theoretical and practical approaches to analyzing convection heat transfer, and we introduced important bulk-stream characteristics for uniform property internal flows.

The theoretical approach to the analysis of convection heat transfer involves the development of mathematical formulations for the fluid flow and energy transfer, which generally take the form of systems of partial differential equations. Representative convection-heat-transfer processes will be analyzed in Chap. 7, which deals with the theory of convection. This treatment provides a theoretical basis for some of the relations for coefficients of friction and heat transfer used in practical engineering analysis, and also provides a framework for dealing with more complex problems.

The practical analysis approach involves the use of relations for coefficients of friction and heat transfer together with fundamental conservation principles. This approach generally results in simple ordinary differential or algebraic equations and offers a means of determining the pressure drop or drag and the heat-transfer performance. Practical engineering analyses are developed for basic forced convection, natural convection, and boiling and condensation processes in Chaps. 8, 9, and 10. This material provides the basis for the evaluation and design of heat exchangers, which is the topic of Chap. 11.

The theoretical and practical approaches to the analysis of convection heat transfer involve the use of a number of important dimensionless parameters. The interpretation of the primary dimensionless groups associated with the analysis of forced convection, natural convection, and boiling and condensation is summarized in Tables 8–9, 9–8, and 10–4.

Finally, our study of convection heat transfer will be restricted to systems involving fluids which are ideal (i.e., ideal gases and incompressible liquids). References 12 and 13 deal with the important topic of compressible flow. Heat and mass transfer in chemically nonhomogeneous fluids is considered in the *mass transfer supplement*.

## REVIEW QUESTIONS

**6–1.** List several important characterizing factors for convection-heat-transfer processes.

**6–2.** Describe the difference between laminar and turbulent flow.

**6–3.** Write defining relations for (a) enthalpy rate $\dot{H}$, (b) mass flow rate $\dot{m}$, and (c) momentum rate $\dot{M}$.

**6–4.** Write defining relations for (a) bulk-stream mass flux $G$ and (b) bulk-stream velocity $U_b$.

**6–5.** Write defining relations for (a) bulk-stream enthalpy rate $\dot{H}_b$ and (b) bulk-stream temperature $T_b$.

**6–6.** List the key elements that characterize the practical and theoretical approaches to the analysis of convection-heat-transfer processes.

**6–7.** Write defining relations for the following dimensionless parameters: (a) Reynolds number $Re_L$, (b) Nusselt number $Nu_L$, and (c) Stanton number $St$.

**6–8.** Define the Prandtl number $Pr$ and state its significance.

**6–9.** List several fluids that are characterized by (a) low values of Prandtl number and (b) moderate values of Prandtl number.

**6–10.** List several advantages and disadvantages associated with the use of the coefficient of heat transfer in the analysis of convection-heat-transfer processes.

## PROBLEMS

**6–1.** One of the methods of determining the friction factor and coefficient of heat transfer involves the use of experimental measurements for the distributions in velocity and temperature. To illustrate, consider the situation in which air at 54°C is flowing over a flat plate with free-stream velocity of 1 m/s and surface temperature of 100°C. Representative measurements for the temperature distribution obtained at an axial location $x_1$ are shown in Fig. P6–1 as a function of distance $y$ from the surface of the plate. Use these data to determine the coefficient of heat transfer $h_x$ at this location.

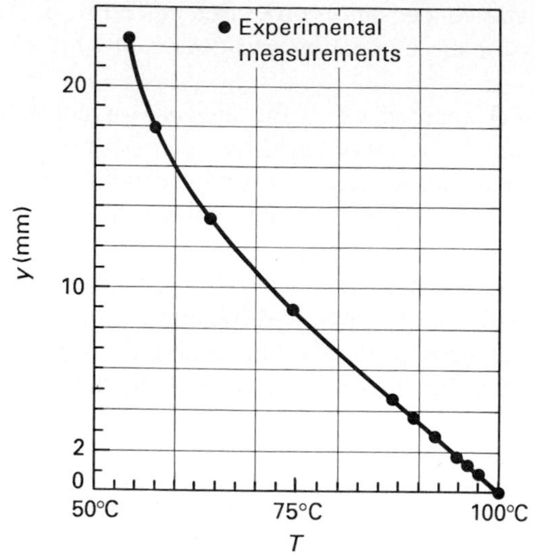


**FIGURE P6–1**

**6–2.** In the theoretical approach to the evaluation of the friction factor and coefficient of heat transfer considered in Chap. 7, solutions are actually developed for the distributions in velocity and temperature across the flow field by applying fundamental principles. For example, an approximate theoretical solution for the temperature distribution for laminar boundary layer flow associated with Prob. 6–1 (see Chap. 7) is given by

$$\frac{T - T_s}{T_\infty - T_s} = 0.331\,\eta\,Pr^{1/3} - 0.00538\,\eta^3\,Pr \qquad \text{for } \eta \leq 4.52\,Pr^{-1/3}$$
$$= 1.0 \qquad \text{for } \eta \geq 4.52\,Pr^{-1/3} \qquad \text{(i)}$$

where $\eta = y/(\nu x/U_\infty)^{1/2}$. (a) Use this relation to show that the coefficient of heat transfer $h_x$ can be represented by

$$Nu_x = \frac{h_x x}{k} = 0.331\,Re_x^{1/2}\,Pr^{1/3} \qquad \text{(ii)}$$

(b) Referring to Fig. P6–1, determine the approximate location $x_1$ at which the data were taken for a free-stream velocity of 1 m/s. (c) Show that Eq. (i) is consistent with the data in Fig. P6–1.

**6–3.** The mean coefficient of heat transfer for natural convection on a vertical plate is a function of the temperature difference $T_s - T_\infty$, the properties $\rho$, $\mu$, $c_P$, $k$, the length $L$, and the buoyancy force, which can be represented by $g(\rho_s - \rho_\infty)$. (a) Show that the mean coefficient of heat transfer $\bar{h}$ can be represented by an expression of the form

$$\overline{Nu_L} = \frac{\bar{h}L}{k} = \text{fn}\,(Gr_L, Pr)$$

where the Grashof number $Gr_L$ is defined by

$$Gr_L = \frac{g(\rho_\infty - \rho_s)L^3}{\rho\nu^2}$$

(b) The bouyancy force for natural convection with small temperature differences is generally represented by

$$g(\rho_s - \rho_\infty) = g\rho\beta(T_\infty - T_s)$$

where the coefficient of thermal expansion $\beta$ is defined by

$$\beta = -\frac{1}{\rho}\frac{\partial\rho}{\partial T}\bigg|_P$$

Use this information to express the Grashof number in terms of $\beta$.

**6–4.** For film condensation on a vertical plate, the coefficient of condensation heat transfer $\bar{h} = q_c''/(T_{sat} - T_s)$ is dependent upon the body force $g(\rho_f - \rho_g)$, which accompanies the large differences in density between the liquid and vapor stages, the latent heat of vaporization relative to the difference between the surface and saturation temperatures $i_{fg}/(T_s - T_{sat})$, the thermophysical properties $\rho$, $\mu$, $c_P$, and $k$ (liquid or vapor), and the length $L$. Use this information together with the dimensional analysis approach to show

$$\overline{Nu}_L = \frac{q_c''}{T_s - T_{sat}}\frac{L}{k} = \text{fn}\,(Pr, Gr_L, Ja)$$

where the Grashof number $Gr_L$ is defined by

$$Gr_L = \frac{g(\rho_f - \rho_g)L^3}{\nu^2}$$

and the Jakob number $Ja$ is

$$Ja = \frac{c_P(T_{sat} - T_s)}{i_{fg}}$$

**6–5.** Because nucleate boiling involves the production of bubbles at nucleation sites, the probler of characterizing this phenomenon is complicated by the fact that the coefficient of boilir heat transfer is a function of the surface tension $\sigma$ as well as the other parameters listed Prob. 6–4. Use this information and the dimensional analysis approach to show

$$Nu_L = \frac{q_c''}{T_s - T_{sat}}\frac{L}{k} = \text{fn}\,(Pr, Gr_L, Ja, Bo_L)$$

for nucleate boiling, where the Bond number $Bo_L$ is defined by

$$Bo_L = \frac{g(\rho_f - \rho_g)L^2}{\sigma}$$

# CHAPTER 7

# CONVECTION HEAT TRANSFER: THEORETICAL ANALYSIS

## 7–1 INTRODUCTION

In this chapter we introduce the theoretical treatment of convection heat transfer. As indicated in Chap. 6, the complete theoretical analysis of convection-heat-transfer problems requires the use of physical laws in the development of mathematical formulations and solutions for the fluid flow and energy transfer within the fluid. The development of differential, integral, or numerical formulations involves (1) the application of the fundamental principles pertaining to mass, momentum, and energy to a control volume within the flow field; and (2) the use of the particular laws pertaining to fluid shear stress and heat flux. These physical laws, which were presented in Chap. 1, are summarized in Table 7–1. Once the mathematical formulations are developed, solutions are obtained for the velocity and temperature distributions and the friction factor and Nusselt number.

In this introductory study attention will be primarily given to developing the fundamentals of laminar flow theory. The treatment of basic laminar convection-heat-transfer theory will be presented in the context of several basic steady one- and two-dimensional internal and external flows and will feature the use of simple analytical solution techniques. Our study of laminar flow theory will be followed by a brief introduction to turbulent flow theory.

## 7–2 LAMINAR FLOW THEORY

The general approach to analyzing laminar convection processes involves the development of (1) mathematical formulations for continuity, momentum, and energy transfer within the fluid; (2) solutions for the velocity profiles $u$ and $v$, and temperature distribution $T$ within the fluid; and (3) predictions for the wall shear stress, wall heat

**TABLE 7–1** Fundamental and particular laws: Summary

1. *Fundamental laws*
   a. Conservation of mass (continuity)
   Rate of creation of mass = 0

$$\Sigma \dot{m}_o - \Sigma \dot{m}_i + \frac{\Delta m_s}{\Delta t} = 0 \quad \text{(i)}$$

   b. Momentum principle (Newton's second law of motion relative to direction $x$)
   Rate of creation of momentum ($RCM_x$) = Sum of forces ($\Sigma F_x$)

$$\Sigma \dot{M}_{o,x} - \Sigma \dot{M}_{i,x} + \frac{\Delta M_{s,x}}{\Delta t} = \Sigma F_x \quad \text{(ii)}$$

   c. Conservation of energy (first law of thermodynamics)
   Rate of creation of energy = 0

$$\Sigma \dot{E}_o - \Sigma \dot{E}_i + \frac{\Delta E_s}{\Delta t} = 0 \quad \text{(iii)}$$

2. *Particular laws*
   a. Newton law of viscosity, $x$ direction†

$$\tau = \mu \frac{\partial u}{\partial y} \qquad \text{rectangular systems} \quad \text{(iv)}$$

$$\tau = \mu \frac{\partial u}{\partial y} = -\mu \frac{\partial u}{\partial r} \qquad \text{cylindrical internal flow system} \quad \text{(v)}$$

   b. Fourier law of conduction
   Axial $x$ direction

$$q_x'' = -k \frac{\partial T}{\partial x} \quad \text{(vi)}$$

   Direction perpendicular to surface

$$q_y'' = -k \frac{\partial T}{\partial y} \qquad \text{rectangular system} \quad \text{(vii)}$$

$$q_y'' = -k \frac{\partial T}{\partial y} = k \frac{\partial T}{\partial r} \qquad \text{cylindrical internal flow system (note that } q_r'' = -q_y'') \quad \text{(viii)}$$

† Equations (iv) and (v) are boundary layer approximations that neglect the contribution of $\partial v/\partial x$. Equations (iv) and (v) actually take the more general forms $\tau = \mu(\partial u/\partial y + \partial v/\partial x)$ and $\tau = -\mu(\partial u/\partial r - \partial v/\partial x)$, respectively.

flux (or wall temperature), and/or coefficients of friction and heat transfer. In our study of this fundamental topic, both differential and integral approaches will be used, with consideration given to the major forced convection/natural convection and internal flow/external flow categories. The basic systems to be studied include two-dimensional flow in tubes and flow over flat plates. The concepts introduced in the study of these classical convection-heat-transfer processes provide a foundation for the theoretical analysis of the more complex problems that are generally encountered in practice.

### 7–2–1 Flow in Tubes

The problem of convection heat transfer for steady two-dimensional laminar flow in a tube with uniform heat flux is considered in this section. This basic problem is illustrated in Fig. 7–1. Although emphasis is placed on circular tubes, the concepts introduced in this section apply to other internal flow systems with uniform cross-sectional area, such as annuli, channels, and parallel plates.

As mentioned in Chap. 6, hydrodynamic and thermal boundary layers develop in the entrance region for internal flow systems, with the flow becoming fully developed in the region downstream where the hydrodynamic and thermal boundary layers are independent of $x$. Recall that fully developed conditions are defined by

$$\frac{\partial u}{\partial x} = 0 \qquad \text{HFD} \tag{7–1}$$

$$\frac{\partial}{\partial x}\left(\frac{T - T_s}{T_b - T_s}\right) = 0 \qquad \text{TFD} \tag{7–2}$$

For these conditions, both the friction factor and the coefficient of heat transfer are independent of $x$ (i.e., $f_x = f$ and $h_x = h$).

In our analysis of this problem, we first develop differential formulations for continuity, momentum, and energy transfer.

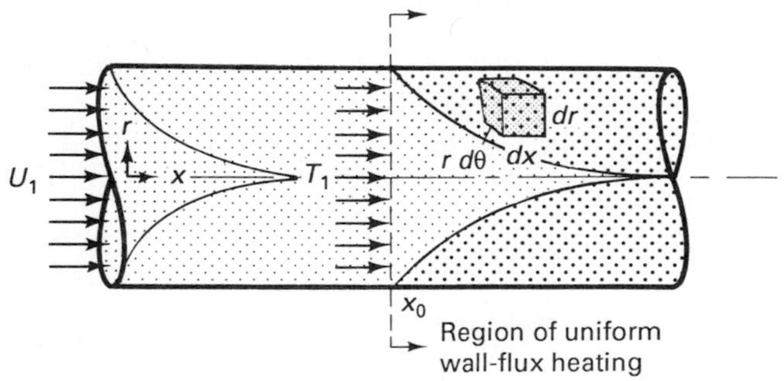


**FIGURE 7–1** Laminar flow in a circular tube with uniform wall-flux heating for $x \geq x_0$.

#### *Mathematical Formulation*

To develop the differential formulation for continuity, momentum, and energy within the entrance and fully developed region, the fundamental laws are applied to the differential volume of fluid $r\,d\theta\,dr\,dx$ shown in Fig. 7–1. The steps required in the development of the differential formulation follow the guidelines established in Chaps. 2 and 3 for conduction heat transfer.

***Continuity*** Applying the principle of conservation of mass to the control volume shown in Figs. 7–1 and 7–2, we obtain

$$\Sigma\,\dot{m}_i = \Sigma\,\dot{m}_o + \cancel{\frac{\Delta m_s}{\Delta t}} \tag{7–3}$$

$$d\dot{m}_x + d\dot{m}_r = d\dot{m}_{x+dx} + d\dot{m}_{r+dr}$$

where $\Delta m_s/\Delta t = 0$ for steady flow, $d\dot{m}_x = \rho ur\,d\theta\,dr$ and $d\dot{m}_r = \rho vr\,d\theta\,dx$. Utilizing the definition of the partial derivative, this expression takes the form

$$\frac{\partial}{\partial x}(d\dot{m}_x)\,dx + \frac{\partial}{\partial r}(d\dot{m}_r)\,dr = 0 \tag{7–4}$$

or

$$\frac{\partial}{\partial x}(\rho u) + \frac{1}{r}\frac{\partial}{\partial r}(\rho r v) = 0 \tag{7–5}$$

For uniform property flow, the continuity equation becomes

$$\frac{\partial u}{\partial x} + \frac{1}{r}\frac{\partial}{\partial r}(rv) = 0 \tag{7–6}$$

The formulation for continuity is completed by writing the boundary condition for $v$,

$$v = 0 \qquad \text{at } r = r_0 \tag{7–7}$$

(The $x$ boundary condition for $u$ will be taken care of in the formulation for momentum transfer, which follows.)

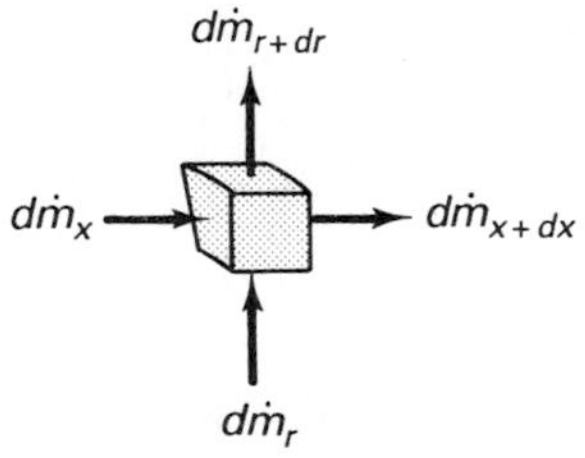


**FIGURE 7–2**
Conservation of mass relative to differential cylindrical control volume; $dA_x = r\,d\theta\,dr$, $dA_r = r\,d\theta\,dx$, $dV = r\,d\theta\,dr\,dx$.

In the HFD region where $\partial u/\partial x = 0$, the continuity equation for incompressible flow reduces to

$$\frac{d}{dr}(rv) = 0 \tag{7–8}$$

By coupling this simple equation with the boundary condition, we find that $v = 0$ in the HFD region.

***Momentum Transfer*** To develop the momentum equation, we apply Newton's second law of motion to the element shown in Figs. 7–1 and 7–3; that is,

$$\sum \dot{M}_{o,x} - \sum \dot{M}_{i,x} + \cancel{\frac{\Delta M_{s,x}}{\Delta t}} = \sum F_x \tag{7–9}$$

The sum of forces is

$$\begin{aligned}\sum F_x &= (\tau\, dA)_r + (P\, dA)_x - (\tau\, dA)_{r+dr} - (P\, dA)_{x+dx} \\ &= (\tau r\, d\theta\, dx)|_r + (Pr\, d\theta\, dr)|_x - (\tau r\, d\theta\, dx)|_{r+dr} - (Pr\, d\theta\, dr)|_{x+dx} \\ &= -\frac{\partial}{\partial r}(r\tau)\, d\theta\, dx\, dr - \frac{\partial P}{\partial x} r\, d\theta\, dr\, dx = -\frac{1}{r}\frac{\partial}{\partial r}(r\tau)\, dV - \frac{\partial P}{\partial x} dV\end{aligned} \tag{7–10}$$

where $dV = r\, d\theta\, dr\, dx$, assuming negligible normal viscous stresses, buoyant forces, and other body forces. The rate of creation of axial momentum is represented by $RCM_x$,

$$\begin{aligned}RCM_x &= d\dot{M}_{x+dx} + d\dot{M}_{r+dr} - d\dot{M}_x - d\dot{M}_r \\ &= \frac{\partial}{\partial x}(d\dot{M}_x)\, dx + \frac{\partial}{\partial r}(d\dot{M}_r)\, dr\end{aligned} \tag{7–11}$$

where $d\dot{M}_x = u\, d\dot{m}_x$ and $d\dot{M}_r = u\, d\dot{m}_r$. Substituting for $d\dot{M}_x$ and $d\dot{M}_r$, $RCM_x$ becomes

$$\begin{aligned}RCM_x &= \frac{\partial}{\partial x}(u\, d\dot{m}_x)\, dx + \frac{\partial}{\partial r}(u\, d\dot{m}_r)\, dr \\ &= \frac{\partial}{\partial x}(u\rho u)\, r\, d\theta\, dr\, dx + \frac{\partial}{\partial r}(u\rho v r)\, d\theta\, dx\, dr \\ &= \frac{\partial}{\partial x}(u\rho u)\, dV + \frac{1}{r}\frac{\partial}{\partial r}(u\rho v r)\, dV\end{aligned} \tag{7–12}$$

Making use of the continuity equation, Eq. (7–5), this expression reduces to

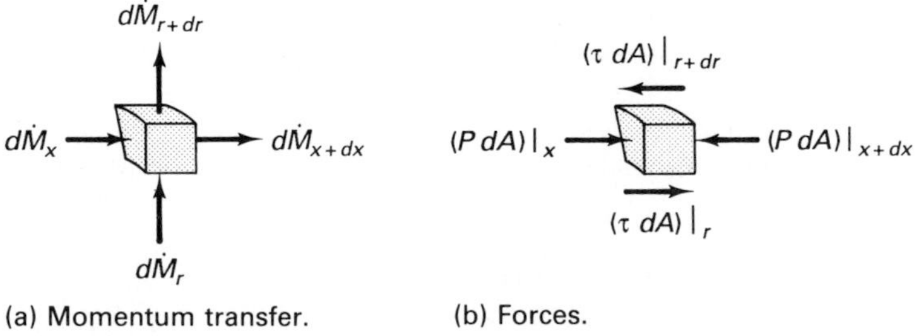


**FIGURE 7–3** Momentum principle relative to differential cylindrical control volume; $x$–direction.

$$
\begin{aligned}
RCM_x &= u\left[\frac{\partial}{\partial x}(\rho u) + \frac{1}{r}\frac{\partial}{\partial r}(\rho v r)\right] dV + \left(\rho u \frac{\partial u}{\partial x} + \rho v \frac{\partial u}{\partial r}\right) dV \\
&= \rho\left(u\frac{\partial u}{\partial x} + v\frac{\partial u}{\partial r}\right) dV
\end{aligned}
\tag{7–13}
$$

Substituting Eqs. (7–10) and (7–13) into Eq. (7–9), we obtain

$$
\rho\left(u\frac{\partial u}{\partial x} + v\frac{\partial u}{\partial r}\right) = -\frac{1}{r}\frac{\partial}{\partial r}(r\tau) - \frac{\partial P}{\partial x} \tag{7–14}
$$

Utilizing Newton's law of viscosity [Eq. (v) in Table 7–1] and assuming that $P$ is essentially a function of $x$ alone, the momentum equation is written as

$$
\rho\left(u\frac{\partial u}{\partial x} + v\frac{\partial u}{\partial r}\right) = \frac{1}{r}\frac{\partial}{\partial r}\left(\mu r\frac{\partial u}{\partial r}\right) - \frac{dP}{dx} \tag{7–15}
$$

or, for constant properties,

$$
u\frac{\partial u}{\partial x} + v\frac{\partial u}{\partial r} = \frac{\nu}{r}\frac{\partial}{\partial r}\left(r\frac{\partial u}{\partial r}\right) - \frac{1}{\rho}\frac{dP}{dx} \tag{7–16}
$$

The differential momentum equation is coupled with boundary conditions for $u$ of the form

$$
u = U_1 \qquad \text{at } x = 0 \tag{7–17}
$$

for a uniform distribution at the entrance,

$$
u = 0 \qquad \text{at } r = r_0 \tag{7–18}
$$

for no slip at the wall, and

$$
\frac{\partial u}{\partial r} = 0 \qquad \text{at } r = 0 \tag{7–19}
$$

because of symmetry.

For HFD flow, $\partial u/\partial x = 0$ and $v = 0$ (based on continuity) such that our differential formulation for momentum transfer in a uniform property fluid is given by

$$
\frac{\nu}{r}\frac{d}{dr}\left(r\frac{du}{dr}\right) - \frac{1}{\rho}\frac{dP}{dx} = 0 \tag{7–20}
$$

and Eqs. (7–18) and (7–19). Referring back to Eq. (7–13), we see that the rate of creation of momentum $RCM_x$ for flow with uniform density is equal to zero for HFD conditions, such that the momentum equation actually represents the force balance $\Sigma F_x = 0$.

## EXAMPLE 7–1

Show that the shear stress varies linearly with $r$ or $y$ for fully developed laminar flow in a circular tube.

### Solution

*Objective* Show that the relation for $\tau$ is linear for HFD flow in a circular tube.

*Schematic* Laminar HFD flow in a circular tube.

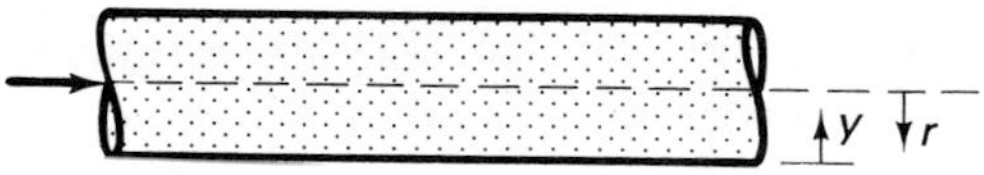


*Assumptions/Conditions*

forced convection
uniform properties
steady-state
Newtonian fluid
nonporous walls
negligible buoyancy and external body forces

*Analysis* To see the behavior of $\tau$ for laminar HFD flow in a circular tube, Eq. (7–20) is first put into the form [see Eq. (7–14)]

$$\frac{1}{r}\frac{d}{dr}(r\tau) + \frac{dP}{dx} = 0 \tag{a}$$

where

$$\tau = \mu \frac{du}{dy} = -\mu \frac{du}{dr}$$

for Newtonian fluids. Noting that $r\tau = 0$ at $r = 0$, Eq. (a) is integrated to obtain

$$r\tau = -\frac{r^2}{2}\frac{dP}{dx} \qquad \text{or} \qquad \tau = -\frac{r}{2}\frac{dP}{dx}$$

Setting $\tau = \tau_s = \tau_0$ at $r = r_0$ to eliminate $dP/dx$, the solution becomes

$$\frac{\tau}{\tau_0} = \frac{r}{r_0} = 1 - \frac{y}{r_0}$$

Thus, the distribution in $\tau$ is indeed linear.

***Energy Transfer*** In analyzing the energy transfer for this problem, we must account for significant components of the molecular conduction heat transfer and energy transported by fluid motion (i.e., advection). To simplify the analysis, we will neglect the effects of kinetic energy, viscous dissipation, body forces, axial conduction, and thermal radiation. (Kinetic energy and/or viscous dissipation are important for viscous fluids such as oil and in high-speed aerodynamic problems, the effects of axial conduction are sometimes significant for liquid metals, and thermal radiation must be accounted for in the analysis of systems involving participating gases and high temperatures.)

Applying the first law of thermodynamics to the control volume $dV$ shown in Fig. 7–4, we write

$$\Sigma \dot{E}_o - \Sigma \dot{E}_i + \cancel{\frac{\Delta E_s}{\Delta t}} = 0$$

$$\underbrace{d\dot{H}_{x+dx} + d\dot{H}_{r+dr} - d\dot{H}_x - d\dot{H}_r}_{\textit{advection}} + \underbrace{dq_{r+dr} - dq_r}_{\textit{molecular conduction}} = 0 \tag{7–21}$$

where $\Delta E_s/\Delta t = 0$ for steady-state conditions, $d\dot{H}_x = i\, d\dot{m}_x$, $d\dot{H}_r = i\, d\dot{m}_r$ and $i$ is the specific enthalpy.

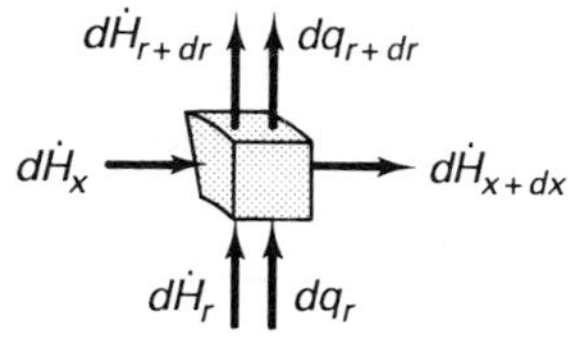


**FIGURE 7–4**
Conservation of energy relative to differential cylindrical control volume.

Utilizing the definition of the partial derivative, it follows that

$$\frac{\partial}{\partial x}(d\dot{H}_x)\, dx + \frac{\partial}{\partial r}(d\dot{H}_r)\, dr + \frac{\partial}{\partial r}(dq_r)\, dr = 0 \tag{7–22}$$

Introducing the Fourier law of conduction [Eq. (viii) in Table 7–1] and substituting for $d\dot{H}_x$ and $d\dot{H}_r$, this equation takes the form

$$\frac{\partial}{\partial x}(\rho u i) + \frac{1}{r}\frac{\partial}{\partial r}(\rho r v i) = \frac{1}{r}\frac{\partial}{\partial r}\left(rk\frac{\partial T}{\partial r}\right) \tag{7–23}$$

or, after making use of the continuity equation, Eq. (7–6),

$$\rho\left(u\frac{\partial i}{\partial x} + v\frac{\partial i}{\partial r}\right) = \frac{1}{r}\frac{\partial}{\partial r}\left(rk\frac{\partial T}{\partial r}\right) \tag{7–24}$$

With $i$ expressed in terms of $c_P$, we obtain

### *Solution*

To solve for the rate of convection heat transfer for laminar flow in a tube, the fluid flow and energy equations must be solved for the velocity and temperature distributions. Once these distributions are known, the friction factor and coefficient of heat transfer can be determined. For problems involving variable property developing flow, the more general continuity, momentum, and energy equations given by Eqs. (7–5), (7–15), and (7–25) must be solved simultaneously. For the somewhat simpler case of uniform property developing flow, Eqs. (7–6) and (7–16) are first solved for the velocity distribution, after which Eq. (7–26) is solved for the temperature profile. Numerical and approximate analytical solutions are available in the literature for these type problems in which the flow is developing [1–12].

Attention is now turned to the development of solutions for the simpler case involving uniform property fully developed laminar flow. An analytical solution is first developed for the fluid-flow aspects of the problem, after which predictions are developed for the temperature distribution and coefficient of heat transfer.

***Fluid Flow—HFD Region*** For HFD flow, $v = 0$ and the momentum equation and boundary conditions are given by

$$\frac{\nu}{r}\frac{d}{dr}\left(r\frac{du}{dr}\right) - \frac{1}{\rho}\frac{dP}{dx} = 0 \qquad (7\text{–}20)$$

and

$$u = 0 \quad \text{at } r = r_0 \qquad \frac{du}{dr} = 0 \quad \text{at } r = 0 \qquad (7\text{–}18,19)$$

To solve this problem, we first develop a solution for the dimensionless velocity distribution.

*Dimensionless Velocity Distribution* Separating the variables in Eq. (7–20) and integrating, we have

$$r\frac{du}{dr} = \frac{1}{\mu}\frac{dP}{dx}\frac{r^2}{2} + C_1 \qquad (7\text{–}31)$$

where $C_1$ is set equal to zero in accordance with Eq. (7–19). Separating the variables and integrating once again, the distribution in velocity $u$ becomes

$$u = \frac{1}{\mu}\frac{dP}{dx}\frac{r^2}{4} + C_2 \qquad (7\text{–}32)$$

Utilizing the no slip condition at the wall, $C_2$ is given by

$$C_2 = -\frac{1}{\mu}\frac{dP}{dx}\frac{r_0^2}{4} \qquad (7\text{–}33)$$

such that our solution for $u$ is

$$u = -\frac{1}{4\mu}\frac{dP}{dx}r_0^2\left[1 - \left(\frac{r}{r_0}\right)^2\right] \tag{7–34}$$

Alternatively, by setting $u$ equal to the centerline velocity $U_c$ at $r$ equal to zero, we have the somewhat more convenient expression,

$$\frac{u}{U_c} = 1 - \left(\frac{r}{r_0}\right)^2 \tag{7–35}$$

Although we now have a theoretical expression for the velocity profile, our solution is incomplete because $U_c$ and $dP/dx$ are as yet unknown.

*Bulk-Stream Velocity and Friction Factor* In order to express these unknown parameters in terms of the bulk-stream velocity $U_b$, which is equal to $U_1$ (see Fig. 7–1), we couple Eq. (7–35) with the defining equation for $U_b$, Eq. (6–34),

$$U_b = \frac{1}{A}\int_A u\, dA \tag{7–36}$$

as follows:

$$\begin{aligned} U_b &= \frac{1}{\pi r_0^2}\int_0^{r_0} u\, 2\pi r\, dr = \frac{2}{r_0^2}\int_0^{r_0} ur\, dr \\ &= \frac{2}{r_0^2}\int_0^{r_0} rU_c\left[1 - \left(\frac{r}{r_0}\right)^2\right] dr = \frac{U_c}{2} = -\frac{1}{8\mu}\frac{dP}{dx}r_0^2 \end{aligned} \tag{7–37}$$

The substitution of this result back into Eq. (7–35) gives

$$u = 2U_b\left[1 - \left(\frac{r}{r_0}\right)^2\right] \tag{7–38}$$

where $U_b = U_1$. This expression is shown in Fig. 7–5 to agree very well with experimental data.

Now that the velocity distribution is fully specified, Newton's law of viscous shear is used to obtain an expression for the wall shear stress $\tau_s = \tau_0$.

$$\tau_0 = \mu\frac{du}{dy}\bigg|_0 = -\mu\frac{du}{dr}\bigg|_{r_0} = \mu 2U_b\left(\frac{2r}{r_0^2}\right)\bigg|_{r_0} = 4\frac{\mu U_b}{r_0} \tag{7–39}$$

It follows that the Fanning friction factor $f$ becomes

$$f = \frac{\tau_0}{\rho U_b^2/2} = \frac{8\mu U_b}{\rho U_b^2 r_0} = 16\frac{\nu}{DU_b} = \frac{16}{Re} \tag{7–40}$$

This equation is compared with experimental data in Fig. 7–6. The agreement between theory and experiment is exceptional for Reynolds numbers below the transitional value of approximately 2000.

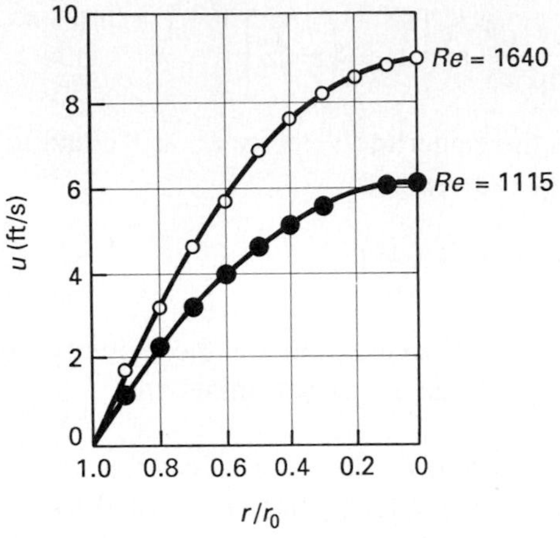


**FIGURE 7–5**
Comparison of Eq. (7–38) with experimental data for $u$—HFD laminar flow in circular tube. (Data from Senecal [13] for flow of air in a 0.75-in.-I.D. tube.)

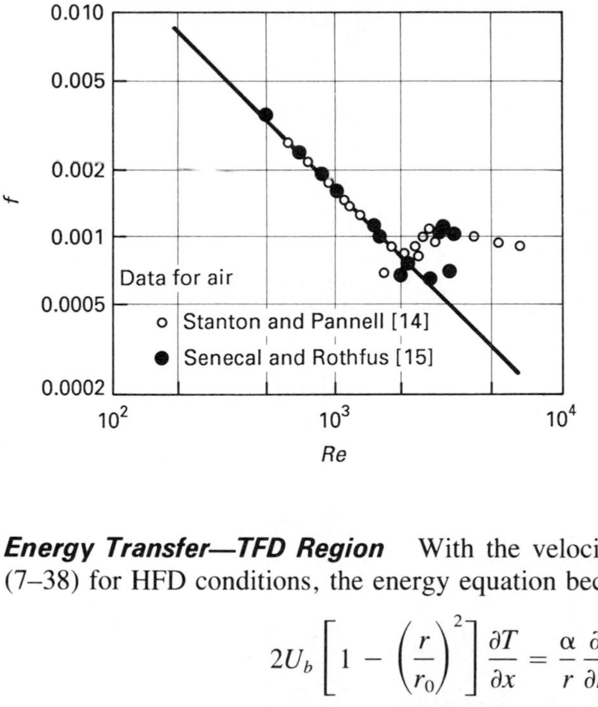


**FIGURE 7–6**
Comparison of Eq. (7–40) with experimental data for $f$—HFD laminar flow in circular tubes.

***Energy Transfer—TFD Region*** With the velocity distribution specified by Eq. (7–38) for HFD conditions, the energy equation becomes

$$2U_b\left[1-\left(\frac{r}{r_0}\right)^2\right]\frac{\partial T}{\partial x}=\frac{\alpha}{r}\frac{\partial}{\partial r}\left(r\frac{\partial T}{\partial r}\right) \tag{7–41}$$

For the TFD region, the term $\partial T/\partial x$ can be specified in terms of $T$, $T_s$, and $T_b$ by utilizing the defining equation for TFD conditions [Eq. (6–47)],

$$\frac{\partial}{\partial x}\left(\frac{T-T_s}{T_b-T_s}\right)=0 \tag{7–2}$$

and by recognizing that the local coefficient of heat transfer $h$ is independent of $x$. For a uniform wall-heat-flux boundary condition, it follows from the general Newton law of cooling,

$$q_c'' = q_0'' = h(T_s - T_b) \tag{7–42}$$

that $T_s - T_b$ is independent of $x$. Thus,

$$\frac{dT_b}{dx} = \frac{dT_s}{dx} \tag{7–43}$$

such that Eq. (7–2) gives

$$\frac{\partial T}{\partial x} = \frac{dT_s}{dx} = \frac{dT_b}{dx} \tag{7–44}$$

Hence, the energy equation takes the one-dimensional form

$$2U_b\left[1 - \left(\frac{r}{r_0}\right)^2\right]\frac{dT_b}{dx} = \frac{\alpha}{r}\frac{d}{dr}\left(r\frac{dT}{dr}\right) \tag{7–45}$$

for uniform wall-flux heating.

Referring to the simple lumped analysis developed in Example 7–2, $dT_b/dx$ is expressed in terms of $q_0''$ by

$$\frac{dT_b}{dx} = \frac{q_0''p}{\dot{m}c_P} = \frac{4q_0''}{\rho c_P U_b D} \tag{7–46}$$

Substituting this relation into Eq. (7–44), the energy equation takes the form

$$\frac{4q_0''}{k}\left(\frac{r}{r_0}\right)\left[1 - \left(\frac{r}{r_0}\right)^2\right] = \frac{d}{dr}\left(r\frac{dT}{dr}\right) \tag{7–47}$$

This equation is coupled with the $r$ boundary conditions

$$\frac{\partial T}{\partial r} = 0 \qquad \text{at } r = 0 \tag{7–28}$$

$$k\frac{\partial T}{\partial r} = q_0'' \qquad \text{at } r = r_0 \tag{7–29}$$

Following the pattern established in the solution of the fluid-flow problem, we first develop a solution for the dimensionless temperature distribution.

*Dimensionless Temperature Distribution* A first integration of Eq. (7–47) from 0 to $r$ and the use of Eq. (7–28) gives

$$r\frac{dT}{dr} = \frac{4q_0''}{r_0 k}\left(\frac{r^2}{2} - \frac{r^4}{4r_0^2}\right) \tag{7–48}$$

[This same result is obtained by integrating from $r_0$ to $r$ with the condition at the wall prescribed by Eq. (7–29).] A second integration yields

$$T = \frac{4q_0''}{r_0 k}\left(\frac{r^2}{4} - \frac{r^4}{16r_0^2}\right) + C_1 \tag{7–49}$$

As just indicated, both boundary conditions satisfy Eq. (7–48), but neither can be utilized to obtain $C_1$ in Eq. (7–49). We circumvent this anomaly by setting $T$ equal to $T_s$ at $r = r_0$. However, it is important to note that $T_s$ is still an unknown function of $x$ that eventually must be evaluated in terms of the specified input $q_0''$. The use of this intermediate step gives

$$C_1 = T_s - \frac{4q_0''}{r_0 k}\frac{3}{16}r_0^2 \tag{7–50}$$

such that the temperature profile becomes

$$T - T_s = -\frac{q_0'' r_0}{k}\left[\frac{3}{4} - \left(\frac{r}{r_0}\right)^2 + \frac{1}{4}\left(\frac{r}{r_0}\right)^4\right] \tag{7–51}$$

To put this expression into a more convenient dimensionless format, $T$ is set equal to $T_c$ at $r$ equal to zero, with the result

$$\frac{T - T_s}{T_c - T_s} = 1 - \frac{4}{3}\left(\frac{r}{r_0}\right)^2 + \frac{1}{3}\left(\frac{r}{r_0}\right)^4 \tag{7–52}$$

where $T_c - T_s = -\frac{3}{4}q_0'' r_0/k$. Whereas the HFD velocity distribution is a second-order polynomial, the temperature distribution is seen to be fourth order.

***Bulk-Stream Temperature and Nusselt Number*** In order to express $T_s$ in terms of $T_b$, we utilize the defining expression for $T_b$ given by Eq. (6–46). This equation reduces to the following form for a circular-tube geometry:

$$T_b = \frac{2}{r_0^2 U_b}\int_0^{r_0} uTr\,dr \tag{7–53}$$

The substitution of the expressions for $u$ and $T$ given by Eqs. (7–38) and (7–51) into this equation results in

$$T_b = T_s - \frac{11}{24}\frac{q_0'' r_0}{k} \tag{7–54}$$

Referring to Example 7–2, because $T_b$ is also given by Eq. (E7–2e), which is based on the simple lumped analysis approach, Eq. (7–54) specifies $T_s$.

Eliminating $T_s$ in Eq. (7–51) by the use of Eq. (7–54), we have a final expression for the temperature distribution, which can be written as

$$T - T_b = -\frac{q_0'' r_0}{k}\left[\frac{7}{24} - \left(\frac{r}{r_0}\right)^2 + \frac{1}{4}\left(\frac{r}{r_0}\right)^4\right] \tag{7–55}$$

To obtain an expression for the Nusselt number $Nu$, we merely rearrange Eq. (7–54), with the result

$$h = \frac{q_0''}{T_s - T_b} = \frac{24}{11}\frac{k}{r_0} \tag{7–56}$$

or

$$Nu = \frac{hD}{k} = \frac{48}{11} = 4.36 \tag{7–57}$$

for uniform wall-heat flux. This equation is consistent with the limiting calculations developed by Sellars et al. [5] for thermal developing flow.

Referring to Example 7–3, we also note that $Nu = 3.66$ for the case of fully developed laminar flow in a circular tube with *uniform wall temperature*. The fact that the solution for uniform wall temperature is 16% lower than the value given by Eq. (7–57) for uniform wall-heat flux indicates the significance of the form of the thermal boundary condition for laminar flow.

As mentioned earlier, the effects of viscous dissipation and kinetic energy should be taken into account for applications involving high-speed flow or viscous fluids. To quantify this point, we note that the relative significance of viscous dissipation and kinetic energy increases with increasing values of the product $Pr\ Ec$, where $Pr$ is the Prandtl number and $Ec = U_F^2/[c_P(T_s - T_F)]$ is the *Eckert number*. Viscous dissipation and kinetic energy effects can be safely neglected for $Pr\ Ec \ll 1$, but should be accounted for when this criterion is not satisfied. Notice that the Eckert number indicates *the significance of the kinetic energy of the flow relative to the enthalpy difference across the boundary layer*.

It should also be noted that the effects of axial conduction generally can be neglected for situations in which the *Peclet number* $Pe = Re\ Pr$ is not small; that is for $Pe \gtrsim 100$. The Peclet number represents the *ratio of enthalpy flow rate to heat conduction rate*. Since $Pe$ is proportional to the Prandtl number $Pr$, we should not be surprised to find that problems in which axial conduction is significant often involve liquid metals.

## EXAMPLE 7–3

Write the mathematical formulation for heat transfer associated with hydrodynamic fully developed laminar flow in a circular tube with uniform wall temperature.

### Solution

*Objective* Write the energy equation and boundary conditions for HFD laminar flow in a circular tube with $T_s = T_0$.

*Schematic* HFD laminar flow in a circular tube—uniform wall temperature $T_0$.

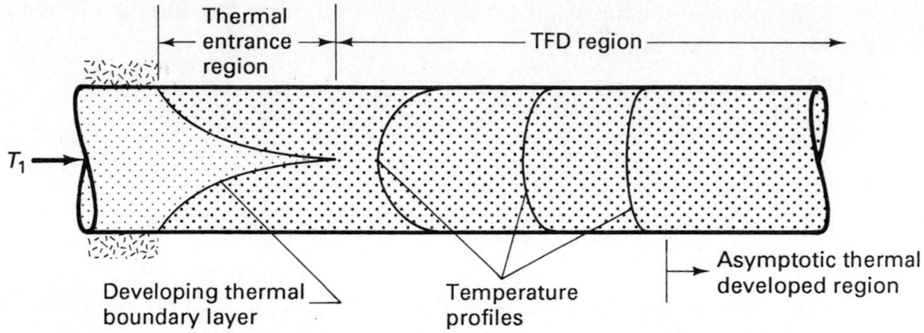
Thermal
entrance
region
TFD region
$T_1$
Developing thermal
boundary layer
Temperature
profiles
Asymptotic thermal
developed region

*Assumptions/Conditions*

forced convection

uniform properties

standard conditions (i.e., steady-state; two-dimensional; ideal, Newtonian, isotropic and single-phase fluid; incompressible flow; no interfacial mass transfer; and negligible buoyancy, kinetic energy, viscous dissipation, external body forces, axial conduction, and thermal radiation)

*Analysis* The differential formulation for energy transfer in the thermal entrance region is represented by Eq. (7–30),

$$u \frac{\partial T}{\partial x} = \frac{\alpha}{r} \frac{\partial}{\partial r}\left(r \frac{\partial T}{\partial r}\right)$$

or, with $u$ given by Eq. (7–38), by

$$2U_b \left[1 - \left(\frac{r}{r_0}\right)^2\right] \frac{\partial T}{\partial x} = \frac{\alpha}{r} \frac{\partial}{\partial r}\left(r \frac{\partial T}{\partial r}\right) \tag{a}$$

with boundary conditions of the form

$$T = T_0 \qquad \text{at } r = r_0 \qquad \frac{\partial T}{\partial x} = 0 \qquad \text{at } r = 0$$

The temperature gradient $\partial T/\partial x$ for TFD flow is expressed in terms of $T$, $T_0$, and $T_b$ by expanding Eq. (7–2); that is

$$\frac{\partial}{\partial x}\left(\frac{T - T_0}{T_b - T_0}\right) = \frac{T_b - T_0}{(T_b - T_0)^2}\left(\frac{\partial T}{\partial x} - \frac{dT_0}{dx}\right) - \frac{T - T_0}{(T_b - T_0)^2}\left(\frac{dT_b}{dx} - \frac{dT_0}{dx}\right) = 0$$

which reduces to

$$\frac{\partial T}{\partial x} = \frac{T - T_0}{T_b - T_0}\frac{dT_b}{dx}$$

Using this relation, Eq. (a) becomes

$$2U_b\left[1 - \left(\frac{r}{r_0}\right)^2\right]\frac{dT_b}{dx}\left(\frac{T - T_0}{T_b - T_0}\right) = \frac{\alpha}{r}\frac{\partial}{\partial r}\left(r\frac{\partial T}{\partial r}\right) \tag{b}$$

Although Eq. (b) does not lend itself to simple analytical treatment, solutions have been developed by an iterative method involving the use of successive approximations for the temperature distribution [11]. The resulting solution for Nusselt number $Nu$ is given by

$$Nu = 3.66$$

for uniform wall temperature conditions.

It should be noted that for this uniform wall-temperature problem the heat flux approaches zero and the bulk-stream temperature $T_b$ approaches $T_0$ as $x$ increases. This region in which $T_b$ no longer changes with $x$ is characterized by $\partial T/\partial x = 0$ and is referred to as *asymptotic thermal developed flow*. Asymptotic thermal developed flows of practical importance commonly occur in annular and Couette flow systems.

## EXAMPLE 7–4

Develop relations for the friction factor, Nusselt number, and distributions in $T_b$ and $T_s$ for fully developed laminar flow between parallel plates with uniform wall-flux heating.

### Solution

*Objective* Develop relations for $f$, $Nu$, $T_b$, and $T_s$ for TFD laminar flow between parallel plates with $q''_c = q''_0$.

*Schematic* TFD laminar flow between parallel plates with uniform wall-heat flux.

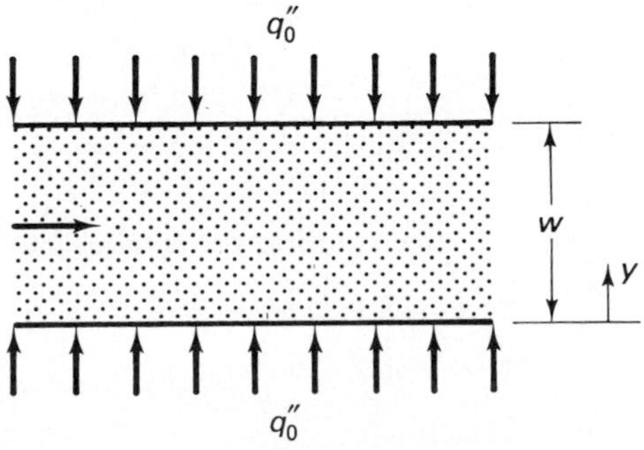

*Assumptions/Conditions*

forced convection
uniform properties
standard conditions

*Analysis* The mathematical formulation for TFD flow between parallel plates is represented by

$$\nu \frac{d^2u}{dy^2} - \frac{1}{\rho}\frac{dP}{dx} = 0 \tag{a}$$

and

$$\rho c_P u \frac{\partial T}{\partial x} = k \frac{\partial^2 T}{\partial y^2} \tag{b}$$

for negligible viscous dissipation, kinetic energy, and other standard conditions, where

$$u = 0 \qquad \text{at } y = 0 \qquad \frac{du}{dy} = 0 \qquad \text{at } y = \frac{w}{2} \tag{c,d}$$

$$-k\frac{\partial T}{\partial y} = q_0'' \qquad \text{at } y = 0 \qquad \frac{\partial T}{\partial y} = 0 \qquad \text{at } y = \frac{w}{2} \tag{e,f}$$

and

$$\frac{\partial T}{\partial x} = \frac{dT_s}{dx} = \frac{dT_b}{dx} \tag{g}$$

for uniform wall-heat flux. Using the lumped/differential formulation approach introduced in Example 7–2, we are also able to write

$$\frac{dT_b}{dx} = \frac{q_0''p}{\dot{m}c_P} = \frac{q_0''2Z}{\rho c_P U_b Z w} = \frac{2q_0''}{\rho c_P U_b w} \tag{h}$$

The velocity distribution is readily obtained by integrating Eq. (a) twice; that is,

$$\frac{du}{dy} = \frac{1}{\mu}\frac{dP}{dx}y + C_1$$

$$u = \frac{1}{\mu}\frac{dP}{dx}\frac{y^2}{2} + C_1 y + C_2$$

Using Eqs. (c) and (d) to evaluate $C_1$ and $C_2$, our solution for $u$ takes the form

$$u = \frac{1}{\mu}\frac{dP}{dx}\left(\frac{y^2}{2} - \frac{yw}{2}\right)$$

To determine the bulk-stream velocity, we write

$$U_b = \frac{1}{w}\int_0^w u\,dy = \frac{1}{w}\int_0^w \frac{1}{\mu}\frac{dP}{dx}\left(\frac{y^2}{2} - \frac{yw}{2}\right)dy$$

$$= \frac{1}{w\mu}\frac{dP}{dx}\left(\frac{y^3}{6} - \frac{y^2 w}{4}\right)\Bigg|_0^w = -\frac{w^2}{12}\frac{1}{w}\frac{dP}{dx}$$

Using this result to express $dP/dx$ in terms of $U_b$, the velocity distribution becomes

$$u = 6U_b\left[\frac{y}{w} - \left(\frac{y}{w}\right)^2\right] \tag{i}$$

Expressions are obtained for the wall-shear stress $\tau_0$ and friction factor $f$ by writing

$$\tau_0 = \mu\frac{du}{dy}\bigg|_0 = \mu 6U_b\left[\frac{1}{w} - 2\frac{y}{w^2}\right]\bigg|_0 = \frac{6\mu U_b}{w} = \rho U_b^2\frac{f}{2}$$

and

$$f = 12\frac{\nu}{wU_b}$$

The hydraulic diameter $D_H$ for flow between parallel plates is given by

$$D_H = \frac{4A}{p} = \frac{4wZ}{2Z} = 2w$$

Thus, the friction factor is expressed in terms of Reynolds number $Re$ by

$$f = 24\frac{\nu}{D_H U_b} = \frac{24}{Re}$$

Substituting Eq. (i) for $u$ into Eq. (b) and using Eqs. (g) and (h) to represent $\partial T/\partial x$, the energy equation for TFD flow with uniform wall-heat flux reduces to a simple second-order differential equation of the form

$$\frac{d^2T}{dy^2} = \frac{12}{w}\frac{q_0''}{k}\left[\frac{y}{w} - \left(\frac{y}{w}\right)^2\right]$$

This equation is integrated twice to obtain

$$\frac{dT}{dy} = \frac{12}{w}\frac{q_0''}{k}\left(\frac{y^2}{2w} - \frac{y^3}{3w^2}\right) + C_1$$

$$T = \frac{12}{w}\frac{q_0''}{k}\left(\frac{y^3}{6w} - \frac{y^4}{12w^2}\right) + C_1 y + C_2$$

Using Eq. (f) to evaluate $C_1$ and representing the unspecified wall temperature at $y = 0$ by $T_s$, the solution for $T$ is put into the form

$$T - T_s = \frac{q_0''w}{k}\left[-\frac{y}{w} + 2\left(\frac{y}{w}\right)^3 - \left(\frac{y}{w}\right)^4\right]$$

This relation is combined with Eq. (i) for $u$ and the defining relation for bulk-stream temperature $T_b$, Eq. (6–46), to obtain

$$T_b = \frac{1}{U_b w}\int_0^w uT\,dy = \frac{1}{U_b w}\int_0^w 6U_b\left[\frac{y}{w} - \left(\frac{y}{w}\right)^2\right]$$
$$\times\left\{T_s + \frac{q_0''w}{k}\left[2\left(\frac{y}{w}\right)^3 - \left(\frac{y}{w}\right)^4 - \frac{y}{w}\right]\right\}dy = T_s - \frac{17}{70}\frac{q_0''w}{k}$$

Rearranging, the Nusselt number $Nu$ becomes

$$Nu = \frac{hD_H}{k} = \frac{q_0''}{T_s - T_b}\frac{2w}{k} = 2\,\frac{70}{17} = 8.24$$

To express $T_b$ in terms of $x$, we simply integrate Eq. (h), with $T_b$ set equal to $T_1$ at $x = 0$, with the result

$$T_b = T_1 + \frac{2q_0''}{\rho c_P U_b}\frac{x}{w} \tag{j}$$

Using this relation, the wall temperature $T_s$ becomes

$$T_s = T_b + \frac{17}{70}\frac{q_0''w}{k} = T_1 + \frac{2q_0''}{\rho c_P U_b}\frac{x}{w} + \frac{17}{70}\frac{q_0''w}{k} \tag{k}$$

in the thermal fully developed region.

As we have noted, this solution is restricted to applications for which viscous dissipation and kinetic energy are negligible. The analysis is generalized to account for these effects in Appendix I and Example 7–5.

**EXAMPLE 7–5**

Develop relations for the Nusselt number $Nu$ and distributions in $T_b$ and $T_s$ for thermal fully developed laminar flow of a highly viscous fluid between parallel plates with uniform wall-flux heating.

**Solution**

*Objective* Develop relations for $Nu$, $T_b$, and $T_s$ for TFD laminar flow of a viscous fluid between parallel plates with $q''_c = q''_0$.

*Schematic* TFD laminar flow between parallel plates with uniform wall-heat flux—highly viscous fluid.

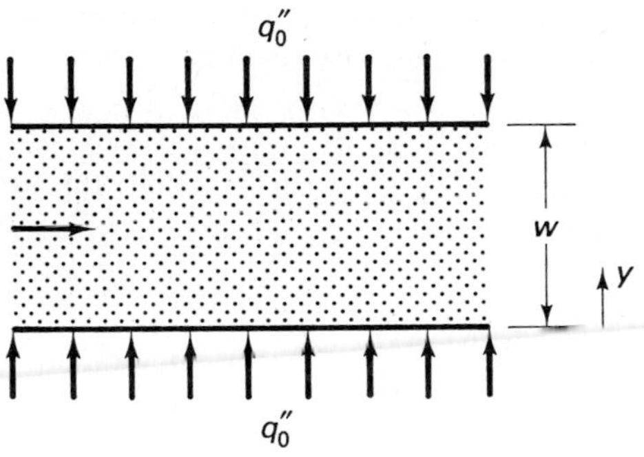


*Assumptions/Conditions*

forced convection
uniform properties
standard conditions, except for significant viscous dissipation effects

*Analysis* As shown in Appendix I, the energy equation for fully developed laminar flow between parallel plates takes the form

$$\rho c_P u \frac{\partial T}{\partial x} = k \frac{\partial^2 T}{\partial y^2} + \mu \left(\frac{du}{dy}\right)^2 \tag{a}$$

where the term $\mu(du/dy)^2$ represents the *energy dissipation* per unit volume, the velocity $u$ is given by Eq. (E7–4i),

$$u = 6U_b \left[\frac{y}{w} - \left(\frac{y}{w}\right)^2\right]$$

and $\partial T/\partial x = dT_b/dx$ for uniform wall-heat flux.

Referring to Fig. E7–5, the lumped/differential formulation for this case is given by

$$\dot{H}_b|_{x+dx} - \dot{H}_b|_x - q''_0 p\, dx - d\dot{E}_\tau = 0$$

or

$$q''_0 p\, dx + d\dot{E}_\tau = \frac{d\dot{H}_b}{dx} dx = \dot{m} c_P \frac{dT_b}{dx} dx \tag{b}$$

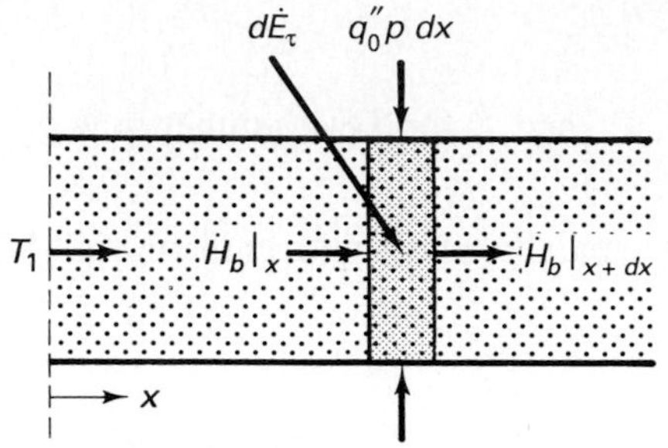


**FIGURE E7–5**
Lumped/differential formulation for energy transfer, including viscous dissipation.

where $d\dot{E}_\tau$ represents the rate of viscous energy dissipation within the lumped/differential element. With $d\dot{E}_\tau$ expressed in terms of the energy dissipation per unit volume $\mu(du/dy)^2$, we obtain

$$d\dot{E}_\tau = \int \mu \left(\frac{du}{dy}\right)^2 dV = Z\,dx \int_0^w 36\,\frac{\mu U_b^2}{w^2}\left[1 - 4\,\frac{y}{w} + 4\left(\frac{y}{w}\right)^2\right] dy$$

$$= 12\,\frac{\mu U_b^2}{w}\,Z\,dx$$

Substituting for $d\dot{E}_\tau$, $\dot{m}$, and the perimeter $p$, Eq. (b) becomes

$$2\,q_0'' + 12\,\frac{\mu U_b^2}{w} = \rho c_P U_b w\,\frac{dT_b}{dx}$$

or

$$\frac{dT_b}{dx} = \frac{1}{\rho c_P U_b w}\left(2q_0'' + 12\,\frac{\mu U_b^2}{w}\right) \tag{c}$$

Combining these results with Eq. (a), we have

$$\frac{6}{w}\left[\frac{y}{w} - \left(\frac{y}{w}\right)^2\right]\left(2q_0'' + 12\,\frac{\mu U_b^2}{w}\right)$$

$$= k\,\frac{d^2T}{dy^2} + 36\,\frac{\mu U_b^2}{w^2}\left[1 - 4\,\frac{y}{w} + 4\left(\frac{y}{w}\right)^2\right]$$

or

$$\frac{d^2T}{dy^2} = \frac{12}{w}\,\frac{q_0''}{k}\left[\frac{y}{w} - \left(\frac{y}{w}\right)^2\right] + 36\,\frac{\mu U_b^2}{kw^2}\left[-1 + 6\,\frac{y}{w} - 6\left(\frac{y}{w}\right)^2\right]$$

with the accompanying boundary conditions given by Eqs. (E7–4e) and (E7–4f). Following the approach used in Example 7–4, the solution for $T$ is given by

$$T - T_s = \frac{q_0''w}{k}\left[-\frac{y}{w} + 2\left(\frac{y}{w}\right)^3 - \left(\frac{y}{w}\right)^4\right]$$
$$+ \frac{\mu U_b^2}{k}\left[-18\left(\frac{y}{w}\right)^2 + 36\left(\frac{y}{w}\right)^3 - 18\left(\frac{y}{w}\right)^4\right]$$

Introducing the defining relation for bulk-stream temperature $T_b$, we obtain

$$T_b = \frac{1}{U_b w}\int_0^w uT\,dy = T_s - \frac{17}{70}\frac{q_0''w}{k} - \frac{54}{70}\frac{\mu U_b^2}{k}$$

or

$$T_s - T_b = \frac{17}{70}\frac{q_0''w}{k} + 0.771\frac{\mu U_b^2}{k} \tag{d}$$

and

$$Nu = \frac{q_0''}{T_s - T_b}\frac{2w}{k} = \frac{140}{17}\left[1 - 0.771\frac{\mu U_b^2}{k(T_s - T_b)}\right]$$
$$= 8.24\,(1 - 0.771\,Pr\,Ec)$$

where $Ec = U_b^2/[c_P(T_s - T_b)]$ is the Eckert number. Notice that the effect of viscous dissipation is to increase the temperature difference $T_s - T_b$ and to decrease $Nu$, with the effect on $Nu$ being less than 1% for this application when $Pr\,Ec <$ 0.013. For a representative temperature difference $T_s - T_b$ equal to 25°C, this criterion is satisfied for air [$Pr = 0.70$, $c_P = 1.01$ kJ/(kg °C)] with $U_b < 21.6$ m/s, for water [$Pr = 5.0$, $c_P = 4.18$ kJ/(kg °C)] with $U_b < 16.5$ m/s, for oil [$Pr = 100$, $c_P =$ 2.4 kJ/(kg °C)] with $U_b < 2.79$ m/s, and for oil [$Pr = 1000$, $c_P = 2.0$ kJ/(kg °C)] with $U_b < 0.805$ m/s. Thus, we would expect the effects of viscous dissipation to be negligible for many applications involving the flow of air or water, but often to be significant for flow of viscous fluids such as oil.

To complete the analysis, we integrate Eq. (c) with $T_b$ set equal to $T_1$ at $x =$ 0, with the result

$$T_b = T_1 + \left(2q_0'' + 12\frac{\mu U_b^2}{w}\right)\frac{x}{\rho c_P U_b w} \tag{e}$$

This result is combined with Eq. (d) to obtain a relation for $T_s$ as a function of $x$ of the form

$$T_s = T_1 + \left(2q_0'' + 12\frac{\mu U_b^2}{w}\right)\frac{x}{\rho c_P U_b w} + \frac{17}{70}\frac{q_0''w}{k} + 0.771\frac{\mu U_b^2}{k} \tag{f}$$

which is applicable to the thermal fully developed region. Notice that Eqs. (e) and (f) reduce to Eqs. (E7–4j) and (E7–4k) for negligible viscous dissipation.

## 7-2-2 Boundary Layer Flow over Plane Surfaces

We now turn our attention to the problem illustrated in Fig. 7-7 of convection heat transfer associated with steady two-dimensional laminar boundary layer flow over a plane surface that is heated or cooled in the region $x \geqslant x_0$. We have already seen in Chap. 6 that hydrodynamic and thermal boundary layers develop over the surface, with $\delta$ and $\Delta$ continually increasing with $x$.

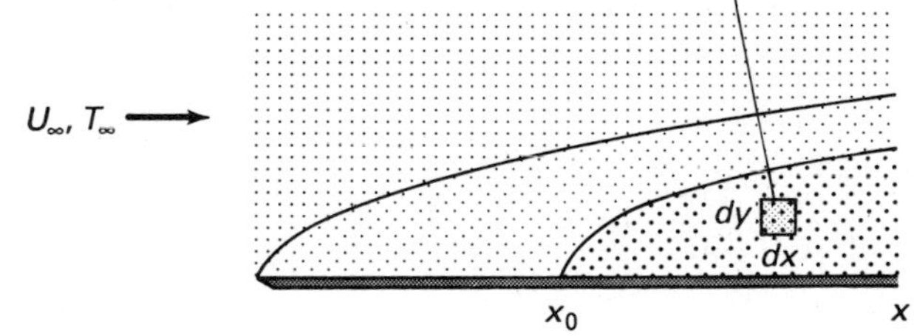


**FIGURE 7-7** Hydrodynamic and thermal boundary layers for flow over a flat plate with heating in the region $x \geq x_0$.

Whereas thc free-stream temperature $T_\infty$ is independent of $x$ for boundary layer flow applications such as this, the free-stream velocity $U_\infty$ may be uniform or may be a function of $x$, depending on the arrangement. For the case of parallel flow over a flat plate such as is shown in Fig. 7-7, $U_\infty$ is uniform. On the other hand, $U_\infty$ follows the power law $U_\infty = Cx^m$ for flow over the wedge-shape body shown in Fig. 7-8. A large number of nonuniform distributions are encountered in practice.

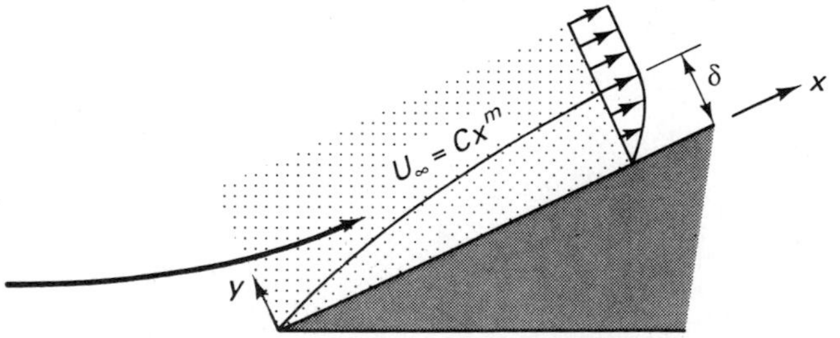


**FIGURE 7-8** Flow over a wedge-shape body.

### *Differential Formulation*

Following the approach developed in the preceding section, the application of the principles of conservation pertaining to mass, momentum, and energy to the differential rectangular control volume shown in Fig. 7-7 leads to a system of boundary layer equations for uniform properties given by

$$\frac{\partial u}{\partial x} + \frac{\partial v}{\partial y} = 0 \tag{7–58}$$

$$u\frac{\partial u}{\partial x} + v\frac{\partial u}{\partial y} = \nu\frac{\partial^2 u}{\partial y^2} - \frac{1}{\rho}\frac{dP}{dx} \tag{7–59}$$

for negligible gravitational and other body force effects,

$$u\frac{\partial T}{\partial x} + v\frac{\partial T}{\partial y} = \alpha\frac{\partial^2 T}{\partial y^2} \tag{7–60}$$

for negligible kinetic energy, viscous dissipation, body forces, axial conduction, and thermal radiation effects. These equations are similar to Eqs. (7–6), (7–16), and (7–26). The seven accompanying boundary conditions are

$$u = U_\infty \quad \text{at } x = 0 \qquad u = U_\infty \quad \text{as } y \rightarrow \infty \tag{7–61,62}$$

$$u = 0 \quad \text{at } y = 0 \qquad v = 0 \quad \text{at } y = 0 \tag{7–63,64}$$

$$T = T_\infty \quad \text{at } x = x_0 \qquad T = T_\infty \quad \text{as } y \rightarrow \infty \tag{7–65,66}$$

and

$$T = T_s \qquad \text{at } y = 0 \tag{7–67a}$$

for specified wall temperature, or

$$-k\frac{\partial T}{\partial y} = q_c'' \qquad \text{at } y = 0 \tag{7–67b}$$

for specified wall-heat flux. By satisfying Eq. (7–59) in the region outside the boundary layer, we are able to express the pressure gradient in terms of the free-stream velocity $U_\infty$ by

$$-\frac{dP}{dx} = \rho U_\infty \frac{dU_\infty}{dx} \tag{7–68}$$

It should also be noted that $v \neq 0$ at $y = 0$ for flow over a porous plate with blowing or suction. The important problem of transpired boundary layer flow is addressed in references 11 and 16–18. The differential formulation for the more general case involving high-speed flow is considered in Appendix I.

Whereas the differential formulation for boundary layer flow generally requires the use of somewhat involved numerical solution techniques, an alternative integral formulation perspective can be used that often lends itself to simpler solution techniques.

### *Integral Formulation*

The integral formulation can be developed by either of two methods. The *indirect method* involves the integration of the differential equations. This approach will be introduced in the context of natural convection flow in Sec. 7–2–3. In the *direct approach* to the development of the integral equations, we apply the fundamental laws pertaining to mass, momentum, and energy transfer to the lumped differential fluid volumes $\delta\ dx\ dz$ and $\Delta\ dx\ dz$. The direct approach is now used to develop the integral formulation for laminar boundary layer flow over plane surfaces.

***Continuity*** Applying the principle of conservation of mass to the control volume $\delta\ dx\ dz$ shown in Fig. 7–9, we obtain

$$\Sigma\, \dot{m}_i = \Sigma\, \dot{m}_o + \cancel{\frac{\Delta m_s}{\Delta t}}$$

$$\dot{m}_x + d\dot{m}_\delta = \dot{m}_{x+dx} \tag{7–69}$$

where $\Delta m_s/\Delta t = 0$ for steady flow, and the mass flow through the wall is zero $[v(x,0) = 0]$; $d\dot{m}_\delta$ is the rate of mass transfer from the free stream into the control volume. The total mass flow rate $\dot{m}_x$ is related to $d\dot{m}_x$ ($= \rho u\ dz\ dy$) by

$$\dot{m}_x = \int d\dot{m}_x = \int_0^\delta (\rho u\ dz)\ dy \tag{7–70}$$

By utilizing the definition of the derivative, Eq. (7–69) reduces to

$$d\dot{m}_\delta = \frac{d\dot{m}_x}{dx}\,dx = \left(\frac{d}{dx}\int d\dot{m}_x\right) dx = \left[\frac{d}{dx}\int_0^\delta (\rho u\ dz)\ dy\right] dx \tag{7–71}$$

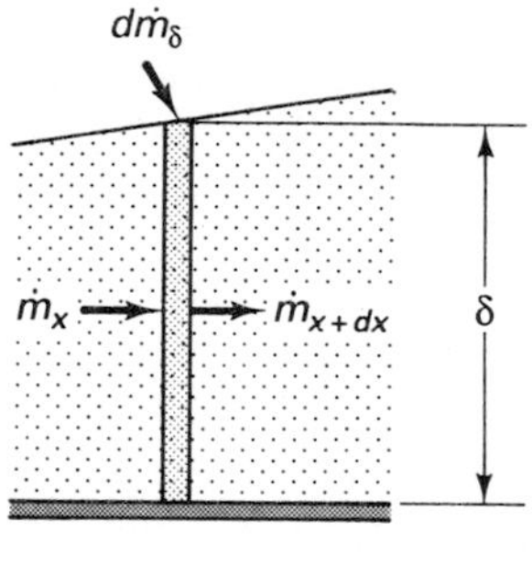


**FIGURE 7–9**
Conservation of mass relative to lumped differential rectangular control volume; $dV = \delta\ dx\ dz$.

Similarly, the differential mass flow rate $d\dot{m}_\Delta$ from the free stream into the thermal boundary layer volume $\Delta\ dx\ dz$ can be represented by

$$d\dot{m}_\Delta = \left(\frac{d}{dx}\int d\dot{m}_x\right) dx = \left[\frac{d}{dx}\int_0^\Delta (\rho u\ dz)\ dy\right] dx \tag{7–72}$$

***Momentum Transfer*** Applying the momentum principle to the lumped differential volume $\delta\ dx\ dz$ shown in Fig. 7–10, we obtain

$$RCM_x = \Sigma\, F_x \tag{7–73}$$

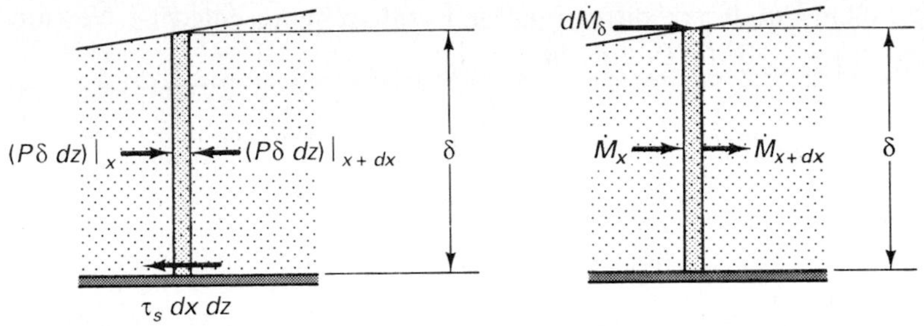


(a) Forces. (b) Momentum transfer.

**FIGURE 7–10** Momentum principle ($x$-direction) relative to lumped differential rectangular control volume; $dV = \delta\, dx\, dz$.

where

$$\Sigma F_x = -\tau_s\, dx\, dz + (P\delta\, dz)|_x - (P\delta\, dz)|_{x+dx} \tag{7–74}$$

and

$$RCM_x = \dot{M}_o - \dot{M}_i + \frac{\Delta M_s}{\Delta t} = \dot{M}_{x+dx} - \dot{M}_x - d\dot{M}_\delta \tag{7–75}$$

with $\Delta M_s/\Delta t = 0$ for steady flow, $d\dot{M}_\delta = U_\infty\, d\dot{m}_\delta$, and

$$\dot{M}_x = \int u\, d\dot{m}_x = \int_0^\delta u(\rho u\, dz)\, dy \tag{7–76}$$

Utilizing the definition of the derivative, Eq. (7–73) reduces to

$$\frac{d\dot{M}_x}{dx}\, dx - d\dot{M}_\delta = -\tau_s\, dx\, dz - \frac{dP}{dx}\,\delta\, dx\, dz \tag{7–77}$$

The substitution of Eq. (7–76) and the defining relationship for $d\dot{M}_\delta$ into this expression gives

$$\left(\frac{d}{dx}\int u\, d\dot{m}_x\right) dx - U_\infty\, d\dot{m}_\delta = -\tau_s\, dx\, dz - \frac{dP}{dx}\,\delta\, dx\, dz \tag{7–78}$$

Substituting for $d\dot{m}_x$ and $d\dot{m}_\delta$, Eq. (7–78) becomes

$$\frac{d}{dx}\int_0^\delta u^2\, dy - U_\infty \frac{d}{dx}\int_0^\delta u\, dy = -\frac{\tau_s}{\rho} - \frac{\delta}{\rho}\frac{dP}{dx} \tag{7–79}$$

for uniform property flow. Using Eq. (7–68) to eliminate $-dP/dx$ in favor of $\rho U_\infty\, dU_\infty/dx$ and rearranging, we obtain

$$U_\infty \frac{d}{dx}\int_0^\delta u\, dy - \frac{d}{dx}\int_0^\delta u^2\, dy + \delta U_\infty \frac{dU_\infty}{dx} = \frac{\tau_s}{\rho} \tag{7–80}$$

With $U_\infty$ taken inside the differential and with $\tau_s$ represented by Newton's law of viscosity, Eq. (7–80) takes the form

$$\frac{d}{dx}\int_0^\delta u(U_\infty - u)\,dy + \frac{dU_\infty}{dx}\int_0^\delta (U_\infty - u)\,dy = \nu\left.\frac{\partial u}{\partial y}\right|_0 = \frac{\tau_s}{\rho} \tag{7–81}$$

which represents the *integral momentum equation.*

For the case of uniform free-stream velocity flow, $dU_\infty/dx = 0$ such that Eq. (7–81) reduces to

$$\frac{d}{dx}\int_0^\delta u(U_\infty - u)\,dy = \nu\left.\frac{\partial u}{\partial y}\right|_0 = \frac{\tau_s}{\rho} \tag{7–82}$$

The boundary conditions that accompany the integral momentum equation are given by Eqs. (7–61)–(7–64). Notice that the boundary condition for $v$ given by Eq. (7–64) has already been satisfied in the development of the integral continuity equation, which itself has been incorporated into the integral momentum equation.

It should be noted that higher-order integral equations for fluid flow can be developed by multiplying the differential continuity and momentum equations by weighing functions and integrating across the boundary layer. For example, the *integral mechanical energy equation* can be obtained by multiplying the momentum equation through by $u$, rearranging and coupling this equation with the continuity equation, and integrating across the boundary layer. Higher-order integral equations such as the mechanical energy equation are used in the development of the most modern multiple-parameter integral methods.

***Energy Transfer*** The application of the first law of thermodynamics to the lumped volume $\Delta\,dx\,dz$ shown in Fig. 7–11 gives (for negligible potential and kinetic energy, viscous dissipation, body force, axial conduction, and thermal radiation effects)

$$\Sigma\,\dot{E}_i = \Sigma\,\dot{E}_o + \cancel{\frac{\Delta E_s}{\Delta t}}$$

$$\dot{H}_x + d\dot{H}_\Delta + dq_c = \dot{H}_{x+dx} \tag{7–83}$$

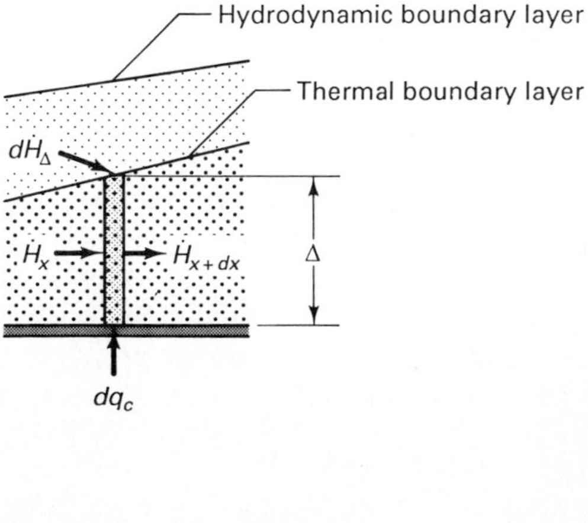
Hydrodynamic boundary layer
Thermal boundary layer
$d\dot{H}_\Delta$
$\dot{H}_x$
$\dot{H}_{x+dx}$
$\Delta$
$dq_c$

**FIGURE 7–11**
Conservation of energy relative to lumped differential rectangular control volume; $dV = \Delta\,dx\,dz$.

where $\Delta E_s/\Delta t = 0$ for steady-state conditions, $d\dot{H}_\Delta = i_\infty \, d\dot{m}_\Delta$, and

$$\dot{H}_x = \int d\dot{H}_x = \int i \, d\dot{m}_x = \int_0^\Delta i(\rho u \, dz) \, dy \tag{7–84}$$

Utilizing the definition of the derivative, Eq. (7–83) is written as

$$d\dot{H}_\Delta + dq_c = \frac{d\dot{H}_x}{dx} \, dx \tag{7–85}$$

With $\dot{H}_x$ given by Eq. (7–84) and with $d\dot{m}_\Delta$ specified on the basis of continuity by Eq. (7–72), we have

$$i_\infty \left( \frac{d}{dx} \int d\dot{m}_x \right) dx + dq_c = \left( \frac{d}{dx} \int i \, d\dot{m}_x \right) dx \tag{7–86}$$

or

$$\frac{d}{dx} \int (i - i_\infty) \, d\dot{m}_x = \frac{dq_c}{dx} \tag{7–87}$$

Replacing $d\dot{m}_x$ by $(\rho u \, dz) \, dy$ and setting $i - i_\infty = c_P(T - T_\infty)$ for flow of ideal fluids with uniform properties, the *integral energy equation* becomes

$$\rho c_P \frac{d}{dx} \int_0^\Delta u(T - T_\infty) \, dy = \frac{dq_c}{dx \, dz} = q_c'' \tag{7–88}$$

The integral energy equation is coupled with the thermal boundary conditions given by Eqs. (7–65) through (7–67). For the case in which a specified wall-heat flux is maintained, $q_c''$ is simply replaced by the specified input. On the other hand, for the case in which the wall temperature is specified, the Fourier law of conduction is utilized, with Eq. (7–88) taking the form

$$\rho c_P \frac{d}{dx} \int_0^\Delta u(T - T_\infty) \, dy = -k \left. \frac{\partial T}{\partial y} \right|_0 = q_c'' \tag{7–89}$$

As in the case of the fluid flow, higher-order integral equations can be developed for energy transfer. However, such higher-order integral energy equations are not presently in common use.

### *Solution*

Because of the nature of the differential formulation given by Eqs. (7–58)–(7–60) and the accompanying boundary conditions, the development of exact solutions for laminar boundary layer flow usually requires the use of numerical-solution techniques [19–21]. The numerical-solution approaches developed in the literature generally involve the use of transformations that provide a basis for achieving efficient numerical calculations. A particularly important feature of the transformation approach is that the transformed equations reduce to ordinary differential equations for the special class of flows for which the velocity and temperature distributions are geometrically similar in the streamwise direction. Boundary layer flow over a flat plate with uniform free-stream velocity and uniform wall temperature or uniform wall-heat flux, and flow over wedge-shape bodies are included in this special category of *similar flows*. Similarity solutions are developed in references 11 and 16–18.

The simpler approximate integral approach involves the solution of ordinary rather than partial differential equations for nonsimilar as well as similar steady two-dimensional boundary layers. In addition, the integral approach provides a frame of reference for the development of the classic transformations used in the numerical-transformation approach. Because of the relative simplicity and versatility of the integral approach, this very popular and practical method will be featured in this section. Because we are dealing with uniform property flow, the fluid-flow aspects of the problem are treated first, after which the energy transfer will be handled. The practical approach presented in this section is extended to the more general case involving nonuniform free-stream velocity and transpiration in references 22 and 23.

We now turn our attention to the classic problem of parallel flow over a heated or cooled flat plate with uniform free-stream velocity. Referring to Eq. (7–68), boundary layer flow with uniform free-stream velocity is characterized by a zero axial pressure gradient; that is,

$$\frac{dP}{dx} = -\rho U_\infty \frac{dU_\infty}{dx} = 0 \tag{7–90}$$

***Integral Solution—Fluid Flow*** The integral momentum equation for boundary layer flow over a flat plate with uniform free-stream velocity,

$$\frac{d}{dx}\int_0^\delta u(U_\infty - u)\,dy = \nu \left.\frac{\partial u}{\partial y}\right|_0 \tag{7–82}$$

involves the one unknown $u$. The boundary layer thickness $\delta$ is of course dependent upon $u$. A simple approximate solution to this integral form of the momentum equation can be obtained by treating $\delta$ as the unknown, with the velocity profile $u$ being approximated in terms of $\delta$ on the basis of the physics of the problem.

*Dimensionless Velocity Distribution* Referring to Fig. 7–12, which depicts the laminar hydrodynamic boundary layer, approximate velocity profiles are easily sketched in that satisfy the boundary conditions

$$u = U_\infty \qquad \text{as } y \to \infty \qquad \text{or at } y \simeq \delta \tag{7–62}$$

$$u = 0 \qquad \text{at } y = 0 \tag{7–63}$$

and the physical requirement

$$\frac{\partial u}{\partial y} = 0 \qquad \text{as } y \to \infty \qquad \text{or at } y \simeq \delta \tag{7–91}$$

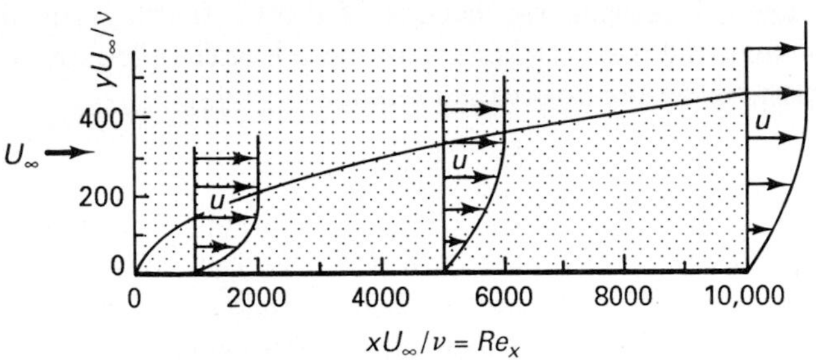


**FIGURE 7–12** Representative velocity distribution for laminar boundary layer flow over a flat plate with uniform free-stream velocity.

Many analytical expressions are available that meet these minimal requirements. One example is the simple second-order polynomial

$$\frac{u}{U_\infty} = 2\frac{y}{\delta} - \left(\frac{y}{\delta}\right)^2 \qquad \text{for } y \leq \delta \tag{7–92}$$

Improvements in our approximation for $u$ can be made by satisfying one or two higher-order requirements, such as

$$\frac{\partial^2 u}{\partial y^2} = 0 \qquad \text{at } y = 0 \tag{7–93}$$

which is indicated by the differential momentum equation, Eq. (7–59) with $dP/dx = 0$, and

$$\frac{\partial^2 u}{\partial y^2} = 0 \qquad \text{at } y = \delta \tag{7–94}$$

in accordance with physical requirements. We can also show that the differential equation must satisfy

$$\frac{\partial^3 u}{\partial y^3} = 0 \qquad \text{at } y = 0 \qquad \text{and} \qquad \text{as } y \to \infty \text{ or at } y \simeq \delta \tag{7–95}$$

plus higher-order derivatives at the outer edge of the boundary layer. The exact solution for $u$ satisfies all these requirements. However, it should be noted that the fulfillment of all these constraints at $y = 0$ and at $y = \delta$ does not guarantee that $u$ satisfies the fundamental differential formulation [Eqs. (7–58) and (7–59)] in the region $0 < y < \delta$.

As already indicated, the second-order profile given by Eq. (7–92) satisfies the primary conditions given by Eqs. (7–62), (7–63), and (7–91). It even satisfies the requirements given by Eq. (7–95), but it fails to satisfy the specifications given by Eqs. (7–93) and (7–94). On the basis of previous studies, more accurate approximations for $u$ have been developed by the use of third-, fourth-, and fifth-order polynomials. The simplest and most convenient of these higher-order approximations is obtained on the basis of the third-order polynomial†

$$u = a_0 + a_1 y + a_2 y^2 + a_3 y^3 \tag{7–96}$$

Since we now have four coefficients, Eqs. (7–62), (7–63), (7–91), and (7–93) can be satisfied. The coupling of Eq. (7–96) with these conditions gives $a_0 = 0$, $2a_2 = 0$, $a_0 + a_1\delta + a_2\delta^2 + a_3\delta^3 = U_\infty$, and $a_1 + 2a_2\delta + 3a_3\delta^2 = 0$. It follows that $a_1 = 3U_\infty/(2\delta)$, and $a_3 = -U_\infty/(2\delta^3)$, such that Eq. (7–96) becomes

$$\frac{u}{U_\infty} = \frac{3}{2}\frac{y}{\delta} - \frac{1}{2}\left(\frac{y}{\delta}\right)^3 \qquad \text{for } y \leq \delta \tag{7–97a}$$

$$= 1 \qquad \text{for } y \geq \delta \tag{7–97b}$$

The velocity profiles shown in Fig. 7–12 are based on this equation. Introducing Newton's law of viscosity, we are able to express the wall-shear stress $\tau_s$ in terms of $\delta$ by

$$\tau_s = \mu \left.\frac{\partial u}{\partial y}\right|_0 = \frac{3}{2}\frac{\mu U_\infty}{\delta} \tag{7–98}$$

As we shall see, this third-order polynomial gives rise to approximate solutions that are within about 3% of the exact solution. Comparable results can be achieved by the use of a third-order polynomial based on the primary constraints together with Eq. (7–94) instead of Eq. (7–93). This level of accuracy is also achieved by the use of a fourth-order polynomial, which satisfies both Eqs. (7–93) and (7–94), and a fifth-order polynomial, which satisfies one of the conditions represented by Eq. (7–95). However, the use of higher-order conditions at the boundaries to establish approximations for the velocity distribution generally proves to be ineffective. Alternative modern methods for achieving more accurate and reliable approximations involve the use of one or more higher-order integral equations. For example, a fourth-order polynomial can be developed with the five coefficients specified in accordance with Eqs. (7–62), (7–63), (7–91), and (7–93) and the integral mechanical energy equation. The use of this two-parameter approach results in solutions that are within 1% of the exact solution. However, this approach is somewhat more involved than the simple one-parameter approach featured in our study. An attractive multiple-parameter approach of a different kind, which involves the use of approximations for the viscous stress in terms of $u$, has been developed by Dorodnitsyn [24], Holt [25], and Fletcher [26].

*Boundary Layer Thickness $\delta$ and Friction Factor* Now that reasonable approximations are available for the velocity profile $u$ in terms of $\delta$, we return to the integral momentum equation in order to develop predictions for $\delta$. Utilizing our third-order polynomial approximation, the integral momentum equation becomes

† An equivalent approach, which provides a basis for extending the analysis to laminar transpired flow and turbulent flow, involves the use of polynomial approximations for shear stress $\tau$ in terms of $y$ [22].

$$U_\infty^2 \frac{d}{dx}\int_0^\delta \left[\frac{3}{2}\left(\frac{y}{\delta}\right) - \frac{1}{2}\left(\frac{y}{\delta}\right)^3\right]\left[1 - \frac{3}{2}\frac{y}{\delta} + \frac{1}{2}\left(\frac{y}{\delta}\right)^3\right] dy = \frac{3}{2}\frac{\nu U_\infty}{\delta} \qquad (7\text{–}99)$$

Performing the integration called for in this equation, we obtain

$$\delta \frac{d\delta}{dx} = \frac{140}{13}\frac{\nu}{U_\infty} \qquad (7\text{–}100)$$

This differential equation in $\delta$ is coupled with the requirement

$$\delta = 0 \qquad \text{at } x = 0 \qquad (7\text{–}101)$$

Equation (7–100) is readily integrated to give

$$\delta^2 = \frac{280}{13}\frac{\nu x}{U_\infty} \qquad \delta = 4.64\sqrt{\frac{\nu x}{U_\infty}} \qquad (7\text{–}102,103)$$

This equation was actually used in the construction of Figs. 7–7 and 7–12.

Substituting this result for $\delta$ into Eq. (7–97a), we obtain an expression for the velocity distribution of the form

$$\frac{u}{U_\infty} = \frac{3}{2}\left[\sqrt{\frac{13}{280}}\frac{y}{\sqrt{x\nu/U_\infty}}\right] - \frac{1}{2}\left[\sqrt{\frac{13}{280}}\frac{y}{\sqrt{x\nu/U_\infty}}\right]^3 \qquad \text{for } y \leqslant \delta$$

$$= 0.323\eta - 0.005\eta^3 \qquad \text{for } \eta \leqslant 4.64 \qquad (7\text{–}104)$$

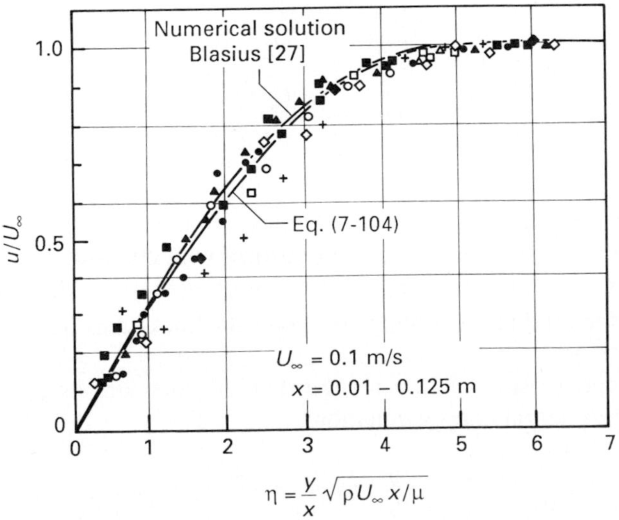


**FIGURE 7–13** Velocity distribution for laminar boundary layer flow over a flat plate with uniform free-stream velocity. (Data from Hansen [2].)

where $\eta = y/\sqrt{x\nu/U_\infty} = \sqrt{280/13}\, y/\delta$. This equation is compared with experimental data and the numerical solution in Fig. 7–13. With $u$ expressed in terms of $\eta$ (or $y/\delta$), we find that the velocity profiles are geometrically similar at all values of $x$. Consequently, $\eta$ is known as a *similarity coordinate*.

To obtain the friction factor $f_x$, Eqs. (7–98) and (7–102) are combined, with the result

$$f_x = \frac{\tau_s}{\rho U_\infty^2/2} = 3\frac{\nu}{U_\infty \delta} = 3\sqrt{\frac{13}{280}}\sqrt{\frac{\nu}{U_\infty x}} = \frac{0.646}{Re_x^{1/2}} \tag{7–105}$$

This expression is only 2.7% below the exact solution,

$$f_x = \frac{0.664}{Re_x^{1/2}} \tag{7–106}$$

which was first developed by Blasius [27] and is shown to be in good agreement with experimental data for laminar flow in Fig. 7–14.

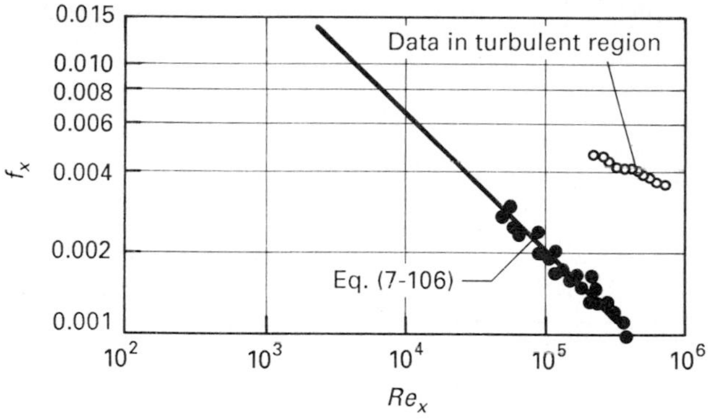


**FIGURE 7–14** Relations for $f_x$—laminar boundary layer flow over a flat plate with uniform free-stream velocity.

***Integral Solution—Energy Transfer*** Consideration is turned to the development of approximate integral solutions for the case of uniform wall heat flux. To simplify the analysis, we will restrict our attention to fluids with moderate to large values of Prandtl number, for which the thermal boundary layer thickness $\Delta$ is less than or of the order of $\delta$.

To develop an approximate solution for the thermal boundary layer thickness $\Delta$, which appears in the integral energy equation,

$$\rho c_P \frac{d}{dx}\int_0^\Delta u(T - T_\infty)\,dy = -k\left.\frac{\partial T}{\partial y}\right|_0 = q_c'' \tag{7–89}$$

we must develop an approximation for the temperature distribution.

*Dimensionless Temperature Distribution* Representative temperature distributions for flow over a flat plate with *uniform wall-heat flux* are illustrated in Fig. 7–15 for cases in which the thermal boundary layer thickness $\Delta$ is less than $\delta$.† Following the pattern of our fluid-flow analysis, $T$ is approximated by the third-order polynomial

$$T = b_0 + b_1 y + b_2 y^2 + b_3 y^3 \qquad (7\text{–}107)$$

where the four coefficients are selected in accordance with the following four conditions, which are analogous to those used in the fluid flow analysis:

$$T = T_\infty \qquad \frac{\partial T}{\partial y} = 0 \qquad \text{as } y \to \infty \text{ or at } y \simeq \Delta \qquad (7\text{–}66,108)$$

$$T = T_s \qquad \frac{\partial^2 T}{\partial y^2} = 0 \qquad \text{at } y = 0 \qquad (7\text{–}67a,109)$$

[Equation (7–109) is obtained by evaluating the differential energy equation, Eq. (7–60), at the wall where $u$ and $v$ are both zero.] Following through with the evaluation of the coefficients $b_n$, we obtain

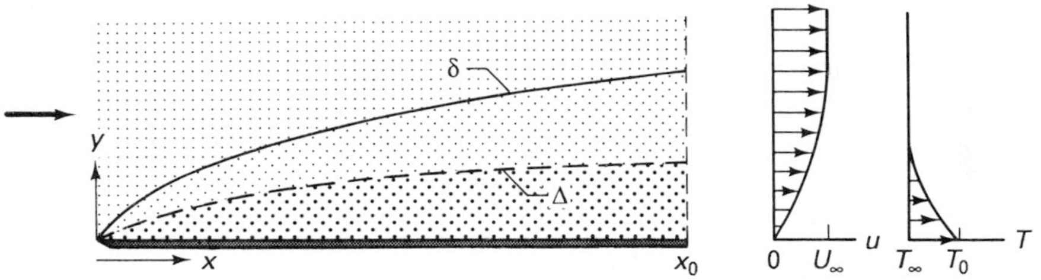


(a) $Pr \cong 4.0$.

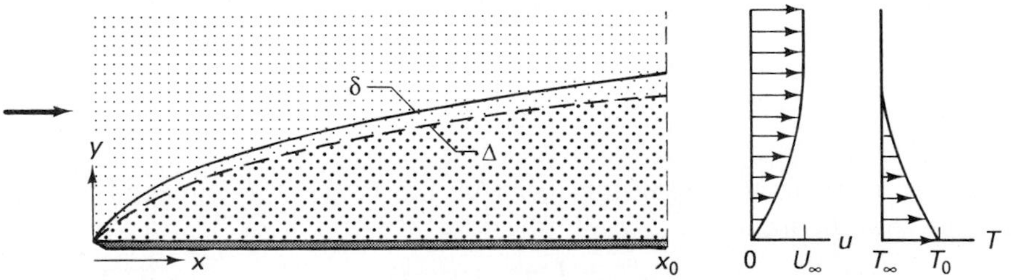


(b) $Pr = 0.72$.

**FIGURE 7–15** Laminar forced convection flow over a flat plate with uniform free-stream velocity and uniform wall-heat flux. (Velocity and temperature profiles shown at $x = x_0$.)

† Concerning the result shown in Fig. 7–15 for $Pr = 0.72$, the integral analysis to follow indicates that $\Delta < \delta$ for uniform wall-heat flux, but $\Delta > \delta$ for uniform wall temperature. This point illustrates the significant effect of the form of the thermal boundary condition on the solution for laminar flow.

$$\frac{T - T_s}{T_\infty - T_s} = \frac{3}{2}\frac{y}{\Delta} - \frac{1}{2}\left(\frac{y}{\Delta}\right)^3 \qquad \text{for } y \leqslant \Delta \tag{7–110a}$$

$$= 1 \qquad \text{for } y \geqslant \Delta \tag{7–110b}$$

which is of the same form as Eq. (7–97) for the dimensionless velocity distribution, except for the important distinction between $\Delta$ and $\delta$. The relationship between wall-heat flux $q''_c$, wall temperature $T_s$, and the thermal boundary layer thickness $\Delta$ is developed by applying the Fourier law of conduction; that is,

$$q''_c = -k\frac{\partial T}{\partial y}\bigg|_0 = \frac{3}{2}\frac{k(T_s - T_\infty)}{\Delta} \tag{7–111}$$

Using Eq. (7–102) for $\delta$, this equation gives rise to an expression for Nusselt number $Nu_x$ of the form

$$Nu_x = \frac{q''_c}{T_s - T_\infty}\frac{x}{k} = \frac{3}{2}\frac{x/\delta}{r} = \frac{3}{2}\sqrt{\frac{13}{280}}\frac{Re_x^{1/2}}{r} = 0.323\frac{Re_x^{1/2}}{r} \tag{7–112}$$

where $r = \Delta/\delta$.

*Thermal Boundary Layer Thickness $\Delta$ and Nusselt Number* To evaluate the integral appearing in the integral energy equation, $T$ is approximated by Eq. (7–110), with $u$ approximated by Eq. (7–97) for $\Delta \lesssim \delta$. Following through with this integration, we obtain

$$\begin{aligned}
&\int_0^\Delta u(T - T_\infty)\,dy \\
&= \int_0^\Delta \left\{U_\infty(T_s - T_\infty)\left[\frac{3}{2}\frac{y}{\delta} - \frac{1}{2}\left(\frac{y}{\delta}\right)^3\right]\left[1 - \frac{3}{2}\frac{y}{\Delta} + \frac{1}{2}\left(\frac{y}{\Delta}\right)^3\right]\right\} dy \\
&= \frac{3}{2}U_\infty(T_s - T_\infty)\delta\left[\frac{1}{10}\left(\frac{\Delta}{\delta}\right)^2 - \frac{1}{140}\left(\frac{\Delta}{\delta}\right)^4\right] = \frac{3}{20}U_\infty(T_s - T_\infty)\delta\left(r^2 - \frac{r^4}{14}\right)
\end{aligned} \tag{7–113}$$

for $r \leqslant 1$. Using this result together with Eq. (7–111), the integral energy equation is put into the form

$$\begin{aligned}
q''_c &= \rho c_P \frac{d}{dx}\left[\frac{3}{20}U_\infty(T_s - T_\infty)\,\delta\left(r^2 - \frac{r^4}{14}\right)\right] \\
&= \frac{1}{10}\frac{d}{dx}\left[\frac{U_\infty q''_c \delta^2}{\alpha}\left(r^3 - \frac{r^5}{14}\right)\right]
\end{aligned} \tag{7–114}$$

Setting $\Delta = 0$ at the location at which heating or cooling is initiated, the solution to this equation becomes

$$\frac{1}{10}\frac{U_\infty q_c'' \delta^2}{\alpha}\left(r^3 - \frac{r^5}{14}\right) = \int_0^x q_c''\, dx = q_0'' x \qquad (7\text{–}115)$$

for uniform wall-heat flux. Using Eq. (7–102) to eliminate $\delta^2$ and rearranging, we have

$$r^3 - \frac{r^5}{14} = \frac{13}{28\,Pr} \qquad (7\text{–}116)$$

where $r \leqslant 1$ indicates that the solution applies to fluids for which $Pr \geqslant 0.5$. Notice that calculations can be obtained for $r$ versus $Pr$ by taking $r$ as the independent variable. As an alternative, a convenient explicit expression can be obtained for $r$ by neglecting the fifth-order term in Eq. (7–116); that is,

$$r \simeq \left(\frac{13}{28\,Pr}\right)^{1/3} = \frac{0.774}{Pr^{1/3}} \qquad (7\text{–}117)$$

which differs from Eq. (7–116) by less than 2.5%. This equation indicates that $\Delta \lesssim \delta$ (i.e., $r \lesssim 1$) for $Pr \gtrsim 0.464$, such that the solution is applicable to common gases and liquids, but does not apply to liquid metals.

Substituting Eq. (7–117) into Eq. (7–112), the solution for Nusselt number $Nu_x$ associated with uniform wall-heat flux becomes

$$Nu_x = 0.417\, Re_x^{1/2}\, Pr^{1/3} \qquad \text{for } Pr \geq 0.464 \qquad (7\text{–}118)$$

This equation is within 5.6% of the numerical solution results reported by Levy [28]†; the exact solution is correlated by

$$Nu_x = 0.442\, Re_x^{1/2}\, Pr^{1/3} \qquad (7\text{–}119)$$

for moderate values of $Pr$.

As shown in Example 7–6, this approach is readily adapted to uniform wall temperature and arbitrary wall-heat flux conditions. For the case of uniform wall temperature, the integral solution for $Nu_x$ is given by

$$Nu_x = 0.331\, Re_x^{1/2}\, Pr^{1/3} \qquad \text{for } Pr \geq 0.928 \qquad (7\text{–}120)$$

which is within 0.3% of the exact solution for moderate values of $Pr$.

It should be emphasized that these solutions are restricted to situations for which $r \leqslant 1$, which requires $Pr \geqslant 0.464$ for uniform wall heat flux and $Pr \geq 0.928$ for uniform wall temperature. In this connection, the idea that $\Delta = \delta$ for $Pr \approx 1$ holds true for uniform wall-temperature conditions, but is invalid for uniform wall-heat flux. The development of integral solutions for $r > 1$ (i.e. $\Delta > \delta$) is suggested as an exercise.

† The comparable fourth-order solution for uniform wall-heat flux is given by

$$Nu_x = 0.450\, Re_x^{1/2}\, Pr^{1/3}$$

which is within 2% of the exact solution for moderate values of $Pr$.

## EXAMPLE 7–6

Use the integral solution approach to develop approximate analytical solutions for the Nusselt number associated with laminar boundary layer flow over a flat plate with uniform free-stream velocity and (a) arbitrary wall-heat flux, or (b) uniform wall temperature.

### Solution

*Objective* Develop approximate integral solutions for $Nu_x$.

*Schematic* Laminar flow over a flat plate with uniform free-stream velocity $U_\infty$.

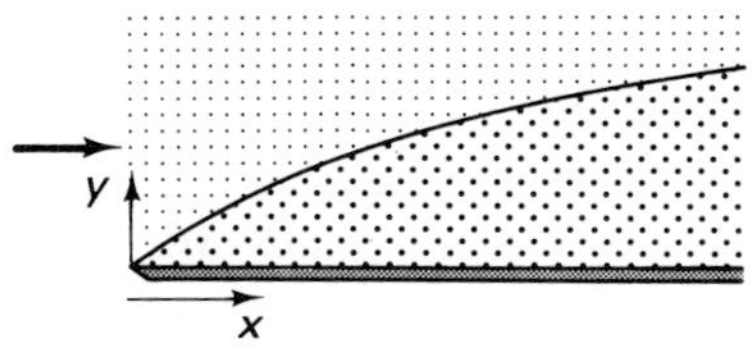


Condition at wall:
(a) Specified wall-heat flux, or
(b) Uniform wall temperature

*Assumptions/Conditions*

forced convection
uniform properties
standard conditions

*Analysis* Referring to the preceding formulation, the use of the third-order polynomial approximations for velocity and temperature permits us to express the integral energy equation in the form of Eq. (7–114),

$$\frac{1}{10}\frac{d}{dx}\left[\frac{U_\infty q_c''\delta^2}{\alpha}\left(r^3 - \frac{r^5}{14}\right)\right] = q_c'' \tag{a}$$

for applications in which $r \leqslant 1$. Neglecting the fifth-order term, this equation reduces to the more convenient form

$$\frac{1}{10}\frac{d}{dx}\left(\frac{U_\infty q_c''\delta^2 r^3}{\alpha}\right) = q_c'' \tag{b}$$

which is within 2.5% of Eq. (a) and which provides the basis for developing convenient explicit analytical solutions.

*(a) Specified Wall-Heat Flux*

Setting $\Delta = 0$ at $x = 0$, the solution to this equation for the case of arbitrarily specified wall-heat flux is simply

$$\frac{U_\infty q_c''\delta^2 r^3}{\alpha} = 10\int_0^x q_c''\, dx$$

or

$$r^3 = \frac{10\,\alpha}{U_\infty \delta^2} \frac{1}{q_c''} \int_0^x q_c''\, dx$$

Using Eq. (7–102) to eliminate $\delta$, this equation reduces to

$$r = \left( \frac{13}{28\, Pr} \frac{1}{x q_c''} \int_0^x q_c''\, dx \right)^{1/3} = \frac{0.774}{Pr^{1/3}} \left( \frac{1}{x q_c''} \int_0^x q_c''\, dx \right)^{1/3}$$

Substituting this result into Eq. (7–112), the Nusselt number $Nu_x$ becomes

$$Nu_x = 0.417\, Re_x^{1/2}\, Pr^{1/3} \left( \frac{1}{x q_c''} \int_0^x q_c''\, dx \right)^{-1/3} \tag{c}$$

Thus, with $q_c''$ specified $Nu_x$ can easily be obtained.

For example, with the heat flux specified by a power law of the form

$$q_c'' = q_0'' x^N$$

Eq. (c) gives rise to

$$Nu_x = 0.417\, Re_x^{1/2}\, Pr^{1/3}\, (N + 1)^{1/3} \qquad \text{for } Pr \geqslant \frac{0.464}{N + 1} \tag{d}$$

Notice that this equation reduces to Eq. (7–118) for uniform wall heat flux ($N = 0$). To see the nature of the distribution in wall temperature indicated by Eq. (d), we write

$$\begin{aligned} T_s - T_\infty = \frac{q_c'' x}{Nu_x k} &= \frac{q_0'' x^{N+1}}{0.417\, k\, Re_x^{1/2}\, Pr^{1/3} (N + 1)^{1/3}} \\ &= \frac{q_0'' x^{N+1/2}}{0.417\, k (U_\infty/\nu)^{1/2}\, Pr^{1/3} (N + 1)^{1/3}} \end{aligned} \tag{e}$$

such that the wall temperature also follows a power-law distribution.

### (*b*) *Uniform Wall Temperature*

To develop solutions for cases in which the wall temperature is specified rather than the wall heat flux, we make use of Eq. (7–111),

$$q_c'' = \frac{3}{2} \frac{k(T_s - T_\infty)}{\Delta}$$

Using this relation to eliminate $q_c''$ in favor of $T_s$, Eq. (b) becomes

$$\frac{1}{10} \frac{d}{dx} \left[ \frac{U_\infty (T_s - T_\infty) \delta r^2}{\alpha} \right] = \frac{T_s - T_\infty}{\Delta}$$

or, for the case of uniform wall temperature ($T_s = T_0$),

$$r\frac{d}{dx}(\delta r^2) = 10\frac{\alpha}{U_\infty \delta}$$

This equation can be solved by writing appropriate homogeneous and particular solutions.

Alternatively, we can make use of the solution given by Eq. (d) for a power law distribution in heat flux. In referring to Eq. (e), we observe that $N = -1/2$ for uniform wall temperature. It follows that the Nusselt number for uniform wall temperature is obtained by setting $N = -1/2$ in Eq. (d); that is,

$$Nu_x = 0.417\, Re_x^{1/2}\, Pr^{1/3}\, 0.5^{1/3} \qquad \text{for } Pr > \frac{0.464}{N+1} = 0.928$$

$$= 0.331\, Re_x^{1/2}\, Pr^{1/3}$$

This solution is within 0.3 to 2.4% of the exact solution, which can be represented by

$$Nu_x = 0.332\, Re_x^{1/2}\, Pr^{1/3} \qquad \text{for } 0.6 < Pr < 10$$

and

$$Nu_x = 0.339\, Re_x^{1/2}\, Pr^{1/3} \qquad \text{for } 10 < Pr$$

## 7–2–3 Natural Convection on Vertical Surfaces

We now turn our attention to the fundamental problem of natural convection flow over a vertical heated or cooled surface that is immersed in an extensive quiescent fluid with ambient temperature $T_\infty$ and density $\rho_\infty$. According to the well-known *Archimedes principle*, a body with density $\rho$ immersed in a fluid with density $\rho_\infty$ experiences an *upward buoyancy force* equal to the weight of the displaced fluid. The net effect of the body weight and the buoyancy force will cause the body to rise if $\rho$ is *less* than $\rho_\infty$ or to sink if $\rho$ is *greater* than $\rho_\infty$. It is the buoyancy force $f_B$, which is caused by temperature-induced density variations in the region near a heated or cooled surface, that provides the driving mechanism for natural convection. Consequently, a heated vertical surface placed in a quiescent fluid will produce a region of warm, less dense, upward-moving fluid near the wall. Similarly, a region of cool, more dense, downward-moving fluid will be sustained near a cooled vertical surface.

Representative distributions in temperature and velocity associated with these two situations are shown in Fig. 7–16. Comparing these profiles with those shown in Fig. 7–15 for forced convection, we observe two important differences. The one very obvious difference is in the nature of the velocity profile. Very much unlike the velocity distribution for forced convection, for this natural convection flow $u$ increases

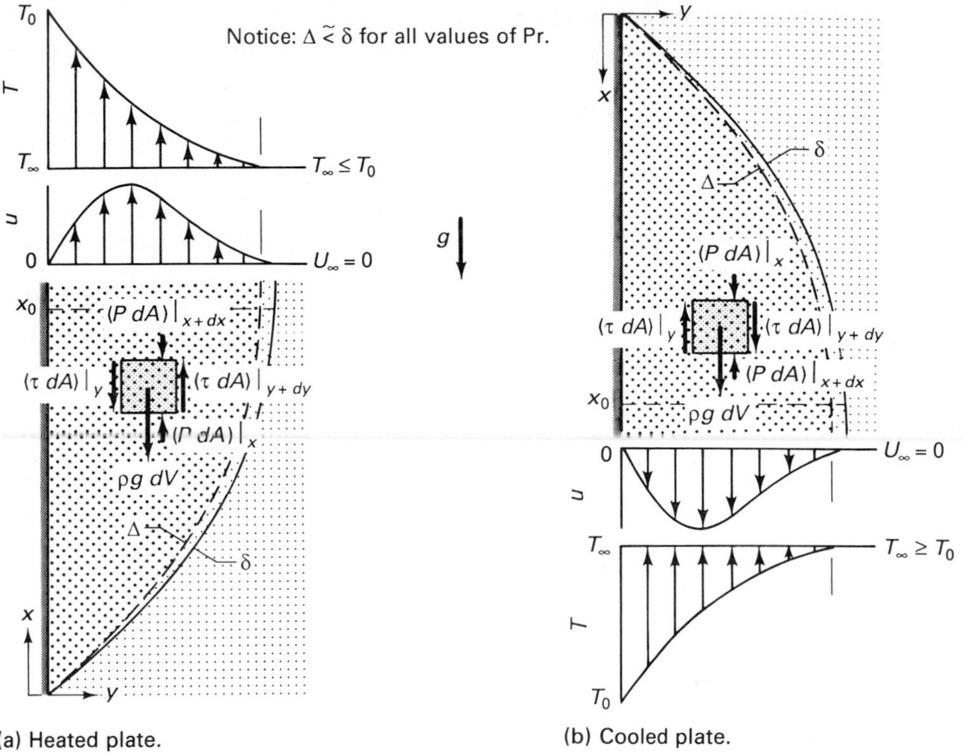


**FIGURE 7–16** Laminar natural convection flow over a vertical plate at uniform temperature $T_0$ in a quiescent fluid at temperature $T_\infty$. (Velocity and temperature profiles are shown at $x = x_0$.)

from zero at the wall to a maximum value, and then decreases to zero at the outer edge of the boundary layer. On the other hand, we observe that the temperature distributions are quite similar, except for the somewhat subtle fact that the thermal boundary layer generally lies within the hydrodynamic boundary layer (i.e., $r < 1$), *regardless* of the value of the Prandtl number *Pr*.† The fact that Δ is less than or of the order of δ for small values of *Pr* is a consequence of the motion-producing buoyancy forces acting within the entire thermal boundary layer. For moderate to large values of *Pr*, the thermal boundary layer lies well within the hydrodynamic boundary layer, as in the case of forced convection, since $\nu >> \alpha$. In this case, the buoyancy force produces a strong fluid shear layer that drives the outer flow field. Thus, for buoyant flows such as this, motion is generally produced across or beyond the entire thermal boundary layer.

† Strictly speaking, Δ/δ does slightly exceed unity for fluids with very low values of *Pr*. For example, $\Delta/\delta \simeq 1.11$ for $Pr = 0.01$ [29].

### *Mathematical Formulation*

The presence of a significant imbalance in the density-related gravitational body force acting on the fluid is the feature that distinguishes natural convection from forced convection. Because the effects of the forces acting on the fluid are accounted for by the momentum equation for standard boundary layer flows, it follows that the continuity and energy equations for natural convection with moderate temperature differences are of the same form as for uniform property forced convection; that is,

$$\frac{\partial u}{\partial x} + \frac{\partial v}{\partial y} = 0 \tag{7–121}$$

$$u\frac{\partial T}{\partial x} + v\frac{\partial T}{\partial y} = \alpha\frac{\partial^2 T}{\partial y^2} \tag{7–122}$$

To develop the momentum equation for natural convection, we simply include the gravitation body force $\rho g_x\, dV$ in the force balance, as shown in Fig. 7–16. To simplify the formulation we will focus attention on the heated plate [Fig. 7–16(a)] for which $g_x = -g$. {The resulting formulation is adapted to the cooled plate [Fig. 7–16(b)] by merely changing the sign associated with terms involving the gravitational acceleration $g$.}

Applying Newton's second law of motion to the differential element shown in Fig. 7–16(a), we write

$$RCM_x = \Sigma F_x \tag{7–123}$$

where

$$\begin{aligned}\Sigma F_x &= (\tau\, dx\, dz)|_{y+dy} - (\tau\, dx\, dz)|_y + (P\, dy\, dz)|_x - (P\, dy\, dz)|_{x+dx} - \rho g\, dx\, dy\, dz \\ &= \left(\frac{\partial \tau}{\partial y} - \frac{\partial P}{\partial x} - \rho g\right) dV\end{aligned} \tag{7–124}$$

and

$$\begin{aligned}RCM_x &= d\dot{M}_{x+dx} + d\dot{M}_{y+dy} - d\dot{M}_x - d\dot{M}_y \\ &= \frac{\partial}{\partial x}(d\dot{M}_x)\, dx + \frac{\partial}{\partial y}(d\dot{M}_y)\, dy\end{aligned} \tag{7–125}$$

Substituting $d\dot{M}_x = u\, d\dot{m}_x$ and $d\dot{M}_y = u\, d\dot{m}_y$ and using the continuity equation, Eq. (7–121), $RCM_x$ becomes

$$\begin{aligned}RCM_x &= \left[\rho\left(u\frac{\partial u}{\partial x} + v\frac{\partial u}{\partial y}\right) + \rho u\left(\frac{\partial u}{\partial x} + \frac{\partial v}{\partial y}\right)\right] dV \\ &= \rho\left(u\frac{\partial u}{\partial x} + v\frac{\partial u}{\partial y}\right) dV\end{aligned} \tag{7–126}$$

Substituting these results for $RCM_x$ and $\Sigma F_x$ into Eq. (7–123), the momentum equation becomes

$$\rho\left(u\frac{\partial u}{\partial x} + v\frac{\partial u}{\partial y}\right) = \mu\frac{\partial^2 u}{\partial y^2} - \frac{dP}{dx} - \rho g \tag{7–127}$$

for uniform viscosity. Except for the gravitational force term $-\rho g$, this equation is identical to the momentum equation for forced convection. As $y \to \infty$, Eq. (7 – 127) reduces to

$$-\frac{dP}{dx} = \rho_\infty g \tag{7 – 128}$$

Substituting this result back into Eq. (7–127), the momentum equation takes the form

$$\rho\left(u\frac{\partial u}{\partial x} + v\frac{\partial u}{\partial y}\right) = \mu\frac{\partial^2 u}{\partial y^2} + g(\rho_\infty - \rho) \tag{7–129}$$

where the term $g(\rho_\infty - \rho)$ represents the driving buoyancy force $f_B$. (Notice that $f_B$ is the difference between the two body forces $-\rho g$ and $-\rho_\infty g$.) The effect of buoyancy is determined by accounting for the dependence of the gravitational term $\rho g$ on temperature. For isothermal conditions, $\rho g$ is constant $(= \rho_\infty g)$ and no flow occurs.

In this formulation for moderate temperature differences, the properties associated with the nonbuoyancy terms are generally evaluated at the free-stream temperature $T_\infty$ or film temperature $T_f = (T_s + T_\infty)/2$. The dependence of $\rho$ on $T$ is often expressed in terms of the *coefficient of thermal expansion* $\beta$, which is defined by

$$\beta = \frac{1}{V}\frac{\partial V}{\partial T}\bigg|_P = -\frac{1}{\rho}\frac{\partial \rho}{\partial T}\bigg|_P \tag{7–130}$$

For ideal gases, $\beta$ is expressed in terms of the absolute temperature $T$ by $\beta = 1/T$. This property is listed in Table A–C–3 for common liquids. Expanding the defining equation for $\beta$, we obtain (for $P$ = constant)

$$\int_{\rho_\infty}^{\rho} d\rho \simeq -\int_{T_\infty}^{T} \rho\beta\, dT = -\overline{\rho\beta}(T - T_\infty) \tag{7–131}$$

or

$$\rho - \rho_\infty = -\overline{\rho\beta}(T - T_\infty) \tag{7–132}$$

where

$$\overline{\rho\beta} = \frac{1}{T - T_\infty}\int_{T_\infty}^{T} \rho\beta\, dT \tag{7–133}$$

Substituting this result for $\rho$ into Eq. (7–129), we obtain

$$\rho\left(u\frac{\partial u}{\partial x} + v\frac{\partial u}{\partial y}\right) = \mu\frac{\partial^2 u}{\partial y^2} + g\,\overline{\rho\beta}(T - T_\infty) \tag{7–134}$$

For situations in which the temperature differences are moderate, the term $\overline{\rho\beta}$ can be approximated by $\rho\beta$, where $\rho$ and $\beta$ are evaluated at $T_\infty$ or $T_f$, such that Eq. (7–134) reduces to

$$u\frac{\partial u}{\partial x} + v\frac{\partial u}{\partial y} = \nu\frac{\partial^2 u}{\partial y^2} + g\beta(T - T_\infty) \qquad (7\text{–}135)$$

The evaluation of the fluid properties at the free-stream or film temperature and the use of $\overline{\rho\beta} = \rho\beta$ is known as the *Boussinesq approximation* [30]. In effect, this simplification involves the assumption of uniform properties, except for variable density in the buoyancy term.

To complete the differential formulation, we write the boundary conditions that accompany the continuity, momentum, and energy equations as follows:

$$u = 0 \quad \text{at } x = 0 \qquad u = 0 \quad \text{as } y \to \infty \qquad (7\text{–}136,137)$$

$$u = 0 \quad \text{at } y = 0 \qquad v = 0 \quad \text{at } y = 0 \qquad (7\text{–}138,139)$$

$$T = T_\infty \quad \text{at } x = x_0 \qquad T = T_\infty \quad \text{as } y \to \infty \qquad (7\text{–}140,141)$$

and

$$T = T_s \quad \text{or} \quad -k\frac{\partial T}{\partial y} = q''_c \quad \text{at } y = 0 \qquad (7\text{–}142\text{a,b})$$

As in the case of forced convection, practical approximate integral methods can be effectively used in the analysis of natural convection boundary layer flow. Following the indirect approach, to develop the integral momentum equation for laminar natural convection boundary layer flow we integrate the differential momentum equation across the boundary layer. First, Eq. (7–135) is rewritten as follows:

$$\frac{\partial u^2}{\partial x} - u\frac{\partial u}{\partial x} + \frac{\partial}{\partial y}(uv) - u\frac{\partial v}{\partial y} = \nu\frac{\partial^2 u}{\partial y^2} + g\beta(T - T_\infty) \qquad (7\text{–}143)$$

Utilizing the continuity equation, Eq. (7–121), this expression reduces to the form

$$\frac{\partial u^2}{\partial x} + \frac{\partial}{\partial y}(uv) = \nu\frac{\partial^2 u}{\partial y^2} + g\beta(T - T_\infty) \qquad (7\text{–}144)$$

Integrating each term of this equation with respect to $y$, we obtain

$$\int_0^y \frac{\partial u^2}{\partial x}\,dy + \int_0^y \frac{\partial}{\partial y}(uv)\,dy = \int_0^y \nu\frac{\partial^2 u}{\partial y^2}\,dy + \int_0^y g\beta(T - T_\infty)\,dy \qquad (7\text{–}145)$$

or

$$\frac{\partial}{\partial x}\int_0^y u^2\,dy + (uv)\Big|_0^y = \nu\frac{\partial u}{\partial y}\Big|_0^y + \int_0^y g\beta(T - T_\infty)\,dy \qquad (7\text{–}146)$$

Setting $y$ equal to $\delta$ and noting that $u = 0$ at $y = \delta$, the integral momentum equation reduces to

$$\frac{d}{dx}\int_0^\delta u^2\,dy = -\nu\frac{\partial u}{\partial y}\bigg|_0 + \int_0^\delta g\beta(T - T_\infty)\,dy \qquad (7\text{–}147)$$

The integral formulation is completed by writing the integral energy equation, which is identical in form to Eq. (7–89); that is,

$$\rho c_P\frac{d}{dx}\int_0^\Delta u(T - T_\infty)\,dy = q_c'' = -k\frac{\partial T}{\partial y}\bigg|_0 \qquad (7\text{–}148)$$

### Solution

Because the momentum equation for natural convection involves distributions in both velocity $u$ and temperature $T$, the development of a solution to this problem generally requires that the fluid flow and energy equations be solved simultaneously. As in the case of forced convection, both numerical/transformation and approximate integral methods are commonly used to solve these equations [29]†. In this connection, results based on the transformation analysis approach indicate that the hydrodynamic and thermal boundary layers are geometrically similar for laminar natural convection flow over a vertical surface with uniform wall-heat flux, uniform wall temperature, and power law distributions in $T_s$ and $q_c''$. This is also found to be the case for forced convection. To develop a practical solution to the problem at hand, we again turn to the approximate integral method, with attention focused on the case of uniform wall temperature conditions.

***Integral Solution—Fluid Flow and Energy Transfer*** The integral momentum and energy equations for natural convection flow over a vertical surface with mild to moderate temperature differences are represented by Eqs. (7–147) and (7–148). Following the approach used in the development of integral solutions for forced convection, the integral equations can be solved by the use of approximations for $u$ and $T$ in terms of the boundary layer thicknesses $\delta$ and $\Delta$.

*Dimensionless Velocity and Temperature Distributions* To maintain consistency with the analysis developed for forced convection in the preceding section, we will approximate the distributions in velocity $u$ and temperature $T$ by third-order polynomials; that is,

$$u = a_0 + a_1 y + a_2 y^2 + a_3 y^3 \qquad (7\text{–}149)$$

and

$$T = b_0 + b_1 y + b_2 y^2 + b_3 y^3 \qquad (7\text{–}150)$$

To evaluate these eight coefficients, the boundary conditions for $u$ and $T$ given by Eqs. (7–136) through (7–142) are coupled with the physical requirements

† An approximate solution for laminar film condensation on a vertical surface, which represents a natural convection flow process involving two phases, is developed in Appendix J.

$$\frac{\partial u}{\partial y} = 0 \qquad \text{at } y = \delta \qquad \frac{\partial T}{\partial y} = 0 \qquad \text{at } y = \Delta \tag{7–151,152}$$

Two additional conditions are written by satisfying Eqs. (7–135) and (7–122) at the wall.

$$\frac{\partial^2 u}{\partial y^2} = -\frac{g\beta(T_s - T_\infty)}{\nu} \qquad \frac{\partial^2 T}{\partial y^2} = 0 \qquad \text{at } y = 0 \tag{7–153,154}$$

The evaluation of the coefficients $a_n$ and $b_n$ in accordance with these eight conditions gives

$$u = \frac{U_x}{4}\left[\frac{y}{\delta} - 2\left(\frac{y}{\delta}\right)^2 + \left(\frac{y}{\delta}\right)^3\right] \tag{7–155}$$

where $U_x = g\beta(T_s - T_\infty)\,\delta^2/\nu$, and

$$\frac{T - T_s}{T_\infty - T_s} = \frac{3}{2}\frac{y}{\Delta} - \frac{1}{2}\left(\frac{y}{\Delta}\right)^3 \qquad \text{for } y \leq \Delta \tag{7–156a}$$

$$= 1 \qquad \text{for } y \geq \Delta \tag{7–156b}$$

which is identical to the approximation given by Eq. (7–110) for forced convection. It also follows that

$$q_c'' = -k\frac{\partial T}{\partial y}\bigg|_0 = \frac{3}{2}\frac{k\,(T_s - T_\infty)}{\Delta} \tag{7–157}$$

and

$$Nu_x = \frac{q_c''}{T_s - T_\infty}\frac{x}{k} = \frac{3}{2}\frac{x}{\Delta} \tag{7–158}$$

*Boundary Layer Thickness and Nusselt Number* Following the approach taken in the analysis of forced convection boundary layer flow, these equations can be substituted into the integral momentum and energy equations, Eqs. (7–147) and (7–148), to produce two equations in the unknowns $\delta$ and $\Delta$. However, because both Eqs. (7–147) and (7–148) involve $u$, $T$, $\delta$, and $\Delta$, the resulting expressions appear to be somewhat intimidating. Consequently, analysts have developed a variety of approximate approaches in order to avoid difficulties. These approaches generally involve the use of the simplifying assumption $\delta \simeq \Delta$, which proves to be quite reasonable for natural convection flow of fluids with moderate to small values of Prandtl number.

The simplest approach of this type is a one-parameter method in which the single unknown $\Delta$ is obtained by the solution of either the integral momentum equation or the integral energy equation. To develop this practical approach, Eqs. (7–155) and (7–156) are used to evaluate the integral appearing in the integral energy equation, with the result

$$\int_0^\Delta u(T - T_\infty)\,dy$$
$$= \int_0^\Delta \frac{U_x}{4}(T_s - T_\infty)\left[\frac{y}{\Delta} - 2\left(\frac{y}{\Delta}\right)^2 + \left(\frac{y}{\Delta}\right)^3\right]\left[1 - \frac{3}{2}\frac{y}{\Delta} + \frac{1}{2}\left(\frac{y}{\Delta}\right)^3\right] dy \qquad (7\text{–}159)$$
$$= \frac{U_x(T_s - T_\infty)\Delta}{105} = \frac{1}{105}\frac{g\beta(T_s - T_\infty)^2\Delta^3}{\nu}$$

Substituting this result into the integral energy equation, Eq. (7–148), and using Eq. (7–157) to express $T_s - T_\infty$ in terms of $q_c''$, we obtain

$$\frac{1}{105}\frac{d}{dx}\left[\frac{g\beta(T_s - T_\infty)^2\Delta^3}{\nu}\right] = \frac{3}{2}\frac{\alpha}{\Delta}(T_s - T_\infty) \qquad (7\text{–}160)$$

for specified wall temperature $T_s$.

With $\Delta$ set equal to zero at $x = 0$, this equation can be solved for the distribution in $\Delta$ for arbitrarily specified inputs for $T_s$.

For uniform wall temperature, Eq. (7–160) is put into the form

$$\Delta\frac{d\Delta^3}{dx} = 3\Delta^3\frac{d\Delta}{dx} = \frac{3}{4}\frac{d\Delta^4}{dx} = \frac{315}{2}\frac{\alpha\nu}{g\beta(T_0 - T_\infty)} \qquad (7\text{–}161)$$

Setting $\Delta = 0$ at $x = 0$, the solution to this equation is

$$\Delta^4 = \frac{4}{3}\frac{315}{2}\frac{\alpha\nu x}{g\beta(T_0 - T_\infty)} \qquad (7\text{–}162)$$

or

$$\frac{\Delta}{x} = 3.81\left[\frac{\nu^2}{g\beta x^3(T_0 - T_\infty)}\frac{\alpha}{\nu}\right]^{1/4} = \frac{3.81}{(Gr_x\,Pr)^{1/4}} = \frac{3.81}{Ra_x^{1/4}} \qquad (7\text{–}163)$$

where the *Grashof number* $Gr_x$ and *Rayleigh number* $Ra_x$ are defined by

$$Gr_x = \frac{g\beta(T_0 - T_\infty)x^3}{\nu^2} \qquad (7\text{–}164)$$

and

$$Ra_x = Gr_x\,Pr = \frac{g\beta(T_0 - T_\infty)x^3}{\alpha\nu} \qquad (7\text{–}165)$$

with properties evaluated at the free-stream temperature $T_\infty$. The form of this equation suggests that the Rayleigh number $Ra_x$ is the primary characterizing parameter for this natural convection process. The Rayleigh number represents the *ratio of bouyancy forces to change in viscous forces in the boundary layer*. Whereas the Grashof number is comparable to the Reynolds number for forced convection (since both parameters pertain to momentum transfer), the Rayleigh number characterizes the coupling through buoyancy of momentum and energy. It so happens that $Ra_L$ is the key parameter for most natural convection flows. In this connection, experimental evidence indicates that natural convection flow over a vertical surface is generally laminar for $Ra_x \lesssim 10^9$ and turbulent for $Ra_x \gtrsim 10^9$.

To establish the range of conditions for which this solution is appropriate, the integral momentum equation can be solved with the assumption $\delta = \Delta$. This step leads to a solution for $\Delta$ of the form $\Delta/x = 3.60/Gr_x^{1/4}$. This relation is equivalent to Eq. (7–163) for $Pr = 1.25$ and is within about 10% for $0.8 \lesssim Pr \lesssim 1.8$. Therefore, the analysis should be reasonably accurate over this important range of Prandtl numbers, which includes common gases. But we would expect this simple result to degenerate for values of $Pr$ much outside this range.

With $\delta$ and $\Delta$ approximated by Eq. (7–163), the velocity and temperature profiles can be obtained from Eqs. (7–155) and (7–156). To compare the solution with numerical results available in the literature, Eq. (7–155) is put into the form

$$\frac{ux}{2\nu\, Ra_x^{1/2}} = \frac{1.81}{Pr}\left[\frac{y}{\Delta} - 2\left(\frac{y}{\Delta}\right)^2 + \left(\frac{y}{\Delta}\right)^3\right] \tag{7–166}$$

with $y/\Delta$ represented by

$$\frac{y}{\Delta} = \frac{y}{x}\frac{x}{\Delta} = \frac{1}{3.81}\left(\frac{y}{x} Ra_x^{1/4}\right) \tag{7–167}$$

Calculations obtained from Eqs. (7–156), (7–166), and (7–167) are shown together with numerical solution results obtained by Ostrach [31] and experimental data for air (with $Pr = 0.73$) in Fig. 7–17. The approximate integral solutions for the distributions in velocity and temperature are observed to be quite respectable.

In connection with this result, it should be noted that the form of the integral solution for $\Delta/x$ given by Eq. (7–163) and the form of Eqs. (7–156) and (7–166) indicate that the distributions in dimensionless velocity $ux/(\nu\, Ra_x^{1/2})$ and temperature $(T - T_s)/(T_\infty - T_s)$ are geometrically similar when expressed in terms of $(y/x)Ra_x^{1/4}$. This result provides a basis for similarity variables that are commonly employed in numerical approaches.

To obtain an expression for Nusselt number $Nu_x$, we substitute Eq. (7–163) for $\Delta/x$ into Eq. (7–158), with the result

$$Nu_x = 0.394\, Ra_x^{1/4} \tag{7–168}$$

As shown in Chap. 9, this equation gives rise to a relation for mean Nusselt number $\overline{Nu}_L$ that is in good agreement with experimental data for air. However, we expect the solution to be unreliable for fluids with values of $Pr$ much different from unity. To check this point, Eq. (7–168) is compared with numerical solution results for $Nu_x/Ra_x^{1/4}$ versus $Pr$ in Fig. 7–18. This figure indicates that the simple one-parameter integral solution is within 5% of the numerical solution for $Pr$ between 0.6 and 1.4, but degenerates for values of $Pr$ outside this range.

Whereas this one-parameter method is incapable of fully accounting for the effect of $Pr$ on natural convection flow, two-parameter methods that satisfy both the

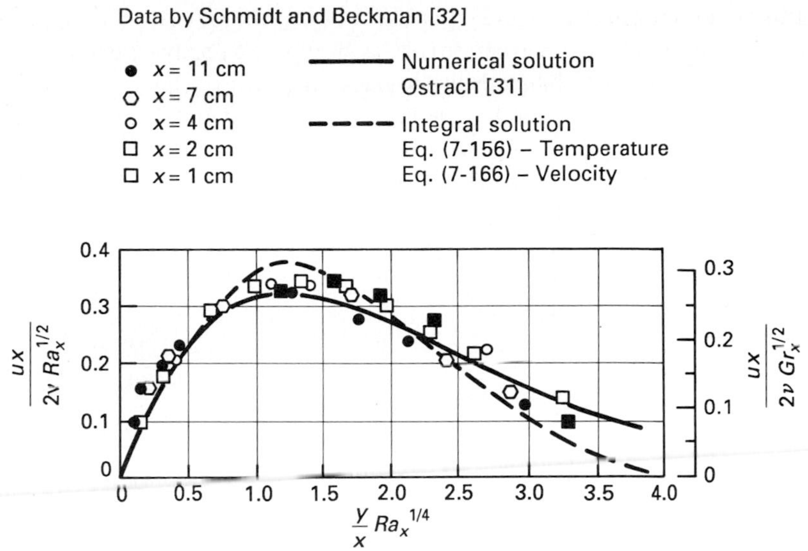


(a) Velocity distribution.

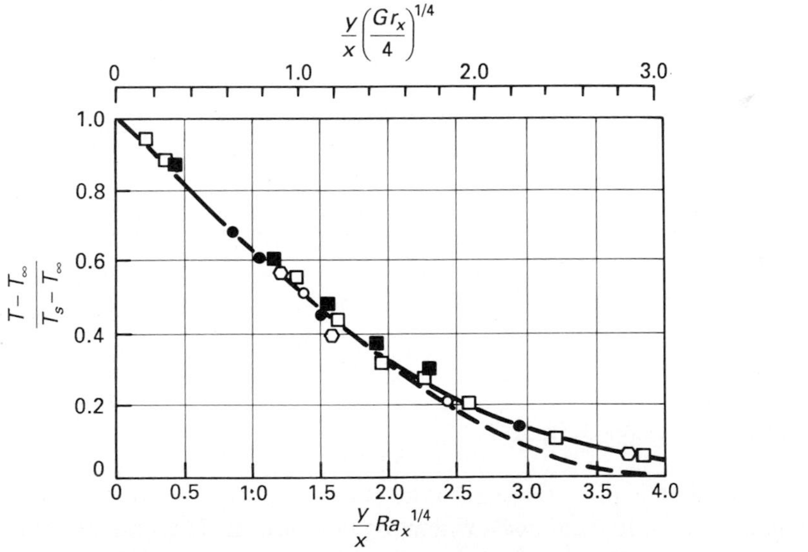


(b) Temperature distribution.

**FIGURE 7–17** Distributions in (a) velocity and (b) temperature for laminar natural convection flow of air ($Pr$ = 0.73) over a vertical plate with uniform wall temperature.

integral momentum equation and integral energy equation perform quite satisfactorily. Two-parameter methods are more involved but prove to be manageable. Calculations for the Nusselt number obtained on the basis of a two-parameter method are shown in Fig. 7–18 to be in quite good agreement with the numerical calculations. The error in the calculations for $Nu_x$ ranges from about 2% for large $Pr$ to less than 0.5% for moderate $Pr$, but reaches an order of 15% for small values of $Pr$. In this connection, the numerical solution is represented to within 2% over the entire range of $Pr$ by the following correlation by Ede [33]:

$$\frac{Nu_x}{Ra_x^{1/4}} = \frac{3}{4}\left(\frac{Pr}{2.5 + 5\sqrt{Pr} + 5\,Pr}\right)^{1/4} \tag{7–169}$$

This equation is also shown in Fig. 7–18.

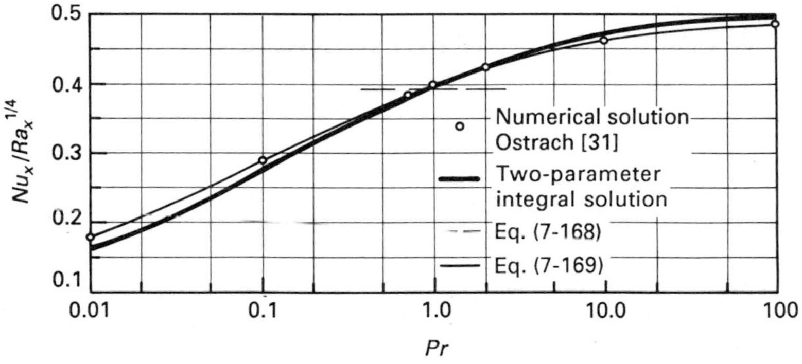


**FIGURE 7–18** Nusselt number for laminar natural convection flow over a vertical plate with uniform wall temperature.

## 7–3 TURBULENT FLOW THEORY

As indicated in Chap. 6, turbulent flow is characterized by nonstreamline unsteady random motion of fluid elements within the flow stream. This chaotic churning action is brought about by minute disturbances within the system, which become unstable for values of the Reynolds number above a certain critical value. For Reynolds numbers below this value, disturbances still occur but are stabilized or dampened by the viscous properties of the fluid, such that laminar conditions prevail.

In the following sections, attention is given to general characteristics of turbulent convection processes and to the classical approach to modeling turbulence. In addition, brief consideration is given to solution results for basic turbulent convection processes.

## 7–3–1 Characteristics of Turbulent Convection Processes

### *The Unsteady Nature of Turbulent Flow*

Extensive experimental flow visualization and anemometer studies have been conducted over the past few years that provide us with a qualitative description of the mechanism associated with turbulent convection processes. First, these studies demonstrate the unsteady character of the entire flow field, including the region very close to the wall. This point is illustrated by Fig. 7–19, in which fluctuations in the instantaneous axial velocity $u$ at a point very close to the wall are shown for turbulent flow over a flat plate.

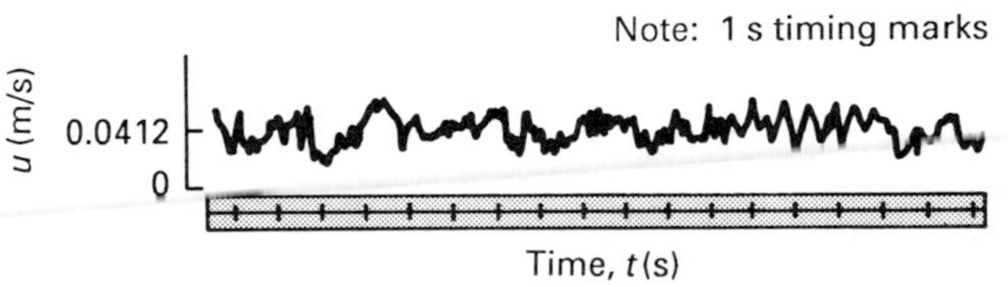


**FIGURE 7–19** Measurements of instantaneous axial velocity $u$ (at $y$ = 0.00127 m; $y^+$ = 8) for fully turbulent boundary layer flow over a flat plate with $U_\infty$ = 0.131 m/s (Runstadler et al. [34]).

A fairly realistic description of the turbulent transport mechanism is shown in Fig. 7–20, which pictures a burst process that involves the intermittent exchange of fluid between the turbulent core and wall region. As mentioned by Kays and Crawford [11], relatively large elements of low-velocity fluid adjacent to the wall surface periodically lift off the surface and eventually break up in the turbulent core. Of course, continuity considerations require that the fluid ejected from the wall region by the burst or lift-off process must be replaced by inrushing fluid with properties, such as axial velocity and temperature, which are associated with the turbulent core.

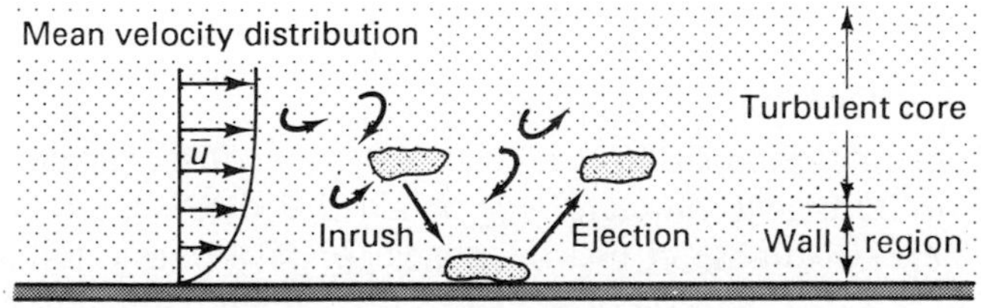


**FIGURE 7–20** Turbulent burst process.

Because turbulence is characterized by fluctuating transport properties, the time-variant velocity profile $u$ is often expressed in terms of mean $\bar{u}$ and fluctuating $u'$ values; that is,

$$u = \bar{u} + u' \tag{7–170}$$

where

$$\bar{u} = \frac{1}{t}\int_0^t u\,dt \tag{7–171}$$

and

$$\overline{u'} = \frac{1}{t}\int_0^t u'\,dt = 0 \tag{7–172}$$

The flow is said to be steady on a time-average basis if $\partial\bar{u}/\partial t = 0$. $u$, $\bar{u}$, and $u'$ are shown in Fig. 7–21 for a situation in which the flow undergoes a change from laminar to turbulent conditions. We see that $u' = 0$ for laminar flow and $u' \neq 0$ for turbulent flow. For developing flows, we have $\partial\bar{u}/\partial x \neq 0$, whereas HFD internal flow is defined for turbulent conditions by

$$\frac{\partial\bar{u}}{\partial x} = 0 \tag{7–173}$$

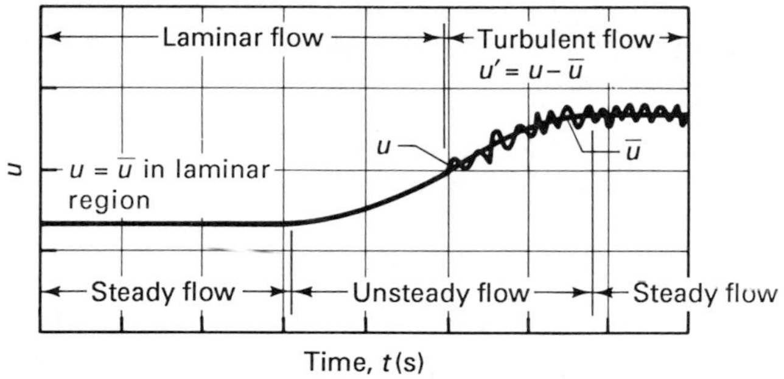


**FIGURE 7–21** Representative instantaneous axial velocity $u$ for process that undergoes a change from laminar to turbulent flow.

The characteristics $v$, $T$, and $P$ are also expressed in terms of mean and fluctuating values. In regard to the temperature profile, TFD conditions are defined for turbulent flow in terms of the time-average temperature distribution $\bar{T}$ by

$$\frac{\partial}{\partial x}\left(\frac{\bar{T} - T_s}{T_b - T_s}\right) = 0 \tag{7–174}$$

Unlike the problem of laminar flow, predictions for the instantaneous unsteady transport properties for turbulent flow cannot be obtained. However, calculations can be developed for the mean transport characteristics for many basic turbulent flow processes by the use of existing theoretical approaches and certain empirical inputs.

The fundamental and particular laws utilized in the analysis of laminar convection processes also apply to turbulent flow. Hence, the principles of continuity, momentum, and energy, together with the Newton law of viscosity and the Fourier law of conduction, will provide the backbone for our study of turbulent flow. But, it is the unsteady and time-average forms of these principles that must be used in the analysis of turbulence.

### *The Consequences of Turbulent Flow*

The practical approach developed in Chap. 8 applies to turbulent flow as well as to laminar flow. However, the appropriate coefficients of friction and heat transfer for turbulent flow must be utilized in these rather simple analyses. Based on the physical picture of turbulence portrayed by Fig. 7–20, we expect that the macroscopic exchange of fluid between the wall and turbulent core regions will bring about a larger transport of momentum and heat than for laminar flow. The coefficients of friction and heat transfer given in Chap. 8 for laminar and turbulent flow substantiate this qualitative observation. This point is reinforced by Figs. 7–6 and 7–14 in which experimental measurements for the friction factors associated with laminar and turbulent HFD flow in circular tubes and flow over flat plates are shown. Notice the increase in friction factor as the flow transcends from laminar to turbulent conditions.

The consequence of this unsteady exchange of fluid between the wall and core regions is also vividly portrayed by the velocity profiles shown in Fig. 7–22 for channel flow. Although the Reynolds number is equal to 4000 for both these profiles, one is laminar and the other is (transitional) turbulent. Notice that much larger wall-region velocities are found in turbulent flow than in laminar flow. Hence, the transport of momentum from the bulk stream to the wall region is indeed most effective for turbulent flow.

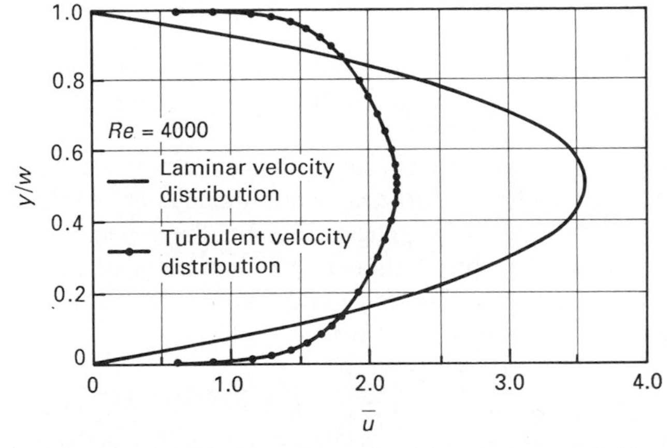


**FIGURE 7–22** Velocity distributions for both laminar and (transitional) turbulent HFD channel flow. (From Knudsen and Katz [35]. Used with permission.)

Similarly, convection heat and mass transfer are much more effective for turbulent flow than for laminar flow. Therefore, industrial convection heat and mass transfer equipment is often operated in the turbulent flow regime in order to achieve high transfer rates.

## 7–3–2 Classical Approach to Modeling Turbulence

The key modeling concepts involved in the classical analysis of turbulent convection processes are introduced in this section in the context of the differential formulation approach for forced convection over plane surfaces. In addition, the very useful integral formulation will be presented.

### *Differential Formulation*

***Fluid Flow*** In the classical approach to turbulence, the continuity and momentum equations are first written in terms of the instantaneous unsteady velocity distribution, after which equations are developed for the time-average velocity distribution.

*Instantaneous Equations* Applying the principles of continuity and momentum to an unsteady fluctuating uniform property flow field, we have

*Continuity*

$$\frac{\partial u}{\partial x} + \frac{\partial v}{\partial y} = 0 \tag{7–175}$$

*Momentum (x-direction)*

$$\rho\left(\frac{\partial u}{\partial t} + u\frac{\partial u}{\partial x} + v\frac{\partial u}{\partial y}\right) = \frac{\partial}{\partial y}\left(\mu\frac{\partial u}{\partial y}\right) - \frac{\partial P}{\partial x} \tag{7–176}$$

The development of these equations follows step by step the formulation of the differential continuity and momentum equations for unsteady laminar boundary layer flow.

Because of the complex random unsteady nature of turbulence, Eqs. (7–175) and (7–176) cannot be solved for the instantaneous fluctuating profiles $u$ and $v$. However, these equations provide a theoretical basis for the development of time-average equations which, when coupled with certain empirical inputs, can be solved for the important mean velocity distributions $\bar{u}$ and $\bar{v}$. Once we know $\bar{u}$ and $\bar{v}$, the mean wall-shear stress $\bar{\tau}_s$ can be determined from

$$\bar{\tau}_s = \mu\left.\frac{\partial \bar{u}}{\partial y}\right|_0 \tag{7–177}$$

and calculations can eventually be obtained for the mean temperature distribution $\bar{T}$ and Nusselt number by solving the time-average energy equation. (For variable property conditions, the time average fluid flow and energy equations must be solved simultaneously.)

*Time-Average Equations* To develop the time-average continuity equation, $u$ and $v$ in Eq. (7–175) are replaced by their mean and fluctuation components $u = \bar{u} + u'$, $v = \bar{v} + v'$; that is,

$$\frac{\partial \bar{u}}{\partial x} + \frac{\partial u'}{\partial x} + \frac{\partial \bar{v}}{\partial y} + \frac{\partial v'}{\partial y} = 0 \qquad (7\text{–}178)$$

Then, taking the time average of each term, we write

$$\overline{\frac{\partial \bar{u}}{\partial x}} + \overline{\frac{\partial u'}{\partial x}} + \overline{\frac{\partial \bar{v}}{\partial y}} + \overline{\frac{\partial v'}{\partial y}} = 0 \qquad (7\text{–}179)$$

or simply,

$$\frac{\partial \bar{u}}{\partial x} + \frac{\partial \bar{v}}{\partial y} = 0 \qquad (7\text{–}180)$$

which is identical in form to Eq. (7–58) for laminar flows.

To develop the time-average momentum equation we first rewrite Eq. (7–176) as

$$\rho\left[\frac{\partial u}{\partial t} + \frac{\partial u^2}{\partial x} + \frac{\partial}{\partial y}(uv) - u\left(\frac{\partial u}{\partial x} + \frac{\partial v}{\partial y}\right)\right] = \frac{\partial}{\partial y}\left(\mu\frac{\partial u}{\partial y}\right) - \frac{\partial P}{\partial x} \qquad (7\text{–}181)$$

By introducing the mean and fluctuating properties into this equation and taking the time average of each term, we obtain (for $\partial\bar{u}/\partial t = 0$)

$$\rho\left[\frac{\partial \bar{u}^2}{\partial x} + \frac{\partial}{\partial y}(\overline{\bar{u}\bar{v}}) + \frac{\partial \overline{u'^2}}{\partial x} + \frac{\partial}{\partial y}(\overline{u'v'})\right] = \frac{\partial}{\partial y}\left(\mu\frac{\partial \bar{u}}{\partial y}\right) - \frac{\partial \bar{P}}{\partial x} \qquad (7\text{–}182)$$

Although the fluctuation terms on the left-hand side of this equation actually represent the transport of momentum by turbulent eddy motion, this equation is traditionally coupled with the continuity equation and put into the form

$$\rho\left(\bar{u}\frac{\partial \bar{u}}{\partial x} + \bar{v}\frac{\partial \bar{u}}{\partial y}\right) = \frac{\partial}{\partial y}\left(\mu\frac{\partial \bar{u}}{\partial y} - \rho\,\overline{u'v'}\right) - \frac{\partial}{\partial x}(\bar{P} + \rho\,\overline{u'^2}) \qquad (7\text{–}183)$$

where $\partial\overline{u'^2}/\partial x$ is generally assumed to be negligible on the basis of experience. Because the term $\mu\,\partial\bar{u}/\partial y$ represents the mean shear stress associated with molecular transport, the term $-\rho\,\overline{u'v'}$ has come to be called the *apparent turbulent shear stress* $\bar{\tau}_t$; that is,

$$\bar{\tau}_t = -\rho\,\overline{u'v'} \qquad (7\text{–}184)$$

$\bar{\tau}_t$ is also referred to as the *Reynolds stress* in honor of the man who first introduced this concept [36]. The actual mean shear stress, $\mu\,\partial\bar{u}/\partial y$, is designated by $\bar{\tau}_y$. Based on this convention, the two terms $\bar{\tau}_t$ and $\bar{\tau}_y$ taken together can be thought of as an *apparent total shear stress* which we shall denote by $\bar{\tau}$.

$$\bar{\tau} = \bar{\tau}_y + \bar{\tau}_t = \mu\frac{\partial \bar{u}}{\partial y} - \rho\,\overline{u'v'} \qquad (7\text{–}185)$$

Based on these definitions for $\bar{\tau}$ and $\bar{\tau}_t$ and assuming negligible pressure gradients in the $y$ direction, the time-average momentum equation, Eq. (7–183), takes the compact and useful form

$$\rho\left(\bar{u}\frac{\partial \bar{u}}{\partial x} + \bar{v}\frac{\partial \bar{u}}{\partial y}\right) = \frac{\partial \bar{\tau}}{\partial y} - \frac{d\bar{P}}{dx} \tag{7–186}$$

With the Reynolds stress term equal to zero for laminar conditions, this equation is seen to reduce to the momentum equation for laminar boundary layer flow, Eq. (7–59).

The differential formulation for turbulent fluid flow is completed by writing the boundary conditions and by specifying the turbulence parameter $\bar{\tau}_t$. The boundary conditions are similar in form to the boundary conditions for laminar flow. For example, we have the no-slip condition at the wall

$$\bar{u} = 0 \qquad \text{at} \quad y = 0 \tag{7–187}$$

*Specification of Reynolds Stress* The distinctive feature of these turbulent fluid-flow equations is the fact that the Reynolds stress $\bar{\tau}_t$ must be specified by the use of theoretical and/or empirical inputs. Although $\bar{\tau}_t$ can be evaluated on the basis of experimental measurements for $u'$ and $v'$, the usual analytical approach, known as the mean-field or Boussinesq [37] method, is to relate this turbulence parameter to the local mean axial velocity $\bar{u}$ by

$$\bar{\tau}_t = \mu_t \frac{\partial \bar{u}}{\partial y} \tag{7–188}$$

where the eddy viscosity $\mu_t$ must be specified; $\nu_t$ ($= \mu_t/\rho$) is the *eddy diffusivity* or *eddy kinematic viscosity*. (The symbol $\epsilon_M$ is often used instead of $\nu_t$ in the turbulence literature.) Notice the similarity between this equation and Newton's law of viscosity, $\tau_y = \mu\ \partial u/\partial y$ or $\bar{\tau}_y = \mu\ \partial \bar{u}/\partial y$. However, unlike $\mu$ which is a property, $\mu_t$ is a turbulence characteristic that varies across the flow field. With the Reynolds stress $\bar{\tau}_t$ expressed in terms of the eddy viscosity $\mu_t$, the apparent total shear stress becomes

$$\bar{\tau} = \bar{\tau}_y + \bar{\tau}_t = (\mu + \mu_t)\frac{\partial \bar{u}}{\partial y} \tag{7–189}$$

or

$$\frac{\bar{\tau}}{\rho} = (\nu + \nu_t)\frac{\partial \bar{u}}{\partial y} \tag{7–190}$$

The differential time-average momentum equation can be written in terms of $\mu_t$ as

$$\rho\left(\bar{u}\frac{\partial \bar{u}}{\partial x} + \bar{v}\frac{\partial \bar{u}}{\partial y}\right) = \frac{\partial}{\partial y}\left[(\mu + \mu_t)\frac{\partial \bar{u}}{\partial y}\right] - \frac{d\bar{P}}{dx} \tag{7–191}$$

Representative measurements for $\nu_t$ are shown in Fig. 7–23 for fully turbulent flow in terms of the dimensionless distances $y^+$ and $y/\delta$; the dimensionless distance $y^+$ is given by

$$y^+ = \frac{U^* y}{\nu} \tag{7–192}$$

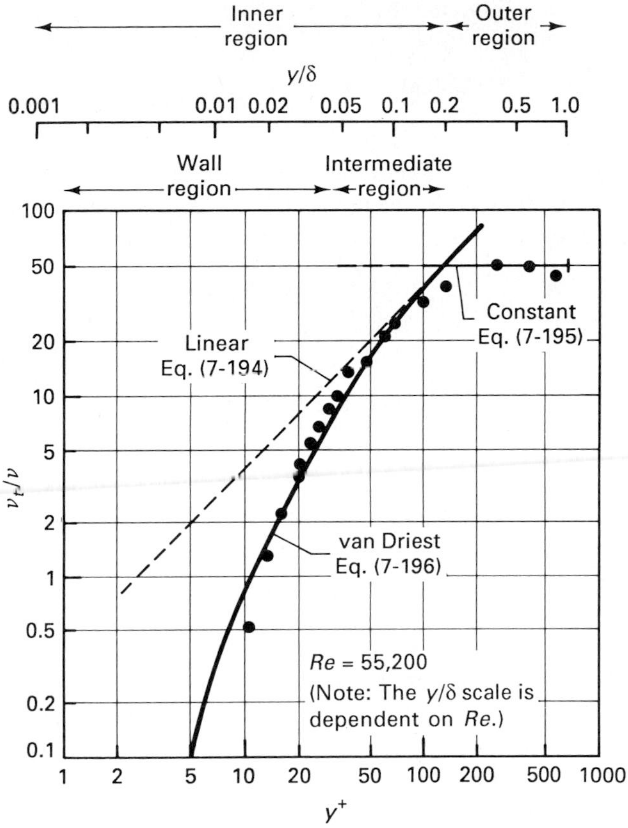


**FIGURE 7–23** Correlations and experimental data for turbulent eddy diffusivity $\nu_t$. (Data for channel flow by Hussain and Reynolds [38].)

The *friction velocity* $U^*$ is defined by

$$U^* = \sqrt{\frac{\overline{\tau}_s}{\rho}} \tag{7–193}$$

and δ represents the channel half-width, tube radius, or boundary layer thickness. Although these data are for channel flow, data for fully turbulent flow in tubes and concentric-circular tube annuli and over surfaces for which no abrupt changes occur follow the same pattern, such that the same correlations for $\nu_t$ can generally be applied to basic fully turbulent flow fields with mild to moderate pressure gradients and transpiration rates.

For practical purposes, a fully turbulent flow field can be broken into the following zones:

| | |
|---|---|
| Inner region | $y/\delta \lesssim y_1/\delta$ |
| Wall region | $y^+ \lesssim 50$ |
| Intermediate region | $y^+ \gtrsim 50$ |
| Outer region | $y/\delta \gtrsim y_1/\delta$ |

where $y_1/\delta$ generally lies between 0.15 and 0.2. The domain outside the wall region is also known as the turbulent core. Referring to Fig. 7–23, we approximate $\nu_t/\nu$ in the intermediate and outer regions (i.e., in the turbulent core) by expressions of the form

$$\frac{\nu_t}{\nu} = \kappa y^+ \qquad \begin{array}{l}\text{Intermediate region}\\ y^+ \gtrsim 50, \quad y/\delta \lesssim y_1/\delta\end{array} \tag{7–194}$$

and

$$\frac{\nu_t}{\nu} = \alpha_1 \delta^+ \qquad \begin{array}{l}\text{Outer region}\\ y/\delta \gtrsim y_1/\delta\end{array} \tag{7–195}$$

where the empirical constant $\kappa$ is approximately 0.41 for fully turbulent flow and $\alpha_1$ ($= \kappa y_1/\delta$) is usually between 0.06 and 0.08.† In the wall region, $\nu_t/\nu$ is clearly nonlinear.

One of the objectives in the analysis of turbulence is to minimize the number of empirical inputs by employing reasonable theoretical models of the actual turbulent transport process. These efforts have been most productive in treating the region adjacent to the wall. Among the various mathematical models for the behavior of $\nu_t$ within the wall region for fully turbulent flow, the formulation by van Driest [40] is perhaps the most popular. This classical analysis gives rise to an expression for $\nu_t/\nu$ within the entire inner region of the form

$$\frac{\nu_t}{\nu} = (\kappa y^+)^2 \left[1 - \exp\left(-\frac{y^+}{a^+}\right)\right]^2 \frac{du^+}{dy^+} \tag{7–196}$$

where $u^+ = \bar{u}/U^*$ and $a^+$ is a damping parameter that is approximately equal to 25 for mild pressure gradients. This equation reduces to Eq. (7–194) as $y^+$ approaches a value of the order of 50. Equation (7–196) is shown to correlate the experimental data in Fig. 7–23 quite well within both the wall and intermediate regions. This basic approach has been extended to boundary layer flows with transpiration and moderate to strong pressure gradients by Kays and Moffat [41], Cebeci and Smith [42], and others.

Other empirical correlations for $\nu_t$ within the inner region have been developed by Rotta [43], Reichardt [44], Deissler [45], Spalding [46], and others. An alternative

† As an alternative approach to characterizing turbulent flow, $\bar{\tau}_t$ is represented by

$$\bar{\tau}_t = \rho \ell^2 \left(\frac{\partial \bar{u}}{\partial y}\right)^2$$

where the *mixing length* $\ell$ is imagined to be the small transverse distance that fluid elements move during a turbulent fluctuation. The mixing length concept was first proposed in 1910 by Prandtl [39].

nonclassical approach to modeling wall turbulence has recently been developed [47–50] which is somewhat akin to the van Driest approach.

It should be noted that approaches have also been proposed for characterizing more complex-type flows associated with transitional turbulence, sudden changes in the boundary conditions, recirculation, and other factors. However, no approach has been devised that is universally applicable to the wide range of conditions encountered in practice [51,52].

***Energy Transfer*** Following the pattern established in the development of the fluid-flow equations, the time-average energy equation for standard conditions can be written as

$$\rho c_P\left(\bar{u}\frac{\partial \bar{T}}{\partial x} + \bar{v}\frac{\partial \bar{T}}{\partial y}\right) = -\frac{\partial \overline{q''}}{\partial y} \tag{7–197}$$

where the *apparent total heat flux* $\overline{q''}$ is defined in terms of the mean heat flux $\overline{q''_y}$ and the *apparent turbulent heat flux* $\overline{q''_t}$ by

$$\overline{q''} = \overline{q''_y} + \overline{q''_t} \tag{7–198}$$

with

$$\overline{q''_y} = -k\frac{\partial \bar{T}}{\partial y} \qquad \text{mean heat flux} \tag{7–199}$$

$$\overline{q''_t} = \rho c_P \overline{v'T'} \qquad \text{apparent turbulent heat flux} \tag{7–200}$$

Notice that Eq. (7–197) reduces to the energy equation for laminar flow, Eq. (7–60), as the turbulent fluctuation term approaches zero.

The time-average energy equation must, of course, be coupled with appropriate boundary conditions. As an example, for turbulent flow over a surface with uniform wall-flux heating, we have

$$-k\frac{\partial \bar{T}}{\partial y} = \overline{q''_0} \qquad \text{at } y = 0 \tag{7–201}$$

Finally, our formulation is closed by specifying the turbulence parameter $\overline{q''_t}$. The apparent turbulent heat flux is generally expressed in terms of $\bar{T}$ by

$$\overline{q''_t} = -k_t\frac{\partial \bar{T}}{\partial y} \qquad \text{or} \qquad \frac{\overline{q''_t}}{\rho c_P} = -\alpha_t\frac{\partial \bar{T}}{\partial y} \tag{7–202,203}$$

where $k_t$ is the *eddy thermal conductivity* and $\alpha_t$ is the *eddy thermal diffusivity*. (The symbol $\epsilon_H$ is often used in place of $\alpha_t$ in the turbulence literature.) It follows that the apparent total heat flux is represented by

$$\overline{q''} = -(k + k_t)\frac{\partial \bar{T}}{\partial y} \tag{7–204}$$

or

$$\frac{\overline{q''}}{\rho c_P} = -(\alpha + \alpha_t)\frac{\partial \overline{T}}{\partial y} \tag{7–205}$$

With $\overline{q''_t}$ expressed in this mean-field format, the differential time-average energy equation takes the form

$$\rho c_P\left(\overline{u}\frac{\partial \overline{T}}{\partial x} + \overline{v}\frac{\partial \overline{T}}{\partial y}\right) = \frac{\partial}{\partial y}\left[(k + k_t)\frac{\partial \overline{T}}{\partial y}\right] \tag{7–206}$$

Although $\nu_t$ and $\alpha_t$ are not fluid properties, they are often expressed in the form of a dimensionless ratio called the *turbulent Prandtl number* $Pr_t$,

$$Pr_t = \frac{c_P \mu_t}{k_t} = \frac{\nu_t}{\alpha_t} \tag{7–207}$$

As early as 1874, Reynolds postulated that $Pr_t$ should be approximately unity because $\overline{q''_t}$ and $\overline{\tau}_t$ both result from the same mechanism; that is, $\alpha_t \simeq \nu_t$ or $Pr_t \simeq 1$. This simple assumption has been utilized with quite good results for basic fully turbulent flow of moderate to high-Prandtl-number fluids. However, recent experimental evidence for moderate $Pr$ fluids indicates that $Pr_t$ is greater than unity in at least a portion of the wall region, falls to a value of the order of 0.9 in the intermediate region out to about $y/\delta = 0.4$, and eventually falls to very small values in the outermost part of the boundary layer [53,54]. The net effect of this $Pr_t$ variation with $y^+$ is evidently balanced out across the boundary layer, such that the $Pr_t = 1$ approximation provides reasonable engineering predictions. However, the turbulent Prandtl number cannot be assumed to be unity for low-Prandtl-number liquid metals. Formulations have been developed for $Pr_t$ for liquid metals by Azer and Chao [55], Jenkins [56], and others.

### *Integral Formulation*

Following the pattern established in Sec. 7–2–3, we integrate the differential time-average continuity, momentum, and energy equations to obtain the integral formulation for turbulent developing flow. The resulting time-average integral momentum and energy equations take the form

$$\frac{d}{dx}\int_0^\delta \overline{u}(U_\infty - \overline{u})\,dy + \frac{dU_\infty}{dx}\int_0^\delta (U_\infty - \overline{u})\,dy = \frac{\overline{\tau}_s}{\rho} \tag{7–208}$$

or, for uniform free-stream velocity,

$$\frac{d}{dx}\int_0^\delta \overline{u}(U_\infty - \overline{u})\,dy = \frac{\overline{\tau}_s}{\rho} \tag{7–209}$$

and

$$\frac{d}{dx}\int_0^\Delta \overline{u}(\overline{T} - T_\infty)\,dy = \frac{\overline{q''_c}}{\rho c_P} \tag{7–210}$$

for basic steady-flow processes with uniform properties.

The boundary conditions for the integral formulation are identical to those utilized in the differential formulation. Notice that these integral equations do *not* involve the turbulent fluctuation terms $\overline{\tau}_t$ and $\overline{q''_t}$! Although the integral formulation for turbulent flow is of the same form as for laminar flow, it should be observed that these modeling equations for turbulence involve $\overline{u}$, $\overline{T}$, $\overline{\tau}_s$, and $\overline{q''_c}$ rather than $u$, $T$, $\tau_s$, and $q''_c$. Thus, the turbulent fluctuating momentum and energy transport mechanism that the parameters $\overline{\tau}_t$ and $\overline{q''_c}$ represent still manages to play its role in the formulation. In this regard, the rather well-behaved nature of the profiles for laminar flow can be attributed to the fact that the momentum and energy transfer mechanisms operate evenhandedly across the entire flow field. On the other hand, the fact that the mean fluctuation momentum and energy transfer (or apparent turbulent shear stress and heat flux, $\overline{\tau}_t$ and $\overline{q''_t}$) associated with turbulent flow vary with $y$ in a nonlinear fashion produces turbulent profiles that possess quite distinctive and different characteristics in the wall region, the intermediate region, and the outer region. Nevertheless, approximate relations for the dimensionless mean velocity and temperature are available [57] which can be used in the development of integral solutions for the local mean wall-shear stress and heat transfer.

## 7–3–3 Solutions

The classical time-average differential continuity, momentum, and energy equations for turbulent convection heat transfer can be solved by exact analytical, numerical, and approximate integral techniques. These methods of solution follow the patterns established in the analysis of laminar flow. In this section, we will introduce the classic Reynolds analogy and we will consider representative solution results for basic turbulent convection processes.

### *The Reynolds Analogy*

The *Reynolds analogy* provides a practical relation between heat and momentum transfer for turbulent boundary layer flow over isothermal surfaces of fluids with Prandtl number of the order of unity. To develop this relation we first combine Eqs. (7–189) and (7–204), with the result

$$\frac{\overline{q''}}{\overline{\tau}} = -\frac{(k + k_t)\,\partial\overline{T}/\partial y}{(\mu + \mu_t)\,\partial\overline{u}/\partial y} = -\frac{k + k_t}{\mu + \mu_t}\frac{d\overline{T}}{d\overline{u}} \tag{7–211}$$

or

$$\frac{\overline{q''}}{\overline{\tau}} = -c_P\left(\frac{k/k_t + 1}{Pr\,k/k_t + Pr_t}\right)\frac{d\overline{T}}{d\overline{u}} \tag{7–212}$$

Since $k_t >> k$ throughout most of the boundary layer, this equation can be approximated by

$$\frac{\overline{q''}}{\overline{\tau}} = -\frac{c_P}{Pr_t}\frac{d\overline{T}}{d\overline{u}} \tag{7–213}$$

Furthermore, since the eddy diffusivity $\nu_t$ and eddy thermal diffusivity $\alpha_t$ both result from the same mechanism of transverse fluctuation, $\alpha_t \simeq \nu_t$ such that $Pr_t \simeq 1$ and Eq. (7–213) reduces to

$$\frac{\overline{q''}}{\overline{\tau}} = -c_P\frac{d\overline{T}}{d\overline{u}} \tag{7–214}$$

Finally, assuming that the ratio $\overline{q''}/\overline{\tau}$ is essentially constant across the boundary layer, Eq. (7–214) is integrated to obtain

$$\frac{\overline{q''_c}}{\overline{\tau}_s} = -c_P\frac{(T_F - T_s)}{U_F} \tag{7–215}$$

or

$$h_x = \frac{\overline{q''_c}}{T_s - T_F} = \frac{\overline{\tau}_s c_P}{U_F} \tag{7–216}$$

and

$$Nu_L = \frac{h_x L}{k} = \frac{f_x}{2} Re_L\, Pr \tag{7–217}$$

which is a general form of the Reynolds analogy. Equation (7–217) can be combined with relations for friction factor to determine the Nusselt number for standard situations involving fluids with moderate values of Prandtl number.

As shown in reference 11, the solution of the momentum equation for fully turbulent HFD internal flow with the turbulent viscosity $\nu_t$ approximated by relations of the type introduced in the previous section gives rise to calculations for friction factor $f$ which can be approximated by the empirical correlation

$$f = 0.046\, Re^{-0.2} \tag{7–218}$$

Equation (7–218) is shown in Fig. 7–24 to compare very well with experimental

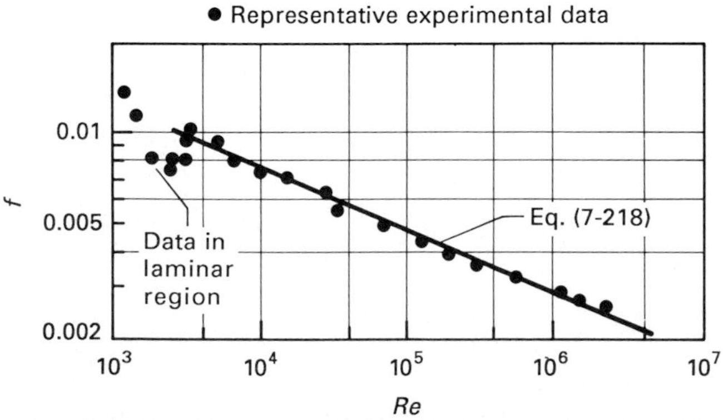


**FIGURE 7–24** Correlation for $f$—HFD turbulent flow in circular tubes.

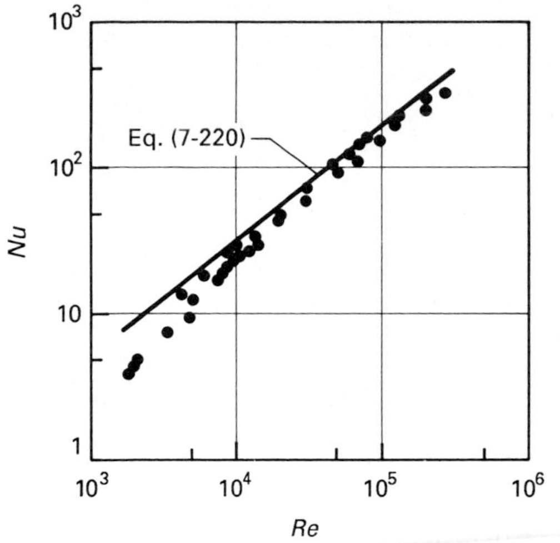


**FIGURE 7–25**
Comparison of calculations for Nusselt number with representative experimental data for air. (Data from Sams and Desmon [58], Deissler and Eian [59], and Barnes and Jackson [60].)

data. Using Eq. (7–218), the Reynolds analogy is written as

$$Nu = 0.023\, Re^{0.8}\, Pr \tag{7–219}$$

Whereas the Reynolds analogy compares favorably with data for gases with $Pr$ of the order of unity, the accuracy of the Reynolds analogy decreases as the difference $|Pr - 1|$ increases. For example, the error in the Reynolds analogy reaches 10% for $0.85 \gtrsim Pr \gtrsim 1.2$ and 20% for $0.72 \gtrsim Pr \gtrsim 1.3$. In this connection, with $Pr_t$ set equal to unity, the theoretical solution (to the energy equation) results in calculations for Nusselt number that are in agreement with the Petukhov and Krillov correlation [61],

$$Nu = \frac{(f/2)\, Re\, Pr}{1.07 + 12.7\sqrt{f/2}\,(Pr^{2/3} - 1)} \tag{7–220}$$

which is applicable for moderate to high values of $Pr$. This equation is compared with experimental data for air in Fig. 7–25.

For the case of fully turbulent boundary layer flow over a flat plate, the integral solution approach yields calculations for friction factor $f_x$ that are approximated by empirical correlations of the form

$$f_x = 0.0592\, Re_x^{-0.2} \tag{7–221}$$

for $10^5 \lesssim Re_x \lesssim 10^7$, and

$$f_x = \frac{0.455}{\ln^2 (0.06\, Re_x)} \tag{7–222}$$

for the broader range $10^5 \lesssim Re_x \lesssim 10^9$. These equations are shown together with experimental data in Fig. 7–26. Using Eq. (7–221), the Reynolds analogy for this application becomes

$$Nu_x = 0.0296\, Re_x^{0.8}\, Pr \tag{7–223}$$

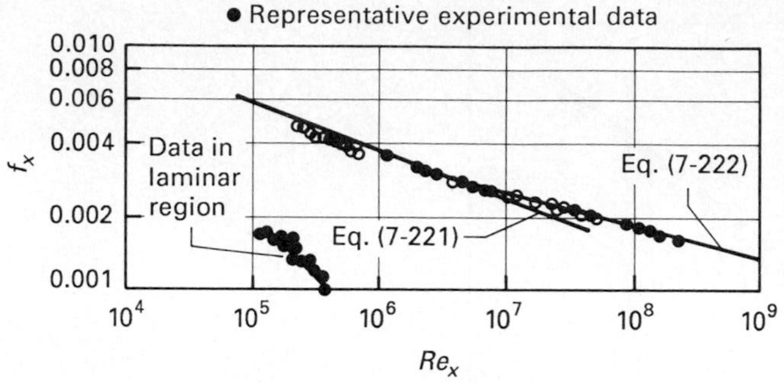


**FIGURE 7–26** Correlations for $f_x$—turbulent boundary layer flow over a flat plate with uniform free-stream velocity.

which is shown to be in good agreement with the data for air in Fig. 7–27. However, as in the case of tube flow, the error in the Reynolds analogy for flow over a flat plate exceeds 20% for values of $Pr$ outside the range 0.72 to 1.3. The theoretical solution for Nusselt number $Nu_x$ can be represented by a correlation developed by White [16],

$$Nu_x = \frac{(f_x/2)\, Re_x\, Pr}{1 + 12.7\sqrt{f_x/2}\,(Pr^{2/3} - 1)} \tag{7–224}$$

which is of the form of Eq. (7–220) and applicable to a wide range of $Pr$.

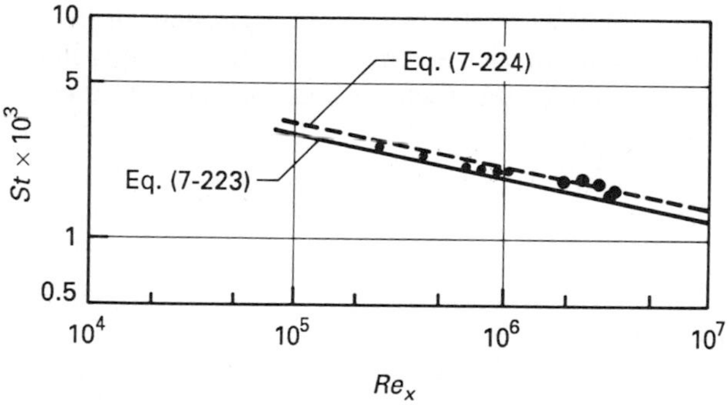


**FIGURE 7–27** Comparison of calculations for Stanton number with experimental data for air (Reynolds et al. [62]).

Recognizing the deficiency in the Reynolds analogy and making use of experimental information, Colburn [63] developed a useful analogy of the form

$$Nu_x = \frac{f_x}{2} Re_x\, Pr^{1/3} \tag{7–225}$$

The *Colburn analogy* proves to be in good agreement with experimental data for fluids with moderate values of $Pr$.

## 7–4 SUMMARY

For purposes of analysis, the two primary categories in convection-heat-transfer systems include *laminar flow* and *turbulent flow* conditions. Whereas the mathematical formulations for laminar flow systems are generally quite straightforward, the solution of the resulting equations can be quite complex. For the simplest problems, as typified by the fully developed tube flow problem introduced in Sec. 7–2–1, analytical solutions can be readily obtained. Analytical solutions to a number of simple laminar convection problems are presented in textbooks by Kays and Crawford [11], White [16], Burmeister [17], and others. Somewhat more complex laminar flow problems involving hydrodynamic and/or thermal development, such as the forced and natural convection boundary layer flows considered in Secs. 7–2–2 and 7–2–3, can be fairly easily solved by means of the integral approach. Analytical solutions can even be developed for some of these types of problems by introducing similarity coordinates. These problems of intermediate complexity are also sometimes solved numerically. The most complex problems involving compressibility effects, variable properties, axial conduction, strong adverse and favorable pressure gradients, mass transfer through a wall, unsteady conditions, and other complications are generally handled by more advanced integral and/or numerical methods.

The analysis of turbulent convection heat transfer involves the conceptual problem of how to mathematically model the complex turbulent transport process itself. The classical mean-field approach introduced in this chapter is the best-known method of attack. Once the eddy diffusivity $\nu_t$ (or mixing length $\ell$) and eddy thermal diffusivity $\alpha_t$ (or turbulent Prandtl number $Pr_t$) are known, the solution of turbulent convection-heat-transfer problems follows the same basic pattern as for laminar flow. Analytical techniques can be utilized for the simplest problems. But numerical and integral techniques are generally called upon to handle the more complex problems, with numerical approaches becoming more and more dominant.

## ■ REVIEW QUESTIONS

**7–1.** Write the fundamental laws that are used in the analysis of convection heat transfer.

**7–2.** List the steps that are involved in the theoretical analysis of laminar convection processes.

**7–3.** What is the difference between incompressible flow and uniform property flow?

**7–4.** Write expressions for the mass transfer rate $d\dot{m}_x$, the momentum transfer rate $d\dot{M}_x$, and the enthalpy flow rate $d\dot{H}_x$ in terms of the primary variables.

**7–5.** Write the differential formulation for fluid flow and energy transfer associated with laminar fully developed flow in a circular tube with uniform wall-heat flux.

**7–6.** Write the differential formulation for fluid flow and energy transfer associated with laminar boundary layer flow over a flat plate with uniform wall temperature.

**7–7.** Write the integral formulation for fluid flow and energy transfer associated with laminar boundary layer flow over a flat plate with uniform wall temperature.

**7–8.** Write the Bernoulli equation and indicate its significance in the analysis of boundary layer flow.

**7–9.** Explain why the thermal boundary layer generally extends well beyond the hydrodynamic boundary layer for laminar forced convection flow of a liquid metal over a flat plate.

**7–10.** Explain why the thermal boundary layer does not extend too far beyond the hydrodynamic boundary layer for laminar natural convection flow of a liquid metal over a vertical plate.

**7–11.** What is the Archimedes principle and how does it relate to natural convection?

**7–12.** What is the physical significance of the Rayleigh number?

**7–13.** What are the main differences between the velocity and temperature distributions for laminar forced convection vs. natural convection?

**7–14.** What is the Boussinesq approximation?

**7–15.** Write the differential formulation for fluid flow and energy transfer associated with laminar natural convection flow over a vertical plate.

**7–16.** Write the integral formulation for fluid flow and energy transfer associated with laminar natural convection flow over a vertical plate.

**7–17.** Illustrate the difference between the shape of the velocity distributions for laminar and turbulent boundary layer flow. Why does this difference occur?

**7–18.** The instantaneous temperature $T$ for turbulent flow is generally expressed in terms of mean $\overline{T}$ and fluctuating $T'$ components. Write the relationship among $T$, $\overline{T}$, and $T'$.

**7–19.** Define (a) Reynolds stress $\overline{\tau}_t$ and (b) apparent total shear stress $\overline{\tau}$.

**7–20.** Definc eddy viscosity $\mu_t$ and explain how it differs from the dynamic viscosity $\mu$.

**7–21.** Define (a) friction velocity $U^*$, (b) dimensionless velocity $u^+$, and (c) dimensionless distance $y^+$.

**7–22.** Define the mixing length.

**7–23.** Define the turbulent Prandtl number $Pr_t$ and explain how it differs from the Prandtl number $Pr$.

**7–24.** What is the significance of the wall region for turbulent boundary layer flow?

**7–25.** What are the capabilities and limitations of the Reynolds analogy?

## PROBLEMS

**7–1.** Develop the continuity and momentum equations for laminar developing flow between parallel plates, assuming variable properties. Also write the limiting forms of these equations for uniform property fully developed flow.

**7–2.** Develop the continuity, momentum, and energy equations for laminar flow between parallel plates for unsteady conditions.

**7–3.** Write the energy equation for developing and thermal fully developed laminar flow between parallel plates for uniform wall temperature $T_0$. Also write the energy equation for asymptotic thermal developed flow associated with uniform wall temperature.

**7–4.** Referring to Prob. 7–3, determine the temperature distribution for asymptotic thermal developed flow with uniform wall temperatures $T_0$ and $T_w$.

**7–5.** Develop a solution for the temperature distribution and Nusselt number for fully developed laminar flow between parallel plates for the case in which one plate is insulated and the other is heated uniformally with $q''_c = q''_0$.

**7–6.** Develop a solution for the temperature distribution and Nusselt number for fully developed laminar flow between parallel plates with uniform wall heat flux into the fluid specified as $q''_0$ at $y = 0$ and $2q''_0$ at $y = w$.

**7–7.** The temperature distribution for TFD flow in a circular tube with uniform wall heat flux is given by Eq. (7–51). Use the defining equation for bulk-stream temperature to show that the Nusselt number can be represented by Eq. (7–57).

**7–8.** Develop a solution for the temperature distribution and wall heat fluxes for asymptotic thermal developed flow between parallel plates with uniform wall temperatures $T_0$ and $T_w$.

**7–9.** Develop a solution for the velocity distribution in laminar fully developed plane Couette flow, for which case $u = 0$ at $y = 0$, $u = U_w$ at $y = w$, and $dP/dx = 0$. Also obtain an expression for the Fanning friction factor $f$.

**7–10.** Develop a solution for the temperature distribution and Nusselt number for thermal fully developed laminar plane Couette flow with uniform wall heat flux into the fluid specified by $q''_w$ at the moving surface and $q''_0 = 0$ at the stationary surface.

**7–11.** Develop a solution for the temperature distribution and Nusselt number for thermal fully developed laminar plane Couette flow with uniform wall heat flux into the fluid specified by $q''_0$ at the stationary surface and $q''_w = 0$ at the moving surface.

**7–12.** Reconsider Prob. 7–10 for plane Couette flow of a highly viscous fluid. Also show that $Pr\,Ec < 0.01$ provides a reasonable criterion for neglecting the effects of viscous dissipation, where $Ec = U_b^2/[c_P(T_s - T_b)]$ is the Eckert number.

**7–13.** Reconsider Prob. 7–11 for plane Couette flow of a highly viscous fluid. Also show that $Pr\,Ec < 0.01$ provides a reasonable criterion for neglecting the effects of viscous dissipation, where $Ec = U_b^2/[c_P(T_s - T_b)]$ is the Eckert number.

**7–14.** Develop a solution for the temperature distribution and wall heat fluxes for asymptotic thermal developed flow of highly viscous fluid between parallel plates with uniform wall temperatures $T_0$ and $T_w$. Also show that the effect of viscous dissipation on the wall heat fluxes is less than 1% for $Pr\,Ec^+ < 0.00166$, where $Ec^+ = U_b^2/[c_P(T_0 - T_w)]$ represents a *modified* Eckert number.

**7–15.** Develop a solution for the temperature distribution and wall heat flux for laminar asymptotic thermal developing plane Couette flow of a highly viscous fluid with uniform wall temperatures $T_0$ and $T_w$. Also show that the effect of viscous dissipation on the wall heat fluxes is less than 1% for $Pr\,Ec^+ < 0.005$, where $Ec^+ = U_b^2/[c_P(T_0 - T_w)]$ is a *modified* Eckert number.

**7–16.** Develop a solution for the temperature distribution and heat flux for laminar asymptotic thermal developing plane Couette flow of a highly viscous fluid with uniform temperature $T_w$ at the moving surface and $q''_0 = 0$ at the stationary surface. Also show that the maximum temperature within the fluid differs from $T_w$ by less than 1% for $Pr\,Ec^* < 0.005$, where $Ec^* = U_b^2/(c_P T_w)$ is a *modified* Eckert number.

**7–17.** Develop a solution for the temperature distribution and wall heat flux for asymptotic thermal developed flow of highly viscous fluid in a circular tube with uniform wall temperature. Also show that the maximum temperature within the fluid differs from $T_0$ by less than 1% for $Pr\,Ec^* < 0.01$, where $Ec^* = U_b^2/(c_P T_0)$ represents a *modified* Eckert number.

**7–18.** Consider a lightly loaded journal bearing lubricated with oil. Develop a relation for the maximum oil temperature in terms of the journal rotational speed $U_J$ and the journal and bearing temperatures $T_J$ and $T_B$.

**7–19.** Consider a lightly loaded journal bearing lubricated with oil. Assuming that no heat is transferred into the journal and that the bearing is maintained at $T_B$, determine the temperature distribution, journal temperature, rate of heat transfer from the bearing, and power required to rotate the journal.

**7–20.** Show that the velocity distribution for laminar fully developed flow in a concentric circular annulus is given by

$$u = -\frac{r_i^2}{4\mu}\frac{dP}{dx}\left[1 - \left(\frac{r}{r_i}\right)^2 + \frac{(r_o/r_i)^2 - 1}{\ln(r_o/r_i)}\ln\frac{r}{r_i}\right]$$

or

$$\frac{u}{U_{\max}} = \frac{(r/r_i)^2 - 1 - 2(r_{\max}/r_i)^2 \ln(r/r_i)}{(r_{\max}/r_i)^2 - 2(r_{\max}/r_i)^2 \ln(r_{\max}/r_i)}$$

where

$$r_{\max}^2 = \frac{r_o^2 - r_i^2}{2\ln(r_o/r_i)}$$

Compare this equation with the experimental data shown in Fig. P7–20. Then outline the steps required to express $u$ in terms of $U_b$ and $f$ in terms of $Re$ and $D_i/D_o$, where $f$ represents the average friction factor over both the inner and outer surfaces.

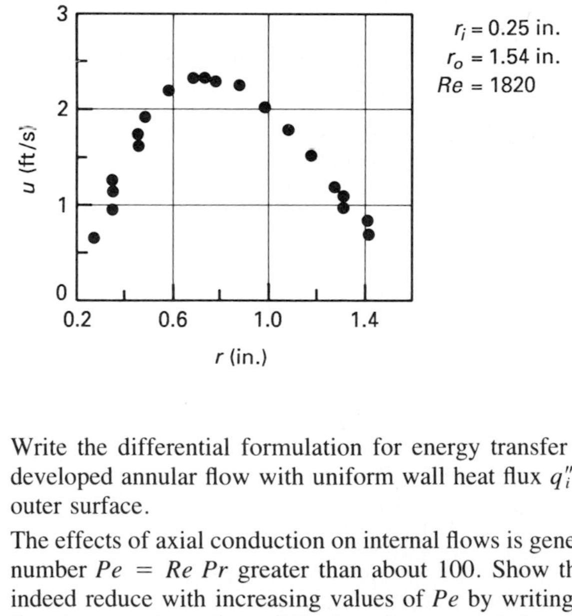
3
2
1
0
u (ft/s)
0.2
0.6
1.0
1.4
r (in.)
$r_i$ = 0.25 in.
$r_o$ = 1.54 in.
Re = 1820

**FIGURE P7–20**
Data for axial velocity distribution—HFD laminar flow in concentric annulus (Rothfus [64]).

**7–21.** Write the differential formulation for energy transfer associated with laminar thermal fully developed annular flow with uniform wall heat flux $q_i''$ at the inner surface and $q_o'' = 0$ at the outer surface.

**7–22.** The effects of axial conduction on internal flows is generally negligible for values of the Peclet number $Pe = Re\,Pr$ greater than about 100. Show that the effect of axial conduction does indeed reduce with increasing values of $Pe$ by writing the energy equation for laminar HFD flow between heated parallel plates in terms of $y/w$ and $u/U_b$ and a suitable dimensionless $x$ variable.

**7–23.** The effect of natural convection on internal forced convection flow between heated or cooled vertical plates is negligible for small values of the ratio $Gr_w/Re^2$, where the Grashof number $Gr_w$ can be represented by

$$Gr_w = \frac{g(\bar{\rho} - \rho)w^3}{\rho\nu^2}$$

and the mean density is defined by $\bar{\rho} = \int_0^w \rho\, dy/w$. Show that the effect of natural convection on the velocity distribution does indeed reduce with decreasing values of $Gr_w/Re$ by developing the momentum equation for fully developed flow between parallel plates with uniform heat flux $q_0''$ in terms of $u/U_b$ and $y$.

**7–24.** Develop a solution for the distributions in velocity and temperature for laminar asymptotic thermal developed flow of an ideal gas between parallel plates with uniform temperatures $T_0$ and $T_w$ for moderate to large values of the ratio $Gr_w/Re$.

**7–25.** Determine the velocity and temperature distributions for asymptotic thermal developed laminar flow between parallel porous plates with uniform wall temperatures $T_0$ and $T_w$ and injection $v_0$ at $y = 0$ and suction $v_w = v_0$ at $y = w$.

**7–26.** Derive Eqs. (7–58), (7–59), and (7–60) for laminar boundary layer flow.

**7–27.** Write the differential formulation for laminar boundary layer flow over a flat plate with uniform free-stream velocity, uniform wall heat flux, and variable property conditions. How does the formulation change for nonuniform free-stream velocity?

**7–28.** Write the continuity, momentum, and energy equations for high-speed laminar boundary layer flow over a flat plate with uniform free-stream velocity, uniform wall temperature, and variable properties.

**7–29.** Develop the integral momentum equation for laminar boundary layer flow given by Eq. (7–81) by integrating the underlying differential equation.

**7–30.** Develop the integral energy equation for laminar boundary layer flow given by Eq. (7–89) by integrating the underlying differential equation.

**7–31.** Develop a fourth-order polynomial approximation for the velocity distribution associated with laminar boundary layer flow over a flat plate with uniform free-stream velocity by satisfying the following five conditions: $u = 0$ and $\partial^2u/\partial y^2 = 0$ at $y = 0$; and $u = U_\infty$, $\partial u/\partial y = 0$, and $\partial^2u/\partial y^2 = 0$ at $y = \delta$. Then develop an approximate integral solution for the friction factor $f_x$.

**7–32.** Develop a fourth-order polynomial approximation for the temperature distribution associated with laminar boundary layer flow over a flat plate by satisfying the following five conditions: $T = T_s$ and $\partial^2T/\partial y^2 = 0$ at $y = 0$; and $T = T_\infty$, $\partial T/\partial y = 0$, and $\partial^2T/\partial y^2 = 0$ at $y = \Delta$. Show that this approximation leads to a relation between wall heat flux $q_c''$, wall temperature $T_s$, and thermal boundary layer thickness $\Delta$ of the form $q_c'' = 2k(T_s - T_\infty)/\Delta$.

**7–33.** Referring to Probs. 7–31 and 7–32, use fourth-order polynomial approximations for velocity and temperature to develop an integral solution for the Nusselt number $Nu_x$ associated with specified wall heat flux of the form $q_c'' = q_0''x^N$. Use this result to obtain relations for $Nu_x$ for uniform wall heat flux and uniform wall temperature.

**7–34.** Using third-order polynomial approximations for $u$ and $T$, develop an integral solution for the Nusselt number $Nu_x$ for laminar flow over a flat plate with uniform free-stream velocity and an unheated starting length followed by uniform wall flux heating.

**7–35.** Using third-order polynomial approximations for $u$ and $T$, develop an integral solution for the Nusselt number $Nu_x$ for laminar flow over a flat plate with uniform free-stream velocity and an unheated starting length followed by uniform wall temperature heating.

**7–36.** Referring to Example 7–6, develop an approximate integral solution for $Nu_x$ by retaining the fifth-order term $r^5/14$ in Eq. (a).

**7–37.** Referring to the integral analysis for laminar boundary layer flow over a flat plate featured in Example 7–6, explain why the solutions are restricted to fluids with $Pr \geq 0.464$ for uniform wall heat flux and $Pr \geq 0.928$ for uniform wall temperature. How can the solution be extended to low values of $Pr$?

**7–38.** The similarity coordinate $\eta = y\,[U_\infty/(\nu x)]^{1/2}$ was established by means of the integral approach for laminar boundary layer flow over a flat plate with uniform free-stream velocity. Show that the differential continuity and momentum equations, Eqs. (7–58) and (7–59), can be transformed into the single ordinary differential equation

$$\frac{d}{d\eta}\left(\frac{u''}{u'}\right) + \frac{1}{2}\frac{u}{U_\infty} = 0 \qquad \text{or} \qquad \zeta''' + \frac{1}{2}\zeta\zeta'' = 0 \qquad \text{(a,b)}$$

where $u' = du/d\eta$, $u'' = d^2u/d\eta^2$ and $\zeta' = u/U_\infty$. Also show that the three boundary conditions for $u$ in $x$ and $y$ reduce to the following two conditions in $\eta$: $u(0) = \zeta'(0) = 0$ and $u(\infty) = \zeta'(\infty) = U_\infty$.

**7–39.** Use the defining relations for stream function $\psi$ ($u = \partial\psi/\partial y$, $v = -\partial\psi/\partial x$) to transform Eq. (7–59) directly into Eq. (P7–38b). Then show that the energy equation given by Eq. (7–60) can be transformed to

$$\theta'' + \frac{Pr}{2}\,\theta' = 0$$

where $\theta = (T - T_0)/(T_\infty - T_0)$ and the primes designate differentiation with respect to $\eta$.

**7–40.** The ordinary differential equation for laminar flow over a flat plate with uniform free-stream velocity given in Prob. 7–38 has been solved by Blasius [27] and others. Based on the classical solution by Blasius, $\zeta''(0) = 0.332$. Representative numerical solution results for the energy equation given in Prob. 7–39 for uniform wall temperature are listed as follows [11]:

| $Pr$ | 0.5 | 0.7 | 1.0 | 7.0 | 10.0 |
|---|---|---|---|---|---|
| $\theta'(0)$ | 0.259 | 0.292 | 0.332 | 0.645 | 0.730 |

Use these results to develop the exact solutions for $f_x$ and $Nu_x$.

**7–41.** Whereas the effects of buoyancy are negligible for internal forced convection flow between vertical parallel plates for $Gr_w/Re \ll 1$, the condition for which natural convection can be neglected for external forced convection flow over a vertical plate is represented by $Ri_L \ll 1$, where $Ri_L = Gr_L/Re_L^2$ is the *Richardson number*. Develop this criterion by rewriting the momentum equation using simple dimensionless variables of the form $U = u/U_\infty$, $V = v/U_\infty$, $X = x/L$, and $Y = y/L$.

**7–42.** Develop a one-parameter integral solution for laminar natural convection flow over a vertical plate with uniform wall temperature using fourth-order polynomial approximations for velocity and temperature.

**7–43.** Develop a one-parameter integral solution for laminar natural convection flow over a vertical plate with uniform wall heat flux using third-order polynomial approximations for velocity and temperature.

**7–44.** Show that the coefficient of thermal expansion can be represented in terms of absolute temperature for ideal gases by $\beta = 1/T$.

**7–45.** Use the results of Appendix J to obtain a relation for the mean Nusselt number for laminar film condensation on a vertical plate.

**7–46.** Following the pattern established in the development of the classical mean momentum equation for turbulent flow, Eq. (7–186), develop the mean energy equation given by Eq. (7–197).

**7–47.** Write the mean continuity, momentum, and energy equations for developing, thermal fully developed, and asymptotic thermal developed turbulent flow between parallel plates for uniform wall temperature.

**7–48.** Referring to Prob. 7–47, sketch the form of the velocity and temperature distributions for asymptotic thermal developing flow with uniform wall temperatures given by $T_0$ and $T_w$.

**7–49.** Write the mean energy equation for developing and thermal fully developed turbulent flow between parallel plates for uniform wall heat flux.

**7–50.** Write the mean momentum and energy equations for fully developed turbulent flow in a circular tube with uniform wall heat flux.

**7–51.** Write the mean energy equation for asymptotic thermal developed high speed turbulent flow in a circular tube with uniform wall temperature.

**7–52.** Write the mean transport equations for turbulent boundary layer flow over a flat plate with nonuniform free-stream velocity and uniform wall heat flux. How does the formulation change for high speed conditions?

**7–53.** Write the mean transport equations for turbulent natural convection flow over a vertical flat plate with uniform wall temperature.

**7–54.** With the eddy diffusivity $\nu_t$ approximated by Eq. (7–194), show that the distribution in mean velocity within the intermediate region for turbulent fully developed flow between parallel plates can be approximated by

$$u^+ = \frac{\bar{u}}{U^*} = \frac{1}{\kappa} \ln y^+ + C \tag{P7–54}$$

where $\kappa = 0.41$ and $C$ is a constant. [Comparisons of this famous logarithmic law with experimental data indicate that the constant $C$ can be set equal to 5.0. Although theoretically restricted to the intermediate region, this equation also provides a fairly good approximation to the data in the outer region.]

**7–55.** With the eddy diffusivity $\nu_t$ approximated by Eq. (7–194) and the turbulent Prandtl number $Pr_t$ set equal to unity for fluids with moderate values of Prandtl number $Pr$, show that the distribution in mean temperature within the intermediate region for turbulent fully developed flow between parallel plates with uniform wall temperature can be approximated by

$$T^+ = \frac{T_s - \bar{T}}{q_0''/(\rho c_P U^*)} = \frac{1}{\kappa} \ln y^+ + B \tag{P7–55a}$$

where $B$ is a constant for a given fluid. [Comparisons of this logarithmic law with experimental data indicate that the constant $B$ can be approximated by

$$B = 12.7\, Pr^{2/3} - 7.7 \tag{P7–55b}$$

Although theoretically restricted to the intermediate region, Eq. (P7–55a) also provides a good approximation to the data in the outer region.]

**7–56.** Show that the logarithmic law given by Eq. (P7–54) is applicable to the intermediate region for turbulent fully developed plane Couette flow. Also sketch the shape of the velocity distribution that should be anticipated.

**7–57.** Demonstrate that the logarithmic laws given in Probs. 7–54 and 7–55 also apply to the intermediate region for fully developed flow in circular tubes.

**7–58.** Similar to the internal flows considered in Probs. 7–54 through 7–57, the inner region for turbulent flow over a flat plate with uniform free-stream velocity is characterized by $\tau \approx \tau_s$ and $q'' \approx q''_c$. Use these approximations to show that the logarithmic laws given in Probs. 7–54 and 7–55 also apply to this classic external turbulent boundary layer flow.

**7–59.** The analogy approach is perhaps the simplest way of developing calculations for the Nusselt number for turbulent boundary layer flow over a flat plate. In this approach, the turbulent Prandtl number $Pr_t$ is set equal to unity and the thermal boundary layer thickness $\Delta$ is approximated by the hydrodynamic boundary layer thickness $\delta$. By utilizing the analogy concept with $u^+$ and $T^+$ approximated by the logarithmic laws given in Probs. 7–54 and 7–55, show that an expression can be developed for $Nu_x$ for uniform wall-temperature heating and fully turbulent flow along the entire length of the plate of the form

$$Nu_x = \frac{(f_x/2)\, Re_x Pr}{1 + \sqrt{f_x/2}(B - C)}$$

Show that this result can be combined with the inputs for $C$ and $B$ given in Probs. 7–54 and 7–55 to obtain

$$Nu_x = \frac{(f_x/2)\, Re_x Pr}{1 + 12.7\sqrt{f_x/2}\,(Pr^{2/3} - 1)}$$

**7–60.** Use Eq. (7–196) and the defining equation for mixing length $\ell$

$$\bar{\tau}_t = \rho \ell^2 \left(\frac{du}{dy}\right)^2$$

to develop an expression for $\ell$ in terms of $y$.

# CHAPTER 8

# CONVECTION HEAT TRANSFER: PRACTICAL ANALYSIS—FORCED CONVECTION

## 8–1 INTRODUCTION

As we have seen, convection processes are categorized according to the geometry of the heat-transfer surface. Three basic categories of forced-convection systems include (1) internal flow in circular tubes, annuli, and noncircular tubular passages with uniform cross-sectional area; (2) external flow over surfaces such as flat plates, cylinders, and spheres; and (3) flow across tube banks. The practical thermal and hydraulic analysis of each of these types of forced-convection processes will be considered in this chapter, with emphasis given to ideal fluids.

## 8–2 INTERNAL FLOW

As indicated in Chap. 6, the practical analysis approach involves the use of coefficients of friction, pressure drop, and heat transfer. Relations for the convection coefficients that are used in the practical analysis of flow through tubular passages are considered in Sec. 8–2–1. The practical thermal analysis approach, which is used in the evaluation and design of heat-transfer equipment, is developed in Sec. 8–2–2, after which brief consideration is given to hydrodynamic aspects in Sec. 8–2–3. The concepts presented in these sections provide a frame of reference for the practical analysis of flow through tube banks and heat-exchanger cores, which is considered in Sec. 8–4 and Chap. 11.

### 8–2–1 Coefficients of Friction, Pressure Drop, and Heat Transfer

A representative heat-transfer core consisting of a tubular passage connected to inlet and exit headers is illustrated in Fig. 8–1. The *contraction ratio* $A/A_1$ is designated by $\sigma$. To characterize appropriately the hydrodynamic and thermal performance of

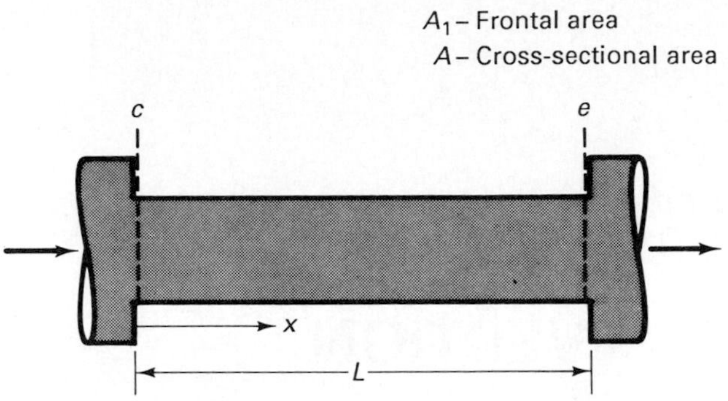
$A_1$ – Frontal area
$A$ – Cross-sectional area
c
e
x
L

**FIGURE 8–1** Passage with abrupt contraction entrance and abrupt expansion exit.

heat-transfer cores such as this, we must account for the effects of flow in the developing and fully developed regions and at the exit.

The coefficient of heat transfer ($h$, $h_x$, $\bar{h}$), which is generally expressed in terms of the Nusselt number,

$$Nu = \frac{hD_H}{k} \qquad (Nu)_x = \frac{h_x D_H}{k} \qquad \overline{Nu} = \frac{\bar{h}D_H}{k} \tag{8–1}$$

and the friction factor ($f$, $f_x$, $\bar{f}$) provide important practical information pertaining to developing and fully developed internal flows.

Because the pressure drop is the primary hydrodynamic dependent variable of interest in the evaluation and design of heat-transfer cores, coefficients for pressure drop are also in common use. The pressure drop over a region extending from the entrance at $x = 0$ (where the pressure is represented by $P_c$) to $x$ is sometimes represented by

$$P_c - P = \frac{px}{A}\frac{f_{\text{app}}}{2}\rho U_b^2 = \frac{2x}{D_H} f_{\text{app}}\,\rho U_b^2 \tag{8–2}$$

where $f_{\text{app}}$ is referred to as the *apparent friction factor*. This approach was introduced in 1942 by Langhaar [1]. An alternative approach to characterizing the pressure drop, which is generally used in modern heat exchanger design methods, involves a relationship of the form

$$P_c - P = \frac{\rho U_b^2}{2}\left(K_c + \frac{4x}{D_H} f\right) \tag{8–3}$$

where $K_c$ is the *entrance-loss coefficient* and $f$ is the friction factor for fully developed flow. For heat-transfer cores with exit headers, we must also account for the pressure loss $\Delta P_{\text{loss}}$ associated with the irreversible free expansion and momentum changes following an abrupt expansion. This pressure loss is generally expressed in terms of the *expansion-loss coefficient* $K_e$ by

$$\Delta P_{\text{loss}} = K_e \frac{\rho U_b^2}{2} \tag{8–4}$$

The practical thermal and hydraulic analysis approaches considered in Secs. 8–2–2 and 8–2–3 feature the use of the heat-transfer coefficients $h$, $h_x$, and $\bar{h}$ and the hydrodynamic coefficients $f$, $K_c$, and $K_e$. We now turn our attention to relations that are available in the literature for these and related coefficients. We shall refer to these theoretical and empirical relations as *convection correlations*. Extensive surveys of convection correlations are available in *Engineering Sciences Data* [2], *Handbook of Heat Transfer* [3], *Compact Heat Exchangers* [4], and *Heat Exchanger Design Handbook* [5].

### Convection Correlations

Convection correlations for Nusselt number, friction factor, and entrance/expansion loss coefficients appearing in the literature for flow in tubular passages are generally expressed in terms of the Reynolds number ($Re = U_b D_H/\nu$), Prandtl number ($Pr = \mu c_P/k$), dimensionless distance $x/D_H$, contraction ratio ($\sigma = A/A_1$), geometry, and thermal boundary conditions by relations of the form

$$\begin{aligned} \textit{Nusselt number} &= \text{fn}\,(Re, Pr, x/D_H, \textit{geometry}, \textit{thermal boundary conditions}) \\ f &= \text{fn}\,(Re, x/D_H, \textit{geometry}) \\ K_c &= \text{fn}\,(Re, x/D_H, \sigma, \textit{geometry}) \\ K_e &= \text{fn}\,(Re, L/D_H, \sigma, \textit{geometry}) \end{aligned} \tag{8–5}$$

and are categorized according to whether the flow is laminar or turbulent. The basis for the theoretical relations presented in this section is developed in Chap. 7.

***Laminar Flow*** For practical purposes, hydrodynamic and thermal fully developed flow can be said to occur for [6]

$$\frac{x}{D_H} \gtrsim 0.05\, Re \qquad \text{HFD} \tag{8–6a}$$

$$\frac{x}{D_H} \gtrsim 0.05\, Re\, Pr \qquad \text{TFD} \tag{8–6b}$$

For HFD and TFD laminar flow with uniform properties, $f$ and $Nu$ are given by

$$f = \frac{C_1}{Re} \qquad Nu = \frac{hD_H}{k} = C_2 \tag{8–7,8}$$

where the constants $C_1$ and $C_2$ are dependent upon geometry, and $C_2$ is dependent upon the thermal boundary conditions. Representative values of $C_1$ and $C_2$ are given

**TABLE 8–1** HFD and TFD laminar flow: Coefficients for Eqs. (8–7) and (8–8)

| *Geometry* | *Friction factor* $f\,Re = C_1$ | *Nusselt Number* $Nu = C_2$ — *Uniform wall temperature* | | *Nusselt Number* $Nu = C_2$ — *Uniform wall heat flux* | |
|---|---|---|---|---|---|
| Square tube | 14.2 | 2.98 | | 3.61 | |
| Circular tube | 16.0 | 3.66 | | 4.36 | |
| Infinite parallel plates | 24.0 | 7.54 | | 8.24 | |
| Circular annulus | | | | | |
| $D_i/D_o$ | | $Nu_i$[a] | $Nu_o$[b] | $Nu_{ii}$[c] | $Nu_{oo}$[d] |
| 0 | 16.0 | $\infty$ | 3.66 | $\infty$ | 4.364 |
| 0.05 | 21.57 | 17.46 | 4.06 | 17.81 | 4.792 |
| 0.1 | 22.34 | 11.56 | 4.11 | 11.91 | 4.834 |
| 0.2 | 23.09 | | | 8.499 | 4.833 |
| 0.25 | | 7.37 | 4.23 | | |
| 0.4 | 23.68 | | | 6.583 | 4.979 |
| 0.5 | | 5.74 | 4.43 | | |
| 0.6 | 23.90 | | | 5.912 | 5.099 |
| 0.8 | 23.98 | | | 5.58 | 5.24 |
| 1.0 | 24.0 | 4.86 | 4.86 | 5.385 | 5.385 |

[a] $Nu_i$–uniform temperature at inner surface; outer surface insulated.
[b] $Nu_o$–uniform temperature at outer surface; inner surface insulated.
[c] $Nu_{ii}$–uniform heat flux at inner surface; outer surface insulated.
[d] $Nu_{oo}$–uniform heat flux at outer surface; inner surface insulated.

in Table 8–1 and in Fig. 8–2. Note that $C_1 = 16$ for flow in a circular tube, with $C_2 = 3.66$ for uniform wall-temperature heating, and $C_2 = 4.36$ for uniform wall-flux heating. Thus, for this particular geometry, the Nusselt number for uniform wall-heat flux is nearly 20% greater than for uniform wall temperature.

It should be noted that because the convection coefficients vary around the perimeter for rectangular passages and other noncircular tubes, $f$ and $h$ are taken as averages over the perimeter.

In the case of fully developed laminar flow in annuli, Table 8–1 and Fig. 8–2 provide correlations for Fanning friction factor $f$ (averaged over both surfaces) and Nusselt number as a function of the diameter ratio $D_i/D_o$. It should be noted that the correlations for Nusselt number represented by $Nu_{ii}$, $Nu_i$, $Nu_{oo}$, and $Nu_o$ pertain to four basic boundary conditions for which one of the two surfaces is insulated; these include $Nu_{ii}$—uniform flux at inner surface; $Nu_i$—uniform temperature at inner surface; $Nu_{oo}$—uniform flux at outer surface; and $Nu_o$—uniform temperature at outer surface. Correlations are available in references 3 to 6 that express the Nusselt numbers at the inner and outer surfaces in terms of these basic solutions for situations in which both surfaces are heated or cooled.

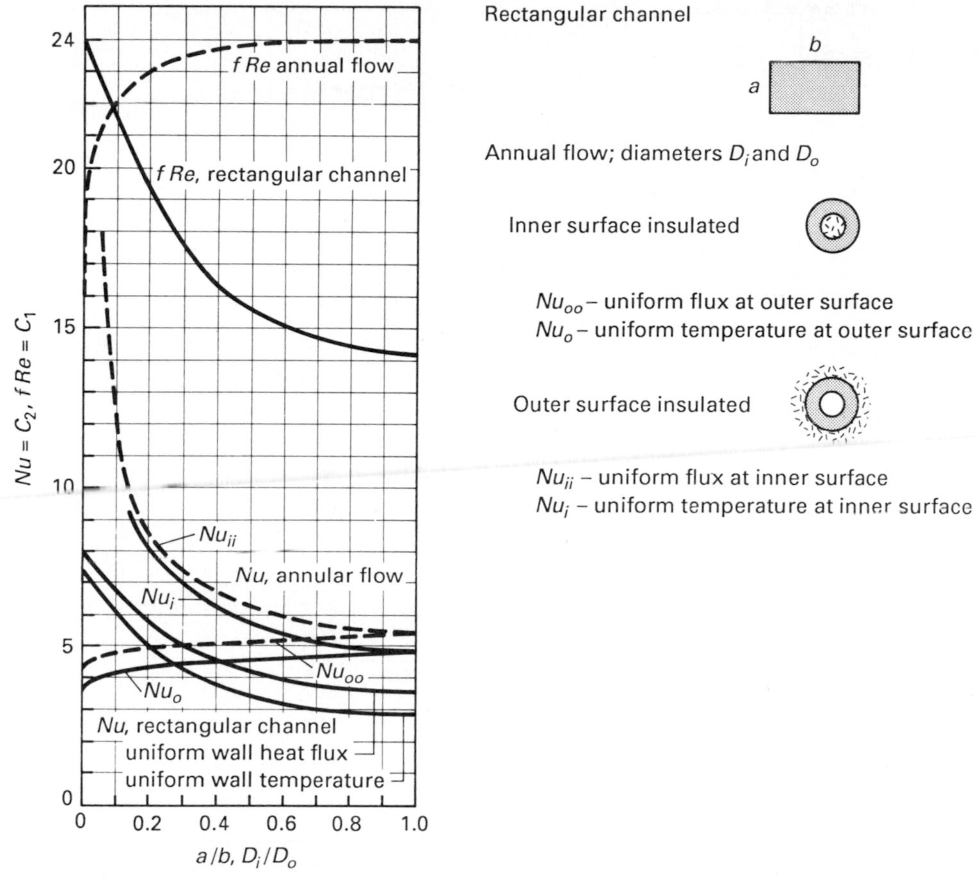


**FIGURE 8–2** Nusselt number and Fanning friction factor for laminar fully developed flow (Kays and Clark [7] and Lundberg et al. [8]).

Turning to hydrodynamic developing flow, for the case in which the frontal area $A_1$ is much greater than the cross-sectional area $A$ of the tubular passage (i.e., $\sigma \simeq 0$), the velocity distribution within the passage develops from an essentially uniform profile at $x = 0$ to a fully developed profile. This problem has been studied for the case of laminar flow in a circular tube by Langhaar [1], Sparrow et al. [9–11], and others. Solution results obtained by Langhaar for $f_x$, $\bar{f}$, and $f_{app}$ are shown in Fig. 8–3. This solution indicates that $f_x$ asymptotically falls toward the HFD value $f$ as $(x/D)/Re$ approaches a value of 0.05. This result provides the basis for the criterion for HFD conditions given by Eq. (8–6). The entrance-loss coefficient $K_c$ for this case is given in Fig. 8–4. Entrance-loss coefficients have also been obtained for flow in circular tubes with nonzero values of the contraction ratio $\sigma$. Correlations are shown in Fig. 8–5 for $K_c$ as a function of $\sigma$ and $Re$ for flow in one or more circular tubes.

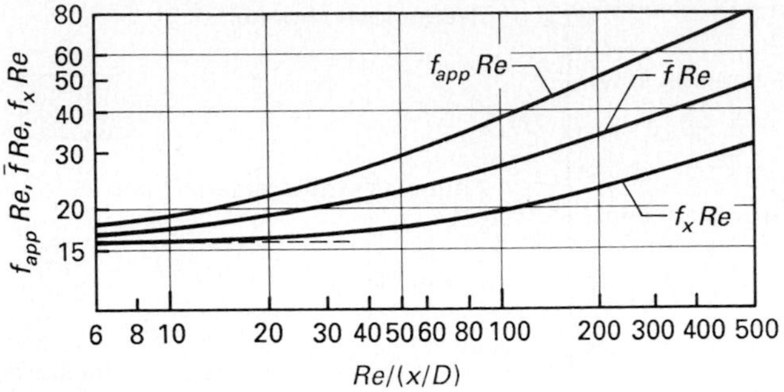


**FIGURE 8–3** Calculations for coefficients of friction by Langhaar [1]—hydrodynamic developing laminar flow in a circular tube.

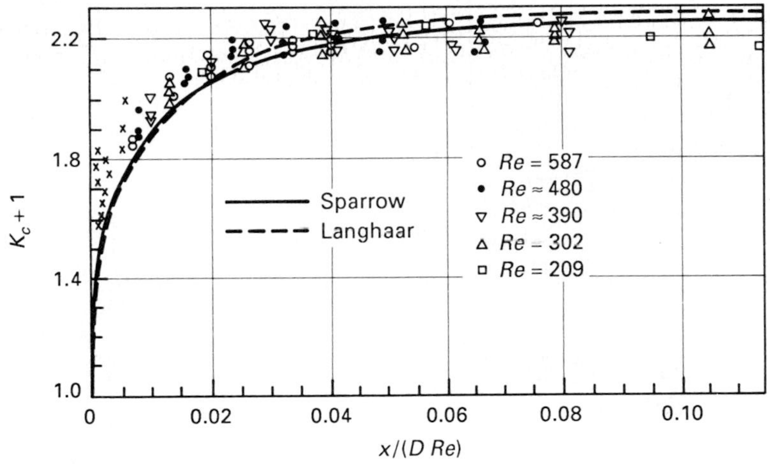


**FIGURE 8–4** Entrance-loss coefficient $K_c$ for laminar flow in a circular tube with $\sigma = 0$ (McComas and Eckert [12]).

This figure also shows correlations for the expansion-loss coefficient $K_e$. Entrance-loss and expansion-loss coefficients associated with other cross-sectional geometries are given in references 4 and 12 through 14.

Theoretical and empirical correlations are available in the literature for the coefficient of heat transfer for thermal developing laminar flow in circular tubes and other geometries. Typical of the design equations that are available is the following popular correlation by Hausen [15] for HFD flow in a circular tube with uniform wall-temperature heating:

$$\overline{Nu} = 3.66 + \frac{0.0668\,\dfrac{Re\,Pr}{x/D}}{1 + 0.04\left(\dfrac{Re\,Pr}{x/D}\right)^{2/3}} \tag{8–9}$$

Hydrodynamic and thermal developed turbulent flow are
shorter distances than for laminar flow. For example, HFD cor
greater than a value of about 10.† TFD conditions occur over ab
for gases such as air, with the length of the thermal entrance
the Prandtl number $Pr$ increases, and vice versa.

The Fanning friction factor for fully developed and fully turb
tubes is well represented by the Prandtl/Nikuradse equation [20]

$$\sqrt{\frac{2}{f}} = 2.46 \ln \left( Re \sqrt{\frac{f}{2}} \right) + 0.292$$

This implicit equation has been shown to be very closely approxim
formula [21]

$$f = (1.58 \ln Re - 3.28)^{-2}$$

Equation (8–15) is compared with experimental data in Fig. 8–7.

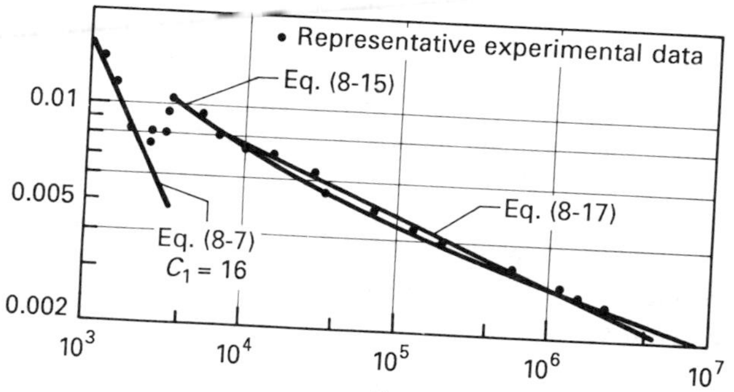


**FIGURE 8–7** Correlations and experimental data for friction factor $f$—HFD flow in circular tubes.

For the Nusselt number, the following correlation by Petukhov
[22] and White [23] is recommended:

$$Nu = \frac{(f/2)\, Re\, Pr}{1.07 + 12.7 \sqrt{f/2}\, (Pr^{2/3} - 1)}$$

This equation is reported to have an accuracy of 5 to 6% in the ranges 1
$5 \times 10^6$ and $0.5 \lesssim Pr \lesssim 200$, and 10% for the same Reynolds number
for $200 \lesssim Pr \lesssim 2000$ [24]. Equation (8–16) is shown in Fig. 8–8, with
Eq. (8–15). Notice that the Nusselt number is much larger for turbulent
for laminar flow.

† Shah and Bhatti [19] suggest the following criterion for the turbulent hydrodynamic entrance le

$$\frac{x}{D} \gtrsim 1.36\, Re^{0.25} \qquad \text{for } Re > 10^4$$



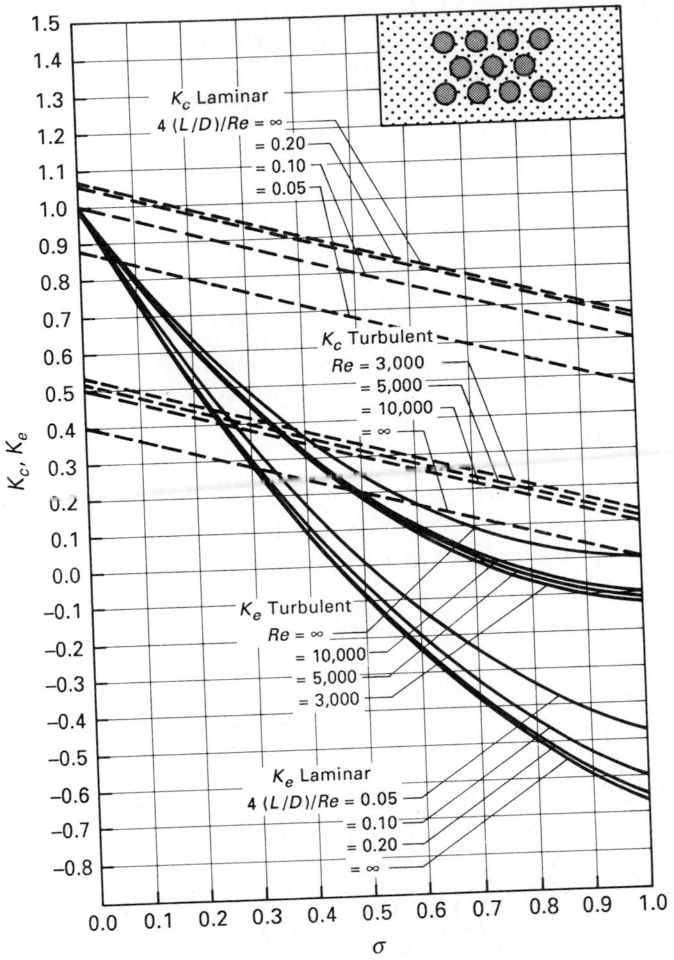


**FIGURE 8–5** Entrance-loss and expansion-loss coefficients for a single or multiple-circular tube heated core with abrupt-contraction entrance and abrupt-expansion exit. (From Kays and London [4]. Used with permission.)

For small values of $x/D$, this equation reduces to

$$\overline{Nu} = 1.67 \left( \frac{Re\, Pr}{x/D} \right)^{1/3} \qquad \text{for} \quad \frac{x/D}{Re\, Pr} \lesssim 0.01 \tag{8–10}$$

These equations are shown in Fig. 8–6. For thermal developing HFD laminar flow with uniform wall-flux heating in short tubes, the local Nusselt number $(Nu)_x$ can be approximated by [16]

$$(Nu)_x = 1.30 \left( \frac{Re\, Pr}{x/D} \right)^{1/3} \qquad \text{for} \quad \frac{x/D}{Re\, Pr} \lesssim 0.01 \tag{8–11}$$

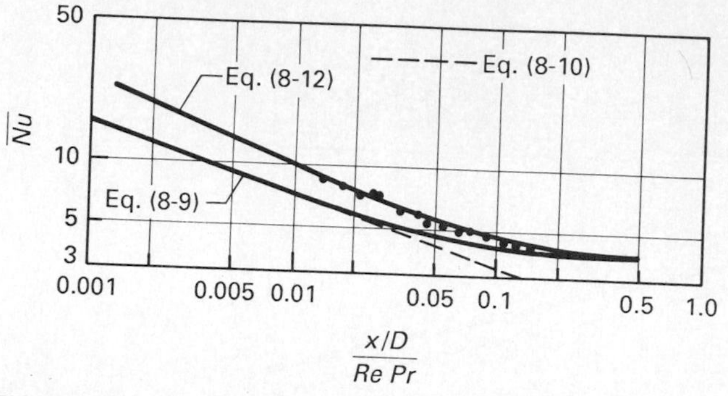


**FIGURE 8–6** Correlations and experimental data for thermal developing flow with uniform wall-temperature heating. (Data for air ($Pr$ = 0.72) from Kays [17].)

Correlations are also available for combined hydrodynam
oping laminar flows. For example, Kays [17] has developed a
oping laminar flow of air in a circular pipe with uniform wall te

$$\overline{Nu} = 3.66 + \frac{0.104 \dfrac{Re\ Pr}{x/D}}{1 + 0.016\left(\dfrac{Re\ Pr}{x/D}\right)^{0.8}}$$

This equation is compared with Eq. (8–9) and with experimental
flow in Fig. 8–6. Note that the Nusselt number is larger for hydrod
flow than for HFD flow. Another correlation for developing flow
temperature, which has been widely used for fluids with $Pr \gtrsim 0.$

$$\overline{Nu} = 1.86\left(\frac{Re\ Pr}{x/D}\right)^{1/3} \qquad \frac{x/D}{Re\ Pr} \lesssim 0.01$$

It should be noted that Eqs. (8–9) to (8–13) do not account
axial conduction. Consequently, these equations are generally restri
in which the *Peclet number* $Pe = Re\ Pr$, which represents the *r*
*flow rate to heat conduction rate*, is not small; that is, $Pe \gtrsim 100$
$Re\ Pr/(x/D) = Pe/(x/D)$ appearing in these equations is the *Gra*
Equations (8–10), (8–11), and (8–13) are most accurate for values of
$x/D \lesssim 0.01\ Re\ Pr$). However, for preliminary design calculations i
with $Pe > 100$, these equations can be extended to $Gz \gtrsim 20$ (i.e
$Re\ Pr$), and $Nu$ can be approximated by the TFD value for $Gz \lesssim 2$
$0.05\ Re\ Pr$).

***Turbulent Flow*** Fully turbulent flow generally occurs for $Re \gtrsim 10^4$
being transitional turbulent in the range $2000 \lesssim Re \lesssim 10^4$.



encountered in heat exchanger applications extends from 500 to 15,000, Kays and London [4] provide performance curves that account for transitional turbulence effects. However, as a first approximation, the Fanning friction factor for HFD transitional turbulent flow is sometimes approximated by

$$f = 0.079\ Re^{-0.25} \tag{8–20}$$

and Eq. (8–16) or Eq. (8–18) is utilized for $Nu$.

The friction factor and Nusselt number are dependent on $x/D_H$, $Re$ and the entrance configuration for developing turbulent flow. Performance curves based on experimental data for air are given by Kays and London [4] for several types of tubes with abrupt-contraction entrances. A convenient correlation for thermal developing flow in short smooth tubes is given by [3]†

$$\overline{Nu} = Nu\left(1 + \frac{C}{x/D}\right) \qquad \text{for} \quad \frac{x_c}{D} \lesssim \frac{x}{D} \lesssim 60 \tag{8–21a}$$

and

$$\overline{Nu} = Nu\,\frac{1.11\ Re^{0.2}}{(x/D)^{0.8}} \qquad \text{for} \quad x \lesssim x_c \tag{8–21b}$$

where $x_c/D = 0.625\ Re^{0.25}$, $C = 1.4$ for the case in which HFD conditions exist at the entrance and $C = 6$ when no hydrodynamic calming section is used. This correlation is sometimes used in preliminary design calculations for heat exchangers.

To provide a practical means of determining the pressure drop $\Delta P$ for developing turbulent flow in heat exchanger cores with abrupt entrance contraction and exit expansion, Kays [14] has developed relations for the entrance-loss coefficient $K_c$ and the expansion-loss coefficient $K_e$ for a number of standard cross sections. Correlations for $K_c$ and $K_e$, which are applicable to cores consisting of single or multiple-circular tubes, are given in Fig. 8–5.

For liquid metals ($0.002 \lesssim Pr \lesssim 0.05$), large amounts of scatter exist in the published data for Nusselt number. A correlation developed by Subbotin et al. [33] and Seban and Shimazaki [34], which is commonly used for design purposes for TFD flows, takes the form

$$Nu = 5.0 + 0.025\ (Re\ Pr)^{0.8} \tag{8–22}$$

where $Re\ Pr \gtrsim 100$ and $L/D \gtrsim 30$. This equation represents the mean of most of the data in the literature for uniform wall-heat flux. Other correlations commonly used for liquid metals include those developed by Skupinski et al. [35] and Labarsky and Kaufman [36].

† Although this equation was developed for turbulent flow of air, it is also often used as a first approximation for other fluids, except for liquid metals.

## EXAMPLE 8–1

Determine the friction factor and the coefficient of heat transfer for fully developed flow of air in a 10-cm-diameter circular tube for flow rates of 0.2 m/s and 2.0 m/s, assuming approximate isothermal conditions at 27°C and atmospheric pressure.

### Solution

*Objective* Determine $f$ and $h$.

*Schematic* Fully developed flow of air in a circular tube.

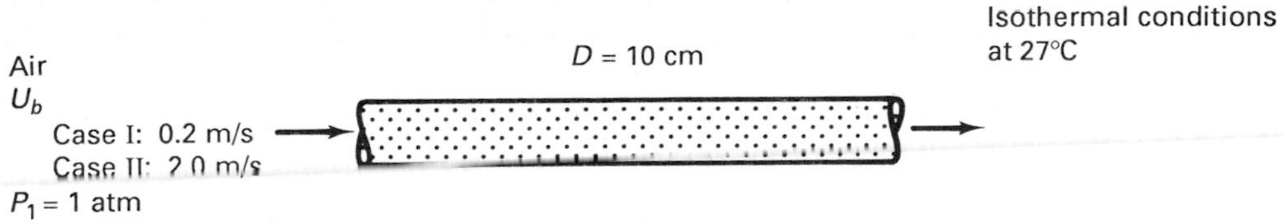
Isothermal conditions
at 27°C
D = 10 cm
Air
U_b
Case I: 0.2 m/s
Case II: 2.0 m/s
P_1 = 1 atm

*Assumptions/Conditions*

forced convection
uniform properties
standard conditions

*Properties* Air at 27°C (Table A–C–5): $\rho = 1.18$ kg/m$^3$, $\mu = 1.85 \times 10^{-5}$ kg/(m s), $\nu = 1.57 \times 10^{-5}$ m$^2$/s, $k = 0.0262$ W/(m °C), $Pr = 0.708$.

*Analysis* The Reynolds numbers for these two flow rates are

$$Re_{\mathrm{I}} = \frac{DU_b}{\nu} = \frac{0.1 \text{ m } (0.2 \text{ m/s})}{15.7 \times 10^{-6} \text{ m}^2/\text{s}} = 1270 \qquad \text{for } U_b = 0.2 \text{ m/s}$$

$$Re_{\mathrm{II}} = 12{,}700 \qquad \text{for } U_b = 2.0 \text{ m/s}$$

Hence, the flow is laminar for the first case and turbulent for the second. The mass flow rates associated with these two situations are

$$\dot{m}_{\mathrm{I}} = \rho A U_b = 1.18 \frac{\text{kg}}{\text{m}^3} \frac{\pi(0.1 \text{ m})^2}{4}\left(0.2 \frac{\text{m}}{\text{s}}\right) = 0.00185 \text{ kg/s}$$

$$\dot{m}_{\mathrm{II}} = 0.0185 \text{ kg/s}$$

For laminar HFD tube flow to occur, $x/D$ must be greater than $0.05\, Re = 63.5$. On the other hand, HFD turbulent flow can occur for $x/D \lesssim 10$. The Fanning friction factors are calculated for fully developed conditions by writing

$$f_{\mathrm{I}} = \frac{16}{Re_{\mathrm{I}}} = \frac{16}{1270} = 0.0126 \qquad \text{from Eq. (8–7)}$$

$$f_{\mathrm{II}} = (1.58 \ln Re_{\mathrm{II}} - 3.28)^{-2} = 0.00737 \qquad \text{from Eq. (8–15)}$$

For short tubes that do not satisfy these $x/D$ requirements, relationships that apply to the developing region must be used.

To achieve TFD tube flow, $x/D$ must be greater than $0.05\ Re\ Pr = 45.0$ for laminar conditions and 10 for turbulent flow. To obtain the coefficients of heat transfer, we first calculate the Nusselt numbers.

$$Nu_{\mathrm{I}} = 3.66 \qquad \text{from Eq. (8–8)}$$

$$Nu_{\mathrm{II}} = \frac{(f/2)\ Re\ Pr}{1.07 + 12.7\sqrt{f/2}\,(Pr^{2/3} - 1)} \qquad \text{from Eq. (8–16)}$$

$$= \frac{(0.00737/2)(12{,}700)(0.708)}{1.07 + 12.7\sqrt{0.00737/2}\,(0.708^{2/3} - 1)} = 36.4$$

The coefficients $h_{\mathrm{I}}$ and $h_{\mathrm{II}}$ are now written as

$$h_{\mathrm{I}} = Nu_{\mathrm{I}}\frac{k}{D} = 3.66\,\frac{0.0262\ \mathrm{W/(m\ ^\circ C)}}{0.1\ \mathrm{m}} = 0.960\ \mathrm{W/(m^2\ ^\circ C)}$$

$$h_{\mathrm{II}} = Nu_{\mathrm{II}}\frac{k}{D} = 9.54\ \mathrm{W/(m^2\ ^\circ C)}$$

## EXAMPLE 8–2

Determine the mean coefficient of heat transfer for flow of air in a circular tube of 5-m length and 10-cm diameter. The mean fluid temperature is about 27°C, the wall temperature is uniform, and the entering velocity and pressure are 0.2 m/s and 0.85 atm.

### Solution

*Objective* Determine $\bar{h}$ over the length $L$.

*Schematic* Flow of air in a circular tube.

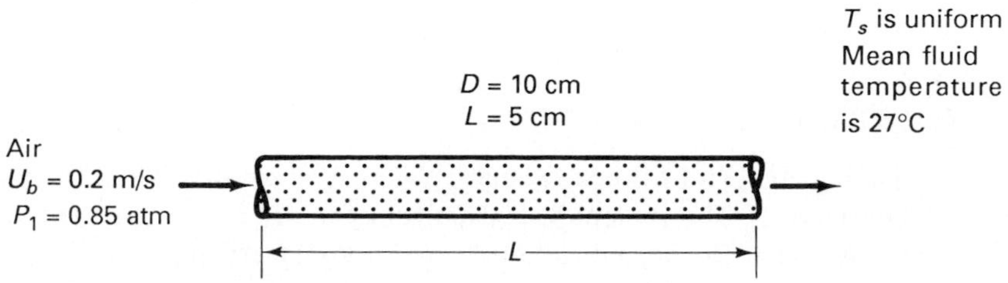
$T_s$ is uniform
Mean fluid temperature is 27°C
D = 10 cm
L = 5 cm
Air
$U_b$ = 0.2 m/s
$P_1$ = 0.85 atm
L

*Assumptions/Conditions*

forced convection
uniform properties

hydrodynamic and thermal developing flow occurs in the entrance region
ideal gas behavior and other standard conditions

*Properties* Because the values of $\mu$, $k$, and $Pr$ for air are fairly insensitive to pressure, these properties can simply be obtained from Table A–C–5; the properties at 300 K are $\mu = 1.85 \times 10^{-5}$ kg/(m s), $k = 0.0262$ W/(m °C), $Pr = 0.708$, and $\rho = 1.18$ kg/m$^3$. Using the ideal gas law to account for the effect of pressure on density, we obtain

$$\rho = 0.85\left(1.18\,\frac{\text{kg}}{\text{m}^3}\right) = 1.00\ \text{kg/m}^3$$

such that

$$\nu = \frac{\mu}{\rho} = \frac{1.85 \times 10^{-5}\ \text{kg/(m s)}}{1.00\ \text{kg/m}^3} = 1.85 \times 10^{-5}\ \text{m}^2\text{/s}$$

*Analysis* To calculate the Reynolds number, we write

$$Re = \frac{DU_b}{\nu} = \frac{0.1\ \text{m}\ (0.2\ \text{m/s})}{1.85 \times 10^{-5}\ \text{m}^2\text{/s}} = 1080$$

which indicates that the flow is laminar.

Assuming that combined hydrodynamic and thermal developing flow occurs in the entrance region, the mean Nusselt number over the length of the tube can be approximated by use of Eq. (8–12),

$$\overline{Nu} = 3.66 + \frac{0.104\,\dfrac{Re\ Pr}{L/D}}{1 + 0.016\left(\dfrac{Re\ Pr}{L/D}\right)^{0.8}}$$

where

$$\frac{L/D}{Re\ Pr} = \frac{5\ \text{m}/(0.1\ \text{m})}{1080(0.708)} = 0.0654$$

Substituting for the term $(L/D)/(Re\ Pr)$, we obtain

$$\overline{Nu} = 3.66 + 1.36 = 5.02$$

or

$$\bar{h} = \overline{Nu}\,\frac{k}{D} = 5.02\,\frac{0.0262\ \text{W/(m °C)}}{0.1\ \text{m}} = 1.32\ \text{W/(m}^2\ \text{°C)}$$

Notice that this result for $\overline{Nu}$ is 38% greater than the value for TFD flow, $Nu$ = 3.66, which oxcurs for $(x/D)/(Re\ Pr) \gtrsim 0.05$ (i.e., $x \gtrsim 3.8$ m).

### *Effects of Property Variation*

The significance of property variation caused by temperature should always be kept in mind when analyzing convection-heat-transfer processes. The two primary effects that are brought about by temperature-induced property change are (1) direct effects caused by temperature variations (a) over the cross section and (b) in the direction of flow; and (2) indirect effects involving the creation of buoyancy forces by variations in the density (or concentration) with temperature. The second of these two effects pertains to combined forced and natural convection, which is discussed in Chap. 9.

The effects of property variation on the local momentum and heat-transfer characteristics that result from the variation in temperature over the cross-section are generally accounted for by utilizing correction factors for the friction factor and Nusselt number. For liquids, where variations in the viscosity are generally the main consideration, these correction factors take the form

$$\frac{f}{f_{cp}} = \left(\frac{\mu_s}{\mu_b}\right)^m \qquad \frac{Nu}{Nu_{cp}} = \left(\frac{\mu_s}{\mu_b}\right)^n \tag{8–23,24}$$

where the subscript $cp$ represents the constant property characteristics, $\mu_s$ is evaluated at the wall temperature $T_s$, the other properties are evaluated at the bulk-stream temperature $T_b$, and the exponents $m$ and $n$ are dependent upon the conditions. On the other hand, the variations in density and thermal conductivity as well as viscosity are usually the most important factors for gases. These three properties are related to the *absolute* temperature for gases in such a way as to permit the use of correction factors of the form

$$\frac{f}{f_{cp}} = \left(\frac{T_s}{T_b}\right)^m \qquad \frac{Nu}{Nu_{cp}} = \left(\frac{T_s}{T_b}\right)^n \tag{8–25,26}$$

over the range $0.33 \lesssim T_s/T_b \lesssim 3.0$. Representative values of $m$ and $n$ are given in Table 8–2 for liquid and gas flow with various conditions. The relationships for friction factor and Nusselt number that are presented throughout this chapter for ideal isothermal conditions may be adjusted to account for mild to moderate property variations by merely multiplying by the appropriate viscosity or temperature-ratio correction factor.

The variation in properties over the length of a tube is often more pronounced than the variation over the cross section. The effect of property variation in the direction of flow on the coefficients of friction and heat transfer are generally accounted for in design work by evaluating the properties at the arithmetic average of the inlet and outlet bulk-stream temperatures, with the Reynolds number defined by Eq. (6–35b). According to Kays and London [4], this approach is adequate for cases involving gas flow with the absolute temperature variation along the tube being less than 2 to 1. When this approach is used, the resulting average value of $T_b$ can also be used to represent $T_b$ in Eqs. (8–23)–(8–26), which are used to correct for local property variation effects over the cross section. For situations in which the temperature variation in the flow direction is large, Kays and London [4] suggest that analyses

**TABLE 8-2** Variable property conditions—coefficients associated with Eqs. (8-23) through (8-26)

| *Fluid* | *Condition* | *m* | *n* | *References* |
|---|---|---|---|---|
| Liquid | Laminar | | | |
| | Cooling | 0.50 | −0.14 | 18,26 |
| | Heating | 0.58 | −0.14 | 18,26 |
| | Turbulent[a] | | | |
| | Cooling | 0.25 | −0.25 | 6,24 |
| | Heating | 0.25 | −0.11 | 6,24,37 |
| Gas | Laminar | 1.0 | 0 | 6 |
| | Turbulent | | | |
| | Cooling | −0.1 | 0 | 6 |
| | Heating[b] | −0.1 | −0.5 | 6 |

[a] The *Nu* correction for turbulent flow of liquids is recommended by Petukhov [24] for the ranges $0.08 \lessapprox \mu_s/\mu_b \lessapprox 40$; $2 \lessapprox Pr \lessapprox 140$; and $10^4 \lessapprox Re \lessapprox 1.25 \times 10^5$.

[b] Sleicher and Rouse [38] recommend

$$n = 0.3 - \left(\log \frac{T_s}{T_b}\right)^{0.25}$$

be developed for separate sections of the tube or heat exchanger over which the temperature variation is not excessive.

### *Enhancement of Heat Transfer*

The correlations for heat transfer and friction presented earlier in this section only apply to passages with smooth surfaces. The heat transfer in industrial applications is often enhanced by the use of fins, rough surfaces, and other techniques. A review of the various methods of enhancement has been developed by Bergles and associates [39–42]. For example, heat transfer for turbulent tube flow has been increased by as much as 400% by the use of rough surfaces. However, large increases in the friction factor have also been found to occur, such that care must be taken to utilize surfaces that are efficient as far as both heat transfer and pressure drops are concerned.

Thermal and hydraulic analyses of systems employing enhancement features such as rough surfaces or fins require the use of special design correlations. Empirical correlations for friction and heat transfer are generally provided by the commercial manufacturers of finned tubes or rough surfaces.

For tubes with rough surfaces characterized by roughness elements with average height $\epsilon$, $f$ is commonly evaluated by use of the well-known *Moody chart*†, which

† The Moody friction factor chart is traditionally presented in terms of the Moody (or Darcy-Weissback) friction factor $\lambda$,

$$\lambda = -\frac{(dP/dx)D_H}{\rho U_b^2/2}$$

where $dP/dx$ represents the pressure gradient for HFD flow. It should be noted that $\lambda = 4f$. Unfortunately, the symbol $f$ is sometimes used in the literature to represent the Moody friction factor as well as the Fanning factor. Therefore, care must be taken to distinguish between these two types of friction factors.

is shown in Fig. 8–10. Typical values of the *roughness* ϵ are given in Table 8–3. However, Fig. 8–10 indicates that the primary characterizing factor for flow in rough tubes is the *relative roughness* ϵ/*D* (and Reynolds number *Re*) rather than the average roughness height ϵ itself.

**TABLE 8–3** Representative values of roughness ϵ

| *Pipe or tube material* | ϵ (μm) |
|---|---|
| Commercial steel or wrought iron | 46 |
| Cast iron | 250 |
| Riveted steel | 900–9000 |
| Asphalted cast iron | 120 |
| Galvanized surface | 150 |
| Drawn tubing | 1.5 |
| Concrete | 300–3000 |
| Wood stave | 180–900 |

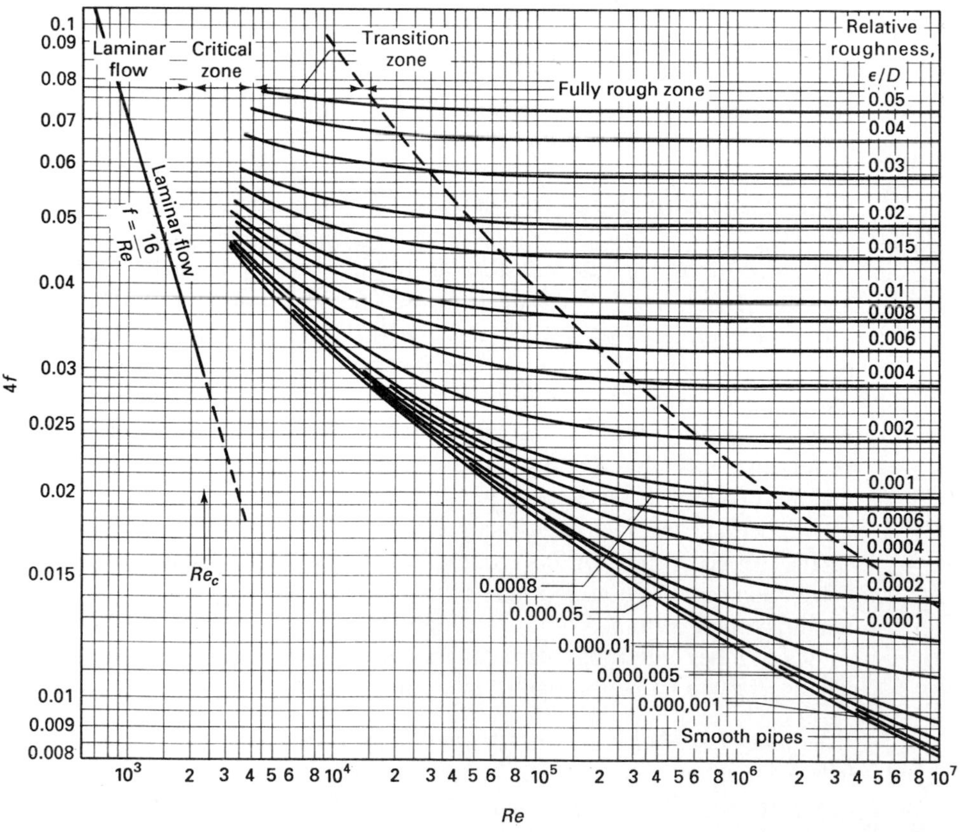


**FIGURE 8–10** The Moody friction factor chart for rough surfaces (Moody [43]).

### *Other Factors*

The coefficients of friction and heat transfer are also sometimes influenced by other factors such as unsteady operation, nonuniform distributions in wall temperature or heat flux, viscous dissipation, phase change (see Chap. 10), and mass transfer through the wall. As indicated earlier, when the convection process is complicated by factors such as these, their influence on the coefficients should be accounted for. Consequently, one must be prepared to search the literature for appropriate correlations, or even to undertake or commission an experimental or theoretical study.

## EXAMPLE 8–3

Show the effect of temperature-induced property variation on the shape of the velocity distribution for laminar flow of (a) gas in a heated tube and (b) liquid in a cooled tube.

### Solution

*Objective* Sketch velocity distributions for uniform and variable property conditions.

*Assumptions/Conditions*

forced convection
variable properties
standard conditions

*Analysis* Referring to Tables A–C–3 through A–C–5, we observe that the viscosity is directly proportional to temperature for gases and inversely proportional for liquids. As a result, when a gas is heated or when a liquid is cooled, the fluid near the wall is more viscous than the fluid further away from the wall. Consequently, we would expect the actual velocity distribution for these conditions to be lower relative to the isothermal velocity profile in the region near the wall and larger in the region away from the wall. This perspective is shown in Fig. E8–3 for laminar tube flow. Of course, the opposite result would be expected for the cooling of a gas or heating of a liquid.

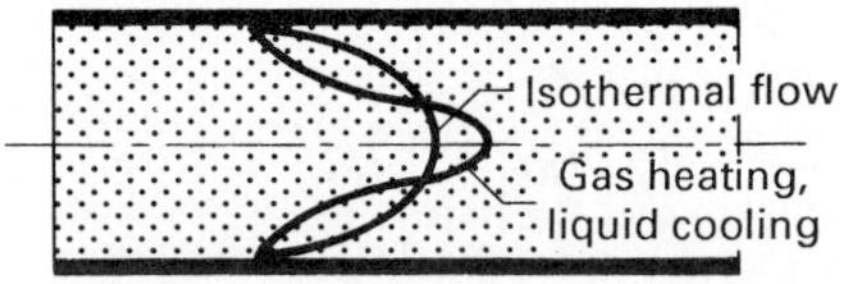


**FIGURE E8–3**
Influence of property variation on velocity distribution for laminar tube flow.

## EXAMPLE 8–4

Air at 11.4°C and 1 atm pressure enters a 10-cm-diameter tube with a mass flow rate of 0.0185 kg/s. The tube wall is maintained at 175°C and the outlet temperature is 42.6°C. Determine the friction factor and the coefficient of heat transfer for approximate fully developed conditions.

### Solution

*Objective* Determine $f$ and $h$ for fully developed conditions.

*Schematic* Fully developed flow of air in a circular tube—uniform wall temperature.

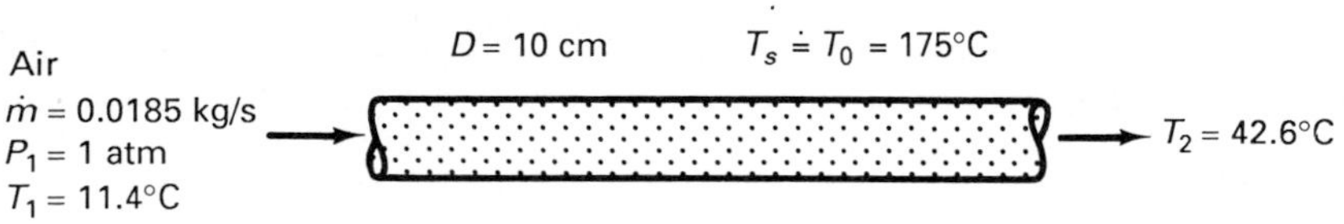
Air
$\dot{m}$ = 0.0185 kg/s
$P_1$ = 1 atm
$T_1$ = 11.4°C
$D$ = 10 cm
$T_s \doteq T_0$ = 175°C
$T_2$ = 42.6°C

*Assumptions/Conditions*

forced convection
moderate property variation
standard conditions

*Properties* Air at 27°C (Table A–C–5): $\rho = 1.18$ kg/m³, $\mu = 1.85 \times 10^{-5}$ kg/(m s), $k = 0.0262$ W/(m °C), $Pr = 0.708$.

*Analysis* To account (approximately) for the effect of property variations on the coefficients of friction and heat transfer, we will evaluate the fluid properties at the arithmetic average bulk-stream temperature and we will employ the correction factors given by Eqs. (8–25) and (8–26),

$$\frac{f}{f_{cp}} = \left(\frac{T_0}{T_b}\right)^m \qquad \frac{Nu}{Nu_{cp}} = \left(\frac{T_0}{T_b}\right)^n$$

To compute the mass flux $G$ and Reynolds number $Re$, we write

$$G = \frac{\dot{m}}{A} = \frac{0.0185 \text{ kg/s}}{\pi(0.1 \text{ m})^2/4} = 2.36 \text{ kg/(m}^2 \text{ s)}$$

and

$$Re = \frac{GD}{\mu} = \frac{[2.36 \text{ kg/(m}^2 \text{ s)}](0.1 \text{ m})}{1.85 \times 10^{-5} \text{ kg/(m s)}} = 12{,}800$$

which is essentially the same as the result obtained in Example 8–1 for isothermal conditions at 27°C. It follows from Example 8–1 that $f_{cp} = 0.00737$ and $Nu_{cp} =$

36.4. Referring to Table 8–2, we find that $m = -0.1$ and $n = -0.5$ for turbulent flow and heating. Setting $T_0 = 448$ K and $T_b = 300$ K, we have

$$f = 0.00737\left(\frac{448\text{ K}}{300\text{ K}}\right)^{-0.1} = 0.00737(0.985) = 0.00708$$

and

$$Nu = 36.4\left(\frac{448\text{ K}}{300\text{ K}}\right)^{-0.5} = 36.4(0.818) = 29.8$$

Therefore, the coefficient of heat transfer is

$$h = Nu\frac{k}{D} = 29.8\frac{0.0262\text{ W/(m °C)}}{0.1\text{ m}} = 7.81\text{ W/(m}^2\text{ °C)}$$

## 8–2 2 Practical Thermal Analysis

Practical lumped analyses are developed in this section for the overall heat-transfer performance of steady internal flow in passages with uniform cross-sectional area, as illustrated in Fig. 8–11. The inlet bulk-stream temperature, fluid properties, mass flow rate, and system dimensions (or outlet bulk-stream temperature $T_2$) are assumed to be known. The main objective of these analyses is to predict the overall rate of heat transfer $q_c$ and the outlet temperature $T_2$ (or the system dimensions) in terms of the specified system parameters and boundary conditions of interest. In addition, we are generally interested in knowing the distribution in bulk-stream temperature $T_b$ and the wall temperature $T_s$ (for cases in which the heat flux is specified).

To develop expressions for the two primary parameters $q_c$ and $T_2$, two independent equations are developed by applying the first law of thermodynamics, Eq. (1–1),

$$\Sigma\,\dot{E}_o - \Sigma\,\dot{E}_i + \frac{\Delta E_s}{\Delta t} = 0 \qquad (8\text{–}27)$$

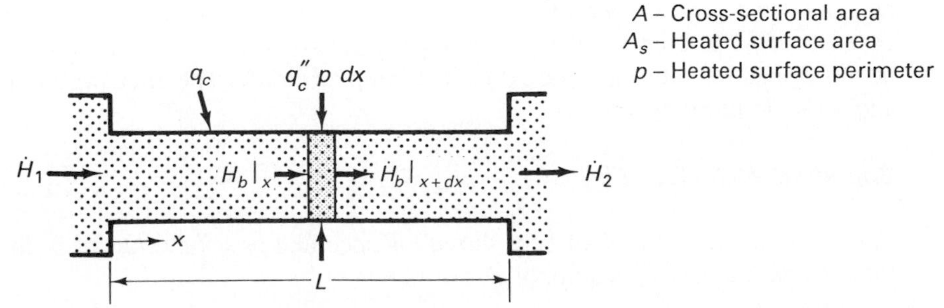


**FIGURE 8–11** Bulk-stream enthalpy and heat-transfer rate associated with flow through lumped volume $AL$ and lumped/differential volume $A\,dx$ within a tubular passage.

to (1) the lumped volume $AL$, and (2) the lumped/differential volume $A\,dx$ (see Fig. 8–11). Applying the first law of thermodynamics to the fluid volume $AL$, we obtain

$$\dot{H}_2 - \dot{H}_1 - q_c = 0 \tag{8–28}$$

or

$$q_c = \dot{H}_2 - \dot{H}_1 = \int_{\dot{H}_1}^{\dot{H}_2} d\dot{H}_b \tag{8–29}$$

Using Eq. (6–45) to express $d\dot{H}_b$ in terms of $T_b$ and $c_P$ for ideal fluids, this equation takes the form

$$q_c = \dot{m} \int_{T_1}^{T_2} c_P \, dT_b \tag{8–30}$$

for uniform $\dot{m}$, which reduces to

$$q_c = \dot{m} c_P (T_2 - T_1) \tag{8–31}$$

for uniform specific heat. Similarly, the application of the first law of thermodynamics to the fluid volume $A\,dx$ gives

$$\dot{H}_b\big|_{x+dx} - \dot{H}_b\big|_x - q_c''\,p\,dx = 0 \tag{8–32}$$

or

$$q_c'' p = \frac{d\dot{H}_b}{dx} \tag{8–33}$$

where $p$ is the perimeter of the heated surface. Substituting for $\dot{H}_b$, this equation becomes

$$q_c'' p = \dot{m} c_P \frac{dT_b}{dx} \tag{8–34}$$

The boundary condition that accompanies this first-order ordinary differential equation is given by

$$T_b = T_1 \qquad \text{at } x = 0 \tag{8–35}$$

Attention is now turned to the solution of these equations for specified wall-heat flux and uniform wall-temperature boundary conditions. These solutions are developed in the context of basic uniform property conditions. (Solutions can also be developed for more general cases involving arbitrarily specified wall temperature and variable property flows.)

### *Specified Wall-Heat Flux*

For cases in which the wall heat flux $q_c''$ is specified as a function of $x$, the overall rate of heat transfer $q_c$ is simply

$$q_c = \int_{A_s} q_c'' \, dA_s = p \int_0^L q_c'' \, dx \tag{8–36}$$

This equation reduces to

$$q_c = q_0''A_s = q_0''pL \tag{8–37}$$

for uniform wall-heat flux (i.e., $q_c'' = q_0''$). With $q_c$ known, the outlet bulk-stream temperature $T_2$ for uniform property flow is obtained from Eq. (8–31),

$$T_2 = T_1 + \frac{q_c}{\dot{m}c_P} \tag{8–38}$$

To obtain an expression for $T_b$, we integrate Eq. (8–34) over the length 0 to $x$, with the result

$$T_b - T_1 = \frac{p}{\dot{m}c_P}\int_0^x q_c''\,dx \tag{8–39}$$

For uniform wall-flux heating, this equation yields

$$T_b - T_1 = \frac{q_0''px}{\dot{m}c_P} = \frac{4q_0''D_H}{k}\frac{p}{p_w}\frac{x/D_H}{Re\,Pr} \tag{8–40}$$

This expression is shown in Fig. 8–12 for flow in a circular tube (with $p = p_w = \pi D$). Notice that the bulk-stream temperature $T_b$ is linear over the length of the tube. In addition, because $T_b$ is independent of the coefficient of heat transfer $h_x$, the predictions for $T_b$ are the same for laminar and turbulent flow.

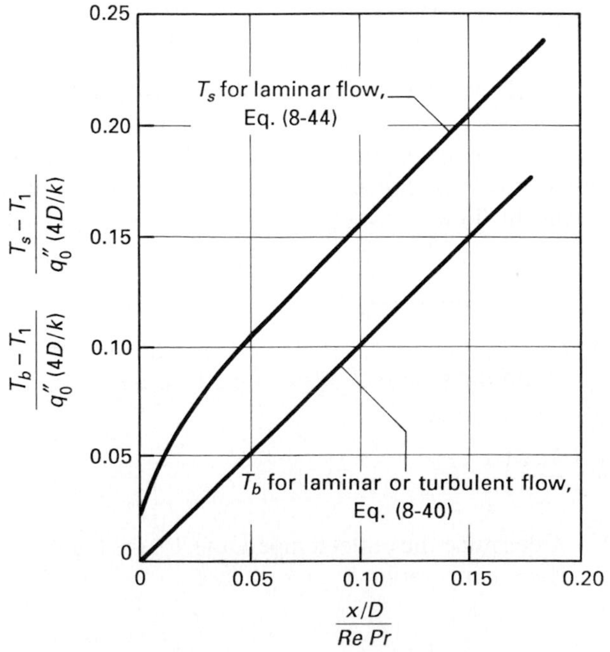


**FIGURE 8–12** Bulk-stream and wall temperatures for flow in a circular tube with uniform wall-heat flux.

An expression is obtained for the unknown surface temperature $T_s$ by combining Eq. (8–39) with the defining equation for the coefficient of heat transfer (the general Newton law of cooling), Eq. (6–48); that is,

$$q_c'' = h_x(T_s - T_b) \tag{8–41}$$

$$T_s - T_1 = \frac{q_c''}{h_x} + \frac{p}{\dot{m}c_P}\int_0^x q_c''\,dx \tag{8–42}$$

For a uniform wall-heat flux, our prediction for $T_s$ takes the form

$$T_s - T_1 = \frac{4q_0''D_H}{k}\left(\frac{1}{4\,Nu_x} + \frac{x/D_H}{Re\,Pr}\frac{p}{p_w}\right) \tag{8–43}$$

Note that $T_s$ is a function of the coefficient of heat transfer. Accordingly, to obtain calculations for $T_s$, $h_x$ or $Nu_x$ must be specified. For example, for HFD laminar thermal developing flow in a circular tube, $Nu_x$ can be approximated by Eq. (8–11) in the entrance region and by Eq. (8–8) ($Nu = 4.36$) in the TFD region, such that Eq. (8–43) becomes

$$\frac{T_s - T_1}{4q_0''D/k} = 0.192\left(\frac{x/D}{Re\,Pr}\right)^{1/3} + \frac{x/D}{Re\,Pr} \qquad \frac{x/D}{Re\,Pr} \lessapprox 0.05 \tag{8–44a}$$

$$= 0.0573 + \frac{x/D}{Re\,Pr} \qquad \frac{x/D}{Re\,Pr} \gtrapprox 0.05 \tag{8–44b}$$

This equation is shown in Fig. 8–12. The wall temperature $T_s$ is seen to be nonlinear in the thermal entrance region, but is linear in the TFD region. Similar calculations can be made for turbulent flow. Because the Nusselt number for turbulent flow is much larger than for laminar flow, we conclude on the basis of Eq. (8–43) that the wall will be cooler for turbulent flow.

## EXAMPLE 8–5

Air at 27°C and atmospheric pressure enters a 10-cm-diameter tube of 1 m length. The bulk-stream velocity within the tube is 2 m/s and the heat flux from the surface of the tube is specified by

$$q_c'' = q_0''\left[1 - \cos\left(\frac{4\pi x}{L}\right)\right]$$

where $q_0'' = 0.541\ \text{kW/m}^2$. Determine the outlet temperature $T_2$, the local bulk-stream temperature $T_b$, and the surface temperature $T_s$ at $x = \frac{6}{8}L$.

### Solution

***Objective*** Determine $T_2$, the distribution in $T_b$, and $T_s$ at $x = x_1 = \frac{6}{8}L$.

*Schematic* Flow of air in a circular tube—nonuniform wall-heat flux.

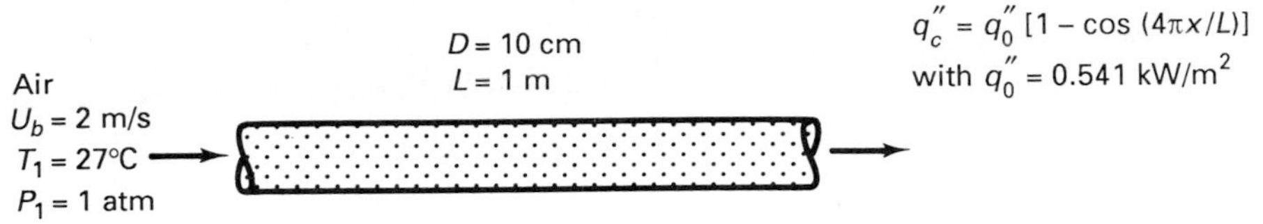
Air
$U_b$ = 2 m/s
$T_1$ = 27°C
$P_1$ = 1 atm
D = 10 cm
L = 1 m
$q''_c = q''_0\,[1 - \cos(4\pi x/L)]$
with $q''_0$ = 0.541 kW/m²

*Assumptions/Conditions*

forced convection
moderate property variation
no hydrodynamic calming section
standard conditions

*Properties* Air at 27°C (Table A–C–5): $\rho = 1.18$ kg/m³, $\mu = 1.85 \times 10^{-5}$ kg/(m s), $c_P = 1.01$ kJ/(kg °C), $k = 0.0262$ W/(m °C), $Pr = 0.708$.

*Analysis* To obtain the outlet temperature for approximate uniform property conditions, we utilize Eq. (8–31), which was obtained by performing an energy balance on the lumped control volume $AL$; that is,

$$q_c = \dot{m}c_P(T_2 - T_1)$$

where

$$q_c = p\int_0^L q''_c\,dx = q''_0 p\int_0^L \left[1 - \cos\left(\frac{4\pi x}{L}\right)\right] dx$$

$$= q''_0 pL = 0.541\,\frac{\text{kW}}{\text{m}^2}\,\pi(0.1\text{ m})(1\text{ m}) = 0.170\text{ kW}$$

Calculating the mass flow rate, we have

$$\dot{m} = \rho A U_b = 1.18\,\frac{\text{kg}}{\text{m}^3}\,\frac{\pi(0.1\text{ m})^2}{4}\,\frac{2.0\text{ m}}{\text{s}} = 0.0185\text{ kg/s}$$

Following through with the calculation for $T_2$, we write

$$T_2 = \frac{q_c}{\dot{m}c_P} + T_1 = \frac{0.170\text{ kW}}{(0.0185\text{ kg/s})[1.01\text{ kJ/(kg °C)}]} + 27°\text{C} = 36.1°\text{C}$$

Referring to Table A–C–5, we see that the variations in properties of air over the temperature range from 27°C to 36.1°C are very small. Consequently, our use of a uniform property analysis with the properties evaluated at 27°C is quite reasonable.

The bulk-stream temperature $T_b$ is obtained from Eq. (8–39), which was developed by applying the first law of thermodynamics to the lumped/differential element $A\,dx$.

$$T_b - T_1 = \frac{p}{\dot{m}c_P} \int_0^x q_c'' \, dx = \frac{q_0'' pL}{\dot{m}c_P} \left[ \frac{x}{L} - \frac{1}{4\pi} \sin \left( \frac{4\pi x}{L} \right) \right]$$

$$= \frac{\pi(0.1 \text{ m})(1 \text{ m})(0.541 \text{ kW/m}^2)}{(0.0185 \text{ kg/s})[1.01 \text{ kJ/(kg °C)}]} \left[ \frac{x}{L} - \frac{1}{4\pi} \sin \left( \frac{4\pi x}{L} \right) \right]$$

$$T_b = 27°\text{C} + 9.10°\text{C} \left[ \frac{x}{L} - \frac{1}{4\pi} \sin \left( \frac{4\pi x}{L} \right) \right] \qquad \text{(a)}$$

This equation is shown in Fig. E8–5. For the purpose of comparison, $T_b$ is also shown for a uniform heat flux of 0.541 kW/m$^2$.

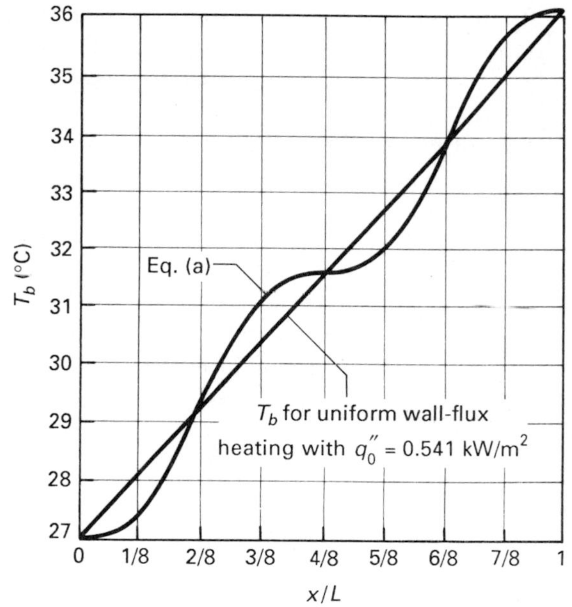


**FIGURE E8–5**
Calculations for bulk-stream temperature $T_b$.

To estimate the local wall temperature $T_s$ at $x = x_1$, we utilize the general Newton law of cooling given by Eq. (8–41),

$$q_c'' = h_x(T_s - T_b)$$

Referring to Example 8–1, the flow is turbulent with $Re = 12{,}700$ and $h = 9.54$ W/(m$^2$ °C). Assuming no calming section, Eq. (8–21a) is used to compute $h_x$ at $x = x_1$, with the result

$$h_{x_1} = h\left(1 + \frac{C}{x_1/D}\right) = 9.54 \frac{\text{W}}{\text{m}^2 \text{ °C}} \left(1 + \frac{6}{7.5}\right) = 17.2 \text{ W/(m}^2 \text{ °C)}$$

It follows that the surface temperature at $x_1$ can be approximated by

$$T_{s,1} = \frac{q''_{c,1}}{h_{x_1}} + T_b = \frac{2(0.541 \times 10^3\ \text{W/m}^2)}{17.2\ \text{W/(m}^2\ \text{°C)}} + 33.8\text{°C} = 96.7\text{°C}$$

Due to the nature of the distribution in $q''_c$, $T_{s,1} = 96.7$°C represents the maximum surface temperature for this problem. Notice that the surface temperature at the tube outlet is equal to $T_b$ since $q''_c = 0$ at $x = L$.

### *Uniform Wall Temperature*

For uniform wall-temperature heating, the heat flux $q''_c$ is expressed in terms of the general Newton law of cooling, Eq. (8–41). Using this representation for $q''_c$, Eq. (8–34) becomes

$$h_x p(T_0 - T_b) = \dot{m}c_P \frac{dT_b}{dx} \tag{8–45}$$

Rearranging, this equation is put into the form

$$\frac{d(T_b - T_0)}{T_b - T_0} = -\frac{ph_x}{\dot{m}c_P}\,dx \tag{8–46}$$

which is integrated to obtain

$$\ln \frac{T_b - T_0}{T_1 - T_0} = -\frac{p}{\dot{m}c_P}\int_0^x h_x\,dx = -\frac{\bar{h}px}{\dot{m}c_P} \tag{8–47}$$

where the mean coefficient of heat transfer $\bar{h}$ is defined by Eq. (6–49),

$$\bar{h} = \frac{1}{x}\int_0^x h_x\,dx \tag{8–48}$$

Thus, the solution for $T_b$ becomes

$$\frac{T_b - T_0}{T_1 - T_0} = \exp\left(-\frac{\bar{h}px}{\dot{m}c_P}\right) \tag{8–49}$$

or

$$\frac{T_b - T_1}{T_0 - T_1} = 1 - \exp\left(-\frac{\bar{h}px}{\dot{m}c_P}\right) \tag{8–50}$$

As shown in Fig. 8–13, this equation indicates that $T_b$ asymptotically approaches $T_0$ as $x$ increases.

To obtain a relation for the outlet bulk-stream temperature $T_2$ we set $x$ equal to $L$ in Eq. (8–50), with the result

$$\frac{T_2 - T_1}{T_0 - T_1} = 1 - \exp\left(-\frac{\bar{h}A_s}{\dot{m}c_P}\right) \tag{8–51}$$

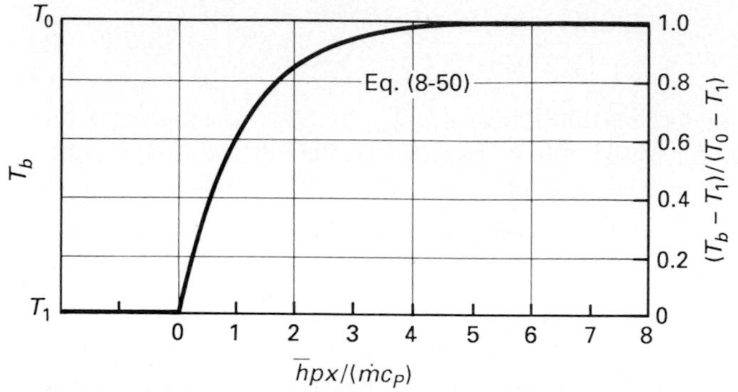


**FIGURE 8–13** Bulk-stream temperature for flow with uniform wall temperature.

where $A_s = pL$. Finally, the coupling of Eqs. (8–31) and (8–51) gives rise to an expression for $q_c$ of the form

$$q_c = \dot{m}c_P(T_0 - T_1)\left[1 - \exp\left(-\frac{\bar{h}A_s}{\dot{m}c_P}\right)\right] \tag{8–52}$$

In the limit, as the dimensionless parameter $\bar{h}A_s/(\dot{m}c_P)$ increases, the outlet temperature $T_2$ approaches $T_0$ and the total rate of heat transfer $q_c$ approaches a maximum value of $\dot{m}c_P(T_0 - T_1)$. This thermodynamic limit occurs at a value of $\bar{h}A_s/(\dot{m}c_P)$ of the order of 4 to 5.

To obtain calculations for $q_c$, $T_2$, and $T_b$ for a given flow condition, we must utilize appropriate correlations for the mean coefficient of heat transfer.

Although the theoretical aspects of the analysis for uniform wall-temperature heating are complete, the predictions for $q_c$ are now put into the effectiveness and log mean temperature difference (*LMTD*) formats, both of which are in common use.

***Effectiveness*** The *effectiveness* $\epsilon$ of a convection heat transfer process is defined by

$$q_c = \epsilon q_{c,\max} \tag{8–53}$$

where the maximum possible rate of heat transfer $q_{c,\max}$ for uniform wall-temperature heating is

$$q_{c,\max} = \dot{m}c_P(T_0 - T_1) \tag{8–54}$$

Introducing Eq. (8–31), we see that the effectiveness also represents the temperature ratio $(T_2 - T_1)/(T_0 - T_1)$. Referring to Eq. (8–51), $\epsilon$ is written as

$$\epsilon = \frac{T_2 - T_1}{T_0 - T_1} = 1 - \exp\left(-\frac{\bar{h}A_s}{\dot{m}c_P}\right) \tag{8–55}$$

This equation is shown in Fig. 8–14 in terms of $\bar{h}A_s/(\dot{m}c_P)$.

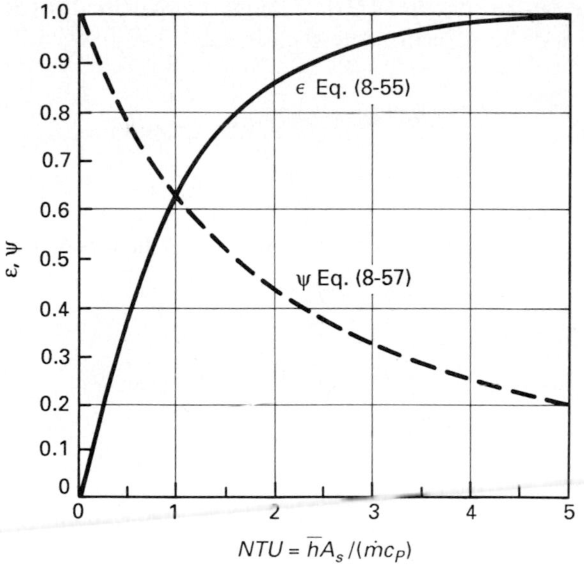


**FIGURE 8–14**
Effectiveness and efficiency for uniform wall-temperature heating.

The dimensionless parameter $\bar{h}A_s/(\dot{m}c_P)$ is sometimes referred to as the *number of transfer units NTU*. The *NTU* provides a comparison between the thermal capacity $\bar{h}A_s$ of the system and the capacity rate $\dot{m}c_P$, and provides an indication of the physical size of the system. The larger the *NTU*, the larger the surface area $A_s$ and the closer the heat-transfer system comes to approaching its thermodynamic limit. However, because $\epsilon$ increases very gradually with *NTU* for $\epsilon \gtrsim 0.8$, the value of *NTU* is generally maintained in the range 0.5 to 2.5 in order to achieve effective use of the surface area.

The significance of the magnitude of *NTU* is reinforced by putting Eq. (8–52) into the form

$$
\begin{aligned}
q_c = \dot{m}c_P(T_0 - T_1)\epsilon &= \bar{h}A_s(T_0 - T_1)\frac{\epsilon}{\bar{h}A_s/(\dot{m}c_P)} \\
&= \bar{h}A_s(T_0 - T_1)\psi
\end{aligned}
\tag{8–56}
$$

where the *efficiency* $\psi$ is defined by

$$\psi = \frac{\epsilon}{NTU} = \frac{1 - \exp(-NTU)}{NTU} = \frac{1 - \exp[-\bar{h}A_s/(\dot{m}c_P)]}{\bar{h}A_s/(\dot{m}c_P)} \tag{8–57}$$

As shown in Fig. 8–14, $\psi$ is in the range 1 to 0.5 for $NTU \lesssim 1.5$, but is less than 0.5 for $NTU \gtrsim 1.5$ and approaches zero as *NTU* becomes large. Because $\psi$ is proportional to the total heat flux $q_c/A_s$, $\psi$ provides a clear indication of the effectiveness of the surface in transferring heat for specified values of $\bar{h}$ and $T_0 - T_1$.

***Log Mean Temperature Difference (LMTD)*** To put the calculations for $q_c$ into the *LMTD* form, we replace the capacity rate $\dot{m}c_P$ in Eq. (8–31) by utilizing Eq. (8–51); that is,

$$\dot{m}c_P = -\frac{\bar{h}A_s}{\ln\dfrac{T_2 - T_0}{T_1 - T_0}} = \frac{\bar{h}A_s}{\ln\dfrac{T_0 - T_1}{T_0 - T_2}} \tag{8–58}$$

This substitution gives rise to the famous *LMTD* equation for heat transfer,

$$q_c = \bar{h}A_s\,LMTD \tag{8–59}$$

where

$$LMTD = \frac{T_2 - T_1}{\ln\dfrac{T_0 - T_1}{T_0 - T_2}} = \frac{\Delta T_1 - \Delta T_2}{\ln\dfrac{\Delta T_1}{\Delta T_2}} \tag{8–60}$$

and $\Delta T_1 = T_0 - T_1$ and $\Delta T_2 = T_0 - T_2$. Comparing Eqs. (8–57) and (8–60), we find that *LMTD* and $\psi$ are related by

$$LMTD = (T_0 - T_1)\psi \tag{8–61}$$

The *LMTD* format is most useful for making design calculations in order to size a heat-transfer system (i.e., to determine the necessary thermal capacity $\bar{h}A_s$) when the inlet, outlet, and surface temperatures are specified. However, for situations in which the outlet temperature is not known, use of the *LMTD* approach to evaluate $T_2$ involves time-consuming iterations. For such cases, the more efficient approach generally is to simply utilize Eqs. (8–51) and (8–52) directly, or to use the effectiveness concept.

## EXAMPLE 8–6

Water at 20°C and 1 atm enters a 10-cm-diameter tube of 10-m length. The mass flow rate of the water is 10 kg/s and the wall temperature is 80°C. Determine the outlet temperature of the water.

### Solution

*Objective* Determine $T_2$.

*Schematic* Flow of water in a circular tube—uniform wall temperature.

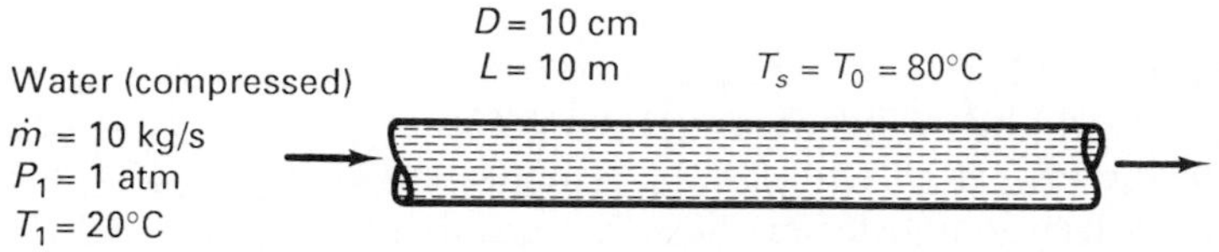
D = 10 cm
L = 10 m
Ts = T0 = 80°C
Water (compressed)
ṁ = 10 kg/s
P1 = 1 atm
T1 = 20°C

*Assumptions/Conditions*

forced convection
moderate property variation
negligible entrance effects on $\bar{h}$
standard conditions

*Properties*

Water at $T = 20°\text{C} = 293$ K (Table A–C–3): $\rho = 1/v_f = 1000$ kg/m$^3$, $\mu = 1.01 \times 10^{-3}$ kg/(m s), $c_P = 4.18$ kJ/(kg °C), $k = 0.603$ W/(m °C), $Pr = 7.01$.

Water at $T = 80°\text{C} = 353$ K: $\mu = 3.54 \times 10^{-4}$ kg/(m s).

*Analysis* Because the outlet temperature is not known, we will utilize the effectiveness approach. Calculating $q_{c,\max}$, we obtain

$$q_{c,\max} = \dot{m}c_P(T_0 - T_1) = 10\,\frac{\text{kg}}{\text{s}}\left(4.18\,\frac{\text{kJ}}{\text{kg °C}}\right)(80°\text{C} - 20°\text{C}) = 2510\text{ kW}$$

Thus, the actual rate of heat transfer is given by

$$q_c = \epsilon q_{c,\max} = \epsilon\, 2510\text{ kW} \tag{a}$$

where the effectiveness $\epsilon$ is given by Eq. (8–55) or Fig. 8–14. To obtain $\epsilon$, we must evaluate the parameter $\bar{h}A_s/(\dot{m}c_P)$, which necessitates the calculation of $\bar{h}$.

The Reynolds number $Re$ is given by

$$Re = \frac{U_bD}{\nu} = \frac{\dot{m}D}{\mu A} = \frac{\dot{m}4}{\mu\pi D} = \frac{4(10\text{ kg/s})}{\pi(0.1\text{ m})[1.01 \times 10^{-3}\text{ kg/(m s)}]} = 126{,}000$$

such that the flow is turbulent. Utilizing Eqs. (8–15) and (8–16) to approximate $Nu$ for isothermal conditions, we have

$$f_{cp} = (1.58\ln 126{,}000 - 3.28)^{-2} = 0.00429$$

$$Nu_{cp} = \frac{(0.00429/2)(126{,}000)(7.01)}{1.07 + 12.7\sqrt{0.00429/2}\,(7.01^{2/3} - 1)} = 719$$

To correct for nonuniform viscosity, we use Eq. (8–24).

$$Nu = Nu_{cp}\left(\frac{\mu_0}{\mu_b}\right)^{-0.11} = 719\left(\frac{3.54 \times 10^{-4}}{1.01 \times 10^{-3}}\right)^{-0.11} = 807$$

Since $x/D = 100$, we approximate $\bar{h}$ by $h$ to obtain

$$\bar{h} = Nu\,\frac{k}{D} = 807\,\frac{0.603\text{ W/(m °C)}}{0.1\text{ m}} = 4860\text{ W/(m}^2\text{ °C)}$$

The number of transfer units $\bar{h}A_s/(\dot{m}c_P)$ is then found to be

$$\frac{\bar{h}A_s}{\dot{m}c_P} = \frac{[4.86\ \text{kW/(m}^2\ {}^\circ\text{C)}]\ \pi(0.1\ \text{m})(10\ \text{m})}{(10\ \text{kg/s})[4.18\ \text{kJ/(kg}\ {}^\circ\text{C)}]} = 0.365$$

Using Eq. (8–55), we determine that $\epsilon \simeq 0.305$.

Returning to Eq. (a), $q_c$ is

$$q_c = 0.305(2510\ \text{kW}) = 766\ \text{kW}$$

To obtain the outlet temperature $T_2$, we substitute into Eq. (8–31), with the result

$$q_c = \dot{m}c_P(T_2 - T_1)$$

$$T_2 = \frac{766\ \text{kW}}{(10\ \text{kg/s})[4.18\ \text{kJ/(kg}\ {}^\circ\text{C)}]} + 20^\circ\text{C} = 18.3^\circ\text{C} + 20^\circ\text{C} = 38.3^\circ\text{C}$$

Referring to Table A–C–3, we find that the specific heat of water at 38.3°C is equal to the value at 20°C to within three significant figures. Therefore, our assumption of uniform specific heat is quite adequate. However, we note that the viscosity $\mu$ changes considerably over this temperature range. Therefore, our analysis can be refined by approximating the properties at the arithmetic average of the inlet and outlet temperatures. This refinement is suggested as an exercise.

## EXAMPLE 8–7

Air at 1 atm flowing at a rate of 0.1 kg/s is to be cooled from 400 to 300 K in a 10-cm-diameter tube with uniform wall temperature of 250 K. Determine the length of the tube.

### Solution

*Objective* Determine the length $L$ required to bring the outlet temperature $T_2$ to 300 K.

*Schematic* Flow of air in a circular tube—uniform wall temperature.

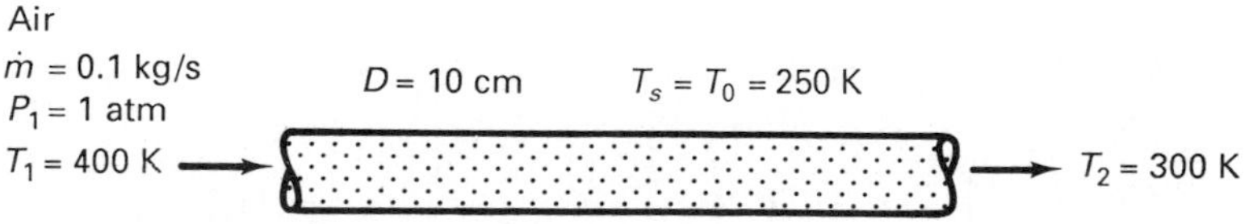
Air
$\dot{m}$ = 0.1 kg/s
$P_1$ = 1 atm
$T_1$ = 400 K
$D$ = 10 cm
$T_s$ = $T_0$ = 250 K
$T_2$ = 300 K

*Assumptions/Conditions*

forced convection
moderate property variation
negligible entrance effects on $\bar{h}$
standard conditions

*Properties* Air at the average inlet and outlet temperature of 350 K (Table A–C–5): $\rho = 0.998$ kg/m$^3$, $\mu = 2.08 \times 10^{-5}$ kg/(m s), $\nu = 2.08 \times 10^{-5}$ m$^2$/s, $c_P = 1.01$ kJ/(kg K), $k = 0.030$ W/(m K), $Pr = 0.697$.

*Analysis* Noting that the specific heat is essentially uniform over the temperature range from 300 to 400 K, we utilize Eq. (8–31) to calculate $q_c$.

$$q_c = \dot{m}c_P(T_2 - T_1) = 0.1\,\frac{\text{kg}}{\text{s}}\left(1.01\,\frac{\text{kJ}}{\text{kg K}}\right)(300\text{ K} - 400\text{ K}) = -10.1\text{ kW}$$

Because the inlet and outlet temperatures are known, we can easily utilize the *LMTD* relationship

$$q_c = \bar{h}A_s\,LMTD \tag{8–59}$$

where

$$LMTD = \frac{\Delta T_1 - \Delta T_2}{\ln\dfrac{\Delta T_1}{\Delta T_2}} = \frac{(250\text{ K} - 400\text{ K}) - (250\text{ K} - 300\text{ K})}{\ln\dfrac{250\text{ K} - 400\text{ K}}{250\text{ K} - 300\text{ K}}} = -91\text{ K}$$

To obtain $\bar{h}$, we calculate the Reynolds number.

$$Re = \frac{DU_b}{\nu} = \frac{\dot{m}D}{A\mu} = 0.1\,\frac{\text{kg}}{\text{s}}\,\frac{4}{\pi(0.1\text{ m})[2.08\times10^{-5}\text{ kg/(m s)}]} = 61{,}200$$

Thus, the flow is turbulent. Utilizing Eqs. (8–15) and (8–16) to calculate $Nu$ for uniform property conditions, we have

$$f_{cP} = (1.58\ln 61{,}200 - 3.28)^{-2} = 0.00501$$

$$Nu_{cP} = \frac{(0.00501/2)(61{,}200)(0.697)}{1.07 + 12.7\sqrt{0.00501/2}\,(0.697^{2/3} - 1)} = 114$$

The property effect associated with the wall-to-bulk-stream temperature difference is corrected for by using Eq. (8–26),

$$Nu = 114\left(\frac{250}{350}\right)^{-0.36} = 129$$

Approximating $\bar{h}$ by $h$, we obtain

$$\bar{h} = Nu\,\frac{k}{D} = 129\,\frac{0.030\text{ W/(m °C)}}{0.1\text{ m}} = 38.7\text{ W/(m}^2\text{ °C)}$$

The length can now be calculated from Eq. (8–59).

$$A_s = \frac{q_c}{\bar{h}\,LMTD} = \frac{-10.1\text{ kW}}{[38.7\text{ W/(m}^2\text{ °C)}](-91\text{ K})} = 2.87\text{ m}^2$$

$$L = \frac{2.87\text{ m}^2}{\pi(0.1\text{ m})} = 9.13\text{ m}$$

Because the resulting value of $L/D$ (= 91.3) is fairly large, the approximation used for $\bar{h}$ (i.e., $\bar{h} \simeq h$) is judged to be reasonable.

## EXAMPLE 8–8

Water at 50°F and 20 psi is heated in a 0.787-in.-diameter tube of 9 ft length. The wall temperature is 80°F, the mass flow rate is 0.69 $\text{lb}_\text{m}$/s, and the velocity distribution at the entrance is approximately uniform. Determine the total rate of heat transfer and the outlet bulk-stream temperature.

### Solution

*Objective* Determine the total rate of heat transfer $q_c$ and the outlet temperature $T_2$.

*Schematic* Flow of water in circular tube.

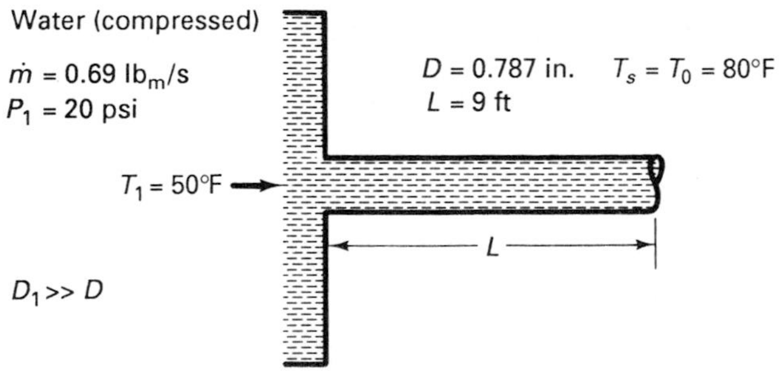
Water (compressed)
ṁ = 0.69 lbm/s
P1 = 20 psi
T1 = 50°F
D1 >> D
D = 0.787 in.
Ts = T0 = 80°F
L = 9 ft
L

*Assumptions/Conditions*

forced convection
moderate property variation
uniform entering velocity
negligible entrance effects on $\bar{h}$
standard conditions

*Properties*

Water at $T = 50°F = 283.3$ K (Table A–C–3): $Pr = 9.29$,

$$\rho = 1000 \frac{kg}{m^3} \frac{0.0624\ lb_m/ft^3}{1\ kg/m^3} = 62.4\ lb_m/ft^3$$

$$\mu = 1.29 \times 10^{-3} \frac{kg}{m\ s} \frac{0.672\ lb_m/(ft\ s)}{1\ kg/(m\ s)} = 8.67 \times 10^{-4}\ lb_m/(ft\ s)$$

$$\nu = \mu/\rho = 1.39 \times 10^{-5}\ ft/s$$

$$c_P = 4.19 \frac{kJ}{kg\ °C} \frac{0.239\ Btu/(lb_m\ °F)}{1\ kJ/(kg\ °C)} = 1\ Btu/(lb_m\ °F)$$

$$k = 0.587 \frac{W}{m\ °C} \frac{0.578\ Btu/(h\ ft\ °F)}{1\ W/(m\ °C)} = 0.339\ Btu/(h\ ft\ °F)$$

Water at $T = 80°F = 300$ K:

$$\mu = 8.55 \times 10^{-4} \frac{kg}{m\ s} \frac{0.672\ lb_m/(ft\ s)}{1\ kg/(m\ s)} = 5.75 \times 10^{-4}\ lb_m/(ft\ s)$$

*Analysis* Assuming that the density and bulk-stream velocity are essentially uniform over the length of the tube, we compute $U_b$ by writing

$$U_b = \frac{\dot{m}}{\rho A} = \frac{0.69\ lb_m/s}{(62.4\ lb_m/ft^3)[\pi(0.787\ ft/12)^2/4]} = 3.27\ ft/s$$

Using this result, the Reynolds number $Re$ becomes

$$Re = \frac{U_b D}{\nu} = \frac{(3.27\ ft/s)(0.787\ ft/12)}{1.39 \times 10^{-5}\ ft^2/s} = 15{,}400$$

such that the flow is judged to be turbulent.

Since the tube-length-to-diameter ratio $L/D$ (= 137) is quite large, we will approximate the mean coefficient of heat transfer $\bar{h}$ by the value for TFD flow, $h$. Using Eqs. (8–15) and (8–16) to compute $f$ and $Nu$ for uniform property conditions, we obtain

$$f = [1.58 \ln 15{,}400 - 3.28]^{-2} = 0.00700$$

and

$$Nu = \frac{(f/2)\, Re\, Pr}{1.07 + 12.7\sqrt{f/2}\,(Pr^{2/3} - 1)}$$

$$= \frac{(0.00700/2)(15{,}400)(9.29)}{1.07 + 12.7\sqrt{0.00700/2}\,(9.29^{2/3} - 1)} = 138$$

To correct for the effect of property variation on Nusselt number, we make use of Eq. (8–24),

$$\frac{Nu}{Nu_{cp}} = \left(\frac{\mu_0}{\mu_b}\right)^{-0.11} = \left(\frac{5.75 \times 10^{-4}}{8.67 \times 10^{-4}}\right)^{-0.11} = 1.05$$

such that the corrected value of $Nu$ becomes

$$Nu = 138(1.05) = 144$$

Thus, we are able to write

$$\bar{h} = h = Nu\frac{k}{D} = 144\frac{0.339\ \text{Btu/(h ft °F)}}{0.787\ \text{ft}/12} = 744\ \text{Btu/(h ft}^2\ \text{°F)}$$

The total rate of heat transfer $q_c$ can now be computed by use of Eq. (8–52),

$$q_c = \dot{m}c_P(T_0 - T_1)\left[1 - \exp\left(-\frac{\bar{h}A_s}{\dot{m}c_P}\right)\right]$$

where the number of transfer units $NTU$ is

$$NTU = \frac{\bar{h}A_s}{\dot{m}c_P} = \frac{[744\ \text{Btu/(h ft}^2\ \text{°F)}]\ \pi(0.787\ \text{ft}/12)(9\ \text{ft})}{(0.69\ \text{lb}_\text{m}/\text{s})[1\ \text{Btu/(lb}_\text{m}\ \text{°F)}](3600\ \text{s/h})} = 0.555$$

Substituting for $NTU$ and the other parameters, $q_c$ becomes

$$q_c = 0.69\frac{\text{lb}_\text{m}}{\text{s}}\left(1\frac{\text{Btu}}{\text{lb}_\text{m}\ \text{°F}}\right)(80\text{°F} - 50\text{°F})[1 - \exp(-0.555)]$$

$$= 8.82\ \text{Btu/s} = 9300\ \text{W}$$

The outlet bulk-stream temperature $T_2$ is obtained by use of Eq. (8–51).

$$\frac{T_2 - T_1}{T_0 - T_1} = 1 - \exp\left(-\frac{\bar{h}A_s}{\dot{m}c_P}\right) = 1 - \exp(-0.555) = 0.426$$

$$T_2 = 50\text{°F} + 0.426\,(80\text{°F} - 50\text{°F}) = 62.8\text{°F}$$

Finally, we note that the analysis could be refined by evaluating the properties at the average inlet and outlet bulk-stream temperatures. However, because the temperature rise over the length of the tube is only 12.8°F, this correction is not necessary.

### 8–2–3 Hydraulic Considerations

The pressure drop $\Delta P$, which must be overcome by pumps or blowers for flow through heat-transfer cores such as the tubular passage shown in Fig. 8–15, can be an important design consideration. The total pressure drop $\Delta P_{1-2}$ $(= P_1 - P_2)$ across a core such as this includes an inlet core pressure drop $\Delta P_{1-c}$ $(= P_1 - P_c)$ caused by the abrupt contraction, a pressure drop $\Delta P_{c-e}$ $(= P_c - P_e)$ due to friction and momentum rate change within the core, and a core exit pressure rise $\Delta P_{2-e}$ $(= P_2 - P_e)$ resulting from the increase in cross-sectional area at the outlet.

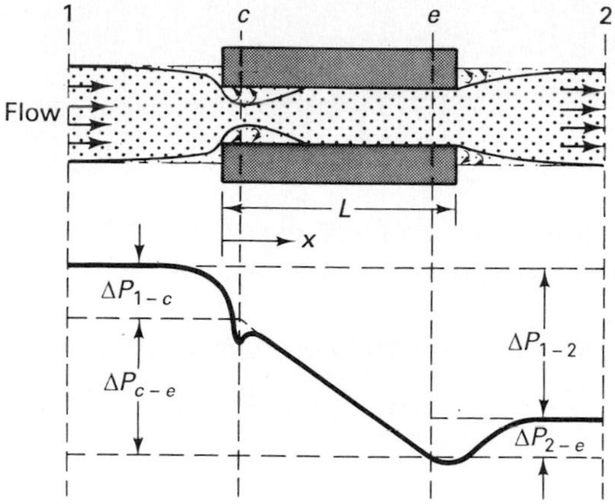


**FIGURE 8–15**
Entrance, core, and exit components of pressure drop for flow through a tubular passage. (From Shah [44]. Used with permission.)

We now turn our attention to the practical hydraulic analysis approach to establishing the pressure drop $\Delta P_{1-2}$ and the related pumping power $\dot{W}_p$ for steady incompressible flow through tubular heating cores. Although the flow will be treated as incompressible [i.e., $\rho \neq \text{fn}\ (P)$], we will account for changes in bulk-stream density caused by differences in temperature $T_b$ over the length of the heated core.

Relations are presented in references 4, 5, and 44 for the total pressure drop $\Delta P_{1-2}$ in terms of $f$, $K_c$, $K_e$, and $\sigma$ for both uniform and variable property flows. The total pressure drop for the case of uniform property flow is given by†

$$\Delta P_{1-2} = \frac{\rho U_b^2}{2}\left(K_c + \frac{4L}{D_H} f + K_e\right) \tag{8–62}$$

Notice that this equation reduces to

$$\Delta P_{1-2} = \frac{\rho U_b^2}{2} \frac{4L}{D_H} f \tag{8–63}$$

for negligible entrance and exit losses. For situations involving gas flow with large temperature differences, $\Delta P_{1-2}$ is represented by

† Equation (8–62) is obtained from the more general form [44]

$$\Delta P_{1-2} = \frac{\rho U_b^2}{2}\left[1 - \sigma^2 + K_c + \frac{4L}{D_H} f - (1 - \sigma^2 - K_e)\right]$$

For the case in which $K_e = 0$ and $\sigma = 1$ at the exit, $\Delta P_{1-2}$ becomes

$$\Delta P_{1-2} = \frac{\rho U_b^2}{2}\left(1 - \sigma^2 + K_c + \frac{4L}{D_H} f\right)$$

$$\Delta P_{1-2} = \frac{G_1^2 v_1}{2}\left[(1 - \sigma^2 + K_c) + \frac{4L}{D_H} f \frac{v_m}{v_1} + 2\left(\frac{v_2}{v_1} - 1\right) - (1 - \sigma^2 - K_e)\frac{v_2}{v_1}\right] \tag{8–64}$$

which accounts for the variation in specific volume over the length of the core; the mean specific volume $v_m$ is generally approximated by the arithmetic average of the inlet and exit values.

The pumping power $\dot{W}_p$, which is required to overcome a pressure drop $\Delta P$, is

$$\dot{W}_p = \frac{\dot{m}\,\Delta P}{\rho} \tag{8–65}$$

for uniform density, or

$$\dot{W}_p = \dot{m}\,\Delta P\, v_m \tag{8–66}$$

for nonuniform properties.

To examine the nature of the pumping power, we consider HFD flow with uniform properties. Using Eq. (8–63) to represent $\Delta P$ for HFD flow, we obtain

$$\dot{W}_p = \frac{(\rho A U_b)\left(\dfrac{4L}{D_H}\dfrac{f}{2}\right)}{\rho} = \frac{\mu^3}{\rho^2}\frac{p_w L}{D_H^3}\frac{f}{2} Re \tag{8–67}$$

With the friction factor represented by Eq. (8–7) for laminar flow and by Eq. (8–17) for turbulent flow, Eq. (8–67) reduces to

$$\dot{W}_p = \frac{C_1}{2}\frac{\mu^3}{\rho^2}\frac{p_w L}{D_H^3} Re^2 = \frac{C_1}{8}\frac{\mu}{\rho^2}\frac{\dot{m}^2 p_w^2 L}{A^3} \tag{8–68}$$

for laminar flow, and

$$\dot{W}_p = 0.023 \frac{\mu^3}{\rho^2}\frac{p_w L}{D_H^3} Re^{2.8} = \frac{0.023}{4^{0.2}}\frac{\mu^{0.2}}{\rho^2}\frac{\dot{m}^{2.8} p_w^{1.2} L}{A^3} \tag{8–69}$$

for turbulent flow. These equations indicate that, among other things, the pumping power is inversely proportional to the square of the fluid density for both laminar and turbulent flow. Therefore, the pumping power is generally small for liquids, but can be quite significant for gases and other low-density fluids. In this connection, it is quite possible to expend as much mechanical energy in pumping low-density fluids as is transferred as heat. Because mechanical energy is generally valued at 4 to 10 times as much as its equivalent in heat transfer for most thermal power systems [4], the hydraulic considerations are often critical in the development of economical heat exchanger designs. As shown in reference 44, the pumping–heating power ratio $\dot{W}_p/q_c$ is approximately proportional to the square of the bulk-stream velocity, $U_b^2$ {i.e.,

$[\dot{m}/(\rho A)]^2\}$. Thus, for applications in which the pumping power is likely to be important, the designer can increase the cross-sectional area in the heat exchanger core by increasing the number of flow passages or the hydraulic diameter.

## EXAMPLE 8–9

Determine the pressure drop and pumping power for flow in the tube of Example 8–8, assuming negligible exit losses.

### Solution

*Objective* Determine $\Delta P_{1-2}$ and $\dot{W}_p$ over the length $L$.

*Schematic* Flow of water in circular tube.

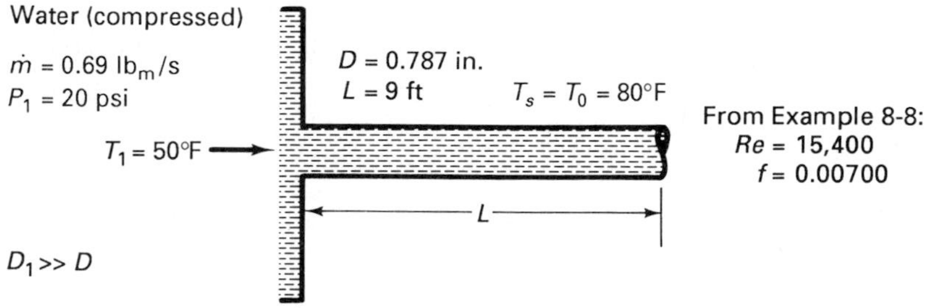

Water (compressed)
ṁ = 0.69 lbm/s
P1 = 20 psi
T1 = 50°F
D1 >> D
D = 0.787 in.
L = 9 ft
Ts = T0 = 80°F
L
From Example 8-8:
Re = 15,400
f = 0.00700


*Assumptions/Conditions*

forced convection
moderate property variation
uniform entering velocity
negligible exit losses
standard conditions

*Analysis* Assuming that the velocity distribution is uniform at the entrance (i.e., $\sigma = 0$) and $\Delta P_{2-e} = 0$, the pressure drop $\Delta P_{1-2}$ is given by (see footnote on page 499)

$$\Delta P_{1-2} = \frac{\rho U_b^2}{2}\left(1 + K_c + \frac{4L}{D_H} f\right) \tag{a}$$

Referring to Example 8–8, $f_{cp} = 0.00700$. Using Eq. (8–23) to correct for the effect of property variation on the friction factor, we obtain

$$f = f_{cp}\left(\frac{\mu_0}{\mu_b}\right)^{0.25} = 0.00700\left(\frac{5.75 \times 10^{-4}}{8.67 \times 10^{-4}}\right)^{0.25} = 0.00700(0.902)$$

$$= 0.00632$$

The entrance-loss coefficient $K_c$ indicated by Fig. 8–5 is approximately 0.48. Substituting into Eq. (a), the pressure drop becomes

$$\Delta P_{1-2} = \frac{62.4}{2}\frac{\text{lb}_\text{m}}{\text{ft}^3}\left(3.27\,\frac{\text{ft}}{\text{s}}\right)^2\left[1.48 + \frac{4(9\text{ ft})(0.00632)}{0.787\text{ ft}/12}\right]$$

$$= 334\,\frac{\text{lb}_\text{m}}{\text{ft s}^2}(1.48 + 3.47) = 1650\,\frac{\text{lb}_\text{m}}{\text{ft s}^2}\,\frac{\text{lb}_\text{f}\text{ s}^2}{32.2\text{ lb}_\text{m}\text{ ft}}$$

$$= 51.2\text{ lb}_\text{f}/\text{ft}^2 = 0.356\text{ psi} = 2460\text{ Pa}$$

where 1 pascal (Pa) = 1 N/m$^2$. Notice that with the flow assumed to be fully developed over the entire length of the tube, the pressure drop obtained from Eq. (8–63) is

$$\Delta P_{1-2} = \frac{\rho U_b^2}{2}\frac{4L}{D_H}f = 334\,\frac{\text{lb}_\text{m}}{\text{ft s}^2}(3.47) = 36.0\text{ lb}_\text{f}/\text{ft}^2 = 0.250\text{ psi}$$

which is 70% of the result which accounts for the entrance effect.

The pumping power associated with flow through the core is given by Eq. (8–65),

$$\dot{W}_p = \frac{\dot{m}\,\Delta P_{1-2}}{\rho}$$

Substituting into this relation, we obtain

$$\dot{W}_p = \frac{(0.69\text{ lb}_\text{m}/\text{s})(51.2\text{ lb}_\text{f}/\text{ft}^2)}{62.4\text{ lb}_\text{m}/\text{ft}^3} = 0.566\text{ ft lb}_\text{f}/\text{s} = 0.767\text{ W}$$

Referring to Example 8–8, we find that the pumping power is very much smaller than the total rate of heat transfer ($q_c$ = 9300 W), which is generally the case for applications involving incompressible fluids.

## 8–3 EXTERNAL FLOW

We now turn our attention to the practical analysis of external flow over surfaces such as flat plates, cylinders, and spheres. As indicated in Sec. 6–2, external flows are characterized by the development of boundary layers that continue to grow in the streamwise direction. The flow is generally laminar for small values of $x$, but develops into turbulence beyond some critical value of $x$ ($= x_c$) for large enough flow rates. These flows are generally characterized in engineering analysis by setting $U_F = U_\infty$ and $T_F = T_\infty$. For uniform property analyses, the properties for external flows are generally evaluated at the free-stream temperature $T_\infty$.

Because of significant differences between external longitudinal flow over flat plates and cylinders and flow across cylinders and spheres, these two basic situations will be discussed separately.

## 8–3–1 Flow over Flat Plates and Cylinders

Boundary layer flow over a flat plate heated in the region $x > x_0$ is shown in Fig. 8–16. An external flow such as this is characterized by uniform free-stream velocity $U_\infty$ and zero pressure gradient. The reference length $L$ for this type flow is commonly set equal to $x$, such that the Reynolds number $Re_x$ becomes

$$Re_x = \frac{U_\infty x}{\nu} \tag{8–70}$$

The flow is usually laminar for $Re_x \lesssim 10^5$, transitional turbulent for $10^5 \lesssim Re_x \lesssim 5 \times 10^5$, and fully turbulent for $5 \times 10^5 \lesssim Re_x$.

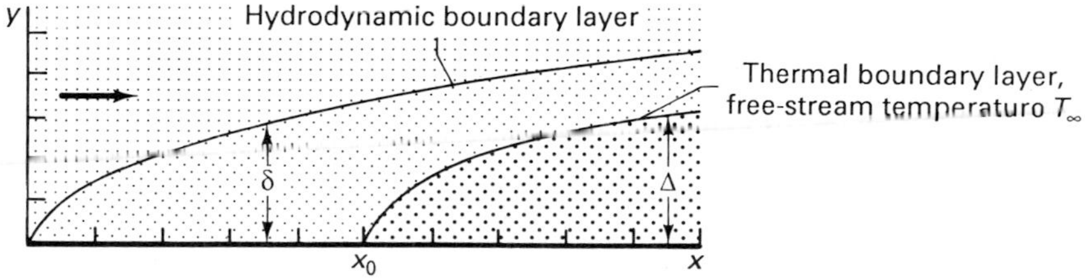


**FIGURE 8–16** Hydrodynamic and thermal boundary layer flow over a flat plate with uniform free-stream velocity and heating in the region $x \geq x_0$.

### *Coefficients of Friction and Heat Transfer*

The definitions for the Fanning friction factor $f_x$ and coefficient of heat transfer $h_x$ for flow over flat plates and cylinders are given by Eqs. (6–20) and (6–22) with $U_F = U_\infty$ and $T_F = T_\infty$; that is,

$$\tau_s = \rho U_\infty^2 \frac{f_x}{2} \qquad q_c'' = h_x(T_s - T_\infty) \tag{8–71,72}$$

The standard definitions for $\bar{f}$ and $\bar{h}$ given by Eqs. (6–21) and (6–23) also apply. The coefficients $h_x$ and $\bar{h}$ for external flows are usually expressed in terms of the Nusselt number ($Nu_x = h_x x/k$, $\overline{Nu} = \bar{h}L/k$) or Stanton number [$St_x = h_x/(\rho c_P U_\infty)$, $\overline{St} = \bar{h}/(\rho c_P U_\infty)$].

***Laminar Flow*** The exact solution for the friction factor $f_x$ for laminar boundary layer flow with uniform free-stream velocity is given by

$$f_x = \frac{0.664}{Re_x^{1/2}} \tag{8–73}$$

after Blasius [45]. This equation is compared with experimental data in Fig. 8–17. To obtain an expression for the mean friction factor $\bar{f}$ over the length $L$, this equation is integrated in accordance with the defining equation, Eq. (6–21), with the result

These equations are also shown together with experimental data and the laminar flow results in Fig. 8–17. The local Fanning friction factor has not been universally characterized for transitional turbulent boundary layer flow. Therefore, correlations such as Eqs. (8–85) and (8–86) are generally utilized in both the transitional and fully turbulent regions.

One of the more simple and useful correlations for Nusselt number for fully turbulent boundary layer flow with uniform free-stream velocity is given by [6]

$$Nu_x = \frac{f_x}{2} Re_x\, Pr^n \qquad \text{or} \qquad Nu_x = 0.0296\, Re_x^{0.8}\, Pr^n \tag{8–87,88}$$

where $n = 0.5$ for $0.5 \lesssim Pr \lesssim 5.0$ and $n = \frac{1}{3}$ for $Pr \gtrsim 5.0$. [Note the similarity between this correlation and the correlation for TFD fully turbulent tube flow given by Eqs. (8–18) and (8–19).] For improved accuracy, the following relationship developed by White [23] is recommended for $0.5 \lesssim Pr$:

$$Nu_x = \frac{(f_x/2)\, Re_x\, Pr}{1 + 12.7\sqrt{f_x/2}\,(Pr^{2/3} - 1)} \tag{8–89}$$

As in the case of internal flow, the effect of the thermal boundary condition on turbulent external boundary layer flow is generally small, except for abrupt changes or low values of $Pe$, such that Eqs. (8–87)–(8–89) can be used for uniform wall temperature or uniform wall-heat flux.† For all practical purposes, this equation is equivalent to the Petukhov-Kirillov correlation, Eq. (8–16), which was recommended for TFD fully turbulent tube flow. Notice that Eq. (8–89) reduces to Eq. (8–87) as the Prandtl number approaches unity.

Practical integral solutions can also be developed for arbitrarily specified wall flux. For the case of an unheated starting length followed by uniform wall-heat flux, $Nu_x$ can be approximated by

$$Nu_x = \frac{\sqrt{f_x/2}\, Re_x\, Pr}{2.17 \ln[(1 - x_0/x)\sqrt{f_x/2}\, Re_x] + 12.7\, Pr^{2/3} - 13.9} \qquad \text{for } 0.5 \lesssim Pr \tag{8–90}$$

which reduces to

$$Nu_x = \frac{\sqrt{f_x/2}\, Re_x\, Pr}{2.17 \ln(\sqrt{f_x/2}\, Re_x) + 12.7\, Pr^{2/3} - 13.9} \qquad \text{for } 0.5 \lesssim Pr \tag{8–91}$$

for uniform wall-heat flux. This equation lies approximately 5% above Eq. (8–89). Because the heat transfer is less sensitive to the exact form of the boundary condition for turbulent flow than for laminar flow, Eq. (8–90) can also be used for step wall-temperature heating.

† The Nusselt number for turbulent boundary layer flow with uniform wall-heat flux is sometimes approximated by

$$Nu_x = 0.0308\, Re_x^{0.8}\, Pr^{1/3}$$

which is only 4% above Eq. (8–88) for $n = \frac{1}{3}$.

For the general case in which transition occurs at a point sufficiently removed from the leading or trailing edges, the mean coefficients $\bar{f}$ and $\bar{h}$ over the length $L$ are effected by conditions within both the laminar and turbulent zones. Under these circumstances, the mean coefficients $\bar{\Lambda}$ (i.e., $\bar{f}/2$, $\bar{h}$) are expressed in terms of the local coefficients $\Lambda$ (i.e., $f_x/2$, $h_x$) by

$$\bar{\Lambda} = \frac{1}{L}\left[\int_0^{x_c} \Lambda_{\text{lam}}\, dx + \int_{x_c}^{L} \Lambda_{\text{turb}}\, dx\right] \tag{8–92}$$

Using Eqs. (8–73) and (8–85) to represent $f_x$ in the laminar and turbulent boundary layers, Eq. (8–92) gives

$$\frac{\bar{f}}{2} = \frac{1}{L}\left[0.332\sqrt{\frac{\nu}{U_\infty}}\int_0^{x_c} x^{-0.5}\, dx + 0.0296\left(\frac{\nu}{U_\infty}\right)^{0.2}\int_{x_c}^{L} x^{-0.2}\, dx\right]$$

$$= \frac{1}{L}\left[2(0.332)\sqrt{\frac{\nu}{U_\infty}}\,x_c^{0.5} + \frac{0.0296}{0.8}\left(\frac{\nu}{U_\infty}\right)^{0.2}(L^{0.8} - x_c^{0.8})\right] \tag{8–93}$$

$$= 0.037\, Re_L^{-0.2} + (0.664\, Re_{x_c}^{-0.5} - 0.037\, Re_{x_c}^{-0.2})\frac{x_c}{L}$$

Similarly, using Eqs. (8–75) and (8–88) to represent the local coefficient $h_x$, the mean coefficient of heat transfer $\bar{h}$ is represented in terms of Nusselt number by

$$\overline{Nu}_L = [0.037\, Re_L^{0.8} + (0.664\, Re_{x_c}^{0.5} - 0.037\, Re_{x_c}^{0.8})]\, Pr^{1/3} \tag{8–94}$$

With $Re_{x_c}$ set equal to a typical value of $5 \times 10^5$, these equations become

$$\frac{\bar{f}}{2} = 0.037\, Re_L^{-0.2} - 0.00174\frac{x_c}{L} \tag{8–95}$$

and

$$\overline{Nu}_L = (0.037\, Re_L^{0.8} - 871)\, Pr^{1/3} \tag{8–96}$$

Notice that for cases in which the boundary layer is tripped at the leading edge or in which $L$ is much greater than $x_c$, Eqs. (8–93) and (8–94) reduce to

$$\bar{f} = 0.074\, Re_L^{-0.2} \qquad \overline{Nu}_L = 0.037\, Re_L^{0.8}\, Pr^{1/3} \tag{8–97,98}$$

***Further Considerations*** Correction factors of the type introduced for tube flow are often utilized to account for the effect of property variations on the friction factor and coefficient of heat transfer. As a matter of fact, Eqs. (8–23) and (8–24) and the values of $m$ and $n$ given in Table 8–2 are recommended by Kays and Crawford [6] for boundary flow of liquids, with the reference temperature set equal to the free-stream temperature $T_\infty$. Similarly, Eqs. (8–25) and (8–26) are sometimes used for gases with $m$ and $n$ being dependent upon the situation. Alternatively, Eckert [48] has suggested that the effect of property variation for boundary layer flow of gases be accounted for by evaluating the properties at the film temperature $T_f = (T_s + T_\infty)/2$.

As in the case of internal flow, the heat transfer is often enhanced for flow over flat plates and cylinders by the use of rough surfaces or fins. But for such surfaces, the manufacturers' design correlations for $\bar{h}$ and $\bar{f}$ must be employed.

Once again, we should be reminded that the coefficients of friction and heat transfer are also sometimes influenced by other factors, such as nonuniform free-stream velocity, curved surfaces, viscous dissipation, phase change, and unsteady operation. When the convection process is complicated by factors such as these, their influence on the coefficients should be accounted for.

### *Practical Analyses*

The coefficients of friction and heat transfer can be used to determine the total force $F_L$ acting on a plate and the total rate of heat transfer $q_c$ or the distribution in wall temperature $T_s$.

***Momentum Transfer*** Based on the definition for wall-shear stress $\tau_s$, the differential shear force $dF_x$, which is caused by fluid flowing over the surface of a flat plate or cylinder at any axial location $x$, is given by

$$dF_x = \tau_s \, dA_s = \tau_s p \, dx \tag{8–99}$$

It follows that the total shear force $F_L$ acting on the body becomes

$$F_L = \int_0^L \tau_s p \, dx \tag{8–100}$$

Introducing the Fanning friction factor, we obtain

$$F_L = \int_0^L \rho U_\infty^2 \frac{f_x}{2} p \, dx = \rho \frac{U_\infty^2}{2} p \int_0^L f_x \, dx = \rho U_\infty^2 \frac{\bar{f}}{2} A_s \tag{8–101}$$

for uniform free-stream velocity and uniform properties.

***Heat Transfer*** To determine the total convection-heat-transfer rate from the surface of a flat plate or cylinder of length $L$, we first express the rate of heat transfer from a differential surface area $dA_s$ $(= p \, dx)$ in terms of the local heat flux $q_c''$.

$$dq_c = q_c'' \, dA_s = q_c'' p \, dx \tag{8–102}$$

Thus, the total rate of convection heat transfer over the entire surface is

$$q_c = p \int_0^L q_c'' \, dx \tag{8–103}$$

where the evaluation of the integral depends on the form of the thermal boundary condition.

*Specified Wall-Heat Flux* For the case in which the heat flux is specified, Eq. (8–103) can be integrated directly to obtain $q_c$. For example, for a uniform wall-heat flux ($q_c'' = q_0''$), $q_c$ is simply

$$q_c = q_0'' pL = A_s q_0'' \tag{8–104}$$

To obtain the local wall temperature, we introduce the coefficient of heat transfer $h_x$ by utilizing the general Newton law of cooling, Eq. (8–72),

$$q_c'' = h_x(T_s - T_\infty) \tag{8–105}$$

Solving for $T_s$, we have

$$T_s - T_\infty = \frac{q_c''}{h_x} = q_c'' \frac{x/k}{Nu_x} \tag{8–106}$$

To calculate $T_s$, Eq. (8–106) must be coupled with the appropriate correlations for $Nu_x$, as illustrated in Example 8–10. Notice that the wall operates at much lower temperatures for turbulent flow than for laminar flow (assuming negligible viscous dissipation effects).

*Specified Wall Temperature* With the wall temperature specified as a function of $x$, $h_x$ is immediately introduced into Eq. (8–103) by utilizing the general Newton law of cooling; that is,

$$q_c = p \int_0^L h_x(T_s - T_\infty)\, dx \tag{8–107}$$

For a uniform wall-temperature boundary condition ($T_s = T_0$), Eq. (8–107) gives rise to an expression for $q_c$ in terms of the mean coefficient of heat transfer $\bar{h}$ of the form

$$q_c = (T_0 - T_\infty)p \int_0^L h_x\, dx = \bar{h} A_s (T_0 - T_\infty) \tag{8–108}$$

This equation is recognized as the simplified form of the Newton law of cooling introduced in Chap. 1.

## EXAMPLE 8–10

Air at 27°C and 1 atm flows at a rate of 3 m/s over a plate of 1-m length. The plate is heated electrically, with a uniform wall flux of 100 W/m$^2$. Determine the approximate temperature along the plate.

### Solution

*Objective* Determine the distribution in wall temperature $T_s$.

*Schematic* Air flow over a flat plate—uniform wall-heat flux.

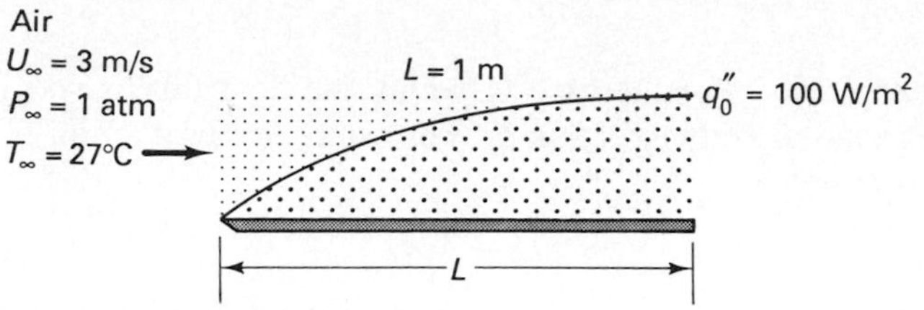


*Assumptions/Conditions*

forced convection
moderate property variation
standard conditions

*Properties* Air at 27°C (Table A–C–5): $\rho = 1.18\ \text{kg/m}^3$, $\mu = 1.85 \times 10^{-5}$ kg/(m s), $\nu = 1.57 \times 10^{-5}\ \text{m}^2\text{/s}$, $k = 0.0262$ W/(m °C), $Pr = 0.708$.

*Analysis* The local surface temperature $T_s$ of the plate is given by Eq. (8–106),

$$T_s = T_\infty + \frac{q_c''}{k}\frac{x}{Nu_x} \tag{a}$$

where $Nu_x$ is dependent on location $x$ and on whether the flow is laminar or turbulent. To see whether the flow becomes turbulent, we calculate $Re_L$.

$$Re_L = \frac{LU_\infty}{\nu} = \frac{1\ \text{m}\ (3\ \text{m/s})}{1.57 \times 10^{-5}\ \text{m}^2\text{/s}} = 1.91 \times 10^5$$

Because $Re_L$ is only slightly greater than $10^5$, the flow may or may not be turbulent.

Using Eq. (8–79) to approximate $Nu_x$ for laminar flow and uniform property conditions, $T_s$ becomes

$$\begin{aligned} T_s &= T_\infty + \frac{q_c''\nu}{0.442\ kU_\infty}\frac{Re_x^{1/2}}{Pr^{1/3}} \\ &= 27°\text{C} + \frac{(100\ \text{W/m}^2)(1.57 \times 10^{-5}\ \text{m}^2\text{/s})}{0.442[0.0262\ \text{W/(m °C)}](3\ \text{m/s})(0.708)^{1/3}} Re_x^{1/2} \\ &= 27°\text{C} + 0.0507\ Re_x^{1/2}\ °\text{C} \end{aligned} \tag{b}$$

This equation is shown in Fig. E8–10. Notice that $T_s$ increases with $x$ by the factor $x^{1/2}$.

If the flow is tripped such that turbulence occurs over most of the plate, $Nu_x$ can be approximated by Eq. (8–88). For this situation $T_s$ is given by

$$T_s = T_\infty + \frac{q_c''}{0.0296}\frac{\nu}{kU_\infty}\frac{Re_x^{0.2}}{Pr^{0.5}} = 27°\text{C} + 0.802\ Re_x^{0.2}\ °\text{C} \tag{c}$$

for uniform property conditions. This equation is also shown in Fig. E8–10.

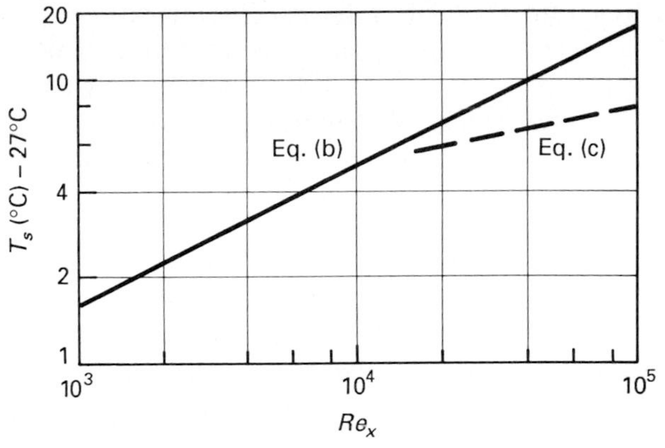


**FIGURE E8–10** Calculations for wall temperature.

Because the calculations for wall temperature increase by as much as 20°C, we would expect moderate property variation effects. Referring to Eq. (a), by evaluating $k$ at the film temperature and using a correction of the form of Eq. (8–26) for $Nu_x$, we find that the maximum wall temperature decreases by about 3% for laminar flow and increases by less than 1% for turbulent flow.

## 8–3–2 Flow Across Cylinders and Spheres

Referring to Fig. 8–18, flow normal to the axis of a circular cylinder and over a sphere is characterized by a diversion of the flow at the *forward stagnation point* where the fluid is brought to rest, and the variation of the free-stream velocity $U_\infty$ and pressure $P_\infty$ and growth of boundary layers in the streamwise direction $x$ along the surface of the body. The boundary layer develops under the influence of a *favorable pressure gradient* ($dP/dx < 0$) and *accelerating* free-stream velocity ($dU_\infty/dx > 0$) over the forward part of the body until the point of minimum pressure $P_\infty$ and maximum

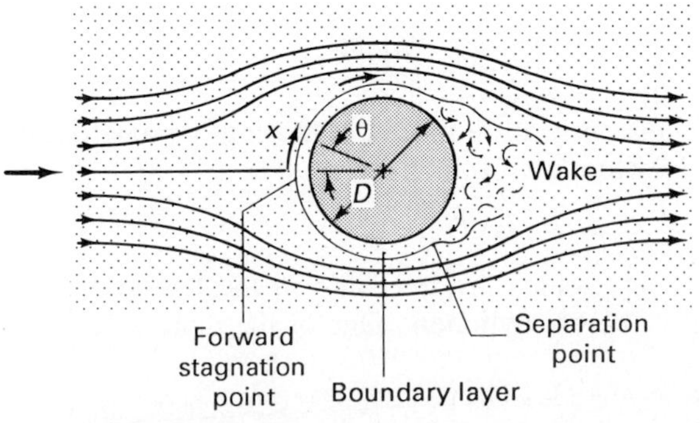


**FIGURE 8–18**
Boundary layer flow over a circular cylinder in crossflow.

velocity $U_\infty$. Beyond this point the boundary layer develops under the complicating influence of *adverse pressure gradient* ($dP/dx > 0$) and *decelerating* flow ($dU_\infty/dx < 0$) which generally produces a region of flow reversal followed by an increase in the growth of the boundary layer and the development of a wake flow involving irregular vortex type motion which persists well downstream of the body. The location beyond which the flow reversal occurs, known as the *separation point*, is characterized by $\partial u/\partial y|_0 = 0$ (i.e., $\tau_s = 0$), where the coordinate $y$ is normal to the surface at all points along the surface. The specific features of the flow are strongly dependent upon the value of the Reynolds number and on whether the flow is laminar or turbulent. Descriptions of these details are provided by Schlichting [49] and White [23].

Although a local Reynolds number is sometimes written for flow normal to a cylinder or sphere in terms of the distance along the curved surface, a more convenient overall Reynolds number is given by

$$Re_{D_N} = \frac{U_\infty D_N}{\nu} \tag{8–109}$$

where the reference length $D_N$ is the height of the body as seen by the onrushing fluid. For example, for circular cylinders and spheres, $D_N$ is simply equal to the diameter $D$.

### Coefficients of Drag and Heat Transfer

The net force acting on a body in crossflow is the result of both the total wall-shear stress and overall pressure drop. Because it is this drag force $F_D$ that is generally needed in design, a *drag coefficient* is defined as

$$F_D = \frac{\rho U_\infty^2}{2} C_D A_N \tag{8–110}$$

where $A_N$ is the area of the body normal to the direction of flow. Figure 8–19 shows the coefficients of drag for flow across cylinders and spheres.

Although empirical correlations are available for the local coefficient of heat transfer $h_\theta$ for flow across cylinders and spheres, practical considerations have resulted in the use of a mean coefficient $\bar{h}$, which is defined by

$$\bar{h} = \frac{1}{2\pi}\int_0^{2\pi} h_\theta \, d\theta \tag{8–111}$$

Accordingly, for standard uniform wall-temperature heating, the rate of heat transfer $q_c$ is given by the simple Newton law of cooling,

$$q_c = \bar{h} A_s (T_s - T_\infty) \tag{8–112}$$

Numerous empirical correlations are available in the literature for the mean Nusselt number $\overline{Nu}_D$ for flow across cylinders. One of the more recently developed

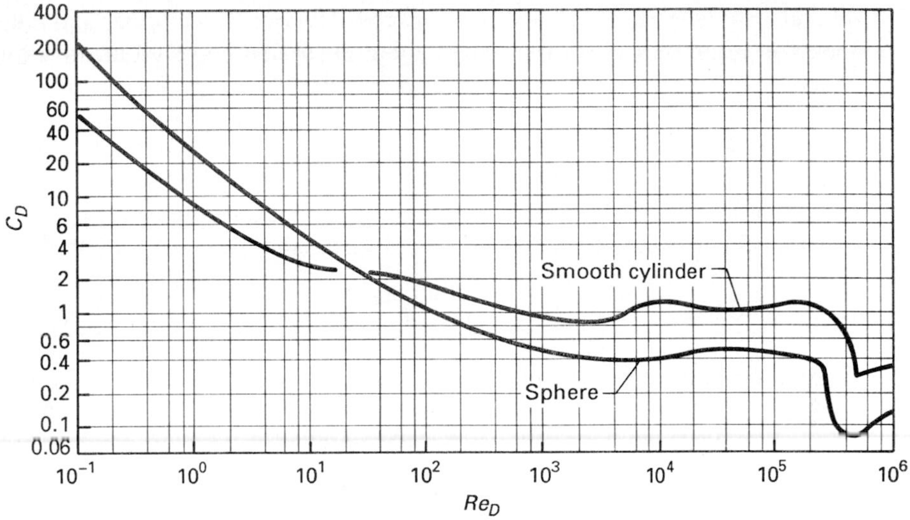


**FIGURE 8-19** Representative drag coefficients for circular cylinder in crossflow and for flow over a sphere. (From Schlichting [49]. Adapted with permission.)

correlations for gas or liquid flow over a circular cylinder with uniform wall temperature is given by [50]

$$\overline{Nu}_D = (0.4\,Re_D^{0.5} + 0.06\,Re_D^{2/3})Pr^{0.4}\left(\frac{\mu_\infty}{\mu_0}\right)^{0.25} \tag{8-113}$$

where $\overline{Nu}_D = \bar{h}D/k$, $10 \lesssim Re_D \lesssim 10^5$, $0.67 \lesssim Pr \lesssim 300$, $0.25 \lesssim \mu_\infty/\mu_0 \lesssim 5.2$, and the properties other than $\mu_0$ are evaluated at $T_\infty$. This correlation has been reported by Whitaker [51] to lie within $\pm 25\%$ of the experimental data.

A similar correlation has been recommended by Whitaker [50] for flow over a sphere which takes the form

$$\overline{Nu}_D = 2 + (0.4\,Re_D^{0.5} + 0.06\,Re_D^{2/3})\,Pr^{0.4}\left(\frac{\mu_\infty}{\mu_0}\right)^{0.25} \tag{8-114}$$

where $3.5 \lesssim Re_D \lesssim 7.6 \times 10^4$, $0.71 \lesssim Pr \lesssim 380$, $1 \lesssim \mu_\infty/\mu_0 \lesssim 3.2$, and all properties except $\mu_0$ are evaluated at $T_\infty$. For higher Reynolds numbers, the following correlation by Achenbach [52] is recommended for gases:

$$\overline{Nu}_D = 430 + a\,Re_D + b\,Re_D^2 + c\,Re_D^3 \tag{8-115}$$

where $a = 5 \times 10^{-3}$, $b = 0.025 \times 10^{-9}$, $c = -3.1 \times 10^{-17}$, $4 \times 10^5 \lesssim Re_D \lesssim 5 \times 10^6$, $Pr \approx 0.71$, and the properties are evaluated at the film temperature $T_f$.

Correlations are available for flow of liquid metals over circular cylinders and spheres in references 53 and 54 and for crossflow over noncircular and nonspherical bodies in references 2, 3, and 55. In addition, design correlations are available for finned tubes which are widely employed in industry.

### Practical Analyses

The use of Eqs. (8–110) and (8–112) directly give the total drag and rate of convection heat transfer from bodies with uniform surface temperature in crossflow.

## EXAMPLE 8–11

Air at atmospheric pressure and 35°C flows across a 1-cm-diameter wire at a velocity of 100 m/s. The power generated by electric current per unit length of the wire is 2000 W/m. Estimate the mean surface temperature of the wire.

### Solution

*Objective* Estimate the mean surface temperature $T_0$.

*Schematic* Air flow across a wire with uniform energy generation.

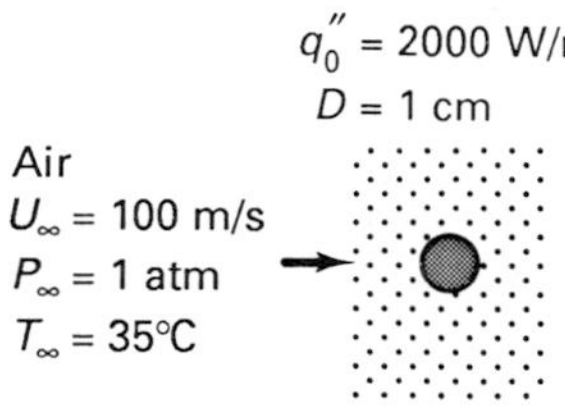
$q_0''$ = 2000 W/m
$D$ = 1 cm
Air
$U_\infty$ = 100 m/s
$P_\infty$ = 1 atm
$T_\infty$ = 35°C

*Assumptions/Conditions*

forced convection
moderate property variation
standard conditions

*Properties* Air at 35°C: $\rho = 1.16$ kg/m$^3$, $\mu = 1.88 \times 10^{-5}$ kg/(m s), $\nu = 1.62 \times 10^{-5}$ m$^2$/s, $k = 0.0268$ W/(m °C), $Pr = 0.706$.

*Analysis* Although the heat-transfer correlations found in the literature were developed for uniform wall-temperature heating, these correlations can be used to obtain an approximate solution for the surface temperature for this uniform heat-flux condition. Calculating the Reynolds number $Re_D$, we have

$$Re_D = \frac{U_\infty D}{\nu} = \frac{(100 \text{ m/s})(0.01 \text{ m})}{1.62 \times 10^{-5} \text{ m}^2\text{/s}} = 61{,}700$$

To obtain the mean Nusselt number $\overline{Nu}_D$, we use Eq. (8–113),

$$\overline{Nu}_D = (0.4\,Re_D^{0.5} + 0.06\,Re_D^{2/3})\,Pr^{0.4}\left(\frac{\mu_\infty}{\mu_0}\right)^{0.25}$$

Approximating $\mu_\infty/\mu_0$ by unity, we obtain

$$\overline{Nu}_D = 168$$

or

$$\bar{h} = \overline{Nu}_D\,\frac{k}{D} = 168\,\frac{0.0268\ \text{W/(m °C)}}{0.01\ \text{m}} = 450\ \text{W/(m}^2\ \text{°C)}$$

An estimate can now be made for the surface temperature by using the Newton law of cooling.

$$T_0 - T_\infty = \frac{q_c}{\bar{h}A_s} = \frac{q_c}{L}\frac{1}{\bar{h}p} = \frac{2000\ \text{W/m}}{[450\ \text{W/(m}^2\ \text{°C)}]\pi(0.01\ \text{m})}$$

$$T_0 = 141\text{°C} + 35\text{°C} = 176\text{°C}$$

To refine the calculation for wall temperature, $\mu_0$ in Eq. (8–113) can be evaluated at 176°C. This step gives rise to $T_0 = 186$°C, which represents a 5.7% change. Further refinement in the value of $\mu_0$ is not justified because of the approximate nature of the heat transfer correlation.

## 8–4 FLOW ACROSS TUBE BANKS

As we shall see in Chap. 11, flow across tube bundles is extremely important in heat-exchanger applications. To introduce this important subject, we consider the basic geometric and bulk-stream characteristics associated with flow through unfinned tube banks in Sec. 8–4–1, present the practical thermal analysis approach in Sec. 8–4–2, and consider hydraulic aspects in Sec. 8–4–3.

### 8–4–1 Geometric and Bulk-Stream Characteristics

The flow and heat-transfer characteristics for unfinned tube banks are dependent upon the tube diameter $D$, the longitudinal pitch $S_L$ and transverse pitch $S_T$ measured between tube centers, number of tube rows crossed $N_L$ and tube columns $N_T$, arrangement of the tubes in the bank, the flow rates, and the fluid. Standard in-line and staggered tube bank arrangements generally used by manufacturers are shown in Fig. 8–20.

Following the convention used by Kays and London [4], the tube bank length $L$ represents an equivalent flow length measured from the leading edge of the first

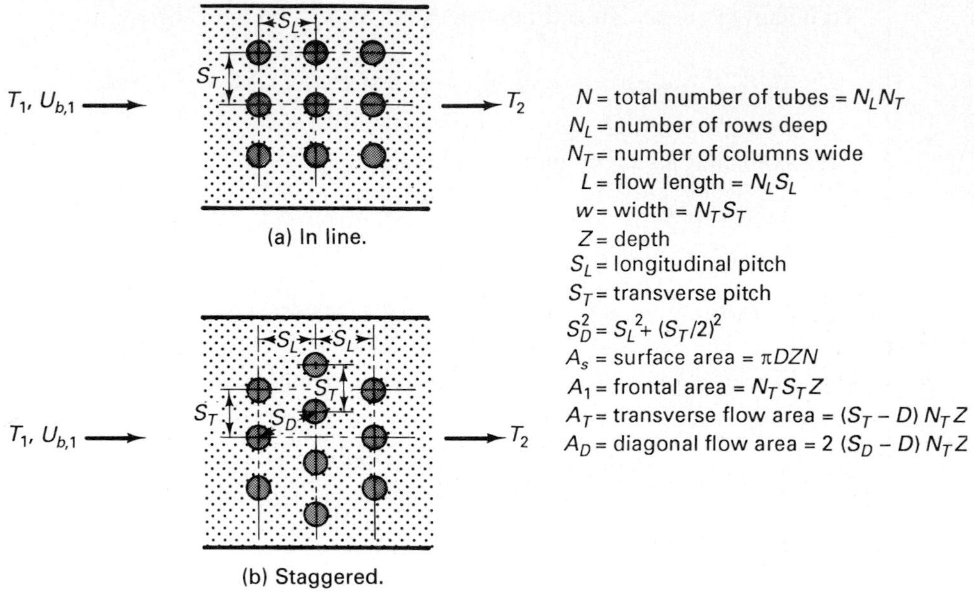
$S_L$
$S_T$
$T_1, U_{b,1}$
$T_2$
(a) In line.
$S_L$ $S_L$
$S_T$
$S_T$
$S_D$
$T_1, U_{b,1}$
$T_2$
(b) Staggered.
$N$ = total number of tubes = $N_L N_T$
$N_L$ = number of rows deep
$N_T$ = number of columns wide
$L$ = flow length = $N_L S_L$
$w$ = width = $N_T S_T$
$Z$ = depth
$S_L$ = longitudinal pitch
$S_T$ = transverse pitch
$S_D^2 = S_L^2 + (S_T/2)^2$
$A_s$ = surface area = $\pi DZN$
$A_1$ = frontal area = $N_T S_T Z$
$A_T$ = transverse flow area = $(S_T - D) N_T Z$
$A_D$ = diagonal flow area = $2 (S_D - D) N_T Z$

**FIGURE 8–20** Tube-bank arrangements.

tube row to the leading edge of a tube row that would follow the last tube row, were another tube row present; that is,

$$L = N_L S_L \tag{8–116}$$

The frontal area $A_1$ and volume $V$ of the tube bank are represented by

$$A_1 = N_T S_T Z \tag{8–117}$$

and

$$V = LA_1 = NS_L S_T Z \tag{8–118}$$

and the total surface area $A_s$ from which heat is convected is given by

$$A_s = \pi DNZ \tag{8–119}$$

where $Z$ is the depth. In practice the surface area $A_s$ is often expressed in terms of the *surface-area density* $\beta$,

$$\beta = \frac{A_s}{V} = \frac{\pi D}{S_L S_T} \tag{8–120}$$

Figure 8–21 shows representative flow patterns for water flowing over a tube bank. The flow exhibits complex features such as separation, recirculation (i.e., flow reversal), and vortex motion that are characteristic of external flow across cylinders. However, because the flow is confined by this multiple tube geometry, it is classified as internal flow for purpose of analysis.

It follows that bulk-stream characteristics are used in the development of the

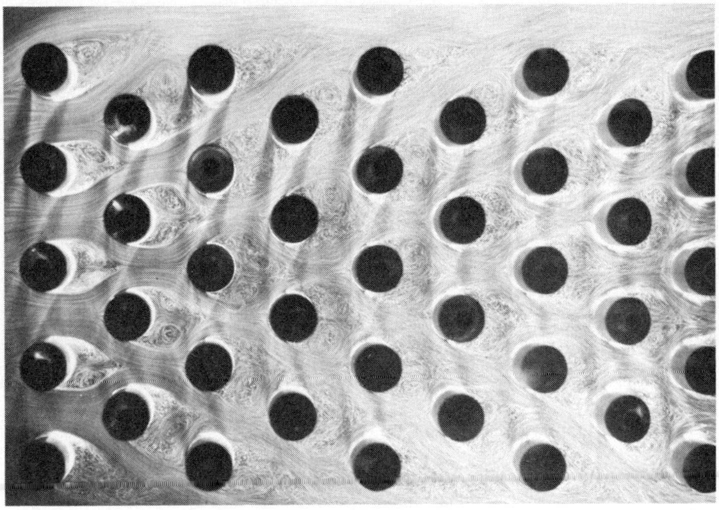

**FIGURE 8–21** Flow pattern for water flowing across staggered array of tubes. (Courtesy of H. Werle, ONERA, Paris.)

practical lumped analysis approach for flow across tube banks. However, because this particular application involves $N$ separate heating surfaces and multiple-flow passages with a nonuniform cross-sectional flow area $A$, the actual distributions in bulk-stream velocity $U_{b,\text{act}}$ and bulk-stream temperature $T_{b,\text{act}}$ within the tube bank are not easily characterized. To overcome this complexity, the tube-bank core is modeled by a number of uniform and continuous flow channels, as illustrated in Fig. 8–22. In this approach, the cross-sectional area is set equal to the *minimum free-flow area* $A_{\min}$ available for fluid flow within the tube bank, and the *effective heat transfer surface perimeter* $p$ is given by

$$p = \frac{A_s}{L} = \frac{\pi DNZ}{L} = \frac{\pi DN_T Z}{S_t} \tag{8–121}$$

which is equal to the *effective wetted perimeter* $p_w$.

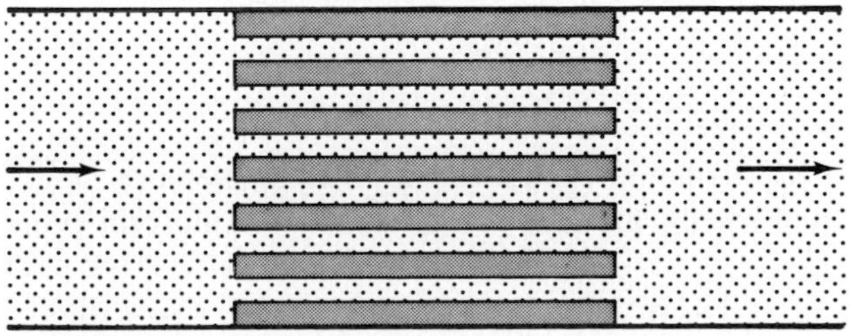

**FIGURE 8–22** Tube-bank core model for practical lumped analysis; $G$ and $U_b$ based on the minimum free-flow area $A_{\min}$ in the core. (From Kays and London [4]. Used with permission.)

Referring to Fig. 8–20, $A_{min} = (S_T - D)N_TZ$ for in-line arrangements. For staggered arrays, the minimum free-flow area may occur between adjacent tubes in a row or between diagonally opposed tubes. Hence, $A_{min}$ is the smaller of the two values $(S_T - D)N_TZ$ and $2(S_D - D)N_TZ$, where the factor 2 is a consequence of the fluid dividing as it moves from the transverse plane through the diagonal planes. These relations for $A_{min}$ are represented by

$$A_{min} = C_A N_T D Z \qquad (8\text{–}122)$$

where the *area coefficient* $C_A$ is

$$C_A = \frac{S_T}{D} - 1 \qquad \text{for in-line arrays, or staggered arrays with } S_D \geq (S_T + D)/2 \qquad (8\text{–}123a)$$

$$C_A = 2\left(\frac{S_D}{D} - 1\right) \qquad \text{for staggered arrays with } S_D < (S_T + D)/2 \qquad (8\text{–}123b)$$

and the *diagonal pitch* $S_D$ is

$$S_D = \left[S_L^2 + \left(\frac{S_T}{2}\right)^2\right]^{1/2} \qquad (8\text{–}124)$$

It follows that the contraction ratio $\sigma$ is given by

$$\sigma = \frac{A_{min}}{A_1} = \frac{C_A D}{S_T} \qquad (8\text{–}125)$$

To specify the hydraulic diameter $D_H$ for flow across tube banks, it is customary to write

$$D_H = \frac{4A_{min}}{p_w} = \frac{4C_A S_L}{\pi} \qquad (8\text{–}126)$$

or

$$\frac{D_H}{4L} = \frac{A_{min}}{A_s} = \frac{C_A}{\pi N_L} \qquad (8\text{–}127)$$

The hydraulic diameter can also be expressed in terms of $\beta$ and $\sigma$ by

$$D_H = 4\frac{A_{min}}{A_s}\frac{V}{A_1} = 4\frac{\sigma}{\beta} \qquad (8\text{–}128)$$

The bulk-stream hydrodynamic characteristics within the core model are represented by $U_b$ and $G$ and satisfy the following relation for uniform property conditions:†

$$\dot{m} = \rho A_{min} U_b = A_{min} G \qquad (8\text{–}129)$$

† The properties $\rho$, $\mu$, and $\nu$ are replaced by the mean values $\rho_b$, $\mu_b$, and $\nu_b$ and $c_P$ is taken as a function of $T_b$ for a variable property analysis. The bulk-stream hydrodynamic characteristics within the core model are also commonly represented by $U_{b,max}$ and $G_{max}$.

The principle of conservation of mass also permits us to write

$$\dot{m} = \rho A_1 U_1 = A_1 G_1 \tag{8–130}$$

It follows that $U_b$ is related to the entering velocity $U_1$ by

$$U_b = \frac{A_1}{A_{\min}} U_1 = \frac{U_1}{\sigma} = \frac{S_T}{C_A D} U_1 \tag{8–131}$$

Similarly, we see that

$$G = \frac{A_1}{A_{\min}} G_1 = \frac{G_1}{\sigma} = \frac{S_T}{C_A D} G_1 \tag{8–132}$$

It also follows that the Reynolds number $Re$ is defined by

$$Re = \frac{U_b D_H}{\nu} = \frac{G D_H}{\mu} \tag{8–133}$$

Based on this definition for the Reynolds number, the flow has been reported to be laminar for $Re$ less than a value of the order of 200, and the flow is transitional turbulent for $200 \lesssim Re \lesssim 6000$.

To complete the picture, the bulk-stream thermal characteristics within the core model are represented by $\dot{H}_b$ and $T_b$, which satisfy Eq. (6–45),

$$d\dot{H}_b = \dot{m} c_P \, dT_b \tag{8–134}$$

## 8–4–2 Practical Thermal Analysis

To adapt the practical thermal analysis approach developed in Sec. 8–2–2 to flow across tube banks, we define the local coefficient of heat transfer $h_x$ by

$$q_c'' = h_x (T_s - T_b) \tag{8–135}$$

where $q_c''$ represents an *effective local heat flux* transferred across the core model (Fig. 8–22) surface area $p\,dx$. Using this definition for $h_x$, the relations developed in Sec. 8–2–2 for flow in tubes with specified wall-heat flux and uniform wall-temperature heating are applicable to flow through tube banks and heat-exchanger cores.

For example, the total heat-transfer rate $q_c$ and exit bulk-stream temperature $T_2$ for the case of uniform wall temperature with uniform properties are given by Eq. (8–31),

$$q_c = \dot{m} c_P (T_2 - T_1) \tag{8–136}$$

or Eq. (8–52),

$$q_c = \dot{m} c_P (T_0 - T_1) \left[ 1 - \exp\left( - \frac{\bar{h} A_s}{\dot{m} c_P} \right) \right] \tag{8–137}$$

and Eq. (8–51)

$$\frac{T_2 - T_1}{T_0 - T_1} = 1 - \exp\left( - \frac{\bar{h} A_s}{\dot{m} c_P} \right) \tag{8–138}$$

where $\bar{h}$ is the mean coefficient of heat transfer. As we have seen, these equations can be combined to obtain

$$q_c = \bar{h}A_s\, LMTD \tag{8–139}$$

where the log mean temperature difference *LMTD* is defined by Eq. (8–60),

$$LMTD = \frac{T_2 - T_1}{\ln \dfrac{T_0 - T_1}{T_0 - T_2}} = \frac{\Delta T_1 - \Delta T_2}{\ln \dfrac{\Delta T_1}{\Delta T_2}} \tag{8–140}$$

Correlations have been developed for the mean coefficient of heat transfer $\bar{h}$ on the basis of experimental data for flow over tube banks with approximate uniform wall temperature by Kays and London [4], Whitaker [50], Grimison [56], Zukauskas [57,58], Bergelin et al. [59], the Babcock and Wilcox Co. [60], and others. Data obtained by Kays and London [4] for a specific tube bank are shown by Fig. 8–23

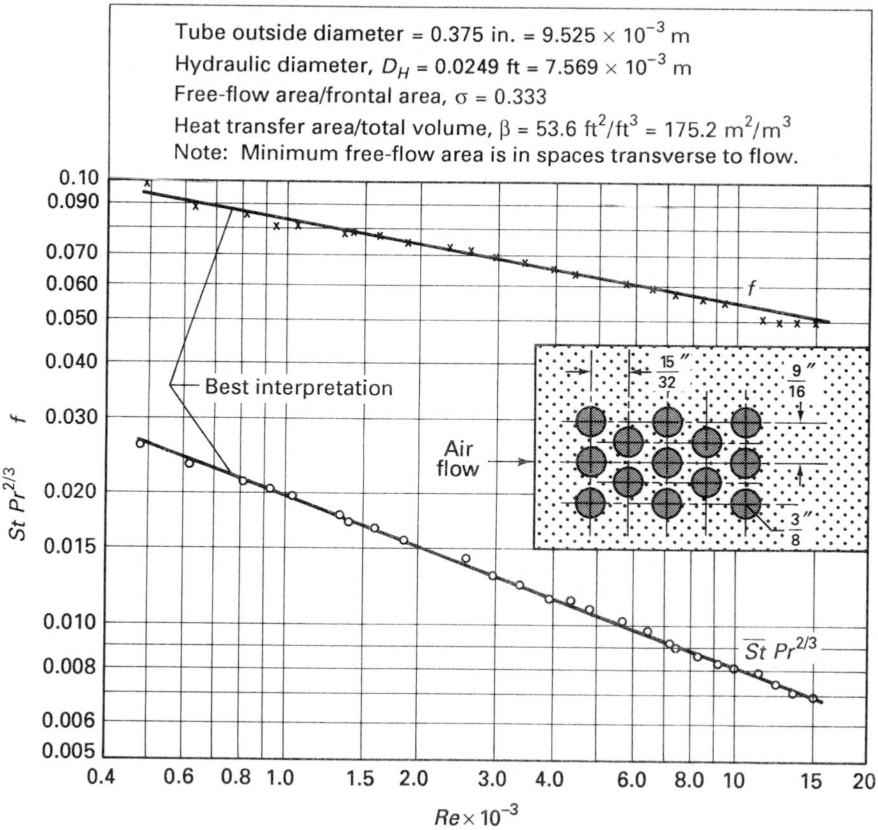


**FIGURE 8–23** Flow normal to a staggered tube bank; surface S1.50–1.25. (From Kays and London [4]. Used with permission.)

in terms of $\overline{St}$ [$= \bar{h}/(\rho c_P U_b)$]. A general correlation developed by Kays and London [4] for staggered tube-bank arrangements of infinite extent is shown in Fig. 8–24. This correlation is used in conjunction with Fig. 8–25 to account for the effect on $\bar{h}$ of a finite number of tube rows in a bank. Although the experimental data on which these curves are based were obtained for air, these correlations can also be used with reasonable confidence for liquids with moderately high values of Prandtl number.

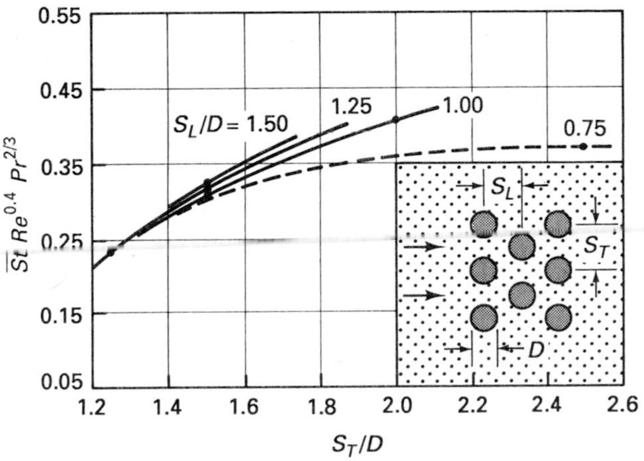


**FIGURE 8–24** Mean Stanton number correlation—gas flow normal to an infinite bank of staggered circular tubes; $300 \leqslant Re \leqslant 15{,}000$. (From Kays and London [4]. Used with permission.)

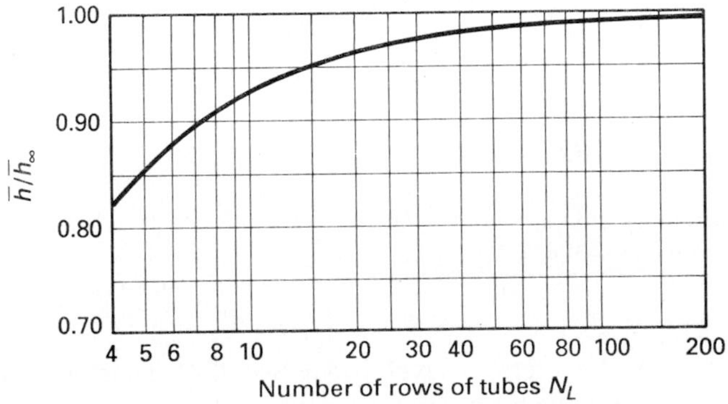


**FIGURE 8–25** Overall influence of row-to-row variations on the heat transfer coefficient for tube banks. (From Kays and London [4]. Used with permission.)

The correlation developed for $\bar{h}$ by Zukauskas [57] is of the form

$$\overline{Nu}_D = C\, Re_D^m\, Pr^n \left(\frac{Pr}{Pr_s}\right)^{0.25} F_c \tag{8–141}$$

where the constants $C$, $m$, and $n$ are given in Table 8–4, and the tube-row correction factor $F_c$ is given in Fig. 8–26.

**TABLE 8–4** Constants of Eq. (8–141) for flow across a tube bank

| *Configuration* | *Range of* $Re_D$ | $C$ | $m$ | $n$ |
|---|---|---|---|---|
| In-line | $10^0$–$10^2$ | 0.9 | 0.4 | 0.36 |
| | $10^2$–$10^3$ | 0.52 | 0.5 | 0.36 |
| | $10^3$–$2 \times 10^5$ | 0.27 | 0.63 | 0.36 |
| | $2 \times 10^5$–$2 \times 10^6$ | 0.033 | 0.8 | 0.4 |
| Staggered | | | | |
| | $10^0$–$5 \times 10^2$ | 1.04 | 0.4 | 0.36 |
| | $5 \times 10^2$–$10^3$ | 0.71 | 0.5 | 0.36 |
| | $10^3$–$2 \times 10^5$ | $0.35(S_T/S_L)^{0.2}$ | 0.6 | 0.36 |
| | $2 \times 10^5$–$2 \times 10^6$ | $0.031(S_T/S_L)^{0.2}$ | 0.8 | 0.36 |

*Source*: From Zukauskas [57].

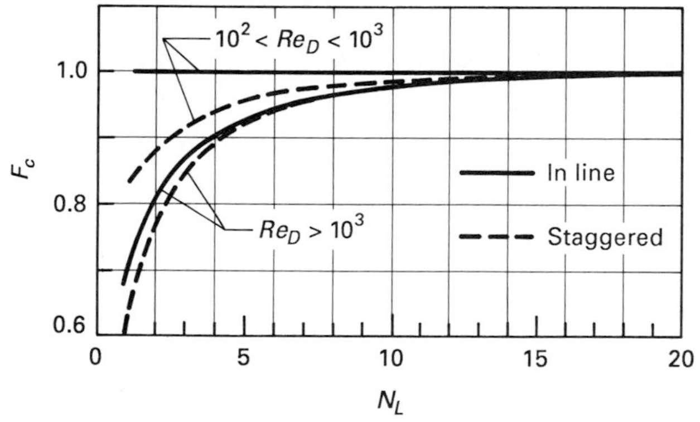


**FIGURE 8–26** A correction factor to account for the tube-row effect for heat transfer in flow normal to bare tube banks (Zukauskas [57]).

It should be noted that the solutions developed in Sec. 8–2–2 and other references for arbitrary wall flux, arbitrary wall temperature, and variable property internal flows also apply to flow across tube banks. However, because of the complex nature of the actual process, little information pertaining to the distribution in the local coefficient of heat transfer $h_x$, which appears in these solutions, is available in the literature. Therefore, to develop first-order solutions for cases involving such arbitrary conditions, $h_x$ is generally approximated by correlations for $\bar{h}$ which are based on uniform wall temperature and uniform properties.

### 8–4–3 Hydraulic Considerations

The practical lumped approach to the analysis of flow through tube banks and heat-exchanger cores developed by Kays and London [4] involves the use of a Fanning friction factor $f$ defined by

$$\tau_s = \rho U_b^2 \frac{f}{2} \qquad (8\text{–}142)$$

where $\tau_s$ represents an effective local wall shear stress acting on the core model surface area $p\ dx$ (see Fig. 8–22) that accounts for the effects of viscous shear and form drag within the actual core, assuming fully developed conditions. In this general approach, $\Delta P_{1-2}$ is given in terms of the entrance-loss and expansion-loss coefficients $K_c$ and $K_e$ by Eq. (8–62),

$$\Delta P_{1-2} = \frac{\rho U_b^2}{2}\left(K_c + \frac{4L}{D_H} f + K_e\right) \qquad (8\text{–}143)$$

for uniform property conditions. For specific applications involving tube banks, $K_c$ and $K_e$ have traditionally been set equal to zero, such that $\Delta P_{1-2}$ is represented by

$$\Delta P_{1-2} = \frac{4L}{D_H}\frac{f}{2}\rho U_b^2 = \frac{\pi N_L}{C_A}\frac{f}{2}\rho U_b^2 \qquad (8\text{–}144)$$

with the entrance and exit losses accounted for by $f$. For situations in which significant changes in density occur over the length of the tube bank, $\Delta P_{1-2}$ is obtained from Eq. (8–63) (with $K_1 = K_2 = 0$); that is,

$$\Delta P_{1-2} = \frac{G^2 v_1}{2}\left[(1 + \sigma^2)\left(\frac{v_2}{v_1} - 1\right) + \frac{4L}{D_H} f \frac{v_m}{v_1}\right] \qquad (8\text{–}145)$$

where $4L/D_H = \pi N_L/C_A$.

Experimental data for $f$ are presented by Kays and London [4] for a number of in-line and staggered tube-bank arrangements. Direct test data obtained for a specific tube bank (staggered arrangement with $D = 0.009525$ m, $S_L = 0.01191$ m, $S_T = 0.01429$ m) are shown in Fig. 8–23. Correlations developed on the basis of data for all of the tube banks with staggered array considered by Kays and London are shown in Fig. 8–27. Similar correlations have been developed by Grimison [56] for both in-line and staggered tube-bank arrangements.

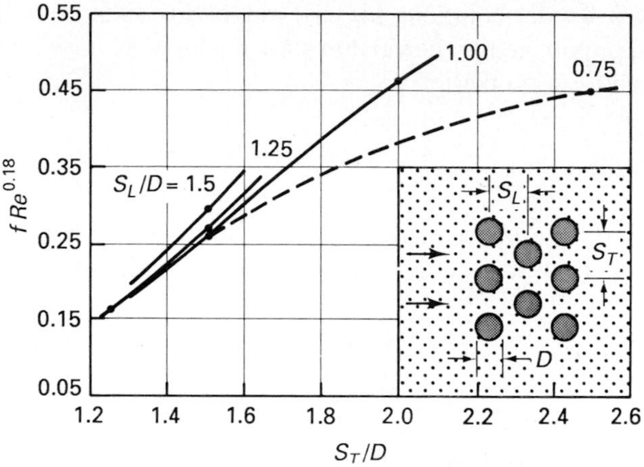


**FIGURE 8–27** Friction factor correlation—flow normal to an infinite bank of staggered circular tubes; $300 \leqslant Re \leqslant 15{,}000$. (From Kays and London [4]. Used with permission.)

In this connection, it is important to note that somewhat different approaches to the correlation of pressure drop for flow across tube banks appear in the literature [57–63]. For example, the recent correlation developed by Zukauskas [57] can be written as

$$\Delta P_{1-2} = 4N_L \frac{f'}{2} \rho U_b^2 \qquad (8\text{–}146)$$

Correlations developed by Zukauskas [57] for $f'$ are shown in terms of $4f'/\chi$ versus $Re_D$ by Fig. 8–28 where $Re_D = U_b D/\nu$; $\chi$ is a correction factor which is generally a function of $Re_D$ and both $S_L/D$ and $S_T/D$. By comparing Eqs. (8–143) and (8–145), we observe that $f'$ and $f$ are related by

$$f = \frac{4C_A f'}{\pi} \qquad (8\text{–}147)$$

It follows that the term $\pi N_L f/C_A$ in Eqs. (8–143) and (8–144) can be replaced by $4N_L f'$.

With $\Delta P$ known, the pumping power can be obtained by use of Eq. (8–65),

$$\dot{W}_p = \frac{\dot{m}\,\Delta P}{\rho} \qquad (8\text{–}148)$$

for uniform property flow, or Eq. (8–66),

$$\dot{W}_p = \dot{m}\,\Delta P\, v_m \qquad (8\text{–}149)$$

for variable property conditions.

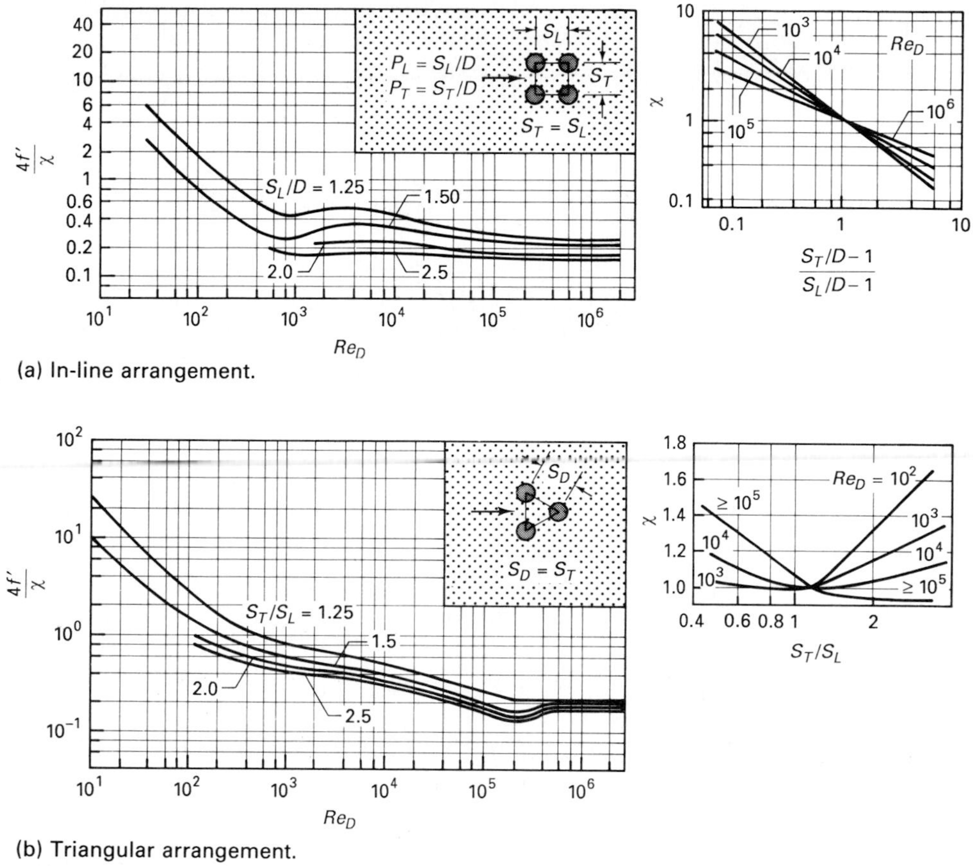


**FIGURE 8–28** Friction factor $f'$ and correction factor $\chi$ for flow across tube banks (Zukauskas [57]).

## EXAMPLE 8–12

The basic geometric parameters listed for the tube-bank surface S1.50–1.25 shown in Fig. 8–23 are

$$D = 0.375 \text{ in.} \quad S_L = 15/32 \text{ in.} \quad S_T = 9/16 \text{ in.}$$

Confirm the values for hydraulic diameter $D_H$, contraction ratio $\sigma$, and surface-area density $\beta$, which are given in the figure.

### Solution

*Objective* Show that $D_H = 0.0249$ ft, $\sigma = 0.333$, and $\beta = 53.6\ \text{ft}^{-1}$.

*Schematic* Tube-bank surface S1.50–1.25.

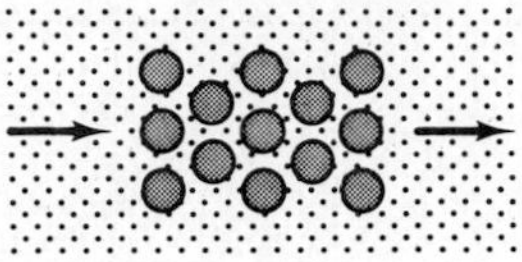

$D = 0.375$ in.
$S_L = 15/32$ in.
$S_T = 9/16$ in.

*Analysis* To provide a basis for evaluating $D_H$, $\sigma$, and $\beta$ for this staggered arrangement, we compute

$$S_D = \left[S_L^2 + \left(\frac{S_T}{2}\right)^2\right]^{1/2} = \left[\left(\frac{15}{32}\text{ in.}\right)^2 + \left(\frac{9/16}{2}\text{ in.}\right)^2\right]^{1/2} = 0.547 \text{ in.}$$

and

$$\frac{S_T + D}{2} = \frac{9/16 \text{ in.} + 0.375 \text{ in.}}{2} = 0.469 \text{ in.}$$

Because $S_D > (S_T + D)/2$, $A_{\min}$ is taken as the area in spaces transverse to the flow; that is,

$$C_A = \frac{S_T}{D} - 1 = \frac{9/16}{0.375} - 1 = 0.5$$

Thus, using Eq. (8–122), $A_{\min}$ is given by

$$\frac{A_{\min}}{N_T Z} = C_A D = 0.5(0.375 \text{ in.}) = 0.188 \text{ in.} = 0.0156 \text{ ft} = 0.00476 \text{ m}$$

We are now able to compute $D_H$ from Eq. (8–126),

$$D_H = \frac{4A_{\min}}{p_w} = \frac{4C_A S_L}{\pi} = \frac{4(0.5)(15/32 \text{ in.})}{\pi}$$

$$= 0.298 \text{ in.} = 0.0249 \text{ ft} = 0.00758 \text{ m}$$

$\sigma$ from Eq. (8–125),

$$\sigma = \frac{A_{\min}}{A_1} = \frac{C_A D}{S_T} = \frac{0.5(0.375 \text{ in.})}{0.562} = 1/3$$

and $\beta$ from Eq. (8–120),

$$\beta = \frac{A_s}{V} = \frac{\pi D}{S_L S_T} = \frac{\pi(0.375 \text{ in.})}{(15/32 \text{ in.})(9/16 \text{ in.})} = 4.47/\text{in.} = 53.6/\text{ft} = 176/\text{m}$$

These results for $D_H$, $\sigma$, and $\beta$ are consistent with the values given by Fig. 8–23.

## EXAMPLE 8–13

A tube bank is used in a commercial heat pump unit to heat air at 1 atm and 15°C. The tube bank consists of eighty 9.52-mm-diameter tubes in a staggered array with 10 rows, 2.5 m depth, 1.19 cm longitudinal pitch, and 1.43 cm transverse pitch. The air enters with a velocity of 6.8 m/s and the tubes are maintained at 65°C. Determine the temperature of the exiting air, total rate of heat transfer, pressure drop, and pumping power.

### Solution

*Objective* Determine $T_2$, $q_c$, $\Delta P$, and $\dot{W}_p$.

*Schematic* Air flow over a tube bank—uniform wall temperature: $T_0 > T_1$.

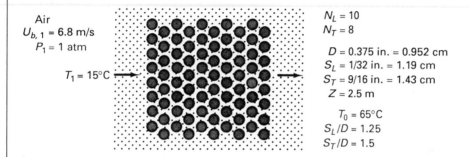

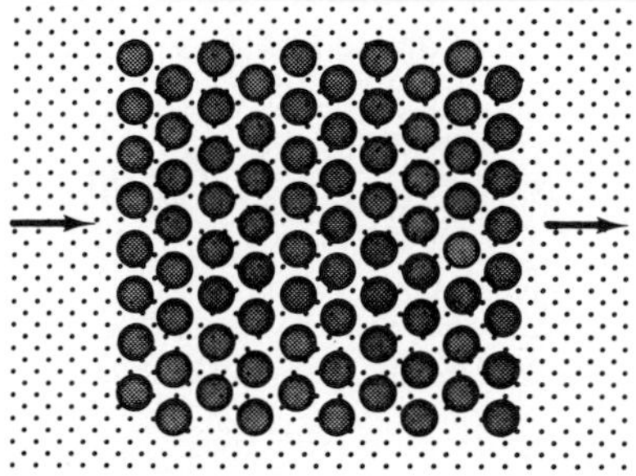
Air
$U_{b,1}$ = 6.8 m/s
$P_1$ = 1 atm
$T_1$ = 15°C
$N_L$ = 10
$N_T$ = 8
$D$ = 0.375 in. = 0.952 cm
$S_L$ = 1/32 in. = 1.19 cm
$S_T$ = 9/16 in. = 1.43 cm
$Z$ = 2.5 m
$T_0$ = 65°C
$S_L/D$ = 1.25
$S_T/D$ = 1.5

*Assumptions/Conditions*

forced convection
moderate property variation
standard conditions

*Properties*

Air at 15°C (Table A–C–5): $\rho = 1.23$ kg/m$^3$, $\mu = 1.76 \times 10^{-5}$ kg/(m s), $\nu = 1.43 \times 10^{-5}$ m$^2$/s, $c_P = 1.01$ kJ/(kg °C), $k = 0.0253$ W/(m °C), $Pr = 0.711$.

Air at 65°C: $Pr = 0.700$.

*Analysis* The entrance area $A_1$ and core surface area $A_s$ are given by

$$A_1 = N_T S_T Z = 8(0.0143 \text{ m})(2.5 \text{ m}) = 0.286 \text{ m}^2$$

and

$$A_s = \pi D N Z = \pi(0.00952 \text{ m})(80)(2.5 \text{ m}) = 5.98 \text{ m}^2$$

This particular arrangement is identical to the tube-bank surface S1.50–1.25, which was considered in Example 8–12. Therefore, we also have $C_A = 0.5$, $D_H = 0.00758$ m $= 0.758$ cm, and $\sigma = 1/3$. It follows that

$$A_{\text{min}} = C_A N_T D Z = 0.5(8)(0.00952\text{ m})(2.5\text{ m}) = 0.0953\text{ m}^2$$

Turning to the hydraulic aspects, the mass flow rate $\dot{m}$ is

$$\dot{m} = \rho A_1 U_{b,1} = 1.23\,\frac{\text{kg}}{\text{m}^3}\,(0.286\text{ m}^2)\left(6.8\,\frac{\text{m}}{\text{s}}\right) = 2.39\text{ kg/s}$$

and the core model bulk-stream velocity $U_b$ is

$$U_b = \frac{A_1}{A_{\text{min}}}\,U_{b,1} = \frac{U_{b,1}}{\sigma} = \frac{6.8\text{ m/s}}{1/3} = 20.4\text{ m/s}$$

These results can be used to compute the Reynolds number $Re$ (and/or $Re_D$) and to evaluate the friction factor $f$ and coefficient of heat transfer $\bar{h}$. The Reynolds number $Re$ is

$$Re = \frac{U_b D_H}{\nu} = \frac{(20.4\text{ m/s})(0.00758\text{ m})}{1.43 \times 10^{-5}\text{ m}^2\text{/s}} = 10{,}800$$

Using Figs. 8–24 and 8–27, we obtain

$$f = 0.27\,Re^{-0.18} = 0.27(10{,}800)^{-0.18} = 0.0507$$

and

$$\overline{St}_\infty\,Pr^{2/3} = 0.32\,Re^{-0.4} = 0.00779$$

or, since $\overline{St} = \overline{Nu}/(Re\,Pr)$,

$$\overline{Nu}_\infty = \frac{\bar{h}_\infty D_H}{k} = 0.32\,Re^{0.6}Pr^{1/3} = 0.32(10{,}800)^{0.6}\,(0.711)^{1/3} = 75.1$$

which gives

$$\bar{h}_\infty = \overline{Nu}_\infty\frac{k}{D_H} = 75.1\,\frac{0.0253\text{ W/(m}^2\text{ °C)}}{0.00758\text{ m}} = 251\text{ W/(m}^2\text{ °C)}$$

Using Fig. 8–25 to correct for the number of rows, $\bar{h}$ becomes

$$\bar{h} = 0.93\,\bar{h}_\infty = 0.93\left(251\,\frac{\text{W}}{\text{m}^2\text{ °C}}\right) = 233\text{ W/(m}^2\text{ °C)}$$

Finally, to approximately correct for the property variation over the cross section, we use Eqs. (8–25) and (8–26); that is,

$$\frac{f}{f_{cp}} = \left(\frac{T_0}{T_b}\right)^m = \left(\frac{338}{288}\right)^{-0.1} = 0.984$$

$$f = 0.0507(0.984) = 0.0499$$

and

$$\frac{\bar{h}}{\bar{h}_{cp}} = \frac{\overline{Nu}}{\overline{Nu}_{cp}} = \left(\frac{T_0}{T_b}\right)^n = \left(\frac{338}{288}\right)^{-0.5} = 0.923$$

$$\bar{h} = 233\,\frac{\text{W}}{\text{m}^2\,°\text{C}}\,(0.923) = 215\ \text{W/(m}^2\,°\text{C)}$$

The outlet temperature $T_2$ and heat transfer rate are given by Eqs. (8–138),

$$\frac{T_2 - T_1}{T_0 - T_1} = 1 - \exp\left(-\frac{\bar{h}A_s}{\dot{m}c_P}\right)$$

and Eq. (8–136),

$$q_c = \dot{m}c_P(T_2 - T_1)$$

To evaluate $T_2$, we first compute the number of transfer units $NTU$.

$$NTU = \frac{\bar{h}A_s}{\dot{m}c_P} = \frac{[215\ \text{W/(m}^2\,°\text{C)}](5.98\ \text{m}^2)}{(2.39\ \text{kg/s})[1.01\ \text{kJ/(kg}\,°\text{C)}]} = 0.533$$

Substituting this value into Eq. (8–138), $T_2$ becomes

$$T_2 = 15°\text{C} + (65°\text{C} - 15°\text{C})[1 - \exp(-0.533)] = 35.7°\text{C}$$

Using Eq. (8–136), $q_c$ is

$$q_c = 2.39\,\frac{\text{kg}}{\text{s}}\left(1.01\,\frac{\text{kJ}}{\text{kg}\,°\text{C}}\right)(35.7°\text{C} - 15°\text{C}) = 50.0\ \text{kW}$$

[This result is also obtained by use of the *LMTD* equation for $q_c$, Eq. (8–139).]

This calculation for $T_2$ indicates a moderate variation in fluid properties over the length of the tube bank. The refinement of the analysis by use of average fluid properties (based on this value of $T_2$) is suggested as an exercise.

To continue with the analysis, the pressure drop is given by Eq. (8–144),

$$\Delta P_{1-2} = \frac{\pi N_L}{C_A}\frac{f}{2}\rho U_b^2 = \frac{\pi(10)}{0.5}\frac{0.0499}{2}\left(1.23\,\frac{\text{kg}}{\text{m}^3}\right)(20.4\ \text{m})^2$$

$$= 802\ \text{N/m}^2 = 0.802\ \text{kPa}$$

Using Eq. (8–148), the pumping power is

$$\dot{W}_p = \frac{\dot{m}\,\Delta P_{1-2}}{\rho} = \frac{(2.39\ \text{kg/s})(802\ \text{N/m}^2)}{1.23\ \text{kg/m}^3} = 1560\ \text{W} = 1.56\ \text{kW}$$

Comparing $q_c$ and $\dot{W}_p$, we find that the pumping-heating power ratio is only 0.032. However, with the velocity increased by a factor of 2, the pumping-heating power ratio increases by a factor of about 4.

In connection with the correlations for $f$ and $\overline{St}$ given in Figs. 8–27 and 8–24, Kays and London [4] established these relations on the basis of experimental data obtained for a number of tube-bank arrangements. Direct data taken for the particular arrangement considered in this example are shown in Fig. 8–23. Using this figure, $f = 0.053$ and $\overline{St}_\infty\, Pr^{2/3} = 0.008$, which are within 4% and 1% of the results obtained using Figs. 8–27 and 8–24. To use the alternative correlations developed by Zukauskas [57], we compute the Reynolds number $Re_D$.

$$Re_D = \frac{DU_b}{\nu} = \frac{D}{D_H} Re = \frac{0.00952 \text{ m}}{0.00758 \text{ m}} 10{,}800 = 13{,}600$$

Using this input together with Figs. 8–26 and 8–28 and Eq. (8–141), we obtain

$$F_c = 0.975 \qquad \frac{4f'}{\chi} = 0.38 \qquad \chi = 1.0$$

and

$$\overline{Nu}_D = C\, Re_D^m\, Pr^n \left(\frac{Pr}{Pr_0}\right)^{0.25} F_c = 0.35 \left(\frac{S_T}{S_L}\right)^{0.2} Re_D^{0.6}\, Pr^{0.36} \left(\frac{Pr}{Pr_0}\right)^{0.25} F_c$$

$$= 0.35 \left(\frac{1.43}{1.19}\right)^{0.2} (13{,}600)^{0.6}\, (0.711)^{0.36} \left(\frac{0.711}{0.700}\right)^{0.25} (0.975) = 94.5$$

Thus,

$$f = \frac{4C_A f'}{\pi} = \frac{4(0.5)\,[0.38\,(1.0)/4]}{\pi} = 0.0605$$

and

$$\bar{h} = 94.5 \frac{0.0253 \text{ W/(m °C)}}{0.00952 \text{ m}} = 241 \text{ W/(m}^2\text{ °C)}$$

These results are within about 19% and 10% of the values obtained using the correlations of Kays and London [4].

## EXAMPLE 8–14

A tube bank with tubes maintained at approximately 58°F is to be used to cool air at atmospheric pressure from 260°F to 80°F. The entering velocity is 10 ft/s and $\frac{3}{8}$-in.-diameter tubes are to be placed in a staggered array with 10 columns, $\frac{3}{8}$ in. longitudinal pitch, and 0.75 in. transverse pitch. Determine the number of tubes that are required, the pumping power per unit depth, and the pumping-heating power ratio.

## Solution

*Objective* Determine the required number of rows $N_L$, the pumping power per unit depth $\dot{W}_p/Z$, and the pumping-heating power ratio $\dot{W}_p/q_c$.

*Schematic*† Air flow over a tube bank—uniform wall temperature: $T_0 < T_1$.

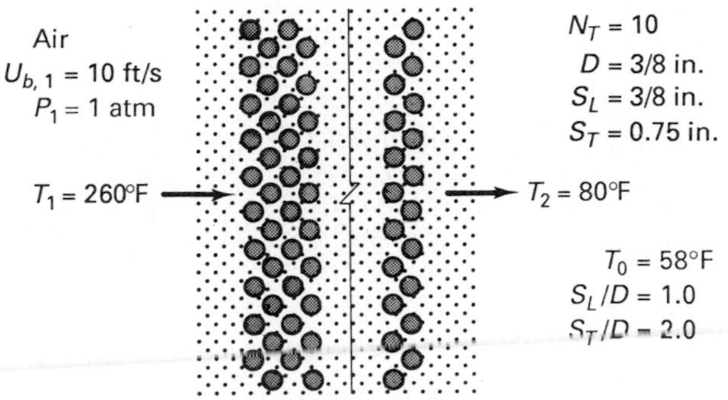
Air
$U_{b,1}$ = 10 ft/s
$P_1$ = 1 atm
$T_1$ = 260°F
$N_T$ = 10
$D$ = 3/8 in.
$S_L$ = 3/8 in.
$S_T$ = 0.75 in.
$T_2$ = 80°F
$T_0$ = 58°F
$S_L/D$ = 1.0
$S_T/D$ = 2.0

*Assumptions/Conditions*

forced convection
moderate property variation
standard conditions

*Properties* Air at average bulk-stream temperature $T_m$ = 170°F = 630°R = 350 K (Table A–C–5):

$$\rho = 0.998\,\frac{\text{kg}}{\text{m}^3}\,\frac{0.0624\ \text{lb}_\text{m}/\text{ft}^3}{\text{kg/m}^3} = 0.0623\ \text{lb}_\text{m}/\text{ft}^3$$

$$\mu = 2.08 \times 10^{-5}\,\frac{\text{kg}}{\text{m s}}\,\frac{0.672\ \text{lb}_\text{m}/(\text{ft s})}{\text{kg/(m s)}} = 1.40 \times 10^{-5}\ \text{lb}_\text{m}/(\text{ft s})$$

$$c_P = 1.01\,\frac{\text{kJ}}{\text{kg K}}\,\frac{0.239\ \text{Btu}/(\text{lb}_\text{m}\ ^\circ\text{F})}{\text{kJ/(kg K)}} = 0.241\ \text{Btu}/(\text{lb}_\text{m}\ ^\circ\text{F})$$

$$k = 0.030\,\frac{\text{W}}{\text{m}^\circ\ \text{C}}\,\frac{0.578\ \text{Btu/(h ft }^\circ\text{F)}}{\text{W/(m }^\circ\text{C)}} = 0.0173\ \text{Btu/(h ft }^\circ\text{F)}$$

$$Pr = 0.697$$

The density at the inlet and exit temperatures varies from 0.0551 $\text{lb}_\text{m}/\text{ft}^3$ (0.883 $\text{kg/m}^3$) to about 0.0735 $\text{lb}_\text{m}/\text{ft}^3$ (1.18 $\text{kg/m}^3$), which represents a 34% increase. Therefore, the effects of density variation over the length of the tube bank should be considered.

† This particular geometric arrangement is denoted as surface S2.00–1.00 in the work of Kays and London [4].

*Analysis* The entrance area $A_1$ and surface area $A_s$ are expressed in terms of the tube bank depth $Z$ by

$$A_1 = N_T S_T Z = 10\left(\frac{0.750}{12}\text{ ft}\right)Z = 0.625\,Z\text{ ft}$$

and

$$A_s = \pi DNZ = \pi\left(\frac{0.375}{12}\text{ ft}\right)(10)N_L Z = 0.982 N_L Z\text{ ft} \tag{a}$$

The minimum free-flow area $A_{\min}$ is determined by writing

$$S_D = \left[S_L^2 + \left(\frac{S_T}{2}\right)^2\right]^{1/2} = \left[(0.375\text{ in.})^2 + \left(\frac{0.750\text{ in.}}{2}\right)^2\right]^{1/2} = 0.530\text{ in.}$$

and

$$\frac{S_T + D}{2} = \frac{0.750\text{ in.} + 0.375\text{ in.}}{2} = 0.562\text{ in.}$$

In this case $S_D < (S_T + D)/2$, such that $C_A$ is given by Eq. (8–123b),

$$C_A = 2\left(\frac{S_D}{D} - 1\right) = 2\left(\frac{0.530\text{ in.}}{0.375\text{ in.}} - 1\right) = 0.827$$

Using this result, we obtain

$$A_{\min} = C_A N_T DZ = 0.827(10)\left(\frac{0.375}{12}\text{ ft}\right)Z = 0.258\,Z\text{ ft}$$

$$D_H = \frac{4C_A S_L}{\pi} = \frac{4(0.827)(0.375\text{ in.})}{\pi} = 0.395\text{ in.} = 0.0329\text{ ft}$$

$$\sigma = \frac{A_{\min}}{A_1} = \frac{0.258\,Z\text{ ft}}{0.625\,Z\text{ ft}} = 0.413$$

The mass flow rate $\dot{m}$ and core model mass flux $G$ are computed by writing

$$\dot{m} = \rho_1 A_1 U_{b,1} = 0.0551\,\frac{\text{lb}_\text{m}}{\text{ft}^3}(0.625\,Z\text{ ft})\left(10\,\frac{\text{ft}}{\text{s}}\right) = 0.344\,Z\text{ lb}_\text{m}/(\text{ft s})$$

and

$$G = \frac{\dot{m}}{A_{\min}} = \frac{0.344\,Z\text{ lb}_\text{m}/(\text{ft s})}{0.258\,Z\text{ ft}} = 1.33\text{ lb}_\text{m}/(\text{ft}^2\text{ s})$$

With the viscosity evaluated at $T_m$, the Reynolds number $Re$ becomes

$$Re = \frac{GD_H}{\mu} = \frac{[1.33\text{ lb}_\text{m}/(\text{ft}^2\text{ s})]\,(0.0329\text{ ft})}{1.40\times10^{-5}\text{ lb}_\text{m}/(\text{ft s})} = 3130$$

Using this result together with Figs. 8–27 and 8–24, $f$ and $\bar{h}_\infty$ become

$$f = 0.46\, Re^{-0.18} = 0.108$$

and

$$\frac{\bar{h}_\infty D_H}{k} = \overline{Nu}_\infty = 0.46\, Re^{0.6}\, Pr^{1/3} = 51.0$$

or

$$\bar{h}_\infty = \overline{Nu}_\infty \frac{k}{D_H} = 51.0 \frac{0.0173 \text{ Btu/(h ft °F)}}{0.0329 \text{ ft}} = 26.8 \text{ Btu/(h ft}^2 \text{ °F)}$$

Since $N_T$ is unknown, the correction for the effect of the number of rows on $\bar{h}$ will be made later. Correcting for the effect of property variation over the cross section, we obtain

$$\frac{f}{f_{cp}} = \left(\frac{518\text{°R}}{630\text{°R}}\right)^{-0.18} = 1.02$$

$$f = 1.02(0.108) = 0.110$$

and

$$\frac{\bar{h}}{\bar{h}_{cp}} = \left(\frac{518\text{°R}}{630\text{°R}}\right)^{-0.5} = 1.10$$

$$\bar{h} = 1.10\left(26.8 \frac{\text{Btu}}{\text{h ft}^2 \text{ °F}}\right) = 29.5 \text{ Btu/(h ft}^2 \text{ °F)}$$

The rate of heat transfer $q_c$ is given by Eq. (8–136),

$$q_c = \dot{m}c_P(T_2 - T_1) = 0.344\, Z \frac{\text{lb}_\text{m}}{\text{ft s}}\left(0.241 \frac{\text{Btu}}{\text{lb}_\text{m} \text{ °F}}\right)(80\text{°F} - 260\text{°F})$$

$$= -14.9\, Z \text{ Btu/(ft s)} = -51.5\, Z \text{ kW/m}$$

and Eq. (8–139),

$$q_c = \bar{h}A_s\, LMTD$$

where $LMTD$ is given by Eq. (8–140),

$$LMTD = \frac{T_2 - T_1}{\ln \dfrac{T_0 - T_1}{T_0 - T_2}} = \frac{80\text{°F} - 260\text{°F}}{\ln \dfrac{58\text{°F} - 260\text{°F}}{58\text{°F} - 80\text{°F}}} = -81.2\text{°F}$$

Combining these relations and solving for $A_s$, we have

$$A_s = \frac{q_c}{\bar{h}\, LMTD} = \frac{-14.9\, Z \text{ Btu/(ft s)}}{[29.5 \text{ Btu/(h ft}^2 \text{ °F)}]\,(-81.2\text{°F})} = 22.4\, Z \text{ ft}$$

Substituting this result into Eq. (a), $N_L$ is given by

$$N_L = \frac{A_s}{0.982\ Z\ \text{ft}} = \frac{22.4\ Z\ \text{ft}}{0.982\ Z\ \text{ft}} = 22.8$$

Using this result together with Fig. 8–25, the analysis is refined by writing

$$\bar{h} = 0.98\bar{h}_\infty = 0.98\left(29.5\,\frac{\text{Btu}}{\text{h ft}^2\ \text{°F}}\right) = 28.9\ \text{Btu/(h ft}^2\ \text{°F)}$$

$$A_s = \frac{-14.9\ Z\ \text{Btu/(ft s)}}{[28.9\ \text{Btu/(h ft}^2\ \text{°F)}]\,(-81.2\text{°F})} = 22.8\ Z\ \text{ft}$$

and

$$N_L = \frac{22.8\ Z\ \text{ft}}{0.982\ Z\ \text{ft}} = 23.2$$

Rounding off, we conclude that 23 tube rows must be used in this application. Thus, a total of 230 tubes is required to satisfy the heating requirement. Because of this large number of tubes, alternative means of accomplishing this task should be considered. The use of finned tubes and compact heat exchangers for applications such as this with large heating requirements is considered in reference 64.

Turning to the hydraulic analysis, to account for the effects of density variation over the length of the tube bank on the pressure drop, we use Eq. (8–145),

$$\Delta P_{1-2} = \frac{G^2 v_1}{2}\left[(1 + \sigma^2)\left(\frac{v_2}{v_1} - 1\right) + \frac{\pi N_L}{C_A} f \frac{v_m}{v_1}\right]$$

$$= \left(1.33\,\frac{\text{lb}_\text{m}}{\text{ft}^2\ \text{s}}\right)^2 \frac{1}{2(0.0551\ \text{lb}_\text{m}/\text{ft}^3)}\left[(1 + 0.413^2)\left(\frac{0.0551\ \text{lb}_\text{m}/\text{ft}^3}{0.0735\ \text{lb}_\text{m}/\text{ft}^3} - 1\right)\right. \tag{b}$$

$$\left. + \frac{\pi(23)(0.108)}{0.827}\,\frac{0.0551\ \text{lb}_\text{m}/\text{ft}^3}{0.0623\ \text{lb}_\text{m}/\text{ft}^3}\right] = (-4.70 + 134)\,\frac{\text{lb}_\text{m}}{\text{ft s}^2}$$

$$= 129\,\frac{\text{lb}_\text{m}}{\text{ft s}^2}\,\frac{\text{lb}_\text{f}\ \text{s}^2}{32.2\ \text{lb}_\text{m}\ \text{ft}} = 4.01\ \text{lb}_\text{f}/\text{ft}^2 = 0.0279\ \text{psi} = 192\ \text{Pa}$$

This pressure drop is in the appropriate range for conventional fans that develop low static pressure.

This result is substituted into Eq. (8–149) to obtain the pumping power $\dot{W}_p$.

$$\dot{W}_p = \dot{m}\,\Delta P\,v_m = 0.344\ Z\,\frac{\text{lb}_\text{m}}{\text{ft s}}\left(4.01\,\frac{\text{lb}_\text{f}}{\text{ft}^2}\right)\left(\frac{1}{0.0621\ \text{lb}_\text{m}/\text{ft}^3}\right)$$

$$= 22.2\ Z\ \text{ft lb}_\text{f}/(\text{ft s}) = 0.0285\ Z\ \text{Btu/(ft s)}$$

Thus, the pumping-heating power ratio is

$$\frac{\dot{W}_p}{q_c} = \frac{0.0285\ Z\ \text{Btu/(ft s)}}{-14.9\ Z\ \text{Btu/(ft s)}} = -0.00191$$

Concerning the result of the hydraulic analysis, it is interesting to note that despite the large density variation over the length of the tube bank, the effect of acceleration represented by the first term in Eq. (b) is only about 3.5% of the core effect. This is often the case for applications involving tube banks and compact heat exchangers, such that the simpler uniform property relation given by Eq. (8–143) (with $\rho$ set equal to $\rho_m$) often suffices.

## 8–5 SUMMARY

The practical engineering analysis of internal flow convection heat transfer processes involves the use of correlations for friction factor $f$, entrance- and expansion-loss coefficients $K_c$ and $K_e$, and the coefficient of heat transfer $h$ in conjunction with lumped/differential formulations for momentum and energy transfer. In this chapter we have presented correlations for $f$, $K_c$, $K_e$, and $h$ for standard single-phase incompressible internal flows which are commonly encountered in practice. Correlations for laminar and turbulent flow presented in this chapter are summarized in Table 8–5. In addition, mathematical formulations for the fluid flow and heat transfer have been developed for uniform property flows. The resulting practical thermal and hydraulic analysis approach applies to internal flows with both uniform and nonuniform cross-sectional area, such that the heat transfer, pressure drop, and pumping power can be computed for heat-transfer cores involving single or multiple tubular passages as well as flow across tube banks. Relations for the key geometric and flow characteristics developed in Sec. 8–4 for flow across tube banks are summarized in Table 8–6. Because of its importance, the practical analysis approach for internal flows is summarized in Table 8–7. As we shall see in Chap. 11, this lumped analysis approach provides the framework for the evaluation and design of heat exchangers. The concepts presented in Secs. 8–2 and 8–4 for internal flows are adapted to finned-tube systems and variable-property flows in references 4, 5, and 64.

The practical approach to the analysis of external flow features the use of defining relations and correlations for friction factor $f_x$ and the coefficient of heat transfer $h_x$. Representative correlations for friction factor $f_x$ and Nusselt number $Nu_x$ for laminar and turbulent flow over surfaces with various heating conditions are listed in Table 8–8.

Finally, the primary dimensionless parameters used in the analysis of forced convection are summarized in Table 8–9.

**TABLE 8–5** Summary of convection correlations for internal flow

| *Condition* | *Correlation* | *Identification* |
|---|---|---|
| LAMINAR FLOW† | | |
| HFD | $f = C_1/Re$ | Eq. (8–7), Table 8–1<br>Fig. 8–2 |
| TFD | $Nu = C_2$ | Eq. (8–8), Table 8–1<br>Fig. 8–2 |
| Thermal entry<br>$T_s = T_0$ | $\overline{Nu} = 3.66 + \dfrac{0.0668 \dfrac{Re\,Pr}{x/D}}{1 + 0.04 \left(\dfrac{Re\,Pr}{x/D}\right)^{2/3}}$ | Eq. (8–9)<br>Hausen [15] |
| $T_s = T_0$<br>$\dfrac{x/D}{Re\,Pr} \lessdot 0.01$ | $\overline{Nu} = 1.67 \left(\dfrac{Re\,Pr}{x/D}\right)^{1/3}$ | Eq. (8–10)<br>Limiting form of Eq. (8–9) |
| $q''_c = q''_0$<br>$\dfrac{x/D}{Re\,Pr} \lessdot 0.01$ | $(Nu)_x = 1.30 \left(\dfrac{Re\,Pr}{x/D}\right)^{1/3}$ | Eq. (8–11)<br>Sellars et al. [16] |
| Hydrodynamic and thermal entry | Performance curves for $f$, $\bar{f}$, $f_{app}$ | Fig. 8–2<br>Langhaar [1] |
| | Performance curves for $K_c$ and $K_e$ | Figs. 8–4, 8–5<br>Kays and London [4] |
| Gas<br>$T_s = T_0$ | $\overline{Nu} = 3.66 + \dfrac{0.104 \dfrac{Re\,Pr}{x/D}}{1 + 0.016 \left(\dfrac{Re\,Pr}{x/D}\right)^{0.8}}$ | Eq. (8–12)<br>Kays [14] |
| $\dfrac{x/D}{Re\,Pr} \lessdot 0.01$ | $\overline{Nu} = 1.86 \left(\dfrac{Re\,Pr}{x/D}\right)^{1/3}$ | Eq. (8–13)<br>Seider and Tate [18] |

**TABLE 8–5** *(Continued)*

| *Condition* | *Correlation* | *Identification* |
|---|---|---|
| TURBULENT FLOW | | |
| HFD | $f = (1.58 \ln Re - 3.28)^{-2}$ | Eq. (8–15) |
| $2000 \leqslant Re \leqslant 10^4$ | $f = 0.079\, Re^{-0.25}$ | Eq. (8–20) |
| $10^4 \leqslant Re$ | $f = 0.046\, Re^{-0.2}$ | Eq. (8–17) |
| TFD<br>$10^4 \leqslant Re \leqslant 5 \times 10^6$<br>$0.5 \leqslant Pr \leqslant 200$ | $Nu = \dfrac{(f/2)\, Re\, Pr}{1.07 + 12.7\sqrt{f/2}\,(Pr^{2/3} - 1)}$ | Eq. (8–16) |
| $10^4 \leqslant Re$ | $Nu = 0.023\, Re^{0.8}\, Pr^n$<br>$n = 0.5$ for $0.5 \leqslant Pr \leqslant 5.0$<br>$n = 1/3$ for $0.5 \leqslant Pr$ | Eq. (8–19)<br>(related correlations by Dittus and Boelter [28] and Colburn [29] |
| Thermal Entry | | |
| $x_c/D \leqslant x/D \leqslant 60$ | $\overline{Nu} = Nu\left(1 + \dfrac{1.4}{x/D}\right)$ | Eq. (8–21a) |
| $x/D \leqslant x_c/D = 0.625\, Re^{0.25}$ | $\overline{Nu} = Nu\dfrac{1.11\, Re^{0.2}}{(x/D)^{0.8}}$ | Eq. (8–21b)<br>*Handbook of Heat Transfer* [3] |
| TFD, Liquid metals<br>$Re\, Pr \geqslant 100$<br>$L/D \geqslant 30$ | $Nu = 5 + 0.025\,(Re\, Pr)^{0.8}$ | Eq. (8–22)<br>Subbotin et al. [33] |
| Hydrodynamic and thermal entry | Performance curves for $K_c$ and $K_e$ | Eq. 8–5<br>Kays and London [4] |
| $x_c/D \leqslant x/D \leqslant 60$ | $\overline{Nu} = Nu\left(1 + \dfrac{6}{x/D}\right)$ | Eq. (8–21a)<br>*Handbook of Heat Transfer* [3] |
| $x/D \leqslant x_c/D = 0.625\, Re^{0.25}$ | See Eq. (8–21b) | |
| Rough surfaces<br>HFD | Moody friction factor chart | Fig. 8–10<br>Moody [43] |
| | $Nu = \dfrac{(f/2)\, Re\, Pr}{1.07 + 12.7\sqrt{f/2}\,(Pr^{2/3} - 1)}$<br>with $f$ obtained from Moody chart | Eq. (8–16) |
| VARIABLE PROPERTY FLOW | | |
| Gas | $\dfrac{f}{f_{cp}} = \left(\dfrac{T_s}{T_b}\right)^m, \dfrac{Nu}{Nu_{cp}} = \left(\dfrac{T_s}{T_b}\right)^n$ | Eq. (8–25, 26)<br>$m$, $n$ from Table 8–2 |
| Liquid | $\dfrac{f}{f_{cp}} = \left(\dfrac{\mu_s}{\mu_b}\right)^m, \dfrac{Nu}{Nu_{cp}} = \left(\dfrac{\mu_s}{\mu_b}\right)^n$ | Eqs. (8–23, 24)<br>$m$, $n$ from Table 8–2 |
| TUBE BANKS | Performance curves for $f$ and $\overline{St}$ | Kays and London [4]<br>Zukauskas [58] |

† Correlations for Nusselt number are restricted to $Pr \geqslant 0.5$.

**TABLE 8–6** Flow across tube banks: Relations for geometric and flow characteristics†

| *Characteristic* | *Relation* | *Equation* |
|---|---|---|
| Derived Geometric Characteristics | | |
| Length | $L = N_L S_L$ | (8–116) |
| Frontal area | $A_1 = N_T S_T Z$ | (8–117) |
| Volume | $V = A_1 L = N S_L S_T Z$ | (8–118) |
| Surface area | $A_s = \pi D N Z$ | (8–119) |
| Surface area density | $\beta = \dfrac{A_s}{V} = \dfrac{\pi D}{S_L S_T}$ | (8–120) |
| Heated perimeter | $p = \dfrac{A_s}{L} = \dfrac{\pi D N_T Z}{S_L}$ | (8–121) |
| Wetted perimeter | $p_w = p$ | |
| Minimum free-flow area | $A_{\min} = C_A N_T D Z$ | (8–122) |
| Area Coefficient | | |
| In-line array | $C_A = \dfrac{S_T}{D} - 1$ | (8–123a) |
| Staggered array | $C_A = \dfrac{S_T}{D} - 1 \quad \text{for } S_D \geqslant \dfrac{S_T + D}{2}$ | (8–123a) |
| | $C_A = 2\left(\dfrac{S_D}{D} - 1\right) \quad \text{for } S_D < \dfrac{S_T + D}{2}$ | (8–123b) |
| | $S_D = \left[S_L^2 + \left(\dfrac{S_T}{2}\right)^2\right]^{1/2}$ | (8–124) |
| Contraction ratio | $\sigma = \dfrac{A_{\min}}{A_1} = \dfrac{C_A D}{S_T}$ | (8–125) |
| Hydraulic diameter | $D_H = 4\dfrac{A_{\min}}{p_w} = 4\dfrac{C_A S_L}{\pi}$ | (8–126) |
| | $= 4\dfrac{\sigma}{\beta}$ | (8–128) |
| Flow Characteristics | | |
| Mass flow rate | $\dot{m} = \rho A_{\min} U_b = A_{\min} G$ | (8–129) |
| | $= \rho A_1 U_1 = A_1 G_1$ | (8–130) |
| Bulk-stream velocity | $U_b = \dfrac{U_1}{\sigma} = \dfrac{S_T}{C_A D} U_1$ | (8–131) |
| Bulk-stream mass flux | $G = \dfrac{G_1}{\sigma} = \dfrac{S_T}{C_A D} G_1$ | (8–132) |
| Reynolds number | $Re = \dfrac{U_b D_H}{\nu} = \dfrac{G D_H}{\mu}$ | (8–133) |

† Basic geometric dimensions: $D$ = tube diameter, $S_L$ = longitudinal pitch, $S_T$ = transverse pitch, $Z$ = depth.

**TABLE 8–7** Summary of the practical thermal analysis approach: Internal flow†

| *Control volume* | *Variables and relations* | *Energy equation* $\Sigma E_o - \Sigma E_i + \dfrac{\Delta E_s}{\Delta t} = 0$ |
|---|---|---|
| lumped/differential<br>$dV = A\,dx$<br>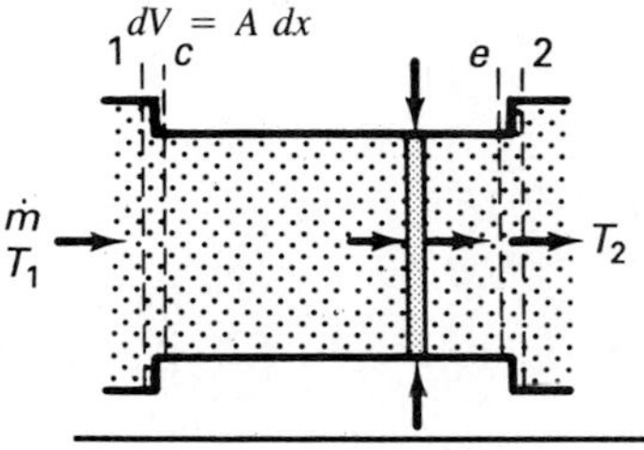 | Independent variables<br>$\dot m$<br>$T_1$, $T_s$ or $q''$<br>Dependent variables<br>$T_b(x)$<br>$q''_s(x)$ or $T_s(x)$<br>Bulk-stream relation<br>$d\dot H_b = \dot m c_P\, dT_b$ | lumped/differential formulation<br>$q''_c p = \dfrac{d\dot H_b}{dx} = \dot m c_P \dfrac{dT_b}{dx}$<br>lumped formulation<br>$q''_c = \dot H_2 - \dot H_1 = \dot m c_P(T_2 - T_1)$ |

| *Coefficients* | *Dimensionless parameters* | *Practical solution results* |
|---|---|---|
| Newton law of cooling (Coefficient of heat transfer $h_x$)<br>$q''_c = h_x(T_s - T_b)$<br>Fanning friction factor $f_x$<br>$\tau_s = \rho U_b^2 \dfrac{f_x}{2}$<br>Entrance-loss coefficient $K_c$<br>$P_c - P = \dfrac{\rho U_b^2}{2}\left(\dfrac{4x}{D_H} f + K_c\right)$<br>Expansion-loss coefficient $K_e$<br>$\Delta P_{\text{loss}} = K_e \dfrac{\rho U_b^2}{2}$ | Reynolds number<br>$Re = \dfrac{GD_H}{\mu} = \dfrac{U_b D_H}{\nu}$<br>Prandtl number<br>$Pr = \dfrac{\mu c_P}{k} = \dfrac{\nu}{\alpha}$<br>Nusselt number<br>$Nu = \dfrac{hD_H}{k}$<br>Stanton number<br>$St = \dfrac{Nu}{Re\,Pr}$ | Solution results given for<br>(1) specified wall-heat flux in Sec. 8-2-2<br>(2) specified wall temperature in [4,5,64]<br>(3) variable properties in [4,5,64]<br>Solution for uniform wall-temperature heating given by<br>$q_c = \dot m c_P(T_0 - T_1)\left[1 - \exp\left(-\dfrac{\bar h A_s}{\dot m c_P}\right)\right]$<br>which is generally represented by<br>$q_c = \bar h A_s\, LMTD$<br>where<br>$LMTD = \dfrac{\Delta T_1 - \Delta T_2}{\ln(\Delta T_1/\Delta T_2)}$<br>or<br>$q_c = \epsilon q_{c,\max}$<br>where $q_{c,\max} = \dot m c_P(T_0 - T_1)$ and<br>$\epsilon = 1 - \exp\left(-\dfrac{\bar h A_s}{\dot m c_P}\right)$<br>Hydraulic: Total core pressure drop<br>$\Delta P_{1-2} = \dfrac{\rho U_b^2}{2}\left(K_c + \dfrac{4L}{D_H} f + K_e\right)$<br>[See Eq. (8–64) for variable density.] |

† See Tables 8–5 and 8–8 for correlations.

**TABLE 8–8** Summary of convection correlations for external flow—flat plate with uniform free-stream velocity

| *Condition* | *Correlation* | *Identification* |
|---|---|---|
| LAMINAR FLOW | | |
| Uniform wall temperature | $f_x = 0.664\, Re_x^{-1/2}$ | Eq. (8–73) |
| $10 \lesssim Pr$ | $Nu_x = 0.339\, Re_x^{1/2}\, Pr^{1/3}$ | Eq. (8–76) |
| $0.6 \lesssim Pr \lesssim 10$ | $Nu_x = 0.332\, Re_x^{1/2}\, Pr^{1/3}$ | Eq. (8–75) |
| $Pr \lesssim 0.3$<br>$Pe \gtrsim 100$ | $Nu_x = \dfrac{0.686\, Re_x^{1/2}}{1 + (1.57/Pr - 7/9)^{1/2}}$ | Eq. (8–78) |
| Uniform Wall-Heat Flux | | |
| $5 \lesssim Pr$ | $Nu_x = 0.451\, Re_x^{1/2}\, Pr^{1/3}$ | Eq. (8–80) |
| $0.3 \lesssim Pr \lesssim 5$ | $Nu_x = 0.442\, Re_x^{1/2}\, Pr^{1/3}$ | Eq. (8–79) |
| $Pr \lesssim 0.3$<br>$Pe \gtrsim 100$ | $Nu_x = \dfrac{0.686\, Re_x^{1/2}}{1 + (0.783/Pr - 7/9)^{1/2}}$ | Eq. (8–81) |
| Step Wall Temperature | | |
| $Pr \gtrsim 0.6[1 - (x_0/x)^{3/4}]$ | $Nu_x = \dfrac{Nu_x\vert_{x0=0}}{[1 - (x_0/x)^{3/4}]^{1/3}}$ | Eq. (8–82) |
| Step Wall-Heat Flux | | |
| $Pr \gtrsim 0.464(1 - x_0/x)$ | $Nu_x = \dfrac{Nu_x\vert_{x0=0}}{(1 - x_0/x)^{1/3}}$ | Eq. (8–83) |
| Mean Coefficients | $\bar{f} = 2 f_L,\ \overline{Nu}_L = 2\, Nu_{x=L}$ | |
| TURBULENT FLOW | | |
| | $f_x = 0.0592\, Re_x^{-0.2}$ | Eq. (8–85) |
| | $f_x = \dfrac{0.455}{\ln^2 (0.06\, Re_x)}$ | Eq. (8–86) |
| Uniform Wall Temperature | | |
| $0.5 \lesssim Pr \lesssim 5.0$ | $Nu_x = 0.0296\, Re_x^{0.8}\, Pr^{0.5}$ | Eq. (8–88) |
| $5.0 \lesssim Pr$ | $Nu_x = 0.0296\, Re_x^{0.8}\, Pr^{1/3}$ | Eq. (8–88) |
| $0.5 \lesssim Pr$ | $Nu_x = \dfrac{(f_x/2)\, Re_x\, Pr}{1.0 + 12.7\sqrt{f_x/2}\, (Pr^{2/3} - 1)}$ | Eq. (8–89) |
| Uniform Wall-Heat Flux | | |
| $0.5 \lesssim Pr$ | $Nu_x = \dfrac{\sqrt{f_x/2}\, Re_x\, Pr}{2.17 \ln(\sqrt{f_x/2}\, Re_x) + B_0}$<br>$B_0 = 12.7\, Pr^{2/3} - 13.9$ | Eq. (8–91) |
| Step Wall-Heat Flux | | |
| $0.5 \lesssim Pr$ | $Nu_x = \dfrac{\sqrt{f_x/2}\, Re_x\, Pr}{2.17 \ln[(1 - x_0/x)\sqrt{f_x/2}\, Re_x] + B_0}$ | Eq. (8–90) |
| Mixed | $\dfrac{\bar{f}}{2} = 0.037\, Re_L^{-0.2} + (0.664\, Re_{x_c}^{-0.5} - 0.037\, Re_{x_c}^{-0.2})\dfrac{x_c}{L}$ | Eq. (8–93) |
| $Pr \gtrsim 0.5$ | $\overline{Nu}_L = [0.037\, Re_L^{0.8} + (0.664\, Re_{x_c}^{0.5} - 0.037\, Re_{x_c}^{0.8})]\, Pr^{1/3}$ | Eq. (8–94) |
| $x_0 = 0$ or $x/x_0$ large | $\bar{f} = 0.074\, Re_L^{-0.2}$ | Eq. (8–97) |
| $Pr \gtrsim 0.5$ | $\overline{Nu}_L = 0.037\, Re_L^{0.8}\, Pr^{1/3}$ | Eq. (8–98) |

**TABLE 8–9** Dimensionless parameters for convection heat transfer: Forced convection

| *Dimensionless group* | *Symbol* | *Definition* | *Interpretation* |
|---|---|---|---|
| Eckert number | $Ec$ | $\dfrac{U_F^2}{c_P(T_s - T_F)}$ | $\dfrac{\text{kinetic energy of flow}}{\text{enthalpy difference across boundary layer}}$ |
| Friction factor | $f_x$ | $\dfrac{\tau_s}{\rho U_F^2/2}$ | $\dfrac{\text{shear force}}{\text{inertia force}}$ |
| Graetz number | $Gz$ | $\dfrac{Re\,Pr}{x/D}$ | $\dfrac{\text{enthalpy flow rate}}{\text{axial heat conduction}}$ |
| Nusselt number | $Nu_L$ or $Nu$ | $\dfrac{hL}{k}$ | $\dfrac{\text{convection for fluid in motion}}{\text{conduction for motionless fluid layer}}$ |
| Prandtl number | $Pr$ | $\dfrac{\nu}{\alpha}$ | relative effectiveness of molecular transport of momentum and energy within the boundary layer |
| Peclet number | $Pe$ | $Re_L\,Pr$ | $\dfrac{\text{enthalpy flow rate}}{\text{heat conduction rate}}$ |
| Reynolds number | $Re_L$ or $Re$ | $\dfrac{U_F L}{\nu}$ | $\dfrac{\text{inertia forces}}{\text{viscous forces}}$ |
| Stanton number | $St$ | $\dfrac{Nu_L}{Re_L\,Pr}$ | $\dfrac{\text{actual convection heat flux}}{\text{enthalpy energy flux capacity}}$ |

## REVIEW QUESTIONS

**8–1.** Define (a) contraction ratio $\sigma$, (b) entrance-loss coefficient $K_c$, and (c) exit-loss coefficient $K_e$.

**8–2.** What are convection correlations?

**8–3.** Write the defining relations for HFD and TFD flow.

**8–4.** Write the general Newton law of cooling.

**8–5.** Write the criterion for the hydrodynamic and thermal entry lengths for internal flow in passages with uniform cross section.

**8–6.** Explain why the friction factor and Nusselt number for flow in a tube are larger if the flow is developing than if the flow is fully developed.

**8–7.** Explain why the friction factor and Nusselt number are larger for turbulent flow than for laminar flow.

**8–8.** How are the effects of property variation on convection generally taken into account in the practical thermal and hydraulic analysis approach?

**8–9.** What is the Moody chart?

**8–10.** Define (a) the number of transfer units *NTU*, (b) the effectiveness $\epsilon$, and (c) the relative effectiveness $\bar{\epsilon}$.

**8–11.** Write the famous *LMTD* equation for heat transfer associated with flow in internal passages and define *LMTD*.

**8–12.** Write an expression for total pressure drop $\Delta P_{1-2}$ for uniform property flow through a passage with abrupt entrance contraction and exit expansion.

**8–13.** What is pumping power? List several cases for which pumping power is an important factor.

**8–14.** Write the Reynolds number criterion for which the flow is generally turbulent for (a) flow in a tube, (b) flow over a flat plate, and (c) flow across a tube bank.

**8–15.** Assuming that the Nusselt number $Nu_x$ for flow over a flat plate is known, explain how the local coefficient of heat transfer $h_x$ can be calculated.

**8–16.** Why is flow over a circular cylinder more complex than flow along a flat plate?

**8–17.** Is flow across a tube bank classified as internal flow or external flow? Explain why.

**8–18.** What is the minimum free-flow area $A_{min}$ for flow across a tube bank?

**8–19.** Show that the hydraulic diameter for flow across a tube bank can be represented by $D_H = 4\sigma/\beta$.

**8–20.** Explain why the solutions for heat transfer developed in Sec. 8–2–2 for flow in tubes can be used in the analysis of flow across tube banks.

## PROBLEMS

**8–1.** The Prandtl/Nikuradse equation for *f*, Eq. (8–14), is implicit. Use this equation to develop an explicit relationship for *Re* in terms of *f*.

**8–2.** Show that Eq. (8–14) can be written as

$$\sqrt{\frac{1}{4f}} = 2 \log (Re \sqrt{4f}) - 0.8$$

**8–3.** Equation (8–15) is sometimes written in the form $4f = (1.82 \log Re - 1.64)^{-2}$. Demonstrate that these two equations are equivalent.

**8–4.** Air enters a square channel with cross-sectional area of 100 cm² at a velocity of 15 m/s and a temperature of 27°C. Determine the friction factor and coefficient of heat transfer for fully developed conditions.

**8–5.** Solve Prob. 8–4 for the case in which the entering velocity is only 0.05 m/s.

**8–6.** Water at 27°C is piped from a supply tank to a heat exchanger using a tube with 2 cm ID and 10 m length. If the mass flow rate is 0.01 kg/s, determine (a) the Reynolds number *Re*, (b) the Fanning friction factor *f*, (c) the entrance-loss coefficient $K_c$, and (d) the coefficient of heat transfer *h* at the exit.

**8–7.** Reconsider Prob. 8–6 for a mass flow rate of 0.1 kg/s.

**8–8.** Engine oil at 60°C flows at a rate of 0.5 kg/s in a 1-m length of 3-cm diameter tubing with an area ratio of 0.5 at the inlet and outlet. Determine (a) the Reynolds number *Re*, (b) the Fanning factor *f*, (c) the entrance-loss coefficient $K_c$, and (d) the mean coefficient of heat transfer $\bar{h}$.

**8–9.** The effect of property variation on the velocity distribution for laminar flow of gas in a hot tube is shown in Example 8–3. Sketch in a representative velocity distribution for the case in which a gas is cooled.

**8–10.** Estimate the friction factor and the coefficient of heat transfer for Prob. 8–4 for the case in which the outlet air temperature is 77°C and the wall temperature is 127°C.

**8–11.** Determine the friction factor and the coefficient of heat transfer for fully developed annular flow of water for radii of 1 cm and 5 cm. The liquid enters at a velocity of 10 cm/s and a temperature of 60°C. The outer surface is insulated and a uniform heat flux is maintained at the inner surface.

**8–12.** Estimate the friction factor and the coefficient of heat transfer for Prob. 8–11 if the outlet water temperature is 14°C and the wall temperature is 7°C.

**8–13.** Solve Example 8–2 for the case in which the wall temperature is 227°C.

**8–14.** Determine the mean coefficient of heat transfer for HFD flow of water in a 10-cm-diameter tube with a short heating section of only 1-m length. The water enters at a velocity of 0.10 m/s and a temperature of 300 K and the section is heated uniformly.

**8–15.** Oil at 50°C flowing at a rate of 0.25 kg/s in a 1-cm-diameter tube enters a 1-m length. The mean wall temperature is 100°C and the oil exits at 70°C. Determine the mean coefficient of heat transfer over the entire length.

**8–16.** Solve Example 8–1 for a tube with relative roughness of 0.01.

**8–17.** Solve Prob. 8–11 for the case in which $U_b$ is equal to 0.1 cm/s.

**8–18.** An annulus with inside and outside radii of 25.4 mm and 89 mm is used to cool water flowing at a rate of 0.1 kg/s from 100°C to 34°C. The surface of the inner tube is maintained at 12°C and the outer surface is insulated. Estimate the friction factor $f$ and the coefficient of heat transfer $h_i$.

**8–19.** The annulus of Prob. 8–18 is modified by incorporating 24 longitudinal fins. The flow area $A$, wetted perimeter $p_w$, and area of finned surface per unit tube length $(A_s)_o/L$ for the finned annulus are given as 3826 mm$^2$, 1275 mm, and 1.035 m. Use the convection correlations shown in Fig. P8–19 to calculate the friction factor $f$ and coefficient of heat transfer $h_i$ for this arrangement.

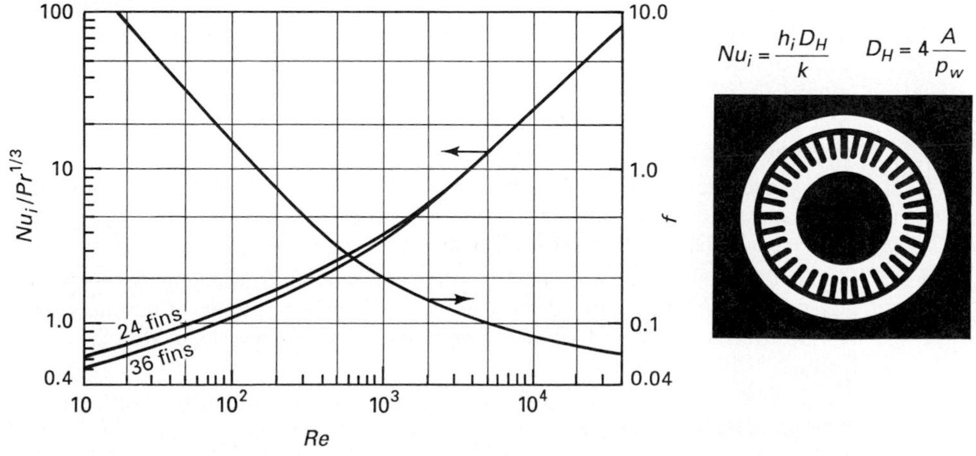


**FIGURE P8–19** Correlations for Nusselt number $Nu_i$ and friction factor $f$—flow in annulus with longitudinal finned tube. (From De Lorenzo and Anderson [65].)

**8–20.** Determine the annual power required to heat 150 gallons per day of water from 55°F to 140°F.

**8–21.** Water at 12°C enters an electrically heated 2-cm-diameter tube of 1 m length. The bulk flow rate is 1 m/s. Determine the bulk-stream temperature $T_b$, outlet temperature $T_2$, and wall temperature $T_s$ at the outlet for uniform wall-flux heating with $q_0'' = 100\ \text{kW/m}^2$.

**8–22.** Referring to Prob. 8–21, determine $T_b$ and $T_2$ for nonuniform wall-flux heating with $q_c'' = 100 \sin(\pi x/L)\ \text{kW/m}^2$.

**8–23.** Referring to Example 8–1, determine the outlet temperature $T_2$ for a length of 5 m for both flow rates, assuming uniform wall-flux heating with $q_0'' = 10\ \text{W/m}^2$.

**8–24.** Water at 50°F is heated in a 4-ft-long annulus with 1-in. and 2-in. radii. A uniform wall flux of 10,000 Btu/(h ft$^2$) is maintained along the inner surface and the outer surface is insulated. Determine the outlet temperatures $T_2$ for bulk flow rates of 0.1 ft/s and 1 ft/s.

**8–25.** Air at 15°C is heated in a 10-m-long 10-cm-square duct with uniform wall flux equal to 100 W/m$^2$. Determine the outlet temperature $T_2$ for a flow rate of 20 m/s.

**8–26.** Solve Prob. 8–25 for a flow rate of 1 m/s and for the case in which only one side is uniformly heated and the other sides are insulated.

**8–27.** Air at 81°F enters a 4-in.-diameter tube with a mass flow rate of 1 kg/s. Determine the length of tube required to bring the air to a temperature of 90°F for nonuniform wall-flux heating of $20x/D$ Btu/(h ft$^2$).

**8–28.** It is desired to preheat the fluid in Prob. 8–6 from 10°C to 40°C by maintaining uniform flux heating of 15 kW/m$^2$ over a short length of the tube wall just preceding the exit. Determine (a) the required length of the heating section, (b) the wall surface temperature at the exit, (c) the total pressure drop, and (d) the pumping power.

**8–29.** Referring to Prob. 8–16, determine the level of uniform wall-flux heating required to increase the water temperature from 27°C to 100°C over a length of 5 m. Also calculate the total pressure drop for this length.

**8–30.** Referring to the tables in Appendix C, we observe that the variation in specific heat over moderate temperature ranges is small for most common fluids. However, $c_P$ for liquid $CO_2$ varies from 1.84 kJ/(kg °C) at −50°C to 36.4 kJ/(kg °C) at 30°C. ($c_P$ can be approximated to within about 1% in the range −20°C to 40°C by $c_P = c_{P1} + a \exp[(T + 20°\text{C})/b]$, where $c_{P1} = 2.05$ kJ/(kg °C), $a = 0.0598$ kJ/(kg °C), and $b = 10.3$°C.) Obviously, a uniform-property analysis for heat transfer to liquid carbon dioxide would be inaccurate, especially for temperatures in the range from 0° to 30°C. To expand upon this point, consider the flow of liquid $CO_2$ in a 2-cm-diameter tube of 1-m length. The mass flow rate is 0.01 kg/s and the inlet temperature is 0°C. Determine the uniform wall flux $q_0''$ required to produce an outlet temperature of 20°C.

**8–31.** Carbon dioxide at −10°C is heated in a 2-cm diameter tube of 1-m length. Determine the outlet temperature for a uniform wall flux of 25 kW/m$^2$ and mass flow rate of 0.1 kg/s. (Note that $c_P$ can be approximated to within about 1% by the relation given in Prob. 8–30.)

**8–32.** Engine oil at 20°C is heated in a 3-cm-diameter tube of 2-m length. Assuming uniform wall-flux heating of 100 kW/m$^3$, determine the outlet temperature $T_2$ if the mass flow rate is 0.2 kg/min. Account for the variation of properties in your analysis by using a linear approximation for $c_P$.

**8–33.** Air at 27°C enters a 10-cm-diameter circular tube of 10-m length with surface at 100°C. Determine the outlet temperature $T_2$ and the total rate of heat transfer for a bulk flow rate of 0.2 m/s.

**8–34.** Refine the solution to Example 8–6 by evaluating the properties at the arithmetic average of the inlet and outlet temperatures.

**8–35.** Water at 10°C enters a 2-cm-diameter 10-m-long tube. The bulk flow rate is 1 m/s. Determine the bulk-stream temperature $T_b$, the outlet temperature $T_2$, and the total rate of heat transfer $q_c$ for a uniform wall temperature of 75°C.

**8–36.** Referring to Example 8–6, estimate the length of tube required to bring the fluid to 35°C.

**8–37.** Water at 40°F is heated in a 20-ft-long annulus with 1-in. and 2-in. radii. The temperature of the inner surface is 150°F and the outer surface is insulated. Estimate the outlet temperature and total rate of heat transfer for a bulk flow rate of 4 ft/s.

**8–38.** Determine the length of tube for the system described in Prob. 8–35 which is required to bring the outlet temperature to 50°C.

**8–39.** For situations such as found in Example 8–6 in which the temperature differences are relatively small, approximate solutions can be developed by setting the fluid temperature $T_F$ in the one-dimensional Newton law of cooling equal to the arithmetic average of the inlet and outlet temperatures. Thus, for a uniform wall temperature condition, $T_0 - T_F$ is approximated by $T_0 - (T_1 + T_2)/2$. Develop an approximate solution to Example 8–6 by utilizing this approach.

**8–40.** Show that the approximate solution approach introduced in Prob. 8–39 for convection heat transfer is within 1% of the *LMTD* solution for $0.75 \lesssim \Delta T_1/\Delta T_2 \lesssim 1.5$.

**8–41.** Solve Example 8–7 by the approximate method introduced in Prob. 8–39.

**8–42.** Water at 20°C with a mass flow rate of 10 kg/s enters a 10-cm-diameter tube with surface at 80°C. Determine the length of tube required to bring the water to 38.3°C. Compare this result with Example 8–6.

**8–43.** Air at 60°F and 1 atm flows through a 6-in-diameter tube of 5 ft length. The surface temperature is maintained at 150°F and the desired outlet temperature is 80°F. Determine the mass flow rate for which these objectives can be achieved.

**8–44.** Air at 27°C and 1 atm enters a 10-cm-diameter tube with a mass flow rate of 0.0025 kg/s. Determine the length of tube required to bring the air to a temperature of 35°C if the surface temperature is 75°C.

**8–45.** Air at 77°C and 1 atm enters a 10-cm-diameter tube with a surface temperature of 27°C. Determine the length of tube required to bring the air temperature to 50°C if the mass flow rate is 0.05 kg/s.

**8–46.** Liquid mercury at 250°C flowing at a rate of 4 kg/s through a 6-cm-ID tube is used to boil water at atmospheric pressure on the outer surface. Assuming that the wall temperature is maintained at about 105°C and that the mercury is to return at no less than 150°C, determine the required length of tube.

**8–47.** Steam at 1.27 MPa condensing on the outer surface of a thin-walled circular tube of 5 cm diameter and 4 m length maintains a uniform surface temperature of 107°C. Water flows through the tube at a rate of 0.5 kg/s, and the inlet and outlet temperatures are 20°C and 60°C. Determine the average convection coefficient for this water flow. Is this value consistent with available convection correlations?

**8–48.** A convenient expression has been developed for the total rate of heat transfer $q_c$ over the length of a tube with uniform wall temperature by combining Eqs. (8–31) and (8–51) [i.e., Eq. (8–52)]. On the other hand, an expression can also be developed for $q_c$ by integrating the Newton law of cooling. Compare these two approaches. Explain why the indirect approach which leads to Eq. (8–52) is preferred.

**8–49.** Demonstrate that the *LMTD* equation [Eq. (8–59)] does not apply to uniform wall-flux heating.

**8–50.** Develop a generalized *LMTD* equation for tube flow with uniform wall temperature and variable specific heat.

**8–51.** Referring to Prob. 8–8, calculate the outlet temperature and the heat transfer rate for a uniform wall temperature of 20°C. Also determine the pressure drop over the length of the tube and the pumping power.

**8–52.** Water at 20°C flowing at a rate of 10 kg/s enters a 10-m-long 10-cm-diameter tube with surface temperature given by $T_s = (20 + 60x/L)$°C. Determine the outlet temperature and the overall rate of heat transfer.

**8–53.** Engine oil flowing through a 1-cm-ID thin-wall copper tube at a rate of 0.5 kg/s is heated from 35°C to 45°C by steam condensing on the outside surface at atmospheric pressure. Develop an approximate solution for the length of tube which is required.

**8–54.** Develop a numerical finite-difference formulation for convection heat transfer in a tube with nonuniform wall temperature and specific heat.

**8–55.** Fluid at 27°C with $Re = 10^4$ flows in a 2-cm-diameter circular tube of 10 m length with $\sigma = 0$ at the entrance and exit. Determine the pressure drop and pumping power for (a) air and (b) water.

**8–56.** Determine the pressure drop and pumping power for Probs. 8–6 and 8–7.

**8–57.** A viscous fluid ($\nu = 3.8 \times 10^{-5}$ m$^2$/s, $\rho = 900$ kg/m$^3$) enters a 1-cm-diameter tube with flow rate of 0.25 kg/s. Assuming an abrupt contraction entrance with $\sigma = 0$, calculate the pressure drop for lengths of (a) 0.1 m, (b) 1 m, and (c) 10 m.

**8–58.** Ammonia at room temperature enters a 1-cm-diameter, 1-m-long tube with flow rate of 0.25 kg/s. Assuming an abrupt contraction entrance with $\sigma = 0$, calculate the pressure drop.

**8–59.** Reconsider Example 8–9 for the case in which $\sigma = 0.5$ at the inlet and exit.

**8–60.** Reconsider Example 8–9 for the case in which the fluid is air with the same value of *Re*.

**8–61.** Air at 27°C and 1 atm flows at a rate of 2 m/s over a flat plate. Determine the mean friction factor $\bar{f}$ and the mean coefficient of heat transfer $\bar{h}$ for a plate length of 25 cm. Also determine the total rate of heat transfer per unit area from the plate if the wall temperature is 100°C.

**8–62.** Solve Prob. 8–61 for the case in which the free-stream velocity is 20 m/s and the plate length is 2.5 m.

**8–63.** Air at 300 K and 1 atm flows over a flat plate at a rate of 35 m/s. The plate is 1 m long, 10 m wide, and is maintained at 350 K. Calculate the mean friction factor $\bar{f}$ and the mean coefficient of heat transfer $\bar{h}$. Also determine the total drag force on the plate and the total rate of heat transfer.

**8–64.** Water at 20°C flows with a free-stream velocity of 1 m/s over a 20-cm-square plate. Determine the total rate of heat transfer from the plate if the surface is at 60°C.

**8–65.** Reconsider Prob. 8–64 for a plate length of 2 m.

**8–66.** Liquid mercury at 20°C flows with a free-stream velocity of 0.05 m/s over a 20-cm-square plate. Determine the total rate of heat transfer from the plate if the surface is at 60°C.

**8–67.** Engine oil at 20°C flows with a free-stream velocity of 1 m/s over a 20-cm-square plate. Determine the total rate of heat transfer from the plate if the surface is at 60°C.

**8–68.** Equations (8–78) and (8–81), which are applicable to laminar boundary layer flow of fluids with low values of Prandtl number, can be developed by use of the approximate integral solution approach. Compare these equations to the following empirical correlations which were developed by Churchill and Ozoe [66] for a wide range in $Pr$:

$$Nu_x = \frac{0.339\, Re_x^{1/2}\, Pr^{1/3}}{[1 + (0.0468/Pr)^{2/3}]^{1/4}} \qquad \begin{array}{l}\text{Uniform wall temperature}\\ Pe \gtrsim 100\end{array}$$

and

$$Nu_x = \frac{0.464\, Re_x^{1/2}\, Pr^{1/3}}{[1 + (0.0207/Pr)^{2/3}]^{1/4}} \qquad \begin{array}{l}\text{Uniform wall-heat flux}\\ Pe \gtrsim 100\end{array}$$

**8–69.** Air at 27°C and 2 atm flows at a rate of 3.5 m/s over a 0.1 m by 1 m glass plate with an electrically conducting film that produces a constant heat flux of 1 kW/m$^2$. Determine the temperature of the plate as a function of $x$. What is the temperature difference between the air and the plate at the trailing edge?

**8–70.** A design is being considered which involves flow of air with free-stream velocity and temperature of 3 m/s and 300 K over a flat plate which consists of ten 3.8 cm by 25.7 cm strip heaters placed side by side and insulated on the bottom. The power provided by each heater is 10 W. According to specifications provided by the manufacturer, the heater surface should not exceed 450 K. Determine whether or not the heater can be safely operated for this application.

**8–71.** Reconsider Prob. 8–70 for the case in which the heated section is to be preceded by an unheated surface of 0.1 m length.

**8–72.** Air at 27°C and 1 atm flows at a velocity of 50 m/s across a 5-mm-diameter wire. The wire temperature is maintained at 100°C. Determine the total rate of heat transfer per unit length of wire.

**8–73.** A 1-cm-diameter wire generates 390 W/m. Determine the flow rate of air required to maintain the wire at 100°C if the air temperature is 27°C.

**8–74.** An electric resistance wire heater with 0.5 mm diameter and 1 m length is placed normal to an air flow. The temperature of the air is 27°C, the free-stream velocity is 10 m/s, and the power output is 20 W. Determine the surface temperature of the wire.

**8–75.** A 0.2-mm-diameter polished-platinum wire of 10 mm length is to be used for a standard hot wire anemometer sensor to measure the velocity of air at 300 K. The temperature of the wire is to be maintained at 500 K by electric resistance heating produced by current controlled by a wheatstone bridge. The electrical resistivity of platinum is 17 μΩ cm. Determine the required electric current for velocities of (a) 1 m/s and (b) 10 m/s.

**8–76.** A heat-transfer correlation has been developed for flow across a cylinder by Churchill and Bernstein [67], which takes the form

$$\overline{Nu_D} = 0.3 + \frac{0.62\, Re_D^{1/2}\, Pr^{1/3}}{[1 + (0.4/Pr)^{2/3}]^{1/4}} \left[1 + \left(\frac{Re_D}{282{,}000}\right)^{5/8}\right]^{4/5}$$

for $Re_D\, Pr \geq 0.2$, with the properties evaluated at the film temperature $T_f$. This correlation provides an uncertainty of within ± 30% for a wide range of conditions, including low to high values of Prandtl number and both uniform wall temperature and uniform wall-heat flux. Use this correlation to solve Prob. 8–72.

**8–77.** Water at 35°C flows at a rate of 4 m/s over a 1-cm-diameter sphere. Calculate the drag and rate of heat transfer if the surface temperature is 75°C.

**8–78.** Air at 35°C flows at a rate of 4 m/s over a 1-cm-diameter sphere. Calculate the rate of heat transfer if the surface temperature is 75°C.

**8–79.** An aluminum sphere of 20 mm diameter is removed from an oven with temperature of 85°C. Estimate the length of time required to cool the sphere to 40°C if it is placed in an air stream at 27°C and 1 atm flowing at 20 m/s.

**8–80.** Referring to Example 8–13, calculate the total heat-transfer rate and pumping power for the case in which the air flow is increased to 10 m/s.

**8–81.** Referring to Example 8–13, what number of rows would be required to bring the exit air temperature to 25°C?

**8–82.** Referring to Example 8–13, what mass flow rate would be required in order to achieve an exit air temperature of 32°C rather than 35.7°C?

**8–83.** The heat pump unit of Example 8–13 is to be used to cool air at 44°C with $T_0 = 10°C$. Determine the temperature of the exiting air and the total rate of heat transfer.

**8–84.** Air at 1 atm and 27°C flows across a bank of 1-in.-diameter tubes with 15 columns and 5 rows with an entering velocity $U_1$ of 7 m/s. The tubes are maintained at 77°C and are arranged in-line with 1.5 in. longitudinal and transverse pitches and 2 m depth. Determine the exit air temperature and the total rate of heat transfer.

**8–85.** Air at 15°C and 1 atm with a velocity of 6 m/s enters a tube bank with $N_L = 7$, $N_T = 8$, $D = 16.4$ mm, $S_L = 34.3$ mm, and $S_T = 32.3$ mm. The tubes are in a staggered array with $T_0 = 70°C$. Determine the heat transfer rate and the pumping power.

**8–86.** Air at 27°C and 1 atm flows at a rate of 5 kg/s over a bank of seventy-five 2.54-cm-diameter tubes of 1.5 m depth. The tubes are in a five-row staggered array with 4-cm pitch and 77°C wall temperature. Determine the total pressure drop, outlet temperature, and rate of heat transfer.

**8–87.** Air at 80°F and 1 atm flows at a rate of 5000 ft$^3$/min over a tube bank, with 200 electrical heating rods with 1-in. diameter and 10 rows. The rods are in a staggered array with 5 ft length and 1.5 in. longitudinal and transverse pitch. Electrical heating produces a uniform flux for each rod of 0.1 kW/ft$^2$. Determine the air temperature and the surface temperature of the rods at the exit.

# CHAPTER 9

# CONVECTION HEAT TRANSFER: PRACTICAL ANALYSIS—NATURAL CONVECTION

## 9–1 INTRODUCTION

As indicated in Chap. 6, flow caused by temperature (or concentration)†-induced density gradients within the fluid is known as *natural convection* (or *free convection*). The most familiar natural-convection flow fields occur as a result of the influence of gravity on fluids in which density gradients have been thermally established. For example, when a vertical cold plate is placed in warm stationary fluid, the temperature of the fluid near the wall will be decreased by conduction heat transfer. As the temperature of the fluid falls, its density will, of course, increase. This difference in density results in the downward flow of the heavier cold fluid near the plate and upward flow of the lighter warm fluid. Similarly, the placement of a hot plate in a cool motionless fluid will result in the upward flow of the light warm fluid near the plate. Examples of natural convection in gravitational force fields include the cooling of electrical devices such as power transistors and transformers, the heating or cooling of building walls on windless days, and the heating of a pan of water.

Another very important type of natural convection flow field occurs in the presence of centrifugal forces that are also proportional to fluid density. This type of natural convection is commonly used to cool rotating components such as turbine blades.

An instrument known as the *Mach–Zehnder interferometer* is often utilized to study natural convection flows. This optical instrument produces interference fringes that are the result of changes in the index of refraction caused by small density differences within the fluid. Consequently, lines of constant temperature can be

† Natural convection caused by concentration-induced density gradients involves interfacial mass transfer which is considered in the *mass transfer supplement*.

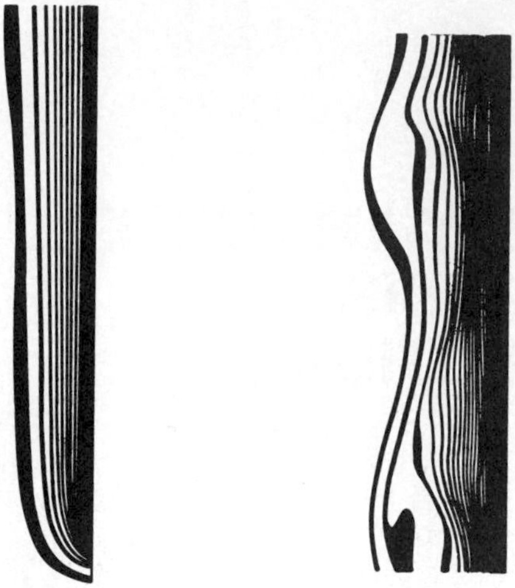

(a) Laminar flow. (b) Turbulent flow.

**FIGURE 9–1**
Interferograms for natural convection flow of air over a vertical flat plate. (Courtesy of E. R. G. Eckert and E. E. Soehngen.)

determined by the use of this instrument. For example, Fig. 9–1 shows photographs of the fringe pattern for natural convection flow over a vertical heated flat plate. This figure clearly indicates the instantaneous flow pattern for this geometry. Because of the streamline pattern observed in Fig. 9–1(a), the flow can be assumed to be laminar over this part of the plate. On the other hand, the pattern observed in Fig. 9–1(b) is clearly irregular, which is characteristic of turbulent flow.

As suggested in Chap. 6, the same classifications found in forced convection pertaining to the geometry of the fluid-solid interface, the nature of the path followed by individual elements of fluid, the type of boundary conditions, and so on, also apply to natural convection systems. Consideration is now given to the characterizing parameters used in the practical analysis of natural convection processes, after which specific attention is given to external natural convection flow, internal natural convection flow, natural convection flow in enclosed spaces, and combined natural and forced convection.

## 9–2 CHARACTERIZING PARAMETERS FOR NATURAL CONVECTION

Because no pumping or blowing is involved in natural convection flow, we will focus attention on the practical thermal analysis approach, which features the use of local and mean coefficients of heat transfer. Theoretical considerations pertaining to natural convection boundary layers are introduced in Sec. 7–2–3 of Chap. 7.

For systems in which the local coefficient of heat transfer $h_x$ can be conveniently obtained, $h_x$ is defined by the general Newton law of cooling,

$$q_c'' = \frac{dq_c}{dA_s} = h_x(T_s - T_F) \tag{9–1}$$

and $\bar{h}$ is defined in terms of $h_x$ by

$$\bar{h} = \frac{1}{A_s}\int_{A_s} h_x\,dA_s \tag{9–2}$$

For more complex systems in which the use of a local coefficient is not practical, $\bar{h}$ is defined by

$$q_c = \bar{h}A_s(\overline{T_s - T_F}) \tag{9–3}$$

where the mean temperature difference $\overline{T_s - T_F}$ depends upon the system geometry and thermal boundary conditions. The local and mean coefficients of heat transfer for natural convection systems are generally expressed in terms of the Nusselt number $Nu_L$. For example, for flow over a flat plate of length $L$, the local Nusselt number is represented by $Nu_x = h_x x/k$ and the mean Nusselt number by $\overline{Nu}_L = \bar{h}L/k$.

The Nusselt number $Nu_L$ is generally expressed in terms of the Rayleigh number $Ra_L$ and the Prandtl number $Pr$ for natural convection processes; that is,

$$Nu_L = \text{fn}\,(Ra_L, Pr) \tag{9–4}$$

The *Rayleigh number* $Ra_L$ is defined by

$$Ra_L = \frac{g\beta(T_s - T_F)\boldsymbol{L}^3}{\alpha\nu} \tag{9–5}$$

where the reference fluid temperature $T_F$ and the reference length $\boldsymbol{L}$ are dependent upon the system geometry. The *coefficient of thermal expansion* $\beta$ is given in Table A–C–3 for several liquids. For ideal gases $\beta$ can be shown to be equal to $1/T$, where $T$ is the absolute temperature of the gas. However, for practical purposes $\beta$ can generally be taken as a constant for moderate temperature differences. Because the Rayleigh number characterizes the coupling through buoyancy of momentum to energy, this dimensionless parameter is the primary variable in natural convection flows. As indicated in Chap. 7, $Ra_L$ represents the product of $Pr$ and the well known *Grashof number* $Gr_L$,

$$Gr_L = \frac{g\beta(T_s - T_F)\boldsymbol{L}^3}{\nu^2} = \frac{Ra_L}{Pr} \tag{9–6}$$

which represents the relative significance of buoyant and viscous effects.†

## 9–3 EXTERNAL NATURAL CONVECTION FLOW

Our attention is first focused on flow over a vertical flat plate with $U_\infty = 0$, which is the classic example of natural convection heat transfer. Consideration will then be

† General defining relations for $Ra_L$ and $Gr_L$, which are applicable to temperature and concentration (including phase change) driven natural convection flows, are given by

$$Ra_L = Gr_L\,Pr = \frac{g(\rho_F - \rho_s)\boldsymbol{L}^3}{\rho\alpha\nu} \qquad Gr_L = \frac{g(\rho_F - \rho_s)\boldsymbol{L}^3}{\rho\nu^2}$$

given to a general class of external flow processes, which includes cylindrical, spherical, and rectangular solid geometries.

The reference temperature $T_F$ for external natural convection flow is equal to the free-stream temperature $T_\infty$, which is uniform for nonstratified conditions. Using the Boussinesq approximation, $\beta$ and the other fluid properties are evaluated at the film temperature $T_f$ or free-stream temperature $T_\infty$.

### 9-3-1 Flow over a Vertical Flat Plate

Similar to the situation in forced-convection boundary layer flow over a flat plate, hydrodynamic and thermal boundary layers develop along the wall of a vertical flat plate in a natural convection flow field. As illustrated in Fig. 9–2 for natural convection flow over a heated plate, the thicknesses of the boundary layers increase with $x$. The flow is laminar toward the front of the plate, but develops into turbulence at a point downstream at which the local Rayleigh number $Ra_x$,

$$Ra_x = \frac{g\beta(T_s - T_\infty)x^3}{\alpha\nu} \tag{9–7}$$

is equal to about $10^9$.

For natural convection flow over a vertical flat plate, Eq. (9–1) applies; that is, the local heat flux $q''_c$ is represented by

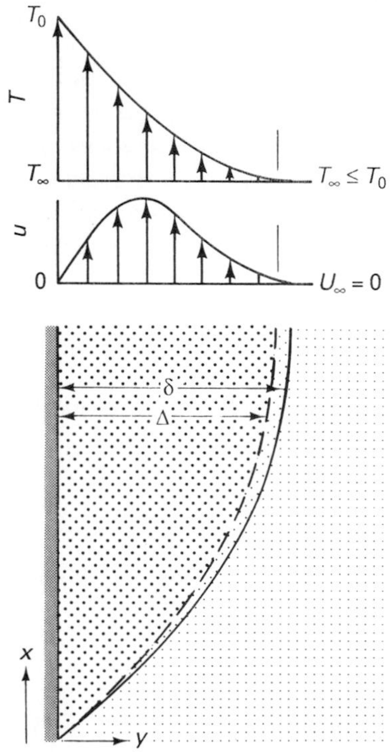


FIGURE 9–2
Representative temperature and velocity profiles for laminar natural convection flow over a heated vertical flat plate.

$$q_c'' = h_x(T_s - T_\infty) \tag{9-8}$$

where $h_x$ is generally expressed in terms of the Nusselt number $Nu_x$,

$$Nu_x = \frac{q_c''}{T_s - T_\infty}\frac{x}{k} = \frac{h_x x}{k} \tag{9-9}$$

Experimental and theoretical studies have been conducted for uniform wall temperature, uniform wall-heat flux, and other boundary conditions. To provide a basis for the practical thermal analysis of natural convection flow over vertical surfaces, we will focus attention on uniform wall-temperature and uniform wall-heat-flux conditions.

### *Uniform Wall Temperature*

For uniform wall-temperature heating ($T_s = T_0 =$ constant), the total heat-transfer rate $q_c$ over a plate with surface area $A_s$ is obtained from Eq. (9–8) by writing

$$q_c = \int_{A_s} h_x(T_s - T_\infty)\, dA_s = (T_0 - T_\infty)\int_{A_s} h_x\, dA_s = \bar{h}A_s(T_0 - T_\infty) \tag{9-10}$$

where $\bar{h}$ is defined by Eq. (9–2).

Theoretical solutions based on boundary layer theory (Chap. 7) for laminar natural convection flow over a vertical plate with uniform wall temperature have been developed by Ostrach [1] and others. Figure 9–3 shows calculations obtained by Ostrach for Nusselt number $Nu_x$ in terms of $Nu_x/Ra_x^{1/4}$ versus $Pr$. These results are represented to within 1% by correlations developed by Ede [2] and Churchill and Usagi [3]. The correlation by Ede takes the form

$$\frac{Nu_x}{Ra_x^{1/4}} = \frac{3}{4}\left(\frac{Pr}{2.5 + 5.0\sqrt{Pr} + 5.0\,Pr}\right)^{1/4} = \text{fn}_1\,(Pr) \tag{9-11}$$

where the right-hand side is represented by $\text{fn}_1\,(Pr)$ and the local Rayleigh number $Ra_x$ is

$$Ra_x = Gr_x\,Pr = \frac{g\beta(T_0 - T_\infty)x^3}{\alpha\nu} \tag{9-12}$$

Equation (9–11) is shown in Fig. 9–3. Using this result, an expression is obtained for the mean Nusselt number $\overline{Nu}_L$ by writing

$$\begin{aligned}\bar{h} &= \frac{1}{L}\int_0^L h_x\, dx = \frac{1}{L}\int_0^L Nu_x\frac{k}{x}\, dx = \frac{k}{L}\int_0^L \text{fn}_1\,(Pr)\frac{Ra_x^{1/4}}{x}\, dx \\ &= \frac{k}{L}\text{fn}_1\,(Pr)\left[\frac{g\beta(T_0 - T_\infty)}{\alpha\nu}\right]^{1/4}\int_0^L x^{-1/4}\, dx \\ &= \frac{k}{L}\text{fn}_1\,(Pr)\left[\frac{g\beta(T_0 - T_\infty)}{\alpha\nu}\right]^{1/4}\frac{L^{3/4}}{3/4}\end{aligned} \tag{9-13}$$

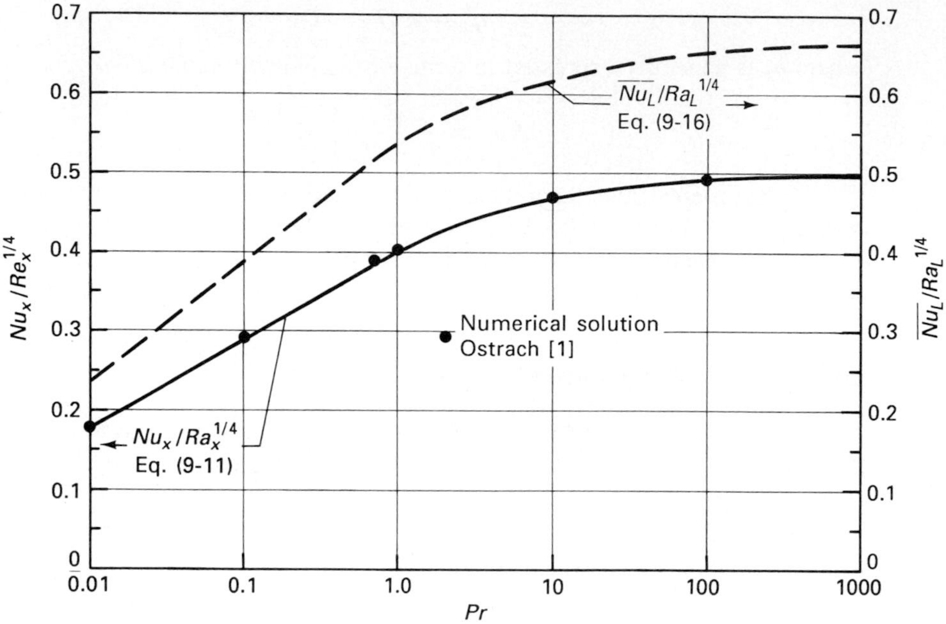


**FIGURE 9–3** Nusselt number $Nu_x$ (and $\overline{Nu_L}$) in terms of Rayleigh number $Ra_x$ (and $Ra_L$) and Prandtl number $Pr$ for vertical surface with uniform wall temperature.

and

$$\overline{Nu_L} = \frac{\bar{h}L}{k} = \frac{4}{3}\,\text{fn}_1\,(Pr)\,Ra_L^{1/4} \tag{9–14}$$

or

$$\overline{Nu_L} = \frac{4}{3} Nu_{x=L} \tag{9–15}$$

Using Eq. (9–10) to represent $Nu_x$, we obtain

$$\frac{\overline{Nu_L}}{Ra_L^{1/4}} = \frac{4}{3}\,\text{fn}_1\,(Pr) = \left(\frac{Pr}{2.5 + 5.0\sqrt{Pr} + 5.0\,Pr}\right)^{1/4} \tag{9–16}$$

This relation, which approximates the exact solution obtained by Ostrach, is compared in Fig. 9–4 with experimental data for gases with values of Prandtl number $Pr$ of the order of unity. The agreement is seen to be quite good in the range $10^5 \lesssim Ra_L \lesssim 10^9$. Equation (9–16) is also shown in Fig. 9–3.

Because of limitations in approximations employed in boundary layer theory, Eqs. (9–11) and (9–16) are restricted to laminar flow with $10^4 \lesssim Ra_x \lesssim 10^9$. However, empirical correlations are available in the literature, which apply in the region for which $Ra_x < 10^4$ as well as in the turbulent zone where $Ra_x > 10^9$. A general

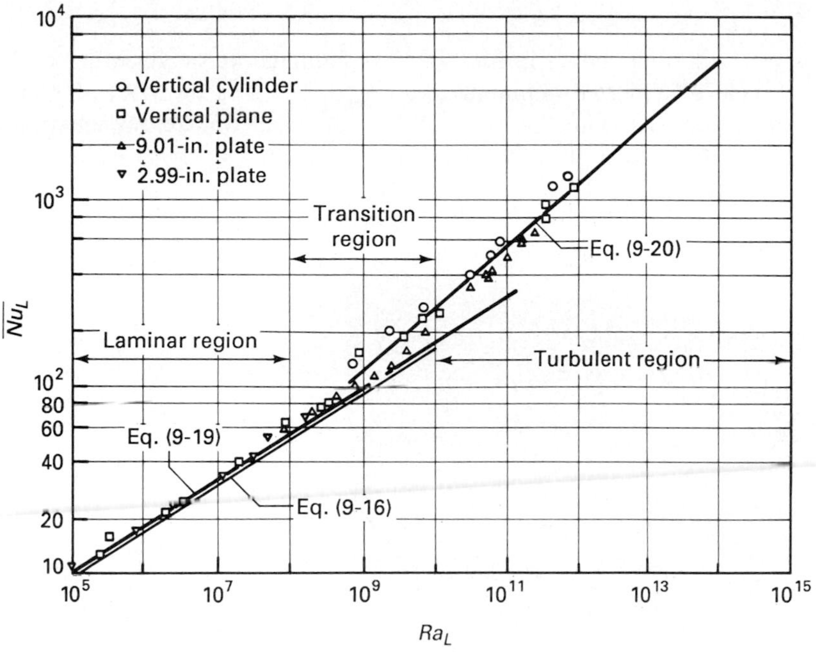


**FIGURE 9–4** Experimental data and correlations for natural convection on a vertical flat plate. (Data assembled by Eckert and Jackson [4].)

correlation developed for $\overline{Nu}_L$ by Churchill and Chu [5] that covers the entire range of $Ra_L$ is of the form

$$\overline{Nu}_L = \left\{0.825 + \frac{0.387\, Ra_L^{1/6}}{[1 + (0.492/Pr)^{9/16}]^{8/27}}\right\}^2 \tag{9–17}$$

For somewhat better accuracy within the laminar flow region, Churchill and Chu [5] suggest the alternative correlation

$$\overline{Nu}_L = 0.68 + \frac{0.670\, Ra_L^{1/4}}{[1 + (0.492/Pr)^{9/16}]^{4/9}} \qquad 0 < Ra_L \lesssim 10^9 \tag{9–18}$$

Other useful relations for uniform wall-temperature heating of gases are given by

$$\overline{Nu}_L = 0.525\, Ra_L^{1/4} \qquad 10^4 \lesssim Ra_L \lesssim 10^9 \tag{9–19}$$

for laminar flow [from Eq. (7–164)]†, and

$$\overline{Nu}_L = 0.13\, Ra_L^{1/3} \qquad 10^9 \lesssim Ra_L \tag{9–20}$$

for turbulent flow [7]. These simple correlations are shown to be in good agreement with experimental data in Fig. 9–4.

† Another popular correlation for laminar flow is given by [6]

$$\overline{Nu}_L = 0.590\, Ra_L^{1/4} \qquad 10^4 \lesssim Ra_L \lesssim 10^9$$

## EXAMPLE 9–1

The power amplifier shown in Fig. E9–1 is mounted vertically in air at 25°C. The case is made of anodized aluminum with a surface area of about 3800 mm$^2$ and a height of 40 mm. Determine the mean coefficient of heat transfer for natural convection cooling, assuming a uniform case temperature of 125°C. Also estimate the power dissipation from the unit.

**FIGURE E9–1**
Power amplifier.

### Solution

*Objective* Determine the mean coefficient of heat transfer $\bar{h}$ and the total rate of heat transfer $q$.

*Schematic* Vertical flat plate representation of power amplifier.

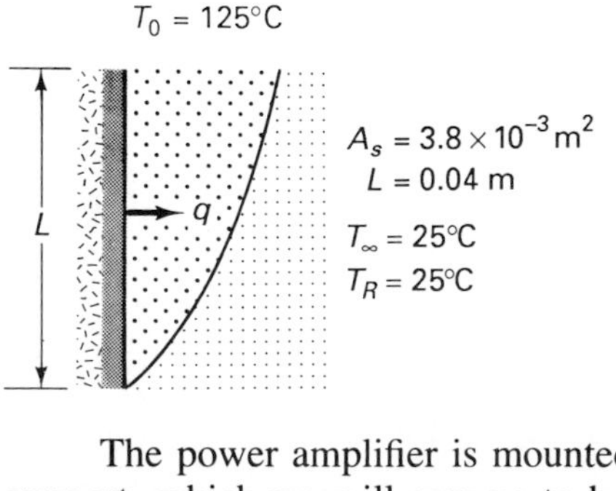

$T_0$ = 125°C
$A_s = 3.8 \times 10^{-3}\,m^2$
$L = 0.04$ m
$T_\infty$ = 25°C
$T_R$ = 25°C
L
q


The power amplifier is mounted vertically with the back surface fastened to a support, which we will assume to be insulated. Because of its small thickness, we will represent the cooling surface of the device by a flat plate.

*Assumptions/Conditions*

natural convection and blackbody thermal radiation cooling
Boussinesq approximation with $\beta$ and other fluid properties evaluated at $T_f$
standard conditions, except for significance of buoyancy and thermal radiation

*Properties* Air at film temperature $T_f = 75°C = 348$ K (Table A–C–5): $\rho = 0.994\ kg/m^3$, $\nu = 2.06 \times 10^{-5}\ m^2/s$, $k = 0.0299$ W/(m °C), $Pr = 0.697$. The coefficient of thermal expansion $\beta$ is approximated by

$$\beta = \frac{1}{T_f} = \frac{1}{348\ K} = 2.87 \times 10^{-3}/K$$

*Analysis* To determine the mean Nusselt number $\overline{Nu}_L$, we first compute the Grashof number $Gr_L$ and Rayleigh number $Ra_L$.

$$Gr_L = \frac{g\beta(T_0 - T_\infty)L^3}{\nu^2}$$

$$= \frac{(9.81\ m/s^2)(2.87 \times 10^{-3}/K)(125°C - 25°C)(0.04\ m)^3}{(2.06 \times 10^{-5}\ m^2/s)^2} = 4.25 \times 10^5$$

$$Ra_L = Gr_L\, Pr = 4.25 \times 10^5\,(0.697) = 2.96 \times 10^5$$

Because the Rayleigh number $Ra_L$ is well below $10^9$, the flow is judged to be laminar. Based on Eq. (9–16) the mean Nusselt number $Nu_L$ is given by

$$\overline{Nu}_L = \left(\frac{Pr}{2.5 + 5\sqrt{Pr} + 5\,Pr}\right)^{1/4} Ra_L^{1/4}$$

$$= \left[\frac{0.697}{2.5 + 5\sqrt{0.697} + 5(0.697)}\right]^{1/4} (2.96 \times 10^5)^{1/4} = 11.9$$

such that the mean coefficient of heat transfer becomes

$$\bar{h} = \overline{Nu}_L \frac{k}{L} = 11.9\,\frac{0.0299\ W/(m\ °C)}{0.04\ m} = 8.90\ W/(m^2\ °C)$$

The total rate of heat transfer $q$ from the surface is

$$q = q_c + q_R$$

where $q_c$ is given by Eq. (9–10). Because the plate is made of black anodized aluminum, we will utilize the blackbody approximation for $q_R$ given by Eq. (1–18), with $T_R$ set equal to the ambient temperature. Substituting for $q_c$ and $q_R$, we have

$$q = \bar{h}A_s(T_0 - T_\infty) + \sigma A_s F_{s-R}(T_0^4 - T_R^4)$$

$$= 8.90\,\frac{W}{m^2\ °C}\,(3.8 \times 10^{-3}\ m^2)(125°C - 25°C)$$

$$+ 5.67 \times 10^{-8} \frac{\text{W}}{\text{m}^2\,\text{K}^4} (3.8 \times 10^{-3}\,\text{m}^2)(1)[(398\text{ K})^4 - (298\text{ K})^4]$$

$$= 3.38\text{ W} + 3.71\text{ W} = 7.09\text{ W}$$

The natural-convection heat transfer accounts for about 48% of the total power dissipated by the power amplifier.

### *Uniform Wall-Heat Flux*

Theoretical solutions have been obtained by Ostrach [1] for the distribution in wall temperature $T_s$ for uniform wall-flux heating (i.e., $q_c'' = q_0'' =$ constant). These results for $T_s$ are generally expressed in terms of the local Nusselt number $Nu_x$ versus the *local flux Rayleigh number* $Ra_x^*$, where

$$Nu_x = \frac{q_0''}{T_s - T_\infty} \frac{x}{k} \tag{9–21}$$

$$Ra_x^* = Nu_x\,Ra_x = Gr_x^*\,Pr = \frac{g\beta q_0'' x^4}{k\alpha\nu} \tag{9–22}$$

and the *local flux Grashof number* $Gr_x^*$ is defined by

$$Gr_x^* = Nu_x\,Gr_x = \frac{g\beta q_0'' x^4}{k\nu^2} \tag{9–23}$$

The solution for $Nu_x$ is shown as a function of $Ra_x^*$ in Fig. 9–5 for $10^4 \lesssim Ra_x \lesssim 10^9$. Fujii and Fujii [8] and Churchill and Ozoe [9] have developed correlations for uniform wall-heat flux. The correlation by Fujii and Fujii is given by

$$\frac{Nu_x}{Ra_x^{*1/5}} = \left(\frac{Pr}{4 + 9\sqrt{Pr} + 10\,Pr}\right)^{1/5} = \text{fn}_2\,(Pr) \tag{9–24}$$

This correlation is within about 1% of the exact solution results shown in Fig. 9–5, and is valid in the range $10^4 \lesssim Ra_x \lesssim 10^9$. As illustrated in Example 9–3, correlations of this type for $Nu_x$ versus $Ra_x^*$ provide an efficient and practical means of calculating the distribution of $T_s - T_\infty$ for a specified wall-heat flux $q_0''$.

Combining Eqs. (9–21) and (9–24), we obtain an expression for the distribution in dimensionless wall temperature of the form

$$\frac{T_s - T_\infty}{q_0''} \frac{k}{L} = \frac{(x/L)^{1/5}}{Ra_L^{*1/5}\,\text{fn}_2\,(Pr)} \tag{9–25}$$

Thus, we see that the distribution in $T_s - T_\infty$ for laminar natural convection in the region away from the leading edge is proportional to $(x/L)^{1/5}$.

To compare the solution results for uniform wall-heat flux and uniform wall temperature, the calculations for $Nu_x$ versus $Ra_x^*$ are expressed in terms of $Nu_x$ versus

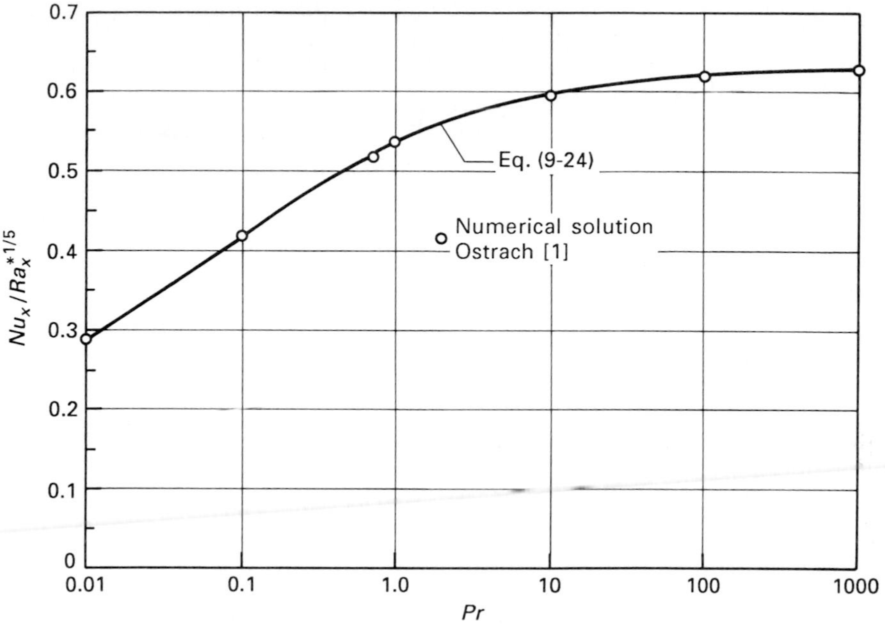


**FIGURE 9–5** Nusselt number $Nu_x$ in terms of flux Rayleigh number $Ra_x^*$ and Prandtl number $Pr$ for vertical surface with uniform wall-heat flux.

$Ra_x$ and shown in Fig. 9–6. Notice that the calculations for uniform wall-heat flux lie approximately 15% above the results for uniform wall temperature. Equation (9–24) is expressed in terms of $Nu_x$ and $Ra_x$ by writing

$$\frac{Nu_x}{(Nu_x\, Ra_x)^{1/5}} = \left(\frac{Pr}{4 + 9\sqrt{Pr} + 10\,Pr}\right)^{1/5} = \text{fn}_2\,(Pr) \tag{9–26}$$

or

$$\frac{Nu_x}{Ra_x^{1/4}} = \left(\frac{Pr}{4 + 9\sqrt{Pr} + 10\,Pr}\right)^{1/4} = [\text{fn}_2\,(Pr)]^{5/4} \tag{9–27}$$

This equation is also shown in Fig. 9–6. In this connection, we also observe that

$$\frac{Nu_x}{Ra_x^{1/4}} = \frac{Nu_x\, Ra_x}{Ra_x^{5/4}} = \frac{Ra_x^*}{Ra_x^{5/4}} \tag{9–28}$$

It follows that the representation of natural convection correlations in terms of $Ra_x^*$ versus $Ra_x$ provides an equivalent and viable alternative to the use of $Nu_x$ versus $Ra_x$ for specified wall temperature and $Nu_x$ versus $Ra_x^*$ for specified wall-heat flux.

Practical relations for natural convection flow of gases over a vertical plate with uniform wall-heat flux have been developed of the forms

$$Nu_x = 0.424\, Ra_x^{1/4} \qquad \text{or} \qquad Nu_x = 0.503\, Ra_x^{*1/5} \tag{9–29a,b}$$

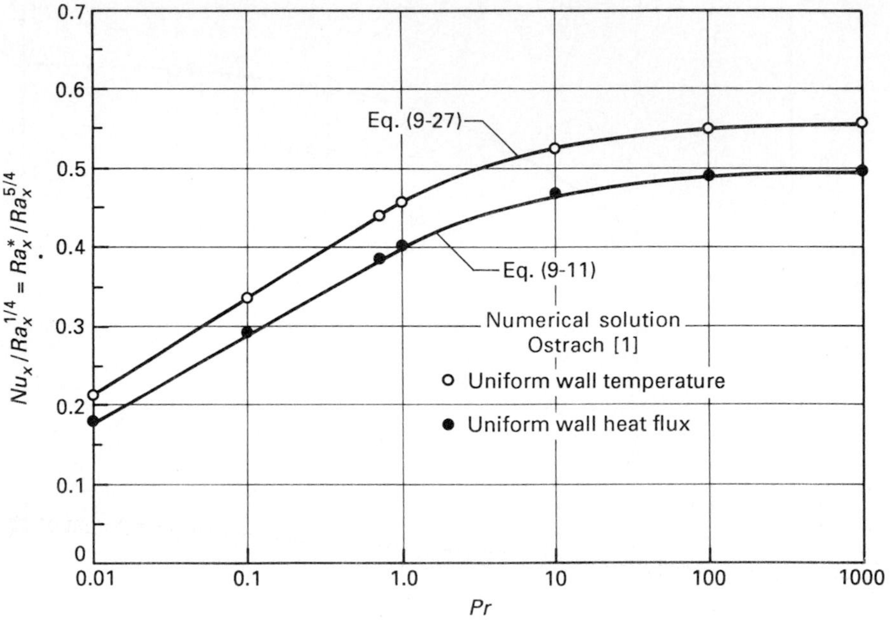


**FIGURE 9–6** Comparison of solution and correlations for natural convection flow over vertical surfaces with uniform wall-temperature and uniform wall-heat-flux conditions.

for laminar flow (see Chap. 7), and

$$Nu_x = 0.0942\, Ra_x^{1/3} \qquad \text{or} \qquad Nu_x = 0.170\, Ra_x^{*1/4} \tag{9–30a,b}$$

for turbulent flow [10].

### *Hydraulic Considerations*

Solutions and correlations are also available in the literature for the friction factor $f_x$ for natural convection flow over vertical plates. However, since no power is expended in mechanical pumping for natural convection flow, further hydraulic considerations are generally not required.

### EXAMPLE 9–2

A vertical plate 10 cm high and 5 cm wide is heated electrically with the heat flux over the surface being uniformly equal to 1.11 kW/m$^2$. Determine the temperature distribution over the plate if both sides are cooled by natural convection with an ambient air temperature of 38°C.

#### Solution

*Objective* Determine the distribution in surface temperature $T_s$.

*Schematic* Vertical flat plate—natural convection heating of air.

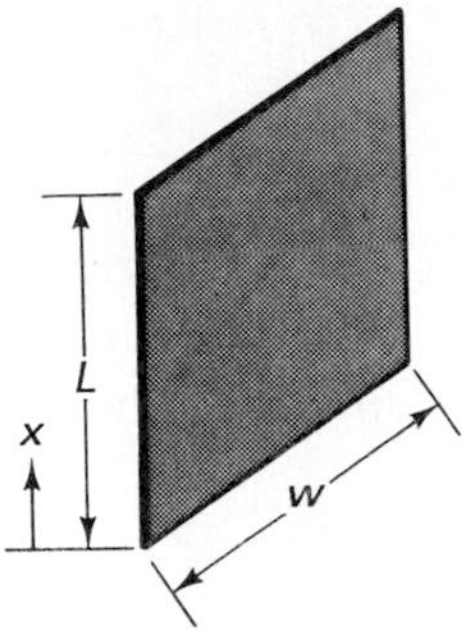


$T_F$ = 38°C
$q_0''$ = 1.11 kW/m$^2$
$L$ = 10 cm
$w$ = 5 cm

*Assumptions/Conditions*

natural convection
Boussinesq approximation with β and other fluid properties evaluated at $T_\infty$
standard conditions, except for significance of buoyancy

*Properties* Because the surface temperature $T_s$ is unknown, we will evaluate the properties of the air at $T$ = 38°C = 311 K. Referring to Table A–C–5, we write $\rho$ = 1.12 kg/m$^3$, $\nu$ = 1.67 × 10$^{-5}$ m$^2$/s, $k$ = 0.0266 W/(m °C), $Pr$ = 0.706, and

$$\beta = \frac{1}{T_\infty} = \frac{1}{311\text{ K}} = 3.22 \times 10^{-3}/\text{K}$$

*Analysis* Because the heat flux is specified in this problem, we first write relations for the flux Grashof number $Gr_x^*$ and the flux Rayleigh number $Ra_x^*$.

$$Gr_x^* = \frac{g\beta q_0'' x^4}{k\nu^2} = \frac{(9.81\text{ m/s}^2)(3.22 \times 10^{-3}/\text{K})(1110\text{ W/m}^2)x^4}{[0.0266\text{ W/(m °C)}](1.67 \times 10^{-5}\text{ m}^2/\text{s})^2}$$

$$= (4.73 \times 10^{12})\left(\frac{x}{\text{m}}\right)^4$$

$$Ra_x^* = Gr_x^*\,Pr = 3.34 \times 10^{12}\left(\frac{x}{\text{m}}\right)^4$$

Assuming that the flow is laminar, we combine this relation for $Ra_x^*$ with the correlation for $Nu_x$ given by Eq. (9–24) (or Fig. 9–6) with the result

$$Nu_x = \frac{q_0''}{T_s - T_\infty}\frac{x}{k} = \left(\frac{Pr}{4 + 9\sqrt{Pr} + 10\,Pr}\right)^{1/5} Ra_x^{*1/5}$$

$$= 0.519\,Ra_x^{*1/5} = 166\left(\frac{x}{\text{m}}\right)^{4/5}$$

or

$$T_s - T_\infty = \frac{q_0''}{166}\frac{x}{k}\left(\frac{\text{m}}{x}\right)^{4/5} = \frac{(1110\text{ W/m}^2)/166}{0.0266\text{ W/(m °C)}}\,x^{1/5}\,\text{m}^{4/5} = 251\left(\frac{x}{\text{m}}\right)^{1/5}\text{ °C}$$

The maximum temperature is at the top of the plate ($x = 0.1$ m).

$$T_{s,\max} - T_\infty = 251\left(\frac{0.1\text{ m}}{\text{m}}\right)^{1/5}\,^\circ\text{C} = 158^\circ\text{C}$$

$$T_{s,\max} = 158^\circ\text{C} + 38^\circ\text{C} = 196^\circ\text{C}$$

It follows that the maximum Grashof number and Rayleigh number become

$$Gr_L = \frac{g\beta(T_s - T_\infty)L^3}{\nu^2} = \frac{(9.81\text{ m/s}^2)(3.22\times10^{-3}/\text{K})(158^\circ\text{C})(0.1\text{ m})^3}{(1.67\times10^{-5}\text{ m}^2/\text{s})^2}$$
$$= 1.79\times10^7$$

and

$$Ra_L = Gr_L\,Pr = 1.79\times10^7\,(0.706) = 1.26\times10^7$$

Because $Ra_L < 10^9$, the flow is indeed laminar.

The solution can be refined by evaluating the properties at a film temperature, with $T_s$ equal to the temperature at $x = L/2$.

## 9-3-2 General External Natural Convection Flows

The rate of heat transfer associated with many standard external natural convection flows encountered in practice can be represented directly by Eq. (9–3) with $T_F$ set equal to $T_\infty$; that is,

$$q_c = \bar{h}A_s(\bar{T}_s - T_\infty) \qquad \text{or} \qquad q_c'' = \bar{h}(\bar{T}_s - T_\infty) \tag{9–31a,b}$$

where $q_c''$ ($= q_c/A$) is the mean heat flux over the surface and $\bar{T}_s$ is the mean surface temperature. The mean coefficient of heat transfer $\bar{h}$ is generally represented by correlations for mean Nusselt number $\overline{Nu}_L$,

$$\overline{Nu}_L = \frac{\bar{h}L}{k} \tag{9–32}$$

in terms of the Rayleigh number $Ra_L$,

$$Ra_L = Gr_L\,Pr = \frac{g\beta(\bar{T}_s - T_\infty)L^3}{\nu\alpha} \tag{9–33}$$

where $\beta$ and other fluid properties are evaluated at the free-stream or film temperature. The characteristic length $L$ used in the defining relations for $\overline{Nu}_L$ and $Ra_L$ is generally taken to approximate the length of travel of the fluid in the boundary layer.

Churchill [11] has developed a general convection correlation that is applicable to a variety of natural convection flows for which the primary buoyant driving force is directed tangential to the surface. This correlation is given by

$$\overline{Nu_L} = (a + 0.331\, b\, Ra_L^{1/6})^2 \tag{9–34}$$

where

$$b = \frac{1.17}{[1 + (0.5/Pr)^{9/16}]^{8/27}} \tag{9–35}$$

The coefficient $a$ and the characteristic length $\boldsymbol{L}$ are listed in Table 9–1 for various geometries. This correlation has been reported to be in good agreement with experimental data for both isothermal and uniformly heated surfaces over the entire range of $Ra_L$ and all $Pr$. Alternative natural convection correlations for these geometries are given in references 6, 7, 12, 13, and others.

Referring to Fig. 9–7, the driving force for natural convection flow over a horizontal plate is entirely due to buoyant forces acting in the direction normal to the surface. This normal buoyant force results in direct vertical motion and in an axial pressure gradient that indirectly drives the flow along the plate. Natural convection correlations have been developed for isothermal horizontal plates of the general form

$$\overline{Nu_L} = b\, Ra_L^m \tag{9–36}$$

The characteristic length $\boldsymbol{L}$ and coefficients $b$ and $m$ are given in Table 9–2 for two basic arrangements. Numerous other convection correlations are available in the literature for horizontal plates that account for various vertical edge conditions.

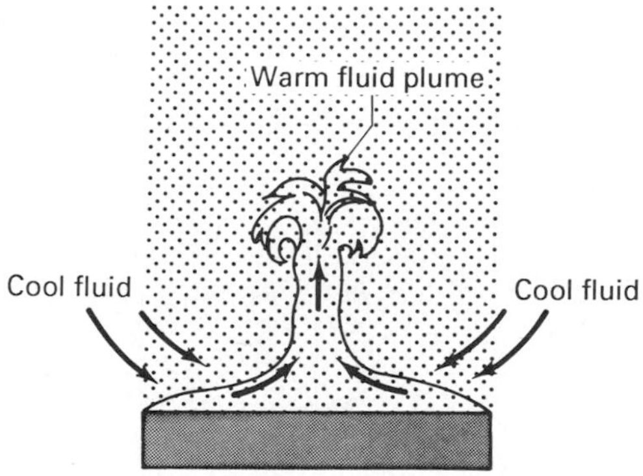


**FIGURE 9–7**
Illustration of natural convection flow over a hot horizontal surface facing up. (Adapted from Gebhart et al. [17]. Used with permission.)

Applications involving the geometries indicated in Tables 9–1 and 9–2 include steam and water pipes, air ducts, room walls, floors and ceilings, and cryogenic containers. For applications involving specified heat-flux conditions, the correlations for $\overline{Nu_L}$ given in this section can be written and plotted in terms of the *flux* Rayleigh number $Ra_L^*$ in order to compute the temperature difference $\overline{T}_s - T_\infty$ without iteration. Further information is available in the literature on topics such as natural convection associated with rotating bodies and finned surfaces. For comprehensive reviews of external natural convection flow, papers by Churchill [11] and Gebhart [15,16] and the reference text by Gebhart and associates [17] are recommended.

**TABLE 9–1** Natural convection: Vertical or inclined surfaces—characteristic length ***L*** and coefficients for Eq. (9–34)

| *Geometry* | ***L*** | *a* | *Comments* |
|---|---|---|---|
| Vertical Surfaces | | | |
| Plate 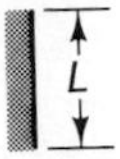  | $L$ | 0.825 | |
| Cylinder 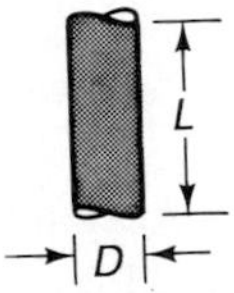  | $L$ | 0.825 | $D/L \gtrsim 35/Gr_L^{1/4}$ See [12, 13] for smaller values of $D/L$ |
| Inclined Surface | | | |
| Heated surface facing down 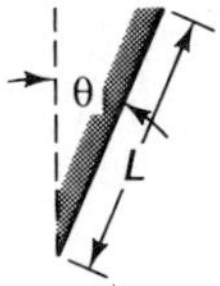  | $L$ | 0.825 | Replace $g$ by $g \cos \theta$ in $Ra_L$ See [14] for heated surface up or cooled surface down |
| Cooled surface facing up   | | | |
| Horizontal Cylinder 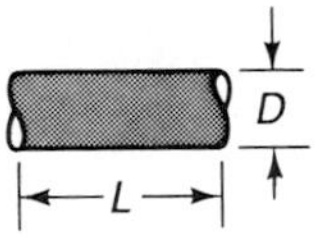  | $\pi D$ | 1.06 | $L \gg D$ |
| Sphere 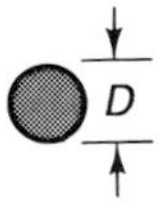  | $\frac{\pi}{2} D$ | 1.77 | |
| Vertical cone 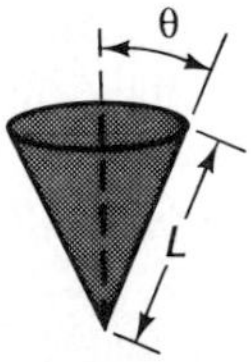  | $\frac{4}{5} L$ | 0.735 | Replace $g$ by $g \cos \theta$ in $Ra_L$ |
| Other Surfaces | | | |
| Inclined disk | $\frac{9}{11} D$ | 0.748 | Replace $g$ by $g \cos \theta$ in $Ra_L$ |
| Spherelike surface with area $A_s$ and volume $V$ | $3\pi V/A_s$ | $A_s^{3/2}/(6V)$ | |

*Source*: Churchill [11].

**TABLE 9–2** Natural convection: Horizontal surfaces—characteristic length ***L*** and coefficients for Eq. (9–36)

| *Horizontal plane surfaces with surface area $A_s$ and perimeter p* | $Ra_L$ | *b* | *m* | *References and comments* |
|---|---|---|---|---|
| $L = A_s/p$ | | | | [18,19] |
| Hot surface up (or cold surface down) | $10^4$–$10^7$ | 0.54 | $\frac{1}{4}$ | Laminar |
| | $10^7$–$10^{11}$ | 0.15 | $\frac{1}{3}$ | Turbulent |
| 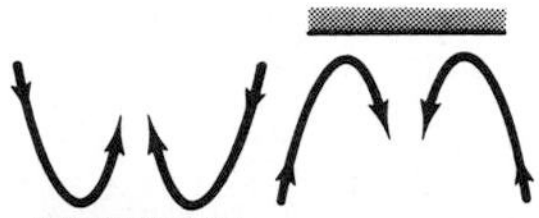 | | | | |
| Hot surface down (or cold surface up) | $10^5$–$10^{11}$ | 0.27 | $\frac{1}{4}$ | Laminar |
| 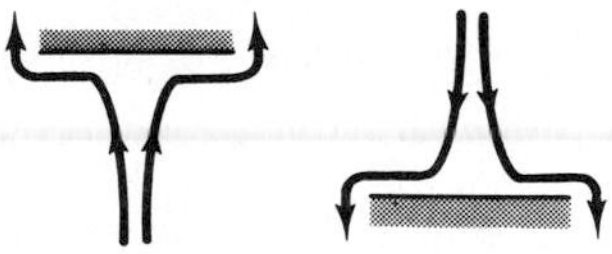 | | | | |

## EXAMPLE 9–3

Estimate the mean coefficient of heat transfer for the power amplifier of Example 9–1 if it is mounted horizontally.

### Solution

*Objective* Determine the effect on $\bar{h}$ of mounting a surface horizontally rather than vertically.

*Schematic* Horizontal flat plate representation of power amplifier.

$A_s = 3800$ mm$^2$ $T_\infty = 25°$C From Example 9-1

$L = 40$ mm $T_0 = 125°$C $Ra_L = 2.96 \times 10^5$

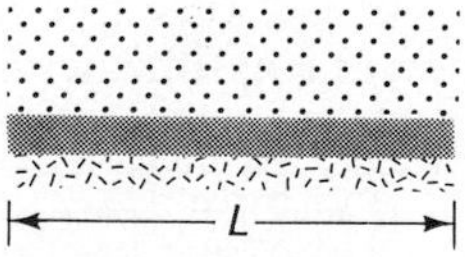


*Assumptions/Conditions*

natural convection

Boussinesq approximation with β and other fluid properties evaluated at $T_f$

standard conditions, except for significance of buoyancy

*Analysis* The mean Nusselt number $\overline{Nu}_L$ for horizontally mounted surfaces can be obtained from Eq. (9–36) with the characteristic length $\boldsymbol{L}$ and coefficients given in Table 9–2; that is,

$$\overline{Nu}_L = b\,Ra_L^m \qquad \boldsymbol{L} = \frac{A_s}{p}$$

and $m = \frac{1}{4}$ or $m = \frac{1}{3}$, depending on the conditions. To determine the perimeter $p$, we write

$$w = \frac{A_s}{L} = \frac{3800\ \text{mm}^2}{40\ \text{mm}} = 95\ \text{mm}$$

$$p = 2(w + L) = 2(95\ \text{mm} + 40\ \text{mm}) = 270\ \text{mm}$$

$$\boldsymbol{L} = \frac{A_s}{p} = \frac{3800\ \text{mm}^2}{270\ \text{mm}} = 14.1\ \text{mm} = 0.0141\ \text{m}$$

Using this value for $\boldsymbol{L}$ and the result from Example 9–1 for $Ra_L$, we obtain

$$Ra_{\boldsymbol{L}} = Ra_L\left(\frac{\boldsymbol{L}}{L}\right)^3 = 2.96 \times 10^5\left(\frac{14.1}{40}\right)^3 = 1.3 \times 10^4$$

Thus the flow is laminar.

For the case in which the device is mounted with the hot surface up, we obtain

$$\overline{Nu}_{\boldsymbol{L}} = 0.54\,Ra_{\boldsymbol{L}}^{1/4} = 0.54(1.3 \times 10^4)^{1/4} = 5.77$$

and

$$\bar{h} = \overline{Nu}_{\boldsymbol{L}}\frac{k}{\boldsymbol{L}} = 5.77\,\frac{0.0299\ \text{W/(m °C)}}{0.0141\ \text{m}} = 12.2\ \text{W/(m}^2\ \text{°C)}$$

On the other hand, with the hot surface facing down, we obtain

$$\overline{Nu}_{\boldsymbol{L}} = 0.27\,Ra_{\boldsymbol{L}}^{1/4} = 2.88$$

and

$$\bar{h} = 2.88\,\frac{0.0299\ \text{W/(m °C)}}{0.0141\ \text{m}} = 6.11\ \text{W/(m}^2\ \text{°C)}$$

Clearly, one should avoid mounting the power amplifier with the hot surface facing down, if at all possible.

## EXAMPLE 9–4

Determine the rate of heat transfer by natural convection from the surface of the cabinet shown in Fig. E9–4. The back of the cabinet is mounted on a vertical wall. Its surface temperature is 125°C and the ambient temperature is 25°C.

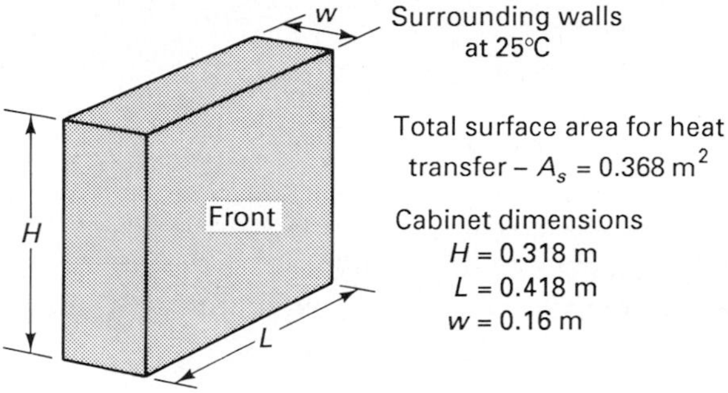
W
Surrounding walls
at 25°C
Total surface area for heat
transfer – $A_s$ = 0.368 m²
Front
H
Cabinet dimensions
H = 0.318 m
L = 0.418 m
w = 0.16 m
L

**FIGURE E9–4**
Cabinet mounted on a vertical wall.

## Solution

*Objective* Determine the rate of convection heat transfer $q_c$.

*Assumptions/Conditions*

natural convection
Boussinesq approximation with $\beta$ and other fluid properties evaluated at $T_f$
standard conditions, except for significance of buoyancy

*Properties* Air at $T_f = 75°C = 348$ K (Table A–C–5): $\rho = 0.994$ kg/m³, $\nu = 2.06 \times 10^{-5}$ m²/s, $k = 0.0299$ W/(m °C), $Pr = 0.697$, and

$$\beta = \frac{1}{T_f} = \frac{1}{348\text{ K}} = 2.87 \times 10^{-3}/\text{K}$$

*Analysis* As a first approximation, we will treat each surface independently. To provide a convenient means of computing the coefficient of heat transfer for each vertical and horizontal surface, the Grashof and Rayleigh numbers are represented in terms of $\boldsymbol{L}$ by

$$Gr_L = \frac{(9.81\text{ m/s}^2)(2.87 \times 10^{-3}/\text{K})(125°\text{C} - 25°\text{C})}{(2.06 \times 10^{-5}\text{ m}^2/\text{s})^2}\boldsymbol{L}^3 = 6.63 \times 10^9 \left(\frac{\boldsymbol{L}}{\text{m}}\right)^3$$

$$Ra_L = Gr_L\,Pr = 4.62 \times 10^9 \left(\frac{\boldsymbol{L}}{\text{m}}\right)^3$$

The characteristic length $\boldsymbol{L}$ and surface areas for the vertical and horizontal surfaces are specified as follows:

*Vertical*

$$A_{s,wH} = 0.16\text{ m }(0.318\text{ m}) = 0.0509\text{ m}^2$$

$$A_{s,LH} = 0.418\text{ m }(0.318\text{ m}) = 0.133\text{ m}^2$$

$$\boldsymbol{L}_v = H = 0.318\text{ m}$$

*Horizontal*

$$A_{s,wL} = 0.16\text{ m }(0.418\text{ m}) = 0.0669\text{ m}^2$$

$$p = 2(0.16\text{ m})(0.418\text{ m}) = 0.578\text{ m}$$

$$L_h = \frac{A_{s,wL}}{p} = \frac{0.0669\text{ m}^2}{0.578\text{ m}} = 0.116\text{ m}$$

We are now in a position to evaluate $Ra_L$, $\overline{Nu}_L$, $\bar{h}$, and $q_c$ for each surface.

*Vertical*

$$Ra_{L,v} = 4.62 \times 10^9 (0.318)^3 = 1.49 \times 10^8$$

$$\overline{Nu}_{L,v} = \left(\frac{Pr}{2.5 + 5\sqrt{Pr} + 5\,Pr}\right)^{1/4} Ra_{L,v}^{1/4} = 56.5 \qquad \text{from Eq. (9–16)}$$

$$\bar{h}_v = 56.5\,\frac{0.0299\text{ W/(m °C)}}{0.318\text{ m}} = 5.27\text{ W/(m}^2\text{ °C)}$$

$$q_{c,v} = \bar{h}A_{s,v}(T_0 - T_\infty)$$

$$= 5.27\,\frac{\text{W}}{\text{m}^2\text{ °C}}\,[2(0.0509\text{ m}^2) + 0.133\text{ m}^2](125°\text{C} - 25°\text{C}) = 124\text{ W}$$

*Horizontal*

$$Ra_{L,h} = 4.62 \times 10^9 (0.116)^3 = 7.21 \times 10^6$$

*Top*

$$\overline{Nu}_{L,\text{top}} = 0.54\,Ra_{L,h}^{1/4} = 28 \qquad \text{from Eq. (9–36), Table 9–2}$$

$$\bar{h}_{\text{top}} = 28\,\frac{0.0299\text{ W/(m °C)}}{0.116\text{ m}} = 7.22\text{ W/(m}^2\text{ °C)}$$

$$q_{c,\text{top}} = 7.22\,\frac{\text{W}}{\text{m}^2\text{ °C}}\,(0.116\text{ m}^2)(125°\text{C} - 25°\text{C}) = 83.8\text{ W}$$

*Bottom*

$$\overline{Nu}_{L,\text{bottom}} = 0.27\,Ra_{L,h}^{1/4} = 14 \qquad \text{from Eq. (9–36), Table 9–2}$$

$$\bar{h}_{\text{bottom}} = 14\,\frac{0.0299\text{ W/(m °C)}}{0.116\text{ m}} = 3.61\text{ W/(m}^2\text{ °C)}$$

$$q_{c,\text{bottom}} = 3.61\,\frac{\text{W}}{\text{m}^2\text{ °C}}\,(0.116\text{ m}^2)(125°\text{C} - 25°\text{C}) = 41.9\text{ W}$$

Summing up the contributions of the vertical, top, and bottom surfaces, the total rate of heat transfer is

$$q_c = q_{c,v} + q_{c,\text{top}} + q_{c,\text{bottom}} = 124\text{ W} + 83.8\text{ W} + 41.9\text{ W} = 250\text{ W}$$

As an aside, the mean coefficient of heat transfer over the entire cabinet can be obtained by writing

$$\bar{h} = \frac{q_c}{A_s(T_0 - T_\infty)} = \frac{250 \text{ W}}{0.368 \text{ m}^2 \, (125°\text{C} - 25°\text{C})} = 6.79 \text{ W/(m}^2 \, °\text{C)}$$

## 9–4 INTERNAL NATURAL CONVECTION FLOW

Internal natural convection flows are often encountered in heat-transfer applications involving fin units, fireplaces, and solar heating units, as well as other systems. To illustrate, the schematic of a natural circulation solar water heater is shown in Fig. 9–8. With the storage tank located above the solar collector, water circulates by natural convection when solar energy is captured by the collector.

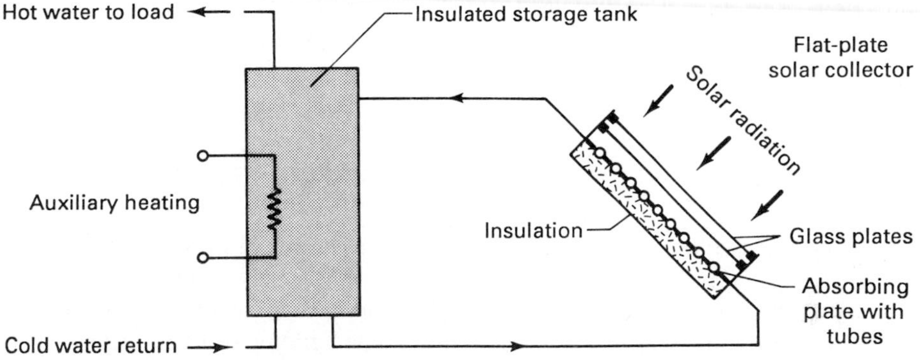


**FIGURE 9–8** Schematic of a natural-convection solar heating system.

Internal natural convection flow systems are complicated by the fact that the bulk-stream temperature $T_b$ generally varies with axial location $x$. The flow pattern depends upon the relative values of the wall temperatures and the temperature of entering fluid $T_1$. For example, for flow between vertical parallel plates with uniform surface temperatures $T_0$ and $T_w$, fluid rises in the vicinity of the warmer wall and falls in the region of the cooler surface for $T_0 > T_1 > T_w$. On the other hand, with $T_0$ and $T_w$ equal and greater than $T_1$, fluid rises throughout the entire system. These two situations are illustrated in Fig. 9–9(a) and (b) for laminar flow conditions. Such parallel-plate geometries are often used to approximate cooling fins in transformers, radiators, and other industrial devices, and in natural-circulation solar flat-plate collectors.

Correlations have been developed on the basis of Eq. (9–3) for internal natural convection flow with uniform wall temperatures $T_0$ and $T_w$, where† $\overline{T_s - T_F} = \overline{T}_s - T_1 = (T_0 + T_w)/2 - T_1$; that is,

$$q_c = \bar{h}A_s(\overline{T}_s - T_1) \tag{9–37}$$

† Correlations are also found in the literature with $\overline{T_s - T_F}$ set equal to $T_0 - T_w$.

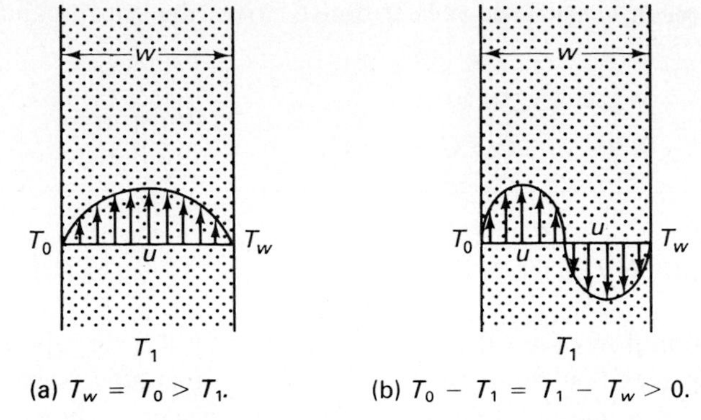


**FIGURE 9–9**
Representative internal natural-convection flow patterns for parallel-plate geometry.

Heat-transfer correlations for other cases involving internal natural convection flow are available in the literature. For example, correlations have been developed for natural convection flow between inclined parallel plates by Tabor [20] and by Dropkin and Somerscales [21].

## 9–5 NATURAL CONVECTION FLOW IN ENCLOSED SPACES

A typical enclosed natural-convection flow system and flow pattern is shown in Fig. 9–10. This type of natural circulation is particularly important in the cooling of electronic devices. In addition, the development of natural-convection flow in enclosures is a factor in the use air gaps to insulate building walls and cryogenic chambers, and in the design of solar flat-plate collectors.

Correlations have been developed for natural-convection heat transfer in enclosures with surfaces of length $H$ maintained at uniform temperatures $T_1$ and $T_2$. These correlations for $\bar{h}$ are based on Eq. (9–3) with $T_s - T_F = T_1 - T_2$; that is,

$$q_c = \bar{h}A_s(T_1 - T_2) \tag{9–38}$$

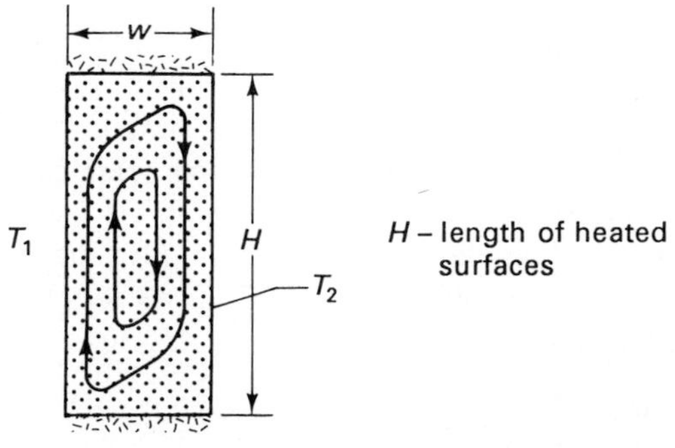


**FIGURE 9–10**
Representative natural-convection flow pattern for an enclosed vertical space; $L = w$, $T_1 > T_2$.

**TABLE 9–3** Natural convection: Characteristics for enclosed spaces with uniform wall temperatures $T_1$ and $T_2$

| *Geometry* | $L$ | $A_s$ | $R_c = L/(\overline{Nu}_L\, kA_s)$ |
|---|---|---|---|
| Rectangular enclosure; width $w$, length of heated surfaces $H$ | $w$ | $A_1 = A_2$ | $\dfrac{w}{A_1 k\, \overline{Nu}_L}$ |
| Vertical concentric annuli; length $H$, end surfaces insulated | $r_2 - r_1$ | $\dfrac{A_2 - A_1}{\ln (A_2/A_1)}$ | $\dfrac{\ln (r_2/r_1)}{2\pi Hk\, \overline{Nu}_L}$ |
| Concentric spheres | $r_2 - r_1$ | $\sqrt{A_1 A_2}$ | $\dfrac{r_2 - r_1}{4\pi r_1 r_2 k\, \overline{Nu}_L}$ |

The mean coefficient $\bar{h}$ is generally expressed in terms of $Ra_L$, $L$, and $H$ by an empirical expression of the form

$$\frac{\bar{h}L}{k} = \overline{Nu}_L = C\, Ra_L^m \left(\frac{H}{L}\right)^n \tag{9–39}$$

where $Ra_L = g\beta L^3(T_1 - T_2)/(\alpha\nu)$; $L$ and $A_s$ are dependent upon the geometry (see Table 9–3). The properties are generally evaluated at the average temperature $(T_1 + T_2)/2$ for internal flows. The coefficients $C$, $m$, and $n$ are given in Table 9–4 for vertical rectangular and annular enclosures (with the enclosing end surfaces insulated), and concentric spheres with $H/L$ set equal to unity). The works by Raithby and Hollands [22]

**TABLE 9–4** Natural convection: Convection correlations for enclosed spaces—coefficients for Eq. (9–39)†

| *Geometry* | *Fluid* | $Ra_L$ | $Pr$ | $C$ | $m$ | $n$ | $H/L$ |
|---|---|---|---|---|---|---|---|
| Rectangular enclosures with vertical surfaces heated, and vertical concentric annuli | Gas | $<2 \times 10^3$ | 0.5–2 | 1 | 0 | 0 | — |
| | | $2 \times 10^3$–$2 \times 10^5$ | 0.5–2 | 0.197 | $\frac{1}{4}$ | $-\frac{1}{9}$ | 11–42 |
| | Liquid | $2 \times 10^5$–$10^7$ | 0.5–2 | 0.073 | $\frac{1}{3}$ | $-\frac{1}{9}$ | 11–42 |
| | | $10^4$–$10^7$ | $1 - 2 \times 10^4$ | $C_1$ | $\frac{1}{4}$ | $-0.3$ | 10–40 |
| | | $10^6$–$10^9$ | 1–20 | 0.046 | $\frac{1}{3}$ | 0 | 1–40 |
| | Gas or liquid | $<10^{10}$ | $<10^5$ | $C_2$ | 0.28 | $-\frac{1}{4}$ | 2–10 |
| | | $> \dfrac{(0.2 + Pr)\, 10^3}{Pr}$ | $10^{-3}$–$10^5$ | $C_3$ | 0.29 | 0 | 1–2 |
| Concentric spheres, set $H/L = 1$ | Gas or liquid | $10^2$–$10^9$ | $0.7$–$4 \times 10^3$ | 0.228 | 0.226 | — | — |

† Vertical surfaces heated and cooled. $C_1 = 0.42\, Pr^{0.012}$, $C_2 = 0.22\, [Pr/(0.2 + Pr)]^{0.28}$, $C_3 = 0.18\, [Pr/(0.2 + Pr)]^{0.29}$.

*Sources*: Jakob [24,25], Graff and Held [26], Emery et al. [27,28], Catton [29], and Scanlan et al. [30,31].

and Kuehn and Goldstein [23] provide additional useful correlations for concentric annuli and spheres.

In regard to natural convection in rectangular enclosures with horizontal heated surfaces, the flow patterns depend upon whether the hotter plate is on the top or the bottom. In the first case, the low-density fluid lies above the heavier fluid, such that no buoyancy effects occur. For this situation, we have pure conduction heat transfer in the fluid, such that $\overline{Nu}_w = 1$.

For the second case, the heavier fluid is on the top. For values of $Ra_w$ less than a critical value, the buoyancy forces are not large enough to cause the fluid to turn over. The critical value of $Ra_w$ for gases is about 1700. For this case, we have stability with no natural convection currents. The flow pattern for values of $Ra_w$ between 1700 and $3.2 \times 10^5$ is laminar and takes the form of cells of circulating fluid. For larger values of $Ra_w$, the flow becomes turbulent and the cellular pattern no longer exists. An empirical equation of the form of Eq. (9–39) with $n = 0$ is recommended for natural convection in horizontal rectangular enclosures heated from below; that is,

$$\overline{Nu}_w = C\, Ra_w^m \tag{9–40}$$

The values of $C$ and $m$ and the ranges in $Ra_w$ are shown in Table 9–5.

**TABLE 9–5** Natural convection: Convection correlations for enclosed rectangular spaces—coefficients for Eq. (9–40)

| *Fluid* | $Ra_w$ | *Pr* | *C* | *m* | *References* |
|---|---|---|---|---|---|
| Horizontal surfaces heated from below | | | | | |
| Gas | $< 1.7 \times 10^3$ | | 1 | 0 | [24–26] [32–34] |
| | $1.7 \times 10^3$–$7.0 \times 10^3$ | 0.5–2 | 0.059 | 0.4 | |
| | $7.0 \times 10^3$–$3.2 \times 10^5$ | 0.5–2 | 0.212 | $\frac{1}{4}$ | |
| | $3.2 \times 10^5 <$ | 0.5–2 | 0.061 | $\frac{1}{3}$ | |
| Liquid | $< 1.7 \times 10^3$ | | 1 | 0 | [25,33–37] |
| | $1.7 \times 10^3$–$6.0 \times 10^3$ | 1–5000 | 0.012 | 0.6 | |
| | $6.0 \times 10^3$–$3.7 \times 10^4$ | 1–5000 | 0.375 | 0.2 | |
| | $3.7 \times 10^4$–$10^8$ | 1–20 | 0.13 | 0.3 | |
| | $10^8 <$ | 1–20 | 0.057 | $\frac{1}{3}$ | |

For natural convection in inclined enclosures with heating from below such as the one shown in Fig. 9–11, studies by Hollands et al. [22,35,38], Catton et al. [39], and Ayyaswamy and Catton [40] indicate that the Nusselt number correlations for vertical enclosures can be utilized up to a critical tilt angle $\theta^*$, with the gravitational acceleration $g$ replaced by its directional component along the surface of the plate, $g \cos \theta$; $\theta^*$ is shown in Table 9–6 as a function of the aspect ratio $H/w$. Utilizing this substitution for $7 \times 10^3 < Ra_w < 3.2 \times 10^5$, we have

$$\overline{Nu}_w = \overline{Nu}_w\big|_{\theta=0} (\cos \theta)^{1/4} \tag{9–41}$$

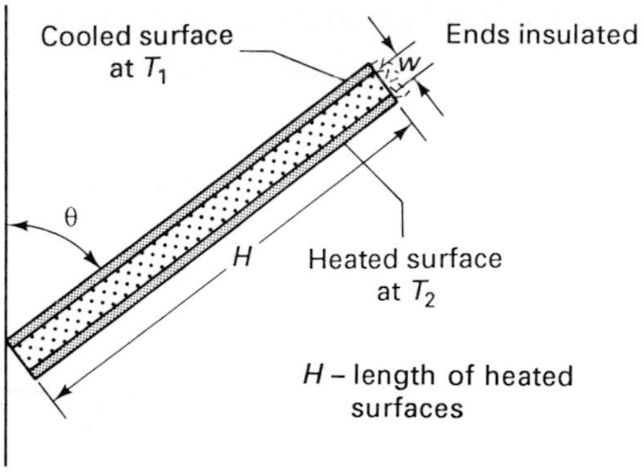
Cooled surface
at $T_1$
Ends insulated
w
θ
H
Heated surface
at $T_2$
H – length of heated
surfaces

**FIGURE 9–11**
Natural convection in an inclined enclosed rectangular space.

This equation has been reported to correlate experimental data in the region $0 \lesssim \theta \lesssim \theta^*$ for all aspect ratios. For tilt angles between $\theta^*$ and $\pi/2$ and large aspect ratios ($H/w \gtrsim 12$), Hollands et al. [41] recommend a correlation of the form

$$\overline{Nu}_w = 1 + 1.44\left[1 - \frac{1708}{Ra_w \sin\theta}\right]^{\bullet}\left\{1 - \frac{1708\,[\sin 1.8(\pi/2 - \theta)]^{1.6}}{Ra_w \sin\theta}\right\} + \left[\left(\frac{Ra_w \sin\theta}{5830}\right)^{1/3} - 1\right]^{\bullet} \tag{9–42}$$

where the notation $[\ ]^{\bullet}$ indicates that the quantity in brackets is to be set equal to zero when its value is less than zero. For situations in which the upper surface of an inclined rectangular enclosure is heated and the lower surface is cooled, Catton [29] recommends the following correlation by Arnold et al. [42] for all aspect ratios:

$$\overline{Nu}_w = 1 + (\overline{Nu}_w|_{\theta=0} - 1)\cos\theta \tag{9–43}$$

It should be noted that the rate of heat transfer $q_c$ is also sometimes expressed in terms of a thermal resistance for natural convection in enclosures as

$$q_c = \frac{T_1 - T_2}{R_c} \tag{9–44}$$

where $R_c = L/(\overline{Nu}_L\, kA_s)$; specific expressions for $R_c$ are given in Table 9–3. Because of the similarity between these expressions for $R_c$ and the expressions developed in

**TABLE 9–6** Critical tilt angle $\theta^*$ for inclined rectangular cavities

| $H/w$ | 1 | 3 | 6 | 12 | >12 |
|---|---|---|---|---|---|
| $\theta^*$ | 65° | 37° | 30° | 23° | 20° |

*Source*: From Hollands et al. [41].

Chap. 2 for $R_k$ for one-dimensional conduction heat transfer, the product $\overline{Nu}_L\, k$ is sometimes referred to as the *apparent thermal conductivity* $k_e$; that is,

$$k_e = \overline{Nu}_L\, k \tag{9–45}$$

For $\overline{Nu}_L = 1$ or $k_e = k$, we have pure conduction heat transfer through the fluid with no natural convection effect.

For more information on the topic of natural convection in enclosures, review articles by Catton [29] and Ostrach [43] are suggested.

## EXAMPLE 9–5

Two concentric blackbody spheres with 2-cm and 5-cm radii are separated by air at 0.5 atm. The surface temperatures are equal to 154°C and 0°C. Determine the total rate of heat transfer in this system.

### Solution

*Objective* Determine the total rate of heat transfer $q$ by natural convection and radiation.

*Schematic* Air contained between concentric blackbody spheres.

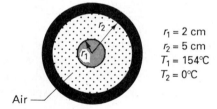

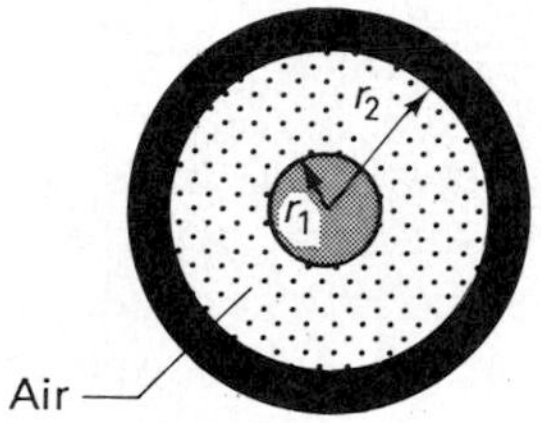


$r_1 = 2$ cm
$r_2 = 5$ cm
$T_1 = 154$°C
$T_2 = 0$°C

*Assumptions/Conditions*

natural convection and blackbody thermal radiation

Boussinesq approximation with β and other fluid properties evaluated at average temperature

standard conditions, except for significance of buoyancy and blackbody thermal radiation

*Properties* Air at average temperature $(T_1 + T_2)/2 = 77$°C $= 350$ K (Table A–C–5): $\rho = 0.998$ kg/m$^3$, $\nu = 2.08 \times 10^{-5}$ m$^2$/s, $k = 0.03$ W/(m °C), $Pr = 0.697$, and

$$\beta = \frac{1}{350\text{ K}} = 2.86 \times 10^{-3}/\text{K}$$

*Analysis* The total rate of heat transfer is given by

$$q = q_c + q_R$$

The rate of heat transfer by natural convection is written as

$$q_c = \bar{h}A_s(T_1 - T_2)$$

where

$$A_s = \sqrt{A_1A_2} = 4\pi r_1 r_2 = 4\pi(0.02 \text{ m})(0.05 \text{ m}) = 0.0126 \text{ m}^2$$

$$L = r_2 - r_1 = 0.05 \text{ m} - 0.02 \text{ m} = 0.03 \text{ m}$$

from Table 9–3,

$$\overline{Nu}_L = 0.228\, Ra_L^{0.226} \tag{9–39}$$

and

$$Gr_L = \frac{g\beta(T_1 - T_2)L^3}{\nu^2}$$

$$= \frac{(9.81 \text{ m/s}^2)(2.86 \times 10^{-3}/\text{K})(154°\text{C} - 0°\text{C})(0.03 \text{ m})^3}{(2.08 \times 10^{-5} \text{ m}^2/\text{s})^2} = 2.7 \times 10^5$$

$$Ra_L = Gr_L\, Pr = 1.88 \times 10^5$$

Substituting this value of $Ra_L$ into Eq. (9–39), we obtain

$$\overline{Nu}_L = 0.228(1.88 \times 10^5)^{0.226} = 3.55$$

and

$$\bar{h} = 3.55\,\frac{0.03 \text{ W/(m °C)}}{0.03 \text{ m}} = 3.55 \text{ W/(m}^2\text{ °C)}$$

Thus, we obtain

$$q_c = 3.55\,\frac{\text{W}}{\text{m}^2\text{ °C}}(0.0126 \text{ m}^2)(154°\text{C} - 0°\text{C}) = 6.89 \text{ W}$$

The radiation heat transfer is given by Eq. (1–18),

$$q_R = \sigma A_1 F_{1-2}(T_1^4 - T_2^4)$$

$$= 5.67 \times 10^{-8}\,\frac{\text{W}}{\text{m}^2\text{ K}^4}[4\pi(0.02 \text{ m})^2][(427 \text{ K})^4 - (273 \text{ K})^4] = 7.89 \text{ W}$$

Thus, the total heat-exchange rate is

$$q = q_c + q_R = 6.89 \text{ W} + 7.89 \text{ W} = 14.8 \text{ W}$$

The natural convection accounts for about 47% of this total.

## 9-6 COMBINED NATURAL AND FORCED CONVECTION

Strictly speaking, natural convection occurs in any nonisothermal forced convection system in which gravitational or centrifugal force fields are present. For example, natural convection superimposed upon forced convection in vertical tubes brings about an increase in heat-transfer rate for laminar upward flow and a decrease for laminar downward flow. That is, the heat transfer is enhanced for laminar flow in vertical tubes when the buoyancy forces are in the direction of flow. On the other hand, the opposite has been found to be true for turbulent flow in vertical tubes. In fact, severe deteriorations in the rate of heat transfer have been reported for turbulent upward vertical flow of high-pressure supercritical fluids in nuclear reactors and fossil-fired steam generators [44,45].

As a guide in determining the significance of natural convection in forced flow fields, the buoyancy forces are generally small and can be neglected for situations in which $Gr_L \ll Re_L^2$. However, for cases in which $Gr_L$ and $Re_L^2$ are of the same order or magnitude, both natural convection and forced convection are usually significant. This point is reinforced by the theoretical and experimental results shown in Fig.

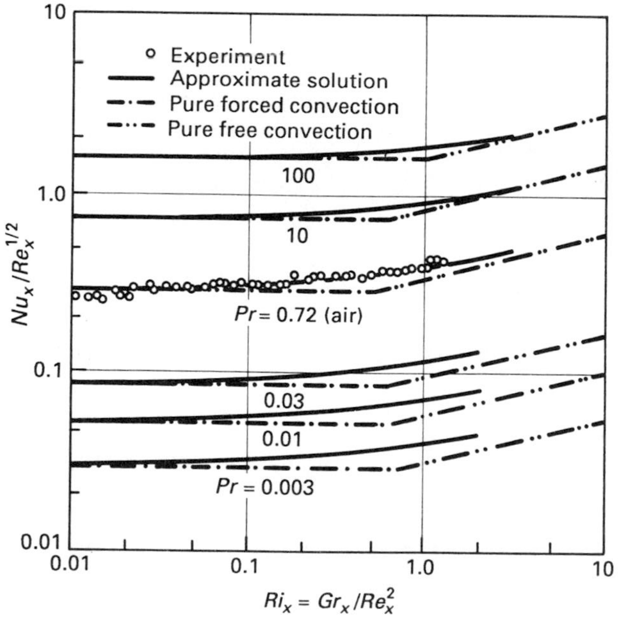


**FIGURE 9–12** Local Nusselt number $Nu_x$ for combined natural and forced convection from an isothermal vertical plate (Lloyd and Sparrow [46]).

9–12 for upward laminar forced convection flow over a vertical heated flat plate with uniform wall temperature. Notice that the theoretical predictions and data for air approach the limiting solution for pure forced convection as the *Richardson number* $Ri_x$,

$$Ri_x = \frac{Gr_x}{Re_x^2} \tag{9–46}$$

falls toward a value of the order of 0.02. The natural convection effect is of the order of 10% for $Ri_x$ equal to about 0.225. The theoretical calculations and experimental data are also seen to approach limiting curves for pure natural convection as $Ri_x$ approaches a value of the order of 10.

Several dimensionless number criteria have been developed to establish the limits between forced and combined convection-heat-transfer regimes. For example, Metais and Eckert [47] published preliminary limit criterion charts in 1964 for vertical and horizontal tube flow of moderate-Prandtl-number fluids. Natural convection limit criteria have also been developed for liquid metals [48] and for supercritical fluids [49].

For information concerning Nusselt number correlations for combined forced and natural convection, one can refer to references 50–52.

## 9–7 SUMMARY

As we have seen, the heat-transfer performance of natural convection flows is generally characterized in terms of the Rayleigh number $Ra_L$ or flux Rayleigh number $Ra_L^*$. Practical natural convection correlations have been presented in this chapter for Nusselt number $Nu_L$ for a number of arrangements involving external flow, internal flow, and flow in enclosures. These natural convection correlations and related dimensionless parameters are summarized in Tables 9–7 and 9–8. With the Rayleigh number known for specified wall temperature, these correlations are readily used to obtain the heat-transfer rate. Similarly, with the flux Rayleigh number known for specified wall-heat flux, the correlation for Nusselt number in terms of flux Rayleigh number provides a straightforward means of obtaining the wall temperature.

Brief consideration has also been given to the important topic of combined natural and forced convection, which becomes important in certain forced-convection flows.

Numerous other important natural convection flows such as thermal plumes, buoyant jets, and flow in stratified ambient media are treated in the literature. The reference textbook by Gebhart et al. [17] provides timely in-depth coverage of a wide range of natural convection flows.

**TABLE 9–7** Summary of convection correlations for natural convection†

| *Condition* | *Correlation* | *Identification* |
|---|---|---|
| Vertical surface | | |
| Isothermal | | |
| Laminar flow $10^4 < Ra_L < 10^9$ | $\dfrac{Nu_x}{Ra_x^{1/4}} = \dfrac{3}{4}\left(\dfrac{Pr}{2.5 + 5\sqrt{Pr} + 5\,Pr}\right)^{1/4}$ | Eq. (9–11) Ede [2] |
| | $\dfrac{\overline{Nu}_L}{Ra_L^{1/4}} = \left(\dfrac{Pr}{2.5 + 5\sqrt{Pr} + 5\,Pr}\right)^{1/4}$ | Eq. (9–16) |
| Laminar and turbulent flow, all $Ra_L$ | $\overline{Nu}_L = \left\{0.825 + \dfrac{0.387\,Ra_L^{1/6}}{[1 + (0.492/Pr)^{9/16}]^{8/27}}\right\}^2$ | Eq. (9–17) Churchill & Chu [5] |
| Laminar flow $0 < Ra_L < 10^9$ | $\overline{Nu}_L = 0.68 + \dfrac{0.670\,Ra_L^{1/4}}{[1 + (0.492/Pr)^{9/16}]^{4/9}}$ | Eq. (9–18) Churchill & Chu [5] |
| Gas Laminar flow $10^4 < Ra_L < 10^9$ | $\overline{Nu}_L = 0.525\,Ra_L^{1/4}$ | Eq. (9–19) McAdams [7] |
| Gas Turbulent flow $10^9 < Ra_L$ | $\overline{Nu}_L = 0.13\,Ra_L^{1/3}$ | Eq. (9–20) McAdams [7] |
| Uniform wall-heat flux | | |
| Laminar flow $10^4 < Ra_x < 10^9$ | $\dfrac{Nu_x}{Ra_x^{*1/5}} = \left(\dfrac{Pr}{4 + 9\sqrt{Pr} + 10\,Pr}\right)^{1/5}$ | Eq. (9–24) Fujii & Fujii [8] |
| | $\dfrac{Nu_x}{Ra_x^{1/4}} = \left(\dfrac{Pr}{4 + 9\sqrt{Pr} + 10\,Pr}\right)^{1/4}$ | Eq. (9–27) Equivalent to Eq. (9–24) |
| Gas Laminar flow | $Nu_x = 0.424\,Ra_x^{1/4}$ | Eq. (9–29a) Chap. 7 |
| | $Nu_x = 0.503\,Ra_x^{*1/5}$ | Eq. (9–29b) Equivalent to Eq. (9–29a) |
| Gas Turbulent flow | $Nu_x = 0.0942\,Ra_x^{1/3}$ | Eq. (9–30a) Vliet & Liu [10] |
| | $Nu_x = 0.170\,Ra_x^{*1/4}$ | Eq. (9–30b) Equivalent to Eq. (9–30a) |
| General external surfaces—Isothermal or uniform heat flux | | |
| Vertical plates, cylinders, inclined surfaces, horizontal cylinders, spheres, vertical cones, and other surfaces all $Ra_L$, $Pr$ | $\overline{Nu}_L = (a + 0.331\,b\,Ra_L^{1/6})^2$ | Eq. (9–34) $a$, $\boldsymbol{L}$—Table 9–1 $b$—Eq. (9–35) Churchill [11] |

† $Gr_L = g\beta(T_s - T_\infty)\boldsymbol{L}^3/\nu^2$, $Ra_L = Gr_L\,Pr = g\beta(T_s - T_\infty)\boldsymbol{L}^3/(\alpha\nu)$, $Ra_L^* = Nu_L\,Ra_L = g\beta q_0''\,\boldsymbol{L}^4/(k\alpha\nu)$, $Gr_L^* = g\beta q_0''\,\boldsymbol{L}^4/(k\nu^2)$.

**TABLE 9–7** *(Continued)*

| *Condition* | *Correlation* | *Identification* |
|---|---|---|
| Horizontal isothermal plates | $\overline{Nu}_L = b\, Ra_L^m$ | Eq. (9–36)<br>$b$, $m$, $\mathbf{L}$—Table 9–2<br>Goldstein et al. [18]<br>Lloyd & Moran [19]<br>$[q_c = \bar{h}A_s(T_1 - T_2)]$ |
| Enclosed spaces<br>Rectangular enclosures, vertical concentric annuli, concentric spheres | $\overline{Nu}_L = C\, Ra_L^m \left(\dfrac{H}{L}\right)^n$ | Eq. (9–39)<br>$\mathbf{L}$, $A_s$—Table 9–3<br>$C$, $m$, $n$, and $Ra_L$ range—Table 9–4<br>various references |
| Horizontal plates heated from below | $\overline{Nu}_w = C\, Ra_w^m$ | Eq. (9–40)<br>$C$, $m$—Table 9–5<br>various references |
| Inclined rectangular space<br>Heated from below cooled from above<br>$0 < \theta < \theta^*$ | $\overline{Nu}_w = (\overline{Nu}_w\vert_{\theta=0})(\cos\theta)^{1/4}$ | Eq. (9–41)<br>$\theta^*$—Table 9–6<br>Hollands et al. [22,35,38] |
| $\theta^* < \theta < \pi/2$ | $\overline{Nu}_w = 1 + 1.44\left[1 - \dfrac{1708}{Ra_w \sin\theta}\right]^{\bullet} \times \left\{1 - \dfrac{1708\,[\sin 1.8(\pi/2-\theta)]^{1.6}}{Ra_w \sin\theta}\right\} + \left[\left(\dfrac{Ra_w \sin\theta}{5830}\right)^{1/3} - 1\right]^{\bullet}$ | Eq. (9–42)<br>$H/w > 12$<br>Hollands et al. [41] |
| Heated from above cooled from below | $\overline{Nu}_w = 1 + (\overline{Nu}_w\vert_{\theta=0} - 1)\cos\theta$ | Eq. (9–43)<br>Arnold et al. [42] |

## ■ REVIEW QUESTIONS

**9–1.** What is natural convection?

**9–2.** What is the general form of convection correlations for natural convection?

**9–3.** Define the Grashof number and state its significance.

**9–4.** Define the Rayleigh number. Why is this parameter important?

**9–5.** Define the flux Rayleigh number and explain its primary use.

**9–6.** Write the criterion that indicates that natural convection flow over a vertical flat plate is likely to be turbulent.

**9–7.** Describe the flow pattern for natural convection flow over a heated horizontal plate.

**TABLE 9–8** Dimensionless parameters for convection heat transfer: Natural convection

| *Dimensionless group* | *Symbol* | *Definition* | *Interpretation* |
|---|---|---|---|
| Grashof number | $Gr_L$ | $\frac{g\beta(T_s - T_\infty)L^3}{\nu^2}$ | Relative significance of buoyant and viscous effects for single phase flow |
| Nusselt number† | $Nu_L$ | $\frac{hL}{k}$ | $\frac{\text{convection for fluid in motion}}{\text{conduction for motionless fluid layer}}$ |
| Prandtl number† | $Pr$ | $\frac{\nu}{\alpha}$ | relative effectiveness of momentum and energy transfer by molecular diffusion within the boundary layer |
| Rayleigh number | $Ra_L$ | $Gr_L\, Pr$ | $\frac{\text{buoyancy forces}}{\text{change in shear forces}}$ |
| Reynolds number† | $Re_L$ | $\frac{U_F L}{\nu}$ | $\frac{\text{inertia forces}}{\text{viscous forces}}$ |
| Flux Rayleigh number | $Ra_L^*$ | $Nu_L\, Ra_L$ | Dimensionless heat flux for natural convection |
| Richardson number | $Ri_x$ | $\frac{Gr_x}{Re_x^2}$ | Buoyant force/inertia force; Relative significance of natural convection and forced convection |

† Dimensionless parameters also used in analysis of forced convection.

**9–8.** Describe the flow pattern for natural convection flow in a horizontal parallel plate enclosure.

**9–9.** Describe the flow pattern for natural convection flow between vertical parallel plates.

**9–10.** What criterion can be used to indicate whether natural convection effects are likely to be significant for forced convection processes?

## PROBLEMS

**9–1.** Determine the Rayleigh number for natural convection flow of air at 50°C and 1 atm over a vertical flat plate 50 cm in length for the following surface temperatures: (a) −46°C; (b) 54°C. Classify the nature of the flow for each of these conditions.

**9–2.** Determine the Rayleigh number for natural convection flow of water at 50°C and 1 atm over a vertical flat plate 5 cm in length for the following surface temperatures: (a) −10°C; (b) 150°C. Classify the nature of the flow for each of these conditions.

**9–3.** Sketch representative hydrodynamic and thermal boundary layers and velocity and temperature distributions for natural convection flow over a cold vertical flat plate for fluids with (a) large $Pr$, (b) moderate $Pr$, and (c) small $Pr$.

**9–4.** Assuming that the power amplifier of Example 9–1 must dissipate 7 W, determine the average surface temperature if the surroundings are at 35°C.

**9–5.** Refine the solution to Example 9–2 by evaluating the properties at the average film temperature.

**9–6.** A plate 10 cm high and 5 cm wide at 175°C is placed vertically in air at 38°C and 1 atm. Determine the average wall-heat flux by natural convection.

**9–7.** Reconsider Prob. 9–6 for the case in which the surrounding fluid is water at 38°C and 1 atm.

**9–8.** Reconsider Prob. 9–6 for the case in which the surrounding fluid is engine oil at 38°C.

**9–9.** A vertical surface 4 m high and 2 m wide exchanges 10 W/m$^2$ of thermal radiation with a nearby furnace. The plate is insulated on the back side. Determine the average surface temperature of the plate if the surrounding air is at 27°C, assuming blackbody surfaces.

**9–10.** A 25-cm-diameter vertical pipe is surrounded by air at 30°C. Determine the rate of natural convection heat transfer for a surface temperature of 200°C and a length of 1 m.

**9–11.** A large thin metal sheet of length $L$ is removed from an electroplating bath and placed in a vertical position to cool by natural convection air flow. Develop a relation for the variation in plate temperature with time $t$ in terms of the initial temperature $T_i$, air temperature $T_\infty$, length $L$, and properties of the plate and fluid.

**9–12.** The surfaces of a 0.5-m by 0.5-m thin plate are maintained at 0°C. Calculate the rate of natural convection heat transfer to air at 90°C and 1 atm if the plate is vertical.

**9–13.** Solve Prob. 9–12 if the plate is horizontal with both sides cooled by convection.

**9–14.** Determine the rate of natural convection heat transfer from two surfaces of a vertical plate with 10-cm height and 4-cm width. The surrounding air temperature is 27°C and the plate surface temperature is 150°C.

**9–15.** A 1-m-long flat plate is inclined at an angle of $\pi/4$ from the horizontal. The air temperature is 40°C and the lower surface of the plate is at 100°C. Determine the mean Nusselt number for natural convection flow.

**9–16.** A 25-cm-diameter horizontal pipe is surrounded by air at 30°C. Determine the rate of natural convection heat transfer for a surface temperature of 200°C and a length of 1 m.

**9–17.** A thin 10-cm-diameter disk with surface temperature of 100°C is situated horizontally in a large container of water at 20°C. Determine the rate of heat input into the plate necessary to balance the natural convection heat transfer from the top and bottom surfaces.

**9–18.** A 5-cm-diameter 2-m-long electrical cable generates 5 W. The surrounding air temperature is 50°C. Estimate the surface temperature of the cable if it is horizontal.

**9–19.** Solve Prob. 9–18 for a vertical orientation.

**9–20.** Determine the rate of natural convection heat transfer from a 10 ft length of 3-in.-O.D. thin-walled horizontal pipe carrying steam at 212°F. The pipe is surrounded by air at 70°F.

**9–21.** Solve Prob. 9–20 if the surrounding fluid is water at 70°F.

**9–22.** A cylindrical heating element 2.54 cm in diameter and 0.5 m long is placed vertically in a container of water at 20°C. The electrically generated heat transfer is 200 W. Estimate the average surface temperature of the element.

**9–23.** Solve Prob. 9–22 for a horizontal orientation.

**9–24.** One surface of a 1-m by 1-m thin plate is maintained at 90°C. The other face is insulated. Calculate the rate of natural convection heat transfer to air at 0°C if the plate is mounted vertically.

**9–25.** Solve Prob. 9–24 for (a) horizontal plate with heated surface up, and (b) horizontal plate with heated surface down.

**9–26.** Air at atmospheric pressure is contained in an enclosure which consists of two vertical plates with 24-cm height and 4-cm width, which are 2 cm apart. These two plates are at 0°C and 50°C, respectively. The four other plates that complete the enclosure are insulated. Determine the rate of heat transfer via natural convection between the two plates.

**9–27.** Determine the total rate of heat transfer between the two plates in Prob. 9–26 by both thermal radiation and natural convection. Assume that all the surfaces are black.

**9–28.** Determine the rate of heat transfer by natural convection in the enclosure of Prob. 9–26 if the 24-cm-high and 2-cm-wide end plates are maintained at 0°C and 50°C, with the other four surfaces insulated.

**9–29.** Determine the rate of heat transfer by natural convection in the enclosure of Prob. 9–26 if the 4-cm by 2-cm top and bottom plates are maintained at 0°C and 50°C, respectively, with the other four surfaces insulated.

**9–30.** A double-pane glass window consists of two 2-mm-thick plates separated by a 5-mm air gap. Determine the rate of natural convection heat transfer across the air gap if the inside surface temperatures of the glass plates are 35°C and 55°C, and the 3-m by 3-m plates are mounted vertically.

**9–31.** Resolve Prob. 9–30 if the temperature of the outer surfaces of the glass plates are specified as 35°C and 55°C.

**9–32.** Reconsider Prob. 2–17, which deals with heat transfer through a vertical exterior wall, by taking into account the effects of natural convection within the air space.

**9–33.** A cryogenic chamber consists of a 2-m-I.D. spherical shell with a 1-cm-thick wall. An outer 2.12-m-I.D. shell surrounds the inner shell, with air at 0.1 atm in the enclosed space. Determine the rate of heat transfer by natural convection if the two surfaces in contact with the enclosed gas are at 40°C and 14°C. (Refer back to Example 1–5 for a related problem.)

**9–34.** A square flat-plate solar collector with a 1-$m^2$ surface area is inclined at an angle of 35° with the horizontal. The bottom collector plate is at a temperature of 180°C and the upper glass plate surface is at 35°C. The glass plate and blackbody collector plates are separated by a distance of 4 cm. The pressure within the enclosure is 0.2 atm. Determine the rate of heat loss across the air gap by natural convection.

**9–35.** Air at 35°C is forced upward along a vertical heated flat plate 1 m high by 10 m wide with a surface temperature of 150°C. Determine the free-stream velocity above which the effects of natural convection can be neglected.

# CHAPTER 10

# CONVECTION HEAT TRANSFER: PRACTICAL ANALYSIS—BOILING AND CONDENSATION

## 10–1 INTRODUCTION

The study of two- and three-phase substances is one of the main topics covered in introductory courses in thermodynamics. Based on our background in thermodynamics, we know that a phase change occurs when a single-phase substance is brought to the saturation state. For example, when the temperature of a subcooled liquid is raised to the saturation temperature $T_{sat}$, vaporization or *boiling* occurs. On the other hand, when the temperature of a superheated vapor is lowered to $T_{sat}$, *condensation* occurs. The thermodynamic study of boiling and condensation as well as melting and freezing is developed in the context of idealistic equilibrium conditions. In practice, however, heat-transfer processes involving phase change occur under nonequilibrium conditions in which the difference between the wall-surface temperature $T_s$ and $T_{sat}$ is not zero. The study of boiling and condensation heat transfer deals with such nonequilibrium liquid-to-vapor and vapor-to-liquid phase-change processes.

Familiar engineering applications of boiling and condensation heat transfer occur in power or refrigeration cycles. In addition, because of the large heat fluxes that can be accomplished in processes that involve phase change, boiling and condensation processes are also used in compact heat exchangers.

It should be mentioned that two-phase heat-transfer processes such as boiling and condensation are considerably more involved than are single-phase convection-heat-transfer processes because of the complicating effects of factors such as property variation, surface tension, latent heat of vaporization, and surface conditions. However, because we will be dealing with the practical thermal analysis of the simplest types of boiling and condensation-heat-transfer processes, the complexities that will be encountered in our introductory treatment of this topic will be minimized.

The general characterizing parameters associated with two-phase heat-transfer processes are introduced in the following section, after which specific attention is given to the phenomena of boiling and condensation in Secs. 10–3 and 10–4.

## 10–2 CHARACTERIZING PARAMETERS FOR TWO-PHASE HEAT-TRANSFER PROCESSES

To characterize two-phase heat-transfer processes, we generally must account for (1) the body force $g(\rho_f - \rho_g)$, which accompanies the large differences in density between the liquid and vapor states; (2) surface tension $\sigma$, for situations involving a liquid-vapor interface with finite curvature or temperature (or concentration) -induced variations in $\sigma$ over the interface;† (3) the latent heat of vaporization relative to the difference between the surface and saturation temperatures $i_{fg}/\Delta T = i_{fg}/(T_s - T_{sat})$; (4) the thermophysical properties of the fluid $\rho$, $\mu$, $c_P$, and $k$ (liquid $f$ or vapor $g$); and (5) a characteristic length $\boldsymbol{L}$.‡ Assuming that the two-phase coefficient of heat transfer $\bar{h} = q_c''/(T_s - T_{sat})$ is dependent upon these parameters, we write

$$\bar{h} = \frac{q_c''}{T_s - T_{sat}} = \text{fn}\,[g(\rho_f - \rho_g), \sigma, i_{fg}/(T_s - T_{sat}), \rho, \mu, c_P, k, \boldsymbol{L}] \tag{10–1}$$

Noting that this relation involves *nine* independent parameters and *four* fundamental dimensions, the dimensional analysis approach indicates *five* independent dimensionless groups and a dimensionless equation of the form

$$\frac{q_c''}{T_s - T_{sat}}\frac{\boldsymbol{L}}{k} = \text{fn}\left[\frac{\mu c_P}{k}, \frac{g(\rho_f - \rho_g)\boldsymbol{L}^3}{\rho \nu^2}, \frac{c_P(T_s - T_{sat})}{i_{fg}}, \frac{g(\rho_f - \rho_g)\boldsymbol{L}^2}{\sigma}\right] \tag{10–2}$$

or

$$\overline{Nu_L} = \text{fn}\,[Pr, Gr_L, Ja, Bo_L] \tag{10–3}$$

where

$$\overline{Nu_L} = \frac{q_c''}{T_s - T_{sat}}\frac{\boldsymbol{L}}{k} = \frac{\bar{h}\boldsymbol{L}}{k} \tag{10–4}$$

is the Nusselt number, $Pr$ is the Prandtl number,

$$Gr_L = \frac{g(\rho_f - \rho_g)\boldsymbol{L}^3}{\rho \nu^2} = \frac{\rho g(\rho_f - \rho_g)\boldsymbol{L}^3}{\mu^2} \tag{10–5}$$

is a generalized form of the *Grashof number*, which represents *the relative significance of buoyant and viscous effects for two-phase flow*,

$$Ja = \frac{c_P(T_s - T_{sat})}{i_{fg}} \tag{10–6}$$

† The surface tension $\sigma$, which results from a net inward force of attraction on molecules at the interface toward molecules within the liquid phase, is defined as the work done in extending the surface of a liquid one unit area. The units commonly used for $\sigma$ are N/m, kg/s$^2$ or dyne/cm.

‡ As discussed in Sec. 10–4–2, additional parameters are required for dropwise condensation.

is the *Jakob number*, which represents *the ratio of the maximum sensible energy absorbed to the latent energy absorbed*, and

$$Bo_L = \frac{g(\rho_f - \rho_g)L^2}{\sigma} \tag{10–7}$$

is the *Bond number*, which represents the *ratio of gravitational forces to surface tension forces*. The surface tension $\sigma$ is a function of both the fluid and temperature, as shown in Tables A–C–3 and A–C–6 of the Appendix. The enthalpy of vaporization $i_{fg}$ is listed in Tables A–C–3(a) and A–C–3(c) for water and several other common fluids. Both $\sigma$ and $i_{fg}$ decrease as the saturation pressure increases.

These and several other dimensionless parameters will be used in the following sections of this chapter to characterize boiling and condensation heat-transfer processes. As we will see, the correlations for boiling and condensation involve the use of properties of saturated liquid ($\rho_f$, $\mu_f$, and $c_{P,f}$) or saturated vapor ($\rho_g$, $\mu_g$, and $c_{P,g}$) in the definition of $Gr_L$ and $Ja$. Unless otherwise indicated, the Nusselt number $Nu_L$ will be expressed in terms of $k_f$. In this connection, $Gr_{L,f}$, $Ja_f$, $Bo_L$, and $Pr_g$ all experience increases and $Pr_f$ decreases with increasing saturation temperature $T_{sat}$ (or saturation pressure $P_{sat}$).

## 10-3 BOILING HEAT TRANSFER

*Boiling heat transfer* occurs when a fluid is exposed to a surface with temperature sufficiently greater than the saturation temperature $T_{sat}$.

The two types of boiling that are found in practice are *pool boiling* and *flow boiling*. Pool boiling occurs in the absence of bulk fluid flow, such as in the boiling of a pan of water. This type of boiling can also be easily produced by submerging an electrically heated wire or a steam-heated tube in a liquid, as shown in Figs. 10–1 and 10–2. Common examples of flow boiling processes include fossil fuel or nuclear steam generators and refrigerant evaporators.

Pool and flow boiling processes are further classified according to whether the bulk-liquid temperature is equal to or less than the saturation temperature. If the bulk temperature of the liquid is below $T_{sat}$, the process is referred to as *subcooled* boiling. Otherwise, the process is known as *saturated* boiling.

### 10-3-1 Pool Boiling

#### *Boiling Curve for Pool Boiling*

For a given fluid, operating pressure, and heating surface, the heat flux $q''_c$ that occurs during pool boiling is dependent upon the excess temperature $\Delta T_e = T_s - T_{sat}$. To illustrate this point, $q''_c$ is plotted in terms of $\Delta T_e$ in Fig. 10–3 for pool boiling of water. This general boiling curve applies to pool boiling of water at any subcritical

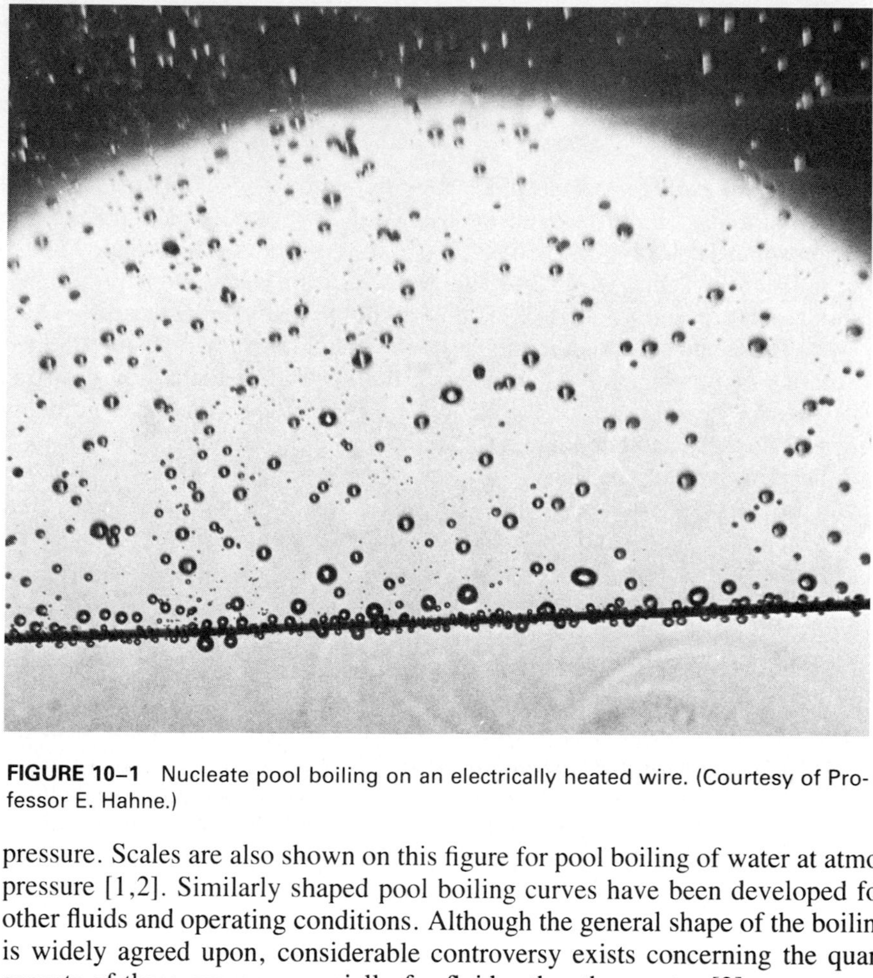

**FIGURE 10–1** Nucleate pool boiling on an electrically heated wire. (Courtesy of Professor E. Hahne.)

pressure. Scales are also shown on this figure for pool boiling of water at atmospheric pressure [1,2]. Similarly shaped pool boiling curves have been developed for many other fluids and operating conditions. Although the general shape of the boiling curve is widely agreed upon, considerable controversy exists concerning the quantitative aspects of these curves, especially for fluids other than water [3].

The temperature of the fluid that is in contact with the wall is $T_s$. Thus, although the bulk-liquid temperature is always less than or equal to $T_{sat}$, the fluid very near the heating surface is superheated, with the degree of superheat being equal to the excess temperature $\Delta T_e$. As we move along the boiling curve from left to right, the degree of superheat of fluid in contact with the heating surface increases.

A boiling curve such as this exhibits rather distinct regimes, which include natural convection, nucleate boiling, transition boiling, and film boiling.

***Natural Convection*** For small values of excess temperature (i.e., $\Delta T_e < \Delta T_{e,A}$) the fluid motion occurs by means of natural convection. The natural convection circulates the slightly superheated liquid from the vicinity of the heating surface to the surface of the pool, where evaporation occurs. No boiling per se occurs in the natural convection regime, such that the heat transfer can be computed by the methods introduced in Chap. 9.

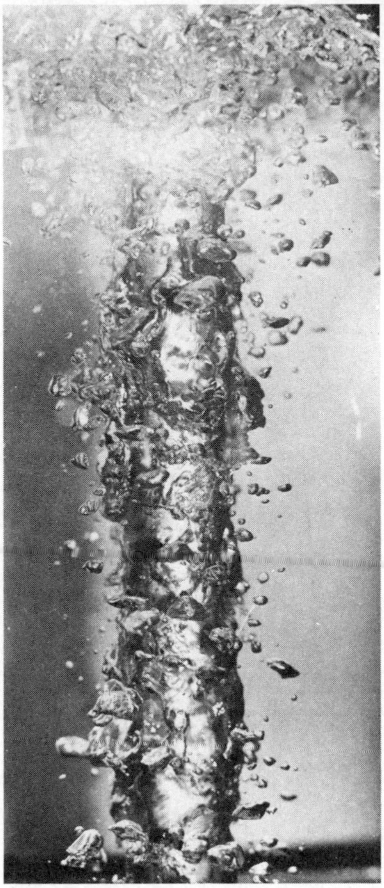

**FIGURE 10–2**
Film pool boiling of methanol on a vertical steam-heated copper tube. (Courtesy of Professor J. W. Westwater.)

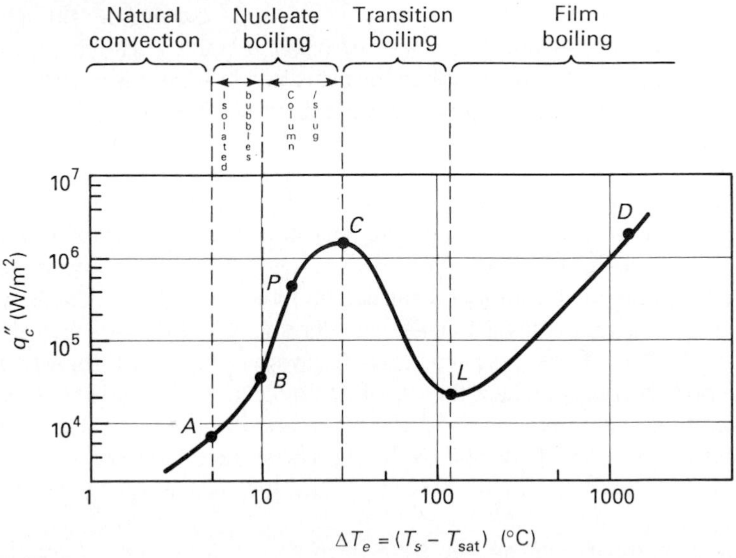


**FIGURE 10–3** Representative boiling curve for pool boiling of water at one atmosphere.

***Nucleate Boiling*** For excess temperatures between $\Delta T_{e,A}$ and $\Delta T_{e,C}$ liquid is transformed into vapor nuclei (i.e., bubbles) at various preferential sites on the heating surface. The nucleate boiling regime is generally characterized according to the behavior of the bubbles.

In the region between the *onset of nucleate boiling* ONB (point $A$) and point $B$, the nucleation sites produce isolated bubbles that eventually break away from the heating surface and rise toward the surface of the pool. The bubble growth and separation result in liquid being circulated and drawn toward the heating surface. It is this intermittent entrainment of cool liquid onto the hot surface which is primarily responsible for the increase in heat flux $q_c''$ in this *isolated bubble-nucleate boiling regime*, with the contribution of evaporation (i.e., energy carried by vapor bubbles) being secondary.

In the region between point $B$ and point $C$ where the heat flux is a maximum, the number of active nucleation sites becomes very large, such that the bubbles coalesce, escape from the surface in the form of jets or columns, and merge into slugs of vapor. The large heat fluxes in this *column/slug-nucleate boiling regime* are produced by the combined effects of liquid entrainment and evaporation. Whereas the effects of evaporation increase with $\Delta T_e$, the contribution of the liquid entrainment reaches a maximum at point $P$. Beyond this point the inflow of cooler liquid toward the heating surface is inhibited by the escaping vapor, with the result that the gradient in $q_c''$ with respect to $\Delta T_e$ decreases until the point is reached at which the surface is no longer continually wetted and the heat flux is a maximum. The *maximum heat flux* $q_{\max}''$ is also commonly referred to as the *critical heat flux* or the *peak heat flux*.

***Transition Boiling*** A dramatic decrease in heat flux occurs in the region immediately beyond the nucleate boiling regime. In this transition boiling regime a vapor film is intermittently built up and partially destroyed, such that partial nucleate boiling continues to occur. However, operation in the transition boiling regime does not occur in practice unless it is possible to control the excess temperature $\Delta T_e$, such as in a tube heated by steam.

***Film Boiling*** A continuous stable vapor film covers the heating surface in the film boiling regime. The location $L$ at which the heat flux is a minimum is known as the *Leidenfrost point*. This point is named for J. G. Leidenfrost, who observed in 1756 that water globules on a very hot metal surface boil away very slowly as they "dance about." Because the vapor has a much lower thermal conductivity than the liquid, very large temperature differences occur across the vapor film. In this high-temperature region of the boiling curve the heat flux is seen to once again increase with increasing excess temperature. This increase in $q_c''$ can be credited to the large driving potential $T_s - T_{\text{sat}}$ and to the effects of thermal radiation. The surface temperature $T_s$ continues to increase with $q_c''$ until the melting point of the material is reached and *burnout* occurs.

Several regimes of pool boiling have been captured by camera for boiling on a copper rod immersed in a fluid as shown in Fig. 10–4. The fin is heated at the wall to the right, such that the excess temperature is a minimum at the left end of

**FIGURE 10–4** Several regimes of pool boiling of isopropanol on a copper fin. (Courtesy of Professor J. W. Westwater.)

the rod and increases as we move toward the wall. Both nucleate boiling regimes are seen to be active over the left one-third of the rod, with transition and film boiling occurring over the remainder of the rod.

The boiling curve shown in Fig. 10–3 can be produced by gradually increasing the surface temperature $T_s$. By slowly increasing $T_s$ we can reach and maintain steady-state boiling at any point along this curve. However, for the case in which the electrical power input $\dot{W}$ to the heater is slowly raised, stable operation between the points $C$ and $D$ cannot be maintained. For this situation, increasing the heat flux just beyond the maximum heat flux, point $C$, results in a rapid often destructive transition to film boiling at point $D$, where the excess temperature may be sufficiently high to cause meltdown of the boiling surface. This point is expanded upon in the following example.

## EXAMPLE 10–1

Utilize the lumped-analysis approach to explain the rapid unstable transition from point $C$ to point $D$ on the boiling curve for the case in which the electrical power input to the heater is gradually increased.

### Solution

*Objective* Explain the sudden unstable increase in $T_s$ for the case in which power $\dot{W}$ to a boiler element is gradually increased.

*Schematic* Boiling of liquid by electric heater.

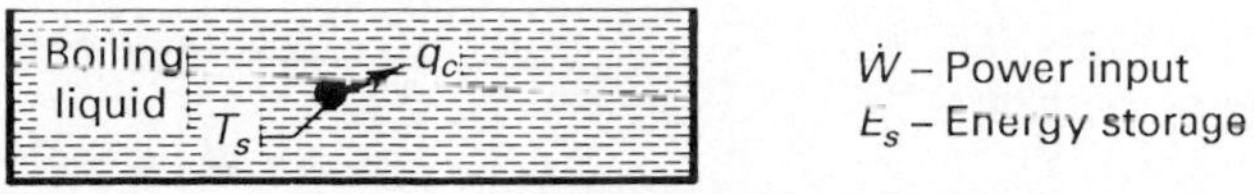
Boiling liquid
$q_c$
$T_s$
$\dot{W}$ – Power input
$E_s$ – Energy storage

*Assumptions/Conditions*

pool boiling

standard conditions, except for unsteady conditions, two-phase fluid, significance of buoyancy and significance of thermal radiation in film boiling regime

*Analysis* To aid in our understanding of this point, we write the lumped energy balance for a boiler element with small Biot number (i.e., negligible resistance to conduction) as follows:

$$\Sigma \dot{E}_i = \Sigma \dot{E}_o + \frac{\Delta E_s}{\Delta t} \qquad \dot{W} = q_c'' A_s + \rho V c_v \frac{dT_s}{dt}$$

Based on this equation, we conclude that the surface temperature $T_s$ will increase with increase in power $\dot{W}$ to the heater. The boiling heat flux $q_c''$ responds to any change in $T_s$ in accordance with the boiling curve. Putting this information together, in the natural convection and nucleate boiling regimes where $q_c''$ increases with increasing $T_s$, a small increase in $\dot{W}$ will produce an increase in $T_s$, which in turn will cause $q_c''$ to increase until a new steady-state surface temperature is reached and $q_c''A_s$ is equal to $\dot{W}$. However, if after reaching the peak heat flux at point $D$, we again increase $\dot{W}$ by a very small amount, an unstable sequence of events is set in motion in which the initial increase in $T_s$ produces a *decrease* in $q_c''$! This degradation in the heat flux $q_c''$ then stimulates the rate of increase in surface temperature $dT_s/dt$. As $T_s$ quickly increase with time at this fixed value of $\dot{W}$, $q_c''$ will change roughly in accordance with the boiling curve until burnout occurs or equilibrium is reached near point $D$. At this point, the rate of heat generation within the heater $\dot{W}$ is again in balance with the rate of heat transfer from the surface $q_c''A_s$. Because point $C$ on the boiling curve is associated with this sometimes destructive event, it is often referred to as the *burnout point*.

### *Convection Correlations for Pool Boiling*

Following the pattern established in the analysis of single-phase heat-transfer processes, the rate of heat transfer associated with the nucleate, transition, and film boiling regimes of pool boiling is given in terms of the *mean coefficient of boiling heat transfer* $h_B$ as†

$$q_c'' = \bar{h}_B (T_s - T_{sat}) \tag{10–8}$$

Note that it is the excess temperature $T_s - T_{sat}$ that is utilized in this defining equation. Unlike the situation encountered in forced convection in which $\bar{h}$ is often independent of $T_s - T_F$, $\bar{h}_B$ is always a function of $T_s - T_{sat}$. Consequently, some experimentalists report their pool boiling heat-transfer data directly in terms of $q_c''$ instead of $\bar{h}_B$.

† The standard coefficient of heat transfer $\bar{h}$ is generally utilized in the correlation of data in the natural convection regime.

Despite the fact that considerable controversy exists concerning the accuracy of available correlations for pool boiling, various design curves have been used to adequately, although perhaps not optimally, design boiling-heat-transfer equipment. Because of the burnout threat in many boiling-heat-transfer applications, the safety of the design comes first and the question of efficiency comes second.

Empirical design correlations for $q_c''$ or $\bar{h}_B$ have been developed for nucleate and film boiling regimes of pool boiling. Some data even exist for the transition boiling regime [4], but general correlations are not presently available for this zone. It should be noted that the geometrical orientation of the surface is relatively unimportant for pool boiling. Therefore, the same correlations are generally used for horizontal, vertical, or inclined surfaces.

Experimental curves for nucleate pool boiling of water on a 0.024-in.-diameter platinum wire are shown in Fig. 10–5 for various saturation pressures. The points at which the maximum heat flux occurs are also shown. The experimental evidence indicates that the heat flux is approximately proportional to $\Delta T_e^3$ throughout most of the nucleate boiling regime. The data have been correlated by Rohsenow [6], Fritz [7], and many others. The popular correlation by Rohsenow takes the form

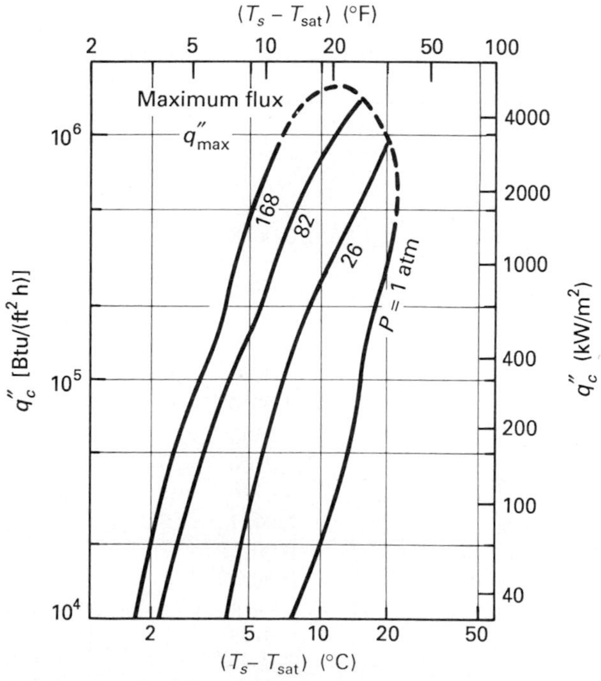


**FIGURE 10–5** Nucleate pool boiling of water at various pressures on electrically heated platinum wire (Addoms [5].

$$\frac{c_{P,f}(T_s - T_{\text{sat}})}{i_{fg}\, Pr_f^s} = C_{sf}\left[\frac{q_c''}{\mu_f i_{fg}}\sqrt{\frac{\sigma}{g(\rho_f - \rho_g)}}\right]^{1/3} \tag{10–9a}$$

or equivalently

$$q_c'' = \mu_f i_{fg}\left[\frac{g(\rho_f - \rho_g)}{\sigma}\right]^{1/2}\left[\frac{c_{P,f}(T_s - T_{\text{sat}})}{C_{sf} i_{fg}\, Pr_f^s}\right]^3 \tag{10–9b}$$

where $C_{sf}$ and $s$ are empirical constants. Table 10–1 gives values of $C_{sf}$ and $s$ for various clean surface-fluid combinations. For example, $C_{sf}$ is equal to 0.013 for a copper-water interface. The Prandtl number exponent $s$ is generally equal to unity for water and 1.7 for other fluids. The uncertainty associated with Rohsenow's correlation is about 25% for $\Delta T_e$, but is as large as 100% for $q_c''$. Thus, $\Delta T_e$ can be computed with far more confidence than $q_c''$.

Equation (10–9) can also be written in the dimensionless form

$$\overline{Nu_L} = \frac{Ja_f^2\, Bo_L^{1/2}\, Pr_f^{1-3s}}{C_{sf}} \tag{10–9c}$$

Due to the nature of this equation and the variations in $Pr_f$, $Ja_f$, and $Bo_L$ with saturation pressure, we conclude that the nucleate boiling heat flux can be enhanced by increasing $P_{\text{sat}}$ or $T_{\text{sat}}$.

**TABLE 10–1** Values of the empirical constants $C_{sf}$ and $s$ in Eq. (10–9) for various liquid-surface combinations

| *Liquid-surface combination* | $C_{sf}$ | $s$ |
|---|---|---|
| Water-copper | | |
| polished | 0.0130 | 1.0 |
| scored | 0.0068 | 1.0 |
| Water-stainless steel | | |
| mechanically polished | 0.0130 | 1.0 |
| ground and polished | 0.0060 | 1.0 |
| chemically etched | 0.0130 | 1.0 |
| Water-brass | 0.0060 | 1.0 |
| Water-nickel | 0.0060 | 1.0 |
| Water-platinum | 0.0130 | 1.0 |
| Benzene-chromium | 0.0100 | 1.7 |
| Carbon tetrachloride-copper | 0.0130 | 1.7 |
| Ethyl alcohol-chromium | 0.0027 | 1.7 |
| Isopropyl alcohol-copper | 0.00250 | 1.7 |
| *n*-Butyl alcohol-copper | 0.00300 | 1.7 |
| *n*-Pentane-copper | | |
| polished | 0.0154 | 1.7 |
| lapped | 0.0049 | 1.7 |
| *n*-Pentane-chromium | 0.015 | 1.7 |

*Source*: From references 5 and 8 through 11.

It should be emphasized that these and other correlations for nucleate pool boiling only apply to that part of the boiling curve for which $\Delta T_e$ is less than the maximum excess temperature $\Delta T_{e,\max}$. The maximum heat flux attainable with nucleate boiling is generally quite crucial to the designer because of the importance of avoiding the burnout dangers of the film boiling regime, particularly in high-performance constant-heat flux systems. The maximum flux for saturated nucleate boiling can be approximated by relations developed by Zuber [12], Rohsenow and Griffith [13], Kutateladze [14], Lienhard and Dhir [15], and others. The equation developed by Zuber on the basis of a hydrodynamic stability analysis is written as

$$q''_{\max} = \frac{\pi}{24}\rho_g i_{fg}\left[\frac{\sigma g(\rho_f - \rho_g)}{\rho_g^2}\right]^{1/4}\left(\frac{\rho_f + \rho_g}{\rho_f}\right)^{1/2} \tag{10–10}$$

Since $\rho_g$ is generally much smaller than $\rho_f$, this equation is commonly approximated by

$$q''_{\max} = \frac{\pi}{24}\rho_g i_{fg}\left[\frac{\sigma g(\rho_f - \rho_g)}{\rho_g^2}\right]^{1/4} \tag{10–11}$$

Although these equations can be utilized as a first approximation for most fluid-surface combinations with relatively large characteristic length, a correction factor must be applied for applications in which the characteristic length of the surface is small relative to the mean bubble diameter [15,16].

According to Eq. (10–11), $q''_{\max}$ can be increased by increasing the gravitational field or by selecting a fluid with a large value of $i_{fg}$. For example, in the selection of a fluid, $i_{fg}$ is higher for water than other common liquids, such that water will sustain the highest maximum heat flux. Further, because the fluid properties are functions of pressure, $q''_{\max}$ can be increased to a maximum for any given fluid by optimizing the pressure. The variation of $q''_{\max}$ with pressure is indicated in Fig. 10–5 for water. The optimum pressure for water is about 100 atm and the maximum heat flux is approximately $3.8 \times 10^6$ W/m$^2$. As an aside, with Eq. (10–11) expressed in dimensionless form (see Table 10–3), we note that $Nu_{L,\max}\, Ja_f = q''_{\max} L c_{P,f}/(i_{fg} k_f)$ is proportional to $Gr_{L,f}^{1/2} Pr_f / Bo_L^{1/4}$.

Zuber [12] has also proposed an equation for the *minimum heat flux* $q''_{\min}$ associated with the Leidenfrost point, which is given by

$$q''_{\min} = 0.09\rho_g i_{fg}\left[\frac{\sigma g(\rho_f - \rho_g)}{(\rho_f + \rho_g)^2}\right]^{1/4} \tag{10–12}$$

The uncertainty of this relation is about 50% for fluids at moderate pressure, but is larger at high pressures. According to this correlation, $q''_{\min} L c_{P,f}/(i_{fg} k_f)$ is proportional to $Gr_{L,f}^{1/2} Pr_f / Bo_L^{1/4}$.

Correlations for film boiling on the surface of a horizontal tube or sphere of diameter $D$ have been developed that take the form

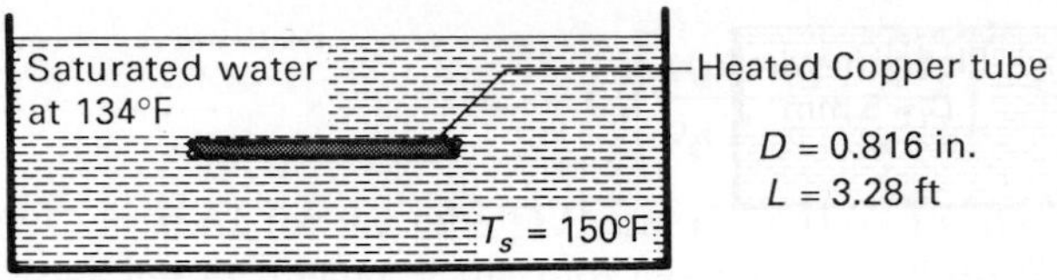
Saturated water
at 134°F
Heated Copper tube
D = 0.816 in.
L = 3.28 ft
T_s = 150°F

*Assumptions/Conditions*

pool boiling
standard conditions, except for two-phase fluid and significance of buoyancy

*Properties* Water at $T = 134°\text{F} = 330$ K (Table A–C–3): $Pr_f = 3.15$,

$$\rho_f = \frac{1}{0.00102}\,\frac{\text{kg}}{\text{m}^3}\,\frac{0.0624\ \text{lb}_\text{m}/\text{ft}^3}{\text{kg/m}^3} = 61.2\ \text{lb}_\text{m}/\text{ft}^3$$

$$\rho_g = \frac{1}{8.82}\,\frac{\text{kg}}{\text{m}^3}\,\frac{0.0624\ \text{lb}_\text{m}/\text{ft}^3}{\text{kg/m}^3} = 0.00707\ \text{lb}_\text{m}/\text{ft}^3$$

$$i_{fg} = 2370\,\frac{\text{kJ}}{\text{kg}}\,\frac{0.4299\ \text{Btu/lb}_\text{m}}{\text{kJ/kg}} = 1020\ \text{Btu/lb}_\text{m}$$

$$c_{P,f} = 4.18\,\frac{\text{kJ}}{\text{kg K}}\,\frac{0.2388\ \text{Btu/(lb}_\text{m}\ °\text{F)}}{\text{kJ/(kg K)}} = 0.998\ \text{Btu/(lb}_\text{m}\ °\text{F)}$$

$$\mu_f = 4.89 \times 10^{-4}\,\frac{\text{kg}}{\text{m s}}\,\frac{0.6720\ \text{lb}_\text{m}/\text{(ft s)}}{\text{kg/(m s)}} = 3.29 \times 10^{-4}\ \text{lb}_\text{m}/\text{(ft s)}$$

$$\sigma = 6.67 \times 10^{-2}\,\frac{\text{N}}{\text{m}}\left(0.2248\,\frac{\text{lb}_\text{f}}{\text{N}}\right)\left(\frac{\text{m}}{3.281\ \text{ft}}\right) = 4.56 \times 10^{-3}\ \text{lb}_\text{f}/\text{ft}$$

*Analysis* The excess temperature is

$$T_s - T_\text{sat} = 150°\text{F} - 134°\text{F} = 16°\text{F}$$

Utilizing the Rohsenow correlation, we calculate the heat flux as follows:

$$q''_c = \mu_f i_{fg}\left[\frac{g(\rho_f - \rho_g)}{\sigma}\right]^{1/2}\left[\frac{c_{P,f}(T_s - T_\text{sat})}{i_{fg}\,Pr_f^s\,C_{sf}}\right]^3$$

$$= 3.29 \times 10^{-4}\,\frac{\text{lb}_\text{m}}{\text{ft s}}\left(1020\,\frac{\text{Btu}}{\text{lb}_\text{m}}\right)\left[\frac{(32.2\ \text{ft/s}^2)(61.2\ \text{lb}_\text{m}/\text{ft}^3)}{4.56 \times 10^{-3}\ \text{lb}_\text{f}/\text{ft}}\right]^{1/2}$$

$$\times \left(\frac{\text{lb}_\text{f}\ \text{s}^2}{32.2\ \text{ft lb}_\text{m}}\right)^{1/2}\left\{\frac{[0.998\ \text{Btu/(lb}_\text{m}\ °\text{F)}](16°\text{F})}{(1020\ \text{Btu/lb}_\text{m})(3.15)(0.013)}\right\}^3$$

$$= 2.17\ \text{Btu/(s ft}^2) = 7820\ \text{Btu/(h ft}^2) = 24.6\ \text{kW/m}^2$$

Thus, the total rate of heat transfer is

$$q_c = q_c''\pi DL = 2.17\,\frac{\text{Btu}}{\text{s ft}^2}\,\pi\left(\frac{0.816\text{ ft}}{12}\right)(3.28\text{ ft}) = 1.52\text{ Btu/s} = 1.60\text{ kW}$$

To calculate the rate of evaporation $\dot{m}$, we write

$$\dot{m} = \frac{q_c}{i_{fg}} = \frac{1.52\text{ Btu/s}}{1020\text{ Btu/lb}_m} = 1.49 \times 10^{-3}\text{ lb}_m\text{/s}$$

### *Enhancement of Pool Boiling*

The heat-transfer correlations given in the previous section apply to smooth surfaces. Many methods have been utilized for producing improvements in pool boiling heat transfer. Perhaps the most common approaches involve the use of rough surfaces and extended surfaces. Several other methods include surface treatment, liquid additives, and mechanical agitation or vibration. A survey of various methods of heat-transfer enhancement is provided by Webb [18]. The main idea in using rough or extended surfaces is to increase the number of nucleation sites. The effect of enhancement techniques such as these on the boiling heat flux is shown in Fig. 10–6. The Thermoexcel-E surface shown in Fig. 10–7 is representative of machined rough surfaces that are available in the marketplace. The use of properly designed fins or rough surfaces such as Thermoexcel-E has also been reported to substantially increase the critical heat flux and delay the onset of transition and film boiling [18,19].

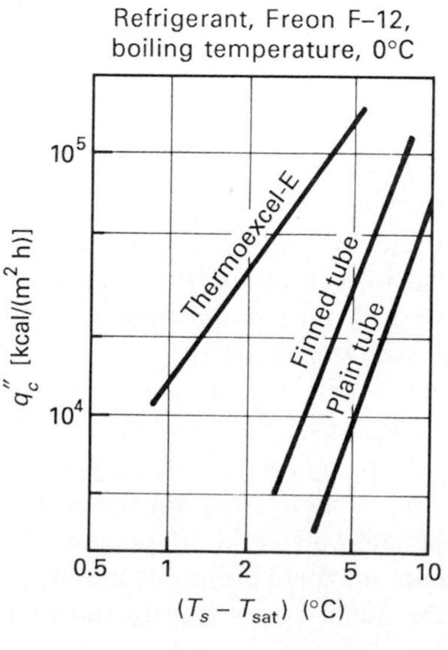


**FIGURE 10–6**
Nucleate pool boiling heat-transfer performance of several surfaces (Arai et al. [19]}.

(a) Plan view.

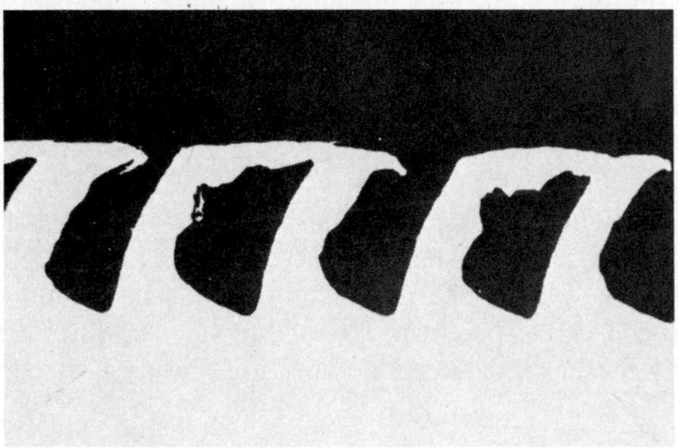

(b) Cross section.

**FIGURE 10–7**
Thermoexcel-E surface for enhanced boiling. (Courtesy of Hitachi, Ltd.)

## 10–3–2 Flow Boiling

### *Physical Description of Flow Boiling*

Flow boiling occurs in internal and external flow forced and natural convection systems. External flow boiling over a cylinder or plate is much like pool boiling because the free-stream temperature and quality are essentially constant with respect to the flow direction. But internal flow boiling is more complex because of changes in the bulk-stream temperature and quality with respect to $x$.

In internal flow boiling processes, liquid (or a liquid-vapor mixture) enters a heating chamber such as the tube shown in Fig. 10–8. Beyond the axial location at which the boiling process is initiated (point a), the quality of the liquid-vapor mixture increases with distance until total vaporization occurs. Thus, for the case in which a subcooled liquid enters the heating chamber, the bulk-fluid temperature increases to a value somewhat less than $T_{sat}$, at which point nucleate boiling is initiated. The bulk temperature then quickly reaches $T_{sat}$ and the quality goes from zero to unity. After

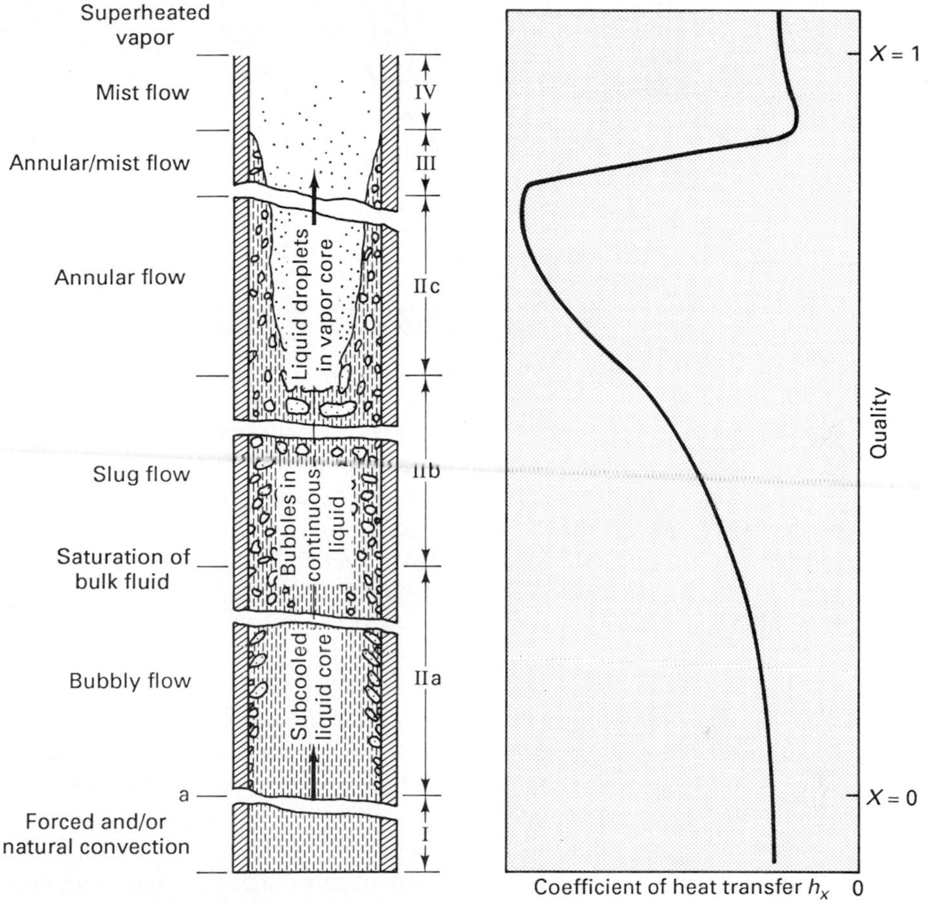


(a) Representative flow patterns. (From Tong [20]. Used with permission.)

(b) Flow boiling curve represented in terms of $h_x$ vs quality $X$.

**FIGURE 10–8** Flow boiling in vertical channel.

all the liquid has been vaporized, the entire bulk fluid becomes superheated. Strictly speaking, $T_{sat}$ varies slightly with $x$ because of pressure losses within the chamber. However, the assumption that $T_{sat}$ is essentially constant is generally quite adequate.

### *Boiling Curve for Flow Boiling*

A representative boiling curve for internal flow boiling in heated tubes or channels is shown in Fig. 10–9. With the excess temperature $\Delta T_e$ represented by $T_s - T_b$ for internal flow and by $T_s - T_{sat}$ for external flow, the boiling curves for flow boiling and pool boiling are of the same general form. Points on the flow boiling curve

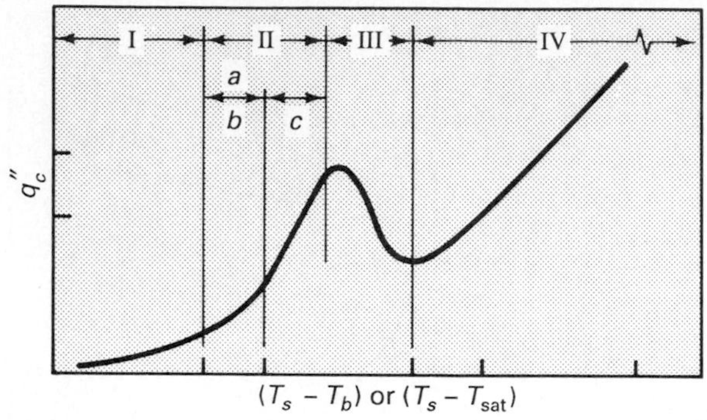


**FIGURE 10–9**
Representative boiling curve for internal flow boiling.

represent the excess temperature $\Delta T_e$ at various axial locations $x$ on the surface versus the local heat flux $q_c''$. The actual numerical values of $q_c''$ and the temperature difference are a function of pressure, geometry, mean velocity, thermal boundary conditions, and degree of subcooling. In this connection, flow boiling curves are also commonly presented in terms of the local coefficient of heat transfer $h_x$ and quality $X$, as illustrated in Fig. 10–8. It should be noted that the local coefficient of heat transfer $h_x$ is defined in the traditional sense; that is,

$$dq_c = h_x\, dA_s\, (T_s - T_F) \tag{10–19}$$

where $T_F = T_b$ for internal flow and $T_F = T_\infty$ for external flow.

The same basic regimes found in pool boiling also occur in flow boiling: forced and/or natural convection (I); nucleate boiling (II); transition boiling (III); and film boiling (IV). However, for internal flow boiling processes, because the fluid is totally confined within the tube or channel with no free interface from which vapor can escape, rather distinctive two-phase flow patterns are associated with each of these regimes. Take, for example, the situation in which a subcooled liquid enters a heated vertical tube in which $T_s$ is greater than $T_{sat}$. Forced and/or natural convection occurs in region I where $T_s - T_b$ is small. But, as in the case of pool boiling, no actual boiling takes place in this zone. When the temperature difference $T_s - T_b$ is sufficiently large to produce regime II nucleate boiling, three basic flow patterns are produced.

First, bubbles form at nucleation sites but do not carry far into the main stream because the local bulk temperature $T_b$ is still below $T_{sat}$. However, as $T_b$ approaches $T_{sat}$, more and more bubbles appear in the bulk stream. Because of the presence of discrete bubbles in the flow stream, regime IIa is sometimes called the *bubbly flow regime*. The bubbles that occur throughout the bulk stream then begin to coalesce into slugs of vapor. Although the quality is quite low in region IIb, the percent of the volume that is occupied by vapor is as large as 50%. Consequently, the bulk flow rate increases. This portion of the nucleate boiling zone is sometimes called the *slug flow regime*. In the third part of the nucleate boiling region, the tube wall is covered by a thin liquid film with vapor and liquid droplets flowing in the center of the tube. Although nucleate boiling occurs in this region, vapor is also believed to be generated by vaporization at the liquid-vapor interface. Large increases in the heat flux continue to occur in this zone, until the heat flux reaches a peak, which is commonly referred to as the *departure from nucleate boiling* (DNB). The coefficients of heat transfer in the vicinity of the DNB can be 20 to 50 times as large as the coefficient for single-phase flow. This part of the nucleate boiling regime (IIc) is also known as the *annular flow regime*.

As in the case of pool boiling, the heat flux in flow boiling processes experiences a drastic decrease in regime III. In this transition boiling region, a wall-drying process occurs in which a vapor film with large thermal resistance gradually develops on the surface. Region III is also sometimes known as the *annular/mist regime*.

In the *film boiling regime*, the wall is completely dried out with liquid droplets being dispersed throughout the bulk of the fluid. Heat is transferred from the wall to the vapor and then from the vapor to the liquid until complete vaporization occurs. Regime IV is also known as the *mist flow regime*.

As in the case of pool boiling, the coefficients of heat transfer are quite high in the nucleate flow boiling regimes. Hence, the wall temperature is usually not too much larger than the bulk-fluid temperature or saturated temperature and the walls are not in danger of meltdown or burnout. However, because of the greatly reduced coefficients of heat transfer in the regimes III and IV, these regions are avoided in certain applications. For example, fossil-fuel-fired boilers or nuclear reactors are usually operated in the nucleate boiling regimes to avoid the onset of DNB. The coefficient of heat transfer attained in these systems is sufficiently high to permit satisfactory operation as long as fouling is prevented by proper water treatment. On the other hand, steam generators for pressurized-water-reactor systems (i.e., water-to-boiling-water heat exchangers) and certain types of process heat exchangers are sometimes operated in regimes III and IV. In these cases, the temperature of the heat source is within the safe operating range of the equipment. In once-through boilers in which superheated steam is produced, the wall-heat flux is generally reduced in the section of the heating chamber where regimes III and IV occur.

The effect of fluid enthalpy (or quality) and heat flux on the wall temperature $T_s$ and bulk-fluid temperature $T_b$ and on the location of the DNB point are indicated in Fig. 10–10 for an upward flow of water in a vertical uniformly heated tube. Notice that $T_b$ increases until the fluid reaches the saturation temperature. $T_b$ then drops slightly in the region where steam is being generated because of a small pressure drop. In the single-phase superheat region, $T_b$ once again increases. Except for the low-heat-flux case, the behavior in surface temperature $T_s$ is quite another matter. For moderate heat flux, $T_s$ remains somewhat above $T_b$ until the DNB point associated with regime III is reached in the region of high steam quality, and a degradation in heat transfer occurs. In this nucleate/film boiling regime, $T_s$ experiences a rather pronounced increase to a mild peak, and then decreases as the steam quality approaches 100%. $T_s$ once again gradually increases in the superheat region. For higher heat fluxes, the DNB point and regime III boiling is reached at lower steam quality and the peak in $T_s$ is higher. As the heat flux increases, the peak in $T_s$ increases and moves in the direction of lower and lower bulk-stream enthalpy, until the metal melts or the tube or channel ruptures.

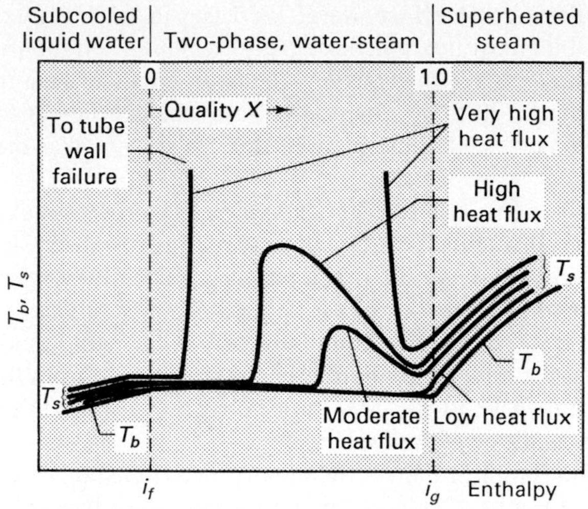


**FIGURE 10–10** Bulk fluid and tube-wall temperature for flow boiling of water. (From *Steam—Its Generation and Use.* Copyright 1978 by Babcock and Wilcox. Used with permission.)

## EXAMPLE 10–4

Sketch a boiling curve for external forced convection flow boiling on an electrically heated wire.

### Solution

*Objective* Show the effect of $U_\infty$ for flow boiling on an electrically heated wire.

*Schematic* Flow boiling on an electrically heated wire.

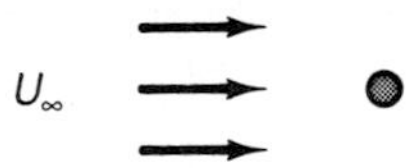


*Assumptions/Conditions*

flow boiling

standard conditions, except for two-phase fluid and significance of buoyancy

*Analysis* For cases in which the free-stream velocity is zero, natural convection and pool boiling control. Therefore, we start with the pool boiling curve shown in Fig. E10–4a. As $U_\infty$ increases, the coefficient of heat transfer in the single-phase natural convection region increases. This effect would be expected to carry over to the nucleate boiling regimes. Thus, boiling curves are sketched in for low, moderate, and high values of $U_\infty$, which should account for the qualitative effects of forced convection.

To reinforce this result, experimental data reported by Lung et al. [21] are shown in Fig. E10–4b. Our external flow boiling curves are in general agreement

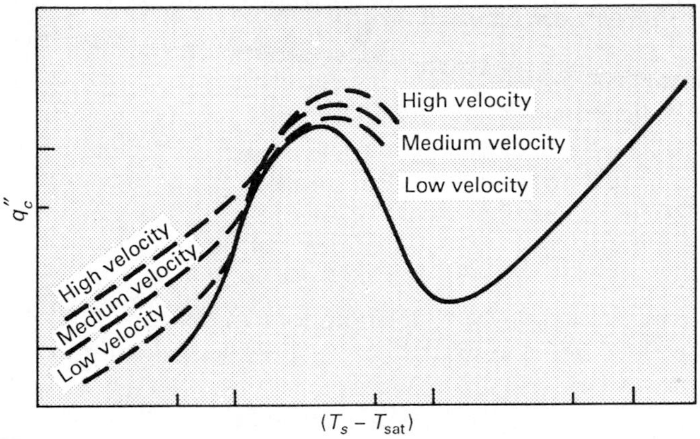


**FIGURE E10–4a** Boiling curve for external flow boiling.

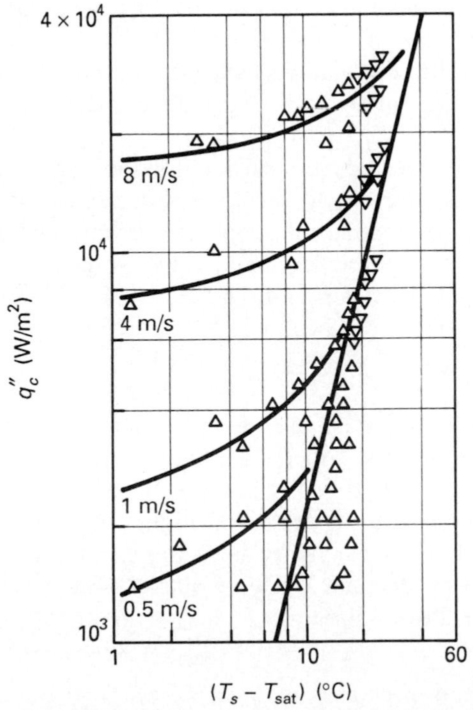


**FIGURE E10–4b**
Experimental data for flow boiling.

with these data. Note that the heat-transfer process is essentially controlled by boiling for larger values of $T_s - T_{sat}$ and by forced convection for smaller values. Experimental and theoretical studies also indicate that the peak heat flux is markedly increased for forced convection.

### *Convection Correlations for Flow Boiling*

Some success has been achieved in developing correlations for the coefficient of heat transfer and the peak heat flux for flow boiling processes. In particular, fairly simple correlations are available for nucleate boiling, in which case the total heat flux $q''_c$ is artificially broken into boiling- and nonboiling-convection components. This idea is reflected in the following equation:

$$q''_c = q''_c|_{\text{boiling}} + q''_c|_{\text{nonboiling}} \tag{10–20}$$

where $q''_c|_{\text{boiling}}$ is approximated by the correlations for two-phase nucleate pool boiling and $q''_c|_{\text{nonboiling}}$ is obtained from the standard correlations for single-phase convection. In regimes IIb and IIc, where the heat-transfer process is essentially governed by the boiling mechanism, Eq. (10–20) can be approximated by

$$q''_c = q''_c|_{\text{boiling}} \tag{10–21}$$

In this region, the heat-transfer flux is essentially independent of the velocity and forced or natural convection effects. These assumptions are consistent with the data shown in Example 10–4.

Empirical burnout or peak heat-flux correlations have been developed for various flow boiling systems. A comprehensive survey of the work done in this area has been presented by Hewitt [22].

Empirical correlations have also been developed for flow boiling heat transfer associated with regimes III and IV. However, the quantitative treatment of this two-phase flow subject is complex and lies beyond the scope of our introductory study. Several references on the subject of two-phase flow include the works by Tong [20], Hsu and Graham [23], Collier [24], and Butterworth and Hewitt [25].

### *Enhancement of Flow Boiling*

Techniques similar to those used in pool boiling are used for improving flow boiling characteristics [18]. One of the more successful methods involves the use of internally ribbed tubes such as the one shown in Fig. 10–11. This type tube has been reported to suppress the onset of DNB and to permit operation at higher heat fluxes than is possible with smooth tubes [25,26]. However, because of the expense of manufacturing internally ribbed tubes, the Babcock and Wilcox Company recommends their use only in high-pressure systems (above 150 atm).

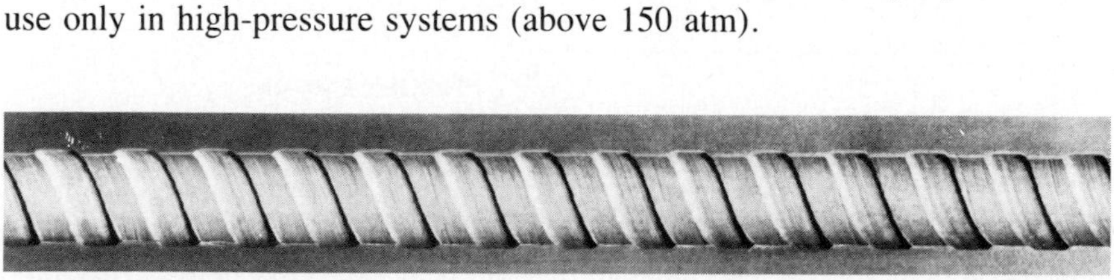

**FIGURE 10–11** Single-lead ribbed tube. (From *Steam—Its Generation and Use.* Copyright 1978 by Babcock and Wilcox. Used with permission.)

## 10–4 CONDENSATION HEAT TRANSFER

*Condensation heat transfer* occurs when a vapor is exposed to a surface with temperature $T_s$ less than the saturation temperature $T_{sat}$. The condensate that is formed flows down the surface under the influence of gravity. If the liquid does not wet the surface, the process is called *dropwise condensation*. Under these conditions, droplets form on the surface as shown in Fig. 10–12. These drops generally flow over the surface along random paths. For those liquids that wet the surface, the condensate forms a film that builds up as it flows down the surface. This *film condensation* process is illustrated in Fig. 10–13. Because of the low velocities associated with the film for laminar flow conditions, the energy transfer across the film is primarily due to conduction, such that the temperature distribution is essentially linear. The existence of such a film, whether in laminar or turbulent flow, increases the resistance to heat transfer, with the result that dropwise condensation is as much as 10 times more effective than film condensation. However, most surfaces become wetted when exposed to a condensing vapor over an extended length of time. Consequently, film condensation is generally encountered in industrial applications and is usually planned for in design work. In this connection, the rate of heat transfer is generally lower when a vapor exists in the presence of a noncondensable gas. Therefore, noncondensable gases are sometimes vented to improve the efficiency of the condensation-heat-transfer process.

Although our attention will be restricted to the surface condensation processes mentioned above, it should be noted that condensation can also occur in the absence of solid boundaries. Examples of these other types of condensation include the formation of fog and the exposure of a vapor to a cool liquid.

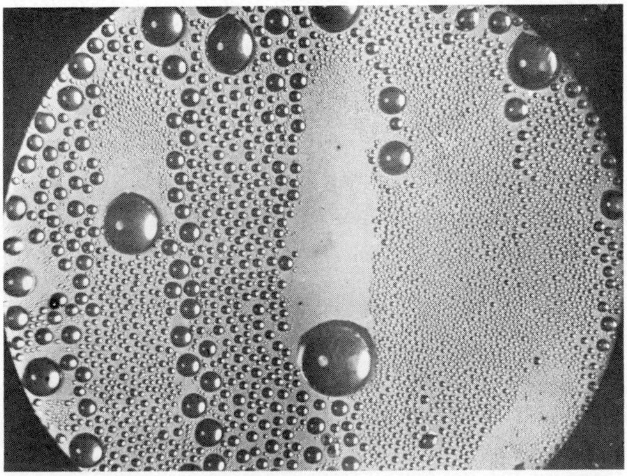

**FIGURE 10–12** Dropwise condensation of steam under ideal conditions. (Courtesy of Professor M. N. Ozisik.)

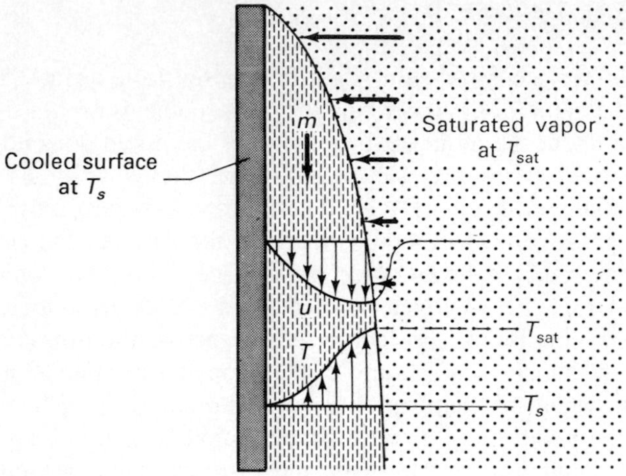
Cooled surface
at $T_s$
$\dot{m}$
Saturated vapor
at $T_{sat}$
$u$
$T$
$T_{sat}$
$T_s$

**FIGURE 10–13** Film of condensate on vertical surface.

### 10–4–1 Film Condensation

According to the work of Shafrin and Zisman [27], a liquid will wet the surface if the surface tension $\sigma$ of the fluid is less than a *critical surface tension* $\sigma_{cr}$, which is characteristic of the surface alone. Aside from certain fluid-surface combinations, such as those involving mercury ($\sigma_{Hg} = 0.47$ N/m at 100°C) or special materials such as Teflon ($\sigma_{cr} = 0.018$ N/m), $\sigma$ is usually less than $\sigma_{cr}$, such that film condensation is most common.

We now turn to the practical analysis of film condensation, which involves the use of a liquid film Reynolds number $Re_f$ and convection correlations for various standard geometries.

#### *Reynolds Number for Film Condensation*

As in single-phase forced-convection processes, the heat transfer in film condensation is dependent upon whether the flow within the film is laminar or turbulent. To provide a criterion by which to characterize the flow, a Reynolds number $Re_f$ for condensate liquid film flow has been defined, which takes the form

$$Re_f = \frac{D_H U_b}{\nu_f} = \frac{4A}{p} \frac{U_b}{\nu_f} \tag{10–22}$$

where $U_b$ is the bulk-velocity flow rate of condensate and $A = p\delta_f$; $\delta_f$ is the local thickness of the film and $p$ is the perimeter. Based on the principle of continuity, $U_b$ can be expressed in terms of the condensate mass flow rate $\dot{m}$ as

$$\dot{m} = \rho_f A U_b \tag{10–23}$$

Thus, $Re_f$ can be written as

$$Re_f = \frac{4\dot{m}}{p\mu_f} \tag{10–24}$$

Notice that the value of $Re_f$ increases as the film grows. For vertical tubes and plates, the critical Reynolds number is approximately 1800.

The properties of the liquid film are generally evaluated at the film temperature $(T_s + T_{sat})/2$.

### Convection Correlations

Consistent with the traditional approach to correlating heat transfer data for single-phase convection processes, the local rate of heat transfer for film condensation is generally expressed in terms of a *coefficient of condensation heat transfer* $h_C$.

$$dq_c = h_C\, dA_s\,(T_{sat} - T_s) \tag{10–25}$$

For a uniform wall-temperature boundary condition, the total rate of heat transfer over the entire surface is given by

$$q_c = \overline{h_C} A_s (T_{sat} - T_s) \tag{10–26}$$

where the mean coefficient of condensation heat transfer is defined by

$$\overline{h_C} = \frac{1}{L}\int_0^L h_C\, dx \tag{10–27}$$

A useful relationship between $\overline{h_C}$ and $Re_f$ can be developed by equating the rate of convection heat transfer from the film to the wall $q_c$, to the rate of energy transfer associated with phase change $\dot{m}i_{fg}$, and sensible heat, which is proportional to $\dot{m}c_{P,f}\,(T_{sat} - T_s)$ (see Appendix J); that is,

$$q_c = \dot{m}i_{fg} + C_1\dot{m}c_{P,f}(T_{sat} - T_s) = \dot{m}\,[i_{fg}(1 + C_1\, Ja_f)] = \dot{m}i'_{fg} \tag{10–28}$$

where $i'_{fg}$ is a modified latent heat similar in form to Eq. (10–14),

$$i'_{fg} = i_{fg}(1 + C_1\, Ja_f) \tag{10–29}$$

and $C_1 = 0.68$ according to the analysis of Rohsenow [28]. It follows that

$$\overline{h_C} = \frac{q_c}{A_s(T_{sat} - T_s)} = \frac{\dot{m}i'_{fg}}{A_s(T_{sat} - T_s)} \tag{10–30}$$

and

$$Re_f = \frac{4\dot{m}}{p\mu_f} = \frac{4q_c}{p\mu_f i'_{fg}} = \frac{4\overline{h_C}A_s(T_{sat} - T_s)}{p\mu_f i'_{fg}} \tag{10–31}$$

Thus, with $\overline{h_C}$ known on the basis of empirical or theoretical correlations, $Re_f$ can be calculated. Several design correlations are now presented for the coefficient of condensation heat transfer associated with external film condensation. Because film condensation involves wetted surfaces with no liquid drop effects, the surface tension $\sigma$ does not appear in the correlations. Referring to the dimensional analysis result introduced in Sec. 10–2, we should expect correlations involving the four dimensionless parameters $Nu_L$, $Pr$, $Gr_L$, and $Ja$.

For information on the more complex topic of film condensation inside tubes and channels, the review by Collier [24] is recommended.

***Vertical Surface*** Referring to Appendix J, Nusselt [29] and Rohsenow [28] developed a theoretical expression for laminar film condensation on an isothermal surface, which can be represented by

$$h_C = 0.707 \left[ \frac{\rho_f g(\rho_f - \rho_g) k_f^3 i'_{fg}}{\mu_f (T_{\text{sat}} - T_s) x} \right]^{1/4} \tag{10–32}$$

This relation has been shown to be applicable for $Pr \gtrsim 0.5$ and $Ja_f \lesssim 1.0$, with high accuracy for $Re_f \lesssim 30$ and with errors within 20% for $Re_f \lesssim 1800$. Based on this famous result for the local coefficient of condensation heat transfer, an expression can be written for the mean coefficient $\overline{h_C}$ of the form

$$\overline{h_C} = 0.943 \left[ \frac{\rho_f g(\rho_f - \rho_g) k_f^3 i'_{fg}}{\mu_f (T_{\text{sat}} - T_s) L} \right]^{1/4} \tag{10–33a}$$

or

$$\overline{Nu_L} = 0.943 \left[ Gr_{L,f} Pr_f \left( \frac{1}{Ja_f} + 0.68 \right) \right]^{1/4} \tag{10–33b}$$

By coupling Eqs. (10–31) and (10–33), the relation for $Re_f$ associated with laminar flow becomes

$$Re_f = 3.77 \left[ \frac{L(T_{\text{sat}} - T_s)}{\mu_f i'_{fg}} \right]^{3/4} \left[ \frac{\rho_f g(\rho_f - \rho_g) k_f^3}{\mu_f^2} \right]^{1/4} \tag{10–34a}$$

or

$$Re_f = 3.77 \left[ Gr_{L,f}^{1/4} Pr_f^{-3/4} \left( \frac{1}{Ja_f} + 0.68 \right)^{-3/4} \right] \tag{10–34b}$$

This equation in either form can be used to calculate $Re_f$. If $Re_f$ is less than 1800, $h_C$ and $\overline{h_C}$ can then be calculated by the use of Eqs. (10–32) and (10–33). Taking this one step further, we couple Eqs. (10–33) and (10–34) to obtain

$$\overline{h_C} = 1.47\, Re_f^{-1/3} \left[ \frac{\rho_f g(\rho_f - \rho_g) k_f^3}{\mu_f^2} \right]^{1/3} \tag{10–35a}$$

or

$$\overline{Nu_L} = 1.47\, Re_f^{-1/3}\, Gr_{L,f}^{1/3} \tag{10–35b}$$

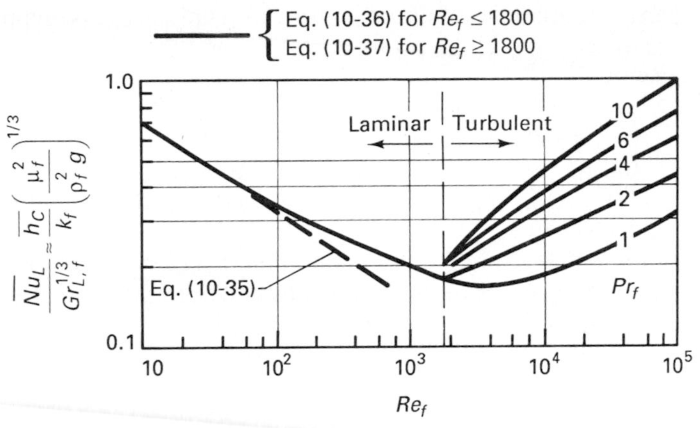


**FIGURE 10–14** Correlations for film condensation heat transfer on a vertical surface.

This expression is shown in Fig. 10–14. It should be noted that since $\rho_f >> \rho_g$, the ordinate in this figure is generally expressed in terms of $\overline{h_C}/k_f\,[\mu_f^2/(\rho_f^2 g)]^{1/3}$, which is essentially equivalent to the more exact term $\overline{Nu_L}/Gr_{L,f}^{1/3}$.

For improved accuracy, which accounts for the effects of waves that form on the condensate film in the range $30 \lesssim Re_f \lesssim 1800$ and for turbulent flow conditions that occur for $Re_f \gtrsim 1800$, the following correlations have been recommended for vertical surfaces:†

$$\overline{h_C} = \frac{Re_f}{1.08\,Re_f^{1.22} - 5.2}\left[\frac{\rho_f g(\rho_f - \rho_g)k_f^3}{\mu_f^2}\right]^{1/3} \tag{10–36a}$$

or

$$\overline{Nu_L} = \frac{Re_f}{1.08\,Re_f^{1.22} - 5.2}\,Gr_{L,f}^{1/3} \tag{1036b}$$

for $30 \lesssim Re_f \lesssim 1800$ by Kutateladze [30], and

$$\overline{h_C} = \frac{Re_f}{8750 + 58\,Pr_f^{-0.5}\,(Re_f^{0.75} - 253)}\left[\frac{\rho_f g(\rho_f - \rho_g)k_f^3}{\mu_f^2}\right]^{1/3} \tag{10–37a}$$

or

$$\overline{Nu_L} = \frac{Re_f}{8750 + 58\,Pr_f^{-0.5}\,(Re_f^{0.75} - 253)}\,Gr_{L,f}^{1/3} \tag{10–37b}$$

for $Re_f \gtrsim 1800$ by Labuntsov [31]. These correlations are also shown in Fig. 10–14.

Equations (10–32) through (10–36) can also be used for film condensation on the outside of vertical tubes if the tube diameter is large compared to the film thickness $\delta$. It should also be noted that because $\rho_f$ is generally so much larger than $\rho_g$, the

† See Table 10–3 for explicit relations for $Re_f$.

term $\rho_f - \rho_g$ appearing in condensation-heat-transfer correlations can be approximated by $\rho_f$ for many situations.

***Inclined Surface*** For plates that are inclined by an angle $\theta$ with the vertical, the equations above can be utilized as a first approximation for film condensation on the upper surface by replacing the gravitational acceleration by its component parallel to the surface, $g \cos \theta$. However, these equations cannot be extended to inclined tubes.

***Horizontal Tube or Sphere*** For laminar film condensation on the outside of a horizontal tube or sphere, $\overline{h_C}$ can be approximated by correlations of the same general form as Eq. (10–33); that is,

$$\overline{h_C} = C\left[\frac{\rho_f g(\rho_f - \rho_g)k_f^3 i'_{fg}}{\mu_f(T_{\text{sat}} - T_s)D}\right]^{1/4} \tag{10–38a}$$

or

$$\overline{Nu_D} = C\left[Gr_{D,f}Pr_f\left(\frac{1}{Ja_f} + 0.68\right)\right]^{1/4} \tag{10–38b}$$

where $C = 0.729$ for the cylinder and $C = 0.815$ for the sphere [16].

For the case of condensation on the outside of a bank of horizontal tubes, the heat flux to the lower tubes is reduced due to condensate falling from above tubes. A conservative estimate can be made for laminar condensation on a vertical row of $N$ tubes by simply replacing $D$ in the correlation for a single tube by $ND$.

### EXAMPLE 10–5

Based on the fact that the temperature distribution within a laminar condensed liquid film on a vertical surface is essentially linear, develop an approximate expression for the thickness of the film in terms of $h_C$.

#### Solution

*Objective* Develop an approximation for $\delta_f$ in terms of $h_C$.

*Schematic* Condensed liquid film.

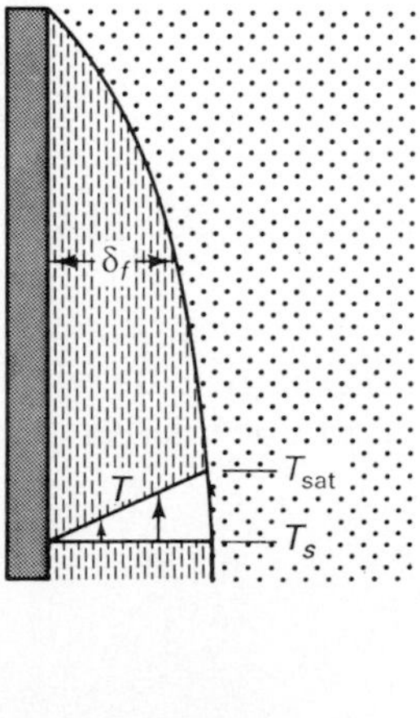

*Assumptions/Conditions*

condensation
wetted surface
laminar flow
uniform properties within liquid phase
standard conditions, except for two-phase fluid and significance of buoyancy

*Analysis* The condensate film thickness $\delta_f$ can be estimated by performing a simple thermal analysis. Assuming as a first approximation that the temperature distribution across the film is linear, we write

$$dq_c = dq_y|_0$$

$$h_C\, dA_s\,(T_s - T_{\text{sat}}) = -k_f\, dA_s \left.\frac{\partial T}{\partial y}\right|_0 = -k_f\, dA_s \frac{T_{\text{sat}} - T_s}{\delta_f}$$

or

$$\delta_f = \frac{k_f}{h_C}$$

Thus, we can approximate $\delta_f$ once relations are available for $h_C$.

**EXAMPLE 10-6**

Steam at a saturation temperature of 57°C condenses on the surface of a 1-m-long 2-cm-diameter vertical tube with a surface temperature of 48°C. Assuming film condensation, determine the total rate of heat transfer and the total mass flow rate of condensate over the entire tube length.

**Solution**

*Objective* Determine $q_c$ and $\dot{m}$.

*Schematic* Film condensation on a vertical tube.

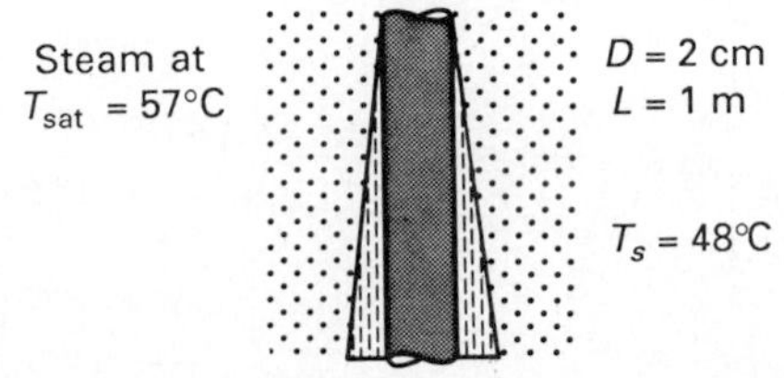

Steam at
$T_{sat}$ = 57°C
$D$ = 2 cm
$L$ = 1 m
$T_s$ = 48°C

*Assumptions/Conditions*

condensation
wetted surface
uniform properties within each phase
standard conditions, except for two-phase fluid and significance of buoyancy
$\rho_g \ll \rho_f$

*Properties* Water at a film temperature of 325 K (Table A–C–3): $\rho_f = 1/v_f = 987$ kg/m$^3$, $i_{fg} = 2380$ kJ/kg, $k_f = 0.645$ W/(m °C), $c_{P,f} = 4.18$ kJ/(kg °C), $\mu_f = 5.28 \times 10^{-4}$ kg/(m s).

*Analysis* Assuming for the moment that the flow is laminar and that the flat plate correlations can be used, the film Reynolds number $Re_f$ is given by Eq. (10–34),

$$Re_f = 3.77\left[\frac{L(T_{\text{sat}} - T_s)}{\mu_f i'_{fg}}\right]^{3/4}\left[\frac{\rho_f g(\rho_f - \rho_g)k_f^3}{\mu_f^2}\right]^{1/4}$$

where

$$Ja_f = \frac{c_{P,f}(T_{\text{sat}} - T_s)}{i_{fg}} = \frac{[4.18\ \text{kJ/(kg °C)}](57°\text{C} - 48°\text{C})}{2380\ \text{kJ/kg}} = 0.0158$$

and

$$i'_{fg} = i_{fg}(1 + C_1\, Ja_f) = 2380\,\frac{\text{kJ}}{\text{kg}}\,[1 + 0.68\,(0.0158)] = 2410\ \text{kJ/kg}$$

Noting that $\rho_f \gg \rho_g$, we write

$$Re_f = 3.77\left\{\frac{1\ \text{m}\,(57°\text{C} - 48°\text{C})}{[5.28 \times 10^{-4}\ \text{kg/(m s)}](2410\ \text{kJ/kg})}\right\}^{3/4}$$
$$\times\left\{\frac{(9.81\ \text{m/s}^2)(987\ \text{kg/m}^3)^2[0.645\ \text{W/(m °C)}]^3}{[5.28 \times 10^{-4}\ \text{kg/(m s)}]^2}\right\}^{1/4} = 161$$

For this value of $Re_f$ the flow is laminar and the wave effect should be relatively small, such that Eq. (10–34) is judged to be reasonable.

The condensate mass flow rate is immediately calculated from Eq. (10–24).

$$\dot{m} = \frac{p\mu_f\, Re_f}{4} = \frac{\pi(0.02\ \text{m})[5.28 \times 10^{-4}\ \text{kg/(m s)}](161)}{4} = 1.34 \times 10^{-3}\ \text{kg/s}$$

To calculate $\bar{h}_C$, we utilize Eq. (10–35a).

$$\bar{h}_C = 1.47\, Re_f^{-1/3}\left[\frac{g\rho_f(\rho_f - \rho_g)k_f^3}{\mu_f^2}\right]^{1/3}$$

$$= 1.47(161)^{-1/3}\left\{\frac{(9.81\ \text{m/s}^2)(987\ \text{kg/m}^3)^2[0.645\ \text{W/(m °C)}]^3}{[5.28\times 10^{-4}\ \text{kg/(m s)}]^2}\right\}^{1/3}$$

$$= 5700\ \text{W/(m}^2\ \text{°C)}$$

The total rate of heat transfer over the tube is

$$q_c = \bar{h}_C A_s(T_{sat} - T_s) = 5700\ \frac{\text{W}}{\text{m}^2\ \text{°C}}\ \pi(1\ \text{m})(0.02\ \text{m})(57\text{°C} - 48\text{°C})$$

$$= 3.22\ \text{kW}$$

As seen in Prob. 10–29 the local film thickness $\delta_f$ for this situation is approximated by

$$\delta_f = 1.52 \times 10^{-4}\ (x/L)^{1/4}\ \text{m}$$

Because $\delta_f$ is much less than the tube diameter $D$, we are justified in utilizing the flat-plate correlation.

Referring back to Example 10–3 on boiling heat transfer, we see that for the same saturation temperature and temperature difference of 9°C (≃16°F), the heat-transfer rate from the vertical surface is approximately 2 times larger for this laminar filmwise condensation than for the boiling. However, at high pressures, boiling heat transfer is much more effective than laminar film condensation.

## EXAMPLE 10–7

Referring back to Example 10–6 in which $1.34 \times 10^{-3}$ kg/s of condensate was produced on a 1-m-long vertical 2-cm-diameter tube, determine the length of a horizontal 2-cm-diameter tube which is required to produce the same condensate mass flow rate.

### Solution

*Objective* Determine the required length $L$.

*Schematic* Film condensation on a horizontal tube.

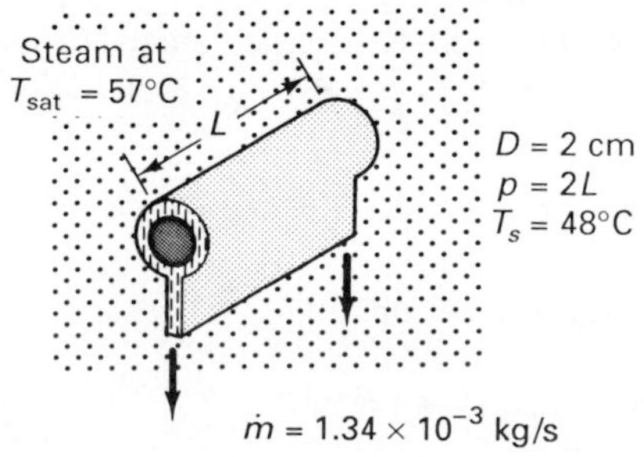
Steam at
Tsat = 57°C
L
D = 2 cm
p = 2L
Ts = 48°C
ṁ = 1.34 × 10⁻³ kg/s

*Assumptions/Conditions*

condensation
wetted surface
uniform properties within each phase
standard conditions, except for two-phase fluid and significance of buoyancy
$\rho_g << \rho_f$

*Properties* Water at a film temperature of 325 K (Table A–C–3): $\rho_f = 1/v_f = 987$ kg/m$^3$, $i_{fg} = 2380$ kJ/kg, $k_f = 0.645$ W/(m °C), $c_{P,f} = 4.18$ kJ/(kg °C), $\mu_f = 5.28 \times 10^{-4}$ kg/(m s).

*Analysis* First, by comparing Eqs. (10–33) and (10–38), we have

$$\frac{\bar{h}_{C,\text{horiz}}}{\bar{h}_{C,\text{vert}}} = \frac{0.729}{0.943}\left(\frac{L}{D}\right)^{1/4} = 0.773\left(\frac{L}{D}\right)^{1/4}$$

Thus, we see immediately that laminar film condensation is more effective for horizontal tubes if the length $L$ is greater than $2.8D$.

Because $\dot{m}$ is known, we will utilize Eq. (10–30) to solve for the length.

$$L = \frac{A_s}{\pi D} = \frac{\dot{m} i'_{fg}}{\pi D \bar{h}_C (T_{\text{sat}} - T_s)} \tag{a}$$

First, assuming laminar flow and setting $i'_{fg} = i_{fg}(1 + C_1\, Ja_f) = 2410$ kJ/(kg °C), Eq. (10–38) is used to calculate $\bar{h}_C$.

$$\bar{h}_C = 0.729\left\{\frac{(9.81\ \text{m/s}^2)(987\ \text{kg/m}^3)^2[0.645\ \text{W/(m °C)}]^3(2410\ \text{kJ/kg})}{[5.28 \times 10^{-4}\ \text{kg/(m s)}](57°\text{C} - 48°\text{C})(0.02\ \text{m})}\right\}^{1/4}$$

$$= 1.17 \times 10^4\ \text{W/(m}^2\ °\text{C)}$$

Notice that this coefficient is over twice as large as the coefficient obtained for the vertical tube! Substituting into Eq. (a), we have

$$L = \frac{(1.34 \times 10^{-3}\ \text{kg/s})(2410\ \text{kJ/kg})}{\pi(0.02\ \text{m})[1.17 \times 10^4\ \text{W/(m}^2\ °\text{C)}](57°\text{C} - 48°\text{C})} = 0.488\ \text{m}$$

To determine whether the flow is actually laminar, we calculate the film Reynolds number. Utilizing Eq. (10–24),

$$Re_f = \frac{4\dot{m}}{p\mu_f}$$

with $p = 2L$ for this horizontal tube geometry, we obtain

$$Re_f = \frac{4(1.34 \times 10^{-3}\ \text{kg/s})}{2(0.488\ \text{m})[5.28 \times 10^{-4}\ \text{kg/(m s)}]} = 10.4$$

Thus, the flow is laminar and Eq. (10–38) does indeed apply.

To conclude, we are able to use a tube that is less than half as long as was required for the vertical arrangement. Consequently, horizontal tube arrangements are often used in condenser design. However, it should be noted that $\bar{h}_C$ can be greater for turbulent film condensation on vertical tubes than for laminar condensation on horizontal tubes.

### *Enhancement of Film Condensation*

The key to improving film condensation heat transfer for a specified fluid, surface temperature $T_s$, and saturation temperature $T_{sat}$ is to decrease the liquid film thickness, which acts as a thermal resistance, and to stimulate turbulence. To accomplish these objectives, rough surfaces and finned tubes have been designed that improve the condensate mass flow $\dot{m}$. Figure 10–15 shows the heat-transfer coefficient for film condensation on horizontal smooth, finned, and rough surfaces. The high heat fluxes obtained by the Thermoexcel-C saw tooth surface have been attributed to a superior condensate dropping ability [32].

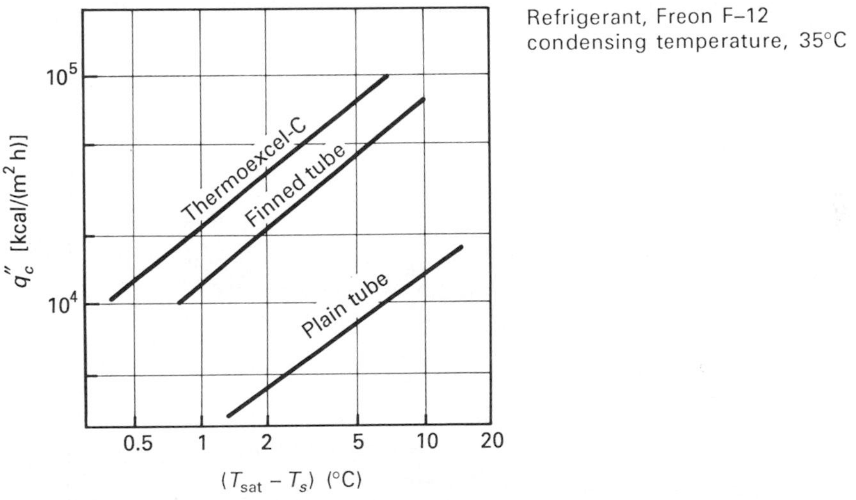


**FIGURE 10–15** Condensation-heat-transfer performance of several surfaces (Arai et al. [19]).

For more information on the enhancement of film condensation for both internal and external flow systems, the review article by Webb [18] is suggested.

## 10–4–2 Dropwise Condensation

To accomplish dropwise condensation, the fluid surface tension $\sigma$ must be greater than the critical surface tension $\sigma_{cr}$. Representative values of $\sigma_{cr}$ are listed in Table 10–2 for several surfaces. Referring to Table A–C–6, we conclude that dropwise

**TABLE 10–2** Critical surface tensions for representative solid materials

| *Material* | $\sigma_{cr} \times 10^3$ N/m |
|---|---|
| Nylon | 46 |
| Polyvinyl chloride | 39 |
| Polystyrene | 33 |
| Polyethylene | 31 |
| Kel-F | 31 |
| Teflon | 18 |
| Platinum with monolayer of | |
| perfluorobutyric acid | 10 |
| perfluorolauric acid | 6 |

*Source*: Sharfin and Zisman [27].

condensation can be achieved on all of these surfaces for water at saturation temperatures below about 160°C.

Experimental data for dropwise condensation of steam are shown in Fig. 10–16. Note the very high coefficients of heat transfer that are achieved. As a matter of fact, dropwise condensation is one of the most effective of all known heat-transfer mechanisms. Consequently, efforts are continuously being made to harness this dy-

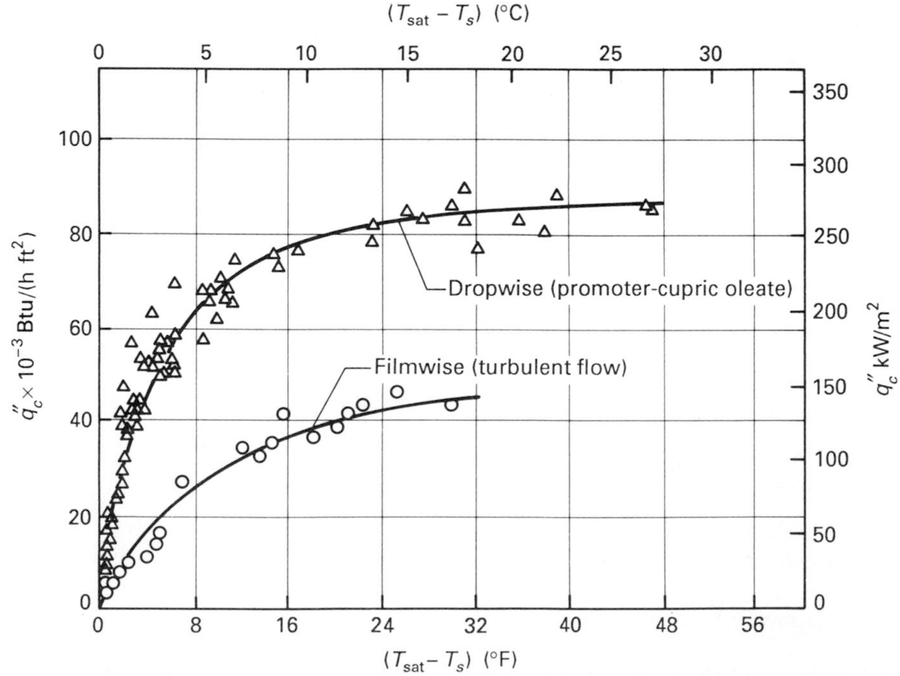


**FIGURE 10–16** Heat flux for condensation of steam at atmospheric pressure on a vertical copper surface (Welch and Westwater [33]).

namic mechanism for industrial use. Various methods have been utilized to attempt to achieve consistent dropwise condensation, such as the use of surface coatings, which reduce the value of $\sigma_{cr}$. Surface coatings and promoters that are sometimes used include Teflon, noble metals, silicones, various waxes, fatty acids (e.g., oleic and stearic), and long-chain hydrocarbons. However, these and most other nonwetting agents do not retain their effectiveness for long lengths of time because of fouling, oxidation, and removal. As an exception, fluoroplastic Emralon 300™ has shown promise as a coating, with a lifetime expectancy of as much as 20 years anticipated [34].

Experimental evidence indicates that the dropwise condensation heat flux can be correlated in terms of $d\sigma/dT$, $T_{sat}$, the properties of the saturated liquid, and other parameters featured in Eq. (10–1); that is,

$$q''_c = \text{fn}\,[g(\rho_f - \rho_g), \sigma, T_s - T_{sat}, i_{fg}, \rho_f, \mu_f, c_{P,f}, k_f, L, |d\sigma/dT|, T_{sat}] \tag{10–39}$$

This equation involves *twelve* parameters and *five* fundamental dimensions. Using dimensional analysis, an equation involving *seven* dimensionless groups is obtained, which is represented by

$$\overline{Nu}_L = \text{fn}\,[Pr_f, Ja_f, Bo_L, S_{L,f}, Ma_{L,f}, T_{sat}/\Delta T_e] \tag{10–40}$$

where $Nu_L$ is the Nusselt number, $Pr_f$ is the Prandtl number, $Ja_f$ is the Jacob number, $Bo_L$ is the Bond number,

$$S_{L,f} = \frac{\sigma}{L\rho_f\, i_{fg}} \tag{10–41}$$

is a *surface energy parameter* that represents the *ratio of surface energy per unit area to the energy flux required to overcome this energy*, and

$$Ma_{L,f} = \frac{L|d\sigma/dT|\,T_{sat}}{\mu_f\,\nu_f} \tag{10–42}$$

is the *modified Marangoni number*, which represents the *ratio of the imbalance in surface tension forces to liquid tangential viscous forces*.

A correlation for dropwise condensation of steam and ethylene glycol that involves six of these seven dimensionless parameters has been proposed by Peterson and Westwater [35], which is represented by

$$\frac{2\sigma T_{sat}\,\bar{h}_C}{i_{fg}\rho_f k_f(T_{sat} - T_s)} = 1.46 \times 10^{-6}\left[\frac{k_f(T_{sat} - T_s)}{\mu_f i_{fg}}\right]^{-1.63} Pr_f^{1/2}\left[\frac{2\sigma|d\sigma/dT|T_{sat}}{\mu_f^2\, i_{fg}}\right]^{1.16} \tag{10–43a}$$

or

$$\overline{Nu}_L = 1.63 \times 10^{-6}\, Pr_f^{2.13}\, Ja_f^{-1.63}\, S_{L,f}^{0.16}\, Ma_{L,f}^{1.16}\, \frac{\Delta T_e}{T_{sat}} \tag{10–43b}$$

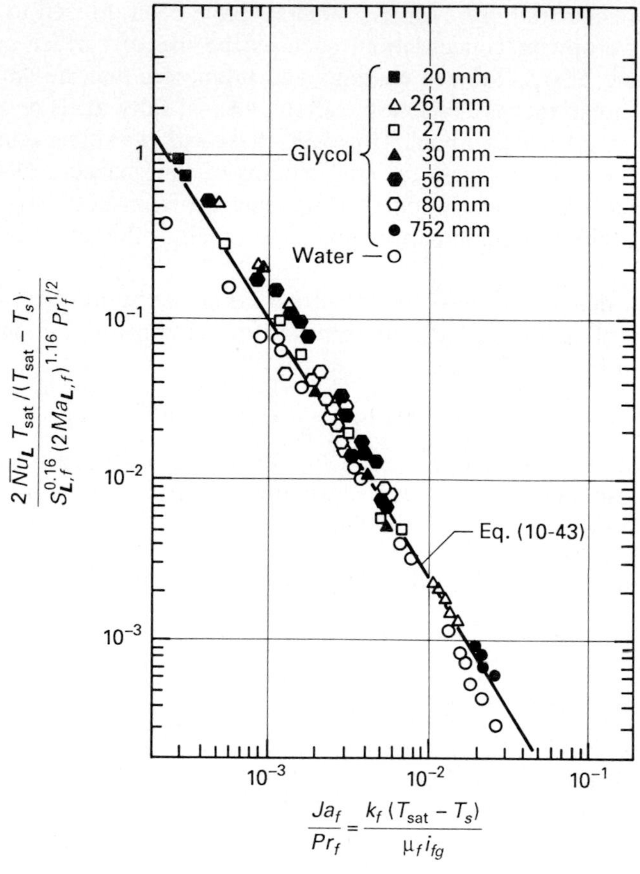


**FIGURE 10–17** Convection correlation for dropwise condensation on vertical surface (Peterson and Westwater [35]).

for $1.65 \lesssim Pr_f \lesssim 23.6$ and $7.8 \times 10^{-4} \lesssim (2\ Ma_{L,f}\,S_{L,f}) \lesssim 2.65 \times 10^{-2}$. This correlation is compared with experimental data in Fig. 10–17. By inspecting this correlation, we note that the modified Marangoni number $Ma_{L,f}$ accounts for the effect of surface tension, with the Bond number $Bo_L$ not appearing, and that the characteristic length $L$ actually cancels out.

## EXAMPLE 10–8

Steam at a saturation temperature of 57°C condenses on the surface of a 1-m-long 2-cm-diameter vertical tube with a surface temperature of 48°C. Assuming that the surface has been coated with an effective nonwetting agent such that dropwise condensation is achieved, determine the total rate of heat transfer over the length of the tube.

## Solution

*Objective* Determine $q_c$, assuming dropwise condensation.

*Schematic* Dropwise condensation on a vertical tube.

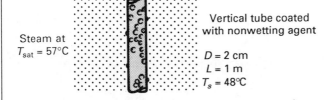

Steam at
$T_{sat}$ = 57°C
Vertical tube coated with nonwetting agent
D = 2 cm
L = 1 m
$T_s$ = 48°C

*Assumptions/Conditions*

condensation
nonwetted surface
uniform properties within each phase
standard conditions, except for two-phase fluid and significance of buoyancy
$\rho_g \ll \rho_f$

*Properties* Water at $T_{sat}$ = 57°C = 330 K (Table A–C–3): $\rho_f = 1/v_f = 984$ kg/m³, $i_{fg}$ = 2370 kJ/kg, $k_f$ = 0.650 W/(m °C), $c_{P,f}$ = 4.18 kJ/(kg °C), $\mu_f = 4.89 \times 10^{-4}$ kg/(m s), $Pr_f$ = 3.15, $\sigma$ = 0.0666 N/m, and

$$\frac{d\sigma}{dT} = \frac{1}{2}\left(\frac{0.0658 - 0.0666}{5} + \frac{0.0666 - 0.0675}{5}\right)$$

$$= -0.17 \times 10^{-3} \text{ N/(m K)}$$

*Analysis* To make use of the Peterson and Westwater correlation for dropwise condensation, Eq. (10–43), we compute the pertinent dimensionless parameters as follows:

*Jakob number*

$$Ja_f = \frac{c_{P,f}(T_{sat} - T_s)}{i_{fg}} = \frac{[4.18 \text{ kJ/(kg °C)}]\,(57°\text{C} - 48°\text{C})}{2370 \text{ kJ/kg}} = 0.0159$$

*Surface energy parameter*

$$S_{L,f} = \frac{\sigma}{L\rho_f\, i_{fg}} = \frac{0.0666 \text{ N/m}}{1 \text{ m }(984 \text{ kg/m}^3)(2370 \text{ kJ/kg})} = 2.86 \times 10^{-11}$$

*Modified Marangoni number*

$$Ma_{L,f} = \frac{\rho_f L |d\sigma/dT| T_{sat}}{\mu_f^2} = \frac{(984\ \text{kg/m}^3)(1\ \text{m})[0.17 \times 10^{-3}\ \text{N/(m K)}](330\ \text{K})}{[4.89 \times 10^{-4}\ \text{kg/(m s)}]^2}$$

$$= 2.31 \times 10^8$$

Since $Pr_f$ and the product $S_{L,f}\,Ma_{L,f} = 2.86 \times 10^{-11}(2.31 \times 10^8) = 0.00661$ both fall in the appropriate range, we substitute into Eq. (10–43) to obtain

$$\overline{Nu}_L = 1.63 \times 10^{-6}\,(3.15)^{2.13}\,(0.0159)^{-1.63}\,(2.86 \times 10^{-11})^{0.16}$$

$$\times\ (2.31 \times 10^8)^{1.16}\left(\frac{9}{330}\right) = 45{,}200$$

It follows that

$$\bar{h}_C = \frac{\overline{Nu}_L}{L/k_f} = \frac{45{,}200}{1\ \text{m}/[0.65\ \text{W/(m °C)}]} = 29{,}400\ \text{W/(m}^2\ \text{°C)}$$

and

$$q_c = \bar{h}_C A_s(T_{sat} - T_s)$$

$$= 29{,}400\ \frac{\text{W}}{\text{m}^2\ \text{°C}}\,[\pi(1\ \text{m})(0.02\ \text{m})](57\text{°C} - 48\text{°C}) = 16.6\ \text{kW}$$

Comparing this result with the value obtained for laminar film condensation in Example 10–6, we find an increase in $\bar{h}_C$ or $q_c$ by a factor of just over 5.

## EXAMPLE 10–9

The *heat pipe* is a device that utilizes evaporation and condensation to transfer heat extremely effectively. A basic heat pipe system is shown in Fig. E10–9a. A heat pipe consists of a closed pipe lined with a wicking material. A condensible fluid is contained within the pipe, with gas filling the hollow core and liquid permeating the wicking material. When one end of the pipe is exposed to a temperature above the

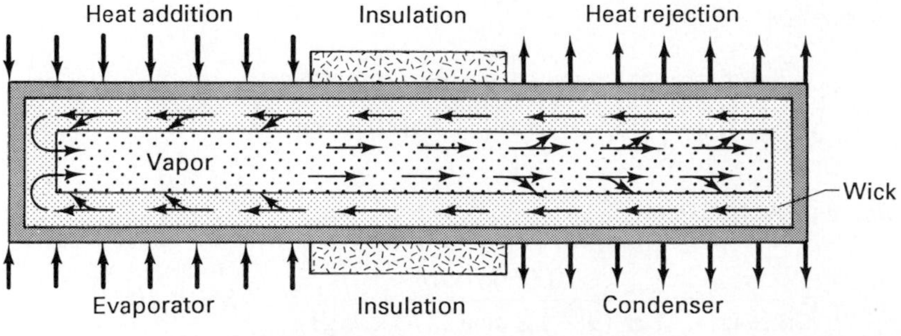


**FIGURE E10–9a** Basic heat pipe.

saturation temperature $T_{sat}$ of the fluid, and the other end is exposed to a temperature below $T_{sat}$, the fluid within the pipe circulates as shown in Fig. E10–9a. This circulation is caused by (1) the evaporation of liquid in the heated portion of the wick, (2) the condensation of fluid back into the cooled section of the wick, and (3) capillary liquid flow within the wick.

Because of the effectiveness of the vaporization and condensation mechanisms and the capillary pumping action, properly designed heat pipes are capable of transferring tens and even hundreds of times as much heat as can be transferred in solid metal bars of equal size. To illustrate the temperature distribution and heat flux in a sodium–stainless steel heat pipe is shown in Fig. E10–9b. Let us compare the heat transfer in this system with conduction heat transfer in a solid stainless steel 1-in.-diameter 18.5-in.-long rod.

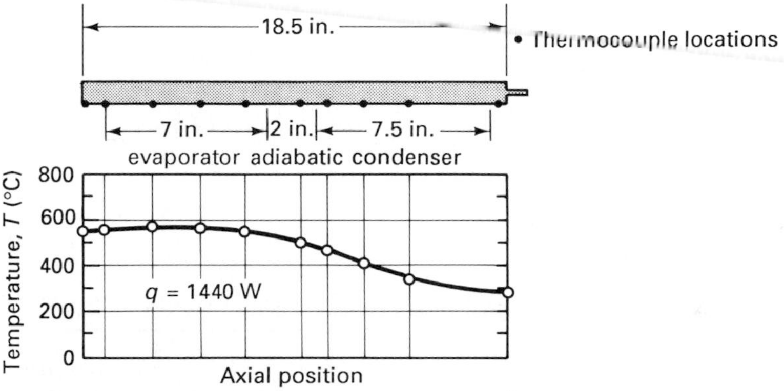


**FIGURE E10–9b** Axial temperature distributions for sodium heat pipe (Dzakowic et al. [36]).

## Solution

*Objective* For purpose of comparison, calculate $q$ in a solid rod with $T_1 = 550°C$ and $T_2 = 300°C$.

*Properties* Stainless steel (AISI 302) at 400°C (Table A–C–1): $k = 17.3$ W/(m °C).

*Analysis* Referring to Fig. E10–9b, the heat pipe transfers 1440 W across a temperature difference of approximately 250°C. The rate of heat transfer in a stainless steel bar with this temperature drop is

$$q = \frac{kA}{L}(T_1 - T_2) = 17.3 \frac{\text{W}}{\text{m °C}} \frac{\pi(1 \text{ in.})^2}{4(18.5 \text{ in.})} (250°\text{C})$$

$$= 184 \frac{\text{W in.}}{100 \text{ cm}} \frac{2.54 \text{ cm}}{\text{in.}} = 4.66 \text{ W}$$

Comparing this rate of heat transfer with the 1440 W transferred in the heat pipe, we have a factor of over 300.

## 10–5 SUMMARY

In this chapter we have dealt with issues pertaining to the identification of the basic physical features associated with boiling and condensation, and with the presentation of characterizing parameters and convection correlations that can be used in the practical analysis approach. The convection correlations and dimensionless parameters featured in this chapter are summarized in Tables 10–3 and 10–4.

As we have seen, large heat fluxes can be established by boiling and condensation processes. However, we have also learned of inherent dangers associated with boiling heat transfer applications involving direct heating by electrical, fossil-fuel, or nuclear energy sources. In such cases, care must be taken to avoid operation beyond the safe nucleate boiling regimes. In the case of condensation heat transfer, we generally expect and design for film condensation, since the surface tension $\sigma$ of most fluids is lower than the critical surface tension $\sigma_{cr}$. However, some success has been achieved in the development of surface coatings and promoters that make possible the decrease in $\sigma_{cr}$ and the establishment of dropwise condensation, which results in particularly high heat fluxes.

**TABLE 10–3** Summary of convection correlations for boiling and condensation

| *Condition* | *Correlation* | *Identification* |
|---|---|---|
| Nucleate boiling | | |
| Heat flux | $q''_c = \mu_f i_{fg}\left[\frac{g(\rho_f - \rho_g)}{\sigma}\right]^{1/2}\left[\frac{c_{P,f}(T_s - T_{sat})}{C_{sf}\, i_{fg}\, Pr_f^s}\right]^3$ | Eq. (10–9b)<br>$C_{sf}$, $s$—Table 10–1<br>Rohsenow [6] |
| | or | |
| | $\overline{Nu}_L = Ja_f^2\, Bo_L^{1/2}\, Pr_f^{1-3s}\, C_{sf}^{-1}$ | Eq. (10–9c) |
| Maximum heat flux | $q''_{max} = \frac{\pi}{24}\rho_g i_{fg}\left[\frac{\sigma g(\rho_f - \rho_g)}{\rho_g^2}\right]^{1/4}\left(\frac{\rho_f + \rho_g}{\rho_f}\right)^{1/2}$ | Eq. (10–10) or<br>Eq. (10–11)<br>Zuber [12] |
| | or | |
| | $\frac{q''_{max}}{i_{fg}}\frac{Lc_{P,f}}{k_f} = \frac{\pi}{24}\frac{Gr_{L,f}^{1/2}\, Pr_f}{Bo_L^{1/4}}\left[\frac{\rho_g(\rho_f + \rho_g)}{\rho_f^2}\right]^{1/2}$ | |
| Minimum heat flux | $q''_{min} = 0.09\,\rho_g i_{fg}\left[\frac{\sigma g(\rho_f - \rho_g)}{(\rho_f + \rho_g)^2}\right]^{1/4}$ | Eq. (10–12)<br>Zuber [12] |
| | or | |
| | $\frac{q''_{min}}{i_{fg}}\frac{L\,c_{P,f}}{k_f} = 0.09\,\frac{Gr_{L,f}^{1/2}\, Pr_f}{Bo_L^{1/4}}\left[\frac{\rho_g^2}{\rho_f(\rho_f + \rho_g)}\right]^{1/2}$ | |

**TABLE 10–3** *(Continued)*

| *Condition* | *Correlation* | *Identification* |
|---|---|---|
| Film Boiling horizontal tube $C = 0.62$ sphere $C = 0.67$ | $\overline{h}_B = C\left[\dfrac{g(\rho_f - \rho_g)\, i_{fg}\,(1 + 0.4\,Ja_g)\,k_g^3}{\nu_g(T_s - T_{sat})D}\right]^{1/4}$ | Eq. (10–13) Zuber [12] |
| | or | |
| Properties of vapor evaluated at film temperature | $\overline{Nu}_D = C\left[Gr_{D,g}\,Pr_g\left(\dfrac{1}{Ja_g} + 0.4\right)\right]^{1/4}$ | Eq. (10–15) |
| with radiation | $\overline{h}_R = \dfrac{\epsilon_s\sigma(T_s^4 - T_{sat}^4)}{T_s - T_{sat}}$ | Eq. (10–16) |
| $\overline{h}_R < \overline{h}_B$ | $\overline{h}_{B,R} = \overline{h}_B + \dfrac{3}{4}\overline{h}_R$ | Eq. (10–17) |
| | $\overline{h}_{B,R} = \overline{h}_B\left(\dfrac{\overline{h}_B}{\overline{h}_{B,R}}\right)^{1/3} + \overline{h}_R$ | Eq. (10–18) |
| Film condensation Vertical surface | | |
| Laminar flow | $h_C = 0.707\left[\dfrac{\rho_f g(\rho_f - \rho_g)k_f^3 i_{fg}(1 + 0.68\,Ja_f)}{\mu_f(T_{sat} - T_s)x}\right]^{1/4}$ | Eq. (10–32) Nusselt [29] |
| | $\overline{h}_C = 0.943\left[\dfrac{\rho_f g(\rho_f - \rho_g)k_f^3 i_{fg}(1 + 0.68\,Ja_f)}{\mu_f(T_{sat} - T_s)L}\right]^{1/4}$ | Eq. (10–33a) |
| | or | |
| | $\overline{Nu}_L = 0.943\left[Gr_{L,f}\,Pr_f\left(\dfrac{1}{Ja_f} + 0.68\right)\right]^{1/4}$ | Eq. (10–33b) |
| | or | |
| | $\overline{h}_C = 1.47\,Re_f^{-1/3}\left[\dfrac{\rho_f g(\rho_f - \rho_g)k_f^3}{\mu_f^2}\right]^{1/3}$ | Eq. (10–35a) |
| | or | |
| | $\overline{Nu}_L = 1.47\,Re_f^{-1/3}\,Gr_{L,f}^{1/3}$ | Eq. (10–35b) |
| $30 < Re_f < 1800$ | $\overline{h}_C = \dfrac{Re_f}{1.08\,Re_f^{1.22} - 5.2}\left[\dfrac{\rho_f g(\rho_f - \rho_g)k_f^3}{\mu_f^2}\right]^{1/3}$ | Eq. (10–36a) Kutateladze [30] |
| | or | |
| | $\overline{Nu}_L = \dfrac{Re_f}{1.08\,Re_f^{1.22} - 5.2}\,Gr_{L,f}^{1/3}$ | Eq. (10–36b) |
| | or | |
| | $Re_f = \left(\dfrac{1}{1.08}\left\{5.2 + \dfrac{4L(T_{sat} - T_s)}{\mu_f\, i'_{fg}} \times \left[\dfrac{\rho_f g(\rho_f - \rho_g)k_f^3}{\mu_f^2}\right]^{1/3}\right\}\right)^{1/1.22}$ | |

**TABLE 10–3** *(Continued)*

| *Condition* | *Correlation* | *Identification* |
|---|---|---|
| Turbulent flow $Re_f > 1800$ | $\bar{h}_C = \dfrac{Re_f}{8750 + 58\, Pr_f^{-1/2}\,(Re_f^{3/4} - 253)} \times \left[\dfrac{\rho_f g(\rho_f - \rho_g)k_f^3}{\mu_f^2}\right]^{1/3}$ | Eq. (10–37a) Labuntsov [31] |
| | or $\overline{Nu}_L = \dfrac{Re_f}{8750 + 58\, Pr_f^{-1/2}\,(Re_f^{3/4} - 253)}\, Gr_{L,f}^{1/3}$ | Eq. (10–37b) |
| | or $Re_f = \left(253 + \dfrac{Pr_f^{1/2}}{58}\left\{\dfrac{4L(T_{sat} - T_s)}{\mu_f i'_{fg}} \times \left[\dfrac{\rho_f g(\rho_f - \rho_g)k_f^3}{\mu_f^2}\right]^{1/3} - 8750\right\}\right)^{4/3}$ | |
| Inclined plates | Use correlations for vertical surfaces with $g$ replaced by $g \cos \theta$ | |
| Horizontal tubes $C = 0.729$ | $\bar{h}_C = C\left[\dfrac{\rho_f g(\rho_f - \rho_g)k_f^3\,(1 + 0.68\, Ja_f)}{\mu_f(T_{sat} - T_s)D}\right]^{1/4}$ | Eq. (10–38a) Lienhard et al. [16] |
| and | or | |
| Spheres $C = 0.815$ | $\overline{Nu}_D = C\left[Gr_{D,f} Pr_f\left(\dfrac{1}{Ja_f} + 0.68\right)\right]^{1/4}$ | Eq. (10–38b) |
| Dropwise condensation | $\overline{Nu}_L = 1.63 \times 10^{-6}\, Pr_f^{2.13}\, Ja_f^{-1.63}\, S_{L,f}^{0.16}\, Ma_{L,f}^{1.16}\, \dfrac{\Delta T_e}{T_{sat}}$ | Eq. (10–43b) Peterson and Westwater [35] |

## REVIEW QUESTIONS

**10–1.** What is (a) boiling and (b) condensation?

**10–2.** What is (a) pool boiling and (b) flow boiling?

**10–3.** Distinguish between subcooled boiling and saturated boiling.

**10–4.** Define excess temperature.

**10–5.** Name the basic regimes for pool boiling.

**10–6.** Name the boiling regime for which radiation heat transfer is most important. Why is this so?

**10–7.** Define (a) critical heat flux, (b) Leidenfrost point, and (c) burnout.

**10–8.** Define the coefficient of boiling heat transfer.

**10–9.** Why are experimental data for boiling heat transfer often presented in terms of the heat flux $q''_c$ rather than the coefficient of boiling heat transfer?

**10–10.** Why is internal flow boiling more complex than external flow boiling?

**10–11.** Define the local coefficient of condensation heat transfer.

**TABLE 10–4** Dimensionless parameters for convection heat transfer: Boiling and condensation

| *Dimensionless group* | *Symbol* | *Definition* | *Interpretation* |
|---|---|---|---|
| Bond number | $Bo_L$ | $\frac{g(\rho_f - \rho_g)L^2}{\sigma}$ | $\frac{\text{gravitational forces}}{\text{surface tension forces}}$ |
| Jakob number | $Ja$ | $\frac{c_P(T_s - T_{sat})}{i_{fg}}$ | $\frac{\text{maximum sensible energy absorbed}}{\text{latent heat absorbed}}$ |
| Grashof number | $Gr_L$ | $\frac{g(\rho_f - \rho_g)L^3}{\rho\nu^2}$ | Relative significance of buoyant and viscous effects for two-phase flow |
| Nusselt number† | $Nu_L$ | $\frac{hL}{k}$ | $\frac{\text{convection for fluid in motion}}{\text{conduction for motionless fluid layer}}$ |
| Prandtl number† | $Pr$ | $\frac{\nu}{\alpha}$ | relative effectiveness of momentum and energy transfer by molecular diffusion within the boundary layer |
| Reynolds number (for condensation film) | $Re_f$ | $\frac{4\dot{m}}{p\mu_f}$ | $\frac{\text{inertia forces}}{\text{viscous forces}}$ |
| Surface energy parameter | $S_{L,f}$ | $\frac{\sigma}{L\rho_f i_{fg}}$ | $\frac{\text{surface energy per unit area}}{\text{energy flux to overcome surface energy}}$ |
| Modified Marangoni number | $Ma_{L,f}$ | $\frac{L\lvert d\sigma/dT\rvert T_{sat}}{\mu_f \nu_f}$ | $\frac{\text{imbalance in surface tension forces}}{\text{liquid tangential viscous forces}}$ |

† Dimensionless parameters also used in analysis of forced convection.

**10–12.** What is the criterion for determining whether film or dropwise condensation will occur?

**10–13.** Define the Reynolds number for film condensation. What is its significance?

**10–14.** Why is the coefficient of condensation heat transfer greater for dropwise condensation than for film condensation?

**10–15.** What is a heat pipe?

## ■ PROBLEMS

**10–1.** The Kutateladze [14] correlation for maximum boiling heat flux is given by $q''_{max} = Ki_{fg}\sqrt{\rho_g}\,[\sigma g(\rho_f - \rho_g)]^{1/4}$; the average value of $K$ is 0.14, but this parameter ranges from 0.13 to 0.19 for various surface conditions. Determine $q''_{max}$ for Example 10–2 by utilizing this correlation and compare the result obtained by the use of the Zuber correlation.

**10–2.** Water at atmospheric pressure is boiled on a mechanically polished stainless steel surface that is heated electrically. Determine the boiling heat flux if the surface temperature is 110°C. Also calculate the maximum excess temperature.

**10–3.** Solve Prob. 10–2 if a scored copper surface is used.

**10–4.** Water at 100 atm is boiled on a chemically etched stainless steel surface. Determine the boiling heat flux if the excess temperature is 10°C.

**10–5.** Water at atmospheric pressure is boiled on an electrically heated copper plate. Determine the surface temperature if the power dissipated per unit surface area is 440 kW/m$^2$.

**10–6.** Water at atmospheric pressure is evaporated from a boiling pan of water at a rate of $5 \times 10^{-4}$ kg/s. The bottom of the copper pan is flat with a 10 cm diameter. Determine the heat flux and the approximate temperature of the bottom surface of the pan.

**10–7.** A 1-cm-diameter 2-m-long copper tube is to be used to boil water at 1 atm. Determine the rate of heat transfer and the rate of evaporation if the surface temperature of the tube is 108°C.

**10–8.** Water at a saturation temperature of 160°C is boiled by a 1-m-long 2-cm-diameter copper tube with surface temperature at 170°C. Determine the rate of heat transfer and the rate of evaporation. Compare these results with predictions for 1 atm and the same excess temperature.

**10–9.** As seen in Fig. 10–5, the boiling heat flux is dependent on pressure. The optimal pressure for water is approximately 100 atm. Show that the maximum heat flux at this pressure is about $4 \times 10^6$ W/m$^2$.

**10–10.** Saturated water at 100 atm is boiled by a 1-m-long 2-cm-diameter copper tube with excess surface temperature of 10°C. Determine the rate of heat transfer and the rate of evaporation. Compare these results with predictions for 1 atm and the same excess temperature.

**10–11.** Compare the maximum heat flux for boiling of water and mercury at 1 atm on a large flat plate.

**10–12.** Use Eq. (10–5) to compute the value of $q''_{\text{min}}$ for pool boiling of water at 1 atm. How does this value compare with the curve shown in Fig. 10–3?

**10–13.** Compute the the minimum boiling heat flux for pool boiling of water at 100 atm.

**10–14.** Combine Eqs. (10–31) and (10–33) to obtain Eq. (10–34).

**10–15.** Combine Eqs. (10–33) and (10–34) to obtain Eq. (10–35).

**10–16.** Use Eq. (10–32) and the defining relation for $h_C$ to obtain Eq. (10–33a).

**10–17.** Steam at a saturation pressure of 14.7 psi condenses on the outside surface of a 1-in. O.D. 12-ft-long vertical tube with surface temperature equal to 200°F. Assuming that filmwise condensation occurs, determine the total rate of heat transfer and the total rate of condensation.

**10–18.** Solve Prob. 10–17 for the case of a horizontal tube.

**10–19.** Determine the length of tube in Prob. 10–17 required to produce a condensate mass flow rate of $3.15 \times 10^{-3}$ lb$_m$/s.

**10–20.** Solve Prob. 10–19 for the case of a horizontal tube.

**10–21.** Ammonia at a saturation temperature of −20°C condenses on the surface of a vertical plate 1 m high by 2 m wide. The surface temperature of the plate is −30°C. Determine the total rate of heat transfer and the total mass flow rate of condensate over the entire surface.

**10–22.** Referring to Prob. 10–21, determine the rate of heat transfer from the upper surface if the plate is inclined at an angle of $\pi/4$ rad with the horizontal.

**10–23.** Freon-12 at a saturation temperature of −10°C condenses on the surface of a 1-m-long 2-cm-diameter vertical tube with a surface temperature of −15°C. Determine the total mass flow rate of condensate.

**10–24.** Determine the condensation rate for Prob. 10–23 if the tube is horizontal.

**10–25.** A 1-cm-diameter 2-m-long vertical tube is to be used to condense steam at 1 atm. Determine the rate of heat transfer and the condensate mass flow rate if the surface temperature of the tube is 90°C. Compare these results to the predictions for boiling heat transfer in Prob. 10–2.

**10–26.** Determine the mass flow rate of condensate for Prob. 10–25 if the tube is horizontal.

**10–27.** The optimal boiling pressure for water is approximately 100 atm. The boiling-heat-transfer rate at this pressure for an excess temperature of 10°C is determined in Prob. 10–10 for a 1-m-long 2-cm-diameter copper tube. For purpose of comparison, determine the rate of film condensation heat transfer for a horizontal tube, at a temperature of 10°C below the saturation temperature for 100 atm.

**10–28.** Referring to Prob. 10–25, calculate $q_c$ for a length of 5.5 m.

**10–29.** Develop an expression for the thickness of laminar film condensation for Example 10–6.

**10–30.** Steam at atmospheric pressure is to be condensed on a vertical 2-cm-diameter tube with a surface temperature of 90°C. Determine the tube length required to produce a condensate mass flow rate of 0.0135 kg/s.

**10–31.** Show that Eq. (10–38) can be represented in terms of $Re_f$ by

$$\bar{h}_C = (2\pi)^{1/3} C^{4/3} Re_f^{-1/3} \left[\frac{\rho_f g(\rho_f - \rho_g)k_f^3}{\mu_f^2}\right]^{1/3}$$

where

$$Re_f = 2\pi C \left[\frac{L(T_{\text{sat}} - T_s)}{\mu_f i'_{fg}}\right]^{3/4} \left[\frac{\rho_f g(\rho_f - \rho_g)k_f^3}{(L/D)^3\mu_f^2}\right]^{1/4}$$

**10–32.** (a) Show that Eq. (10–36) can be put into the alternative form:

$$Re_f = \left(\frac{1}{1.08}\left\{5.2 + \frac{4L(T_{\text{sat}} - T_s)}{\mu_f i'_{fg}}\left[\frac{\rho_f g(\rho_f - \rho_g)k_f^3}{\mu_f^2}\right]^{1/3}\right\}\right)^{1/1.22}$$

(b) Show that Eq. (10–37) can be put into the alternative form:

$$Re_f = \left(253 + \frac{Pr_f^{1/2}}{58}\left\{\frac{4L(T_{\text{sat}} - T_s)}{\mu_f i'_{fg}}\left[\frac{\rho_f g(\rho_f - \rho_g)k_f^3}{\mu_f^2}\right]^{1/3} - 8750\right\}\right)^{4/3}$$

(c) What is the benefit of writing Eqs. (10–36) and (10–37) in these alternative forms?

**10–33.** Calculate the condensate mass flow rate for Example 10–8.

**10–34.** Reconsider Prob. 10–25 for the case in which the tube is coated with an effective wetting agent.

# CHAPTER 11

# CONVECTION HEAT TRANSFER: PRACTICAL ANALYSIS— HEAT EXCHANGERS

## 11–1 INTRODUCTION

*Heat exchangers* are devices that transfer heat between fluids at different temperatures. Heat exchangers are used in a wide range of industrial and commercial applications. Examples are found in the power, air-conditioning, refrigeration, cryogenics, heat recovery, process, automotive, aircraft, marine, and manufacturing industries, as well as in many products available in the marketplace.

In this chapter we will describe the more common types of heat exchangers that are commercially available, discuss the important functions of evaluation, selection, and design, and utilize the principles presented in Chaps. 6 through 10 to develop the practical approach to heat exchanger analysis.

## 11–2 TYPES OF HEAT EXCHANGERS

As indicated by Shah [1], a heat exchanger consists of (1) active heat exchanging elements such as a core or a matrix containing the heat transfer surface, and (2) passive fluid distribution elements such as headers, manifolds, tanks, inlet and outlet nozzles or pipes, and seals. The surface of the exchanger core, which is in direct contact with fluids and through which heat is transferred, is referred to as the *heat transfer surface*. The various types of heat exchangers that are commercially available may be categorized according to the geometric configuration of the heat transfer surface and flow arrangement as well as certain other considerations.

## 11–2–1 Double-Pipe Heat Exchangers

The *double-pipe* module shown in Fig. 11–1 is the simplest type of heat exchanger. In the hairpin unit shown in this figure, two double-pipes are joined at one end by a U-bend and a return bend housing on the shell side. As illustrated in Fig. 11–2, double-pipe heat exchangers involve combinations of tube and annular flow, with the tube and annular flows being in the same direction for *parallel-flow* (concurrent) arrangements and in opposing directions for *counterflow*. Whereas devices of this type were originally used as small-size classical counterflow heat exchangers, present

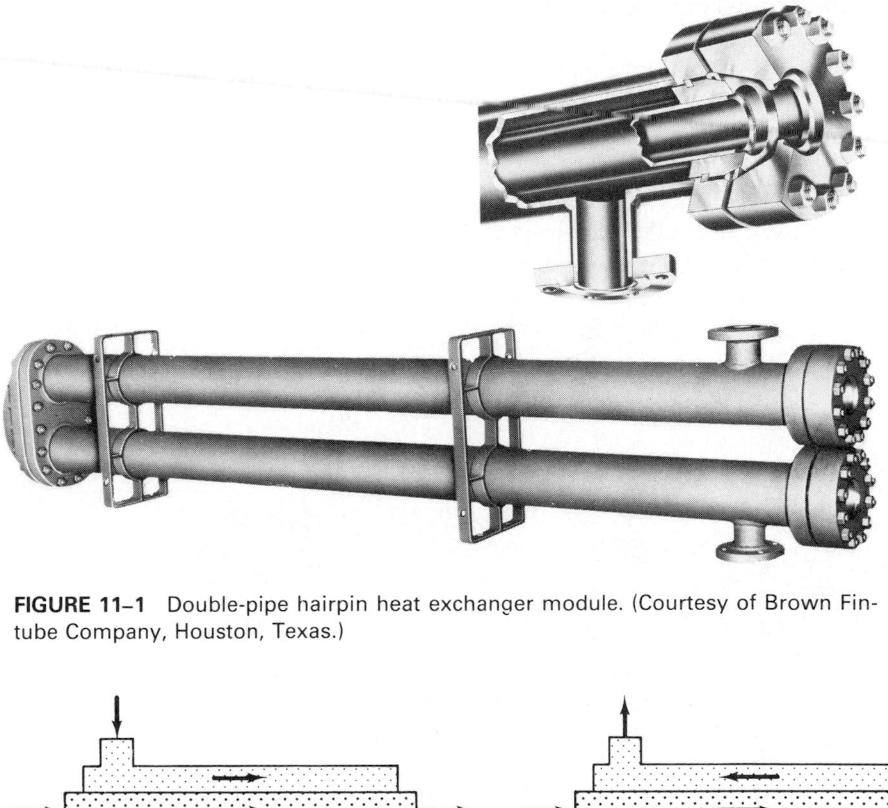

**FIGURE 11–1** Double-pipe hairpin heat exchanger module. (Courtesy of Brown Fintube Company, Houston, Texas.)

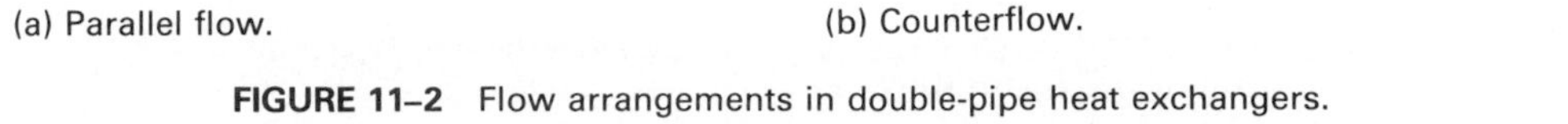

(a) Parallel flow. (b) Counterflow.

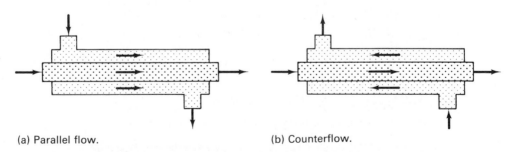

**FIGURE 11–2** Flow arrangements in double-pipe heat exchangers.

applications involve the use of multiple double-pipe hairpin units with radial or longitudinal finned tubes. Nevertheless, because of the relatively small surface area and cross-sectional flow area, double-pipe heat exchangers are generally reserved for applications that require low to moderate heat transfer rates. As a rule of thumb, applications for which double-pipe heat exchangers are economical normally involve heat transfer surface areas less than 50 $m^2$ and heat transfer rates less than 1000 kW.

In applications requiring the transfer of high rates of heat, other kinds of heat exchangers are utilized. Two types that provide sufficiently large surface areas are shell-and-tube heat exchangers and crossflow heat exchangers.

### 11–2–2 Shell-and-Tube Heat Exchangers

Typical shell-and-tube heat exchangers are shown in Figs. 6–3 and 11–3. Heat exchangers of this type are generally classified according to the number of tube and shell passes. The shell-and-tube heat exchanger shown in Fig. 6–3 has one shell pass and one tube pass, whereas the heat exchangers pictured in Fig. 11–3 are constructed with a single shell pass and double tube pass. The standard shell type is TEMA E, with entry and outlet nozzles at the opposite ends in single shell arrangements.† Heat exchangers with multiple shell and tube passes are often employed in industry. However, because of their relatively large size and weight, shell-and-tube heat exchangers are generally not used in aircraft, automotive, and marine applications.

Baffles are generally used to increase the heat-transfer performance in com-

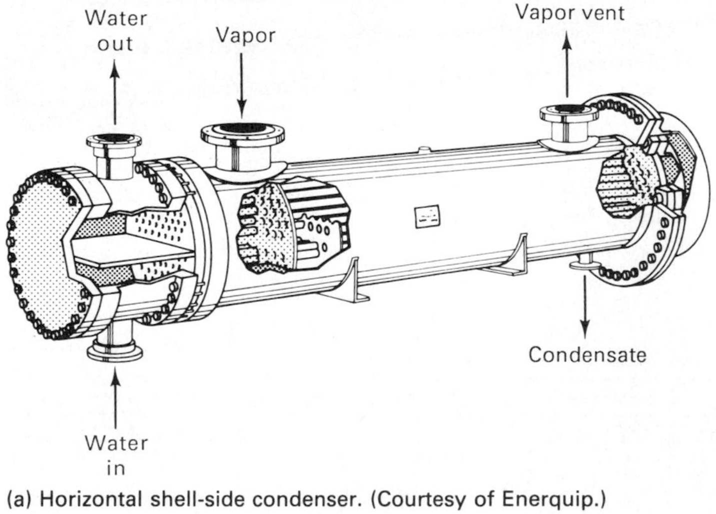


(a) Horizontal shell-side condenser. (Courtesy of Enerquip.)

**FIGURE 11–3** Typical shell-and-tube heat exchangers—single pass on shell side, two passes on tube side.

† TEMA stands for Tubular Exchanger Manufacturers Association.

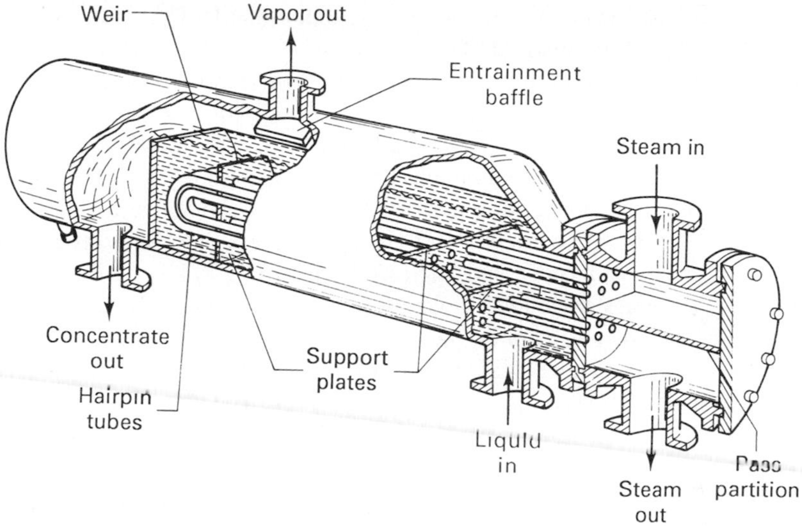


(b) Horizontal shell-side evaporator with hairpin tubes. (From *Heat Exchanger Design Handbook* [2]. Used with permission.)

**FIGURE 11–3 (continued)**

mercial shell-and-tube exchangers by diverting the flow on the shell side across the tubes. The use of baffles also provides a means of strengthening the mechanical structure of the exchanger. Two of the most common baffle arrangements are illustrated in Fig. 11–4. Segmental baffles are characterized by the ratio of the height of the

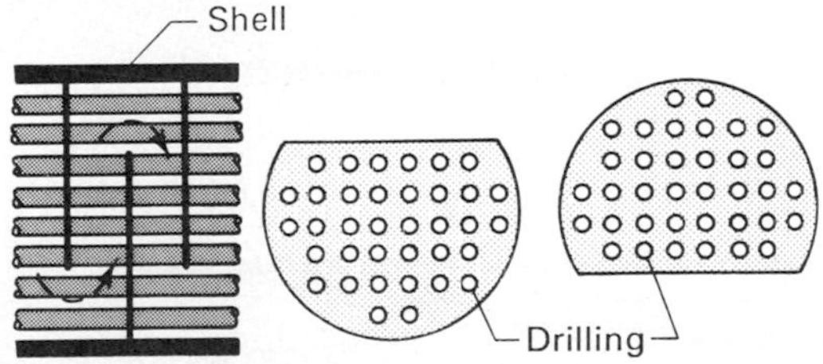


(a) Segmental baffles (25% cut).

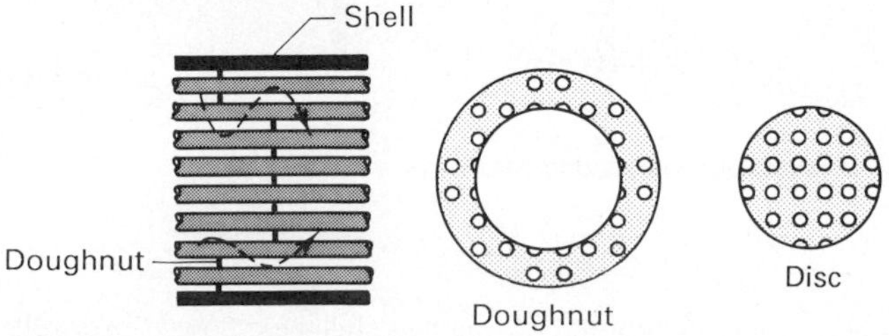


(b) Disc- and doughnut baffles.

**FIGURE 11–4**
Baffles used in shell-and-tube heat exchangers. (From Kern [3]. Used with permission.)

cut-out segment to the inside diameter $D_{i,s}$ of the shell. The optimum heating-to-pumping power performance has been found to occur for a baffle cut of about 20 percent [4]. The *baffle spacing*, also known as *baffle pitch*, is designated by $B$. The baffle spacing is generally maintained between $\frac{1}{5}$ to 1 times $D_{i,s}$.

Referring to Fig. 11–5, baffled shell-side flow is very complex. As indicated by Taborek [5], only part of the fluid takes the desired path B through the tube bundle, whereas a potentially substantial portion flows through the leakage areas between the tubes and baffle (path A) and between shell and baffle (path E) and through the bypass area between the tube bundle and shell wall (path C). For situations in which the bundle-to-shell bypass clearance exceeds a value of about 30 mm, sealing strips attached to the baffles and running along the length of the shell are commonly used to force the bypass stream back into the tube bundle. Typical flow fractions of the various streams for well-designed baffles are listed in Table 11–1 for laminar and turbulent flows. Whereas these baffle arrrangements do improve the overall heat-transfer performance because of the mixing and approximate cross-flow pattern that is established, low heat-transfer rates occur in the regions of relative stagnant flow and leakage.

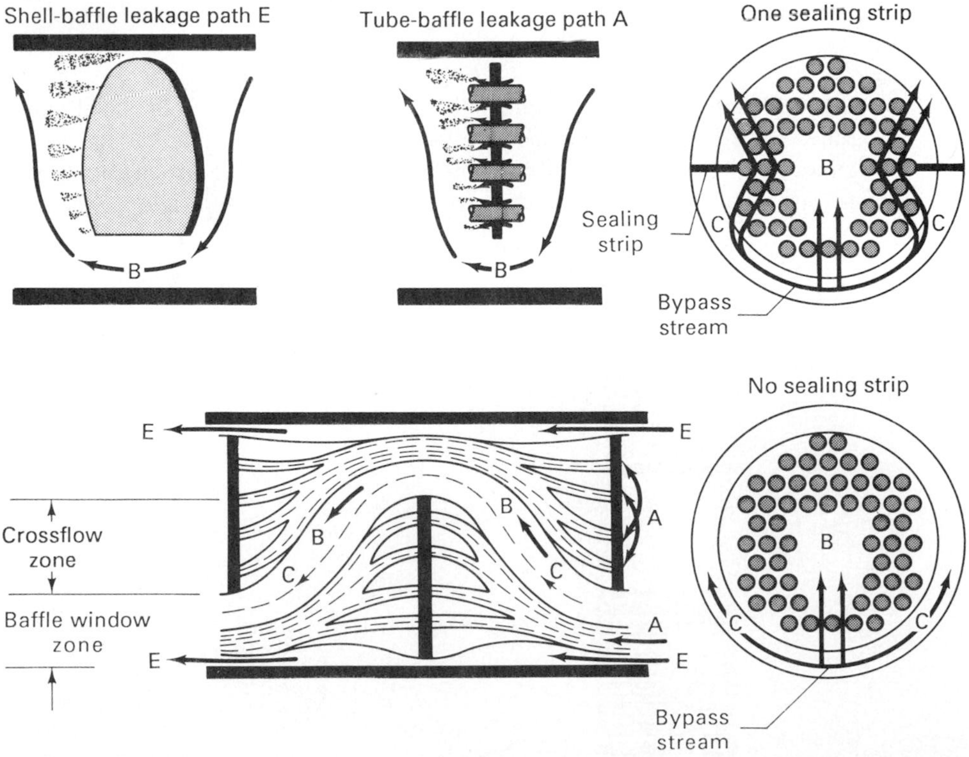


**FIGURE 11–5** Schematic of flow distribution for baffled shell-side flow (Tinker [6]); crossflow path B across tube bundle, bypass path C between shell and tube bundle.

**TABLE 11–1** Typical flow fractions of various streams for baffled shell-side flow

| *Stream* | *Laminar* (%) | *Turbulent* (%) |
|---|---|---|
| Main crossflow (B) | 25–50 | 40–70 |
| Bundle bypass (C) | 20–30 | 15–20 |
| Shell-to-baffle leakage (E) | 6–40 | 6–20 |
| Tube-to-baffle leakage (A) | 4–10 | 9–20 |

*Source*: Taborek [5].

Another type of baffle known as *rodbaffle* consists of vertical and horizontal rods that provide support and containment of the tubes. In addition to reducing flow-induced vibration, the rodbaffle system has been reported to produce uniform free-flowing turbulence, thereby eliminating dead or low flow areas in the bundle and increasing the effective heat-transfer area [7].

The three standard types of tube arrangements used in shell-and-tube heat exchangers are shown in Fig. 11–6. The triangular and square arrangements are similar to the staggered and inline tube bank arrays introduced in Sec. 8–4. These arrangements are characterized by tube pitch $S_t$, tube outside diameter $d_o$, and shell inside diameter $D_{i,s}$. As pointed out by Kern and Kraus [8], these various layouts permit some selectivity on the shell side between heat transfer and allowable pressure drop. In applications in which the pressure drop is not a primary consideration, such as high pressure operation with liquids, it may be desirable to pack as much surface as possible into a given size shell. On the other hand, a less dense arrangement that provides lower resistance to the flow may be required for gases at moderate pressures.

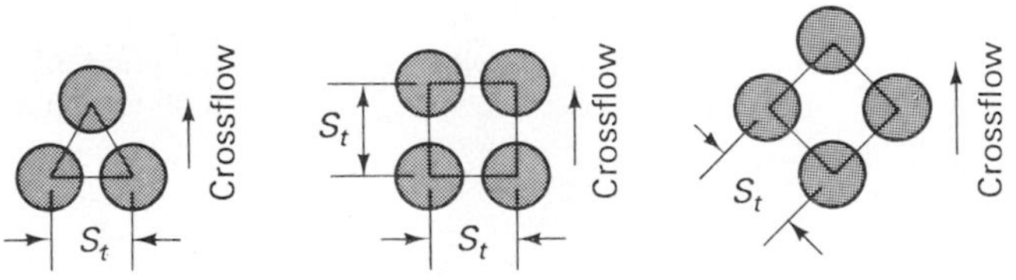


(a) Triangular. (b) Square. (c) Rotated square.

**FIGURE 11–6** Typical tube arrangements used for shell-and-tube heat exchangers.

## 11–2–3 Crossflow Heat Exchanger

Crossflow heat exchangers consist of a number of interconnected tubes or passageways that are separated by one or more channels, as illustrated in Fig. 11–7. Heat exchangers of this type are often used when one or both of the fluids is gas. In this connection, heat exchangers designed for the purpose of cooling a hot fluid stream by passing air in crossflow over one or more plane or finned tubes are known as *air coolers*.

Crossflow heat exchangers and solution results are categorized according to whether the fluid streams can be considered to be *mixed* or *unmixed*. Mixing of a

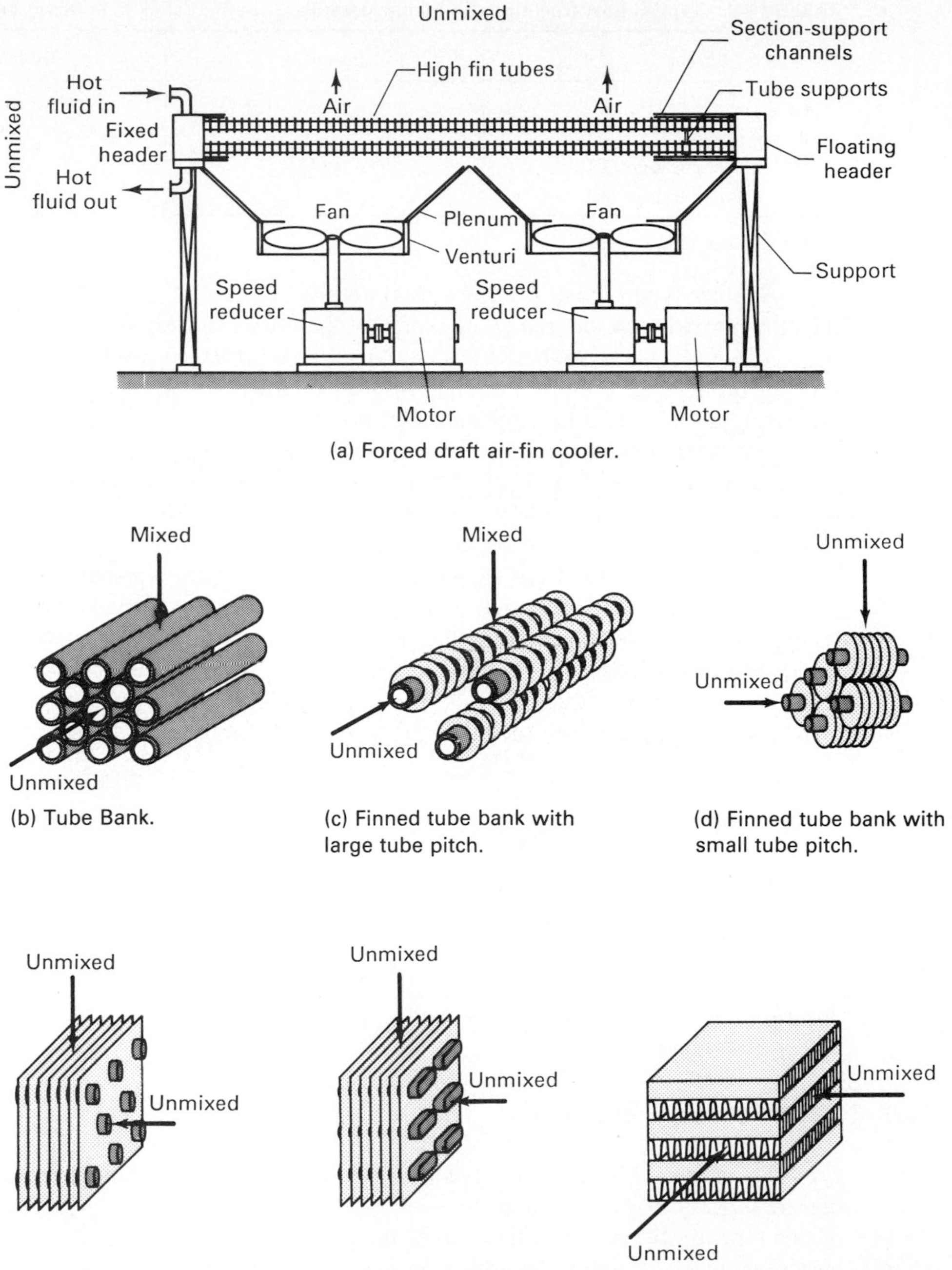


(a) Forced draft air-fin cooler.

(b) Tube Bank.

(c) Finned tube bank with large tube pitch.

(d) Finned tube bank with small tube pitch.

(e) Tube-fin heat exchanger.

(f) Flat tube-fin heat exchanger.

(g) Plate-fin heat exchanger.

**FIGURE 11–7** Crossflow heat exchangers. (From Kern and Kraus [8] and Kays and London [9]. Used with permission.)

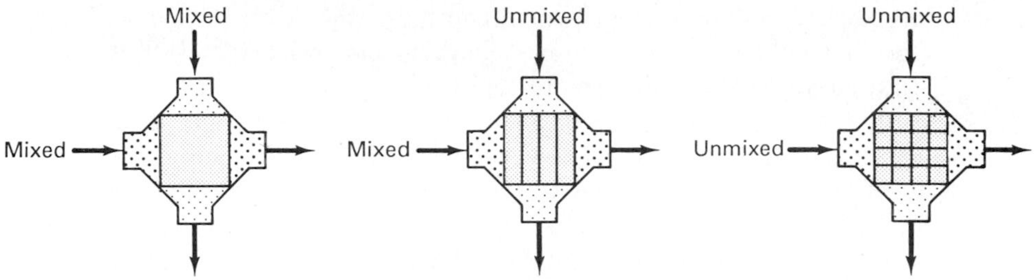


(a) Both streams mixed.
(b) One stream mixed, one stream unmixed.
(c) Both streams unmixed.

**FIGURE 11–8** Mixed and unmixed categories of crossflow heat exchangers.

stream implies that its flow is unrestricted in the lateral direction (i.e., in the direction of the other stream). Conversely, unmixed flow implies that the flow of a stream is restricted in the lateral direction. The three possible combinations of mixed and unmixed streams are represented schematically in Fig. 11–8. As we will see in Sec. 11–5, the heat transfer performance of crossflow heat exchangers is better for unmixed streams than for mixed streams. Although the categorization of a stream is sometimes clear cut, as in the cases shown in Fig. 11–7e–g, the actual flow is better described as being partially mixed in many applications. For example, the degree of mixing for crossflow over the high-fin tubes shown in Fig. 11–7c,d ranges from mixed for wide spacing to unmixed for close spacing. However, in spite of this reality, solution approaches available in the literature require that the stream be designated by one extreme or the other (i.e., mixed or unmixed).

The plate heat exchanger illustrated in Fig. 11–9 is another important type of

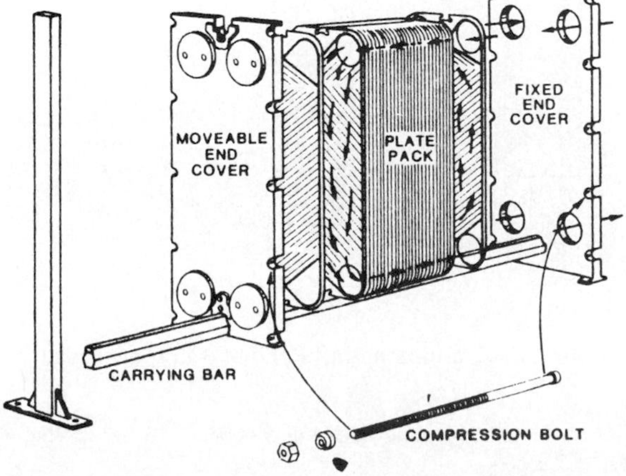


**FIGURE 11–9**
Gasketed plate heat exchanger—exploded view. (From *Heat Exchanger Design Handbook* [2]. Used with permission.)

unmixed crossflow arrangement. Plate heat exchangers are built of thin plates that are clamped together to form passages. As shown in Fig. 11–9, the plates are fitted with gaskets, which are used to seal the unit and to direct the two liquids countercurrently through the narrow passages between alternate pairs of heat transfer plates. Heat exchangers of this type are easily dismantled for cleaning, inspection, or alteration and are widely used in the dairy, beverage, food, pharmaceutical, and synthetic rubber industries [1]. However, because of limitation in gasket material, plate heat exchangers are generally restricted to pressures below 10 to 15 atm and temperatures below 150°C.

## 11–2–4 Regenerators

In the types of heat exchangers described above, two fluids are separated by a thin wall through which heat is transferred.† These direct transfer heat exchangers are commonly called *recuperators*. Another type of heat exchanger known as the *regenerator* involves the alternate use of the same flow passages by both fluids. In such storage-type heat exchangers, energy is first transferred by convection from one fluid and stored in the matrix wall, and later transferred from the matrix to the other fluid.

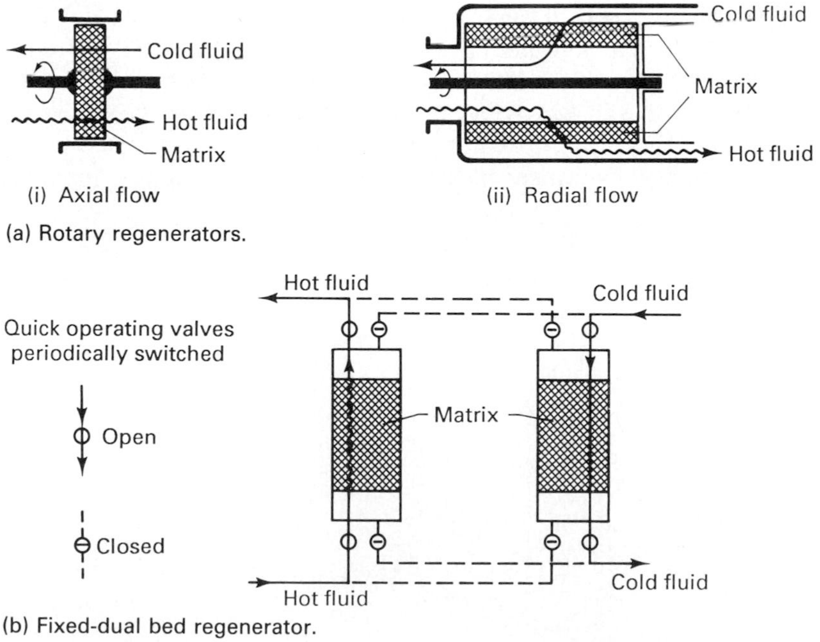


**FIGURE 11–10** Regenerators. (From Kays and London [9]. Used with permission.)

† In certain types of heat exchangers two immiscible fluids are in direct contact. Water cooling towers are an example of such *direct contact* heat exchangers.

Thus, energy is periodically stored and rejected by the matrix wall. Examples of storage-type heat exchangers include the rotary regenerator and the fixed dual-bed regenerator shown in Fig. 11–10. Representative matrix surfaces are shown in Fig. 11–11. In addition to providing a very compact heat transfer surface, the surface of a regenerator matrix tends to be self-cleaning because of the periodic flow reversals and lack of permanent flow-stagnation regions. This self-cleaning feature has been demonstrated by the Ljungstrom air preheaters used in TVA power plants, which burn various grades of coal [9].

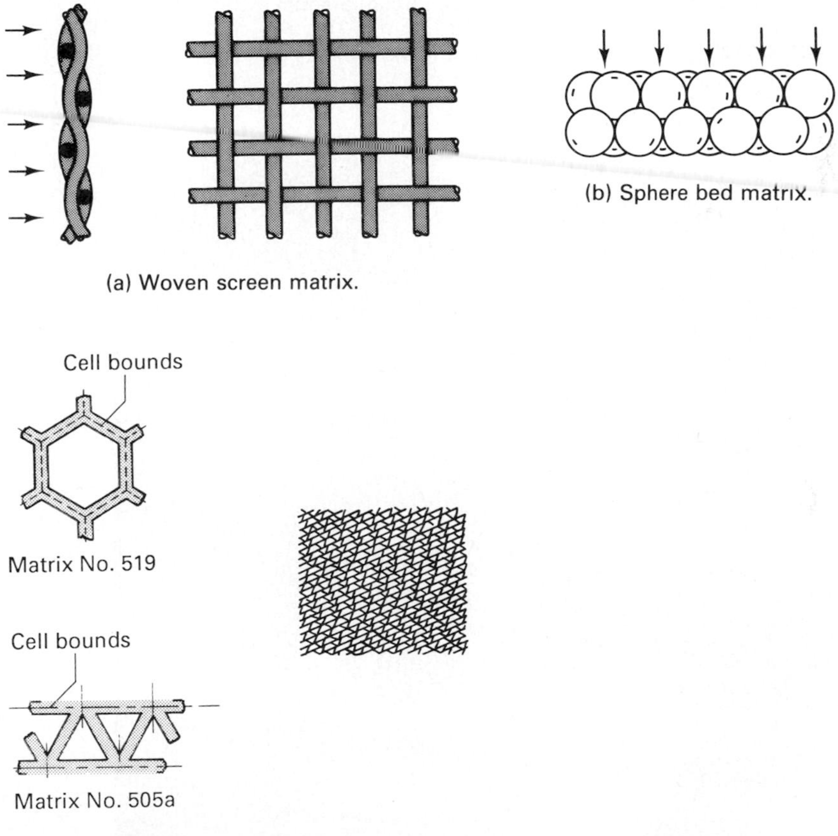
(b) Sphere bed matrix.
(a) Woven screen matrix.
Cell bounds
Matrix No. 519
Cell bounds
Matrix No. 505a
(c) Glass ceramic matrix.

**FIGURE 11–11** Regenerator matrix surfaces. (From Kays and London [9]. Used with permission.)

## 11–2–5 Heat Exchanger Performance/Size

High-performance heat exchangers are characterized by high heating-pumping power ratios and small volumes. Various methods are routinely used to enhance the performance and reduce the size of heat exchangers. For example, the rate of heat transfer

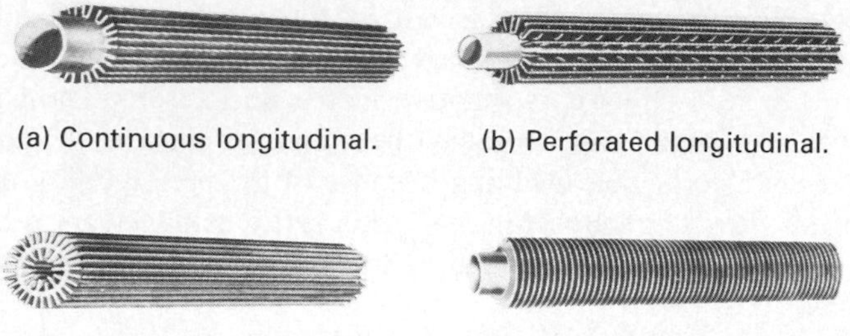

(a) Continuous longitudinal. (b) Perforated longitudinal.

(c) Internal-external. (d) Circumferential.

**FIGURE 11–12** Fintubes. (Courtesy of Brown Fintube Company, Houston, Texas.)

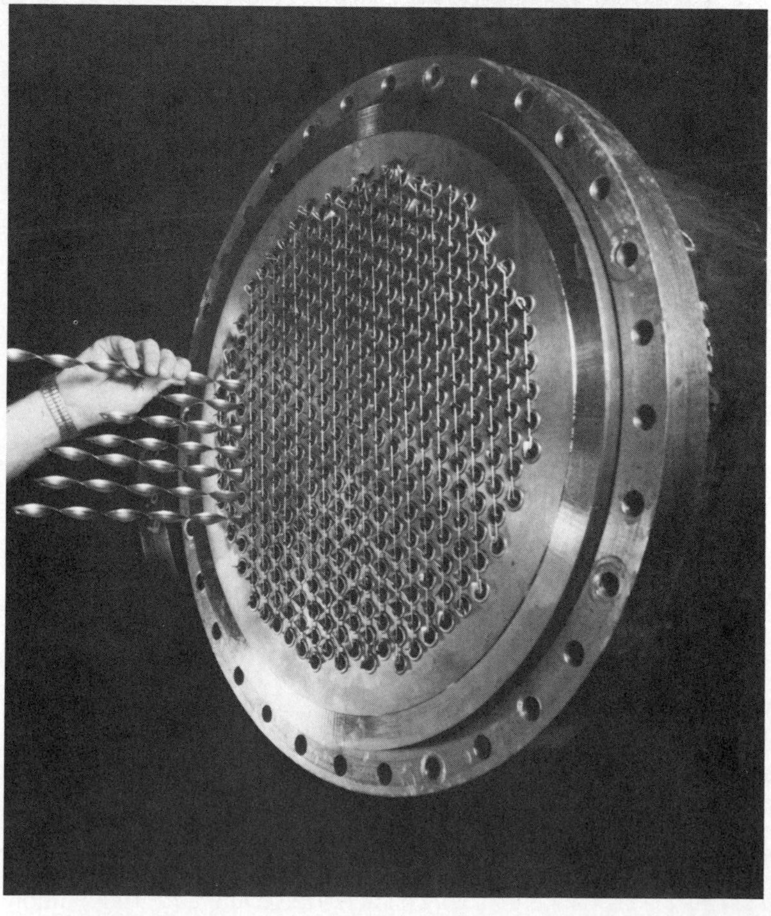

**FIGURE 11–13** Turbulator tube insert. (Courtesy of Brown Fintube Company, Houston, Texas.)

to fluids flowing in tubular heat exchangers is often enhanced by the use of longitudinal, circumferential or spiraled fins, spines, ribs or grooves. Several types of finned tubes are shown in Fig. 11–12. Similarly, tube-fin and plate-fin heat exchangers are commonly constructed of corrugated (wavy) or interrupted (strip, louvered or perforated) fins. Louvered fins are formed by cutting small strips in the fin and lifting the strips above the base fin surface.

Other methods of enhancing the heat-transfer performance of tubular heat exchangers involve the use of various types of inserts that promote mixing. The *turbulator* tube insert shown in Fig. 11–13 has been reported to provide significant increases in heat-transfer rate and discourages the buildup of fouling deposits on the heat-transfer surface. Another concept features the use of a series of fluted and solid spheres mounted on connecting rods, as shown in Fig. 11–14, with the fluted spheres inserted into each tube and the solid spheres arranged around the tubes. In addition to enhancing the heating-pumping power ratio and reducing fouling, the *sphere matrix* inserts are reported to eliminate tube vibration in shell-and tube heat exchangers. Other devices such as coiled tubes and mechanical aids that stir the fluid or scrape the surface are also found in commercial practice.

Heat exchangers with *surface area density* $\beta$ (i.e., ratio of surface area to volume $A_s/V$) greater than about 700 $m^2/m^3$ are referred to as *compact heat exchangers*. Because of their smaller size and weight, compact heat exchangers are prevalent in

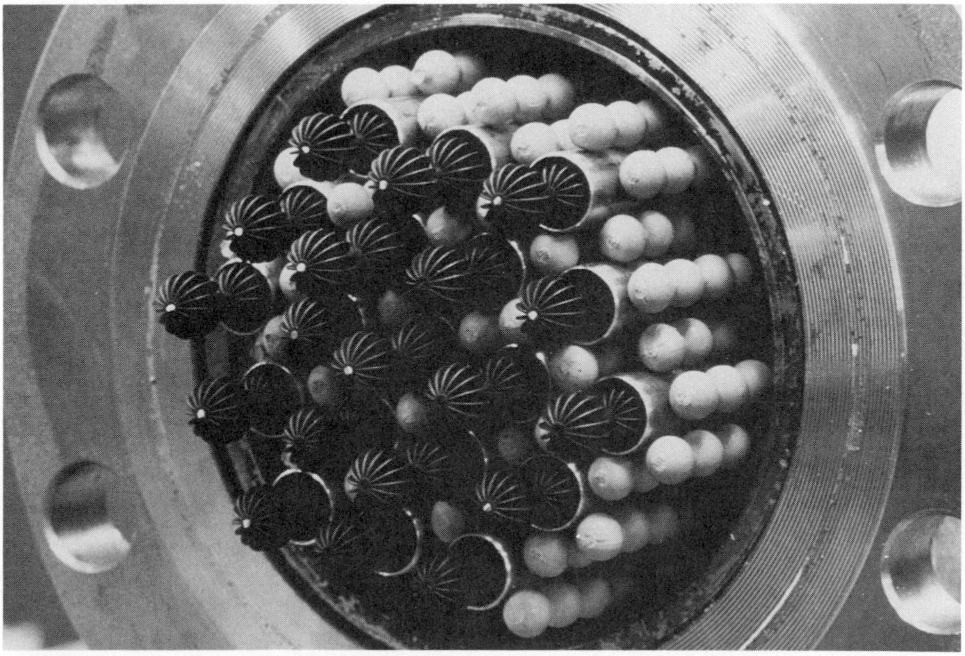

**FIGURE 11–14** Sphere matrix system featuring fluted sphere inserts and solid sphere rod supports. (Courtesy of Vapor Corporation.)

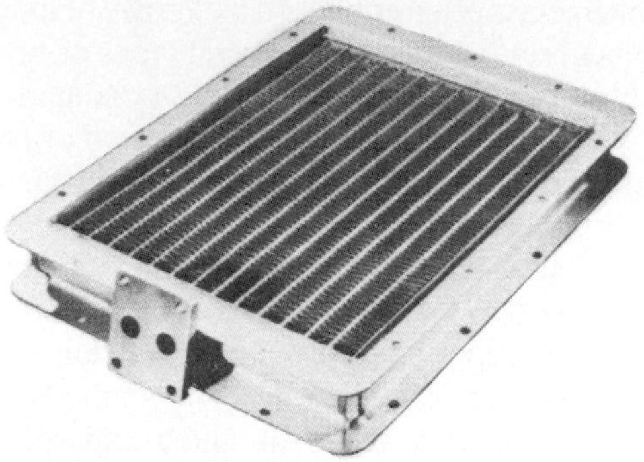

(a) Plate-fin air-oil cooler.

(b) Air-oil temperature regulator.

**FIGURE 11–15** Compact heat exchangers for use in aircraft. (Courtesy of AiResearch Division—Allied Signal Aerospace Company.)

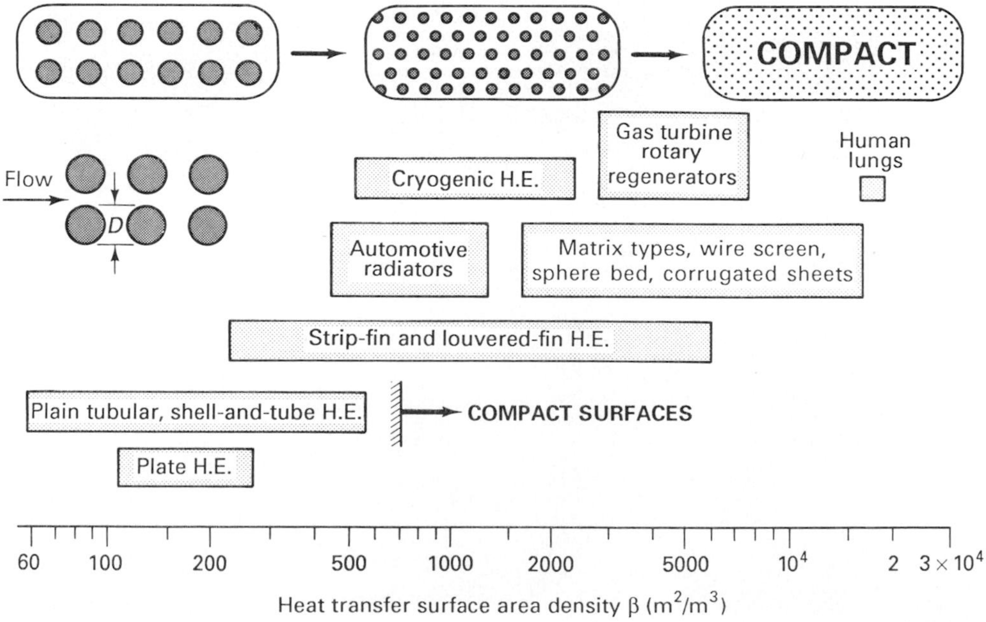


**FIGURE 11–16** Surface area density spectrum of heat exchangers. (From Shah [1]. Used with permission.)

aircraft, vehicular, and marine systems. Two compact crossflow heat exchangers designed for use in aircraft are shown in Fig. 11–15. A spectrum of surface area density prepared by Shah [1] for representative types of heat exchangers is shown in Fig. 11–16. A typical shell-and-tube heat exchanger with 25.4 mm (1 in.) diameter tubes, which is commonly used in power plant condensers, will have a value of $\beta$ of about 130 $m^2/m^3$, such that it would be considered to be noncompact.† On the other hand, modern automobile radiators with 5.5 fins/cm (14 fins/in.) qualify as compact heat exchangers since they generally have an area density of the order of 1100 $m^2/m^3$, which is approximately equivalent to 3-mm-diameter tubes. Interestingly, the human lungs are an amazingly compact heat and mass transfer system, having a surface area density of about 17,500 $m^2/m^3$, which is equivalent to about 0.19 mm tubes.

## 11–3 EVALUATION AND DESIGN

As suggested in Chap. 1, the objective of the thermal and hydraulic evaluation function is to determine the total rate of heat transfer, outlet temperatures, and pressure drop that can be produced by an existing system under given operating conditions (i.e., fluids, mass flow rates, and inlet temperatures). The evaluation or rating of heat-exchanger performance provides the basis of (1) prescribing changes in the operation, and (2) determining when an existing unit must be cleaned, overhauled, modified, or replaced.

In its broadest sense, heat exchanger design encompasses (1) component selection, (2) thermal and hydraulic design, (3) mechanical and metallurgical design, (4) architectural design, and (5) operation and maintenance considerations, all of which should be performed in the context of the overall plant or system, with interaction between the requirements of each of these individual design facets. The objective of the thermal and hydraulic design function is to determine the surface area required to transfer a specified rate of heat with acceptable pressure drop for given fluids, mass flow rates, and terminal (i.e., inlet and exit) temperatures.

The practical lumped analysis approach provides the basis for thermal-hydraulic evaluation and design of heat exchangers. Practical thermal and hydraulic analyses will be developed momentarily for double-pipe, shell-and tube, and crossflow heat exchangers. However, as a preliminary to the development of the practical thermal analysis of heat exchangers, we must first consider the concept of the overall coefficient of heat transfer.

## 11–4 OVERALL COEFFICIENT OF HEAT TRANSFER

Unlike the single fluid processes studied in the previous chapters, the surface temperatures or heat fluxes are not specified for heat exchangers. This complication is overcome by use of the overall coefficient of heat transfer.

† The relation between $D_H$ and $\beta$ used by Shah [1] to approximately represent typical heat exchanger configurations is given by $\beta = 3333/D_H$.

### 11–4–1 Local Overall Coefficient of Heat Transfer

Referring to the counterflow arrangement shown in Fig. 11–17, the differential rate of heat transfer $dq$ across a wall of thickness $\delta$, length $dx$, and uniform perimeters $p_c$ and $p_h$ can be written in terms of the local thermal resistance $R'$ as†

$$dq = \frac{T_h - T_c}{R'_h + R'_k + R'_c} = \frac{T_h - T_c}{\dfrac{1}{(h\,dA_s)_h} + R'_k + \dfrac{1}{(h\,dA_s)_c}} \tag{11–1}$$

where $T_h$ and $T_c$ represent the local bulk-stream temperatures of the hot and cold streams, respectively, the heat-transfer surface areas are represented by $dA_{s,h} = p_h\,dx$ and $dA_{s,c} = p_c\,dx$; $R'_k = \delta/(k\,dA_s)$ for plane walls and $R'_k = \ln(d_o/d_i)/(2\pi k\,dx)$ for cylindrical tube walls with inside and outside diameters $d_i$ and $d_o$. For counterflow (or parallel-flow) arrangements such as this with uniform inlet temperatures $T_{h,i}$ and $T_{c,i}$, the distributions in $T_h$ and $T_c$ are generally functions of $x$. The defining equation for the *local overall coefficient of heat transfer U* is

$$dq = U\,dA_s\,(T_h - T_c) \tag{11–2}$$

where the reference surface area $dA_s$ is normally set equal to $dA_{s,h}$ or $dA_{s,c}$. Combining Eqs. (11–1) and (11–2), we obtain an expression for the local overall coefficient of heat transfer of the form

$$\frac{1}{U\,dA_s} = \frac{1}{(U\,dA_s)_h} = \frac{1}{(U\,dA_s)_c} = \frac{1}{(h\,dA_s)_h} + R'_k + \frac{1}{(h\,dA_s)_c} \tag{11–3}$$

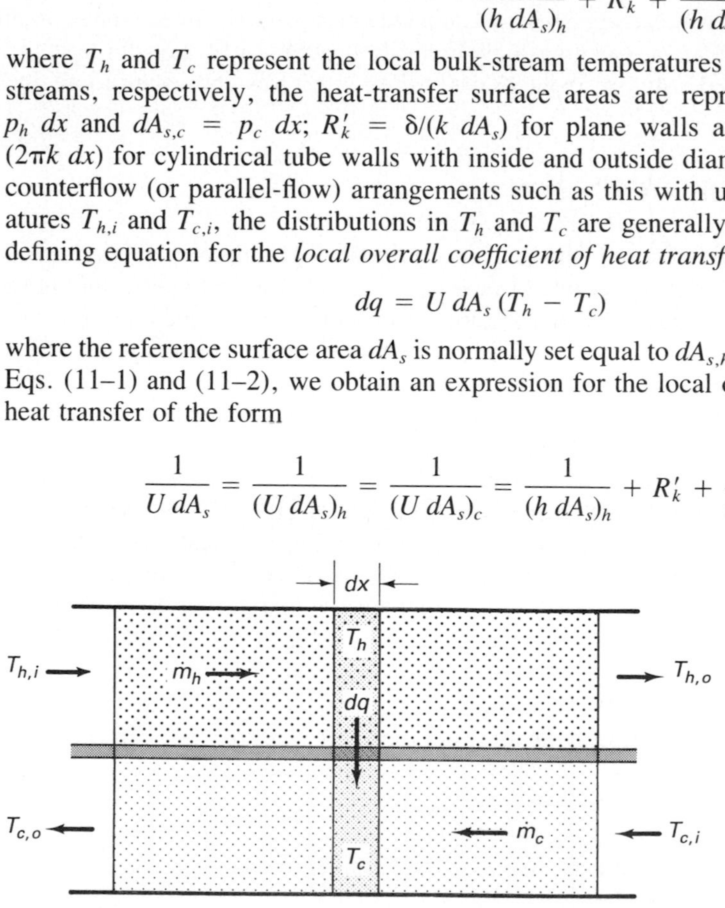


**FIGURE 11–17** Sketch for rate of heat transfer $dq$ between two counterflowing fluids separated by a wall.

† Alternative representations commonly found in the literature for shell-and-tube and double-pipe heat exchangers are of the form

$$dq = \frac{T_s - T_t}{R'_s + R'_k + R'_t} \quad \text{or} \quad dq = \frac{T_o - T_i}{R'_o + R'_k + R'_i}$$

where the shell or annular side is represented by $s$ or $o$ and the tube side by $t$ or $i$.

Since the perimeters are uniform, this equation can also be written as†

$$\frac{1}{UA_s} = \frac{1}{(UA_s)_h} = \frac{1}{(UA_s)_c} = \frac{1}{(hA_s)_h} + R_k + \frac{1}{(hA_s)_c} \tag{11–4}$$

where $A_s = pL$, $A_{s,h} = p_hL$, $A_{s,c} = p_cL$, and $L$ is the length of the heat exchanger wall; $R_k = \delta/(kA_s)$ for plane walls, and $R_k = \ln(d_o/d_i)/(2\pi kL)$ for cylindrical tube walls. From these equations we see that $U_h$ and $U_c$ are geometrically related by $UA_s = U_hA_{s,h} = U_cA_{s,c}$. We also observe that $U$ is nonuniform for situations in which $h_h$ and/or $h_c$ are nonuniform.

The local coefficients of heat transfer $h_h$ and $h_c$ are generally represented by the use of correlations for Stanton number $St_h$ and $St_c$ (or Nusselt number $Nu_h$ and $Nu_c$) in terms of hydraulic diameter $D_H$ and Reynolds number $Re$ of the type introduced in Chaps. 8–10 for internal flow through passages with smooth, finned, rough, or otherwise modified surfaces.

## 11-4-2 Mean Overall Coefficient of Heat Transfer

Similar to the defining equation for $\bar{h}$, the *mean overall coefficient of heat transfer* $\bar{U}$ for a wall with surface area $A_s$ is given by

$$\bar{U} = \frac{1}{A_s}\int_{A_s} U\, dA_s \tag{11–5}$$

As we will see, it is $\bar{U}$ that is generally called for in the analysis of heat exchangers.

In design work $\bar{U}$ is computed by the use of Eq. (11–5) together with general (design) correlations for the coefficients of heat transfer of the hot and cold fluids. For the two limiting cases in which either the hot-side or cold-side thermal resistance controls, $\bar{U}$ can be approximated by $\bar{h}_h$ or $\bar{h}_c$, respectively (for clean unfinned surfaces). On the other hand, for cases in which these thermal resistances are of the same order of magnitude, both $h_h$ and $h_c$ must be used in the evaluation of $\bar{U}$. However, in practice $U$ is often treated as a constant, such that $\bar{U} \simeq U$. In applications for which this is permissible, $\bar{U}$ is sometimes calculated by evaluating the coefficients $h_h$ and $h_c$ at the midpoint of the heat exchanger. $\bar{U}$ is also commonly approximated by setting $h_h$ and $h_c$ in Eq. (11–4) equal to the mean values $\bar{h}_h$ and $\bar{h}_c$ at the outlets; that is,

$$\frac{1}{\bar{U}A_s} = \frac{1}{(\bar{h}A_s)_h} + R_k + \frac{1}{(\bar{h}A_s)_c} \tag{11–6}$$

† This equation commonly appears in the literature as

$$\frac{1}{U_h} = \frac{1}{h_h} + \frac{\delta}{k(A_k/A_{s,h})} + \frac{1}{h_c(A_{s,c}/A_{s,h})}$$

based on surface area $A_{s,h}$, or

$$\frac{1}{U_c} = \frac{1}{h_h(A_{s,h}/A_{s,c})} + \frac{\delta}{k(A_k/A_{s,c})} + \frac{1}{h_{s,c}}$$

based on surface area $A_{s,c}$, where $A_k$ is the effective area of the wall; $A_k = A_s$ for plane walls and $A_k = (d_o - d_i)L/\ln(d_o/d_i)$ for cylindrical tube walls.

Whereas design work involves the use of Eq. (11–5) together with generalized correlations for $h_h$ and $h_c$ to calculate $\overline{U}$, the evaluation function is commonly undertaken with the aid of specific (rating) correlations for $\overline{U}$, which are established on the basis of the analysis of previously recorded performance data (i.e., terminal temperatures, fluid properties, and mass flow rates) for the unit in question. The use of the practical thermal analysis approach in calculating $\overline{U}$ will be considered in Sec. 11–5–1.

Representative values of $\overline{U}$ are given in Table 11–2 for various situations found in practice.

**TABLE 11–2** Overall heat-transfer coefficients—representative values

| | $\overline{U}$ | |
|---|---|---|
| *Application* | Btu/(h ft² °F) | W/(m² °C) |
| Steam condenser | 200–$10^3$ | $10^3$–6 × $10^3$ |
| Freon-12 condenser with water coolant | 50–200 | 300–$10^3$ |
| Water-to-water | 100–300 | 600–2 × $10^3$ |
| Water-to-oil | 20–60 | 100–400 |
| Water-to-gasoline | 60–90 | 300–500 |
| Steam-to-light fuel oil | 30–60 | 200–400 |
| Steam-to-heavy fuel oil | 10–30 | 60–200 |
| Steam-to-gasoline | 50–200 | 300–$10^3$ |
| Finned-tube heat exchanger; water in tubes, air over tubes | 5–10† <br> 70–150‡ | 30–60† <br> 420–840‡ |
| Finned-tube heat exchanger; steam in tubes, air over tubes | 5–50† <br> 70–700‡ | 30–300† <br> 420–4200‡ |

† Based on area of finned surface.
‡ Based on area of unfinned surface.
*Source*: Based on [10].

### 11–4–3 Effect of Fouling

It should be noted that the overall coefficient of heat transfer is often reduced by fouling deposits such as dirt that accumulate on the heat-exchanger walls. For example, Fig. 11–18 shows severe fouling on the super-heater tubes of a boiler. The effect of such deposits is usually accounted for by the *fouling factor* $F_f$, which is defined in terms of the coefficient of heat transfer $h$ by

$$F_f = \frac{1}{h_f} - \frac{1}{h} \qquad (11\text{–}7)$$

**FIGURE 11–18** Ash deposit fouling on secondary superheater tubes. (From *Steam—Its Generation and Use*. Copyright 1978 by Babcock and Wilcox. Used with permission.)

where $h_f$ represents the coefficient of heat transfer after fouling has occurred. It follows that the overall coefficient of heat transfer $U$ for a wall with fouling on both surfaces is represented by

$$\frac{1}{UA_s} = \frac{1}{(hA_s)_h} + \left(\frac{F_f}{A_s}\right)_h + R_k + \left(\frac{F_f}{A_s}\right)_c + \left(\frac{1}{hA_s}\right)_c \tag{11–8}$$

for unfinned surfaces. Representative fouling factors are given in Table 11–3. A review of the literature on fouling in heat exchangers is provided in references 11 and 12.

Corrosion in heat exchangers is another problem that is encountered in practice, especially in the chemical industry. To overcome this problem, glass and plastic tubes and shells are often used. For example, the double-pipe tube bank shown in Fig. 11–19 is constructed of glass-lined steel pipe. The excellent thermal conduction properties of borosilicate materials make the use of glass quite attractive in applications involving corrosive chemicals.

**TABLE 11–3** Fouling factors—representative values

| *Type fluid* | $F_f$ (h ft² °F/Btu) | $F_f$ (m² °C/W) |
|---|---|---|
| Seawater: | | |
| Below 50°C | $5 \times 10^{-4}$ | $9 \times 10^{-5}$ |
| Above 50°C | $1 \times 10^{-3}$ | $2 \times 10^{-4}$ |
| Treated boiler feedwater above 50°C | $1 \times 10^{-3}$ | $2 \times 10^{-4}$ |
| Fuel oil | $5 \times 10^{-3}$ | $9 \times 10^{-4}$ |
| Quenching oil | $4 \times 10^{-3}$ | $7 \times 10^{-4}$ |
| Alcohol vapors | $5 \times 10^{-4}$ | $9 \times 10^{-5}$ |
| Steam, non-oil-bearing | $5 \times 10^{-4}$ | $9 \times 10^{-5}$ |
| Industrial air | $2 \times 10^{-3}$ | $4 \times 10^{-4}$ |
| Refrigerating liquid | $1 \times 10^{-3}$ | $2 \times 10^{-4}$ |

*Source*: Based on [13].

Other factors that influence the overall coefficient of heat transfer include fins, baffles, multiple tube passes, crossflow, and two-phase flow. For example, the overall coefficient of heat transfer for finned surfaces is given by

$$\frac{1}{UA_s} = \frac{1}{(\eta_o h A_s)_h} + (R_{tc})_h + \left(\frac{F_f}{\eta_o A_s}\right)_h + R_k + \left(\frac{F_f}{\eta_o A_s}\right)_c + (R_{tc})_c + \frac{1}{(\eta_o h A_s)_c} \tag{11–9}$$

where the net surface efficiency $\eta_o$ is defined by Eq. (2–127) and $R_{tc}$ represents the thermal contact resistance between the fin and wall.

**FIGURE 11–19**
Double-pipe tube bank constructed of glass-lined steel pipes. (Courtesy of De Dietrich, Inc.)

## EXAMPLE 11–1

Freon F-12 at $-20°C$ and 0.4 MPa flows in the annulus of a small double-pipe heat exchanger at a rate of 0.265 kg/s. Hot water at 98°C and 0.2 MPa passes through the tube with a mass flow rate of 0.035 kg/s. The heat exchanger is constructed of thin-walled copper tubing with 2-cm inside diameter, 3-cm outside diameter, and 3-m length. Determine the approximate mean overall coefficient of heat transfer.

### Solution

*Objective* Determine $\overline{U}$.

*Schematic* Double-pipe heat exchanger: counterflow arrangement.

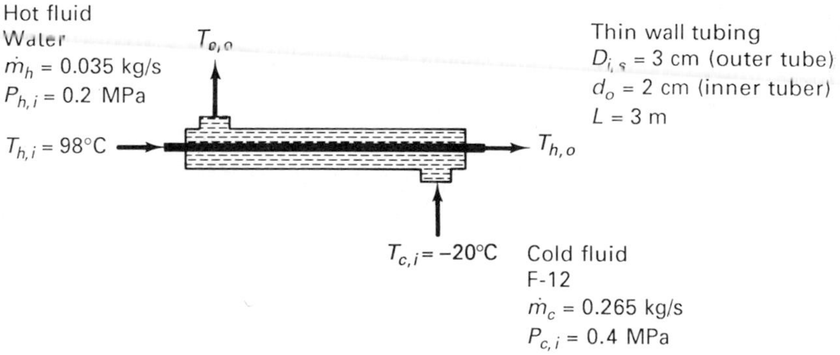
Hot fluid
Water
$\dot{m}_h$ = 0.035 kg/s
$P_{h,i}$ = 0.2 MPa
$T_{h,i}$ = 98°C
$T_{c,o}$
Thin wall tubing
$D_{i,s}$ = 3 cm (outer tube)
$d_o$ = 2 cm (inner tuber)
$L$ = 3 m
$T_{h,o}$
$T_{c,i}$ = −20°C
Cold fluid
F-12
$\dot{m}_c$ = 0.265 kg/s
$P_{c,i}$ = 0.4 MPa

*Assumptions/Conditions*

negligible conduction resistance $R_k$
negligible fouling
negligible entrance effects
uniform properties
standard conditions

*Properties* Assuming that both fluids are compressed, the properties are approximated by use of Table A–C–3 for saturated liquid.

Hot water at $T = 98°C = 371$ K (Table A–C–3): $\rho = 1/v_f = 961$ kg/m$^3$, $\mu = 2.86 \times 10^{-4}$ kg/(m s), $\nu = 2.98 \times 10^{-7}$ m$^2$/s, $c_P = 4.21$ kJ/(kg °C), $k = 0.680$ W/(m °C), $Pr = 1.78$.

Freon F-12 at $-20°C$ (Table A–C–3):† $\rho = 1460$ kg/m$^3$, $\nu = 2.35 \times 10^{-7}$ m$^2$/s, $\mu = \rho\nu = 3.43 \times 10^{-4}$ kg/(m s), $c_P = 0.907$ kJ/(kg °C), $k \simeq 0.071$ W/(m °C), $Pr \simeq 4.4$.

† $k$ and $Pr$ are given to within two significant figures.

*Analysis* The Reynolds number of the hot water flowing in the tube is calculated first by writing

$$D_{H,h} = d_i = d_o = 2 \text{ cm}$$

$$Re_h = \left(\frac{D_H U_b}{\nu}\right)_h = \left(\frac{D_H}{\mu}\frac{\dot{m}}{A}\right)_h = \frac{d_i \dot{m}_h}{\mu_h \pi d_i^2/4}$$

$$= \frac{4\dot{m}_h}{\mu_h \pi d_i} = \frac{4(0.035 \text{ kg/s})}{[2.86 \times 10^{-4} \text{ kg/(m s)}]\ \pi\ (0.02 \text{ m})} = 7790$$

Thus, the flow is transitional turbulent. Because $L/D$ (= 150) is much greater than 10, we assume that the flow is fully developed with negligible entrance effects. To approximate the coefficient of heat transfer $h_h$, we utilize Eqs. (8–16) and (8–20).

$$f_h = 0.079(7790)^{-0.25} = 0.00841$$

$$Nu_h = \frac{(0.00841/2)(7790)(1.78)}{1.07 + 12.7\sqrt{0.00841/2}\,(1.78^{2/3} - 1)} = 40.0$$

$$h_h = Nu_h\left(\frac{k}{D_H}\right)_h = 40.0\,\frac{0.68 \text{ W/(m °C)}}{0.02 \text{ m}} = 1360 \text{ W/(m}^2\text{ °C)}$$

For the Freon flowing in the annulus, we write

$$D_{H,c} = D_{i,s} - d_o = 1 \text{ cm}$$

$$Re_c = \left(\frac{D_H}{\mu}\frac{\dot{m}}{A}\right)_c = \frac{\dot{m}_c(D_{i,s} - d_o)}{\mu_c \pi (D_{i,s}^2 - d_o^2)/4} = \frac{4\dot{m}_c}{\mu_c \pi (D_{i,s} + d_o)}$$

$$= \frac{4(0.265 \text{ kg/s})}{[3.43 \times 10^{-4} \text{ kg/(m s)}]\ \pi\ (0.05 \text{ m})} = 19{,}700$$

Referring to Fig. 8–9 and noting that $d_o/D_{i,s} = D_i/D_o = 0.666$, we find that $Nu_c = Nu_{ii} \approx Nu/0.97$. Thus, for this application $Nu_c$ can be set equal to $Nu$ as a reasonable conservative approximation. Using Eqs. (8–15) and (8–16) to calculate $Nu$, we obtain

$$f_c = (1.58 \ln 19{,}700 - 3.28)^{-2} = 0.00656$$

$$Nu_c = \frac{(0.00656/2)(19{,}700)(4.4)}{1.07 + 12.7\sqrt{0.00656/2}\,(4.4^{2/3} - 1)} = 124$$

$$h_c = Nu_c\left(\frac{k}{D_H}\right)_c = 124\,\frac{0.071 \text{ W/(m °C)}}{0.01 \text{ m}} = 880 \text{ W/(m}^2\text{ °C)}$$

The mean overall coefficient of heat transfer $\overline{U}$ is now approximated by using Eq. (11–6) with $A_s = A_{s,h} \simeq A_{s,c}$, $\overline{h}_h \simeq h_h$, and $\overline{h}_c \simeq h_c$,

$$\frac{1}{\overline{U}} = \frac{1}{h_h} + \frac{1}{h_c} = \left(\frac{1}{1360} + \frac{1}{880}\right)\frac{\text{m}^2\ {}^\circ\text{C}}{\text{W}}$$

$$\overline{U} = 534\ \text{W/(m}^2\ {}^\circ\text{C)}$$

If the outlet temperatures and wall temperatures are known, our calculations for $h_h$ and $h_c$ can be refined by evaluating the properties of each fluid at the arithmetic average of its inlet and outlet temperatures and by utilizing the correction factors introduced in Chap. 8.

It should be noted that the use of the somewhat less conservative approximation for $Nu_c$ indicated by Fig. 8–9 gives rise to $Nu_c = Nu/0.97 = 128$, $h_c = 907$ W/(m² °C), and $\overline{U} = 544$ W/(m² °C). Thus, this refinement results in a change of only 1.9% in the calculation for $\overline{U}$.

## EXAMPLE 11–2

A parallel-flow double-pipe heat exchanger is constructed of 0.113-in.-thick steel (0.5% carbon) tubing with 0.824-in.-I.D. inner tube and 2-in.-I.D. outer tube. The tube-side and shell-side local coefficients of heat transfer for a certain application are reported to be approximated by 200 Btu/(h ft² °F) and 1000 Btu/(h ft² °F), respectively, and the tube-side fouling factor is $5.67 \times 10^{-4}$ (h ft² °F)/Btu. Calculate the local overall coefficient of heat transfer.

### Solution

*Objective* Determine $U$.

*Schematic* Double-pipe heat exchanger: parallel-flow arrangement.

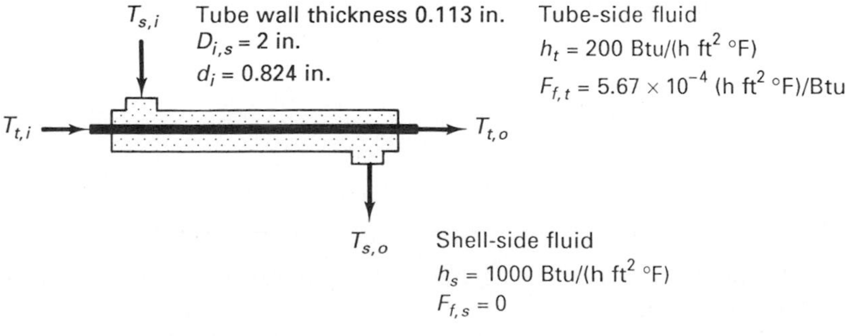
$T_{s,i}$
Tube wall thickness 0.113 in.
$D_{i,s}$ = 2 in.
$d_i$ = 0.824 in.
Tube-side fluid
$h_t$ = 200 Btu/(h ft² °F)
$F_{f,t}$ = 5.67 × 10⁻⁴ (h ft² °F)/Btu
$T_{t,i}$
$T_{t,o}$
$T_{s,o}$
Shell-side fluid
$h_s$ = 1000 Btu/(h ft² °F)
$F_{f,s}$ = 0

*Assumptions/Conditions*

uniform properties

standard conditions

*Properties* Carbon steel (0.5% C) (Table A–C–1):

$$k \simeq 54 \frac{\text{W}}{\text{m °C}} \frac{0.5778 \text{ Btu/(h ft}^2 \text{ °F)}}{\text{W/(m °C)}} = 31.2 \text{ Btu/(h ft}^2 \text{ °F)}$$

*Analysis* The local overall coefficient of heat transfer for this situation is represented by Eq. (11–8),

$$\frac{1}{UA_s} = \frac{1}{(hA_s)_t} + \left(\frac{F_f}{A_s}\right)_t + R_k + \left(\frac{F_f}{A_s}\right)_s + \frac{1}{(hA_s)_s}$$

where the subscripts $t$ and $s$ are used in place of $h$ and $c$, and

$$R_k = \frac{\ln(d_o/d_i)}{2\pi k L}$$

Setting $A_s = A_{s,t} = \pi d_i L$, the local overall coefficient of heat transfer relative to the tube-side surface area is given as

$$\frac{1}{U_t} = \frac{1}{h_t} + F_{f,t} + \frac{d_i}{2k} \ln\left(\frac{d_o}{d_i}\right) + F_{f,s} \frac{d_i}{d_o} + \frac{1}{h_s} \frac{d_i}{d_o}$$

Substituting for the various parameters, we obtain

$$\frac{1}{U_t} = \left[\frac{1}{200} + 5.67 \times 10^{-4} + \frac{0.824/12}{2(31.2)} \ln\left(\frac{1.05}{0.824}\right) + 0 + \frac{1}{1000}\left(\frac{0.824}{1.05}\right)\right] \frac{\text{h ft}^2 \text{ °F}}{\text{Btu}}$$

$$= (0.005 + 0.000567 + 0.000267 + 0.000785) \frac{\text{h ft}^2 \text{ °F}}{\text{Btu}}$$

$$= 0.00664 \text{ h ft}^2 \text{ °F/Btu}$$

or

$$U_t = 151 \text{ Btu/(h ft}^2 \text{ °F)}$$

## EXAMPLE 11–3

Air at 15°C and 1 atm with entering velocity of 6.8 m/s is to be heated in a crossflow heat exchanger with eighty 9.52-mm-diameter 2.5-m-long thin wall tubes. The tubes are placed 10 tubes deep in a staggered array with longitudinal and transverse pitches of 1.19 cm and 1.43 cm, respectively. Compressed water at 65°C enters the tubes with a mass flow rate of 0.25 kg/s. Determine the mean overall coefficient of heat transfer for this application.

**Solution**

*Objective* Determine $\overline{U}$.

*Schematic* Crossflow heat exchanger.

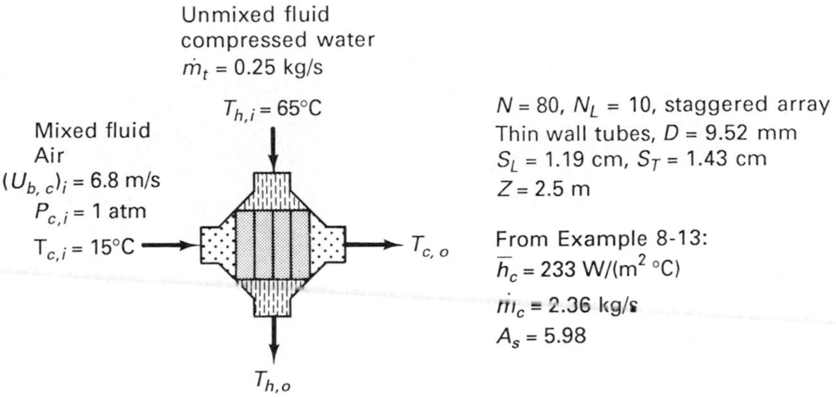
Unmixed fluid
compressed water
$\dot{m}_t = 0.25$ kg/s
$T_{h,i} = 65°C$
Mixed fluid
Air
$(U_{b,c})_i = 6.8$ m/s
$P_{c,i} = 1$ atm
$T_{c,i} = 15°C$
$T_{c,o}$
$T_{h,o}$
$N = 80$, $N_L = 10$, staggered array
Thin wall tubes, $D = 9.52$ mm
$S_L = 1.19$ cm, $S_T = 1.43$ cm
$Z = 2.5$ m
From Example 8-13:
$\bar{h}_c = 233$ W/(m$^2$ °C)
$\dot{m}_c = 2.36$ kg/s
$A_s = 5.98$

*Assumptions/Conditions*

negligible conduction resistance $R_k$
negligible fouling
uniform properties
standard conditions

*Properties* Water at $T = 65°C = 338$ K (Table A–C–3): $\rho = 1/v_f = 981$ kg/m$^3$, $\mu = 4.33 \times 10^{-4}$ kg/(m s), $\nu = \mu/\rho = 0.442 \times 10^{-6}$ m$^2$/s, $c_P = 4.19$ kJ/(kg °C), $k = 0.658$ W/(m °C), $Pr = 2.75$.

*Analysis* Referring back to Example 8–13, the mean coefficient of heat transfer $\bar{h}_c$ for flow of air over the tube bank is approximately 233 W/(m$^2$ °C) (neglecting variable property effects).

To determine the mean coefficient of heat transfer $\bar{h}_h$ for the tubular water flow, we must calculate the Reynolds number. The bulk-stream velocity is given by

$$U_{b,h} = \left(\frac{\dot{m}}{\rho A}\right)_h = \frac{0.25 \text{ kg/s}}{(981 \text{ kg/m}^3)\,[80\,\pi\,(0.00952 \text{ m})^2/4]} = 0.0447 \text{ m/s}$$

such that $Re_h$ becomes

$$Re_h = \left(\frac{DU_b}{\nu}\right)_h = \frac{(0.00952 \text{ m/s})(0.0447 \text{ m/s})}{0.442 \times 10^{-6} \text{ m}^2\text{/s}} = 963$$

Thus, the flow is laminar and the mean Nusselt number can be approximated from Eq. (8–12),

$$\overline{Nu}_h = 3.66 + \frac{0.104 \dfrac{Re\,Pr}{x/D}}{1 + 0.016 \left(\dfrac{Re\,Pr}{x/D}\right)^{0.8}}$$

Setting $x = Z$, we have

$$\left(\frac{x/D}{Re\,Pr}\right)_h = \frac{2.5/0.00952}{963(2.75)} = 0.0992$$

such that the mean Nusselt number for the water at the outlet of the heat exchanger becomes

$$\overline{Nu}_h = 3.66 + \frac{0.104/0.0992}{1 + 0.016(1/0.0992)^{0.8}} = 4.61$$

Thus, $\bar{h}_h$ is

$$\bar{h}_h = \left(\overline{Nu}\,\frac{k}{D}\right)_h = 4.61\,\frac{0.658\ \text{W/(m °C)}}{0.00952\ \text{m}} = 319\ \text{W/(m}^2\ \text{°C)}$$

Assuming that the thermal resistance of the tube wall and fouling are negligible, the mean overall coefficient of heat transfer is approximated by

$$\frac{1}{\overline{U}} = \frac{1}{\bar{h}_h} + \frac{1}{\bar{h}_c} = \left(\frac{1}{319} + \frac{1}{233}\right)\frac{\text{m}^2\ \text{°C}}{\text{W}}$$

or

$$\overline{U} = 135\ \text{W/(m}^2\ \text{°C)}$$

As in the previous example, this approximation for $\overline{U}$ can be refined to account for property variation with temperature once the outlet temperatures and the wall temperatures are known.

## 11–5 HEAT EXCHANGER ANALYSIS

The situations that must be dealt with in the practical analysis of direct transfer heat exchangers (recuperators) range from relatively simple unfinned double-pipe modules involving basic counterflow or parallel flow with uniform cross-sectional area to more complex finned-tube double-pipe, baffled shell-and-tube, and crossflow arrangements with combinations of mixed and unmixed crossflow, counterflow and parallel flow, finned and unfinned tubes, and uniform and nonuniform cross-sectional areas.

To develop a general approach that is applicable to this variety of flow arrangements, we model the heat exchanger core by uniform and continuous flow channels through which the fluids pass in the appropriate relative counterflow, parallel flow, and crossflow paths. Following the approach in Sec. 8–4, the cross-sectional area of the core model is represented by the minimum free-flow area $A_{min}$ for flow passages with nonuniform cross section, such as crossflow over tube banks or flow over finned tubes.

To provide a general framework for analysis, the conditions of the thoroughly mixed entering and exiting streams will primarily be specified by the use of subscripts with reference to the hot ($h$) and cold ($c$) fluids according to the schedule listed in Table 11–4. An alternative perspective that is commonly used in the literature is based on the shell ($s$) and tube ($t$) sides, as is also indicated in Table 11–4.

**TABLE 11–4** Subscripts used to designate mixed entering and exiting conditions of the fluid stream†

| | *Flow direction* | | *Station* | | |
|---|---|---|---|---|---|
| *Reference* | *inlet* | *outlet* | 1 | 2 | *n* |
| hot fluid | $h,i$ | $h,o$ | $h,1$ | $h,2$ | $h,n$ |
| cold fluid | $c,i$ | $c,o$ | $c,1$ | $c,2$ | $c,n$ |
| shell (annulus) | $s,i$ | $s,o$ | $s,1$ | $s,2$ | $s,n$ |
| tube | $t,i$ | $t,o$ | $t,1$ | $t,2$ | $t,n$ |

† Other conventions are sometimes used in the literature. Examples include in for $i$, out for $o$, $i$ for $t$, and $o$ for $s$. In addition, cap $T$ and lower case $t$ are sometimes used to differentiate between the hot and cold (or shell and tubular) fluids.

The general aspects associated with the practical thermal and hydraulic analysis of direct transfer heat exchangers that are applicable to double-pipe, shell-and-tube, and crossflow heat exchangers are developed in this section. The practical analysis of regenerators is presented by Kays and London [9].

### 11–5–1 Practical Thermal Analysis

The practical thermal analysis of heat exchangers features the representation of the heat transfer rate in terms of the local overall coefficient of heat transfer $U$ by Eq. (11–2),†

$$dq = U\,dA_s\,(T_h - T_c) \tag{11–2}$$

and, referring to Fig. 11–20, is based on lumped and lumped/differential formulations of the first law of thermodynamics of the general forms

$$q = \dot{H}_{h,i} - \dot{H}_{h,o} \qquad q = \dot{H}_{c,o} - \dot{H}_{c,i} \tag{11–10,11}$$

$$dq = -d\dot{H}_h \qquad dq = d\dot{H}_c \tag{11–12,13}$$

† As discussed in reference 2, Eqs. (11–2), (11–12), and (11–13) are put into a somewhat different form for shell-and-tube heat exchangers.

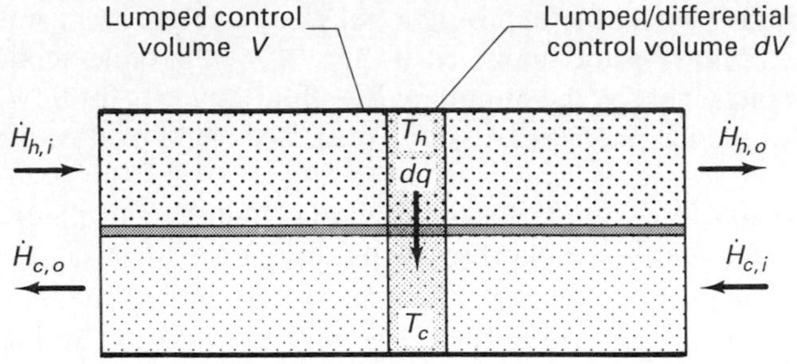


**FIGURE 11–20** General perspective for lumped and lumped/differential formulation presented in context of a basic counterflow arrangement.

which represent the total and differential rates of heat transfer from the hot fluid [Eqs. (11–10) and (11–12)] and to the cold fluid [Eqs. (11–11) and (11–13)]. The solution of Eqs. (11–2) and (11–10)–(11–13) depends on the nature of the distributions in bulk-stream temperatures $T_h$ and $T_c$ and on whether the fluids are single-phase or two-phase. Whereas the bulk-stream temperatures are both functions of $x$ alone for counterflow and/or parallel-flow in double-pipe and well mixed shell-and-tube heat exchanger units, two independent variables (e.g., $x$ and $z$) are required to describe the distributions in $T_h$ and $T_c$ for flow in crossflow heat exchangers.

For single-phase fluids, Eqs. (11–10)–(11–13) become

$$q = (\dot{m}c_P)_h(T_{h,i} - T_{h,o}) \quad (11\text{–}14)$$

and

$$q = (\dot{m}c_P)_c(T_{c,o} - T_{c,i}) \quad (11\text{–}15)$$

for uniform specific heats, and

$$dq = -(\dot{m}c_P)_h\, dT_h \qquad dq = (\dot{m}c_P)_c\, dT_c \quad (11\text{–}16,17)$$

for one-dimensional distributions in bulk-stream temperature $T_h(x)$ and $T_c(x)$, such as occur in double-pipe or well baffled shell-and-tube heat exchangers.† The products $(\dot{m}c_P)_h$ and $(\dot{m}c_P)_c$ are known as the *capacity rates* and are commonly designated by $C_h$ and $C_c$, respectively. These equations provide the basis for relating the total rate of heat transfer $q$ and the outlet temperatures $T_{h,o}$ and $T_{c,o}$ to the overall coefficient of heat transfer, surface area $A_s$, and other system parameters (i.e., fluid properties, mass flow rates $\dot{m}_h$ *and* $\dot{m}_c$, inlet fluid temperatures $T_{h,i}$ and $T_{c,i}$, and geometry and flow arrangement). For basic uniform property flows, the solutions to these equations are expressed in terms of the effectiveness $\epsilon$ or the log mean temperature difference *LMTD*, with both solution formats involving the mean overall coefficient of heat transfer $\overline{U}$.

† Equations (11–16) and (11–17) take a more general form for two-dimensional systems involving crossflow [2].

### *Effectiveness Method*

The *effectiveness* $\epsilon$ is defined by Eq. (8–53),

$$q = q_{max}\,\epsilon \tag{11–18}$$

where the maximum rate of heat transfer $q_{max}$ corresponds to the thermodynamic limiting case in which the outlet temperature of the fluid with minimum capacity rate approaches the inlet temperature of the other fluid; that is,

$$q_{max} = C_{min}(T_{h,i} - T_{c,i}) \tag{11–19}$$

with the minimum capacity rate $(\dot{m}c_P)_{min}$ represented by $C_{min}$. Combining these relations with Eqs. (11–14) and (11–15), we obtain

$$q = C_{min}(T_{h,i} - T_{c,i})\,\epsilon \tag{11–20}$$

or

$$\epsilon = \frac{C_h}{C_{min}}\left(\frac{T_{h,i} - T_{h,o}}{T_{h,i} - T_{c,i}}\right) = \frac{C_c}{C_{min}}\left(\frac{T_{c,o} - T_{c,i}}{T_{h,i} - T_{c,i}}\right) \tag{11–21}$$

Thus,

$$\epsilon = \frac{T_{h,i} - T_{h,o}}{T_{h,i} - T_{c,i}} \tag{11–22}$$

for $C_{min} = C_h < C_c$, or

$$\epsilon = \frac{T_{c,o} - T_{c,i}}{T_{h,i} - T_{c,i}} \tag{11–23}$$

for $C_{min} = C_c < C_h$.

As shown in the context of the analysis of double-pipe heat exchangers (Example 11–4), the effectiveness $\epsilon$ can be expressed in terms of the *capacitance ratio* $C^*$,

$$C^* = \frac{C_{min}}{C_{max}} \tag{11–24}$$

and the *number of transfer units NTU*,

$$NTU = \frac{\overline{U}A_s}{C_{min}} \tag{11–25}$$

which provides a dimensionless index of the size of the heat exchanger; the term $\overline{U}A_s$ represents the heat exchanger *thermal capacity*. In general, $\epsilon$ is a function of $C^*$, $NTU$, and the geometry and flow arrangement; that is,

$$\epsilon = \epsilon(C^*, NTU) \qquad \text{or} \qquad NTU = NTU\,(C^*, \epsilon) \tag{11–26a,b}$$

for a specific geometry and flow arrangement, with the solution results generally expressed in form of analytical relations and/or charts for double-pipe, shell-and-tube, and crossflow heat exchangers. Whereas relations of the form of Eq. (11–26a) are generally used for evaluation of heat exchanger performance, the alternative form given by Eq. (11–26b) is more convenient for design work.

Representative analytical relations for $\epsilon$-*NTU* are given in Tables 11–5a and 11–5b for several standard double-pipe, shell-and-tube, and crossflow heat exchangers. In addition, typical effectiveness-*NTU* charts are given in Figs. 11–21 to 11–23. (An extensive listing of analytical expressions and charts for heat exchanger effectiveness is presented in references 2 and 9.) In applying these relationships and charts, the hot and cold reference temperatures $T_h$ and $T_c$ are interchangeable, except for applications involving certain types of shell-and-tube heat exchangers, and crossflow heat

**TABLE 11–5a** Relations for heat exchanger effectiveness of the form of Eq. (11–26a)

| *Heat exchanger type* | *Effectiveness* | |
|---|---|---|
| All exchangers for $C^* = 0$ | $\epsilon = 1 - \exp(-NTU)$ | (a) |
| Double-pipe | | |
| parallel flow | $\epsilon = \dfrac{1 - \exp[-NTU(1 + C^*)]}{1 + C^*}$ | (b) |
| counterflow | $\epsilon = \dfrac{1 - \exp[-NTU(1 - C^*)]}{1 - C^* \exp[-NTU(1 - C^*)]}$ | (c) |
| counterflow, $C^* = 1$ | $\epsilon = \dfrac{NTU}{1 + NTU}$ | (d) |
| Shell-and-tube | | |
| one shell (TEMA E) 2, 4, . . . . . , tube passes | $\epsilon = \dfrac{2}{(1 + C^*) + (1 + C^{*2})^{1/2}(1 + e^{-\Gamma})/(1 - e^{-\Gamma})}$ | (e) |
| | $\Gamma = NTU(1 + C^{*2})^{1/2}$ | |
| Crossflow | | |
| one fluid mixed, one fluid unmixed | | |
| $C_{\min} = C_{\text{mixed}} = C_s$ | $\epsilon = \dfrac{T_{s,i} - T_{s,o}}{T_{s,i} - T_{t,i}} = 1 - \exp\left(-\dfrac{\Gamma}{C^*}\right)$ | (f) |
| | $\Gamma = 1 - \exp(-C^* NTU)$ | |
| $C_{\min} = C_{\text{unmixed}} = C_t$ | $\epsilon = \dfrac{T_{t,o} - T_{t,i}}{T_{s,i} - T_{t,i}} = \dfrac{1}{C^*}[1 - \exp(-\Gamma C^*)]$ | (g) |
| | $\Gamma = 1 - \exp(-NTU)$ | |
| both fluids unmixed | $\epsilon = 1 - \exp\left[\dfrac{\exp(-\Gamma C^* NTU) - 1}{\Gamma C^*}\right]$ | (h) |
| | $\Gamma = NTU^{-0.22}$ | |
| both fluids mixed | $\epsilon = \left[\dfrac{1}{1 - \exp(-NTU)} + \dfrac{C^*}{1 - \exp(-C^* NTU)} - \dfrac{1}{NTU}\right]^{-1}$ | (i) |

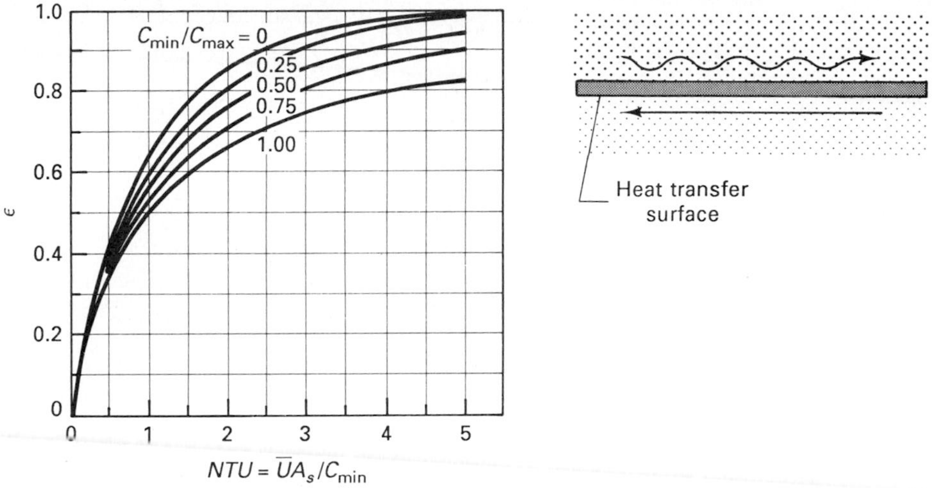


(a) Counterflow.

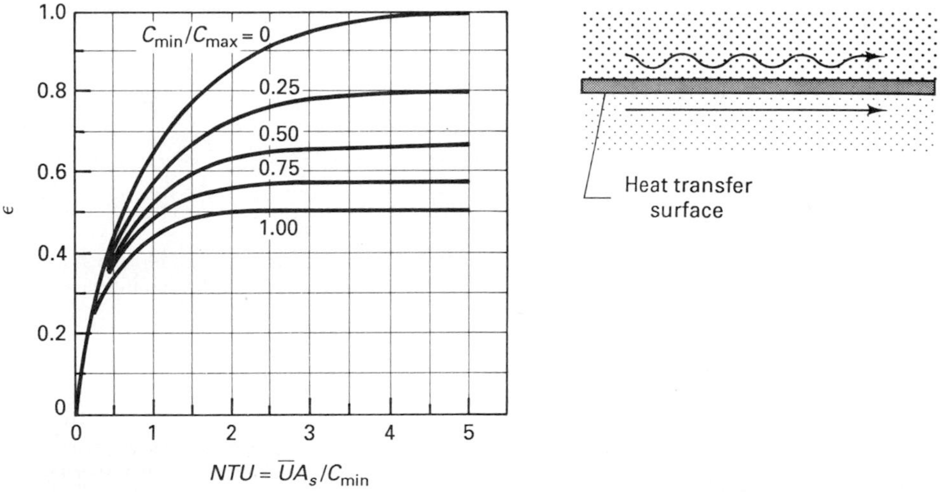


(b) Parallel flow.

**FIGURE 11–21** Effectiveness for double-pipe heat exchangers. (From Kays and London [9]. Used with permission.)

exchangers with one stream mixed and one stream unmixed. To identify and deal with these special arrangements in which the reference temperatures used to establish the effectiveness are not interchangeble, we will designate the shell or mixed side by subscript $s$ and the tube or unmixed side by subscript $t$. Equations (11–19)–(11–

**TABLE 11–5b** Relations for heat exchanger effectiveness of the form of Eq. (11–26b)

| *Heat exchanger type* | *NTU* | |
|---|---|---|
| All exchangers for $C^* = 0$ | $NTU = -\ln(1-\epsilon)$ | (a) |
| Double-pipe | | |
| parallel flow | $NTU = -\dfrac{\ln[1-\epsilon(1+C^*)]}{1+C^*}$ | (b) |
| counterflow | $NTU = \dfrac{1}{C^*-1}\ln\dfrac{\epsilon-1}{\epsilon C^*-1}$ | (c) |
| counterflow, $C^* = 1$ | $NTU = \dfrac{\epsilon}{1-\epsilon}$ | (d) |
| Shell-and-tube | | |
| one shell (TEMA E) 2, 4, . . . , tube passes | $NTU = -\dfrac{1}{(1+C^{*2})^{1/2}}\ln\dfrac{E-1}{E+1}$ | (e) |
| | $E = \dfrac{2/\epsilon-(1+C^*)}{(1+C^{*2})^{1/2}}$ | |
| Crossflow one fluid mixed one fluid unmixed | | |
| $C_{\min} = C_{\text{mixed}} = C_s$ | $NTU = -\dfrac{1}{C^*}\ln[C^*\ln(1-\epsilon)+1]$ | (f) |
| $C_{\min} = C_{\text{unmixed}} = C_t$ | $NTU = -\ln\left[1+\dfrac{1}{C^*}\ln(1-\epsilon C^*)\right]$ | (g) |

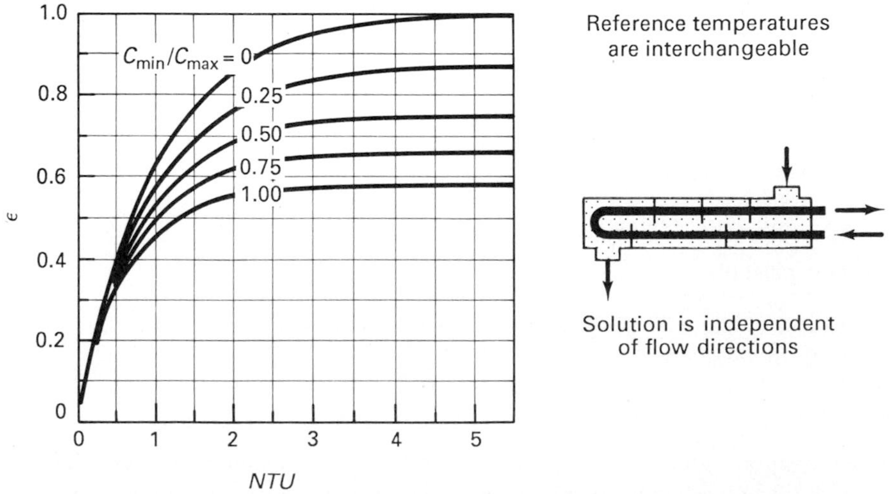

1.0
0.8
0.6
0.4
0.2
0
ε
0
1
2
3
4
5
NTU
$C_{min}/C_{max} = 0$
0.25
0.50
0.75
1.00
Reference temperatures are interchangeable
Solution is independent of flow directions


**FIGURE 11–22** Effectiveness for one shell and 2, 4, . . . tube pass heat exchanger; well baffled TEMA E type shell. (From Kays and London [9]. Used with permission.)

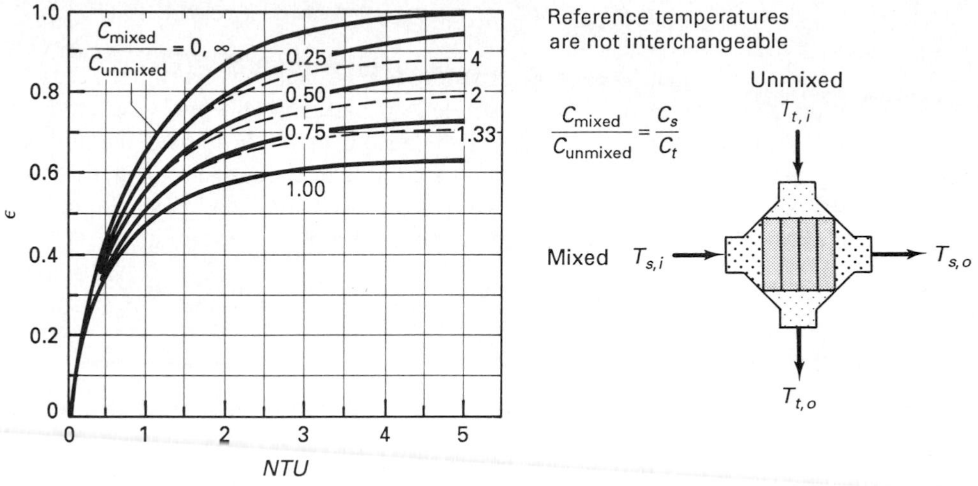


(a) One fluid mixed and one fluid unmixed.

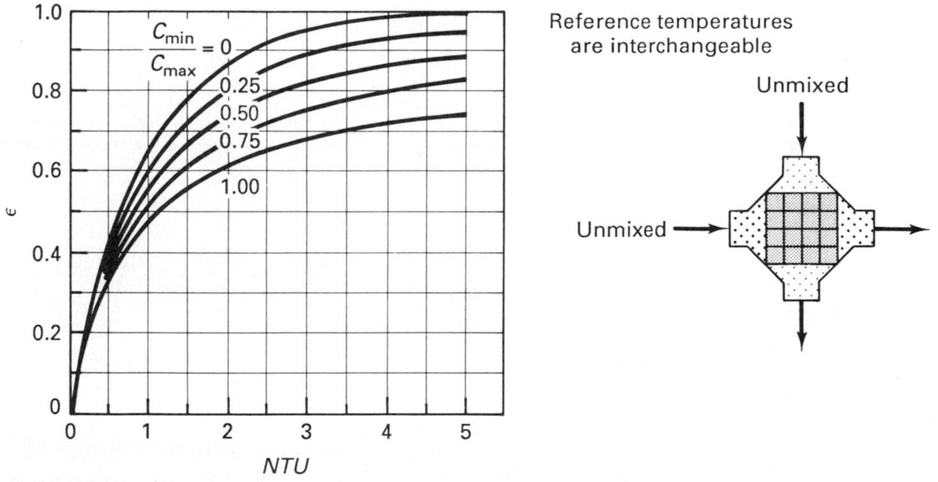


(b) Both fluids unmixed.

**FIGURE 11–23** Effectiveness for crossflow heat exchangers. (From Kays and London [9]. Used with permission.)

23) can be adapted to this alternative reference scheme by merely replacing $T_h$ by $T_s$ and $T_c$ by $T_t$ and by specifying whether $C_{\min} = C_s$ or $C_{\min} = C_t$. With $C^*$ set equal to zero for applications involving two-phase flow in boilers or condensers, the relations in Table 11–5 reduce to

$$\epsilon = 1 - \exp(-NTU) \qquad \text{or} \qquad NTU = -\ln(1 - \epsilon) \tag{11–27a,b}$$

which is similar in form to Eq. (8–55). Notice that for this important type of application, the relation for $\epsilon$ is independent of the geometry and flow arrangement.

As illustrated in Examples 11–5 through 11–9, the $\epsilon$-$NTU$ method proves to be very useful in both evaluation and design of heat exchangers. Note that although this method provides a basis for solving some evaluation and design problems without iteration, iterative solution techniques are commonly required.

## EXAMPLE 11–4

Develop expressions for effectiveness which are applicable to parallel flow and counterflow in double-pipe heat exchangers.

### Solution

*Objective* Develop expressions for $\epsilon$ that are applicable to double-pipe modules.

*Schematic* Double-pipe heat exchangers.

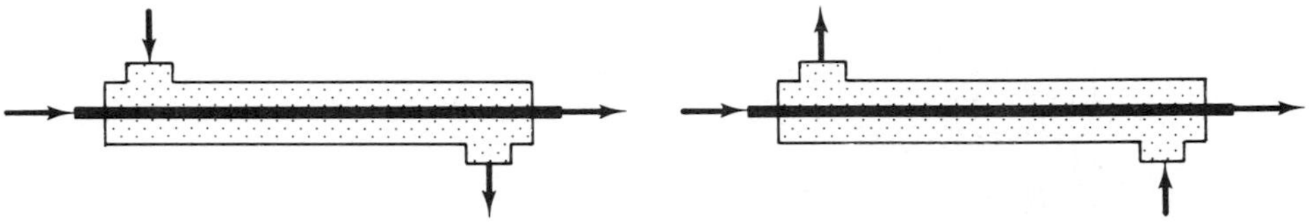

(a) Parallel flow (b) Counterflow

*Assumptions/Conditions*

uniform properties
standard conditions

*Analysis* Focusing attention on single-phase flow with uniform properties, the practical thermal analysis involves the solution of the lumped formulation given by Eqs. (11–14) and (11–15),

$$q = (\dot{m}c_P)_h(T_{h,i} - T_{h,o}) \qquad q = (\dot{m}c_P)_c(T_{c,o} - T_{c,i}) \tag{11–14,15}$$

and the lumped/differential formulation given by Eqs. (11–16) and (11–17),

$$dq = -(\dot{m}c_P)_h\, dT_h \qquad dq = (\dot{m}c_P)_c\, dT_c \tag{11–16,17}$$

with the differential heat transfer rate $dq$ expressed in terms of the local overall coefficient of heat transfer $U$ by Eq. (11–2),

$$dq = U\, dA_s\,(T_h - T_c) \tag{11–2}$$

The control volumes associated with these lumped/differential formulations are shown in Fig. E11–4. As shown in this figure, we designate parallel-flow and counterflow arrangements relative to a coordinate system with $x = 0$ at the inlet of the hot fluid (station 1) and $x = L$ at the other end (station 2) by writing

$$(\dot{m}c_P)_h = C_h \qquad (\dot{m}c_P)_c = C_c \qquad \text{for parallel flow}$$

$$(\dot{m}c_P)_h = C_h \qquad (\dot{m}c_P)_c = -C_c \qquad \text{for counterflow}$$

which accounts for the fact $\dot{m}_h$ and $\dot{m}_c$ both have positive streamwise directions for parallel flow and have opposing streamwise directions for counterflow.

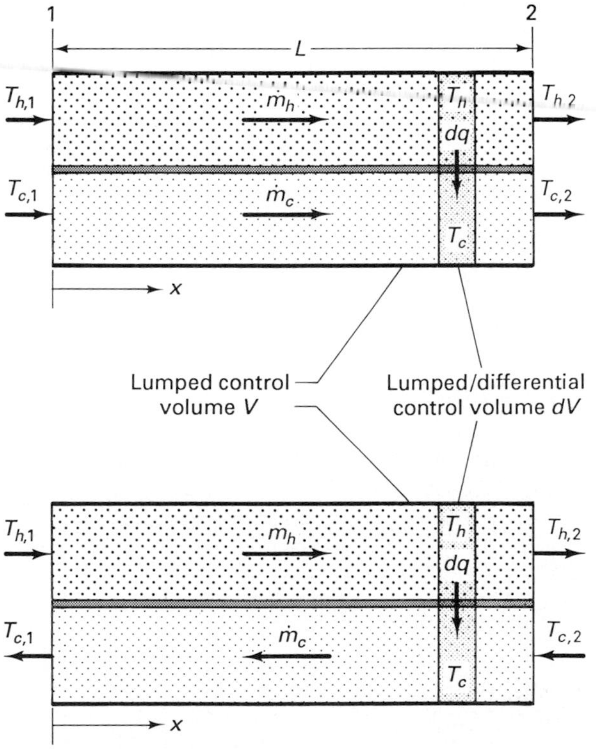


(a) Parallel-flow arrangement.

(b) Counterflow arrangement.

**FIGURE E11–4**
Coordinate system for double-pipe heat exchanger module.

Following this perspective with $dq$ represented by Eq. (11–2) and $dA_s$ set equal to $p\,dx$, Eqs. (11–14)–(11–17) become

$$q = C_h(T_{h,1} - T_{h,2}) \qquad \frac{dT_h}{T_h - T_c} = -\frac{Up}{C_h}\,dx \qquad \text{hot fluid} \qquad \text{(a,b)}$$

$$q = \pm C_c(T_{c,2} - T_{c,1}) \qquad \frac{dT_c}{T_h - T_c} = \frac{Up}{\pm C_c}\,dx \qquad \text{cold fluid} \qquad \text{(c,d)}$$

where parallel flow and counterflow correspond to the positive and negative signs, respectively.

To simplify matters, Eqs. (b) and (d) are combined to form a single differential equation for the dependent variable $T_h - T_c$; that is,

$$\frac{d(T_h - T_c)}{T_h - T_c} = -\left(\frac{1}{C_h} \pm \frac{1}{C_c}\right) Up\, dx$$

or, for convenience,

$$\frac{d(T_h - T_c)}{T_h - T_c} = -\zeta Up\, dx \tag{e}$$

where

$$\zeta = \frac{1}{C_h} \pm \frac{1}{C_c} \tag{f}$$

Referring to Fig. E11–4, the boundary condition for $T_h - T_c$ is designated relative to stations 1 and 2 by

$$T_h - T_c = T_{h,1} - T_{c,1} \qquad \text{at } x = 0$$

for both parallel-flow and counterflow arrangements. Using this boundary condition, the solution to Eq. (e) is simply

$$\ln \frac{T_h - T_c}{T_{h,1} - T_{c,1}} = -\int_0^x \zeta Up\, dx = -\zeta \overline{U} px$$

for uniform perimeter $p$ and uniform properties, where $\overline{U}$ is the mean overall coefficient of heat transfer over the length $x$,

$$\overline{U} = \frac{1}{x} \int_0^x U\, dx$$

Rearranging, the solution is put into the form

$$\frac{T_h - T_c}{T_{h,1} - T_{c,1}} = \exp\left(-\zeta \overline{U} px\right) \tag{g}$$

Setting $T_h - T_c = T_{h,2} - T_{c,2}$ at $x = L$, this equation reduces to

$$\frac{T_{h,2} - T_{c,2}}{T_{h,1} - T_{c,1}} = \exp\left(-\zeta \overline{U} A_s\right) \tag{h}$$

where $\overline{U}$ is evaluated over the length $L$ and the product $\overline{U}A_s$ represents the thermal capacity of the heat exchanger.

Equations (a), (c), and (h) provide the basis for developing solutions for any three dependent variables for parallel-flow and counterflow double-pipe heat exchangers. We now combine these equations to express the solutions in terms of the effectiveness. The effectiveness $\epsilon$ is defined by Eq. (11–18),

$$\epsilon = \frac{q}{q_{max}} = \frac{q}{C_{min}(T_{h,i} - T_{c,i})}$$

which takes the form

$$\epsilon = \frac{q}{C_{min}(T_{h,1} - T_{c,1})} \tag{i}$$

for parallel flow, and

$$\epsilon = \frac{q}{C_{min}(T_{h,1} - T_{c,2})} \tag{j}$$

for counterflow. Therefore, Eqs. (a), (c), and (h) are combined to obtain an expression for $q/(T_{h,i} - T_{c,i})$. This is done by substituting Eqs. (a) and (c) into Eq. (h) to eliminate (1) $T_{h,2}$ and $T_{c,2}$ for parallel flow with $(\dot{m}c_P)_c = C_c$, and (2) $T_{h,2}$ and $T_{c,1}$ for counterflow with $(\dot{m}c_P)_c = -C_c$. Taking these steps, we obtain

$$q = \frac{T_{h,1} - T_{c,1}}{\zeta_p}[1 - \exp(-\zeta_p \overline{U}A_s)] \tag{k}$$

where

$$\zeta_p = \frac{1}{C_h} + \frac{1}{C_c} \tag{l}$$

for parallel flow, and

$$q = \frac{(T_{h,1} - T_{c,2})[1 - \exp(-\zeta_c \overline{U}A_s)]}{1/C_h - (1/C_c)\exp(-\zeta_c \overline{U}A_s)} \tag{m}$$

where

$$\zeta_c = \frac{1}{C_h} - \frac{1}{C_c} \tag{n}$$

for counterflow. Substituting Eq. (k) into Eq. (i), we have

$$\epsilon = \frac{(T_{h,1} - T_{c,1})[1 - \exp(-\zeta_p \overline{U}A_s)]}{\zeta_p C_{min}(T_{h,1} - T_{c,1})} = \frac{1 - \exp(-\zeta_p \overline{U}A_s)}{1 + C^*} \tag{o}$$

for parallel flow. Similarly the substitution of Eq. (m) into Eq. (j) gives

$$\epsilon = \frac{1 - \exp(-\zeta_c \overline{U}A_s)}{C_{min}/C_h - (C_{min}/C_c)\exp(-\zeta_c \overline{U}A_s)} \tag{p}$$

for counterflow.

The parameters $\zeta_p \overline{U}A_s$ and $\zeta_c \overline{U}A_s$ appearing in Eqs. (o) and (p) can be expressed in terms of the capacitance ratio $C^*$ and number of transfer units $NTU$,

$$C^* = \frac{C_{min}}{C_{max}} \qquad NTU = \frac{\overline{U}A_s}{C_{min}}$$

by writing

$$\zeta_p \overline{U}A_s = \left(\frac{1}{C_h} + \frac{1}{C_c}\right)\overline{U}A_s = \frac{\overline{U}A_s}{C_{\min}}\left(1 + \frac{C_{\min}}{C_{\max}}\right) = NTU\,(1 + C^*)$$

for parallel flow, and

$$\zeta_c \overline{U}A_s = \left(\frac{1}{C_h} - \frac{1}{C_c}\right)\overline{U}A_s = \frac{\overline{U}A_s}{C_{\min}}\left(\frac{C_{\min}}{C_h} - \frac{C_{\min}}{C_c}\right) = NTU\,(1 - C^*)$$

for counterflow with $C_{\min} = C_h$, or

$$\zeta_c \overline{U}A_s = NTU\,(C^* - 1)$$

for counterflow with $C_{\min} = C_c$. Substituting these relations into Eqs. (o) and (p), we obtain

$$\epsilon = \frac{1 - \exp\left[-NTU\,(1 + C^*)\right]}{1 + C^*} \tag{q}$$

for parallel flow, and

$$\epsilon = \frac{1 - \exp\left[-NTU\,(1 - C^*)\right]}{1 - C^* \exp\left[-NTU\,(1 - C^*)\right]} \tag{r}$$

for counterflow with $C_{\min} = C_h$, or

$$\epsilon = \frac{1 - \exp\left[-NTU\,(C^* - 1)\right]}{C^* - \exp\left[-NTU\,(C^* - 1)\right]} \tag{s}$$

for counterflow with $C_{\min} = C_c$. A careful comparison of Eqs. (r) and (s) reveals that these two equations are equivalent.

## EXAMPLE 11–5

Freon F-12 at −20°C and 0.4 MPa flowing at a rate of 0.265 kg/s is heated in a double-pipe heat exchanger. Hot water with a mass flow rate of 0.035 kg/s enters the tubes at a temperature of 98°C and 0.2 MPa. The heat exchanger is constructed of thin-walled copper tubing with 2-cm inside diameter, 3-cm outside diameter and 3-m length. Determine the total rate of heat transfer and the distributions in bulk-stream and wall temperature for a parallel-flow arrangement.

### Solution

*Objective* Determine $q$ and the distributions in $T_h$, $T_c$, and $T_s$ for parallel flow.

*Schematic* Double-pipe heat exchanger: parallel flow arrangement.

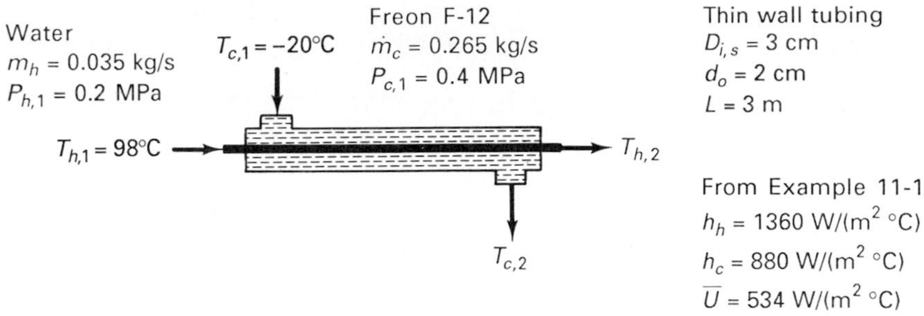
Water
$m_h = 0.035$ kg/s
$P_{h,1} = 0.2$ MPa
$T_{h,1} = 98°C$
$T_{c,1} = -20°C$
Freon F-12
$\dot{m}_c = 0.265$ kg/s
$P_{c,1} = 0.4$ MPa
$T_{h,2}$
$T_{c,2}$
Thin wall tubing
$D_{i,s} = 3$ cm
$d_o = 2$ cm
$L = 3$ m
From Example 11-1:
$h_h = 1360$ W/(m$^2$ °C)
$h_c = 880$ W/(m$^2$ °C)
$\overline{U} = 534$ W/(m$^2$ °C)

*Assumptions/Conditions*

negligible conduction resistance $R_k$
negligible fouling
negligible entrance effects
uniform properties
standard conditions

*Properties*

Water at $T = 98°C = 371$ K (Table A–C–3): $c_P = 4.21$ kJ/(kg °C).

F-12 at $-20°C$ (Table A–C–3): $c_P = 0.907$ kJ/(kg °C).

*Analysis* The heat transfer surface area $A_s$ and thermal capacity $\overline{U}A_s$ are obtained by writing

$$A_s = \pi d_o L = \pi(0.02 \text{ m})(3 \text{ m}) = 0.188 \text{ m}^2$$

and

$$\overline{U}A_s = 534 \frac{\text{W}}{\text{m}^2\ °\text{C}} (0.188 \text{ m}^2) = 100 \text{ W/°C}$$

Because the inlet temperatures and the flow rates are given, the effectiveness approach will be utilized to determine the rate of heat transfer and the outlet temperatures. Therefore, calculations are first obtained for the parameters $C_h$, $C_c$, and $NTU$.

$$C_h = \dot{m}_h c_{P,h} = 0.035 \frac{\text{kg}}{\text{s}} \left(4.21 \frac{\text{kJ}}{\text{kg °C}}\right) = 0.147 \text{ kW/°C}$$

$$C_c = \dot{m}_c c_{P,c} = 0.265 \frac{\text{kg}}{\text{s}} \left(0.907 \frac{\text{kJ}}{\text{kg °C}}\right) = 0.24 \text{ kW/°C}$$

$$C_{\min} = C_h = 0.147 \text{ kW/°C} \qquad C_{\max} = C_c = 0.24 \text{ kW/°C}$$

$$C^* = \frac{C_{\min}}{C_{\max}} = \frac{0.147}{0.24} = 0.613$$

$$NTU = \frac{\overline{U}A_s}{C_{\min}} = \frac{100\ \text{W/°C}}{0.147\ \text{kW/°C}} = 0.68$$

$$q_{\max} = C_{\min}(T_{h,i} - T_{c,i}) = 0.147\ \frac{\text{kW}}{\text{°C}}\ [98\text{°C} - (-20\text{°C})] = 17.3\ \text{kW}$$

For this parallel-flow arrangement, the effectiveness $\epsilon$ is given by Eq. (b) in Table 11–5a or by Fig. 11–21b. For better accuracy, we utilize the analytical relation to obtain

$$\epsilon = \frac{1 - \exp\left[-NTU\,(1 + C^*)\right]}{1 + C^*} = \frac{1 - \exp\left[-0.68(1.61)\right]}{1.61} = 0.413$$

Utilizing Eq. (11–18), we write

$$q = \epsilon q_{\max} = 0.413(17.3\ \text{kW}) = 7.14\ \text{kW}$$

The outlet temperatures are calculated as follows:

$$T_{c,2} = \frac{q}{C_c} + T_{c,1} = \frac{7.14\ \text{kW}}{0.24\ \text{kW/°C}} - 20\text{°C} = 9.75\text{°C}$$

$$T_{h,2} = T_{h,1} - \frac{q}{C_h} = 98\text{°C} - \frac{7.14\ \text{kW}}{0.147\ \text{kW/°C}} = 49.4\text{°C}$$

To get a better feel for the problem, expressions are developed for the bulk-stream temperature distributions. To obtain predictions for $T_h$ and $T_c$, we first substitute Eq. (E11–4g) into Eq. (E11–4b).

$$dT_h = -(T_{h,1} - T_{c,1})\left[\exp(-\zeta\overline{U}px)\right]\frac{Up}{C_h}\,dx$$

Integrating from 0 to $x$ with $\overline{U}$ and $U$ set equal to 534 W/(m² °C) as a first approximation, we obtain

$$\frac{T_h - T_{h,1}}{T_{h,1} - T_{c,1}} = \frac{1}{\zeta C_h}\left[\exp(-\zeta\overline{U}px) - 1\right]$$

$$T_h = T_{h,1} + \frac{T_{h,1} - T_{c,1}}{\zeta C_h}\left[\exp\left(-\zeta C_{\min}\, NTU\,\frac{x}{L}\right) - 1\right] \qquad \text{(a)}$$

where $NTU = 0.68$ and

$$\zeta C_h = \zeta C_{\min} = 1 + C^* = 1.61$$

Similarly, the coupling of Eqs. (E11–4g) and (E11–4d) gives rise to an expression for $T_c$ of the form

$$T_c = T_{c,1} + \frac{T_{h,1} - T_{c,1}}{\zeta(\dot{m}c_P)_c}\left[1 - \exp\left(-\zeta C_{\min}\, NTU\,\frac{x}{L}\right)\right] \qquad \text{(b)}$$

where we take the positive sign [i.e., $(\dot{m}c_P)_c = C_c$] for parallel flow and

$$\zeta C_c = \frac{C_c}{C_h} + 1 = 2.63$$

Thus, we have

$$T_h = 98°\text{C} + \frac{98°\text{C} + 20°\text{C}}{1.61}\left\{\exp\left[-1.61(0.68)\frac{x}{L}\right] - 1\right\} \tag{c}$$

and

$$T_c = -20°\text{C} + \frac{98°\text{C} + 20°\text{C}}{2.63}\left\{1 - \exp\left[-1.61(0.68)\frac{x}{L}\right]\right\} \tag{d}$$

These equations are plotted in Fig. E11–5.

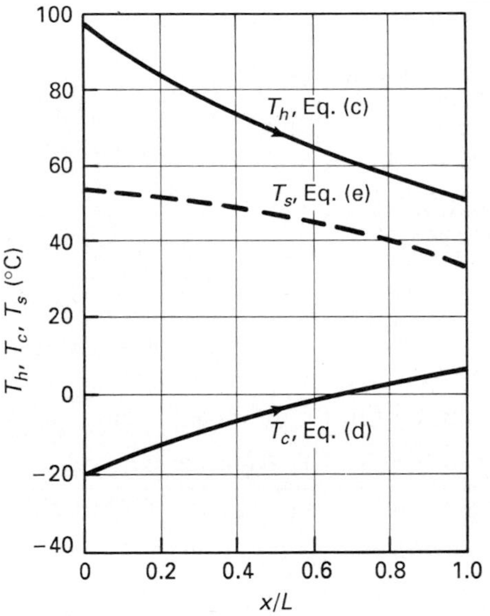


**FIGURE E11–5**
Calculations for bulk-stream and wall temperatures $T_h$, $T_c$, and $T_s$ for parallel-flow arrangement.

To approximate the wall temperature $T_s$, we write

$$U(T_h - T_c) = h_h(T_h - T_s)$$

Approximating $h_h$ by 1360 W/(m$^2$ °C) and $U$ by 534 W/(m$^2$ °C), we obtain

$$\frac{T_h - T_s}{T_h - T_c} = \frac{U}{h_h} \simeq \frac{534}{1360} = 0.393 \tag{e}$$

Calculations for $T_s$ obtained by the use of this equation are shown in Fig. E11–5. The bulk-stream temperatures and wall temperature at the mid-point are estimated to be $T_h = 79.1°\text{C}$, $T_c = -1.09°\text{C}$, and $T_s = 47.6°\text{C}$.

Now that the outlet fluid temperature, the bulk-stream temperatures, and the wall temperatures have been estimated, our analysis can be refined by approximating the properties at the arithmetic average of the inlet and outlet temperatures and by utilizing the property correction factors for $h_h$ and $h_c$ given by Eq. (8–24). This refinement is suggested as an exercise.

## EXAMPLE 11–6

Reconsider Example 11–5 for counterflow.

### Solution

*Objective* Determine $q$ and the distributions in $T_h$, $T_c$, and $T_s$ for counterflow.

*Schematic* Double-pipe heat exchanger: counterflow arrangement.

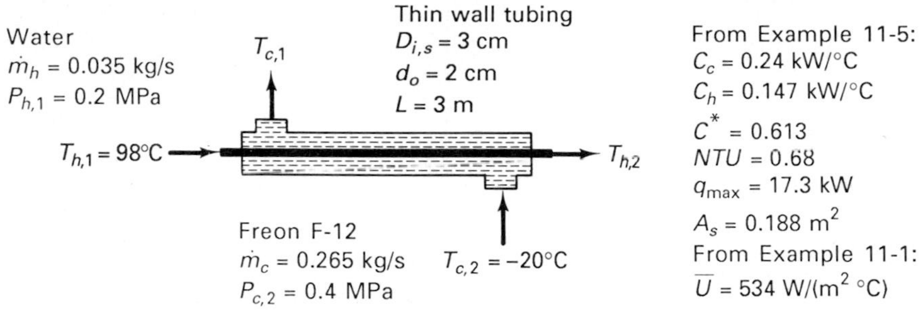
Thin wall tubing
$D_{i,s}$ = 3 cm
$d_o$ = 2 cm
L = 3 m
Water
$\dot{m}_h$ = 0.035 kg/s
$P_{h,1}$ = 0.2 MPa
$T_{c,1}$
$T_{h,1}$ = 98°C
$T_{h,2}$
Freon F-12
$\dot{m}_c$ = 0.265 kg/s
$P_{c,2}$ = 0.4 MPa
$T_{c,2}$ = −20°C
From Example 11-5:
$C_c$ = 0.24 kW/°C
$C_h$ = 0.147 kW/°C
$C^*$ = 0.613
NTU = 0.68
$q_{max}$ = 17.3 kW
$A_s$ = 0.188 m²
From Example 11-1:
$\overline{U}$ = 534 W/(m² °C)

*Analysis* For counterflow, $\epsilon$ is given by Eq. (c) in Table 11–5a,

$$\epsilon = \frac{1 - \exp[-NTU(1 - C^*)]}{1 - C^* \exp[-NTU(1 - C^*)]} = \frac{1 - \exp[-0.68(0.387)]}{1 - 0.613\exp[-0.68(0.387)]} = 0.438$$

Calculating $q$, we have

$$q = 0.438(17.3 \text{ kW}) = 7.58 \text{ kW}$$

Continuing with the calculation of the outlet temperatures $T_{c,1}$ and $T_{h,2}$, we write

$$q = C_c(T_{c,2} - T_{c,1}) = C_h(T_{h,1} - T_{h,2})$$

$$T_{c,1} = \frac{q}{C_c} + T_{c,2} = \frac{7.58 \text{ kW}}{0.24 \text{ kW/°C}} - 20\text{°C} = 11.6\text{°C}$$

$$T_{h,2} = T_{h,1} - \frac{q}{C_h} = 98\text{°C} - \frac{7.58 \text{ kW}}{0.147 \text{ kW/°C}} = 46.4\text{°C}$$

Notice that $q$ is about 6% greater for counterflow than for parallel flow. Referring to Fig. 11–21, we observe that even greater benefit occurs from counterflow operation for larger values of *NTU*. (As a double check, our calculations can be confirmed by utilizing the *LMTD* approach.)

The bulk-stream temperature distributions are given by Eqs. (E11–5a) and (E11–5b), with $(\dot{m}c_P)_c$ set equal to $-C_c$ and

$$\zeta = \frac{1}{C_h} - \frac{1}{C_c}$$

such that

$$\zeta C_h = \zeta C_{\min} = 1 - \frac{C_h}{C_c} = 1 - 0.613 = 0.387$$

and

$$-\zeta C_c = -\frac{C_c}{C_h} + 1 = -0.633$$

Substituting into Eqs. (E11–5a) and (E11–5b), we have

$$T_h = 98°\text{C} + \frac{98°\text{C} - 11.6°\text{C}}{0.387}\left\{\exp\left[-0.387(0.68)\frac{x}{L}\right] - 1\right\} \tag{a}$$

and

$$T_c = 11.6°\text{C} + \frac{98°\text{C} - 11.6°\text{C}}{-0.633}\left\{1 - \exp\left[-0.387(0.68)\frac{x}{L}\right]\right\} \tag{b}$$

These distributions are shown in Fig. E11–6.

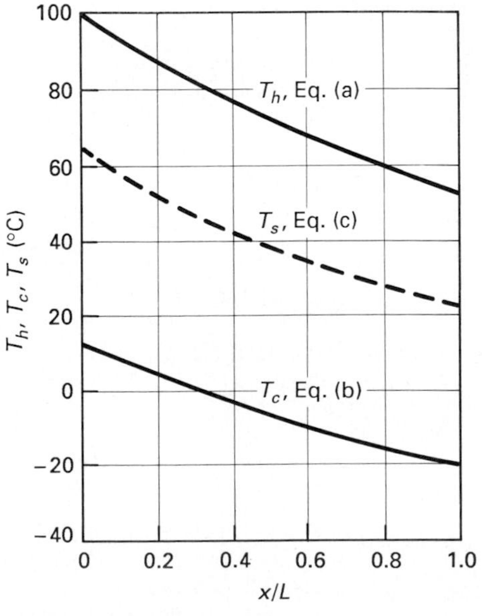


**FIGURE E11–6**
Calculations for bulk-stream and wall temperatures $T_h$, $T_c$, and $T_s$ for counterflow arrangement.

As shown in Example 11–5, the wall temperature $T_s$ can be approximated by

$$\frac{T_h - T_s}{T_h - T_c} = \frac{U}{h_h} \simeq 0.393 \tag{c}$$

This expression is shown in Fig. E11–6.

## EXAMPLE 11–7

The evaluation function sometimes requires that the mass flow rates of two streams be determined for an existing heat exchanger with specified fluids and terminal temperatures. Outline the steps that would be required to determine $\dot{m}_h$ and $\dot{m}_c$ for the double-pipe heat exchanger application of Example 11–6 with the outlet temperatures specified as $T_{c,o} = 11.6°\text{C}$ and $T_{h,o} = 46.4°\text{C}$.

### Solution

*Objective* Set up a computational strategy for evaluation of mass flow rates $\dot{m}_h$ and $\dot{m}_c$ for an existing double-pipe heat exchanger with specified fluids and terminal temperatures.

*Schematic* Double-pipe heat exchanger: counterflow arrangement.

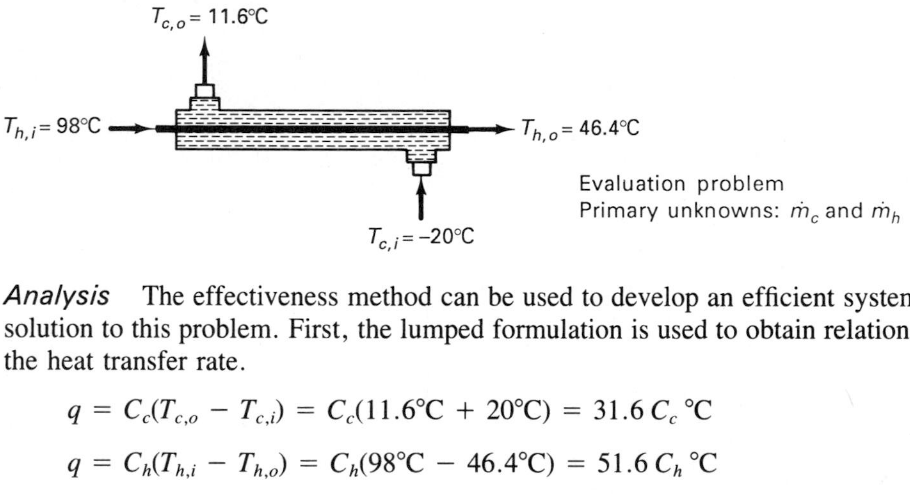


*Analysis* The effectiveness method can be used to develop an efficient systematic solution to this problem. First, the lumped formulation is used to obtain relations for the heat transfer rate.

$$q = C_c(T_{c,o} - T_{c,i}) = C_c(11.6°\text{C} + 20°\text{C}) = 31.6\, C_c\, °\text{C}$$

$$q = C_h(T_{h,i} - T_{h,o}) = C_h(98°\text{C} - 46.4°\text{C}) = 51.6\, C_h\, °\text{C}$$

from which we obtain

$$C^* = \frac{C_h}{C_c} = \frac{31.6}{51.6} = 0.612$$

Thus,

$$C_{\min} = C_h$$

Therefore, the effectiveness is calculated by writing

$$\epsilon = \frac{q}{q_{\max}} = \frac{C_h(T_{h,i} - T_{h,o})}{C_{\min}(T_{h,o} - T_{c,i})} = \frac{98°\text{C} - 46.4°\text{C}}{98°\text{C} + 20°\text{C}} = 0.437$$

Knowing $\epsilon$ and $C^*$, the number of transfer units $NTU$ can be obtained by using Fig. 11–21 or Eq. (c) in Table 11–5b. Using the analytical approach, we obtain

$$NTU = \frac{1}{C^* - 1} \ln \frac{\epsilon - 1}{\epsilon C^* - 1} = \frac{1}{0.612 - 1} \ln \frac{0.437 - 1}{0.437(0.612) - 1} = 0.679$$

To complete the analysis, we must evaluate $\overline{U}$, $C_h$, and $C_c$. $\overline{U}$ and $C_h$ are related by

$$NTU = \frac{\overline{U}A_s}{C_{\min}} = \frac{\overline{U}A_s}{C_h}$$

or

$$C_h = \frac{\overline{U}A_s}{NTU} = \frac{\overline{U}(0.188 \text{ m}^2)}{0.679} = 0.277\overline{U} \text{ m}^2 \qquad \text{(a)}$$

The following iterative solution approach can now be used for evaluating $\dot{m}_h$ and $\dot{m}_c$:

(1) Set first approximation $C_h = C_h^{(i)}$ $(i = 1)$;

(2) Calculate $C_c$, $\dot{m}_h$, and $\dot{m}_c$ from

$$C_c = C_h^{(i)} C^* \qquad \dot{m}_h = \frac{C_h^{(i)}}{c_{P,h}} \qquad \dot{m}_c = \frac{C_c}{c_{P,c}}$$

(3) Calculate $Re_h$ and $Re_c$ from (see Example 11–1)

$$Re_h = \frac{4\dot{m}_h}{\mu_h \pi d_i} \qquad Re_c = \frac{4\dot{m}_c}{\mu_c \pi (D_{i,s} + d_o)}$$

(4) Evaluate $\overline{Nu}_h$ and $\overline{Nu}_c$ using appropriate correlations and calculate $\overline{h}_h$, $\overline{h}_c$, and $\overline{U}$;

(5) Use Eq. (a) to obtain new value of $C_h$,

$$C_h^{(i+1)} = 0.277\overline{U} \text{ m}^2$$

(6) To obtain final solutions for $\dot{m}_h$ and $\dot{m}_c$, repeat steps 2–5 until satisfactory convergence is achieved.

This iterative approach is typical of the type of analysis that must be used in evaluating the mass flow rate in existing heat exchangers with specified fluids and terminal temperature.

## EXAMPLE 11–8

A 1–2 pass shell-and-tube heat exchanger with surface area of 62.5 $m^2$ is to be designed for cooling oil at 80°C with specific heat of 2.5 kJ/(kg °C) and mass flow rate of 5.0 kg/s. Water at 15°C flowing at 9.94 kg/s is used as the cooling fluid. Determine the heat transfer and outlet temperatures, assuming an overall coefficient of heat transfer equal to 300 W/($m^2$ °C).

### Solution

*Objective* Determine $q$, $T_{h,o}$, and $T_{h,i}$.

*Schematic* Shell-and-tube heat exchanger: 1–2 pass arrangement.

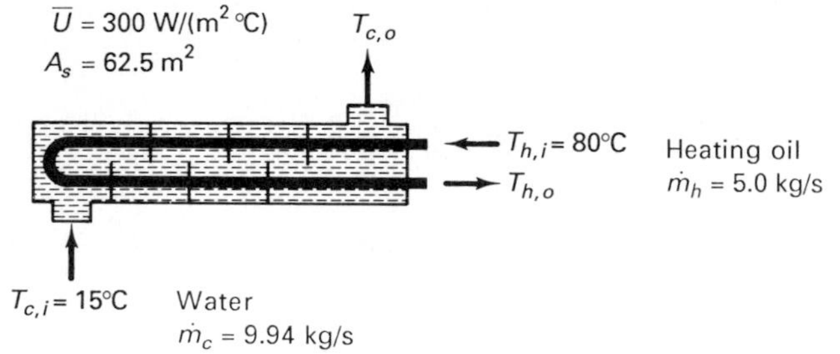


*Assumptions/Conditions*

uniform properties
standard conditions

*Properties*

Water at $T = 15°C = 288$ K (Table A–C–3): $c_P = 4.19$ kJ/(kg °C).

Hot oil: $c_P = 2.5$ kJ/(kg °C).

*Analysis* To employ the effectiveness approach, we first calculate the capacitance rates and $NTU$, assuming uniform property conditions.

$$C_h = (\dot{m}c_P)_h = 5\,\frac{\text{kg}}{\text{s}}\left(2.5\,\frac{\text{kJ}}{\text{kg °C}}\right) = 12.5\ \text{kW/°C}$$

$$C_c = (\dot{m}c_P)_c = 9.94\,\frac{\text{kg}}{\text{s}}\left(4.19\,\frac{\text{kJ}}{\text{kg °C}}\right) = 41.6\ \text{kW/°C}$$

Thus,

$$C_{\min} = C_h \qquad C^* = \frac{12.5}{41.6} = 0.3$$

and

$$NTU = \frac{\overline{U}A_s}{C_{\min}} = \frac{[300 \text{ W/(m}^2\text{ °C)}](62.5 \text{ m}^2)}{12.5 \text{ kW/°C}} = 1.5$$

Substituting these results into Eq. (e) in Table 11–5a, we are able to calculate $\epsilon$.

$$\Gamma = NTU\,(1 + C^{*2})^{1/2} = 1.5\,(1 + 0.3^2)^{1/2} = 1.57$$

and

$$\epsilon = \frac{2}{(1 + C^*) + (1 + C^{*2})^{1/2}\,(1 + e^{-\Gamma})/(1 - e^{-\Gamma})} = 0.691$$

The heat transfer rate $q$ is obtained by writing

$$q = \epsilon q_{\max} = \epsilon C_{\min}(T_{h,i} - T_{c,i})$$

$$= 0.691\left(12.5\,\frac{\text{kW}}{\text{°C}}\right)(80\text{°C} - 15\text{°C}) = 561 \text{ kW}$$

It follows that

$$q = C_h(T_{h,i} - T_{h,o}) = C_c(T_{c,o} - T_{c,i})$$

$$T_{h,o} = T_{h,i} - \frac{q}{C_h} = 80\text{°C} - \frac{561 \text{ kW}}{12.5 \text{ kW/°C}} = 35.1\text{°C}$$

$$T_{c,o} = T_{c,i} + \frac{q}{C_c} = 28.5\text{°C}$$

These results are observed to be consistent with the calculations obtained by means of the *LMTD* approach in Example 11–14 for the corresponding design problem.

Finally, we note that the variation in $c_P$ for water entering at 15°C and exiting at 28.5°C is less than 1%, such that the uniform property analysis should be reasonable, providing that the variation in $c_P$ of the heating oil is not too large.

## EXAMPLE 11–9

Air at 15°C and 1 atm with entering velocity of 6.8 m/s is to be heated in a crossflow heat exchanger with eighty 9.52-mm-diameter thin wall tubes of 2.5 m length. The tubes are placed 10 tubes deep in a staggered array with longitudinal and transverse pitches of 1.19 cm and 1.43 cm, respectively. Hot water at 65°C enters the tubes with a mass flow rate of 0.25 kg/s. Determine the rate of heat transfer and the outlet temperatures.

### Solution

*Objective* Determine $q$, $T_{t,o}$, and $T_{s,o}$.

*Schematic* Crossflow heat exchanger.

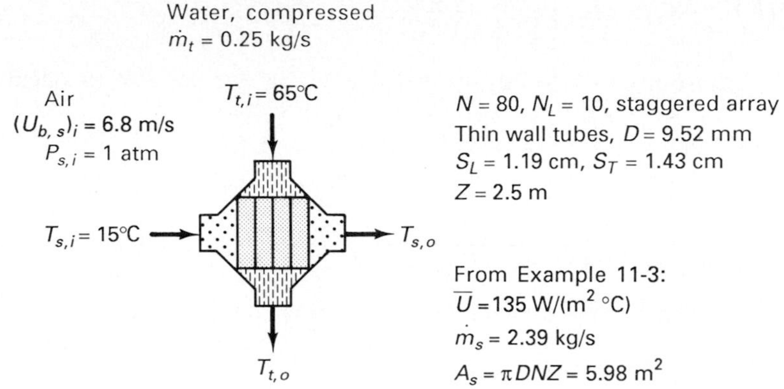
Water, compressed
$\dot{m}_t = 0.25$ kg/s
Air
$(U_{b,s})_i = 6.8$ m/s
$P_{s,i} = 1$ atm
$T_{t,i} = 65°C$
$T_{s,i} = 15°C$
$T_{s,o}$
$T_{t,o}$
$N = 80$, $N_L = 10$, staggered array
Thin wall tubes, $D = 9.52$ mm
$S_L = 1.19$ cm, $S_T = 1.43$ cm
$Z = 2.5$ m
From Example 11-3:
$\overline{U} = 135$ W/(m$^2$ °C)
$\dot{m}_s = 2.39$ kg/s
$A_s = \pi DNZ = 5.98$ m$^2$

*Assumptions/Conditions*

negligible conduction resistance $R_k$
negligible fouling
uniform properties
standard conditions

*Properties*

Air at $T$ = 15°C = 288 K (Table A–C–5): $\rho$ = 1.23 kg/m$^3$, $\mu$ = 1.76 × 10$^{-5}$ kg/(m s), $c_P$ = 1.01 kJ/(kg °C), $k$ = 0.0253 W/(m °C), $Pr$ = 0.711.

Water at $T$ = 65°C = 338 K (Table A–C–3): $\rho$ = 981 kg/m$^3$, $\mu$ = 4.33 × 10$^{-4}$ kg/(m s), $c_P$ = 4.19 kJ/(kg °C), $k$ = 0.658 W/(m °C), $Pr$ = 2.75.

*Analysis* To utilize the effectiveness approach, we calculate the capacity rates and *NTU*.

$$C_t = (\dot{m}c_P)_t = 0.25\,\frac{\text{kg}}{\text{s}}\left(4.19\,\frac{\text{kJ}}{\text{kg °C}}\right) = 1.05\text{ kW/°C}$$

$$C_s = (\dot{m}c_P)_s = 2.39\,\frac{\text{kg}}{\text{s}}\left(1.01\,\frac{\text{kJ}}{\text{kg °C}}\right) = 2.41\text{ kW/°C}$$

$$C_{\min} = C_t = 1.05\text{ kW/°C} \qquad C^* = \frac{C_t}{C_s} = \frac{1.05}{2.41} = 0.436$$

$$\overline{U}A_s = 135\,\frac{\text{W}}{\text{m}^2\text{ °C}}(5.98\text{ m}^2) = 0.807\text{ kW/°C}$$

$$NTU = \frac{\overline{U}A_s}{C_{\min}} = \frac{0.807\text{ kW/°C}}{1.05\text{ kW/°C}} = 0.769$$

These results are substituted into Eq. (g) in Table 11–5a to obtain

$$\Gamma = 1 - \exp(-NTU) = 1 - \exp(-0.769) = 0.537$$

$$\epsilon = \frac{1}{C^*}[1 - \exp[-C^*\Gamma)] = \frac{1}{0.436}\{1 - \exp[-0.436(0.537)]\} = 0.479$$

(Figure 11–23 can be used to check this result.)

The total rate of heat transfer $q$ is calculated as follows:

$$q_{\max} = C_{\min}(T_{t,i} - T_{s,i}) = 1.05\,\frac{\text{kW}}{°\text{C}}\,(65°\text{C} - 15°\text{C}) = 52.5\text{ kW}$$

$$q = \epsilon q_{\max} = 0.479\,(52.5\text{ kW}) = 25.1\text{ kW}$$

The outlet fluid temperatures are now calculated.

$$q = C_t(T_{t,i} - T_{t,o})$$

$$T_{t,o} = T_{t,i} - \frac{q}{C_t} = 65°\text{C} - \frac{25.1\text{ kW/°C}}{1.05\text{ kW/°C}} = 41.1°\text{C}$$

$$q = C_s(T_{s,o} - T_{s,i})$$

$$T_{s,o} = \frac{q}{C_s} + T_{s,i} = \left(\frac{25.1\text{ kW}}{2.41\text{ kW/°C}} + 15°\text{C}\right) = 25.4°\text{C}$$

The effect of fluid property variation with temperature can be approximately accounted for by resolving the problem with the thermal properties of the air and water evaluated at the arithmetic average of the inlet and outlet bulk-stream temperatures. In addition, the wall temperature at the midpoint of the heat exchanger can be estimated, such that $\bar{h}_t$ and $\bar{h}_s$ can be corrected for effects of property variation.

Note that our solution for $q$ of 25.1 kW lies well below the value of 50.0 kW obtained in Example 8–13 for air flow over a tube bank with tubes maintained at constant temperature of 65°C. This difference is attributed to the lower tube-wall temperature.

### *Log Mean Temperature Difference (LMTD) Method*

As shown in Example 11–10, the solution for the heat transfer rate $q$ in double-pipe heat exchanger modules with counterflow or parallel flow can be expressed in the form

$$q = \overline{U}A_s\,LMTD \tag{11–28}$$

where the *log mean temperature difference LMTD* is defined relative to the frame of reference shown in Fig. 11–24 by

$$LMTD = \frac{(T_{h,1} - T_{c,1}) - (T_{h,2} - T_{c,2})}{\ln \dfrac{T_{h,1} - T_{c,1}}{T_{h,2} - T_{c,2}}} = \frac{\Delta T_1 - \Delta T_2}{\ln(\Delta T_1/\Delta T_2)} \tag{11–29}$$

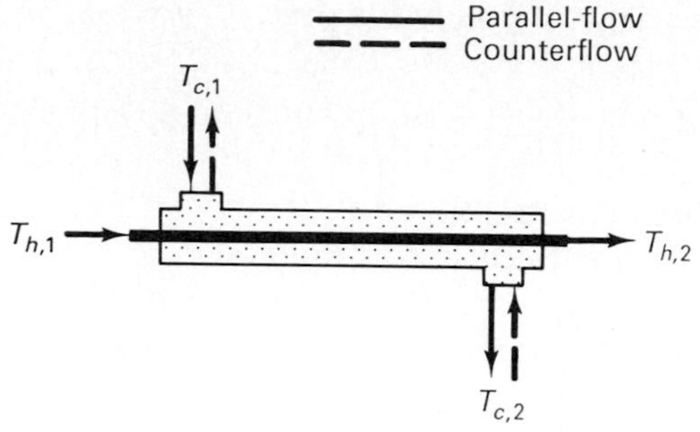


**FIGURE 11–24**
Double-pipe heat exchanger.

with $\Delta T_1 = T_{h,1} - T_{c,1}$ and $\Delta T_2 = T_{h,2} - T_{c,2}$.

Practical solution results for more complex heat exchangers are commonly expressed in the generalized *LMTD* format

$$q = \overline{U}A_s(F\ LMTD) \tag{11–30}$$

where *LMTD* now represents the log mean temperature difference associated with an equivalent counterflow single-pass arrangement with the same hot and cold stream entering and exiting temperatures,

$$LMTD = \frac{(T_{h,i} - T_{c,o}) - (T_{h,o} - T_{c,i})}{\ln \dfrac{T_{h,i} - T_{c,o}}{T_{h,o} - T_{c,i}}} \tag{11–31}$$

and $F$ is a *correction factor* that accounts for the extent to which the actual flow differs from ideal counterflow. Notice that $F = 1$ for counterflow or parallel flow in double-pipe heat exchanger units. The value of $F$ in properly designed shell-and-tube and crossflow heat exchangers generally lies between 0.75 and unity, with very much lower values indicating poor performance. The correction factor $F$ is a function of the geometry and flow arrangement and temperature ratios represented by $P$ and $R$; that is,

$$F = F(P, R) \tag{11–32}$$

for a specific geometry and flow arrangement. The temperature ratios $P$ and $R$ can be represented by

$$P = \frac{T_{c,o} - T_{c,i}}{T_{h,i} - T_{c,i}} \qquad R = \frac{T_{h,i} - T_{h,o}}{T_{c,o} - T_{c,i}} \tag{11–33a,b}$$

or equivalently,

$$P = \frac{T_{t,o} - T_{t,i}}{T_{s,i} - T_{t,i}} \qquad R = \frac{T_{s,i} - T_{s,o}}{T_{t,o} - T_{t,i}} \tag{11–34a,b}$$

Representative charts for $F$ are shown in Figs. 11–25 and 11–26 for shell-and-tube and crossflow heat exchangers. It should be noted that the reference temperatures used to determine $P$, $R$, and $F$ for the *LMTD* method are generally interchangeable. A more extensive listing of charts is provided by Taborek [14]. The standard *LMTD* method proves to be convenient for solving basic heat exchanger design problems, but involves iterative calculations when used in the evaluation of an existing heat exchanger with unspecified outlet temperatures.

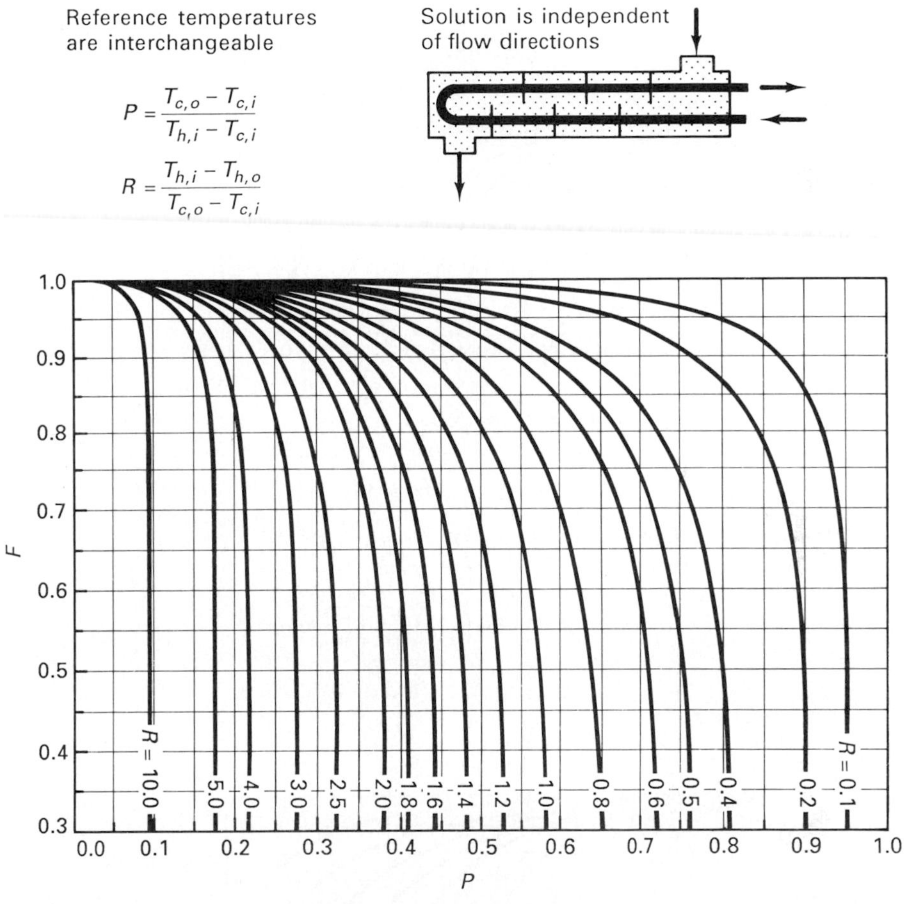


**FIGURE 11–25** Correction factor for one shell and 2, 4, . . . tube pass heat exchanger; well baffled TEMA E type shell (Bowman et al. [15]).†

† The correction factor $F$ for one shell and an even number of tube passes is given by Bowman [15] as

$$F = \frac{(R^2 + 1)^{1/2}}{R - 1} \log \frac{1 - P}{1 - PR} \left[ \log \frac{2/P - 1 - R + (R^2 + 1)^{1/2}}{2/P - 1 - R - (R^2 + 1)^{1/2}} \right]^{-1}$$

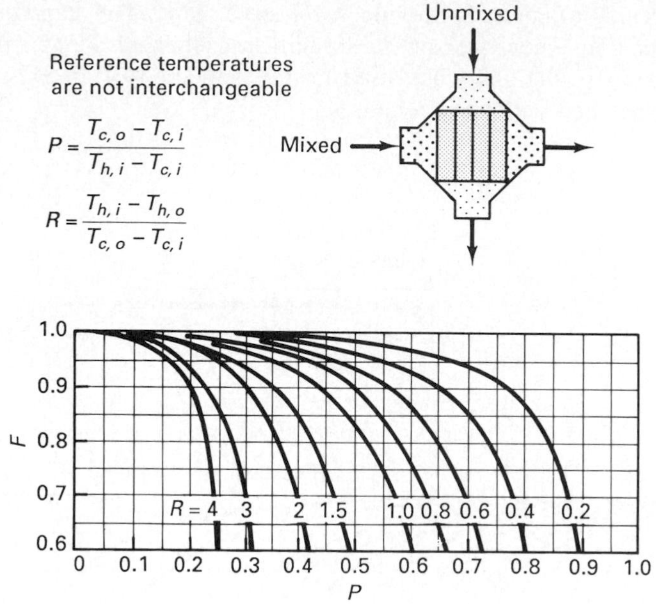
Unmixed
Reference temperatures are not interchangeable
$P = \frac{T_{c,o} - T_{c,i}}{T_{h,i} - T_{c,i}}$
$R = \frac{T_{h,i} - T_{h,o}}{T_{c,o} - T_{c,i}}$
Mixed
F
1.0
0.9
0.8
0.7
0.6
R = 4
3
2
1.5
1.0
0.8
0.6
0.4
0.2
0 0.1 0.2 0.3 0.4 0.5 0.6 0.7 0.8 0.9 1.0
P

(a) One fluid mixed and one fluid unmixed.

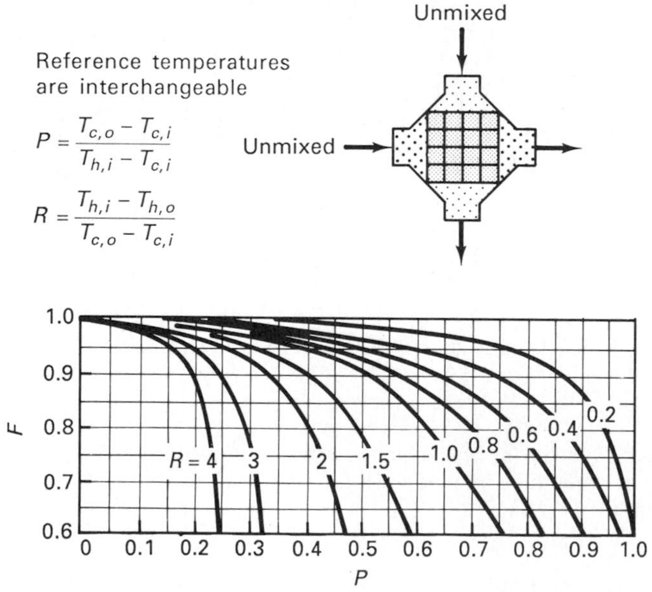
Unmixed
Reference temperatures are interchangeable
$P = \frac{T_{c,o} - T_{c,i}}{T_{h,i} - T_{c,i}}$
$R = \frac{T_{h,i} - T_{h,o}}{T_{c,o} - T_{c,i}}$
Unmixed
F
1.0
0.9
0.8
0.7
0.6
R = 4
3
2
1.5
1.0
0.8
0.6
0.4
0.2
0 0.1 0.2 0.3 0.4 0.5 0.6 0.7 0.8 0.9 1.0
P

(b) Both fluids unmixed.

**FIGURE 11–26** Correction factors for crossflow heat exchangers (Bowman et al. [15]).

### EXAMPLE 11–10

Develop the famous *LMTD* equation for parallel flow or counterflow double-pipe heat exchangers.

**Solution**

*Objective* Develop Eq. (11–28),

$$q = \overline{U}A_s\, LMTD$$

*Assumptions/Conditions*

uniform properties
standard conditions

*Analysis* Equations (a), (c), and (h) of Example 11–4 provide the basis for developing the *LMTD* equation for double-pipe heat exchangers; that is,

$$q = C_h(T_{h,1} - T_{h,2}) \qquad q = \pm C_c(T_{c,2} - T_{c,1})$$

and

$$\frac{T_{h,2} - T_{c,2}}{T_{h,1} - T_{c,1}} = \exp(-\zeta \overline{U}A_s)$$

where

$$\zeta = \frac{1}{C_h} \pm \frac{1}{C_c}$$

To express the solutions in the *LMTD* format, we first put these equations into the form

$$C_h = \frac{q}{T_{h,1} - T_{h,2}} \qquad \pm C_c = \frac{q}{T_{c,2} - T_{c,1}} \tag{a,b}$$

and

$$\zeta \overline{U}A_s = \left(\frac{1}{C_h} + \frac{1}{\pm C_c}\right)\overline{U}A_s = -\ln\frac{T_{h,2} - T_{c,2}}{T_{h,1} - T_{c,1}}$$

or

$$\frac{1}{C_h} \pm \frac{1}{C_c} = \frac{1}{\overline{U}A_s}\ln\frac{T_{h,1} - T_{c,1}}{T_{h,2} - T_{c,2}} \tag{c}$$

Substituting Eqs. (a) and (b) into Eq. (c), the solution for $q$ becomes

$$\frac{T_{h,1} - T_{h,2}}{q} + \frac{T_{c,2} - T_{c,1}}{q} = \frac{1}{\overline{U}A_s}\ln\frac{T_{h,1} - T_{c,1}}{T_{h,2} - T_{c,2}}$$

which reduces to

$$q = \overline{U}A_s\, LMTD \tag{d}$$

where

$$LMTD = \frac{(T_{h,1} - T_{c,1}) - (T_{h,2} - T_{c,2})}{\ln \dfrac{T_{h,1} - T_{c,1}}{T_{h,2} - T_{c,2}}} = \frac{\Delta T_1 - \Delta T_2}{\ln (\Delta T_1/\Delta T_2)}$$

with $\Delta T_1 = T_{h,1} - T_{c,1}$ and $\Delta T_2 = T_{h,2} - T_{c,2}$. This result applies to both parallel-flow and counterflow arrangements.

Comparing Eq. (d) with the more general log mean temperature difference equation given by Eq. (11–30),

$$q = \overline{U}A_s(F\, LMTD)$$

the correction factor $F$ for parallel flow or counterflow in a double-pipe heat exchanger module is simply equal to unity.

## EXAMPLE 11–11

A double-pipe heat exchanger is used to cool 55 $\text{lb}_\text{m}$/min of oil with a specific heat of 0.525 Btu/($\text{lb}_\text{m}$ °F) from 122°F to 104°F. A cooling fluid enters the annulus of the exchanger at 68°F and exits at 77°F. The mean overall coefficient of heat transfer $\overline{U}$ is 88 Btu/(h ft$^2$ °F). Determine the heat-exchanger surface area $A_s$ for both parallel flow and counterflow.

### Solution

*Objective* Determine $A_s$ for parallel flow and counterflow.

*Schematic* Double-pipe heat exchangers with $\overline{U} = 88$ Btu/(h ft$^2$ °F).

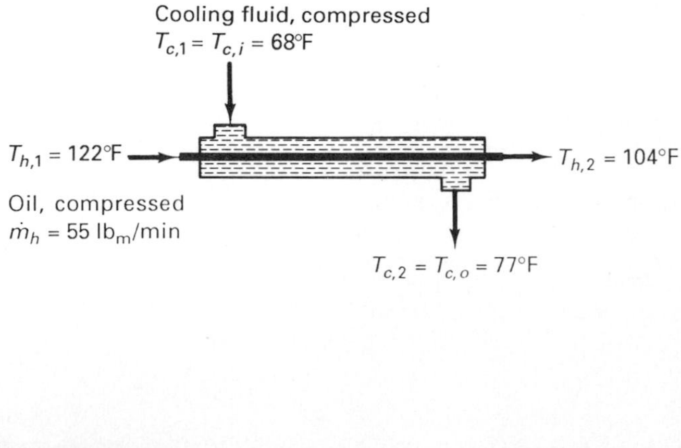
Cooling fluid, compressed
$T_{c,1} = T_{c,i} = 68$°F
$T_{h,1} = 122$°F
$T_{h,2} = 104$°F
Oil, compressed
$\dot{m}_h = 55$ $\text{lb}_\text{m}$/min
$T_{c,2} = T_{c,o} = 77$°F

(a) Parallel-flow arrangement

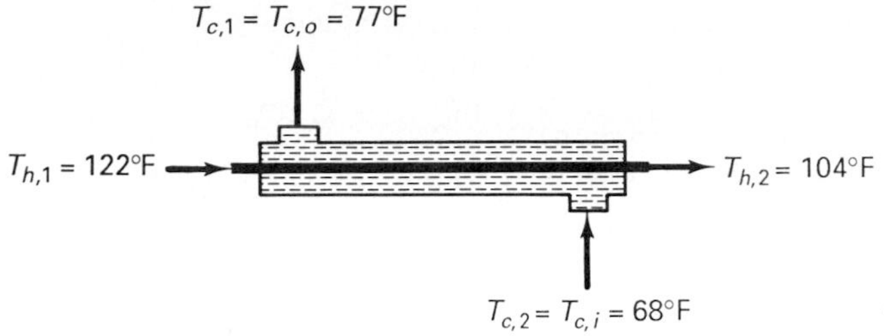

$T_{c,1} = T_{c,o} = 77°F$
$T_{h,1} = 122°F$
$T_{h,2} = 104°F$
$T_{c,2} = T_{c,i} = 68°F$


(b) Counterflow arrangement

*Assumptions/Conditions*

uniform properties
standard conditions

*Properties* Hot oil: $c_P = 0.525$ Btu/(lb$_m$ °F).

*Analysis* Because the exit temperatures are given, we can conveniently use the *LMTD* approach. The total rate of heat transfer $q$ is

$$q = (\dot{m}c_P)_h\,\Delta T_h = (\dot{m}c_P)_c\,\Delta T_c = 55\,\frac{\text{lb}_\text{m}}{\text{min}}\left(0.525\,\frac{\text{Btu}}{\text{lb}_\text{m}\,°\text{F}}\right)(122°\text{F} - 104°\text{F})$$

$$= 520 \text{ Btu/min} = 31{,}200 \text{ Btu/h}$$

*Parallel flow*

$$LMTD = \frac{(122°\text{F} - 68°\text{F}) - (104°\text{F} - 77°\text{F})}{\ln \dfrac{122°\text{F} - 68°\text{F}}{104°\text{F} - 77°\text{F}}} = 39°\text{F}$$

$$q = \overline{U}A_s\,LMTD$$

$$A_s = \frac{q}{\overline{U}\,LMTD} = \frac{31{,}200 \text{ Btu/h}}{[88 \text{ Btu/(h ft}^2\,°\text{F)}](39°\text{F})} = 9.09 \text{ ft}^2$$

*Counterflow*

$$LMTD = \frac{(122°\text{F} - 77°\text{F}) - (104°\text{F} - 68°\text{F})}{\ln \dfrac{122°\text{F} - 77°\text{F}}{104°\text{F} - 68°\text{F}}} = 40.3°\text{F}$$

$$A_s = \frac{q}{\overline{U}\,LMTD} = 8.8 \text{ ft}^2$$

**EXAMPLE 11–12**

A double-pipe heat exchanger is utilized to heat water with mass flow rate of 10 kg/s from 15°C to 33°C. The heating fluid enters at 75°C with a capacity rate of 25 kW/°C and the mean overall coefficient of heat transfer is 1570 W/(m$^2$ °C). Determine the necessary surface area for parallel-flow and counterflow operation.

**Solution**

*Objective* Determine $A_s$ for parallel flow and counterflow.

*Schematic* Double-pipe heat exchangers with $\overline{U} = 1570$ W/(m² °C).

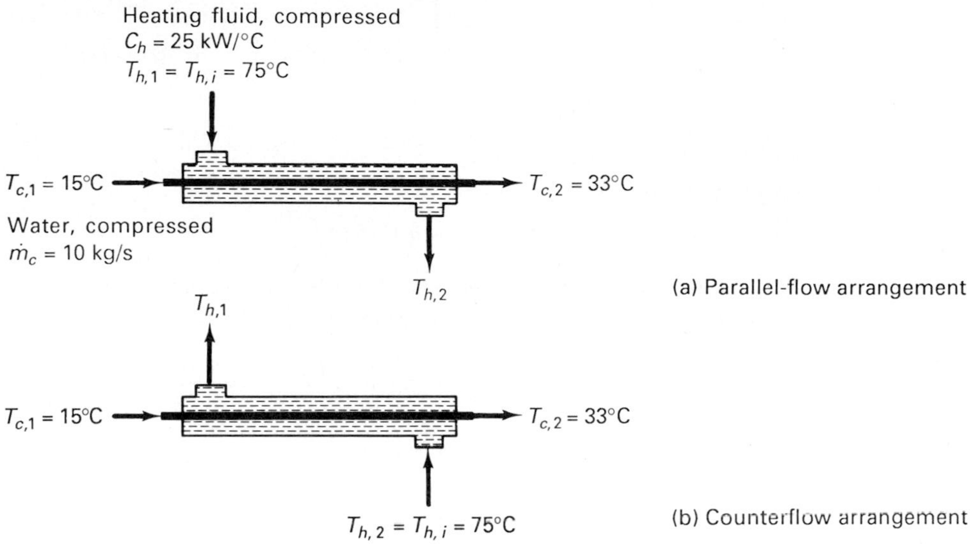


*Assumptions/Conditions*

uniform properties
standard conditions

*Properties* Water at $T = 15°C = 288$ K (Table A–C–3): $c_P = 4.18$ kJ/(kg °C).

*Analysis* Utilizing the *LMTD* equation, we have

$$A_s = \frac{q}{\overline{U}\, LMTD} \tag{a}$$

for both parallel-flow and counterflow arrangements. The capacity rate of the water and the rate of heat transfer $q$ are given by

$$C_c = (\dot{m}c_P)_c = 10\,\frac{\text{kg}}{\text{s}}\left(4.18\,\frac{\text{kJ}}{\text{kg °C}}\right) = 41.8\ \text{kW/°C}$$

$$q = C_c(T_{c,2} - T_{c,1}) = 41.8\,\frac{\text{kW}}{\text{°C}}\,(33°\text{C} - 15°\text{C}) = 752\ \text{kW}$$

The outlet temperature $T_{h,o}$ of the heating fluid is calculated as

$$q = C_h(75°\text{C} - T_{h,o})$$

$$T_{h,o} = 75°\text{C} - \frac{752 \text{ kW}}{25 \text{ kW/°C}} = 44.9°\text{C}$$

Thus, we can calculate the *LMTD*.

*Parallel flow*

$$LMTD = \frac{(75°\text{C} - 15°\text{C}) - (44.9°\text{C} - 33°\text{C})}{\ln \dfrac{75°\text{C} - 15°\text{C}}{44.9°\text{C} - 33°\text{C}}} = 29.7°\text{C}$$

*Counterflow*

$$LMTD = \frac{(44.9°\text{C} - 15°\text{C}) - (75°\text{C} - 33°\text{C})}{\ln \dfrac{44.9°\text{C} - 15°\text{C}}{75°\text{C} - 33°\text{C}}} = 35.6°\text{C}$$

Substituting these inputs into Eq. (a), we obtain

$$A_s = \frac{752 \text{ kW}}{[1.57 \text{ kW/(m}^2\text{ °C)}]\ LMTD}$$

$$= 16.1 \text{ m}^2 \qquad \text{for parallel flow}$$

$$= 13.5 \text{ m}^2 \qquad \text{for counterflow}$$

Thus, the counterflow arrangement offers an approximate 20% savings on surface area. However, because of the rather large surface area required for this application, more compact-type heat exchangers should be considered.

## EXAMPLE 11–13

Saturated water at 1 atm with a quality of 0.1 and mass flow rate of 12 kg/min is to be cooled to 80°C. Cooling water at 15°C and 20 kg/min mass flow rate is available. Determine the surface area required for a counterflow double-pipe heat exchanger if the mean overall coefficient of heat transfer is 400 W/(m$^2$ °C), with the cooling fluid flowing in the annulus.

### Solution

***Objective*** Determine $A_s$ for counterflow.

***Schematic*** Double-pipe heat exchanger: counterflow arrangement.

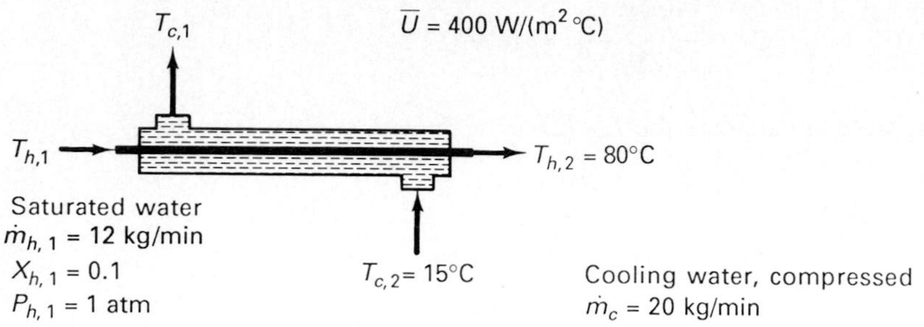


*Assumptions/Conditions*

uniform properties in each phase
standard conditions, except for phase change

*Properties*

Saturated water at 1 atm (*Steam Tables*): $i_f = 419$ kJ/kg, $i_{fg} = 2260$ kJ/kg.

Saturated liquid water at 80°C (*Steam Tables*): $i_{h,2} = 335$ kJ/kg.

Water at $T = 15°C = 288$ K (Table A–C–3): $c_P = 4.19$ kJ/(kg °C).

*Analysis* To calculate the enthalpy of the entering saturated water we write

$$X_{h,1} = 0.1 = \frac{i_{h,1} - i_f}{i_{fg}}$$

$$i_{h,1} = 0.1\left(2260\,\frac{\text{kJ}}{\text{kg}}\right) + 419\,\frac{\text{kJ}}{\text{kg}} = 645\text{ kJ/kg}$$

The rate of heat transfer $q_{\text{II}}$ required to completely condense 12 kg/min of this two-phase water is

$$q_{\text{II}} = \dot{m}_h(i_{h,1} - i_f) = 12\,\frac{\text{kg}}{\text{min}}\,(645 - 419)\,\frac{\text{kJ}}{\text{kg}} = 2710\text{ kJ/min} = 45.2\text{ kW}$$

The energy required to subcool the saturated liquid water to 80°C is calculated as follows:

$$q_{\text{I}} = \dot{m}_h(i_f - i_{h,2}) = 12\,\frac{\text{kg}}{\text{min}}\,(419 - 335)\,\frac{\text{kJ}}{\text{kg}} = 1010\text{ kJ/min} = 16.8\text{ kW}$$

Hence, the total rate of heat transfer $q$ is

$$q = q_{\text{I}} + q_{\text{II}} = (1010 + 2710)\,\frac{\text{kJ}}{\text{min}} = 3720\text{ kJ/min} = 62\text{ kW}$$

With this information and noting that the specific heat of water is about 4.19 kJ/(kg °C), we can obtain the cooling fluid outlet temperature $T_{c,1}$ by writing

$$C_c = (\dot{m}c_P)_c = 20\,\frac{\text{kg}}{\text{min}}\left(4.19\,\frac{\text{kJ}}{\text{kg °C}}\right)\left(\frac{\text{min}}{60\text{ s}}\right) = 1.40\text{ kW/°C}$$

$$q = C_c(T_{c,1} - T_{c,2})$$

$$T_{c,1} = \frac{q}{C_c} + T_{c,2} = \frac{62\text{ kW}}{1.40\text{ kW/°C}} + 15\text{°C} = 59.3\text{°C}$$

Similarly, we can write the following expression for the temperature of the cooling water at the point at which the quality of the saturated fluid is zero:

$$q_{\text{I}} = C_c(T_a - T_{c,2})$$

$$T_a = \frac{q_{\text{I}}}{C_c} + T_{c,2} = \frac{16.8\text{ kW}}{1.40\text{ kW/°C}} + 15\text{°C} = 27.0\text{°C}$$

The bulk temperatures of the two fluids are sketched in Fig. E11–13.

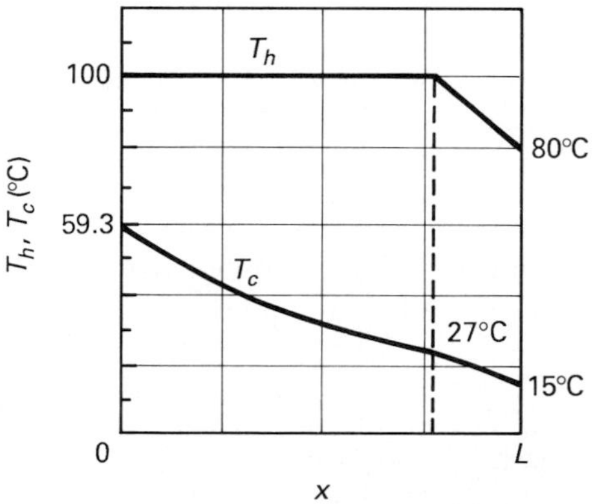


**FIGURE E11–13**
Calculations for bulk-stream temperatures $T_h$ and $T_c$—counterflow arrangement.

The surface area $A_{s,\text{I}}$ required to subcool the hot water is determined as follows:

$$LMTD_{\text{I}} = \frac{(100\text{°C} - 27.0\text{°C}) - (80\text{°C} - 15\text{°C})}{\ln\dfrac{100\text{°C} - 27.0\text{°C}}{80\text{°C} - 15\text{°C}}} = 68.9\text{°C}$$

$$q_{\text{I}} = \overline{U}A_{s\text{I}}\,LMTD_{\text{I}}$$

$$A_{s\text{I}} = \frac{q_{\text{I}}}{\overline{U}\,LMTD_{\text{I}}} = \frac{16{,}800\text{ W}}{[400\text{ W/(m}^2\text{ °C)}](68.9\text{°C})} = 0.610\text{ m}^2$$

Similarly, to determine the area $A_{s\text{II}}$ required to condense the vapor, we write

$$LMTD_{\text{II}} = \frac{(100\text{°C} - 59.3\text{°C}) - (100\text{°C} - 27.0\text{°C})}{\ln\dfrac{100\text{°C} - 59.3\text{°C}}{100\text{°C} - 27.0\text{°C}}} = 55.3\text{°C}$$

$$q_{II} = \overline{U}A_{sII}\, LMTD_{II}$$

$$A_{sII} = 2.04 \text{ m}^2$$

The total surface area of the heat exchanger is

$$A_s = A_{sI} + A_{sII} = 0.610 \text{ m}^2 + 2.04 \text{ m}^2 = 2.64 \text{ m}^2$$

It should be noted that for two-phase problems such as this, one cannot calculate the surface area $A_s$ directly from the overall *LMTD*. This is because of the change in the shape of the temperature difference curve in the two-phase and single-phase regions (see Fig. E11–13).

## EXAMPLE 11–14

A shell-and-tube heat exchanger with one shell pass and two tube passes heats water at 15°C with a mass flow rate of 9.94 kg/s. Heating oil [$c_P$ = 2.5 kJ/(kg °C)] enters the tubes at 80°C and exits at 35°C with a mass flow rate of 5.0 kg/s. Determine the surface area of the heat exchanger if the mean overall coefficient of heat transfer is 300 W/(m² °C).

### Solution

*Objective* Determine $A_s$.

*Schematic* Shell-and-tube heat exchanger: 1–2 pass arrangement.

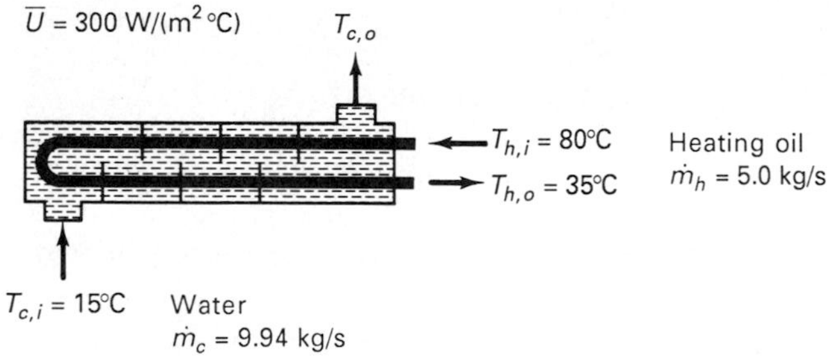

$\overline{U}$ = 300 W/(m² °C)
$T_{c,o}$
$T_{h,i}$ = 80°C
$T_{h,o}$ = 35°C
Heating oil
$\dot{m}_h$ = 5.0 kg/s
$T_{c,i}$ = 15°C
Water
$\dot{m}_c$ = 9.94 kg/s


*Assumptions/Conditions*

uniform properties
standard conditions

*Properties*

Water at $T$ = 15°C = 288 K (Table A–C–3): $c_P$ = 4.19 kJ/(kg °C).

Heating oil: $c_P$ = 2.5 kJ/(kg °C).

*Analysis* The heat transfer $q$ and outlet temperature $T_{c,o}$ of the water are obtained by writing

$$q = (\dot{m}c_P)_h\,(T_{h,i} - T_{h,o}) = 5.0\,\frac{\text{kg}}{\text{s}}\left(2.5\,\frac{\text{kJ}}{\text{kg °C}}\right)(80°\text{C} - 35°\text{C}) = 562\ \text{kW}$$

and

$$q = (\dot{m}c_P)_c\,(T_{c,o} - T_{c,i})$$

$$T_{c,o} = \frac{q}{(\dot{m}c_P)_c} + T_{c,i} = \frac{562\ \text{kW}}{(9.94\ \text{kg/s})\,[4.19\ \text{kJ/(kg °C)}]} + 15°\text{C} = 28.5°\text{C}$$

Now that both outlet temperatures are known, we are able to calculate the *LMTD* for ideal counterflow conditions.

$$LMTD = \frac{(T_{h,i} - T_{c,o}) - (T_{h,o} - T_{c,i})}{\ln \dfrac{T_{h,i} - T_{c,o}}{T_{h,o} - T_{c,i}}}$$

$$= \frac{(80°\text{C} - 28.5°\text{C}) - (35°\text{C} - 15°\text{C})}{\ln \dfrac{80°\text{C} - 28.5°\text{C}}{35°\text{C} - 15°\text{C}}} = 33.3°\text{C}$$

To utilize Fig. 11–25, $P$ and $R$ are calculated.

$$P = \frac{T_{c,o} - T_{c,i}}{T_{h,i} - T_{c,i}} = \frac{28.5°\text{C} - 15°\text{C}}{80°\text{C} - 15°\text{C}} = 0.208$$

$$R = \frac{T_{h,i} - T_{h,o}}{T_{c,o} - T_{c,i}} = \frac{80°\text{C} - 35°\text{C}}{28.5°\text{C} - 15°\text{C}} = 3.33$$

Based on Fig. 11–25, we obtain a correction factor $F$ of about 0.9. To calculate $A_s$, we employ Eq. (11–30).

$$A_s = \frac{q}{\overline{U}\,(F\,LMTD)} = \frac{562\ \text{kW}}{[300\ \text{W/(m}^2\ \text{°C)}]\,[0.9\,(33.3°\text{C})]} = 62.5\ \text{m}^2$$

This design problem can also be efficiently solved by use of the effectiveness method. Following this approach, we write

$$\epsilon = \frac{T_{h,i} - T_{h,o}}{T_{h,i} - T_{c,i}} = \frac{80°\text{C} - 35°\text{C}}{80°\text{C} - 15°\text{C}} = 0.692$$

$$C_h = (\dot{m}c_P)_h = 12.5\ \text{kW/°C} \qquad C_c = (\dot{m}c_P)_c = 41.6\ \text{kW/°C}$$

$$C_{\min} = C_h = 12.5\ \text{kW/°C} \qquad C^* = 0.3$$

$NTU = 1.5$ from Fig. 11–22 [or Eq. (e) in Table 11–5b]

$$A_s = NTU\,\frac{C_{\min}}{\overline{U}} = 1.5\,\frac{12.5\ \text{kW/°C}}{300\ \text{W/(m}^2\ \text{°C)}} = 62.5\ \text{m}^2$$

which is consistent with the result obtained by the *LMTD* method.

## EXAMPLE 11–15

An unmixed crossflow heat exchanger is to be constructed that will heat 2.5 kg/s of air at 1 atm from 15°C to 30°C. Hot water enters at 52.5°C. The mean overall coefficient of heat transfer is 300 W/(m² °C). Determine the required surface area to produce an outlet water temperature of 24°C.

### Solution

*Objective* Determine required heat transfer surface area $A_s$.

*Schematic* Crossflow heat exchanger: both fluids unmixed.

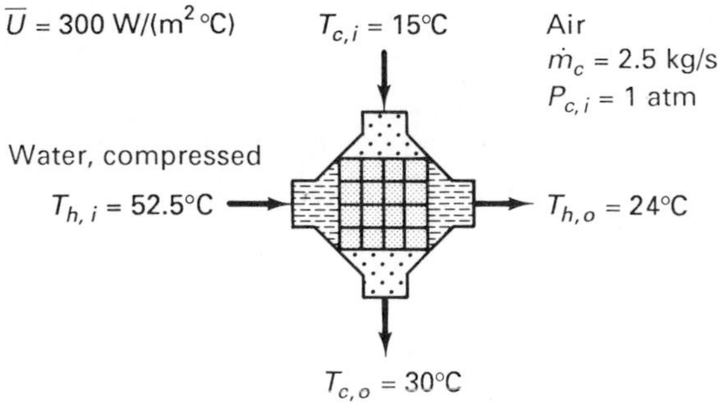
$\overline{U}$ = 300 W/(m² °C)
$T_{c,i}$ = 15°C
Air
$\dot{m}_c$ = 2.5 kg/s
$P_{c,i}$ = 1 atm
Water, compressed
$T_{h,i}$ = 52.5°C
$T_{h,o}$ = 24°C
$T_{c,o}$ = 30°C

*Assumptions/Conditions*

uniform properties
standard conditions

*Properties* Air at $T = 15°C = 288$ K (Table A–C–5): $c_P = 1.01$ kJ/(kg °C).

*Analysis* To determine the unknown surface area $A_s$, we utilize Eq. (11–30),

$$A_s = \frac{q}{\overline{U}(F\,LMTD)}$$

where

$$q = (\dot{m}c_P)_c\,(T_{c,o} - T_{c,i})$$

$$= 2.5\,\frac{\text{kg}}{\text{s}}\left(1.01\,\frac{\text{kJ}}{\text{kg °C}}\right)(30°\text{C} - 15°\text{C}) = 37.9\text{ kW}$$

Calculating the log mean temperature difference, we have

$$LMTD = \frac{(52.5°\text{C} - 30°\text{C}) - (24°\text{C} - 15°\text{C})}{\ln\dfrac{52.5°\text{C} - 30°\text{C}}{24°\text{C} - 15°\text{C}}} = 14.7°\text{C}$$

To utilize Fig. 11–26, we calculate $P$ and $R$.

$$P = \frac{T_{c,o} - T_{c,i}}{T_{h,i} - T_{c,i}} = \frac{30°\text{C} - 15°\text{C}}{52.5°\text{C} - 15°\text{C}} = 0.4$$

$$R = \frac{T_{h,i} - T_{h,o}}{T_{c,o} - T_{c,i}} = \frac{52.5°\text{C} - 24°\text{C}}{30°\text{C} - 15°\text{C}} = 1.9$$

Using these inputs, the correction factor $F$ is approximately 0.84.

Substituting into Eq. (11–30), we obtain

$$A_s = \frac{37.9 \text{ kW}}{[300 \text{ W/(m}^2\ °\text{C)}]\,[0.84(14.7°\text{C})]} = 10.2 \text{ m}^2$$

## *Specification of $\overline{U}$*

In design work the mean overall coefficient of heat transfer $\overline{U}$ is generally specified by the use of design correlations for the mean (or local) coefficients of heat transfer $\overline{h}_h$ and $\overline{h}_c$ (or $h_h$ and $h_c$) represented by Nusselt number $\overline{Nu}$ $[(\overline{h}D_H/k)_h, (\overline{h}D_H/k)_c]$ or Stanton number $\overline{St}$ $[\overline{h}_h/(\rho c_P U_b)_h, \overline{h}_c/(\rho c_P U_b)_c]$ in terms of Reynolds number $Re$ $[(D_H U_b/\nu)_h, (D_H U_b/\nu)_c]$ and Prandtl number $Pr$ $[(\mu c_P/k)_h, (\mu c_P/k)_c]$, and inputs for resistance associated with fouling (and fins), using the relations presented in Sec. 11–4. The nature of the correlations for $\overline{h}_h$ and $\overline{h}_c$ is dependent on the geometry, flow arrangement, and on whether the flow is laminar or turbulent. Correlations of the type introduced in Chap. 8 for $\overline{h}$ can generally be used for double-pipe and crossflow heat exchangers. However, more complex design correlations are generally used to characterize $\overline{h}$ for baffled shell-side flow. Information pertaining to the evaluation of $\overline{U}$ for shell-and-tube and crossflow heat exchangers is introduced in references 2 and 9.

In the thermal analysis of existing heat exchangers, correlations can be readily established for $\overline{U}$ on the basis of recorded measurements for the terminal temperatures and flow rates by the use of solutions for effectiveness $\epsilon$ or the correction factor $F$. For example, using the effectiveness approach $\epsilon$ can be computed from Eq. (11–22) or Eq. (11–23) and $C^*$ can be obtained from the defining relation, Eq. (11–24). Knowing $\epsilon$ and $C^*$, the $\epsilon(C^*, NTU)$ curve, which corresponds to the geometry and flow arrangement of interest, can be used to evaluate the number of transfer units $NTU$, after which $\overline{U}$ can be computed by putting Eq. (11–25) into the form

$$\overline{U} = \frac{C_{\min}}{A_s} NTU \tag{11–35}$$

Similarly, following the $LMTD$ approach, Eqs. (11–30), (11–32), and (11–33) can be used to compute $LMTD$, $P$, and $R$, after which $F$ can be obtained from the appropriate $F(P, R)$ curve. Finally, $\overline{U}$ can be computed from Eq. (11–30) and Eq. (11–14) or Eq. (11–15),

$$\overline{U} = \frac{q}{A_s(F\,LMTD)} = \frac{C_h(T_{h,i} - T_{h,o})}{A_s(F\,LMTD)} = \frac{C_c(T_{c,o} - T_{c,i})}{A_s(F\,LMTD)} \tag{11–36}$$

Calculations for $\overline{U}$ obtained in either of these ways account for all of the various complicating factors that may be present, such as fouling, fins, and baffles. By specifying the mean coefficient of heat transfer on one side (say the tube side) the calculations obtained for $\overline{U}$ over a range of operating conditions can be used to develop a correlation for the mean coefficient of heat transfer for the other side (say the shell-side) in terms of *Nu* or *St* versus *Re* and *Pr*.

### 11–5–2 Hydraulic Considerations

Referring to Figs. 8–15 and 8–22 and to Secs. 8–2–3 and 8–4–3, the total pressure drop $\Delta P_{1-2}$ (relative to inlet and outlet stations 1 and 2) for incompressible flow on one side of a heat exchanger is generally represented by Eq. (8–62),

$$\Delta P_{1-2} = \frac{\rho U_b^2}{2}\left(K_c + \frac{4L}{D_H}f + K_e\right) \tag{11–37}$$

for uniform density, or Eq. (8–64),

$$\Delta P_{1-2} = \frac{G^2 v_1}{2}\left[1 - \sigma^2 + K_c + \frac{4L}{D_H}f\frac{v_m}{v_1} + 2\left(\frac{v_2}{v_1} - 1\right) - (1 - \sigma^2 - K_e)\frac{v_2}{v_1}\right] \tag{11–38}$$

for situations in which significant variations in density are caused by heating or cooling. Whereas Eq. (11–37) is generally applicable to liquids or to gases with low to moderate rates of heating, Eq. (11–38) should be used for gas flow with moderate to strong heating. In design work, the effective Fanning friction factor $f$, entrance-loss coefficient $K_c$, and expansion-loss coefficient $K_e$ are specified in accordance with design correlations for flow through tubes, annuli, tube banks, and other heat exchanger cores. In this connection, the entrance and exit losses are accounted for by $f$ with $K_c$ and $K_e$ set equal to zero for flow across tube banks (and through matrix surfaces), such that Eqs. (11–37) and (11–38) reduce to

$$\Delta P_{1-2} = \frac{\rho U_b^2}{2}\left(\frac{4L}{D_H}f\right) \tag{11–39}$$

for uniform density, and

$$\begin{aligned}\Delta P_{1-2} &= \frac{G^2 v_1}{2}\left[1 - \sigma^2 + \frac{4L}{D_H}f\frac{v_m}{v_1} + 2\left(\frac{v_2}{v_1} - 1\right) - (1 - \sigma^2)\frac{v_2}{v_1}\right] \\ &= \frac{G^2 v_1}{2}\left[(1 + \sigma^2)\left(\frac{v_2}{v_1} - 1\right) + \frac{4L}{D_H}f\frac{v_m}{v_1}\right]\end{aligned} \tag{11–40}$$

for variable density.

As pointed out by Kays and London [9], losses in return bends or headers must be accounted for separately, as must any losses in inlet and exit headers and nozzles and associated valves and ducting. The additional losses associated with such components are generally represented by

$$\Delta P_f = K_f \frac{\rho U_b^2}{2} \tag{11–41}$$

where $K_f$ is the *component-loss coefficient*. Values for $K_f$ are given in references 2, 16, and 17 for a number of common fittings, valves, and cross sections. Representative values are listed in Table 11–6 for several types of heat exchanger components. For example, combining Eqs. (11–37) and (11–41), the tube-side or shell-side pressure drop in standard heat exchangers for uniform density is given by

$$\Delta P = \frac{\rho U_b^2}{2}\left(K_c + \frac{4L}{D_H} f + K_e + K_f\right) \tag{11–42}$$

**TABLE 11–6** Loss coefficients $K_f$ for heat exchanger components

| *Component* | $K_f$ |
|---|---|
| 180° tube bend | 1.5 |
| Double-pipe return housing for annular flow | 0.5 |
| Shell-and-tube return header for tube flow | 4.0 |

It should be noted that similar to the situation involved in the calculation of the mean overall coefficient of heat transfer, the specification of the friction factor $f$ in the evaluation of existing heat exchangers can be established by the use of rating correlations that are based on the analysis of specific performance data. This approach is particularly useful in the evaluation of shell-and-tube heat exchangers for which the baffled shell-side design correlations for $f$ are quite involved. In this approach, measurements for the pressure drop $\Delta P_{1-2}$ together with standard inputs for $K_c$, $K_e$, and $K_f$ are substituted into Eq. (11–42) to compute $f$ for isothermal conditions. By varying the flow rate over the range of conditions of interest, a rating correlation is readily developed for $f$ in terms of $Re$; that is,

$$f = \frac{D_H}{4L}\left(\frac{2}{\rho U_b^2}\Delta P_{1-2} - K_c - K_e - K_f\right) \tag{11–43}$$

Specific attention is given to the evaluation of pressure drop and pumping power in double-pipe, shell-and-tube, and crossflow heat exchangers in references 2 and 9.

## EXAMPLE 11–16

A 30° API petroleum distillate oil with mass flow rate of 2.18 kg/s is to be cooled from 114°C to 66°C using cooling-tower water with an available inlet to outlet temperature range of 26°C to 50°C. Standard double-pipe hairpin heat exchanger modules of 7.62 m length constructed of 1.5 and 3 in. schedule 40 wrought steel pipes are to be used. The allowable pressure drop for each stream is approximately 0.0828 MPa (12 psi) and the required minimum fouling factors are specified as 0.00015 m$^2$ °C/W for the water and 0.0007 m$^2$ °C/W for the oil. Estimate the number of modules required and how they should be arranged, using a simple uniform property analysis and assuming that the resulting outlet temperatures must be within 10% of the specified values.

## Solution

*Objective* Determine the number and arrangement of double-pipe modules required to achieve

$$T_{h,o} = 66°\text{C} \qquad T_{c,o} = 50°\text{C}$$

with

$$T_{h,i} = 114°\text{C} \qquad T_{c,i} = 26°\text{C}$$

and pressure drops less than 0.0828 MPa.

*Schematic* Double-pipe hairpin heat exchanger module.

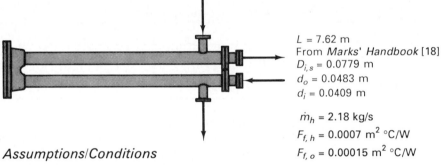

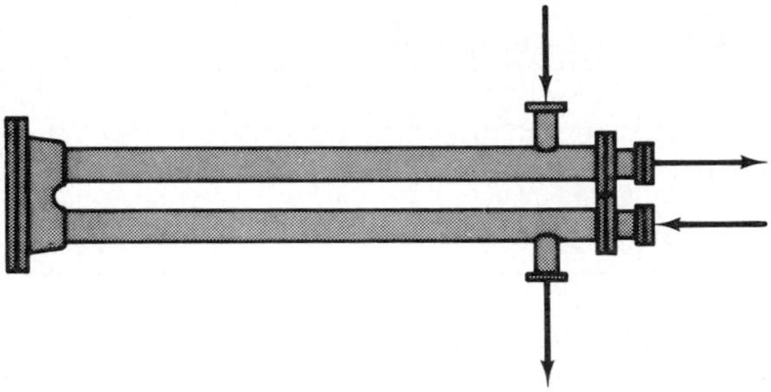

$L$ = 7.62 m
From *Marks' Handbook* [18]
$D_{i,s}$ = 0.0779 m
$d_o$ = 0.0483 m
$d_i$ = 0.0409 m

$\dot{m}_h$ = 2.18 kg/s
$F_{f,h}$ = 0.0007 m$^2$ °C/W
$F_{f,o}$ = 0.00015 m$^2$ °C/W

*Assumptions/Conditions*

moderate property variation
standard conditions

*Properties*

30° API oil at 90°C:† $\rho$ = 820 kg/m$^3$, $\mu$ = 0.00245 kg/(m s), $c_P$ = 2.17 kJ/(kg °C), $k$ = 0.128 W/(m °C), $Pr$ = 41.5.

Water at $T$ = 38°C = 311 K (Table A–C–3): $\rho$ = 993 kg/m$^3$, $\mu$ = 0.000682 kg/(m s), $c_P$ = 4.17 kJ/(kg °C), $k$ = 0.63 W/(m °C), $Pr$ = 4.51.

Wrought steel pipe (Table A–C–1): $k$ = 59 W/(m °C).

*Analysis* The cross-sectional areas and hydraulic diameters of the two passages are calculated as follows:

*Tube*

$$A_t = \frac{\pi d_i^2}{4} = \frac{\pi (0.0409\ \text{m})^2}{4} = 0.00131\ \text{m}^2$$

$$D_{H,t} = d_i = 0.0409\ \text{m}$$

† The properties of various petroleum fractions are available in TEMA [13].

*Annulus*

$$A_a = \frac{\pi}{4}(D_{i,s}^2 - d_o^2) = \frac{\pi}{4}(0.0779^2 - 0.0483^2)\ \text{m}^2 = 0.00293\ \text{m}^2$$

$$D_{H,a} = \frac{4A}{p_w} = \left(\frac{D_{i,s}^2 - d_o^2}{D_{i,s} + d_o}\right) = D_{i,s} - d_o = (0.0779 - 0.0483)\ \text{m} = 0.0296\ \text{m}$$

To determine the heat transfer rate and mass flow rate of the cooling water we employ the lumped energy formulation.

$$q = (\dot{m}c_P)_h(T_{h,i} - T_{h,o}) = 2.18\ \frac{\text{kg}}{\text{s}}\left(2.17\ \frac{\text{kJ}}{\text{kg °C}}\right)(114°\text{C} - 66°\text{C}) = 227\ \text{kW}$$

$$q = (\dot{m}c_P)_c(T_{c,o} - T_{c,i})$$

$$\dot{m}_c = \frac{q}{c_{P,c}(T_{c,o} - T_{c,i})} = \frac{227\ \text{kW}}{[4.17\ \text{kJ/(kg °C)}](50°\text{C} - 26°\text{C})} = 2.27\ \text{kg/s}$$

To provide a basis for establishing which fluid is to pass through the annulus, which has a little over twice the cross-sectional area of the tube, we will compute the bulk-stream velocities and Reynolds numbers for the two possible combinations, using a series arrangement in order to maximize $\overline{U}$.

*Oil*: $$U_{b,h} = \frac{\dot{m}_h}{\rho_h A} = \frac{2.18\ \text{kg/s}}{(820\ \text{kg/m}^3)A} = \frac{0.00266\ \text{m}^3/\text{s}}{A}$$

*Tube* $A = A_t$

$$U_{b,h} = \frac{0.00266\ \text{m}^3/\text{s}}{0.00131\ \text{m}^2} = 2.03\ \text{m/s}$$

$$Re_h = \frac{(D_H U_b)_h}{\nu_h} = \frac{0.0409\ \text{m}\ (2.03\ \text{m/s})}{[0.00245\ \text{kg/(m s)}]/(820\ \text{kg/m}^3)} = 27{,}800$$

*Annulus* $A = A_a$

$$U_{b,h} = \frac{0.00266}{0.00293}\ \frac{\text{m}}{\text{s}} = 0.908\ \text{m/s}$$

$$Re_h = \frac{0.0296\ (0.908)}{0.00245/820} = 9000$$

*Water*: $$U_{b,c} = \frac{\dot{m}_c}{\rho_c A} = \frac{2.27\ \text{kg/s}}{(993\ \text{kg/m}^3)A} = \frac{0.00229\ \text{m}^3/\text{s}}{A}$$

*Tube* $A = A_t$

$$U_{b,c} = \frac{0.00229}{0.00131}\ \frac{\text{m}}{\text{s}} = 1.75\ \text{m/s}$$

$$Re_c = \frac{(D_H U_b)_c}{\nu_c} = \frac{0.0409\,(1.75)}{0.000682/993} = 104{,}000$$

*Annulus* $A = A_a$

$$U_{b,c} = \frac{0.00229}{0.00293}\,\frac{\text{m}}{\text{s}} = 0.781 \text{ m/s}$$

$$Re_c = \frac{0.0296\,(0.781)}{0.000682/993} = 33{,}700$$

These calculations indicate that the oil will be in transitional turbulent flow if placed in the annulus. On the other hand, if the oil is placed in the tube, the flow of both fluids will be fully turbulent. Therefore, this arrangement is tentatively selected with the intention of maximizing the overall coefficient of heat transfer.

The friction factors, Nusselt numbers, and coefficients of heat transfer are now computed for fully developed conditions using Eqs. (8–15) and (8–16).

*Oil in tube*

$$f_h = (1.58 \ln 27{,}800 - 3.28)^{-2} = 0.00602$$

$$Nu_h = \frac{(f/2)\,Re\,Pr}{1.07 + 12.7\sqrt{f/2}\,(Pr^{2/3} - 1)}$$

$$= \frac{(0.00602/2)(27{,}800)(41.5)}{1.07 + 12.7\sqrt{0.00602/2}\,(41.5^{2/3} - 1)} = 398$$

$$h_h = Nu_h \frac{k_h}{D_{H,t}} = 398\,\frac{0.128 \text{ W/(m °C)}}{0.0409 \text{ m}} = 1250 \text{ W/(m}^2\text{ °C)}$$

*Water in annulus*—Referring to Fig. 8–9 and noting that $d_o/D_{i,s} = D_i/D_o = 0.62$, we conclude that $Nu_c = Nu_{ii} = Nu/0.96$. Thus, we may simply set $Nu_c = Nu$ as a conservative estimate, and write

$$f_c = (1.58 \ln 33{,}700 - 3.28)^{-2} = 0.00575$$

$$Nu_c = \frac{(0.00575/2)(33{,}700)(4.51)}{1.07 + 12.7\sqrt{0.00575/2}\,(4.51^{2/3} - 1)} = 194$$

$$h_c = 194\,\frac{0.63 \text{ W/(m °C)}}{0.0296 \text{ m}} = 4130 \text{ W/(m}^2\text{ °C)}$$

Using these results, the overall coefficient of heat transfer based on the outer surface area of the tube $A_s = A_{s,o}$ becomes (neglecting the wall resistance)

$$\frac{1}{\overline{U}} \simeq \frac{1}{h_h}\frac{d_o}{d_i} + F_{f,h}\frac{d_o}{d_i} + F_{f,c} + \frac{1}{h_c}$$

$$= \left[\frac{1}{1250}\frac{48.3}{40.9} + 0.0007\frac{48.3}{40.9} + 0.00015 + \frac{1}{4130}\right]\frac{\text{m}^2\ ^\circ\text{C}}{\text{W}}$$

$$= (0.000945 + 0.000827 + 0.00015 + 0.000242)\frac{\text{m}^2\ ^\circ\text{C}}{\text{W}}$$

or

$$\overline{U} = 462\ \text{W/(m}^2\ ^\circ\text{C)}$$

with the prescribed fouling, and

$$\overline{U} = 843\ \text{W/(m}^2\ ^\circ\text{C)}$$

for no fouling. The wall resistance given by

$$R_k = \frac{\ln (r_o/r_i)}{2\pi(kL)_{\text{wall}}}$$

cannot be formally computed until the total length of the system is estimated. However, by substituting for $r_o$, $r_i$, and $k_{\text{wall}}$, we find that $R_k$ for one module is an order of magnitude smaller than $R_c$ and $R_h$. At this point it should be mentioned that the overall coefficient of heat transfer for clean conditions is about 28% lower for the case in which the oil flows in the annulus instead of the tube.

The *LMTD* approach can now be used to compute the required surface area for the ideal counterflow arrangement in which the tube and annulus of each module are placed in series. The calculation proceeds as follows:

$$LMTD = \frac{(114^\circ\text{C} - 50^\circ\text{C}) - (66^\circ\text{C} - 26^\circ\text{C})}{\ln \dfrac{114^\circ\text{C} - 50^\circ\text{C}}{66^\circ\text{C} - 26^\circ\text{C}}} = 51.1^\circ\text{C}$$

$$q = \overline{U}A_s\, LMTD$$

$$A_s = \frac{q}{\overline{U}\, LMTD} = \frac{227\ \text{kW}}{[462\ \text{W/(m}^2\ ^\circ\text{C)}](51.1^\circ\text{C})} = 9.62\ \text{m}^2$$

The surface area per unit is given by

$$A_{s,\text{module}} = 2L\pi d_o = 2(7.62\ \text{m})\ \pi\ (0.0483\ \text{m}) = 2.31\ \text{m}^2$$

Dividing $A_s$ by $A_{s,\text{module}}$, we obtain a value of 4.16, such that 4 modules are estimated to be required, providing that the pressure drops are within specifications. Using the effectiveness method, this number of modules results in calculations for the outlet temperatures that are within 1.6% of the specified values.

The pressure drops for the two streams are represented by Eq. (11–42) with $K_c = K_e = 0$. The total lengths of the tubular and annular passageways for $M$ modules in series are given by

$$L_t = 2(7.62\ \text{m})M = 15.2M\ \text{m} \qquad L_a = 15.2M\ \text{m}$$

Substituting into these equations, we obtain

*Hot tubular fluid (oil)*

$$(\Delta P_t)_{1-2} = \left(\frac{\rho U_b^2}{2}\right)_t \left[\left(\frac{4L}{D_H}f\right)_t + 1.5M\right]$$

$$= \frac{1}{2}\left(820\ \frac{\text{kg}}{\text{m}^3}\right)\left(2.03\ \frac{\text{m}}{\text{s}}\right)^2 \left[\frac{4(15.2\ \text{m})(0.00602)}{0.0409\ \text{m}} + 1.5\right]M$$

$$= 17{,}700M\ \text{kg/(m s}^2)$$

$$= 0.0708\ \text{MPa} \qquad \text{for } M = 4$$

*Cold annular fluid (water)*

$$(\Delta P_a)_{1-2} = \left(\frac{\rho U_b^2}{2}\right)_a \left[\left(\frac{4L}{D_H}f\right)_a + 0.5M\right]$$

$$= \frac{1}{2}\left(993\ \frac{\text{kg}}{\text{m}^3}\right)\left(0.781\ \frac{\text{m}}{\text{s}}\right)^2 \left[\frac{4(15.2)(0.00575)}{0.0296} + 0.5\right]M$$

$$= 3730M\ \text{kg/(m s}^2)$$

$$= 0.0149\ \text{MPa} \qquad \text{for } M = 4$$

This result indicates that the cold stream is well within the pressure drop restrictions, but that the hot fluid is very near the maximum allowable. If 4 units are used and the actual pressure drop on the oil side is found to exceed the acceptable value, one or two of the streams can be arranged in parallel in order to reduce $(\Delta P_t)_{1-2}$.

Because the heat transfer rate for this application (227 kW) is considerably below 1000 kW, the use of a double-pipe heat exchanger is quite appropriate. For much larger heat-transfer rates, shell-and-tube heat exchangers should be considered.

## 11–6 SUMMARY

As we have seen, heat exchangers are classified according to geometry, flow arrangement, size, and certain other factors. The standard types of indirect contact recuperators include double-pipe, shell-and-tube, and crossflow configurations. Regenerators and direct contact heat exchangers are also in common use. The flow arrangements are further classified according to the number of passes, whether the fluid streams are in parallel flow, counterflow, crossflow, or combination of these, and, in the case of crossflow heat exchangers, whether the streams should be designated as mixed or unmixed. We have also seen that various means are employed in order to enhance the performance and reduce the size of heat exchangers. In addition to baffling of shell-side fluids, surface modifications such as fins (plain, corrugated, louvered, strip, perforated, longitudinal, circumferential, etc.), and inserts are often employed. It is with the aid of such devices that the design of compact heat exchangers ($\beta \gtrsim 700\ m^2/m^3$) have become possible.

In this chapter we have adapted the practical analysis approach presented in Chap. 8 to the evaluation and design of heat exchangers, with emphasis placed on recuperators. As we have seen, this approach involves the use of correlations for the coefficients of heat transfer $h$ and friction $f$ and the concept of the overall coefficient of heat transfer $U$ together with lumped and lumped/differential formulations.

The practical thermal analysis approach gives rise to solution results for heat transfer that are conveniently expressed in terms of the effectiveness $\epsilon$ and *LMTD* for steady-state uniform property flow with uniform bulk-stream temperatures and overall coefficient of heat transfer over the cross section of the heat exchanger. Relations for the effectiveness $\epsilon$ and correction factor $F$ have been presented for standard double-pipe, shell-and-tube, and crossflow arrangements. (The practical thermal analysis approach can be adapted to processes that involve nonuniform properties, unsteady operation, nonuniform bulk-stream temperature, and nonuniform overall coefficient of heat transfer by the use of computer techniques [19–21].) In addition, consideration has been given to the specification of the mean overall coefficient of heat transfer.

The analysis of regenerators, direct contact heat exchangers, and fired heat exchangers with significant levels of thermal radiation involve the application of the basic principles introduced in this and previous chapters, but is generally more involved. Once an individual has achieved a thorough understanding of the principles introduced in this chapter, which pertain to the analysis of basic recuperators, the literature that deals with the more complex type heat exchangers can be more readily followed. Works that deal with the analysis of regenerators, direct contact heat exchangers, and fired heat exchangers include references 2, 9, 19, 22–24.

## REVIEW QUESTIONS

**11–1.** What are the primary reasons for using baffles in a heat exchanger?

**11–2.** What is a recuperator?

**11–3.** What is the difference between a mixed and unmixed crossflow arrangement?

**11–4.** What are the objectives of the evaluation function?

**11–5.** List several ways in which the heat-transfer performance can be improved for tubular heat exchangers.

**11–6.** What is a compact heat exchanger?

**11–7.** Define the surface area density $\beta$.

**11–8.** Define the local overall coefficient of heat transfer $U$.

**11–9.** Define the fouling factor $F_f$.

**11–10.** Define the log mean temperature difference *LMTD*.

**11–11.** For what type of heat-exchanger problems is the effectiveness method preferable to the *LMTD* method of analysis?

**11–12.** Describe how experimental measurements can be used together with the *LMTD* approach to determine the mean overall coefficient of heat transfer $\overline{U}$.

**11–13.** Write an expression for the pressure drop $\Delta P_{1-2}$ in terms of the effective Fanning friction factor $f$ for (a) flow in a bundle of tubes with abrupt changes in area at the entrance and exit, and (b) flow across a tube bank.

**11–14.** Explain how double-pipe heat exchanger units can be arranged in order to achieve significantly smaller overall pressure drop.

**11–15.** Explain how the number of tube passes affects the tube side pressure drop in a shell-and-tube heat exchanger.

## PROBLEMS

**11–1.** Using the appropriate criterion, classify the following heat exchanger surfaces as compact or noncompact: (a) double-pipe surface of Example 11–1, (b) shell-and-tube surface of Example 11–8 for $V = 0.5\ \text{m}^3$, (c) tube bank of Fig. 8–23, and (d) plate-fin surface of Fig. P11–1.

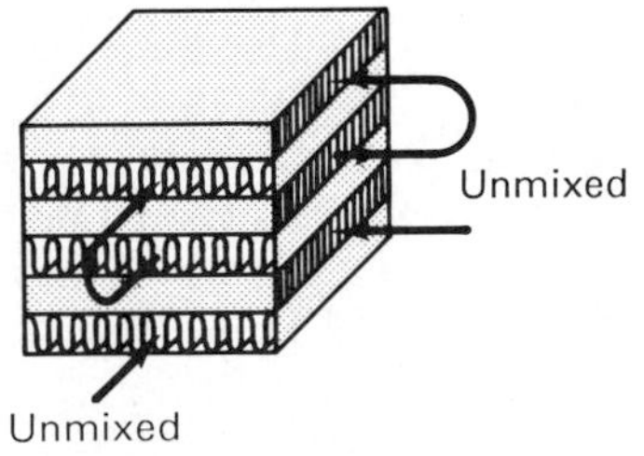


Fin pitch = 782 per m
Plate spacing = 6.35 mm
$D_H$ = 1.875 mm
Fin metal thickness = 0.152 mm
$\beta = 1841\ \text{m}^2/\text{m}^3$
$A_F/A_o = 0.849$

**FIGURE P11–1** Multipass plate-fin heat exchanger surface 19.86. (From Kays and London [9]. Used with permission.)

**11–2.** Determine the overall coefficient of heat transfer for a thin-walled double-pipe heat exchanger with inside pipe diameter of 10 cm, outer pipe diameter of 12 cm, and length 1.5 m. Air enters the tube at 250 K with a velocity of 10 m/s and water enters the annulus at 10°C with a velocity of 1 m/s.

**11–3.** Reconsider Prob. 11–2 for the case in which F-12 at 250 K flows in the tube with a velocity of 0.15 m/s.

**11–4.** To provide a more conservative estimate for the overall coefficient of heat transfer for Prob. 11–2, assume a net fouling factor of $2.5 \times 10^{-4}$ m$^2$ °C/W. Is this level of fouling significant?

**11–5.** To provide a more conservative estimate for the overall coefficient of heat transfer for Prob. 11–3, assume a net fouling factor of $2.5 \times 10^{-4}$ m$^2$ °C/W. Is this level of fouling significant?

**11–6.** Reconsider Example 11–2 for the case in which the fouling factors are given by $2 \times 10^{-3}$ m$^2$ °C/W for each fluid.

**11–7.** Referring to Table 11–2, estimate the values of net fouling factor which would generally be significant (i.e., cause a change in $U$ of 10% or more) for the following heat exchanger applications: (a) steam condenser, (b) F-12 condenser, (c) water to oil, (d) steam to heavy fuel oil, and (e) finned-tube heat exchanger with water in tubes and air over tubes.

**11–8.** Resolve Example 11–1 for the case in which the double-pipe heat exchanger is constructed of 2-mm-thick stainless steel AISI 302 tubing. The inside tube diameters are 2 cm and 3 cm, respectively.

**11–9.** Reconsider Example 11–1 for the case in which $d_i = d_o = 6$ mm.

**11–10.** Hot water at 98°C flows at a rate of 25 cm/s through a horizontal stainless-steel pipe with 5 cm I.D. and 5 mm wall thickness. The exterior of the pipe is exposed to air at 20°C and 1 atm. Determine the mean overall heat-transfer coefficient with respect to the inside surface area of the pipe.

**11–11.** Solve Example 11–3 for the case in which the mass flow rate of the water is 2.5 kg/s instead of 0.25 kg/s.

**11–12.** Referring to Example 11–5, the outlet temperatures for the Freon and water are 49.4°C and 9.75°C, respectively, and the wall temperature at the midpoint is 47.6°C. Use this information to refine the calculations for $\overline{U}$.

**11–13.** Develop expressions for $T_{c,2}$ and $T_{h,2}$ in a parallel-flow double-pipe heat exchanger for large values of $L$.

**11–14.** Develop an expression for $q$ in a counterflow double-pipe heat exchanger for the case in which $|C_c| = C_h$.

**11–15.** Demonstrate that Eqs. (r) and (s) of Example 11–4 are equivalent.

**11–16.** Check the solution to Example 11–5 by computing $q_c$, $T_{c2}$, and $A_s$ by the *LMTD* approach, assuming that $T_{c2} = 9.75$°C.

**11–17.** Refine the solution of Example 11–5 by evaluating the properties of the fluids at the average temperatures $(T_{c1} + T_{c2})/2$ and $(T_{h1} + T_{h2})/2$, and by utilizing the property correction factor for $h_c$ and $h_h$ given by Eq. (8–24). (See Prob. 11–12.)

**11–18.** To check the predictions for $T_{c1}$ and $T_{h2}$ in Example 11–6, utilize the *LMTD* approach to calculate $q$.

**11–19.** A double-pipe heat exchanger with thin walls and $D_{i,s} = 3$ cm, $d_o = 2$ cm, and $L = 3$ m is to be used to heat ethylene glycol at 20°C flowing at a rate of 1.25 kg/s. The heating fluid is water at 98°C flowing at a rate of 0.035 kg/s. Estimate the mean overall coefficient of heat transfer with the tubular fluid taken as the (a) water and (b) ethylene glycol.

**11–20.** Referring to Prob. 11–19, determine the core pressure drop of the viscous fluid for (a) annular flow and (b) tubular flow.

**11–21.** Reconsider the evaluation of heat transfer rate and the outlet temperatures associated with Example 11–5 for a heat exchanger length of 10 m.

**11–22.** Reconsider the evaluation of heat transfer rate and the outlet temperatures associated with Example 11–6 for a heat exchanger length of 10 m.

**11–23.** Water at 20°C flows in the annulus of a double-pipe heat exchanger at a rate of 2 kg/s. Hot water at 90°C flows through the tube at a rate of 0.2 kg/s. The heat exchanger is constructed of 1-mm-thick copper tubing with inside diameters of 3 cm and 5 cm, and 5 m length. Determine the outlet temperatures and the rate of heat transfer for a counterflow arrangement.

**11–24.** Use the effectiveness approach to solve Example 11–11 for parallel flow.

**11–25.** Use the effectiveness approach to solve Example 11–12 for counterflow.

**11–26.** Use the *LMTD* approach to determine the heat transfer rate for Example 11–5.

**11–27.** Water is heated in a double-pipe heat exchanger from 15°C to 40°C. Oil [$c_P$ = 2.5 kJ/(kg °C)] with a mass flow of 0.03 kg/s and inlet and outlet temperatures of 80°C and 35°C serves as the heating fluid. Utilize the *LMTD* approach to determine the required surface area for a counterflow arrangement with $\overline{U}$ = 300 W/(m$^2$ °C).

**11–28.** Water is heated in a counterflow double-pipe heat exchanger from 35°C to 85°C by an oil with a specific heat of 1.5 kJ/(kg °C) and mass flow rate of 50 kg/min. The oil is cooled from 215°C to 180°C and the overall coefficient of heat transfer is 400 W/(m$^2$ °C). Determine the rate of heat transfer, the mass flow rate of the water, and the surface area $A_s$ of the heat exchanger.

**11–29.** A double-pipe heat exchanger cools water with a mass flow rate of 0.5 kg/s from 50°C to 5°C. A refrigerant enters at −25°C with a capacity rate of 5.0 kW/°C. The mean overall coefficient of heat transfer of the heat exchanger is 1200 W/(m$^2$ °C). Determine the surface area required for parallel flow and counterflow operation.

**11–30.** A double-pipe heat exchanger is utilized to heat oil with a capacity rate of 15 kW/°C from 30°C to 80°C. Hot water enters at 125°C and 2 atm and exits at 95°C. Determine the surface area and rate of heat transfer if $\overline{U}$ is equal to 250 W/(m$^2$ °C) for counterflow.

**11–31.** Water flowing at a rate of 6 kg/min is heated from 35°C to 75°C by oil which undergoes a change from 130°C to 100°C. The overall coefficient of heat transfer is 400 W/(m$^2$ °C). Determine the capacity rate of the oil and the surface area of a double-pipe heat exchanger for counterflow.

**11–32.** Calculate and plot the bulk-stream temperature distributions for the parallel-flow arrangement of Example 11–11.

**11–33.** Calculate and plot the bulk-stream temperature distributions for the counterflow arrangement of Example 11–11.

**11–34.** Calculate and plot the bulk-stream temperature distributions for the counterflow arrangement of Example 11–12.

**11–35.** Following the pattern established in the approximate solution developed for the uniform wall temperature boundary condition in Prob. 8–39, the heat transfer in a heat exchanger can be approximated by

$$q_c = \overline{U}A_s \frac{\Delta T_2 + \Delta T_1}{2}$$

As in the analysis of Chap. 8, the substitution of $(\Delta T_2 + \Delta T_1)/2$ for the *LMTD* relationship for a double-pipe heat exchanger leads to less than 1% error for $0.75 \leqslant \Delta T_1/\Delta T_2 \leqslant 1.5$. The coupling of this expression with Eqs. (11–14) and (11–15) provides a simple direct means of approximating $q$ and the outlet temperatures. Utilize this approximate approach to estimate the rate of heat transfer and outlet temperature for Example 11–5.

**11–36.** Utilize the approximate approach of Prob. 11–35 to estimate the rate of heat transfer and outlet temperature for Example 11–6.

**11–37.** Solve Example 11–13 by the effectiveness approach.

**11–38.** Referring to Example 11–5, calculate the rate of heat transfer for the case in which the heating water is saturated with a quality of 0.05 and a temperature of 98°C.

**11–39.** Demonstrate that Eq. (8–55) is applicable to double-pipe heat exchangers for situations in which $C^* \rightarrow 0$.

**11–40.** Exhaust from diesel engines used for generating electricity is available at 600 K and mass flow rate of 180 kg/h. It is desired to heat air at 27°C with mass flow rate of 100 kg/h using a double-pipe heat exchanger unit with inside tube diameters of 7.38 cm and 9.8 cm, wall thickness (copper) of 2.77 mm, and length of 2 m. Determine the outlet temperature of each stream, assuming a counterflow arrangement with air flow in the tube, using the properties of carbon dioxide to approximate the properties of the exhaust gas.

**11–41.** Water flowing at a mass rate of 10 kg/s is to be heated from 15°C to 33°C. The heating fluid enters a 1–2 pass shell-and-tube heat exchanger at 75°C with a capacity rate of 25 kW/°C. Assuming that the heat exchanger is rated at an overall coefficient of heat transfer of 1700 W/(m$^2$ °C) for these conditions, determine the required surface area $A_s$.

**11–42.** A heat exchanger with one shell pass and two tube passes heats 20,000 lb$_m$/h of water in the tubes from 100°F to 250°F. The heating water enters as saturated liquid at 500°F and exits at 300°F. Determine the rate of heat transfer and the surface area of the heat exchanger if $\overline{U} = 1200$ Btu/(h ft$^2$ °F).

**11–43.** Water is heated in a shell-and-tube heat exchanger from 35°C to 85°C by an oil with a specific heat of 1.5 kJ/(kg °C) and mass flow rate of 50 kg/min. The oil is to be cooled from 215°C to 180°C and the overall coefficient of heat transfer is 400 W/(m$^2$ °C). The water makes one shell pass and the oil makes two tube passes. Estimate the rate of heat transfer, the mass flow rate of the water, and the surface area $A_s$ of the heat exchanger.

**11–44.** Outline the steps required in solving Example 11–8 by the *LMTD* method.

**11–45.** From a test on a single-shell, two-tube-pass heat exchanger, the following data are available: oil [$c_P$ = 2 kJ/(kg °C)] in turbulent flow inside the tubes entered at 70°C at a rate of 0.6 kg/s and exited at 40°C; water flowing on the shell side entered at 15°C and exited at 25°C. A change in service conditions requires the cooling of the oil from an initial temperature of 65°C but at $\frac{3}{2}$ of the flow rate used in the performance test. Estimate the outlet temperature of the oil for the same water mass flow rate and inlet temperature.

**11–46.** A well baffled 1–2 shell-and-tube heat exchanger is used to cool a viscous oil [$c_P$ = 2.10 kJ/(kg °C)] from 340 K to 310 K. The oil flows in the tubes at a rate of 0.1 kg/s, with the flow pattern being laminar. The water flowing on the shell side is heated from 290 K to 300 K. A change in service conditions requires that the mass flow rate of the oil be increased by 25%. Estimate the outlet temperatures of the oil and water for the same inlet temperatures and mass flow rate of water, assuming that the oil flow remains laminar.

**11–47.** Solve Example 11–14 by the effectiveness approach.

**11–48.** A heat exchanger using 0.5 kg/s of steam at the exhaust of a turbine with a temperature of 50°C and quality of 90% is to be used to heat 3 kg/s of seawater [$c_P$ = 3.98 kJ/(kg °C)] from 20°C to 40°C. The heat exchanger is to be sized for one shell pass for the steam and four tube passes with 20 parallel tube circuits of 2.50-cm-I.D. and 2.75-cm-O.D. copper tubing. For the clean heat exchanger, the mean heat-transfer coefficients of the steam and water sides are approximately 10,000 W/(m$^2$ °C) and 1500 W/(m$^2$ °C), respectively. Calculate the tube length required for long-term service.

**11–49.** A 1–4 shell-and-tube heat exchanger with 42 in. ID shell, 1144 $\frac{3}{4}$-in. OD by 16 ft long 16 BWG tubes ($d_o$ = 0.0191 m, $d_i$ = 0.0166 m) on 1 in. square pitch, and twenty-two 20% cut baffles spaced 8.33 in. apart is to be used to cool 30 API oil [$c_P$ = 2.17 kJ/(kg °C)] with a temperature of 114°C flowing at a rate of 43.6 kg/s. Cooling water flowing at a rate of 45.4 kg/s is to enter the shell side at a temperature of 26°C. According to design calculations that involve the use of standard convection correlations which account for the effect of baffles, the mean overall coefficient of heat transfer is about 436 W/(m$^2$ °C) for no fouling. Use this information to determine the heat transfer rate and the outlet temperatures of the two fluids. (See Example 11–16 for the properties of 30 API oil.)

**11–50.** Referring to Prob. 11–49, the baffle spacing $B$ is observed to be equal to $0.2D_{i,s}$. According to TEMA standards, the baffle spacing generally lies within $0.2D_{i,s}$ to $D_{i,s}$ with the overall coefficient of heat transfer usually being higher for smaller spacing. To expand upon this point, an analysis has been developed to see the effect of reducing the number of baffles to four, with $B = 0.912D_{i,s}$. The analysis indicates that the mean overall coefficient of heat transfer for this application using only four baffles is 391 W/(m$^2$ °C), which is about 9% lower than the value obtained using 22 baffles. Calculate the heat-transfer rate and the outlet temperatures of the two fluids for the case in which only four baffles are used.

**11–51.** Water at 100°F flowing at a rate of 30,000 lb$_m$/h is heated in a 1–2 shell-and-tube heat exchanger with 72 tubes of 0.75 in. diameter and 5.4 ft length. Water at 200°F flowing in the tubes at a rate of 15,000 lb$_m$/h is used as the heating fluid. The overall heat-transfer coefficient is 250 Btu/(h ft$^2$ °F). Determine the heat transfer rate and the outlet temperatures of both fluids.

**11–52.** Water enters the tubes of a single-pass heat exchanger at 5°C and leaves at 55°C. On the shell side, 60 kg/min of steam condenses at 100°C. Calculate the overall heat transfer coefficient and the required flow rate of water if the surface area of the exchanger is 12 m$^2$.

**11–53.** Water in two tube passes is used to cool oil [$c_P$ = 2.35 kJ/(kg °C)] in one shell pass. The surface area of the heat exchanger is 25.3 m$^2$. The water enters at 20°C and leaves at 70°C. The oil enters at 180°C and 2.67 kg/s and leaves at 80°C. Determine the heat-transfer rate, the flow rate of water, and the overall coefficient of heat transfer. Also sketch the variation in bulk-stream temperatures in the tube and annulus as a function of $x$.

**11–54.** A 1–4 shell-and-tube heat exchanger containing 80 thin walled 25-mm diameter tubes must be designed to heat 2.5 kg/s of water from 15°C to 85°C. The heating is to be accomplished by passing hot engine oil [$c_P$ = 2.35 kJ/(kg °C)], which is available at 160°C, through the shell side of the exchanger. The oil is known to provide an average convection coefficient $\bar{h}$ of 400 W/(m$^2$ °C) on the outside of the tubes. Determine the oil flow rate if the oil leaves the exchanger at 100°C. How long must the shell be to accomplish the desired heating. (See Incropera and DeWitt [25], Example 11–2 for a similar problem.)

**11–55.** Air at 15°C and 1 atm with a mass flow rate of 2.37 kg/s is to be heated in a crossflow heat exchanger with eighty 9.52-mm-diameter 2-m-long tubes. The tubes are placed in a staggered array with 10 rows, 2.5 m depth, 1.19 cm longitudinal pitch, and 1.43 cm transverse pitch.

Hot water at 65°C enters the tubes with a mass flow rate of 1.66 kg/s. Estimate the overall coefficient of heat transfer, the rate of heat transfer, and the exit temperatures of the air and water.

**11–56.** A crossflow heat exchanger is used to heat 0.6 kg/s of air (mixed) at 15°C. Hot water enters the tubes at 45°C with a mass flow rate of 0.5 kg/s. The overall coefficient of heat transfer is 300 W/($m^2$ °C) and the total surface area of the heat exchanger is 5 $m^2$. Estimate the heat-transfer rate and the exit bulk-stream temperatures.

**11–57.** Solve Example 11–9 for the case in which the mass flow rate of the heating water is 2.5 kg/s instead of 0.25 kg/s.

**11–58.** Referring to Example 11–9, determine the depth of the tube bank which is required to bring the exiting water temperature to 35°C.

**11–59.** Air at 260°F with entering velocity of 10 ft/s is cooled in the tube bank of Example 8–14 with 23 tube rows and 6 ft depth. The cooling fluid is water at 58°F with mass flow rate of 45 $lb_m$/s. Determine the outlet temperatures of the fluids.

**11–60.** An unmixed crossflow heat exchanger is used to heat 1.25 kg/s of air from 15°C to 30°C. Hot water enters at 52.5°C. The mean overall coefficient of heat transfer is 300 W/($m^2$ °C) and the total surface area of the heat exchanger is 10 $m^2$. Determine the heat-transfer rate, exit-water temperature, and mass flow rate of water.

**11–61.** Solve Example 11–15 by the effectiveness method.

**11–62.** Air flowing at a rate of 2 kg/s is heated in a crossflow heat exchanger from 27°C to 50°C. Water enters the tubes at 90°C and exits at 75°C. Estimate the mass flow rate of the water, the rate of heat transfer, and the surface area of the heat exchanger if $\overline{U}$ is equal to 450 W/($m^2$ °C).

**11–63.** An unmixed crossflow heat exchanger is to be used to heat air with the hot exhaust gases from a turbine. The air enters with a temperature of 27°C and a mass flow rate of 0.7 kg/s. The temperature, specific heat, and flow rate of the hot gases are 900°C, 1 kJ/(kg °C), and 0.6 kg/s. Determine the surface area of the heat exchanger required to produce an air temperature of 150°C if $\overline{U}$ is equal to 300 W/($m^2$ °C).

# APPENDIXES

A MATHEMATICAL CONCEPTS, 705

B DIMENSIONS, UNITS, AND SIGNIFICANT FIGURES, 707

C THERMOPHYSICAL PROPERTIES, 710

D EARTH TEMPERATURE DATA FOR SELECT U.S. CITIES, 735

E ANALYTICAL SOLUTIONS FOR CONDUCTION HEAT TRANSFER, 735

F NUMERICAL COMPUTATIONS, 744

G MATHEMATICAL FUNCTIONS, 745

H TABLE FOR RADIATION FUNCTIONS, 746

I VISCOUS DISSIPATION:
MATHEMATICAL FORMULATION FOR LAMINAR FLOW, 747

J FILM CONDENSATION:
APPROXIMATE SOLUTION FOR LAMINAR FLOW, 750

# A MATHEMATICAL CONCEPTS

## A–1 The Calculus

The importance of the calculus in the study of heat transfer cannot be overstated. Therefore, we want briefly to review fundamental concepts pertaining to the calculus.

### *The Derivative*

For cases in which the dependent variable $\psi$ is a function of only one independent variable such as $x$, we are reminded of the following definition for the derivative:

$$\frac{d\psi}{dx} = \lim_{\Delta x \to 0} \frac{\psi(x + \Delta x) - \psi(x)}{\Delta x} \tag{A–1}$$

Because $dx$ itself is infinitesimal, this equation can be written in the form

$$\frac{d\psi}{dx} = \frac{\psi(x + dx) - \psi(x)}{dx} \tag{A–2}$$

Hence, $\psi(x + dx)$ can be written in terms of $\psi(x)$ and $d\psi/dx$ as

$$\psi(x + dx) = \psi(x) + \frac{d\psi}{dx} dx \tag{A–3}$$

This equation will be found to be very important in the analysis of one-dimensional conduction heat transfer in Chap. 2.

If $\psi$ is a function of more than one independent variable such as $x$, $y$, $z$ and $t$, the partial derivative of $\psi$ with respect to $x$ at the location $x,y,z$ and at the instant $t$ is defined as

$$\frac{\partial\psi}{\partial x} = \lim_{\Delta x \to 0} \frac{\psi(x + \Delta x) - \psi(x)}{\Delta x} \quad \text{at} \quad x,y,z \text{ and } t \tag{A–4}$$

or

$$\frac{\partial\psi}{\partial x} = \frac{\psi(x + dx) - \psi(x)}{dx} \quad \text{at} \quad x,y,z \text{ and } t \tag{A–5}$$

For this more general situation, $\psi(x + dx)$ is expressed in terms of $\psi(x)$ and the partial derivative; that is,

$$\psi(x + dx) = \psi(x) + \frac{\partial\psi}{\partial x} dx \tag{A–6}$$

Similar expressions can be written which involve partial derivatives with respect to the other independent variables. For example, the partial derivative of $\psi$ with respect to $y$ can be written as

$$\frac{\partial\psi}{\partial y} = \frac{\psi(y + dy) - \psi(y)}{dy} \quad \text{at} \quad x,y,z \text{ and } t \tag{A–7}$$

### *Differential Equations*

Both ordinary and partial differential equations are encountered in the study of heat transfer. For situations in which the equations are *linear*, analytical solution techniques often can be used. Second-order linear ordinary differential equations take the form

$$\psi + a(x)\frac{d\psi}{dx} + b(x)\frac{d^2\psi}{dx^2} = c(x) \qquad \text{(A–8)}$$

where the coefficients $a$, $b$, and $c$ are functions of the independent variable $x$ only. This equation is *nonlinear* if any one of the coefficients $a$, $b$, or $c$ is a function of $\psi$. Similarly, partial differential equations which involve coefficients that are functions of only the independent variables are linear.

The number of boundary conditions in an independent variable required in the solution of any ordinary or partial differential equation is equal to the highest-order differential in that variable. Similar to the definition for a linear differential equation, linear boundary conditions take the general form

$$\psi + d(x)\frac{d\psi}{dx} + e(x)\frac{d^2\psi}{dx^2} = f(x) \qquad \text{at} \quad x = x_1 \qquad \text{(A–9)}$$

A differential equation or boundary condition that is satisfied by a function $\psi$ is said to be *homogeneous* if it is also satisfied by $C\psi$, where $C$ is an arbitrary constant. Thus, the linear ordinary differential equation given by Eq. (A–8) is homogeneous if $c(x)$ is zero and is *nonhomogeneous* if $c(x) \neq 0$. Similarly, the boundary condition given by Eq. (A–9) is homogeneous if $f(x)$ is zero.

As an example of a problem involving linear nonhomogeneous equations, consider Eq. (A–8) with boundary conditions of the form

$$\psi + d_1(x)\frac{d\psi}{dx} = f_1(x) \qquad \text{at} \quad x = x_1 \qquad \text{(A–10)}$$

$$\psi + d_2(x)\frac{d\psi}{dx} = f_2(x) \qquad \text{at} \quad x = x_2 \qquad \text{(A–11)}$$

These types of boundary conditions are often encountered in heat-transfer problems. Notice that this problem involves three nonhomogeneous terms, $c(x)$, $f_1(x)$, and $f_2(x)$. The solution to this problem takes the form

$$\psi = \psi_0 + C_1\psi_1 + C_2\psi_2 \qquad \text{(A–12)}$$

where $\psi_0$ is the particular solution of the nonhomogeneous differential equation and $C_1\psi_1 + C_2\psi_2$ is the solution of the homogeneous differential equation. The constants of integration $C_1$ and $C_2$ are evaluated on the basis of the boundary conditions.

## A–2 Coordinate Systems

The foregoing review of basic mathematical concepts has been presented in the context of the *Cartesian coordinate system* shown in Fig. A–A–1(a). Our study will also

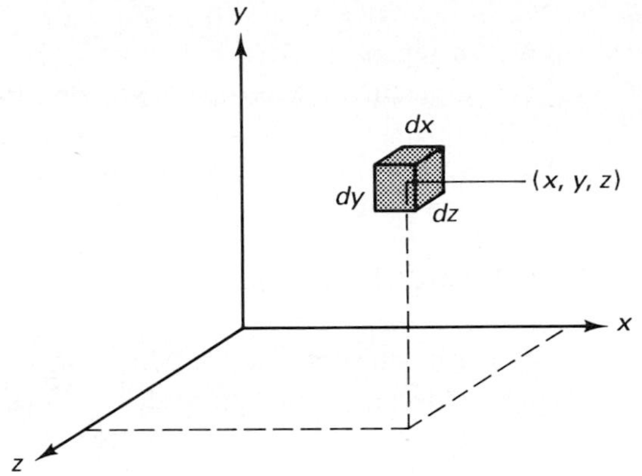

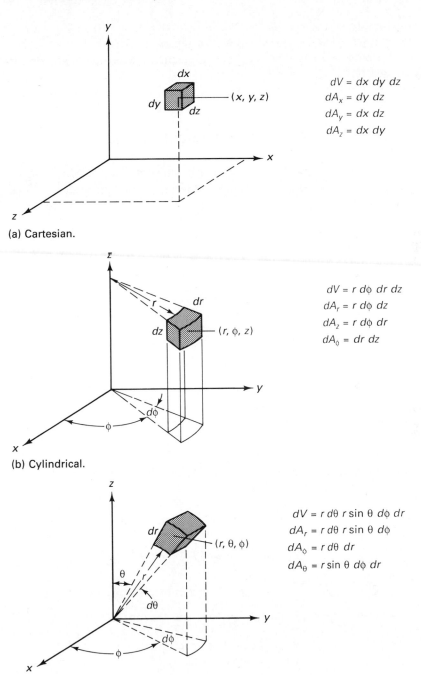

$$dV = dx\,dy\,dz$$
$$dA_x = dy\,dz$$
$$dA_y = dx\,dz$$
$$dA_z = dx\,dy$$

(a) Cartesian.

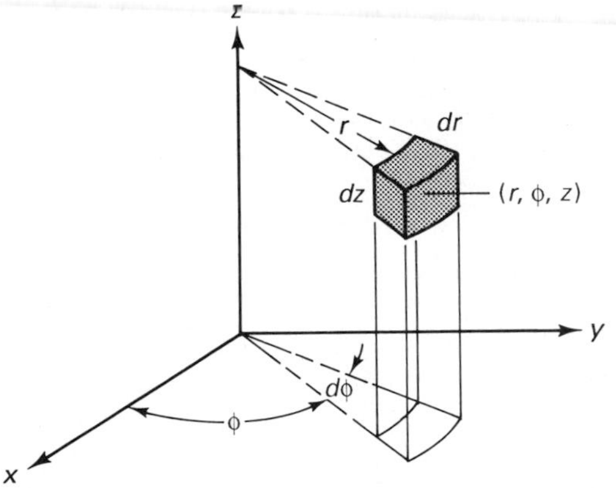


$$dV = r\,d\phi\,dr\,dz$$
$$dA_r = r\,d\phi\,dz$$
$$dA_z = r\,d\phi\,dr$$
$$dA_\phi = dr\,dz$$

(b) Cylindrical.

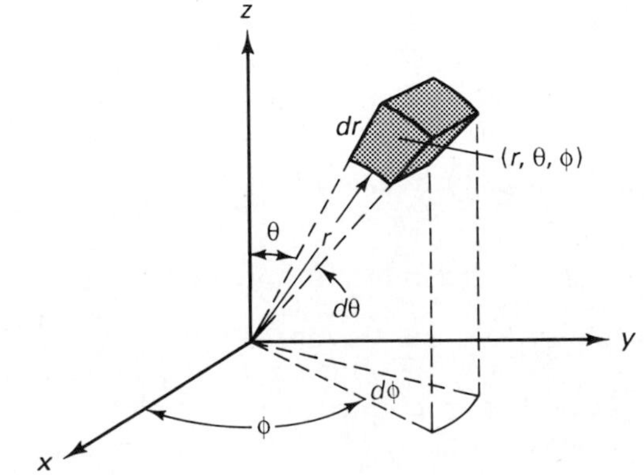


$$dV = r\,d\theta\,r\sin\theta\,d\phi\,dr$$
$$dA_r = r\,d\theta\,r\sin\theta\,d\phi$$
$$dA_\phi = r\,d\theta\,dr$$
$$dA_\theta = r\sin\theta\,d\phi\,dr$$

(c) Spherical.

**FIGURE A–A–1** Coordinate systems.

require the use of the *cylindrical coordinate system* shown in Fig. A–A–1(b), and the *spherical coordinate system* shown in Fig. A–A–1(c).

The temperature distribution can be a function of one, two, or three spatial coordinates (e.g., $x$, $y$, and $z$) and a single time variable $t$. Systems in which the temperature is a function of time will be referred to as unsteady. Steady-state conditions prevail when the temperature distribution within a system is independent of time.

## B DIMENSIONS, UNITS, AND SIGNIFICANT FIGURES

Until the transition from English to metric (the International System, SI) units is complete, it will be necessary for us to be conversant with both systems. Consequently, both SI and English units will be used in our study, with emphasis given to the SI system.

The units for the four key fundamental dimensions—*time*, *length*, *mass*, and *temperature*—in these systems are summarized as follows:

| *System* | *Time, t* | *Length, L* | *Mass, m* | *Temperature, T* |
|---|---|---|---|---|
| SI | second, s | meter, m | kilogram, kg | Kelvin, K<br>Celsius, °C |
| English | second, s | foot, ft | pound mass, $lb_m$ | Rankine, °R<br>Fahrenheit, °F |

Focusing attention on the units for temperature, in the SI system the *Kelvin* K is used for the absolute temperature scale† and the *degree Celsius* °C is used for the Celsius temperature scale. These two SI units for temperature are related by

$$T(\text{K}) = T(°\text{C}) + 273.2°\text{C} \tag{B–1}$$

Note that an increment °C is equal to an increment of 1 K; i.e.,

$$°\text{C} = 1\ \text{K} \tag{B–2}$$

The Rankine °R and Fahrenheit °F units for temperature scale in the English system are related by

$$T(°\text{R}) = T(°\text{F}) + 459.7°\text{F} \tag{B–3}$$

and

$$°\text{F} = 1°\text{R} \tag{B–4}$$

The units for the SI and English systems are related by

$$T(°\text{F}) = \frac{9°\text{F}}{5°\text{C}}\, T(°\text{C}) + 32°\text{F} \tag{B–5}$$

$$°\text{R} = \frac{5}{9}\,\text{K} \qquad \text{or} \qquad °\text{F} = \frac{5}{9}\,°\text{C} \tag{B–6,7}$$

† Note that the SI unit for the absolute temperature scale is Kelvin K and not degree Kelvin °K.

We are also reminded of the following units for the derived dimensions *force*, *energy*, and *power*:

| *System* | *Force, F* | *Energy, E* | *Power, P* |
|---|---|---|---|
| SI<br>English | newton, N<br>pound force, $lb_f$ | joule, J<br>British thermal unit, Btu | watt, W<br>Btu/h |

By definition, we have

$$1\text{ N} = 1\text{ kg m/s}^2 \qquad 1\text{ J} = 1\text{ N m} \qquad 1\text{ W} = 1\text{ J/s} \qquad \text{(B–8,9,10)}$$

Symbols and basic units for parameters used in the text are summarized in the Nomenclature. For convenience, conversion factors are listed on the inside back cover page.

The relation between the Newton *N* and the units for mass, length, and time is of course based on *Newton's second law*, which takes the form

$$F = ma \qquad \text{(B–11)}$$

for an object of mass *m* undergoing an acceleration *a*. Newton's second law of motion is given in the context of a fluid control volume by Eq. (1–3). Newton's other laws of motion and gravity include the *first law of motion* (or *law of inertia*), which states that any object in a state of rest or of uniform linear motion will remain in such a state unless acted upon by an external force; the *third law of motion* (or *law of action and reaction*), which states that every force (or action) gives rise to an opposing force (or reaction) of equal strength but opposite direction; and the *universal law of gravitation*, which states that two masses $m_1$ and $m_2$ separated by a distance *r* will be attracted toward one another by a force *F*, which can be represented by

$$F = \frac{Gm_1m_2}{r^2} \qquad \text{(B–12)}$$

where $G = 6.673 \times 10^{-11}\text{ N m}^2/\text{kg}^2$ is the *universal gravitational constant*. Notice that Eq. (B–12) can be put into the form of Eq. (B–11) by setting $m = m_1$ and $g = Gm^2/r^2$.

In regard to heat transfer, the following units generally will be utilized in this text:

*q*—heat-transfer rate, W

*Q*—total heat transfer, J

The total heat transfer over an increment of time *Q* is related to *q* by the equation

$$q = \frac{\partial Q}{\partial t} \qquad \text{(B–13)}$$

Hence, *Q* can be obtained from *q* by writing

$$Q = \int_0^t q\,dt \tag{B–14}$$

at any point in space.

Three significant figures will be maintained for most of our calculations.

## C THERMOPHYSICAL PROPERTIES

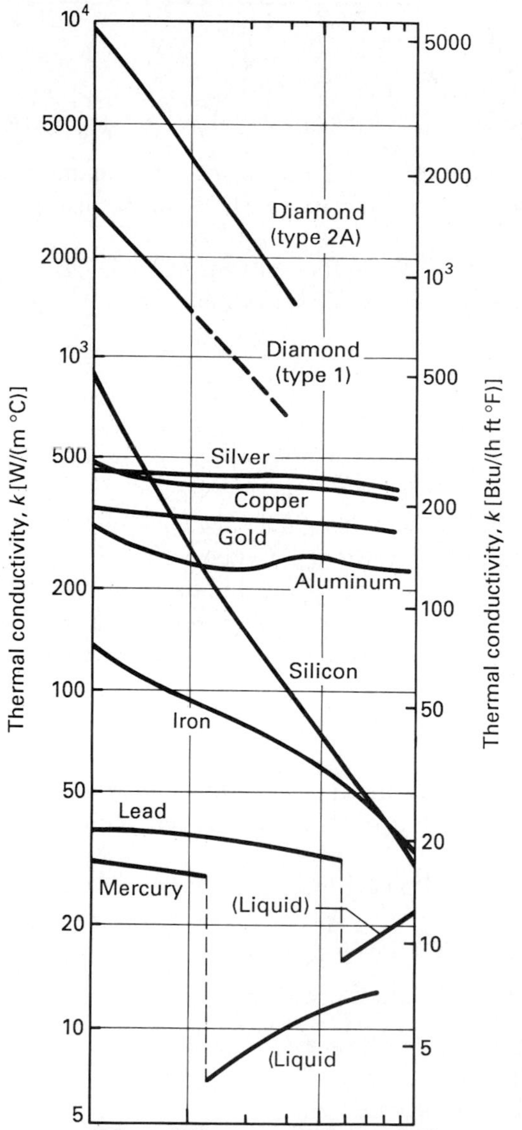


**FIGURE A–C–1**
Dependence of thermal conductivity on temperature: metals and other good heat conductors—moderate temperature zone. (From Touloukian et al. [1,2]. Used with permission.)

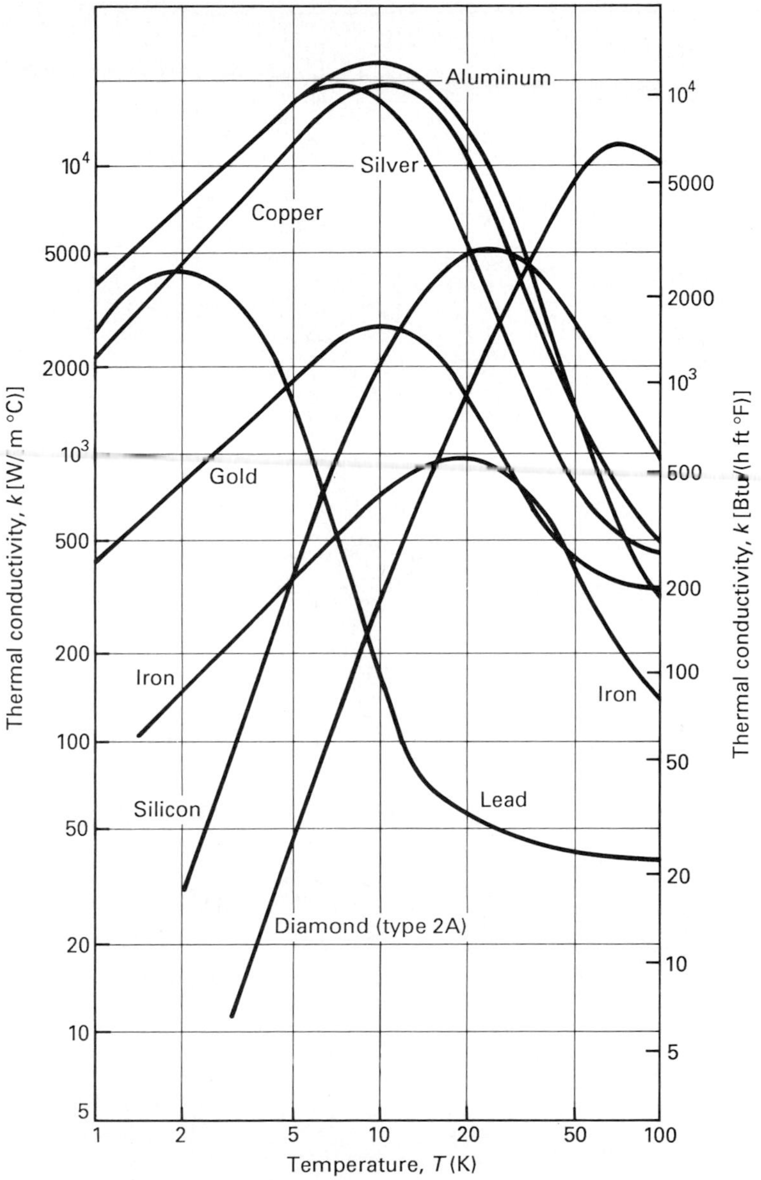


**FIGURE A–C–2** Dependence of thermal conductivity on temperature: metals and other good heat conductors—cryogenic temperature zone. (From Touloukian et al. [1,2]. Used with permission.)

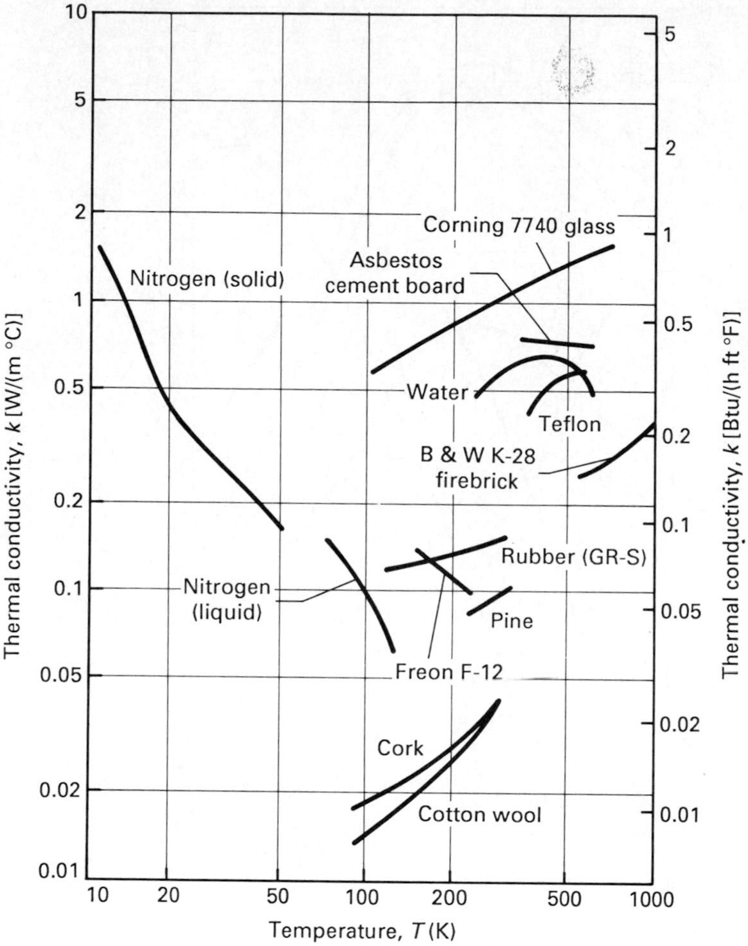


**FIGURE A–C–3** Dependence of thermal conductivity on temperature: nonmetallic solids and saturated liquids. (From Touloukian et al. [2,3]. Used with permission.)

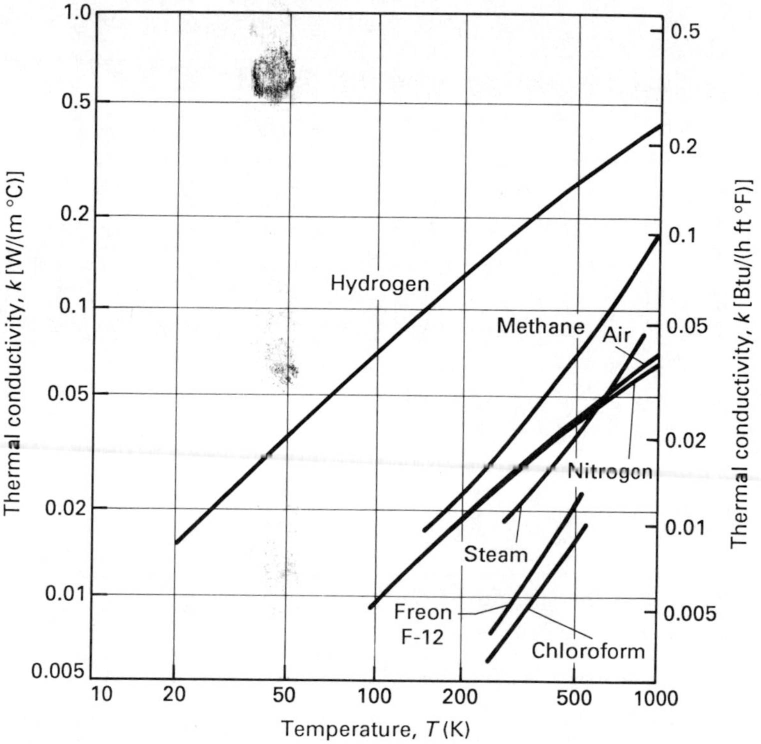


**FIGURE A–C–4** Dependence of thermal conductivity on temperature: gases. (From Touloukian et al. [3]. Used with permission.)

**TABLE A–C–1(a)** Property values of solid metals: metallic elements

| Element | Melting point (K) | *Properties at 293 K* ρ (kg/m³) | $c_v$ [kJ/(kg °C)] | k [W/(m °C)] | α × 10⁶ (m²/s) | *Variation of properties with temperature T(K)*† k [W/(m °C)]/$c_v$ [kJ/(kg °C)] 100 | 200 | 400 | 600 | 800 | 1000 | 1200 |
|---|---|---|---|---|---|---|---|---|---|---|---|---|
| Aluminum | 933 | 2702 | 0.896 | 236 | 97.5 | 301 | 237 | 240 | 232 | 220 | | |
| | | | | | | **0.480** | **0.795** | **0.945** | **1.02** | **1.140** | | |
| Antimony | 904 | 6684 | 0.208 | 24.6 | 17.7 | | 30.2 | 21.2 | 18.2 | 16.8 | | |
| Beryllium | 1550 | 1850 | 1.75 | 205 | 63.3 | 990 | 301 | 161 | 126 | 107 | 89 | 73 |
| | | | | | | **0.195** | **1.07** | **2.10** | **2.50** | **2.70** | **2.90** | |
| Bismuth | 545 | 9780 | 0.124 | 7.9 | 6.51 | 16.5 | 9.7 | 7.0 | | | | |
| | | | | | | **0.114** | **0.122** | **0.129** | | | | |
| Boron | 2573 | 2500 | 1.047 | 28.6 | 10.9 | 211 | 52.5 | 18.7 | 11.3 | 8.1 | 6.3 | 5.2 |
| | | | | | | **0.121** | **0.567** | **1.38** | **1.79** | **2.04** | **2.21** | |
| Cadmium | 594 | 8650 | 0.231 | 97 | 48.5 | 203 | 99.3 | 94.7 | | | | |
| | | | | | | **0.198** | **0.222** | **0.242** | | | | |
| Cesium | 302 | 1873 | 0.230 | 36 | 83.6 | | 36.8 | | | | | |
| Chromium | 2118 | 7160 | 0.440 | 91.4 | 29.0 | 159 | 111 | 87.3 | 80.5 | 71.3 | 65.3 | 62.4 |
| | | | | | | **0.188** | **0.376** | **0.474** | **0.530** | **0.570** | **0.605** | **0.670** |
| Cobalt | 1765 | 8862 | 0.389 | 100 | 29.0 | 167 | 122 | 84.8 | 67.4 | 58.2 | 52.1 | 49.3 |
| | | | | | | **0.220** | **0.350** | **0.420** | **0.465** | **0.510** | **0.580** | **0.680** |
| Copper | 1356 | 8950 | 0.383 | 386 | 113 | 467 | 400 | 379 | 370 | 359 | 345 | 331 |
| | | | | | | **0.250** | **0.355** | **0.395** | **0.415** | **0.430** | **0.450** | **0.480** |
| Germanium | 1211 | 5360 | 0.322 | 59.9 | 34.7 | 232 | 96.8 | 43.2 | 27.3 | 19.8 | 17.4 | 17.4 |
| | | | | | | **0.190** | **0.290** | **0.337** | **0.348** | **0.357** | **0.375** | **0.395** |
| Gold | 1336 | 19300 | 0.129 | 316 | 127 | 327 | 323 | 312 | 304 | 292 | 278 | 262 |
| | | | | | | **0.109** | **0.124** | **0.131** | **0.135** | **0.140** | **0.145** | **0.155** |
| Hafnium | 2495 | 13280 | | 23.1 | | | 24.4 | 22.3 | 21.3 | 20.8 | 20.7 | 20.9 |
| Indium | 430 | 7300 | | 82.2 | | | 89.7 | 74.5 | | | | |
| Iridium | 2716 | 22500 | 0.134 | 147 | 48.8 | 172 | 153 | 144 | 138 | 132 | 126 | 120 |
| Iron | 1810 | 7870 | 0.452 | 81.1 | 22.8 | 134 | 94 | 69.4 | 54.7 | 43.3 | 32.6 | 28.2 |
| | | | | | | **0.218** | **0.388** | **0.495** | **0.580** | **0.688** | **0.985** | **0.615** |
| Lead | 601 | 11340 | 0.129 | 35.3 | 24.1 | 39.7 | 36.6 | 33.8 | 31.2 | | | |
| | | | | | | **0.118** | **0.125** | **0.132** | **0.142** | | | |
| Lithium | 454 | 534 | 3.39 | 77.4 | 42.7 | | 88.1 | 72.1 | | | | |
| Magnesium | 923 | 1740 | 1.02 | 156 | 88.2 | 169 | 159 | 153 | 149 | 146 | | |
| | | | | | | **0.645** | **0.927** | **1.06** | **1.16** | **1.26** | | |
| Manganese | 1517 | 7290 | 0.486 | 7.78 | 2.2 | | 7.17 | | | | | |

| | | | | | | | | | | | | |
|---|---|---|---|---|---|---|---|---|---|---|---|---|
| Mercury | 234 | 13546 | | | | | 28.9 | | | | | |
| Molybdenum | 2883 | 10240 | 0.251 | 138 | 53.7 | 179 | 143 | 134 | 126 | 118 | 112 | 105 |
| | | | | | | **0.141** | **0.224** | **0.261** | **0.275** | **0.285** | **0.295** | **0.308** |
| Nickel | 1726 | 8900 | 0.446 | 91 | 22.9 | 164 | 106 | 80.1 | 65.5 | 67.4 | 71.8 | 76.1 |
| | | | | | | **0.233** | **0.385** | **0.487** | **0.595** | **0.532** | **0.564** | **0.597** |
| Niobium | 2741 | 8570 | 0.270 | 53.6 | 23.2 | 55.2 | 52.6 | 55.2 | 58.2 | 61.3 | 64.4 | 67.5 |
| | | | | | | **0.192** | **0.254** | **0.279** | **0.288** | **0.298** | **0.307** | **0.316** |
| Palladium | 1825 | 12020 | 0.244 | 71.8 | 24.5 | 76.5 | 71.6 | 73.6 | 79.7 | 86.9 | 94.2 | 102 |
| | | | | | | **0.168** | **0.227** | **0.251** | **0.261** | **0.271** | **0.281** | **0.291** |
| Platinum | 2042 | 21450 | 0.133 | 71.4 | 25.0 | 77.5 | 72.4 | 71.6 | 73.0 | 75.5 | 78.6 | 82.6 |
| | | | | | | **0.100** | **0.125** | **0.136** | **0.141** | **0.146** | **0.152** | **0.157** |
| Potassium | 337 | 860 | 0.741 | 103 | 162 | | 104 | 52 | | | | |
| Rhenium | 3453 | 21100 | 0.137 | 48.1 | 16.6 | 58.9 | 51 | 46.1 | 44.2 | 44.1 | 44.6 | 45.7 |
| | | | | | | **0.097** | **0.127** | **0.139** | **0.145** | **0.151** | **0.156** | **0.162** |
| Rhodium | 2233 | 12450 | 0.248 | 150 | 48.6 | 186 | 154 | 146 | 136 | 127 | 121 | 115 |
| | | | | | | **0.150** | **0.225** | **0.258** | **0.280** | **0.299** | **0.317** | **0.334** |
| Rubidium | 312 | 1530 | 0.348 | 58.2 | 109 | | 58.9 | | | | | |
| Silicon | 1685 | 2330 | 0.703 | 153 | 93.4 | 884 | 264 | 98.9 | 61.9 | 42.2 | 31.2 | 25.7 |
| | | | | | | **0.256** | **0.549** | **0.780** | **0.856** | **0.900** | **0.934** | **0.955** |
| Silver | 1234 | 10500 | 0.234 | 427 | 174 | 444 | 430 | 425 | 412 | 390 | 379 | 361 |
| | | | | | | **0.187** | **0.225** | **0.239** | **0.250** | **0.262** | **0.277** | **0.292** |
| Sodium | 371 | 971 | 0.121 | 133 | 114 | | 138 | | | | | |
| Tantalum | 3269 | 16600 | 0.138 | 57.5 | 25.1 | 59.2 | 57.5 | 57.8 | 58.6 | 59.4 | 60.2 | 61 |
| | | | | | | **0.108** | **0.131** | **0.142** | **0.144** | **0.147** | **0.150** | **0.153** |
| Tin | 505 | 5750 | 0.227 | 67.0 | 51.3 | 85.2 | 73.3 | 62.2 | | | | |
| | | | | | | **0.188** | **0.215** | **0.243** | | | | |
| Titanium | 1953 | 4500 | 0.522 | 21.9 | 9.32 | 30.5 | 24.5 | 20.4 | 19.4 | 19.7 | 20.7 | 22 |
| | | | | | | **0.300** | **0.465** | **0.551** | **0.591** | **0.633** | **0.675** | **0.620** |
| Tungsten | 3653 | 19300 | 0.134 | 179 | 69.2 | 214 | 197 | 162 | 139 | 128 | 121 | 115 |
| | | | | | | **0.088** | **0.124** | **0.139** | **0.144** | **0.147** | **0.150** | **0.154** |
| Uranium | 1407 | 19070 | 0.113 | 27.4 | 12.7 | 21.7 | 25.1 | 29.6 | 34 | 38.8 | 43.9 | 49 |
| | | | | | | **0.092** | **0.105** | **0.122** | **0.142** | **0.171** | **0.175** | **0.157** |
| Vanadium | 2192 | 6100 | 0.502 | 31.4 | 10.3 | 35.8 | 31.5 | 32.1 | 34.2 | 36.3 | 38.6 | 41.2 |
| | | | | | | **0.265** | **0.441** | **0.528** | **0.554** | **0.578** | **0.612** | **0.662** |
| Zinc | 693 | 7140 | 0.385 | 121 | 44.0 | 122 | 123 | 116 | 105 | | | |
| | | | | | | **0.294** | **0.363** | **0.398** | **0.432** | | | |
| Zirconium | 2125 | 6750 | 0.272 | 22.8 | 12.8 | 33.2 | 25.2 | 21.6 | 20.7 | 21.6 | 23.7 | 25.7 |
| | | | | | | **0.200** | **0.258** | **0.294** | **0.315** | **0.335** | **0.354** | **0.336** |

† Values for $c_v$ shown in **bold** typeset.
*Sources*: Touloukian et al. [1,2], Eckert and Drake [4], and Raznjevic [5].

**TABLE A–C–1(b)** Property values of solid metals: alloys

| Metal | Composition (%) | Properties at 293 K | | | | Variation of properties with temperature $T$(K)† | | | | | | |
|---|---|---|---|---|---|---|---|---|---|---|---|---|
| | | $\rho$ (kg/m$^3$) | $c_v$ [kJ/(kg °C)] | $k$ [W/(m °C)] | $\alpha \times 10^5$ (m$^2$/s) | $k$ [W/(m °C)]/$c_v$ [kJ/(kg °C)] 100 | 200 | 400 | 600 | 800 | 1000 | 1200 |
| Aluminum | | | | | | | | | | | | |
| Duralumin | 94–96 Al. 3–5 Cu. trace Mg | 2787 | 0.883 | 164 | 6.676 | | 163<br>**0.787** | 186<br>**0.925** | 186<br>**1.042** | | | |
| Silumin | 87 Al, 13 Si | 2659 | 0.871 | 164 | 7.099 | | 165 | | | | | |
| Copper | | | | | | | | | | | | |
| Aluminum Bronze | 95 Cu, 5 Al | 8666 | 0.410 | 83 | 2.330 | | | | | | | |
| Bronze | 75 Cu, 25 Sn | 8666 | 0.343 | 26 | 0.859 | | | | | | | |
| Red brass | 85 Cu, 9 Sn, 6 Zn | 8714 | 0.385 | 61 | 1.804 | | | | | | | |
| Brass | 70 Cu, 30 Zn | 8522 | 0.385 | 111 | 3.412 | 75 | 95<br>**0.360** | 137<br>**0.395** | 149<br>**0.425** | | | |
| German silver | 62 Cu, 15 Ni, 22 Zn | 8618 | 0.394 | 24.9 | 0.733 | | | | | | | |
| Constantan | 60 Cu, 40 Ni | 8922 | 0.410 | 22.7 | 0.612 | 17<br>**0.237** | 19<br>**0.362** | | | | | |
| Iron | | | | | | | | | | | | |
| Cast iron | ≈4 C | 7272 | 0.420 | 52 | 1.702 | | | | | | | |
| Wrought iron | 0.5 C | 7849 | 0.460 | 59 | 1.626 | | | 56 | 47 | 39 | 34 | 33 |
| Steel | | | | | | | | | | | | |
| Carbon steel | 0.5 C | 7833 | 0.465 | 54 | 1.474 | | | 51 | 44 | 43 | 32 | 30 |
| | 1 C | 7801 | 0.473 | 43 | 1.172 | | | 43 | 39 | 34 | 30 | 28 |
| | 1.5 C | 7753 | 0.486 | 36 | 0.970 | | | 36 | 35 | 32 | 29 | 28 |

| | | | | | | | | | | | | |
|---|---|---|---|---|---|---|---|---|---|---|---|---|
| Chrome steel | 1 Cr | 7865 | 0.460 | 61 | 1.665 | | | 54 | 46 | 38 | 33 | 33 |
| | 5 Cr | 7833 | 0.460 | 40 | 1.110 | | | 38 | 35 | 31 | 29 | 29 |
| | 10 Cr | 7785 | 0.460 | 31 | 0.867 | | | 31 | 29 | 28 | 28 | 28 |
| Chrome | 15 Cr, 10 Ni | 7865 | 0.460 | 19 | 0.526 | | | | | | | |
| nickel steel | 20 Cr, 15 Ni | 7833 | 0.460 | 15.1 | 0.415 | | | | | | | |
| Nickel steel | 10 Ni | 7945 | 0.460 | 26 | 0.720 | | | | | | | |
| | 20 Ni | 7993 | 0.460 | 19 | 0.526 | | | | | | | |
| | 40 Ni | 8169 | 0.460 | 10 | 0.279 | | | | | | | |
| | 60 Ni | 8378 | 0.460 | 19 | 0.493 | | | | | | | |
| Nickel- | 80 Ni, 15 Cr | 8522 | 0.460 | 17 | 0.444 | | | | | | | |
| chrome steel | 40 Ni, 15 Cr | 8073 | 0.460 | 11.6 | 0.305 | | | | | | | |
| Inconel X-750 | 73 Ni, 15 Cr, 6.7 Fe | 8510 | 0.439 | 11.7 | 0.31 | 8.7 | 10.3 | 13.5 | 17 | 20.5 | 24 | 27.6 |
| | | | | | | | **0.372** | **0.473** | **0.510** | **0.546** | **0.626** | |
| Manganese steel | 1 Mn | 7865 | 0.460 | 50 | 1.388 | | | | | | | |
| | 5 Mn | 7849 | 0.460 | 22 | 0.637 | | | | | | | |
| Silicon steel | 1 Si | 7769 | 0.460 | 42 | 1.164 | | | | | | | |
| | 5 Si | 7417 | 0.460 | 19 | 0.555 | | | | | | | |
| Stainless steel | AISI 302 | 8055 | 0.480 | 15.1 | 0.391 | | | 17.3 | 20 | 22.8 | 25.4 | |
| | | | | | | | | **0.512** | **0.559** | **0.585** | **0.606** | |
| | AISI 304 | 7900 | 0.477 | 14.9 | 0.395 | | | 16.6 | 19.8 | 22.6 | 25.4 | 28 |
| | | | | | | | | **0.515** | **0.557** | **0.582** | **0.611** | **0.640** |
| | AISI 316 | 8238 | 0.468 | 13.4 | 0.348 | | | 15.2 | 18.3 | 21.3 | 24.2 | |
| | | | | | | | | **0.504** | **0.550** | **0.576** | **0.602** | |
| | AISI 347 | 7978 | 0.480 | 14.2 | 0.371 | | | 15.8 | 18.9 | 21.9 | 24.7 | |
| | | | | | | | | **0.513** | **0.559** | **0.585** | **0.606** | |
| Tungsten steel | 1 W | 7913 | 0.448 | 66 | 1.858 | | | | | | | |
| | 5 W | 8073 | 0.435 | 54 | 1.525 | | | | | | | |

† Values for $c_v$ shown in **bold** typeset.
*Sources*: Eckert and Drake [4], Desai et al. [6], and others.

**TABLE A–C–2** Property values of solid nonmetals

| *Substance* | $T$ (°C) | $\rho$ (kg/m$^3$) | $c_v$ [kJ/(kg °C)] | $k$ [W/(m °C)] | $\alpha \times 10^7$ (m$^2$/s) |
|---|---|---|---|---|---|
| Structural and heat-resistant materials | | | | | |
| Asphalt | 20–55 | 2120 | | 0.74–0.76 | |
| Brick | | | | | |
| Building brick | 20 | 1600 | 0.84 | 0.69 | 5.2 |
| Carborundum brick | 600 | | | 18.5 | |
| (50% SiC) | 1400 | | | 11.1 | |
| Chrome brick | 200 | 3000 | 0.84 | 2.32 | 9.2 |
| | 550 | | | 2.47 | 9.8 |
| | 900 | | | 1.99 | 7.9 |
| Fireclay brick, burnt 2426°F | 500 | 2000 | 0.96 | 1.04 | 5.4 |
| | 800 | | | 1.07 | |
| | 1100 | | | 1.09 | |
| Burnt 2642°F | 500 | 2300 | 0.96 | 1.28 | 5.8 |
| | 800 | | | 1.37 | |
| | 1100 | | | 1.40 | |
| Magnesite | 200 | | 1.13 | 3.81 | |
| (50% MgO) | 650 | | | 2.77 | |
| | 1200 | | | 1.90 | |
| Masonry | 20 | 1700 | 0.837 | 0.658 | |
| Silica (95% $SiO_2$) | 20 | 1900 | | 1.07 | |
| Zirconia (62% $ZrO_2$) | 20 | 3600 | | 2.44 | |
| Cement, mortar | 23 | | | 1.16 | |
| Concrete | | | | | |
| Cinder | 23 | | | 0.76 | |
| Stone 1–2–4 mix | 20 | 1900–2300 | 0.88 | 1.37 | 8.2–6.8 |
| Glass | | | | | |
| Window | 20 | 2700 | 0.84 | 0.78 (avg) | 3.4 |
| Borosilicate | 30–75 | 2200 | | 1.09 | |
| Plaster, gypsum | 20 | 1440 | 0.84 | 0.48 | 4.0 |
| Plexiglas | 20 | 1180 | | 0.195 | |
| Plywood | 20 | 590 | 1.22 | 0.109 | |
| Sand | | | | | |
| dry | 20 | | | 0.582 | |
| moist | 20 | 1640 | | 1.13 | |
| Stone | | | | | |
| Granite | | 2640 | 0.82 | 1.73–3.98 | 8–18 |
| Limestone | 100–300 | 2500 | 0.90 | 1.26–1.33 | 5.6–5.9 |
| Marble | | 2500–2700 | 0.80 | 2.07–2.94 | 10–13.6 |
| Sandstone | 40 | 2160–2300 | 0.71 | 1.83 | 11.2–11.9 |
| Wood (across the grain): | | | | | |
| Balsa | 30 | 140 | | 0.055 | |
| Cypress | 30 | 460 | | 0.097 | |
| Fir | 23 | 420 | 2.72 | 0.11 | 0.96 |
| Maple or oak | 30 | 540 | 2.4 | 0.166 | 1.28 |
| Yellow pine | 23 | 640 | 2.8 | 0.147 | 0.82 |
| White pine | 30 | 430 | | 0.112 | |

**TABLE A–C–2** (continued)

| *Substance* | $T$ (°C) | $\rho$ (kg/m$^3$) | $c_v$ [kJ/(kg °C)] | $k$ [W/(m °C)] | $\alpha \times 10^7$ (m$^2$/s) |
|---|---|---|---|---|---|
| | | | Insulating material | | |
| Asbestos† | 0 | 470–570 | 0.816 | 0.154 | 3.3–4 |
| Cardboard, corrugated | | | | 0.064 | |
| Corkboard | 30 | 160 | | 0.043 | |
| Cork: | | | | | |
| Regranulated | 32 | 45–120 | 1.88 | 0.045 | 2–5.3 |
| Ground | 32 | 150 | | 0.043 | |
| Diatomaceous earth (Sil-o-cel) | 0 | 320 | | 0.061 | |
| Felt: | | | | | |
| Hair | 30 | 130–200 | | 0.036 | |
| Wool | 30 | 330 | | 0.052 | |
| Fiber, insulating board | 20 | 240 | | 0.048 | |
| Glass fiber | 0 | 220 | | 0.035 | |
| Glass wool | 23 | 24 | 0.7 | 0.038 | 22.6 |
| Kapok | 30 | | | 0.035 | |
| Magnesia, 85% | 38 | 270 | | 0.067 | |
| | 93 | | | 0.071 | |
| | 150 | | | 0.074 | |
| | 204 | | | 0.080 | |
| Rock wool | 32 | 160 | | 0.040 | |
| Loosely packed | 150 | 64 | | 0.067 | |
| | 260 | | | 0.087 | |
| Sawdust | 23 | | | 0.059 | |
| Silica aerogel | 32 | 140 | | 0.024 | |
| Wood shavings | 23 | | | 0.059 | |
| | | | Soils and rocks‡ | | |
| Heavy soil, saturated | 30 | 2300 | 0.838 | 2.42 | 9.04 |
| Heavy soil, damp solid masonry | 30 | 2100 | 0.963 | 1.30 | 6.45 |
| | | 2300 | 0.879 | | |
| Heavy soil, dry | 30 | 2000 | 0.838 | 0.865 | 5.16 |
| Light soil, damp | | 1600 | 1.05 | | |
| Light soil, dry | 30 | 1440 | 0.838 | 0.346 | 2.80 |
| Dense rock | 30 | 2300 | 0.838 | 3.46 | 12.9 |
| Average rock | 30 | 2800 | 0.838 | 2.42 | 10.3 |

† Well-documented health hazards are associated with long-term exposure to asbestos. Therefore, asbestos insulating material is no longer recommended.

‡ *Source*: Adapted with permission from ASHRAE Handbook of Fundamentals [7].

**TABLE A–C–2** (continued)

| *Substance* | *T* (°C) | ρ (kg/m³) | $c_v$ [kJ/(kg °C)] | *k* [W/(m °C)] | $\alpha \times 10^7$ (m²/s) |
|---|---|---|---|---|---|
| | | Other materials | | | |
| Coal | | | | | |
| Anthiarite | 27 | 1350 | 1.26 | 0.26 | |
| Bituminous in situ | | 1300 | | 0.60 | 3.5 |
| Cotton | 27 | 80 | 1.30 | 0.06 | |
| Diamonds | | | | | |
| Type I | 0 | | | 1000 | |
| Type IIa | 0 | 3500 | 0.51 | 2650 | 14800 |
| Type IIb | 0 | | | 1510 | |
| Foods | | | | | |
| Apple, red (75% $H_2O$) | 27 | 840 | 3.6 | 0.513 | |
| Banana (76% $H_2O$) | 27 | 840 | 3.60 | 0.513 | |
| Beef | 25 | | | | 1.35 |
| Chicken, white meat | −75 | | | 1.60 | |
| (74% $H_2O$) | −20 | | | 1.35 | |
| | 0 | | | 0.472 | |
| | 20 | | | 0.489 | |
| Egg white | | | | | 1.37 |
| Human tissue | | | | | |
| skin | 27 | | | 0.37 | |
| fat layer (adipose) | 27 | | | 0.2 | |
| muscle | 27 | | | 0.41 | |
| Ice | 0 | 913 | 1.83 | 2.22 | 0.124 |
| Leather (sole) | 27 | 998 | | 0.159 | |
| Linoleum | 20 | 535 | | 0.081 | |
| Paper | 27 | 930 | 1.34 | 0.18 | |
| Rubber, vulcanized | | | | | |
| soft | 27 | 1100 | 2.01 | 0.13 | |
| hard | 27 | 1190 | | 0.16 | |
| Snow | | | | | |
| loose | 0 | 110 | | 0.05 | |
| packed | 0 | 500 | | 0.19 | |

*Sources*: *International Critical Tables* [8], Incropera and DeWitt [9], and others.

**TABLE A–C–3(a)** Property values of saturated water

| Temperature, T (K) | Pressure P (bar)[a] | Heat of Vaporization $i_{fg}$ (kJ/kg) | Specific Volume (m³/kg) $v_f \times 10^3$ | Specific Volume (m³/kg) $v_g$ | Specific Heat [kJ/(kg K)] $c_{P,f}$ | Specific Heat [kJ/(kg K)] $c_{P,g}$ | Thermal Conductivity [W/(m K)] $k_f \times 10^3$ | Thermal Conductivity [W/(m K)] $k_g \times 10^3$ | Viscosity [kg/(m s)] $\mu_f \times 10^6$ | Viscosity [kg/(m s)] $\mu_g \times 10^6$ | Prandtl Number $Pr_f$ | Prandtl Number $Pr_g$ | Surface Tension $\sigma_f \times 10^3$ (N/m) | Expansion Coefficient $\beta_f \times 10^6$ (K$^{-1}$) | Temperature T (K) |
|---|---|---|---|---|---|---|---|---|---|---|---|---|---|---|---|
| 273.15 | 0.00611 | 2502 | 1.000 | 206.3 | 4.217 | 1.854 | 569 | 18.2 | 1750 | 8.02 | 12.99 | 0.815 | 75.5 | −68.05 | 273.15 |
| 275 | 0.00697 | 2497 | 1.000 | 181.7 | 4.211 | 1.855 | 574 | 18.3 | 1652 | 8.09 | 12.22 | 0.817 | 75.3 | −32.74 | 275 |
| 280 | 0.00990 | 2485 | 1.000 | 130.4 | 4.198 | 1.858 | 582 | 18.6 | 1422 | 8.29 | 10.26 | 0.825 | 74.8 | 46.04 | 280 |
| 285 | 0.01387 | 2473 | 1.000 | 99.4 | 4.189 | 1.861 | 590 | 18.9 | 1225 | 8.49 | 8.81 | 0.833 | 74.3 | 114.1 | 285 |
| 290 | 0.01917 | 2461 | 1.001 | 69.7 | 4.184 | 1.864 | 598 | 19.3 | 1080 | 8.69 | 7.56 | 0.841 | 73.7 | 174.0 | 290 |
| 295 | 0.02617 | 2449 | 1.002 | 51.49 | 4.181 | 1.868 | 606 | 19.5 | 959 | 8.89 | 6.62 | 0.849 | 72.7 | 227.5 | 295 |
| 300 | 0.03531 | 2438 | 1.003 | 39.13 | 4.179 | 1.872 | 613 | 19.6 | 855 | 9.09 | 5.83 | 0.857 | 71.7 | 276.1 | 300 |
| 305 | 0.04712 | 2426 | 1.005 | 27.90 | 4.178 | 1.877 | 620 | 20.1 | 769 | 9.29 | 5.20 | 0.865 | 70.9 | 320.6 | 305 |
| 310 | 0.06221 | 2414 | 1.007 | 22.93 | 4.178 | 1.882 | 628 | 20.4 | 695 | 9.49 | 4.62 | 0.873 | 70.0 | 361.9 | 310 |
| 315 | 0.08132 | 2402 | 1.009 | 17.82 | 4.179 | 1.888 | 634 | 20.7 | 631 | 9.69 | 4.16 | 0.883 | 69.2 | 400.4 | 315 |
| 320 | 0.1053 | 2390 | 1.011 | 13.98 | 4.180 | 1.895 | 640 | 21.0 | 577 | 9.89 | 3.77 | 0.894 | 68.3 | 436.7 | 320 |
| 325 | 0.1351 | 2378 | 1.013 | 11.06 | 4.182 | 1.903 | 645 | 21.3 | 528 | 10.09 | 3.42 | 0.901 | 67.5 | 471.2 | 325 |
| 330 | 0.1719 | 2366 | 1.016 | 8.82 | 4.184 | 1.911 | 650 | 21.7 | 489 | 10.29 | 3.15 | 0.908 | 66.6 | 504.0 | 330 |
| 335 | 0.2167 | 2354 | 1.018 | 7.09 | 4.186 | 1.920 | 656 | 22.0 | 453 | 10.49 | 2.88 | 0.916 | 65.8 | 535.5 | 335 |
| 340 | 0.2713 | 2342 | 1.021 | 5.74 | 4.188 | 1.930 | 660 | 22.3 | 420 | 10.69 | 2.66 | 0.925 | 64.9 | 566.0 | 340 |
| 345 | 0.3372 | 2329 | 1.024 | 4.683 | 4.191 | 1.941 | 668 | 22.6 | 389 | 10.89 | 2.45 | 0.933 | 64.1 | 595.4 | 345 |
| 350 | 0.4163 | 2317 | 1.027 | 3.846 | 4.195 | 1.954 | 668 | 23.0 | 365 | 11.09 | 2.29 | 0.942 | 63.2 | 624.2 | 350 |
| 355 | 0.5100 | 2304 | 1.030 | 3.180 | 4.199 | 1.968 | 671 | 23.3 | 343 | 11.29 | 2.14 | 0.951 | 62.3 | 652.3 | 355 |
| 360 | 0.6209 | 2291 | 1.034 | 2.645 | 4.203 | 1.983 | 674 | 23.7 | 324 | 11.49 | 2.02 | 0.960 | 61.4 | 697.9 | 360 |
| 365 | 0.7514 | 2278 | 1.038 | 2.212 | 4.209 | 1.999 | 677 | 24.1 | 306 | 11.69 | 1.91 | 0.969 | 60.5 | 707.1 | 365 |
| 370 | 0.9040 | 2265 | 1.041 | 1.861 | 4.214 | 2.017 | 679 | 24.5 | 289 | 11.89 | 1.80 | 0.978 | 59.5 | 728.7 | 370 |
| 373.15 | 1.0133 | 2257 | 1.044 | 1.679 | 4.217 | 2.029 | 680 | 24.8 | 279 | 12.02 | 1.76 | 0.984 | 58.9 | 750.1 | 373.15 |
| 375 | 1.0815 | 2252 | 1.045 | 1.574 | 4.220 | 2.036 | 681 | 24.9 | 274 | 12.09 | 1.70 | 0.987 | 58.6 | 761 | 375 |
| 380 | 1.2869 | 2239 | 1.049 | 1.337 | 4.226 | 2.057 | 683 | 25.4 | 260 | 12.29 | 1.61 | 0.999 | 57.6 | 788 | 380 |
| 385 | 1.5233 | 2225 | 1.053 | 1.142 | 4.232 | 2.080 | 685 | 25.8 | 248 | 12.49 | 1.53 | 1.004 | 56.6 | 814 | 385 |
| 390 | 1.794 | 2212 | 1.058 | 0.980 | 4.239 | 2.104 | 686 | 26.3 | 237 | 12.69 | 1.47 | 1.013 | 55.6 | 841 | 390 |
| 400 | 2.455 | 2183 | 1.067 | 0.731 | 4.256 | 2.158 | 688 | 27.2 | 217 | 13.05 | 1.34 | 1.033 | 53.6 | 896 | 400 |
| 410 | 3.302 | 2153 | 1.077 | 0.553 | 4.278 | 2.221 | 688 | 28.2 | 200 | 13.42 | 1.24 | 1.054 | 51.5 | 952 | 410 |
| 420 | 4.370 | 2123 | 1.088 | 0.425 | 4.302 | 2.291 | 688 | 29.8 | 185 | 13.79 | 1.16 | 1.075 | 49.4 | 1010 | 420 |

**TABLE A–C–3(a)** (continued)

| *Temperature, T* (K) | *Pressure* $P$ (bar)[a] | *Heat of Vaporization* $i_{fg}$ (kJ/kg) | *Specific Volume* (m³/kg) | | *Specific Heat* [kJ/(kg K)] | | *Thermal Conductivity* [W/(m K)] | | *Viscosity* [kg/(m s)] | | *Prandtl Number* | | *Surface Tension* $\sigma_f \times 10^3$ (N/m) | *Expansion Coefficient* $\beta_f \times 10^6$ (K⁻¹) | *Temperature T* (K) |
|---|---|---|---|---|---|---|---|---|---|---|---|---|---|---|---|
| | | | $v_f \times 10^3$ | $v_g$ | $c_{P,f}$ | $c_{P,g}$ | $k_f \times 10^3$ | $k_g \times 10^3$ | $\mu_f \times 10^6$ | $\mu_g \times 10^6$ | $Pr_f$ | $Pr_g$ | | | |
| 430 | 5.699 | 2091 | 1.099 | 0.331 | 4.331 | 2.369 | 685 | 30.4 | 173 | 14.14 | 1.09 | 1.10 | 47.2 | — | 430 |
| 440 | 7.333 | 2059 | 1.110 | 0.261 | 4.36 | 2.46 | 682 | 31.7 | 162 | 14.50 | 1.04 | 1.12 | 45.1 | — | 440 |
| 450 | 9.319 | 2024 | 1.123 | 0.208 | 4.40 | 2.56 | 678 | 33.1 | 152 | 14.85 | 0.99 | 1.14 | 42.9 | — | 450 |
| 460 | 11.71 | 1989 | 1.137 | 0.167 | 4.44 | 2.68 | 673 | 34.6 | 143 | 15.19 | 0.95 | 1.17 | 40.7 | — | 460 |
| 470 | 14.55 | 1951 | 1.152 | 0.136 | 4.48 | 2.79 | 667 | 36.3 | 136 | 15.54 | 0.92 | 1.20 | 38.5 | — | 470 |
| 480 | 17.90 | 1912 | 1.167 | 0.111 | 4.53 | 2.94 | 660 | 38.1 | 129 | 15.88 | 0.89 | 1.23 | 36.2 | — | 480 |
| 490 | 21.83 | 1870 | 1.184 | 0.0922 | 4.59 | 3.10 | 651 | 40.1 | 124 | 16.23 | 0.87 | 1.25 | 33.9 | — | 490 |
| 500 | 26.40 | 1825 | 1.203 | 0.0766 | 4.66 | 3.27 | 642 | 42.3 | 118 | 16.59 | 0.86 | 1.28 | 31.6 | — | 500 |
| 510 | 31.66 | 1779 | 1.222 | 0.0631 | 4.74 | 3.47 | 631 | 44.7 | 113 | 16.95 | 0.85 | 1.31 | 29.3 | — | 510 |
| 520 | 37.70 | 1730 | 1.244 | 0.0525 | 4.84 | 3.70 | 621 | 47.5 | 108 | 17.33 | 0.84 | 1.35 | 26.9 | — | 520 |
| 530 | 44.58 | 1679 | 1.268 | 0.0445 | 4.95 | 3.96 | 608 | 50.6 | 104 | 17.72 | 0.85 | 1.39 | 24.5 | — | 530 |
| 540 | 52.38 | 1622 | 1.294 | 0.0375 | 5.08 | 4.27 | 594 | 54.0 | 101 | 18.1 | 0.86 | 1.43 | 22.1 | — | 540 |
| 550 | 61.19 | 1564 | 1.323 | 0.0317 | 5.24 | 4.64 | 580 | 58.3 | 97 | 18.6 | 0.87 | 1.47 | 19.7 | — | 550 |
| 560 | 71.08 | 1499 | 1.355 | 0.0269 | 5.43 | 5.09 | 563 | 63.7 | 94 | 19.1 | 0.90 | 1.52 | 17.3 | — | 560 |
| 570 | 82.16 | 1429 | 1.392 | 0.0228 | 5.68 | 5.67 | 548 | 76.7 | 91 | 19.7 | 0.94 | 1.59 | 15.0 | — | 570 |
| 580 | 94.51 | 1353 | 1.433 | 0.0193 | 6.00 | 6.40 | 528 | 76.7 | 88 | 20.4 | 0.99 | 1.68 | 12.8 | — | 580 |
| 590 | 108.3 | 1274 | 1.482 | 0.0163 | 6.41 | 7.35 | 513 | 84.1 | 84 | 21.5 | 1.05 | 1.84 | 10.5 | — | 590 |
| 600 | 123.5 | 1176 | 1.541 | 0.0137 | 7.00 | 8.75 | 497 | 92.9 | 81 | 22.7 | 1.14 | 2.15 | 8.4 | — | 600 |
| 610 | 137.3 | 1068 | 1.612 | 0.0115 | 7.85 | 11.1 | 467 | 103 | 77 | 24.1 | 1.30 | 2.60 | 6.3 | — | 610 |
| 620 | 159.1 | 941 | 1.705 | 0.0094 | 9.35 | 15.4 | 444 | 114 | 72 | 25.9 | 1.52 | 3.46 | 4.5 | — | 620 |
| 625 | 169.1 | 858 | 1.778 | 0.0085 | 10.6 | 18.3 | 430 | 121 | 70 | 27.0 | 1.65 | 4.20 | 3.5 | — | 625 |
| 630 | 179.7 | 781 | 1.856 | 0.0075 | 12.6 | 22.1 | 412 | 130 | 67 | 28.0 | 2.0 | 4.8 | 2.6 | — | 630 |
| 635 | 190.9 | 683 | 1.935 | 0.0066 | 16.4 | 27.6 | 392 | 141 | 64 | 30.0 | 2.7 | 6.0 | 1.5 | — | 635 |
| 640 | 202.7 | 560 | 2.075 | 0.0057 | 26 | 42 | 367 | 155 | 59 | 32.0 | 4.2 | 9.6 | 0.8 | — | 640 |
| 645 | 215.2 | 361 | 2.351 | 0.0045 | 90 | — | 331 | 178 | 54 | 37.0 | 12 | 26 | 0.1 | — | 645 |
| 647.3 [b] | 221.2 | 0 | 3.170 | 0.0032 | ∞ | ∞ | 238 | 238 | 45 | 45.0 | ∞ | ∞ | 0.0 | — | 647.3[b] |

[a] 1 bar = $10^5$ N/m²; 1 atm = 1.0133 bar = 0.10133 MPa.
[b] Critical temperature.

*Sources*: Adapted with permission from Incropera and DeWitt [9] and Liley [10].

**TABLE A-C-3(b)** Property values of saturated liquids

| $T$ (°C) | $\rho$ (kg/m$^3$) | $c_P$ [kJ/(kg °C)] | $k$ [W/(m °C)] | $\alpha \times 10^7$ (m$^2$/s) | $\nu \times 10^6$ (m$^2$/s) | $Pr$ | $\beta \times 10^3$ (1/°C) |
|---|---|---|---|---|---|---|---|
| | | | Ammonia, $NH_3$ | | | | |
| −50 | 703.69 | 4.463 | 0.547 | 1.742 | 0.435 | 2.60 | |
| −40 | 691.68 | 4.467 | 0.547 | 1.775 | 0.406 | 2.28 | |
| −30 | 679.34 | 4.476 | 0.549 | 1.801 | 0.387 | 2.15 | |
| −20 | 666.69 | 4.509 | 0.547 | 1.819 | 0.381 | 2.09 | |
| −10 | 653.55 | 4.564 | 0.543 | 1.825 | 0.378 | 2.07 | |
| 0 | 640.10 | 4.635 | 0.540 | 1.819 | 0.373 | 2.05 | 2.2 |
| 10 | 626.16 | 4.714 | 0.531 | 1.801 | 0.368 | 2.04 | 2.3 |
| 20 | 611.75 | 4.798 | 0.521 | 1.775 | 0.359 | 2.02 | 2.5 |
| 30 | 596.37 | 4.890 | 0.507 | 1.742 | 0.349 | 2.01 | |
| 40 | 580.99 | 4.999 | 0.493 | 1.701 | 0.340 | 2.00 | |
| 50 | 564.33 | 5.116 | 0.476 | 1.654 | 0.330 | 1.99 | |
| | | | Carbon dioxide, $CO_2$ | | | | |
| −50 | 1,156.3 | 1.84 | 0.0855 | 0.4021 | 0.119 | 2.96 | |
| −40 | 1,117.8 | 1.88 | 0.1011 | 0.4810 | 0.118 | 2.46 | |
| −30 | 1,076.8 | 1.97 | 0.1116 | 0.5272 | 0.117 | 2.22 | |
| −20 | 1,032.4 | 2.05 | 0.1151 | 0.5445 | 0.115 | 2.12 | |
| −10 | 983.38 | 2.18 | 0.1099 | 0.5133 | 0.113 | 2.20 | |
| 0 | 926.99 | 2.47 | 0.1045 | 0.4578 | 0.108 | 2.38 | |
| 10 | 860.03 | 3.14 | 0.0971 | 0.3608 | 0.101 | 2.80 | |
| 20 | 772.57 | 5.0 | 0.0872 | 0.2219 | 0.091 | 4.10 | 14.0 |
| 30 | 597.81 | 36.4 | 0.0703 | 0.0279 | 0.080 | 28.7 | |
| | | | Sulfur dioxide, $SO_2$ | | | | |
| −50 | 1,560.8 | 1.3595 | 0.242 | 1.141 | 0.484 | 4.24 | |
| −40 | 1,536.8 | 1.3607 | 0.235 | 1.130 | 0.424 | 3.74 | |
| −30 | 1,520.6 | 1.3616 | 0.230 | 1.117 | 0.371 | 3.31 | |
| −20 | 1,488.6 | 1.3624 | 0.225 | 1.107 | 0.324 | 2.93 | |
| −10 | 1,463.6 | 1.3628 | 0.218 | 1.097 | 0.288 | 2.62 | |
| 0 | 1,438.5 | 1.3636 | 0.211 | 1.081 | 0.257 | 2.38 | |
| 10 | 1,412.5 | 1.3645 | 0.204 | 1.066 | 0.232 | 2.18 | |
| 20 | 1,386.4 | 1.3653 | 0.199 | 1.050 | 0.210 | 2.00 | 1.94 |
| 30 | 1,359.3 | 1.3662 | 0.192 | 1.035 | 0.190 | 1.83 | |
| 40 | 1,329.2 | 1.3674 | 0.185 | 1.019 | 0.173 | 1.70 | |
| 50 | 1,299.1 | 1.3683 | 0.177 | 0.999 | 0.162 | 1.61 | |

**TABLE A–C–3(b)** (continued)

| $T$ (°C) | $\rho$ ($kg/m^3$) | $c_P$ [kJ/(kg °C)] | $k$ [W/(m °C)] | $\alpha \times 10^7$ ($m^2/s$) | $\nu \times 10^6$ ($m^2/s$) | $Pr$ | $\beta \times 10^3$ (1/°C) |
|---|---|---|---|---|---|---|---|
| Methyl chloride, $CH_3Cl$ | | | | | | | |
| −50 | 1,052.6 | 1.4759 | 0.215 | 1.388 | 0.320 | 2.31 | |
| −40 | 1,033.4 | 1.4826 | 0.209 | 1.368 | 0.318 | 2.32 | |
| −30 | 1,016.5 | 1.4922 | 0.202 | 1.337 | 0.314 | 2.35 | |
| −20 | 999.39 | 1.5043 | 0.196 | 1.301 | 0.309 | 2.38 | |
| −10 | 981.45 | 1.5194 | 0.187 | 1.257 | 0.306 | 2.43 | |
| 0 | 962.39 | 1.5378 | 0.178 | 1.213 | 0.302 | 2.49 | |
| 10 | 942.36 | 1.5600 | 0.171 | 1.166 | 0.297 | 2.55 | |
| 20 | 923.31 | 1.5860 | 0.163 | 1.112 | 0.293 | 2.63 | |
| 30 | 903.12 | 1.6161 | 0.154 | 1.058 | 0.288 | 2.72 | |
| 40 | 883.10 | 1.6504 | 0.144 | 0.996 | 0.281 | 2.83 | |
| 50 | 861.15 | 1.6890 | 0.133 | 0.921 | 0.274 | 2.97 | |
| Dichlorodifluoromethane (Freon F-12), $CCl_2F_2$ | | | | | | | |
| −50 | 1,546.8 | 0.8750 | 0.067 | 0.501 | 0.310 | 6.2 | |
| −40 | 1,518.7 | 0.8847 | 0.069 | 0.514 | 0.279 | 5.4 | |
| −30 | 1,489.6 | 0.8956 | 0.069 | 0.526 | 0.253 | 4.8 | 1.9 |
| −20 | 1,460.6 | 0.9073 | 0.071 | 0.539 | 0.235 | 4.4 | |
| −10 | 1,429.5 | 0.9203 | 0.073 | 0.550 | 0.221 | 4.0 | |
| 0 | 1,397.4 | 0.9345 | 0.073 | 0.557 | 0.214 | 3.8 | 3.1 |
| 10 | 1,364.3 | 0.9496 | 0.073 | 0.560 | 0.203 | 3.6 | |
| 20 | 1,330.2 | 0.9659 | 0.073 | 0.560 | 0.198 | 3.5 | |
| 30 | 1,295.10 | 0.9835 | 0.071 | 0.560 | 0.194 | 3.5 | |
| 40 | 1,257.13 | 1.0019 | 0.069 | 0.555 | 0.191 | 3.5 | 4.4 |
| 50 | 1,216.0 | 1.0216 | 0.067 | 0.545 | 0.190 | 3.5 | |
| Eutectic calcium chloride solution, 29.9% $CaCl_2$ | | | | | | | |
| −50 | 1,319.8 | 2.608 | 0.402 | 1.166 | 36.35 | 312.0 | |
| −40 | 1,315.0 | 2.6356 | 0.415 | 1.200 | 24.97 | 208.0 | |
| −30 | 1,310.2 | 2.6611 | 0.429 | 1.234 | 17.18 | 139.0 | |
| −20 | 1,305.5 | 2.688 | 0.445 | 1.267 | 11.04 | 87.1 | |
| −10 | 1,300.7 | 2.713 | 0.459 | 1.300 | 6.96 | 53.6 | |
| 0 | 1,296.1 | 2.738 | 0.472 | 1.332 | 4.39 | 33.0 | |
| 10 | 1,291.4 | 2.763 | 0.485 | 1.363 | 3.35 | 24.6 | |
| 20 | 1,286.6 | 2.788 | 0.498 | 1.394 | 2.72 | 19.6 | |
| 30 | 1,282.0 | 2.814 | 0.511 | 1.419 | 2.27 | 16.0 | |
| 40 | 1,277.2 | 2.839 | 0.523 | 1.445 | 1.92 | 13.3 | |
| 50 | 1,272.5 | 2.868 | 0.535 | 1.468 | 1.65 | 11.3 | |

**TABLE A–C–3(b)** (continued)

| $T$ (°C) | $\rho$ (kg/m$^3$) | $c_P$ [kJ/(kg °C)] | $k$ [W/(m °C)] | $\alpha \times 10^7$ (m$^2$/s) | $\nu \times 10^6$ (m$^2$/s) | $Pr$ | $\beta \times 10^3$ (1/°C) |
|---|---|---|---|---|---|---|---|
| | | | Glycerin, $C_3H_5(OH)_3$ | | | | |
| 0 | 1,276.0 | 2.261 | 0.282 | 0.983 | 8310 | 84,700 | |
| 10 | 1,270.1 | 2.319 | 0.284 | 0.965 | 3000 | 31,000 | |
| 20 | 1,264.0 | 2.386 | 0.286 | 0.947 | 1180 | 12,500 | 0.50 |
| 30 | 1,258.1 | 2.445 | 0.286 | 0.929 | 500 | 5,380 | |
| 40 | 1,252.0 | 2.512 | 0.286 | 0.914 | 220 | 2,450 | |
| 50 | 1,245.0 | 2.583 | 0.287 | 0.893 | 150 | 1,630 | |
| | | | Ethylene glycol, $C_2H_4(OH)_2$ | | | | |
| 0 | 1,130.8 | 2.294 | 0.242 | 0.934 | 57.53 | 615 | |
| 20 | 1,116.6 | 2.382 | 0.249 | 0.939 | 19.18 | 204 | 0.65 |
| 40 | 1,101.4 | 2.474 | 0.256 | 0.939 | 8.69 | 93 | |
| 60 | 1,087.7 | 2.562 | 0.260 | 0.932 | 4.75 | 51 | |
| 80 | 1,077.6 | 2.650 | 0.261 | 0.921 | 2.98 | 32.4 | |
| 100 | 1,058.5 | 2.742 | 0.263 | 0.908 | 2.03 | 22.4 | |
| | | | Engine oil (unused) | | | | |
| 0 | 899.12 | 1.796 | 0.147 | 0.911 | 4280 | 47,100 | |
| 20 | 888.23 | 1.880 | 0.145 | 0.872 | 900 | 10,400 | 0.70 |
| 40 | 876.05 | 1.964 | 0.144 | 0.834 | 240 | 2,870 | |
| 60 | 864.04 | 2.047 | 0.140 | 0.800 | 83.9 | 1,050 | |
| 80 | 852.02 | 2.131 | 0.138 | 0.769 | 37.5 | 490 | |
| 100 | 840.01 | 2.219 | 0.137 | 0.738 | 20.3 | 276 | |
| 120 | 828.96 | 2.307 | 0.135 | 0.710 | 12.4 | 175 | |
| 140 | 816.94 | 2.395 | 0.133 | 0.686 | 8.0 | 116 | |
| 160 | 805.89 | 2.483 | 0.132 | 0.663 | 5.6 | 84 | |
| | | | Mercury, Hg | | | | |
| 0 | 13,628 | 0.1403 | 8.20 | 42.99 | 0.124 | 0.0288 | |
| 20 | 13,579 | 0.1394 | 8.69 | 46.06 | 0.114 | 0.0249 | 0.182 |
| 50 | 13,506 | 0.1386 | 9.40 | 50.22 | 0.104 | 0.0207 | |
| 100 | 13,385 | 0.1373 | 10.51 | 57.16 | 0.0928 | 0.0162 | |
| 150 | 13,264 | 0.1365 | 11.49 | 63.54 | 0.0853 | 0.0134 | |
| 200 | 13,145 | 0.1570 | 12.34 | 69.08 | 0.0802 | 0.0116 | |
| 250 | 13,026 | 0.1357 | 13.07 | 74.06 | 0.0765 | 0.0103 | |
| 315.5 | 12,847 | 0.134 | 14.02 | 81.5 | 0.0673 | 0.0083 | |

*Source*: Adapted with permission from Eckert and Drake [4].

**TABLE A–C–3(c)** Properties of saturated liquid-vapor

| *Fluid* | $T_{sat}$ (°C) | $i_{fg}$ (kJ/kg) | $\rho_f$ (kg/m³) | $\rho_g$ (kg/m³) |
|---|---|---|---|---|
| Ethanol | 78† | 846 | 757 | 1.44 |
| Ethylene Glycol | 197† | 812 | 1111 | — |
| Glycerin | 290† | 974 | 1260 | — |
| Mercury | 127 | 301.9 | 13370 | — |
| | 227 | 298.9 | 13130 | — |
| | 357† | 301 | 12740 | 3.90 |
| Refrigerant F-12 | −30† | 165 | 1488 | 6.32 |
| | 27 | 136.9 | 1305 | — |
| | 47 | 123.8 | 1229 | — |

† 1 atmosphere

*Sources*: Incropera and DeWitt [9] and others.

**TABLE A–C–4** Property values of common liquid metals

| *Metal* | *Melting point* (°C) | *Normal boiling point* (°C) | $T$ (°C) | $\rho \times 10^{-3}$ (kg/m³) | $c_P$ [kJ/(kg °C)] | $k$ [W/(m °C)] | $\mu \times 10^3$ [kg/(m s)] | $Pr$ |
|---|---|---|---|---|---|---|---|---|
| Bismuth | 271 | 1480 | 316 | 10.01 | 0.144 | 16.4 | 1.62 | 0.014 |
| | | | 760 | 9.47 | 0.165 | 15.6 | 0.79 | 0.0084 |
| Lead | 327 | 1740 | 371 | 10.5 | 0.159 | 16.1 | 2.40 | 0.024 |
| | | | 704 | 10.1 | 0.155 | 14.9 | 1.37 | 0.016 |
| Lithium | 179 | 1320 | 204 | 0.51 | 4.19 | 38.1 | 0.60 | 0.065 |
| | | | 982 | 0.44 | 4.19 | | 0.42 | |
| Mercury | −38.9 | 357 | 10 | 13.6 | 0.138 | 8.1 | 1.59 | 0.027 |
| | | | 316 | 12.8 | 0.134 | 14.0 | 0.86 | 0.0084 |
| Potassium | 63.9 | 760 | 149 | 0.81 | 0.796 | 45.0 | 0.37 | 0.0066 |
| | | | 704 | 0.67 | 0.754 | 33.1 | 0.14 | 0.0031 |
| Sodium | 97.8 | 883 | 204 | 0.90 | 1.34 | 80.3 | 0.43 | 0.0072 |
| | | | 704 | 0.78 | 1.26 | 59.7 | 0.18 | 0.0038 |
| Sodium—potassium: 22% Na | 19 | 826 | 93.3 | 0.848 | 0.946 | 24.4 | 0.49 | 0.019 |
| | | | 760 | 0.69 | 0.883 | | 0.146 | |
| 56% Na | −11.1 | 784 | 93.3 | 0.89 | 1.13 | 25.6 | 0.58 | 0.026 |
| | | | 760 | 0.74 | 1.04 | 28.9 | 0.16 | 0.058 |
| Lead—bismuth: 44.5% Pb | 125 | 1670 | 288 | 10.3 | 0.147 | 10.7 | 1.76 | 0.024 |
| | | | 649 | 9.84 | | | 1.15 | |

*Source*: Adapted with permission from Knudsen and Katz [11].

**TABLE A–C–5(a)** Property values of gases at atmospheric pressure†

| $T$ (K) | $\rho$ (kg/m$^3$) | $c_P$ [kJ/(kg °C)] | $k$ [W/(m °C)] | $\alpha \times 10^4$ (m$^2$/s) | $\mu \times 10^5$ [kg/(m s)] | $\nu \times 10^6$ (m$^2$/s) | $Pr$ |
|---|---|---|---|---|---|---|---|
| | | | | Air | | | |
| 100 | 3.6010 | 1.0266 | 0.009246 | 0.02501 | 0.6924 | 1.923 | 0.770 |
| 150 | 2.3675 | 1.0099 | 0.013735 | 0.05745 | 1.0283 | 4.343 | 0.753 |
| 200 | 1.7684 | 1.0061 | 0.01809 | 0.10165 | 1.3289 | 7.490 | 0.739 |
| 250 | 1.4128 | 1.0053 | 0.02227 | 0.13161 | 1.488 | 9.49 | 0.722 |
| 300 | 1.1774 | 1.0057 | 0.02624 | 0.22160 | 1.846 | 15.68 | 0.708 |
| 350 | 0.9980 | 1.0090 | 0.03003 | 0.2983 | 2.075 | 20.76 | 0.697 |
| 400 | 0.8826 | 1.0140 | 0.03365 | 0.3760 | 2.286 | 25.90 | 0.689 |
| 450 | 0.7833 | 1.0207 | 0.03707 | 0.4222 | 2.484 | 28.86 | 0.683 |
| 500 | 0.7048 | 1.0295 | 0.04038 | 0.5564 | 2.671 | 37.90 | 0.680 |
| 550 | 0.6423 | 1.0392 | 0.04360 | 0.6532 | 2.848 | 44.34 | 0.680 |
| 600 | 0.5879 | 1.0551 | 0.04659 | 0.7512 | 3.018 | 51.34 | 0.680 |
| 650 | 0.5430 | 1.0635 | 0.04953 | 0.8578 | 3.177 | 58.51 | 0.682 |
| 700 | 0.5030 | 1.0752 | 0.05230 | 0.9672 | 3.332 | 66.25 | 0.684 |
| 750 | 0.4709 | 1.0856 | 0.05509 | 1.0774 | 3.481 | 73.91 | 0.686 |
| 800 | 0.4405 | 1.0978 | 0.05779 | 1.1951 | 3.625 | 82.29 | 0.689 |
| 850 | 0.4149 | 1.1095 | 0.06028 | 1.3097 | 3.765 | 90.75 | 0.692 |
| 900 | 0.3925 | 1.1212 | 0.06279 | 1.4271 | 3.899 | 99.3 | 0.696 |
| 950 | 0.3716 | 1.1321 | 0.06525 | 1.5510 | 4.023 | 108.2 | 0.699 |
| 1000 | 0.3524 | 1.1417 | 0.06752 | 1.6779 | 4.152 | 117.8 | 0.702 |
| 1100 | 0.3204 | 1.160 | 0.0732 | 1.969 | 4.44 | 138.6 | 0.704 |
| 1200 | 0.2947 | 1.179 | 0.0782 | 2.251 | 4.69 | 159.1 | 0.707 |
| 1300 | 0.2707 | 1.197 | 0.0837 | 2.583 | 4.93 | 182.1 | 0.705 |
| 1400 | 0.2515 | 1.214 | 0.0891 | 2.920 | 5.17 | 205.5 | 0.705 |
| 1500 | 0.2355 | 1.230 | 0.0946 | 3.262 | 5.40 | 229.1 | 0.705 |
| 1600 | 0.2211 | 1.248 | 0.100 | 3.609 | 5.63 | 254.5 | 0.705 |
| 1700 | 0.2082 | 1.267 | 0.105 | 3.977 | 5.85 | 280.5 | 0.705 |
| 1800 | 0.1970 | 1.287 | 0.111 | 4.379 | 6.07 | 308.1 | 0.704 |
| 1900 | 0.1858 | 1.309 | 0.117 | 4.811 | 6.29 | 338.5 | 0.704 |
| 2000 | 0.1762 | 1.338 | 0.124 | 5.260 | 6.50 | 369.0 | 0.702 |
| 2100 | 0.1682 | 1.372 | 0.131 | 5.715 | 6.72 | 399.6 | 0.700 |
| 2200 | 0.1602 | 1.419 | 0.139 | 6.120 | 6.93 | 432.6 | 0.707 |
| 2300 | 0.1538 | 1.482 | 0.149 | 6.540 | 7.14 | 464.0 | 0.710 |
| 2400 | 0.1458 | 1.574 | 0.161 | 7.020 | 7.35 | 504.0 | 0.718 |
| 2500 | 0.1394 | 1.688 | 0.175 | 7.441 | 7.57 | 543.5 | 0.730 |

**TABLE A-C-5(a)** (continued)

| $T$ (K) | $\rho$ (kg/m$^3$) | $c_P$ [kJ/(kg °C)] | $k$ [W/(m °C)] | $\alpha \times 10^4$ (m$^2$/s) | $\mu \times 10^5$ [kg/(m s)] | $\nu \times 10^6$ (m$^2$/s) | $Pr$ |
|---|---|---|---|---|---|---|---|
| | | | Ammonia, $NH_3$ | | | | |
| 300 | 0.6894 | 2.158 | 0.0247 | 0.166 | 1.015 | 14.7 | 0.887 |
| 320 | 0.6448 | 2.170 | 0.0272 | 0.194 | 1.09 | 16.9 | 0.870 |
| 340 | 0.6059 | 2.192 | 0.0293 | 0.221 | 1.165 | 19.2 | 0.872 |
| 360 | 0.5716 | 2.221 | 0.0316 | 0.249 | 1.24 | 21.7 | 0.872 |
| 380 | 0.5410 | 2.254 | 0.0340 | 0.279 | 1.31 | 24.2 | 0.869 |
| 400 | 0.5136 | 2.287 | 0.0370 | 0.315 | 1.38 | 26.9 | 0.853 |
| 420 | 0.4888 | 2.322 | 0.0404 | 0.356 | 1.45 | 29.7 | 0.833 |
| 440 | 0.4664 | 2.357 | 0.0435 | 0.396 | 1.525 | 32.7 | 0.826 |
| 460 | 0.4460 | 2.393 | 0.0463 | 0.434 | 1.59 | 35.7 | 0.822 |
| 480 | 0.4273 | 2.430 | 0.0492 | 0.474 | 1.665 | 39.0 | 0.822 |
| 500 | 0.4101 | 2.467 | 0.0525 | 0.519 | 1.73 | 42.2 | 0.813 |
| 520 | 0.3942 | 2.504 | 0.0545 | 0.552 | 1.80 | 45.7 | 0.827 |
| 540 | 0.3795 | 2.540 | 0.0575 | 0.597 | 1.865 | 49.1 | 0.824 |
| 560 | 0.3708 | 2.577 | 0.0606 | 0.634 | 1.93 | 52.0 | 0.827 |
| 580 | 0.3533 | 2.613 | 0.0638 | 0.691 | 1.995 | 56.5 | 0.817 |
| | | | Carbon dioxide, $CO_2$ | | | | |
| 220 | 2.4733 | 0.783 | 0.010805 | 0.05920 | 1.111 | 4.490 | 0.818 |
| 250 | 2.1657 | 0.804 | 0.012884 | 0.07401 | 1.259 | 5.813 | 0.793 |
| 300 | 1.7973 | 0.871 | 0.016572 | 0.10588 | 1.496 | 8.321 | 0.770 |
| 350 | 1.5362 | 0.900 | 0.02047 | 0.14808 | 1.721 | 11.19 | 0.755 |
| 400 | 1.3424 | 0.942 | 0.02461 | 0.19463 | 1.932 | 14.39 | 0.738 |
| 450 | 1.1918 | 0.980 | 0.02897 | 0.24813 | 2.134 | 17.90 | 0.721 |
| 500 | 1.0732 | 1.013 | 0.03352 | 0.3084 | 2.326 | 21.67 | 0.702 |
| 550 | 0.9739 | 1.047 | 0.03821 | 0.3750 | 2.508 | 25.74 | 0.685 |
| 600 | 0.8938 | 1.076 | 0.04311 | 0.4383 | 2.683 | 30.02 | 0.668 |
| | | | Carbon monoxide, CO | | | | |
| 220 | 1.55363 | 1.0429 | 0.01906 | 0.11760 | 1.383 | 8.903 | 0.758 |
| 250 | 1.36530 | 1.0425 | 0.02144 | 0.15063 | 1.540 | 11.28 | 0.750 |
| 300 | 1.13876 | 1.0421 | 0.02525 | 0.21280 | 1.784 | 15.67 | 0.737 |
| 350 | 0.97425 | 1.0434 | 0.02883 | 0.2836 | 2.009 | 20.62 | 0.728 |
| 400 | 0.85363 | 1.0484 | 0.03226 | 0.3605 | 2.219 | 25.99 | 0.722 |
| 450 | 0.75848 | 1.0551 | 0.0436 | 0.4439 | 2.418 | 31.88 | 0.718 |
| 500 | 0.68223 | 1.0635 | 0.03863 | 0.5324 | 2.606 | 38.19 | 0.718 |
| 550 | 0.62024 | 1.0756 | 0.04162 | 0.6240 | 2.789 | 44.97 | 0.721 |
| 600 | 0.56850 | 1.0877 | 0.04446 | 0.7190 | 2.960 | 52.06 | 0.724 |

**TABLE A-C-5(a)** (continued)

| $T$ (K) | $\rho$ (kg/m$^3$) | $c_P$ [kJ/(kg °C)] | $k$ [W/(m °C)] | $\alpha \times 10^4$ (m$^2$/s) | $\mu \times 10^5$ [kg/(m s)] | $\nu \times 10^6$ (m$^2$/s) | $Pr$ |
|---|---|---|---|---|---|---|---|
| | | | Helium, He | | | | |
| 200 | 0.2435 | 5.200 | 0.1177 | 0.9288 | 1.566 | 64.38 | 0.694 |
| 255 | 0.1906 | 5.200 | 0.1357 | 1.3675 | 1.817 | 95.50 | 0.70 |
| 366 | 0.13280 | 5.200 | 0.1691 | 2.449 | 2.305 | 173.6 | 0.71 |
| 477 | 0.10204 | 5.200 | 0.197 | 3.716 | 2.750 | 269.3 | 0.72 |
| 589 | 0.08282 | 5.200 | 0.225 | 5.215 | 3.113 | 375.8 | 0.72 |
| 700 | 0.07032 | 5.200 | 0.251 | 6.661 | 3.475 | 494.2 | 0.72 |
| 800 | 0.06023 | 5.200 | 0.275 | 8.774 | 3.817 | 634.1 | 0.72 |
| 900 | 0.05286 | 5.200 | 0.298 | 10.834 | 4.136 | 781.3 | 0.72 |
| | | | Hydrogen, $H_2$ | | | | |
| 30 | 0.84722 | 10.840 | 0.0228 | 0.02493 | 0.1606 | 1.895 | 0.759 |
| 50 | 0.50955 | 10.501 | 0.0362 | 0.0676 | 0.2516 | 4.880 | 0.721 |
| 100 | 0.24572 | 11.229 | 0.0665 | 0.2408 | 0.4212 | 17.14 | 0.712 |
| 150 | 0.16371 | 12.602 | 0.0981 | 0.475 | 0.5595 | 34.18 | 0.718 |
| 200 | 0.12270 | 13.540 | 0.1282 | 0.772 | 0.6813 | 55.53 | 0.719 |
| 250 | 0.09819 | 14.059 | 0.1561 | 1.130 | 0.7919 | 80.64 | 0.713 |
| 300 | 0.08185 | 14.314 | 0.182 | 1.554 | 0.8963 | 109.5 | 0.706 |
| 350 | 0.07016 | 14.436 | 0.206 | 2.031 | 0.9954 | 141.9 | 0.697 |
| 400 | 0.06135 | 14.491 | 0.228 | 2.568 | 1.086 | 177.1 | 0.690 |
| 450 | 0.05462 | 14.499 | 0.251 | 3.164 | 1.178 | 215.6 | 0.682 |
| 500 | 0.04918 | 14.507 | 0.272 | 3.817 | 1.264 | 257.0 | 0.675 |
| 550 | 0.04469 | 14.532 | 0.292 | 4.516 | 1.348 | 301.6 | 0.668 |
| 600 | 0.04085 | 14.537 | 0.315 | 5.306 | 1.429 | 349.7 | 0.664 |
| 700 | 0.03492 | 14.574 | 0.351 | 6.903 | 1.589 | 455.1 | 0.659 |
| 800 | 0.03060 | 14.675 | 0.384 | 8.563 | 1.740 | 569 | 0.664 |
| 900 | 0.02723 | 14.821 | 0.412 | 10.217 | 1.878 | 690 | 0.676 |
| 1000 | 0.02451 | 14.968 | 0.440 | 11.997 | 2.016 | 822 | 0.686 |
| 1100 | 0.02227 | 15.165 | 0.464 | 13.726 | 2.146 | 965 | 0.703 |
| 1200 | 0.02050 | 15.366 | 0.488 | 15.484 | 2.275 | 1107 | 0.715 |
| 1300 | 0.01890 | 15.575 | 0.512 | 17.394 | 2.408 | 1273 | 0.733 |
| 1333 | 0.01842 | 15.638 | 0.519 | 18.013 | 2.444 | 1328 | 0.736 |
| | | | Nitrogen, $N_2$ | | | | |
| 100 | 3.4808 | 1.0722 | 0.009450 | 0.025319 | 0.6862 | 1.971 | 0.786 |
| 200 | 1.7108 | 1.0429 | 0.01824 | 0.10224 | 1.295 | 7.568 | 0.747 |
| 300 | 1.1421 | 1.0408 | 0.02620 | 0.22044 | 1.784 | 15.63 | 0.713 |
| 400 | 0.8538 | 1.0459 | 0.03335 | 0.3734 | 2.198 | 25.74 | 0.691 |
| 500 | 0.6824 | 1.0555 | 0.03984 | 0.5530 | 2.570 | 37.66 | 0.684 |
| 600 | 0.5687 | 1.0756 | 0.04580 | 0.7486 | 2.911 | 51.19 | 0.686 |
| 700 | 0.4934 | 1.0969 | 0.05123 | 0.9466 | 3.213 | 65.13 | 0.691 |
| 800 | 0.4277 | 1.1225 | 0.05609 | 1.1685 | 3.484 | 81.46 | 0.700 |
| 900 | 0.3796 | 1.1464 | 0.06070 | 1.3946 | 3.749 | 91.06 | 0.711 |
| 1000 | 0.3412 | 1.1677 | 0.06475 | 1.6250 | 4.400 | 117.2 | 0.724 |
| 1100 | 0.3108 | 1.1857 | 0.06850 | 1.8591 | 4.228 | 136.0 | 0.736 |
| 1200 | 0.2851 | 1.2037 | 0.07184 | 2.0932 | 4.450 | 156.1 | 0.748 |

**TABLE A–C–5(a)** (continued)

| $T$ (K) | $\rho$ (kg/m$^3$) | $c_P$ [kJ/(kg °C)] | $k$ [W/(m °C)] | $\alpha \times 10^4$ (m$^2$/s) | $\mu \times 10^5$ [kg/(m s)] | $\nu \times 10^6$ (m$^2$/s) | $Pr$ |
|---|---|---|---|---|---|---|---|
| | | | | Oxygen, $O_2$ | | | |
| 100 | 3.9918 | 0.9479 | 0.00903 | 0.023876 | 0.7768 | 1.946 | 0.815 |
| 150 | 2.6190 | 0.9178 | 0.01367 | 0.05688 | 1.149 | 4.387 | 0.773 |
| 200 | 1.9559 | 0.9131 | 0.01824 | 0.10214 | 1.485 | 7.593 | 0.745 |
| 250 | 1.5618 | 0.9157 | 0.02259 | 0.15794 | 1.787 | 11.45 | 0.725 |
| 300 | 1.3007 | 0.9203 | 0.02676 | 0.22353 | 2.063 | 15.86 | 0.709 |
| 350 | 1.1133 | 0.9291 | 0.03070 | 0.2968 | 2.316 | 20.80 | 0.702 |
| 400 | 0.9755 | 0.9420 | 0.03461 | 0.3768 | 2.554 | 26.18 | 0.695 |
| 450 | 0.8682 | 0.9567 | 0.03828 | 0.4609 | 2.777 | 31.99 | 0.694 |
| 500 | 0.7801 | 0.9722 | 0.04173 | 0.5502 | 2.991 | 38.34 | 0.697 |
| 550 | 0.7096 | 0.9881 | 0.04517 | 0.6441 | 3.197 | 45.05 | 0.700 |
| 600 | 0.6504 | 1.0044 | 0.04832 | 0.7399 | 3.392 | 52.15 | 0.704 |
| | | | | Water vapor (steam) | | | |
| 380 | 0.5863 | 2.060 | 0.0246 | 0.2036 | 1.271 | 21.6 | 1.060 |
| 400 | 0.5542 | 2.014 | 0.0261 | 0.2338 | 1.344 | 24.2 | 1.040 |
| 450 | 0.4902 | 1.980 | 0.0299 | 0.307 | 1.525 | 31.1 | 1.010 |
| 500 | 0.4405 | 1.985 | 0.0339 | 0.387 | 1.704 | 38.6 | 0.996 |
| 550 | 0.4005 | 1.997 | 0.0379 | 0.475 | 1.884 | 47.0 | 0.991 |
| 600 | 0.3652 | 2.026 | 0.0422 | 0.573 | 2.067 | 56.6 | 0.986 |
| 650 | 0.3380 | 2.056 | 0.0464 | 0.666 | 2.247 | 66.4 | 0.995 |
| 700 | 0.3140 | 2.085 | 0.0505 | 0.772 | 2.426 | 77.2 | 1.000 |
| 750 | 0.2931 | 2.119 | 0.0549 | 0.883 | 2.604 | 88.8 | 1.005 |
| 800 | 0.2739 | 2.152 | 0.0592 | 1.001 | 2.786 | 102 | 1.010 |
| 850 | 0.2579 | 2.186 | 0.0637 | 1.130 | 2.969 | 115 | 1.019 |

† The values of $\mu$, $k$, $c_P$, and $Pr$ are fairly insensitive to pressure for He, $H_2$, $O_2$, $N_2$, and air. The effect of pressure on $\rho$, $\nu$, and $\alpha$ can generally be accounted for by use of the ideal gas law.

*Source*: Adapted with permission from Eckert and Drake [4].

**TABLE A–C–5(b)** Molecular weight and specific heat ratio of common gases‡

| *Gas* | *Chemical formula* | *Molecular weight* $\mathcal{M}$ | *Specific heat ratio* $\gamma$ |
|---|---|---|---|
| Air | — | 28.97 | 1.400 |
| Ammonia | $NH_3$ | 17.03 | |
| Carbon dioxide | $CO_2$ | 44.01 | 1.289 |
| Carbon monoxide | CO | 28.01 | 1.400 |
| Helium | He | 4.003 | 1.667 |
| Hydrogen | $H_2$ | 2.016 | 1.409 |
| Nitrogen | $N_2$ | 28.01 | 1.400 |
| Oxygen | $O_2$ | 32.00 | 1.393 |
| Water vapor | $H_2O$ | 18.02 | 1.327 |

‡ Values for specific heat ratio $\gamma = c_P/c_v$ are given at 300 K.

**TABLE A–C–6** Liquid-vapor surface tension of various fluids

| *Liquid* | *Saturation temperature* (°C) | *Surface tension*† $\sigma \times 10^3$ (N/m) |
|---|---|---|
| Ammonia | −70 | 42.4 |
| | −50 | 37.9 |
| | −40 | 35.4 |
| Benzene | 10 | 30.2 |
| | 50 | 25.0 |
| | 70 | 22.4 |
| Carbon dioxide | −30 | 10.1 |
| | −10 | 6.14 |
| | 10 | 2.67 |
| | 30 | 0.07 |
| Ethyl alcohol | 10 | 23.2 |
| | 50 | 19.9 |
| | 100 | 15.7 |
| Hydrogen | −258 | 2.80 |
| | −255 | 2.29 |
| | −253 | 1.95 |
| Mercury | 10 | 488 |
| | 50 | 480 |
| | 100 | 470 |
| | 200 | 450 |
| Methane | 90 | 18.9 |
| | 100 | 16.3 |
| | 115 | 12.3 |
| Nitrogen | 78 | 8.75 |
| | 85 | 7.17 |
| | 90 | 6.04 |
| Water | 0 | 75.6 |
| | 15.6 | 73.2 |
| | 37.8 | 69.7 |
| | 93.3 | 60.1 |
| | 100 | 58.8 |
| | 160 | 46.1 |
| | 227 | 31.9 |
| | 293 | 16.2 |
| | 360 | 1.46 |
| | 374 | 0 |

† 1 N/m = $10^3$ dyne/cm

**TABLE A–C–7** Emissivities of various surfaces†

*Metallic substances*

| *Surface* | *Temperature* | | | | | | | |
|---|---|---|---|---|---|---|---|---|
| | 100 K | 300 K | | 600 K | | 1000 K | | |
| Aluminum | | | | | | | | |
| Highly polished, film | 0.02 | 0.04 | | 0.06 | | | | |
| Foil | 0.06 | 0.07 | | | | | | |
| Anodized | | | (400 K) | | | | | |
| | | 0.82 | 0.76 | | | | | (1500 K) |
| Oxidized | | | | **0.69** | | **0.55** | | **0.41** |
| Brass | | | (329 K) | | (573 K) | | | |
| Polished | | **0.05** | | | **0.032** | | | |
| Oxidized | | | | **0.22** | | | | |
| Tarnished | | | **0.20** | | | | | |
| Chromium | | | | | | | | |
| Polished or plated | **0.05** | **0.10** | | **0.14** | | | | |
| Copper | | | | | | | | |
| Highly polished, film | | 0.03 | | 0.04 | | 0.04 | | |
| Lightly tarnished | | **0.037** | | | | | | |
| Oxidized | | **0.78** | | | | 0.86 | | |
| Gold | | | | | | | | |
| Highly polished, film | 0.01 | 0.03 | | 0.04 | | 0.06 | | |
| Foil | 0.06 | 0.07 | | | | | | |
| Iron | | | (398 K) | | (698 K) | | | |
| Polished | | | | | **0.14** | | | |
| Oxidized | | | **0.78** | | | | | |
| Lead | | | (403 K) | | | | | |
| Polished | | | **0.056** | | | | | |
| Oxidized | | **0.28** | | | | | | |
| Nickel | | | | | | | (1200 K) | |
| Polished | | | | 0.09 | | 0.14 | 0.17 | |
| Oxidized | | | | 0.40 | | 0.57 | | |
| Silver | | | | | (800 K) | | (1100 K) | |
| Polished | | 0.02 | | 0.03 | 0.05 | | 0.08 | |
| Stainless steels | | | | | | | | (1400 K) |
| Typical, polished | | **0.17** | | **0.19** | | **0.30** | | |
| Typical, lightly oxidized | | | | | **0.33** | **0.40** | (1200 K) | |
| Typical, highly oxidized | | | | | **0.67** | **0.70** | **0.70** | |
| AISI type 310 | | | | | 0.26 | 0.29 | | 0.29 |
| AISI type 310 oxidized | | | (360 K) | (500 K) | (760 K) | | | |
| at 1255 K | | | 0.46 | 0.36 | 0.64 | | | |
| Tin, bright | | **0.07** | | | | | | |
| Zink | | | | (530 K) | | | | (1600 K) |
| Polished | | 0.02 | | 0.03 | | | | 0.06 |
| Tarnished | | **0.25** | | | | | | |

† Values for normal emissivity $\epsilon_0$ shown in **bold** typeset.

**TABLE A–C–7** (continued)

| *Surface* | *Temperature* 600 K | 800 K | 1000 K | 1200 K | 1500 K | 2000 K | 2500 K | 3300 K |
|---|---|---|---|---|---|---|---|---|
| Molybdenum | | | | | | | | |
| Polished | 0.06 | | 0.10 | | 0.15 | 0.21 | 0.26 | |
| Rough | 0.25 | | 0.31 | | 0.42 | | | |
| Oxidized | 0.80 | 0.82 | | | | | | |
| Platinum | | | | | | | | |
| Polished | | 0.10 | 0.13 | | 0.18 | | | |
| Tantalum | | | | | | | | |
| Polished | | | | 0.11 | 0.17 | 0.23 | 0.28 | |
| Tungsten | | | | | | | | |
| Polished | | | **0.10** | | **0.18** | **0.25** | **0.29** | **0.39** |

*Nonmetallic substances*

| *Surface* | *Temperature* 300 K | | | 800 K | | | |
|---|---|---|---|---|---|---|---|
| Asphalt | 0.93 | | | 0.90 | | | |
| Building materials | | | | | | | |
| Brick-fire clay | 0.90 | | | 0.70 | (1600 K)<br>0.75 | | |
| Gypsum or plaster board | 0.90–0.92 | | | | | | |
| Masonry, plastered | **0.93** | | | | | | |
| Plaster, lime white, rough | **0.93** | | | | | | |
| Wood | | | | | | | |
| Beech, planed | | (343 K)<br>**0.94** | | | | | |
| Oak, planed | **0.88** | | | | | | |
| Carbon (graphite) | | | | | (2000 K)<br>**0.65** | (2300 K)<br>**0.69** | (2600 K)<br>**0.74** |
| Cloth | 0.75–0.90 | | | | | | |
| Concrete | 0.88–0.93 | | | | | | |
| Glass | | | | | | | |
| Window | 0.90–0.95 | | | | | | |
| Corning | | (139 K)<br>0.68 | (260 K)<br>0.82 | | | | |
| Pyrex | | (82 K)<br>**0.85** | (420 K)<br>**0.88** | | (1100 K)<br>**0.75** | | |
| Ice | | | | | | | |
| Smooth | **0.97** | | | | | | |
| Rough | **0.985** | | | | | | |
| Paints | | | | | | | |
| Aluminum bronze | | | (373 K)<br>**0.20–0.40** | | | | |
| Aluminum enamel (rough) | **0.39** | | | | | | |
| Aluminum heated to 325°C | | | (423–588 K)<br>**0.35** | | | | |

## TABLE A–C–7 (continued)

*Nonmetallic substances*

| *Surface* | *Temperature* | | | | | | |
|---|---|---|---|---|---|---|---|
| | 300 K | | | 800 K | | | |
| Bakelite enamel | | | (353 K) **0.935** | | | | |
| Black, Parsons | 0.98 | | (460 K) 0.98 | | | | |
| Enamel | | | | | | | |
| White, rough | **0.90** | | | | | | |
| Black, bright | **0.88** | | | | | | |
| Lampblack | 0.96 | | (530 K) 0.97 | | | | (1600 K) 0.97 |
| Oil | | (273–473 K) **0.88** | | | | | |
| Shellac, black | | | | | | | |
| Bright | **0.82** | | | | | | |
| Dull | | (348–418 K) **0.91** | | | | | |
| White, acrylic | 0.90 | | | | | | |
| White, zink oxide (ZnO) | 0.92 | | | | (1100 K) 0.82 | (1200 K) 0.81 | (1300 K) 0.82 |
| Paper, white | 0.92–0.97 | | | | | | |
| Porcelain, glazed | **0.93** | | | | | | |
| Pyrex | **0.82** | | (600 K) **0.80** | | (1000 K) **0.71** | **0.62** | |
| Pyroceram | **0.85** | | **0.78** | | **0.69** | | (1500 K) **0.57** |
| Quartz, fuzed (rough) | **0.93** | | | | | | |
| Refractories (furnace liners) | | | | | | (1400 K) | (1600 K) |
| Alumina brick | | | | **0.40** | **0.33** | **0.28** | **0.33** |
| Magnesia brick | | | | **0.45** | **0.36** | **0.31** | **0.40** |
| Kaolin insulating brick | | | | **0.70** | (1200 K) **0.57** | **0.47** | **0.53** |
| Rubber | | | | | | | |
| Soft, gray | **0.86** | | | | | | |
| Hard, black (rough) | **0.95** | | | | | | |
| Sand | 0.90 | | | | | | |
| Silicon carbide | | | **0.87** | | (1000 K) **0.87** | | (1500 K) **0.85** |
| Skin | 0.95 | | | | | | |
| Snow | 0.82–0.90 | | | | | | |
| Soil | 0.93–0.96 | | | | | | |
| Rocks | 0.88–0.95 | | | | | | |
| Teflon | 0.85 | (400 K) 0.87 | (500 K) 0.92 | | | | |
| Vegetation | 0.92–0.96 | | | | | | |
| Water | 0.96 | | | | | | |

*Sources*: Raznjevic [5], Touloukian et al. [12–14], Mallory [15], Gubareff et al. [16], and Kreith and Kreider [17].

## D EARTH TEMPERATURE DATA FOR SELECT U.S. CITIES

**TABLE A–D–1** Earth temperature data for select U.S. cities†

| | *City* | $T_M$ (°F) | $\Delta T_s$ (°F) | $t_0$ (day) |
|---|---|---|---|---|
| AL | Birmingham | 65 | 19 | 31 |
| AZ | Phoenix | 73 | 23 | 33 |
| AR | Little Rock | 64 | 21 | 32 |
| CA | Los Angeles | 64 | 7 | 54 |
| CO | Denver | 52 | 22 | 37 |
| DC | Washington | 57 | 22 | 36 |
| FL | Jacksonville | 71 | 14 | 32 |
| GA | Atlanta | 62 | 19 | 32 |
| ID | Boise | 53 | 21 | 34 |
| IL | Chicago | 51 | 25 | 37 |
| | Urbana | 53 | 26 | 35 |
| IN | Indianapolis | 55 | 24 | 34 |
| IA | Des Moines | 52 | 28 | 35 |
| KS | Topeka | 56 | 26 | 35 |
| KY | Louisville | 60 | 22 | 33 |
| LA | New Orleans | 70 | 15 | 32 |
| MA | Portland | 48 | 22 | 39 |
| MS | Plymouth | 51 | 21 | 43 |
| MI | Detroit | 50 | 25 | 39 |
| MN | Minneapolis | 47 | 29 | 35 |
| MS | Columbus | 65 | 19 | 32 |
| MO | Kansas City | 58 | 26 | 35 |
| MT | Billings | 49 | 23 | 37 |
| NB | Lincoln | 53 | 28 | 34 |
| NV | Las Vegas | 69 | 23 | 32 |
| NJ | Trenton | 55 | 22 | 38 |
| NM | Albuquerque | 59 | 22 | 31 |
| NY | Albany | 50 | 25 | 38 |
| NC | Greensboro | 60 | 20 | 31 |
| ND | Bismarck | 44 | 31 | 33 |
| OH | Akron | 52 | 23 | 37 |
| | Columbus | 55 | 22 | 34 |
| OK | Ok. City | 62 | 23 | 34 |
| | Tulsa | 62 | 23 | 34 |
| OR | Portland | 54 | 13 | 37 |
| PA | Philadelphia | 55 | 22 | 34 |
| | Pittsburgh | 52 | 23 | 36 |
| SC | Charleston | 66 | 16 | 32 |
| SD | Rapid City | 50 | 25 | 38 |
| TN | Knoxville | 61 | 21 | 31 |
| | Nashville | 60 | 21 | 32 |
| TX | Ft. Worth | 68 | 21 | 34 |
| | Houston | 71 | 16 | 33 |
| UT | Salt Lake | 53 | 24 | 35 |
| VT | Burlington | 46 | 26 | 37 |
| VA | Richmond | 60 | 19 | 33 |
| WA | Seattle | 53 | 12 | 36 |
| | Spokane | 49 | 21 | 32 |
| WV | Charleston | 58 | 20 | 33 |
| WI | Madison | 49 | 27 | 36 |
| WY | Cheyenne | 48 | 21 | 39 |

† $T_M$—Annual mean earth temperature. (Essentially constant to depth of 200 ft.)
$\Delta T_s$—Amplitude of annual change in ground surface temperature.
$t_0$—Phase constant (i.e., day of the year at which ground surface temperature is a minimum).
*Source:* Adapted with permission from ASHRAE Handbook of Fundamentals [7].

## E ANALYTICAL SOLUTIONS FOR CONDUCTION HEAT TRANSFER

### E–1 Steady Two-Dimensional Conduction in Rectangular Solids

The separation of variables approach is utilized in this section to solve for the temperature distribution in the steady two-dimensional rectangular plate shown in Fig. 3–5. The differential formulation is given by

$$\frac{\partial^2 \psi}{\partial x^2} + \frac{\partial^2 \psi}{\partial y^2} = 0 \tag{3–30}$$

$$\psi = 0 \quad \text{at} \quad x = 0 \qquad \psi = 0 \quad \text{at} \quad x = L \tag{3–31,32}$$

$$\psi = 0 \quad \text{at} \quad y = 0 \qquad \psi = F(x) \quad \text{at} \quad y = w \tag{3–33,34}$$

where $\psi = T - T_1$ and $F(x) = f(x) - T_1$. Because only one nonhomogeneous condition appears in this system of equations, we are ready to employ the separation of variables approach.

We now assume a product solution of the form

$$\psi(x,y) = X(x)\, Y(y) \tag{E–1–1}$$

where $X(x)$ and $Y(y)$ are functions of $x$ and $y$, respectively. Substituting this assumed product solution into Eqs. (3–30) through (3–34), we obtain

$$Y \frac{d^2X}{dx^2} + X \frac{d^2Y}{dy^2} = 0 \tag{E–1–2}$$

$$X(0)\, Y(y) = 0 \qquad X(L)\, Y(y) = 0 \tag{E–1–3,4}$$

$$X(x)\, Y(0) = 0 \qquad X(x)\, Y(w) = F(x) \tag{E–1–5,6}$$

Rearranging Eq. (E–1–2) with the purpose of separating the variables, we obtain

$$\frac{1}{X}\frac{d^2X}{dx^2} = -\frac{1}{Y}\frac{d^2Y}{dy^2} \tag{E–1–7}$$

The left-hand side of this equation is a function of $x$ alone and the right-hand side is only a function of $y$. It follows that in order to satisfy Eq. (E–1–7), both sides of the equation must be equal to a constant; that is,

$$\frac{1}{X}\frac{d^2X}{dx^2} = -\frac{1}{Y}\frac{d^2Y}{dy^2} = \gamma_n \tag{E–1–8}$$

where eigenvalues $\gamma_n$ represent all constants for which Eqs. (E–1–2) through (E–1–6) are satisfied. The value of $\gamma_n$ must be selected on the basis of a consideration of the boundary conditions.

Equation (E–1–8) gives us two equations of the forms

$$\frac{d^2X}{dx^2} - \gamma_n X = 0 \tag{E–1–9}$$

and

$$\frac{d^2Y}{dy^2} + \gamma_n Y = 0 \tag{E–1–10}$$

Both of these equations are satisfied by exponential functions. The assumptions

$$X = Ae^{ax} \qquad \text{and} \qquad Y = Be^{by} \tag{E–1–11,12}$$

lead to values of $a$ and $b$ of the forms

$$a^2 = +\gamma_n \qquad a = \pm\sqrt{\gamma_n} \tag{E–1–13a,b}$$

$$b^2 = -\gamma_n \qquad b = \pm\sqrt{-\gamma_n} \tag{E–1–14a,b}$$

Hence, the solutions become

$$X = A_1 e^{\sqrt{\gamma_n}\,x} + A_2 e^{-\sqrt{\gamma_n}\,x} \tag{E–1–15}$$

$$Y = B_1 e^{\sqrt{-\gamma_n}\,y} + B_2 e^{-\sqrt{-\gamma_n}\,y} \tag{E–1–16}$$

The constants $A_1$, $A_2$, $B_1$, $B_2$, and $\gamma_n$ are determined by use of the boundary conditions. We observe that the boundary conditions for $X$ are both homogeneous. Thus, $x$ is the homogeneous coordinate in this problem and $y$ is the nonhomogeneous coordinate. The boundary conditions in the homogeneous coordinate are considered first. Equations (E–1–3) and (E–1–4) give

$$A_1 + A_2 = 0 \tag{E–1–17}$$

and

$$A_1 e^{\sqrt{\gamma_n}\,L} + A_2 e^{-\sqrt{\gamma_n}\,L} = 0 \tag{E–1–18}$$

or

$$A_1 (e^{\sqrt{\gamma_n}\,L} - e^{-\sqrt{\gamma_n}\,L}) = 0 \tag{E–1–19}$$

If $A_1$ is taken as zero, we have the trivial solution $\psi = 0$. Therefore, the term in parentheses must be zero; that is,

$$e^{\sqrt{\gamma_n}\,L} - e^{-\sqrt{\gamma_n}\,L} = 0 \tag{E–1–20}$$

Our problem is to find all values of $\gamma_n$ for which this equation is satisfied. To simplify matters, we introduce the substitution

$$\gamma_n = \pm \lambda_n^2 \tag{E–1–21}$$

For positive values of $\gamma_n$, Eq. (E–1–20) becomes

$$e^{\lambda_n L} - e^{-\lambda_n L} = 0 \qquad \text{or} \qquad \sinh(\lambda_n L) = 0 \tag{E–1–22,23}$$

This equation is only satisfied for $\lambda_n L = 0$, or $\lambda_n = 0$. The substitution of $\lambda_n = 0$ (or $\gamma_n = 0$) into Eqs. (E–1–15) and (E–1–16) is seen to produce the trivial result that $X$ is independent of $x$ and $Y$ is independent of $y$. Therefore, we conclude that $\gamma_n$ is not positive.

For negative values of $\gamma_n$, Eq. (E–1–20) takes the form

$$e^{i\lambda_n L} - e^{-i\lambda_n L} = 0 \qquad \text{or} \qquad \sin(\lambda_n L) = 0 \tag{E–1–24,25}$$

Because

$$\sin(n\pi) = 0 \qquad \text{for} \qquad n = 0,1,2,\ldots \tag{E–1–26}$$

we conclude that Eq. (E–1–20) is satisfied for

$$\lambda_n L = n\pi \qquad \text{for} \qquad n = 0,1,2,\ldots \tag{E–1–27}$$

Thus, $\gamma_n$ is given by

$$\gamma_n = -\lambda_n^2 = -\left(\frac{n\pi}{L}\right)^2 \qquad \text{for} \qquad n = 0,1,2, \ldots \tag{E–1–28}$$

and our solution for $X$ and $Y$ for any integer $n$ takes the form

$$X_n = A_1 e^{i(n\pi/L)x} + A_2 e^{-i(n\pi/L)x} \tag{E–1–29}$$

or (since $A_2 = -A_1$)

$$X_n = 2A_1 i \sin \frac{n\pi x}{L} \tag{E–1–30}$$

and

$$Y_n = B_1 e^{(n\pi/L)y} + B_2 e^{-(n\pi/L)y} \tag{E–1–31}$$

We now utilize the $y$-boundary conditions. Setting $Y = 0$ at $y = 0$ in accordance with Eq. (E–1–5), we have

$$0 = B_1 + B_2 \tag{E–1–32}$$

Thus, Eq. (E–1–31) takes the form

$$Y_n = B_1[e^{(n\pi/L)y} - e^{-(n\pi/L)y}] \qquad \text{or} \qquad Y_n = 2B_1 \sinh \frac{n\pi y}{L} \tag{E–1–33,34}$$

Substituting this result together with Eq. (E–1–30) into Eq. (E–1–1) gives

$$\psi_n = X_n Y_n = \left(2A_1 i \sin \frac{n\pi x}{L}\right)\left(2B_1 \sinh \frac{n\pi y}{L}\right) = C_n \sinh \frac{n\pi y}{L} \sin \frac{n\pi x}{L} \tag{E–1–35}$$

for any integer $n$, where $C_n = 2iA_1B_1$. Because $\psi_n$ represents a solution for each integer, the full solution is given by

$$\psi = \sum_{n=1}^{\infty} \psi_n = \sum_{n=1}^{\infty} C_n \sinh \frac{n\pi y}{L} \sin \frac{n\pi x}{L} \tag{E–1–36}$$

The coefficient $C_n$ is evaluated by use of the final nonhomogeneous $y$-boundary condition by substituting Eq. (E–1–36) into Eq. (E–1–6); that is,

$$F(x) = \sum_{n=1}^{\infty} C_n \sinh \frac{n\pi w}{L} \sin \frac{n\pi x}{L}$$

$$F(x) = \sum_{n=1}^{\infty} c_n \sin \frac{n\pi x}{L} \tag{E–1–37}$$

where $c_n = C_n \sinh (n\pi w/L)$. This equation is recognized as *the Fourier sine series* expansion of the function $F(x)$, where $c_n$ can be expressed as

$$c_n = \frac{2}{L} \int_0^L F(x) \sin \frac{n\pi x}{L}\, dx \tag{E–1–38}$$

Thus, our final solution for the temperature distribution takes the form

$$T - T_1 = \psi = \sum_{n=1}^{\infty} c_n \frac{\sinh(n\pi y/L)}{\sinh(n\pi w/L)} \sin\frac{n\pi x}{L} \tag{E–1–39}$$

## E–2 Unsteady One-Dimensional Conduction with Convective Cooling

Exact solutions are listed in this section for the temperature distribution in a plane wall, infinite cylinder, and sphere for unsteady one-dimensional convective cooling. The solutions are developed in reference 18 by means of the separation of variables approach.

### *Plane Wall*

The mathematical formulation for convective cooling of a plane wall of thickness $2L$ is given by

$$\frac{\partial^2 T}{\partial x^2} = \frac{1}{\alpha}\frac{\partial T}{\partial t}$$

$$T = T_i \qquad \text{at } t = 0$$

$$-k\frac{\partial T}{\partial x} = h(T - T_F) \qquad \text{at } x = L \tag{3–56}$$

and

$$\frac{\partial T}{\partial x} = 0 \qquad \text{at } x = 0$$

or

$$-k\frac{\partial T}{\partial x} = h(T_F - T) \qquad \text{at } x = -L$$

The solution to this equation is given by Schneider [18] as

$$\Theta = \frac{T - T_F}{T_i - T_F} = \sum_{n=1}^{\infty} C_n \exp(-\gamma_n^2 Fo) \cos\left(\gamma_n \frac{x}{L}\right) \tag{E–2–1}$$

where the coefficient $C_n$ is given by

$$C_n = \frac{4 \sin \gamma_n}{2\gamma_n + \sin(2\gamma_n)} \tag{E–2–2}$$

and the eigenvalues $\gamma_n$ are positive roots of the transcendental equation

$$\gamma_n \tan \gamma_n = Bi_0 \tag{E–2–3}$$

The first five roots of this equation are given in Table A–E–2–1. As pointed out in Sec. 3–9–3, for values of the Fourier number $Fo$ greater than about 0.2, Eq. (E–2–1) can be approximated by truncating second and higher-order terms. The re-

**TABLE A–E–2–1** First five roots $\gamma_n$: $\gamma_n \tan \gamma_n = Bi_0$

| $Bi_0$ | $\gamma_1$ | $\gamma_2$ | $\gamma_3$ | $\gamma_4$ | $\gamma_5$ |
|---|---|---|---|---|---|
| 0.000 | 0.0000 | 3.1416 | 6.2832 | 9.4248 | 12.5664 |
| 0.002 | 0.0447 | 3.1422 | 6.2835 | 9.4250 | 12.5665 |
| 0.004 | 0.0632 | 3.1429 | 6.2838 | 9.4252 | 12.5667 |
| 0.006 | 0.0774 | 3.1435 | 6.2841 | 9.4254 | 12.5668 |
| 0.008 | 0.0893 | 3.1441 | 6.2845 | 9.4256 | 12.5670 |
| 0.010 | 0.0998 | 3.1448 | 6.2848 | 9.4258 | 12.5672 |
| 0.020 | 0.1410 | 3.1479 | 6.2864 | 9.4269 | 12.5680 |
| 0.040 | 0.1987 | 3.1543 | 6.2895 | 9.4290 | 12.5696 |
| 0.060 | 0.2425 | 3.1606 | 6.2927 | 9.4311 | 12.5711 |
| 0.080 | 0.2791 | 3.1668 | 6.2959 | 9.4333 | 12.5727 |
| 0.100 | 0.3111 | 3.1731 | 6.2991 | 9.4354 | 12.5743 |
| 0.200 | 0.4328 | 3.2039 | 6.3148 | 9.4459 | 12.5823 |
| 0.300 | 0.5218 | 3.2341 | 6.3305 | 9.4565 | 12.5902 |
| 0.400 | 0.5932 | 3.2636 | 6.3461 | 9.4670 | 12.5981 |
| 0.500 | 0.6533 | 3.2923 | 6.3616 | 9.4775 | 12.6060 |
| 0.600 | 0.7051 | 3.3204 | 6.3770 | 9.4979 | 12.6139 |
| 0.700 | 0.7506 | 3.3477 | 6.3923 | 9.4983 | 12.6218 |
| 0.800 | 0.7910 | 3.3744 | 6.4074 | 9.5087 | 12.6296 |
| 0.900 | 0.8274 | 3.4003 | 6.4224 | 9.5190 | 12.6375 |
| 1.000 | 0.8603 | 3.4256 | 6.4373 | 9.5293 | 12.6453 |
| 1.500 | 0.9882 | 3.5422 | 6.5097 | 9.5801 | 12.6841 |
| 2.000 | 1.0769 | 3.6436 | 6.5783 | 9.6296 | 12.7223 |
| 3.000 | 1.1925 | 3.8088 | 6.7040 | 9.7240 | 12.7966 |
| 4.000 | 1.2646 | 3.9352 | 6.8140 | 9.8119 | 12.8678 |
| 5.000 | 1.3138 | 4.0336 | 6.9096 | 9.8928 | 12.9352 |
| 6.000 | 1.3496 | 4.1116 | 6.9924 | 9.9667 | 12.9988 |
| 7.000 | 1.3766 | 4.1746 | 7.0640 | 10.0339 | 13.0584 |
| 8.000 | 1.3978 | 4.2264 | 7.1263 | 10.0949 | 13.1141 |
| 9.000 | 1.4149 | 4.2694 | 7.1806 | 10.1502 | 13.1660 |
| 10.000 | 1.4289 | 4.3058 | 7.2281 | 10.2003 | 13.2142 |
| 15.000 | 1.4729 | 4.4255 | 7.3959 | 10.3898 | 13.4078 |
| 20.000 | 1.4961 | 4.4915 | 7.4954 | 10.5117 | 13.5420 |
| 30.000 | 1.5202 | 4.5615 | 7.6057 | 10.6543 | 13.7085 |
| 40.000 | 1.5325 | 4.5979 | 7.6647 | 10.7334 | 13.8048 |
| 50.000 | 1.5400 | 4.6202 | 7.7012 | 10.7832 | 13.8666 |
| 100.000 | 1.5552 | 4.6658 | 7.7764 | 10.8871 | 13.9981 |

sulting relations are summarized in Table 3–5 and form the basis for the famous Heisler charts given in Fig. 3–12.

### *Infinite Cylinder*

The mathematical formulation for convective cooling of an infinite cylinder with radius $r_0$ is given by

$$\frac{1}{r}\frac{\partial}{\partial r}\left(r\frac{\partial T}{\partial r}\right) = \frac{1}{\alpha}\frac{\partial T}{\partial t} \tag{3-63}$$

$$T = T_i \qquad \text{at } t = 0$$

$$\frac{\partial T}{\partial r} = 0 \qquad \text{at } r = 0 \tag{3-65}$$

and

$$-k\frac{\partial T}{\partial r} = h(T - T_F) \qquad \text{at } r = r_0$$

The solution to this equation is given by Schneider [18] as

$$\Theta = \frac{T - T_F}{T_i - T_F} = \sum_{n=1}^{\infty} C_n \exp\left(-\gamma_n^2 Fo\right) J_0\left(\gamma_n \frac{r}{r_0}\right) \tag{E-2-4}$$

**TABLE A–E–2–2** First five roots $\gamma_n$: $\gamma_n J_1(\gamma_n) = Bi_0 J_0(\gamma_n)$

| $Bi_0$ | $\gamma_1$ | $\gamma_2$ | $\gamma_3$ | $\gamma_4$ | $\gamma_5$ |
|---|---|---|---|---|---|
| 0.00 | 0.0000 | 3.8317 | 7.0156 | 10.1735 | 13.3237 |
| 0.02 | 0.1995 | 3.8369 | 7.0184 | 10.1754 | 13.3252 |
| 0.04 | 0.2814 | 3.8421 | 7.0213 | 10.1774 | 13.3267 |
| 0.06 | 0.3438 | 3.8473 | 7.0241 | 10.1794 | 13.3282 |
| 0.08 | 0.3960 | 3.8525 | 7.0270 | 10.1813 | 13.3297 |
| 0.10 | 0.4417 | 3.8577 | 7.0298 | 10.1833 | 13.3312 |
| 0.20 | 0.6170 | 3.8835 | 7.0440 | 10.1931 | 13.3387 |
| 0.30 | 0.7465 | 3.9091 | 7.0582 | 10.2029 | 13.3462 |
| 0.40 | 0.8516 | 3.9344 | 7.0723 | 10.2127 | 13.3537 |
| 0.50 | 0.9408 | 3.9594 | 7.0864 | 10.2225 | 13.3611 |
| 0.60 | 1.0184 | 3.9841 | 7.1004 | 10.2322 | 13.3686 |
| 0.70 | 1.0873 | 4.0085 | 7.1143 | 10.2419 | 13.3761 |
| 0.80 | 1.1490 | 4.0325 | 7.1282 | 10.2516 | 13.3835 |
| 0.90 | 1.2048 | 4.0562 | 7.1421 | 10.2613 | 13.3910 |
| 1.00 | 1.2558 | 4.0795 | 7.1558 | 10.2710 | 13.3984 |
| 2.00 | 1.5994 | 4.2910 | 7.2884 | 10.3658 | 13.4719 |
| 3.00 | 1.7887 | 4.4634 | 7.4103 | 10.4566 | 13.5434 |
| 4.00 | 1.9081 | 4.6018 | 7.5201 | 10.5423 | 13.6125 |
| 5.00 | 1.9898 | 4.7131 | 7.6177 | 10.6223 | 13.6786 |
| 6.00 | 2.0490 | 4.8033 | 7.7039 | 10.6964 | 13.7414 |
| 7.00 | 2.0937 | 4.8772 | 7.7797 | 10.7646 | 13.8008 |
| 8.00 | 2.1286 | 4.9384 | 7.8464 | 10.8271 | 13.8566 |
| 9.00 | 2.1566 | 4.9897 | 7.9051 | 10.8842 | 13.9090 |
| 10.00 | 2.1795 | 5.0332 | 7.9569 | 10.9363 | 13.9580 |
| 15.00 | 2.2509 | 5.1773 | 8.1422 | 11.1367 | 14.1576 |
| 20.00 | 2.2880 | 5.2568 | 8.2534 | 11.2677 | 14.2983 |
| 30.00 | 2.3261 | 5.3410 | 8.3771 | 11.4221 | 14.4748 |
| 40.00 | 2.3455 | 5.3846 | 8.4432 | 11.5081 | 14.5774 |
| 50.00 | 2.3572 | 5.4112 | 8.4840 | 11.5621 | 14.6433 |
| 100.00 | 2.3809 | 5.4652 | 8.5678 | 11.6747 | 14.7834 |

where the coefficient $C_n$ is given by

$$C_n = \frac{2}{\gamma_n} \frac{J_1(\gamma_n)}{J_0^2(\gamma_n) + J_1^2(\gamma_n)} \qquad \text{(E–2–5)}$$

and the eigenvalues $\gamma_n$ are positive roots of the transcendental equation

$$\gamma_n J_1(\gamma_n) = Bi_0 J_0(\gamma_n) \qquad \text{(E–2–6)}$$

The Bessel functions $J_0$ and $J_1$ are listed in Table A–G–2 and the first five roots of this equation are given in Table A–E–2–2. These results provide the basis for the approximate Heisler relations and charts given in Table 3–6 and Fig. 3–13.

### *Sphere*

The mathematical formulation for convective cooling of a sphere of radius $r_0$ is given by

$$\frac{1}{r^2}\frac{\partial}{\partial r}\left(r^2 \frac{\partial T}{\partial r}\right) = \frac{1}{\alpha}\frac{\partial T}{\partial t} \qquad \text{(3–64)}$$

$$T = T_i \qquad \text{at } t = 0$$

$$\frac{\partial T}{\partial r} = 0 \qquad \text{at } r = 0 \qquad \text{(3–65)}$$

and

$$-k\frac{\partial T}{\partial r} = h(T - T_F) \qquad \text{at } r = r_0$$

The solution to this equation is given by Schneider [18] as

$$\Theta = \frac{T - T_F}{T_i - T_F} = \sum_{n=1}^{\infty} C_n \exp(-\gamma_n^2 Fo) \frac{\sin(\gamma_n r/r_0)}{\gamma_n r/r_0} \qquad \text{(E–2–7)}$$

where the coefficient $C_n$ is given by

$$C_n = \frac{4(\sin \gamma_n - \gamma_n \cos \gamma_n)}{2\gamma_n - \sin(2\gamma_n)} \qquad \text{(E–2–8)}$$

and the eigenvalues $\gamma_n$ are positive roots of the transcendental equation

$$1 - \gamma_n \cot \gamma_n = Bi_0 \qquad \text{(E–2–9)}$$

The first five roots of this equation are given in Table A–E–2–3. These results provide the basis for the approximate Heisler relations and charts given in Table 3–7 and Fig. 3–14.

**TABLE A–E–2–3** First five roots $\gamma_n$: $1 - \gamma_n \cot \gamma_n = Bi_0$

| $Bi_0$ | $\gamma_1$ | $\gamma_2$ | $\gamma_3$ | $\gamma_4$ | $\gamma_5$ |
|---|---|---|---|---|---|
| 0.000 | 0.0000 | 4.4934 | 7.7253 | 10.9041 | 14.0662 |
| 0.005 | 0.1224 | 4.4945 | 7.7259 | 10.9046 | 14.0666 |
| 0.010 | 0.1730 | 4.4956 | 7.7265 | 10.9050 | 14.0669 |
| 0.020 | 0.2445 | 4.4979 | 7.7278 | 10.9060 | 14.0676 |
| 0.030 | 0.2991 | 4.5001 | 7.7291 | 10.9069 | 14.0683 |
| 0.040 | 0.3450 | 4.5023 | 7.7304 | 10.9078 | 14.0690 |
| 0.050 | 0.3854 | 4.5045 | 7.7317 | 10.9087 | 14.0697 |
| 0.060 | 0.4217 | 4.5068 | 7.7330 | 10.9096 | 14.0705 |
| 0.070 | 0.4551 | 4.5090 | 7.7343 | 10.9105 | 14.0712 |
| 0.080 | 0.4860 | 4.5112 | 7.7356 | 10.9115 | 14.0719 |
| 0.090 | 0.5150 | 4.5134 | 7.7369 | 10.9124 | 14.0726 |
| 0.100 | 0.5423 | 4.5157 | 7.7382 | 10.9133 | 14.0733 |
| 0.200 | 0.7593 | 4.5379 | 7.7511 | 10.9225 | 14.0804 |
| 0.300 | 0.9208 | 4.5601 | 7.7641 | 10.9316 | 14.0875 |
| 0.400 | 1.0528 | 4.5822 | 7.7770 | 10.9408 | 14.0946 |
| 0.500 | 1.1656 | 4.6042 | 7.7899 | 10.9499 | 14.1017 |
| 0.600 | 1.2644 | 4.6261 | 7.8028 | 10.9591 | 14.1088 |
| 0.700 | 1.3525 | 4.6479 | 7.8156 | 10.9682 | 14.1159 |
| 0.800 | 1.4320 | 4.6696 | 7.8284 | 10.9774 | 14.1230 |
| 0.900 | 1.5044 | 4.6911 | 7.8412 | 10.9865 | 14.1301 |
| 1.000 | 1.5708 | 4.7124 | 7.8540 | 10.9956 | 14.1372 |
| 1.500 | 1.8366 | 4.8158 | 7.9171 | 11.0409 | 14.1724 |
| 2.000 | 2.0288 | 4.9132 | 7.9787 | 11.0856 | 14.2075 |
| 3.000 | 2.2889 | 5.0870 | 8.0962 | 11.1727 | 14.2764 |
| 4.000 | 2.4557 | 5.2329 | 8.2045 | 11.2560 | 14.3434 |
| 5.000 | 2.5704 | 5.3540 | 8.3029 | 11.3349 | 14.4080 |
| 6.000 | 2.6537 | 5.4544 | 8.3914 | 11.4086 | 14.4699 |
| 7.000 | 2.7165 | 5.5378 | 8.4703 | 11.4773 | 14.5288 |
| 8.000 | 2.7654 | 5.6078 | 8.5406 | 11.5408 | 14.5847 |
| 9.000 | 2.8044 | 5.6669 | 8.6031 | 11.5994 | 14.6374 |
| 10.000 | 2.8363 | 5.7172 | 8.6587 | 11.6532 | 14.6870 |
| 11.000 | 2.8628 | 5.7606 | 8.7083 | 11.7027 | 14.7335 |
| 16.000 | 2.9476 | 5.9080 | 8.8898 | 11.8959 | 14.9251 |
| 21.000 | 2.9930 | 5.9921 | 9.0019 | 12.0250 | 15.0625 |
| 31.000 | 3.0406 | 6.0831 | 9.1294 | 12.1807 | 15.2380 |
| 41.000 | 3.0651 | 6.1311 | 9.1987 | 12.2688 | 15.3417 |
| 51.000 | 3.0801 | 6.1606 | 9.2420 | 12.3247 | 15.4090 |
| 101.000 | 3.1105 | 6.2211 | 9.3317 | 12.4426 | 15.5537 |

**TABLE A–G–2** Bessel functions of the first kind

| $x$ | $J_0(x)$ | $J_1(x)$ | $x$ | $J_0(x)$ | $J_1(x)$ | $x$ | $J_0(x)$ | $J_1(x)$ |
|---|---|---|---|---|---|---|---|---|
| 0.0 | 1.0000 | 0.0000 | 1.0 | 0.7652 | 0.4400 | 2.0 | 0.2239 | 0.5767 |
| 0.1 | 0.9975 | 0.0499 | 1.1 | 0.7196 | 0.4709 | 2.1 | 0.1666 | 0.5683 |
| 0.2 | 0.9900 | 0.0995 | 1.2 | 0.6711 | 0.4983 | 2.2 | 0.1104 | 0.5560 |
| 0.3 | 0.9776 | 0.1483 | 1.3 | 0.6201 | 0.5220 | 2.3 | 0.0555 | 0.5399 |
| 0.4 | 0.9604 | 0.1960 | 1.4 | 0.5669 | 0.5419 | 2.4 | 0.0025 | 0.5202 |
| 0.5 | 0.9385 | 0.2423 | 1.5 | 0.5118 | 0.5579 | 2.5 | −0.0484 | 0.4971 |
| 0.6 | 0.9120 | 0.2867 | 1.6 | 0.4554 | 0.5699 | 2.6 | −0.0968 | 0.4708 |
| 0.7 | 0.8812 | 0.3290 | 1.7 | 0.3980 | 0.5778 | 2.7 | −0.1424 | 0.4416 |
| 0.8 | 0.8463 | 0.3688 | 1.8 | 0.3400 | 0.5815 | 2.8 | −0.1850 | 0.4097 |
| 0.9 | 0.8075 | 0.4059 | 1.9 | 0.2818 | 0.5812 | 2.9 | −0.2243 | 0.3754 |

## H TABLE FOR RADIATION FUNCTIONS

**TABLE A–H–1** Radiation functions

| $\lambda T_s$ (μm K) | $\frac{E_{b,0\to\lambda}}{E_b}$ | $\frac{E_{b\lambda}}{E_b}$ | $\lambda T_s$ | $\frac{E_{b,0\to\lambda}}{E_b}$ | $\frac{E_{b\lambda}}{E_b}$ |
|---|---|---|---|---|---|
| 200 | 0.000000 | $1.1782 \times 10^{-27}$ | 6,200 | 0.754140 | $0.78453 \times 10^{-4}$ |
| 400 | 0.000000 | $1.5404 \times 10^{-13}$ | 6,400 | 0.769234 | 0.72566 |
| 600 | 0.000000 | $3.2687 \times 10^{-8}$ | 6,600 | 0.783199 | 0.67163 |
| 800 | 0.000016 | $3.1137 \times 10^{-7}$ | 6,800 | 0.796129 | 0.62206 |
| 1,000 | 0.000321 | $3.7229 \times 10^{-5}$ | 7,000 | 0.808109 | 0.57659 |
| 1,200 | 0.002134 | 1.6460 | 7,200 | 0.819217 | 0.53487 |
| 1,400 | 0.007790 | 4.2226 | 7,400 | 0.829527 | 0.49660 |
| 1,600 | 0.019718 | 7.8266 | 7,600 | 0.839102 | 0.46147 |
| 1,800 | 0.039341 | $1.1799 \times 10^{-4}$ | 7,800 | 0.848005 | 0.42921 |
| 2,000 | 0.066728 | 1.5502 | 8,000 | 0.856288 | 0.39956 |
| 2,200 | 0.100888 | 1.8524 | 8,500 | 0.874608 | 0.33543 |
| 2,400 | 0.140256 | 2.0699 | 9,000 | 0.890029 | 0.28320 |
| 2,600 | 0.183120 | 2.2032 | 9,500 | 0.903085 | 0.24044 |
| 2,800 | 0.227897 | 2.2627 | 10,000 | 0.914199 | 0.20523 |
| 3,000 | 0.273232 | 2.2627 | 10,500 | 0.923710 | 0.17609 |
| 3,200 | 0.318102 | 2.2179 | 11,000 | 0.931890 | 0.15184 |
| 3,400 | 0.361735 | 2.1411 | 11,500 | 0.939959 | 0.13155 |
| 3,600 | 0.403607 | 2.0433 | 12,000 | 0.945098 | 0.11448 |
| 3,800 | 0.443382 | 1.8328 | 13,000 | 0.955139 | $0.87794 \times 10^{-5}$ |
| 4,000 | 0.480877 | 1.8160 | 14,000 | 0.962898 | 0.68373 |
| 4,200 | 0.516014 | 1.6977 | 15,000 | 0.969981 | 0.53993 |
| 4,400 | 0.548796 | 1.5810 | 16,000 | 0.973814 | 0.43174 |
| 4,600 | 0.579280 | 1.4682 | 18,000 | 0.980860 | 0.28533 |
| 4,800 | 0.607559 | 1.3606 | 20,000 | 0.985602 | 0.19582 |
| 5,000 | 0.633747 | 1.2592 | 25,000 | 0.992215 | $0.86857 \times 10^{-6}$ |
| 5,200 | 0.658970 | 1.1642 | 30,000 | 0.995340 | 0.44130 |
| 5,400 | 0.680360 | 1.0758 | 40,000 | 0.997967 | 0.14889 |
| 5,600 | 0.701046 | 0.99392 | 50,000 | 0.998953 | $0.63336 \times 10^{-7}$ |
| 5,800 | 0.720158 | 0.91829 | 75,000 | 0.999713 | 0.13151 |
| 6,000 | 0.737818 | 0.84861 | 100,000 | 0.999905 | $0.42648 \times 10^{-8}$ |

# I VISCOUS DISSIPATION: MATHEMATICAL FORMULATION FOR LAMINAR FLOW

To analyze viscous (and high speed) internal or external boundary layer flows, we must account for the effects of viscous dissipation (and kinetic energy).

Referring to the internal flow illustrated in Fig. A–I–1, the viscous shearing forces acting on the surfaces $dA_y$ and $dA_{y+dy}$ produce rates of work, the magnitudes of which are represented by

$$|d\dot{W}_\tau|_y = |\tau u \, dA|_y \tag{I–1}$$

and

$$|d\dot{W}_\tau|_{y+dy} = |\tau u \, dA|_{y+dy} \tag{I–2}$$

Depending upon the location of the element in the flow field for this particular geometry, the stresses $\tau_y$ and $\tau_{y+dy}$ may be operating in either direction and $|d\dot{W}_\tau|_{y+dy}$ may be greater than, less than, or even equal to $|d\dot{W}_\tau|_y$. The *net* rate of work $d\dot{W}_\tau$

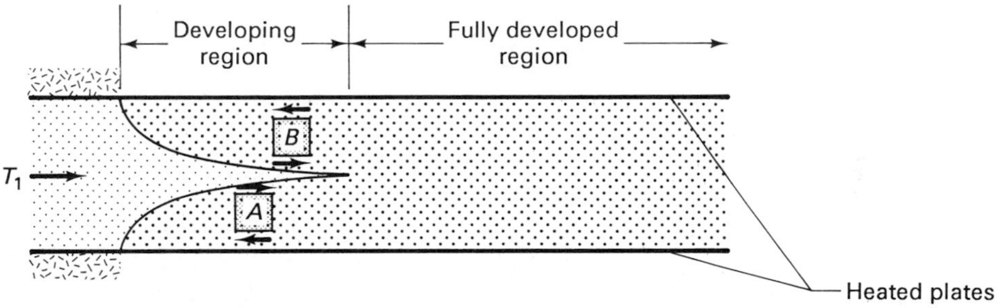


(a) Control volumes within flow field.

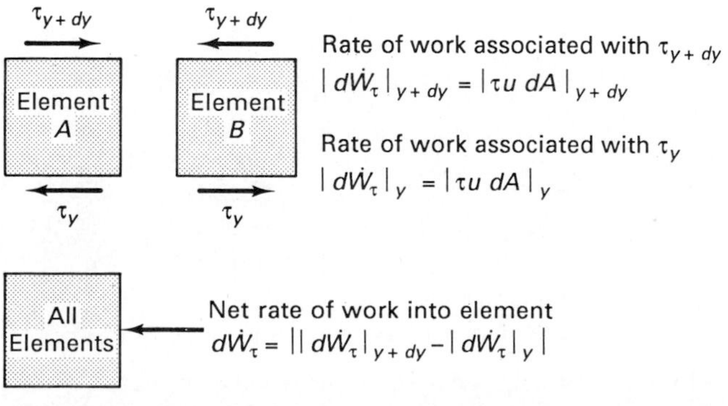


(b) Rates of work associated with shearing stresses acting on fluid elements.

**FIGURE A–I–1** Viscous or high speed laminar flow between parallel plates.

that is actually transferred into the element as a result of viscous shearing action is equal to the absolute difference between these two rates of work; that is,†

$$d\dot{W}_\tau = \left| |d\dot{W}_\tau|_{y+dy} - |d\dot{W}_\tau|_y \right| = \left| |\tau u\, dA|_{y+dy} - |\tau u\, dA|_y \right|$$
$$= \frac{\partial}{\partial y}(\tau u\, dA_y)\, dy = \frac{\partial}{\partial y}(\tau u)\, dV \tag{I–3}$$

which is greater than (or equal to) zero.

The viscous energy generation term $d\dot{W}_\tau$ is shown together with the rates of kinetic energy [$d\dot{KE}_x = (u^2/2)\, d\dot{m}_x$ and $d\dot{KE}_y = (u^2/2)\, d\dot{m}_y$], enthalpy, and heat transfer entering and exiting a control volume in Fig. A–I–2. By incorporating each of these terms into the first law of thermodynamics, we obtain

$$d\dot{H}_x + d\dot{KE}_x + d\dot{H}_y + d\dot{KE}_y + dq_y + d\dot{W}_\tau$$
$$= d\dot{H}_{x+dx} + d\dot{KE}_{x+dx} + d\dot{H}_{y+dy} + d\dot{KE}_{y+dy} + dq_{y+dy} \tag{I–4}$$

where $d\dot{H}_x = i\, d\dot{m}_x$ and $d\dot{H}_y = i\, d\dot{m}_y$, or

$$\frac{\partial}{\partial x}(d\dot{H}_x)\, dx + \frac{\partial}{\partial x}(d\dot{KE}_x)\, dx + \frac{\partial}{\partial y}(d\dot{H}_y)\, dy$$
$$+ \frac{\partial}{\partial y}(d\dot{KE}_y)\, dy + \frac{\partial}{\partial y}(dq_y)\, dy = d\dot{W}_\tau \tag{I–5}$$

Substituting for the various terms and introducing the Fourier law of conduction, Eq. (I–5) becomes (assuming $u >> v$)

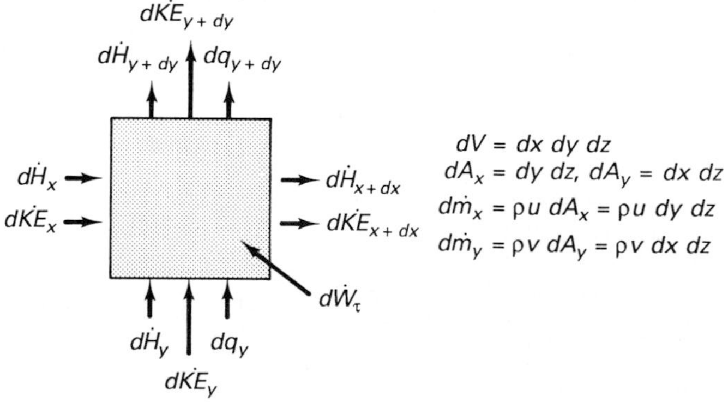


**FIGURE A–I–2** Conservation of energy relative to differential control volume for high speed laminar flow.

† Designating the two work rates by $|d\dot{W}_\tau|_{\text{greater}}$ and $|d\dot{W}_\tau|_{\text{lesser}}$, we note that $|d\dot{W}_\tau|_{\text{greater}} = d\dot{W}_\tau + |d\dot{W}_\tau|_{\text{lesser}}$, where $d\dot{W}_\tau$ is transferred into the fluid element and $|d\dot{W}_\tau|_{\text{lesser}}$ is transmitted through the element and into the adjacent layer of fluid.

$$\frac{\partial}{\partial x}[d\dot{m}_x(i + u^2/2)]\,dx + \frac{\partial}{\partial y}[d\dot{m}_y(i + u^2/2)]\,dy$$
$$- \frac{\partial}{\partial y}\left(k\frac{\partial T}{\partial y}\,dA_y\right)dy = \frac{\partial}{\partial y}(\tau u\,dA_y)\,dy \tag{I–6}$$

or

$$\frac{\partial}{\partial x}[\rho u(i + u^2/2)] + \frac{\partial}{\partial y}[\rho v(i + u^2/2)] = \frac{\partial}{\partial y}(\tau u) + \frac{\partial}{\partial y}\left(k\frac{\partial T}{\partial y}\right) \tag{I–7}$$

The continuity and momentum equations for viscous and high speed flows are of the same form as for fluids with moderate to low viscosity and speed. For two-dimensional laminar flow over plane surfaces, these equations are readily shown to take the forms

$$\frac{\partial}{\partial x}(\rho u) + \frac{\partial}{\partial y}(\rho v) = 0 \tag{I–8}$$

and

$$\rho\left(u\frac{\partial u}{\partial x} + v\frac{\partial u}{\partial y}\right) = \frac{\partial \tau}{\partial y} - \frac{dP}{dx} \tag{I–9}$$

or

$$\rho\left(u\frac{\partial u}{\partial x} + v\frac{\partial u}{\partial y}\right) = \frac{\partial}{\partial y}\left(\mu\frac{\partial u}{\partial x}\right) - \frac{dP}{dx} \tag{I–10}$$

These equations are similar in form to Eqs. (7–5), (7–14), and (7–15) for flow in circular tubes.

The continuity and momentum equations are often used to put the energy equation into more convenient forms. To see this point, the continuity equation is combined with Eq. (I–7) to obtain

$$\rho\left[u\frac{\partial}{\partial x}(i + u^2/2) + v\frac{\partial}{\partial y}(i + u^2/2)\right] = \frac{\partial}{\partial y}(\tau u) + \frac{\partial}{\partial y}\left(k\frac{\partial T}{\partial y}\right) \tag{I–11}$$

Taking a step further, we write

$$\frac{\partial}{\partial y}(\tau u) = \tau\frac{\partial u}{\partial y} + u\frac{\partial \tau}{\partial y} \tag{I–12}$$

and use the momentum equation to obtain

$$\begin{aligned}\frac{\partial}{\partial y}(\tau u) &= \tau\frac{\partial u}{\partial y} + u\frac{dP}{dx} + \rho u\left(u\frac{\partial u}{\partial x} + v\frac{\partial u}{\partial y}\right)\\ &= \tau\frac{\partial u}{\partial y} + u\frac{dP}{dx} + \rho\left[u\frac{\partial}{\partial x}(u^2/2) + v\frac{\partial}{\partial y}(u^2/2)\right]\end{aligned} \tag{I–13}$$

Substituting this result into Eq. (I–11), we have

$$\rho\left(u\frac{\partial i}{\partial x}+v\frac{\partial i}{\partial y}\right)=\tau\frac{\partial u}{\partial y}+u\frac{dP}{dx}+\frac{\partial}{\partial y}\left(k\frac{\partial T}{\partial y}\right) \tag{I–14}$$

Setting $\tau = \mu\ \partial u/\partial y$ in accordance with Newton's law of viscous stress, our final result is

$$\rho\left(u\frac{\partial i}{\partial x}+v\frac{\partial i}{\partial y}\right)=\mu\left(\frac{\partial u}{\partial y}\right)^2+u\frac{dP}{dx}+\frac{\partial}{\partial y}\left(k\frac{\partial T}{\partial y}\right) \tag{I–15}$$

or, in terms of specific heat $c_P$,

$$\rho c_P\left(u\frac{\partial T}{\partial x}+v\frac{\partial T}{\partial y}\right)=\mu\left(\frac{\partial u}{\partial y}\right)^2+u\frac{dP}{dx}+\frac{\partial}{\partial y}\left(k\frac{\partial T}{\partial y}\right) \tag{I–16}$$

The term $\mu(\partial u/\partial y)^2$ represents the energy dissipation per unit volume. This term is commonly designated by $\mu\Phi$, where the *dissipation function* $\Phi$ is defined by

$$\Phi=\left(\frac{\partial u}{\partial y}\right)^2 \tag{I–17}$$

for this basic two-dimensional flow. Notice that $\Phi$ is always positive. Except for very high speed flows, the pressure energy term $u\ dP/dx$ is generally quite small. For the usual case in which the viscous dissipation term $\mu\Phi$ and the pressure gradient energy are small, Eq. (I–16) reduces to

$$\rho c_P\left(u\frac{\partial T}{\partial x}+v\frac{\partial T}{\partial y}\right)=\frac{\partial}{\partial y}\left(k\frac{\partial T}{\partial y}\right) \tag{I–18}$$

This equation is similar in form to Eq. (7–25), which was developed for flow in a circular tube. In this connection, the corresponding equation for laminar flow in a circular tube, which is applicable to viscous and high speed flows, is given by

$$\rho c_P\left(u\frac{\partial T}{\partial x}+v\frac{\partial T}{\partial r}\right)=\mu\left(\frac{\partial u}{\partial r}\right)^2+u\frac{dP}{dx}+\frac{1}{r}\frac{\partial}{\partial r}\left(kr\frac{\partial T}{\partial r}\right) \tag{I–19}$$

## J FILM CONDENSATION: APPROXIMATE SOLUTION FOR LAMINAR FLOW

Film condensation on a cool vertical surface that is exposed to a saturated vapor represents a natural convection flow process involving two phases. The velocity and temperature distributions across a thin film of liquid condensate that is formed on a vertical surface at temperature $T_0 < T_{sat}$ is illustrated in Fig. A–J–1. Notice that the velocity reaches a maximum value near the liquid-vapor interface where the viscous stress approaches zero and that the temperature rises from $T_0$ at the wall to $T_{sat}$ at the edge of the condensate film. Thus, the thickness of the thermal boundary layer $\Delta$

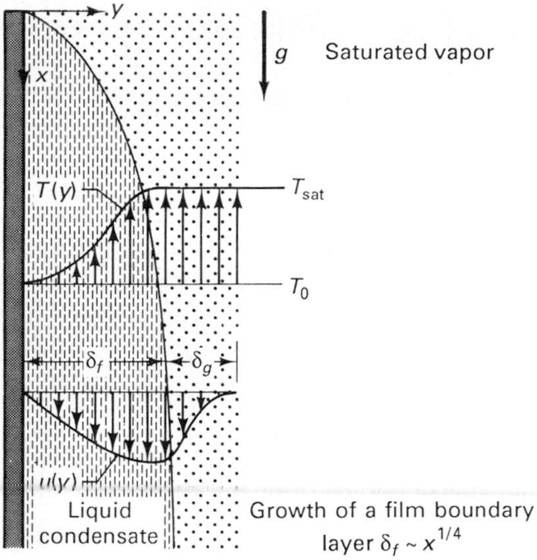


**FIGURE A–J–1** Laminar film condensation on a vertical surface; $g_x = g$.

and condensate film $\delta_f$ are equivalent, but the hydrodynamic boundary layer thickness $\delta = \delta_f + \delta_g$ actually extends beyond the liquid film.

Assuming that (1) the thicknesses $\delta_f$ and $\delta_g$ are small, (2) the flow within the condensate and overlying vapor is laminar, and (3) the properties in each phase are approximately uniform, the transport equations applicable to each phase are represented by

$$\frac{\partial u}{\partial x} + \frac{\partial v}{\partial y} = 0 \tag{J–1}$$

from Eq. (7–121),

$$\rho\left(u\frac{\partial u}{\partial x} + v\frac{\partial u}{\partial y}\right) = \mu\frac{\partial^2 u}{\partial y^2} - g(\rho_\infty - \rho) \tag{J–2}$$

from Eq. (7–129) [with the sign associated with the gravitational acceleration $g$ changed since $g_x$ is in the direction of flow (i.e., $g_x = g$)], and

$$\rho c_P\left(u\frac{\partial T}{\partial x} + v\frac{\partial T}{\partial y}\right) = k\frac{\partial^2 T}{\partial y^2} \tag{J–3}$$

from Eq. (7–122), where the properties for saturated liquid are used in the condensate film and the properties for saturated vapor are used in the vapor phase. The boundary conditions that accompany these equations include

$$u = 0 \qquad v = 0 \qquad T = T_0 \qquad \text{at } y - 0$$
$$u = 0 \qquad T = T_{\text{sat}} \qquad \text{as } y \rightarrow \infty \tag{J–4}$$

plus interfacial conditions that assure continuity in the distributions in velocity, temperature, stress, and energy transfer. This formal mathematical formulation has been solved by Koh, Sparrow, and Hartnett [19].

A very useful approximate solution to this problem was first proposed by Nusselt [20], which involves the simplifying assumption that the advection terms in Eqs. (J–2) and (J–3) are negligible for the low velocity associated with laminar flow, such that the momentum and energy equations within the condensate film reduce to

$$\frac{d^2u}{dy^2} = -\frac{g(\rho_f - \rho_g)}{\mu_f} \qquad \frac{d^2T}{dy^2} = 0 \tag{J–5,6}$$

Arguing further that the shear stress at the liquid-vapor interface must be small, Nusselt was able to eliminate the momentum equation for the vapor film by assuming

$$\frac{\partial u}{\partial y} = 0 \qquad \text{at } y = \delta_f \tag{J–7}$$

Equation (J–5) is readily integrated to obtain

$$u = -\frac{g(\rho_f - \rho_g)y^2}{2\mu_f} + C_1 y + C_2 \tag{J–8}$$

Evaluating the constants of integration $C_1$ and $C_2$ in accordance with the two boundary conditions $u = 0$ at $y = 0$ and $du/dy = 0$ at $y = \delta_f$, the solution for $u$ takes the form

$$u = \frac{g(\rho_f - \rho_g)\delta_f^2}{2\mu_f}\left[2\frac{y}{\delta_f} - \left(\frac{y}{\delta_f}\right)^2\right] \tag{J–9}$$

This result is used to obtain an expression for the condensate mass flow rate $\dot{m}_x$ for a plate depth $b$ by writing†

$$\begin{aligned} \dot{m}_x &= b\int_0^{\delta_f} \rho_f u\, dy = b\int_0^{\delta_f}\left\{\frac{\rho_f g(\rho_f - \rho_g)\delta_f^2}{2\mu_f}\left[2\frac{y}{\delta_f} - \left(\frac{y}{\delta_f}\right)^2\right]\right\} dy \\ &= \frac{b\rho_f g(\rho_f - \rho_g)\delta_f^3}{3\mu_f} \end{aligned} \tag{J–10}$$

Similarly, the solution to Eq. (J–6), which satisfies the conditions $T = T_0$ at $y = 0$ and $T = T_{\text{sat}}$ at $y = \delta_f$ is given by

$$T = T_0 + (T_{\text{sat}} - T_0)\frac{y}{\delta_f} \tag{J–11}$$

† The condensate mass flow rate $\dot{m}_x$ is represented by $\dot{m}$ in Chap. 10.

Using this linear relation for $T$, the heat flux *into* the surface is given by

$$q_c'' = k_f \frac{\partial T}{\partial y}\bigg|_0 = \frac{k_f(T_{sat} - T_0)}{\delta_f} \tag{J–12}$$

such that the local coefficient of condensation heat transfer $h_C$ becomes

$$h_C = \frac{q_c''}{T_{sat} - T_0} = \frac{k_f}{\delta_f} \tag{J–13}$$

To evaluate the film thickness $\delta_f$, which appears in the solutions for $u$, $\dot{m}$, $T$, and $h_C$, we consider mass and energy balances on the lumped/differential element $\delta_f\, dx$ shown in Fig. A–J–2.

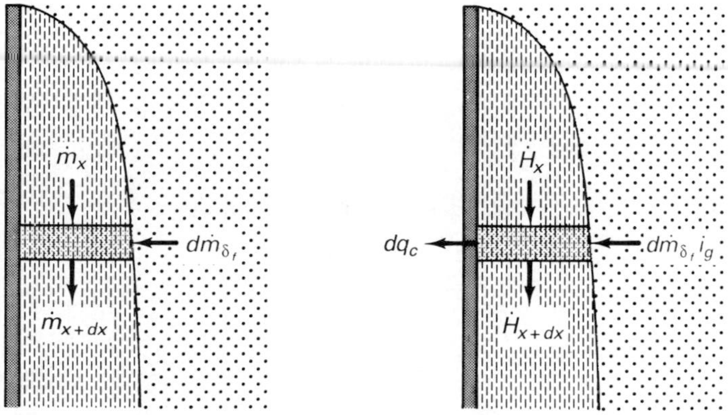


(a) Mass balance. (b) Energy balance.

**FIGURE A–J–2** Lumped/differential control volume within the condensate film.

The mass balance indicates

$$d\dot{m}_{\delta_f} = \dot{m}_{x+dx} - \dot{m}_x = \frac{d\dot{m}_x}{dx}\, dx = d\dot{m}_x \tag{J–14}$$

where

$$d\dot{m}_x = \frac{b\rho_f g(\rho_f - \rho_g)}{\mu_f}\, \delta_f^2\, d\delta_f \tag{J–15}$$

from Eq. (J–10).

Following through with the energy balance, we obtain

$$dq_c = d\dot{m}_{\delta_f}\, i_g - (\dot{H}_{x+dx} - \dot{H}_x) = d\dot{m}_x\, i_g - d\dot{H}_x \tag{J–16}$$

where

$$dq_c = b\,dx\,q_c'' = \frac{b\,dx\,k_f(T_{\text{sat}} - T_0)}{\delta_f} \tag{J–17}$$

from Eq. (J–12). To approximate the condensate enthalpy flow rate $\dot{H}_x$, we neglect the *sensible heat* absorbed due to the temperature within the film falling below $T_{\text{sat}}$; that is,

$$\dot{H}_x \approx \dot{m}_x i_f \tag{J–18}$$

Combining Eqs. (J–16) and (J–18), we obtain

$$dq_c = d\dot{m}_x\, i_g - d(\dot{m}_x i_f) = d\dot{m}_x\, i_{fg} \tag{J–19}$$

where $i_{fg}$ is the *latent heat of vaporization*. This equation also indicates

$$q_c = \dot{m}_x i_{fg} \qquad \text{or} \qquad \dot{m}_x = \frac{q_c}{i_{fg}} \tag{J–20,21}$$

A relation is obtained for $\delta_f$ by substituting for $d\dot{m}_x$ and $dq_c$ in Eq. (J–19).

$$\delta_f^3\, d\delta_f = \frac{k_f\, \mu_f(T_{\text{sat}} - T_0)\, dx}{\rho_f g(\rho_f - \rho_g) i_{fg}} \tag{J–22}$$

Setting $\delta_f = 0$ at $x = 0$, Eq. (J–22) is integrated to obtain

$$\delta_f - \left[\frac{4\mu_f k_f(T_{\text{sat}} - T_0)x}{\rho_f g(\rho_f - \rho_g) i_{fg}}\right]^{1/4} \tag{J–23}$$

Substituting this result into Eq. (J–13), the relation for $h_C$ takes the form

$$h_C = \frac{k_f}{\delta_f} = \left[\frac{\rho_f g(\rho_f - \rho_g) i_{fg}\, k_f^3}{4\mu_f(T_{\text{sat}} - T_0)x}\right]^{1/4} \tag{J–24}$$

or, in terms of Nusselt number $Nu_x$,

$$Nu_x = \frac{h_C x}{k_f} = \left[\frac{\rho_f g(\rho_f - \rho_g) i_{fg}\, x^3}{4k_f \mu_f(T_{\text{sat}} - T_0)}\right]^{1/4} \tag{J–25}$$

It should be noted that this well-known result is usually refined by accounting for the effects of sensible heat in the relation for $\dot{H}_x$. Following the approach developed in Sec. 6–4–2 for internal flows, $\dot{H}_x$ can be represented by

$$\dot{H}_x = b\rho_f \int_0^{\delta_f} [c_{P,f}(T - T_{\text{sat}}) + i_f]u\, dy \tag{J–26}$$

which reduces to Eq. (J–18) when the sensible heat is neglected. Using Eqs. (J–9) and (J–11) for $u$ and $T$, this equation becomes

$$\dot{H}_x = \dot{m}_x[i_f + C_1\, c_{P,f}(T_{\text{sat}} - T_0)] \tag{J–27}$$

where $C_1 = \frac{3}{8}$, and the analysis gives rise to relations for $q_c$ and $Nu_x$ of the form

$$q_c = \dot{m}_x \, i'_{fg} \qquad \text{or} \qquad \dot{m}_x = \frac{q_c}{i'_{fg}} \tag{J–28,29}$$

and

$$Nu_x = \left[ \frac{\rho_f g(\rho_f - \rho_g) i'_{fg} \, x^3}{4k_f \mu_f (T_{\text{sat}} - T_0)} \right]^{1/4} \tag{J–30}$$

where the *modified latent heat of vaporization* $i'_{fg}$ is represented by

$$i'_{fg} = i_{fg} + C_1 \, c_{P,f}(T_{\text{sat}} - T_0) \tag{J–31}$$

As a result of a higher-order analysis of the problem, Rohsenow [21] obtained a similar result with $C_1 = 0.68$. The Rohsenow solution has been found to be in excellent agreement with experimental data for low condensate mass flow rates. However, it should be noted that the simpler result represented by Eqs. (J–20) through (J–25) is generally within 2% of the refined solution.

The results of this analysis are used in Sec. 10–4, which deals with the practical analysis of condensation heat transfer. In addition, more accurate relations are given that apply to higher condensate mass flow rates associated with wavy laminar flow and turbulent flow.

# NOMENCLATURE

## FUNDAMENTAL UNITS†

### SI Units

| | |
|---|---|
| A | ampere (electric current $I$) |
| Å | angstrom ($10^{-10}$ m; wavelength $\lambda$) |
| cm | centimeter ($10^{-2}$ m; length $L$) |
| °C | degree centigrade (temperature $T$) |
| J | joule (N m; energy $E$) |
| kg | kilogram (mass $m$) |
| kJ | kilojoule ($10^3$ J; energy $E$) |
| kW | kilowatt ($10^3$ W; power $\dot{W}$) |
| K | kelvin (absolute temperature $T$) |
| m | meter (length $L$) |
| mm | millimeter ($10^{-3}$ m; length $L$) |
| nm | nanometer ($10^{-9}$ m; wavelength $\lambda$) |
| N | newton (kg m/s$^2$; force $F$) |
| Pa | Pascal (N/m$^2$; pressure $P$, stress $\tau$) |
| rad | radian (plane angle) |
| s | second (time $t$) |
| sr | steradian (solid angle $\omega$) |
| V | volt |
| W | watt (J/s, power $\dot{W}$) |
| μm | micrometer ($10^{-6}$ m; wavelength $\lambda$) |
| Ω | ohm |

### English Units

| | |
|---|---|
| Btu | British thermal unit (energy $E$) |
| deg | degree (plane angle) |
| ft | foot (length $L$) |
| h | hour (time $t$) |
| in. | inch (length $L$) |
| $\text{lb}_\text{f}$ | pound force (32.2 $\text{lb}_\text{m}$ ft/s$^2$; force $F$) |
| $\text{lb}_\text{m}$ | pound mass (mass $m$) |
| °F | degree Fahrenheit (temperature $T$) |
| °R | degree Rankine (absolute temperature $T$) |

## PARAMETERS††

### English Symbols

| | |
|---|---|
| $A$ | cross-sectional area, m$^2$ or ft$^2$ |
| $A_{\min}$ | minimum free-flow area in tube bank or heat exchanger core, m$^2$ or ft$^2$ |
| $A_F$ | area of fin surface, m$^2$ or ft$^2$ |

† Roman typeset is used throughout the text for units.

†† *Italic* typeset is used throughout the text for parameters.

| | |
|---|---|
| $A_N$ | area of body normal to direction of flow, $m^2$ or $ft^2$ |
| $A_o$ | area of finned surface, $m^2$ or $ft^2$ |
| $A_p$ | area of prime (unfinned) surface, $m^2$ or $ft^2$ |
| $A_s$ | surface area, $m^2$ or $ft^2$ |
| $A_1$ | frontal area for tube bank or heat exchanger core, $m^2$ or $ft^2$ |
| $B$ | baffle pitch, m or ft |
| $c$ | speed of light, m/s or ft/s |
| $c_0$ | speed of light in vacuum, m/s or ft/s |
| $c_P$ | specific heat at constant pressure, J/(kg °C) or Btu/($lb_m$ °F) |
| $c_v$ | specific heat at constant volume, J/(kg °C) or Btu/($lb_m$ °F) |
| $C$ | thermal capacitance, J/°C or Btu/°F; absolute capacity rate for flow in heat exchanger core ($\dot{m}c_P$), kW/°C or Btu/(h °F) |
| $C_A$ | area coefficient for tube bank [defined by Eq. (8–123)] |
| $C_D$ | drag coefficient [defined by Eq. (8–110)] |
| $C_e$ | electrical capacitance, Farad (F) |
| $C(r,t)$ | dimensionless temperature for infinite cylinder $[(T - T_F)/(T_i - T_F)]$ |
| $C^*$ | capacitance ratio ($C_{min}/C_{max}$) |
| $C_{min}$ | minimum absolute capacity rate for flow in heat-exchanger core, kW/°C or Btu/(h °F) |
| $C_{max}$ | maximum absolute capacity rate for flow in heat-exchanger core, kW/°C or Btu/(h °F) |
| $d$ | diameter for inner tube of double-pipe arrangement, m or ft |
| $D$ | diameter, m or ft |
| $D_{i,s}$ | inside diameter of shell for shell-and-tube heat exchanger, m or ft |
| $D_H$ | hydraulic diameter ($4A/p_w$ or $4A_{min}/p_w$), m or ft |
| $D_N$ | height of body for crossflow, m or ft |
| $e$ | energy per unit mass, J/kg or Btu/$lb_m$; the number 2.7182818 |
| $E$ | energy, J or Btu; total emissive power, $W/m^2$ or Btu/(h $ft^2$); substitution used in Table 11–5b |
| $E_b$ | blackbody total emissive power, $W/m^2$ or Btu/(h $ft^2$) |
| $E_e$ | electrical voltage, V |
| $E_\lambda$ | monochromatic emissive power [defined by Eq. (5–4)], $W/m^2$ or Btu/(h $ft^2$) |
| $E_{0\to\lambda}$ | subtotal emissive power [defined by Eq. (5–5)], W/($m^2$ μm) or Btu/(h $ft^2$ μm) |
| $f$ | Fanning friction factor |
| $f'$ | friction coefficient for tube bank [defined by Eq. (8–146)] |
| $F$ | force, N or $lb_f$; correction factor for *LMTD* [defined by Eq. (11–30)] |
| $F_c$ | correction factor for tube row effect [see Eq. (8–141)] |
| $F_D$ | drag force, N or $lb_f$ |
| $F_f$ | fouling factor [defined by Eq. (11–7)], $m^2$ °C/W or $ft^2$ °F h/Btu |
| $F_{s-R}$ | radiation shape factor |
| $\mathcal{F}_{s-R}$ | radiation factor |
| $g$ | gravitational acceleration, $m/s^2$ or $ft/s^2$ |
| $G$ | irradiation, $W/m^2$ or Btu/(h $ft^2$); mass flux [defined by Eq. (6–33)], kg/(s $m^2$) or $lb_m$/(s $ft^2$) |
| $h$ | convection heat-transfer coefficient, W/($m^2$ °C) or Btu/(h $ft^2$ °F) |
| $h_f$ | convection heat-transfer coefficient with fouling, W/($m^2$ °C) or Btu/(h $ft^2$ °F) |
| $h_{tc}$ | thermal contact coefficient, W/($m^2$ °C) or Btu/(h $ft^2$ °F) |
| $H$ | enthalpy, J or Btu; height, m or ft |
| $\dot{H}$ | enthalpy rate, W or Btu/h |
| $i$ | specific enthalpy, J/kg or Btu/$lb_m$ |
| $i_{fg}$ | latent heat of vaporization, J/kg or Btu/$lb_m$ |
| $i'_{fg}$ | corrected or modified latent heat of vaporization (Chap. 10), J/kg or Btu/$lb_m$ |
| **i, j, k** | unit vectors |
| $I$ | radiation intensity [defined by Eq. (5–14)], W/($m^2$ sr) or Btu/(h $ft^2$ sr) |
| $I_e$ | electric current, A |
| $J$ | radiosity, $W/m^2$ or Btu/(h $ft^2$) |
| $k$ | thermal conductivity, W/(m °C) or Btu/(h ft °F) |

| | |
|---|---|
| $k_e$ | effective thermal conductivity for natural convection in an enclosed space, W/(m °C) or Btu/(h ft °F) |
| $k_t$ | eddy thermal conductivity for turbulent flow, W/(m °C) or Btu/(h ft °F) |
| $K_c$ | entrance-loss coefficient |
| $K_e$ | expansion-loss coefficient |
| $KE$ | kinetic energy, J or Btu |
| $L$ | length, m or ft |
| $\mathbf{L}$ | characteristic length, m or ft |
| $\ell$ | characteristic length ($V/A_s$), m or ft; turbulent mixing length, m or ft |
| $LMTD$ | log mean temperature difference [defined by Eq. (11–29)], °C or °F |
| $m$ | mass, kg or $lb_m$; convective fin parameter $\{[\bar{h}p/(kA)]^{1/2}\}$, $m^{-1}$ or $ft^{-1}$ |
| $\dot{m}$ | mass flow rate, kg/s or $lb_m$/s |
| $m_R$ | radiative fin parameter $\{[\sigma p F_{s-R}/(kA)]^{1/2}\}$, $m^{-1}$ or $ft^{-1}$ |
| $M$ | momentum, kg m/s or $lb_m$ ft/s; number of heat transfer lanes in a curvilinear sketch (Sec. 3–8) |
| $\dot{M}$ | momentum flow rate, N or $lb_f$ |
| $N$ | number of tubes; number of fins; number of temperature increments in a curvilinear sketch (Sec. 3–8) |
| $N_L$ | number of tubes in longitudinal direction |
| $N_T$ | number of tubes in transverse direction |
| $NTU$ | number of transfer units for heat exchanger $[\bar{h}A_s/(\dot{m}c_P)$ or $\bar{U}A_s/C_{min}]$ |
| $p$ | perimeter, m or ft |
| $P$ | pressure, $N/m^2$ or $lb_f/ft^2$; heat exchanger parameter [defined by Eq. (11–33a) or Eq. (11–34a)] |
| $P(x,t)$ | dimensionless temperature for infinite plate $[(T - T_F)/(T_i - T_F)]$ |
| $PE$ | potential energy, J or Btu |
| $Q$ | energy transfer, J or Btu |
| $q$ | heat-transfer rate, W or Btu/h |
| $\dot{q}$ | rate of energy generation per unit volume, $W/m^3$ or Btu/(h $ft^3$) |
| $q''$ | heat flux, $W/m^2$ or Btu/(h $ft^2$) |
| $q''_{max}$ | maximum heat flux, $W/m^2$ or Btu/(h $ft^2$) |
| $\overline{q''}$ | *apparent total mean heat flux for turbulent flow* $(\overline{q''_y} + \overline{q''_t})$, $W/m^2$ or Btu/(h $ft^2$) |
| $\overline{q''_t}$ | apparent turbulent mean heat flux for turbulent flow $(\rho c_P \overline{v'T'})$, $W/m^2$ or Btu/(h $ft^2$) |
| $r, \phi, z$ | cylindrical coordinates |
| $r, \theta, \phi$ | spherical coordinates |
| $r_c$ | critical radius, m or ft |
| $r_0$ | cylinder or sphere radius, m or ft |
| $R$ | thermal resistance, °C/W or °F h/Btu; heat exchanger parameter [defined by Eq. (11–33b) or Eq. (11–34b)] |
| $R'$ | thermal resistance associated with differential area, °C/W or °F h/Btu |
| $RCM$ | rate of creation of momentum, N or $lb_f$ |
| $R_{tc}$ | thermal contact resistance, °C/W or °F h/Btu |
| $R_e$ | electrical resistance, Ω |
| $R_F$ | fouling factor, $m^2$ °C/W or $ft^2$ °F h/Btu |
| $S$ | conduction shape factor, m or ft |
| $S_D$ | diagonal pitch for tube bank, m or in. |
| $S_L$ | longitudinal pitch for tube bank, m or in. |
| $S_T$ | transverse pitch for tube bank, m or in. |
| $S(x,t)$ | dimensionless temperature for semi-infinite solid $[(T - T_F)/(T_i - T_F)]$ |
| $t$ | time, s or h |
| $T$ | temperature, °C, K, °F, or °R |
| $T_b$ | bulk-stream temperature [see Eq. (6–46)] |
| $u, v, w$ | components of fluid velocity, m/s or ft/s |
| $u^+$ | dimensionless velocity used in analysis of turbulent flow, m/s or ft/s |
| $U$ | velocity, m/s or ft/s; internal energy, J or Btu; overall coefficient of heat transfer [defined by Eq. (11–2)], W/($m^2$ °C) or Btu/(h $ft^2$ °F) |
| $U_b$ | bulk-stream velocity [defined by Eq. (6–34)] |
| $U^*$ | friction velocity $[(\bar{\tau}_s/\rho)^{1/2}]$, m/s or ft/s |
| $v$ | specific volume, $m^3$/kg or $ft^3/lb_m$; component of fluid velocity normal to wall, m/s or ft/s |
| $V$ | volume, $m^3$ or $ft^3$ |
| $\dot{V}$ | volume flow rate, $m^3$/s or $ft^3$/s |
| $w$ | width, m or ft |

$\dot{W}$ power, W or Btu/h

$\dot{W}_p$ pumping power [see Eq. (8–65)], W or Btu/h

$\dot{W}_\tau$ viscous energy dissipation rate [see Appendix I], W or Btu/h

$x, y, z$ rectangular coordinates, m or ft

$x_c$ length beyond which flow is turbulent

$X$ quality for two-phase flow

$y^+$ dimensionless distance from wall ($yU^*/\nu$)

$Z$ depth, m or ft; number of unknown nodal temperatures in a finite-difference formulation

$Z_s$ total number of nodes in a finite-difference formulation

## □ Greek Symbols

*alpha*

$\alpha$ thermal diffusivity, $m^2/s$ or $ft^2/s$; absorptivity

$\alpha_0$ temperature coefficient of resistance, $°C^{-1}$ or $°F^{-1}$

$\alpha_t$ eddy thermal diffusivity for turbulent flow, $m^2/s$ or $ft^2/s$

*beta*

$\beta$ coefficient of thermal expansion, 1/°C or 1/°F; surface-area density of heat-exchanger core ($A_s/V$), $m^{-1}$ or $ft^{-1}$

$\beta_T$ temperature coefficient of thermal conductivity, $°C^{-1}$ or $°F^{-1}$

*gamma*

$\gamma$ specific-heat ratio ($c_P/c_v$)

$\gamma_n$ eigenvalues (see Appendix E)

$\Gamma$ substitution used in Table 11–5a

*delta*

$\delta$ hydrodynamic boundary-layer thickness, m or ft; wall thickness, m or ft

$\Delta$ thermal boundary-layer thickness, m or ft; difference between two values

*epsilon*

$\epsilon$ emissivity; surface roughness, m or in.; heat-exchanger effectiveness; small increment

$\epsilon_H$ eddy thermal diffusivity for turbulent flow (often used instead of $\alpha_t$), $m^2/s$ or $ft^2/s$

$\epsilon_M$ eddy diffusivity for turbulent flow (often used instead of $\nu_t$), $m^2/s$ or $ft^2/s$

$\epsilon_\theta$ directional emissivity [defined by Eq. (5–17)]

*zeta*

$\zeta$ defined by Eq. (E11–4f)

*eta*

$\eta$ similarity parameter

$\eta_F$ fin efficiency

$\eta_o$ net surface efficiency (or temperature effectiveness)

*theta*

$\theta$ zenith angle, rad

$\Theta$ dimensionless temperature distribution $[(T - T_F)/(T_i - T_F)]$

*kappa*

$\kappa$ turbulence coefficient

*lambda*

$\lambda$ wavelength, μm; Blasius friction factor ($4f$)

$\lambda_n$ defined by Eq. (E–1–21)

$\Lambda$ general representation of fanning friction factor or coefficient of heat transfer

*mu*

$\mu$ viscosity, kg/(m s) or $lb_m$/(ft s)

$\mu_t$ eddy viscosity for turbulent flow, kg/(m s) or $lb_m$/(ft s)

*nu*

$\nu$ kinematic viscosity, $m^2/s$ or $ft^2/s$; radiation frequency, $s^{-1}$

$\nu_t$ eddy diffusivity for turbulent flow, $m^2/s$ or $ft^2/s$

*xi*

$\xi$ general coordinate direction (can represent $x$, $y$, $z$, or $r$), m or ft

*pi*

$\pi$ the number 3.14159265

*rho*

$\rho$ density ($1/v$), $kg/m^3$ or $lb_m/ft^3$; reflectivity

$\rho_e$ electrical resistivity, m Ω or ft Ω

*sigma*

| | |
|---|---|
| $\sigma$ | Stefan-Boltzmann constant, W/($m^2$ $K^4$) or Btu/(h $ft^2$ °$R^4$); surface tension, N/m or $lb_f$/ft; contraction ratio ($A_{min}/A_1$) |

*tau*

| | |
|---|---|
| $\tau$ | shear stress, N/$m^2$ or $lb_f/ft^2$; transmissivity |
| $\bar{\tau}$ | apparent total mean shear stress for turbulent flow ($\bar{\tau}_y + \bar{\tau}_t$), N/$m^2$ or $lb_f/ft^2$ |
| $\bar{\tau}_t$ | apparent turbulent mean shear stress for turbulent flow ($-\rho\overline{u'v'}$), N/$m^2$ or $lb_f/ft^2$ |

*phi*

| | |
|---|---|
| $\phi$ | azimuthal angle, rad |
| $\Phi$ | viscous dissipation function, $s^{-2}$ |

*chi*

| | |
|---|---|
| $\chi$ | correction for friction factor associated with flow across tube bank (see Fig. 8–28) |

*psi*

| | |
|---|---|
| $\psi$ | Substitution $T - T_1$, °C; efficiency |

*omega*

| | |
|---|---|
| $\omega$ | solid angle ($d\omega = dA_r/r^2$), sr |
| $\Omega$ | ohm |

## ☐ Subscripts

| | |
|---|---|
| *a* | annulus |
| *b* | blackbody; bulk-stream conditions |
| *B* | boiling |
| *c* | centerline; convection; cold fluid; contraction entrance; counterflow |
| *C* | condensation |
| *cp* | constant properties |
| *D* | based on diameter |
| *e* | expansion exit; electrical |
| *f* | saturated liquid conditions; condensate film |
| *F* | reference fluid condition |
| *g* | saturated vapor conditions; graybody in thermal radiation |
| *h* | hot fluid |
| *HS* | heat sink |
| *i* | influx; inside diameter; tube side; initial condition |
| *k* | conduction |
| *L* | based on length |
| ***L*** | based on characteristic length |
| *m* | *x* node in finite-difference formulation |
| *n* | *y* node in finite-difference formulation |
| *o* | efflux; outside diameter; annular or shell side |
| 0 | uniform wall condition |
| *p* | parallel flow |
| *r* | radial direction |
| *R* | reference; radiation |
| *s* | surface conditions; storage; shell side; mixed flow in crossflow heat exchanger |
| sat | saturated conditions |
| *ss* | steady state |
| *t* | turbulent; tube side; unmixed flow in crossflow heat exchanger |
| *T* | transistor |
| *w* | wetted surface |
| *x* | local condition on a surface relative to the *x* coordinate; *x*-direction |
| *y* | *y*-direction |
| $\lambda$ | monochromatic radiation properties |
| $\infty$ | free-stream conditions |
| $\xi$ | general coordinate direction |

## ☐ Superscripts

| | |
|---|---|
| ′ | turbulent fluctuating quantity |
| ″ | per unit area (flux) |
| · | per unit time (rate) |
| *k* | iteration index number |
| *m,n* | coefficients in empirical correlations |
| $\tau$ | finite-difference index for time |

## ☐ Overbar

| | |
|---|---|
| — | surface average conditions; time mean |

## Dimensionless Groups

$Bi$ Biot number ($h\ell/k$ for convection)
$Bo_L$ Bond number $[g(\rho_f - \rho_g)L^2/\sigma]$
$Ec$ Eckert number $\{U_b^2/[c_P(T_s - T_F)]\}$
$Fo$ Fourier number $(\alpha t/\ell_0^2)$
$Gr_L$ Grashof number [defined by Eq. (9–6) or Eq. (10–5)]
$Gr_L^*$ flux Grashof number $(Nu_L\, Gr_L)$
$Gz$ Graetz number $[Re\, Pr/(x/D)]$
$Ja$ Jakob number $[c_P(T_s - T_{sat})/i_{fg}]$
$Ma_{L,f}$ modified Marangoni number [defined by Eq. (10–42)]
$Nu$ Nusselt number for internal flow passage $(hD_H/k)$
$Nu_L$ Nusselt number $(hL/k)$
$Pe$ Peclet number $(Re\, Pr)$
$Pr$ Prandtl number $(\nu/\alpha)$
$Ra_L$ Rayleigh number $(Gr_L\, Pr)$
$Ra_L^*$ flux Rayleigh number $(Nu_L\, Ra_L)$
$Re$ Reynolds number for internal flow passage $(U_b D_H/\nu)$
$Re_L$ Reynolds number $(U_F L/\nu)$
$Re_f$ Reynolds number for film condensation $[4\dot{m}/(p\mu_f)]$
$Ri_x$ Richardson number $(Gr_x/Re_x^2)$
$S_{L,f}$ surface energy parameter $(\sigma/(L\rho_f i_{fg}))$
$St$ Stanton number $[Nu_L/(Re_L\, Pr)$ or $Nu/(Re\, Pr)]$

## MATHEMATICAL OPERATIONS AND FUNCTIONAL RELATIONS†

fn general functional relation
$\text{fn}_i$ specific functional relation; $i = 1, 2$
log logarithm to the base 10
ln logarithm to the base $e$

† Roman typeset is used throughout text for mathematical operations and functional relations.

# GLOSSARY

## ■ GENERAL

| | |
|---|---|
| Heat transfer $q$ | The transfer of energy across a system boundary caused solely by a temperature difference. |
| Finite-difference energy balance method | The development of nodal equations for energy transfer by the use of finite-difference approximations for the components of energy transfer. |
| Finite-difference discretization method | The use of finite-difference approximations in the transformation of the differential formulation into nodal equations. |
| Finite-difference discretization error or truncation error | The error associated with the finite-difference increment. |
| Finite-difference round-off error | The error associated with the use of a finite number of significant figures. |
| Finite-difference nodal network | System of subvolumes and nodes associated with finite-difference formulation. |
| Gauss-Seidel method | An effective iterative scheme for solving systems of linear or nonlinear algebraic equations. |
| Laws, fundamental | |
| Conservation of mass | Given by Eq. (1–2), $$\Sigma\, \dot{m}_o - \Sigma\, \dot{m}_i + \frac{\Delta m_s}{\Delta t} = 0$$ |
| First law of thermodynamics | *Conservation of energy* given by Eq. (1–1), $$\Sigma\, \dot{E}_o - \Sigma\, \dot{E}_i + \frac{\Delta E_s}{\Delta t} = 0$$ |

Newton's second law — *Momentum principle* given by Eq. (1–3),

$$\Sigma \dot{M}_{o,x} - \Sigma \dot{M}_{i,x} + \frac{\Delta M_{s,x}}{\Delta t} = \Sigma F_x$$

Temperature $T$ — A property that is an index of the kinetic energy possessed by molecules, atoms, and subatomic particles of a substance.

Thermal resistance $R$ or $R'$ — Defined by Eq. (1–23),

$$q = \frac{\Delta T}{R} \quad \text{or} \quad dq = \frac{\Delta T}{R'}$$

## ■ CONDUCTION

Conduction heat transfer — The transfer of energy caused by physical interaction between molecules, atoms, and subatomic particles of a substance at different temperatures (level of kinetic energy).

Conduction shape factor $S$ — Defined by Eq. (1–12),

$$q = kS(T_1 - T_2)$$

Critical radius $r_c$ — The radius at which the heat transfer rate in a cylindrical or spherical body is maximum for the case in which the surface is exposed to convection or radiation.

Laws, particular

Fourier law of conduction — One-dimensional form given by

$$q_\xi = -k\frac{\partial T}{\partial \xi}$$

where $\xi$ represents a space dimension.

General Fourier law of conduction — Multidimensional form given by Eq. (3–1),

$$q'' = q_x''\,\mathbf{i} + q_y''\,\mathbf{j} + q_z''\,\mathbf{k}$$

Superconductors — Substances under low-temperature conditions that have extremely high thermal conductivities.

Temperature coefficient of resistance $\alpha_0$ — Pertaining to the electrical resistivity $\rho_e$ and defined by Eq. (2–68),

$$\rho_e = \rho_0(1 + \alpha_0 T)$$

Temperature coefficient of thermal conductivity $\beta_T$ — Pertaining to substances with temperature dependent thermal conductivity $k(T)$ and defined by Eq. (2–60),

$$k(T) = k_0(1 + \beta_T T)$$

Thermal conductivity $k$ — A thermophysical property of the conducting medium which represents the rate of conduction heat transfer per unit area for a temperature gradient of 1 °C/m (or 1 °F/ft).

Thermal contact coefficient $h_{tc}$ — Pertaining to heat transfer at the interface between two substances denoted by I and II and defined by Eq. (2–28),

$$q_{tc} = h_{tc}A[T_{\text{I}}(0) - T_{\text{II}}(0)]$$

Thermal diffusivity $\alpha$ — Thermophysical property of a substance defined by

$$\alpha = \frac{k}{\rho c_v}$$

Thermal resistance $R_k$ — Defined by Eq. (1–25),

$$R_k = \frac{1}{kS}$$

## RADIATION

Absorption — The process of conversion of radiation incident on matter to internal energy.

Absorptivity $\alpha$ — The fraction of thermal radiation incident on a surface which is absorbed. Modifiers: directional, hemispherical, monochromatic, total.

Blackbody — An object that absorbs all the thermal radiation reaching its surface ($\alpha = 1$). Denoted by subscript *b*.

Black light — Ultraviolet radiation in the range (0.32 − 0.40 μm). Also known as UVA radiation.

Configuration factor — Equivalent to *shape factor*.

Diffuse — Modifier that indicates the same intensity or irradiation in all directions.

Directional — Modifier that refers to a particular direction or angle.

Electromagnetic radiation — Energy that is produced as a result of charged particles which undergo acceleration.

Electromagnetic spectrum — Various types of electromagnetic radiation that are characterized according to wavelength $\lambda$ or frequency $\nu$.

Emissive power $E$ — Rate of thermal radiation emitted by a body per unit surface area. Modifiers: blackbody, monochromatic, total.

Emissivity $\epsilon$ — Ratio of the radiation emitted by a surface to the radiation emitted by a blackbody at the same temperature. Modifiers: directional, hemispherical, monochromatic, total.

Frequency $\nu$ — The number of characteristic cycles per unit time of electromagnetic radiation, which is related to the *wavelength* $\lambda$ and *propagation speed c* by

$$\nu = c/\lambda$$

Graybody — Surfaces for which $\epsilon = \epsilon_\lambda$. Denoted by subscript *g*.

Greenhouse effect — The warming of an environment which is heated by solar radiation as a result of the transmitting medium being essentially transparent to the low-wavelength incoming thermal radiation (mainly visible and near infrared) and opaque to the longer-wavelength infrared radiation (mainly middle and far IR) emitted from the surface.

Hemispherical — Modifier which refers to all directions in the space above a surface.

| | |
|---|---|
| Index of refraction $n$ | Equal to the ratio $c_0/c$. |
| Infrared radiation (IR) | Electromagnetic radiation in the wavelength range 0.76 to 1000 μm. Hot bodies are a primary source of infrared radiation. This part of the thermal radiation region includes *near* IR ($0.76 < \lambda < 1.50$ μm), *middle* IR ($1.50 < \lambda < 5.60$ μm), and *far* IR ($5.60 < \lambda < 1000$ μm). |
| Intensity $I$ | The total rate of thermal radiation emitted per unit solid angle $d\omega$ and per unit area normal to the direction $\phi$, $\theta$. Modifiers: monochromatic, total. |
| Irradiation $G$ | The incoming thermal radiation flux. Modifier: diffuse, monochromatic, total. |
| Laws, particular | |
| Kirchhoff's law | Relation between emission and absorption properties for blackbody thermal radiation in an isothermal enclosure. (See Example 5–5.) |
| Stefan-Boltzmann law | Emissive power of a blackbody given by Eq. (5–2), $E_b = \sigma T_s^4$ where $\sigma = 5.67 \times 10^{-8}$ W/(m$^2$ K$^4$) is the *Stefan-Boltzmann constant.* |
| Planck's law | Monochromatic distribution of emission from a blackbody given by Eq. (5–7). |
| Quantum theory | Theory developed by Planck which relates the photon energy $e$ to the frequency $\nu$ and *Planck's constant* $h = 6.625 \times 10^{-34}$ J s by $e = h\nu = \frac{h}{\lambda}$ |
| Reciprocity law | Relation between shape factors represented by [see Eq. (5–29)] $A_s F_{s-R} = A_R F_{R-s}$ |
| Wien's law | Locus of wavelength associated with peak emission by a blackbody given by Eq. (5–8), $T_s \lambda_{max} = 2898$ μm K |
| Microwave radiation | Electromagnetic radiation in the wavelength range $10^3$ to $10^5$ μm. High-power microwaves which are produced by special types of electron tubes known as magnetrons are used in the cooking of food. (Low-power microwaves which are produced by electron tubes such as klystrons and traveling wave-tubes are used in radar and telecommunications.) |
| Monochromatic | Modifier that refers to single wavelength radiation component. |
| Monochromatic distribution | Refers to variation of radiation with wavelength. |
| Nonparticipating medium | Substances that are completely transparent to thermal radiation. |
| Participating medium | Substances that absorb and emit significant amounts of thermal radiation. |
| Photons (or quanta) | Discrete packets of energy associated with the emission and absorption of electromagnetic radiation by changes in atomic energy state. |

**Radiation** — Equivalent to electromagnetic radiation.

**Radiation factor $\mathcal{F}_{s-j}$** — Defined by Eq. (5–72),

$$q_{s-j} = A_s\mathcal{F}_{s-j}\left(\frac{\epsilon_s}{\alpha_s}E_{bs} - \frac{\epsilon_j}{\alpha_j}E_{bj}\right)$$

**Radiation heat transfer $q_R$** — The transfer of energy across a system boundary by means of an electromagnetic mechanism which is caused solely by a temperature difference.

**Radiation shape factor $F_{s-R}$** — The fraction of thermal radiation leaving a diffuse surface $A_s$ that passes through a nonparticipating medium to surface $A_R$.

**Radiosity $J_s$** — The rate of thermal radiation emitted and reflected per unit area from a surface $A_s$. $J_s$ is expressed in terms of $E_s$, $\rho_s$, and $G_s$ by Eq. (5–49),

$$J_s = E_s + \rho_s G_s$$

Modifiers: monochromatic, total.

**Rayleigh effect** — Molecular scattering in the atmosphere which results in the redirection of much of the incoming solar radiation. According to J. W. S. Rayleigh, the degree of scattering is inversely proportional to $\lambda^4$.

**Reflection** — The process in which radiation incident on a surface is redirected.

**Reflectivity $\rho$** — The fraction of thermal radiation incident on a surface which is reflected. Modifiers: directional, hemispherical, monochromatic, total.

**Reradiating surfaces** — An insulated surface with temperature which is totally governed by thermal radiation.

**Scattering** — Redirection of thermal radiation as a result of reflection from molecules of gas, dust, and other particles in the atmosphere.

**Semitransparent** — Refers to a medium which transmits a portion of the incident radiation.

**Spectral** — Equivalent to monochromatic.

**Specular reflection** — The condition for which the angle of reflection is equal to the angle of incidence.

**Solar radiation** — Electromagnetic radiation emitted from the sun which is concentrated in the low wavelength region ($0.3 - 2.5$ μm) of the thermal spectrum, with the peak occurring at about 0.50 μm. The spectral distribution approximates that of a blackbody at 5762 K.

**Thermal radiation** — Electromagnetic energy in the intermediate wavelength range 0.1 to 1000 μm that is associated with the temperature of the emitting surface. Thermal radiation is generally considered to consist of ultraviolet, visible, and infrared radiation.

**Thermal radiation networks** — Analogous electrical circuits which involve the use of thermal resistances of the form

$$R_{s-j} = \frac{1}{A_s F_{s-j}}$$

Thermal radiation shields — Thin plates or shells of highly reflective materials separated by evacuated spaces. (Radiation shields are used in superinsulative composite walls.)

Thermal resistance $R_R$ — Defined by Eq. (1–26),

$$R_R = \frac{1}{\sigma A_s F_{s-R}(T_s + T_R)(T_s^2 + T_R^2)}$$

Total — Modifier which refers to all wavelengths.

Transmission — The process of radiation passing through matter.

Transmissivity $\tau$ — The fraction of thermal radiation incident on a surface which is transmitted.

Ultraviolet radiation (UV) — Electromagnetic radiation in the wavelength range 0.01 to 0.40 μm. The part of the ultraviolet region that is classified as thermal radiation includes UVA ($0.32 < \lambda < 0.40$ μm), UVB ($0.29 < \lambda < 0.32$ μm), UVC ($0.20 < \lambda < 0.29$ μm), and vacuum UV ($0.1 < \lambda < 0.2$ μm).

Visible radiation (light) — Electromagnetic radiation in the wavelength range 0.40 – 0.76 μm). The visible region includes *violet* ($0.40 < \lambda < 0.44$ μm), *blue* ($0.44 < \lambda < 0.49$ μm), *green* ($0.49 < \lambda < 0.54$ μm), *yellow* ($0.54 < \lambda < 0.60$ μm), *orange* ($0.60 < \lambda < 0.63$ μm), and *red* ($0.63 < \lambda < 0.76$ μm).

View factor — Equivalent to *shape factor*.

Wavelength $\lambda$ — The characteristic length of electromagnetic radiation which is related to the *frequency* $\nu$ and *propagation speed* $c$ by

$$\lambda = c/\nu$$

White light — The summation of all visible wavelengths.

## ■ CONVECTION†

Advection — The process by which energy is transferred by fluid motion.

Apparent turbulent shear stress $\overline{\tau_t}$ — Defined by Eq. (7–184),

$$\overline{\tau_t} = -\rho\overline{u'v'}$$

Apparent turbulent heat flux $\overline{q_t''}$ — Defined by Eq. (7–200),

$$\overline{q_t''} = \rho c_P \overline{T'v'}$$

Boundary layer

- hydrodynamic — The region of a flow field near a surface in which the viscous effects are significant.
- thermal — The region of a flow field near a surface in which the effects of molecular conduction are significant.

† The definition and interpretation of the primary dimensionless groups associated with the analysis of forced convection, natural convection, and boiling and condensation is summarized in Tables 8–9, 9–8, and 10–4.

Boundary layer approximations — Approximations proposed by Prandtl which pertain to the relative magnitude of velocity and velocity gradients within the boundary layer. (See Sec. 6–2.)

Bulk-stream enthalpy rate $\dot{H}_b$ — Defined by Eq. (6–40),

$$\dot{H}_b = \int_{\dot{m}} i \, d\dot{m} = \int_A i\rho u \, dA$$

Bulk-stream mass flux $G$ — Mass flow rate per unit cross-sectional area; that is,

$$G = \frac{\dot{m}}{A}$$

Bulk-stream temperature $T_b$ — Defined by Eq. (6–46),

$$T_b = \frac{\rho}{\dot{m}} \int_A Tu \, dA$$

for uniform property flow.

Bulk-stream velocity $U_b$ — Defined by Eq. (6–34),

$$U_b = \frac{\dot{m}}{\rho A} = \frac{G}{\rho} = \frac{1}{A} \int_A u \, dA$$

for uniform density.

Continuity — Equivalent to fundamental principle of *conservation of mass*.

Convection — The transfer of heat from a surface to a moving fluid.

Coefficient of heat transfer,

local $h_x$ — Defined by Eq. (1–21),

$$dq_c = h_x \, dA_s \, (T_s - T_F)$$

mean $\overline{h}$ — Defined by Eq. (1–22),

$$\overline{h} = \frac{1}{A_s} \int_{A_s} h_x \, dA_s$$

Darcy friction factor $\lambda$ — Related to *Fanning friction factor f* by

$$\lambda = 4f$$

Eddy diffusivity $\nu_t$ (or $\epsilon_M$) — Defined by

$$\nu_t = \mu_t/\rho$$

Eddy thermal conductivity $k_t$ — Used in the mean-field method to relate the *apparent turbulent heat flux* to the *mean temperature* according to Eq. (7–202),

$$\overline{q_t''} = -k_t \frac{\partial \overline{T}}{\partial y}$$

Eddy thermal diffusivity $\alpha_t$ (or $\epsilon_H$) — Defined by

$$\alpha_t = k_t/(\rho c_P)$$

Eddy viscosity $\mu_t$ — Used in the mean-field method to relate the *apparent turbulent shear stress* $\overline{\tau}_t$ to the *mean axial velocity* $\overline{u}$ according to Eq. (7–188),

$$\overline{\tau}_t = \mu_t \frac{\partial \overline{u}}{\partial y}$$

Fanning friction factor

local $f_x$ — Defined by Eq. (6–20),

$$\tau_s = \rho U_F^2 \frac{f_x}{2}$$

**mean $\bar{f}$** — Defined by Eq. (6–21),

$$\bar{f} = \frac{1}{A_s}\int_{A_s} f_x \, dA_s$$

**Fin efficiency $\eta_F$** — Pertaining to a convecting fin and defined by Eq. (2–122),

$$\eta_F = \frac{q_F}{q_{\max}} = \frac{q_F}{\bar{h}A_F(T_0 - T_F)}$$

**Friction velocity $U^*$** — A characteristic velocity generally used in the analysis of turbulent flow, which is defined by Eq. (7–193),

$$U^* = \sqrt{\frac{\tau_s}{\rho}}$$

**Hydraulic diameter $D_H$** — A characteristic length for internal-flow systems, which is defined by Eq. 6–27),

$$D_H = 4\frac{A}{p_w}$$

for tubes, or by Eq. (8–126),

$$D_H = 4\frac{A_{\min}}{p_w}$$

for tube banks and other heat-exchanger surfaces.

**Incompressible flow** — Flow processes in which pressure induced changes in density are negligible.

**Isotropic** — Describes the condition in which properties are independent of direction.

**Laminar flow** — Flow in which individual elements of fluid follow smooth streamline paths.

**Laws, particular**

- **Newton law of cooling** — Given by Eq. (1–20),

  $$q_c = \bar{h}A_s(T_s - T_F)$$

  for uniform surface and fluid temperatures.

- **Newton law of cooling, general** — Defining relation for the local coefficient of heat transfer $h_x$, given by Eq. (1–21),

  $$dq_c = h_x \, dA_s(T_s - T_F)$$

- **Newton law of viscous stress** — One-dimensional, steady-state form given by Eq. (1–19),

  $$\tau = \mu\frac{du}{dy}$$

  where $\mu$ is the fluid viscosity and $\tau$ is the viscous stress acting in the streamwise $x$-direction.

**Mixing length $\ell$** — An alternative mean-field representation for $\bar{\tau}_t$ proposed by Prandtl which takes the form

$$\bar{\tau}_t = \rho\ell^2\left(\frac{\partial\bar{u}}{\partial y}\right)^2$$

**Net surface efficiency $\eta_o$** — Pertaining to a convecting fin and defined by Eq. (2–126),

$$q_o = \eta_o\bar{h}A_o(T_s - T_F)$$

**Newtonian fluids** — Fluids which satisfy the Newton law of viscous stress (i.e., stress is proportional to velocity gradient).

Reynolds stress $\overline{\tau}_t$ — Equivalent to *apparent turbulent shear stress.*

Similar flow — Boundary layer flows for which the velocity and temperature distributions are geometrically similar in the streamwise $x$-direction. Similar flow in tubes is referred to as *fully developed.*

Temperature effectiveness — Equivalent to *net surface efficiency.*

Thermal resistance $R_c$ — Defined by Eq. (1–27),

$$R_c = \frac{1}{\bar{h}A_s}$$

Turbulent flow — Flow in which the movement of individual elements of fluid is unsteady and random in nature (i.e., fluctuating).

Turbulent Prandtl number $Pr_t$ — Defined by

$$Pr_t = \nu_t/\alpha_t = c_P\mu_t/k_t$$

Viscous dissipation — The rate at which energy is converted into thermal energy by shearing forces. (See Appendix I).

## ☐ Forced Convection

Colburn analogy — An empirical correlation for turbulent boundary layer flow of the form

$$Nu_x = \frac{f_x}{2} Re_x Pr^{1/3}$$

which is applicable to forced convection flow of fluid with moderate values of Prandtl number.

Contraction ratio $\sigma$ — Pertaining to internal-flow passages and defined by

$$\sigma = A_{min}/A_1$$

Convection correlations — Theoretical and empirical relations for coefficients of friction, pressure drop, and heat transfer.

Drag coefficient $C_D$ — Pertaining to external flow and defined by Eq. (8–110),

$$F_D = \frac{\rho U_\infty^2}{2} C_D A_N$$

Effectiveness $\epsilon$ — Pertaining to internal-flow passages and defined by Eq. (8–53),

$$q_c = \epsilon q_{c,\max}$$

Efficiency $\psi$ — Pertaining to internal-flow passages and defined by Eq. (8–57),

$$\psi = \frac{\epsilon}{NTU}$$

Entrance-loss coefficient $K_c$ — Pertaining to internal-flow passages and defined by Eq. (8–3),

$$P_c - P = \frac{\rho U_b^2}{2}\left(K_c + \frac{4x}{D_H} f\right)$$

Expansion-loss coefficient $K_e$ — Pertaining to internal flow passages and defined by Eq. (8–4),

$$\Delta P_{\text{loss}} = K_e \frac{\rho U_b^2}{2}$$

Forced convection — Fluid flow that is caused by mechanical devices such as pumps, fans, or compressors.

Fully developed flow

hydrodynamic (HFD) — Internal flow in which the velocity distribution is geometrically similar in the streamwise $x$-direction. HFD flow in an impermeable tube is also characterized by a constant value of the friction factor; that is, $f_x = f$.

thermal (TFD) — Internal flow in which the temperature distribution is geometrically similar in the streamwise $x$-direction. TFD flow in an impermeable tube is also characterized by a constant value of the coefficient of heat transfer; that is, $h_x = h$.

Log mean temperature difference *LMTD* — Pertaining to internal flow passages and defined by Eq. (8–60),

$$LMTD = \frac{\Delta T_1 - \Delta T_2}{\ln (\Delta T_1/\Delta T_2)}$$

Number of transfer units *NTU* — Defined by

$$NTU = \bar{h}A_s/(\dot{m}c_P)$$

for tube flow, and

$$NTU = UA_s/C_{\min}$$

for heat exchanger cores.

Pumping power $\dot{W}_p$ — Power required to overcome pressure drop $\Delta P$ for internal flow, defined by Eq. (8–65),

$$\dot{W}_p = \frac{\dot{m}\,\Delta P}{\rho}$$

for uniform density.

Reynolds analogy — A theoretical relation for turbulent forced convection boundary layer flow of the form

$$Nu_x = \frac{f_x}{2} Re_x\, Pr$$

which is based on the similarity between momentum and energy transfer for turbulent flow of fluid with Prandtl number of the order of unity.

Surface-area density $\beta$ — Pertaining to flow in heat-exchanger cores and defined by

$$\beta = A_s/V$$

## ☐ Natural Convection

Archimedes principle — A body immersed in a fluid experiences an upward buoyancy force equal to the weight of the displaced fluid.

Boussinesq approximation — Pertaining to natural convection, the evaluation of the fluid properties at the free-stream temperature.

Coefficient of thermal expansion $\beta$ — Defined by Eq. (7–130),

$$\beta = -\frac{1}{\rho}\frac{\partial \rho}{\partial T}\bigg|_P$$

Natural convection — Fluid flow that is caused by temperature (or concentration) induced density gradients within the fluid.

## Boiling and Condensation

| | |
|---|---|
| Burnout | The condition associated with *film boiling* in which the temperature of the surface reaches the melting point. |
| Dropwise condensation | A relatively uncommon condensation process in which liquid droplets form on the surface and flow under the influence of gravity. The *non-wetting* conditions associated with dropwise condensation occur when the surface tension $\sigma$ of the fluid is greater than a critical value $\sigma_{cr}$ which is characteristic of the surface material. |
| Excess temperature $\Delta T_e$ | Defined by $\Delta T_e = T_s - T_{\text{sat}}$ |
| Film condensation | A relatively common condensation process in which a liquid film forms on the surface and flows under the influence of gravity. The wetting conditions associated with film condensation occur when the surface tension $\sigma$ of the fluid is less than a critical value $\sigma_{cr}$ which is characteristic of the surface material. |
| Film boiling | Boiling regime in which a stable vapor film covers the heating surface. The heat flux $q''_c$ in this high-temperature region of the boiling curve increases with increasing excess temperature. |
| Heat pipe | A device consisting of a closed pipe lined with wicking material which utilizes evaporation and condensation to transfer heat effectively. |
| Latent heat of vaporization $i_{fg}$ | Equivalent to the *enthalpy of vaporization*. |
| Latent heat of vaporization, corrected (or modified) $i'_{fg}$ | Defined by $i'_{fg} = i_{fg} + C_1 c_P (T_s - T_{\text{sat}}) = i_{fg}(1 + C_1\, Ja)$ where $Ja$ is the *Jakob number*. |
| Leidenfrost point | The point of minimum heat flux on the pool-boiling curve. |
| Maximum heat flux (or critical heat flux, peak heat flux) $q''_{\max}$ | The point on the pool-boiling curve at which the heat flux reaches a maximum. |
| Nucleate boiling | Boiling regime in which liquid is transformed into vapor nuclei (i.e., bubbles) at various preferential sites on the heating surface. |
| Surface tension $\sigma$ | The work done in extending the surface of a liquid one unit area. The surface tension of a fluid results from a net inward force of attraction on molecules at the interface toward molecules within the liquid phase. |
| Surface tension, critical $\sigma_{cr}$ | A characteristic surface tension of materials which forms the basis for the following surface wetting criterion: $\sigma < \sigma_{cr}$ liquid *will* wet surface; $\sigma > \sigma_{cr}$ liquid *will not* wet surface |
| Transition boiling | The region between the maximum and minimum heat fluxes in which a vapor film is intermittently built up and partially destroyed. This region is characterized by a decrease in heat flux $q''_c$ with increasing excess temperature. |

## Heat Exchangers

| | |
|---|---|
| Capacity rate $C$ | Defined by $$C = \dot{m}c_P$$ and usually subscripted to identify the fluid stream. |
| Component loss coefficient $K_f$ | Pertaining to pressure drop in return bends, headers and other components, defined by Eq. (11–41), $$\Delta P_f = K_f \frac{\rho U_b^2}{2}$$ |
| Design function, thermal and hydraulic | Determination of the surface area $A_s$ required to transfer a specified rate of heat with acceptable pressure drop for given fluids, mass flow rates, and terminal temperatures. |
| Evaluation (or rating) function, thermal and hydraulic | Determination of the total rate of heat transfer $q_c$, outlet temperatures, and pressure drop that can be produced by an existing system under given operating conditions. |
| Fouling | Deposits on a surface that retard the transfer of heat. |
| Fouling factor $F_f$ | Defined by Eq. (11–7), $$F_f = \frac{1}{h_f} - \frac{1}{h}$$ where $h_f$ represents the coefficient of heat transfer *after* fouling has occurred. |
| Heat exchangers | Devices that transfer heat between fluids at different temperatures. |
| Compact | Heat exchangers in which the *surface area density* $\beta$ is greater than about 700 $m^2/m^3$. |
| Direct contact | Heat exchangers in which two immiscible fluids are in direct contact. |
| Recouperators | Heat exchangers in which two fluids are separated by a thin wall through which heat is transferred. |
| Regenerators | Heat exchangers in which the same flow passages are alternately used by both fluids. |
| Mixed stream | Pertaining to crossflow heat exchangers, a flow stream which is unrestricted in the lateral direction (i.e., in the direction of the other stream). |
| Overall coefficient of heat transfer | |
| local $U$ | Defined by Eq. (11–2), $$dq = U\,dA_s\,(T_h - T_c)$$ |
| mean $\overline{U}$ | Defined by Eq. (11–5), $$\overline{U} = \frac{1}{A_s}\int_{A_s} U\,dA_s$$ |
| Unmixed stream | Pertaining to crossflow heat exchangers, a flow stream which is restricted in the lateral direction (i.e., in the direction of the other stream). |

# REFERENCES

## CHAPTER 1


1. Fourier, J. B., *Théorie analytique de la Chaleur*. Paris, 1822. English translation by A. Freeman, Dover Publications, Inc., New York, 1955.
2. Touloukian, Y. S., et al., *Thermophysical Properties of Matter*; 13 vols. plus index. New York: Plenum Publishing Co., 1970–1977.
3. Bolz, R. E., and G. L. Tuve, eds., *Handbook of Tables for Applied Engineering Science*, 2nd ed. Boca Raton, Fla.: CRC Press, 1973.
4. American Society of Heating, Refrigeration and Air Conditioning Engineers, *Handbook of Fundamentals*, Chaps. 17 and 31. New York: ASHRAE, 1985.
5. Powell, R. W., C. Y. Ho, and P. E. Liley, "Thermal Conductivity of Selected Materials," *NSRDS-NBS 8*. Washington, D.C.: U.S. Department of Commerce, National Bureau of Standards, 1966.
6. Ho, C. V., R. W. Powell, and P. E. Liley, *Thermal Conductivity of Elements*, Vol. 1, *First Supplement to Journal of Physical and Chemical Reference Data*. Washington, D.C.: American Chemical Society, 1972.
7. Eckert, E. R. G., and R. M. Drake, *Analysis of Heat and Mass Transfer*. New York: McGraw-Hill Book Company, 1972.
8. Kays, W. M., and M. E. Crawford, *Convective Heat and Mass Transfer*, 2nd ed. New York: McGraw-Hill Book Company, 1980.
9. Bird, R. B., W. E. Stewart, and E. N. Lightfoot, *Transport Phenomena*. New York: John Wiley & Sons, 1960.

## CHAPTER 2


1. Kreyszig, E., *Advanced Engineering Mathematics*, 4th ed. New York: John Wiley & Sons, 1979.
2. Barzelay, M. E., K. N. Tong, and G. F. Holloway, "Effect of Pressure on Thermal Conductance of Contact Joints," *NACA Tech. Note 3295*, May 1955.
3. Clausing, A. M., "Heat Transfer at the Interface of Dissimilar Metals—The Influence of Thermal Strain," *Int. J. Heat Mass Transfer*, 9, 1966, 791.
4. Moore, C. J., Jr., H. A. Blum, and H. Atkins, "Classification Bibliography for Thermal Contact Resistance Studies," *ASME Paper 68-WA/HT-18*, December 1968.
5. Arpaci, V. S., *Conduction Heat Tranfer*. Reading, Mass.: Addison-Wesley Publishing Co., 1966.
6. Schneider, P. J., *Conduction Heat Transfer*. Reading, Mass.: Addison-Wesley Publishing Co., 1955.
7. Gardner, K. A., "Efficiency of Extended Surfaces," *Trans. ASME*, 67, 1945, 621.
8. Sparrow, E. M., and D. K. Hennecke, "Temperature Depression at the Base of a Fin." *J. Heat Transfer*, 92, 1970, 204.


## CHAPTER 3


1. Kreyszig, E., *Advanced Engineering Mathematics*, 4th ed. New York: John Wiley & Sons, 1979.
2. Wylie, C. R., and L. C. Barrett, *Advanced Engineering Mathematics,* 5th ed. New York: McGraw-Hill Book Company, 1982.
3. Ozisik, M. N., *Heat Conduction*. New York: John Wiley & Sons, 1980.
4. Arpaci, V. S., *Conduction Heat Transfer*. Reading, Mass.: Addison-Wesley Publishing Co., 1966.
5. Carslaw, H. S., and J. C. Jaeger, *Conduction of Heat in Solids*. London: Oxford University Press, 1947.
6. Schneider, P. J., *Conduction Heat Transfer*. Reading, Mass.: Addison-Wesley Publishing Co., 1955.
7. Bewley, L. V., *Two-Dimensional Fields in Electrical Engineering*. New York: Macmillan Publishing Co., 1948.
8. Kreith, F., and W. Z. Black, *Basic Heat Transfer*. New York: Harper & Row, Publishers, 1980.
9. Langmuir, I., E. O. Adams, and F. A. Meikle, "Flow of Heat through Furnace Walls," *Trans. Am. Electrochem. Soc.*, 24, 1913, 53.
10. Rudenberg, R., "Die Ausbreitung der Luft-und Erdfelder um Hochspannungsleitungen besonders bei Erd-und Kurzschlüssen," *Electrotech. Z*, 46, 1945, 1342.
11. Andrews, R. V., "Solving Conductive Heat Transfer Problems with Electrical-Analogue Shape Factors," *Chem. Eng. Progr.*, 5, 1955, 67.

12. Hahne, E., and U. Grigull, "Formfaktor und Formwiderstand der stationären mehrdimensionalen Wärmeleitung," *Int. J. Heat Mass Transfer*, 18, 1975, 75.
13. American Society of Heating, Refrigeration and Air Conditioning Engineers, *Handbook of Fundamentals*. New York: ASHRAE, 1985.
14. Heisler, M. P., "Temperature Charts for Induction and Constant Temperature Heating," *Trans. ASME*, 69, 1947, 227.
15. Grober, H. S., and U. Grigull, *Fundamentals of Heat Transfer*. New York: McGraw-Hill Book Company, 1961.
16. Eckert, E. R. G., and R. M. Drake, *Analysis of Heat and Mass Transfer*. New York: McGraw-Hill Book Company, 1972.
17. Holman, J. P., *Heat Transfer*, 6th ed. New York: McGraw-Hill Book Company, 1986.
18. Kreith, F., and M. S. Bohn, *Principles of Heat Transfer*, 4th ed. New York: Harper & Row, Publishers, 1986.
19. Lienhard, J. H., *A Heat Transfer Textbook*, 2nd ed. Englewood Cliffs, N.J.: Prentice-Hall, 1987.
20. Janna, W. S., *Engineering Heat Transfer*. Boston: PWS Publishers, 1986.

## CHAPTER 4

1. *Handbook of Numerical Heat Transfer*, edited by W. J. Minkowycz, E. M. Sparrow, G. E. Schneider, and R. H. Pletcher. New York: John Wiley & Sons, 1988.
2. Desai, C. S., *Elementary Finite Element Method*. Englewood Cliffs, N.J.: Prentice-Hall, 1979.
3. Kreyszig, E., *Advanced Engineering Mathematics*, 4th ed. New York: John Wiley & Sons, 1979.
4. Patankar, S. V., *Numerical Heat Transfer and Fluid Flow*. New York: McGraw-Hill/Hemisphere, 1980.
5. Mitchell, A. R., and D. F. Griffith, *Finite Difference Methods in Partial Differential Equations*. New York: John Wiley & Sons, 1980.
6. Myers, G. E., *Analytical Methods in Conduction Heat Transfer*. New York: McGraw-Hill Book Company, 1971.
7. Fox, L., *Numerical Solution of Ordinary and Partial Differential Equations*. Reading, Mass.: Addison-Wesley Publishing Co., 1962.
8. Smith, G. D., *Numerical Solution of Partial Differential Equations with Exercises and Worked Solutions*. London: Oxford University Press, 1965.
9. Jacob, M., *Heat Transfer*. New York: John Wiley & Sons, 1949.
10. Baliga, B. R., and S. V. Patankar, "Elliptic Systems: Finite Element Method II," Chap. 11 in *Handbook of Numerical Heat Transfer*. New York: John Wiley & Sons, 1988.
11. Holman, J. P., *Heat Transfer*, 6th ed. New York: McGraw-Hill Book Company, 1986.
12. Incropera, F. P., and D. P. DeWitt, *Fundamentals of Heat Transfer*, 2nd ed. New York: John Wiley & Sons, 1985.
13. Kreith, F., and W. Z. Black, *Basic Heat Transfer*. New York: Harper & Row, Publishers, 1986.

## CHAPTER 5


1. Touloukian, Y. S., and D. P. DeWitt, *Thermophysical Properties of Matter*, Vol. 7, *Thermal Radiative Properties—Metallic Elements and Alloys*. New York: Plenum Publishing Co., 1970.
2. Touloukian, Y. S., and D. P. DeWitt, *Thermophysical Properties of Matter*, Vol. 8, *Thermal Radiative Properties—Nonmetallic Solids*. New York: Plenum Publishing Co., 1970.
3. Touloukian, Y. S., D. P. De Witt, and R. S. Hernicz, *Thermophysical Properties of Matter*, Vol. 9, *Thermal Radiative Properties—Coatings*. New York: Plenum Publishing Co., 1970.
4. Gubareff, G. G., J. E. Jansen, and R. H. Torborg, *Thermal Radiation Properties Survey*, Honeywell Research Center, Honeywell Regulator Company, Minneapolis, Minn., 1960.
5. Sparrow, E. M., and R. D. Cess, *Radiation Heat Transfer*. Washington, D.C.: Hemisphere Publishing Co., 1978.
6. Planck, M., *The Theory of Heat Radiation*. New York: Dover Publications, Inc., 1959.
7. Dunkle, R. V., J. T. Gier, and co-workers, "Snow Characteristics Project, Progress Report," University of California, Berkeley, 1953.
8. Seban, R. A., "The Emissivity of Transition Metals in the Infrared," *J. Heat Transfer*, C87, 1965, 173.
9. Sieber, W., "Zusammensetzung der von Werk- und Baustoffen zurückgeworfenen Wärmestrahlung," *Z. Tech. Physik*, 22, 1941, 130.
10. Schmidt, E., and E. Eckert, "Über die Richtungsverteilung der Wärmestrahlung," *Forsch. Ing. Wes.*, 6, 1935, 175.
11. Siegel, R., and J. R. Howell, *Thermal Radiation Heat Transfer*, 2nd ed. New York: McGraw-Hill Book Company, 1981.
12. Dietz, A. G. H., "Diathermanous Materials and Properties of Surfaces," in *Space Heating with Solar Energy* by R. W. Hamilton. Cambridge, Mass.: The MIT Press, 1954.
13. Seal, M., "The Increasing Applications of Diamond as an Optical Material and in the Electronics Industry," *Ind. Diamond Rev.*, April 1978, 130.
14. Williams, D. A., T. A. Lappin, and J. A. Duffie, "Selective Radiation Properties of Particular Coatings," *J. Engr. Power*, 95A, 1963, 213.
15. Edwards, D. K., "Radiation Interchange in a Nongray Enclosure Containing an Isothermal Carbon-dioxide–Nitrogen Gas Mixture," *J. Heat Transfer*, c84, 1962, 1.
16. Hottel, H. C., and R. S. Egbert, "Radiant Heat Transmission from Water Vapor," *AIChE Trans.*, 38, 1942.
17. Hottel, H. C., "Radiant Heat Transmission," Chap. 4 in *Heat Transmission*, 3rd ed., by W. H. McAdams. New York: McGraw-Hill Book Company, 1954.
18. Eckert, E. R. G., and R. M. Drake, *Analysis of Heat and Mass Transfer*. New York: McGraw-Hill Book Company, 1972.
19. Mackey, C. O., L. T. Wright, Jr., R. E. Clark, and N. R. Gay, "Radiant Heating and Cooling, Part *I*," *Cornell Univ. Eng. Expt. Sta. Bull.*, 32, 1943.
20. Hottel, H. C., "Radiant Heat Transmission," *Mech. Eng.*, 52, 1930, 699.
21. Howell, J. R., *A Catalog of Radiation Configuration Factors*. New York: McGraw-Hill Book Company, 1982.
22. Hamilton, D. C., and W. R. Morgan, "Radiant Interchange Configuration Factors," *NACA TN 2836*, 1952.

23. Kreyszig, E., *Advanced Engineering Mathematics*, 4th ed. New York: John Wiley & Sons, 1979.
24. Gebhart, B., *Heat Transfer*, 2nd ed. New York: McGraw-Hill Book Company, 1971.
25. Duffie, J. A., and W. A. Beckman, *Solar Energy Thermal Processes*. New York: John Wiley & Sons, 1974.
26. Thekaekara, M. P., "Data on Incident Solar Radiation," *Suppl. Proc. 20th Ann. Meeting Inst. Environ. Sci.*, 21, 1974.
27. Daniels, F., Jr., "Physical Factors in Sun Exposure," *Arch. Dermatol.*, 85, 1962, 358.
28. Thekaekara, M. P., and A. J. Drummond, "Standard Values for the Solar Constant and Its Special Components," *Nat. Phys. Sci.*, 229, 1971, 6.
29. Anderson, E. E., *Solar Energy Fundamentals for Designers and Engineers*. Reading, Mass: Addison-Wesley Publishing Co., 1982.
30. Howell, J. R., R. B. Bannerot, and G. C. Vliet, *Solar-Thermal Energy Systems Analysis and Design*. New York: McGraw-Hill Book Company, 1982.

## CHAPTER 6

1. Keenan, J. H., F. G. Keyes, P. G. Hill, and J. G. Moore, *Steam Tables*. New York: John Wiley & Sons, 1978.
2. Prandtl, L., "Über Flüssigkeitsbewegung bei sehr kleiner Reibung," *Proc. 3rd Int. Math. Kong. Heidelberg*, 1904.
3. White, F. M., *Heat Transfer*, Reading, Mass.: Addison-Wesley Publishing Co., 1984.
4. Adiutori, E. F. "Origins of the Heat Transfer Coefficient," *Mechanical Engineering*, 112, 1990, 46.
5. Buckingham, E., "On Physically Similar Systems: Illustrations of the Use of Dimensional Analysis," *Phys. Rev.*, 4, 1914, 345.
6. Bridgeman, P. W., *Dimensional Analysis*. New Haven, Conn.: Yale University Press, 1931.
7. Langhaar, H. L., *Dimensional Analysis and Theory of Models*. New York: John Wiley & Sons, 1951.
8. Kreith, F., and M. S. Bohn, *Principles of Heat Transfer*, 4th ed. New York: Harper & Row, Publishers, 1986.
9. Lienhard, J. H., *A Heat Transfer Textbook*, 2nd ed. Englewood, Cliffs, N.J.: Prentice Hall, 1987.
10. Fox, R. W., and A. T. McDonald, *Introduction to Fluid Mechanics*. New York: John Wiley & Sons, 1978.
11. Arpaci, V. S., and P. S. Larsen, *Convection Heat Transfer*. Englewood Cliffs, N.J.: Prentice Hall, 1984.
12. John, E. A., *Gas Dynamics*. Englewood Cliffs, N.J.: Prentice Hall, 1984.
13. Benedict, R. P., *Fundamentals of Gas Dynamics*. New York: John Wiley & Sons, 1983.

## CHAPTER 7

1. Graetz, L., "Über die Wärmeleitfähigkeit von Flüssigkeiten," *Ann. Phys. Chem.*, 25, 1885, 337.

2. Hansen, M., "Velocity Distribution in the Boundary Layer of a Submerged Plate," *NACA TM* 585, 1930.
3. Langhaar, H. L., "Steady Flow in the Transition Length of a Straight Tube," *J. Appl. Mech.*, 9, 1942, A55.
4. Kays, W. M., "Numerical Solution for Laminar Flow Heat Transfer in Circular Tubes," *Trans. ASME*, 77, 1955, 1265.
5. Sellars, J. R., M. Tribus, and J. S. Klein, "Heat Transfer to Laminar Flows in a Round Tube or Flat Conduit, the Graetz Problem Extended," *Trans. ASME*, 78, 1956, 441.
6. Seigel, R., E. M. Sparrow, and T. M. Hallman, "Steady Laminar Heat Transfer in Circular Tube with Prescribed Wall Heat Flux," *Appl. Sci. Res.*, A7, 1958, 386.
7. Goldberg, P., "A Digital Computer Solution for Laminar Flow Heat Transfer in Circular Tubes," M. S. Thesis, Mechanical Engineering Department, Massachusetts Institute of Technology, 1958.
8. Heaton, H. S., W. C. Reynolds, and W. M. Kays, "Heat Transfer in Annular Passages, Simultaneous Development of Velocity and Temperature Fields in Laminar Flow," *Int. J. Heat Mass Transfer*, 7, 1964, 763.
9. Sparrow, E. M., S. H. Lin, and T. Lundren, "Flow Development in the Hydrodynamic Entrance Region of Tubes and Ducts," *Phys. Fluids*, 7, 1964, 338.
10. McComas, S. T., and E. R. G. Eckert, "Laminar Pressure Drop Associated with the Continuum Entrance Region and for Slip Flow in a Circular Tube," *J. Appl. Mech.*, 32, 1965, 765.
11. Kays, W. M., and M. E. Crawford, *Convection Heat and Mass Transfer*, 2nd ed. New York: McGraw-Hill Book Company, 1980.
12. *Handbook of Heat Transfer Fundamentals*, 2nd ed., edited by W. M. Rohsenow, J. P. Hartnett, and E. N. Ganic. New York: McGraw-Hill Book Company, 1985.
13. Senecal, V. E., "Characteristics of Transition Flow in Smooth Tubes," Ph.D. Thesis, Carnegie Institute of Technology, 1952.
14. Stanton, T. E., and J. R. Pannell, "Similarity of Motion in Relation to the Surface Friction of Fluids," *Trans. Roy. Soc. (London)*, A214, 1914, 199.
15. Senecal, V. E., and R. R. Rothfus, "Transition Flow of Fluids in Smooth Tubes," *Chem. Eng. Progr.*, 49, 1953, 533.
16. White, F. M., *Viscous Fluid Flow*. New York: McGraw-Hill Book Company, 1974.
17. Burmeister, L. C., *Convective Heat Transfer*. New York: John Wiley & Sons, 1983.
18. Schlichting, H., *Boundary Layer Theory*, 7th ed. New York: McGraw-Hill Book Company, 1979.
19. Patankar, S. V., *Numerical Heat Transfer and Fluid Flow*. Washington, D.C.: Hemisphere Publishing Co., 1980.
20. *Handbook of Numerical Heat Transfer*, edited by W. J. Minkowycz, E. M. Sparrow, G. E. Schneider, and R. H. Pletcher. New York: John Wiley & Sons, 1988.
21. Anderson, D. A., J. C. Tannehill, and R. H. Pletcher, *Computational Fluid Mechanics and Heat Transfer*. New York: Hemisphere Publishing Corp., 1984.
22. Thomas, L. C., and W. L. Amminger, "A Practical One-Parameter Integral Method for Laminar Incompressible Boundary Layer Flow with Transpiration," *J. Appl. Mech.*, 110, 1988, 474.
23. Thomas, L. C., and W. L. Amminger, "A Two-Parameter Integral Method for Laminar Transpired Thermal Boundary Layer Flow," *AIAA J.*, 28, 1990, 205.

24. Dorodnitsyn, A. A., *Advances in Aeronautical Sciences*, Vol. 3, New York: Pergamon Press, 1960.
25. Holt, M., *Numerical Methods in Fluid Dynamics*. Berlin: Springer-Verlag, 1984.
26. Fletcher, C. A. J., *Computational Galerkin Methods*. Berlin: Springer-Verlag, 1984.
27. Blasius, H., "Grenzschichten in Flüssigkeiten mit kleiner Reibung," *Z. Math. Phys.* 56, 1908, 1; English translation in *NACA TM* 1256.
28. Levy, S., "Heat Transfer to Constant-Property Laminar-Boundary Layer-Flows with Power-Function Free-Stream Velocity and Wall Temperature Variation," *J. Aeronaut. Sci.*, 19, 1952, 341.
29. Gebhart, B., Y. Jaluria, R. L. Mahajan, and B. Sammakia, *Buoyancy-Induced Flows and Transport*. Washington, D.C.: Hemisphere Publishing Co., 1988.
30. Boussinesq, J., *Théorie Analytique de la Chaleur*, Vol. 2. Paris: Gauthier-Villars, 1903.
31. Ostrach, S., "An Analysis of Laminar Free-Convection Flow and Heat Transfer about a Plate Parallel to the Direction of the Generating Body Force," *NACA Rept.* 1111, 1953.
32. Schmidt, E., and W. Beckman, "Das Temperatur-und Geschwindig keitsfeld von einer Wärme-abgebenden, senkrechten Platte bei natürlicher Konvektion, *Forsch-Ing. Wes*, 1, 1930, 391.
33. Ede, A. J., "Advances in Free Convection," in *Advances in Heat Transfer*, 4, 1967, 1.
34. Runstadler, P. W., S. J. Kline, and W. C. Reynolds, "An Experimental Investigation of the Flow Structure of the Turbulent Boundary Layer," *Rept. MD-8*, Stanford University, 1963.
35. Knudsen, J. D., and D. L. Katz, *Fluid Dynamics and Heat Transfer*. New York: McGraw-Hill Book Company, 1958.
36. Reynolds, O., *Scientific Papers of Osborne Reynolds*, Vol. II. London: Cambridge University Press, 1901.
37. Boussinesq, J., "Théorie de l'écoulement tourbillant," *Mem. Pres. Acad. Sci.* (Paris), 23, 1877, 46.
38. Hussain, A. K. M. F., and W. C. Reynolds, "Measurements in Fully Developed Turbulent Channel Flow," *J. Fluids Eng.*, 97, 1975, 569.
39. Prandtl, L., "Eine Beziehung zwischen Wärmeaustausch and Strömungswiderstand der Flüssigkeiten," *Phys. Zeit.*, 11, 1910, 1072.
40. van Driest, E. R., "Turbulent Boundary Layer in Compressible Fluids," *J. Aero. Sci.*, 18, 1951, 145.
41. Kays, W. M., and R. J. Moffat, *Studies in Convection*, Vol. 1, London: Academic Press, 1975, 213.
42. Cebeci, T., and A. M. O. Smith, *Analysis of Turbulent Boundary Layers*. New York: Academic Press, 1974.
43. Rotta, J., "Das in Wandnähe gültige Geschwindigkeitsgesetz turbulenter Strömungen," *Ing. Arch.*, 18, 1950, 277.
44. Reichardt, H., "Vollständige Darstellung der turbulenten Geschwindigkeitsverteilung in glatten Leitungen," *ZAMM*, 31, 1951, 208.
45. Deissler, R. G., "Analysis of Turbulent Heat Transfer, Mass Transfer, and Friction in Smooth Tubes at High Prandtl Numbers," *NACA TN* 3145, 1954.
46. Spalding, D. B., "A Single Formula for the Law of the Wall," *J. Appl. Mech.*, 28, 1961, 455.
47. Danckwerts, P. V., "Significance of Liquid-Film Coefficients in Gas Absorption," *Ind. Eng. Chem.*, 43, 1951, 1460.

48. Einstein, H. A., and H. L. Li, "The Viscous Sublayer along a Smooth Boundary," *ASCE, J. Mech. Div.*, 82, 1956, 293.
49. Hanratty, T. J., "Turbulent Exchange of Mass and Momentum with a Boundary," *AIChE J.* 2, 1956, 1.
50. Thomas, L. C., "The Surface Rejuvenation Model of Wall Turbulence: Inner Laws for $u^+$ and $T^+$," *Int. J. Heat Mass Transfer*, 23, 1980, 1097.
51. Kline, S. J., D. J. Cockrell, M. V. Morkovin, and G. Sovran, *Proc. Comput. Turbulent Boundary Layer*, Vol. I. Department of Mechanical Engineering, Stanford University, 1968.
52. Bradshaw, P., T. Cebeci, and J. Whitelaw, *Engineering Calculational Methods for Turbulent Flow*. New York: Academic Press, 1981.
53. Bradshaw, P., "Compressible Turbulent Shear Layers," in *Annual Review of Fluid Mechanics*, Vol. 9. Palo Alto, Calif.: Annual Reviews, Inc., 1977.
54. Simpson, R. L., D. G. Whitten, and R. J. Moffat, "An Experimental Study of the Turbulent Prandtl Number of Air with Injection and Suction," *Int. J. Heat Mass Transfer*, 13, 1970, 125.
55. Azer, N. Z., and B. T. Chao, "Turbulent Heat Transfer in Liquid Metals—Fully Developed Pipe Flow with Constant Wall Temperature," *Int. J. Heat Mass Transfer*, 3, 1961, 77.
56. Jenkins, R., "Variation of Eddy Conductivity with Prandtl Modulus and Its Use in Predictions of Turbulent Heat Transfer Coefficients," *Heat Transfer and Fluid Mechanics Institute*, Stanford University, 1951.
57. Thomas, L. C., and S. M. F. Hasani, "Supplementary Boundary Layer Approximations for Turbulent Flows," *J. Fluids Eng.*, 111, 1989, 420.
58. Sams, E. W., and L. G. Desmon, "Heat Transfer from High Temperature Surfaces to Fluids," *NACA Memo* E9D12, 1949.
59. Deissler, R. G., and C. S. Eian, "Analytical and Experimental Investigation of Heat Transfer with Variable Fluid Properties," *NACA TN* 2629, 1952.
60. Barnes, J. F., and J. D. Jackson, "Heat Transfer to Air, Carbon Dioxide and Helium Flowing through Smooth Circular Tubes under Conditions of Large Surface/Gas Temperature Ratio," *J. Mech. Eng. Sci.*, 3, 1961, 303.
61. Petukhov, B. S., and V. V. Kirillov, "Heat Exchange for Turbulent Flow of Liquid in Tubes," *Teploenerg.*, 4, 1958.
62. Reynolds, W. C., W. M. Kays, and S. J. Kline, "Heat Transfer in the Turbulent Incompressible Boundary Layer—III: Arbitrary Wall Temperature and Heat Flux," *NASA Memo* 12-3-58W, 1958.
63. Colburn, A. P., "A Method of Correlating Forced Convection Heat Transfer Data and Comparison with Fluid Friction," *Trans. AIChE*, 29, 1933, 1974.
64. Rothfus, R. R., "Velocity Distribution and Fluid Friction in Concentric Annuli," PhD Thesis, Carnegie Institute of Technology, 1948.

## CHAPTER 8

1. Langhaar, H. L., "Steady Flow in the Transition Length of a Straight Tube," *J. Appl. Mech.*, 9, 1942, A55.
2. *Engineering Sciences Data*, London: Heat Transfer Subsciences, Technical Editing and Production Ltd., 1970.

3. *Handbook of Heat Transfer*, edited by W. M. Rohsenow and J. P. Hartnet. New York: McGraw-Hill Book Company, 1973.
4. Kays, W. M., and A. L. London, *Compact Heat Exchangers*, 3rd ed. New York: McGraw-Hill Book Company, 1984.
5. *Heat Exchanger Design Handbook*, Washington, D.C.: Hemisphere Publishing Co., 1983.
6. Kays, W. M., and M. E. Crawford, *Convective Heat and Mass Transfer*, 2nd ed. New York: McGraw-Hill Book Company, 1980.
7. Kays, W. M., and S. H. Clark, TR No. 17, Mechanical Engineering Department, Stanford University, 1953.
8. Lundberg, R. E., W. C. Reynolds, and W. M. Kays, ''Heat Transfer with Laminar Flow in Concentric Annuli with Constant and Variable Wall Temperature and Heat Flux,'' *NASA TN*-1972, Washington, D.C., 1963.
9. Lundgren, T. S., E. M. Sparrow, and J. B. Starr, ''Pressure Drop Due to the Entrance Region in Ducts of Arbitrary Cross-Section,'' *J. Basic Eng.*, 86, 1964, 620.
10. Sparrow, E. M., T. S. Chen, and V. K. Johnson, ''Laminar Flow and Pressure Drop in Internally Finned Annular Ducts,'' *Int. J. Heat Mass Transfer*, 7, 1964, 583.
11. Sparrow, E. M., and A. Haji-Sheikh, ''Flow and Heat Transfer in Ducts of Arbitrary Shape with Arbitrary Thermal Boundary Conditions,'' *J. Heat Transfer*, 88, 1966, 351.
12. McComas, S. T., and E. R. G. Eckert, ''Laminar Pressure Drop Associated with the Continuum Entrance Region and for Slip Flow in a Circular Tube,'' *J. Appl. Mech.*, 32, 1965, 765.
13. Burmeister, L. C., *Convective Heat Transfer*. New York: John Wiley & Sons, 1983.
14. Kays, W. M., ''Loss Coefficients for Abrupt Changes in Flow Cross Section with Low Reynolds Number Flow in Single and Multiple Tube Systems,'' *Trans. ASME*, 72, 1950, 91.
15. Hausen, H., ''Darstellung des Wärmeüberganges in Rohren durch verallgemeinerte Potenzbeziehungen,'' *Z. Ver. Deut. Ing.*, 4, 1943, 91.
16. Sellars, J. R., M. Tribus, and J. S. Klein, ''Heat Transfer to Laminar Flow in a Round or Flat Conduit—The Graetz Problem Extended,'' *Trans. ASME*, 78, 1956, 441.
17. Kays, W. M., ''Numerical Solutions for Laminar-Flow Heat Transfer in Circular Tubes,'' *Trans. ASME*, 77, 1955, 1265.
18. Seider, E. M., and C. E. Tate, ''Heat Transfer and Pressure Drop of Liquids in Tubes,'' *Ind. Eng. Chem.*, 28, 1936, 1429.
19. Shah, R. K., and M. S. Bhatti, ''Assessment of Test Techniques and Correlations for Single-Phase Heat Exchangers,'' 6th NATO Advanced Study Institute, Thermal/Hydraulic Fundamentals and Design of Two-Phase Heat Exchangers, Povadevarzin, Portugal, July 1987.
20. Nikuradse, J., ''Wärmeübergang in Rohrleitungen,'' *Forsch. Ing. Wes.*, 1932, 356.
21. Petukhov, B. S., and V. N. Popov, *Teplofiz. Vysok. Temperatur* (High Temperature Heat Physics), 1, 1963.
22. Petukhov, B. S., and V. V. Kirillov, ''Heat Exchange for Turbulent Flow of Liquid in Tubes,'' *Teploenerg.*, 4, 1958.
23. White, F. M., *Viscous Fluid Flow*. New York: McGraw-Hill Book Company, 1974.
24. Petukhov, B. S., ''Heat Transfer and Friction in Turbulent Pipe Flow with Variable Physical Properties,'' in *Advances in Heat Transfer*. New York: Academic Press, 1970, 504.

25. Barnes, J. F., and J. D. Jackson, ''Heat Transfer to Air, Carbon Dioxide and Helium Flowing through Smooth Circular Tubes under Conditions of Large Surface/Gas Temperature Ratio,'' *J. Mech. Eng. Sci.*, 3, 1961, 303.
26. Deissler, R. G., and C. S. Eian, ''Analytical and Experimental Investigation of Heat Transfer with Variable Fluid Properties,'' *NACA TN* 2629, 1952.
27. Sams, E. W., and L. G. Desmon, ''Heat Transfer from High Temperature Surface to Fluids,'' *NACA Memo E9D12*, 1949.
28. Dittus, F. W., and L. M. K. Boelter, ''Heat Transfer in Automobile Radiators of the Tubular Type,'' Univ. of California–Berkeley, *Pub. Eng.* 2, 1930, 443.
29. Colburn, A. P., ''A Method of Correlating Forced Convection Heat Transfer Data and a Comparison with Fluid Friction,'' *Trans. AIChE*, 29, 1933, 1974.
30. Kays, W. M., and E. Y. Leung, ''Heat Transfer in Annular Passages: Hydrodynamically Developed Turbulent Flow with Arbitrarily Prescribed Heat Flux,'' *Int. J. Heat Mass Transfer*, 6, 1963, 537.
31. Patel, V. C., and M. R. Head, ''Some Observations on Skin Friction and Velocity Profiles in Fully Developed Pipe and Channel Flow,'' *J. Fluid Mech.*, 38, 1969, 181.
32. Lawn, C. J., ''Turbulent Heat Transfer at Low Reynolds Numbers,'' *J. Heat Transfer*, 91, 1969, 532.
33. Subbotin, V. I., A. K. Papovyants, P. L. Kirillov, and N. N. Ivanovskii, ''A Study of Heat Transfer to Molten Sodium in Tubes,'' *Soviet J. Atomic Energy*, 13, 1962, 380.
34. Seban, R. A., and T. T. Shimazaki, ''Heat Transfer to Fluid Flowing Turbulently in a Smooth Pipe with Walls at Constant Temperature,'' *Trans. ASME*, 73, 1951, 803.
35. Skupinski, E., J. Tortel, and L. Vautrey, ''Determination des Coefficients de Convection d'un Alliage Sodium-Potassium dans un Tube Circulative,'' *Int. J. Heat Mass Transfer*, 8, 1965, 937.
36. Lubarsky, B., and S. J. Kaufman, ''Review of Experimental Investigations of Liquid-Metal Heat Transfer,'' *NASA TN* 3336, 1955.
37. Allen, R. W., and E. R. G. Eckert, ''Friction and Heat-Transfer Measurements to Turbulent Pipe Flow of Water ($Pr = 7$ and 8) at Uniform Wall–Heat Flux,''*J. Heat Transfer*, 86, 1964, 301.
38. Sleicher, C. A., and M. W. Rouse, ''A Convenient Correlation for Heat Transfer to Constant and Variable Property Fluids in Turbulent Pipe Flow,'' *Int. J. Heat Mass Transfer*, 18, 1975, 677.
39. Bergles, A. E., ''Enhancement of Heat Transfer,'' *Sixth Int. Heat Transfer Conf.*, Toronto, 6, 1978, 89.
40. Schlager, L. M., M. B. Pate, and A. E. Bergles, ''A Survey of Refrigerant Heat Transfer and Pressure Drop Emphasizing Oil Effects and In-Tube Augmentation, *ASHRAE Trans.*, 93, 1987.
41. Bergles, A. E., V. Nirmalan, G. H. Junkhan, and R. L. Webb, Bibliography on Augmentation of Convection Heat and Mass Transfer—II, Rept. HTL-31, ISU-ERI-Ames—84221, Iowa State University, Ames, Iowa, 1983.
42. Bergles, A. E., ''Techniques to Augment Heat Transfer,'' in *Handbook of Heat Transfer*. New York: McGraw-Hill Book Company, 1973.
43. Moody, L. F., ''Friction Factors for Pipe Flow,'' *Trans. ASME*, 66, 1944, 671.

44. Shah, R. K., "Classification of Heat Exchangers," in *Heat Exchangers—Thermal-Hydraulic Fundamentals and Design*. New York: McGraw-Hill Book Company, 1981.
45. Blasius, H., "Das Ähnlichkeitsgesetz bei Reibungsvorgängen in Flüssigkeiten," *Forsch. Ing. Wes.*, 131, 1913.
46. Liepmann, H. W., and S. Dhawan, "Direct Measurements of Local Skin Friction in Low-Speed and High-Speed Flow," *Proc. First U.S. Nat. Cong. Appl. Mech.*, 1951.
47. Wieghardt, K., and W. Tillmann, "On the Turbulent Friction Layer for Rising Pressure," *NACA TM* 1314, 1951.
48. Eckert, E. R. G., "Engineering Relations for Friction and Heat Transfer to Surfaces in High Velocity Flow," *J. Aero. Sci.*, 22, 1955, 585.
49. Schlichting, H., *Boundary Layer Theory*, 7th ed. New York: McGraw-Hill Book Company, 1979.
50. Whitaker, S., *Elementary Heat Transfer Analysis*. New York: Pergamon Press, 1976.
51. Whitaker, S., "Forced Convection Heat Transfer Correlations for Flow in Pipes, Past Flat Plates, Single Cylinders, Single Spheres and for Flow in Packed Beds and Tube Bundles," *AIChE J.*, 18, 1972, 361.
52. Achenbach, E., "Heat Transfer from Spheres up to Re $= 6 \times 10^6$," *Sixth Int. Heat Transfer Conf.*, Toronto, 5, 1978, 341.
53. Ishiguro, R., K. Sugiyama, and T. Kumada, "Heat Transfer around a Circular Cylinder in a Liquid-Sodium Crossflow," *Int. J. Heat Mass Transfer*, 22, 1979, 1041.
54. Witte, L. C., "An Experimental Study of Forced-Convection Heat Transfer from a Sphere to Liquid Sodium," *J. Heat Transfer*, 90, 1968, 9.
55. Jacob, M., *Heat Transfer*, Vol. 1. New York: John Wiley & Sons, 1949.
56. Grimison, E. D., "Correlation and Utilization of New Data on Flow of Gases over Tube Banks," *Trans. ASME*, 59, 1937, 538.
57. Zukauskas, A., "Heat Transfer in Banks of Tubes in Crossflow of Fluid," *'Mintis' Vilnius*, 1968, 124.
58. Zukauskas, A., "Convective Heat Transfer in Crossflow," in *Handbook of Single-Phase Convection Heat Transfer*, edited by S. Kakac, R. K. Shah, and W. Aung. New York: John Wiley & Sons, 1987.
59. Bergelin, O. P., G. A. Brown, and S. C. Doberstein, "Heat Transfer and Fluid Friction during Flow across Banks of Tubes," *Trans. ASME*, 74, 1952.
60. *Steam—Its Generation and Use*, 38th ed. New York: The Babcock and Wilcox Company, 1975.
61. Jacob, M., "Heat Transfer and Flow Resistance in Crossflow of Gases Over Tube Banks," *Trans. ASME*, 60, 1938, 384.
62. Gaddis, E. S., and V. Gnielinski, "Pressure Drop in Crossflow Across Tube Bundles," *Int. Chem. Eng.*, 25, 1985, 1.
63. Gnielinski, V., A. Zukauskas, and A. Shrinkska, "Banks of Plane and Finned Tubes," in *Heat Exchanger Design Handbook*. Washington, D.C.: Hemisphere Publishing Co., 1983.
64. *Heat Exchangers–Thermal-Hydraulic Fundamentals and Design*. New York: McGraw-Hill Book Company, 1981.
65. DeLorenzo, B., and E. D. Anderson, "Heat Transfer and Pressure Drop of Liquids in Double-Pipe Fin-Tube Heat Exchangers," *Trans. ASME*, 67, 1945, 697.

66. Churchill, S. W., and H. Ozoe, "Correlations for Laminar Forced Convection in Flow over an Isothermal Flat Plate and in Developing and Fully Developed Flow in an Isothermal Tube," *J. Heat Transfer*, 95, 1973, 46.

67. Churchill, S. W., and M. Bernstein, "A Correlating Equation for Forced Convection from Gases and Liquids to a Circular Cylinder in Crossflow," *J. Heat Transfer*, 99, 1977, 300.

## CHAPTER 9

1. Ostrach, S., "An Analysis of Laminar Free-Convection Flow and Heat Transfer about a Flat Plate Parallel to the Direction of the Generating Body Force," *NACA Rept.*, 1953, 1111.
2. Ede, A. J., in *Advances in Heat Transfer*, Vol. 4. New York: Academic Press, 1967, 1.
3. Churchill, S. W., and R. Usagi, "A General Expression for the Correlation of Rates of Transfer and Other Phenomena," *AIChE J.*, 18, 1972, 1121.
4. Eckert, E. R. G., and T. W. Jackson, "Analysis of Turbulent Free-Convection Boundary Layer on a Flat Plate," *NACA Rept.* 1015, 1951.
5. Churchill, S. W., and H. H. S. Chu, "Correlating Equations for Laminar and Turbulent Free Convection from a Vertical Plate," *Int. J. Heat Mass Transfer*, 18, 1975, 1323.
6. Bayley, F. J., J. M. Owen, and A. B. Turner, *Heat Transfer*. London: Thomas Nelson & Sons Ltd., 1972.
7. McAdams, W. H., *Heat Transmission*, 3rd ed. New York: McGraw-Hill Book Company, 1954.
8. Fujii, T., and M. Fujii, The Dependence of Local Nusselt Number on Prandtl Number in the Case of Free Convection along a Vertical Surface with Uniform Heat Flux," *Int. J. Heat Mass Transfer*, 19, 1976, 121.
9. Churchill, S. W., and H. Ozoe, "A Correlation for Laminar Free Convection from a Vertical Plate," *J. Heat Transfer*, 95, 1973, 540.
10. Vliet, G. C., and C. K. Liu, "An Experimental Study of Turbulent Natural Convection Boundary Layers," *J. Heat Transfer*, 91, 1969, 517.
11. Churchill, S. W., "Free Convection around Immersed Bodies," in *Heat Exchanger Design Handbook*. Washington, D.C.: Hemisphere Publishing Co., 1983.
12. Cebeci, T., Laminar-Free-Convection Heat Transfer from the Outer Surface of a Vertical Slender Circular Cylinder," *Proc. Fifth Int. Heat Transfer Conf.*, NC 1.4, 1974, 15.
13. Minkowycz, W. J., and E. M. Sparrow, "Local Nonsimilar Solutions for Natural Convection on a Vertical Cylinder," *J. Heat Transfer*, 96, 1974, 178.
14. Fuji, I., and H. Imura, "Natural Convection Heat Transfer from a Plate with Arbitrary Inclination," *Int. J. Heat Mass Transfer*, 15, 1972, 755.
15. Gebhart, B., *Heat Transfer*. New York: McGraw-Hill Book Company, 1971.
16. Gebhart, B., "Natural Convection Flows and Stability," in *Advances in Heat Transfer*. New York: Academic Press, 1973.
17. Gebhart, B., Y. Jaluria, R. L. Mahajan, and B. Sammakia, *Buoyancy-Induced Flows and Transport*. Washington, D.C.: Hemisphere Publishing Co., 1988.
18. Goldstein, R. J., E. M. Sparrow, and D. C. Jones, "Natural Convection Mass Transfer Adjacent to Horizontal Plates," *Int. J. Heat Mass Transfer*, 16, 1973, 1025.

19. Lloyd, J. R., and W. R. Moran, "Natural Convection Adjacent to Horizontal Surface of Various Planforms," *ASME Paper No. 74-WA/HT-66*, 1974.

20. Tabor, H., "Radiation, Convection and Conduction Coefficients in Solar Collection," *Bull. Res. Council Israel*, 6C, 1958, 155.

21. Dropkin, D., and E. Somerscales, "Heat Transfer by Natural Convection in Liquids Confined by Two Parallel Plates which are Inclined at Various Angles with Respect to the Horizontal," *J. Heat Transfer*, 87, 1965, 77.

22. Raithby, G. D., and K. G. T. Hollands, "A General Method of Obtaining Approximate Solutions to Laminar and Turbulent Free Convection Problems," in *Advances in Heat Transfer*, Vol. 11. New York: Academic Press, 1975, 265.

23. Kuehn, T. H., and R. J. Goldstein, "Correlating Equations for Natural Convection Heat Transfer between Horizontal Circular Cylinders," *Int. J. Heat Mass Transfer*, 19, 1976, 1127.

24. Jakob, M., "Free Convection through Enclosed Plane Gas Layers," *Trans. ASME*, 68, 1946, 189.

25. Jakob, M., *Heat Transfer*, Vol. 1. New York: John Wiley & Sons, 1949.

26. Graff, J. G. A., and E. F. M. Van der Held, "The Relation between the Heat Transfer and Convection Phenomena in Enclosed Plane Air Layers," *Appl. Sci. Res.*, 3, 1952, 393.

27. MacGregor, R. K., and A. P. Emery, "Free Convection through Vertical Plane Layers: Moderate and High Prandtl Number Fluids," *J. Heat Transfer*, 91, 1969, 391.

28. Emery, A., and N. C. Chu, "Heat Transfer across Vertical Layers," *J. Heat Transfer*, 87, 1965, 110.

29. Catton, I., "Natural Convection in Enclosures," *Sixth Int. Heat Transfer Conf.*, Toronto, 6, 1978, 13.

30. Weber, N., R. E. Rowe, E. H. Bishop, and J. A. Scanlan, "Heat Transfer by Natural Convection between Vertically Eccentric Spheres," *ASME Paper 72-WA/HT-2*, 1972.

31. Scanlan, J. A., E. H. Bishop, and R. E. Rowe, "Natural Convection Heat Transfer between Concentric Spheres," *Int. J. Heat Mass Transfer*, 13, 1970, 1857.

32. O'Toole, J., and P. L. Silveston, "Correlation of Convective Heat Transfer in Confined Horizontal Layers," *Chem. Eng. Progr. Symp.*, 57, 1961, 81.

33. Goldstein, R. J., and T. Y. Chu, "Thermal Convection in a Horizontal Layer of Air," *Progr. Heat Mass Transfer*, 2, 1969, 55.

34. Globe, S., and D. Dropkin, "Natural-Convection Heat Transfer in Liquids Confined by Two Horizontal Plates and Heated from Below," *J. Heat Transfer*, 81, 1959, 24.

35. Hollands, K. G. T., G. D. Raithby, and L. Konicek, "Correlation Equations for Free Convection Heat Transfer in Horizontal Layers of Air and Water," *Int. J. Heat Mass Transfer*, 18, 1975, 879.

36. Schmidt, E., "Free Convection in Horizontal Fluid Spaces Heated from Below," *Proc. Int. Heat Transfer Conf., Boulder, Colo., ASME*, 1961.

37. Clifton, J. V., and A. J. Chapman, "Natural Convection on a Finite Size Horizontal Plate," *Int. J. Heat Mass Transfer*, 12, 1969, 1573.

38. Hollands, K. G. T., S. E. Unny, and G. D. Raithby, "Free Convective Heat Transfer across Inclined Air Layers, *ASME Paper 75-HT-55*, 1975.

39. Catton, I., P. S. Ayyaswamy, and R. M. Clever, "Natural Convection in a Finite, Rectangular Slot Arbitrarily Oriented with Respect to the Gravity Vector," *Int. J. Heat Mass Transfer,* 17, 1974, 173.

40. Ayyaswamy, P. S., and I. Catton, "The Boundary-Layer Regime for Natural Convection in a Differentially Heated, Tilted Rectangular Cavity," *J. Heat Transfer*, 95, 1973, 543.
41. Hollands, K. G. T., S. E. Unny, G. D. Raithby, and L. Konicek, "Free-Convection Heat Transfer across Inclined Air Layers," *J. Heat Transfer*, 98, 1976, 189.
42. Arnold, J. N., P. N. Bonaparte, I. Catton, and D. K. Edwards, "Experimental Investigation of Natural Convection in a Finite Rectangular Region Inclined at Various Angles from 0° to 180°," *Proc. HTFMI*, Stanford, Calif.: Stanford University Press, 1974.
43. Ostrach, J., "Natural Convection in Enclosures," in *Advances in Heat Transfer*, New York: Academic Press, 1972.
44. Shitsman, M. E., "Natural Convection Effect on Heat Transfer to Turbulent Water Flow in Intensively Heated Tubes at Supercritical Pressure," Symp. Heat Transfer and Fluid Dynamics of Near Critical Fluids, *Proc. Inst. Mech. Eng.*, 182, Part 31, 1968.
45. Jackson, J. D., and K. Evans-Lutterodt, "Impairment of Turbulent Forced Convection Heat Transfer to Supercritical Pressure $CO_2$ Caused by Buoyancy Forces," University of Manchester, England, *Res. Rept. N-E-2*, 1968.
46. Lloyd, J. R., and E. M. Sparrow, "Combined Forced and Free Convection Flow on Vertical Surfaces," *Int. J. Heat Mass Transfer*, 13, 1970, 434.
47. Metais, B., and E. R. G. Eckert, "Forced, Mixed and Free Convection Regimes," *J. Heat Transfer*, 86, 1964, 295.
48. Buhr, H. D., A. D. Carr, and R. R. Balzhiser, "Temperature Profiles in Liquid Metals and the Effects of Superimposed Free Convection in Turbulent Flow," *Int. J. Heat Mass Transfer*, 11, 1968, 641.
49. Shiralkar, B., and P. Griffith, "The Effect of Swirl, Inlet Conditions, Flow Direction, and Tube Diameter on the Heat Transfer to Fluids at Supercritical Pressure," *J. Heat Transfer*, 92, 1970, 465.
50. Oosthuizen, P. H., and S. Madan, "Combined Convective Heat Transfer from Horizontal Cylinders in Air," *J. Heat Transfer*, 92, 1970, 194.
51. Gebhart, B., T. Audunson, and L. Pera, "Forced, Mixed and Natural Convection from Long Horizontal Wires, Experiments at Various Prandtl Numbers," *Fourth Int. Heat Transfer Conf., Paris*, IV, Sec. 3.2, 1970.
52. Brown, C. K., and W. H. Gauvin, "Combined Free and Forced Convection," Pts. I and II, *Can. J. Chem. Eng.*, 43, 1965, 306.

## CHAPTER 10

1. Farber, E. A., and R. L. Scorah, "Heat Transfer to Boiling Water under Pressure," *Tran ASME*, 70, 1948, 369.
2. Nukyiyama, S., "Maximum and Minimum Values of Heat Transmitted from a Metal to Boil Water under Atmospheric Pressure," *Japan Soc. Mech. Eng.*, 37, 1934, 367.
3. Cooper, M. G., "Nucleate Boiling," *Sixth Int. Heat Transfer Conf., Montreal*, 6, 1978,
4. Hahne, E., and U. Grigull, *Heat Transfer in Boiling*. Washington, D.C.: Hemisphere lishing Co., 1977.
5. Addoms, J. N., "*Heat Transfer at High Rates to Water Boiling Outside Cylinders*," Thesis, Department of Chemical Engineering, Massachusetts Institute of Technology,

6. Rohsenow, W. M., "A Method of Correlating Heat Transfer Data for Surface Boiling Liquids," *Trans. ASME*, 74, 1952, 969.
7. Fritz, W., "Grundlagen der Wärmeübertragung beim Verdampfen von Flüssigkeiten," *Chem. Eng. Tech.*, 11, 1963, 753.
8. Piret, E. L., and H. S. Isbin, "Natural Circulation Evaporation Two-Phase Heat Transfer," *Chem. Eng. Prog.*, 50, 1954, 305.
9. Cichelli, M. T., and C. F. Bonilla, "Heat Transfer to Liquids Boiling under Pressure," *Trans. AIChE*, 41, 1945, 755.
10. Cryder, D. S., and A. C. Finalbargo, "Heat Transmission from Metal Surfaces to Boiling Liquids: Effect of Temperature of the Liquid on Film Coefficient," *Trans. AIChE*, 33, 1937, 346.
11. Vachon, R. I., G. H. Nix, and G. E. Tanger, "Evaluation of Constants for the Rohsenow Pool-Boiling Correlation," *J. Heat Transfer*, 90, 1968, 239.
12. Zuber, N., "On the Stability of Boiling Heat Transfer," *Trans. ASME*, 80, 1958, 711.
13. Rohsenow, W., and P. Griffith, "Correlations of Maximum Heat Transfer Data for Boiling of Saturated Liquids," *Chem. Eng. Progr. Symp. Ser.*, 52, 1956, 47.
14. Kutateladze, S. S., "Heat Transfer in Condensation and Boiling," *USAEC Rept. AEC–tr–3770*, 1952.
15. Lienhard, J. H., and V. K. Dhir, "Extended Hydrodynamic Theory of the Peak and Minimum Pool Boiling Heat Fluxes," *NASA*, Rept. CR-2270, July 1973.
16. Lienhard, J. H., V. K. Dhir, and D. M. Riherd, "Peak Pool Boiling Heat Flux Measurements on Finite Horizontal Flat Plates," *J. Heat Transfer*, 95, 1973, 477.
17. Bromley, L. A., "Heat Transfer in Stable Film Boiling," *Chem. Eng. Progr.*, 46, 1950, 221.
18. Webb, R. L., "The Evolution of Enhanced Surface Geometries for Nucleate Boiling," *Heat Transfer Eng.*, 2, 1981, 46; and "Nucleate Boiling on Porous Coated Surfaces," *Heat Transfer Eng.*, 4, 1983, 71.
19. Arai, N., T. Fukushima, A. Arai, T. Nakajima, K. Fujie, and Y. Nakayama, "Heat Transfer Tubes Enhancing Boiling and Condensation in Heat Exchangers of a Refrigeration Machine," *ASHRAE J.*, 83, 1977, 58.
20. Tong, L. S., *Boiling Heat Transfer and Two-Phase Flow*. New York: John Wiley & Sons, 1965.
21. Lung, H., K. Latsch, and H. Rampf, "Boiling Heat Transfer to Subcooled Water in Turbulent Annular Flow," in *Heat Transfer in Boiling*, edited by E. Hahne and U. Grigull. Washington, D.C.: Hemisphere Publishing Co., 1977.
22. Hewitt, G. F., "Critical Heat Flux in Flow Boiling," *Sixth Int. Heat Transfer Conference, Toronto*, 6, 1978, 143.
23. Hsu, Y. Y., and R. W. Graham, *Transport Processes in Boiling and Two-Phase Systems*. New York: McGraw-Hill Book Company, 1976.
24. Collier, J. G., *Convective Boiling and Condensation*. New York: McGraw-Hill Book Company, 1972.
25. Butterworth, D., and G. F. Hewitt, *Two-Phase Flow and Heat Transfer*. Oxford: Oxford University Press, 1977.
26. Ackerman, J. W., "Pseudoboiling Heat Transfer to Supercritical Pressure Water in Smooth and Ribbed Tubes," *J. Heat Transfer*, 92, 1970, 490.
27. Shafrin, E. G., and W. A. Zisman, "Constitutive Relations in the Wetting of Low Energy

Surface and the Theory of the Retraction Method of Preparing Monolayers,'' *J. Phys. Chem.* 64, 1960, 516.

28. Rohsenow, W. M., ''Heat Transfer and Temperature Distribution in Laminar Film Condensation,'' *Trans. ASME*, 78, 1956, 1645.
29. Nusselt, W., ''Die Oberflächenkondensation des Wasserdampfes,'' *Z. Ver. Deut. Ing.*, 60, 1916, 541.
30. Kutateladze, S. S., *Fundamentals of Heat Transfer*. New York: Academic Press, 1963.
31. Labuntsov, D. A., ''Heat Transfer in Film Condensation of Pure Steam on Vertical Surface and Horizontal Tubes,'' *Teploenergetika*, 4, 1957, 72.
32. Nakayama, W., T. Daikoku, H. Kuwahara, and K. Kakizaki, ''High Performance Heat Transfer Surface Thermoexcel,'' *Hitachi Rev.*, 24, 1975, 329.
33. Welch, J. F., and J. W. Westwater, ''Microscopic Study of Dropwise Condensation,'' Int. Developments in Heat Transfer, *Proc. of the Int. Heat Transfer Conference*, University of Colorado, Part II, 1961, 302.
34. Desmond, R. M., and B. V. Karlekar, ''Experimental Observations of a Modified Condenser Tube Design to Enhance Heat Transfer in a Steam Condenser,'' *ASME* Paper 80-HT-53, 1980.
35. Peterson, A. C., and J. W. Westwater, ''Dropwise Condensation of Ethylene Glycol,'' *Chem. Eng. Progr., Symp.* Ser. 62, 1966, 135.
36. Dzakowic, G. S., et al., ''Experimental Study of Vapor Velocity Limit in a Sodium Heat Pipe,'' *ASME Paper 69-HT-21, National Heat Transfer Conference, Minneapolis, Minn.*, 1969.

## CHAPTER 11

1. Shah, R. K., ''Classification of Heat Exchangers,'' in *Heat Exchangers—Thermal—Hydraulic Fundamentals and Design*. New York: McGraw-Hill Book Company, 1981.
2. *Heat Exchanger Design Handbook*. Washington, D.C.: Hemisphere Publishing Co., 1984.
3. Kern, D. Q., *Process Heat Transfer*. New York: McGraw-Hill Book Company, 1950.
4. Lohrisch, F. W., *Hydrocarbon Process Petrol. Refiner*, 42, 1963, 177.
5. Taborek, J., ''Shell-and-Tube-Heat Exchangers: Single-Phase Flow,'' in *Heat Exchanger Design Handbook*. Washington, D.C.: Hemisphere Publishing Co., 1984.
6. Tinker, T., ''General Discussion on Heat Transfer,'' *Proc. Inst. Mech. Eng. London*, 1951, 89.
7. Small, W. M., and R. K. Young, ''The Rod Baffle Heat Exchanger,'' *Heat Transfer Eng.*, 1979, 21.
8. Kern, D. Q., and A. D. Kraus, *Extended Surface Heat Transfer*. New York: McGraw-Hill Book Company, 1972.
9. Kays, W. M., and A. L. London, *Compact Heat Exchangers*, 3rd ed. New York: McGraw-Hill Book Company, 1984.
10. Mueller, A. C., ''Thermal Design of Shell-and-Tube Heat Exchangers for Liquid-to-Liquid Heat Transfer,'' *Eng. Bull., Res. Ser., 121*, Purdue University Engineering Experiment Station, 1954.
11. Epstein, N., ''Fouling in Heat Exchangers,'' *Sixth Int. Heat Transfer Conf.*, Toronto, 6, 1978, 235.
12. Somerscales, E. F. C., and J. G. Knudsen (eds.), *Fouling of Heat Transfer Equipment*. Washington, D.C.: Hemisphere Publishing Co., 1981.

13. *Standards of Tubular Exchanger Manufacturers Association*, 6th ed. New York: Tubular Exchanger Manufacturers Association, Inc., 1978.
14. Taborek, J., ''Charts for Mean Temperature Difference in Industrial Heat Exchanger Configurations,'' in *Heat Exchanger Design Handbook*. Washington, D.C.: Hemisphere Publishing Co., 1984.
15. Bowman, R. A., D. C. Mueller, and W. M. Nagle, ''Mean Temperature Differences in Heat Exchanger Design,'' *Trans. ASME*, 62, 1940, 283.
16. Benedict, R. P., *Fundamentals of Pipe Flow*. New York: John Wiley & Sons, 1980.
17. Miller, D. S., *Internal Flow Systems*. Bedford, UK: British Hydromechanics Research Association (BHKA) Fluid Engineering, 1978.
18. *Marks' Standard Handbook for Mechanical Engineers*, 9th ed, edited by E. A. Avallone and T. Baumeister III. New York: McGraw-Hill Book Company, 1986.
19. *Engineering Data Book*. Allen Park, Mich.: UOP–Wolverine Tube Division, 1967.
20. Breber, G., ''Computer Programs for Design of Heat Exchangers,'' in *Heat Transfer Equipment Design*, edited by R. K. Shah, E. C. Subbarao, and R. A. Mashelkar. New York: Hemisphere Publishing Co., 1988.
21. Spalding, D. B., ''Computational Fluid Dynamics Applied to the Prediction of the Heat Transfer, Mechanical and Thermal-Stream Behavior of Heat Exchangers,'' in *Heat Transfer Equipment Design*, edited by R. K. Shah, E. C. Subbarao, and R. A. Mashelkar. New York: Hemisphere Publishing Co., 1988.
22. *Steam—Its Generation and Use*, 39th ed. New York: The Babcock and Wilcox Company, 1978.
23. Rohsenow, W. M., and J. P. Hartnett, *Handbook of Heat Transfer*, (*Applications*), 2nd ed. New York: McGraw-Hill Book Company, 1985.
24. Fraas, A. P., and M. N. Ozisik, *Heat Exchanger Design*. New York: John Wiley & Sons, 1965.
25. Incropera, F. P., and D. P. DeWitt, *Fundamentals of Heat and Mass Transfer*, 2nd ed. New York: John Wiley & Sons, 1985.

## APPENDIXES

1. Touloukian, Y. S., R. W. Powell, C. Y. Ho, and P. G. Klemens, *Thermophysical Properties of Matter*, Vol. 1: *Thermal Conductivity—Metallic Elements and Alloys*. New York: Plenum Publishing Co., 1970.
2. Touloukian, Y. S., R. W. Powell, C. Y. Ho, and P. G. Klemens, *Thermophysical Properties of Matter*, Vol. 2: *Thermal Conductivity—Nonmetallic Solids*. New York: Plenum Publishing Co., 1970.
3. Touloukian, Y. S., P. E. Liley, and S. C. Saxena, *Thermophysical Properties of Matter*, Vol. 3: *Thermal Conductivity—Nonmetallic Liquids and Gases*. New York: Plenum Publishing Co., 1970.
4. Eckert, E. R. G., and R. M. Drake, *Analysis of Heat and Mass Transfer*. New York: McGraw-Hill Book Company, 1972.
5. Raznjevic, K., *Handbook of Thermodynamic Tables and Charts*, 3rd ed. New York: McGraw-Hill Book Company, 1976.
6. Desai, P. D., T. K. Chu, R. H. Bogaard, M. W. Ackermann, and C. Y. Ho, ''Part I: Thermophysical Properties of Stainless Steels,'' *CINDAS Special Report*, Purdue University, West Lafayette, Ind., September 1976.

7. American Society of Heating, Refrigerating and Air Conditioning Engineers, *ASHRAE Handbook of Fundamentals*. New York: ASHRAE, 1989.
8. *International Critical Tables*. Washington, D.C.: National Academy of Sciences, 1978.
9. Incropera, F. P., and D. P. DeWitt, *Fundamentals of Heat Transfer*, 2nd ed. New York: John Wiley & Sons, 1985.
10. Liley, P. E., Steam Tables in SI Units, private communication, School of Mechanical Engineering, Purdue University, West Lafayette, Ind., 1989.
11. Knudsen, J. G., and D. L. Katz, *Fluid Dynamics and Heat Transfer*. New York: McGraw-Hill Book Company, 1958.
12. Touloukian, Y. S., and D. P. DeWitt, *Thermophysical Properties of Matter*, Vol. 7: *Thermal Radiative Properties—Metallic Elements and Alloys*. New York: Plenum Publishing Co., 1970.
13. Touloukian, Y. S., and D. P. DeWitt, *Thermophysical Properties of Matter*, Vol. 8: *Thermal Radiative Properties—Nonmetallic Solids*. New York: Plenum Publishing Co., 1970.
14. Touloukian, Y. S., D. P. DeWitt, and R. S. Hernicz, *Thermophysical Properties of Matter*, Vol. 9: *Thermal Radiative Properties—Coatings*. New York: Plenum Publishing Co., 1970.
15. Mallory, J. F., *Thermal Insulation*. New York: Van Nostrand Reinhold, 1969.
16. Gubareff, G. G., J. E. Janssen, and R. H. Torborg. *Thermal Radiation Properties Survey*. Minneapolis: Minneapolis-Honeywell Regulator Company, 1960.
17. Kreith, F., and J. F. Kreider, *Principles of Solar Energy*. Washington, D.C.: Hemisphere Publishing Co., 1978.
18. Schneider, P. J., *Conduction Heat Transfer*. Reading, Mass.: Addison-Wesley Publishing Co., 1955.
19. Koh, J. C., E. M. Sparrow, and J. P. Hartnett, "The Two-Phase Boundary Layer in Laminar Film Condensation," *Int. J. Heat Mass Transfer*, 2, 1961, 69.
20. Nusselt, W., "Die Oberflächenkondensation des Wasserdampfes," *Z. Ver. Deut. Ing.*, 60, 1916, 541.
21. Rohsenow, W. M., "Heat Transfer and Temperature Distribution in Laminar Film Condensation," *Trans. ASME*, 78, 1956, 1645.

# INDEX


Absorptivity, 13, 273
  monochromatic, 277
Absorptivity approximations:
  graybody, 279
  isothermal, 279
  nonisothermal, 280
Advection, 16
Air cooler, 633
Analog field plotter, 141
Analogy:
  heat-electrical, 25
  heat-momentum (Reynolds analogy), 451
Anemometer, 368
Anodized aluminum heat-sink fin units, 82, 84, 88
Approaches to the analysis of:
  conduction, 10
  convection, 374
  practical, 296, 375
  radiation, 296
  theoretical, 375
Archimedes principle, 430
Area coefficient, tube bank, 518
Assumptions/Conditions, standard conditions for convection, 406
Asymptotic thermal developed flow, 407

Baffles, 630
  disc and doughnut, 631
  pitch (spacing), 632
  rod baffle, 633
  segmental, 631
BASIC programs, 215, 238, 744
Beam length, mean, 283, 285
Bessel functions, 163
Biot number, 95
  convection, 78
  radiation, 296, 331
Blackbody, 13, 262, 275
Blackbody thermal radiation, 297
Black light, 260
Blasius solution, 423
Boiling (see also Convection, practical analysis of boiling and condensation):
  burnout, 588
  characteristic parameters, 584
  coefficient of boiling heat transfer, 590
  departure from nucleate boiling (DNB), 601
  dimensionless groups, summary, 625
  flow boiling, 598
    boiling curve for, 599
    convection correlations for, 604
    enhancement of, 604
    physical description of, 598
  heat transfer, 585
  Leidenfrost point, 588, 593
  maximum heat flux (critical heat flux or peak heat flux), 588
  minimum heat flux, 588, 593
  onset of nucleate boiling (ONB), 588
  pool boiling, 585
    boiling curve for, 587

convection correlations for, 590, 622
enhancement of, 594
regimes of flow boiling:
annular, 601
annular/mist, 601
bubbly, 600
forced and/or natural convection, 600
mist, 601
slug, 600
regimes of pool boiling:
natural convection, 586
nucleate, 588
column/slug, 588
film, 588
isolated bubbles, 588
transition, 588
saturated, 585
subcooled, 585
Bond number, 389, 585
Boundary conditions, 38
composite walls, 47
convection, 368
other types, 56
standard, 39
table, 41
Boundary layer:
approximations, 373
hydrodynamic, 373
thermal, 373
thickness, 373
Boundary layer flow, 373, 414
Boussinesq approximation, 434
Buckingham pi theorem, 377
Bulk-stream:
enthalpy rate, 383
mass flux, 382
temperature, 384, 404
velocity, 382, 401
Buoyancy, 430
Burnout, 588

Calculus, 705
Capacity rate, 654
minimum, 655
Capacity ratio, 655
Cartesian coordinate system (see Rectangular coordinate system)
Chips (see Compact integrated circuits)
Coefficient of friction, 375
Coefficient of heat transfer, 17, 375
mean, 376
Colburn analogy, 454, 472, 504
Combined modes of heat transfer, 19, 346
Compact heat exchangers, 639
Compact integrated circuits (ICs or chips), 27, 48, 88
Composite walls, conduction heat transfer within, 47
combined series-parallel arrangement, 53
criteria for one-dimensional analysis, 54
parallel arrangement, 52
series arrangement, 51
Condensation (see also Convection, practical analysis of boiling and condensation):
characterizing parameters, 584
dimensionless groups, 625
dropwise, 605, 615, 623
film, 605
coefficient of condensation heat transfer, 607
convection correlations, 607, 622
enhancement of, 615
Reynolds number for, 606
thickness, 606
Conduction, 5
analytical solutions for:
steady two-dimensional conduction in rectangular solid, 31, 735
unsteady one-dimensional conduction with convective cooling, 157, 739
infinite cylinders, 163, 740
plane walls, 159, 739
semi-infinite solids, 150
spheres, 164, 742
in composite walls, 47
in cylindrical sections, steady, one-dimensional, 62
in plane walls, steady (see Heat transfer, one-dimensional)
in radial systems, steady, 58
internal energy sources, with, 71
one-dimensional (see One-dimensional heat transfer)
variable thermal conductivity, with, 67
Conduction, Multidimensional, 118–87
analogy approaches, 140–43
electrical analogy, 140
analog field plotter, 141
analytical approaches, 131–40
separation-of-variables method, 131
differential formulation, 120–30
Cartesian coordinate system, 120
radial coordinate systems, 126
general Fourier law of conduction, 118, 119
graphical approaches, 143–46
steady two-dimensional systems, 143
introduction, 118
mathematical formulations, 119, 120

Conduction, Multidimensional (*Continued*)
practical solution results, 146–80
steady multidimensional heat-transfer systems, 146
unsteady one-dimensional heat transfer in:
flat plates, circular cylinders and spheres, 157
semi-infinite solids, 150
unsteady two-and-three-dimensional heat transfer systems, 173
solution techniques, 130, 131
summary, 180, 181
Conduction, numerical approach, 188–256
finite-difference approximations, 190–95
spatial derivatives, 190
summary, 192
time derivatives, 192
introduction, 188
nodal network, 189, 190
steady systems, analysis of, 195–224
finite-difference formulation, 195
accuracy, 199
exterior nodal equations, 197
interior nodal equations, 196
summary of nodal equations, 199
finite-difference solutions, 203
direct methods, 223
iterative methods, 209
unsteady analysis method, 224
summary, 247, 248
unsteady systems, analysis of, 224–47
explicit method, 226
numerical solutions, 233
stability considerations, 229
implicit method, 240
nodal equations, 224
R/C network formulation, 243
Conduction shape factor, 11, 146
Configuration factor (see Radiation shape factor)
Conservation principle:
energy, 4
mass, 4
Continuity equation, 393
Contraction-loss coefficient (see Entrance-loss coefficient)
Contraction ratio, 463
Convection correlations:
boiling, pool, 590–94
summary, 622, 623
combined natural and forced convection, 576
condensation:
dropwise, 617
film, 607–9
summary, 623, 624
forced convection, 17, 465
circular tube, 466
concentric tube annulus, 466, 473
cylinder in cross flow, 512
external flow,
laminar, 503
turbulent, 505
finned annulus, 543
flat plate, 503
summary, 540
internal flow:
laminar, 465
summary, 536
turbulent, 470
transitional, 473
liquid metals, 474
noncircular tubes, 466
property variation, effect of, 478
spheres, 512
tube bank, 520–22, 524
natural convection, 17
enclosures, 570–73
external, general, 562–64
internal flow, 569
summary, 578, 579
vertical plate:
uniform wall-heat flux, 558–60
uniform wall temperature, 553–55
Convection, introduction, 364–89
approaches to analysis of, 374–81
dimensional analysis, 377
practical, 375
coefficients of friction and heat transfer, 375
theoretical, 375
boundary layer, 373, 374
characteristics of internal flow, 381–86
hydrodynamic entrance and fully developed regions, 382
thermal entrance and fully developed regions, 383
factors associated with characterization of, 364–73
boundary conditions, 368
forced and natural convection, 364
internal and external flow, 365
laminar and turbulent flow, 367
miscellaneous, 372
time and space dimensions, 367
type of fluid, 369
summary, 386, 387
Convection, practical analysis of boiling and condensation, 583–627
boiling heat transfer, 585–604
flow boiling, 598

boiling curve for flow boiling, 599
convection correlations for flow boiling, 604
enhancement of flow boiling, 604
physical description of flow boiling, 598
pool boiling, 585
boiling curve for pool boiling, 585
film boiling, 588
natural convection, 586
nucleate boiling, 588
transition boiling, 588
convection correlations for pool boiling, 590
enhancement of pool boiling, 594
characterizing parameters for two-phase heat-transfer processes, 584, 585
condensation, 605–21
dropwise, 615
film, 606
approximate solution for laminar flow, 750
convection correlations, 607
horizontal tube or sphere, 610
inclined surface, 610
vertical surface, 608
enhancement of film condensation, 615
Reynolds number for film condensation, 606
introduction, 583, 584
summary, 622–25
Convection, practical analysis of forced convection, 463–548
external flow, 502–15
flow across cylinders and spheres, 511
coefficients of drag and heat transfer, 512
practical analyses, 514
flow over flat plates and cylinders, 503
coefficients of friction and heat transfer, 503
further considerations, 507
laminar flow, 503
turbulent flow, 505
practical analyses, 508
heat transfer, 508
specified wall-heat flux, 509
specified wall temperature, 509
momentum transfer, 508
flow across tube banks, 515–35
geometric and bulk-stream characteristics, 515
hydraulic considerations, 523
practical thermal analysis, 519
internal flow, 463–502
coefficients of friction, pressure drop, and heat transfer, 463
convection correlations, 465
laminar flow, 465
turbulent flow, 470
effects of property variations, 478
enhancement of heat transfer, 479
other factors, 481
hydraulic considerations, 498
introduction, 463
practical thermal analysis, 483
specified wall heat flux, 484
uniform wall temperature, 489
effectiveness, 490
log mean temperature difference, 492
summary, 535–41
Convection, practical analysis of heat exchangers, 628–703
evaluation and design, 641
heat exchanger analysis, 652–96
hydraulic considerations, 690
practical thermal analysis, 653
effectiveness method, 655
log mean temperature difference method, 675
specification of $\overline{U}$, 689
introduction, 628
overall coefficient of heat transfer, 641–52
effect of fouling, 644
local overall coefficient of heat transfer, 642
mean overall coefficient of heat transfer, 643
summary, 696, 697
types of heat exchangers, 628–41
crossflow heat exchangers, 633
double-pipe heat exchangers, 629
performance/size, 637
regenerators, 636
shell-and-tube heat exchangers, 630
Convection, practical analysis of natural convection, 549–82
characterizing parameters for natural convection, 550, 551
combined natural and forced convection, 576, 577
external natural convection flow, 551–63
flow over a vertical flat plate, 552
hydraulic considerations, 560
uniform wall-heat flux, 558
uniform wall temperature, 553
general external natural convection flows, 562
internal natural convection flow, 569
introduction, 549, 550
natural convection flow in enclosed spaces, 570–75
summary, 577–80
Convection, theoretical analysis of, 390–462
introduction, 390
laminar flow theory, 390–440
boundary layer flow over plane surfaces, 414
differential formulation, 414

Convection, theoretical analysis of (*Continued*)
integral formulation, 415
continuity, 416
energy transfer, 418
momentum transfer, 416
solution, 419
integral solution—energy transfer, 423
dimensionless temperature distribution, 425
thermal boundary layer thickness and Nusselt number, 426
integral solution-fluid flow, 420
boundary layer thickness and friction factor, 422
dimensionless velocity distribution, 420
flow in tubes, 392
mathematical formulation, 392
continuity, 393
energy transfer, 397
momentum transfer, 394
solution, 400
energy transfer—TFD region, 402
bulk-stream temperature and Nusselt number, 404
dimensionless temperature distribution, 403
fluid flow—HFD region, 410
bulk-stream velocity and friction factor, 401
dimensionless velocity distribution, 400
natural convection on vertical surfaces, 430
mathematical formulation, 432
solution, 435
integral solution—fluid flow and energy transfer, 435
boundary layer thickness and Nusselt number, 436
dimensionless velocity and temperature distributions, 435
summary, 455
turbulent flow theory, 440–54
characteristics of turbulent convection processes, 441
consequences of turbulent flow, 443
unsteady nature of turbulent flow, 441
classical approach to modeling turbulence, 444
differential formulation, 444
energy transfer, 449
fluid flow, 444
instantaneous equations, 444
specification of Reynolds stress, 446
time-average equations, 444
integral formulation, 450
solutions, 451
Reynolds analogy, 451
Critical heat flux, 588
Critical radius, 63
Crossflow heat exchangers, 633
mixed streams, 633
unmixed streams, 633
Cryogenic applications, 19, 21
Curvilinear squares, 144
Cylinders:
conduction heat transfer in, 59
differential formulation, 126
Heisler charts, 165, 166
Heisler relations, 163
internal energy sources, with, 74
radial, 58
steady, 58
thermal resistance, 59
unsteady, 160
convection heat transfer:
forced convection, 511
natural convection, 564
Cylindrical coordinate system, 126, 707
Cylindrical section, conduction heat transfer in, 62

Damping parameter, turbulent, 448
Darcy friction factor, 376
Design and evaluation, functions of, 2, 641
Differential approach:
multidimensional conduction, 118
one-dimensional conduction, 34
Diffuse emitters, 271
Diffuse reflection, 280
Dimensional analysis, 377
Dimensional continuity, principle of, 4
Dimensionless parameters:
table of, for convection:
boiling and condensation, 625
forced convection, 541
natural convection, 580
table of, general, 761
Dimensions, units, and significant figures, 708
Dissipation (also see Viscous dissipation)
energy, 411
function, 750
viscous, 411
Double-pipe heat exchangers, 629
counterflow, 629
hairpin unit, 691
parallel flow (concurrent flow), 629
Drag, coefficient of, 512
Dropwise condensation, 615

Earth temperature:
amplitude of annual variation in ground surface temperature, 32, 156
annual mean, 32, 149, 156
data for select cities, 735
phase constant, 32, 156
Eckert number, 405, 457, 541
modified, 457
Eddy diffusivity, 446
Eddy kinematic viscosity, 446
Eddy thermal conductivity, 449
Eddy thermal diffusivity, 449
Effectiveness:
heat exchanger, 655–59
relative, 491
tube flow, 490
Electrical analogy, 140
method of (see Analogy)
thermal radiation, 300
unsteady lumped capacitance system, 105
Electrical motors, heat transfer in, 114
Electrical resistivity, 71
Electromagnetic radiation, 12, 257
Electromagnetic spectrum, 13, 258
Emissive power, 4
directional, 271
monochromatic, 265
subtotal, 264
total, 13, 262
Emissivity, 14, 262
directional, 271
monochromatic, 267
various surfaces, of, 263
Energy:
conservation of (see First law of thermodynamics)
dissipation, 411
equation, 398
generation, 71
kinetic, 748
storage, 4
Enhancement of heat transfer, 479
Enthalpy, 5, 370
Enthalpy of vaporization (see Latent heat of vaporization)
Enthalpy rate, 370
Entrance-loss coefficient, 464
Entrance region:
hydrodynamic, 382
thermal, 383
Equations of state, 4
Error function, 150
complementary, 150
table of, 745
Evaluation and design, functions of, 2, 641
Evaporation (see *mass transfer supplement*), 372
Excess temperature, 586
Expansion-loss coefficient, 464
Extended surfaces, 76–95
analysis, 78
Biot number, 78
circumferential fins, 85
effective perimeter, 88
equivalent surface area, 88
fin efficiency, 82
fin resistance, 82
finned area fraction, 88
finned systems, 87
local heat flux per fin, 88
net surface efficiency, 87
prime surface area, 87
rectangular and triangular fins, 85
spines, 85
External flow:
general description, 365
practical analysis of, 502–15
theoretical analysis of:
forced convection, 414
natural convection, 430

Fanning friction factor, 376
Film condensation, 606
coefficient of condensation heat transfer for, 607
enhancement of, 615
Reynolds number for, 606
Film temperature, 508
Fin efficiency, 82
charts, 85
Finite difference approximations, 190
spatial derivatives, 190
accuracy (error), 191–94, 199
backward, 190
central, 191
forward, 190
three-point formula, 249
time derivatives, 192
backward, 192, 225
forward, 192, 225
stability, 192, 229
Finite-difference formulation, 120
accuracy, 199
discretization method, 196
energy balance method, 196, 198
error:
discretization (truncation), 199
round-off, 199
R/C networks, 243

Finite-difference formulation (*Continued*)
steady systems, 195
unsteady systems, 224
explicit, 226
implicit, 240
stability, 229
Finite-difference solutions, steady, 203
direct methods, 223
Gaussian elimination, 223
matrix, 223
iterative methods, 209
Gauss-Seidel, 209
relaxation, 210
unsteady analysis method, 224
Finite-difference solutions, unsteady:
explicit, 233–40
implicit, 241
Finite-element method, 120, 188, 248
Finned area fraction, 88
Fins (see Extended surfaces)
Fin tubes (also see Extended surfaces), 638
corrugated, 639
louvered, 639
First law of thermodynamics, 4
Flame, 285
Flat plate (see Plane wall)
Flat plate collector, 356
Flow boiling (see Boiling)
Forced convection, 17
combined with natural convection, 576
convection correlations, 465
laminar flow, 465
property variation, effect of, 478
summary:
external flow, 540
internal flow, 536
turbulent flow, 470
practical analysis (see Convection, practical analysis of)
summary of practical thermal analysis approach for internal flow, 539
theoretical analysis (see Convection, theoretical analysis)
Fouling factor, 644
representative values, 646
Fourier equation, 121
Fourier law of conduction, 6
finite-difference approximation, 194
general, 7, 118
tensor form, 119
vector form, 119
Fourier number, 158
Free convection (see Natural convection)
Freezing line depth, 156
Frequency, electromagnetic waves, 258
Friction, coefficient of, 376
flow across tube banks, 520
forced convection flows:
external, 503
internal, 465
Friction factor, 376, 541
apparent, 464
Darcy, 376
Fanning, 376
mean, 376
Friction velocity, 447
Fully developed flow:
hydrodynamic, 382
thermal, 385
Fundamental laws, 4, 391

Gaussian elimination, 223
Gauss-Seidel iteration method, 209, 242
General Fourier law of conduction, 7, 118
General Newton law of cooling, 79
Graetz number, 470, 541
Graphical approaches to analysis of conduction, 143
Grashof number, 388, 437, 551, 584
Grashof number, flux, 558
Graybodies, 269
Greenhouse effect, 278
Grober charts (see Heisler charts)

Heat exchangers (see also Convection, practical analysis of heat exchangers):
charts for:
correction factors, 677, 678
effectiveness, 657–59
coefficient of heat transfer, overall, 641
compact, 639
direct contact, 636
evaluation and design, 641
practical thermal analysis, 653
effectiveness method, 655
*LMTD* method, 675
pressure drop, 690
pumping power, 691
rating, 641
regenerator matrix surfaces, 637
thermal capacity, 655
tube inserts, 639
types of, 628
crossflow, 633
double-pipe, 629
regenerators, 636
shell-and-tube, 630

Heat-flow lines, 144
Heat flux, apparent total, 449
Heat pipe, 620
Heat pump, ground source, 33
Heat pump water heater (HPWH), 31
Heat sink, 82
Heat transfer, 1
  accumulative, 97, 154, 160, 162, 165, 168
  analogy between electric current flow and, 140
  basic modes of (see Mechanisms, basic transport)
  conduction (see Conduction)
  convection (see Convection)
  energy-related applications, 1–3
  one-dimensional (see Heat transfer, one-dimensional)
  radiation (see Radiation heat transfer)
Heat transfer, coefficient of
  boiling:
    flow boiling, 599
    pool boiling, 590
  condensation, 607
  flow across tube banks, 520
  forced convection
    external flow, 503
    internal flow, 465
  heat exchangers, overall, 641
  local, 376
  mean, 376
  natural convection, 550
  typical values, 18
Heat-transfer correlations (see Convection correlations)
Heat transfer, introduction, 1–33
  analogy between heat transfer and the flow of electric current, 25–28
  basic transport mechanisms and particular laws, 5–24
    combined modes of heat transfer, 19
    conduction, 5
      analysis of conduction, 10
        conduction shape factor, 11
      Fourier law of conduction, 6
        thermal conductivity, 7
    convection, 16
      Newton law of cooling, 16
        coefficient of heat transfer, 17
      Newton law of viscous stress, 16
    radiation, 12
      radiation-heat-transfer rate, 14
      Stefan-Boltzmann law, 13
  fundamental laws, 4, 5
  summary, 28, 29
Heat transfer, one-dimensional, 34–117
  conduction in a plane wall, 34–58
    boundary conditions, 38
      composite walls, 47
        combined series-parallel arrangements, 53
        parallel arrangements, 52
        series arrangements, 51
      other types, 56
      standard, 39
    differential formulation, 34
    differential formulation/solution, summary, 57
    short method, 57
    solution, 36
  conduction in radial systems, 58–67
    analysis, 59
    critical radius, 63
  extended surfaces, 76–95
    analysis, 78
    fin efficiency, 82
    finned systems, 87
    fin resistance, 82
    net surface efficiency, 87
  internal energy sources, 71–76
    analysis, 72
  introduction, 34
  practical solution results, 105–8
  summary, 108
  unsteady heat transfer systems, lumped capacitance, 95–105
    analysis, 96
    electrical analogy, 105
  variable thermal conductivity, 67–71
    analysis, 68
Heisler charts, 161, 162, 164, 165, 167, 168
Heisler relations, 159, 163, 166
High speed flow, 747
Hydraulic diameter, 381
  tube bank, 518
Hydraulic radius, 381
Hydrodynamically fully developed flow, 382, 392, 442
Hydrodynamic boundary layer, 373
Hydrodynamic entrance region, 382

Ideal fluids, 371
Ideal gas law, 286
Incompressible liquids, 371
Index of refraction, 258
Infrared radiation, 261
Insulation, 10
  critical radius, 63
Integral equations for boundary layer flow:
  continuity, 416

Integral equations for boundary layer flow (*Continued*)
  energy, 419
  mechanical energy, 418
  momentum, 417
Integral formulation, 120, 415
Integral methods, for boundary layer flow, 422
Integral solutions:
  energy transfer:
    specified wall-heat flux, 428
    uniform wall-heat flux, 426
    uniform wall temperature, 429
  fluid flow, uniform free-stream velocity, 420
  natural convection, 435
Intensity, 270
Interferometer, Mach-Zehnder, 549
Internal energy, 5
Internal energy sources, 71–76
  maximum temperature, 73, 107
Internal flow:
  characteristics of, 381–86
  general description, 365
  practical analysis of, 463–502
    natural convection, 563
    tube banks, 515–35
  summary of practical thermal analysis approach, 539
  theoretical analysis of, 392
Irradiation, 273
  monochromatic, 278
Isotherms, 144
Isotropic, 7, 370

Jakob number, 389

Kirchhoff's law, 275, 277

Lambert cosine law, 271
Laminar flow:
  general description, 367
  theory:
    condensation, film, 750
    forced convection, 390
    natural convection, 430
    viscous high speed flow, 749
Laplace equation, 121
Latent heat of vaporization, 371
  corrected, 594
  modified, 607, 755
Leidenfrost point, 588, 593
Liquid metals:
  convection correlations for, 474, 504
  property values of, 726
Liquid petroleum gas (LPG), 21
Liquids (saturated), property values of, 721–26
Log mean temperature difference, 492
  correction factor, 676
  heat exchangers, 675
  temperature ratios, 676
Loss coefficient:
  entrance, 464
  expansion, 464
  heat exchanger component, 691

Magnetron, 261
Marangoni number, modified, 617
Mass:
  conservation of, 4
  flow rate, 370
Mass transfer (see *mass transfer supplement*), 372
Mathematical concepts, 705
Mathematical functions:
  Bessel functions of the first kind, 163
  error functions, 150
Matrix methods, 223
Mechanisms, basic transport, 5
Melanoma, 33
Metals, property values of:
  liquid, 726
  solid, 714–17
Methodology (see Approaches to the analysis of)
Metric system, 708
Microwave oven, 23
Microwave radiation, 261
Minimum free-flow area, 517
Mixing-cup temperature (bulk-stream temperature),
Mixing length, 448, 462
Modes of heat transfer (also see Mechanisms, basic transport)
  combined, 19
Molecular weight, table of, 730
Momentum equation, 395
Momentum rate, 370
Momentum, rate of creation of axial, 391, 394, 416
Monochromatic (spectral):
  absorptivity, 277
  emissive power, 265
  emissivity, 267
  irradiation, 278
  reflectivity, 277
  transmissivity, 277
Monte Carlo method, 357
Moody friction factor chart, 479

Natural convection (see also Convection, practical analysis of natural convection):
  characteristic parameters, 550

combined with forced convection, 576
concentration driven (see also *mass transfer supplement*), 364
convection correlations:
enclosed spaces, 571, 574
external flows, general, 562–65
vertical flat plate, 553–55, 558–60
dimensionless parameters, summary, 579
enclosed spaces, flow in, 570
external flow, 551
integral solution, 435
internal flow, 563
introduction, 17, 364, 549
theoretical analysis, 430
Net surface efficiency, 87
Newtonian fluid, 370
Newton-Raphson method, 45, 110
Newton's law of cooling, 16, 76
general, 79, 376
Newton's law of viscous stress, 16
Newton's laws of motion:
first law (or law of inertia), 709
second law, 709
third law (or law of action and reaction), 709
universal law of gravitation, 709
Newton's second law of motion, 4
Nodal network, 189
Nodal points, 189
Nonmetals, property values, 718–20
Nonparticipating fluids, 13
Number of transfer units, 491, 655
Numerical approach (see Conduction, numerical approach)
Nusselt number, 376, 386, 541

One-dimensional heat transfer (see Heat transfer, one-dimensional)
Overall coefficient of heat transfer, 641
finned surfaces, for, 646
local, 642
mean, 643
representative values, 644

Participating gases, 13, 282
absorptivity, 283
emissivity, 283
radiation heat transfer through, 334
Particular laws, 4, 5, 391
Peclet number, 405, 541
Perimeter:
effective heat transfer surface, 517
effective wetted, 517
Photons, 258
Photosphere, 351
Photosynthesis, 261
Planck law, 265
Plane wall:
conduction in, 34, 157
Heisler charts, 160, 161
Heisler relations, 159
internal energy generation, with, 72
steady, one-dimensional (see also Heat transfer, one-dimensional), 34–58
thermal resistance, 37
unsteady, 157
convection:
forced, 503
natural, 564
Poisson equation, 121
Pool boiling (see Boiling)
Practical solution results:
conduction, multidimensional, 146–80
conduction, one-dimensional, 105
tables, 106, 107
unsteady lumped capacitance systems, 108
Prandtl number, 371, 374
Pressure drop:
flow in heat exchangers, 690
flow in tube banks, 523
flow in tubes, 498
Product solution, 132, 177, 179
Property variation effects, convection, 478
Pumping power, 500, 524, 691

Quantum theory, 258

Radial systems, conduction in, 58
analysis, 59
critical radius, 63
Radiation factor, 324
table, 325
Radiation functions, table of, 746
Radiation heat transfer, 257–363
combined with other modes, 346
introduction, 257
physical mechanism, 257–61
electromagnetic spectrum, 258
thermal radiation, 259
infrared (IR), 261
microwaves, 261
ultraviolet (UV), 260
visible (light), 261
practical analysis of, 296–351
blackbody thermal radiation, 297
bisurface systems, 297
multisurface systems, 298

Radiation heat transfer (*Continued*)
thermal radiation networks, 300
nonblackbody thermal radiation, 306
bisurface systems, 310
multisurface systems, 317
reradiating surfaces, 322
thermal radiation shields, 318
radiation factor, 324
radiosity, 307
systematic solution approach, 326
transparent medium, 334
participating gases, 334
blackbody enclosures, 335
gray gas/multiple-surface enclosure, 336
gray gas/single-surface enclosures, 335
near-blackbody enclosures, 338
transparent solids, 338
blackbody bisurface systems, 338
effects of diffuse nonblackbody surfaces and reflecting medium, 342
unsteady systems, 330
radiation shape factor, 287–95
design curves, 289
law of reciprocity, 288
summation principles, 288
solar radiation, 351
solar-energy systems, 355
solar-thermal systems, 355
solar resource, 351
summary, 357
thermal radiation properties, 261–87
of gases, 282
surface emission properties, 262
blackbody thermal radiation, 265
monochromatic emissive power, 265
subtotal emissive power, 264
thermal radiation emitted from real surfaces, 267
thermal radiation intensity, 270
total emissive power, 262
surface irradiation properties, 273
directional effects, 280
monochromatic irradiation properties, 277
total irradiation properties, 273
Radiation heat transfer, introduction, 12
Radiation heat transfer rate, 14
generalized equation, 93
Radiation shape factor, 14, 287
design curves, 289
law of reciprocity, 288
summation principles, 288
Radiosity, 307
Rating and sizing (see Evaluation and design, function of)
Rayleigh effect, 353
Rayleigh number, 437, 551
Rayleigh number, flux, 558
R/C network approach, 141
Reciprocity, law of, 288, 299, 363
Rectangular coordinate system, 706
Rectangular solid, conduction in, 120
differential formulation, 120
numerical finite-difference formulation, 189
separation of variables method, 131
Recuperators, 636
Reflecting medium, 342
Reflectivity, 13, 273
monochromatic, 277
Regenerators, 636
Relaxation method, 210
Reradiating surfaces, 322
Resistance (see Thermal resistance)
Reynolds analogy, 451
Reynolds number, 374
film condensation, for, 606
Reynolds stress, 445
Richardson number, 460, 541, 577
Roughness, 480
relative, 480

Saturated liquids, property values of, 721–26
Scatter, radiation, 353
Second law of thermodynamics, 4, 6, 229, 231
Semi-infinite solid, conduction in, 150
Separation-of-variables method, 131, 736
Shape factor (see Radiation shape factor)
Shear stress, 375, 396
apparent total, 445
Shell-and-tube heat exchangers, 630
Short method, one-dimensional conduction, 57, 59
Silicon heat-sink compound (grease), 49
Similar flow, 365, 420
Similarity coordinate, 423, 460
Similarity solutions, 420
Sizing and rating (see Design and evaluation, function of)
Solar collectors, 356
Solar constant, 352
Solar energy, 2
Solar irradiation:
extraterrestrial, 352, 354
monochromatic, 352
terrestrial, 353
Solar radiation, 259, 267, 351–56
solar-energy systems, 355

solar resource, 351
solar-thermal systems, 355
Solid angle, 271
Solids, property values of:
metals, 714–17
nonmetals, 718–20
Specific heat:
constant pressure, 5, 371
constant volume, 5, 371
ratio, 730
Spectral thermal radiation properties (see Monochromatic)
Specular reflection, 280
Spheres:
conduction in, 59, 160
differential formulation, 128
internal energy sources, with, 74
Heisler charts, 167, 168
Heisler relations, 164
steady, 59
thermal resistance, 59
unsteady, 160
convection:
forced convection, 511
natural convection, 564
Spherical coordinate system, 128, 707
Spines, 77
Stability criterion, finite-difference method, 230
Stanton number, 376, 386, 541
Stefan-Boltzmann law, 13, 262
Summation principles, 288
Sunburn radiation, 260
Superconductors, 7
Superinsulations, 19, 22, 349
Superposition, principle of, 138
Surface-area density, 516, 639
spectrum, 640
Surface energy parameter, 617
Surface tension, 389, 584
critical, 606, 616
tables of, 722, 731
Symbols (see Nomenclature)

Taylor series expansion, 190, 192
Temperature, 5
bulk-stream (see Bulk-stream temperature)
coefficient of resistance, 71
coefficient of thermal conductivity, 67
effectiveness (see Net surface efficiency)
excess, 586
film, 508
free-stream, 374, 502
units of, 708
Thermal boundary layer, 373
Thermal circuits, 25
Thermal conductivity, 6, 7
apparent, 449, 574
variable, 67, 710–13
Thermal contact coefficient, 48
Thermal diffusivity, 121
Thermal entrance region, 383
Thermal expansion, coefficient of, 433, 551
Thermal fully developed flow, 385, 392, 442
Thermal radiation, 258
Thermal radiation heat transfer (see Radiation heat transfer)
Thermal radiation network elements, 308, 310
Thermal radiation networks, 300
Thermal radiation shape factor (see Radiation shape factor)
Thermal radiation shields, 318
Thermal resistance, 25
fin, 82
tables, 83, 84
local, 25, 107
steady three-dimensional systems, 148
steady two-dimensional systems, 147
Thermal time constant, 97
Thermodynamics, 3
first law of, 4
second law of, 4, 6, 229, 231
Thermophysical properties, 710–34
Time constant, thermal, 97
Total emissive power, 13, 262
Transistors (see Compact integrated circuits)
Transmissivity, 13, 273
monochromatic, 277
Transparent medium, radiation heat transfer through, 334
participating gases, 334
transparent reflecting solid, 342
transparent solid, 338
Transpiration (see *mass transfer supplement*)
Transport mechanisms and particular laws, 5–24
combined modes of heat transfer, 19
conduction, 5
analysis of, 10
conduction shape factor, 11
Fourier law of conduction, 6
thermal conductivity, 7
convection, 16
Newton law of cooling, 16
coefficient of heat transfer, 17
Newton law of viscous stress, 16
radiation, 12

Transport mechanisms and particular laws (*Continued*)
radiation heat transfer rate, 14
Stefan-Boltzmann law, 13
Tube banks, flow across:
core model, 517
correlations for pressure drop and heat transfer, 520–22, 524, 525
effective local heat flux, 519
hydraulic considerations, 523
hydraulic diameter, 518
in-line, 516
minimum free-flow area, 517
pitch:
longitudinal, 515
transverse, 515
practical thermal analysis of, 519
staggered, 516
summary of relations for geometric and flow characteristics, 538
surface-area density, 516
Tubes, convection in, 463
Turbulent burst phenomenon, 441
Turbulent flow:
characteristics of, 367
theory (see Convection, theorctical analysis)
Turbulent heat flux, apparent, 449
Turbulent Prandtl number, 450
Turbulent shear stress, apparent, 445

Ultraviolet radiation, 260
Units, 708
Unsteady heat transfer systems, one-dimensional, 95–105
electrical analogy, 105
Unsteady thermal radiation systems, 330

Variable thermal conductivity, 67
View factor (see Radiation shape factor)
Viscous dissipation (see also Dissipation):
mathematical formulation for laminar flow, 747
Visible radiation (light), 261

Water bed, heated, 31
Wavelength, 258
Wave number, 282
White light, 261
Wien's displacement law, 266

## PHYSICAL CONSTANTS

| | |
|---|---|
| Ideal gas constant | $\bar{R}$ = 8.3144 kJ/(kgmole K)<br>= 0.08205 $m^3$ atm/(kgmole K)<br>= 1545.3 ft $lb_f/(lb_{mole}$ °R)<br>= 1.986 Btu/($lb_{mole}$ °R) |
| Speed of light in vacuum | $c_0$ = 2.998 × $10^8$ m/s<br>= 186280 mile/s |
| Stefan-Boltzmann constant | $\sigma$ = 5.6697 × $10^{-8}$ W/($m^2$ $K^4$)<br>= 1.7122 × $10^{-9}$ Btu/(h $ft^2$ °$R^4$) |
| Standard gravitational acceleration (sea level) | $g$ = 9.8066 m/$s^2$<br>= 32.174 ft/$s^2$ |
| Standard atmospheric pressure | $P_a$ = 0.10133 MPa<br>= 1.0133 bar<br>= 14.696 $lb_f$/in.$^2$<br>= 760 mm Hg |

## CONVERSION FACTORS

### Fundamental Dimensions

| *Quantity* | *Conversion* | *Other Equivalents* |
|---|---|---|
| Length | 1 m = 3.2808 ft | 1 m = $10^2$ cm<br>= $10^3$ mm = $10^6$ μm = $10^{10}$ Å<br>1 mile = 5280 ft = 1.6093 km |
| Time | 1 s = h/3600 | |
| Mass | 1 kg = 2.2046 $lb_m$ | 1 kg = $10^3$ g |
| Temperature | $T$(°C) = [$T$(°F) − 32°F]5/9<br>$T$(K) = [$T$(°R)]5/9 | $T$(K) = $T$(°C) + 273.15°C<br>$T$(°R) = $T$(°F) + 459.67°F |
| Temperature difference | 5°C = 9°F<br>5 K = 9°R | 1 K = 1°C<br>1°R = 1°F |